Universal Access in HCI:

Towards an Information Society for All

HUMAN FACTORS AND ERGONOMICS

Gavriel Salvendy, Series Editor

Bullinger, H.-J., and Ziegler, J. (Eds.) : *Human–Computer Interaction: Ergonomics and User Interfaces*

Bullinger, H.-J., and Ziegler, J. (Eds.) : *Human–Computer Interaction: Communication, Cooperation, and Application Design*

Stephanidis, C. (Ed.) : *User Interfaces for All: Concepts, Methods, and Tools*

Smith, M. J., Salvendy, G., Harris, D., and Koubeck, R. J. (Eds.) : *Usability Evaluation and Interface Design: Cognitive Engineering, Intelligent Agents and Virtual Reality*

Smith, M. J., and Salvendy, G. (Eds.) : *Systems, Social and Internationalization Design Aspects of Human–Computer Interaction*

Stephanidis, C. (Ed.) : *Universal Access in HCI: Towards an Information Society for All*

Stanney, K. (Ed.) : *Handbook of Virtual Environments Technology: Design, Implementation, and Applications*

Meister, D., and Enderwick, T. : *Human Factors in System Design, Development, and Testing*

Universal Access in HCI:

Towards an Information Society for All

Volume 3 of the Proceedings of HCI International 2001
9th International Conference on Human–Computer Interaction
Symposium on Human Interface (Japan) 2001
4th International Conference on Engineering Psychology
and Cognitive Ergonomics
1st International Conference on Universal Access
in Human–Computer Interaction
August 5–10, 2001 New Orleans, Louisiana, USA

Edited by
Constantine Stephanidis
ICS FORTH and University of Crete

2001

LAWRENCE ERLBAUM ASSOCIATES, PUBLISHERS
Mahwah, New Jersey   London

Lawrence Erlbaum Associates, Inc., Publishers
10 Industrial Avenue
Mahwah, NJ 07430

Library of Congress Cataloging-in-Publication Data

Universal Access in HCI: Towards an Information Society
for All / edited by Constantine Stephanidis
        p. cm.
Includes bibliographical references and index.
ISBN 0-8058-3609-8 (cloth : alk. paper) (Volume 3)

ISBN 0-8058-3607-1 (Volume 1)
ISBN 0-8058-3608-X (Volume 2)
ISBN 0-8058-3610-1 (Set)

2001

Books published by Lawrence Erlbaum Associates are printed on
acid-free paper, and their bindings are chosen for strength and durability.

Printed in the United States of America
10   9   8   7   6   5   4   3   2   1

Table of contents

Part 2
Adaptive and Intelligent Interfaces

Part 4
Multimodal, Continuous and Ubiquitous Interaction

Part 6
Human Factors, Ergonomics, Guidelines and Standards

Part 7
Access to information

Part 8
Applications

Part 9
Assistive Technologies

Part 10
Cultural, Legal, Ethical and Social Issues

Preface

The Ninth International Conference on Human-Computer Interaction, HCI International 2001, is held in New Orleans, Louisiana U.S.A., 5-10 August 2001, jointly with the Symposium on Human Interface (Japan) 2001, the 4th International Conference on Engineering Psychology and Cognitive Ergonomics, and the 1st International Conference on Universal Access in Human-Computer Interaction. A total of 2,738 individuals from industry, academia, research institutes, and governmental agencies from 37 countries submitted their work for presentation, and only those submittals that were judged to be of high scientific quality were included in the program. These papers address the latest research and application in the human aspects of design and use of computing systems. The papers accepted for presentation thoroughly cover the entire field of human-computer interaction, including the cognitive, social, ergonomic, and health aspects of work with computers. These papers address major advances in knowledge and effective use of computers in a variety of diversified application areas, including offices, financial institutions, manufacturing, electronic publishing, construction, and health care.

We are most grateful to the following cooperating organizations:

Chinese Academy of Sciences
Human Factors and Ergonomics Society
Institute of Industrial Engineers

International Ergonomics Association
Japan Ergonomics Society
Japan Management Association

The 229 papers contributing to this book cover the following areas, related to Universal Access in Human-Computer Interaction:

Universal Design
Adaptive and Intelligent Interfaces
Architectures and Tools
Multimodal, Continuous and Ubiquitous
  Interaction
User Diversity and User Participation

Human Factors, Ergonomics, Guidelines
  and Standards
Access to Information
Applications
Assistive Technologies
Cultural, Social, Ethical and Legal Issues

The selected papers on other HCI topics are presented in the two companion books, one edited by M. J. Smith, G. Salvendy, D. Harris and R. J. Koubek, and the other by M. J. Smith and G. Salvendy.

We wish to thank the following Board members who so diligently contributed to the success of the conference and to the direction of the content of the three books. The Conference Board members include:

Engineering Psychology and Cognitive Ergononmics
Chris Babar, *UK*
Ken Boff, *USA*
Guy Boy, *France*
Carlo Cacciabue, *Italy*
Judy Edworthy, *UK*
Arthur Fisk, *USA*
Margie Galer-Flyte, *UK*
Michael W. Haas, *USA*
Craig Harvey, *USA*
Erik Hollnagel, *Sweden*
Kenji Itoh, *Japan*
Peter Jorna, *Netherlands*
Masaaki Kurosu, *Japan*
Kenneth R. Laughery, Sr., *USA*
David Morrison, *Australia*
Sundaram Narayanan, *USA*
Reiner Onken, *Germany*
Eduardo Salas, *USA*
Neville Stanton, *UK*
Eric Tang, *ROC*
Christopher D. Wickens, *USA*

Universal Access in Human-Computer Interaction
Demosthenes Akoumianakis, *Greece*
Elisabeth Andre, *Germany*
David Benyon, *UK*
Noelle Carbonell, *France*
Jan Ekberg, *Finland*
Jon Gunderson, *USA*
Seppo Haataza, *Finland*
Ilias Iakovidis, *Belgium*
Julie Jacko, *USA*
Arthur Karshmer, *USA*
Alfred Kobsa, *Germany*
Pier Luigi Emiliani, *Italy*
Harry Murphy, *USA*
Michael Pieper, *Germany*
Christian Stary, *Austria*
Hirotada Ueda, *Japan*
Jean Vanderdonckt, *Belgium*
Gregg Vanderheiden, *USA*
Annika Waern, *Sweden*
Gerhard Weber, *Germany*
Michael Wilson, *UK*
Toshiki Yamoka, *Japan*

This conference could not have been held without the diligent work of Kim Gilbert, the conference administrator and Myrna Kasdorf, the program administrator, who were both invaluable in the completion of these books. Also, a special thanks goes to Dr. Xiaowen Fang, the student liaison, for all his outstanding efforts.

Michael J. Smith
University of Wisconsin-Madison
Madison, Wisconsin 53706
USA

Gavriel Salvendy
Purdue University
West Lafayette, Indiana 47907
USA

Constantine Stephanidis
ICS-FORTH and
University of Crete
GREECE

Richard J. Koubek
Wright State University
Dayton, Ohio 45435 USA

Don Harris
Cranfield University
Cranfield, Bedford MK43 OAL
UK

June 2001

HCI International 2003

The Tenth International Conference on Human-Computer Interaction, HCI International 2003, will take place jointly with the Symposium on Human Interface (Japan) 2003, the 5th International Conference on Engineering Psychology and Cognitive Ergonomics, and the 2nd International Conference on Universal Access in Human-Computer Interaction, in the island of Crete, Greece, 22-27 June 2003. The conference will cover a broad spectrum of HCI-related themes, including theoretical issues, methods, tools and processes for HCI design, new interface techniques and applications. The conference will offer a pre-conference program with tutorials and workshops, parallel paper sessions, panels, and exhibitions. For more information please visit the URL address: http://hcii2003.ics.forth.gr

General Chair:

Prof. Constantine Stephanidis
ICS-FORTH and University of Crete
Heraklion, Crete, Greece
Telephone: +30-81-391741
Fax: +30-81-391740
Email: cs@ics.forth.gr

The proceedings will be published by Lawrence Erlbaum and Associates.

PART 1

UNIVERSAL DESIGN

Bridging the Gap between Design for All and Assistive Devices

Julio Abascal*, Antón Civit**

*Laboratory of Human-Computer Interaction for Special Needs
The University of the Basque Country, Donostia, Spain
E-mail: julio@si.ehu.es

**Robotics and Computer Technology for Rehabilitation Laboratory
University of Seville, Seville, Spain
E-mail: civit@icaro.fie.us.es

Abstract

The "Design for all" strategy requires not only a clear commitment from the designers, but also the development of adequate methodologies and tools. Interoperability among the diverse devices that are used simultaneously is a key factor for universal access. Many users have physical or cognitive limitations to handle several different devices at the same time. On the other hand, some disabled people need specific devices to be able to use mainstream technology. The introduction of the Bluetooth protocol can solve many of these problems through wireless communication and interoperation. The objective of this paper is to show the important role that Bluetooth can play in granting accessibility to mainstream equipment for people who require using specific assistive devices.

1. Introduction

The huge social change that wireless technologies in general, and mobile telephony in particular, has meant for modern society is well known. For many disabled and older people mobile communications offer a promising new world of security, personal communication and autonomous life. But it is not only opportunities that are foreseen. If these devices are not designed for all, the promises could become barriers for social integration (Abascal et al., 2000).

It is also known that some disabled people need special equipment to gain access to products designed for all. For instance a deaf or hard of hearing person may need an audio prosthesis to use a telephone set. The latter can be considered as designed for all if it has inductive coupling, to allow the audio prosthesis to capture the signal. Consequently, the features of this specific equipment are crucial to the success of universal design concepts. Currently most of this equipment works independently or is connected through point to point cabling or other types of ad hoc connections. Although a great effort has been made to develop wired LANs for rehabilitation technology applications, these have only become widespread for wheelchair applications.

Nevertheless, wireless networks which are smaller than wide area networks (WAN) are becoming increasingly popular. Different technologies for wireless local area networks (WLAN) and wireless personal area networks (WPAN) are appearing on the market. Specifically, personal area networks can become very important to disabled users, due to their need for equipment interconnection and interoperation. For instance, a frequent requirement is the connection of assistive technology equipment to standard mainstream market devices and services. Proprietary protocols and non standard connectors make this task especially difficult. However Bluetooth (an open standard developed among others by Ericsson, IBM, Intel, Nokia, Toshiba, 3com, Lucent, Microsoft and Motorola) provides an open wireless cable replacement profile that could help greatly in improving this situation[1].
The objective of this paper is to show the important role that the Bluetooth protocol can play in bringing accessibility to "designed for all" equipment for people using specific assistive devices.

---

[1] The first Bluetooth devices have already appeared on the market. A list of over 100 products can be currently (06/02/01) found in http://www.bluetooth.com. Further proliferation into other products is expected before the end of 2001.

2. Bluetooth Technology

Bluetooth is a new wireless technology oriented towards the creation of Personal Area Networks (PAN). A very good introduction to Bluetooth can be found in (Compaq, 2000). A PAN, called a *piconet* in Bluetooth terminology, covers a 10m bubble around a central master device. Up to 8 devices can actively participate in a *piconet* while up to 256 can remain "parked" so that they can replace an active device very quickly if they are needed. Up to 10 different *piconets* can exist inside any 10m bubble. Any device can, in principle, participate in various *piconets*.

The bandwidth of a Bluetooth personal network is 1Mb/s. The supported channel configurations are up to 3 simultaneous full duplex 64kb/s voice channels, and 432kb/s symmetric or 721/56kb/s asymmetric asynchronous data connections.

The Bluetooth specification establishes a set of profiles that define different usage modes. Some of these profiles are related to generic services like *General Access* or *Service Discovery* while others are directly related to usage models. Among these we can find:
- *Serial Port*: defines a serial port emulation for devices that traditionally use point to point connections. This profile is directly related to the cable replacement usage mode and will be, at least initially, one of the most useful for RT related applications.
- *Cordless Telephony*: establishes the way in which a cell phone can use Bluetooth to act as a cordless phone.
- *Headset*: defines the transmission and reception of voice data over a Bluetooth link.
- *Dial-up networking*: used as a replacement for the computer / cell phone link.
- *LAN access*: set up a personal network using point to point protocol.
- *Object exchange, push, file transfer and synchronization*: these profiles define the way in which different objects like files, business cards, appointments, etc., are exchanged over Bluetooth PANs.

Among the possible applications, some of the most common will be "use anywhere" Internet bridges, automatic cordless PDA synchronizers, interactive conference file transfer, instant postcards, etc.

3. Scenarios

In this section some situations in which wireless personal networks may be useful to the disabled user will be described. The need for interoperable wireless connected equipment will be also shown. The analysis of these scenarios will allow us to show the needs, procedures for use, barriers, etc., that people with disabilities and elderly people might find and how these can be overcome. The first two scenarios are related to mobile/wireless communications while the rest are related to other aspects of the users' daily life.

3.1 Voice Call Scenarios

This scenario is related to the most frequent use of mobile communications. The user simply wants to communicate by voice with somebody. This is, apparently, a very simple situation but when it is more deeply analyzed, it can be found that certain people in specific environments can experience serious difficulties. In this scenario the user has to locate the phone, hold it, somehow dial the number, talk and hear during the conversation, hang the phone up and, probably, replace it somewhere. Due to their difficulty in performing one or more of these tasks, some users hardly fit into this scenario.

Hard of hearing users also experience important problems when using GSM mobile phones or DECT wireless terminals (Hansen et al., 1996). These are due to the interference between the phone RF transmissions and the hearing aids. Although these effects can be minimized if they are considered in the design phase of both the phone and the hearing aid, it is also true that a Bluetooth enabled hearing aid, working together with a Bluetooth enabled microphone, could act as a Bluetooth headset. In this way the sound output of the phone could be directly fed into the user's hearing aid greatly increasing the system usability. As an added benefit, in the same hearing aid a microphone could be used with a fixed Bluetooth enabled phone and with a mobile. Later on other possible applications for the Bluetooth enabled hearing aid will analyzed.

Deaf users are not the only ones who have access problems in the simple phone call scenario. As a further example, the case of users with severe mobility restrictions can be considered. These users make up one of the groups for which mobile and wireless phones can be, in principle more beneficial. For those users whose only means of moving around is an electric wheelchair, most of the devices required for everyday living must be installed in the chair. For them a mobile/wireless phone can be the most natural element to communicate with their friends or relatives or to contact emergency services in case of need. In this case, the biggest problem is not the conversation itself, but the possibility of starting and ending it. Most of these users have very limited control over their hands, which makes it impossible to pick up the terminal, hold it and dial in the usual way. Thus, if a terminal is to be used in this situation, there are some new requirements that would allow complete hands-free operation:

- External microphone and speaker connections should be available.
- It should be possible to answer an incoming call without using the hands.
- It should be possible to place a call without using the hands.

Those requirements are not only valid for the wheelchair user scenario but they are also good for the car driver scenario. This is probably the reason why the first two requirements are met by most currently available terminals. The third one is technically more complex and some solutions that are valid for the wheelchair scenario may not be adequate for the car driver and vice versa.

Most current mobile phone terminals include an interface connector that usually allows an external device full control over almost every phone function. If this is the case, the best solution is probably to have a single device acting as user interface for all other devices resident in the wheelchair. This is the main idea behind wheelchair buses like M3S and DX. To the best of our knowledge, interfaces to connect mobile terminals to these buses are not currently available. Unfortunately, in most cases, the terminal interface is not publicly documented and this restricts the number of devices that can implement this solution. A natural solution would be to use a Bluetooth enabled phone (it could be mobile and wireless at the same time) and control it through Bluetooth from the users standard wheelchair controller. This would require a device to connect the wheelchair bus to Bluetooth. This device is currently not available. More applications of the Bluetooth bridge for wheelchair buses will be discussed later on.

People with severe confusion problems could also benefit greatly from mobile/wireless communications but, in this case, the user interface must be kept much simpler than in all the cases above. Here a very simple user assistant device could help the user in his or her everyday tasks. This type of device is currently available. It is possible, in principle, that this device could control a fixed or mobile phone through Bluetooth. The system must probably decide with little user participation whom to call.

Blind users experience difficulties in locating the terminals. Bluetooth-enabled mobiles, that can act as wireless terminals when the user is at his/her home or office, can act as a single personal terminal thus being much easier to operate[2].

3.2   Text message scenarios

Text messaging in mobile networks has become increasingly popular in recent years. But text messages have their own accessibility problems. These problems are very well known because they are just a variation of the computer access problems that were common before the multimedia era.

It is interesting to note that some ideas originally developed for users with strong mobility restrictions like *word prediction* (Garay et al., 1997) and *reduced keyboard disambiguation* (Lesher et al., 1998) have come to the mainstream market through their application in mobile telephony. The size requirement of an ordinary mobile phone rules out standard QWERTY keyboards for text input and thus the standard 9 or 12 keyed keyboard must be used for this propose. When this is done all users enter a functional disability situation and must employ techniques that users with little control over their mobility have been using for years. These means that all the approaches invented to reduce the number of keystrokes required to write a word have become important to the public in general. This process will be beneficial for disabled users in the long run as it will bring about an improvement in word prediction and disambiguation techniques.

---

2 Dual mode GSM / DECT terminals, which are currently available, are another possible alternative.

6

The main problem with messages for users with restricted mobility is being able to write them. Even if the terminal incorporates a word predictor it will not be directly usable by most users with important mobility restrictions. For this purpose, it is essential that the terminal can be controlled externally from a device that can easily be commanded be the user. This was already mentioned in the "voice call" scenario and Bluetooth is a very good alternative for this purpose.

Blind users have two different problems. First, they must be able to read the messages. Considering that these messages are usually very short, text-to-speech conversion is the best alternative for retrieving them. Text-to-speech is a mature technology but to the authors' best knowledge, is not available in current terminals. There is no doubt that future terminals will be able to read text messages.

The second problem is more difficult and more interesting, as it differs from the standard computer access problem. With ordinary computer keyboards blind users don't have great difficulties for inputting text. As has been mentioned above, the ordinary keyboard is almost ruled out for mobile designs and word prediction and reduced keyboard disambiguation methods are becoming commonplace. With this type of input device, blind users experience much greater difficulties than with keyboards with unambiguous mapping. This is due to the fact that with word prediction the user must be ready to see the different suggested alternatives and choose among them[3]. Thus a problem arises: how can this be implemented in a way that is natural for a blind user? A simple solution could be to stick to QWERTY keyboards or at least to unambiguous keyboard mappings, thus small Bluetooth enabled QWERTY keyboards could be used for inputting text to the mobile phone.

3.3 E-mail Scenario

This is one of the original scenarios that led to the design of Bluetooth and it is especially interesting for the case of disabled users. Currently, for instance, a wheelchair user who writes an e-mail needs to connect to a fixed network, or use a mobile modem, to send it. The case for reception is obviously analogous. If the user's computer (or whatever device is used to send and receive e-mail) is Bluetooth enabled, the user can write and e-mail and send it as soon as he or she is near enough to the Bluetooth modem or network connection. This procedure is also valid for accessing the Internet or printing files. Bluetooth can be a good solution in the case of blind users who experience difficulties in making connections to fixed networks too.

3.4 Environment Control Scenario

Many wheelchair users require some form or another of environmental control. The Bluetooth specification defines three classes of radios. Current products use class 3 where the transmission power is limited to 1mW and the range to 10m. Environment control applications will require the use of class 1 radios that limit the output power to 100mW. There are several projects considering the use of Bluetooth for environment control applications but, as Bluetooth's main scope is personal networks, other alternatives (e.g. *HomeRF*) may be more suited to this purpose.

The wheelchair bus / Bluetooth bridge, already discussed in section 3.1 would be very useful for this application. Using this device it would be possible to control the equipment linked to the smart home network using the user specific wheelchair controller. If the smart home network is not based on Bluetooth another device would be required to act as a bridge between Bluetooth and the existing network.

3.5 Short Range Environment Control

Under this title we consider several environment-related applications that do not require the use of class 1 radios. Some example applications are door openers, in which a Bluetooth enabled door opens when a wheelchair user wants to pass through it and Bluetooth light switches that can be operated directly from a wheelchair nearby. Of course, some of these applications may become mass-market applications and became especially accepted by elderly users.

---

[3] For example, in the T9 algorithm when the user likes the first alternative he or she will just hit the spacebar once after inputting the word; if the second alternative is desired the spacebar must be hit twice, etc. Other actions are necessary to input a word that is not included in the dictionary.

3.6 Short Range Payment Related Applications

Bluetooth has the possibility of establishing several secure modes including link level unfrocked security. In this mode authentication is required before using a device. This mode together with the four key based encryption mechanisms could, in principle, provide enough security for using Bluetooth in commercial transactions. Thus a wheelchair or blind user could, for example, pay for a bus ticket without having to take out notes and coins or inserting a card into a specific machine. Contactless transponder based cards can also be used for this purpose but the available "intelligence" in these devices is much more limited.

4. Good interface design

The interoperability among devices allowed by Bluetooth can change drastically the accessibility to mainstream design for all devices. But there is another requirement. It is necessary to avoid the hypothetical lack of ability of the users to handle complex devices and also their acceptability. Both lack of ability and rejection of the technology, if they exist, are frequently due to the low quality of the interface. Thus, good design of interfaces is needed to avoid these two important problems. Moreover, badly designed interfaces make use difficult, not only by disabled and elderly people, but by everyone.

5. Conclusions

There are sectors of the population unable to use the systems that are designed for all without specific equipment. So it is essential that products and services are designed in such a way that, when necessary, they are open to possible adaptation for specific user needs. But it is also essential to open these systems to interoperability. Bluetooth, as an open wireless interface, can be very helpful in permitting users unable to use the equipment directly to use it through other devices that are better adapted to their needs.

Bluetooth has great potential to change the lives of disabled and elderly people. Industry is starting to take into account this population as a potential market, mainly due to three reasons: the rising proportion of disabled people in western society, the possibility of government-subsidized prices for people with disabilities and the potential introduction of some of these devices in the mainstream market. However the advantages offered to disabled and elderly people can only be useful if the design takes into account the real needs and requirements of elderly and disabled users.

References
Abascal, J., Civit, A. (2000). *ERCIM*
Compaq Commercial Portables Product Marketing, White paper: Bluetooth Technology Overview, November 2000. Available at:
    http://www.compaq.com/products/wireless/wpan/files/WhitePaper_BluetoothTechnologyOverview-QA.pdf
Garay, N., Abascal, J.G. (1997). Intelligent Word-Prediction to Enhance Text Input Rate. In Moore J. et al. (eds.) *Proceedings of the 1997 International Conference on Intelligent User Interfaces (IUI-97)*. ACM Press, pp. 241-4.
Hansen M.Ø., Poulsen, T. (1996). *Evaluation of noise in hearing instruments caused by GSM and DECT mobile telephones*. Scand Audiol; 25: 227-232.
Lesher, G,W., et al. (1998). Optimal Character Arrangements for Ambiguous Keyboards. *IEEE Transactions on Rehabilitation Engineering*, Vol. 6, No. 4, pp 415-23.

Re-thinking HCI in terms of universal design

Demosthenes Akoumianakis[1], Constantine Stephanidis[1, 2]

[1]Institute of Computer Science
Foundation for Research and Technology-Hellas
Science &Technology Park of Crete, Heraklion, Crete, GR-71110, Greece
Email: cs@ics.forth.gr

[2]Department of Computer Science, University of Crete

Abstract

This paper sets out to investigate some of the challenges posed to Human-Computer Interaction (HCI) by the recent rise of interest on universal design and universal access. Specifically, we briefly review the premises of universal design, and examine the implications on HCI prevailing practice and research agenda. It is argued that HCI stands to gain from appreciating and respecting the base lines of universal design. In particular, the primary benefit for HCI, is in expanding its scope and underlying methodological ground to cope with diversity in the emerging Information Society.

1. Background

Architects have long recognised the need for a design code of practice, which would respect, value and accommodate diverse human abilities. The normative perspective of such a code of practice is that: "*Instead of responding only to the minimum demands of laws, which require a few special features for disabled people, it is possible to design most manufactured items and building elements to be usable by a broader range of human beings, including children, elderly people, people with disabilities, and people of different sizes.*" Encyclopaedia of Architecture, Design, Engineering and Construction, 1989, p. 754.

Progressively, such a commitment was translated to engineering recommendations, standard rules and legislative clauses. The term used to refer to this new engineering thinking is universal design. In the 1990's, universal design obtained a broader connotation and attracted considerable interest in the field of Information Technology and Telecommunications (IT&T). Several projects were carried out in an attempt to address technical issues, raise awareness and advance an understanding of the challenges pertaining to the appropriation of the benefits of universal design in IT&T.

In the field of HCI, the rise of interest on universal design is a recent development. It results from the need to address the shortcomings in traditional HCI design practices to cope with the different dimensions of diversity (Stephanidis, Salvendy, et al., 1998; Stephanidis, Salvendy, et al., 1999; Stephanidis, 2001). To this end, universal design has been the focal issue of concern in the context of an International Scientific Forum[1] and several research and development projects over the past decade (for a review see Stephanidis & Emiliani, 1999). These experiences have created the compelling need to establish on a more formal basis a wider, interdisciplinary research community to consolidate recent experiences on universal design, explore new challenges relevant to HCI and provide the tools needed to appropriate its benefits.

In this paper, we set out to investigate the implications of universal access and universal design on HCI research and practice agendas. Specifically, we examine the traditional assumptions which have withhold HCI research in the past

---

[1] The International Scientific Forum "Towards an Information Society for All' was launched in 1997, as an international ad hoc group of experts sharing common visions and objectives, namely the advancement of the principles of Universal Access in the emerging Information Society. The Forum held three workshops to establish interdisciplinary discussion, exchange of knowledge, dissemination, and international co-operation. The 1st workshop took place in San Francisco, USA, August 29, 1997, and was sponsored by IBM. The 2nd took place in Crete, Greece, June 15-16, 1998. The 3rd took place in Munich, Germany, August 22-23, 1999. The latter two events were partially funded by the European Commission. The Forum has produced two White Papers (Stephanidis, Salvendy et al., 1998; Stephanidis, Salvendy et al., 1999).

and how they are re-shaped by the emergence of an Information Society and the compelling need to accommodate diversity in HCI design. Then, we extrapolate to address key question for HCI theory, methodology and engineering practice, which need to be re-considered to provide the ground for inclusive and universal design. The paper ends with a summary of key findings and conclusions.

2. Human Computer Interaction in transition

In recent years, several research communities have tried to make projections as to the future type and scale of research needed to advance theory and engineering practice within HCI. This becomes evident when one considers the number of new journals which progressively appear and compete for a share of the market in HCI archival publications. Furthermore, it is also evidenced from the focus of special issues in existing HCI journals (see for example the special issues of the journals, Human Computer Interaction, ACM Transactions on Computer-Human Interaction and the User Modelling and User-adapted Interaction). In the majority of these cases, the common conclusion derived is the need to revisit the basics of HCI in the light of recent advances and foreseen trends (see Table 1). One such trend, which is increasingly gaining in interest and research attention, is universal access. There are several reasons for this. Some of them are related to the appreciation of the universal access challenges to HCI, while others may be associated with the failure of traditional research strands to influence the on-going shift in the technological paradigm.

Table 1: Contrasting paradigms for HCI

	Paradigms	
	Traditional	**Emerging**
Design & Development model	Waterfall	Iterative (Spiral)
Design goal	Meet well defined user requirements	Understanding the global execution context of tasks; coping with changing requirements, not known apriori
Implementation goal	Comply to well-defined architectural system flows	Platform independence Scale-ability Interoperability
Content management	Storage and representation	Personalization, transcoding and delivery

It is frequently argued that the hardest target to achieve en route to the information age, often referred to as the Information Society, is the design of new computer-embodied artefacts to facilitate the broad range of emerging activities (Stephanidis et al., 1998a; Stephanidis et al., 1998b). However, for this to be attained, HCI should revisit some of the basic assumptions which have shaped recent work and progress. Some of these are briefly discussed below.

The "average" typical user. One of the basic guidelines for HCI design is to study the user. Many recent methodological frames of reference introduce this guideline as a milestone for subsequent stages in a user-centred design process. Additionally, there has been a wide of variety of instruments, for capturing requirements and studying users. Notwithstanding the relative merits of each method or tool, it is important to mention that their usefulness should be carefully judged. This is not due to the validity of the techniques, but rather due to the changing conception of who are the users. More specifically, in the context of the emerging distributed and communication-intensive information society, users are no longer the computer-literate, skilled and able-bodied workers driven by performance-oriented motives. Nor do users constitute a homogeneous mass of information seeking actors with standard abilities, similar interests and common preferences with regards to information access and use. Instead, our conception of users should accommodate all potential citizens, including the young and the elderly, residential users, as well as those with situational or permanent disability. Consequently, it becomes increasingly complex for designers to know the users of their products in advance and ever more compelling to design for the broadest possible end user communities. This raises implications on design methodology and instruments, as well as the technical and user perceived qualities to be delivered.

The business environment. Another assumption which has influenced HCI throughout its short history has been the context of use in which computer-based designs are encountered. Due to the unlimited business demand for information processing, the HCI community has progressively acquired a bias and habitual tendency towards outcomes (i.e. theories, methods and tools) which satisfy the business requirements and demonstrate performance improvements and productivity gains. However, since the early 1990s, analysts have been concerned with the increasing residential demand for information, which is now anticipated to be much higher than its business counterpart. Consequently, designers should progressively adapt their thinking to facilitate a shift from designing tools for productivity improvement to designing computer-mediated environments of use. This leads to the next point, which addresses the need to extend the prevailing metaphors for interaction design to suite the changing requirements.

The desktop computer-embodiment. The desktop embodiment of current computers is perhaps the most prominent innovation delivered by the user interface software industry, including the tool development sector. However, the diffusion of the Internet as an information highway and the proliferation of advanced interaction technologies (e.g. mobile devices, network attachable equipment, etc), signify that many of the tasks to be performed by humans in the information age will no longer be bound to the visual desktop. New metaphors are likely to prevail as design catalysts of the emerging virtual spaces and the broader type and range of computer-mediated human activities. Arguably, these metaphors should encapsulate an inherently social and communication-oriented character in order to provide the guiding principles and underlying theories for designing more natural and intuitive computer embodiments. Consequently, the challenge lies within the scope of finding powerful themes and design patterns to shape the construction of novel communication spaces. At the same time it is more than likely that no single design perspective, analogy or metaphor will suffice as a panacea for all, potential users or computer-mediated human activities. Design will increasingly entail the articulation of diverse concepts, deeper knowledge and more powerful representations to describe the broader range and scope of interactive patterns and phenomena. To this end, developmental theories of human communication and action may provide useful insights towards enriched and more effective methodological strands to interaction design.

To the above needed revisions, universal design offers its early design focus on accessibility, which in practical terms entails understanding and designing for the global execution context of tasks. A task (i.e. a piece of functionality) can be executed in different contexts, which arise due to different machine environments, terminals or user conditions (i.e. location, preferences). Due to differences in information processing between users and machines, designers create symbolic representations through which users perceive a task. Increasingly, designers will have to device alternative suitable symbolic representations to cope with the diversity in the potential execution contexts of tasks. These images need to be use-adapted and tailorable manifestations of non-interactive functions (tasks). Nevertheless, the question is how to design them so as to enable anyone, anytime, anywhere access.

3. Implications

3.1 Underlying theory and science base

In the light of the above, the prime theoretical challenge that arises is on the availability of a suitable scientific frame of reference to provide the grounds for understanding and designing for the global execution context of tasks. At the core of this question is the capability of a frame of reference to study context, something which has attracted the attention of HCI researchers in the recent past. To this end, the HCI community seems to appreciate the shortcomings of contemporary approaches, such as human factors evaluation and cognitive science, and recognises the need for (Carroll, 1991):

- Better utilisation of the existing knowledge and science base, by utilising the experimental ground of social and human sciences (e.g. anthropology, sociology)
- Broadening the range and/or extending the scope of information processing psychology with concepts from developmental approaches to HCI, such as activity theory (Bødker, 1991), language/action theory (Winograd, 1988), situated action model (Suchman, 1987) and distributed cognition (Norman, 1993).

The normative perspective adopted in these efforts is that interactions between humans and information artefacts should be studied in specific social contexts (as opposed to laboratories or otherwise controlled settings) and should take into account the distinctive properties that characterise them. Despite this common commitment to the study of context, the above alternatives differ with respect to at least three dimensions (Nardi, 1996), namely the unit of analysis in studying context, the categories offered to support a description of context and the extend to which each treats actions as structured prior or during activity.

3.2 Underlying methodological ground

Finally, the above theoretical and scientific debate raises implications on underlying methodology and engineering HCI practice. In terms of HCI methodology, the universal design movement has had a moderate impact. The vast majority of the available universal design materials remain in the shadow of the Human Factors Evaluation tradition. For instance, the universal design principles (Story, 1998) constitute general context-independent recommendations, which require substantial interpretation on behalf of designers. Also universal design lacks the solid methodological ground to enable early assessments of quality attributes pertinent to universal access, comprehensive evaluation techniques or formal methods to specify what universal access implies in certain application areas. Consequently, at the present time, any methodological transfers between HCI and Universal Design are likely to have the universal design community at the receiving end.

3.3 Engineering practice

In terms of engineering practice, the situation is slightly different. Specifically, there have been several proposals indicating desirable attributes of an engineering methodology to advance a universal design perspective in HCI. Some of these proposals cover the area of user interface architectures, while others emphasise the need for a new class of user interface development techniques. In particular, popular user interface architectural models, are considered inadequate for a number of reasons. First of all, they do not explicitly account for the notion of accessibility of an interactive application, as considered by universal design advocates. Thus, they do not address a range of issues, which recent research indicates as cornerstones for universal design in HCI. Examples of such issues include: multiple platform environments, toolkit integration, platform abstraction, etc., which arise from the proliferation of novel interaction platforms and diversity in usage patterns. Secondly, these models offer no account of user interface adaptation (Stephanidis, 2001), which is a central theme in design for all in HCI. As a result, key decisions such as when and what to adapt are not addressed, while the components that are needed to drive adaptations at the level of the user interface are totally missing. Thirdly, existing architectural models offer implementation-oriented views of user interface architectures, thus delimiting the role of design and not addressing how design knowledge can be propagated to development and implementation phases. As a result, the application of these models in current HCI practices leads to re-implementations (reactive approach) rather than instantiation of an alternative design (proactive approach).

Apart from the architectural challenge, universal design inspired efforts into HCI reveal shortcomings in prevalent user interface development techniques. Specifically, since the key ingredient of universal design in HCI is the capability to encapsulate alternative interactive behaviours through abstraction, it follows that development methods closer to the physical level of interaction are inappropriate (see Figure 1). This renders inappropriate a large and popular collection of development techniques, which include presentation-based approaches, such as VisualBasic™ and TAE plus™ (TAE Plus, 1998[2]), physical-task based methods, such as UAN (Hartson et al., 1990) and demonstration-based techniques, such as those in Peridot (Myers, 1995). On the other hand, techniques, which focus on higher-level dialogue properties and offer mechanisms for articulating alternative interactive components stand a better chance and could be considered as candidates for an engineering platform suitable for

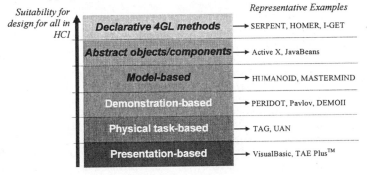

Figure 1: User interface development techniques: Ranking and suitability index

universal design in HCI. These techniques include declarative 4GL methods, typically found in some UIMSs such as SERPENT (Bass et al., 1990) and HOMER (Savidis et al., 1995), abstract objects / components, such as meta-

---

[2] http//www.cen.com/tae/

widgets (Blattner et al., 1992) or component-ware technologies, such as Active X^{TM} by Miscrosoft and JavaBeans^{TM} by SunSoft, and some model-based techniques such as those in MATSERMIND (Szekely et al., 1995).

4. Conclusions

This paper has attempted to reveal that universal design is neither a utopia nor "wishful thinking" by socially oriented researchers. It poses several challenges to the fields in which it has been applied, and this is the case with HCI as well as the emerging Information Society. Nevertheless, universal design offers several advances to the fields, which succeed to adopt, internalise and appropriate its benefits. This is now evident in engineering disciplines such as interior design (e.g., housing, interiors and the design of household appliances and controls. In the area of HCI, the concept and principles of universal design are quite new. Nevertheless, there have been several projects, which have demonstrated both the technical feasibility as well as the demands posed to the practice of HCI. This paper, has briefly reviewed some of these efforts and identified the implications for the HCI research agenda. The main conclusion drawn is that for HCI, universal design offers new research directions, which expand the scope of studying computer-mediated human activities. Specifically, by appreciating and respecting the base line of universal design, HCI stands to benefit and extend its experimental and empirical ground towards a more thorough treatment of diversity in the emerging Information Society.

References

Bass, L., Hardy, E., Little, R., Seacord, R. (1990). Incremental development of user interfaces. In G. Cockton (Ed.), *Engineering for Human-Computer Interaction.* (pp. 155-173). The Netherlands: North-Holland.

Blattner, M., Glinert, E., Jorge, J., Ormsby, G. (1992). Metawidgets: Towards a theory of multimodal interface design. In *Proceedings of the COMPSAC'92 Conference*, Chicago. IEEE Computer Society Press.

Bødker, S. (1991). *Through the Interface: A Human Activity Approach to User Interface Design.* Hillsdale: Lawrence Erlbaum Associates.

Carroll, J. (Ed.) (1991). *Designing Interaction: Psychology at the HCI.* Cambridge: Cambridge University Press.

Hartson, H.R., Siochi, A., Hix, D. (1990). The UAN: A User-Oriented Representation for Direct Manipulation Interface Design. *ACM Transactions on Information Systems*, 8(3), 289-320.

Myers, B. (1995). User Interface Software Tools. *ACM Transactions on Human-Computer Interaction*, 2(1), 64-103.

Nardi, B. (1996). *Context and consciousness: Activity Theory and HCI.* Cambridge MA: MIT Press.

Norman, D. (1993). *Things That Make Us Smart.* Reading, MA: Addison-Wesley Publishing Company.

Savidis, A., Stephanidis, C. (1995). Developing Dual User Interfaces for Integrating Blind and Sighted Users: The HOMER UIMS. In *Proceedings of the Conference on Human Factors in Computing Systems (CHI '95)*, Denver, Colorado (pp. 106-113). New York: ACM Press.

Stephanidis, C. (Ed.) (2001). *User Interfaces for All - concepts, methods and tools.* Mahwah, NJ: Lawrence Erlbaum Associates (ISBN 0-8058-2967-9).

Stephanidis, C., Emiliani, P-L. (1999). Connecting to the information society: a European perspective. *Technology and Disability*, 10(1), 21-44.

Stephanidis, C., Salvendy, G., et al., (1998). Toward an Information Society for All: An International R&D Agenda. *International Journal of Human-Computer Interaction*, 10(2), 107-134.

Stephanidis, C., Salvendy, G., et al., (1999). Toward an Information Society for All: HCI challenges and R&D recommendations. *International Journal of Human-Computer Interaction*, 11(1), 1-28.

Story, M.F. (1998). Maximising Usability: The Principles of Universal Design. *Assistive Technology*, 10, 4-12.

Suchman, L.A. (1987). *Plans and Situated Actions: The problem of human machine communication.* Cambridge: Cambridge University Press.

Szekely, P., Sukaviriya, P., Castells, P., Muthukumarasamy, J., Salcher, E. (1995). Declarative interface models for user interface construction tools: the Mastermind approach. In Bass L., Unger C. (Eds.), *Engineering for Human-Computer Interaction, Proceedings of EHCI'95* (pp. 120-150). London: Chapman & Hall.

Winograd, T. (1988). A language/action perspective on the Design of Co-operative Work. *Human Computer Interaction*, 3(1), 3-30.

Designing multiuser voice control interface for home entertainment systems

Shuo-Hsiu Hsu*, Lin-Lin Chen*, Y. C. Chu[†]

*National Taiwan University of Science and Technology, Taipei, Taiwan
[†]Philips Research East Asia – Taipei, Taipei, Taiwan

Abstract

An interface is proposed to provide multiuser Chinese speech control of a home entertainment system (a video player), and to facilitate resolution of conflicts in control. The multi-user speech interface allows designation of speech input to only one or to all users. Experiments were then conducted to assess usability of the proposed interface. Tasks were designed to stimulate cooperative and competitive behaviors under time constraints. Our results indicate that even for competitive tasks, conflicts among users can be resolved within at most two to three turns of interaction. We also found that recognition performance is a critical influential factor for the usage behavior. In addition, most subjects preferred the master control mode to the hands-free shared control mode due to better recognition rates, indicating that the hands-free advantage for voice control is beneficial only if the recognition rate reaches a satisfactory level.

1. Introduction

Human communications are multimodal in nature. People communicate with each other by languages, gestures, mimics and nonlinguistic sounds. Instead of expressing a message in only one way, people often use several communicative channels in parallel to express their meaning more clearly [3].

The use environment for any home entertainment system is inherently multi-user and multimodal. In designing a speech interface for such a system, one of the important issues is the combination of multiple input and output modalities, including remote control, voice, visual display, and other modes of interaction. Seamless integration of these modalities helps create a natural interface that best matches the users' expectations and the input/output attributes' perceptual structure. In addition, it is necessary to provide means for multiple users to interact with the system, to encourage social dialogue, and to facilitate conflict resolution.

Previous studies on dual-mode interfaces showed that a dual-mode interface provides better efficiency than a single mode interface does [2]. A successful case is the interface combined with speech and keyboard in a specific mixture level, where the speech interface coexists with other modalities rather than replaces them [2]. Another synergistic example is the multimodal interface used in the electric map system by combining speech, typing and handwriting [1]. In this interface, speech is known for one of the natural way of human interaction and typing is the common tool for fast information input to systems. In addition, handwriting is sometimes useful for specific tasks that are difficult for speech [8,9].

Conflicts are one of the critical issues because a group of users is expected to use the system at the same time. For a home entertainment system with a remote control, two or more users may fight for the ownership of the remote control to win the right to issue commands. Resolution of the conflicts relies on the social behavior patterns developed in this group. For speech interfaces in a multi-user environment, people are even more likely to engage in a conflict situation. Studies at Philips categorized different conflicts into eight scenarios. These scenarios were further classified into three types: conflicts with issuing commands, conflicts with talking on the phone, and conflicts with the environment. The first conflict type occurs either when users issue commands at will or when they disagree on what they want to do. When the second conflict type occurs, users tend to either be forced to use the remote control or hang up the phone to eliminate the disturbance. For conflicts with the environment, users can only give up voice control and use the remote control instead. It is not clear how people will choose to resolve different types of conflicts and how long it will take for them to do so.

This paper reports the design of a multiuser speech interface for a video player, and the results of experiments conducted to assess the proposed interface. We describe the interface design in the next section by analyzing the use environment and describing the design concept. Details of the experiment design and the method to analyze data obtained in different phases are discussed in the third section. This is followed by findings from the experiments concerning the control behavior, subject preference, influential factors, and general opinions of the subjects.

2. Interface design

To facilitate the design of speech interfaces for home entertainment systems, we divide the use environment into two areas according to different levels of access to control. We call these two areas "remote area" and "local area" as shown in Figure 1. Different design concepts were then derived by placing input/output devices such as voice receiving units, voice recognition units and feedforward or feedback units in different areas. With the combination of local voice receiving unit, remote voice recognition unit, and remote/local feedback, one of the concepts developed is a remote controller with the speech recognition capability that provides users with obvious differentiation of authority when operating the speech interface in a group.

To accept speech input from multiple users, we implemented an omni-directional microphone in the main set (the video player). This microphone accepts speech

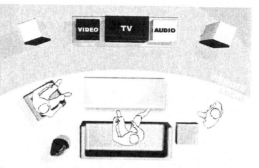

Figure 1 Remote and Local Areas

input from every user in the room whenever the master controller activates it. To receive commands from the master user only, the master controller is equipped with a unidirectional microphone. The unidirectional microphone allows the master controller to accept speech input from its holder while neglects input from other speakers. The received speech will be transmitted to the main set for further processing (because the speech recognizer is located in the main set). Whoever holds the master controller becomes the master user for voice control.

A red button on the master controller is designed to designate speech input in a multi-user environment. User can easily switch between three different designation scenarios by pressing the red button: accepting speech input from master user only, accepting speech input from everyone in the room, or rejecting all speech input. By pressing the red button, the master user can activate the microphone on the master controller while disable the one on the main set. In this case, only verbal commands from the master user will be recognized. Commands from other speakers will be neglected. Similarly, the master user can switch the input channel to the main set, allowing every user in the room to issue verbal commands. Pushing the button again can disable all speech input completely. Each time the red button is pressed, the device will switch from one scenario to the next. Feedback is provided both on the main set and on the hand-held master controller. Due to the limitation of not being able to implement any graphic output with the video player (so that feedback can be shown directly on TV screen), an LED display set on the video player's panel is used for indicating whether the omni-directional microphone is activated. On the master controller, a red light near the unidirectional microphone is turned on when the microphone is activated. Since the master controller is designed to be a hand-held device, the grand "hands-free" advantage of voice control is no longer valid in this case. It is then crucial for us to evaluate in the following experiments to see if the users can accept the master controller. In our experiment, two modes of control offered by the master controller will be used: a *master control mode*, in which only the user holding the master controller can issue voice commands; and a *shared control mode*, in which all users can issue voice commands to the home entertainment system directly.

3. Experiment design

3.1 Participants

People tend to enjoy entertainment activities in the home environment with either their family members or friends. To achieve our experimental purposes, a group with fewer than three people may not stimulate enough interactions/conflicts. Thus we decided to have three subjects in each group. A total of twelve groups are tested in the experiments. Each group consists of either all friends or all members of the same family, aging from eighteen to forty-five.

3.2 Overall procedure and experiment phases

To observe the behavior of subjects interacting with the interface and with each other in the group, experiments are divided into different phases. In each phase of the experiments, tasks are designed to stimulate interactions between subjects and also reflect the actual usage scenarios for each mode of interface. Experiments are designed with cooperation and competition in mind. The experiments consist of four phases: phase A (shared control phase), phase

B (master control phase), phase C (mixed control phase), and phase D (mixed control phase with remote control). The first three phases use different modes of voice control, while a remote controller is used together with the master controller in the forth phase.

During phase A, all users are allowed to control the home entertainment system by issuing commands directly to the system. In phase B, the master controller is introduced and only one user is selected and allowed to be the master user in charge of issuing voice commands by speaking to the master controller. In phase C, the users are allowed to switch between master control and shared control modes. Anyone in the group can use the master controller to become the master user, or they may switch to the shared control mode. During phase D, users are given an additional option of controlling the system with a remote control.

3.3 Experimental tasks and procedure

In order to observe different patterns of interactions during the experiments, tasks are designed with two stimuli: time limit and emphasis of cooperation or competition. Typical tasks in four phases are that subjects receive several questions and are asked to find the answers by playing videos. In phases A and *B (shared control phase and master control phase)*, competitive tasks are assigned and allowed to finish without time limits. Each subject gets a different picture-searching task. The subjects must execute the tasks together (by issuing commands directly to the system unit in phase A, or by asking the master user to issue commands in phase B). The subjects then take a five-minute break to lessen the influence of the former phases, before moving to the third phase. In phase C (*mixed control phase*), time limit is placed on cooperative tasks to stimulate more interactions in the group. Subjects have to finish a common task together by choosing either the master or the shared modes. As the previous case, subjects again take a five-minute break before moving to phase D. In phase D (*mixed control phase with remote control*), tasks are designed for subjects to compete for awards. Within a time limit, the subject finding more answers than others gets an extra reward. In addition to voice control, subjects can use the remote control to execute the task.

3.4 Data analysis

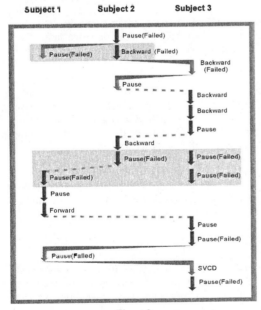

Figure 2

To trace the control behavior of subjects in different phases, we developed a novel diagram, called *control flow diagram*, to graphically represent the process recorded on video. A control flow diagram consists of several columns, each representing a subject, and a number of rows, each corresponding to one turn of interaction. Different symbols were used to express various control actions taken by the subjects, so that all relevant events occurred during experiments can be shown in the diagram.

A typical control flow diagram is illustrated in Figure 2. This diagram has three columns (for the three subjects) and 17 rows (for the 17 turns of interaction). It can be seen quite clearly who owns the control authority, when the control switches, and when conflicts occur. We found this kind of graphic representation an excellent way to summarize observations of interactions among subjects. The use of the control flow diagram also makes it easier to calculate the recognition rate of subjects, and to examine how subjects deal with conflicts.

4. Findings

4.1 User Preference

In phase C (*mixed control phase without remote control*), we found that subjects often chose to use the master controller, rather than the shared mode. Only two of the twelve groups switched to the share mode at some points of

16

the experiments. In the mixed control mode with remote control, we found that seven out of the twelve groups executed tasks using both remote and voice control in parallel. The other five groups chose to use remote control exclusively. We also found that four of twelve groups use the master controller to issue the command when a particular function cannot be found on the remote controller immediately.

4.2 Influence of recognition rate on usage behavior

In phase A (*shared control phase*), we observed a large number of control switches from one subject to another. To analyze the primary cause for these switches, we calculated the frequencies of control switches due to recognition error against the total number of control switches. In total, 49% of the control switches occurred when a subject encountered a recognition error.

Subjects' preference of the master mode over the shared mode in phase C is another indication of how recognition performance dominates the user behavior. To substantiate this point, we conducted a separate experiment to compare the recognition rates in these two modes. The recognition rates in the master mode were clearly better than those in the shared mode due to a much closer talking distance. The preference of the master mode over the hands-free shared mode due to better recognition rates indicates that the hands-free advantage for voice control is beneficial only if the recognition rate reaches a satisfactory level.

4.3 Conflicts

In our experiments, we concentrated on the first type of conflicts, i.e. conflicts with issuing commands. We recognized it as a *conflict* whenever two or more subjects issue (same or different) commands at the same time. For the twelve groups of subjects, we observed a total number of 29 conflicts. We considered a conflict to be over, when one of the subjects gains control for issuing commands.

We found that, out of the 29 cases of conflicts observed, thirteen cases ended up with no change in control authority. For the rest of sixteen cases when the control authority did change hand, nine cases ended with control taken over by a subject *not* involved in the conflict, and only seven cases with control taken over by a subject involved in the conflict.Observations in these experiments show that conflicts can be resolved rather quickly. Of the total 29 conflicts, all except three were resolved within eight seconds. The other three conflicts last 21, 24, 31 seconds, respectively. For example, the two conflicts in Figure 5 last three seconds and seven seconds respectively.

During our experiments, conflicts did not seem to cause any problem to the subjects in executing their tasks. Even though the tasks were designed to stimulate competitive behavior, behavior patterns developed in all four phases are rather *cooperative*.

4.4 Other influential factors

In all four phases of experiments, a subject's personal character also influenced how actively they tried to issue commands. In phase A (*shared control phase*), we found that subjects with a high recognition rate did not necessarily issue commands all the time and dominate the control in the group. Subjects with a low recognition rate may try hard to issue commands. In these cases, a subject's character affects the control behavior during the experiments. Comparing groups of friends with groups of family members, subjects of family members seem to be more apparent in their status than those of friends. The status in the family is transformed into dominating authority. We found that elder brothers or sisters directed the younger ones to issue commands for them even though they themselves have a better recognition rate. In phase B (*master control phase*), subjects without the master controller usually solicited help from the master user by directly naming the commands, indirectly expressing what they wish to achieve, or by using gestures.

4.5 General opinions from subjects

Comparing the shared mode with the master mode, only seven of the thirty-six subjects regarded the master mode unnecessary because of its similarity to remote control. For conflict resolution, most subjects agreed that the use of the master controller could help solve the problem of people fighting for control authority via voice. Some subjects even suggested that the system should recognize specific users in the shared mode for conflict resolution. Lower recognition rates makes the shared mode difficult to use due to a longer talking distance. The ambient noise can easily get into the microphone and consequently affect the recognition rate. This is especially critical when one subject issues the command with others chatting on the side. A few subjects indicated that the master control mode provides some kind of privacy for them to issue commands while they have to do that publicly in the shared mode.

4.6   Impact of remote control

In the mixed control phase with remote control, seven of thirty-six subjects insisted on using the remote controller to execute the tasks exclusively. Four of the thirty-six subjects indicated that using remote control and voice control in parallel makes it convenient for user to choose the suitable mode for specific purposes. In general subjects found remote control a natural fall back to voice control. However, due to the lack of a proper introduction, most subjects found the remote controller difficult to use. Usually subjects spent the first half of the phase in getting familiar with the remote controller. Thus the impact of the remote control on task execution often started in the 2nd half of the phase.

5. Conclusion

User behavior and preference for voice control in a multi-user environment was examined systematically via usability studies. Twelve groups, each consisting of three subjects who are either friends or members of the same family, were tested. A graphic representation of the control flows has been developed for process analysis. The results reveal that even for competitive tasks, conflicts among users were resolved within at most two to three turns. The majority of conflicts (26 out of 29, or 90%) ended in 8 seconds. Recognition performance was found to be a dominating factor in control switching and in choosing different control modes. In total 344 control switches from one subject to another, 168 switches (49%) were caused by recognition errors. Evidence also shows that most subjects preferred the master mode to the hands-free shared mode due to better recognition rates. This is a clear indication that the hands-free advantage for voice control is beneficial only if satisfactory recognition performance can be reached.

References

1. Cheyer, A., Julia, L. (1998). Multimodal Maps: An Agent-Based Approach. In *Multimodal Human Computer Communication: Systems, Techniques, and Experiments*, Berlin, Springer, Lecture Notes in Artificial Intelligence 1374, pp. 111-121.
2. Murata, A. (1998) "Effectiveness of Speech Response Under Dual-task Situations," *International Journal of Human-Computer Interaction*, v.10, n. 3, p. 283-292.
3. Bunt, H., Ahn, R., Beun, R.J., Borghuis, T., van Overveld, K. (1998). Multimodal Cooperation with the Denk System. In *Multimodal Human Computer Communication: Systems, Techniques, and Experiments*, Berlin, Springer, Lecture Notes in Artificial Intelligence 1374, pp. 39-67.
4. Higgins, E.T., Kruglanski, A.W. (1996). *Social Psychology: Handbook of Basic Principles*, New York, The Guilfor Press, pp. 745-776.
5. Siroux, J., Guyomard, M., Multon, F., Remondeau, C. (1998). Modeling and Processing of Oral and Tactile Activities in the GEORAL System. In *Multimodal Human Computer Communication: Systems, Techniques, and Experiments*, Berlin, Springer, Lecture Notes in Artificial Intelligence 1374, pp. 101-110.
6. Jouke, R., Pete, D., Paul, K. (1998). Using Automatic Speech Recognition in Consumer Products. Philips Nat Lab Report 7033.
7. Martin, J.C., Veldman, R., Béroule, D. (1998). Developing Multimodal Interfaces: A Theoretical Framework and Guided Propagation Networks. *Multimodal Human Computer Communication: Systems, Techniques, and Experiments*, pp. 158-187.
8. Oviatt, S. (1994). Toward Empirically-Based Design of Multimodal Dialogue Sytems. *In Proceedings of AAAI'94-IM4S*, Stanford, pp.30-36.
9. Oviatt, S., Olsen, E. (1994) Integration Themes in Multimodal Human-Computer Interaction. In *Proceedings of ICSLP'94*, Yokohama, p. 551-554.
10. Vogten, L.L.M., Kaufholz, P., Bekker, M.M., de Ridder, H. (1998). Voice Control of an Audio Set, Phillips Nat Lab Report 7042.

A practical approach to design for Universal Access: the Information Point case study

Simeon Keates[a], Patrick Langdon[a], P. John Clarkson[a], Peter Robinson[b]

[a]Engineering Design Centre, Department of Engineering, University of Cambridge, Trumpington Street, Cambridge CB2 1PZ. United Kingdom

[b]Computer Laboratory, University of Cambridge, New Museums Site, Pembroke Street, Cambridge CB2 3QG. United Kingdom

Abstract

It is known that many products are not accessible to large sections of the population. Designers instinctively design for able-bodied users and are either unaware of the needs of users with different capabilities, or do not know how to accommodate their needs into the design cycle. The aim of this paper is to present a 7-level design approach for implementing design for universal access. A summary of the principal methods for designing for users with different capabilities is given and a case study highlights the use of the design approach.

1. Introduction

There are several existing approaches for designing more inclusive interfaces. However, there are shortcomings of each of these approaches that prevent each of them from being used to provide the definitive design approach that designers can use in *all* circumstances. The principal weaknesses stem from the targeted nature of the approaches. The existing design approaches are often targeted at specific population groups or impairment types. For example, Transgenerational Design [1] focuses on design for the elderly. Alternatively, they focus on specific impairment types, as for Rehabilitation Design [2]. They can also be targeted at specific cultures. For instance, Universal Design [3] dominates US/Japanese approaches to inclusive design, whereas Europe has generally tended to develop other methods, such as the User Pyramid Approach [4]. The prescribed methods of application of the existing methods are often vague. For example, Universal Design is more of an ethos than a rigorous, systematic design approach. There are very few structured descriptions of the implementation of Universal Design in more detail than broad design objectives [3]. Consequently, while combined the existing approaches may offer complete coverage of the population needs, individually they do not. Therefore, there is a need for a new approach that draws on the strengths of the existing inclusive design approaches and offers practical and measurable design criteria.

2. The 7-level design approach

To meet the need for a new design approach, the 7-level approach has been developed based on the known stages of interaction [5] and usability heuristic evaluations [6]. Developing an interface for universal access involves understanding the fundamental nature of the interaction. Typical interaction with an interface consists of the user perceiving an output from the system, deciding a course of action and the implementing the response. These steps can be explicitly identified as perception, cognition and motor actions [5] and relate directly to the user's sensory, cognitive and motor capabilities respectively.

To produce a new design approach, these interaction components have to be combined with the 3 basic stages of design: (1) define the problem; (2) - develop a solution; and (3) evaluate the solution. Initially, a 5-level design approach was developed that divided Stage 2 of the design process into three constituent steps that address each of the interaction steps [7]. The 5-level approach has been successfully applied to the design of a software interface for the control of an assistive technology robot [8]. However, further resolution of the approach is possible by separating the problem definition into two steps: defining the user wants/desires and defining the user needs, i.e. the required functionality of the product interface. To reflect these separate objectives, the evaluation procedure similarly needs to become two stage: verifying that the required functionality is provided and validating that the system satisfies the user wants/desires. This generates the 7-level design approach shown in Figure 1.

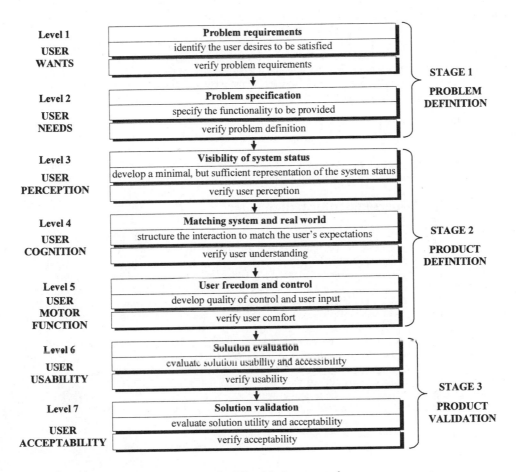

Figure 1. The 7-level design approach.

2.1 Applying the 7-level approach

The 7-level approach addresses each of the system acceptability goals identified by Nielsen [6]. The approach has been applied to a number of case studies including the design of a software interface for an interactive robot [7] and the review of an information point described later.

Level 1 defines the user needs, that is the social motivation for designing the product. This can be identified through softer, sociological assessment methods. Questionnaires and interviews are effective for identifying such user needs.

Level 2 focuses on specifying the required utility of the product. Traditional engineering requirements capture techniques can be used, as can task analysis [5, 6]. Alternatively, functional assessments of rival products or observation of existing methods can provide insight into the necessary functionality.

Levels 3 to 5 focus on the stages of interaction. Usability and accessibility techniques can be applied directly to these levels, as can anthropometric and ergonomic data and standards. Prototypes of varying fidelity play a key role in these levels.

Level 3 addresses how the user perceives information from the system. This involves assessing the nature and adjustability of the media used, their appropriateness for the utility, and the physical layout. Anthropometric data are important to ensure that the output is in a position that the user can perceive it. Ergonomic and empirical data from trials are also necessary to ensure that the stimuli are intense enough to be perceived. Ideally, environmental conditions, such as lighting and noise, also need to be identified and modeled.

Level 4 assesses the matching of the system contents and behavior to the user mental model. Once the output channels are defined, the content/utility can be added to the system and evaluated because the functionality for monitoring the system is in place. Literally the user can see/hear/etc. the data. Common techniques to map the user system behavior to user expectations include cognitive walkthroughs.

Level 5 focuses on the user input to the system. As with level 3, this involves assessing the nature and adjustability of the media, their appropriateness for the utility, and the physical layout. Again anthropometric measures are important to ensure that the input media are within the operating range of the user. Ideally, empirical data from user trials needs to be gathered to evaluate the effectiveness of the input solutions. These can be supported by adopting user modeling techniques. Where user trials are impossible, suitably calibrated user models can be used to provide design data.

Level 6 involves the evaluation of the complete system to ensure satisfactory utility, usability and accessibility. Formal user trials and usability/accessibility assessments are essential at this point, before the design can progress to the final level, 7.

Level 7 assesses the resultant system against the user needs. This mirrors Nielsen's social acceptability requirement. Softer, more qualitative approaches are generally needed, such as surveys, interviews and questionnaires.

3. Case study: the personal information point

The Post Office (TPO) is one of the UK's largest employers, directly employing 220,000 people. As with all large UK companies, TPO is expected to comply with the Disability Discrimination Act, which stipulates that all companies offering services to the public should not bar use of those services on the grounds of disability. The age profile of TPO customers is biased towards the over-65's, with a large proportion of customers visiting post offices to collect their pensions. With the increased variability in capability associated with aging, the incentive for inclusive design is very great. TPO is aiming to introduce automated customer information points (IPs) into post offices to augment the current range of services. TPO is also very keen to ensure that the product is as usable and accessible as possible, not simply for DDA compliance, but also to engender customer satisfaction. An initial concept system, Figure 2, has been developed by a design company for review as a possible solution to TPO's requirements. The concept IP consists of three display heads mounted on a central free-standing pedestal of fixed height. Each unit head has a small LCD panel mounted at a fixed angle, with three buttons on either side of the screen for input and a telephone handset for audio output. The output is primarily recorded video footage with accompanying soundtrack. The purpose of the case study here is to examine the usability of the prototype and how this would have been affected by the use of the 7-level approach.

Figure 2. The concept Information Point

The 7-level design approach can be divided into the following steps for the IP:

Level 1	What is the aim of the IP?
Level 2	What are the IP system requirements?
Level 3	How does the user receive information from it?
Level 4	Does the user understand what is happening?
Level 5	How does the user enter information and what are the physical demands?
Level 6	Does the IP meet the functionality and usability requirements?
Level 7	Does the IP meet the stated aim?

3.1 Level 1 - System aims

The aims of the IP are more sociological than functional and consequently open to interpretation. One of the aims is to provide TPO customers with an introduction to modern technology as a precursor to post office transactions becoming increasingly automated. the subsidiary aim is to reduce queue length for the counter service in a cost-effective manner. These aims must be achieved in such a way that meets legislated accessibility requirements. No explicit usability or performance criteria were stipulated.

3.2 Level 2 - System requirements

The target user of the new Information Point is the typical post office customer. Owing to the age profile of typical post office customers, the real target person is likely to be older and have at least one major impairment, most likely visual, auditory or motor. The user are likely to be more resistant to change and to new technology.

The concept IP was designed to be bright, open and attractive to tempt customers to use it. However, to be of real use, the IP also needs to offer easy access to significant content, with sufficient bandwidth for the interaction. For maximum accessibility and legislative compliance, the design process should be entirely driven by user needs and product objectives. The concept IP was developed to meet the broad aims identified in Level 1. This implies that the concept IP may not offer the necessary functionality or usability because these issues have not been explicitly defined by TPO at this stage.

3.3 Level 3 - User Output

The concept IP used a combination of video footage displayed on the screen and accompanying soundtrack on the telephone handset. Neither output mode was sufficiently complete to provide all the information on its own. Consequently, the 1.7 million adults in the UK with visual impairments [9] and the 3 million hearing impaired adults would be unable to use the system successfully. This does not include those who are environmentally impaired through background glare or noise. Using user-aware design, a mix of output measures would provide the best population coverage. For instance, the inclusion of text subtitles to supplement audio output would reduce the number of potential users being excluded.

The other major concern when addressing user output is that the output solution chosen is positioned in an accessible manner. The number of people excluded by the fixed height and angle of the concept IP display is a good example of this. The display viewing height is set rather high at 1610 mm, with a lower bound on the viewing height of 1451 mm. This excludes half of women of 65 and approximately a quarter of all adult women, rendering 6 million women unable to view the display [9].

3.4 Level 4 - User understanding/cognition

With the concept IP not having any real content at this stage, detailed evaluation of the user understanding of the system was not possible. However, it was possible to estimate the likelihood of good user understanding being attainable by the IP. The structure of the interaction, and hence of the interface, is partly dependent upon input and output modes chosen. Under the 7-level approach, only the output modes will have been specified by this stage. Looking only at the output capabilities, the small IP screen limits the richness of content that could be achieved by a larger display. It also requires many more levels of searching to reach a target goal because fewer options and less information can be displayed simultaneously. However, with the focus on the technology in the design of the concept IP and the decision to include only six buttons for input, the functionality of the software interface is arbitrarily limited to selection from a list. However, if the list is well designed, then the system has the potential at least to be easy to learn.

3.5 Level 5 - User input

The prototype IP only offers button input through a limited number of buttons. Text entry is impossible without the use of voice recognition technology or scanning input to a keyboard emulator. Some text entry requirements could be circumvented, though, by the use of smart cards that contained information about the user and could be inserted into the IP. This is comparatively restrictive. The other issue of concern is the physical positioning of the input for user comfort. The free-standing design of the prototype is likely to affect anyone who has difficulty standing for extended periods of time. The requirement that both hands be used eliminates anyone who needs to use a walking stick for support, some 57% of the over 60's who make up the majority to TPO's customers. Across the whole UK population, 10.7 million people have difficulty standing and would be unable to use the prototype for any extended period of time. The associated reaching and dexterity requirements for operating the buttons and the handset also excludes a further 3 million people [9].

3.6   Levels 6 & 7 - Evaluation

This review process represents the first stage of the evaluation of the concept IP. The IP is currently undergoing a re-design to take into consideration the accessibility flaws revealed by this analysis. The first evaluation prototypes of the new design should be out later this year for trial periods in three local post offices.

4.   Conclusions

Having applied the 7-level approach as a review, a number of pointers for the re-design of the prototype have been identified. Levels 1 and 2 show that the existing design does not offer sufficient functionality to meet the requirements. Levels 3 to 5 illustrate the importance of being aware of the user needs throughout the design process. Approximately 21 million potential users could have been excluded from using the IP because of the failure to consider different user physical capabilities at the outset. That is 45% of the total UK adult population and does not include those who may be environmentally impaired.

The scale of exclusion identified in the case study shows how important it is to implement a structured approach to user-aware design. By examining the interaction in terms of perception, cognition and motor functions, the required user sensory cognitive and motor capabilities can be identified. It also illustrates how arbitrary design choices, such as requiring the user to stand, can exclude large numbers of users. Fortunately, potential, and often simple, remedies can also be identified. For instance placing the unit at desk height and having chairs nearby would remove almost all of the difficulties for the 17.8 million people with lower motor capabilities and physical attributes from the designer's target user.

4.1   Further work

This paper has shown the level of exclusion that can be avoided through the application of a methodical user centered design approach such as the 7-level approach, and highlights areas where the greatest benefit can be achieved. Although only one case study was presented, the approach has been applied to a number of other case studies and has proven to be of benefit in all so far. However, one constant is that the approach was being applied by usability experts. It is recognized that the 7-level approach is a high level approach and requires usability expertise for its interpretation and application in its current form.

Consequently, the next stage of development is to develop the approach so that it can be applied by people without exclusively usability expertise, such as designers. The intention is to explicitly identify the different usability assessment methods that can be used in conjunction with specific levels of the approach and allow the designer to choose from whichever method is most suitable for their abilities, engineering discipline, resources and the task in hand. The ultimate aim is to populate the approach with modular steps that are focused towards progressing through the 7 levels and that will allow the approach to be applied to a wide range of products, whether hardware or software.

References

1.   Pirkl, J. (1993)Transgenerational Design: Products for an aging population, Van Nostrand Reinhold, NY.
2.   Hewer, S., et al. (1995) The DAN teaching pack: Incorporating age-related issues into design courses. RSA, London.
3.   Bowe, F.G.(2000). Universal Design in Education, Bergin & Gavey.
4.   Benktzon, M. (1993). Designing for our future selves: the Swedish experience. Applied Ergonomics 24, 1, 19-27.
5.   Card, S.K., Moran, T.P., Newell, A. (1983) The Psychology of Human-Computer Interaction. LEA, Inc.
6.   Nielsen, J. (1993). Usability Engineering. Morgan Kaufmann Publishers, Inc, San Francisco, CA.
7.   Keates S, Clarkson PJ, Robinson P. (1999). Designing a usable interface for an interactive robot. Proceedings of the 6th Int'l Conference on Rehabilitation Robotics. 156-162.
8.   Keates S, Clarkson PJ, Harrison LJ, Robinson P. (2000). Towards a practical inclusive design approach. Proceedings of the 1st ACM Conference on Universal Usability, 45-52.
9.   ADULTDATA - The handbook of adult anthropometric and strength measurements (1998). Department of Trade and Industry.

Creating a new approach to design and a new vision for networked appliances and net-based services

Naomi Kihara

R & D Department, Corporate Design Center
Matsushita Electric Industrial Co., Ltd..
14thFloor, Twin21 National Tower
Shiromi, Chuo-ku, Osaka, Japan

Abstract

In this report, we introduced our unique activity to create a new vision for networked appliances from designers point of view. Which includes new design philosophies based on human-oriented and universal perspectives, with some ideas, and visual works.

1. Background

Let me start by asking a question. What kind of non-PC device would be ideal for using the net-based services in this wired society which is awash with information? And do you have a clear image of what our new networked lifestyle might look like in future? In this research project, we tried to offer a convincing vision of a new lifestyle and to set up a new design philosophy based on a more human-oriented and universal perspective.

In response to the accelerating trend toward a fully networked world, Matsushita Electric Industrial Co., Ltd. is undertaking a wide range of projects to provide IT-based products and services to help make consumers' lives more convenient and comfortable. Part of our effort is focused on developing networked appliances for home and mobile networks.

Don't you feel a little uncomfortable with the idea of "networked" appliances? If so, it's probably because we are concerned that home appliances might undergo the same rapid development as personal computers did.

People's uneasiness about the impersonal nature of technology revolution urged us to launch this project, which is aimed at crafting a new approach to designing Networked appliances based on completely new thinking. Our aim is to contribute to realizing a more relaxed lifestyle through human-centered technologies.

2. Course of action

In this project, we teamed up with an independent design office, AXIS, for the following two reasons.

(1) We needed a proposal that could impress Matsushita's originality on people both inside and outside the company. So we thought we would be better advised to seek third-party evaluation.

(2) We wanted a partner remote from us in both geographical and specialization terms that could complement our expertise so that we could exchange information that is new to each other.

We spent half a year shaping and refining ideas for new Networked appliances and Net-based services, drawing on the outcomes of our market research.

Eventually, our ideas boiled down to three basic concepts.

We believe we have come up with innovative design concepts based on unique and intriguing ideas for developing products that are easy and simple to use, and which add more fun to life.

Matsushita's design section has compiled a guide for our universal design standards, which includes the following requirements.

- Matsushita's design should make it easier to lead an independent life.
- Matsushita's design should help people to enjoy a more human-centered life.

The concepts we have developed are designed to fulfill these requirements.

24

3. Concept: a doorknob

3.1 Story

A doorknob is the handle of a door.
It announces that we are facing a door.
Every door has a doorknob...
...and anybody can open the door by using the doorknob.
When we grip a doorknob to open a door, we do so by our own choice.
We believe our future networked appliances and net-based services need to be "as simple and easy to use as a doorknob."
An appliance that warns us before it suffers a breakdown, for instance, would make us feel more relaxed and comfortable with the idea of using network-enabled versions of familiar appliances.
We would feel very easy and comfortable with an appliance if we could
- connect it to the Internet by simply lifting a handset
- repair it or have it repaired by following the instructions given.
- break the connection by simply putting down the handset
We can develop universal, human-centered designs for both products and services by trying to make them easy to get a grip on, open to decisions by users with little technological knowledge, and comfortable to use.

3.2 Concept (concept for networking)

We have adopted the concept of a "doorknob" as a symbol of simplicity and operability and of the promise of a more relaxed lifestyle. The "doorknob" symbolizes a positive action and a fascination in something that will be gained from the action.
A "doorknob" can also be interpreted to represent the analog world, as symbolized by the familiar "switch."

3.3 Supporting Concepts: "Discommunication" (a concept relating to products) and "Identification" (a concept relating to customers)"

In our lexicon, "discommunication" means that we have a system within ourselves that makes us cut or find another way of communicating when they become too much of a burden for us.
The explosive growth of high-tech-based communications systems and the mind-boggling amount of information traveling around the globe at the speed of light have started provoking a negative reaction from us. We need a new mode of communication. One solution to the problem is to enable discommunication.
By creating an alternative path, discommunication, in products and services to liberate users from current communication that requires too-much effort to gain information, they will experience first-ever operations.
The aim of discommunication is to urge customers to experience a renewed awareness of their minds and bodies, to remind them of past experiences, and to stimulate their intuition.
This will become a system that can offer a new option (relaxation) as an alternative to the current communications environment.

4. Achievements

We have already introduced some of these concepts and ideas to people both inside and outside our company in a film showing our images of "a comfortable lifestyle with networked appliances." We have also reported our activities to top executives through two in-house exhibitions.
We have filed patent applications for four of the ideas that have been born out of this project.
We are going to continue our effort to advertise our new universal approach to design to audiences both inside and outside the company and to develop and offer innovative solutions for both products and services based on our unique "doorknob" concept and our special designing and R&D perspectives.

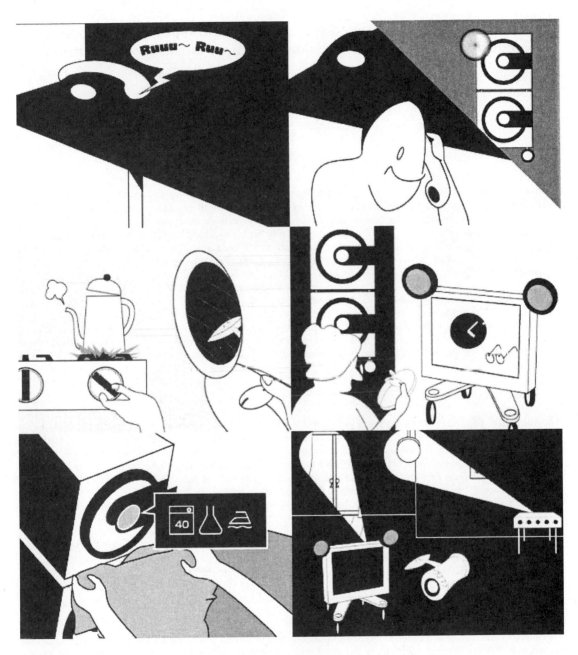

(Images from the film)
Presented by CDC R&D Knob Project Team + AXIS Design Dept.

The marriage of activity theory and human-computer interaction: understanding practice to develop computer systems for workgroups

Jayne Klenner-Moore

King's College, Department of Mass Communications and
Media Technologies Wilkes-Barre, PA USA

Abstract

Activity theory is the basis for understanding human learning through the discovery and observation of how humans develop through the use and creation of tools within their culture. Activity theory and cultural psychology give us a way to look at how we learn with the cultural tools in the context of the work we do - everyday living.

In today's organizational society, this goal may be achieved in a virtual collaboration of workers who are not necessarily working at either the same time or the same place to achieve a goal. The cognitive, collaborative tool-the computer allows us to work in a new way that exploits the constraints of place and time.

To understand the nature of Human-Computer Interaction (HCI) we need to understand the framework in which the system is employed. We also need to ask ourselves how the fundamental definition of work has changed.

Sylvia Scribner describes practice as "a socially-constructed activity organized around some common object" (Scribner, 1997 in Tobach et al) Scribner encourages us to look at the practices themselves in order to understand the way that we acquire knowledge and skills as we work to become experts in these areas. It is important that we look at work practices as the main area in which adults are active. This paper will look at how the computer acts as the cultural tool for mediated action and how understanding practice can help us to develop better systems for workgroups.

1. Introduction

The purpose of this paper is to support the notion that Activity Theory research can in fact help us to build better interactive computer systems for workgroups and understand the changing nature of work with computers. Traditionally the division of labour consisted of multiple people working toward a common goal. For example Leont'ev's story of primeval collective hunt which includes the beaters performing a integral part of the hunt, if not actually killing the prey. In today's organizational society, this goal may be achieved in a virtual collaboration of workers who are not necessarily working at either the same time or the same place to achieve a goal. The cognitive, collaborative tool-the computer allows us to work in a new way that exploits the temporal constraints of time and place.

Kutti asks the question "why now" when talking about the social-cultural investigation that is blossoming at this point in human history. She answers this question this way, " The relationship between work and information technology is nowadays very close- in fact, the implementation of IS (Information Systems) may be the most common means of changing the way work is done in organizations" (Kuutti, 1999). Can you think of any job that does not, currently involve the use of a computer? This then at the very least becomes an artefact or tool worthy of exploration.

To understand the nature of Human-Computer Interaction (HCI) we need to understand the framework in which the system is employed. We also need to ask ourselves how the fundamental definition of work has changed. Do the definitions of work and division of labour that come from Marx still hold true? How our activities are mediated with the computer is evident even in our daily language. For instance, in the 1970's for example our parents might have had on their business cards a business phone number, a home phone number, and if lucky maybe a fax number for contacting them. Today most business cards hold all that information plus an email and in many cases a URL. When you say goodbye to students for holiday, do you ask them to write you a letter or send an email? Even collaboration on conference papers can be done virtually with email and file sharing.

2. Practice

Sylvia Scribner describes practice as "a socially-constructed activity organized around some common object" (Scribner, 1997 in Tobach et al) Scribner encourages us to look at the practices themselves in order to understand the way that we acquire knowledge and skills as we work to become experts in these areas. It is important that we look at work practices as the main area in which adults are active. In Scribner's research, it was important that the study of labour activity be analyzed. She does this by looking at the two mutually dependent aspects of mediation in labour activity as derived from Marx and Engels, these are the use and making of tools and that the labour process is performed in the joint collective activity. As Engels said " Labour begins with the making of tools" (Scribner 1997 in Tobach p. 4).

One way to look at how the practice of work has changed is to look at how we structure the knowledge of an organization. The introduction of knowledge-based systems into organizational structures has an impact on the success of the organization. The central characteristic that sets Knowledge-Based Systems (KBS) from other Information Technologies (IT) is that its focus is on knowledge, not information (Hendriks, 1999). Hendriks defines KBS as "computer systems that use a formal mechanism to represent or simulate specific aspects of human knowledge, and to apply these representations in actual problem situations". Knowledge-based systems are an interesting source of mediated activity as they are a prime example of how we can use the computer as a part of a larger system, which involves a way of change in the way we do work.

One other interesting point that Hendriks mentions is that "Organizational capabilities rather than product-market combinations have become the lasting source of advantage". This is a demonstration of how our "global economy" has moved from the Industrial Age to that of the Information Age and how the use of computers has helped to implement that change. Would this be a true transformation as Davydof defines it? The evolution has also changed the way we need to look at the division of labour. If we agree with Scribner then, in that the study of the practice of work will help us to understand the developmental process of the individual and the culture of that work, then it would follow that we can not ignore the inclusion of the computer in that study.

This use of the mediating tool of calculator, computer, intranets, chat rooms give us a solid physical protocol or means for studying human-computer interaction under the principals of activity theory. It presents us with Wertsch's agent and agency (Wertsch, 1998) as the unit of analysis.

The importance of using AT to further study work in the context of HCI gives us a richer description of the practice of work, the understanding of the culturally situated individual with the group and a deeper understanding of the user and the interface. As Aboulafia, Gould and Spyrou describe, "Activity Theory on the other hand, takes a broader view of the 'technisation' of human operations and places HCI within this wider framework" (Aboulafia, Gould, Spyrou, 1995).

3. Describing Activity

It would be pertinent to include at this point a brief overview of Leont'ev's three-level model of activity. The hierarchy of the model is that the top level of activity is object-related goals, the middle level is driven by a goal, and the bottom level of operations is driven by a goal, which is driven by the tools, and objects that exist in the context of the activity.

Table 1 Three level hierarchy adapted from http://www.marxists.org/archive/leontev/index.htm

Level	Oriented Toward	Carried Out By
Activity	Object/Motive	Community
Action	Goal	Individual or Group
Operation	Conditions	Routinized Human or Machine

This is a very fitting area for discussing the computer as tool, for it also acts as mediator between the individual, group and community and a storage house of symbols and communication. Table 1 also makes an important addition, that of the outcome. How in fact does the use of the computer act on or change the outcome of the activity, if in fact it does? Think of your own work practice what are the tools; what is the object, what is the outcome of your work practice? This area as well can be looked at when we are studying the interaction of computer and human. If you want to get into an even more interesting conversation of HCI, you could look at a very artificial environment such as the one we use with artificial agents. The agent acts as a section of the division of labour, they interact in a virtual community to achieve a goal-oriented outcome based on rules and defined by the domain.

4.1 Computer Supported Collaborative Work

We have mentioned previously, the computer can play an important role in the activity of work. The area that we will be looking at here, allowing that there are myriads of areas to look at this problem such as Computer Supported Collaborative Learning (CSCL), situated cognition, among many others. Computer supported collaborative work is the study of how we work together with the computer as a tool.

If you download MSN's latest browser you are automatically "hooked into" a community of users. This new browser allows us to work together in many ways the structure of the MSN browser offers a new opening interface, which allows multiple users to sign on if you have a hotmail account.

This service allows for a server space in which students can upload files, photos, carry on live chats, leave threaded messages on a bulletin board and create web pages or links of interest to the community. In this instance, the computer and its online system have become an additional community space, workspace and social space.

For another example of the use of computer software that enables the function of work, look at the latest version of Dream Weaver. This version of the software allows for the ability of multiple workers (developers) to work on a site by the use of a system that tells the group when a page is out being worked on and when its back thereby reducing the mistakes usually made by versions being updated remotely by members of a team.

For another example of the use of computer software that enables the function of work, look at the latest version of Dream Weaver. This version of the software allows for the ability of multiple workers (developers) to work on a site by the use of a system that tells the group when a page is out being worked on and when its back thereby reducing the mistakes usually made by versions being updated remotely by members of a team.

4.2 Workspace analysis for collaborative settings

Following along the work of Alan Dix and Devina Ramundy we can look at better designing the interfaces that we use when developing interfaces and systems for remote use." The rapid growth in world-wide communications has enabled cooperative users to collaborate and access shared resources when distributed remotely" (Ramundy, 2000). Remote work may involve looking at cross-cultural ethnographies to better understand the roles of worker in different contexts and cultures. We need to look at the practices from multiple angles to better understand the development of creating shared workspaces.

If we follow that memory and thinking in daily life are not separate from, but are part of, doing (Scribner in Tobach, 1997), then how we work at work and how our work is mediated by the tools we use is the best way to understand the nature of the activity. If that work involves the use of the computer, then how we work with the machine also should be looked at to understand the how in how we work. In our Dream Weaver example, development teams will often work on multiple pages of Web site make revision and in the end need to bring together a finished product. By having the system monitor where things are and what the current version it is, distributes some of the cognitive load away from the user and onto the system.

Engeström and Middleton talk about the sociology of work and they give us two broad traditions the first is the macro level discussions of the impact of technological development of the skills and organization of work, the second is the micro sociological analysis of locally constructed and negotiated work activities (Engeström and Middleton, 1998).

Practice becomes an interesting starting point for the study of work itself. The study of practice can involve many fields of expertise, not only the Activity Theory researcher. Engeström and Middleton discuss the issue of human agency in the study of work and practice and call for a new framework of concepts that speak directly to practice (Engeström, Y. and Middleton, D., 1998).

In many ways, the isolation of the computer from human activity does not allow for a complete understanding of the interaction between the two. "Many of the problems associated with the information processing model of cognitive science can be related to incompleteness and the narrow base of this approach" (Aboulafia, Gould, and Spyrou, 2000).

5.1 Discussion

Bødker and Grønbeck propose that we study situations to help us develop computer technology for the workplace. Their premise is that we should develop using collaboration between the user (worker) and the designer to help design in the development of prototypes. They suggest four categories of situations for analysis.

1. Situations where the future work situation with a new computer application is simulated to investigate the future work activity.
2. Situations where the prototype is manipulated and used as a basis for idea exploration.
3. Situations focusing on the designers' learning about the users' work practice.
4. Situations where the prototyping tool or the design activity as such becomes the focus (Bødker, Grønboek, 1998).

This area is a key part of the research in adaptive graphical user interfaces and how we can develop them to work in context with the user. This are in particular is ripe for further development within the scope of Activity Theory.

5.2 Multiple Perspectives In Research

As Davydof (In Engestrom, 1999) mentions one problem of activity theory is the inderdiciplinarity and organizing the study of human action. We need to conduct research that involves the different perspectives of the different disciplines, such as psychologist and sociologists, educators and HCI professionals. We can look at research involving human-computer interaction and its implications for work practice among other areas.

6. Conclusion

The use of AT as a theoretical background and way of looking at work can give us a better picture of the actual practice humans perform in everyday mundane tasks. We can therefore utilize the different observational and dialectical approaches inherent in AT to look at the way in which humans interact with computers. It is important also that we look at the way the tool is developed to support a network of users and networks in general. Activity theory and cultural psychology give us a multi level view of how the infrastructure works physically, technically, and culturally. Because of the many new uses of technology it is imperative that we begin to look at those areas such as HCI, CSCW as well in a collaboration with the social sciences and work together as Davydof suggests to find a better picture of the culture of work.

References
Aboulafia, A., Gould, E., Spyrou, T. (1995). *Activity Theory vs Cognitive Science in the Study of Human-Computer Interaction*. IRIS Information systems research seminar in Scandinavia, Gjern, Denmark, http://iris.informatik.gu.se/conference.
Bødker, S., Grønboek, Kaj (1998). Users and designers in mutual activity: An analysis of cooperative activities in systems design. *Cognition and Communication at Work*. Y. a. M. Engström, D. Cabridge, The Press Syndicate of The University of Cambridge: 130-139.
Bødker, S., Grønboek, Kaj (1998). *Users and designers in mutual activity: An analysis of cooperative activities in systems design.*

Engeström, Y. (1999). Activity theory and individual and social transformation. *Perspectives on Activity Theory*. Y. Engeström. Cambridge, The Press Syndicate of the University of Cambridge: 3.

Engström, Y. a. M., D., Ed. (1998). *Introduction: Studying work as mindful practice*. Cognition and Communication at Work. Cambridge, The Press Syndicate of the University of Cambridge.

Hendriks, P. H. (1999). "The organisatioanl impact of knowledge-based systems: a knowledge perspective." *Knowledge-Based Systems* **12**(199): 159 - 169.

Kutti, K. (1999). Activity Theory, transformation of work, and information systems design. *Perspectives on Activity Theory*. Y. Engeström. Cambridge, The Press Syndicate of The University of Cambridge: 360 -376.

Ramundy-Ellis, D. (2000*). Temporal Interface Issues and Software Architecture for Remote Cooperative Work*, http://www.comp.lancs.ac.uk/computing/users/devina/phd.htm.

Tobach, E., Falmagne, R.J., Parlee, M.B., et al, Ed. (1997). *Mind in action: A functional approach to thinking. Mind and Social Practice Selected Writings of Sylvia Scribner*. Cambridge, The Press Syndicate of the University of Cambridge.

Wertsch, J. V. (1998). *Mind as Action*. Oxford, Oxford University Press.

Designing Internet-based Systems and Services for All: Problems and Solutions

Panayiotis Koutsabasis*, Jenny S. Darzentas*, Julio Abascal **, Thomas Spyrou *, John Darzentas *

* University of the Aegean, Department of Product and Systems Design, Ermoupolis, Syros, Greece, GR-84100, Email: {kgp, jennyd, tsp, idarz} @ aegean.gr
** Laboratory of Human-Computer Interaction for Special Needs, The University of the Basque Country, Donostia, Spain, E-mail: julio@si.ehu.es

Abstract

Most current electronic services do not address the breadth of design issues necessary to comply with 'design-for-all' concepts, despite the fact that various design-for-all tools are available. One important reason for this gulf seems to be that most of the aforementioned work has not been provided to designers and Information Technology (IT) industry in a form that can enable them to easily include it in their design processes. This paper argues that design-for-all practical tools and methods need to be presented to designers, in an easily understandable and applicable manner, which will be generic enough to cover a wide range of requirements. The paper illustrates some of the obstacles preventing designers of Internet based systems and services from designing for all and suggests the need for a design aid environment that would bring the results from this work within the reach of designers.

1. Introduction

In the area of Internet-based services, various tools have been developed aiming at aiding designers in designing applications that can be universally accessed or designed for all. For example, at the level of requirements capture, methodologies and techniques aim to assist designers to trace user needs and characteristics in order to achieve a problem formulation that can lead to universally accessible services. At the level of design and modelling, recent advances in areas, such as adaptive user interfaces and software agents, provide solutions to conceptual and engineering issues related to design-for-all. At the level of policy, various recommendations, guidelines and standards for accessible design are available and try to address the problem of designing for all in a generic manner. Finally, at the level of development, a few tools are available, which implement part of existing recommendations.

However the extent and impact of this work in Internet-based services has not yet been widely seen. Typically, most designs still tend to address "average" persons' needs. As identified in the main report of CEN/ISSS project on Design for All and Assistive Technology (available at: http://www.ict.etsi.fr/activities/Design_for_All/INDEX.htm), *'few people represent the average person, with the consequence that if a product is designed for the average person, it might be uncomfortable or impossible for most people to use it'*. 'Design for all' has been described as the concept of addressing the needs of all potential users (Stephanidis et al, 1999). The need for designed for all products, is fuelled by the rapidly increasing demand for universal access covering a broad range of user requirements, reflecting the changing nature of human activities, the variety of contexts of use, the increasing availability and diversification of information and the proliferation of technological platforms.

One significant reason to explain the gulf between the existence of various design-for-all tools and the fact that most electronic services do not address the breadth of design issues necessary to comply with 'design-for-all' concepts, seems to be that most of relevant work has not been provided to designers in a form that can enable them to easily include it into their design processes. From the perspective of the IT professional, the process of designing for an inclusive information society requires awareness and to-the-point guidance concerning these design-for-all tools.

The paper argues that design-for-all practical tools and methods need to be presented to designers, in an easily understandable and applicable manner, which will be generic enough to cover a wide range of requirements. The paper illustrates some of the obstacles that prevent designers of Internet based systems and services from designing for all, in terms of guidance, applicability and take-up of existing work. Furthermore the need for a design aid environment that would bring the relevant results from this work within the reach of designers is suggested and a first view regarding its basic concepts is discussed.

The work presented in this paper is partially funded by the European Commission in terms of project IST-2000-26211 IRIS (Incorporating the Requirements of People with Special Needs or Impairments to Internet-based Systems and Services). The authors would like to acknowledge the contributions of their colleagues from EUROPEAN DYNAMICS

S.A., University of Aegean, GMD-Forschungszentrum Informationstechnik GmbH, University of Basque Country and ISdAC (Information Society disAbilities Challenge).

2. 'Design-for-All' aids

A considerable portion of previous and existing work on design for all includes aids related to approaches for requirements capture of a wide range of characteristics of users relative to requirements for access to computer-based systems; the design and development of system models related to the adaptation to user abilities, characteristics, preferences and tasks; and a number of standards, guidelines and recommendations that include the aforementioned requirements and models in computer-based systems and services and aim to guide designers to incorporate those into their designs. This section demonstrates the need for further extensions / adaptations of this existing work so that it captures the breadth of design-for-all requirements, by describing some examples of these aids in the aforementioned areas.

2.1 Approaches for user requirements in Design for All

In many traditional software applications the potential users are either known in advance (for example for the case of a specific in-house application), or the metaphors to be employed into the functional design are well known (for example in the case of logistics applications). However, with the advent of the Internet, services have to address user requirements, characteristics, cultures at a worldwide level. In this sense, these services need to address the requirements of neither individual, nor small groups of users but rather large, diverse and multicultural user communities (Stephanidis et al, 1999). In addition, the already established metaphors of the office desktop may not be appropriate for the case of multi-user, distributed environments and interactions. Specifically in the area of assistive technology, user requirements capture usually had as a starting point the classification of users according to their disabilities, mainly aiming at developing individual solutions for each type of disability.

These types of approaches currently are being overcome by approaches that take into account user abilities and hence are much closer to "design for all" concepts. These approaches view the use of technology under "special circumstances" rather than viewing users "with special needs", which also allows for the target population to be much wider. For instance, people using a browser while driving a car may have similar access needs that blind people. In this way not only the potential market is increased, but also disabled people's needs are considered in an inclusive manner. This inclusive approach is also in accordance with the needs of people with disabilities, who are not interested in very specific products that cover their needs, which are also hard to afford (besides the psychological factors related to very specialised solutions), but they are interested to ensure that mainstream applications are widely accessible and incorporate their requirements. It is known that people with disabilities may need special equipment to access technology, but once they can, they should not have further accessibility problems. Thus the designer is committed to the objective to avoid added barriers.

Given the above problematique, an approach towards user requirements can be to use a methodology that helps to analyse, in a uniform manner, the special abilities, characteristics and constraints required by each user, in a given environment to design for a specific (set of) task(s). Userfit is a methodology oriented to requirements capture and specification for Assistive Technology design that can be described as user centred, system oriented and promoting iterative design (Poulson et al, 1996). The objective of Userfit is to help in the collection and processing of the information related to the user, by means a set of summary tools.

Furthermore, in order to accommodate the requirement for the collection and organisation of a wide range of knowledge about users, there is a need for tools that can aid the analyst/designer to better organise this knowledge and lessen the effort for timing tasks. For instance, for the case of Userfit the filling of the required summaries for a complex application is quite tedious. Therefore a tool that can significantly ease analysts' tasks is being developed (Figure 1).

2.2 Designing for adaptive system behaviour

As illustrated above, the design-for-all concept calls for the incorporation of a wide range of user requirements into service designs. However these requirements may result into conflicting or not simultaneously interoperable technical solutions. For example, the use of small fonts is not appropriate for users with sight impairments. Thus, in order to design for all, it is important that the system can provide alternative modalities, which are often related to human sensory cues.

Furthermore, even if a system supports multiple states and can cover a wide range of requirements, a given system state may not be at all accessible by users with requirements that totally conflict with the given system state, with the consequence that these users cannot change the given system state at all. For example, a default 'small fonts' setting cannot be altered by users with vision impairments, even if the system is configurable regarding its font size, because it is not at all accessible by them at the given default state. Therefore, besides supporting multimodalities, it has been realised that an approach towards achieving design for all is that of designing for system adaptivity, i.e. the system inherent property to trace and automatically adapt to user requirements.

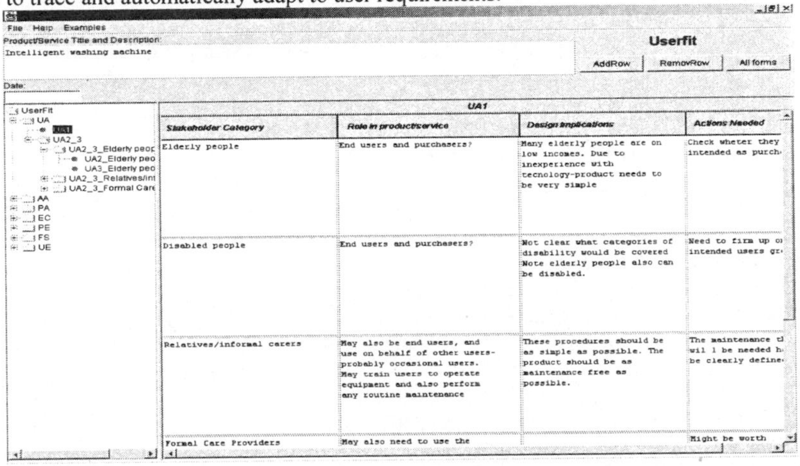

Figure 1: A snapshot of the user interface of a tool that can assist designers towards the use of the Userfit methodology.

A widely discussed paradigm for adaptive user interfaces is that of interface agents. Interface agents 'watch over the shoulder' user actions and can collaborate with them to ease the performance of a task or even undertake tasks on their behalf. Interface agents have been designed for a number of application domains. For the case of ACTS project GAIA (Generic Architecture for Information Availability) (Hands et al 2001), an interface agent was designed to provide various types of assistance to users of Internet electronic brokerage services in two contexts: the user interface context, to assist users to comprehend and manipulate the user interface: and the domain of application context, to provide users with information and advice according to their preferences (Koutsabasis et al 1999).

Despite the fact that interface agents are a promising paradigm for adaptive system design, there is still work to be done in a number of areas. For example, work on interface agents has not fully taken advantage of recent advances on user modelling research (Koutsabasis et al 2001) or mixed initiative interaction theories (Cohen et al 1998), which are highly relevant to their design and can enhance it towards behaviours and actions that are more predictable and accurate from the user point of view. Furthermore, there is a need for tools and libraries that can be used directly by designers in order to design such adaptive systems, thus contributing to design for all. For example, despite the fact that there is active research on interface agents in the last few years, there are still no specific development environments (a review of software agent developer environments, see Agentbuilder Web site, Agent construction tools: http://www.agentbuilder.com/AgentTools/index.html). Interface agent development environments should aim to ease the integration of interface agents into existing environments, ensuring a degree of adaptation to user characteristics as well as compliance with existing application requirements, such as those related to tasks to be performed and media to be employed for effective user-system interaction.

2.3   Applying accessibility guidelines to user interface design

In order to provide designers with guidance regarding the application of design for all principles, a number of recommendations, specifications and standards have emerged (Nicolle & Abascal, 2001). These types of tools aim to assist designers to reduce the large number of design options into those that follow accessibility and usability principles. These initiatives broadly fall under the following classes of work: a) general standards for usability; b) technical accessibility guidelines; and c) domain specific specifications for user profiles.An example of a general standard for usability is ISO 9241 'Ergonomic Requirements for office work with visual display terminals (VDTs)'. This standard describes ergonomic requirements on several issues such as: task, visual display, keyboard, environment, colours and dialogue and it may be a basis for Web accessibility certification. The fact that it is general in scope, results that it may require further interpretations in order to be used further.. An example of an elaboration of ISO 9241 for the evaluation of user interfaces is that of Gappa et al (1997) who have developed a guideline oriented expert-based evaluation method that prepares the requirements of the standard to be tested in about 450 test items. However, the task of placing general guidelines into a particular context is not easy and may result into nearly exhaustive enumerations of properties and characteristics. Furthermore, such tasks, related to the interpretation of general recommendations, may discourage designers to consider the use of such recommendations into their design processes.

A large number of technical guidelines are available to support the design of usable and accessible Web sites. However, as Vanderdonckt (1999) remarks *'most of these guidelines come from multiple guideline sources with various trust levels so that their application does not guarantee any improvement of either the accessibility or the usability of Web sites'.* Furthermore the extent to which each set of guidelines addresses technical issues, as well as the degrees of freedom according to which each set of guidelines can be used to aid the design process significantly vary. Stephanidis and Akoumianakis (1999) demonstrate that different engineering perspectives in the implementation of guidelines can lead to different interpretations and can influence the quality of the final products.

Work in specific application domains has also led to the production of specifications related to adaptations of services to user characteristics according to user profiles. Within the context of learning technology environments, there is much effort on the one hand, at defining metadata for educational content, and on the other, at specifying learner user profiles. For example in the distance learning sector, specification such as IMS Enterprise (available at: http://www.imsproject.org/enterprise/index.html) and IEEE LTSC LOM (Learning Object Metadata, available at: http://ltsc.ieee.org/doc/wg12/LOM_WD4.htm), aim to develop frameworks for interoperable, personalized services in this area. However there is still work to be done, for example towards joint efforts for defining metadata schemata and user profiles due to the complementarity of these two types of specifications (Konstantopoulos et al, 2001).

The establishment of the aforementioned types of recommendations requires tools, which can automatically aid designers to apply such guidelines at design time. For a few of these recommendations, such tools are already available. For example, Bobby (available at: http://www.cast.org/bobby) is a Web-based tool that analyses Web pages for their accessibility to people with disabilities following the W3C.WAI Web content accessibility guidelines (available at: http://www.w3.org/TR/WCAG20) and extensions to this tools already exist to cover the needs of very large Web sites, i.e. more that 10.000 pages (Cooper et al, 1999).

However these tools address very specific needs and have not been yet well situated into the design process for Internet-based design. From the perspective of the IT professional, the process of designing and developing for an inclusive information society requires awareness and to-the-point guidance with respect to design-for-all tools. In order to provide effective guidance to designers, a general-to-specific aiding approach is required that will organise and elaborate the bewildering variety of strands of existing work into into a format that can deliver practical results.

3. Towards a design aid environment to assist designers to design for all

IST-2000-26211 project IRIS aims to contribute towards addressing the above requirements. The main objectives of the IRIS project are to:

- Encapsulate into a design aid environment, work on design-for-all tools and methods; user modelling theories and methods including users with special needs; guidelines, recommendations and results from work about hypermedia, enrolment and accessibility; and
- Use this environment to redesign and enhance existing services in the areas of teleworking / on-line learning and electronic commerce, guided by rigorous user testing and evaluation.

The IRIS project will identify the suitability of a range of existing tools and methods for design for all, such as those related to guidelines, standards and recommendations for usability and accessibility, user modelling and profiles, delivering a range of media and content formats. It will identify and incorporate diverse strands of work, from multiple sources: i.e. open fora, such as W3C; industry, such as the Sun Accessibility Initiative, and standardisation organisations, ISO and CEN-ISSS, as well as other research projects. The project will also work on elaborating models of user requirements relevant to media, by involving large and international groups of users with varying abilities, with the aim of translating these models into technical characteristics of communication channels so that services may be configured to these characteristics. By building upon this body of existing work, as well as the research with users, IRIS will work towards extensions of existing design tools and methods, so that these are in accordance with 'Design for All' concepts, as well as the technical development of new tools that can easily be incorporated into the design process and can assist designers. This work will contribute towards the first objective of the IRIS project, the design of a designer aid environment that can assist designers to design for all. The specification, design and development of the information infrastructure (e.g. such as user models – profiles and content descriptions) and of user centred techniques for adaptation of media and content has started employing the technologies of directory services and software agents respectively, as part of the design aid environment.

The IRIS design aid environment will be used to further develop existing Internet services, currently deployed in a commercial context, in the selected areas of electronic commerce and teleworking/on-line learning. The use of the IRIS designer aid environment and the application of principles for inclusive design to the redesign of those services can significantly enhance their impact and target market. The evaluation of the redesigned Internet services will be achieved in a user centred manner, with the participation of large international groups of users with special needs, which will enable IRIS to make the best use of their varying requirements and insight.

A diagrammatic illustration of the IRIS approach for aiding designers to design for all is shown in Figure 2.

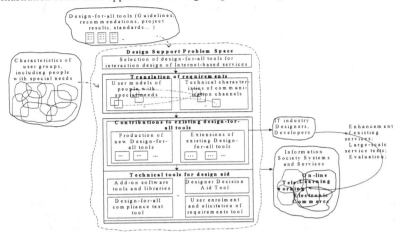

Figure 2: A conceptual view of the IRIS design aid environment context and impact.

References

Cohen, R. Allaby, C. Cumbaa, C. Fitzgerald, M. Ho, K. Hui, B. Latulipe, C. Lu, F. Moussa, N. Pooley, D. Qian, A. Siddiqi, S. (1998) What is Initiative? *User Modeling and User-Adapted Interaction 8: pp. 171–214,Kluwer Academic Publishers.*

Cooper, M. Limbourg, Q. Mariage, C., Vanderdonckt, J. (1999) Integrating Universal Design into a Global Approach for Managing Very Large Web Sites, *Proceedings of the 5th ERCIM Workshop on User Interfaces for All*, available at: http://ui4all.ics.forth.gr/UI4ALL-99/proceedings.html, Dagstuhl, Germany, 28th Nov.-1st Dec.

Gappa, H. Oppermann, R. Pieper, M. (1997) Certifying Web Accessibility for the Handicapped by ISO 9241 Conformance Testing, *Proceedings of the 3rd ERCIM Workshop on User Interfaces for All*, available at: http://ui4all.ics.forth.gr/UI4ALL-97/proceedings.html, Obernai, France, 3-4 November 1997

Hands, J. Darzentas, J. Darzentas, J. S. Spyrou, T. Koutsabasis, P. Smith, R. (2001) GAIA: A Generic Architecture and Applicable Platform for Information Brokerage, *to appear in: Electronic Commerce, Research and Applications*, Elsevier.

Konstantopoulos M., Darzentas J.S., Koutsabasis P., Spyrou T., Darzentas J. (2001) Towards Integration of Learning Objects Metadata and Learner Profiles Design: Lessons Learnt from GESTALT, to appear in *Interactive Learning Environments*.

Koutsabasis P. Darzentas J.S. Spyrou T. Darzentas J. (1999) Facilitating User – System Interaction: the GAIA Interaction Agent, *Proceedings of 32th Hawaiian Conference on system Sciences (HICSS-32)*, IEEE, pp. 322, Maui Hawaii, 5-8 Jan. 1999.

Koutsabasis, P. Darzentas, J. S. Spyrou, T. Darzentas, J. (2001), Modeling Agents for Providing User Interface Assistance: The Design of the GAIA Interaction Agent, *to appear in: Electronic Commerce, Research and Applications*, Elsevier.

Nicolle C., Abascal J. (editors) (2001) *Inclusive Guidelines for HCI.* Taylor & Francis

Poulson, D. Ashby, M. Richardson, S. (editors) (1996) Userfit: A practical handbook on user-centered design for Assistive Technology, European Commission TIDE 1062 USER project.

Stephanidis, C., Akoumianakis, D. (1999) Accessibility guidelines and scope of formative HCI design input: Contrasting two perspectives, *Proceedings of the 5th ERCIM Workshop on User Interfaces for All*, available at: http://ui4all.ics.forth.gr/UI4ALL-99/proceedings.html, Dagstuhl, Germany, 28th Nov.-1st Dec. 1999.

Stephanidis, C. Salvendy, G. Akoumianakis, D. Arnold, A. Bevan, N. Dardailler, D. Emiliani, P. L. Iakovidis, I. Jenkins, P. Karshmer, A. Korn, P. Marcus, A. Murphy, H. Oppermann, C. Stary, C. Tamura, H. Tscheligi, M. Ueda, H. Weber, G. Ziegler, J. (1999, November) Toward an Information Society for All: HCI challenges and R&D recommendations, *International Journal of Human-Computer Interaction*, Vol. 11(1), 1999, pp. 1-28.

Vanderdonckt, J. (1999) Towards a Corpus of Validated Web Design Guidelines, *Proceedings of 5th ERCIM Workshop on User Interfaces for All*, available at: http://ui4all.ics.forth.gr/UI4ALL-99/proceedings.html, Dagstuhl, Germany, 28thNov-1stDec.

Case study: designing universal access for intelligent household facilities

Heidi Krömker, Klaus Kuhn, Nina Sandweg

Siemens AG, Corporate Technology, Information and Communication, User Interface Design,
Otto-Hahn-Ring 6, D-81730 München, Germany
heidi.kroemker@mchp.siemens.de, nina.sandweg@mchp.siemens.de,
klaus.kuhn@mchp.siemens.de

Abstract

Intelligent household devices will only be successful if every member of the family, from children to grandparents, likes them and can use them quickly. The case study for the Home Set illustrates how User Interface Designers create operating design concepts for singles, senior citizens and young families within a systematic usability life cycle. Windows, light and many other things can be operated automatically in a postcard-sized display. This study points out those tasks in the usability-engineering life cycle which are especially important for universal access.

1. Introduction

The objective of the Home Set project was to develop an integrated, modular system which automates routine domestic chores for apartment and house owners. The system consists of a variable number of actuators and sensors which have a wireless link to a central base station. The electronic devices (such as lights, window openers, washing machines or stereo equipment) can be configured and controlled via the base station. The system can be plugged into a telephone connection, allowing remote control and monitoring via a telephone or mobile device.

Home Set is designed to permit universal access. This means it should do justice to the requirements of various kinds of households (e.g. working couples, families with children, senior citizens), while considering the differing demands of European and United States markets. The entire system was designed to be configured and operated by means of a postcard-sized touch screen display located at the central base station. Interaction with the display was to not assume familiarity with use of a PC, in order to allow system operation by those without PC experience.

2. Usability engineering life cycle for universal access

Before development began, the operating design concept was created in a user- and task-oriented process (Krömker, 98).[1].

The design starts with requirement analyses (Mayhew, 1999). For universal access, the following steps in the usability life cycle are especially important:

- User profiles
  Universal access means defining the profiles with regard to capabilities and the contexts in which the device is used.
- Task analysis
  The tasks must be derived from the various contexts of the various user profiles.
- General design principles
  These principles must be expanded, in order to do justice to the differing capabilities of the users. There are much greater variations in sensory abilities, vision and cognitive processes.
- Usability goals
  These goals must reflect the entire range of the preceding analyses.

---

[1] The operating design concept was for the EWW NP at Siemens AG, Vienna, Austria.. This project was supervised by Michael Burmester, Marc Hassenzahl (currently employed by User Interface Design GmbH), Axel Platz, Klaus Kuhn and Nina Sandweg.

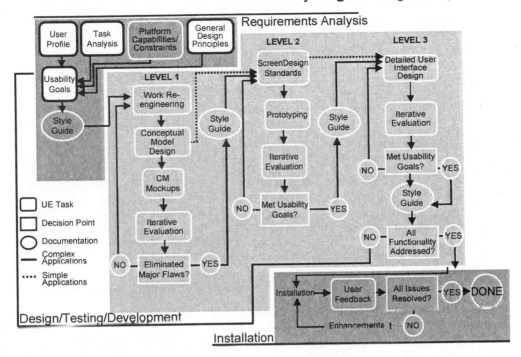

Fig. 1: Usability engineering tasks: the white fields are especially important for universal access (based on Mayhew, 1999)

Focus groups were used to gather the requirements. This qualitative technique permits in-depth access to the various user profiles and tasks.

In every step of the subsequent design work, the designs were visualized. Solutions sketched on paper and simple slide shows illustrate initial ideas. A high-fidelity simulation presents a mature design and a detailed view of the features. Based on these visualizations, we were able to evaluate the degree to which adherence to design rules had lead to universal access.

This is necessary, because requirement analyses and rule-based design can only approximate the actual user requirements. The evaluation was based on usability testing, because this technique was equally suitable for all user profiles.

3. Requirements analysis

In Germany and the U.S., focus groups were assembled in order to obtain fundamental and target group-specific requirements for the functional model and the user interface. Authentic users of the various target groups participated in the analysis (Kuhn, 2000).

A study by the Institute for Social Research in Berlin specified various household types (Meyer et al., 1999): employed singles (living alone), working couples (with and without children), traditional families, couples whose children have already left home, senior citizens living alone.

The user profiles were derived from these households. Small offices with up to five employees (such as offices for attorneys, M.D.s or architects) were added as a further target group. Based on these household types, persons were invited to participate in the focus groups. Eleven focus groups were formed (five in Germany, six in the U.S.) comprising a total of 75 participants, 34 female and 41 male. The youngest participant was 19, the oldest 82.

In these focus groups, lists of typical domestic chores were created. Functional ideas for Home Set were then derived and assessed. A detailed analysis of the focus group results revealed differences in user demands for the product's features:

Dual-career couples, dual-career families:
- Acceptance of merchandise (mail, packages, etc.) when away
- Access for craftsmen when away
- Remote access, remote control
- Electronic replacement for key ring
  (e.g. biometric access)
- Automatic ventilation
- Shopping service
- Cleaning service

Families:
- Watching after children (e.g. monitoring sleeping or ill children)

Empty nest, senior citizens:
- Reminder function (e.g. to-do list, appointment book)
- Mainly safety and comfort functions

Small Offices:
- Central locking (for file cabinets, etc., in addition to doors)
- Central control of PCs (PCs are switched on in the proper sequence, , servers first)

The results of the German and U.S. focus groups were analyzed separately. We discovered that 70% of the expected features were the same in both target markets. The German participants stressed security-related functions, such as a burglar alarm or checking if safety-related loads are switched off (e.g. irons, stoves). The American groups preferred comfort-related features, such as automatic door openers or monitoring of lights.

A very significant result of all discussion groups was that the system could handle many chores; however, the user wants to always have control of the overall system.

4. Designing the operating design concept

The first step was to establish usability goals for universal access. This would allow those involved in the design process to base their work on a common objective:

- The system assumes responsibility for major tasks, relieving the user. It is a tool which must prove to be useful and usable in the household. The user is primarily concerned about the chores; the system should be as "invisible" as possible. The system should perform the chores imperceptibly.
- A novice user is not necessarily aware that sensors, actuators and interconnections are required to perform a chore. Such persons need advice and ideas for experiments. Experienced users want to have the system perform novel, creative tasks.
- The user wants to have control and direct access. In particular, she wants to be able to quickly undo tasks performed by the system.
- Everyday life is not routine. That is, the user wants to decide ad hoc if or how chores will be performed. At a high level of abstraction, everyday life is routine. Many people leave the house in the morning and return in the evening. However, when analyzing this routine it becomes apparent that this does not hold true on all days (weekends, holidays, etc.). Also, the times when events occur or the conditions vary ("Today, the roof window can stay open because it will not rain"). In addition, some activities are uniform for a certain time and then change, making it necessary to permanently adapt the procedures. The technology suddenly becomes "visible" again, but users clearly reject this. Whenever more than one person lives in a household, the differing daily rhythms of the inhabitants break the routine. There is effectively no more routine.
- The rooms of an apartment or house provide the most natural way of characterizing it. However, features such as "Motion switches on light" might apply to several rooms. The user is mostly concerned with features in the house. The rooms are seen as secondary.

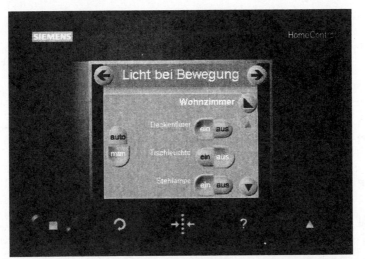

Fig. 2: User interface for home appliance control. The round shape, colors and 2 ½-dimensional design of the touch screen's elements indicate where the user must touch the screen to perform an activity.

To achieve the goal of universal access, we took special consideration of the following variations in the users' capabilities (Lohrum, 2000):

Sensory processes

- Vision
  As a person's visual system grows old, distance vision, light sensitivity and perception of depth, color and details are affected. For this reason, the screen accentuates these differences. Large letters are used, for enhanced readability.
- Hearing
  One's hearing also changes. Feedback signals were therefore limited to frequencies between 800 and 1,000 Hz.
- Tactile abilities
  When designing the button size, we took into consideration that touch sensitivity and fine motor abilities change with age.

Cognitive functions

- Reaction speed and memory
  The information is presented without timeouts. The user can decide how long she wants the information to be displayed. Nothing must be remembered to perform tasks.
- Information processing
  Several design features are included to do justice to changes in perceptual flexibility. The screen displays only information required for the next step. Animated sequences are used to clarify screen changes. When the context changes, the change is illustrated by moving the window's contents out of view, revealing the new contents.

5. Technical conditions

One of the major challenges in designing the user interface for Home Set was the extremely small size of the display. Merely 8.95 cm by 11.4 cm, it is about the size of a postcard. Also, users without PC experience and senior citizens should be able to use the user interface intuitively. They should even be able to configure and install it.

When designing displays on touch screens, the touchable elements must be made large enough to allow users to invoke activities. In addition, it must be obvious which parts of the user interface will cause an event. For Home Set, therefore, all touchable elements are round like fingers. Also, these elements are orange, a "warm" color, while all other parts of the screen are in "cool" colors (gray and blue). A "2½-dimension" effect is used for touchable elements, thus differentiating them from the background and inviting users to touch them (cf. Fig. 2).

6. Usability tests

In Germany and in the U.S., usability tests were conducted with two iterations. These tests were performed to verify the usability of the operating design concept during the design and development process for the Home Set. Potential for optimizing the design of specific components was explored and used during further development. In addition to usability, the attractiveness of the visual design was examined.

The evaluation objectives and questions resulted directly from the development of the operating design concept. To conduct the tests, a simulator tailored specifically to these issues was designed in order to verify the relevant components. The usability tests were performed in Germany and the U.S. with a total of 17 test persons in the first iteration and 12 test persons in the second. The participants were invited on the basis of age, computer experience and household type.

7. Conclusion

A very high degree of usability and acceptance for the Home Automation System was achieved by basing the design strictly on the usability process and focusing on the requirements. In both Germany and the U.S, the operating design concept described above has been shown to be very simple and easy to understand. Considering the complexity and novelty of the product, it is especially noteworthy that even users over 50 and persons with very little computer experience had little difficulty in operating the product without instructions. The results of the project demonstrated that the established usability-engineering life cycle suffices for the design of universal access. However, with regard to the user profiles, tasks and general design principles it is important to systematically take into account consideration the greater degree of variance in requirements.

References

Krömker, H. (1998). Gestaltung der Mensch-Maschine Schnittstelle in der industriellen Forschung und Entwicklung. In: ITG-Fachbericht 154, Technik für den Menschen, Gestaltung und Einsatz benutzungsfreundlicher Produkte, Vorträge der ITG-Fachtagung am 26. Und 27. Oktober 1998 an der katholischen Universität Eichstätt, VDE-Verlag GmbH Berlin, pp. 171-179.

Kuhn, K. (2001). Problems and Benefits of Requirement Gathering with Focus Groups: A Case Study. In: Krömker, H. (Special Issue Editor): Ease and Joy of Use for Complex Systems at Siemens. International Journal of Human-Computer Interaction (IJHCI), Lawrence Erlbaum Associates, Inc., Publishers, Mahwah, New Jersey, pp. 309-326.

Meyer, S., Böhm, U. & Fischer, B. (1999). Siemens in every home – Smart home for all. Unpublisched manuscript, Institute for Social Research, Berlin, Germany

Mayhew, D. J. (1999). The usability life cycle, Morgan Kaufmann Publishers, Inc., San Francisco, pp. 34.

Lohrum, P. (2000). Nicht alle Kunden sind jung – intergenerative Produktgestaltung. In: Meyer-Hentschel, G. and H. (Editor), Handbuch Seniorenmarketing, Deutscher Fachverlag, Frankfurt am Main, S. 379-419.

Study on menu usability and structural analysis

Takuo Matsunobe*, Haruhiko Sato[†]

*Wakayama University, [†]Kyushu Institute of Design

1. Introduction

The introduction of GUI-based operating systems like Windows and Macintosh has made PCs prevalent in Japanese homes over the last few years, adding to the number of computer beginners. At the same time, a number of software houses are marketing a wide variety of titles. But not a few users have difficulty using a PC, since they "don't understand software terminology," and want to "use a computer for a specific purpose but don't know how." Software houses have addressed this by explaining their terminologies and operating procedures in manuals and "help" functions.

One of the things hindering ease of use is that each software title uses a different set of terminology and procedures for the same operations. If terminology and procedures were standardized, there wouldn't be any difficulty understanding how to navigate a software title you had never used before. Each software house has unified these things to some extent among its own different titles, but a common standard has yet to be applied among different manufacturers and software categories. In the worst of cases, there might be a title you could launch but not be able to quit because of differences in procedures. At present, standardization of operating procedures among different manufacturers would be difficult.

From the point of view of the beginner, it would be desirable to be able to use many different types of software from the outset if only they learn some initial basic terminology and procedures. Unfortunately, however, this is not the case. Each OS manufacturer has its own set of guidelines1,2, and none of these have been created for a specifically Japanese environment. Rather, they are Japanese versions of English environments.

This being the case, it is important to know how users sort each operation on the menu so we can begin standardizing it. In this research project, we used a cluster analysis of how PC users sort each operation into which items to find out how they correlate each operation with the menu items used in software. In our survey, we used menu items selected from existing major software titles.

2. Study method

To survey the correlation between operations users want to perform and menu items, we carried out a questionnaire on web browsers and word processors, the two most frequently used types of software program.

2.1 Questionnaire

The questionnaire had three sections: 1) individual history of PC use; 2) the relationship between web browser operations and menu items; and 3) the relationship between word processor operations and menu items.

1) Survey of individual history of PC use

Questions included length of experience using PCs, purposes, OS used, software used and average use time.

2, 3) Questionnaire on operations of web browsers and word processors, and menu items

For our survey of 2) and 3), we prepared a questionnaire by sampling operations and menu items from the software (Table 1). Operations that overlap in different titles were combined. The term "menu items" usually refers to the options within a menu, but in this study we use it to refer to the titles that are constantly displayed across the top of the window when the software is in use (in Windows, the menu bar and drop down menu). Word processor software menu items were sampled from the twelve titles shown in the

Table 1 software list

Japanese
MS-WORKS(wordprocessor
MS-
ichitar
claris works(wordprocessor
wordperfe
wordpa
macwrite
EG
Macwor
Solo
Web browser
Internet
Netscape
Netscape

table. We succeeded in isolating 10 menu items (Table 2) and 80 operations for web browsers, and 16 menu items (Table 2) and 501 operations for word processing programs. Since we believed beginners might have difficulty selecting menu items in web browsers, some of the terminology being unique to that type of software, here we added the five items of "home page," "page," "option," "tool" and "setting." Two entries, "meaning unknown" and "not applicable," were added to the menu item lists for both. Respondents were asked to choose the menu items they select before they execute operations. They were also instructed to select "not applicable" when they believed there were no applicable menu items, and "meaning unknown" when they didn't understand what a specific operation meant. Also, when they didn't understand a specific term in the operating instructions, they were asked to circle it.

2.2 Survey method
The questionnaire was given to 22 university and graduate school students (14 males, 8 females) with an average age of 23.3 (from 21 to 29 years old).

2.3 Method of Analysis
To sort the operations, we counted the number of respondents who selected each menu item for each operation and carried out a cluster analysis. Similarly, we did a cluster analysis of the menu items alone. For our analysis, we used STATSOFT's VisualSTAT.

3. Results

3.1 Sorting of the Respondents
On the length experience using a PC, the average number of years was 2.8 (from 0 to 10 years). Likewise, the average number of hours of use per day was 2.8 (from 0 to 10 hours). All of the respondents had used PCs in some way or another, and there were no beginners. Of the operating systems used (multiple answers were allowed), 16 had experience with Windows95, 4 with Windows 3.1, 6 with DOS, 11 with MacOS and 6 with UNIX. Next, we counted the number of "meaning unknown" responses concerning operations (Fig. 1). Since we believe the number of "meaning unknown" responses reflects the level of software experience, which is the single most important survey item, the respondents were divided into an "Advanced User Group (respondents A-Q)" and a "Beginner Group (respondents R-V), according to the number of "meaning unknown" answers.

3.2 Advanced User Group
1) On the basis of the answers of the Advanced User Group, the number of respondents who selected specific menu items for each operation was tallied. After grouping the operations using the menu items that were selected by the largest number of respondents, we surveyed which other menu items were selected within each operation cluster. We found that for some operation clusters, almost all the respondents selected the same menu item, and for others a wide variety of items were selected (Fig. 2 shows an example of an operation cluster where diverse menu items were selected). Operation clusters indicated by nearly all the respondents included "file," "edit" and "help" for both web browsers and word processors.

Table 2 menu items

Web browser

japanese	english
file	file
hensyu	edit
hyoji	display
ido	move
jump	jump
okiniiri	my favorites
bookmark	bookmark
directory	directory
window	windows
help	help

wordprocessor

japanese	english
file	file
hensyu	edit
hyoji	display
sonyu	insertion
size	size
sosyoku	clip-art
syotai	fonts
syoshiki	formatting
tool	tools
dogu	instruments
sakuhyo	making tables
keisen	rules
outline	outlines
spell	spelling
window	windows
help	help

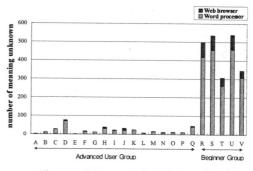

Figure 1 Number of "meaning unknown"

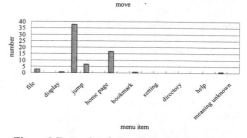

Figure 2 Example of operation clusters where

2) Cluster Analysis

To learn how users sort each operation into menu items, we performed a cluster analysis based on the number of respondents who selected each menu item for each operation. Agglutination was employed for a cluster analysis, the Euclidean distance for measuring distance and furthest neighbor for binding.3

Fig. 3 shows the dendrogram obtained by our cluster analysis of web browser operations. By referring to the progress of the binding, operations were grouped at a Euclidean distance of 30. As a result, the groups were characterized by "edit," "explanations and information display," "file," "setting," "move homepage," "display" and "others."

Similarly, we produced a dendrogram for word processors. Operations were grouped at a Euclidean distance of 5. The groups were characterized by "file," "edit," "explanations and help," "display," "display preferences," "window views," "setting formats," "setting character formats," "special functions," "insertion," "tables," "tables and rules" and "others."

Next, cluster analysis was applied to the names used for menu items to learn which items the respondents were liable to confuse. Figures 4 and 5 show the dendrograms produced. On the basis of these diagrams, with menu items for web browsers grouped at a Euclidean distance of 20, it was found that respondents often confuse "options" with "tools," and that there is some confusion among the terms "home pages," "book marks," "my favorites," "windows," "directory," "pages" and "jump."

Next, word processor menus were grouped at a Euclidean distance of 80. Clear-cut clusters were not formed for "making tables," "tools," "help," "fonts," "rules," "windows," "clip-art," "spelling," "outlines," "instruments" and "size."

On the basis of the above analysis, a correlation was obtained of the menu items and operations for web browsers and word processors. The results are shown in Tables 3 and 4. Operations and menu items are related to each other for the menu items "file," "edit," "help" and "display" in web browsers and word processors.

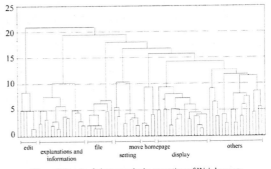

Figure 3 Result of cluster analysis, operation of Web browser

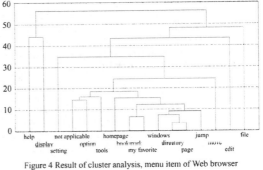

Figure 4 Result of cluster analysis, menu item of Web browser

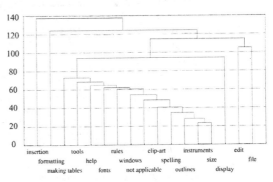

Figure 5 Result of cluster analysis, menu item of Word processor

3.3 Answers from the Beginner Group

We tallied the answers from the Beginner Group and sorted them using cluster analysis, but found no big clusters. This is presumably because most of the answers were "meaning unknown." Therefore we investigated the terms that caused the respondents to respond "meaning unknown."

3.4 Terms whose "meaning is unknown"

Terms whose meaning respondents didn't understand when they answered were clustered under: terms unique to the Internet, such as "Upload" and "Web"; terms unique to word processors, such as "Shading" and "Margin"; PC terms such as "Short Cut" and "Property"; and others that didn't belong to any of the above. We concluded that the chief reason was that they

Table 3
Web browser

operation cluster	menu item
file	file
edit	edit
explanations and information	help
about display	display
about setting	setting, option
move	bookmark, my favorite
URL	dorectory, jump, move
upload	homepage
load	page
others	tools, windows

didn't understand the technical terms: the "other" category included "Delete," whose meaning they must have known but didn't understand in the context of the operating manual.

We then classified the "meaning unknown" terms from the point of view of the Japanese language into: "terminology only in Japanese" such as "Seigyo-moji" (Control Letters), "Bundan-Kinshi" (Prevent Separation), "Shoki-Settei" (Default) and "Henko-Rireki" (Change History); "combinations of katakana and kanji characters" such as "System Settei" (preferences), "Version Joho (Information)," "Shoshiki (Formatting) Bar," and "Link-Shu (Links) on the Website"; "katakana terms" such as "File," "Folder," "Macro," "Scroll," "Security," and "Support"; and English terms such as "Plug-in," "Java," and "Web." More than half of the "meaning unknown" terms came under the grouping "English terms which included terms in katakana."

4. Consideration

Operations were classified into the following two groups by menu items selected:

A group whose menu items selected agreed; and
A group whose menu items selected didn't agree.

The first group might have been following manufacturers' guidelines,1,2) which helped unify correlation between operations and menu item labels. As a result, the group is united by the menu item labels, rather than the characteristics of the operations: "File," which includes such operations as "Save Text as a File," "Quit (this program)" and "Print Document"; "Help," which includes such operations as "About Help," "View Software Version" and "View Software Instructions"; and "Edit," which includes such operations as "Copy Selection," "Cut Selection" and "Redo Last Operation."

Table 4
word processor

operation cluster	menu item
file	file
edit	edit
explanations and information	help
format setting	formatting
about insertion	insertion
about display	display
display setting	windows
window display	
font setting	font,size,clip-art
special functions	spelling,outline
table	making table
rules	rules
others	tools,instruments

With regard to the second group, cluster analysis showed that the respondents had a problem with operations which overlap under more than one menu item label and those which do not have appropriate labels in the menu items. Take web browsers, for example. Operations which overlap with more than one entry among the labels of the menu items were about "Move Home Page," with labels for corresponding menu items including "Home," "Bookmarks," "My Favorites," "Directory," "Page," "Jump" and "Move." Similarly, "operations related to settings" overlap with such menu item labels as "Preferences" and "Options." A closer look at "operations related to preferences" showed a different overlapping pattern. Besides "Preferences" and "Options," menu items labels related to each setting operation were selected. For example, for the operation of "setting tool bars that are to be displayed on the screen," some selected "Preferences" and others, "View." As a result, they selected different items for operations related to "Preferences." In such a case, we believe those operations should be included both in a menu item that includes all the "operations related to preferences" and an item under "View," so as to meet users' mental models. Similarly, "Set Tool Bars" (Tools and Preferences), "Set Paper Size and Margins for Printing" (File and Preferences), "Set Font Size for Home Page" (Preferences and View), "Show Thumbnail Window with Outline of the Text" (Outline and Window) were found to overlap among groups whose menu item names are different. It was believed these operations are better included under both menu items.

Cluster analysis also found combinations of menu item labels that can be easily confused between "Options" and "Tools," and among "Home," "Bookmarks," "My Favorites," "Window," "Directory," "Page" and "Jump." Among these menu item names, "Move Home Page" and "Record Location of Home Page" overlapped. It might therefore be better to separate them into two menu item names, "Move Home Page" and "Record Location of Home Page." We believe the chief rationale for this to be that, though it depends on the number of operations included under certain items, item labels for home pages have yet to be fixed at this point. Some items didn't form any definite clusters, probably because only a few respondents selected those item names. We therefore believe it worthwhile to review the necessity of those item names and, if necessary, menu item names that correspond to the classes of the operations should be proposed.

From the findings of the survey, we verified that the standardization of menu items such as "File," "Edit," "View" and "Help," promoted under manufacturers' guidelines, should benefit experienced users. Since beginners found these difficult to interpret, however, we believe that the terminology, including menu item names, needs to be examined.4

We also believe there is reason to make proposals on new terminology for those terms whose meaning was unknown, in which katakana terms should be expressed in kanji characters as much as possible. We feel this to be important because some users didn't even understand what "files" and "folders" are, the terms currently in use. Use of kanji ideographic characters, unlike katakana phonetic symbols, should help users understand the meanings, as reported by Matsunobe, et al.5 Technical terms also need to be replaced by terms which general users can understand. One of the most important considerations when coining terms is, we think, whether they will be understandable to beginners. We believe we should start by examining terms in future studies so we can make expressions for operations easily comprehensible, leading to a standardization of menus an examination from the point of view of the beginner. The terms which were found to make operations difficult should ideally be replaced with Japanese terminology. However, users are advised to memorize those terms which are commonly used to avoid unnecessary confusion. Everything considered, it would be desirable to begin with standard default menus that individuals could customize, for the sake of operability for existing users.

Now that the survey findings have shown how operations should be classified, we believe it possible to create standard Japanese menus or provide an index for Japanese menus when the terms and structures of menu items are being discussed. Along these lines, menu item labels should be examined taking into account those where users were confused in making their selections of classifications, and those which users could classify without any confusion. On the basis of the above, the structural survey of menus by cluster analysis was found to be effective in identifying user intentions. When new software titles are developed, the creation of menus should be made easier by investigating how users correlate terms with menu items and applying cluster analysis to the findings.

5. Conclusion

In this study, we surveyed Japanese menus of PC software by correlating operations with menu items and evaluating the outcome by cluster analysis. The following are our conclusions.

(1) We grouped operations and menu items on the basis of users' selection. It was found that the unification between operations and such menu items as "File, "Edit" and "Help" had seen progress.

(2) Some operations should not be included under more than one menu item.

(3) Operations that overlap among more than one menu item and those for which appropriate menu item labels are not assigned make classification complex.

(4) Because some terms which users didn't understand made operations difficult, technical terms in each software title should be simplified. Beginners found English and its katakana transliteration difficult to understand, thus hindering operations.

(5) Survey of menus by cluster analysis is effective in building menu structures.

If we examine terms and menu structures further on the basis of the above findings, we believe we will be able to suggest more user-friendly menus.

References

1   Microsoft Corporation: Windows User Interface Design Guide, Askii, Tokyo, 1995.

2   Apple Computer, Inc.: Human Interface Guidelines: The Apple Desktop Interface (Japanese version), Addison Wesley Publishers Japan, Tokyo, 1989.

3   Kinoshita E. (1995). A Guide to Multivariate Analysis Using Easy Mathematical Models, 89-103, Kindai Kagaku-sha, Tokyo.

4   (Compled by) Yasuyuki Kikuchi, et al. (1995). GUI Design Guidebook, Kaibun-do, Tokyo.

5   Matsunobe T., Yoshimura K., Oda Y, Sato H. (1999). On Terms Used in Japanese Menus of PC Software, Proceedings for the Human Ergology Society's 34th Convention.

How to Utilize Task Taxonomies for the Design of Web Applications 'for All'

Chris Stary

University of Linz, Department of Business Information Systems
Freistädterstraße 315, A-4040 Linz, Austria
stary@ce.uni-linz.ac.at - www.ce.uni-linz.ac.at

Abstract

Although there exists a variety of frameworks to develop web applications based on usability-engineering principles, few approaches exist that guide developers on how to put these guidelines to practice. In this paper we suggest to utilize task taxonomies for web application design. Several taxonomies are discussed, for browsing-oriented interaction tasks, and for generic, but domain-oriented work tasks in distributed information spaces. The embodiment of taxonomy-driven design in task-based interface engineering is discussed. The need of providing a variety of interaction modalities at the interface as well as of the flexibility to accomplish tasks in different ways is elaborated.

1. Introduction

Several factors have made the Web increasingly available to millions of users (around the world). The goal of many developers is to provide Web-based services to broader markets. Designing effective and usable Web applications has also become increasingly important to survival in the e-commerce market. In order to achieve widespread utility and usage of their solutions developers need to understand the users, their tasks, and how to make the (envisioned) services accessible to a variety of users (e.g., Preece et al., 1994). However, the design of Web applications is still considered to be a primarily artistic endeavor (Spool et al., 1999). On one hand, some empirical findings about pitfalls to avoid or success stories to follow might direct developers in their strive for usable interfaces. On the other hand, frameworks have been developed to lead way to 'successful' user interfaces.

There is an implicit assumption behind most of the Web-application developments. They are assumed to be accessible for 'all', since once provided in an intranet or once accessible via the Internet a diversity of users might access the facilities and information provided, either without or within a particular task or situational context. Although web application development should be driven by these contexts, recent empirical results, such as provided by Selvidge (1999), show that web designers are still typically developing web applications for marketing appeal rather than for usability issues. If usability deficiencies cannot be removed, neither the acceptability by users nor the sustainable diffusion of web applications in e-markets seems very likely (Silberer et al., 2000).

In this paper we suggest to enhance the design process of web applications through task frameworks rather than novel heuristics and results from successful design solutions, as e.g., proposed by Nielsen (1996). Recognizing the importance of the nature of Web-related tasks we propose the pro-active embodiment of that knowledge in a structured design process, both at the level of domain- as well as interaction-specification. In the following, we first deal with traditional frameworks in Web-usability engineering capturing (Web-related) tasks, before we demonstrate the benefits of generic task modeling. These benefits mainly stem from the implementation-independent representation of design knowledge as well as from building flexible systems that serve a variety of users instead of sticking to particular application domains and interaction facilities.

2. From taxonomies to taskonomies in web engineering

The categorization of tasks has a long tradition in user interface design. For instance, Carter (1985) has grouped user-oriented system functions according to their generic (task) capabilities in interaction. He distinguished functions for control in/output from those for data in/output. The group of functions for control contains execution functions and information functions. It comprises activities that are session-specific (execution functions) and situation-specific, such as user guidance, help, and tutoring (information functions). The group of functions for data manipulation contains so-called data functions and information functions. Data functions comprise all functions for the manipulation of data structures and data values, such as insert. Information functions as part of data functions address user task-related activities, such as searching for information.

This framework is extremely helpful for designers, since it provides a level of abstraction that allows on one hand to concretize design knowledge for a particular application domain. On the other hand it supports reuse of design knowledge and the specification of commonalties that are required for framework definition and refinement at an implementation-independent level. Such a level of abstraction seems to be inevitable for structured design.

The information functions as part of data functions can further be refined to particular sub task types. According to Ye (1991) information search can be classified into two major categories: known-item and unknown-item search. In a known-item search, users know precisely the objects for which they are looking for, whereas in an unknown-item search they do not know exactly the items they are looking for. Both types are relevant for web applications accessible for 'all', such as Internet-based e-commerce applications (see, e.g., Kalakota et al., 1996). In a typical unknown-item search web applications suggest a set of terms that is assumed to be relevant for (a variety of) users or in a particular application domain - see, for instance www.netscape.com for the first approach. However, unknown-item search poses a greater challenge to the user-interface designer than known-item search, in particular in cases where a search task requires users to traverse multiple search spaces with a limited knowledge about the application domain. That challenge requires to rethink the representation and navigation through the search and response space.

While performing a task analysis for web applications Byrne et al. (1999) came up with a web-specific taxonomy of user tasks. Based on research in the field of information finding and termed 'taskonomy' it comprises six generic tasks: Use information, locate on page, go to page, provide information, configure browser, and react to environment. Each task contains a set of sub tasks related to the task. For instance, 'provide information' captures provision activities such as search string, shipping address and survey response. Since web applications are increasingly becoming standard in public information spaces and thus, accessible by 'all', such a taskonomy can be utilized in the course of structured design and the construction of flexible user interface architectures. It also serves as reference model for evaluation, since the tasks to be performed are understood from the perspective of users and usage rather than from the perspective of technology and implementation.

3.   Embedding taskonomies in task-based web design

How can such taxonomies be embedded in Web-application design? In the following a solution is proposed that results from our experiences with the model-based TADEUS-environment and design methodology (Stary, 1999). Following a multi-perspective idea user interface development in TADEUS is based on four different models: Task, problem domain data, user, and interaction (domain) model. These models have to be refined and finally, migrated by the designer in the course of development to an application model, in order to enable task-based implementations of the development case at hand. Each of the models is composed of a structural and behavior specification. All models are specified using the same object-oriented notation facilitating the migration process.

Taskonomies for Web applications can be embedded at several stages of development and, related to that, at different levels of specification, namely at the level of task specification, and the level of interaction specification. At the level of task specifications generic task descriptions can be put into the context of actual user tasks. Assume a reservation task for a course offered by a virtual school. At that level several possibilities can be specified for the accomplishment of that task. In Figure 1 a typical design space is shown at a generic level for search tasks. In that case two ways to accomplish the search are captured, either search via newsgroups or search via the Web. In the first case a request is posted to a newsgroup, in the latter case an URL and options are provided.

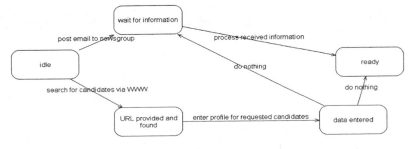

Figure 1. Generic behavior specification of a search task

48

This example demonstrates one of the core activities in designing user interfaces for 'all'. Although there might be standardized procedures for particular tasks to follow in public information spaces, such as the Web, design of user interfaces for 'all' requires to capture a certain flexibility at the generic task (content) level. This holds for work situations as well as for public use of applications. Since users might have different views on the accomplishment of tasks not only a variety of different procedures might lead to identical (work) results, but also the arrangement of information and the navigation at the user interface might differ significantly from user to user. This brings us to the second level of way utilizing taskonomies for user interfaces 'for all': the level of interaction. Figure 2 shows a generic structure description of interaction facilities as being used in TADEUS. In this case GUI elements and browsing facilities have been conceptually integrated. The boxes denote classes, the black triangles aggregations, the white ones specializations.

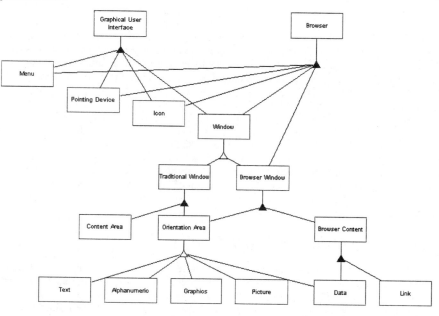

Figure 2. Generic structure specification of GUIs and browsers

Such an integrated specification allows to assign generic navigation and interaction tasks from taskonomies, such as proposed by Byrne et al. (1999). Some of the entries might become part of an application specification. Figure 3 shows the reservation case based on selected items from the browser taxonomy.

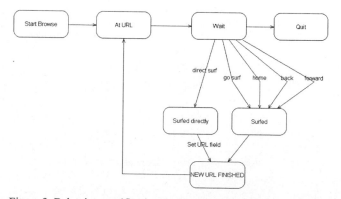

Figure 3. Behavior specification derived from a taskonomy of web tasks

Specification techniques based on taskonomies allow to embed empirical results, such as provided by Hu et al. (1999), either in terms of guidelines or pre-programmed behavior. For instance, the finding that lists should be better designed through ranking entries than alphabetical ordering in case of low familiarity with the task, can either be encoded in a particular method or be assigned to the class list box through a comment field for the designer.

Finally, switching between modalities of interaction is facilitated through generic task specification at the interaction level. Assume the user interface should be able to provide access through browsing for visually impaired and visually enabled persons. In that case taskonomies of browsing systems could capture mapping of function keys to the basic interaction features for browser-based interaction, according to empirical data, e.g., as provided by Zajicek et al. (1998): F1 - Load a URL etc. Using that kind of mechanism, modality- or style switching can be lifted to a more conceptual, since implementation-independent level of abstraction, and does not remain implicitly (since hard coded). This way, utmost flexibility of user interface technology can be achieved at design time, enriching the interaction space for a variety of users.

References

1. Byrne, M.D. ; John, B.E. ; Wehrle, N.S. ; Crow, D.C. (1999).: The Tangled Web We Wove : A Taskonomy of WWW Use, *CHI'99*, ACM, 544-561.
2. Carter, J.A., Jr. (1985). A Taxonomy of User-Oriented Data Processing Functions, *Trends in Ergonomics / Human Factors II*, eds : Eberts, R.E., Eberts, C.G., 335-342, Elsevier.
3. Hu, P.J.-H. ; Ma, P.Ch. ; Chau, P.Y.K. (1999). Evaluation of User Interface Designs for Information Retrieval Systems : A Computer-Based Experiment, *Decision Support Systems*, Vol. 27, 125-143.
4. Kalakota, R. ; Whinston, A.B. (1996). *Frontiers of Electronic Commerce*, Addison Wesley, Reading, Massachusetts.
5. Nielsen, J. (1996). Top Ten Mistakes in Web Design, *http://www.sun.com/columns/alertbox/ 9605.html*.
6. Preece, J.; Rogers, Y.; Sharp, H.; Benyon, D.; Holland, S.; Carey, T. (1994). Human-Computer Interaction, Addison-Wesley, Wokingham.
7. Selvidge, P. (1999). Reservations about the Usability of Airline Web Sites, *CHI'99* Extended Abstracts, ACM, 306-307.
8. Silberer, G. ; Fischer, L. (2000). Acceptance, Implications and Success of Kiosk Systems (in German), *Multimediale Kioskterminals*, eds : Silberer, G. ; Fischer, L., Gabler, Wiesbaden, 221-245.
9. Spool, J.M ; Scanlon, T. ; Snyder, C. ; Schroeder, W. (1999). Measuring Web Usability, *CHI'99* Extended Abstracts, ACM, 354.
10. Stary, Ch. (1999). Toward the Task-Complete Development of Activity-Oriented User Interfaces, *Int. J. Human-Computer Interaction*, 11(2), 153-182.
11. Stary, Ch. (2001). User Diversity and Design Representations. Towards Increased Effectiveness in Design for All, *Universal Access in the Information Society*, 1(1), 2001.
12. Ye , M.M. (1991).: System Design and Cataloging meet the Users: User Interface Design to Online Public Access Catalogs, *Journal of American Society for Information Science*, 42(2), 78-98.
13. Zajicek, M.; Powell, C.; Reeves, C. (1998). A Web Navigation Tool for the Blind, *3rd International Conference on Assistive Technologies*, ACM, 1998.

IS4ALL:
Promoting Universal Design in Healthcare Telematics

Constantine Stephanidis[1,2]

[1] Institute of Computer Science
Foundation for Research and Technology - Hellas
Science and Technology Park of Crete
GR-71110, Heraklion, Crete, Greece
Email: cs@ics.forth.gr

[2] Department of Computer Science, University of Crete

Abstract

IS4ALL (Information Society for All) is a EC-funded Thematic Network (Working Group) aiming to advance the principles and practice of Universal Access in Information Society Technologies, by establishing a wide, interdisciplinary and closely collaborating network of experts to provide the European IT&T industry in general, and Health Telematics in particular, with a comprehensive code of practice on how to appropriate the benefits of universal design. This paper outlines the objectives and methodological approach of IS4ALL.

1. Introduction

The Information Society has the potential to improve the quality of life of citizens, the efficiency of our social and economic organisation, and to reinforce cohesion. However, as with all major technological changes, it can also have disadvantages, introducing new barriers, human isolation and alienation (the so-called "digital divide"), if the diverse requirements of all potential users are not taken seriously into consideration and if an appropriate "connection" to computer applications and services is not guaranteed. It is in this context that the notion of Universal Access becomes critically important for ensuring social acceptability of the emerging Information Society. Universal access implies the accessibility and usability of Information Society Technologies (IST) by anyone, anywhere, anytime. In the context of the emerging Information Society, Universal Access becomes predominantly an issue of design, and the important issue arises of how is it possible to design systems that permit systematic and cost-effective approaches to accommodating all users. Universal Design in the Information Society has been defined as the conscious and systematic effort to proactively apply principles, methods and tools, in order to develop IST products and services that are accessible and usable by all citizens, thus avoiding the need for a posteriori adaptations, or specialised design. The rationale behind the deployment Universal Design in IST is grounded on the claim that designing for the "typical" or "average" user, as the case has been with "conventional" design of Information Technology and Telecommunications (IT&T) applications and services, leads to products which do not cater for the needs of the broadest possible population, thus excluding categories of users by limitng the context of use and restricting the choice of technology or access medium to a predefiend and narrow range. Proactive strategies entail a purposeful effort to build-in universal access features into a product, starting from the early stages of product design.

As a proactive strategy, Universal Design is already practised in several engineering disciplines, such as, for example, civil engineering and architecture, with many applications in interior and workplace design and housing. While existing knowledge may be considered sufficient to address the accessibility of physical spaces, this is not the case with Information Society Technologies, where Universal Design is still posing a major challenge.

In the recent past, R&D work has demonstrated the feasibility of Design for all in the field of HCI [Stephanidis & Emiliani, 1999]. Efforts towards universal accessibility of the interactive components of Information Society Technologies have met wide appreciation by an increasing proportion of the international research community, thus leading to the foundation of working groups and scientific forums. The International Scientific Forum "Towards an Information Society for All' was launched in 1997, as an international ad hoc group of experts sharing common visions and objectives, namely the advancement of the principles of Universal Access in the emerging Information

Society. The Forum held three workshops[1] to establish an interdisciplinary discussion comunity, a common vocabulary to facilitate exchange and dissemination of knowledge, and to promote international co-operation. The Forum has produced two White Papers (Stephanidis et al., 1998 and Stephanidis et al., 1999 reporting on an evolving international R&D agenda focusing on the development of an Information Society acceptable to all citizens. The proposed agenda addresses technological and user-oriented issues, application domains, and support measures. The Forum has also elaborated on the proposed agenda by identifying challenges in the field of human-computer interaction, and clusters of concrete recommendations for international collaborative R&D activities. Moreover, the Forum has addressed the concept of *accessibility* beyond the traditional fields of inquiry (e.g., assistive technologies, housing, etc), in the context of selected mainstream Information Society Technologies, and important application domains with significant impact on society as a whole (e.g., Healthcare).

Based on the success of its initial activities, the Forum has proposed IS4ALL (Information Society for All) – a new EC-funded thematic network aiming to advance the principles and practice of Universal Access towards the wider IST community. In particular, the project focuses on the area of Healthcare Telematics, a critical Information Society application domain, and on emerging technologies shaping the nature and contents of this domain. IS4ALL therefore establishes a wide, interdisciplinary and closely collaborating "network of experts" (Working Group) to provide the European Healthcare industry with a comprehensive information package detailing how to appropriate the benefits of universal design.

2. Concepts, principles and objectives

The primary focus of the proposed activities of IS4ALL is on the impact of advanced desktop and mobile interaction technologies on emerging Healthcare products and services. The Health Telematics domain has been selected on the grounds of being a critical service sector, catering for the population at large, and at the same time involving a variety of diverse target user groups (e.g., doctors, nurses, administrators, patients). These characteristics render it a complex domain, due to inherent diversity, and an ideal "testbed" for exemplifying the principles of Universal Access and assessing both the challenges and the opportunities in the context of the emerging Information Society. By emerging interaction platforms we mean primarily advanced desktop-oriented environments (e.g., advanced GUIs, 3D graphical toolkits, visualisers), and mobile platforms (e.g., palmtop devices) enabling ubiquitous access to electronic data from anywhere, and at anytime. Such technologies are expected to bring about radical improvements in the type and range of Health Telematics services. In this context, accounting for the accessibility, usability and acceptability of these technologies at an early stage of their development is likely to improve their market impact as well as the actual usefulness of the end products.

The specific technological / scientific objectives to be attained by IS4ALL can be summarised as follows:

- Consolidate existing knowledge on Universal Access in the context of Information Society Technologies, which is currently dispersed across different international sites and actors, into a comprehensive code of design practice (e.g., enumeration of methods, process guidelines, etc).
- Translate the consolidated wisdom to concrete recommendations for emerging technologies (e.g., emerging desktop and mobile platforms) in Health Telematics.
- Demonstrate the validity and applicability of the recommendations in the context of concrete scenarios drawn from an experimental regional Health Telematics network.
- Promote the Universal Access principles and practice in Healthcare Telematics through a mix of outreach activities, which include seminars, and participation in major international conferences, concertation meetings, and project clustering events.

3. Technical approach and expected outcomes

To attain the above stated objectives, IS4ALL proceeds in the three phases, namely scenario analysis, consolidation and outreach.

---

[1] The 1st workshop took place in San Francisco, USA, August 29, 1997, and was sponsored by IBM. The 2nd took place in Crete, Greece, June 15-16, 1998. The 3rd took place in Munich, Germany, August 22-23, 1999. The latter two events were partially funded by the European Commission.

3.1   Scenario analysis

IS4ALL seeks to collect data by:

- defining an appropriate set of instruments to elicit and document best practice and experience in the area of Universal Access; some of these instruments may be reference case studies on how Universal Access is being practised and state of the art reports on technologies and application areas, etc.
- extracting and developing scenarios (relevant to a regional Healthcare Telematics network as well as the to the activities of the industrial participants) to demonstrate the validity and applicability of such a code of practice; these scenarios are formulated around an agreed common theme, namely electronic healthcare records [Tsiknakis et al., 1997]

The methodological approach of the project links with recent developments in scenario-based analysis (Carroll, 1995; Carroll, 2001). Scenarios are narrative descriptions acting both as design vocabularies and validation platforms. Used as design vocabularies, scenarios provide a "user" oriented language for understanding human activities. As an approach to evaluation, scenarios help define, focus and articulate an evaluation to suit specific requirements and intended objectives. In IS4ALL scenarios will be employed for both these purposes. Figure 1 depicts precisely this dual role of scenarios in the context of systems development.

Specifically, as empirical tools, scenarios help the analyst to understand the existing structure and organization of tasks, and to unfold prevailing patterns of use in a user community. At the same time, as analytical instruments, scenarios provide useful insights, which help re-engineer a process or artifact so as to make users aware (i.e. through prototypes) of the envisioned situation and the corresponding usage patterns.

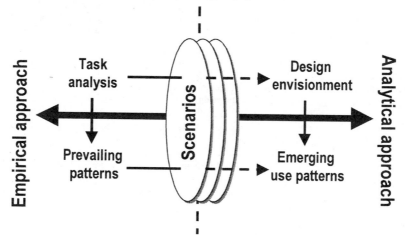

Figure 1: Scenarios provide a common design vocabulary and a minimal evaluation context

In the context of IS4ALL, scenarios are perceived as narrative descriptions of computer-mediated human activities in a Healthcare Telematics environment. The social setting of a Healthcare Telematics environment may be bound to a clinic within the hospital, a ward within a clinic or even to the end users business or residential environment. The scope of such scenarios is intended to be narrow and focused on very specific issues. As an example, a possible scenario could be focused on patient's access to the electronic health record from home, using platforma such as desktop PCs, or while on the move, using a portable device which runs Windows CE. Such a scenario will play a dual role. In the first place, it will describe a clear and unambiguous design resource. Secondly, it will offer a reference point for validating IS4ALL instruments. As a design resource, it may be used to extract parameters, which should appear in, for instance, the Universal Access Context Description interview. As a validation platform, the scenario may be used to assess the compiled Universal Access Description interview. Reference scenarios of IS4ALL will be characterised by the following quality characteristics:

- They will unfold the diversity that characterises the use of EHRs; as already pointed out this could be a complex endeavor, therefore the focus will be on a few well defined cases;
- They will reveal current and anticipated use of EHRs; in this context, anticipated use refers to studying a forthcoming situation which is not supported by current practices, such as, for example, devising a suitable

metaphoric representation and style of interaction for accessing an HER from a mobile device with limited display capabilities.

- They will provide a context for design rationale; this implies that scenarios should provide sufficient details of interactive episodes in real contexts of use so that analysis of desired/undesired user consequences can be undertaken
- They will be complete; a scenario is complete when it provides all the information needed for further refinement.

3.2 Consolidation

In a subsequent phase, based on the data gather through scenario analysis the project will consolidate these findings into a code of Healthcare Telematics universal design practice. Schematically, the consolidation phase will follow the iterative cycle outlined in Figure 2.

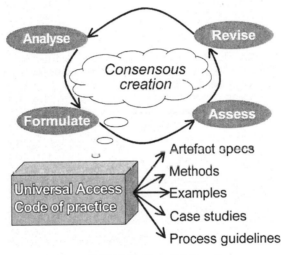

Figure 2. IS4ALL Consolidation phase

The concrete outcomes of the consolidation phase will be:

- A process-oriented code of practice for introducing universal access into Healthcare Telematics product / service lifecycle. Such a code of practice will provide European industries with state of the art reference material and support services on how to approach, internalise and exploit the benefits of universal design in the development of Healthcare Telematics products and services.
- Examples and case studies of good practice
- A handbook of design methods, including principles and guidelines, design techniques and evaluation methods.

3.3 Outreach

IS4ALL plans to reach a wide community within Healthcare Telematics, but also other sectors of the IST inductry. To this effect, the project will undertake several activities. Firstly, IS4ALL will organise seminars, targeted to mainstream IT&T industry, to be held in different European countries. These seminars will help IS4ALL to reach a wide community of potential participants and make them aware of the principles and practice of Universal Access in Healthcare Telematics. Secondly, IS4ALL will target relevant international standards organisations to facilitate updates in draft international standards, or the introduction of new work items which accommodate the project's results. Of primary importance will be standardisation at an international (i.e. ISO) level, and in particular the new Work Item on Accessibility under ISO9241/SC4/WG5. Finally, IS4ALL will participate in a variety of conferences and international workshops, offering presentations and review of relevant results.

4. Concluding remarks

At present, existing guidelines on Universal Access are of a high level of abstraction. This renders them impractical or even unsuitable to be used by industry designers without major efforts. Neither the resources (regarding the number of personnel, the amount of time and money involved) nor the expertise needed to apply these high level guidelines to the needs of individual sectors are available within companies. As a result, despite the fact that the basic wisdom about Universal Design in many cases may exist, the available guidelines remain unused. This gap has been identified as one of the main drawbacks that prevent European industries from applying the principles of Universal Access. IS4ALL constitutes an effort to overcome such a gap in the filed of Healthcare Telematics.

It is important to acknowledge that universal access to Information Society Technologies is a long-term research target. It is just beginning to appear on a more systematic basis on the research agendas of non-market institutions (e.g., NSF in the USA and EC in the EU) and national research funding agencies. In the current stage it is important for the research community to establish a solid R&D agenda, before other support measures can be initiated. The pre-requisite for this is a clear and common vocabulary, which will identify the issues at stake. Following this, new R&D agendas can be discussed.

Acknowledgements

IS4ALL comprises ICS-FORTH, Greece, as the co-ordinating partner, and the following member organisations: including Microsoft Healthcare Users Group Europe (MS-HUGe), European Health Telematics Association (EHTEL), Consiglio Nazionale delle Ricerche – Istituto di Ricerca sulle Onde Elettromagnetiche (CNR-IROE), Forschungszentrum Informationstechnik GmbH (GMD), Institut National de Recherche en Informatique et Automatique – Laboratoire lorrain de recherche en informatique et ses applications (INRIA) and Fraunhofer-Gesellschaft zur Foerderung der angewandten Forschung e.V. - Institut fur Arbeitswirtschaft und Organisation (FhG-IAO). Several other organisations are subcontracted as affiliated organisations of the Network (http://is4all.ics.forth.gr/).

References

Carroll, J. (Ed.) (1995). *Scenario-Based Design: Envisioning Work and Technology in System Development*. New York: Wiley.

Carroll, J. (2001). *Making Use: Scenario-Based Design Human-Computer Interactions*. Cambridge: MIT Press.

Stephanidis, C., Emiliani, P-L. (1999). "Connecting" to the information society: a European perspective. *Technology and Disability*, 10 (1), 21-44.

Stephanidis, C., Salvendy, G., Akoumianakis, D., Bevan, N., Brewer, J., Emiliani, P. L., Galetsas, A., Haataja, S., Iakovidis, I., Jacko, J., Jenkins, P., Karshmer, A., Korn, P., Marcus, A., Murphy, H., Stary, C., Vanderheiden, G., Weber, G., Ziegler, J. (1998). Toward an Information Society for All: An International R&D Agenda. *International Journal of Human-Computer Interaction*, 10 (2), 107-134.

Stephanidis, C., Salvendy, G., Akoumianakis, D., Arnold, A., Bevan, N., Dardailler, D., Emiliani, P. L., Iakovidis, I., Jenkins, P., Karshmer, A., Korn, P., Marcus, A., Murphy, H., Oppermann, C., Stary, C., Tamura, H., Tscheligi, M., Ueda, H., Weber, G., Ziegler, J. (1999). Toward an Information Society for All: HCI challenges and R&D recommendations. *International Journal of Human-Computer Interaction*, 11 (1), 1-28.

Tsiknakis, M., Chronaki, C.E., Kapidakis, S., Nikolaou, C., Orphanoudakis, S.C. (1997). An integrated architecture for the provision of health telematic services based on digital library technologies. *International Journal on Digital Libraries, Special Issue on "Digital Libraries in Medicine"*, 1 (3), 257-277.

Making products user-friendly and charming using Human Design Technology

Toshiki Yamaoka, Takuo Matsunobe

Faculty of System Engineering, Wakayama University, Wakayama, Japan

Abstract

This paper describes how to make products user-friendly and charming using 70 design items and processes of Human Design Technology (HDT). 70 design items are classified into: 1) Usability and user interface design items (29 items); 2) Kansei (sensitivity) design items (9 items); 3) Universal design items (9 items); 4) Product liability (PL) design items (6 items); 5) Robust design items (5 items); 6) Maintenance items (2 items); 7) Ecological design items (5 items); 8) Other (5 items). HDT is consist of 6 steps: (1) Gathering user requirements; (2) Grasping current circumstances; (3) Formulating product concepts; (4) Designing (synthesizing); (5) Evaluating the design; (6) Surveying usage conditions. As one example, an alarm clock was designed based on HDT.

1. What is Human design technology?

Human design technology (Yamaoka, 2000) integrates fields like marketing research, ergonomics, cognitive science, industrial design, usability evaluation and statistics (multiple variable analysis) in order to design user friendly products that have broad popular appeal. It is defined then as technology that scientifically analyzes human beings and uses various information related to humans (i.e. physiology, psychology, cognition and behavior) as design conditions.

The technology not only applies to all processes ranging from product planning to design and evaluation, but its logical, quantitative approach can also be applied all the way through the upstream end of goods production.

In the past we relied on intuition as we reviewed processes as analytically and quantitatively as possible to make sure there were no design condition oversights in the drive to produce sound goods based on user needs.

The situation that precipitated human design technology was the fractionalization of work at upstream stages of product development where individual supervisors focused solely on their own specific area entirely for show and thus crippled their ability to generate products with broad-based appeal. Then came a change in thinking in goods production from an emphasis on material requirements to an emphasis on user requirements that necessitated an ability to grasp the overall situation.

2. HDT process

The following steps comprise the HDT process: (1) Gathering user requirements; (2) Grasping current circumstances; (3) Formulating product concepts; (4) Designing (synthesizing); (5) Evaluating the design; (6) Surveying usage conditions

The range of studies includes gathering and analyzing user requirements, formulating product concepts and then condensing and evaluating the designs. The next step is production and sales which are followed up by surveys on usage conditions for the consumer product.

2.1 Gathering user requirements

The very first step is to extract user requirements using various methodologies like task analysis, group interviews and direct observations. Task analysis and direct observations are stressed in HDT because they can uncover latent user requirements by analyzing unconscious user behavior.

(1) Direct observation method

The following describes direct observation in more detail.

1) Observe based on five aspects of the human machine interface (physical aspect, information aspect, temporal aspect, environmental aspect and organizational aspect).

2) Find vestiges of user behavior because incompatibilities between users and system will always turn up in some form or another.

3) Consider possible operational and behavioral cues.

4) Consider recognizable differences.

5) Examine limitations on user operation and behavior imposed by the system.

(2) 3P(point) Task analysis method

Consider a typical scenario where the product being surveyed will be used. Write down the tasks that will be performed in each scenario in the order that they come up. If a task is comprised of subtasks, then enter these subtasks in the task column as well. Write down real and anticipated user problems in the data processing sequence consisting of effective acquirement of information --> ease of understanding and judgment -->comfortable operation. The cues when you are looking at problem areas are 1) emphasis, 2) simplicity and 3) consistency for effective acquirement of information and then terminology, 1) term, 2) cues, 3) mapping, 4) consistency, 5) feedback and 6) system structure (whether the operating principle of the device is understood) for ease of understanding and judgment and then 1) posture, 2) fit, 3) torque for comfortable operation (Yamaoka, 2000). Finally consider proposals for resolving the real and anticipated problems that were extracted.

It is a good idea to refer to the resolution proposals when they come up because they present opportunities for working out solutions. A rough design proposal can be developed by condensing the resolution proposals. Since the product concept is still undecided at this point however, they will be used strictly for reference.

2.2   Grasping current circumstances

This step determines how users perceive products currently on the market. Make note of the results and use them to come up with remedies. The very simple correspondence analysis method is generally used here.

2.3   Formulating product concepts

This step takes the user requirements that have been gathered and condenses them down into 10 items or less based on like functions. List the condensed user needs (defined as user requirements) appropriately at the top or bottom to form a product concept (Fig1).

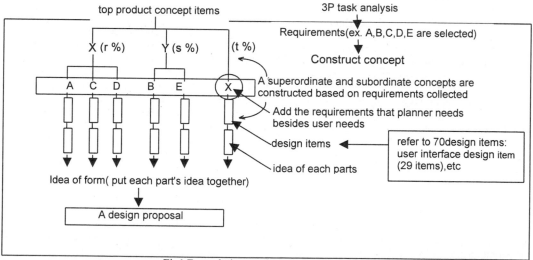

Fig1 Formulating product concept

This method is sufficient in most cases, but it does not reflect the cognitive pattern of users. If you want to look at the patterns systematically in more detail based on real users, ask target user groups of 10 to 20 people to rank the requirements you have extracted and condensed. The analytic hierarchy process (AHP) is a good way to rank (that is to weight) items. Compare the item groups ranked at the top and bottom, and then ask the target users why they ranked the items as they did (this is called laddering). Continue asking the target users why they answered as they did until they run out of answers. Use the same procedure to ask about the item ranked number 2 and all subsequent item groupings until you reach the item ranked at the bottom.

This data represents the cognitive pattern of the target users and you organize it to look at it systematically. Since it is merely a summary of the cognitive patterns of users toward the requirement groupings at this level, the goal of planners must be added to it along with additional items representing purpose. Appropriate design items are added to the items at the very bottom of these system views to complete the system view for a product concept.

The user interface, Kansei(sensitive engineering), universal and physical design items are condensed into these design items for your convenience. Use the AHP process to weight each item in the system view. This weighted value will also be used for the production cost ratio, so items weighted at the low end can be cut if the cost is not right.

2.4   Designing (synthesizing)

Condense the design items at the very bottom of the product concept system view as design proposals on par with the top product concept items. Since item groupings have already been established as shown below for design study areas like the user interface and universal design, use these design items (Yamaoka, Nishimura, et al., 2000)

1) Usability and user interface design items (29 items) (Yamaoka, et al., 1998,Yamaoka, Suzuki, et al., 2000)
2) Kansei (sensitivity) design items (9 items) (Yamaoka, et al., 2000)
3) Universal design items (9 items) (Yamaoka, 1999)
4) Product liability (PL) design items (6 items)
5) Robust design items (5 items)
6) Maintenance items (2 items)
7) Ecological design items (5 items)
8) Other (5 items)

These design items were extracted for the product concept from a viewpoint of user requirements, so create a visual representation, using words if visual representation proves difficult, for each design item as you did with the top item. Bring together all the tentative design items that were visually represented and use them to form a design proposal (Fig1). Design items are usually narrowed down at this stage, but if that is not the case, then propose designs that incorporate undecided product attribute levels, let target users rank the proposals, and then apply conjoint analysis to narrow down the design items. Since you can calculate the rate of contribution for an attribute and the utility value of a level, you should select a design item with a high score.

2.5   Evaluating the design

At this step, let target users evaluate the design proposals using the AHP method in order to study the feasibility of the proposals (meaning that you verify the specifications). You then validate the efficacy of the proposals by applying task or other analysis to a product mockup (product model). You should compare the proposals with rival products monitored in the market using a design rendering and mockup, and you might want to analyze the proposal using the correspondence analysis mentioned earlier as well.

If you want to study the relationship of design items to the overall evaluation of design proposals, then you can give out questionnaires to target users and perform multiple regression analysis on the results to calculate the relative importance of the design items as a weighted value.

2.6   Surveying usage conditions

The usage condition survey is based on the concept of the post occupant evaluation (POE) survey for architects and owners in the building industry. The method determines how a person uses a product once it has been purchased and it allows the user and the person taking the survey to jointly extract problem areas. It therefore relies heavily on direct observation and task analysis, but questionnaires are used as well.

3. Design Items

Design is done using following design items which include the definition and its examples as database. These items are cue to design and make products user-friendly and charming.

3.1   User Interface design items (29 items)

(1) Construction of user-oriented UI system: 1) Flexibility, 2) Customization for different user levels, 3) User protection, 4) Universal design, 5) Application to different cultures
(2) Encouragement of the user's motivation: 6) Provision of user enjoyment, 7) Provision of sense of accomplishment, 8) The user's leadership, 9) Reliability
(3) Construction of effective interaction:
3-1 Effective acquirement of information: 10) Clue, 11) Simplicity, 12) Ease of information retrieval, 13) At a glance interface, 14) Mapping, 15) Identification

3-2 Ease of understanding and judgment: 16) Consistency, 17) Mental model, 18) Presentation of various information, 19) Term/Message, 20) Minimization of users' memory load

3-3 Comfortable operation: 21) Minimization of physical load, 22) Sense of operation, 23) Efficiency of operation

(4) Common keywords: 24) Emphasis, 25) Affordance, 26) Metaphor, 27) System Structure, 28) Feedback, 29) Help

For more information see: (Yamaoka, et al., 1998; Yamaoka, 2000; Yamaoka, Suzuki, et al., 2000)

3.2 Kansei (sensitivity) design items (5 items)

1) Color: Color elements like peaceful colors and unconventional colors.
2) Fit: A sense of human and machine integration such as comfortable shape or an enveloping sense.
3) Shape: Elements like a simple shapes or smart shapes.
4) Functionality and convenience: Elements related to function and convenience such as good functions and usability.
5) Sense of material: Elements that have a sense of material such as the richness of a material or the novel use of a material.
6) Design images: Design image elements like contemporary, nostalgic and chic.
7) Ambiance: Elements like nice interior and relaxing atmosphere
8) New combinations: The effect of completely new combinations such as image and audio combinations or harmonizing contradictory items.
9) Unexpected application: Although this is closely tied to new combinations, it is a basic item that evokes a Kansei reaction.

For more information see: (Yamaoka, et al., 2000)

3.3 Universal design items (9 items)

1) Adjustability, 2) Redundancy, 3) Understanding function and feature at a glance, 4) Feedback, 5) Error tolerance, 6) Acquisition of information, 7) Understanding and judgment of information, 8) Operation, 9) Continuity of information and operation

For more information see: (Yamaoka,1999)

3.4 Product liability (PL) design items (6 items)

1) Elimination of risk. 2) Fool proof. 3) Tamper proof. 4) Guard. 5) Interlock. 6)Warning label.

3.5 Robust design items (5 items)

1) Strong material. 2) Examining shape. 3) Strong structure. 4) Design reduced or avoided stress. 5) Design for unconscious behavior

3.6 Maintenance items (2 items)

1) Keeping space. 2) Easy operation

3.7 Ecological design items (5 items)

1) Durability. 2) Recycling, 3) Very few materials. 4) Most suitable materials. 5) Flexible design.

3.8 Other (5 items)

Five aspects of the human machine interface
1) Physical aspect. 2) Information aspect. 3) Temporal aspect. 4) Environmental aspect. 5) Organizational aspect.

4. Designing an alarm clock based on HDT

For HDT application, an alarm clock was designed as one example (Yamaoka, Nishimura, et al., 2000).
(1) Gathering user requirements: 3P(point) Task analysis was done to collect user requirements.
    Extracted user requirements are good operation, compact size, pretty shape, fresh color and so on.
(2) Grasping current circumstances
    The nine typical alarm clocks on the market were selected to search how users perceive these clocks.
    These clocks are evaluated by 29 persons using 11 adjective like pretty, modern, simple, decorative and so on.

Collected data was analyzed by correspondence analysis method.

(3) Formulating product concepts: Product concept was structured from user requirements. Main concept is "Everybody in a family can operate alarm clock easily"
Sub concepts are 1) fashionable (10%), 2) easy to use(15%), 3) easy to operate(30%), 4) changeable alarm sound(20%), 5) good form(15%), 6) in harmony with indoor(10%).   Each sub concepts are weighted by the percentile values. 70 design items related to each sub concepts were selected to visualize the concept.

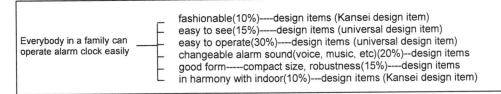

Fig.2 Product concept of an alarm clock

(4) Designing (synthesizing)
A new alarm clock was designed based on design items related to items at the bottom of the product concept.

Fig.3 A new alarm clock design

(5 Evaluating the design
A new design plan of an alarm clock was evaluated compared with the nine typical alarm clocks on the market. Evaluation was done by using AHP(Analytic Hierarchy Process) method.   Subjects are 7 consumers.
As result, this new design plan was evaluated highly.

5. Conclusion

In the paper, I described the concept and design method underlying HDT for making products user-friendly and charming.   The design point in HDT is to provide designs that apply design items to all design keywords at the bottom of a systematic product concept.

References

(1) Yamaoka,T.(2000). Primer of Design Information Science,pp272-298 Japan Standard Association
(2)Yamaoka,T. et al. (2000). International Sensibility Ergonomics Symposium,282-289
(3) Yamaoka,T. et al. (1998). Structured user interface design & evaluation method,
    user interface design committee, Research Institute of Human Engineering for Quality Life
(4) Yamaoka,T.,Nishimura,M.,Takami,I.Hotta,H. (2000). Concept and method of Human Design Technology(1), Proceedings of the 47th Annual Conference of JSSD
(5) Yamaoka,T. (1999). Constructing universal design method,40th Japan ergonomics Conference,16I1-1
(6) Yamaoka,T.,Okada,A. (1999). User interface design in actual practice, Kaibundo Publishing
(7) Yamaoka,T.,Suzuki,K.,Fujiwara,Y. (ed.) (2000). Structured user interface design and evaluation, Kyoritu Publishing Co.

PART 2

ADAPTIVE AND INTELLIGENT INTERFACES

Web-Based Characters For Universal Access: Experience With Different Player Technologies

Elisabeth André

German Research Center for Artificial Intelligence GmbH
Stuhlsatzenhausweg 3, D-66123 Saarbrücken, Germany
email: andre@dfki.de

Abstract

A growing number of web-based applications makes use of lifelike characters or agents as a metaphor for highly personalized human-machine communication. Based on either cartoon drawings, recorded video images of persons, or 3D body models, lifelike characters enrich the repertoire of available communication styles and promise to make the visit of a web site a more enjoyable experience. Unfortunately, a number of characters are inaccessible to a larger group of users since they require the installation of additional software packages or can be started in specific computing environments only. In this paper, we will discuss three different player technologies that are currently being used in various DFKI projects and analyze them with respect to accessibility and usability.

1. Introduction

With the advent of web browsers that are able to execute programs embedded in web pages, the use of animated characters for the presentation of information over the web has become possible. Instead of surfing the web on their own, users can join a tour, ask the lifelike character for assistance or even delegate a complex search task to it.

Lifelike characters promise to increase the accessibility of web-based applications since they allow for the emulation of communication styles common in human-human dialogue. Especially, inexperienced computer users benefit from a more natural style of communication which releases them from the burden to learn and familiarize themselves with less native interaction techniques. Furthermore, lifelike characters provide new opportunities for users with sensory impairments. SigningAvatar[TM] [9] is developing avatars that are able to communicate with deaf users via sign language. Massaro and colleages [6] make use of a talking head to visualize the appropriate articulation for children with hearing loss. Last but not least, there is the entertaining and affective function of an animated character. An empirical study by Lester and colleagues [5] revealed that an animated pedagogical may have a strong positive effect on the students' perception of the learning experience. Mulken and colleagues [8] conducted a study to compare presentations with and without a Persona. It turned out that subjects experienced learning tasks presented by the Persona as being less difficult than those without a life-like character.

To increase the acceptance and usability of a character, its audio-visual appearance, personality, and behaviors need to be tailored to the anticipated user group and application. For instance, users with auditory impairments could benefit from a character that allows for a clear illustration of articulation while the quality of voice is decisive for users with visual impairments who won't notice subtleties in the mimics. Accessibility also requires that a character can be used across a broad range of platforms. In the following, we will report on our experience with three different player technologies for characters that have been employed in various DFKI projects and analyze them with respect to accessibility and usability.

2. The AiA Personas: Java-Based Animated Web-Presenters

In the AiA project (Adaptive Communication Assistant for Effective Infobahn Access), we developed a number of personalized information assistants that facilitate user access to the Web [1] by providing orientation assistance in a dynamically expanding navigation space. These assistants are characterized by their ability to retrieve relevant information, reorganize it, encode it in different media (such as text, graphics, and animation), and present it to the user as a multimedia presentation.

64

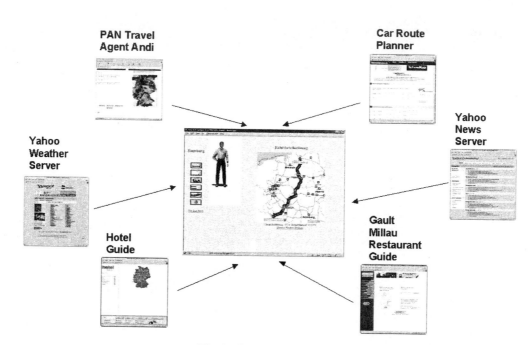

Fig. 1: The AiA Travel Agent

Fig. 1 shows one of our applications, which is a personalized travel agent. Suppose the user wants to travel to Hamburg and is starting a query for typical travelling information. To comply with the user's request, the AiA system retrieves information about Hamburg from various web servers, e.g. a weather, a restaurant and a hotel server, selects relevant units, restructures them and uses an animated character to present them to the user. The novelty of AiA is that the presentation scripts for the characters and the hyperlinks between the single presentation parts are not stored in advance, but generated automatically from pre-authored documents fragments and items stored in a knowledge base using a plan-based approach.

The AiA personas have been realized with DFKI's Java-based player technology. To view a Persona presentation, the user does not need to install any software on his or her local machine. Instead the presentation engine is downloaded as a Java-applet. To support the integration of animated agents into web interfaces, our group has developed a toolkit called PET (Persona-Enabling Toolkit). PET provides an XML-based language for the specification of Persona commands within conventional HTML-pages. These extended HTML-pages are then automatically transformed into a down-loadable Java-based runtime environment which drives the presentation on standard web browsers. PET may be used in two different ways. First of all, it can be used by a human author for the production of multimedia presentations which include a lifelike character. Second, we have the option to automate the complete authoring process by making use of our presentation planning component to generate web pages that include the necessary PET-commands.

3. The Inhabited Market Place: A Presentation Team Based on the Microsoft Agent™ Package

The objective of the Inhabited Market Place is to investigate sketches, given by a team of lifelike characters, as a new form of sales presentation [2]. The basic idea is to communicate information by means of simulated dialogues that are observed by an audience.

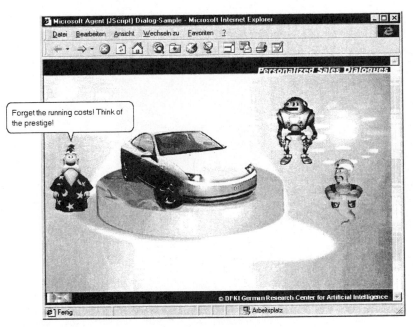

Fig. 2: Inhabited Market Place

Fig. 2 shows a dialogue between the Microsoft characters Merlin as a car seller and Genie and Robby as buyers. Genie has uttered some concerns about the high running costs which Merlin tries to play down. The presentation is not just a mere enumeration of the plain facts about the car. Rather, the facts are presented along with an evaluation under consideration of the observer's interest profile.

To automatically create scripts for the presentation team, we use the same planning approach as for the AiA personas. The outcome of the authoring process is an HTML file that includes control sequences for the characters. For the animation of the characters, we use the Microsoft Agent[TM] package [7]. In contrast to the PET approach, the agents' runtime engine is not part of the web page. Rather, to view the presentation with the Microsoft characters, the Microsoft Agent package has to be installed on the user's local machine.

4. The Virtual News Agency: An Animated Presentation Team Based on SMIL

In the applications described above, the characters' presentations only included static material, such as written text or images. However, a web presentation may also comprise dynamic media objects, such as video and audio, all of which have to be displayed in a spatial and temporally coordinated manner and need to be synchronized with the agents' communicative gestures. Fig. 3 shows a typical application scenario with two animated agents - a newscaster and a technician - that watch and comment on a video while it is displayed on a screen.

Since the other two player technologies do not include any facilities for controlling the timing of other dynamic media, we decided to rely on the Synchronized Multimedia Integration Language (SMIL) in this application scenario. SMIL is a standard recommended by the W3 consortium with the aim of adding synchronization to the Web (see http://www.w3.org/TR/REC-smil/). As in the other two applications, we use our presentation planning technology to generate expressions in the SMIL language that are played within a web browser using a SMIL player that supports streaming over the web, such as Real Player 8 Basic (see http://service.real.com/).

Fig. 3: Virtual News Agency

5. Conclusion

In this paper, we sketched three applications that are based on the same planning technology, but exploit different player technologies for the visualization of the Personas. Our experience has shown that there is not yet a general-purpose player technology which meets the needs of all our applications.

The Microsoft Agent™ toolkit includes a number of useful software packages that support the creation of engaging character applications, such as animation components and components for the recognition and synthesis of natural language. Due to its comfortable application programmer's interface, this toolkit is an excellent choice for rapid prototyping. Being free of charge, it is also a low-cost entry point for new character projects. However, its accessibility and usability is limited by the fact that the Microsoft package needs to be installed on the user's local machine to view a presentation with Microsoft Agent characters. Furthermore, HTML-pages including control sequences for Microsoft Agent characters have to be played within the Microsoft Internet Explorer.

The objective behind the development of the PET toolkit was to support the accessibility and usability of character-based web applications. Being implemented as a Java applet, the PET player runs on any standard web browser. In addition, the user is not required to download a dedicated plug-in or to install other (third-party) software. These characteristics are in particular of interest for commercial web sites. By making use of the HTML event model, PET also fulfills the necessary prerequisites for the creation of interactive Personas. On the other hand, there is a tradeoff between the player functionality and the overall download time of the applet. Therefore, the approach is less suitable for applications that require a rich repertoire of character behaviors and complex media combinations.

SMIL provides a declarative language for the specification of the spatial and temporal layout and in combination with players, such as RealPlayer 8, sophisticated streaming and synchronization technology. These features are in particular useful if a character's communicative actions have to be coordinated with other dynamic media, such as video and audio. However, SMIL does not support the incremental design of presentations and thus does not allow for the integration of interactive Personas. Furthermore, specifying plan operators for SMIL presentations requires some training since it presumes a basic understanding of spatial and temporal constraints.

6. Some Future Directions

Currently, most character developers focus on the Web site owner, e.g. an E-Commerce company, who owns the character. Even if such characters adapt themselves to the specific user, it is still the provider who has the final control over the characters' behavior which is essentially determined by the provider's and not the user's goals. A great challenge for future research is the development of personalized user-owned characters. Whereas provider-owned characters inhabit a specific web site to which they are specialized, user-owned characters may take the user to unknown places, make suggestions, direct her attention to interesting information or simply have a chat with her about the place they are jointly visiting. Clearly, the degree to which an animated agent is able to make an unknown web page accessible to the user will depend to a large extent on the information gathered from a web site. There are several approaches to tackling this issue. One direction is to rely on sophisticated methods for information retrieval and extraction. However, we are still far from robust approaches capable of analyzing arbitrary Web pages consisting of heterogeneous media objects, such as text, images, and video. Another approach uses so-called annotated environments [3] which provide the knowledge that agents need to appropriately perform their tasks. These annotations can be compared to markups of a Web page. Our hope is that with the increasing popularity of agents, a standard for such annotations will be developed that will significantly ease the characters' work.

References

1. André, E., Rist, T., Müller, J. (1999). Employing AI Methods to Control the Behavior of Animated Interface Agents. *Applied Artificial Intelligence* 13:415-448.
2. André, E., Rist, T. , van Mulken, S., Klesen, M., Baldes, S. (2000). The Automated Design of Believable Dialogues for Animated Presentation Teams. In: Cassell et al. (eds.): *Embodied Conversational Agents*, 220-255, Cambridge, MA: MIT Press.
3. Doyle, P., Hayes-Roth, B. (1998). Agents in Annotated Worlds. Proc. of the Third International Conference on *Autonomous Agents*, 173-180, New York: ACM Press.
4. Elliott, C., Brzezinski, J. (1998). Autonomous Characters as Synthetic Characters. In *AI Magazine*, 19(2):13-30.
5. Lester, J.C., Converse, S.A., Kahler, S.E., Barlow, S.T., Stone, B.A., Bhogal, R.S. (1997). The persona effect: Affective impact of animated pedagogical agents. In S. Pemberton, ed., *Human factors in computing systems, CHI'97 conference proceedings*, 359–366. New York: ACM Press.
6. Massaro, D.W., Cohen, M.M., Beskov, J., Cole, R.A. (2000). Developing and Evaluating Conversational Agents. In: Cassell et al. (eds.): *Embodied Conversational Agents*, 287-318, Cambridge, MA: MIT Press.
7. Microsoft (1999). Microsoft Agent: Software Development Kit. Redmond, Wash.: Microsoft Press.
8. Van Mulken, S., André, E., Müller, J. (1998). The persona effect: How substantial is it? In H. Johnson, L. Nigay and C. Roast, eds., *People and Computers XIII* (Proceedings of HCI-98), 53-66. Berlin: Springer.
9. Wideman, C.J. Popson, S.J. (2000). Sign Language Assistive Technology Offers Access to Digital Media, in: Proc. of CSUN'S Sixteenth Annual International Conference "Technology and Persons with Disabilities", http://www.csun.edu/cod/conf2001/proceedings/index.html

Maximizing educational opportunity for every type of learner: adaptive hypermedia for Web-based education

Peter Brusilovsky

School of Information Sciences, University of Pittsburgh
135 North Bellefield Avenue, Pittsburgh, PA 15260

Abstract

The paper discusses the problems of applying adaptive hypermedia techniques, namely adaptive navigation support in Web-based educational courseware. Adaptive navigation support techniques provide support mechanisms, which are tailored to accommodate the current knowledge, background and learning goal of an individual user. Web-based education with adaptive guidance is a way to maximize educational opportunity for every type of learner.

1. Introduction

Web-based education is still far from achieving its main goal - reaching wide distance audience and, in particular, students from underrepresented groups. The experience shows that Web-based courses work relatively well only for well-prepared and well-organized students who know what to learn and can manage their learning. We think that one of the bottlenecks of Web-based education is the course material, which currently comes in various forms - lectures, tutorials, examples, quizzes, and assignments. In all current Web-based courses, the course material is in most cases still implicitly oriented for a traditional on-campus audience - reasonably homogeneous, reasonably well-prepared and well-motivated students. However, Web-based courses are to be used by a much wider variety of users than any campus-based courses. These users may have very different goals, backgrounds, knowledge levels, and learning capabilities. A Web-based course, which is designed with a particular class of users in mind (as it is usually done for on-campus courses) may not suit other users. The only way to proceed is to make the course material richer and more flexible so that different students can get the personalized content and the personalized order of its presentation. Current Web-based courses are not flexible. In the best case the course material is a network of static hypertext pages with some media enhancement. Neither the teacher nor the delivery system can adapt the course presentation to different students. As a result, some students waste their time learning non-relevant or known material and some students fail to understand (or just misunderstand) the material and overload a distance teacher with multiple questions and requests for additional information.

The solution is to develop Web-based courses, which can adapt to the users with very different backgrounds, starting knowledge on the subject and learning goals. Currently, several research groups are investigating various ways to approach this solution (Brusilovsky, 1999). At the moment, the most promising set of relevant techniques can be found the area of adaptive hypermedia (Brusilovsky, 1996). Our own work in this direction is centered around applying various adaptive navigation support techniques to develop adaptive guidance mechanisms which are specially tailored to accommodate the current knowledge, learning goals, and information seeking tasks of that individual user. Guidance in this context addresses the problem of a user's unproductive wandering, refocusing them on their learning objectives, suggesting logical next steps to inform them about the knowledge structure of the hyperspace, or re-sequencing materials according to their demonstrated knowledge of content. Adaptive guidance is especially important for Web-based courses because in many cases the user is "alone" working with it (probably from home). That is why "external" guidance that a colleague or a teacher typically provides adaptively in a normal classroom situation, is not available.

To explore the problems of adaptive guidance in educational context we have developed *InterBook* (Brusilovsky, Eklund, & Schwarz, 1998) - a system for authoring and delivering adaptive electronic textbooks (AET) on the WWW . InterBook uses a content metadata (prerequisite concepts and outcome concepts) about every courseware object and a overlay model of student knowledge to provide every student with several kinds of adaptive navigation support and guidance - adaptive page sequencing, adaptive help, and adaptive link annotation.

The paper reviews adaptive navigation support functionality provided by InterBook, provides a brief review of similar works in this area, and advocates the use of this technology in large-scale Web-based education.

2. Knowledge representation and content structuring

The key to adaptivity in an adaptive textbook is knowledge about its domain represented in the form of domain model and knowledge about individual students represented in the form of individual student models. The domain model serves as a basis for structuring the content of an adaptive ET. We distinguish two content parts in each AET: a glossary and a textbook . This section provides some minimal information about knowledge representation and content structuring. Some more information can be found in (Brusilovsky, et al., 1998).

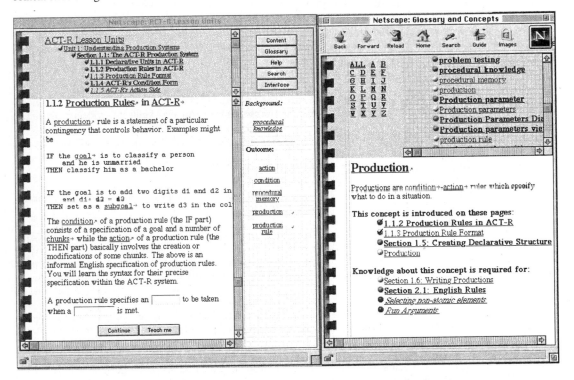

Figure 1. A section of an electronic textbook (left) and a glossary page (right) as presented by InterBook. Links to all book pages are annotated with colored bullets. Links to glossary concept pages on the right panel of the textbook window are annotated with checkmarks.

AET uses the simplest form of domain model: a set of domain concepts. By concepts we mean elementary pieces of knowledge for the given domain identified by a domain expert. The domain model provides a structure for an overlay student model which is a representation of the student's knowledge of the subject. For each domain model concept, an individual student's knowledge model stores some value which is an estimation of the student knowledge level of this concept (for example, "unknown", "learned" and "well-learned").

In InterBook, each AET is hierarchically structured into units of different level: chapters, sections, and subsections. To make AET "more intelligent" and to connect it to the glossary, we have to let the system know what each unit of the textbook is about. It is done by indexing of textbook units with domain model concepts. Several books on the same subject form a bookshelf. All books from the same bookshelf are indexed with the same set of domain model concepts. Each terminal unit has an attached list of related concepts (we call this list spectrum of the unit). For each involved concept, the spectrum represents the name and the role of the concept in the unit (each concept can be either a outcome concept or a prerequisite concept). The system has an option to show all outcome and background concepts for the current section on a page border to the right of the section content (Figure 1)

The Glossary is, in fact, a visualized domain network. Each node of the domain network is represented by a node of the hyperspace, while the links between domain network nodes constitute main paths between hyperspace nodes. The links between domain model concepts constitute navigation paths between glossary entries. Thus, the structure of the glossary resembles the pedagogic structure of the domain knowledge. In addition to providing a description of a concept, each glossary entry provides links to all book sections which introduce or require the concept. This means that the glossary integrates traditional features of an index and a glossary.

3. Student modeling and adaptive guidance

To support the student navigating through the course, the system uses adaptive annotation, adaptive sorting, and direct guidance technologies. Adaptive annotation means that the system uses visual cues (icons, fonts, colors) to show the type and the educational state of each link. Direct guidance means that the system can suggest to the student the next part of the material to be learned.

The key to all adaptive functionality is student modeling. The system maintains an up-to-date model of individual student knowledge on the subject. The student modeling mechanism accepts two kinds of evidence of student knowledge of a concept:

- a student has visited a page which presents some information about a concept (i.e., the page has this concept among outcome concepts)
- a student correctly answers a question which checks the knowledge of this concept

The latter evidence is stronger, so no "well-learned" grade can be given to a concept unless the student confirms his or her knowledge by answering a test.

Using the student model, it is possible to distinguish several educational states for each unit of an electronic book: the content of a unit can be known to the student (all outcome concepts have been already learned), ready to be learned, or not ready to be learned (the latter example means that some prerequisite knowledge is not yet learned). The icon and the font of each link presented to the student are computed dynamically from the individual student model. They always inform the student about the type and the educational state of the unit behind the link. In InterBook, red means not ready to be learned, green means ready and recommended, and white means no new information. A checkmark is added for already visited units (Figure 1). The same mechanism can be used to distinguish and show several levels of student knowledge of the concepts shown on the concept bar. In InterBook, no annotation means "unknown", a small checkmark means "known" (learning started), a medium checkmark means "learned" and a big checkmark means "well-learned" (Figure 1). For many students, adaptive guidance provides enough support to make a navigation decision. Those who hesitate to make a choice could push the button "Teach me" and the system will apply several heuristics to select the most suitable node among those ready to be learned.

Another type of guidance that the system can provide is goal-based learning. The system knowledge about the course material comprises knowledge about what the prerequisite concepts are for any unit of the textbook. Often, when students have problems with understanding some explanation or example or solving a problem, the reason is that some prerequisite material is not understood well. In that case they can request prerequisite-based help (using a special button) and, as an answer to help request, the system generate a list of links to all sections which present some information about background concepts of the current section. This list is adaptively sorted according to the student's knowledge represented in the student model: more "helpful" sections are listed first (Figure 2). Here "helpful" means how informative the section is to learn about the background concepts. For example, the section which presents information about an unknown background concept is more informative than a section presenting information about a known concept. The section which presents information about two unknown background concepts is more informative than a section presenting information about one concept.

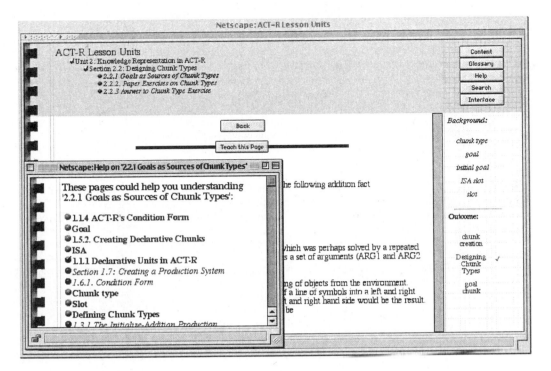

Figure 2. Prerequisite-based help is provided using adaptive link sorting technology

4. Similar works and prospects

A number of works suggest the use of adaptive navigation support technologies for educational applications. Adaptive link annotation is currently the most popular technology. Most often these annotations are provided in the form of visual cues, i.e. different icons, font colors, or font types. In Web context It was used first in ELM-ART and since that applied in all descendants of ELM-ART such as InterBook and many other systems (Brusilovsky, Schwarz, & Weber, 1996; De Bra, & Calvi, 1998; Henze, Naceur, Nejdl, & Wolpers, 1999). ELM-ART and InterBook also use adaptive navigation support by sorting. Another popular technology is hiding and disabling. The options here are either to remove the link, make the link completely non-functional, or mask it. This technology was explored in several systems such as AHA (De Bra, et al., 1998).

We think that adaptive navigation support technologies can be successfully used to provide adaptive guidance in the hyperspace of learning material in the context of Web-based education thus maximizing educational opportunity for every type of learner. Our own work and the work of other researchers brought promising results. In particular, empirical studies have shown that adaptive navigation support can significantly reduce navigation overhead, encourage students to use non-linear navigation and spend more time working with the system, and improve quality of learning (Brusilovsky, 1998). Our own approach based on prerequisite-outcome indexing of learning material looks particularly attractive. It is relatively simple, scalable, yet powerful enough to guide the students. We have already used it for implementing AET for Web-based courses. Currently we are extending the approach to handle large-scale Web-based courses. At the moment the price of indexing a course may look too big for a benefit of being able to provide an adaptive guidance. However, we think that in the future concept-based guidance will be an important part of any courseware engineering system. The emerging standards on Learning Objects Metadata (IEEE LTCS WG12, 2000) are quite compatible with our prerequisite-outcome approach to course indexing. As soon as larger and larger amounts of indexed course material will become available, the use of use of concept-based adaptive guidance will become cost-efficient.

References

Brusilovsky, P. (1996). Methods and techniques of adaptive hypermedia. *User Modeling and User-Adapted Interaction*, **6** (2-3), 87-129.

Brusilovsky, P. (1998). Adaptive Educational Systems on the World-Wide-Web: A Review of Available Technologies. *Proceedings of Workshop "WWW-Based Tutoring" at 4th International Conference on Intelligent Tutoring Systems (ITS'98)*, August 16-19, 1998. San Antonio, TX, - pp. , available online at http://www-aml.cs.umass.edu/~stern/webits/itsworkshop/brusilovsky.html.

Brusilovsky, P. (1999). Adaptive and Intelligent Technologies for Web-based Education. *Künstliche Intelligenz*, (4), 19-25, available online at http://www.contrib.andrew.cmu.edu/~plb/papers/KI-review.html.

Brusilovsky, P., Eklund, J., & Schwarz, E. (1998). Web-based education for all: A tool for developing adaptive courseware. *Computer Networks and ISDN Systems*, **30** (1-7), 291-300.

Brusilovsky, P., Schwarz, E., & Weber, G. (1996). ELM-ART: An intelligent tutoring system on World Wide Web. In C. Frasson, G. Gauthier, & A. Lesgold (Ed.), *Third International Conference on Intelligent Tutoring Systems, ITS-96* (Vol. 1086, pp. 261-269). Berlin: Springer Verlag.

De Bra, P., & Calvi, L. (1998). AHA! An open Adaptive Hypermedia Architecture. *The New Review of Hypermedia and Multimedia*, **4** 115-139.

Henze, N., Naceur, K., Nejdl, W., & Wolpers, M. (1999). Adaptive hyperbooks for constructivist teaching. *Künstliche Intelligenz*, (4), 26-31.

IEEE LTCS WG12. (2000). *LOM: Working Draft Document v3.6*. Learning Object Metadata Working Group of the IEEE Learning Technology Standards Committee,

Interface design for mass-customization in e-commerce

Martin G. Helander

Nanyang Technological University
Singapore 639798
mahel@ntu.edu.sg

Halimahtun M. Khalid

Universiti Malaysia Sarawak
94300 Kota Samarahan, Sarawak, Malaysia
mkmahtun@idea.unimas.my

Abstract

A model for design of a mass-customization interface is presented. Information on Customer Needs is used to predict Functional Requirements, which in turn determine the Design Parameters of the merchandize. Customers will then interact with the web page and design their product and submit their order. Once products have been sold, the sales will be analyzed by the Marketing, and the resulting information will be used to update Customer Needs. There are several critical human factors issues, including the assessment of customer needs, the design of a platform of products for mass-customization and the design of the web page for input of customers design options and specifications.

1. Introduction

The objective of mass -customization is to produce highly customized products rapidly and cost-effectively. This strategy involves more than the manufacturing, logistics system, and marketing. It requires that manufacturers can understand customer needs, so that customer preferences can be incorporated in product design, see Table 1.

Several companies, such as Dell Computers and Oracle offer web-based design of artifacts. For an overview see Helander and Khalid (1999). In Dell's case, a customer may select various functionalities and components to compose the computer of his choice. This task is fairly simple – it can be finished without consideration to esthetics or component location. The case of Oracle is different - a customer will design a network server. Typically a professional working for a company performs this task. It may take extended time, since the components must be carefully selected to optimize the performance of the system.

A company that offers mass-customized products has a difficult planning and design task. This is not only an issue of designing a single product, but rather a product platform that can present so many varieties of a design, that a majority of customers can find a satisfactory combination of product features.

In the past, the Do-It-Yourself (DIY) concept implied that a customer assembled a product designed by a company. Here the roles are reversed: A customer designs while the company assembles. Hence we need to understand why the customer/designer would like to engage in this activity and what the underlying motivation is.

There are several human factors issues: Motivation - why do customers like to design? How can the design task be presented so that the customer can easily understand available options? From a procedural perspective - how do customers go about design? Are certain design procedures more usable than others – such as top-down design or bottom-up design?

Table 1: Characteristics of Contemporary and Customized Manufacturing

Contemporary Manufacturing		Customized Manufacturing
Mass production	→	Mass customization
Standard specifications	→	Customized specifications
Design for the customers	→	Design by the customer
Customer-designer-manufacturer link	→	Customer/designer-manufacturer
Vendor delivery	→	Direct delivery
Off-line communication	→	Web-based communication

2. Objective

There is one major objective of this paper - to present a framework that can be used for predicting and modeling user needs, as they are expressed in DIYD for mass customization. Such information can be used for design of a product platform as well as a user interface where customers may easily express their needs both with respect to the selected product features and for the purpose of useful design procedures.
A secondary objective is to present some priorities for future research.

3. A Model of Mass Customization through DIYD Design

A model of mass customization is presented in Figure 1. It is necessary to assess customer needs, since these are fundamental to the design of the product as well as the web presentation. However, initially an e-commerce outfit for customized products will have a poor appreciation of customer needs. This is because the commodity is new, and mass customization is also new. Over time, as experience is gained, and the company understands customer preferences, it will be easier to predict customer needs.

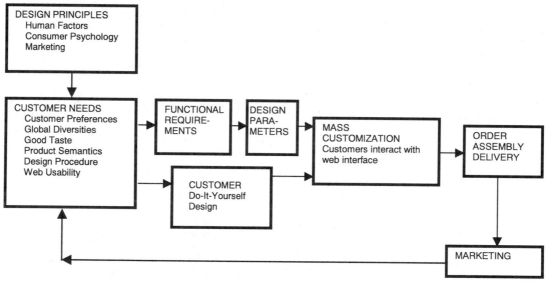

Figure 1. Estimating Customer Needs for Do-it-Yourself Design

In order to facilitate mass customization one must also identify interface design features that make the Web site more usable, thereby assisting the customer during the various phases of Do-It-Yourself-Design (DIYD) process. From the estimation of customer needs we may formulate Functional Requirements (FRs) and Design Parameters (DPs) (Suh, 1990). The FRs identify the goals/purposes of the product at a high level and the DPs identify the Product/Parts Catalogue. An example is given below.

3.1 *Customer motivation.* Why do people like to design their own products?

There are tow major issues - the ability to design and the motivation to design.

Research suggests that design ability is possessed by everyone (Cross, 1995). Through design, individuals have the opportunity to express their personality. The potential benefits can be far reaching, such as maximizing customer values and dealing with actual customer needs.

In order to reflect individuality, people have become increasingly preoccupied with "designer labels"; designed objects are regarded as having status and value. Taste reflects our personality and also reveals part of our personality to others (Abelson, 1986). Through design, individuals have the opportunity to express their personality. Dilnot (1997) went even further in claiming that: "Objects (help) make us. Making (and designing) are moments of making (and designing) ourselves ... The object is the substitute for ourselves in the special sense that things work to provide us artificially what nature "neglected" to bestow us."

To summarize, opportunities to express their needs directly via a Web-based system may enhance customer satisfaction. Given the opportunity, customers will want to design themselves and purchase their own design – the individualized artifacts. In the future customers will no longer be passive recipients of goods and services; they will like to be involved.

3.2 Estimating Customer Needs.

For a new product it can be very difficult to estimates customer needs. Goel and Pirelli (1992) observed that the results of market investigations for new products are not always informative – the true market needs are not divulged from a post survey, they can only be estimated after customers have actually bought the product.

Lacking good estimates of market needs one is then limited to past sales experiences and principles of good taste. Psychological research can help in clarifying some of the controversies of what constitutes good taste. Research in Gestalt psychology emphasizes the equilibrium and over all impression of objects. Objects with visual balance have, for centuries, been said to be particularly pleasing (Zee, 1986). Objects that are symmetrical and have even weight distribution are regarded as more "balanced" and preferred over other objects (Locher et al., 1992). Although these findings seem to support the idea of simple design over complex, Berlyne (1971) claimed that people prefer complexity: they look more at complex figures than simple figures. He also mentioned that people prefer objects that are moderately familiar to objects that are over-familiar. Martindale and Uemura (1983) suggested that aesthetic preference is linked to the physiological arousal potential of artifacts and that preference diminishes with increased familiarity. Folk psychology is ambivalent in this respect: "old favorites" versus "familiarity breeds contempt".

3.3 Design procedures.

There are potential problems in designing or building products. One major problem is the limitations of the human information processing system including the working memory. This makes it difficult to build or specify products with more than 10 items. In setting up a Web-based interface for design and mass customization, a significant challenge is to help customers formulate their implicit requirements and make them explicit to the manufacturer. In this process one cannot overload the customer with seemingly "trivial" questions. Instead, one has to assume how certain product details can be designed by drawing from principles of "good taste", that is generic preferences that have emerged in human factors and marketing research. These principles can be implemented in a product design and thereby offer the potential customer a better opportunity to design a satisfactory product.

The purpose of this research is therefore to make options easy to understand and easy to incorporate into mass customization design. There are several potential ways to perform mass customization, including:
* Selecting from a list of alternative items from one or several menus;
* Selecting items one-by-one either bottom-up or top-down;
* Assembling sequentially;
* Building subassemblies first, which are then assembled in turn;
* Substituting items in a finished design; or
* Putting together a puzzle of parts presented visually.

The optimal strategy for mass customization depends on what object is being designed (i.e. Apparel, consumer products, computers.

In a separate study (Helander and Khalid, 1999) experiments were conducted to investigate the preferred design procedures on the web, using the design of a watch as an example. The task involved the selection of several different design elements including: type of watch, casing, dials, strap, numerals, color, and so forth. The watch is presented at www.idtown.com. The study concluded that that top down presentation of design items is preferred.

The results may seem a little surprising since Guindon (1990) showed that designers do not follow a top-down design procedure. However, our results indicate that when designers need to use design elements presented on the web, the logical expectation is that they must be organized top-down.

4. Discussion

4.1 Customer Needs

This article has summarized several human factors design problems for interface design for mass customization. There is much evidence to suggest that mass customization will actually take off. Dell Computers and Oracle provide much support for this. In order to make mass customization palpable to the customer we must solve the problems of design of a product platform. The design process turns out to be much more problematic in mass customization than for design of single products. This is because a product platform must be proposed that can cater to differences in customer needs. There are global and ethnic differences, which must be understood. Research has not yet produced much guidance in this respect. Our understanding of global differences in shopping patterns remains superficial.

In order to design the product platform it is necessary to estimate Customer needs. This is not easy, and for new products it may be close to impossible. AS result companies will therefore often introduce a new product without much though to customer needs, but with the expectation that the market introduction will furnish them with viable date so that the product design can be quickly updated. This has become common policy for example in software design, where the version 1.0 is expected to be replaced immediately.

Lacking information about customer needs, a designer must then resort to principle of "good taste". Some psychological research is available that can guide a designer in this respect.

There is a great need for research to model customer needs. The present literature provides only a few rudimentary models.

4.2 Design of the Web Environment

In conventional shopping customers rely on sales personnel to help with their product selection. In the same way, manufacturers have relied on sales and marketing people to predict the preference of customers. With e-shopping, however, there is no middle-person because customers communicate directly with the manufacturer. This may be problematic. Similar situations occur with catalog shopping. However, the problem with mass customization is worse because while the former only requires the customer to select from a limited list, the latter aims to provide what the customer wants.

A presentation of an exhaustive list may not be possible and would, in fact, confuse customers. As a consequence, it is important to aid customer's Decision-Making Process. We must therefore understand how to present the product elements for mass customization. The optimal presentation format depends on the type of customer (Kahn and Huffman, 1998). For example, servers manufactured by Oracle are difficult to specify since they require deep understanding of computers. This is a task for a computer expert employed by a company and this customer is much different from the casual customer that we have in mind in this paper. To buy a Dell computer is a much less demanding task, since there are nnot so many design elements to specify.

There is hardly any research in this important area, so any initiative would be important. It is, for example important study the Decision-Making Process (DMP), customer satisfaction levels, and search effectiveness change as functions of searching methods (e.g., tree, keywords, visual inspection), task complexity (e.g., number of choices presented on screen), and customer variables (e.g., accountability associated with the decision, time pressure) (Coombs and Avrunin, 1977).

It is also important to develop one or several cognitive models of customer decision behavior. Some models have a theoretical basis in decision making research, such as Expected Utility Theory, Prospect Theory, Multi-Attribute Value Models, the Lens Model, Preference Trees, Elimination by Aspect, Framing, and Regret Theories.

References

Abelson, R.P. (1986). Beliefs are like possessions. *Journal for the Theory of Social Behavior*, 16, 223-50.

Coombs, C., Avrunin, G.S. (1977). Single peaked preference functions and theory of preference. *Psychological Review*, 84 (2), 216-230.

Cross, N. (1995). Discovering design ability. In R. Buchanan & V. Margolin (Eds), *Discovering Design*. Chicago: The University of Chicago Press.

Crozier, R. (1994). *Manufactured Pleasures. Psychological Responses to Design*. Manchester: Manchester University Press.

Dilnot, C. (1997). The gift. In V. Margolin & R. Buchanan (eds.), *The Idea of Design*. Cambridge, MA: MIT Press.

Goel, V., & Pirolli, P. (1992). The Structure of Design Problem Spaces. *Cognitive Science, 16*, 395-429.

Guindon, R. (1990). Designing the Design Process: Exploiting Opportunistic Thoughts. *Human-Computer Interaction, 5*, 305-344.

Helander, M.G., & Khalid, H.M. (1999). Modelling Customer Needs in Web-based Do-it-Yourself Design. In Abesekwera, J., Lonnroth, E-R, Piamonte, D.P.T., & Shanavaz, H. (Eds.), *Proceedings of the 10th Year Anniversary Ergonomics Conference*, (pp. 9-13). Sweden: Lulea University of Technology.

Kahn, B., & Huffman, C. (1998). Variety for sale: Mass customization or mass confusion? *Journal of Retailing*, 74(4), 491-513.

Locher, P., Smets, G., & Overbeeke, K. (1992). The contribution of stimulus attributes, viewer expertise and task requirements on visual and haptic perception of balance. In *Proceedings of the 12th International Congress of the International Association for Empirical Esthetics*, Hochschule der Kunste, Berlin.

Martindale, C., & Uemura, A. (1983). Stylistic change in European music. *Leonardo*, **16**, 225-8.

Suh, N. (1990). *Principles of Design*. Oxford: Oxford University Press.

Zee, A. (1986). Fearful Symmetry: The Search for Beauty in Modern Physics. New York: Macmillan.

Towards Automatic Adaptation of Data Interfaces

Claire Knight, Malcolm Munro

Visualisation Research Group, Research Institute in Software Evolution, Department of Computer Science, University of Durham, Durham, DH1 3LE.

Abstract

Personalisation and user interface technologies are now at the stage where user preference can be inferred with some degree of success. Such approaches are not necessarily always the best approach as user autonomy is important for some. There is also potential for component based data interfaces. Such interfaces can be loaded and applied on demand with (distributed) component technologies and running code can be replaced on the fly. This theoretically enables multi-metaphor data interfaces to be taken a step further from the direct customisation level at which most currently work. This paper examines these future moves from the point of view of a directly customisable multi-metaphor visualisation tool. It currently allows visualisations and data sources to be linked and provides much of the connectivity for the user; enabling non-expert use. The automatic adaptation of such multi-metaphor tools can stem from this work. This can only help towards the goal of information access for all.

1. Introduction

The goal of an Information Society for All requires that interfaces to software are changed to be able to accommodate the wide range of expected users. Because no two users are alike, and all users have their own preferences and ways of working then flexible interfaces that provide choices in the representation of data, and allow for expansion, provide a way of overcoming the subjective nature of interfaces. Unfortunately this goal of having an easily personalised and customised user interface is easier to perceive than to create. Many issues exist with technology constraints, and also with what should be expected of a user. Obvious changes such as the colour scheme used can be inferred from the windowing system defaults; indeed these may be enforced by the window managing software. What about the case where such defaults are not required, or cause problems with some other information display? Something that seems trivial has many possibilities.

There is also the maintainability concept where users all have their own displays, layouts, and favourite ways of using the software. There has to be some way in which this can be reconciled for software updates and versioning; and all the better if these can maintain user preferences through the change process. Software is not a trivial object; it is complex on its own, and that complexity increases with the number of interactions that it participates in.

The research presented in this paper puts forward a framework in which the ideals of an Information Society for All, and therefore catering for a wide range of diversity, can start to be achieved. It has been reached through the desire for more customisation and flexibility with visualisations of large abstract data sources, and these provide examples of the concepts. Future work that derives from this is the automatic inference of aspects of the interface. This can help both novice users who need an entry route to the system, and advanced users who already have ways of working with the system and expect changes in interfaces to reflect this prior experience.

2. Personalisation & customisation

Personalisation is something that is hard to quantify because the term is used to represent many aspects of a customisable and flexible interface (Kramer et al.). It is important to remember that the key aspect of a personalisable interface (whatever that may mean!) is that it needs to bring some form of *value-add* to the user of that interface. There is always the problem, especially with new technology, of falling into the feature trap so beloved of software applications developers; the addition of fancy features, mainly to act as selling points, or as something that their competitor's products do not have. It is necessary to ensure that this does not happen with any interface, but personalisable features should be included for a justifiable reason.

An interface will be deemed successful if it is able to support the user in achieving their current goal. This means that the personalisation aspects should be related to the tasks expected to be supported. Obviously domain knowledge is also of importance with any personalisation. If something is always constant in a given domain, making that a frequently configurable parameter is likely to cause users, particularly experienced ones, irritation. Conversely, some editing of "constants" is necessary for the once every x years that the value changes.

There is also a need to empower users through supporting them in the tasks and activities that they need to carry out, but not to restrict them too much in the way this is done. A flexible interface is much more likely to be able to adapt to slight changes in working practices than a rigid one. Clearly large changes may force some interface redesign, but in cases like this there is usually a system redesign as well. The impression of empowerment is an important one; it supports and promotes the view of a usable interface, and it allows for the creation of intelligence amplifying tools. Tools that help with the completion of real tasks become pervasive (at least in the minds of the users) and can therefore be considered to be a success (Karat et al.). Intelligence amplification does not seek to replace users in the way that much of artificial intelligence does. Instead it looks to support users through aiding them with tasks that a computer is more than capable of doing, and then allowing the insight and experience of the user to be utilised to actually achieve the end result. Intelligence amplification also supports the view the tools should support the humans, and not control the activities that they are able to carry out.

Personalisation in many ways means that an interface feels like it was designed for that user (or at worst, that class of users). Customisation deals more with the configurable items that an interface has to enable any user to fine-tune certain aspects of it. With current technologies and algorithms there is some cross over between these two concepts, but this distinction is appropriate for this work.

Much of computer science remains an art, despite the scientific endeavour and theory behind it. Those facets of computer science that deal with graphics and interfaces (in any combination) come even closer to the view of feeding information into a magic black box and then the perfect result appearing at the other end. Those within the discipline appreciate that this is not the case and that engineering and scientific principles can be applied. This research is an attempt to achieve this in a flexible way.

3. Visualisations

Visualisations are unusual in that there are often two interfaces involved for interaction with the user. This can even be the case even with three-dimensional visualisations located in virtual environments composed of sensors and panel displays. There is the more usual GUI that is used to start the application (if on a desktop display) and often for the configuration of the visualisation with, for example, the data to be utilised and the representations and metaphors required. There is then the visualisation itself, which from a usability engineering (Neilson) point of view should be able to directly interacted with, at least at a rudimentary level such as navigation.

Visualisation is also subject to many of the same problems as interfaces over what is considered to be aesthetically pleasing, easy to use, easy to learn, necessary displays for tasks and so on. Because of the high level of subjectivity, and in many ways more so than for GUIs because of the freedom of display dimensions and metaphor, the ability to personalise and customise visualisations is seen as vital in ensuring their acceptance in standard interfaces and from usability engineering perspectives.

It has been suggested that representation (at an interface level rather than for visualisations, in which case the metaphor is the guiding factor) ultimately affects the presentation and interaction with a service or product (Pednault). It is because the representations used limits the extent to which personalisation and customisation can be applied, and also the information that can be captured and utilised towards furthering this aspect of the interface. Some of this is obviously visible to the user through the display presented to them, and in the range of interaction devices that can be utilised with it. Other aspects of it relate to the structures and code that exist in that product or service at the programming and architectural side of things. Because so many of these issues are dependent on the code and the requirements upon which it was based, then it is hard to say what these should be for all systems. A step in the right direction would be something like XML (XML) whereby data exchange could be made common through the use of embedded descriptions of that data structure. This issue is included here because it is similar to the problem experienced by many visualisations. It is technically possible to factor and/or transform most data sets, especially given a range of interfaces to data stores, but whether that is meaningful for the visualisation is a different issue! (Knight)

Based on all of these issues the concept of a visualisation and interface framework has been developed. At the moment it relies on standard GUI technologies for the initial display and much of the control of the prototype, but the concepts still apply with smaller devices and other advances in technology. It provides a framework into which various data sources and visualisations can be registered, and attempts to broker between those the user wishes to use. This gives great flexibility and allows for an environment that can be customised (as far as the available plug-ins) for the user and task.

4. Flexible interface framework

This section describes the framework that has been developed, and is being refined through experience, for developing flexible visual analysis and understanding systems through the integration of visualisations and data sources. Such a framework also has a place in interface technology. For the same reason that you need to give users both an easy route into using a system (through default displays) and also the ability to configure to their way of working as they gain more experience. This empowering of users means that they are more likely to accept the tool. From both an interface and a visualisation perspective this type of framework goes some way towards making the goal of information access for all a reality.

Similar research has been done for visualizations, such as the work on creating a database schemata dependent framework. *Snap-Together Visualization* (North) is a model that creates explicit interaction links between multiple views, and is based on relations and actions. Interactions with objects in one view cause the views at the other end of the relationships to also look that data up based on the action defined for that link, and thus changing an object in one view will update all linked views. This work differs in that it seeks to be more flexible. It does not rely on the presence or use of database technologies. It also aims to work at a more automated level for the linking and translation between data sources and visualisations. *Snap* requires that links are made explicitly which means that users need to be aware of databases and the use of keys to identify related data.

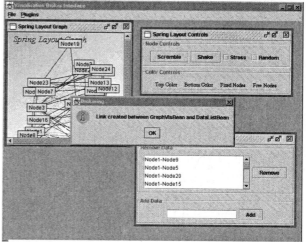

Figure 1 - Plug-in interface

The framework is based on a plug-in technology that allows several types of plug-ins to be registered with the system. Once these plug-ins are known about then the system can utilise them in combination. At the moment this is under user direction, and there is very little in the way of data brokering (from a visualisation perspective) to ensure the images meet the data needs of the information. The next step is to utilise some degree of automatic personalisation, but still allow full user control should they wish. The concept of brokering can also be brought into play with users controlling actual linking of visualisations to data. This allows a given data source to know which of the registered plug-ins would be suitable for displaying it, and also for any visualisation to be aware of the actual data sources that are usable in its display.

Figure 1 shows a view of the prototype in action. This has two components plugged in. The top two windows are connected with the visualisation shown in the top left, whilst the bottom right displays the data source it is currently working with. It is appreciated that this interface is rudimentary and a standard GUI from an interface perspective, but it is theory that is important to consider. The fully specified framework will allow much more plug-in flexibility, to the extent of interacting with other plug-ins, the framework, and so on. It will also have some degree of automation to deal with personalisation and brokering of data. All of these are necessary for a flexible, personalisable interface to allow access for all.

5. Further work

There are many ways in which this work can develop. The main direction is to incorporate linking with distributed provisions such as lookup and directory services to allow for utilisation of remote and/or third-party interface objects. There are obviously issues to do with licensing, actual abilities of the remote code, and security that need to also be addressed. There is also the concept of service provision which industrial initiatives such as .NET (Microsoft), J2EE (Sun – Java 2 Enterprise Edition) and JINI (Sun – JINI) are considering now. These are (in various guises) frameworks for developing distributed component based services and may provide mechanisms to solve some of the problems, such as security, which would be an issue for remote plug-ins.

6. Conclusions

This interface has presented a plug-in framework that allows for user freedom in the utilisation of visualisations and data sources to enable them to carry out analysis. Such a framework is necessary if the ideal of information for all is to be achieved because of the huge variability in user preferences. The flexibility to utilise different displays on different devices to suit the user and task is important. Empowering the user is a good way of achieving user satisfaction, and interface acceptance.

References

Karat, J., Karat, C., and Ukelson, J. (2000). Affordances, Motivation, and the Design of User Interfaces, Communications of the *ACM, Vol. 43, No. 8*, August 2000.

Knight, C. (2000). *Virtual Software in Reality*, Ph.D. Thesis, Department of Computer Science, University of Durham, June 2000.

Kramer, J., Noronha, S., and Vergo, J. (2000). A User-Centered Design Approach to Personalization, Communications of the *ACM, Vol. 43, No. 8*, August 2000.

Microsoft (2001). *Microsoft .NET Home Page*, current page available on the world wide web with last known update in early 2001: http://www.microsoft.com/net/

Neilson, J. (1993). *Usability Engineering*, San Francisco, Academic Press Professional Publishing.

North C., and Shneiderman, B. (2000). Snap-Together Visualization: A User Interface for Coordinating Visualizations via Relational Schemata, *Advanced Visual Interfaces (AVI) 2000*, May 2000.

Pednault, E. P. D. (2000). Representation is Everything, *Communications of the ACM, Vol. 43, No. 8*, August 2000.

Sun (2001). *Java™ 2 Platform, Enterprise Edition*, current page available on the world wide web with last known update on 13th March 2001: http://java.sun.com/j2ee/?frontpage-javaplatform

Sun (2001). *Jini™ Network Technology*, current page available on the world wide web with last known update in February 2001: http://www.sun.com/jini/

XML (2001). *XML.ORG – The XML Industry Portal*, current page available on the world wide web with last known update on 12th March 2001: http://www.xml.org/

Intelligent Affective Interfaces:
A User-Modeling Approach for Telemedicine

Christine L. Lisetti[1], Michael Douglas, Cynthia LeRouge

Information Systems and Decision Sciences,
University of South Florida
4202 East Fowler Avenue CIS 1040
Tampa, Florida, USA*
{lisetti,mdouglas,clerouge}@coba.usf.edu
http://coba.usf.edu/lisetti

Abstract

We address some of the current challenges in intelligent interfaces for universal access and present the design of an intelligent interface which is aimed at 1) input processing the user's sensory modalities (or modes) via various media, 2) building (or encoding) a model of the user's emotions (MOUE) and 3) adapting its multimedia output to provide the user with an easier and more natural technology access and interaction. We identify key research issues relating to one particularly rich application of MOUE, namely home health care provided via telemedicine.

1. Emotions in computer-mediated communication

With the mass appeal of Internet-centered applications, it has become obvious that the digital computer is no longer viewed as a machine whose main purpose is to compute, but rather as a machine (with its attendant peripherals and networks) that provides new ways for human interaction. Indeed, until recently information was conveyed from the computer to the user mainly via the visual modality, whereas inputs from the user to the computer transpired through the keyboard and pointing devices via the user's motor mode.

The recent emergence of multi-modal interfaces, as our everyday tools, restores a better balance between our physiology and sensory/motor skills, and impacts (for the better we hope) the richness of routine activities. Given recent progress in user-interface primitives composed of gesture, speech, context and affect, it seems feasible to design environments that provide mass accessibility as well as a natural (versus a technologically imposing) feel.

In particular, as we reintroduce the use of all of our senses within our modern computer tools via multi-media devices -- or at least the **Visual, Kinesthetic, and Auditory (V, K, A)** – we liberate the possibility to incorporate both the user's cognition and emotions while designing computer interfaces. Computers are rapidly encroaching areas of our lives that typically involve socio-emotional content. Hence it is important to build user models not only of the physical characteristics of users, but also of their emotions. Since emotional representation in humans involves physiology, beliefs and goals, they can be perceived at the input analysis level, encoded into the dialog control and user model, and responded to appropriately at the output generation level as described in (Lisetti and Bianchi, 2001) (see Figure 1).

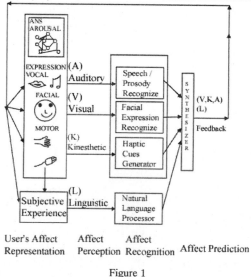

User's Affect   Affect   Affect
Representation   Perception Recognition   Affect Prediction

Figure 1

[1] This research has been partially funded by Intel Corp, Interval Research Corp, and Office of Naval Research.

2. The computer mediated communication paradox

CMC technologies – electronic mail, computer conferencing, distribution lists, multi-user dungeons (MUDs), and chat rooms – are all intended to facilitate communication. However, ambiguity regarding whether these technologies create richer and more meaningful communication exist (Kraut, 1998): such is the computer-mediated communication paradox.

Early CMC theories (before the emergence of digital cameras) suggested that electronic (text only) communication led to communication that was impersonal with limited exchange of *socio-emotional content*[2]. Several studies have concluded that text-only CMC is competitive with audio and video conferencing for *task-dimensional content* but did not rate as appropriate for building relationships, bargaining, and possibly for providing compassionate care to isolated patients.

The basic assumption of these CMC theories – characterized as the "cues-filtered out" approaches to technology effects – was that CMC does not transmit the same level of vibrancy and interaction that exists in face-to-face interaction. The nature of a small group is indeed to embody face-to-face, textually rich (broadband) communication. Facial expressions, bodily gestures, tone of voice, accent and speech rhythm (often culturally styled) provide richness of communication that, outside of the laboratory, has not been reproduced electronically. Therefore, users of CMC systems are thought to exhibit fewer of their natural communication behaviors. Studies show that as bandwidth narrowed, media allowed less "social presence", and communication tended to be less friendly, emotional, or personal, but more businesslike and task-oriented.

Other studies that discuss the effect on social involvement and psychological well-being of using the Internet, describe results showing significant declines of communication with family members in household, and decline in social circle circumference, accompanied with an increase in depression and loneliness (Kraut, 1998). These findings render it increasingly important for technology developers to take a *socio-technological approach* to their design, by developing technology that adapts (or co-adapts) to social processes, which are at the center of human lives.

3. Computer mediated communication in telemedicine

Emotions are particularly important in telemedicine where patients with various illnesses are treated or monitored remotely by medical professionals using multiple media devices. Whereas it might be *desirable* that interfaces acknowledge and respond to user's emotions in a natural and enjoyable manner for most users, it is *crucial* that they do so for the large majority of novice users and techno-phobic people (who can experience fear, frustration, or anger during HCI) and it is *vital* that they do so for remote and isolated patients who cannot travel because of physical immobility (who need tele-presence and care) and particularly for patients with emotional disorders (who experience depression, anxiety attacks, or post-traumatic disorder). Hence, we raise the question – how can a model of user's emotions be used to enhance the world of telemedicine within the context of computer-mediated communication.

The John Hopkins University School of Medicine defines telemedicine as the use of telecommunications to facilitate health care. Phone lines, fax machines, electronic mail, video-conferencing, and highly complex workstations that communicate instantaneously between two or more remote sites are all technologies engaged in telemedicine. Direct patient care via telemedicine currently has two main forms: 1) high-end real-time videoconferencing with medical peripherals used for patient examinations and peer consultations among participants at remote medical facilities and 2) low-end home health care monitoring systems connecting remote and isolated patients with medical personnel via relatively inexpensive media.

We are currently interested in *home health care* telemedicine services, which typically exist in three different forms:
1. the patient has a small medical monitoring device with an LCD monitor and a camera . The device can be used for assessing blood pressure, blood glucose levels and interactive videoconferencing with the consultant at the central workstation. The camera can also be used to transmit pictures of external afflictions, like scars or wounds;

---

[2] Socio-emotional content is defined as interactions that show solidarity, compassion, tension relief, agreement, antagonism, tension, and disagreement, whereas task-dimensional content is defined as interactions that ask for or give factual information or opinions.

2.   the patient has a health buddy interactive device that connects to a web site; each day the device prompts the patient with objective (yes/no) questions and uploads the answers to the web site; a nurse reads the answers and determines the gravity of the patient's state based upon patient responses and calls the patient if deemed necessary;

3.   the patient interacts strictly via email access with a nurse.

4. MOUE: an intelligent affective interface

We propose a Model Of User's Emotions (MOUE) architecture that can take as input both mental and physiological components associated with a particular emotion. It is illustrated in Figure 2.

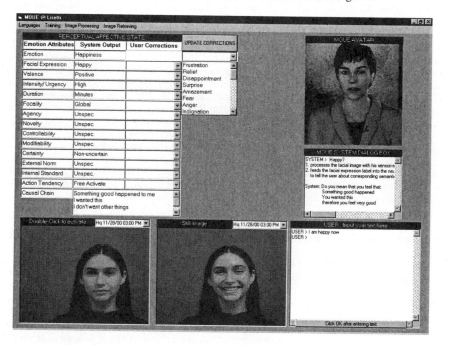

Figure 2: Model Of User's Emotions (MOUE) architecture

Using the same terminology as Maybury and Wahlster (1998), physiological components are to be identified and collected from observing the user via receiving sensors or *medium*: camera, mouse, microphone through the human senses employed to express emotion, i.e. the different *modalities* or *modes* which refer to: Visual, Kinesthetic, and Auditory (V, K, A)[3]. The system is also intended to receive input from Linguistic tools (L) in the form of linguistic terms for emotion concepts, which describe the subjective experience associated with a particular emotion.

The output of the system is given in the form of a synthesis for the most likely emotion concept corresponding to the sensory observations. This synthesis constitutes a descriptive feedback to the user about his and her current state, derived from the user's ongoing video stream, and a selected sequence of still images. The system is extensible by providing appropriate multi-modal feedback to the user depending upon his/her current state. Feedback adjustments involve varying aspects of the interface artificial agent, the avatar. Variations may be made to voice intonation, voice speed and pitch, face skin color, gender and facial expression (via animation).

---

[3] We limit ourselves to the three modalities (V, K, A) because we currently have found more emotion-relevant literature on those modalities than on the other two Olfactory and Taste (O, T).

5. Socio-technical research questions for moue in telemdecine

5.1   Patient Interface

Through a Wizard of Oz type of experiment (Dahlback, 1998), we investigate how MOUE will affect the patient. Will it annoy, or intimidate the patient, will it make the patient feel more comfortable, and cared for, will it help the patient's recovery; what type of avatar's face is most appropriate to use for such an application (male, female, black, white, young, etc.), what kind of voice, tone of voice, facial expressions, etc? What demographic class of patients is more positively responsive (male, female, young, elderly, etc.)?

5.2   Health-Provider Interface

What is a suitable design interface for emotional reading software to provide readings to home health care service providers? Suitable representation should provide an adequate scope of information and elicit user confidence in MOUE capabilities. Hence, additional questions arise:
*   What mode of MOUE feedback do health care service providers find most useful in a telemedicine home health care situation?
*   What is the most useful form of MOUE feedback among individual and combined options of: (1) Text; (2) Patient picture; (3) Figure (e.g. chart, graph, or table) representation.
*   Does the presence of a patient picture increase confidence in MOUE feedback?

Additionally, though few would argue a patient's emotional state is a factor of consideration in all forms of medical care, future research may discern that patient MOUE readings are more useful for home monitoring of certain types of health conditions.

References

Dahlback, N. et al. (1998). Wizard of Oz Studies – Why and How. In Maybury, M. and Wahlster, W, (Eds) *Readings in Intelligent User Interfaces*, San Francisco, CA: Morgan Kaufman Publishers.

Kraut, R. et al. (1998). Social Impact of the Internet: What does it Mean?. *Communications of the ACM*, 41(12): 21-22, 1998.

Lisetti, C.L., Bianchi, N. (2001, to appear). Modeling Multimodal Expression of User's Affective Subjective Experience. *User Modeling and User Adapted Interaction*.

Maybury, M., Wahlster, W. (Eds.) (1998). *Readings in Intelligent User Interfaces*, Morgan Kaufman Press.

Intelligent interfaces for universal access: challenges and promise

Mark T. Maybury

Information Technology Division
The MITRE Corporation
202 Burlington Road
Bedford, MA 01730, USA
maybury@mitre.org
http://www.mitre.org/resources/centers/it

Abstract

This article outlines some challenges and opportunities in the area of universal access to intelligent interfaces. Technological advances in computing and communication together with increasingly available connectivity provide both new challenges and new opportunities for disadvantaged individuals to become more full participants in society. It is also the case that careful exploitation of these advances can make advantaged workers that much more efficient or effective, for example, when they find themselves in communications and/or computing disadvantaged environments such as while mobile or in remote areas.

1. Universal interfaces

Growing volumes of information, shrinking product creation cycles, and increasing global competition demand optimal utilization of scare expert human resources. Effective human machine interfaces and information services promise to increase access and productivity for all. Furthermore, governments are moving more actively to promoting universal access. For example, in the United States, Section 508 of the Workforce Reinvestment Act, passed in November 1998, requires that people with disabilities have equity in use of electronic and information technology (Peet 2001).

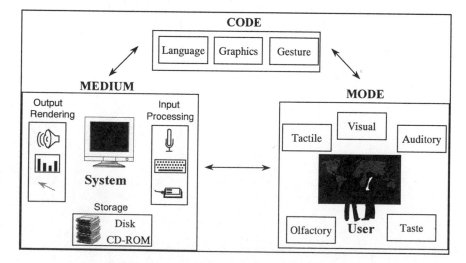

Figure 1. Medium, Mode and Code

To frame our intelligent user interface analysis, we first define some terminology (see Figure 1). Following Maybury and Wahlster (1998), we distinguish the terms medium, mode, and code. By *mode* or *modality* we refer primarily to the human senses employed to process incoming information, i.e., vision, audition, olfaction, touch, and taste. In contrast, we use *medium* to refer to the material object or carrier used for saving or presenting information including computer input/output devices (e.g., microphone, speaker, screen, pointer). As illustrated in Figure 1, medium is material-centered whereas modality is human-centered. We use the term *code* to refer to a system of symbols, syntax, and semantics (e.g., natural language, pictorial language, gestural language) used to encode and transmit information, often via multiple media which may support multiple modes. For example, a natural language code might use typed/written text or speech, which would rely upon visual or auditory modalities and associated media (e.g., keyboard, microphone). In the remainder of this paper we consider challenges and opportunities with multimedia/mode/code interfaces for all.

While today's computer-user interfaces have advanced beyond command line interfaces, they remain principally direct manipulation also called WIMP (windows, icons, menus, and pointing) interfaces. As Figure 2 illustrates, the typical architecture of these systems consists of three primary components. Presentation refers to the widgets – including windows, icons, and menus – that represent the application and are the primary mechanisms of communication and control between the user and system. Dialogue control refers to the data and flow control logic that manages the sensing and action from user interactions with the presentation. Application programming interfaces (APIs) are the routines invoked to perform the underlying system functions (e.g. inserting, deleting, modifying or otherwise manipulating application objects).

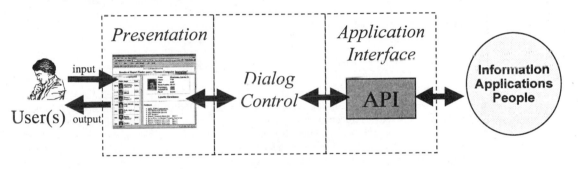

Figure 2. Presentation, Dialog Control and APIs

This is the architecture that is found today's popular graphical user interfaces (e.g., Microsoft Windows, Macintosh, and UNIX windowing systems). Accordingly, today's interface designers have three primary targets for enhancing the universality of human computer interfaces: presentation, dialog control, and APIs. For example, the interface can be customized by providing presentations of dialog control which are tailored to individual user needs (e.g., voice navigation or large, highly contrasted buttons for the visually impaired) or customizing APIs (e.g., limiting available functionality for novice or memory challenged users).

2.   Intelligent interfaces for all

Sophisticated interface tailoring requires modeling and reasoning about the domain, task, user, and dialog, as well as interaction media, modes and codes via interactive devices. A grand challenge of universal interfaces is to represent, reason, and exploit these models to more effectively process input, generate output, and manage the dialogue and interaction between human and machine so that we maximize the efficiency, effectiveness, and naturalness, if not joy, of interacting. These key interface functions form the basis of Table 1 that articulates some grand challenges for each of these functions. Table 1 also indicates benefits that accrue for all users if these challenges are addressed, which we briefly discuss in turn.

88

Table 1. Interface Challenges and Benefits

Interface Function	Grand Challenges	*POTENTIAL BENEFITS*
Input Analysis	Interpretation of imprecise, ambiguous, and/or partial multimodal input	Flexibility and choice in alternative media input, synergistic input
Output Generation	Automated generation of coordinated speech, natural language, gesture, animation, non-speech audio, generation, possibly delivered via interactive, animated life-like agents.	Mixed media (e.g., text, graphics, video, speech and non-speech audio) and mode (e.g., linguistic, visual, auditory) displays tailored to the user and context.
Dialog Control	Mixed initiative natural interaction that deals robustly with context shift, interruptions, feedback, and shift of locus of control.	Ability to tailor flow and control of interactions and facilitate interactions including error detection and correction tailored to individual physical, perceptual and cognitive differences. Motivational and engaging life-like agents.
Agent/User Modeling	Unobtrusive learning, representation, and use of models of agents (including the user).	Enables tracking of user characteristics, skills and goals in order to enhance interaction.
API	Dealing with increasingly broad and complex application functionality.	Simplification of functionality, possibly limited by user and/or context models. Automated task completion. Task help tailored to situation, context, and user. Mobile and substitutable interfaces for disabled users.

Input Analysis. One of the limitations of current user interfaces is the restricted choice, flexibility and expressiveness of input mechanisms. Today these are typically keyboard and mouse. An example of the limited flexibility is that input from these media must occur using a single input device or sequential input from several devices. An example of limited power is that mouse input is typically constrained to 2D input. Now consider that a user in a hands-eyes busy task (e.g., driving, working with machinery, performing medical diagnosis/procedures) may not have the luxury of using their hands or even eyes to provide input. This limitation can be severe for the motor or visually impaired. Universal interfaces should overcome a user's absolute or situational inability to utilize particular modes (e.g., auditory, visual, tactile) and/or interactive devices (e.g., screen, keyboard, mouse). Advances to broaden the available media to perform similar functions can be important, such as the ability to navigate and/or select options using voice or eye gestures instead of hands and keyboard. This furthermore requires codes and/or artificial languages that capture not only the syntax of each media and modality, but also the semantics of various naturally occurring languages (e.g., natural language, graphics, gesture). By providing similar communicative actions across codes, systems can enable users to accomplish the same function across codes (e.g., performing a selection with a typed, spoken, or gestural input). Finally, analysis of the pragmatics (i.e., beliefs, intent, emotions) of inputs can provide even richer models of the user. For example, most speech recognizers ignore intonation associated with signals thus loosing valuable indicators from pitch or intonation patterns of the user's emotional state (e.g., calm, agitated or stressed).

Output Generation. Just as current input devices limit users' expressiveness, so too presentation devices limit the range of expressiveness of computers. Effective presentations require reasoning about communicative intent, selecting content, allocating content to particular codes (e.g., language, graphics), and realizing content on or across particular codes, media and modes. For example, the same intention and propositional content (e.g., a severe weather warning) can be realized in a coordinated fashion in a range of media, modes and codes (e.g., as a non-speech audio siren, as a natural language italicized 12-point Times font email message). Selection can be driven by a range of criteria including but not limited to properties of the message (e.g., importance, urgency, size), the intended recipient (e.g., their physical, perceptual, cognitive, emotional state), the situation (e.g., a small screen public kiosk), and the state of the speaker (e.g., the personality of an animated agent). Limitations in user perceptual modalities (e.g., auditory, visual, tactile) can suggest preferred or required realizations. For example, hearing impaired users often use alarm clocks that vibrate their bed. Microsoft Windows provides user configurable capabilities such as StickyKeys or MouseKeys, ShowSounds, and auditory alerts to support users with tactile, auditory, and visual impairments, respectively. Finally, in perceptual or cognitive overload situations, an effective rendering of information can actually reduce user attention requirements and stress.

Dialog Control & User/Agent Modeling. Even the best input devices and output displays are not useful without control. Whereas current WIMP interfaces primarily support interaction via a standard set of devices including windows, icons, menus and dialogue boxes, these can be tailored to specific user and situational needs. For example, surface display properties of these interactive devices such as their size, or location can enable visual or motor impaired users to perform selections or provide input. In addition, time-outs or displays that change function with time can be sensitive to the reaction times of individual users. For example, elderly users sometimes require more time for query formulation and response generation during human computer dialogues than younger users.

Interfaces with the ability to manage discourse promise more natural interaction that deals robustly with context, interruptions, feedback, and support of mixed-initiative, wherein the locus of control ebbs and flows naturally between system and user. Explicit modeling and management of dialogue would support the ability to tailor the flow and control of interactions and facilitate them using mechanisms such as error detection and correction tailored to individual physical, perceptual and cognitive differences. Doing so requires modeling the user, their (physical, perceptual, and cognitive) characteristics as well as their beliefs, goals, and plans. Representing, reasoning and acting upon these models plays a foundational role in input processing, output generation, and management of dialogue and interaction.

Recent research focusing on life-like animated agents promises more human-like interactions. For example, knowing that the elderly typically have difficulty hearing sounds above 3 kHz can dictate the use of a low-pitched voice in an animated interface agent. While the use of emotion in a life like agent has not yet been shown to increase end user performance, motivational effects have been demonstrated in some applications (André, Rist, and Müller 2000). This could be valuable, for example, for motivation and/or attention disorders. Also, users with time and/or resource limitations may choose to delegate certain functions to an interface agent.

API. Whereas the complexity and breadth of desktop applications can be overwhelming for typical users, they can be disastrous for users with attention deficits. For example, Microsoft Word has over one thousand commands and yet only 20 of these constitute 80% of usage (Linton et al 1999, 2000). Using models of the user and task, it is feasible to limit the range of available functions, to support automated task completion, and to provide help tailored to the user, task and context. In addition, having a standard interface API would support ready substitution for personal interfaces. If used with wireless devices, standard APIs could ease the connectivity of individual devices tailored to the needs of mobility impaired users or those with limited dexterity or control (e.g. Parkinson's patients).

3. Conclusion

Truly universal access for all presents grand challenges to the user interface community. It demands a series of scientific advances that will require moving beyond our current interface architectures. Intelligent dialogue control as well as intelligent processing and management of multiple media, modalities and codes is fundamental to this progress and promises increased flexibility in input, processing, and output. Intelligence can be applied to the interface to enable a range of functions, including but not limited to:

1. Understanding imprecise, ambiguous, and/or partial multimodal/media input
2. Generating coordinated, cohesive, and coherent tailored multimodal/media presentations
3. Interaction management that is sensitive to models of the user, domain, task, and context, providing such functions as semi or automated completion of delegated tasks, recovery from miscommunication, tailored interaction styles, and interface adaptation.

Universality will require new modeling advances, such as languages that incorporate partial representation and reasoning, and geospatial and temporal reasoning. Input devices and rendering mechanisms will need to be tailored to the physical, perceptual and cognitive nature of the user, possibly even within individual sessions. They will also need to adapt to the requirements of the environment (e.g., personal computer, public kiosk, mobile digital assistant). Intelligent and adaptive interfaces promise universal access for all across a broad set of domains for work, learning and play.

References

André, E., Rist, T., Müller, J. (1999). Employing AI Methods to Control the Behavior of Animated Interface Agents. *Applied Artificial Intelligence*. 13: 415-448.

Boykin, S., Merlino, M. (Feb. 2000). A. Machine Learning of Event Segmentation for News on Demand. *Communications of the ACM*. Vol 43(2): 35-41.

Linton, F., Joy, D., Schaefer, H-P. (1999). Building User and Expert Models by Long-Term Observation of Application Usage. In J. Kay (Ed.), UM99: User Modeling: Proceedings of the Seventh International Conference (pp. 129-138). New York: Springer Verlag. [Selected data are accessible from an archive on http://zeus.gmd.de/ml4um/]

Linton, F., Joy, D., Schaefer, H-P., Charron, A. (2000). OWL: A Recommender System for Organization-Wide Learning. Educational Technology and Society. http://ifets.ieee.org/periodical/

Maybury, M., Merlino, A., Morey, D. (1997). Broadcast News Navigation using Story Segments, ACM International Multimedia Conference, Seattle, WA, November 8-14, 381-391.

Maybury, M., Wahlster, W. (eds). (1998). *Readings in Intelligent User Interfaces*. Morgan Kaufmann Press.

Maybury, M. (Feb. 2000). News on demand: Introduction. *Communications of the ACM*. Vol 43(2): 32-34.

Peet, M. (2001). Information Access for the Disabled: the Section 508 Mandate and its Implications for Intelligent Interface Development. In 1st International Conference on Universal Access in HCI. New Orleans, LA .

AVANTI: a universally accessible web browser

Alexandros Paramythis[1], Anthony Savidis[1], Constantine Stephanidis[1, 2]

[1] Institute of Computer Science (ICS)
Foundation for Research and Technology-Hellas (FORTH)
Science and Technology Park of Crete
Heraklion, Crete, GR-71110, Greece
cs@ics.forth.gr

[2] Department of Computer Science, University of Crete

Abstract

This paper reports on the development of a universally accessible web browser, carried out following a generic and systematic methodology for the provision of user interfaces accessible by anybody, anywhere and at anytime. The paper presents an overview of the design, implementation and evaluation phases of the web browser, focusing on the aspects related to universal access. The aim is to present a case study illustrating the applicability of the adopted methodology and techniques in the construction of a large-scale universally accessible user interface.

1. Introduction

The emergence of the Information Society and the increasing use of the Internet and the World Wide Web by the population at large, have introduced new dimensions in Human-Computer Interaction (HCI), necessitating the development of user interfaces which provide accessibility and high-quality interaction to anybody, everywhere, at anytime. The need to proactively address these requirements is becoming increasingly evident, and has recently resulted in increased attention to the concepts of Universal Access and Universal Design. More specifically, the specialization of the principles, practices and goals of Universal Design in the field of HCI has given rise to the concept of User Interfaces for All (Stephanidis, 1995; Stephanidis, 2001a).

This paper presents a case study in the development of a universally accessible web browser, capable to dynamically tailor itself in order to provide accessibility and high-quality interaction to users with different abilities, skills, requirements and preferences, in different contexts of use. The AVANTI web browser has been developed as a front-end to the AVANTI information system (see acknowledgments) (Stephanidis et al., 2001) following the Unified User Interface Development methodology (U^{2}ID). Unified User Interfaces (U^{2}Is) employ adaptability (self-adaptation based on knowledge available prior to the commencement of interaction) and adaptivity (self-adaptation based on knowledge acquired through interaction monitoring) techniques to ensure accessibility and high-quality interaction for all potential users (Stephanidis, 2001b). Adaptive software systems have been considered in a wide range of recent research efforts (see, e.g., Brusilovsky et al., 1998). Until recently, however, adaptive techniques have had limited impact on the issue of Universal Access. In fact, in many cases, adaptive techniques and assistive technologies have shared terminological references (the most prominent being the concept of "adaptation" itself), sometimes with fundamental differences in the interpretations of these terms. The U^{2}ID methodology is the first (and so far the only) documented approach providing support for universal access through the utilisation of user interface adaptations. It provides an integrated development methodology, an architectural framework, and dedicated tools for developing universally accessible interfaces. As a result, universal access can be designed, integrated, and evaluated as an embedded attribute of the user interface of interactive components and applications.

2. Designing for multiple user groups

The AVANTI information system was developed with the objective to address the interaction requirements of individuals with diverse abilities, skills, requirements and preferences (including disabled and elderly people), using web-based multimedia applications and services. AVANTI advocated a new approach to the development of web-based information systems. In particular, it put forward a conceptual framework for the construction of systems that support adaptability and adaptivity at both the content and the user interface levels [Fink et al., 1998; Stephanidis et al, 2001]. The AVANTI framework comprises five main components: (i) a collection of multimedia databases accessed through a common communication interface; (ii) the User Modeling Server (UMS) [Fink et al., 1998], which maintains and updates individual user profiles, as well as user stereotypes; (iii) the Content Model [Fink, 1997], which retains a meta-description of the information available in the system; (iv) the Hyper-Structure Adaptor [Kobsa, 1999], which adapts the information content, according to user characteristics; and (v) the User Interface component. The latter was called upon to provide an accessible and usable interface to a range of user categories, irrespective of physical abilities or technology expertise, and to support various contexts of use. The target end-user groups include: (i) "able-bodied" people; (ii) blind people; and, (iii) motor-impaired peopl, with different degrees of difficulty in employing traditional input devices. Further requirements include: (i) supporting users with any level of computer expertise; (ii) supporting users with or without previous experience in the use of web-based software; (iii) supporting "personal" use (i.e., from the user's own system), as well as "public" use (i.e., from public information kiosks); (iv) continuously support users as their communication and interaction requirements change over time.

From the above, it follows that the design space for the user interface of the AVANTI web browser was rather large and complex, covering a range of diverse user requirements, different contexts of use, and dynamically changing interaction situations. These requirements dictated the development of a new experimental front-end, which would not be based on existing web browser technology. In fact, the accessibility requirements posed by the user categories addressed in AVANTI could not be met by the customisability features supported by commercial web browsers, nor through the use of third-party assistive products. Instead, it was necessary to have full control over both the task structure and the interaction dialogue, including the capability to arbitrarily modify them. The U²ID approach addresses the above requirements [Stephanidis, 2001b].

Following the U²I design method, the design of the user interface follows three main stages [Savidis et al., 2001]: (a) enumeration of different design alternatives to cater for the requirements of the users and the context of use; (b) encapsulation of the design alternatives into appropriate abstractions and integration into a polymorphic task hierarchy and (c) development and documentation of the design rationale.

Having defined the design space as well as the primary user tasks to be supported by the AVANTI browser, different dialogue patterns (referred to as styles) were defined and designed for each task, to cater for the identified interaction requirements. Consider, for example, the design of a web browsing user task, namely that of specifying the URL of a web document (the "open location" task). This is a typical task in which user expertise is of paramount importance. For instance, an experienced user would probably prefer a "command-based" approach, which is the approach followed by most modern browsers, while a novice user might benefit from a simplified approach, such as choosing from a list of pre-selected destinations. When the physical abilities of the user are also considered, the "open location" task may need to be differentiated even further, in order to be accessible by a blind person, or a person with a motor impairment. Moreover, there are cases in which the "open location" task should not be available at all, such as when the interface is used at a public information point.

The second stage in the U²I design process concerns the definition of a task hierarchy for each task. This includes a hierarchical decomposition of each task in sub-tasks and styles, as well as the definition of task operators (e.g., BEFORE, OR, XOR, * and +) that enable the expression of dialogue control flow formulas. The task hierarchy encapsulates the "nodes of polymorphism", i.e., the nodes at which one has to "select" an instantiation style in order to proceed further down the task hierarchy. Polymorphism may concern alternative task sub-hierarchies, alternative abstract instantiations for a particular task, or alternative mappings of a task to physical interactive artifacts. During the design of the AVANTI user interface, the definition and design of alternative instantiation styles has been performed in parallel with the definition of the task hierarchy for each particular task. This entails the adoption of an iterative design approach in the polymorphic task decomposition phase.

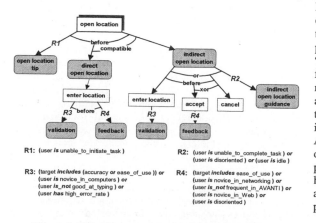

Figure 1: Example polymorphic task decomposition

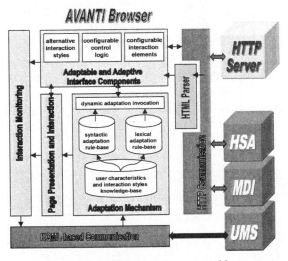

Figure 2: The AVANTI web browser architecture

An initial requirements analysis phase led to the definition of two sets of user-oriented characteristics, which formed the basis of the decision parameters that drove the polymorphic decomposition process. The first set was termed "static" user characteristics, to denote that these characteristics are unlikely to change in the course of a single interaction session. These characteristics are assumed to be known prior to the initiation of interaction (retrieved from the user profile), and comprises: (i) physical abilities; (ii) native language; (iii) familiarity with computing, the web in general, and the AVANTI system itself; (iv) the overall interaction target (speed, comprehension, accuracy, error tolerance); and, (v) user preferences regarding aspects of the application and the interaction. The second set of characteristics was termed "dynamic user states and interaction situations", to denote that the evidence they hold is usually derived at run-time, through interaction monitoring. The following "states" and "situations" were included: (i) user familiarity with specific tasks (i.e., evidence of the user's capability to successfully initiate and complete certain tasks); (ii) ability to navigate; (iii) error rate; (iv) disorientation (i.e., inability to cope with the current state of the system); and, (v) user idle time.

A set of user and usage-context characteristics is associated with each polymorphic alternative during the decomposition process, providing the mechanism for deciding upon the need for, and selecting, different styles or style components. This set of characteristics constitutes the adaptation rationale, which is depicted explicitly on the hierarchical task structure (see Figure 1) in the form of design parameters associated with specific "values", which were then used as "transition attributes", qualifying each branch leading away from a polymorphic node. These attributes were later used to derive the run-time adaptation decision logic of the final interface.

3. From Unified Design to Unified Implementation

The development of the AVANTI browser's architecture (see Figure 2) was based on the U^2I architectural framework [Savidis and Stephanidis, 2001]. This framework proposes a specific way of structuring the implementation of interactive applications by means of independent intercommunicating components, with well-defined roles and behavior. A Unified User Interface is comprised of: (a) the Dialogue Patterns Component; (b) the Decision-Making Component; (c) the User Information Server; and (d) the Context Parameters Server.

In the AVANTI browser architecture, the adaptable and adaptive interface components, the interaction monitoring component, and the page presentation and interaction component are the functional equivalent of the Dialogue Patterns Component (DPC) in the U^2I architecture. They encapsulate the implementation of the various alternative dialogue patterns (interaction / instantiation styles) identified during the design process, and are responsible for their activation / de-activation, applying the adaptation decisions made by the respective module. Moreover, each style implementation has integrated functionality for monitoring user interaction and reporting the interaction sequence back to the user modeling component. The adaptation mechanism directly corresponds to the Decision Making Component (DMC) in the U^2I architecture. It encompasses the logic for deciding upon and triggering adaptations, on the basis of information stored in its knowledge space. The role of the User Information Server (UIS) in the U^2I architecture is played by the UMS of the AVANTI system. The UMS is implemented as a server, usually remotely located on the network, as it offers central, multi-user modeling functionality. The communication between the user interface and the UMS is bilateral: the interface sends messages regarding user actions at the physical and task levels (e.g., "the user pressed reload", or "the user successfully completed the loading of a new document", respectively). The UMS employs a set of stereotypes that store categorized knowledge about the users, their interactions and environment, and a set of rules, to draw inferences on the current state of the user, based on the monitoring information. Finally, the Context Parameters Server in the U^2I architecture does not have an equivalent component in the AVANTI browser's architecture, as the needs of the user interface are limited in that respect and are covered by the UI itself.

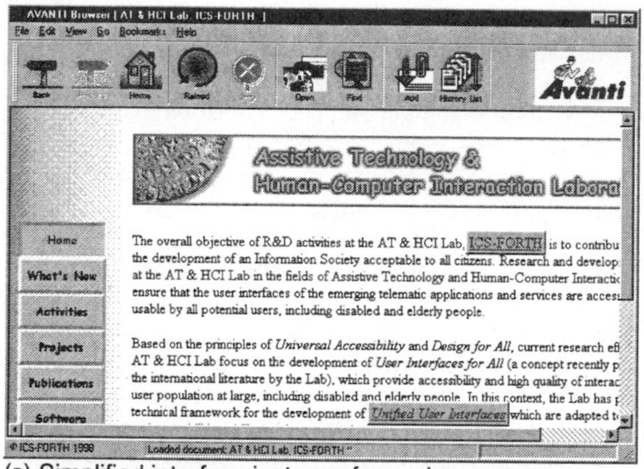

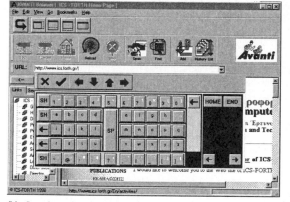

(a) Simplified interface instance for novice users

(b) Interface instance for motor-impaired user, supporting switch-based interaction

Figure 4: Instances of adaptability

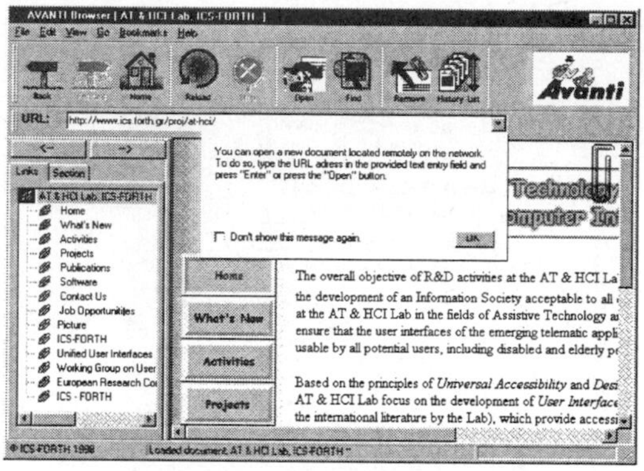

Figure 5: Instance of adaptive prompting

In the context of realizing the polymorphic task hierarchy into a U^2I of the design parameters that guided the introduction of different instantiation styles need to be transforemd into a form appropriate for the construction of a decision-making component. In the AVANTI implementation, the adaptation rationale was transformed into corresponding adaptation rules that are valuated at run-time. The re-valuation of any particular rule is triggered by the addition or modification of inferred user- or context-related parameters in the interface's knowledge space (either from user profiles, or from the UMS).

Figure 4 and 5 illustrate some of the categories of adaptation supported in the AVANTI browser. Specifically, 4(a), presents a simplified instance intended for use by a user unfamiliar with web browsing. In the second instance, 4(b), the interface is adapted for a motor-impaired user (interaction is performed through the "automatic scanning" technique [Stephanidis et al., 2001). The single interface instance in figure 5 illustrates a case of adaptive prompting, occurring when the interface has received an inference from the UMS, indicating that the user is unable to initiate the "open location" task. When this inference is added into the interface's knowledge space, the relevant rules are triggered. These cause the activation of a "tip" dialog notifying the user about the existence of the "open location" functionality and offering some indications of the steps involved in completing the task.

4. Conclusions

This paper has presented the employment of the U^2ID framework in the development of the AVANTI web browser. In particular, it has focused on those aspects of the user interface development process that are characteristic of the unification concept, and on the implications this has in the development life cycle of universally accessible interactive products. U^2ID differs from traditional approaches to developing user interfaces as: (i) it is oriented towards iteratively capturing and satisfying multiple user- and context-oriented requirements; (ii) it employs adaptation-based techniques in order to facilitate and enhance personalized interaction.

The AVANTI unified interface is capable of adapting itself to suit the requirements of able-bodied, blind and motor impaired users in a variety of contexts of use. Adaptability and adaptivity are used extensively to tailor and enhance the interface respectively, in order to effectively and efficiently meet the target of interface individualisation for end users. As the design of the UI has followed the principles of universal design, and the implementation is based on the U^2I architecture, inclusion of additional target user groups and contexts of use is facilitated [Savidis and Stephanidis, 2001]. One of the challenges in the development of the AVANTI browser has been the choice of suitable evaluation methods and techniques to be used for the evaluation of U^2Is in particular, and adaptable and adaptive interfaces in general. Although there have been several attempts in the past to construct both objective and subjective expert- and user-based evaluation methods in this field (for a review, see [Hook, 1998]), the dynamic dimensions of adaptive user interfaces require further investigation as far as evaluation is concerned. The approach taken in the development of the AVANTI browser has been the introduction of a two-fold assessment process, which involved expert evaluation and user-based evaluation [Andreadis et al., 1998]. Expert evaluation activities were intended to assess the appropriateness of the designed styles for the specific interaction context and the particular user characteristics for which they were intended, as well as the adaptation rules. The main goal of usability studies has been to derive some initial results on the overall usability of the information systems. In total, 175 users (from all user categories) were involved in the studies. The results of the evaluation were encouraging both in terms of user acceptance of the adaptation characteristics of the interface, and in terms of the fulfillment of the initial goals that led to the employment of unification in the user interface. The experience acquired has demonstrated the applicability of the methodologies, techniques and tools comprising the U^2ID paradigm in the construction of a large-scale, "real-world" universally accessible user interface.

Acknowledgments

The development of the Unified User Interface development approach has been supported through collaborative research and development projects, partially funded by the European Commission. These projects are: (i) the TIDE-ACCESS TP1001 project (Development Platform for Unified Access to Enabling Environments) of the European Commission (DG XIII); and, (ii) the ACTS-AVANTI AC042 project (Adaptable and Adaptive Interaction in Multimedia Telecommunications Applications) of the European Commission (DG XIII).

References

Andreadis, A., Giannetti, L., Marchigiani, E., Rizzo, A., Schiatti, E., Tiberio, M., Pentila, M., Perala, J., Leikas, J., Suihko, T., Emiliani, P.L., Bini, A., Nill, A., Sabbione, A., Sfyrakis, M., Stary, C., Totter, A. (1998). *Global evaluation of the experiments.* ACTS AC042 AVANTI Project, Deliverable DE030.

Brusilovsky, P., Kobsa, A., Vassilova, J. (eds.) (1998). *Adaptive Hypertext and Hypermedia Systems.* Dordrecht, Netherlands: Kluwer Academic Publishers.

Fink, J. (1997). *Prototype Content Model.* ACTS-AVANTI AC042 Project Deliverable 017.

Fink, J., Kobsa, A., Nill, A. (1998). Adaptable and Adaptive Information Provision for All Users, Including Disabled and Elderly People. *New Review of Hypermedia and Multimedia* 4, 163-188.

Höök, K. (1998). Evaluating the Utility and Usability of an Adaptive Hypermedia System. *Journal of Knowledge-Based Systems*, 10 (5).

Kobsa, A.(1999). Adapting Web Information to Disabled and Elderly Users. In *Proceedings of WebNet-99* (Honolulu, HI).

Savidis, A., Akoumianakis, D., Stephanidis, C. (2001). The Unified User Interface Design Method. In C. Stephanidis (ed.) *User Interfaces for All – Concepts, Methods and Tools* (pp. 417-440). Mahwah, NJ: LEA (ISBN 0-8058-2967-9, 760 pages).

Savidis, A., Stephanidis, C. (2001). The Unified User Interface Software Architecture. In C. Stephanidis (ed.) *User Interfaces for All – Concepts, Methods and Tools* (pp. 389-415). Mahwah, NJ: LEA (ISBN 0-8058-2967-9, 760 pages).

Stephanidis, C. (1995). Towards User Interfaces for All: Some Critical Issues. Panel Session "User Interfaces for All - Everybody, Everywhere, and Anytime". In Y. Anzai, K. Ogawa & H. Mori (Eds.), *Symbiosis of Human and Artifact - Future Computing and Design for Human-Computer Interaction [Proceedings of the 6th International Conference on Human-Computer Interaction (HCI International '95)]*, Tokyo, Japan (vol. 1, pp. 137-142). Amsterdam: Elsevier, Elsevier Science.

Stephanidis, C., Salvendy, G., et al., (1998). Toward an Information Society for All: An International R&D Agenda. *International Journal of Human-Computer Interaction*, 10(2), 107-134.

Stephanidis C. (2001a). User Interfaces for All: New perspectives into HCI. In C. Stephanidis (ed.) *User Interfaces for All – Concepts, Methods and Tools* (pp. 3-17). Mahwah, NJ: LEA (ISBN 0-8058-2967-9, 760 pages).

Stephanidis C. (2001b) The concept of Unified User Interfaces. In C. Stephanidis (ed.) *User Interfaces for All – Concepts, Methods and Tools* (pp. 371-388). Mahwah, NJ: LEA (ISBN 0-8058-2967-9, 760 pages).

Stephanidis, C., Paramythis, A., Sfyrakis M., Savidis A. (2001). A Case Study in Unified User Interface Development: The AVANTI Web Browser. In C. Stephanidis (ed.) *User Interfaces for All – Concepts, Methods and Tools* (pp. 525-568). Mahwah, NJ: LEA (ISBN 0-8058-2967-9, 760 pages).

Information Access for the Disabled: The Section 508 Mandate and its Implications for Intelligent Interface Development

Margot Peet

The MITRE Corporation
1820 Dolley Madison Blvd.
McLean, VA 22102[*]
mpeet@mitre.org

Abstract

Recent groundbreaking legislation is aimed at improving the public's ability to successfully interact with all forms of electronic and information technology. The Information Technology community is responding to these challenges with a number of efforts geared toward providing infrastructure, standards and guidelines for making user interfaces accessible to the disabled. In this paper we review this legislation, Section 508 of the Workforce Rehabilitation Act, and its definition of accessibility. We then discuss its anticipated impacts on the IT community. We look at increased standards activities oriented toward providing users with means to change an interface to accommodate personal preferences and capabilities. We also look at new government activities in standardized testing and validation for accessibility in response to Section 508. We review emerging commercial technologies that focus on personalization in regard to capabilities and preferences. Finally, we attempt to predict how developers of intelligent user interfaces can exploit these new activities and technologies.

1. Introduction

"Curiouser and curiouser", wrote Lewis Carroll. Who would have expected that (lack of) accessibility of user-computer interfaces would sit squarely at ground zero of the controversial presidential election of 2000? The infamous butterfly ballot of Palm Beach County is indeed, just that: a badly flawed user interface. Public response to the confusion it sowed shed an interesting light on the social impact of information technology and user interfaces. The inability of the mostly elderly public to interact with complex human-computer interfaces is generally perceived as evidence of a lack of intelligence.

Recent groundbreaking legislation is aimed at improving the public's ability to successfully interact with all forms of electronic and information technology. The information technology community is responding to this challenge with a number of efforts oriented toward providing infrastructure, standards, and guidelines for making user interfaces accessible. In this paper, we review this legislation – what it calls for and how it defines compliance – and attempt to predict its anticipated impacts on the IT community. We discuss how developers of the intelligent user interface technology of the future can exploit this "first wave" of activity.

2. The Section 508 Mandate

Section 508 of the Workforce Rehabilitation Act, passed in 1998 [1], requires that electronic and information technology (E&IT) developed, procured, maintained, or used by the Federal government be accessible to people with disabilities. Compliance with the law will begin to be enforced in June 2001. The law mandates that

- Individuals with disabilities who are Federal employees have access to and use of information and data that is comparable to the access to and use of the information and data by Federal employees who are not individuals with disabilities; and that

---

[*] This work has been funded by the MITRE Corporation under project 1901M008 AA.

- Individuals with disabilities who are members of the public seeking information or services from a Federal department or agency have access to and use of information and data that is comparable to the access to and use of the information and data by such members of the public who are not individuals with disabilities. This applies in particular to Federal government web pages.

Restrictions to the law state that federal agencies are exempt if that can prove that compliance imposes "undue burden". National security systems are exempt from Section 508 requirements. Finally, Section 508 does not require federal agencies to purchase assistive technology or peripheral devices, such as screen readers or Braille devices, for employees.

On the international scene, the rights of the disabled to information access are addressed in a range of ways. Some countries rely merely on guidelines. Others, notably Australia and the U.K., have passed legislation regarding the rights of the disabled to information access (the Disability Discrimination Acts of the U.K and Australia) [2]. Portugal, too, has legislated that web pages be accessible along the lines of guidelines set out by the World Wide Web Consortium's (W3C) Web Accessibility Initiative (WAI) [3]. Japan's Ministry of Economy, Trade and Industry (METI, formerly MITI) is circulating recommendations for IT accessibility that will eventually become regulations for doing business with the ministries [4].

In the United States there already exist disability rights laws, notably Section 255 of the Telecommunications Act, that call for accessible telecommunications. However, Section 508 differs in an important way from this related civil rights legislation and from international legislation. Section 508 has *teeth*; that is, it is enforceable. Section 508 calls for an "independent Federal agency devoted to accessibility for people with disabilities", the Architectural and Transportation Barriers Compliance Board, or "Access Board", to develop a set of definitions and accessibility standards [5]. These will be incorporated into the Federal Acquisition Regulation (FAR). Should a Federal agency fail to comply with these standards, a complaint or civil action seeking to enforce compliance with the standards may be filed. Accessibility is therefore strictly defined as compliance with the Access Board's standards, and enforcement in mandated by means of government acquisition regulations.

2.1 Access Board Standards for Web-based Intranet and Internet Information and Applications

The Access Board decided to establish their standards as a set of performance criteria, rather than design standards, so as to provide the regulated parties flexibility to meet their objectives in a more cost-effective manner. Their intent is to provide technical requirements as well as functional performance criteria, and to be as descriptive as possible so procurement officials and others would know when Section 508 compliance has been achieved.

Issuing realistic standards proved to be a controversial undertaking, with considerable input from industry, academia, government, and advocacy groups. Final standards were published in the Federal Register on December 21, 2000, setting June 21, 2001 as the date at which they would become effective and complaints could be filed.

Although the Access Board's standards cover all E&IT, web-based intranct and internet information and applications command particular interest for the development of intelligent user interfaces. The Access Board's standards lean heavily on the Web Content Accessibility Guidelines (WCAG) set out by the W3C's Web Accessibility Initiative. Of the sixteen provisions outlined to define compliance of web-based intranet and internet information and applications, nine are borrowed from the WCAG, with slight changes to wording to reflect the regulatory, as opposed to the voluntary, nature of the provisions [6].

3. The first wave: community response to Section 508

Section 508 is a significant piece of legislation for the technology community, as well as the user community. As Maybury [7] and others have noted, "accessible" user interfaces are extensible to able-bodied populations, who under one condition or another, can be considered temporarily disabled (viz., an eyes-busy driver places similar demands on a user interface as a blind user). Developments in the IT community in response to the Section 508 mandate can be viewed as "electronic curb cuts". Just as curb cuts required under the American with Disabilities Act benefited individuals temporarily "disabled" by bicycles or strollers, so too can IT developments generated by

Section 508 compliance be exploited by the intelligent interface developer community. In this section, we discuss developments in the IT community that are the first wave of activity spurred by Section 508, and their implications.

The Growth of Standards

As mentioned above, the W3C has played a prominent role in establishing voluntary guidelines for accessible web content, and many of these guidelines evolved into regulatory language in the Section 508 provisions set forth by the Access Board. These guidelines also form the basis of a number of evaluation tools that assist web page authors in identifying changes to their pages needed for the disabled to more easily use their Web pages, notably Bobby (CAST)[8], and The Wave (Institute on Disabilities, Temple University)[9].

In 1998, the National Committee for Information Technology Standards (NCITS) formed an Information Technology Accommodation Study Group (NCITS-ITA) to develop a set of protocols named the Alternative Interface Access Protocol (AIAP). These protocols would convey information about user preferences and capabilities between a user interface system and another system with which the user intends to interact. Alternative interfaces could be accommodated or constructed, in real time if necessary, to provide fundamental access to computing services and information regardless of any limitation of the user.

In October 2000, the NCITS ITA evolved into the NCITS V2 Committee on Information Technology Access Standards. This committee is an industry consensus standards group whose focus is to develop standards to provide users with a means to change an interface, by either allowing them to control that interface though a different interface device or to allow them to download information or code to the device to change its interface. The first project of this group is to transition the working group's protocol into an Alternate (User) Interface Access Protocol standard [10].

The activities of this group encompass but also extend beyond meeting the challenges of Section 508. Ultimately, the AIAP could provide a standardized means of personalization of information technology by accommodating users' preferences and capabilities in real time.

Standardized Testing and Validation for Accessibility

The General Services Administration (GSA) is establishing a forum to bring together E&IT industry and users to establish an infrastructure to support implementation of Section 508 requirements by vendors, and provide testing and validation processes to help attain quality assurance for these standard specifications and products. The ADIT project (Accessibility for people with Disabilities through standards, Interoperability and Testing) will establish an Accessibility Forum which will, among its tasks, launch a "Proof of Concept" testing project to refine procedures for subsequent application to other user groups and E&IT products [11].

Emerging Commercial Technologies

In anticipation of regulatory changes demanded by enforcement of Section 508, some industry vendors are adapting and developing tools and infrastructure to support accessibility. The X.Org Consortium, steward of the X windows system, has chartered an Accessibility Task Force to ensure that X windows meets all applicable accessibility standards [12]. The GNOME [13] and Athena Operating Systems [14] offer personalization capabilities, and the Java [tm] Accessibility API provides a contract for communicating information between the user-interface components of a Java application and an Assistive Technology [15].

A number of commercial ventures are beginning to exploit XML to make user interfaces more accessible to the general public. Edapta (recently acquired by Reef)[16] has developed software based on the AIAP Protocols to accommodate users' preferences and capabilities in accessing web sites. This technology was originally oriented toward the accessibility market, and is now being extended to electronic commerce applications. In addition, there are a number of emerging commercial transcoding services and servers that allow web developers to convert a site's data and then deliver content suitable to mobile devices. They rely on XML technology to transcode web sites to mobile devices. Built for the able-bodied community, they can potentially be applied to the disabled community.

4. Intelligent user interfaces – the next wave

Section 508 is spurring the development of accessibility standards, infrastructure and tools in the IT community. This session is entitled "Intelligent Interfaces for Universal Access". With the passage of Section 508, it is now an opportune time to flip this title. The intelligent interface developer's community can now think in terms of "Universal Access for Intelligent Interfaces", that is, exploiting the emerging standards, infrastructure and tools generated by Section 508 in the development of intelligent user interfaces.

Three features characterize intelligent user interfaces [17]:

> Tailoring of information to meet an individual's needs
> Multimedia output for presenting information
> Multimodal input for accessing information and for input

In other words, personalization and user interaction through multiple input and output modalities characterize intelligent user interfaces. The accessibility tools, standards, and infrastructure developed in the wake of Section 508 match these characteristics. Section 508 requires the development of user interfaces that utilize more than one input/output modality, and many of the commercial technologies emerging from the accessibility marketplace support the personalization of user interfaces.

Of particular interest to the intelligent interface community is the development of the Alternate Interface (User) Interface Protocol by the NCITS Information Access Standards Committee. This protocol will pave the way for multimodal input to access information. Also, the ADIT activity may result in useful validation protocols for testing intelligent interfaces. Finally, developments in the commercial sector that use XML technology to transcode web sites holds promise, with the caveat of its limitation in the presentation of non-textual forms such as graphics or videos. In these cases, much deeper mechanisms are required such as inter- and intra-media transformation, which may also become central for future intelligent user interface research as well. Nevertheless, as a whole, these activities are the first step in making user interfaces truly accessible to all.

References

[1]   *Online at* http://www.itpolicy.gsa.gov/cita/508.htm
[2]   *Online at* http://www.caslon.com.au/accessibilityguide2.htm
[3]   *Online at* http://www.acessibilidade.net/petition/government_resolution.html
[4]   *Online at* http://www-3.ibm.com/able/hr/law.html
[5]   *About the Board. Online at* http://www.access-board.gov/news/508-final.htm.
[6]   *WAI homepage at* http://www.w3.org/WAI/
[7]   Maybury, M. (2001).   Intelligent Interfaces for Universal Access: Challenges and Promise.   In *First International Conference on Universal Access in HCI, New Orleans, LA, 5-10 August, 2001*
[8]   *Online at* http://www.cast.org/bobby/
[9]   *Online at* http://www.temple.edu/inst_disabilities/piat/wave/
[10] *Online at* http://www.ncits.org/tc_home/v2.htm
[11] *Online at* http://adit.aticorp.org/#http://adit.aticorp.org/
[12] Cheikes, B.A. (2001). Design for Accessibility: Meeting the 'Section 508 Challenge'. In *First International Conference on Universal Access in Human Computer Interaction,* New Orleans, LA, 5-10 August, 2001
[13] *Online at* http://www.sun.com/access/gnome/faqs.html
[14] *Online at* http://www.rocklyte.com/athena/
[15] *Online at* http://migratetoday.sun.com/access/articles/
[16] *Online at* www.edapta.com
[17] Maybury, M., Wahlster, W. (eds) (1998). *Readings in Intelligent User Interfaces.* Morgan Kaufmann Press.

Human-Computer Collaboration for Universal Access

Charles Rich, Candace L. Sidner and Neal Lesh

Mitsubishi Electric Research Laboratories
201 Broadway
Cambridge, Massachusetts, 02139, USA
rich|sidner|lesh@merl.com

Abstract

We describe an approach to universal access based on the idea of making the computer a collaborator, and an application-independent technology for implementing such interfaces.

1. Introduction

As computer professionals, we have all observed the phenomenon that ordinary people tend to anthropomorphize the computer programs they use. Because this tendency inevitably leads to frustration (viewed as people, current programs are terribly unresponsive, uncooperative, and capricious!), our usual reaction is to try to train people to think differently. An alternative response is to accept the fact that people are most used to dealing with other people and, just as the science of ergonomics attempts to adjust the physical properties of devices to suit human physiology, we need to adjust the modes of interaction of computer systems to suit human psychology, not vice versa.

In the Collagen (for *Coll*aborative *agen*t) project, we have focused on the natural human tendency to collaborate. Stated most generally, collaboration is a process in which two or more participants coordinate their actions toward achieving shared goals. Furthermore, most collaboration involves communication (discourse). Human collaborative discourse has been studied both empirically and computationally by researchers in linguistics and psychology. Our goal in this project is to apply what is known about human-human collaboration to the human-computer case.

Two key features of collaborative discourse are communicating about goals and asking for advice. Communicating about goals is much better than just commanding actions not only because it is more natural, but also because the other person (or computer agent) may have access to information about how to satisfy your goals which you do not. For example, suppose you are in a room that is too warm. It is much better to tell your host that you are too warm than to command him to open the window, or turn off the furnace, or turn on the air conditioner, since your host may know, for example, that the window is stuck, the air conditioner, since your host may know, for example, that the window is stuck is stuck, the air conditioner is broken, etc.

To continue the example, your host may not be able to decide between opening the window and turning on the air conditioner without first asking you what your preference is between traffic noise and air conditioner noise. In order to recognize when it needs advice and to explain the difficulty encountered, a collaborative agent must to some degree be able to reason about beliefs and intentions.

At the engineering level, our approach has been to develop an application-independent collaboration manager called Collagen^{TM}. A collaboration manager is a software component that mediates the interaction between a software agent and a user. It is similar to what is often called a discourse manager, except that it keeps track of not only the linguistic and attentional state of a discourse, but also the collaborative intentions of the participants. However, it is less than a fully automated planning system, because it does not by itself decide what the agent should do or say next (though it may provide some candidates); it primarily provides a representation for recording the decisions that the agent has made and communicated. Collagen is implemented in Java using the Java Beans^{TM} architecture.

1 User says "Hello."
2 Agent says "Hello, what's your name?"
3 User says "Candy."
4 Agent says "Hello, Candy!"
5 User says "Let's set the clock." 6
6 Agent says "What is the time?"
7 User says "The time is 11:00 AM."
8 Agent sets the clock to 11:00 AM.
   ...
9 User says "How do I play back a video from my camcorder?"
10 Agent says "First, plug in the camcorder cables."
11 User says "How do I plug in the camcorder cables?"
12 Agent points to where you plug in the camcorder cables.
13 Agent says "The red cable plugs into the red terminal, the black cable plugs into the black terminal, and the yellow cable plugs into the yellow terminal. Make sure the other ends of the cables plug into the corresponding colors on the camcorder."
14 User says "What next?"
15 Agent says "Now press Play on your camcorder. You should see a picture on the TV screen."
16 User says "Ok."
17 Agent says "I don't see a picture. Make sure the camcorder is plugged in or has fresh batteries, and is turned on."
18 User says "Ok."
19 Agent says "Good. It's working."

(a) VCR Agent.

1 Agent says "Please list the other (non-boolean) attributes used in the columns of the logic table."
2 User says "The attributes are: voltage."
3 Agent creates a new user-defined attribute.
4 Agent enters "voltage" as the name of the new attribute.
5 Agent says "Please select the data type of the new attribute."
6 User selects float as the data type of the new attribute.
7 User clicks ok to accept the definition of voltage.
8 Agent says "Please edit the initial setting for the voltage attribute (if necessary) and then click ok."
9 User clicks ok to accept 0.0 as the initial setting for voltage.
10 Agent shows the expression editor for column 1.
11 Agent enters (voltage > 0.0) as the title for column 1.
12 Agent says "Please define an expression for column 1."
... [user goes to lunch] ...
14 User says "Where are we?"
15 Agent says "You are defining an expression for column 1, as part of making a logic table."
16 User enters (voltage >= 1000.0) as the title for column 1.
17 Agent shows the logical operator window.
18 Agent selects the >= operator.
19 Agent drags and drops >= into expression for column 1.
20 Agent selects the voltage attribute.
21 Agent drags and drops voltage into expression for column 1.
22 Agent types "1000.0" into expression for column 1.

(b) Symbol Editor Agent.

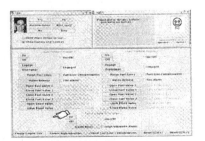

1 Agent says "Let's start the generator."
2 User says "Ok. "
3 Agent says "You take it from here."
4 User presses the alarm reset button.
5 User says "What next?"
6 Agent says "Please press the gas turbine generator on button."
7 Agent points to the gas turbine generator on button.
8 User presses the gas turbine generator button.
9 Agent says "Good. We have successfully started the generator."

(c) Gas Turbine Agent.

1 Agent says "Let's set the temperature in a room."
2 User selects the bathroom.
3 User says "What next?"
4 Agent says "What should the temperature setting be?"
5 User says "18 degrees Centigrade."
6 Agent sets the temperature in the bedroom to 18 degrees.

(d) Thermostat Agent.

Figure 1: Some agents built with Collagen.
(The underlined phrases are examples of collaborative communication.)

Over the past several years, we and our research colleagues outside of Mitsubishi Electric have used Collagen to build prototype collaborative agents in a number of different application domains (see Figure 1). For example, the collaborative agent in Figure 1a helps a user set up and program a video cassette recorder. The agent in Figure 1b, developed in collaboration with the Industrial Electronics and Systems Laboratory of Mitsubishi Electric in Japan, guides a user through the process of achieving a typical task using a sophisticated graphical interface development tool, called the Symbol Editor.

The agent in Figure 1c was developed in collaboration with the Information Sciences Institute of the University of Southern California. This agent teaches a student user how to operate a gas turbine engine and generator configuration using a simple software simulation. This work is part of a larger effort, which also involves the MITRE Corporation (Gertner *et al.* 2000), to incorporate application-independent tutorial strategies into Collagen. Teaching and assisting are best thought of as points on a spectrum of collaboration (Davies *et al.* 2001), rather than as separate capabilities.

Finally, Figure 1d shows an agent being developed at the Delft University of Technology to help people program a home thermostat (Keyson *et al.* 2000). This agent will eventually be able to help people analyze their behavior patterns and construct complicated heating and cooling schedules to conserve energy. It is part of a larger research project at Delft to add intelligence to consumer products.

We have also built agents for air travel planning (Rich & Sidner 1998) and email (Gruen *et al.* 1999). All of these agents are currently research prototypes.

The remainder of this paper is a very brief overview of the theory and methods used to implement collaborative human-computer interfaces such as these. For more detailed information, please refer to the referenced papers.

2. Collaborative Interface Agent

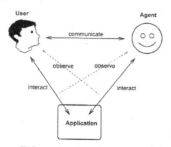

We have taken the approach of adding *a collaborative interface agent* (see figure above) to a conventional direct-manipulation graphical user interface. The name of our software system, Collagen (for *Coll*aborative *agent*), derives from this approach. (Collagen is also a fibrous protein that is the chief constituent of connective tissue in vertebrates.)

The interface agent approach mimics the relationships that typically hold when two humans collaborate on a task involving a shared artifact, such as two mechanics working on a car engine together or two computer users working on a spreadsheet together.

In a sense, our approach is a very literal-minded way of applying collaborative discourse theory to human-computer interaction. We have simply substituted a software agent for one of the two humans that would appear in figure above if it were a picture of human-human collaboration.

Notice that the software agent in this paradigm is able both to communicate with and observe the actions of the user and vice versa. Among other things, collaboration requires knowing when a particular action has been done. In Collagen, this can occur two ways: either by a reporting communication ("I have done *x*") or by direct observation. Another symmetrical aspect of the figure is that both the user and the agent can interact with the application program.

3. Collaborative Discourse Theory

Collaboration is a process in which two or more participants coordinate their actions toward achieving shared goals. Most collaboration between humans involves communication. *Discourse* is a technical term for an extended communication between two or more participants in a shared context, such as a collaboration.

In 1986, Grosz and Sidner, based on the empirical study of natural human collaboration, proposed a tripartite framework for modelling task-oriented discourse structure. The first (*intentional*) component records the beliefs and intentions of the discourse participants regarding the tasks and subtasks ("purposes") to be performed. The second (*attentional*) component captures the changing focus of attention in a discourse using a stack of "focus spaces" organized around the discourse purposes. As a discourse progresses, focus spaces are pushed onto and popped off of this stack. The third (*linguistic*) component consists of the contiguous sequences of utterances, called "segments," which contribute to a particular purpose.

Grosz and Sidner (1990) extended this basic framework with the introduction of SharedPlans, which are a formalization of the collaborative aspects of a conversation. The SharedPlan formalism models how intentions and mutual beliefs about shared goals accumulate during a collaboration. Grosz and Kraus (1996) provided a comprehensive axiomatization of SharedPlans in including extending it to groups of collaborators.

Most recently, Lochbaum (1998) developed an algorithm for discourse interpretation using SharedPlans and the tripartite model of discourse. This algorithm predicts how conversants follow the flow of a conversation based on their understanding of each other's intentions and beliefs.

4. Discourse State

Participants in a collaboration derive benefit by pooling their talents and resources to achieve common goals. However, collaboration also has its costs. When people collaborate, they must usually communicate and expend mental effort to ensure that their actions are coordinated. In particular, each participant must maintain some sort of mental model of the status of the collaborative tasks and the conversation about them-we call this model the *discourse state*.

Among other things, the discourse state tracks the beliefs and intentions of all the participants in a collaboration and provides a focus of attention mechanism for tracking shifts in the task and conversational context. All of this information is used by an individual to help understand how the actions and utterances of the other participants contribute to the common goals.

In order to turn a computer agent into a collaborator, we needed a formal representation of discourse state and an algorithm for updating it. The discourse state representation currently used in Collagen, illustrated in Figure 2, is a partial implementation of Grosz and Sidner's theory of

Figure 2: Example discourse state in VCR agent.

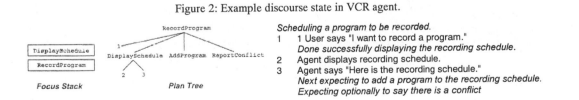

Focus Stack   Plan Tree

Scheduling a program to be recorded.
1   1 User says "I want to record a program."
    Done successfully displaying the recording schedule.
2   Agent displays recording schedule.
3   Agent says "Here is the recording schedule."
    Next expecting to add a program to the recording schedule.
    Expecting optionally to say there is a conflict

collaborative discourse (see Section 3). It consists of a stack of goals, called the *focus stack* (which will soon become a stack of focus spaces to better correspond with the theory), and a *plan tree* for each goal on the stack. The top goal on the focus stack is the "current purpose" of the discourse. A plan tree in Collagen is an (incomplete) encoding of a partial SharedPlan between the user and the agent. For example, Figure 2 shows the focus stack and plan tree immediately following the discourse events numbered 1-3 on the right side of the figure.

5. Related Work

This work lies at the intersection of many threads of related research in artificial intelligence, computational linguistics, and user interface. We believe it is unique, however, in its combination of theoretical elements and implemented technology. Other theoretical models of collaboration (Levesque *et al.* 1990) do not integrate the intentional, attentional and linguistic aspects of collaborative discourse, as SharedPlan theory does. There has been much related work on implementing collaborative dialogues in the context of specific applications, based either on discourse planning techniques (Chu-Carroll & Carberry 1995; Ahn *et al.* 1995; Allen *et al.* 1996; Stein *et al.* 1999)) or rational agency with principles of cooperation (Sadek & De Mori 1997). None of these research efforts, however, have produced software that is reusable to the same degree as Collagen.

References

[1]   Ahn R. *et al.* (1995).The DenK-architecture: A fundamental approach to user-interfaces. *Artificial Intelligence Review*, 8: pp.431-445.

[2]   Allen J. *et al.* (1996). A robust system for natural spoken dialogue. In *Proc. 34th Annual Meeting of the* ACL, pages 62-70, Santa Cruz, CA.

[3]   Chu-Carroll J., Carberry S. (1995). Response generation in collaborative negotiation. In *Proc. 33rd Annual Meeting of the* ACL, pages 136-143, Cambridge, MA, June.

[4]   Davies J., Lesh N., Rich C., Sidner C., A. Gertner, Rickel, J. (2001). Incorporating tutorial strategies into an intelligent assistant. *In Proc. Int. Conf. on Intelligent User Interfaces*, Santa Fe, NM.

[5]   Gertner A., Cheikes B., Haverty L. (2000). Dialogue management for embedded training. In AAAI *Symposium on Building Dialogue Systems for Tutorial Applications, Tech. Report FS-00-01*.

[6]   Grosz B. J., Kraus S. (1996). Collaborative plans for complex group action. *Artificial Intelligence*, 86(2):269-357.

[7]   Grosz B. J., Sidner C. L. (1986). Attention, intentions, and the structure of discourse. *Computational Linguistics*, 12(3): pp.175-204.

[8]   Grosz B. J., Sidner C. L. (1990) Plans for discourse. In P. R. Cohen, J. L. Morgan, and M. E. Pollack, editors, *Intentions and Communication*, pages 417-444. MIT Press, Cambridge, MA.

[9]   Gruen D., Sidner C., Boettner C., Rich C. (1999). A collaborative assistant for email. In *Proc. ACM SIGCHI Conference on Human Factors in Computing Systems*, Pittsburgh, PA.

[10] Keyson D. *et al.* (2000). The intelligent thermostat: A mixed-initiative user interface. In *Proc. A CM SIGCHI Conf. on Human Factors in Computing Systems*, pages 59-60, The Hague, The Netherlands.

[11] Levesque H. J., Cohen P R, Nunes J. H. T. (1990). On acting together. In *Proc. 8th National Conf. on Artificial Intelligence*, pages 94-99, Boston, MA.

[12] Lochbaum K. E. (1998). A collaborative planning model of intentional structure. *Computational Linguistics*, 24(4).

[13] Rich C., Sidner C.(1998). Collagen: A collaboration manager for software interface agents. *User Modeling and User-Adapted Interaction*, 8(3/4): pp.315-350.

[14] Sadek D., De Mori R. (1997). Dialogue systems. In R. De Mori, editor, *Spoken Dialogues with Computers*. Academic Press.

[15] Stein A., Gulla J. A., Thiel U. (1999). User-tailored planning of mixed initiative information seeking dialogues. *User Modeling and User-Adapted Interaction*, 9(1-2):pp.133-166.

Towards Services that Enable Ubiquitous Access to Virtual Communication Spaces

Thomas Rist

DFKI Saarbrücken, Germany. Email: rist@dfki.de

Abstract

The growing-together of telecommunication and information technology in conjunction with the emergence of portable, connected, and ubiquitous computing devices will bring about a new generation of conferencing and telecommunication systems that enable ubiquitous communication by multiple modalities and media. An obvious goal in this context is to achieve device interoperability among heterogeneous *information appliances*. However, in addition to base technology development, new communication support services are needed to enhance the usability of the upcoming generation of computer-mediated communication systems.

1. Motivation and background

Growing together of telecommunication and information technology enables new forms of teleconferencing, collaborative work among geographically dispersed teams, and getting-together in virtual spaces. At the same time, there is an increasing diversification in communication terminals including PC and pocket computers, PDA's, and especially mobile phones with built-in computing capabilities. In order to enable ubiquitous access to shared virtual meeting and workspaces we need to take into account that users are often not equipped on an equal footing in terms of output and input capabilities. In this contribution, we report on work towards communication support services for tele-communication environments with heterogeneous devices.

The work has been conducted in the context of the Magic Lounge[*] project. The project name also stands for a virtual meeting space in which the members of a geographically dispersed community can come together, chat with each other and carry out joint, goal-directed activities. The typical Magic Lounge users are groups of ordinary people with little knowledge on the underlying telecommunication technology, but with the need to collaborate on everyday tasks. Hence, the motivations for entering the Magic Lounge are as diverse as are the interests of the potential users themselves. For instance, one group of system users are people from a number of small Danish islands who wish to meet in a virtual space to exchange and share ideas, experience and knowledge on matters relating to their hobbies, professions, or to problems in their local communities. A French version of the system has been given to the Youth Service of Villejuif, a small team of young women and men involved in the problems of suburbs of large French towns. In what follows, we discuss key features of the Magic Lounge system in more detail.

2. Accessing the magic lounge by mobile devices

An important project goal is to enable access to a virtual meeting space through mobile devices to account for the rapidly growing community of mobile but nevertheless connected users. Our research focus is on issues that arise when trying to integrate mobile devices with limited display and interaction capabilities into a platform for teleconferencing and collaborative work, such as the Magic Lounge system (cf. Figure 1).

Mobile access to the meeting space is partly based on WAP, *(Wireless Application Protocol)* which employs WML *(Wireless Markup Language)*, and partly on a wireless telephone connection. Figure 2 illustrates an interaction sequence using WML pages. Magic Lounge services, such as a chat facility, are listed and can be launched by selecting them (first and second snapshot of Figure 2). Contributions made by other communication partners can be presented as text (third snapshot of Figure 2), or likewise via the audio channel. Messages or notes can also be submitted in text form (fourth snapshot of Figure 2). We use an automated presentation planner to allow for flexible tailoring of both content and layout of WML pages in a way that suits the display restrictions of the target devices.

[*] Magic Lounge is funded under the Esprit Long-Term Research pro-active initiative i3 and involves as partners: DFKI, Germany, Odense University, Denmark; LIMSI-CNRS France, Université de Technologie de Compiègne (UTC), Compiègne, France Siemens AG, Germany, and The Danish Isles - User Community, Denmark.

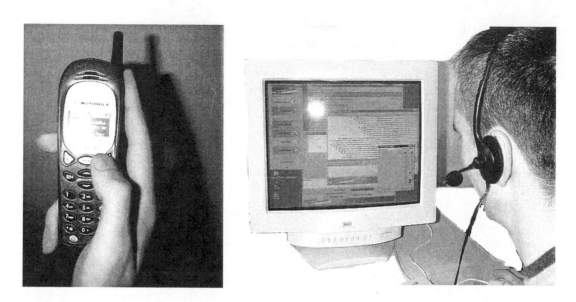

Figure 1: Accessing a virtual meeting space through heterogeneous communication devices. On the right-hand side, a user is accessing the Magic Lounge by an ordinary PC with audio connection, while on the left-hand side a WAP-enabled mobile phone is used by another user to access the same virtual meeting space.

Informal user studies [1] indicate that using a mobile phone to access the Magic Lounge virtual meeting space is well appreciated by the users and is a relatively easy task to perform - though several procedures, such as writing a text message, are yet laborious, and some confusion may arise because of unchangeable device-specific assignments of functions to keys. However, there is hope that interaction design can be significantly improved by switching to upcoming WAP versions.

Figure 2: Magic Lounge WAP interface for a mobile phone (Nokia 7110).

3. A framework for mixed audio/chat communication

With the goal to identify key concepts, services and functions that may drive the development of the next generation of computer-mediated communication systems we were especially interested in the way how ordinary people use technology, such as chat and audio conferencing. User studies carried out by [2] have shown that users are quite willing to give up a fair deal of pre-designed special-purpose structure in their common textual workspace in favor of naturalness and flexibility of interaction if a speech channel is available. There is still an ongoing discussion (also among the Magic Lounge project partners) whether text-based communication, such as chat or instant messaging, will become less important when audio quality via the Internet and mobile nets improves and becomes more dominant and cheaper. However, although speech potentially improves group interaction, a certain degree of text-based workspace structure is still needed. In particular, a highly structured textual *memory* of the interaction is necessary not only to support hardware heterogeneity, but also to implement the permanence of information across meetings. For the realization of a meeting memory and to support a flexible choice between communication modalities, the project developed a communication framework that is based on the notion of referable objects (in the sense of objects to which one can refer to), and communicative acts. Essentially, communicative acts denote activities, such as exchanging audio or chat messages among the communication partners. By treating the communication acts as referable objects, it is possible to reconstruct the flow of activity between all the clients (humans and also system components) at higher levels of abstraction, and to reveal the various relations that may exist between the single acts.

4. A conversation memory for virtual meeting spaces

An extensive study of current teleconferencing and CSCW systems, as well as the project's own participatory design work with user groups [3, 4], revealed a strong need for a meeting memory that can be queried by newcomers or latecomers who want to know what has happened in a meeting so far [5]. Our hypothesis is that such a meeting memory must support the *user's cognitive activity of remembering*. In this view, it are the users who remember while the system only provides assistance in this process, e.g. by providing various views on information that has been observed by the system and which may become critical to multi-party conversational performance. In particular, a temporal meeting browser [6] allows users to navigate back and forth through recorded meetings and to inspect individual contributions in a non-linear manner, whereas a "tree-view" of a conversation emphasises the oftentimes hierarchical structure of the turns in a conversation (cf. Figure 4, middle window).

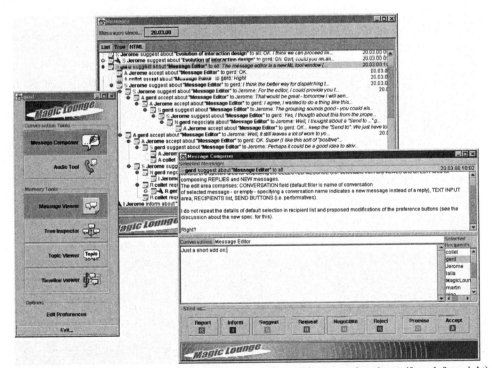

Figure 4: Basic windows of the Magic Lounge PC client interface. This screen shot shows (from left to right) the main tool palette; a message viewer using a tree-like display of the hierarchical structure as detected in a conversation; the message composition window with special send buttons to indicate speech acts.

The contents of the Magic Lounge conversation memory emerges from communication and collaboration acts that have been observed in virtual meeting places. Therefore, the conception of the memory component foresees the recording of spoken and typed utterances as well as other interaction events, such as the mutual exchange of references to electronic documents. As such, we use the term virtual meeting space in a broad sense and consider the fact that such spaces may simultaneously integrate traditional human-computer interaction processes (for instance, database or Web access) and also be related to what could be called a whole group story made of synchronous and asynchronous interactions. Contributions are recorded in the memory together with structural information, such as sender and recipient(s), the intended and declared communicative act, and the conversation to which a contribution belongs.

5. Conclusions

In this contribution we presented work towards ubiquitous access to virtual meeting spaces. The main challenge of this endeavor is to bridge the gap between communication modes that arises when people connect via different devices, such as a PC and a mobile phone. Driven by our observations made in the Magic Lounge project, we predict a high demand for (a) the smooth integration of multiple communication modalities such as voice and text, and (b) the availability of a mixed-modality conversation memory that can be accessed by the users at any time, and in addition, may provide the necessary contextual information for other services.

Acknowledgments

The work presented in this paper is a joint effort of the Magic Lounge project team. I'd like to thank all project members for their valuable contributions.

References

[1] Rist, T., Brandmeier, P., Herzog, G. (2000). Evaluation of the Mobile Access to the Magic Lounge System Magic Lounge Deliverable D7-Y3.

[2] Bernsen, N.O., Dybkjær, L., Luz, S. (2000). Evaluation of the Final Magic Lounge System. Magic Lounge Deliverable D6-Y3.

[3] Masoodian, M., Cleal, B. (1999). User-Centred Design of a Virtual Meeting Environment for Ordinary People. Conference Proceedings of HCI International'99, 8th International Conference on Human-Computer Interaction, Vol. 2, Munich, Germany, August (pp. 528-532).

[4] Henry, C., Vapillon, J., Collet, C., Martin, J. (2000). Supporting Virtual Meetings for Collective Activities: User-Centered Design in the Magic Lounge project. In. Proc. of the i3 Annual Conference, Jönköping, Sweden, September (pp. 57-62).

[5] Rist, T., Martin, J.-C., Néel, F.D., Vapillon, J. (2000). On the Design of Intelligent Memory Functions for Virtual Meeting Places: Examining Potential Benefits and Requirements. In: Journal Le Travail Humain, Vol. 63 No. 3/2000, pp. 203-225.

[6] Luz, S. F., Roy, D.M. (1999). Meeting browser: A system for visualising and accessing audio in multicast meetings. In Proceedings of the International Workshop on Multimedia Signal Processing, pp. 489-494. IEEE Signal Processing Society, September.

The design of an adaptive web tool to support human information seeking and retrieval strategies: some theoretical implications

Francesca Rizzo

University of Siena
Communication Sciences Dept.
Via dei termini, 6 53100 Siena
francy@media.unisi.it

Abstract

Just like the size of the Internet itself, no one knows exactly how many documents there are on the World Wide Web, or how many servers there are providing access to these documents. Finding what you want in this morass of information is no small task. The number of search tools has grown, and while most people make use of these tools, it is safe to say that many don't really understand their functionalities.

This work introduces the concept of adaptability (Rizzo, Marchigiani, Andreadis, 1997) as the fundamental feature, which characterizes technologies that allow people accessing and seeking information they want. In particular the paper describes how the adaptability design issue, encountered during one national project for designing an open multimedia environment to retrieve and organise documents, is facing through an attempt to improve interactivity between end users and electronic system.

1. Information, information seeking and new technologies

Our world is becoming more complex and interconnected largely because of developments in technology and the importance of information to human and technical development (Zuboff, 1988). We live in an information society in which more people have to manage a growing level of interactions, a large quantity of information and data and complex electronic environments. The main consequences of this phenomenon are:

1) Larger volumes of information. We are dealing with more kind of information in many aspects of our lives. More of us are engaged in work process such as generating, managing and communicating information to produce and provide goods and services for the global economy. This in turn requires most people to go beyond the workplace in order to learn new skill and acquire new knowledge to their jobs.

2) New forms and aggregations of information. We are engaged with information in new forms, especially electronic, digital forms that are more abstract, more dynamic, and more easy to handle than in printed or painted form. But information in digital form is both enabling and complicating. On one hand electronic information is more accessible from anywhere in the world. On the other hand it is less available for billion of people unaided by technologies. Moreover, we are dependent on digital representation of information to express "knowledge" in forms that we can understand and perceive. Electronic digital information is simple because it is easily represented by only two elements (bits); but it is complex because of the many sets of coding that are necessary for humans to make sense out of digital information.

3) New tools for working with information. Mainly this means that significant resources (time, financial resources) must be continuously spent in acquiring, learning to use, applying and maintaining electronic tools. Secondarily more cognitive efforts have to be made in understanding information based on digital expression.

It seems clear that many of the new electronic systems to access, manage, seek and retrieve information are still unable to support people to satisfy their information needs.

In order to face this problem in the project described below, our main hypothesis are as follows:
- ❖ we were concerned in designing a system to support people engaged in a solution of an information problem in order to change their state of knowledge;
- ❖ we had design an interface system able to allow users to interact with a digital environment in a way similar to the interaction with human information sources;
- ❖ information seeking is an iterative process (Marchionini, 1995).

2. The project's background

The project presented aims at designing, developing and testing the interface of an electronic system in order to support effective and intelligent information searching processes in the field of clothing and textile industrial districts in Italy. For this reason a regional service centre was created in 1980 offering standard information about clothing and textile market trend and development.

The main purpose of the project is to support working people with different roles in the field of textile and clothing industry, making them able to access, retrieve, modify, and share useful information (Chesi, Rizzo, 2000). The system design process is developed toward the implementation of an adaptivity user interface. The concept of adaptivity is concerned with the interactive characteristics of interface that can support user in a continuous communication flow process with the system, in accordance with the tracking of interaction.

In order to reach the goals described above, we have applied a User Centred Approach (Norman, 1993) which implies the principles of:
  the involvement of real end users in designing and evaluating the application from the beginning of the design process; the development of a series of prototype; the iterative process of design, evaluation and redesign.

2.1 Human strategies to search information

The implementation oh high level of adaptivity in an electronic application depends on the knowledge of the end users needs, skill and preferences as well as the description of task requirements. In order to individualize the users needs and the general interfaces requirements for our interface we have first analysed the process of communication flow, observing the activities of people in one real problem-solving context (Rizzo, Chesi, 2000).

The main result we have obtained from this phase has been the users' action model of information seeking behavior. Our end users have been divided into two groups: users that know what information they need and users that do not know what they are searching for.

In the first case, users' activity could be described as an information retrieval process based on analytical information strategies.

In the second case, information seeking became a problem solving activity (Rumiati, 1990) based on browsing information strategies.

This second case can be briefly described as follows: the first step was characterised for a general awareness of the knowledge of problem domain. In addition, the user often showed that he was not fully aware of his goal. In this phase, a human operator helped the client to understand his problem through a question and answer process that aided the identification of the possible information source required, satisfying the user's need. The human operator could make available a better possible solution for typical problems (i.e. information about spring-summer trend for women), contingent problems (i.e. I need information's about behavior and performance of competitors) or for the identification of obscure problems (i.e. how can I improve in my field?) supporting interactively the process of information problems refinement.

There are many studies about the definition of a model of information seeking behavior (Ellis, 1989) which describes people behavior when using technologies such as search engine. Sheiderman, Byrd and Croft 1998) have individualized four phases. Normally users start a search process defining their information needs and goals; then they decide to employ an electronic system for supporting the four phases: formulation of intention (what search, where search) actions, review of results, refinement.

Information seeking process, as well as described above and from empirical results of our investigation, can be defined as fundamentally interactive. It depends on activities performed by information seeker, feedback received from information environment, goals modification based on this feedback. Normally in everyday life people can manage their interaction with information thanks to the personal knowledge and the cognitive information strategies and tactics developed to use the classical information sources (newspaper, books, library, catalogs etc.).

How to solve the issue of interactivity in an electronic system for supporting information seeking and retrieval? If the term of interactivity is referred at the capabilities of the interface system to dynamically modify their communication characteristics during the human computer interaction then the problem is what levels of adaptivity are effective and efficient (ISO/DIS 9241-11,1997) in ameliorate the problem of information pollution on the net.

2.2 Information seeking and Human Computer Interaction Approach: final notes

In order to develop a robust interaction style for an electronic interface system should start from the good representation of the model of information seeking action. From this point of view the results obtained from human Computer Interaction studies can be critical in order to facing the challenge of interactivity issue on electronic seeking and retrieval system. In particular the Norman's model of human action (Hutchins, Holland and Norman, 1986) provides a very powerful tool.

The Norman's model describes five states (goal, intention, action, perception, evaluation) and three distance (semantic, referential and intereferential the first two are present on the both sides of execution and evaluation of the model). The three forms of distance can help us describing the relationship between the task user has in mind and the way the task can be performed through the interface. But there are many cases in which - during their activities - people have to redefine their goal, the way in which they interact with the context depending either on the action carried out or on the produced results. Bagnara and Rizzo (1989) have defined the distance between two different goals, mediated by a series of action; as scenario distance or issue distance.

Currently the main problem is that user interface for information access are focused on query formulation and output examination, ignoring the problem of correct identification and refinement of information problem and users needs. One possible way to answer at this problem could be the implementation of interfaces able to support the redefinition of users goals, expectation and action's plan.

References
Bagnara S., Rizzo A. (1989). A Methodology for the analysis of errors processes in human computer interaction. *Proceedings of Computers: Organizational, Management and health aspect.* Amsterdam.
Chesi, C., Rizzo F. (2000). Open multimedia system to retrieve and organise documents: an adaptive web based IR system in the field of textile and clothing industry. *Adaptive Hypermedia and Adaptive web-based system. Proceedings of 1st International conference,* Trento: Springer & Verlag.
Ellis , D. (1989). A behavioral Model for information retrieval System. *Journal of information sciences,* 15, 37-59.
Hutchins, E., Holland, j. D., Norman, D. A. (1986). Direct manipulation interfaces. In Norman D. A., Draper S. W. (Eds.), *User centered system design: New prospective on human computer intercation.* Hillsdale, NJ: Erlbaum Associates.
Marchionini, G. (1995). *Information seeking in electronic environments.* Cambridge MA: University Press Cambridge.
Norman, D. A.(1993). *Things that make us smart.* New York: Addison Wesley.
Rizzo, A., Marchigiani E., Andreadis (1997). The AVANTI project: prototyping and evaluation with cognitive walkthrough based on Norman's model of action. *Proceedings of Designing Interactive System,* Amsterdam.
Rumiati, R. (1990). *Giudizio e decisione,* Milan: Il Mulino.
Shneiderman, B., Byrd, D., W. B. Creoft (1998). Sorting out Searching. *Communication of ACM,* 41, 75- 82.
Zuboff, S. (1988). *In the age of smart machine,* New York: Basic Book.

Consensus-based adaptive user interfaces for universal access systems

Janusz Sobecki, Ngoc Thanh Nguyen

Dept. of Information Systems, Wrocław University of Technology
Wyb. Wyspiańskiego 27, 50-370 Wrocław, POLAND
e-mail: {sobecki,thanh}@pwr.wroc.pl

Abstract

Today's universal access systems must be ready to serve different types of users. Population of users is very differentiated, so it is almost impossible to design for every information system a single, equally appropriate, user interface for each user. Instead, we postulate to construct adaptive interfaces that take into account experiences of all the population to build users profile by means of consensus methods.

1. Introduction

Universal access systems are becoming more and more popular. Nowadays it is not necessary to start any computer to be in the situation of interaction with some kind of a system. It could be the TV set (especially with the DVD player or the digital satellite tuner), the banking teller machines or any other automatic selling machines. Besides their functionalities user interaction is the key factor, but quite often it is very difficult to use these systems. The reasons for this are different: they are too complicated, users don't know how systems are functioning, users are not acquainted with this particular interface, systems use national languages that the user doesn't know or what happens quite often nowadays user don't understand vocabulary used in system texts (Sobecki 1999).

User interface plays very important role in today's information systems, especially in so called general access systems. These kind of systems could have even millions of users, from the other hand each user uses even dozens of such a systems every day. That's the reason why the quality of user interfaces i.e. their ability to adapt to the user needs is a critical factor of their functioning. Today in the era of competitive economy, system solutions that are not accepted by the user's population have a little chance to survive.

General access systems are supplied by quite many developers, so the interfaces of such systems are usually very different. They often use quite different interaction styles, i.e. function keys or direct graphical manipulation (Newman 1996). Quite often they assume that there exists a single interface, which is the most suitable for all the users, but users could differ in many aspects: gender, age, physical, social, anthropological, educational background, experience, etc. So, there is no wonder that these differences may result in great problems in using unique interfaces. The designers should also be aware that potential users of their systems have already some experience with some other systems, that have often very close functionality and so the users have already some interaction habits derived from prior experience.

The methods for making systems more appropriate for users are very different and solutions can come from at least several fields of science: HCI (human computer interaction) with Interaction Styles or User Models, AI (Artificial Intelligence) with Natural Language Processing or IR (Information Retrieval) with User Profiles, the domain of Hypermedia and Multimedia Information Systems brings also some ideas in this matter (Sobecki 2000). There are already more multidisciplinary approaches to this problem i.e. Interface Agents as emerging from more general Agent Paradigm (Wolldridge 1995). The results of these works are available now in many commercial applications in form of personal assistants. They are implemented to support the Internet search engines and browsers, or office applications (i.e. MS Office Assistants). HCI, for example, worked out several methods for developing useful systems, these methods however concentrate upon statistically centered interface adoption to so called average user (Peerce 1996).

In this paper the method for constructing consensus based adaptive user interfaces for general access systems is presented. Consensus methods turned out to be a very suitable tool in solving conflicts in distributed systems. One can use them for reconciling the disagreements in knowledge states about the same real world of agents in multi-agent systems (Katarzyniak 2000 & Nguyen 1999). An attempt of using consensus methods to make an agreement between inconsistent profiles for interface agents was presented in works (Nguyen et. al. 2000 & 2001). A very good example for such general access systems is an internet based information system. This system is usually accessed by a population of heterogeneous users, and we assume that it is possible, using consensus base methods, to automatically adapt an interface for each particular user, basing on similar users experiences with this system.

2. Interface Agents

P. Mae defines an interface agent as an agent that acts as a kind of intelligent assistant to a user with respect to some computer application (Wolldridge 1995). The MIT agents act primarily by observing their users, and by application some machine learning mechanisms. For each new situation, the agent computes the distances, between the current state and all appropriate states stored in the memory. Together with these past states corresponding actions taken by the user in the past are stored. Then, for this new situation, the recalled action is selected which state resembles the most, or in other words has the smallest distance between, this new situation (Fleming 1999). The distance function could be constructed in many different ways, for example in form of sum of weighted features describing the particular state.

In some cases adaptive interfaces are composed not from a single but rather set of agents, that usually are simple processes which run 'in the background', to decompose the task of implementing the interface. Generally, interface agents try to help the user by assigning him a particular user model comparing his activity with the population of other users and so offering him appropriate assistance. Some of them, as mentioned above, also take into account the prior user's interaction with this system. Both methods are focused on this specific application and its environment (Keeble 2000).

It is believed that interface agent should meet several requirements, i.e.: adaptivity, autonomy and collaboration (for interface agents collaboration with the user is essential). Some researchers add also other requirements such as robustness, defined as an ability to perform in restricted conditions (Brown 1998). A possible implementation solution of interface agents is Bayesian belief networks used for example in Microsoft Office 97 Office Assistant (Hedberg 1998). Bayesian belief networks can represent cause-and-effect relationships between nodes, which are weighted to express how they affect each other. One of the most interesting aspects of the belief networks is their non-determinism, so they can infer most reasonable solution, given the network of relationships, probabilities and trade-off inherent in the situation.

Interface agents to predict properly the user intents must work on accurate user model, but quite often they can be inaccurate. There could be many different reasons for this, but most important are: differences between users and users' behavior changes over time. It is possible to predict user intent and to model user behavior having defined some metrics and functions, such as: precision, reactive and autonomy metrics, as well as utility function (Brown 1998). Some authors (Fleming 1998) identify such uncertain situations and propose in some cases to consult user to make the decision.

3. Adaptive interface architecture for general access systems

General access systems have usually a dynamic population of users located in distributed environment. It is obvious that general access systems have very different users, but we assume that also the same user's interface preferences changes dynamically over time, i.e. music preferences (top-hits). In our architecture these systems form a multi-agent system that communicates with these users. Agents collect information about the environment, i.e. users, in form of the user profiles. These profiles consists of users experiences with this or other systems, his/her achievements, facts about consulting help information, committed errors, etc. These data are useful not only to classify users, by application of some sort of distant function, but also enables to construct adoptively an interface for each user. User profiles are different among population of agents, because the user's environment is highly dynamic, users have different experiences and usually differs in some aspects between themselves.

Each time a user interacts with any system, he/she uses the interface that is in a particular form. This interface may be more or less convenient for the user. The interface, for each user, is described in form of the interface profile. It is possible, by observing the process of interaction, to evaluate interaction performances, by means of some type of usability function, and decide whether the interface is appropriate for the particular user. Then, for each user we can select interface profiles of users with profiles that resemble the particular one and find the consensus between them. This calculated interface profile gives rise to final interface construction.

There are three major problems in the idea outlined above, i.e. concrete form of a user profile, an interface profile and a usability function. The user profile, like in many internet based systems could contain recently used systems (visited pages), its options, his/her interaction (i.e. entered data) and some additional information that could be used in the usability function. The usability function could be based on the following features: time of completing appropriate functions, committed errors and their corrections, rapid change of options, consulting help information, the number of the user's requests for an assistance, etc. The interface profile from the other hand specifies the interface itself. It is assumed that in adoptive systems interfaces are constructed dynamically for each user. To do so the process has to be highly parameterized, potential interface attributes could be selected raises from quite simple as: fonts, their styles, colors and sizes, background colors or graphics, buttons in form of icons or texts; or more complex as: parameters of used set of fonts (their correlations), used whole interface templates (as for example themes in Microsoft FrontPage) or style of buttons animation.

In the following section some theoretical aspects of user profile determination by means of consensus method is presented. Because of limited length of the paper, the problems of user classification will not be addressed in this paper.

4. Determining of user profiles' consensus

In work (Nguyen 2000a) the author defined a general consensus system for solving conflicts in distributed systems. In this paper this system will be used to determine the consensus for concrete conflict, in which the interface agents have different profiles (as the best in their opinion) for serving of the same class of users. A consensus system is defined as follows:

$$Consensus\_Sys = (X,F,P,Z)$$

where

- X- a finite set of consensus carriers,
- F- a finite set of functions
- P- a finite set of relations on carriers
- Z- a set of first order logic formulas, for which the relation system (X,F,P) is a model.

For our distributed interface system we specify the above notions as follows (the names of carriers will be used as attribute names, if it will not cause the ambiguity):

1. $X = \{Interface,Class,Degree,A_1,A_2,...,A_n\}$,

where:

- $\quad$ *Interface* is the set of interface agents,
- $\quad$ *Class* – the set of user classes,
- $\quad$ *Degree* = [0,1], and
- $\quad$ $A=\{A_1,A_2,...,A_n\}$ is a set of attributes describing the interface profiles for user service.

2. $F = \{Credibility\}$

where *Credibility: Interface→Degree* is a function that assigns to an interface agent a number representing his credibility. In this work we assume that the function is constant.

3. $P = \{Profile^+,Profile^-\}$

where $Profile^+,Profile^- \subseteq Interface \times Class \times A_1 \times A_2 \times...\times A_n$

We interpret a tuple of relation *Profile*$^+$, for example, $<i_1,k_1,a_1,a_2,...,a_n>$ as follows: in opinion of agent i_1 the best profile for serving clients from class k_1 should be the tuple $<A_1:a_1,A_2:a_2,...,A_n:a_n>$ (or in other words utility function for this profile is over an average). A tuple $<i_2,k_2,a'_1,a'_2,...,a'_n>$ belonging to relation *Profile*$^-$ means that according to agent i_1 for users of class k_2 the profile $<A_1:a'_1,A_2:a'_2,...,A_n:a'_n>$ is poor for their service (or in other words utility function for this profile is below an average).

4. Z: Logical formulas representing conditions which have to be specified by the tuples belonging to relations from P.

A consensus situation is then defined as follows:

$$s=(\{Profile^+,Profile^\pm,Profile^-\},Class\to\{A_1,A_2,...,A_n\})$$

where

$$Profile^\pm=Dom(Profile)\backslash(Profile^+\cup Profile^-) \text{ for } Dom(Profile)= Interface\times Class\times A_1\times A_2\times...\times A_n.$$

The set $\{Profile^+,Profile^\pm,Profile^-\}$ is then the basis of consensus, and the relationship $Class\to\{A_1,A_2,...,A_n\}$ is the consensus subject. The elements of consensus basis are called:

- $\quad$ *Profile*$^+$ - positive element,
- $\quad$ *Profile*$^\pm$ - uncertain element, and
- $\quad$ *Profile*$^-$ - negative element.

The interpretations of elements of relations *Profile*$^+$ and *Profile*$^-$ were given above. Notice that relation *Profile*$^\pm$ is the same type as relations *Profile*$^+$ and *Profile*$^-$, its elements should be interpreted as the uncertainty of interface agents. For example, if the tuple $<i_3,k_3,a''_1,a''_2,...,a''_n>$ belongs to relation *Profile*$^\pm$ then it means that agent i_3 does not know (has no foundations to state) if profile $<A_1:a''_1,A_2:a''_2,...,A_n:a''_n>$ is the best profile for users from class k_3 or not.

The distance function δ between tuples $r,r'\in Dom(Profile)_{\{A1,A2,...,An\}}$ may be defined as follows:

$$\delta(r,r')=\sum_{i=1}^n\delta_i(r_{A_i},r'_{A_i})$$

where function δ_i is a distance function between the elements of set $Dom(Profile)_{Ai}$. Notice that if the structure of set $Dom(Profile)_{Ai}$ is homogeneous then it should be possible to define such function. On the other hand, one can prove that if all functions δ_i are metrics, then function δ should also be a metric.

In work (Nguyen 2000b) the author defines 6 postulates for consensus. It was also proved that if it is possible to define a distance function δ between tuples of $\mathrm{Dom}(Profile)_{\{A1,A2,...,An\}}$, then for given domain $<\{P^+,P^\pm,P^-\},\{A \to B\}>$ the following consensus function

$$C(P) = \{c \in \mathrm{Dom}(Profile)_{\{A1,A2,...,An\}}: (c_A=a) \Rightarrow (\Sigma_{x \in P+}\delta(c_B,x) = \min_{y \in Dom(P)} \Sigma_{x \in P+}\delta(y,x))\}$$

should satisfy all the postulates.

The algorithms for consensus determining are able to be worked out if the structures of elements of set $\mathrm{Dom}(Profile)$ and function δ_i are known. Below we present an algorithm for the case when the elements of set $\mathrm{Dom}(Profile)$ are intervals of numbers. This structure is very popular in representing uncertain and incomplete information referring to many parameters of user profiles. The distance function between real number intervals $r=[r_*,r^*]$ and $r'=[r'_*,r'^*]$ is defined as:

$$\delta(r,r') = |r_* - r'_*| + |(r^*-r_*)-(r'^*-r'_*)|$$

and the algorithm which determines a consensus for given intervals is the following:

Given: Relation T consisting of n real number intervals $tp_j=[tp_{j*},tp_j^*]$ for $i=1,2,...,n$

Result: Consensus $tp=[tp_*,tp^*]$ such that $\sum_{i=1}^{n}\delta(tp,tp_i) = \min_{tp' \in T} \sum_{i=1}^{n}\delta(tp',tp_i)$

Procedure:
```
BEGIN
    if n=1 then
                begin
                    tp:=tp₁; goto END;
                end
    else
      begin /*Creating sets with repetitions */
          X₁:=(tpᵢ* | i=1,2,...,n); /*of lower values */
          X₂:=(tpᵢ* | i=1,2,...,n); /*of upper values */
      end;
    sort sets X₁ and X₂ in increasing order;
    k:=⌊(n+1)/2⌋; k':= ⌊n/2⌋+1; /*where ⌊x⌋ is the greatest
                                   integer not greater than x*/
    for x in X₁ do set integer tp* such that xₖ'*≥tp*≥xₖ*;
    for x in X₂ do set integer tp* such that xₖ'*≥tp*≥xₖ* and
    tp*≥tp*;
    tp:=[tp*,tp*]
END.
```

5. Conclusions

The method for adaptive user interface construction presented here, as in many prior works in the HCI domain also has its statistical context, but its goal is to build automatically an appropriate interface for particular user (or class of users) instead of building one interface for so called average user. The consensus based method for adoptive interface construction is based on the assumption that it is possible to describe interface attributes values in terms of ordered values. Unfortunately for some postulated in the 3rd paragraph interface profile attributes, i.e. interface templates, this property could be quite difficult to satisfy. In such occasions more complex procedures should be applied.

References

1. Brown S.M. (et. al.) (1998), Using Explicit Requirements and Metrics for Interface Agent User Model for Interface Agent User Model Correction. Autonomous Agents, Minneapolis.1998
2. Fleming M., Cohen R. (1999), User Modeling in the Design of Interactive Interface Agents. UM99 User Modeling. Proceedings of the Seventh International Conference. Springer. 1999, pp.67-76. Wien, Austria.
3. Francisco-Revilla L., Shipman F.M. (2000), Adaptive Medical Information Delivery Combining User, Task ans Situation Models. Intelligent User Interfaces New Orleans LA USA 2000, pp. 94-97.
4. Hedberg S.R. (1998), Is AI going mainstream at last? A look inside Microsoft Research. IEEE Intelligent Systems, March/April 1998, pp. 21-25.

5. Katarzyniak R., Nguyen N.T. (2000), Using consensus methods to determine encapsulated profiles of the world state distributed in multiagent systems. S. Buchanan (Ed.), Proceedings of 14th Int. Conference on Systems Engineering, ICSE'2000, Coventry, UK, September 2000, pp. 300-304.

6. Keeble R.J., Macredie R.D., Williams D.S. (2000), User Environments and Individuals: Experience with Adaptive Agents. Cognition, Technology & Work 2000, vol. 2, pp 16-26.

7. Newman W.M., Lamming M.G. (1996), Interactive System Design. Harlow: Addison-Wesley 1996.

8. Nguyen N.T. (2000), Using consensus methods for solving conflicts of data in distributed systems. M. Batosek et al. (Eds.), Proceedings of 27th International Conference SOFSEM, Lecture Notes on Computer Science vol. 1963, Springer –Verlag, 2000, pp. 409-417.

9. Nguyen N.T. (1999), A computer-based multiagent system for building and updating models of dynamic populations located in distributed environments. Proc. of the 5th Conference on Computers in Medicine, Łódź 1999, vol. 2, pp. 133-137.

10. Nguyen N.T., Sobecki J. (2001), Consensus-Based Methods Applied to the Intelligent User Interface Development. To appear in: Proceedings of 1st International Conference on Information System Modelling, Czech Republic, 2001.

11. Nguyen N.T., Sobecki J. (2000), Making agreement of user service profile in the model of distributed interfaces. Proceedings of 2nd National Conference MISSI'2000, Wroclaw Univ. of Technology, 2000 (in polish).

12. Nguyen N.T. (2000), Using consensus methods for determining the representation of expert information in distributed systems. Cerri (Ed.): Proceedings of 9th International Conference on Artificial Intelligence AIMSA'2000, Lecture Notes on Artificial Intelligence vol. 1904, Springer –Verlag, 2000, pp. 11-20.

13. Peerce J. (et. al.) (1996), Human-Computer Interaction. Harlow: Addison-Wesley 1996.

14. Sobecki, J. (1999), Personal Agents for Public Access Interactive Systems. Conference Materials Second International Conference on Research for Information Society 14-16 October 1999, volume B, pp. 7/1-7/7.

15. Sobecki, J. (2000), Interactive Multimedia Information Planning. Valiharju T. (ed.), Digital Media in Networks 1999. Perspectives to Digital World. University of Tampere 2000, pp. 38-44.

16. Wooldridge M., Jennings N.R. (1995), Intelligent Agents: Theory & Practice. Knowledge Engineering Review 1995 100(2), pp. 115-152.

Adaptable and Intelligent User Interfaces to Heterogeneous Information

Maximilian Stempfhuber

Social Sciences Information Centre (IZ), Lennéstr. 30, 53113 Bonn, Germany
E-mail: st@bonn.iz-soz.de

Abstract

Information nowadays is available to a large number of users over the Internet. Along with the ease of access at the technical level comes the need to support a wide range of users having different skills and information needs. Users' skills may vary at the level of computer literacy - knowing how to operate a computer and the software to access the information - and at the level of task or domain knowledge which is necessary to find relevant information. In existing information systems, there often is a sharp distinction between user interfaces and functions for beginners and for experienced users, making a smooth transition nearly impossible. Difficulties add up, if multiple heterogeneous information sources have to be accessed to satisfy the user's information needs. On the basis of Visual Formalisms and Query Previews we propose a user interface design for information systems which visualizes semantic dependencies between different information sources and the indexing vocabularies used by them. It exploits human's information processing capabilities and avoids the often incomplete transfer of knowledge involved with metaphors. At the same time it allows for user-controlled adaptation of the information presented on the screen, showing the inexperienced user only the information he can understand while allowing the experienced information seeker to stepwise add information necessary for advanced search strategies. Intelligent components transform search terms between the vocabularies of the information sources, a process that is transparent to the beginner but can be controlled by the expert at a very detailed level.

1. Introduction

With the growth of the internet during the last years, everyone now may slip into the role of an information producer or consumer at any time and nearly any place in the world. Along with it comes a new diversity of users and information sources, for which standards do neither exist nor can be enforced.

Current activities in information science develop standards for metadata (e.g. Dublin Core), indexing vocabularies or document structures and try to support their activities with methods for automatic metadata extraction, vocabulary switching or schema mapping. Their success is limited by the fact that standards normally can not be enforced on the Web and that information providers may have plenty of good reasons for being distinct from others. In addition, the average user seeking for information on the Internet is normally not aware of the internals of search engines and portals or the differences in quality of information sources.

Looking at human-computer interaction, many advances have been made since the first graphical user interfaces (GUI) appeared. But still there is no single model or framework for designing user interfaces in a way that would ensure best usability for every potential user. Too often, the demands of users at one extreme of the scale are left out because they seem too few to be a market or too demanding to meet their needs without neglecting the others. The result will be a user interface that is too simple for a professional or too complex for a beginner. Often both interfaces are combined within one system without providing a smooth transition between them. A good example arc current search engines on the Internet with the radically different user interface for the beginner (normally a single line to enter terms) and for the expert (e.g. a facility to enter complex expressions in a formal language).

For the domain of information retrieval we tried to develop a concept for designing user interfaces that uses a very restricted graphical syntax and suffices several - often contrary - goals:

- Equally well suited for beginners and experts.
- Minimal effort for comprehending the user interfaces basics.
- Adaptable to different task and domain knowledge.
- Adaptable to simple and complex information needs.

2. Detecting users' needs

To detect the users' needs in the domain of information retrieval we conducted a questionnaire about their experiences with and expectations to the retrieval systems they normally use. Since we were primarily interested in the maximum range of functions that an information system should provide, 14 professional information brokers were interviewed. For 13 of them the retrieval of documents was a regular part of their work and they had on average 15.4 years of experience in document retrieval.

The first group of questions dealt with preferences in user interfaces (graphical vs. command line), retrieval models (Boolean vs. statistical) and functions/information supporting query formulation and review of results. The alternatives had to be rated on a scale from 1 to 5, with 1 meaning "never used" or "unimportant feature" and 5 meaning "used nearly exclusive" or "very important feature". We found out that graphical entry forms (on average 3.6 out of 5 points) in combination with Boolean logic (4.7) are the preferred mode of interaction and retrieval model. Also six entry fields had to be rated according to their importance for query formulation (search terms, free text, title, author, year of publication, classification). They all got between 3.1 (classification) and 4.9 (search terms) points on average, with five additional fields being suggested by the users. When asking for important features that support query formulation, the full range of Boolean operators, means to manually enter a Boolean expression (in a separate window and within every field) and a flexible combination of search fields with Boolean operators were rated 4.5 or better. Five additional features were suggested by the users and got between 2.7 and 3.9 points. Only one feature - an entry form with a simplified, restricted Boolean logic - was rated significantly lower (1.9). All ten types of information, that could be displayed for the query results were rated between 2.5 (graphical display of distribution of hits per search term) and 4.9 (number of documents found).

With the second part of the questionnaire we tried to find out if entry forms for queries imply a single interpretation as a Boolean query through their layout (e.g. title AND author vs. title OR author), since many information systems combine them with AND without letting the user change the operator. Five different screen layouts had to be evaluated and the subjects wrote down the most reasonable Boolean expression to combine (some of) them. They specified also the default operator that should be used within fields to combine two or more search terms. As a default for combining fields, the AND operator was suggested, but seven of the subjects explicitly noted that every operator should be available. Within fields the preferred default was the OR operator. The use of AND as a default between fields is in contrast to the most frequently used combinations of fields, which were also produced by the subjects: The most frequently used fields ('search terms' and 'free text' from part 1 of the questionnaire) were exclusively combined with OR. An explanation for this might be that some subjects thought of relevant search strategies, whereas the others only chose the operator which – independent of a strategy - narrowed the search.

The last part contained questions to detect the necessary features for searching in multiple heterogeneous databases at the same time. The major problems are differences between the databases in structure (i.e. fields), content (i.e. document types) and indexing (i.e. vocabularies, intellectual vs. automatic indexing). Many systems today support this 'cross search', but often only with a limited set of features compared to a search in only one database (e.g. no thesaurus). To detect important features, a list containing 19 of them had to be rated according to their usefulness (e.g. thesaurus support, changing databases during query formulation, live display of number of hits, ranking results by relevance, iterative retrieval). All of the features were rated between 2.9 (display of differences between thesauruses) and 4.7 (show if a search term is valid in any of the databases).

In summary, the participating experts had very similar expectations in regard to the features they rated important in information systems. Nearly every feature, regardless if existing in current systems or not, was judged to be at least of medium importance. No suggestion to simplify the system or to reduce functionality was accepted. For user interface design this means that tailoring the system only to novice users may reject the experts, while the functionality needed by the experts is far too complex for the beginner.

To resolve this contradiction, we use Visual Formalisms and Query Previews as a means to visualize complex semantics of the underlying data in a way that is well known and easy to comprehend. At the same time it allows the user to add complexity at a very fine grained level to make the user interface suit his specific needs.

3. Design idea

The domain of our research, the simultaneous retrieval of documents from multiple information sources, raises problems because of the heterogeneity of the information units - the documents - and the collections - the databases - which contain them. Even text documents from one source can be heterogeneous in *structure*, thinking of bibliographic data of monographs and journal articles (e.g. author, title, year of publications) in contrast to

descriptions of research projects (e.g. title, duration, funding). Both document types are contained in the databases SOLIS (literature) and FORIS (research projects and gray literature) of the Social Science Information Centre.
One other source of heterogeneity in *semantics* are the vocabularies used for indexing documents. Many information providers, like libraries, scientific information centres or commercial database publishers use specialized indexing vocabularies (e.g. thesauruses or classifications) to describe the relevant topics of the documents stored in their databases. These vocabularies normally differ in the level of detail (libraries have a large, information centres a very narrow scope of relevant literature), semantics (the same word may have different meanings in different domains) and language. Table 1 shows typical differences - which rise from heterogeneity - and the components of an information system which can handle them.

Table 1: Levels of heterogeneity and system components to handle them

Differences	System component
Structure (different fields, e.g. author, title, editor, year of publication)	user interface
Content (different document types, e.g. monographs, articles, project descriptions)	user interface
Indexing vocabulary (different thesauruses and/or languages)	cross-concordances, statistical transfer modules, user interface

It has been shown in our questionnaire, that the specific needs of beginners and experts may be very different. Table 2 lists the needs of both groups in the context of user interface complexity and system features.

Table 2: Different needs of novices and experts

Novice	Expert
simple user interface	complete set of features
no training	willing to spend time to learn advanced features
few additional information	detailed information to determine search direction and to analyze results
short query, immediate presentation of results	complex queries, frequent reformulation
predefined structure of the query	flexible structure of query
information if/how often a search term was found	search term is valid in which databases, search term is found how often in which databases, transformation of search terms between databases
information about size of the result set (total number of hits)	hits per search term and database, hits per database, total number of hits

Dynamic Queries (Shneiderman et al. 1992, Ahlberg et al. 1992) and *Query Previews* (Plaisant 1999) are two techniques to build user interfaces for database systems that allow interaction and give immediate feedback on the changes in the query result. Dynamic Queries support fast and reversible modifications of queries by using direct manipulation (e.g. with sliders or selection from lists) instead of typing. Query Previews use special metadata about the distribution of attribute values to calculate the size of the result set on-the-fly instead of performing time consuming database searches. This allows the user to see the effects that changes to the query have on the result set before the query is actually sent to the server. It therefore minimizes the chance of an empty result set.
One thing Dynamic Queries do not always handle properly is to show semantic dependencies between the attribute values used for searching and the elements of the result set. A simple example is a query for restaurants where one search attribute is *cuisine* and the other *average entree*. The initial selection of 'French' and 'Italian' cuisine may yield 23 restaurants, the following selection of $20-25 for an average entree could reduce the number of restaurants to 16 without indicating how many of them have French cuisine or if there still are any Italian restaurants that meet the criteria.
To show these semantic dependencies we use a *Visual Formalism* (Nardi&Zarmer 1993). Visual Formalisms - in contrast to metaphors - do not depend on the transfer of knowledge from one domain (e.g. mechanical typewriter) to some other (e.g. computer + word processing software), which is often incomplete and leads to user errors. They use

humans' visual information processing capabilities and rely on general objects, like maps, tables or diagrams. The term *Visual* is used because they are to be generated, comprehended and communicated by humans; they are *Formal* because they are to be manipulated, maintained, and analyzed by computers (Harel 1988). One very important aspect of Visual Formalisms is that they are interactive and therefore are at the same time output (the displayed information) and input (the user can manipulate the display and the data).

3.1 ODIN user interface components

The figures 1 and 2 show the two basic components of the ODIN framework (Object-oriented Dynamic user INterfaces; Stempfhuber 2001): query filters and query attributes. Both use a table to show the dependencies in the data of the underlying information system.

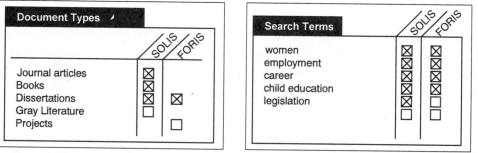

Figure 1: Filter for document types   Figure 2: Control for entering search terms

The filter 'document types' shows the dependencies between each document type (e.g. journal article, book) and database, expressed by the combination of rows (document types) and columns (database name). A cell in the table is empty, if the specific document type does not exist in a certain database and contains a check box control, if the document type exists in the database. By default, all check boxes are selected, so that the search will include all available information. To restrict the search to specific document types and databases, the user opens the filter and de-selects some check boxes, like it is shown in figure 2 (no projects from the database FORIS).

In the same way, a table can display if a search term is valid for a database. Figure 2 shows the combination of five search terms and four databases. In this simplified example, the information system uses automatic indexing and therefore may not have adequate metadata at hand if a search term is valid. But even in this situation it allows the user to control the query at a very detailed level, so that search terms yielding too many hits in certain databases can be excluded only from these databases, but not from others.

3.2 Adapting the user interface

From the questionnaire with expert users of information systems it was clear, that they need much more information to guide their search than novice users, which often are satisfied by browsing through the results immediately after initial query formulation. The adaptive ODIN control for search attributes allows the user interface to be as simple as possible for the novice and can be enhanced with additional information by the expert.

Figure 3 shows the entry field for search terms with the maximum level of additional information. The first column contains Boolean operators to modify the logic of the query by direct manipulation. The columns for the databases have been enriched with the number of hits per search term and database (number within each cell), the hits per database (column sum) / search term (row sum) and the total number of documents (lower right corner). When letting the mouse pointer rest over a cell in the table, a fly-over help will give information about the transformation of query terms between the databases, which takes place automatically in the background.

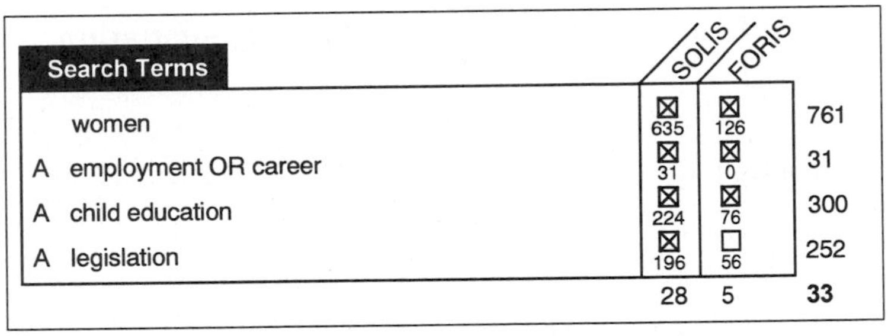

Search Terms	SOLIS	FORIS	
women	☒ 635	☒ 126	761
A employment OR career	☒ 31	☒ 0	31
A child education	☒ 224	☒ 76	300
A legislation	☒ 196	☐ 56	252
	28	5	**33**

Figure 3: Maximum level of information for search terms

4. Conclusion

We have presented a novel design for user interfaces of information systems which allows both novice and expert users to efficiently find information in heterogeneous information sources. The design is based on the Visual Formalism of a table and is able to express the dependencies between search attributes and databases even in complex situations. For the novice, all information can be hidden and the user interface consists only of a multi-line entry field per search attribute. The expert can adapt the interface to its needs and display additional information at the most detailed level possible. This fits the demands of both user groups and allows smooth transition from novice to expert with a single and consistent user interface design.

The ODIN user interface has been implemented as a prototype for text retrieval and as a fully functional information system for geographical information, MURBANDY (Stempfhuber et al. 2001). A heuristic evaluation of the prototype with five experts ensured completeness of features and comprehensibility of the design.

Currently user tests are carried out in a controlled setting with both systems to evaluate ease of use and comprehensibility of the table display. This will also give information about the adequate amount of information for the initial configuration of the screen. While experts explicitly stated that all information (see figure 3) should be displayed permanently because of its usefulness for their normal work, it is clear that this is not appropriate for all users in all situations. Pre-tests showed that the ODIN concept is easily understood because of its simple structure and can be transferred to different domains very well. There were only minor problems with dynamic screen layout, which was heavily used in MURBANDY, but not with text retrieval.

References

Ahlberg, C.; Williamson, C.; Shneiderman, B. (1992). *Dynamic Queries for Information Exploration: An Implementation and Evaluation.* In: Proceedings of CHI'92 Conference on Human Factors in Computing Systems, Monterey, CA, May 3-7, 1992. pp. 619-626.

Harel, D. (1988). *On Visual Formalisms.* In: Communications of the ACM, Vol. 31 (5), May 1988, pp. 514-530.

Nardi, B. A.; Zarmer, C. L. (1993). *Beyond Models and Metaphors: Visual Formalisms in User Interface Design.* In: Journal of Visual Languages and Computing, 1993, 4, pp. 5-33.

Plaisant, C.; Shneiderman, B.; Doan, K.; Bruns, T. (1999). *Interface and data architecture for query preview in networked information systems.* In: ACM Transactions on Information Systems, Vol. 17 , Nr. 3, pp. 320-341.

Shneiderman, B.; Williamson, C.; Ahlberg, C. (1992). *Dynamic Queries: Database Searching by Direct Manipulation.* In: CHI'92 Conference on Human Factors in Computing Systems, Monterey, CA United States, May 3-7, 1992, pp. 669-670.

Stempfhuber, M. (2001). *ODIN. Objektorientierte Dynamische Benutzungsoberflächen. Behandlung struktureller und semantischer Heterogenität in Informationssystemen mit den Mitteln der Softwareergonomie.* Dissertation im Fach Informatik an der Universität Koblenz-Landau (to appear).

Stempfhuber, M.; Hermes, B.; Demicheli, L.; Lavalle, C. (2001). Enhancing Dynamic Queries and Query Previews: Integrating Retrieval and Review of Results within one Visualization. In: Oberquelle, H.; Oppermann, R.; Krause, J. (eds.): Proceedings of Mensch & Computer 2001, Bad Honnef, May 5-8, 2001, pp.317-326.

Distributed User Modelling for Universal Information Access

Julita Vassileva

Computer Science Department, University of Saskatchewan,
1C101 Engineering Bldg., 57 Campus Drive, Saskatoon, S7N 5A9, Canada

Abstract

In a distributed multi-agent based software environment, the traditional monolithic user model ceases to exist and is replaced by user model fragments, developed by the various software agents populating the environment. These model fragments have been developed in a variety of specific contexts to help achieve various goals. User models are thus fragmented, relativized, local, and often quite shallow. They are inherently inconsistent with one another and reflect not only characteristics of the users, but also certain social relationships among them. With the arising proliferation of models and data, the user modelling problem transforms into retrieval and integration of the available user model fragments "just in time" by a particular computational agent to the breadth and depth needed for a specific purpose. In this paper we explore the implications of distributed user models, drawing examples from I-Help, a collaborative system for peer help.

1. Introduction

The next generation of mobile, distributed, autonomous computer applications will allow future computing environments to be accessible from everywhere: not only on desk-top and lap-tops, but also on palm-tops, cell phones; these environments will be worn (data glasses, watches, etc.) and be embedded in everyday devices, home appliances, in the environment. People with various cultural background, goals, knowledge, and disabilities will be interacting with these mobile and ubiquitous computing environments in all imaginable contexts: in vehicles, in meetings, on public transport, while shopping, relaxing, eating, cooking etc... Meeting user needs adequately is crucial for future software applications, since the variety of contexts and devices in which software will be used implies a huge diversity of user needs to be met. The only way of achieving adaptation in distributed software environments is by designing systems that are no longer static stand-alone applications, but dynamic integrative environments that configure themselves according to the individual needs of the user, the context of use, and the platform requirements. Such systems should be able to operate independently of a person's location. In order to achieve such environments, the application functions have to be packaged into small atomic units or agents, which can re-assemble themselves dynamically depending on situational variables. Thus the atomic units will be reused in an appropriate way depending on the current hosting environment, e.g. uniformity and compatibility is achieved. These two qualities are important for two reasons: first, to allow the benefits for interaction and interoperability of software in the networked world, and second, to reduce the load on users learning how to operate a variety of software on different devices.

User modelling is crucial in order to track the context of the user. Due to the variety of contexts ensured by the mobility and ubiquity of the computing devices, a lot of information is needed to capture the aspects that are important for the user's needs, for example, about her exact location, state etc. The more information is available, the more adequate the user model and therefore, the better the adaptation of the functionality towards the individual user will be. Tracking the user's movement, location, state is especially important since the users are mobile.

We believe that conventional methods for software design even though appropriate for distributed applications design are not sufficient for applications supporting universal user access. These methods assume providing a fixed functionality defined at their design time independently of the conditions of their use, the experience or typical tasks of the user. There are techniques ensuring interoperability in distributed software, for example, CORBA, Jini and DCOM. However, with distributed objects, even though objects may run on different platforms, applications generally form a single monolithic entity of tightly bound objects, with hand-coded calls to known methods of pre-existing objects. The problem that the applications may require different functionality in a wide variety of contexts (platforms, locations, user needs etc.) cannot be addressed. There is research on methods allowing software applications to, for example, dynamically constrain the offered functionality when the parameters of the environment do not allow the provision of full functionality (Noble & Satayanaranyanan, 1999). For example, if a web browser is invoked on a laptop using a low-bandwidth network connection (modem or wireless), the browser should automatically adapt to ignore images that slow down the transfer of information. However, this adaptation is only with respect to the hardware and not with respect to the needs of the individual user. For example, for a particular task of the user it might be crucial to get a specific image, even though this image might not be important for 99% of the users. Providing applications with the capability to dynamically configure themselves, as proposed by Martin, Cheyer & Morgan (1999), but so that they meet best the needs of their users, according to the platform and context of use (e.g. location, task etc.) would lead to flexible self-adaptive software applications for universal access. The emerging environments that will dominate the software landscape of the future will be distributed, multi-user, ubiquitous, and mobile. There will be very few monolithic applications – the new applications will be inherently distributed and agent based. Negotiation between agents will become the way software applications function and interact with users; there will be no pre-defined at design time "method calls", but dynamically established links as result of negotiation among software agents. The functionality of the applications will emerge as a result of collaboration between software agents and users.

User modelling (UM) plays a crucial role in this type of environments. Continuous contact between users and ubiquitous information / communication technology allows for very fine-grained tracking of users' activities under different circumstances and by different modelling agents. At any given moment of time there is no consistent model of a user; there are many "snapshots" taken by various agents, in different contexts, containing totally different information. Therefore, in a distributed multi-agent based software environment, the traditional user model is replaced by user model fragments, developed by the various software agents, populating the environment in a variety of specific contexts and with various goals.

2. An example: the I-Help system

An example of such an environment is I-Help (Greer et al., 1998, Vassileva et al., 1999), a multi-agent system that allows users to request, receive and give peer help synchronously and asynchronously. I-Help is fully implemented and has been deployed in different versions with more than 1000 users. Ffor more information about evaluation of I-Help in these deployments, see Greer et al. (2001).

I-Help provides seamless access for students to a variety of distributed help resources (human resources, like peer help and expert advice, as well as electronic resources, like threads in discussion forums, FAQ entries, and web-resources). Users are represented in I-Help by their personal agents; electronic resources or software applications are represented by application agents. A user request for help is sent by the user's personal agent to an agent broker who locates an appropriate other agent (either personal or application agent) using a database of models of the agent's resources. These resources can be the competence of user in the domain or the topic of material covered by an electronic material. I-Help introduces negotiation between agents and payment for help in terms of ICUs (I-Help Currency Units). On an agent-level the I-Help economy helps in regulating the supply and demand of human help resources. On a user-level, the market mechanism helps prevent overloading competent users with help-requests and motivates users to provide high quality help. In this function, it is similar to reputation / ranking based mechanisms used by Web sites like *slashdot.com*, *plastic.com* and *www.thewines.com*. In large multi-user environments, it cannot be expected that users will be intrinsically motivated to help other users. An economic regulating mechanism is important in multi-user environments since otherwise they tend to get invaded by "harvesters" which cause degradation in the performance (Adar & Huberman, 2000).

2.1 Adaptation in I-Help

There are various types of adaptations taking place in I-Help:

- Adaptation to the user's level of competence and specific help request.
- Adaptation of the number of help-requests directed to the user to his/her level of busy-ness, priorities at the moment, general preferences, and interpersonal relationships.
- Adaptation of the type of help-resource and the interface to the physical location of the user and the type of device being used, whether a browser on a PC, a hand held device, or a cell phone.
- Adaptation of the agent's appearance and basic behaviour to the user's wishes.
- Adaptation of the negotiation strategies deployed by the personal agents to the preferences of the user.
- Adaptation of the negotiation strategies deployed by the personal agents to the tactics of other users' agents.

Adaptation in multi-agent multi-user systems, like I-Help, happens all the time. Since there is no predefined behaviour to be "adapted", the behaviour of the system emerges depending on (1) the circumstances, (2) complex interrelationships between users and agents, (3) "just-in-time" generated models, and (4) decision-making techniques to judge the situation at hand and to calculate the utilities of the possible actions. Decision-theoretic approaches have become increasingly popular in the UM community recently (Bohnenberger et al., 2001; Jameson et al., 2000; Mudgal & Vassileva, 2000; Suryadi & Gmytrasiewicz, 1999). They require, however, a numeric or probabilistic representation of the situation (i.e. the user modelling features and other variables describing the context) and numeric evaluations of the payoffs of each possible action. These numbers are hard to invent; they need to be determined experimentally.

Adaptation within the multi-agent I-Help system is based on 1) models of users and other agents maintained by the personal agents, and 2) models of the involved software applications maintained by the application agents. The latter type of models (i.e. models of the resources provided by software applications: web-pages, discussion forum threads, etc.) is an important part of the adaptation of the system, since they are traded on the market by the application agents to satisfy users' needs. However, in this paper we focus on the user and agent models maintained by personal and diagnostic agents.

2.2 Agents developing models of users and other agents

A multitude of user models is created by various agents and with various purposes. *Application agents* (agents of discussion forum threads, web pages, search engines) build their own *user profiles* representing features relevant to the context of the application, *based* on their interaction with the users and using "traditional" UM techniques.

Diagnostic agents (agents representing web-based test items, questionnaires etc.) represent a special type of application agents for a specific goal to create *user models in a particular area* of activity / knowledge and with a particular structure.

Personal agents (agents representing users), maintain user models containing private user characteristics. Examples of such characteristics are *lists with the person's friends and enemies, the user's preferences about negotiation* (how greedy/generous the personal agent should be in negotiation, the subjective importance for the user of certain resources like time or money), and the user's current goal. In addition, personal agents manage the set of competence models/profiles of the user for certain domains. Diagnostic agents create these domain specific user models on user's or personal agent's request. During negotiation with other agents, the personal agent acts as a representative of the user. The negotiation preferences and characteristics that control the agent's behaviour with other agents (for example, egoism, greediness, or generosity) are selected by the user. They reflect the way the user wishes to be perceived by the "world", therefore, indirectly, they represent also a kind of model of the user. During negotiation, the agents try to optimize their actions and to predict the opponent's actions. For this purpose, they create *models of the other agent's "character"* (priorities). Thus each personal agent models the character that the other user wants his/her agent to represent in the agent community.

Matchmaker / broker agents manage *databases of user models* (profiles) for a certain population of users; each broker is specialized to deal with models of certain user characteristics and to perform matchmaking for a specific purpose.

2.3 User modelling process: making sense of fragments

During negotiation, the agents take into account the relationships between the users (if they exist), for example, by changing the negotiation strategy for friends or enemies (offering a discount or an extra high price). After repetitive successful negotiations followed by successful help sessions between the users, the agents offer to add a new relationship between the users in their models, thus increasing the number of "friends" of their users.

The users are notified by their agents only when a deal has been arranged, and they can agree to participate in the help session or they can discard the message. If the user always discards the notifications of his/her agent, it won't be able to earn virtual money. After a failed deal due to an undelivered resource (i.e. the user refused to help), the personal agent of the other user notifies the matchmaker. The matchmaker serves also as a "better business bureau": the personal agent, which breaks a deal, will gain bad reputation, and other agents will start avoiding it.

After a help session, the personal agent of each user presents to the user a brief questionnaire about the usefulness of the help session, and the perceived knowledge of the other party. This information is used to update the models of the users involved in the help session.

In summary, I-Help is an example of a system with many users interacting at any point of time with a varying pool of agents. In such a setting, there is no one monolithic user model associated with each user. Rather the knowledge about the user is distributed among the various agents who interact with the user (both human and software agents). User and agent models are thus fragmented, relativized, local, and quite shallow. They are inherently inconsistent with one another and reflect not only characteristics of the users, but also certain social relationships among them. In addition, depending on who is modelling and who is being modelled in a distributed multi-agent environment, there can be agents modelling users, users modelling agents and agents modelling other agents. With the arising wealth of models and data, the user-modelling problem transforms into retrieval and integration of the available user model fragments "just in time" by a particular computational agent to the breadth and depth needed for a specific purpose. Thus, the need for integrating user model fragments grows in importance, and the ideal of maintaining a single monolithic user model is less desirable (and likely intractable).

3. Discussion: distributed user modelling issues

With distributed applications/environments user modelling becomes the process of assembling and summarizing fragmented user information from potentially diverse sources. The key to making sense of this widely distributed information is the ability to interpret multi-modal information from multiple heterogeneous relevant sources and to integrate this information into a user model of appropriate granularity. The focus is shifted from the model itself to the process of modelling. The model is computed "just in time" (Kay, 1999) and only makes sense in the context in which it is created (time, purpose, the agent creating it, the agent being modelled, the available sources of information). This introduces many new requirements for the user modelling process. The main questions boil down to how to manage all this information:

- How does one locate the agent who has a relevant model given the context and the purpose for which the model is needed?
- How does one make sense of possibly inconsistent and even contradictory data?
- In general, how does one interpret models created by other agents?
- How does one ensure persistency and integrity of user models in such an environment? Can it be expected at all?
- How to ensure privacy?

New techniques will be required in order to carry out distributed, just in time integration of user models, relying heavily on modelling interpersonal relationships, decision making. Changes in the traditional methods for system evaluation will be necessary, since currently there is a lack of methodologies for evaluation of systems with emerging behaviour.

Many privacy and security issues arise in such environments that must be tackled. The idea of delegating the responsibility of user modelling to autonomous (and even worse, economically motivated) agents can be worrisome. Are agents guaranteed to serve their users' best interests? Of course, if agents maintain only distributed context-loaded profiles, a full integration of the information by the "Big Brother" would be so expensive that it would be practically impossible. However, integration of specific data, with a specific purpose would be possible. Even if we trust the personal agents not to reveal some particular data, how do we know which piece of data may be critically integrated with other data and used against us?. The police warning "Anything you say can be used against you" will mean, "any information revealed by you or your agent can potentially be used against you". In this case, the role of personal agents should become similar to the role of lawyers protecting the interests of their users. However, this requires a level of intelligent reasoning of the agents that is unlikely to be feasible in the near future. And even if users instruct their agents not to reveal *any* information about them (at the cost of much functionality and many potential benefits of multi-agent multi-user modelling), they can't prevent other agents from modelling their behaviour, just as people cannot prevent other people from observing their actions, making conclusions and thinking certain things about them. Any person and agent who has encountered the user or his/her agent will be able to develop a model of him / her, which it may give to a third agent or user on request. So "universal access" to information becomes "universal transparency", which makes every user potentially vulnerable.

Another problem is how to avoid spreading negative rumours and ill-informed gossip among agents, with the attendant risk of isolating agents or users from certain services (Bicknell, 2000). Reputation networks and rumours are necessary protection mechanisms in a multi-agent community, but how is it possible to avoid their misuse? Active research in I-Help into these issues is now underway (Winter, 1999).

So the privacy issues of this new type of multi agent / user modelling are many and deep. On the other hand, we are witnessing the appearance of inherently distributed applications, like Gnutella (Gnutella, 2000), and Mojo (Mojonation, 2000), allowing storage of data or computation to happen on demand on any machine in a distributed environment without an all-knowing centralized component, depending completely on the patterns of computer usage of the participants. It seems that the idea of revealing the possibly private information of computer usage and even providing their "sacred" computers with all their data to unknown third parties is not so undesirable for users, when it is not done by a centralized institutions and when it is based on an economic rewarding mechanism. We believe that, as this example shows, people will readily give some of their privacy in exchange for the convenience and benefits provided by the new technology and highly adaptive environment. Yet, these issues have to be tackled, since this technology, as any powerful technology, can be easily and badly misused.

4. Conclusion

These revised ideas about user modelling will shift the user modelling research agenda. Processes such as retrieval, aggregation, and interpretation will be much more important than they have been. Many very interesting research issues surrounding these techniques will have to be explored. In a fragmented, distributed, and universally accessible technological environment, user modelling will increasingly be viewed as essential to building an effective system, but will also increasingly be seen to be tractable as new techniques emerge from these explorations. Nevertheless, as our I-Help experiments have already shown, it will not be necessary to resolve all of these issues in order to usefully user model.

References:

Adar E., Huberman B. (2000) Free Riding on Gnutella. *First Monday.* available on line at: http://www.firstmonday.dk/issues/issue5_10/adar/index.html

Bicknell, C. (2000) Anti-Fraud That's Anti-Consumer, Wired News, July 24, 2000, available on line at: http://www.wired.com/news.print/0,1294,37642,00.htm

Bohnenberger, T., & Jameson, A. (2001) When Policies Are Better Than Plans: Decision-Theoretic Planning of Recommendation Sequences In J. Lester (Ed.), *IUI2001: International Conference on Intelligent User Interfaces. New York: ACM.*

Greer, J., McCalla, G., Cooke, J., Collins, J., Kumar, V., Bishop, A. and Vassileva, J. (1998) The Intelligent HelpDesk: Supporting Peer Help in a University Course. In Goetl B., Half H., Redfield C., Shute V. (Eds.) *Intelligent Tutoring Systems: Proceedings ITS'98.* LNCS No1452. (pp.494-503). Springer Verlag: Berlin.

Greer J., McCalla G., Vassileva J., Deters R., Bull S., Kettel L. (2001) Lessons Learned in Deploying a Multi-Agent Learning Support System: The I-Help Experience, to appear in Moore J (Ed.) *AI and Education: Proceedings of AIED'2001,* San Antonio, Texas, May 19-24, 2001.

Gnutella, 2000; available on line at: http://gnutella.wego.com/

Jameson, A., Großmann-Hutter, B., March, L., Rummer, R., Bohnenberger, T., & Wittig, F. (2000) When Actions Have Consequences: Empirically Based Decision Making for Intelligent User Interfaces. *Knowledge-Based Systems,* 13. In press.

Kay, J. (1999). *A Scrutable User Modelling Shell for User-Adapted Interaction.* Ph.D. Thesis. Basser Department of Computer Science. University of Sydney. Sydney. Australia.

Martin D., Cheyer A., Moran D. (1999). The Open Agent Architecture: A Framework for Building Distributed Software Systems. *Applied Artificial Intelligence.* 13, 91-128.

Mojonation (2000): available on line at: http://www.mojonation.com/

Mudgal, C., Vassileva, J. (2000) An Influence Diagram Model for Multi-Agent Negotiation, In M. Klusch & L. Kerschberg (Eds.) *Cooperative Information Agents: Proceedings of CIA'2000.* LNAI 1860. (pp.107-118). Springer Verlag: Berlin-Heidelberg.

Noble B.D., Satayanaranyanan M. (1999) Experience with adaptive mobile applicatons in Odyssey, *Mobile Networks and Applications,* 4, 4, 245-255.

Suryadi, D., Gmytasiewicz, P. (1999) Learning Models of Other Agents using Influence Diagrams, in Kay J. (Ed.) *User Modelling: Proceedings of the 7th International Conference on User Modelling.* (pp. 223-232). Springer: Wien-New York.

Vassileva J., Greer J., McCalla G., Deters R., Zapata D., Mudgal C., Grant S. (1999) A Multi-Agent Approach to the Design of Peer-Help Environments. In Lajoie S. and Vivet M. (Eds.) *Artificial Intelligence and Education: Proceedings of AIED'99.* (pp.38-45). IOS Press: Amsterdam.

Winter, M. (1999) The Role of Trust and Security Mechanisms in an Agent-Based Peer-Help Environment, , *Proceedings of the Workshop on Deception, Trust, and Fraud in Agent Societies* (pp. 139-149), associated with Autonomous Agents '99, Seattle WA.

Decision-theoretic approaches to user interface adaptation: implications on Universal Access

Vasilios Zarikas[1], Alexandros Paramythis[1], Constantine Stephanidis[1,2]

[1] Institute of Computer Science, Foundation for Research and Technology – Hellas, Science and Technology Park of Crete, GR-71110 Heraklion, Crete, Greece
email: cs@ics.forth.gr

[2] Department of Computer Science, University of Crete, Greece

Abstract

This paper discusses the employment of decision-theoretic approaches in adaptive user interfaces, towards the goal of Universal Access. Specifically, the paper examines ways in which the requirements posed by Universal Access on human-computer interaction can be addressed through automatic user interface adaptation, using decision-theoretic frameworks. Furthermore, the paper discusses some of the implications of employing such adaptation frameworks in user interfaces intended for use by diverse user groups, in differing contexts of use.

1. Introduction

Universal Access in the context of Human-Computer Interaction refers to the conscious and systematic effort to practically apply principles, methods and tools of universal design in order to develop high quality user interfaces, accessible and usable by a diverse user population with different abilities, skills, requirements and preferences, in a variety of contexts of use, and through a variety of different technologies [Stephanidis, 2001b]. Following the preceding definition, the development of universally accessible interactions involves addressing diversity along a number of different dimensions, including [Stephanidis, 2001b] intra- and inter- individual differences; cultural background; interaction technologies; environment and task contexts; etc.

A relatively recent approach to anticipating and addressing diversity in the context of Universal Access is the attempt to develop interactive systems that are capable of automatically adapting to a wide variety of user- and context- related requirements [Stephanidis, 2001a]. Adaptive User Interfaces (AUIs) [Dieterich, et al., 1993] is a research field that has received considerable attention in the past years. Their basic premise lies with the capability to dynamically (i.e., at run-time) determine aspects of the interaction state (mainly with respect to user characteristics, goals, plans, etc.) and respond to changes in these aspects as the need arises [Browne, et al., 1990]. Decision theory offers itself naturally as a methodological approach to the design and implementation of such dynamic behaviour.

[Norvig & Cohn, 1997] describe decision theory as the combination of utility theory (a way of formally representing a user's preferences) and probability theory (a way of representing uncertainty), and argue that it is an appropriate means for designing adaptive software that will operate in an uncertain environment.

The application of decision theoretic approaches in the development of intelligent and adaptive interactive software systems has gained momentum in recent years (see, for example, [Brown, Santos, and Banks, 1997], [Horvitz, 1999], [Murray, and VanLehn, 2000]). This has been mainly due to its employment in popular commercially available products. For instance, the seminal work of the Decision Theory and Adaptive Systems Group of Microsoft Research[1] (e.g., [Horvitz, et al., 1998]) has given rise to some of the best known incarnations of adaptive software components, as part of the Microsoft Office suite of products. The rest of this paper will seek to outline ways in which decision theory-based adaptive user interfaces can facilitate Universal Access, as well as identify some of the implications of employing decision theory in this context.

---

[1] http://www.research.microsoft.com/dtas/

2. Decision Theory in the context of Universal Access

2.1 Employing decision theory in adaptive interaction

As discussed in the previous section, the HCI-oriented challenges posed by Universal Access regard the multiple dimensions involved in rendering an interactive system or service accessible and usable by all its potential users, in different contexts of use, and through a multitude of access devices and platforms. Meeting those challenges through the application of adaptive interaction techniques may involve:

- identifying all those elements of diversity that may impact the interaction between users and the system, and enumerating their potential "values" (thus defining the space of diversity that is to be addressed, and the adaptation "determinants");
- identifying suitable alternative interactions that may cater for these elements of diversity (thus defining the adaptation "constituents") ;
- capturing the rationale that relates the identified alternatives to (combinations of) "values" in the design space (which serves as the adaptation "logic" [Stephanidis, et al, 2001]);
- "detecting" changes in these user-, context-, platform-, etc. related "values" (during, or between interaction sessions);
- deciding upon and applying adaptations (e.g., by de-/activating the alternative interactions), on the basis of the detected changes and the adaptation logic;
- assessing the effects of adaptations to determine whether these are in line with the expected results, as these are represented in the adaptation logic.

It should be noted that the above is not implied to be an either exhaustive, or "required" list of items with respect to the employment of adaptation towards Universal Access. Rather, it is intended as a basis for discussing the potential merits of applying decision theoretic approaches in this direction.

Decision theory recognises that, when a decision is made, the ranking of available alternatives produced by using a decision criterion, has to be consistent with the decision maker's objectives and preferences. The theory offers a rich collection of techniques and procedures to reveal preferences and to introduce them into models of decision. It is not concerned with defining objectives, designing the alternatives or assessing the consequences; it usually considers them as given from outside, or previously determined. Given a set of alternatives, a set of consequences, and a correspondence between those sets, decision theory offers conceptually simple procedures for choice. Thus, using, for example, decision analysis techniques, designers of an adaptive interface can define, without conflicts and incoherence among probabilities-weights and in a "readable" yet "computable" form, design strategies as vectors of objectives, design goals (e.g., performance, accuracy, ease of use) as criteria, and alternatives-constituents (e.g., text vs. graphics vs. audio) as attributes. The uncertainty that is inherent in the design of universally accessible, adaptive user interfaces, can also be treated through proven decision theoretic approaches and instruments, by: applying decision-making criteria from the field of game theory, to effect non trivial selection rules; employing dynamic subjective probabilities for design artefacts with limited empirical support; etc.

In the particular case of employing Bayesian belief networks, the following may also be considered potential benefits in the domain of adaptive user interfaces. Firstly, the bi-directional property of these networks offers the opportunity to learn causal relationships and hence to gain knowledge in order to predict consequences of interaction. Secondly, the structure of Bayesian networks permits the encoding of prior knowledge and data, which usually come in both probabilistic and deterministic form and in causal structure. In the rest of this section a number of existing, decision theory-based systems will be presented as exemplars of how the above relate to adaptive interaction.

Among the most important "determinants" of adaptation are the characteristics of users, including their knowledge, interaction goals and plans, organisational roles, etc. User modelling [Kobsa, 1993] and plan recognition [Waern, 1997] are two areas of research that have explicitly addressed the issue of capturing (and, in some cases, reacting upon) such information. The Bayesian student modelling framework used in ANDES [Conati, et al. 1997] is one example of the contribution of decision theory in these fields. Specifically, ANDES uses a probabilistic framework to perform three kinds of assessment: plan recognition, prediction of student's goals and actions, and long-term assessment of the student's domain knowledge. Other decision theoretic frameworks related to user modelling and plan recognition are described in [Carberry, 1990], [Wu, 1991], [Pynadath and Wellman, 1995], etc.

Employing decision theoretic approaches in adaptive interaction also opens up new possibilities for automatically generating human-comprehensible explanations of what a system decides to adapt and why (see, e.g., [Chajewska, and Draper, 1998]). Such explanations can be of paramount importance in at least two ways. Firstly, they allow users to acquire and maintain an understanding of the system behaviour and the underlying rationale, thus potentially rendering dynamic modifications a system feature that can be anticipated (or even depended upon), while at the same time increasing the user-perceived consistency of the system. Secondly, they enable the elicitation of user feedback regarding the system's model of its "world", which, in turn, can be used to assess and adjust decision making parameters at run-time.

For instance, with regards to decision theoretic learning techniques, one can consider the characteristic example of the intelligent interface agent reported in [Brown, Santos, and Banks, 1997], which is capable of altering the topology of its embedded Bayesian knowledge network, in order to better adapt itself to modelling a particular user.

A rather complementary perspective to learning is demonstrated by the Automated Travel Assistant [Lindens, Hanks, and Lesh, 1997], which differs from traditional decision analytic and planning frameworks, in that a complete model is not available prior to interaction (e.g., elicited, or constructed by a human expert). Rather, the system starts with minimal information about the user's preferences, while the respective model is incrementally inferred by presenting the user with candidate solutions and receiving critique over those solutions. [Wolfman, et al., 2001] take a similar approach, in the employment of a decision-theoretic framework for the selection of interaction modes in the SMARTedit programming by demonstration system.

The systems presented thus far provide some concrete examples of how decision theory can contribute to the development of adaptive interactions, taking into account various user- and interaction context- related dimensions. Although the cited work is not explicitly intended to facilitate Universal Access, its relevance and applicability towards meeting the related HCI challenges should be apparent.

2.2   Implications on Universal Access

Having discussed ways in which decision theoretic approaches can be employed to effect adaptive interaction, we now shift our attention to the implications and impact of decision theory in the development of universally accessible interactive systems and services.

Perhaps the most important contribution of decision theory to Universal Access, is the provision of formal means to represent uncertainty. Uncertainty is an inherent but often neglected element when one attempts to acquire, represent, or use knowledge that concerns as diverse a range of design parameters as Universal Access requires. No solution can be imaginably appropriate for all users, in all circumstances of use (or even for a single user in the course of time). And one can hardly characterise with certainty the precise preconditions for considering a solution "appropriate" for deployment. Under this perspective, decision theory offers the capability to explicitly model and, in fact, capitalise upon the incompleteness of existing design knowledge, when striving for Universal Access.

At a more general level, the employment of decision theory can offer a structured, "prescriptive" approach to the design and implementation of universally accessible systems. Specifically, as has been discussed in the previous section, decision theory is applicable to, and can facilitate, all phases of adaptive software development. Combined with higher-level methodological approaches for capturing and addressing the requirements posed by Universal Access (such as User Interfaces for All [Stephanidis, 2001b]), a decision theoretic framework for adaptation can "guide" the design and implementation of accessible interactions, offering the appropriate instruments to support each step of the process, starting from the initial encoding of related empirical findings and expert knowledge, all the way to supporting dynamic behaviour at run-time.

Concerning the encoding of design knowledge in particular, decision theoretic frameworks have the added benefit of allowing one to use that knowledge at run-time, not indiscriminately (i.e., without questioning its validity), but in a manner that allows for reassessment and "fine-tuning" of that knowledge, given the user and usage context at hand (which is directly related to the potential of decision theory for structured learning). Furthermore, with the advent of evaluation frameworks that explicitly address the derivation of design feedback from adaptive interactions [Paramythis, Totter, and Stephanidis, 2001-submitted], it is foreseen that it will be possible to exploit decision theoretic frameworks in the creation and extraction of new, empirically founded design knowledge, with direct relevance to Universal Access.

Finally, another related benefit stems from the representational power that decision theory affords to the encoding and progressive refinement of universal design knowledge. This power originates from the capacity to model and seamlessly relate the various dimensions that may influence interaction at any given point in time. For example, a designer may initially model separately the design parameters of users, interaction technologies, the application domain, observable environment factors, etc. These modelled parameters can, at a later stage, be combined to form a global execution context (i.e., the combination of any and all parameters that may influence interaction at any given point in time), so that decisions made to facilitate access are not based only on partial, or segmented views of the "world". Seen from a different perspective, creating such a global context enables one to assess the validity of design knowledge that was derived under more constrained, or controlled conditions.

3. Conclusions

This paper has addressed the employment of decision theoretic approaches to the development of universally accessible interactions, focusing on how decision theory can facilitate and inform automatic user interface adaptation in this context. A number of existing decision theory-based adaptive systems have been briefly presented, to demonstrate ways in which decision theory can be used to approach the challenges posed by universal access. Finally, some of the main implications of decision theoretic adaptation on universal access have been discussed.

The authors are currently investigating the utilisation of decision theory-based adaptation frameworks in the user interface of the Nautilus Web browser[2] (a follow-up to the AVANTI Web browser [Stephanidis, et al, 2001]), which is intended for use by a diverse user population (including people with various forms and degrees of disability), in different contexts of use (e.g., on a user's personal computer, at public information points).

Similar approaches are being applied in the case of the PALIO[3] framework for nomadic, location-sensitive and location-based services. Furthermore, a new modular approach to the evaluation of adaptive user interfaces [Paramythis, Totter, and Stephanidis, 2001-submitted] is planned to be applied in both of the above cases, to assess the feasibility of deriving empirical findings on adaptive user interaction that will be reusable across application boundaries, in the context of universal access.

References

[Brown, Santos, and Banks, 1997] Brown, S.M., Santos., E., Banks, S.B. (1997). A Dynamic Bayesian Intelligent Interface Agent, In *Proceedings of the Sixth International Interfaces Conference (Interfaces 97)* (pp. 118-120) Montpellier, France.

[Browne et al., 1990]. Browne, D., Norman, M., Riches, D. (1990). Why Build Adaptive Systems? In D. Browne, P. Totterdell, & M. Norman (eds.), *Adaptive User Interfaces* (pp. 15-57). Academic Press.

[Carberry, 1990] Carberry, S. (1990). Incorporating default inferences into plan recognition. In *Proceedings of the 8th National Conference on Artificial Intelligence* (pp. 471-478).

[Chajewska, and Draper, 1998] Chajewska, U., Draper, D.L. (1998). Explaining Predictions in Bayesian Networks and Influence Diagrams. In Peter Haddawy & Steve Hanks (eds.), *Proceedings of the 1998 AAAI Spring Symposium on "Interactive and Mixed-Initiative Decision-Theoretic Systems"* (pp. 23-31). AAAI Press.

[Conati, et al., 1997] Conati, C., Gertner, A., VanLehn, K., Druzdzel, M. (1997). Online student modeling for coached problem solving using Bayesian networks. In *Proceedings of the Sixth International Conference on User Modeling* (pp.231-242) Sardinia, Italy. User Modeling, Springer-Verlag.

[Dieterich et al., 1993]. Dieterich, H., Malinowski, U., Kühme, T., Schneider-Hufschmidt, M. (1993). State of the Art in Adaptive User Interfaces. In Schneider-Hufschmidt, M., Kühme, T., & Malinowski, U. (eds.), *Adaptive User Interfaces: Principles and Practice* (pp. 13-48). Elsevier Science Publishers BV.

---

[2] Project EPET 98AMEA28 Nautilus, funded by the Hellenic Ministry of Development.

[3] IST-1999-20656 PALIO project "Personalised Access to Local Information and services for tOurists", partially funded by the European Commission.

[Horvitz, 1999] Horvitz, E. (1999). Principles of Mixed-Initiative User Interfaces. In *Proceedings of CHI '99, ACM SIGCHI Conference on Human Factors in Computing Systems*, Pittsburgh, PA.

[Horvitz et al., 1998] Horvitz, E., Breese, J., Heckerman, D., Hovel, D., Rommelse, K. (1998). The Lumiere Project: Bayesian User Modeling for Inferring the Goals and Needs of Software Users. In *Proceedings of the Fourteenth Conference on Uncertainty in Artificial Intelligence* (pp. 256-265) Madison, WI. Morgan Kaufmann: San Francisco.

[Kobsa, 1993]. Kobsa, A. (1993). User Modelling: Recent Work, Prospects and Hazards. In Schneider-Hufschmidt, M., Kühme, T., & Malinowski, U. (eds.), *Adaptive User Interfaces: Principles and Practice* (pp. 111-128). Elsevier Science Publishers BV.

[Linden, Hanks, and Lesh, 1997] Linden, G., Hanks, S., Lesh, N. (1997). Interactive Assessment of User Preference Models: The Automated Travel Assistant. In Jameson, A., Paris, C., & Tasso, C. (Eds.) *User Modeling: Proceedings of the Sixth International Conference, UM97* (pp. 67-78). Vienna, New York: Springer Wien New York.

[Murray, and VanLehn, 2000] Murray, R.C., VanLehn, K. (2000). DT Tutor: A decision-theoretic, dynamic approach for optimal selection of tutorial actions. In Gauthier, Frasson, VanLehn (eds.) *Proceedings of the 5th Int'l Conf "Intelligent Tutoring Systems"* (pp. 153-162). Springer, lecture notes in Computer Science, Montreal, Canada, Vol. 1839.

[Norvig and Cohn, 1997] Norvig, P., Cohn, D. (1997). *Adaptive software.* PC AI Magazine, 11 (1).

[Paramythis, Totter, and Stephanidis, 2001-submitted] Paramythis, A., Totter, A., Stephanidis, C. (submitted, 2001). A modular approach to the evaluation of Adaptive User Interfaces. Submitted for publication in the *Workshop on Empirical Evaluations of Adaptive Systems*, July 2001, Sonthofen, Germany.

[Pynadath, and Wellman, 1995] Pynadath, D.V., Wellman, M.P. (1995). Accounting for context in plan recognition, with application to traffic monitoring. In *Proceedings of the 11h Conference on Uncertainty in Artificial Intelligence*, pp. 472-481.

[Stephanidis, 2001a] Stephanidis, C. (2001). The concept of Unified User Interfaces. In C. Stephanidis (Ed.) *User Interfaces for All - Concepts, Methods, and Tools* (pp. 371-388). Mahwah, NJ: Lawrence Erlbaum Associates (ISBN 0-8058-2967-9, 760 pages).

[Stephanidis, 2001b] Stephanidis, C. (Ed.). (2001). *User Interfaces for All - Concepts, Methods, and Tools.* Mahwah, NJ: Lawrence Erlbaum Associates (ISBN 0-8058-2967-9, 760 pages).

[Stephanidis, et al, 2001] Stephanidis, C., Paramythis, A., Sfyrakis, M., and Savidis, A. (2001). A Case Study in Unified User Interface Development: The AVANTI Web Browser. In C. Stephanidis (Ed.) *User Interfaces for All - Concepts, Methods, and Tools* (pp. 525-568). Mahwah, NJ: Lawrence Erlbaum Associates (ISBN 0-8058-2967-9, 760 pages).

[Waern, 1997] Waern, A. (1997). Local Plan Recognition in Direct Manipulation Interfaces. In *Proceedings of the ACM International conference on Intelligent User Interfaces*, Orlando, USA.

[Wittig, 1999] Wittig, F. (1999). Learning Bayesian Networks With Hidden Variables for User Modeling (UM99) pp. 343-344.

[Wolfman, et al., 2001] Wolfman, S.A., Lau, T., Domingos, P., Weld, D.S. (2001). Mixed Initiative Interfaces for Learning Tasks: SMARTedit Talks Back. In *Proceedings of the Intelligent User Interfaces Conference (IUI'01)*.

[Wu, 1991] Wu, D. (1991). Active Acquisition of User Models: Implications for Decision-Theoretic Dialog Planning and Plan Recognition. *User Modeling and User-Adapted Interaction*, 1, 149-172.

PART 3

ARCHITECTURES AND TOOLS

Handling User Diversity in Task-Oriented Design

Chris Stary

University of Linz, Department of Business Information Systems
Freistädterstraße 315, A-4040 Linz, Austria
stary@ce.uni-linz.ac.at, www.ce.uni-linz.ac.at

Abstract

In this paper a shift from traditional role modeling in the context of task-based design is proposed towards user model-driven design, thus, implementing the pro-active concept of user interfaces for all. It will be shown that traditional models, namely the task, user, problem domain data, and interaction model can be utilized, however, the design process and design representations have to be enriched, in order to meet the demands for user interfaces supporting diverse user groups.

1. Introduction

Universal access to information in the information society targets towards enabling interaction with information technologies for a variety of users and for a variety of domains, i.e. the development of user interfaces and applications 'for all'. In case interests, tasks and needs of people are not considered properly in the course of developing access facilities, it is very likely that the acceptability of these technologies will be low (cf. e.g., Preece et al., 1994). Features for interaction have to take into account the context of interaction, the capability to switch between modalities for interaction, and situational as well as content dynamics.

Unfortunately, current user-interface development technologies lack an integrative consideration of these characteristics. This makes the process of designing user interfaces for all as well as the operation of adaptive solutions an ineffective and time consuming procedure. Hence, in this paper we focus on the nature of knowledge about user diverse user groups, and how that knowledge can be put into the context of task specifications for effective human-centered interface design as well as adaptive interactive systems.

From the software architecture point of view, task specifications and adaptation procedures to user needs and preferences have been put into mutual context through identifying models of tasks, users, and interaction for (automated) adaptation, see e.g., Hoppe (1988); Norcio et al. (1989); Sukaviriya et al. (1993); Thomas et al. (1993). In particular, Hoppe (1988) considers adequate adaptivity to be based on user- *and* task modeling. In case the focus is more on the task and the situation the user is dealing with (diagnosis of the user's current needs based on what he/she is actually doing) adaptivity is called context-adaptivity. Otherwise, adaptivity is termed individualization. The consideration of software architectures for adaptation and adaptability documents the need for software that is 'appropriate to the *conjunction* of *user*, *task*, and *environment*', as to be read in Brooke (1991).

2. Handling user diversity

According to the continuing embodiment of interactive software systems into work organizations and other systems of society, there are several possibilities to categorize users: (organizational) roles, skills for handling information technology, societal needs, and human capabilities. Empirical research has shown that there are substantial differences in the use of computer systems through individuals, either with regard to gender, mental effort, or functional roles of users, e.g., BIT (2000), or due to limitations of continually changing resources, e.g., Jameson et al. (1999). Hence, capturing those items seems to be crucial both, in the course of development, and during the operation of interactive software systems. In addition, computer technology is increasingly becoming an organizational technology, thus, increasingly controlling work and production processes. For instance, through the advent of Workflow Management Systems, interactive software systems determine the content and flow of work, and thus, the way users interact with technology along the process of work (see, e.g., Glasson et al., 1994). Hereby, functional role specifications indicate the involvement of users in a work flow, namely, the relationship between a user and a work activity. In case social aspects of the role are taken into account, a structural role (in addition to the functional role) has to be defined (Dobson et al., 1994). In case both type of roles are specified, a so-called intentional model of a system can be described (Colman et al., 2000).

A first attempt to categorize users with respect to their role they can play in the context of interactive systems and automated information processing has been performed by Cotterman et al. (1989). Their 3D-grid allows to specify value-triples representing user expertise in software construction, operating and management, as well as user involvement in information consumption/production processes in the company. An experienced worker in customer service is then described by the same means like unexperienced staff members in accounting. Other approaches capturing user roles and capabilities stem from business process (re)engineering, as illustrated in the subsequent example (Table 1). It demonstrates role/function modeling of an insurance company (names have been changed). Apparently, this kind of representation does only capture user characteristics marginally, in contrast to models set up after user and/or task analysis, e.g., Hackos et al. (1998). These models target towards in-depth understanding of individual capabilities and needs of users. This content is represented through stereoptypical profiles in traditional user modeling, e.g., McTear (2000).

Table 1: User knowledge 'topography' according to Scheer (1998).

Knowledge in the Field of ... Employee's Name	Product Development	Legal Issues Computing	Skill Development	Customer Relationship Management
Smith	Market analysis Feasibility study	Contracting	Coaching	------
Meyer	Project Management	Claim handling	Lecturing	Binding program

However, from the traditional approaches in user modeling, cognitive engineering, task analysis and business process re-engineering (for an in-depth discussion on these issues see Stary, 2001) we cannot expect a migration or integrated representation of user characteristics acquired through the different approaches for analysis or modeling. Development 'for all', nevertheless, should capture different sources of knowledge in a coherent and comprehensive way. Tasks, procedures for their accomplishment, individual skills and preferences should be represented and processed in an uniform way or at least be provided with well-defined interfaces when specified, in order to design user interfaces open for a diverse set of users. User models should also capture the dynamics of interaction, i.e. adaptation mechanisms and representations that comprise the results of adaptation in the context of human-computer interaction. Hence, we have to look for novel schemes of representation allowing to view knowledge about users and tasks at an accurate level of abstraction for developers. User models at such a level of abstraction have to meet several objectives:

1. They have to allow the use of a *unifying notation for specification*. Meeting this objective is a pre-requisite for integrating or migrating different perspectives, such as functional and structural role descriptions.
2. They have to be usable for *analysis, design, implementation, evaluation and adaptation*. Meeting this objective enables the communication of design ideas as well as results from different phases of development and evaluation among analysts, developers, and users.
3. They have to be *easy to modify* to reflect changes and the results from software adaptation. Meeting this objective enables to reflect changes proposed by the adaptation mechanism at an implementation-independent level of software description.

3. Representation for context-adaptivity

The challenge of developing accurate design representations is well known, either through formulating the need for usable design representations (see, e.g., Agarwal et al., 2000) or through stating their crucial importance in the course of specifying development knowledge (see, e.g., HCI, 1999). In the following a scheme for representing task and user knowledge is discussed. It is based on the results of the TADEUS-project (Stary, 1999). It follows the idea of a task-centered approach to user modeling, and is ontology-driven, in the sense that

(i) the construction of the design representation is initially carried out in a top-down fashion, namely by concretizing (populating) a given ontology for tasks and user interface styles

(ii) the adaptation procedure feeds back data to the concrete application objects and the ontology (bottum-up changes).

TADEUS is a model-based user interface development approach, and implements the recently brought up idea of shared task models (Brazier et al., 2000):

- 'During knowledge acquisition, a shared task model is a means to acquire a common understanding of a task in interaction with experts.
- During design of a system, a shared task model can be used (1) as a basis for user interaction, and also (2) as a basis for the design of a clarification support agent.' (ibid., p.78)

TADEUS is based on four different models that are integrated in the course of development to an application model. Each model is composed of a structural and behavior specification. One of these models is the user model. According to the TADEUS methodology it has to be specified after and in correlation to the task model. The structural part comprises user group definitions as the organization of tasks requires. There are two ways to define user groups from the perspective of an organization, namely the functional and the individual perspective.

For instance, each department of an organization at hand has a particular set of tasks to perform. These tasks are coupled with privileges, such as the privilege to manipulate salary data in case of human resource management (functional perspective). Besides that functional role description each staff member of the organization has also a user profile based on individual skills and preferences, such as accounting and the use of button bars instead of menus (individual perspective). Accurate user modeling in TADEUS requires an integrated view on both perspectives. Hence, a functional description of a user might deviate from an assumed best practice description and lead to individual sub processes for task accomplishment, however, resulting in the requested output of work. The same holds for the features and modalities provided for interaction. In addition, a user might be involved in several work processes. In this case, a TADEUS user model captures all (parts of) work processes where the modeled user plays a functional role in task accomplishment. Assume, a user is working as an accountant as well as a product developer and member of the training department. From the functional perspective this user would integrate several functional roles in his/her specific task and behavior specification.

In TADEUS integration is performed initially at the level of structure specifications, namely in terms of mutually relating Object Relationship Diagrams. TADEUS instantiates and propagates the relationships of tasks to the concerned data (represented in the problem domain data model) and to the involved dialog elements (represented in the interaction model), and finally, to an application model. Coupling the user context with the task context requires the use of particular relationships, namely 'handles'-relationships. A user 'handles' a certain task in a certain functional role. The level of experience and skill development is captured through personal profiles. The latter are also part of the class descriptions of the user model. As a consequence, each user is captured through both, the assignment of tasks (using the 'handles'-relationship) and his/her personal profile. This stereotypical specification of role behavior can directly be encoded, as done in TADEUS through Object Behavior Diagrams (based on the personal profile of a user).

The adaptation of the software system towards the actual behavior of a user can be achieved through sub symbolic representations, since to recognize, classify by tasks, and adjust user patterns in changing environments, traditional representation and analysis techniques cannot be used in a straightforward way, e.g., Sanchez et al. (1992); Chen et al. (1997). Starting point can be any Object Behavior Diagram OBD of the TADEUS knowledge representation. Hence, the starting OBD might either concern the presentation and organization of tasks, data required for task accomplishment, or interaction modalities. Dynamic user modeling this way might lead to a change of paths (state transitions) in OBDs that constitute the life cycles of objects. This specification is then successively changed according to the individual knowledge, experiences, skills, and style of task accomplishment. Typically, a set of transitions (being part of an OBD) describing a certain modality, such as a Graphical User Interface (GUI), is encoded to the sub symbolic representation scheme.

Methodologically, the paths of the OBD selected are converted to sets of paired actions that become entries of a so-called adjacency matrix (for details see Stary, 2001). The permutation of corresponding rows and columns enables the re-arrangement of the relations represented by the entries. The matrix is fed to a Kohonen network (Kohonen, 1989). According to the individual behavior of a user

- data are collected with respect to the selected sequence of state transitions,
- behavioral patterns can be identified through categorizing sequences of state transitions, and
- behavior can be predicted, as soon as the user enters a particular sequence.

In step 1 (Identification of the Starting Point) a particular dynamic model (i.e. a state transition diagram) has to be selected from the TADEUS specifications. It represents more or less directly the default behavior of the intended application from the organization's perspective. In step 2 (Graph-Matrix Transformation) the paths of the state transition diagram selected in step 1 are converted to sets of paired actions and transitions that become entries of a so-called adjacency matrix (McGrew et al., 1992). In step 3 the matrix is fed to a Kohonen network (step 3) which is a self-organizing (i.e. unsupervised) associative memory system. Hence, learning occurs without being biased by any observer (as it has been the case for static user modeling by the analyst or designer), and without indicating desired behavior (output) to given inputs. According to the behavior of a user data are collected with respect to the

138

selected sequence of state transitions. Hence, behavioral patterns can be identified through categorizing sequences of state transitions. To each state a node of the network is assigned (initialization). In the course of learning while the user is interacting with a prototype of the user interface in TADEUS, the states have to be recognized by the network. Those states that are recognized, are marked. In case they cannot be recognized (i.e. assigned to predefined nodes), they will be assigned to separate nodes of the network. As such, predefined sequences will be re-enforced, whereas novel sequences can be detected through newly assigned nodes and weights expressing the strength of connection with other nodes (states). Assume, a user a GUI is fed to the network. In the course of task accomplishment the user switches to the command window (which is a dedicated window) several times. The network might recognize this focussing on a certain window and deliver as an input to the symbolic representation that the entire task has been accomplished through the use of a single window of type command or console.

In step 4 (Extraction) the result of the learning process is re-transferred to a state-transition representation to contrast the learned behavior with the specified default behavior (in OBDs). Now, adaptation can be performed according to individual user needs. Assume the result of neural networking of above. As the command window has been represented in an OBD, the designer might look, whether it is possible to switch to this window exclusively, and omit the other windows for this user. In case the runtime of the application environment is able to react immediately the recognized behavior can be implemented directly at the user interface. However, in order to support the design at the specification level, the result of the learning process has to be re-transferred to an OBD that can be contrasted with the OBD (identified as the starting point for adaptation). As a consequence, regularities in user behavior for task accomplishment as well as particular needs for interaction can be acquired and directly represented through specifications (at an implementation-independent layer). In addition, problem solving activities can be optimized individually, redefining the initially assumed 'best practice' of accomplishing tasks, e.g. through high-level activities (macros). Finally, multiple patterns for accessing the same information can become evident or derived from user behavior. This approach is a promising candidate to enable design 'for all' in the sense of personalized user interfaces, as recently described in CACM (2000).

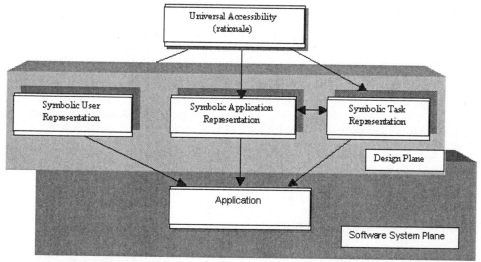

Figure 1. Universal Accessibility and the Role of Design Representations

4. Conclusive summary

Designing user interfaces for a diversity of end users requires the representation of contexts of use, user characteristics, interaction devices and styles, and their mutual interdependencies. The latter are required to support the design process based on different perspectives and/or models when it should lead to a complete specification of an interactive solution. The design knowledge has to be kept in a notation that enables modification, either through designers or through users in the course of providing feedback to the design ideas. For effective design, the plane of representation spanned by the elements listed above, can be supported through a tool such as TADEUS (see Figure 1), in order (i) to create artifacts that are complete with respect to the specification of context, and (ii) to provide a feedback loop in the design process, e.g., through prototyping.

In this paper a shift from traditional role modeling in the context of task-based design has been proposed towards adaptive usage-driven design, thus, implementing a comprehensive, since pro-active and reactive concept of user interfaces for all. In order to achieve adaptation in a task context traditional models, such as task, user, problem domain data, and interaction models can be utilized. However, the representation plane and application plane have to be modified, in order to meet the demands for open user groups design specifications and adaptive user interface behavior.

References

1. Agarwal, R.; Prabuddha, D.; Sinsha, A.P.; Tanniru; M. (2000). On the Usability of OO Representations, *Communications of the ACM*, 43(10), 83-89.
2. BIT: *Behaviour and Information Technology*: Special Issue on Individual Issues in the Use of Computers, 19(4), pp. 283-313, July-August 2000.
3. Brazier, F.M.T.; Jonker, C.M.; Treur, J.; Wijngaards, N.J.E. (2000). On the Use of Shared Task Models in Knowledge Acquisition, Strategic User Interaction and Clarification, *International Journal Human-Computer Interaction*, 52, 77-110.
4. Brooke, J. (1991). Usability, Change, Adaptable Systems and Community Computing, *HCI'91*, Vol. 2, 1093-1097, Elsevier (North Holland).
5. CACM: *Communications of the ACM*: Special Issue on Personalization, 43(8), August 2000.
6. Chen, Q.; Norcio, A.F. (1997). Modeling a User's Domain Knowledge with Neural Networks, *Human-Computer Interaction*, 9(1), 25-40.
7. Colman, A.W.; Leung, Y.K. (2000). Using Intentional Models for the Interface Design of Multi-Level Systems, *International Journal Human-Computer Interaction*, 52, 1007-1029.
8. Cotterman, W.W.; Kumar, K. (1989). User Cube: A Taxonomy of End Users, *Communications of the ACM*, 32(11), 1313-1320.
9. Dobson, J.E.; Blyth, A.; Chudge, J.; Strens, R. (1994). The ORDIT Approach to Organizational Requirements, *Requirements Engineering Social and Technical Issues*, London, Academic Press.
10. Glasson, B.C.; Hawryszkiewycz, I.T., Underwood, B.A.; Weber, R.A. (1994) (eds): *Business Process Re-engineering: Information Systems Opportunities and Challenges*, Elsevier Sciences, Amsterdam.
11. Hackos, J. T.: Redish, J.C. (1998). User and Task Analysis for Interface Design, Wiley, New York.
12. HCI: *Human-Computer Interaction*: Special Issue 'Representations in Interactive Systems Development', 14 (1&2), 1999.
13. Hoppe, H.U. (1988). Task-Oriented Parsing - A Diagnostic Method to be Used by Adaptive Systems. ACM *CHI'88*, pp. 240-247, 1988, Washington.
14. Jameson, A.; Schäfer, R.; Weis, Th.; Berthold, A.; Weyrath, Th. (1999). Making Systems Sensitive to the User's Time and Working Memory Constraints, *International Workshop* on *Intelligent User Interfaces*, ACM, 79-86.
15. Kohonen, T. (1989). *Self-Organization and Associative Memory*, Springer, New York.
16. McGrew, J. (1992). Task Analysis, Neural Nets and Very Rapid Prototyping, *Neural Networks and Pattern Recognition in Human-Computer Interaction*, 91-100, Ellis Horwood, N.Y.
17. McTear, M.F. (2000). Intelligent Interface Technology: From Theory to Reality, *Interacting with Computers*, 12, 323-336.
18. Norcio, A.F.; Stanley, A. (1989). Adaptive Human-Computer Interfaces: A Literature Survey and Perspective, *IEEE Transactions on Systems, Man, and Cybernetics*, 19(2), 399-408.
19. Preece, J.; Rogers, Y.; Sharp, H.; Benyon, D.; Holland, S.; Carey, T. (1994). *Human-Computer Interaction*, Addison-Wesley, Wokingham.
20. Sanchez-Sinencio, E.; Lau, C. (1992). *Artificial Neural Networks: Paradigms, Applications, and Hardware Implementations*, IEEE, New York.
21. Scheer, A.-W. (1998). *ARIS – From Business Processes to Applications* (in German), Springer, Berlin.
22. Stary, Ch. (1999). Toward the Task-Complete Development of Activity-Oriented User Interfaces, *Int. J. Human-Computer Interaction*, 11(2), 153-182.
23. Stary, Ch. (2001): User Diversity and Design Representations. Towards Increased Effectiveness in Design for All, *Universal Access in the Information Society*, 1(1), Springer.
24. Sukavirya, P.N.; Foley, J.D. (1993). Supporting Adaptive Interfaces in a Knowledge-Based User Interface Environment. *International Workshop on Intelligent User Interfaces*, ACM, 107-113, Orlando.
25. Thomas, C.G.; Krogsoeter, M. (1993). An Adaptive Environment for the User Interface of Excel. *International Workshop on Intelligent User Interfaces*, ACM, 123-130, Orlando.

Usability Engineering For Different European Countries

Konrad Baumann

Philips Internet Fax Appliances, Gutheil-Schoder-Gasse 10, A-1102 Vienna, Austria
and
Fachhochschule Joanneum, Alte Poststrasse 149, A-8043 Graz, tel +43 316 5453 8615, fax +43 316 5453 8601, konrad.baumann@fh-joanneum.at

Abstract

This paper describes the product improvements of Philips fax machines that have been implemented following a detailed analysis of customer feedback from several European countries. It also points out that within Europe there are important country-specific differences in usability-related customer problems. Examples are shown in the figures.

1. Introduction

According to [Norman 1998] the usability of a product gets in the center of the attention when the product performance has passed the critical point (see figure 1). Left of this point it is the goal of a manufacturer to improve product performance. Right of this point the technology is "good enough" and therefore irrelevant. Customers do not aim for more and better technology, but for more convenience, reliability, and lower cost.

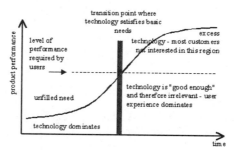

Figure 1: The technology life cycle [Norman 1998]

Fax machines have passed the transition point already a few years ago. For fax machines the life cycle of the whole product category is at the point where sales will start to decrease and manufacturers will face price erosion.

In this phase, however, it gets more difficult to solve user interface design problems because of the price erosion and an expected decrease in sales. Solutions tend to be perceived as good solutions as long as they are cheap or even help the company to save money.

Our main source of information about customer problems with our fax machines is the analysis of call center feedback. Therefore we defined a keydriver which is directly affected by product usability [Baumann 1999] and can be measured in terms of cost savings. Then we identified possible improvement actions related to this keydriver which we sorted according to their impact. The results will be shown in the following.

2. Call center data collection

In the Philips call center in Vienna every customer-triggered event leading to an activity of a call center agent is recorded in a database. A typical call center agent handles between about 50 and 100 events daily. This makes the database grow by a lot of entries per working day.

The problem classification system assigns a name and code to each problem. It derives from the standard repair code system which we had to modify, however, because it originally did not contain any usability-related problem causes.

It is defined like a tree structure now starting at the top level with "customer problem call" as opposed to "call-back", "customer fax" and other types of events.

At the next level of the tree the branch "customer problem call" is split into "usability problem call" and "technical problem call", "marketing call" and "shipment-related call". The definition of a usability problem is that it can be solved by giving advice, while technical problems can only be solved by hardware exchange, repair, software reset, or software update. Another definition of a usability problem is that in this case the product has always performed according to specification.

At the next tree levels the problem areas are defined. These problem areas and all lower tree levels specify the problem in more and more detail.

The output of the problem database is e.g. a report indicating the rate of usability problem calls for each detailed problem or for each problem area. The usability problem call rate is a figure indicating the number of related customer calls divided by the average number of sold products (appliances or service packages) in the same time frame. It only becomes valid after six months of sales of a new product.

3.   Country-specific usability problems

Let us explain a fax-specific typical usability issue. Usability issues are responsible for a considerable part of the calls to the Philips Fax call center. They mainly stem from the installation phase of the product. Most of them arise when the customer connects the product to the same telephone line where also other devices are operated (e.g. cordless phone and answering machine) and when the user changes the setting of the ring count.

Please refer to Figure 2 for a typical distribution of call reasons.

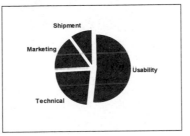

Figure 2: Reasons why customers call the call center for a typical fax machine

The ring count is the number of ringing signals that occur before the fax switch becomes active. After the selected number of ringing signals the fax switch automatically takes the line and detects possible fax sending request signals (CNG signals, i.e. short beeps) on the telephone line. In this case a fax reception is initiated.

In case of no CNG signals the fax machine concludes that the call is a voice call and starts to ring again until the call is taken by somebody or by an internal or external answering machine. The feature that makes the ringing start again is called "active fax switch" and is proprietary to Philips.

There are several reasons why, on the other hand, the fax switch leads to confusion and generates calls to the call center. First, the fax switch is a rather complex part and helps to solve problems which usually not even technically experienced customers are aware of:

1)   the problem that a call has to be taken by the device before fax CNG signals can be detected, and
2)   the problem that ringing would usually stop in that moment, which is not desirable in case of a voice call.

Both these problems are side-effects of the features "automatic fax reception", "silent fax reception", and "active fax switch". It is evident that in most cases complex product features introduce new usability problems to a product.

Like the timer feature included to video cassette recorders, the said features of fax machines are part of an essential functionality of the product and the customers are therefore forced to deal with them. They can not just ignore them like e.g. the organizer function of most mobile phones is ignored by most of the product owners.

The fax switch issue is made again more complex by the country-specific technical details of the telephone systems. For example in parallel connection of devices on a telephone line the feature does not work properly. However it is advertised and therefore expected by many customers. This explains the higher percentage of usability problems we see in France.

Especially there are country-specific cultural standards which generate difficulties. E.g. in China it is a cultural standard to wait for a fax reception signal (CED signal, i.e. a long beep) before sending a fax. In this case the fax machine and the calling person silently wait for the respective tone signal of the other side before hanging up in frustration. This was the main usability problem with fax machines reported in 1999 by our China call center.

Compared with the fax switch, other new and sales-relevant features like "fax via internet" and "detachable document scanner" create much less calls in the Philips fax call center. Let us assume - without evidence - that most customers simply do not use them.

After this look into country-specific usability issues of information appliances let us return to the process of dealing with them in the improvement process.

4. Data processing

The database enables us to compare differences between countries and differences between products. It has been interesting to see that usability problems are more frequent in France (see figure 3). This can be explained by the higher percentage of parallel connections of multiple devices to a single telephone line.

Also we clearly see that more complex products have a higher rate of calls per device sold (see figure 4). This can be explained by the product complexity of course, but also by the fact that office users have less tolerance towards problems and are more demanding regarding the use of sophisticated features than home users who often buy the simpler product version.

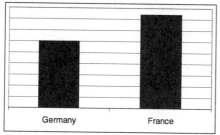

Figure 3: Usability-related calls per sales for the same fax machine in two different countries. The reason for the significant difference is the different connection of multiple devices to a single telephone line. Note: The values are not shown for confidentiality reasons.

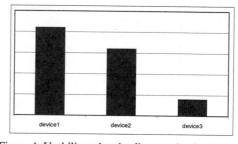

Figure 4: Usability-related calls per sales for three different fax machines. Devices 1 and 2 are for office and home office use, device 3 is for home use.

5. Evaluation method

Detailed improvement proposals can be collected by means of heuristic evaluation, benchmarking of competitors' products, brainstorming, or application of known usability principles to product aspects.

In our case the favoured improvement proposals for the described problem cluster "fax switch" strongly depended on the job of a person (product manager, usability professional, user manual author, software developer, hardware developer, call center agent). They ranged from hiding the fax switch in the user manual (which was already done), creating two ways of setting the ring count for novice and expert users (which was also done) to eliminating the "active fax switch" feature (which was impossible for marketing reasons).

Finding a solution supported by all involved parties was a difficult task. The key idea in this situation was to let the customers decide themselves, which we achieved by the help of the matrix evaluation method. This metod has been presented at HCII99 and is described in detail in [Baumann 1999].

The method closes the overall feedback loop about customer satisfaction and product usability and has been used successfully in the Philips Business Unit Fax as a quality process tool and decision aid. The matrix method is based on a very detailed and schematic estimation procedure involving several people in the company. The improvement proposals are evaluated against the different problem call reasons as they appear in the call center. The result of the matrix method is a prioritized action list showing the financial impact of around 60 improvement actions.

The action list (figure 4) shows the financial impact of potential usability-related improvements per device sold. The figures can be compared directly to the cost prices in the product's bill of material or to other expenses necessary for implementing the improvement.

6. Improvement actions

When specifying and developing the current Philips Fax product line in 1999 and 2000 the following main user interface improvements resulted from the described method have been implemented:

The fax switch has been redesigned. The former two setting methods for novice and expert users were replaced by a single one. It follows the former "expert" setting style by asking for input of the ring count. The term "toll-free rings" (indicating the number of ringing signals that occur before the fax machine takes the line) was replaced by the term "fax rings". The preset ring count combinations named "normal", "quick" etc. were skipped. These relatively small changes have approached the user interface of the fax switch to the mental model generated by the users when they are facing a fax switch for the first time.

Another important change has been to switch off in the default factory setting a few features that are not used by the majority of customers. This is for example the feature "silent fax reception" and the "timer" feature switching automatically between two ring count settings for day and night mode. These features are important to have in a product when the product is evaluated by a comparative test in a magazine. Also they are important for sales reasons. However, in a true "plug-and-play" product they should stay in the background until the customer actively looks for them and switches them on.

Furthermore the operation panel has been enhanced by a button that switches the answering machine on or off. The fact that this function needed to be done by several keypresses in a menu before was also a major reason for customer calls.

The feature "email and internet access" that is offered with both product lines recently developed makes it necessary to have an alphabetic keyboard on the fax machine. Unlike other manufacturers who show the keyboard on the device front Philips decided to hide the keyboard under a flap that carries a few other keys operating the answering machine. This reduces the visible number of keys by around 50%.

The keys of the alphabetic keyboard, however, are only as big as the keys of a mobile phone. This is often perceived as too small for the relatively big appliance but it is a necessary compromise with both financial and technical requirements.

In the same time all keys of the operation panel were grouped to functional blocks. There are functional blocks for the answering machine, for the fax functions, for the telephone, and for the display and the keys related to menu selection. This improves the perceived usability at first sight and gives the product a well-balanced simple appearance.

Finally the acoustic signals were redesigned. New signals were introduced, like a "switch on" and a "switch off" sound (raising respectively falling tone sequence) they give appropriate feedback when a key is pressed that switches something on and off.

For the "detachable scanner" feature some acoustic signals have been created, too, indicating the start and the end of a page, normal operation, memory warning and overflow condition.

As a positive side-effect ten new ringer tones ranging from conservative to video-game style have been composed and implemented.

Please see Figure 5 for a visualization of the financial improvement potential of single improvement actions.

Please see Figure 6 for a picture of the improved fax machine.

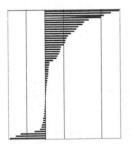

Figure 5: Estimated financial impact of usability improvement proposals on a single product. The highest positive values correspond to around 30 cents.

Figure 6: Operation panel of the Philips i-jet vox fax machine. In the center there is a flap which covers the alphabetic keyboard used for the input of email and internet addresses. On top of this flap there are the keys operating the answering machine. This appliance is for sale only in European countries.

7. Country-specific features

For cost reasons it is desirable not to have big differences in the country versions of our products. Like the manufacturers of cars, Philips tries to eliminate country-specific products or at least to create a fax machine which can be sold in all over Europe in the same version. This is not easy because of the described technical diversity and the user diversity. The remaining country-specific parts of a product are: cables and connectors, the ROM carrying the software, the user manual and other printed information packed with the product.

Many of the mentioned changes made it easier to simplify the user manual which was redesigned to be in a two-column layout at the same time. In total the user manual was shortened by 50% which is an important cost factor in itself. Straight guidelines were included into the user manual's fax switch section saying how to set the two ring counts for specific situations arising in specific countries. Also an additional two-page short installation guide was redesigned and improved.

The help pages the fax machine can print out were transformed to a two-column layout. This was a major challenge having an impact on the translation, the UI specification, and the implementation in software. Instead of two pages there are now seven pages of help information text permanently stored in memory.

Also we invested a lot of time in a careful redesign of the menu options and the feedback texts displayed by the software program. This was done in every one of the 10 language versions in which the appliance is available.

Any further country-specific functional difference of the product was not possible to implement. This would have doubled the effort of field testing, manual writing and call center evaluation for every country. Instead of translating the user manual in every language it would have been necessary to rewrite it or parts of it. The same would have been the case for the software and for the display texts. This would be impossible to do within the financial targets and deadlines that have to be met.

We are convinced that it is best to have a single well-tested product version which offers easy-to-use basic functionality, a lot of well-structured but non-prominent additional features and a good description in country-specific user manuals and help functions. The basic functionality of this product should take into account all country-specific issues and thereby avoid most problems. This is much better than to generate many country versions and thereby making compromises in time, cost and - most important - quality.

8. Conclusions

It can be concluded that usability engineering methods applied at the time of concept creation made it easier for the team to know what to do first and helped to find decisions when necessary. At the same time they do not prevent the team to stay creative and to go for new solutions that have to be tested by different methods. The method helps making usability improvements "hard facts" and therefore helps "selling" them to colleagues in financial and technical departments used to deal with facts and figures. It has been shown that the measures listed above had a positive long-term effect on our keydriver, the usability-related call center calls per sales. After half a year of production of the new product line the planning and implementation efforts of the mentioned improvements are already covered by savings in the call center. Besides this we are quite sure to have influenced sales and customer satisfaction positively.

Acknowledgments

I would like to acknowledge the management and all employees of Philips Internet Fax Appliances in Vienna, Austria, who contributed to the described activities and who made them possible.

References

Baumann, K., Lanz, H.(1998). "Mensch-Maschine-Schnittstellen elektronischer Geraete". Berlin: Springer.
Baumann, K., Thomas, B. (2001). "User Interface Design for Electronic Appliances". London: Taylor & Francis.
Baumann, K. (1999) "Matrix Evaluation Method for Planned Usability Improvements Based on Customer Feedback", In H.-J. Bullinger and J. Ziegler: Human Computer Interaction, London: Lawrence Erlbaum, paper presented at HCII99 in Munich, Germany.
Jordan, P.W., Thomas, B. et al. (1996). "Usability Evaluation in Industry". London: Taylor & Francis.
Norman, D.A. (1998) "The Invisible Computer". Cambridge, MA: The MIT Press.
Stary, Ch. (1994). "Interaktive Systeme", Braunschweig: Vieweg.

Force-Feedback in Computer-Mediated Communication

Scott Brave, Clifford Nass, Erenee Sirinian

Department of Communication, Stanford University, Stanford, CA 94305-2050 USA,
{brave, nass, erenee}@stanford.edu

Abstract

This study investigates the effect of force-feedback in computer-mediated communication. Participants completed a screen-based maze task with an alleged remote participant in a 2 (task characteristics: cooperative vs. competitive) by 2 (modality: haptic/force-feedback vs. visual) balanced, between-participants experiment. There were a number of cross-over interactions. In the competitive task, participants felt more powerful and more positively overall when interacting through force-feedback than when interacting visually. They also liked the other participant more and trusted them more. The opposite results were obtained for the cooperating participants. Implications for including force-feedback in computer-mediated communication are outlined.

1. Introduction

Human-computer interaction has traditionally focused on audio-visual interfaces. Over the past several years, however, the research community has given new consideration to the tactile dimension of mediated interaction. Recent work by Ishii's Tangible Media Group, for example, has explored the use of real-world physical objects as interface elements in what they term Tangible User Interfaces (TUIs) (Ishii & Ullmer, 1997). An alternate approach has been the enhancement of traditional graphics-based interfaces through the use of force-feedback pointing devices. Commercially-available force-feedback devices, such as Immersion's FEELit mouse and Sensible's PHANToM, can serve a dual role as both input device and force display, enabling users to physically interact with onscreen objects and worlds.

Current research into force-feedback (haptic) interfaces falls into three main categories. The first is simulation, where the goal is to enhance the sense of realism in virtual worlds (included in this category are virtual reality applications such as surgery training and gaming) (see Burdea, 1996; Salisbury, Brock, Massie, Swarup, & Zilles, 1995). The second is scientific visualization, where force output serves as an additional channel for exploring complex data sets. Finally, a growing body of literature has begun to investigate the potential for force-feedback to improve the efficiency of conventional GUI interactions (e.g., Rosenberg & Brave, 1996; Miller & Zelenick, 1998; Münch & Dillmann, 1997).

All of the above research focuses exclusively on single-user interactions. However, the tactile dimension is potentially of great value not only in our interactions with the inanimate virtual world, but also in mediated interpersonal interactions. A few projects have begun to create such interactive systems. One of the first attempts at multi-user force-feedback interaction, *Telephonic Arm Wrestling* (White & Back, 1986), provided a basic mechanism to simulate the feeling of arm wresting over a telephone line. More recently, Fogg, Cutler, Arnold, and Eisback (1998) described *HandJive*, a pair of linked hand-held objects for playing haptic games. *InTouch* (Brave, Ishii, & Dahley, 1998) is a desktop device that employs force-feedback to create the illusion of a shared physical object over distance, enabling simultaneous physical manipulation and interaction. (Other examples include Goldberg & Wallace, 1993; Noma & Miyasato, 1996; Oakley, Brewster, & Gray, 2000; Strong & Gaver, 1996) Many of these projects report positive reactions from users based on informal user testing. However, formal studies have not been conducted to evaluate the effects of computer-mediated haptic communication (for an exception, see Basdogan, Slater, Durlach, & Shrinivasan, 1998). This paper presents an initial experimental study intended to help fill this gap.

Although interaction through today's force-feedback devices is a far cry from real physical contact, one conceptual framework to begin thinking about multi-user force-feedback interactions is remote *touch*. Touch is a powerful means of communication—one that offers an immediacy and intimacy unparalleled by words or images. The firm handshake, an encouraging pat on the back, a comforting hug, all demonstrate the profound expressiveness of physical contact. In the real world, touch can further serve as a powerful mechanism for reinforcing trust and establishing group-bonding. Depending on context however, touch can also be utilized to assert dominance, display power, and even cause harm. For this reason, we use both a cooperative and a competitive task in our experiment.

Because real-world touch has been shown to have effects on both affective state (Ashton, 1980; Fisher, Rytting, & Heslin, 1976; Patterson, Powell, & Lenihan, 1986) and interpersonal evaluation (Fisher et al.; Hornick, 1987; Wycoff & Holley, 1990), we measure each as a dependent variable.

2. Method

2.1 Participants

Participants were 48 university undergraduates enrolled in communication classes. Participants were randomly assigned to condition, with gender approximately balanced across conditions. All participants signed informed consent forms, were debriefed at the end of the experiment session, and received class credit for their participation.

2.2 Procedure

The experiment was a 2 (modality: haptic vs. visual) by 2 (valence of intent: benevolent vs. hostile, i.e. cooperative vs. competitive) balanced, between-participants design. Upon arrival to the lab, the participant was told that he or she would be working on a computer-based task with a participant in another room. The participant was then left with a consent form while the experimenter supposedly went to check if the participant and experimenter in the other room were ready to begin. In actuality, there was only one participant and all interaction with the supposed other participant was simulated by the computer (a reverse "Wizard of Oz" experiment).

After returning, the experimenter explained the task, a maze, to the participant. This task was similar to a paper-and-pencil maze, except that it was done on the computer screen. The keyboard's arrow keys controlled the movement of a cursor within the maze; the goal was to exit the maze as quickly as possible. The maze task was described to the participant using a simplified version of the maze. The experimenter then explained that the other participant had already been through the maze several times and would now be observing the actual participants' performance. Depending on the condition, the participant was also told that the other participant would be trying either to help them through their maze (cooperative) or make it more difficult for them to complete the maze (competitive). The assistance or hindrance would come in the form of directional suggestions made either visually or haptically, depending on condition. The experimenter then left the room and the participant began the maze task.

The maze task was chosen for two reasons. First, communication through touch seemed reasonable in the context of finding your way through a maze. Second, the maze task constrains the interaction in such a way that simulating interaction with another person is feasible (all suggestions were in one of four directions and occurred at key points within the maze). After completing the maze task, the participant was instructed to fill out a paper-and-pencil questionnaire asking for his or her assessment of the interaction and the other participant.

2.3 Manipulation

2.3.1 Valence of Intent

To instantiate benevolent intent (cooperation), the participant was told that the other participant's score on the maze task would be a combination of both of their times through the maze. In other words, the other participant wanted the participant to do *well* and would therefore try to help as much as possible. All suggestions in this condition were in the correct direction to reach the exit.

Hostile intent (competition) was instantiated by telling the participant that the other participant would receive a score based on how much better or worse their time was than the participant's. In other words, the other participant wanted the participant to do *poorly* and would therefore try to "mess them up" while working on the maze. To be maximally confusing, half of the suggestions were correct and half were incorrect (if all the responses were incorrect, the participant would know what to do).

2.3.2 Modality

In the visual condition, suggestions came in the form of arrows appearing on the screen next to the maze. During a suggestion, the appropriate directional arrow was shown above, below, and to each side of the maze to ensure visibility. The arrows remained onscreen for 300mS.

In the haptic condition, suggestions were made through "pushes" (directional forces) on a force-feedback joystick (Immersion Corporation's Impulse Engine 2000). Participants in this condition controlled cursor movement in the maze with their dominant hand and held the force-feedback joystick with their other hand to receive the suggestions.

The duration of the push was also 300mS, increasing in intensity from zero to maximum intensity over the first 150mS and then decreasing back to zero in the next 150mS.

2.4 Measures

Attitudinal measures were based on responses to a paper-and-pencil questionnaire. The questionnaire included two kinds of adjectives. The first set of questions asked, "How well do each of the following adjectives describe how you feel," followed by a series of adjectives. Each adjective was associated with a ten-point Likert scale anchored by "Describes Very Poorly" and "Describes Very Well." The second set of questions asked: "How well do each of the following adjectives describe the person you worked with," followed by a series of adjectives. Each adjective was again associated with a ten-point Likert scale anchored by "Describes Very Poorly" and "Describes Very Well." Based on theory and confirmed by factor analysis, we developed four attitudinal measures. All indices were highly reliable.

Feeling of power of the participant was comprised of three items: powerful, dominant, and in control (Cronbach's alpha = .81)

Positive affect of the participant was comprised of seven items: comfortable, good, pleasant, happy, positive, successful, and capable (alpha=.88).

Liking of partner was comprised of five items: friendly, warm, likable, pleasant, compassionate, and attractive (alpha=.87)

Trust of partner was comprised of seven items: honest, not tricky, trustworthy, reliable, sincere, cooperative, and helpful (alpha=.96)

3. Results

3.1 Power feeling

There was a significant cross-over interaction with respect to feeling powerful, $F(1, 44)=10.89$, $p<.002$ (see Figure 1). Touch participants felt more powerful in the competitive condition, but less powerful in the cooperative condition. As an artifact of the extreme value in the haptic-competitive condition, there was a main effect for intent, $F(1,44)=6.44$, $p<.02$.

3.2 Participant Affect

There was a significant cross-over interaction with respect to affective state, $F(1,44)=10.85$, $p<.002$ (see Figure 2). Touch led to much more positive feelings in the competitive than in the cooperative conditions, while leading to slightly more negative affect in the cooperative condition. There was a main effect for modality $F(1, 44)= 5.54$, $p<.02$, which was an artifact of the extremely low liking for the competitive visual case.

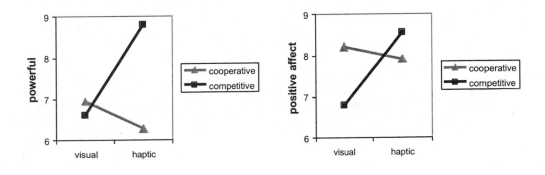

Figure 1. Feeling of Power   Figure 2. Positive Affect

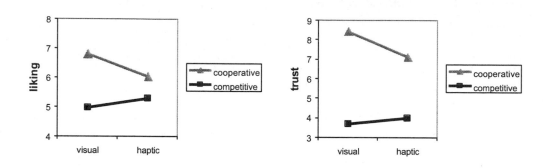

Figure 3. Liking of Ostensible Partner   Figure 4. Trust of Ostensible Partner

3.3   Liking

There was a significant interaction between valence of intent and modality with respect to liking, $F(1,44)=4.81$, $p<.03$ (see Figure 3). Consistent with the previous results, cooperative participants liked the interaction partner better when viewing onscreen arrows as compared to touch, while competitive participants liked their ostensible partner better in the haptic case. Not surprisingly, cooperative participants liked the interaction better than competitive participants, $F(1,44)=26.86$, $p<.001$. There was no main effect for modality.

3.4   Trust

There was a significant interaction between valence of intent and modality with respect to trust, $F(1,44)=4.51$, $p<.04$ (see Figure 4). Contrary to initial expectations, cooperative participants trusted their ostensible partner less when interacting via haptics as compared to visuals, while there was no difference in the competitive conditions. Not surprisingly, cooperative participants trusted their partners more than did competitive participants, $F(1,44)=106.79$, $p<.001$. There was no main effect for modality.

4.   Discussion

Communication via force-feedback had a dramatic effect on users' feelings about the interaction and users' perceptions of their interaction partners. Interestingly, touch did not have a *direct* effect on users' attitudes; instead, modality interacted with the perceived intent of the (ostensible) interaction partner.

The largest effects were in users' feelings of power and their affective state. Namely, in the competitive condition, interaction through touch made users feel more powerful and more positively overall. However, in the cooperative condition, the effect of touch was the opposite: Interaction through touch made users feel less powerful and less positively overall. This effect can be explained by considering the psychology of holding. Holding the joystick gives the participant a feeling of being in command of their partner. This is true because the forces on the joystick— though always felt—were not strong enough to overcome the participant's actions. In other words, the distant partner could influence the joystick, but the participant had final say over the joystick's position and movement. In the competitive condition, this gave the participant a clear sense of power. The participant's goal was to overcome their partner and the joystick served as a medium for establishing and reinforcing their dominance. Such dominance was not possible in the visual modality, since the arrows can only be ignored, not commanded.

In the cooperative condition, on the other hand, being in command of the other participant was useless. The participant was in a position of ineffectual authority, because command over the joystick didn't serve any practical or psychological end. The result was now the opposite: The participant felt powerless. The results seen for affective state now have an obvious explanation. A participant who feels powerful would almost surely feel more positively overall; the opposite would be true for a participant who felt less powerful. It is also reasonable to expect that a participant who felt more powerful and positive would give more positive evaluations of the partner than a participant who feels more powerless and negative would.

5. Conclusion

With the advent of commercial force-feedback devices, researchers are giving serious consideration to the incorporation of haptics into both HCI and CMC applications. The underlying sentiment is often that the added dimension of touch will inherently lead to more efficient and satisfying interfaces. The results of the present study, however, indicate that the effect of mediated touch can be positive *or* negative, depending on context. Much as touch in the real world, mediated touch is a psychologically complex phenomenon, requiring careful consideration before use. More work is clearly necessary before a set of guidelines for applying mediated touch can be developed, but this study will serve as a starting point for future investigation.

References

Ashton, N.L. (1980). Affective reactions to interpersonal distances by friends and strangers. *Bulletin of the Psychoinomic Society*, 15(5), 306-308.

Brave, S., Ishii, H., Dahley, A. (1998). Tangible interfaces for remote collaboration and communication. *Proceedings of CSCW '98: Conference on Computer Supported Cooperative Work*, 169-178.

Burdea, G.C. (1996). *Force and touch feedback for virtual reality*. New York: John Wiley & Sons, Inc.

Fisher, J.D., Rytting, M., Heslin, R. (1976). Hands touching hands: Affective and evaluative effects of an interpersonal touch. *Sociometry*, 49(4), 416-421.

Fogg, B.J., Cutler, L., Arnold, P., Eisback, C. (1998). HandJive: A device for interpersonal haptic entertainment. *Proceedings of CHI '98: Conference on Human Factors in Computing Systems*, 57-64.

Goldberg, K., & Wallace, R. (1993). Denta-Dentata. Visual Proceedings of SIGGRAPH '93: International Conference on Computer Graphics and Interactive Techniques.

Ho, C., Basdogan, C., Slater, M., Durlach, N., Shrinivasan, M.A. (1998) An experiment on the influence of haptic communication on the sense of being together. *Proceedings, BT Presence Workshop*. BT Labs.

Hornick, J. (1987). The effect of touch and gaze upon compliance and interest of interviewees. *The Journal of Social Psychology*, 127(6), 681-683.

Ishii, H., Ullmer, B. (1997). Tangible bits: Towards seamless interfaces between people, bits and atoms. *Proceedings of CHI'97: Conference on Human Factors in Computing Systems*, 234-241.

Miller, T., Zeleznik, R. (1998). An insidious Haptic invasion: adding force feedback to the X desktop. *Proceedings of UIST'98: Symposium on User Interface Software and Technology*, 59 – 64.

Münch, S., Dillmann, R. (1997). Haptic output in multimodal user interfaces. *Proceedings of IUI'97: International Conference on Intelligent User Interfaces*, 105 – 112.

Noma, H. Miyasato. T. (1997). Haptic communication for cooperative object manipulation. *Proceedings of the International Workshop on New Media Technology*, 83-88.

Oakley, I., Brewster, S.A., Gray, P.D. (2000). Communicating with feeling. *Proceedings of the First Workshop on Haptic Human-Computer Interaction*, pp 17-21.

Patterson, M.L., Powell, J.L., Lenihan, M.G. (1986). Touch, compliance, and interpersonal affect. *Journal of Nonverbal Behavior*, 10(1), 41-50.

Rosenberg, L., Brave, S. (1996). Using force feedback to enhance human performance in graphical user interfaces. *Companion Proceedings of CHI'96: Conference on Human Factors in Computing Systems*, 291-292.

Salisbury, K., Brock, D., Massie, T., Swarup, N., Zilles, C (1995). Haptic rendering: Programming touch interaction with virtual objects. *Proceedings of the 1995 Symposium on Interactive 3D Graphics*, 23 – 130.

Strong, R., Gaver, B. (1996). Feather, scent and shaker: Supporting simple intimacy. Videos, Demonstrations, and Short Papers of CSCW'96: Conference on Computer Supported Cooperative Work, 29-30.

Wycoff, E.B., Holley, J.D. (1990). Effects of flight attendants' touch upon airline passengers' perceptions of the attendant and the airline. *Perceptual and Motor Skills, 71*, 932-934.

White, N., and Back D. (1986). Telephonic arm wrestling. Shown at *The Strategic Arts Initiative Symposium* (Salerno, Italy, Spring 1986). See http://www.normill.com/artpage.html

Automated evaluation of accessibility guidelines

Michael Cooper

CAST
39 Cross St
Peabody, MA 01960 USA
+1 978 531 8555
mcooper@cast.org

Abstract

One requirement of Universal Accessibility is that documents be authored in such a way that existing technologies can perform the transformations needed to meet the needs of different users. It is possible for an automated tool to evaluate a document for the presence, absence, or validity of language-specific features relevant to this need. This can help document authors quickly identify usability problems and prepare a plan for repair.

CAST's Bobby™ evaluates HTML documents for conformance to Web Content Accessibility Guidelines 1.0 maintained by the World Wide Web Consortium. The 65 checkpoints are implemented as 92 evaluations, of which 27% can be evaluated automatically. An additional 55% can be evaluated only semi-automatically, either because the guideline is complex or because human judgment is required. Bobby cannot evaluate the remaining 18%. The intention of this implementation of the guidelines is to help authors identify problems that are otherwise hard to find, and to teach authors unfamiliar with the guidelines about the principles in the context of their own page.

Feedback from users shows that the main utility of Bobby is in those items that have fully automated evaluation. Users will know that a problem exists or does not exist, and can usually be given concrete information about the needed repair. Although a semi-automated evaluation can give contextualized information about how to make the determination, users find this confusing. When deciding on repair strategies, in fact, users tend to omit the 76% of partially or fully manual evaluations from consideration altogether. Automated evaluation is, then, an important strategy to ensure universal usability of a document, and its utility is directly tied to the level of automaticity it provides. It should not be used, however, as the sole criterion for evaluating documents for purposes other than targeting repairs.

1. Introduction

Accessibility to persons with disabilities of electronic documents is a key component of Universal Design and is often an indicator of usability. The spread of the World Wide Web brings new opportunities to people with disabilities but it is important that the technology not erect new barriers. This creates the concept of and underscores the importance of accessibility features.

The ability of a document to adapt to the characteristics of the user is a core component of accessibility. This adaptation can include supporting output to and input from multiple types of hardware devices, such as CRT monitors, speech synthesizers, and Braille devices for output, and keyboards, mice, and switches for input. These adaptations support users with sensory and manual disabilities. It can also include subtler adaptations, such as providing the ability to adjust the size, color, and contrast of displayed text, or providing supplementary descriptions for complex images. These adaptations support people with minor sensory disabilities and with learning disabilities, who have impaired cognitive ability in a discrete domain [Meyer & Rose, 1998]. Because all people have individual differences in their areas of strength, this flexibility ultimately can benefit all users.

Often, and usually preferably, adaptation of content is supported by assistive technologies that interact with the primary document. Being specialized devices, assistive technologies, or their software drivers, must interact with document viewers via a public API. Documents formats must provide the requisite information for the viewer to provide via the API.

HTML, viewed by Web browsers, has been structured since its inception with some supports for accessibility. HTML 4.0 [Raggett et al, 1999] provides numerous accessibility features. Guidelines for Web site accessibility have been created by the Web Access Initiative (WAI) at the World Wide Web Consortium (W3C). Building on guidelines originally created by the Trace Research and Development Center (Trace), the WAI released the Web Content Accessibility Guidelines (WCAG) in 1999 [Chisholm et al, 1999]. Since Bobby's release, the specifications for its implementation have been adopted by the WAI as a supportive component of the WCAG [Ridpath & Chisholm, 2000].

Divided into 14 Guidelines and further presented as 65 Checkpoints, the WCAG cover a range of techniques for creating accessible web documents. These techniques focus mostly on HTML but are relevant to other document formats commonly used in web documents, such as images, sound, video, scripts, applets, and plugins. Broadly classified, these guidelines outline requirements to ensure that

- content can be transformed to meet the needs of the user,
- content that cannot be transformed has textual alternatives,
- older technologies can be used to access the content,
- unstandardized technologies are avoided,
- specific accessibility problems with current browsers and access aids are avoided.

2. Automated evaluation with Bobby

While working to use Web technology in educational contexts, CAST discovered the importance of accessibility guidelines. Using the guidelines then maintained by the Trace Center, CAST created Bobby in 1996. There were a number of reasons that the guidelines needed additional support in an automated evaluator:

- the guidelines were technical and difficult to understand by lay users,
- a manual search of a document to find accessibility problems could be time-consuming,
- CAST believed the guidelines could be taught to document authors more effectively if applied within the context of their own work.

The name "Bobby" comes from an American stereotype of the British police, or "bobbies", perceived as friendly neighborhood police looking for trouble before it starts, and in a friendly manner guiding people in a new direction. CAST's intent was that Bobby would help document authors to find accessibility problems and provide concrete suggestions for repair early in the development process before they become prohibitively difficult to fix, ideally before users encounter the problems and leave the site.

Bobby's implementation of the 65 Checkpoints are as 92 discrete evaluations. These evaluations are categorized into 3 levels of support: Full Support, Partial Support, and No Support. In Full Support, Bobby can reliably detect the presence of guideline violations without either failing to detect any violations (false negatives) or erroneously detecting violations (false positives). Partial Support means Bobby can detect that a particular guideline might be applicable, but cannot assure the user that a violation does or does not exist, or Bobby cannot detect all guideline violations. For No Support, Bobby cannot detect anything and simply alerts the user to the guideline for every page, regardless of its applicability. 17 evaluations (18%) have No Support and are simply prompts to the user for every page. 50 evaluations (55%) have Partial Support and require supplementary user evaluation. The remaining 25 evaluations (27%) have Full Support and are fully automated. For most users, Bobby's utility is mainly in the Full Support items.

Bobby evaluates by parsing an HTML page into tokens and examining structural elements of interest. Structural elements can include HTML tags, attributes and their values, as well as simple relationships among HTML tags. No internal document representation is created beyond a flat list of the tokens in source code order. Bobby evaluates each token in turn and applies applicable guidelines. It can also set variable for use when evaluating later tokens, but it cannot change the evaluation of tokens that have already been examined. When a guideline violation is found, it stores the line number in which the violation occurred with the violation type.

After the entire document has been processed Bobby generates a report. Each error type found appears as a short title hyperlinked to a document containing an extended description of the error and how to repair it [CAST, 2001]. The report also shows the HTML source for the line(s) in which the error occurred (Figure 1). The linked description for each error type indicates the reason for the guideline, provides examples of both problem and repaired code, and describes how to perform the repair.

Later versions of Bobby include the ability in the report to redisplay the original web page, annotated with small "Bobby hat" icons that show visually the location of errors and link to appropriate part of the textual report (Figure 2). This is intended to support users who do not code HTML directly and instead understand their pages in terms of objects in the visual layout. A Bobby hat beside an image indicates that there is a problem with that image, and users can easily locate that in their authoring tool and correct the problem. Question marks indicate the location of Partial Support items to distinguish them from the Full Support items that receive the hat icon. In order to prevent these annotated pages from being overwhelmed by these icons, only Priority 1 items are annotated.

1. Provide alternative text for all images. (33 instances)
   Line 256: <IMG SRC="http://service.bfast.com/bfast/serve?bfmid=2
   Line 320: <td valign=top bgcolor=#FFFFFF width=1><img src="http
   width="3" height="1"></td><td valign=top bgcolor=#000033 width
   border="0" width="1" height="1"></td><td valign=top bgcolor=#8(
   src="http://a1636.g.akamai.net/7/1636/797/94fb0c3ed8a8f9/graphic
   width=1><img src="http://a1636.g.akamai.net/7/1636/797/94fb0c3
   bgcolor=#FFFFFF width=1>
   Line 321: <img src="http://a1636.g.akamai.net/7/1636/797/94fb0c:
   Line 342: <NOBR><img src="http://a1636.g.akamai.net/7/1636/79
   HREF="/traffic/11.shtml">Updates</a></NOBR>
   Line 346: <img src="http://a1636.g.akamai.net/7/1636/797/fb370f1
   HREF="/weather">Updates</a>
   Line 359: Martin Luther King Jr.'s dream of racial harmony
   SRC="http://a1636.g.akamai.net/7/1636/797/fb370f15c24533/graphi
   href="/dailyglobe/2015/metro/The_struggle_never_ends+.shtml">St
   Line 360: <img src="http://a1636.g.akamai.net/7/1636/797/fb370f1
   HREF="javascript:MM_openBrWindow

Figure 1: Text of a Bobby Report

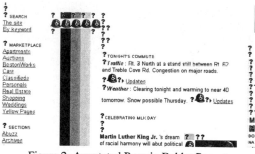

Figure 2: Annotated Page in Bobby Report

3. Analysis of evaluation

There are several aspects of this implementation of evaluation that affects the level of support for automated evaluation that Bobby is able to provide. These include

- reliance on structural features,
- limited ability to evaluate relationships among objects,
- evaluates source code,
- does not examine other languages than HTML,
- complicated heuristics,
- cannot make subjective judgment.

3.1 Structural features

Structural components of a document are easy to evaluate because they can be parsed into specific tokens. In the case of HTML, these structural features are tags, their attributes, and attribute values. In addition, the relationship of tags to each other can have meaning. For instance, a TR tag defines a table row, and one or more TD tags defines the cells within that row.

Many of the guidelines can be aligned directly with structural features. These guidelines can be completely evaluated by such a tool. Checkpoint 1.1 requires that all IMG elements have an alt attribute. When Bobby encounters an IMG element it checks for the presence of an alt attribute and flags a violation if one is not found. Checkpoint 7.2 requires that there be no BLINK elements in the document – if Bobby finds one, it flags a violation.

Guidelines that cannot be tied directly to simple structures are more difficult to evaluate automatically. Checkpoints 13.4 and 13.5 suggest the use of repeated, consistent navigation elements across pages in a collection. These elements may exist as complex structures of tags, but slight differences in structure, even with the same effect for the user, can render recognition impossible without sophisticated algorithms. Checkpoints 5.1 and 5.2 require that table headers be appropriately identified, for instance by use of the TH table header tag instead of the TD table cell tag. The presence of table cells with boldface or other formatting of their textual contents inside them may signal the need for this; however, there are many possible ways (such as FONT or B tags, or style sheets) by which this may be achieved, and it is difficult to anticipate them all.

3.2 Relationships among objects

Elements in a document can be related in several ways, such as proximity in the document structure or in the visual layout, and container or nesting relationships. Checkpoint 13.1 requires that links be clearly identified, in part through the use of good link text. Evaluating link text requires evaluating the textual portion of any element contained by an A tag. Checkpoint 5.2 requires that tables have appropriate header markup; evaluating this involves knowing the container and complex proximity relationships of each cell (TH or TD) in the table.

Guidelines that involve the relationships among multiple elements in a document require the evaluator to ascertain these relationships. Bobby's current flat page model does not make relational data available. Better models are needed for better evaluation. During design discussions of Bobby we have identified the need for a tree-like structural model that can be traversed in an arbitrary manner, such as the W3C's Document Object Model [Apparao et al, 1998] specifies, which provides easy access to container relationships. Another model needed is a visual layout model in order to evaluate the positional relationships of objects.

3.3 Source code

The ability to evaluate the source code of Web documents provides significant advantages for analysis. Without the source code, it would be necessary to evaluate document elements by querying a public API of a browser. Such APIs often do not expose all the items of interest to an evaluator, which needs more information than users generally need. Because Bobby has access to the source code of HTML, it has complete access to the content of an HTML page. This creates the need for it to parse the document, with the issues described above, but gives it the ability to create a document representation appropriate to the unique needs of an automated evaluator.

3.4 Languages supported

While Web pages can generally be assumed to be HTML, they often incorporate elements in other languages. These may be source code languages (parsed by the evaluator from a text stream), such as scripts (e.g., ECMAScript) and style sheets (e.g., Cascading Style Sheets), or they may be unparsed languages, such as images and applets. For each of these languages, there must be additional specialized capacity to process that language: additional parsing modules or the ability to query the API of binary included elements.

Bobby does not currently support any document language other than HTML. This means that guidelines applicable to scripts, style sheets, images, etc., aside from evaluating the HTML code that includes them, cannot be supported. This can significantly reduce the value of the automated component of the evaluation for documents that rich in non-HTML features.

3.5 Judgment

The largest difficulty in automatic evaluation of guidelines is subjective judgments. Guidelines can range from very objective to very subjective [Vanderdonckt, 1999]. The computer cannot make subjective determinations. Checkpoint 14.1 requires that language be as clear and simple as is appropriate. The computer cannot easily determine if the language is clear and simple, and cannot determine if the language complexity is appropriate for the intended use.

We have found it possible in Bobby to divide guidelines into objective and subjective components, increasing the overall level of support for guidelines, at the cost of a more complicated report. Checkpoint 1.1 requires that images (the IMG element) be provided with alternative content, including always an alt attribute, and if needed for further explanation, a longdesc attribute whose value is the URI of a more extended description. Bobby can detect the presence of a longdesc attribute, but if it is not present, it cannot determine if this is a violation of the guideline, or if the image was not needed. If Bobby does not detect a longdesc attribute, then, it issues a query to the user but does not identify a definitive violation. Because the alt attribute is always required, however, that component of the guideline has full support.

3.6 Complicated algorithms

Some objective guidelines, and even some subjective guidelines, could be considered feasible to evaluate given the right algorithm. Some such algorithms, however, could be prohibitively difficult to implement. Checkpoint 4.1 requires that a document identify changes in the human language. Language pattern detection or dictionaries of known languages could conceivably detect unmarked changes in the language. The development and computing to accomplish this, however, is many times the size of most automated evaluators in use today. This factor significantly decreases the level of support low-budget evaluators would be able to provide.

4. Conclusion

Applying guidelines to the design of documents helps to ensure universal accessibility. Automated evaluation of the conformance of documents to such guidelines is important to the creation of well-structured documents. Our experience with Bobby has shown that factors both intrinsic to the guidelines and of the implementation of the evaluation tool determine the possible results of the evaluation. This knowledge may be useful to the design of better evaluation tools. It may also inform the creation of useful guidelines when such guidelines are intended to have, or for usability reasons should have, automatic evaluation support.

References

Apparao, V, Byrne, S., Champion, M., Isaacs, S., Jacobs, I., Le Hors, A., Nicol, G., Robie, J., Sutor, R., Wilson, C., & Wood, L., eds. (1998). DOM Level 1 Specification. http://www.w3.org/TR/REC-DOM-Level-1/

CAST (2001). Bobby Documentation. http://www.cast.org/Bobby/Documentation331.cfm.

CAST (2001). Bobby Report Help Files. http://www.cast.org/Bobby/ReportHelpFiles906.cfm

Chisholm, W., Vanderheiden, G., & Jacobs, I., eds. (1999). Web Content Accessibility Guidelines 1.0. http://www.w3.org/TR/1999/WAI-WEBCONTENT-19990505/, W3C.

Meyer, A. & Rose, D.H. (1998). Learning to read in the computer age. Cambridge, MA: Brookline Books.

McCathieNevile, C., Koivunen, M., & Jacobs, I. (2000). SMIL and SVG - Towards Accessible Multimedia. In Technology and Persons with Disabilities 2000 Proceedings, http://www.csun.edu/cod/conf2000/proceedings/0102McCathieNevile.html

Raggett, D., Le Hors, A., & Jacobs, I., eds. (1999). HTML 4.0 Specification. http://www.w3.org/TR/html4/.

Ridpath, C., & Chisholm, W., eds. (2000). Techniques For Accessibility Evaluation And Repair Tools (draft). http://www.w3.org/TR/AERT.

Vanderdonckt, J. (1999). Developmental milestones towards a tool for working with guidelines. Interacting with Computers, Vol. 12, N°2, November 1999, pp. 81-118.

Modeling Preference for Adaptive User-Interfaces

Jacob Eisenstein

RedWhale Software
277 Town & Country Village, Palo Alto, CA 94303
jacob@redwhale.com

Abstract

The incorporation of plastic and adaptive user-interfaces into the model-based paradigm requires a new, more flexible modeling formalism. Rather than modeling the user-interface as a set of static structures and mappings, the UI should be modeled as a set of design *preferences*. Preferences are frequently many-to-one or many-to-many relationships that elude conventional UI modeling, which has largely focused on one-to-one mappings. In this paper, a spectrum of preference relations is described, and a new syntax for modeling preference is proposed. This spectrum extends from simple one-to-one *bindings* to complex *design guidelines* that can be structured together to implement decision trees. This new representation allows decision trees to be tightly integrated into the user-interface model itself, enhancing their flexibility and power.

1. Introduction

Model-based user-interface development has been characterized as a process of creating *mappings* between elements in various model components [3]. For example, interactor selection can be thought of as finding a set of appropriate mappings between abstract elements in the domain model and widgets from a presentation model. Typically, these mappings are assumed to be "one-to-one", connecting two single elements together. However, many important user-interface design relationships require many-to-one or even many-to-many mappings. While a general-purpose framework for representing such mappings might be desirable for the flexibility that it offers, such a framework would add considerable complexity to the representation language and impose a significant burden on user-interface modeling tools. One specific phenomenon that commonly eludes description by one-to-one mappings is *preference*. Preference relations allow for a more flexible and adaptable specification of the user-interface; rather than specifying exactly how the UI must appear, the designer can specify what *would be* preferred, and in which situation. This freedom is particularly important for user-interfaces that must run on heterogeneous devices [4], since the contexts of use vary and may even change at run-time. Preference modeling is also critical if interaction is to be dynamically customized for the user; preference relations can be used to show how the user-interface should change in relation to the user model. In this paper, a formalism for modeling preference is proposed.

2. A Spectrum of Preference Relations

Preference relations range from simple one-to-one bindings to sophisticated design guidelines. This spectrum of preferences can be divided into two broad classes: concrete preferences and abstract preferences. Concrete preferences specify their targets directly, although an ordered list of targets is permitted. Abstract preferences specify their targets indirectly, based on *criteria* that describe characteristics of either the targets themselves or some other *object*.

	Condition	Target	Criteria	Object
Binding	1	1	None	None
Simple Preference	Any number	1	None	None
Ordered Preference	Any number	Any number	None	None
Abstract Preference	Any number	More than 1	Criteria apply to targets; logical and preferential criteria are allowed	None
Design guideline	Any number	Any number	Criteria apply to object; only logical criteria are allowed	Any number

156

2.1 Concrete Preferences

In any preference relation, there is first a set of *conditions*, under which the preference becomes valid. For example, if user U1 prefers for presentation element P1 (say, a listbox) to represent domain object D1, then U1 and D1 are the conditions. P1 is the *target* of the preference relation—the thing that is preferred. Every preference relation must have at least one condition and one target. A preference relation that involves only one condition and one target will be referred to as a *binding*. A preference relation that involves any number of conditions and a single target will be referred to as a *simple preference*. A designer might also want to enumerate a set of interactors, in order of preference. In this case, there are multiple targets. Preferences that involve multiple conditions and multiple targets, are referred to as *ordered* preferences.

2.2 Abstract Preferences

Bindings, simple preferences, and ordered preferences are all *concrete* preferences, because the targets are specified directly. Rather than specifying the element that is preferred, a designer may wish to instead specify the *characteristics* of the element that is preferred. For example, a designer may prefer whatever presentation element requires the least number of clicks, or whatever dialog structure imposes the least cognitive load. This kind of preference is referred to as an *abstract* preference.

Abstract preferences have an additional feature—a set of *criteria*. The criteria determine which target is selected; in the example above, the number of clicks required is the criterion. For an abstract preference, the targets are possible selections, and criteria determine which target is actually chosen. Needless to say, an abstract preference is required to have more than one target. There are two types of criteria: preferential and logical. Preferential criteria specify those characteristics that cause one target to be *preferred*. For example, a preferential criterion might say to choose the dialog structure with the least complexity. Conversely, logical criteria specify what kinds of targets are *allowed*. A logical criterion might say not to choose any dialog structure with complexity greater than 50. If there is more than one preferential criteria, then each criterion must have a *priority* to indicate how important it is relative to the others. Logical criteria need no priority; if the value of the feature under consideration by a logical criterion is in violation of the criterion, then the associated target is ruled out.

When used in combination, preferential and logical criteria can allow for very complex preference specifications. For example, consider the following specification: "For task model Tm2 and domain model Dm1, user U4 likes presentation elements that require few clicks, take up little screen real estate, and have lots of colors. The criteria of having lots of colors is most important, followed by the number of clicks. But if amount of space occupied is too small, it won't be visible, so disallow it altogether." In this situation, Tm2, Dm1, and U4 are the conditions. The set of presentation elements under consideration are the targets. There are three preferential criteria: number of colors (more is preferred), number of clicks (less is preferred), amount of space occupied (less is preferred). In addition, there is one logical criteria: the amount of space occupied must be greater than some constant. For a table describing this preference relation, see the second scenario in section 3.1.4.

2.3 Design Guidelines

Thus far, it has been assumed that the criteria apply only to the targets. That is, distinctions can be made between the targets only on the basis of characteristics of the targets themselves. It is possible to say, for example, "U4 prefers the least complex task tree," but it is not possible to say, "If U4's experience level is less than *intermediate*, then use presentation element P3." To do this, we need to separate the list of targets from the criteria.

Preference relations where the criteria do not necessarily apply to the targets are called *design guidelines*. Design guidelines are the most abstract and complex preference relations that we will attempt to model. For this class of preference relations, an additional feature must be considered: a list of objects. The criteria refer to features of each object. For design guidelines, all criteria must be logical. If the value of every criteria is true for every object, then a mapping is created between the conditions and each of the targets.

Recall the example above: "If U4's experience level is less than intermediate, then use P3." In this case, U4 is the object. The criterion is the "experience level" attribute. The target is P3. If the experience level meets the criteria of being less than intermediate, then the target P3 is chosen and the mapping is made.

The target of a design rule could itself be another preference relation. For example, suppose we wanted to model the following preference: "if U4's experience level is less than intermediate, then use the task tree with the least amount of complexity." Once again, U4 is the object, and experience level is the criterion. But the target is a preference relation – specifically, an *abstract preference*. That abstract preference says to choose the task tree with the least amount of complexity. So for the abstract preference, the target is a list of task trees and the criterion is the amount of complexity. The condition of the abstract preference is simply the design guideline that targeted it.

By creating a series of *design guidelines* that target each other, we can implement a decision tree to represent complex abstract preferences. Decision trees have already been used for representing UI design guidelines [5]. However, never before has it been possible to integrate a decision tree so tightly into the user interface model. This offers a number of important advantages. Whereas the set of discriminants is static in all known previous applications of decision trees to UI design, this new formalism permits any attribute defined in the user interface model to act as a discriminant. Moreover, rather than limiting use of the decision tree to one particular kind of mapping—e.g., interactor selection, as in [5]—any type of mapping can be performed. Decision trees can just as easily be used to select among alternative dialog structures or even color schemes.

To see how this can work, let us consider an example in which a decision tree selects interactors by first considering the type of a domain element D1, and then the experience level of the user U1. We discriminate based on the type of D1 (e.g. float, integer, string) by creating a set of design guidelines G_1... G_N: one for each possible type. Each such guideline has U1 and D1 as conditions. D1 serves as the object, and the type of D1 is the criterion. For G_1, the criterion is met iff the type of D1 is "integer." For G_2, the criterion is met iff the type of D1 is "string," and so on. If the criterion is met, then D1 and U1 are mapped on to the targets; otherwise, no mapping is made. In this case, the targets are another set of design guidelines. For G_1, those targets are design guidelines $G_{1.1}$... $G_{1.M}$. For G_2, those targets are design guidelines $G_{2.1}$... $G_{2.M}$. Thus, if the type of D1 is "integer," then D1 and U1 are mapped on to the design guidelines $G_{1.1}$... $G_{1.M}$. These design guidelines must then be evaluated.

If the criterion is met for at least one of G_1... G_N, then the evaluation of the decision tree continues. $G_{1.1}$ is structured similarly to G_1. The condition of $G_{1.1}$ is now G_1, and the object is now U1; the target is an interactor. The criterion is the experience level of U1. For $G_{1.1}$, the criterion is met iff the experience level of U1 is "advanced." For $G_{1.2}$, the criteria is met iff the experience level of U1 is "intermediate," and so on. Each of $G_{1.1}$... $G_{1.n}$ represents a user level, and each has a target P, which is an interactor. If the criterion is met, then G_1 is mapped on to P. Mappings of this kind are transitive. If U1 and D1 are mapped onto G_1, and if G_1 is mapped onto the interactor P, then U1 and D1 are mapped onto P. The design guidelines are evaluated in a breadth-first manner; once all guidelines are evaluated, the mappings (if any) are returned. We are then finished evaluating the decision tree.

The tight integration of the decision tree with user-interface model offers significant advantages over previous design guideline implementation strategies. However, it poses a problem of model binding. In the example decision tree, it is assumed that the relevant design guidelines have been created with the appropriate objects and conditions, which are themselves parts of the user-interface model. In reality, this is not likely to be the case, because decision trees of design guidelines are usually created independent of any specific user-interface model, and are intended for reuse among several user-interface models. The solution is to underspecify the decision tree, leaving the objects and conditions empty. When the decision tree is applied to a specific interface model, a binding procedure for linking up the appropriate objects and conditions is necessary. Of course, some kind of binding must be performed whenever a decision tree or any other set of general design rules is applied to a specific user-interface model.

3. Implementing Preference Relations in XIML

XIML, the eXtensible Interaction Markup Language, is an XML-based user-interface modeling language, described in [1,2]. This formalism for modeling preference relations has been incorporated into XIML as part of the Design Model Component. The design model consists of a list of preference elements. A preference element can have four child elements: *conditions, targets, objects* and *criteria*. Each of these elements contains a list of *relation statements*, which indicate a mapping to another element somewhere else in the UI specification. Relation statements are simple one-to-one mappings, with a reference to an element's ID and a semantic definition. Relation statements can also include little bits of information, which are specified in *attribute statements*.

All criteria demand three common attribute statements: the name of the criteria (e.g. "screen space", "user experience level"), the type of the criteria ("preferential" or "logical"), and the behavior. The behavior of preferential criteria can take the following values: *minimize, maximize, approach, retreat*. If either of the latter two values is taken, then a *threshold* must be supplied; this is the value that the designer is trying to approach or retreat from. Preferential criteria also have a *priority*, which specifies the importance of the criterion relative to other criteria. The example in section 3.1.4 should make this clearer. The behavior of logical criteria can take the following values: *greater than, less than, equals, not equals*. All logical criteria must have a *threshold*. Logical criteria do not have a priority; if all criteria are met then the mapping holds; otherwise it does not.

158

3.1 Examples

This section offers examples of how to model various preference phenomena. All examples have also been modeled in XIML code; please contact the author for more information.

3.1.1 Binding
Scenario: "User U1 prefers presentation element P1."

CONDITIONS	TARGETS	OBJECTS	CRITERIA
U1	P1		

3.1.2 Simple Preference
Scenario: "User U1 prefers presentation element P1 for representing domain model Dm1."

CONDITIONS	TARGETS	OBJECTS	CRITERIA
U1, Dm1	P1		

3.1.3 Ordered Preference
Scenario: "User U1 prefers presentation element P1 to presentation element P2, for representing domain model Dm1."

CONDITIONS	TARGETS	OBJECTS	CRITERIA
U1, Dm1	P1 – priority 100 P2 – priority 50		

3.1.4 Abstract Preferences
Scenario: "User U1 prefers the presentation element that occupies the least screen space, but not less than 100 square pixels."

CONDITIONS	TARGETS	CRITERIA		OBJECT
U1	P1, P2,...Pn	Name	Screen Space	
		Type	Preferential	
		Behavior	Minimize	
		Priority	100	
		Name	Screen Space	
		Type	Logical	
		Behavior	Greater Than	
		Threshold	100	

Scenario: "For task model Tm2 and domain model Dm1, user U4 likes presentation elements that require few clicks, take up little screen real estate, and have lots of colors. The criteria of having lots of colors is most important, followed by the number of clicks. But if the amount of space occupied is too small, it won't be visible, so disallow it altogether."

CONDITIONS	TARGETS	CRITERIA		OBJECT
Tm2, Dm1, U4	P1, P2, ... Pn	Name	Screen Space	
		Type	Preferential	
		Behavior	Minimize	
		Priority	25	
		Name	Screen Space	
		Type	Logical	
		Behavior	Greater Than	
		Threshold	100	
		Name	Clicks	
		Type	Preferential	
		Behavior	Minimize	
		Priority	50	
		Name	Number of Colors	
		Type	Preferential	
		Behavior	Maximize	
		Priority	100	

3.1.5 Design Rules

Scenario: "If user U1 is *not* of experience level *expert*, then select the presentation element with the least cognitive complexity."

Design Rule G1

CONDITIONS	TARGETS	CRITERIA		OBJECT
U1	G2	Name	Experience Level	U1
		Type	Logical	
		Behavior	Not Equals	
		Threshold	Expert	

Abstract Preference G2

CONDITIONS	TARGETS	CRITERIA		OBJECT
G1	P1, P2, ... Pn	Name	Cognitive Complexity	
		Type	Preferential	
		Behavior	Minimize	
		Priority	100	

References

1. Eisenstein J.. "XIML: The eXtensible Interaction Markup Language." RedWhale Software internal document. Please contact author for additional information.
2. Puerta, and Eisenstein J. (2001). "A Representational Basis for User-Interface Transformations." 2001 CHI Workshop on Transforming the UI for Anyone, Anywhere. Seattle WA.
3. Puerta and Eisenstein J. (1999). "Towards a General Computational Framework for Model-Based Interface Development Systems." Knowledge-Based Systems, Vol. 12, pp. 433-442.
4. Thevenin D. and Coutaz J. (1999). "Plasticity of User Interfaces: Framework and Research Agenda", in Proceedings of INTERACT'99. Edinburgh: IOS Press.
5. Vanderdonckt J. and Berquin P. (1999) "Towards a Very Large Model-Based Approach for User Interface Development", Proceedings of UIDIS'99. Los Alamitos: IEEE Press, pp. 76-85.

Towards a general guidance and support tool for usability optimization

Christelle Farenc[a], Philippe Palanque[a], Christian Bastien[bc], Dominique Scapin[b], Marco Winckler[a]

[a]L.I.H.S., University of Toulouse I,
place Anatole France, 31042 Toulouse Cedex, France
[b]INRIA, Domaine de Voluceau
B.P. 105, 78153 Le Chesnay Cedex, France
[c]Laboratoire d'Ergonomie Informatique, University of René Descartes
45 rue des Saints-Pères, 75270 Paris Cedex 06, France

Abstract

This paper reports on the work that has been done by the team above within the EvalWeb project. The work presented here describes how web usability guidelines should be refined in order to be suitable for the ergonomic design and evaluation of web sites and to be embedded in a software tool.

1. Introduction

The problem of designing usable interactive applications' user interfaces has been addressed in details for more than a decade. Today, the problem of usability of Web sites is becoming more and more critical as the number of Web users is still increasing in an exponential way. Besides, it seems that usability problems found in WIMP interfaces are still replicated in Web sites.

Web interfaces evaluation versus Wimp User Interfaces evaluation

Numerous evaluation methods exist for Web site evaluation. These methods may be grouped into three categories:

- Classical methods for WIMP user interface that are re-used for Web site without modification (example: end-user testing),
- Methods for WIMP user interface that have been adapted for Web interface (example: the WAMMI (12) questionnaire),
- New methods defined for the particular context of Web interface evaluation. This category brings together all remote evaluation methods (example: Tele-conference (7) supporting evaluation).

These numerous evaluation methods help make Web sites more usable. However, the use of these methods is sometimes considered a hard task due to many factors:

- Project budget and time for Web site development are (most of the time) more limited than for WIMP user interfaces,
- Designers of Web sites are not necessary designers of WIMP user interface. Their skills to conducts usability evaluation and their usability awareness are (usually) poor
- Due to the high frequency of Web site modification, the evaluation methods must be re-applied several times,
- User population of Web sites is expanding in age, in expectations, in information needs, in task types and in user abilities. In this context, most of user centered evaluation methods are difficult to apply.

What kind of methods should be defined?

The reasons presented above justify a evaluation method, that should be cheaper, easier to use for non-expert in usability, easier to integrate in an iterative design and of course more efficient than the current ones.
However, the evaluation is a complex task that requires knowledge and expertise. For usability experts, most of the evaluation methods are "simple". For non-experts, however, evaluation methods are considered "complex" mostly because they appeal to usability knowledge and expertise for their effective application throughout the evaluation process. Usability guidelines may be one way to alleviate this lack of expertise. Research has shown that careful application of guidelines had positive impact on usability (6).

For these reasons, our work is focusing on Web usability guidelines. In order to make these guidelines "easy to use" for non-experts, we propose a tool that integrates usability guidelines and which provides general guidance for non-expert designers during Web sites evaluation and design processes.

The paper is structured as follows: section 2 reports on the usability knowledge contained in Web usability guidelines. Section 3 proposes a summary of different tools for working with guidelines. And finally, in section 4, we present the guidelines organization and the incorporation of guidelines into a new development life cycle dedicated to the construction of Web sites.

2. Web usability guidelines

Many Web usability guidelines are available in the scientific literature and on Web sites. In (16), a description of these different sources of guidelines is proposed.
However, contrary to the WIMP usability guidelines, the process employed to develop Web guidelines is more informal (15). Web usability guidelines are often produced by common sense or observation of good practices. Moreover, Web usability guidelines mainly concern interface look and feel (15) and more specific aspects such as navigation, graphic use, hypertext links, etc.
At the organization level, Web design guidelines are structured along concepts such as usability (8), along design stages (9) or even along specific aspects of the Web (10).

3. Existing tools for working with guidelines

Two categories of tools exist to provide assistance to designers:
1 *Passive tools*: this category only provides designers with powerful access to guidelines, but these guidelines are in no way executable in any sense. Facilities can be offered to gather guidelines, to select them, to build a report for evaluation purposes, or for documentation, illustration or teaching.
2 *Active tools*: this category provides designers with tools capable of some form of processing guidelines, either at design time or at evaluation time. Representative tools belonging to this category are reported in (20). In particular for the Web, there are validation checkers concentrating on the HTML Code (3), such as WebLint (2), HTML Validator (22), but these verifications, although useful, are not related to usability. To our knowledge, only three tools are devoted to some form of automated testing of Web usability guidelines:

- **Bobby** (4), from CAST, automatically checks a Web page or a series of Web pages against accessibility guidelines promoted by the Web Accessibility Initiative (WAI) (21). Clicking on any Bobby hat appearing on the resulting page leads to a reference where the problem is detected and some comment on the guideline. WAI guidelines have been considered for implementation and evaluation through Bobby with three different levels of support: fully automated, partially automated, and manual.
- **WebSAT** (18), from the WebMetrics Suite (17) automatically test a single or a series of Web pages against a predefined set of guidelines that are not necessarily focusing on accessibility. As Bobby, a static analysis of the HTML code is performed and submitted to usability testing by a formal approach. Results are then summarized into six categories: accessibility (e.g., "All images not used as links should contain ALT tags"), form use (e.g., "The form should include a functionality for returning the completed form"), performance, maintainability, navigation, and readability (e.g., "Try to limit the density of the Web page"). Similarly, explanation for each usability defect can be provided, as well as a checklist of them.
- **Design Advisor** (5) automatically critiques a Web page while the designer is modifying its content. The critique is based on a visual hierarchy of perceptual guidelines assuming that page elements are searched according to a priority order: motion, size, images, color, text style, and position. From these guidelines, the tool automatically superimposes a scanning path on the Web page, thus highlighting usability problems. Although this tool is primarily based on these specific guidelines, it is interesting to note here that they have been implemented as Prolog clauses and the attributes of Web elements as facts. Having such an inference engine would be very practical for us, as adding, deleting, or modifying any guideline in the knowledge base would have no effect on their execution. But so far, we are unsure that all guidelines can be restricted to Prolog clauses. A deeper understanding of these possible restrictions, where any, is required. It is not certain that all guidelines can be implemented this way.

162

Our main goal is to produce a tool for automated testing of usability guidelines as far as they can be implemented. For this purpose, a specific tool for working with guidelines is needed. The next section explains how this goal can be reached.

4. Our approach: The EvalWeb project

Integration of guidelines in the design process

Guidelines incorporated into a development cycle are usually used manually or automatically at the evaluation stage. Our approach is both to embody guidelines in a software tool and to integrate it in each design phase.

In order to reach this goal, it is necessary to associate guidelines with different phases during the development cycle in order to present them where relevant.

For this purpose, each guideline is formalized in a systematic way. Guidelines can be decomposed in a "premise" and a "conclusion". For example: "Each personal homepage should contain the company logo at the top of the page". This rule can be redefined like this: "IF the current page is a personal homepage THEN the page should contain the company logo AND the logo should be at the top of the page". The "premise" of guidelines defines the context of validity and the "conclusion" defined actions that should be verified.

Our work consists in determining in which design phases "premises" and "conclusions" should be considered. For this purpose, the following general development life cycle has been proposed.

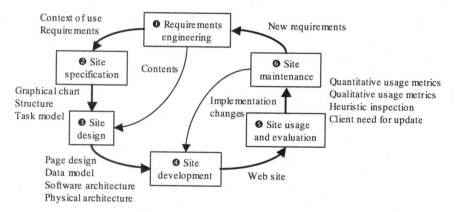

Figure 1: general development life cycle for a Web site

In the previous example:
- During the "requirement engineering" phase, the designer must indicate if pages correspond to personal pages in a company Web site.
- During the "specification" phase, the homepage should contain the logo company and this logo should be at the top of the page.

Guidelines organization

To improve the organizational structure of guidelines, we have also classified guidelines according to ergonomic criteria (defined as usability dimensions by (1) and by index keys.
- The classification by *ergonomic criteria* provides designers with a general draft that embeds ergonomic rules into the design process describing how they interfere with usability. This classification also allows attributing a level of importance to each guideline. For example, a guideline related to "task compatibility" is more important than a guideline related to "grouping items by format".
- The classification by *index keys* allows flexible access to guidelines, but is more adequate while envisioning a solution:
  - For the automatic selection of guidelines for a particular element of Web site;
  - For the selection of relevant guidelines when only few particular elements of Web site have been modified.

5. Present and future work

In this study, three sources containing guidelines have been selected ((14),(11),(19)). These sources do not deal only with Web interfaces. Therefore, we also have integrated a recent compilation of ergonomic guidelines specific to the Web (13).

Due to the large number of guidelines, a selection has been done. We have extracted 466 non-redundant and relevant guidelines in the context of Web site design. Afterwards, each selected guidelines has been classified according to the structuring presented in §4.

Future work will concern tool support for the integration of guidelines in the design process of Web site. This paper presented some of the potential benefits and some shortcomings of automated testing of usability guidelines. A prototype is currently implemented; but, based on these preliminary results, it appears that a more complete and functional version of the tool would be a promising contribution to automated testing of usability guideline.

References

1 Bastien, C. and Scapin, D. (1995). Evaluating a User Interface with Ergonomic Criteria, *International Journal of Human-Computer Interaction*, Vol. 7, pp. 105–121.

2 Bowers, N. (1996) Weblint: Quality Assurance for World-Wide Web, *in Proc. of 5th Int. World Wide Web Conf. WWW'5*, Paris, May 6-10, 1996, Elsevier, Amsterdam. Accessible at http://www5conf.inria.fr/fich_html/papers/P34/Overview.html

3 Clark, D. and Dardailler, D. (1999). Accessibility on the Web: Evaluation & Repair tools to make it possible. Accessible at http://www.dinf.org/csun_99/session0195.html

4 Cooper, M. (1999). Evaluating Accessibility and Usability of Web Pages, *in Proc. of 3rd International Conference on Computer-Aided Design of User Interfaces CADUI'99*, Louvain-la-Neuve, October 21-23, 1999, Kluwer Academics, Dordrecht, pp. 33–42.

5 Faraday, P (1999) Visually Critiquing Web Pages, *in Proc. of IFIP TC. 13 Conf. on Human-Computer Interaction INTERACT'99*, Edinburgh, August 1999, A. Sasse & Ch. Johnson (eds.), IOS Press.

6 Grose E.M., Forsythe C.h. and Ratner, J. (1998). Human Factors and Web development. *Lawrence Erlbaum Associates*, Mahwah.

7 Hartson H.R., Castillo J.C., Kelso J., Kamler J. and Neale W.C. (1996). Remote Evaluation: The Network as an Extension of the Usability Laboratory In *Proc. of ACM Conference on Human Factors in Computing Systems CHI'96*, April 13 - 18, Vancouver Canada, 228-235

8 http://www.eng.buffalo.edu/~ramam_m/

9 http://www.usability.serco.com/nonframe/web.html

10 http://webmaster.info.aol.com/webstyle/index.html

11 ISO Draft International Standard (DIS) 9241-11 (1999), Ergonomic Requirements for office work with visual display terminals, Part 11: Guidance on Usability, International Standards Organization, Geneva.

12 Kirakowski J. and Bozena C. (1998). Measuring the usability of web sites. Human Factors and Ergonomics Society Annual Conference, Chicago

13 Leulier, C., Bastien, C. and Scapin, D. (1998). Compilation of ergonomic guidelines for the design and evaluation of Web sites. *Commerce & Interaction Report*. Rocquencourt, France: Institut National de Recherche en Informatique et en Automatique.

14 Mayhew, D. (1992). *Principles and Guidelines in Software User Interface Design*, Prentice Hall, Englewood Cliffs.

15 Ratner J., Grose E., Forsythe, C. (1996). Characterization and Assessment of HTML Style Guides. In: *Proc. of ACM Conference on Human Factors in Computing Systems CHI'96*. Vol.2, pp. 115–116.

16 Scapin D., Leulier C., Vanderdonckt J., Mariage C., Bastien C., Farenc C., Palanque P. and Bastide, R. (2000). Towards Automated Testing of Web Usability Guidelines In. *Tools for Working with Guidelines* London Springer; pp. 293-304.

17 Scholtz, J.(1998). WebMetrics: A Methodology for Producing Usable Web Sites, *in Proceedings of the 42nd Annual Meeting of Human Factors and Ergonomics Society HFES'98*, Chicago, October 5-9, 1998, Vol. E, Human Factors and Ergonomics Society, Santa Monica, pp. 1612.

18 Scholtz, J. Laskowski, S and Downey, L. (1998). Developing Usability Tools and Techniques for Designing and Testing Web Sites, *in Proceedings of the 4th International Conference on Human Factors and the Web HFWeb'98*, Basking Ridge, June 5, 1998, J. Cantor (ed.). Accessible at http://www.research.att.com/conf/hfweb/proceedings/scholtz/index.html

19   Vanderdonckt, J (1994). *Guide ergonomique des interfaces homme-machine*, Presses Universitaires de Namur, Namur.

20   Vanderdonckt, J.(1999). Development Milestones towards a Tool for Working with Guidelines, *Interacting with Computers*, Vol. 12, N°2, pp. 81-118. Accessible at http://belchi.qant.ucl.ac.be/publi/1999/Milestones.pdf

21   WAI Accessibility Guidelines, Web Accessibility Initiative, World Wide Web Consortium, Geneva, 1998. Accessible at http://www.w3.org/wai

22   Yahoo category on HTML Validation Checkers, 2000. Accessible at http://dir.yahoo.com/Computers_and_Internet/Information_and_Documentation/Data_Formats/ HTML/Validation_and_Checkers/

Software Development and Open User Communities

Peter Forbrig, Anke Dittmar

University of Rostock, Department of Computer Science,
Albert-Einstein-Str. 21, D-18051 Rostock, Germany

Abstract

The request for software systems which can be used by different groups of people under different conditions asks for support by new modularization principles on the modeling level.

This paper follows the idea of combining several submodels, as e.g. task and object models, to specify different views on software systems and their situations of use as known from model-based approaches. But further, a separate specification of the general aspects of a model and the constraints imposed by concrete situations is suggested.

At the example of the adapted action model this specification strategy is deeper explained and it is shown how it can contribute to a more flexible modeling for open user communities.

1. Introduction

Human beings apply interactive software systems to fulfil tasks. In this sense, software is nothing else but a tool for us. We use it like a hammer or a mechanical machine to achieve the goals set by the appropriate tasks. Thanks to the development of new interactive techniques, the way a user can perform his task has become less restrictive. Thus, window systems and direct manipulation allow him to execute several subtasks concurrently.

"There is no point in building a system that is functionally correct or efficient if it doesn't support user's tasks or if users cannot employ the interface to understand how the system will achieve task objects" (Duke & Harrison, 1995). Model-based approaches like (Paterno 2000), (Puerta, Cheng, Ou & Min, 1999) or (Wilson, Johnson, Kelly, Cunningham & Markopoulos, 1993) follow this idea. Task-based and object-oriented techniques are used to derive, for example, user interfaces the user can cope with.

In this paper, it is assumed that a software system has to support tasks in a certain area of application which can be specified by a task model. An example of a task model describing the preparation a meal is given in Fig.1. It demonstrates that task modeling is not restricted to any kind of tasks or tools.

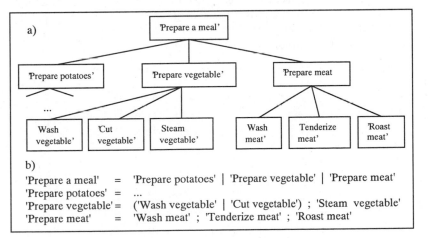

Figure 1. Preparation of a meal

We claim, however, that the existing models are to rigid to specify all requirements of interactive systems for open user communities. Besides different user characteristics different constraints of the user's actual environment have to be considered. Has the user to perform a task alone or does he work in a group? Is he disabled in a certain way? What is his standard of knowledge? Does the user have a large screen or does he use a mobile phone to work with an INTERNET connection?... The term *specific situation of use* refers to such questions.

Let us have a look at the simple example in Fig.1. Basically, the model allows us to prepare potatoes, vegetable and (most important) meat concurrently. There are only general constraints coming from the field of application like washing and tenderizing the meat before roasting it.

But, the model neglects individual preferences or abilities as well as specific constraints in the actual environment. Imagine, for example, that a person can handle at most two pots or pans at the same time. He could prepare the potatoes in one pot, steam the vegetable in a pan and, then, roast the meat 'immediately afterwards' in the same pan to get a hot meal, nevertheless.

Software design for open user communities asks for support by new modularization principles on the model level. We argue that a separation between kernel models which describe the general requirements imposed by the area of interest and additional models is a step in the requested direction. An additional model can be seen as an extension of the appropriate kernel model adapting it to a specific situation of use.

Before in Sect.3 this more flexible specification process is described Sect.2 gives a proposal to modify a 'classical' task model allowing model adaptations. The paper is closed by a summary.

2. From 'classical' task models to model adaptations

A 'classical' or simple task model consists of two parts as to be seen in the abstract example of Fig.2. The *hierarchical description* H decomposes the task T into subtasks T_i until the level of basic tasks. To each basic task T_i an operation op_i is assigned which has to be executed in order to fulfill T_i. The task tree is the most stable part of the task model. It gives an impression what has to be done whereas the *sequential description* S specifies the order in which the operations can be performed to fulfil the whole task. Therefore, in S the task T is considered as a process of a process algebra in the usual way of task-based approaches (e.g. (Paterno, 2000)). S is built up by two types of equations. An equation of the first type describes temporal constraints between those subtasks which have a common parent node in the task tree (equ_T, equ_{T1}, and equ_{T2} in Fig.2). Equations of the second type define the above mentioned mapping between basic tasks and operations of the set OP (equ_{T11}, equ_{T12}, equ_{T21}, and equ_{T22} in the example). For reasons of brevity equations of the last type as well as operations are omitted later on. Temporal dependencies between subtasks can be formulated by well-known operators as used e.g. in (Johnson, Wilson & Markopoulos, 1991). In Fig.2 the temporal operators for parallel ($|$) and sequential execution (;) of subtasks were applied. Thus, T can be fulfilled by executing one of the following 6 sequences of basic tasks: $\langle T_{11},T_{12},T_{21},T_{22}\rangle$, $\langle T_{11},T_{21},T_{12},T_{22}\rangle$, $\langle T_{11},T_{21},T_{22},T_{12}\rangle$, $\langle T_{21},T_{11},T_{12},T_{22}\rangle$, $\langle T_{21},T_{11},T_{22},T_{12}\rangle$, $\langle T_{21},T_{22},T_{11},T_{12}\rangle$.

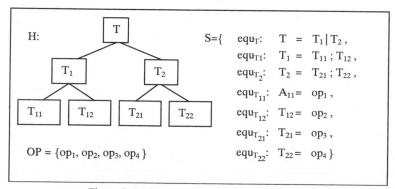

Figure 2. A simple task model for the abstract task T

In this paper a simple task model is taken to describe the *kernel task model*. Its task tree determines which subtasks can be performed in general. The sequential description contains only those temporal relations which come from the area of application. In Fig.2, T_1 and T_2 could be interpreted as tasks decomposed into further subtasks concerning the input of some data to the software system (T_{11} and T_{21}) and the calculation of the appropriate results (T_{12} and T_{22}). Further it is known that the execution of T_{12} and T_{22} takes a long time (even with the best processor). Obviously, T_{11} has to be performed before T_{12} just as T_{21} before T_{22}. But, there are no temporal constraints between T_1 and T_2. A user interface derived from this task model could consists of two windows W_1 and W_2 in parallel, W_1 reflecting T_1, W_2 reflecting T_2. In both windows the calculation is disabled until all input data are given.

Considering open user communities there are also changing constraints arising from the specific situation of use. In our example, a user could work with a small device which can display only one of the windows W_1 and W_2 at the same time. It is blocked until a calculation is finished. The task model could be adapted to this situation by saying that T_1 has to be executed for enabling T_2 because T_1 has a higher priority, for example. (We admit this is only a scholarly example but we hope it is sufficient to explain the idea.)

$$
\begin{aligned}
S \quad =\{ \; & equ_T: \quad T \;\; = T_1 \mid T_2, \\
& equ_{T_1}: \; T_1 = T_{11} \; ; T_{12}, \\
& equ_{T_2}: \; T_2 = T_{21} \; ; T_{22} \; \} \\[6pt]
& equ_c: \;\; C = T_1 ; T_2
\end{aligned}
$$

part of the kernel model

specific constraints

Figure 3. An adaptation of Fig.2

In our approach, a kernel task model is adapted to a specific situation of use by additional temporal constraints noted as equations similar to them in the sequential description of the kernel model. That means that the task hierarchy itself is considered as stable but the set of possible execution sequences can be restricted. The possibility of disabling some of optional subtasks is also included. Fig.3 shows an adaptation of the kernel model in Fig.2.

The above mentioned situation is reflected in the specific constraint equ_C which restricts the 6 possible execution sequences to only one: $(\langle T_{11}, T_{12}, T_{21}, T_{22} \rangle)$.

Fig.4 illustrates an adapted model for the introductory example (a person could handle 2 pots or pans at most). The temporal operator $\langle\langle\;\rangle\rangle$ is used to specify the relation ('Roast meat' *immediately afterwards* 'Steam vegetable') because $\langle\langle T_i \rangle\rangle$ means that the execution of subtask T_i has not to be disturbed by any other subtask of the task model.

$$
\begin{aligned}
S \quad =\{ \; & \text{Prep. meal} & = \;\; & \text{Prep. potatoes} \mid \text{Prep. vegetable} \mid \text{Prep. meat,} \\
& \text{Prep. potatoes} & = \;\; & \ldots\,, \\
& \text{Prep, vegetable} & = \;\; & (\,\text{Wash vegetable} \mid \text{Cut vegetable}\,)\,; \text{Steam vegetable,} \\
& \text{Prep. meat} & = \;\; & \text{Wash meat ; Tenderize meat ; Roast meat} \}
\end{aligned}
$$

equ_c: $\;C = \langle\langle \text{Steam vegetable ; Roast meat} \rangle\rangle$

Figure 4. An adaptation of Fig.1

3. A more flexible specification of software

It is widely accepted that the use of different models during the software development leads to more matured systems because they can catch different aspects of the problem space. Most model-based methods distinguish between a model of the existing task situation and the envisioned one. These models are further subdivided into submodels describing the tasks (task model), the environment (business object model) and the users (user model).

168

They support the derivation of a design model of the interactive system. Deeper discussions of this topic can be found in (Forbrig, 1999).

Although such approaches contribute to a flexible specification of software the submodels itself are often to rigid to describe different user needs. Consequently, we are often confronted with specifications which are either to restrictive for single users or give them too much freedom within the application context.

This paper proposes to separate a submodel into a kernel part and additional ones to allow more flexible specifications. Whereas the kernel part is stable or static (as far as you can consider a model as stable) an additional part depends on the specific situation of use. The general idea of the approach is illustrated in Fig.5. As to be seen the model of the interactive system is derived from the kernel submodels as usual. But in the kernel models only the general constraints given by the field of application are specified. Models of specific constraints can adapt a kernel model to a specific situation of use and change the model of the interactive system dynamically.

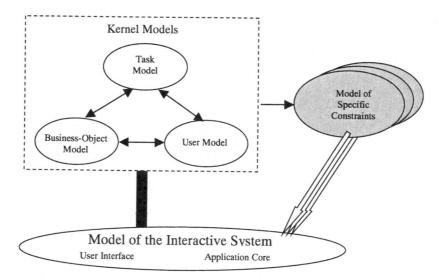

Figure 5. Ontology of the model-based approach

4. Summary

Software systems for open user communities have to reflect the different needs of users. Hence, it is reasonable to distinguish even in the specification phase between two parts of description. Firstly, a general one which is valid in all situations of use. Secondly, a specific one which takes into consideration the characteristics of a single user or group of users.

In Sect.2 a proposal of adapting task models was made. Models of specific temporal constraints adapt a kernel model by restriction without changing the nature of it completely. The idea is based on process algebras which allow a very precise specification of temporal relations. A more detailed explanation is given in (Dittmar 2000). We believe that similar mechanisms of adaptation can also be found for the other submodels within a model-based approach.

Of course, there are many open questions. Sometimes it can be difficult to categorize a constraint as general or specific, for example. Nevertheless, we believe that the principle of adaptation can support a more flexible software development.

References

Duke, D. J., Harrison, M. D. (1995). Mapping user requirements to implementations. Software Engineering Journal, Vol. 1, 13-20.

Johnson, P., Wilson, S., Markopoulos P. (1991). A framework for task based design.

Paterno, F. (2000). Model-Based Design and Evaluation of Interactive Applications. Springer Verlag.

Puerta, A., Cheng, E., Ou, T., Min, J. (1999). MOBILE: User-Centred Interface Building. In: Proceedings of the ACM Conf. on Human Aspects on Computing Systems CHI'99. ACM Press, New York 426-433.

Wilson, S., Johnson, P., Kelly, C. (1993). Cunningham, J., Markopoulos, P.: Beyond hacking: A model based approach to user interface design. In: Alty, J.L., Diaper, D., Guest, S. (eds.): People and Computers VIII, Proceedings of the HCI'93 Conference, 418-423

Dittmar, A. (2000). More precise descriptions of temporal relations within task models, ICSE2000, Workshop Design Specification and Verification of Interactive Systems, LNCS 1946,151-168

Forbrig, P. (1999). Task- and object-oriented development of interactive systems - How many models are necessary?. Proc. DSVIS 99, Braga, 225-237

Appliance independent specification of User Interfaces by XML – A model-based view

Peter Forbrig, Andreas Müller, Clemens H. Cap

University of Rostock
Department of Computer Science
Albert-Einstein-Str. 21, 18051 Rostock, Germany
{pforbrig, xray}@informatik.uni-rostock.de

Abstract

Modern software systems have to be applicable via different devices. Therefore it is important to have a chance to specify a user interface once appliance independent and then to generate interactively different presentations. This paper follows such an approach with a special focus on mobile devices. It presents a concept of device independent user interface design based on the XML-technology. The concept is applied to a small e-commerce example.

1. Motivation

Different devices with different capabilities are used by modern information and communication systems to present the application logic. Modern banking software should for instance provide a HTML-based user interface for a web-based usage of the service as well as a WAP-based mobile communication service. A customer should also be allowed to use the software by special automata (bancomat).

In our approach the three models (user, task and business-object model) are the basis for the development of the application logic and the user interface design. Figure1 represents the main relations between different models.

It allows the computer based development of interactive systems based on the TADEUS approach (Elwert 1995).

This approach allows a reuse of the designed models apart from final representations. To have a flexible development process from an abstract interaction model to a specific representation there was the idea to use the XML technology for representation and transformation features. The specific features of devices are specified by XML-descriptions as well and the transformation process is based on this information. This approach allows a smooth integration of newly developed devices into already existing software. It also allows also a better support of a larger number of devices, which results in a better access to information resources by a lot of users. They can use the available devices and sometimes also those devices they can handle best.

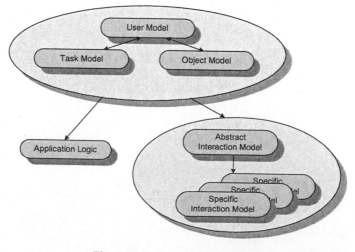

Fig. 1: Model-based user interface design

2. Related work in specifying user interfaces by XML

There are already some approaches to specify user interfaces by markup languages. There is for instance UIML (Abrams 1999) a XML-based language for describing user interfaces. It comes with different rendering programs allowing the generation of user interfaces on different platforms. Up to now the generation for WML, Java AWT and swing are available.
A special renderer for each target-device is needed. This is an important disadvantage The flexibility for selecting target-device features is limited only because there the lack of a mapping concept.

XUL has its focus on window-based graphical user interfaces by XML. Its disadvantage is the limitation to such graphical user interfaces. It is not applicable to interfaces of small mobile devices.
There is no abstraction of interaction functionality available. The device-independent specification of user interfaces was not planned. The intention was the development of the mozilla-project.

3. Specification of user interfaces

A user interface is considered as a simplified version of the MVC model and is separated into a model component and a presentation component. The model component describes the feature of the user interface on an abstract level. It is also called abstract interaction model.
The user-interface objects with their representation are specified in the presentation component. During the development a mapping forming the abstract interaction model to the specific interaction model is necessary. TADEUS has a term representation of both models. The transformation process is described by attribute grammars. For different platforms different grammars are necessary. The development of such grammars for different mobile devices is sometimes time consuming. It would be good to have more support for the transformation process. Especially the integration of the specific features of different devices is difficult. A general support would be appreciated.
The XML technology (Goldfarb 1999) offers such an opportunity. It allows the description of the abstract interaction model, the description of specific characteristics of different devices and the specification of the transformation process, which will be shown later. Let us first have a look at a model of the transformation process forming an abstract interaction model to a specific user interface:

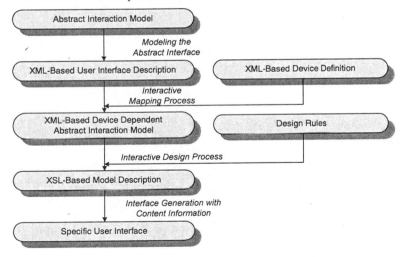

Fig. 2: Creation of a user interface

The following paragraphs will show a specific way of using XML in the proposed manner. The example of an e-shop will be used to show the principles of this approach. From abstract specification to concrete user interfaces the reader is able to follow the development stages.

3.1 XML-Based Abstract Interaction Model

The abstract interaction model is transformed into a notation which is based on a language of the XML family. The basic document type definition (DTD) was introduced in (Müller, Forbrig 2001).

The user interface is at this stage composed of one or more user interface objects (UIO). A UIO can be composed of other UIO's or one or more input-/output values. These values are in an abstract form and describe interaction features of the user interface in an abstract way. Later there will be a mapping process of these abstract UIO's to concrete ones. There are different types of these abstract UIO's. They have different attributes to specify their behaviour. One example to explain is the input_1-n value. Here we have some attributes to specify these abstract UIO. The input_1-n value for instance can be mapped to a slider as a concrete UIO. The range_min attribute specify the minimum-value of input possibilities, the range_max attribute specify the maximum-value. The range_interval value is for fixing the interval within the range. Also we can specify a list of input values. Now lets have a look at an example of a simple user interface of a little e-shop system:

```
<?xml version="1.0" encoding="ISO-8859-1"?> .  .  .  .
<ui ui_name="e-shop">
    <uio name="title"> .  .  .  .
        <output name="title">
            <output_string/>
        </output>
    </uio> .  .  .  .
.  .  .  .
    <uio name="Article">
        <output name="Art_Name">
            <output_string/>
        </output> .  .  .  .
        <input name="Order">
            <input_trigger/>
        </input>
    </uio> .  .  .  .
 </ui>
```

Example 1: Part of description of a simple user interface of an e-shop system

Because of lack of space only a few attributes are mentioned here.

3.2 XML-Based Device Definition

A universal description language for describing properties of various target devices in abstract and comparable matter is necessary to support the transformation process from an abstract interaction model to a device dependent abstract interaction model. Based on (Mundt 2000) such a language was developed. Specifications written in this language are XML documents describing the specific features of devices.

Because a lack of space here the corresponded DTD can be looked at (Müller, Forbrig 2001). The specifications are not only necessary for this transformation they influence the specification of formatting rules as well. Following example shows a very short fragment of a device definition of java.AWT:

```
<?xml version="1.0" encoding="ISO-8859-1"?>
<?xml-stylesheet type="text/xsl" href="?.xsl"?>
<!DOCTYPE device SYSTEM "devdef-1.dtd">

<device id="Advanced Windowing Toolkit">
    <device id="java.awt.Button">
        <service id="?" kind="trigger">
            <name>Location</name>
            <feature>java.awt.Button.setLocation</feature>
        </service>
    </device>
    <vendor>SUN Microsystems</vendor>
```

```
    <version>2.0</vendor>
</device>
```

Example 2: Device definition for java.AWT

3.3 XML-Based Device Dependent Abstract Interaction Model

This model is still on an abstract level but it fulfils already some constraints coming from the device specification. It uses available features and omits not available services. The result of the mapping process is a file in which all abstract UIO's are mapped to concrete UIO's of a specific representation. The structure of this file is based on a XML-DTD which was introduced in (Müller, Forbrig 2001).

This file is specific according to a target device and includes all typical design properties of concrete UIO's like colour, position, size etc. It is a collection of option-value pairs where the content of the values specified later on in the design process and describes a "skeleton" of a representation of a specific user interface. The user interface is designed on the basis of this skeleton during the following design process by designers or ergonomics.

A part (one UIO) of our simple example of an e-shop system mapped as a HTML-representation looks like follows:

```
. . . .
    <uio name="Ok">
        <input name="Ok">
            <input_trigger/>
            <device>
                <type>BUTTON</type>
                <parameter>
                    <option>TYPE</option>
                    <value>submit</value>
                </parameter> . . . .
            </device> . . . .
        </uio> . . . .
</ui_map>
```

Example 3: E-shop abstract user interface specified as abstract device dependent (HTML) interaction model

A part of the same user interface mapped as a java.AWT looks like introduced in the next example:

```
. . . .
    <uio name="Ok">
        <input name="Ok">
            <input_trigger/>
            <device>
                <type>java.awt.Button</type>
                <parameter>
                    <option>java.awt.Button.setLocation</option>
                    <value></value>
                </parameter> . . . .
            </device> . . . .
        </uio> . . . .
</ui_map>
```

Example 4: E-shop abstract user interface specified as abstract device dependent (java.AWT) interaction model

The wml-example is omitted because of its similarity to the HTML document.

3.4 XSL-Based Model Description

The creation of the XSL-based model description is based on the knowledge of available UIO's for specific representations. It is necessary to know which values of properties of a UIO are available in the given context. The XML-based device dependent abstract interaction model (skeleton) and available values of properties are used to create a XSL-based model description specifying a complete user interface.

3.5 Specific User Interface

By XSL transformation a file describing a specific user interface will be generated. Examples for java.AWT and HTML are given in (Müller 2001). Unpublished WML examples were developed as well.

The XSL transformation process consists of two sub-processes:

1. Creation of a specific file representing the user interface (Java.AWT, Swing, WML, VoiceXML, HTML,...) and
2. Integration of content (data base, application) into the user interface.

The generated model is already a specification runable on the target platform. It is constructed by a transformation process which uses as input the device dependent abstract interaction model, which is coded in XML.

There are two different cases. The user interface is generated once like for java.AWT or it has to be generated dynamically several times like for WML. In the first case there is no need for a new generation of the user interface if the contents changes. The content handling is handled within the generated description file.

In the second case the content will be handled by instructions within the XSL-based model description. Each modification of the contents results in a new generation of the generated description file.

4. Results and Future Work

The current work includes the specification of a XML-based language for the description of user interfaces. A concept for the mapping process from abstract UIO's to concrete UIO's was developed. A XML language was developed which allows the mapping of abstract user interfaces to concrete once on mobile devices. This concept is already evaluated by several examples. A tool supporting the mapping and in the design process is under development. It delivers already promising results. Up to now it is not integrated in previous phases of the development process. Our experiments showed that XML is a promising technology for the platform independent generation of user interfaces. Further studies will show whether dialogue sequences of already developed applications can be integrated into the design process. The usage of patterns could enhance the development process.

References

Abrams, M., et al: UIML Tutorial. http://www.harmonia.com

Benyon, D., Green, T., Bental, D. (1999). Conceptual Modelling of User Interface Development. Springer.

Goldfarb, C.F., Prescod, P.(1999). XML-Handbook, Prentice-Hall.

Foley, J.D. (1994). History, Results and Bibliography of the User Interface Design Environment (UIDE), an Early Model-based Systems for User Interface Design and Implementation, in Proc. of DSV-IS'94, Carrara, 8-10 June 1994, pages 3-14.

Forbrig, P. (1999). Task- and Object-Oriented Development of Interactive Systems: How many models are necessary?, DSVIS'99, Braga, Portugal.

Forbrig, P., Stary, C. (1998). "From Task to Dialog: How Many and What Kind of Models do Developers Need ?" CHI Workshop "From Task to Dialogue: Task-Based User Interface Design", Los Angeles.

Elwert, T., Schlungbaum, E. (1995). Modelling and Generation of Graphical User Interfaces in the TADEUS Approach , In: P. Palanque, R. Bastide (Eds.): Designing, Specification, and Verification of Interactive Systems. Wien, Springer, S. 193-208.

Harrison , M.D.; J.C. Torres (Eds.. (1997). Proceedings of the 4th Eurographics Workshop on Design, Specification and Verification of Interactive Systems. University of Granada, p. 427-440.

Makopoulos, P.; Johnson, P. (eds). (1998). Design, Specification and Verification of Interactive Systems '98. Springer-Verlag, Wien New York.

Markus, A.,Smilonich, N., Thompson, L. (1995). The Cross-GUI-Handbook. Addison-Wesley.

Müller, A., Forbrig, P. (2001). Using XML for Model - based User Interface Design. CHI Workshop "Transforming UI Anyone. Anywhere", Seattle.

Müller, A. (2001). Spezifikation von Benutzerschnittstellen durch eine XML-basierte Markup-Sprache. Rostocker Informatik-Berichte 25.

Mundt, Th. (2000). Gerätedefinitionssprache DEVDEF. http://wwwtec.informatik.uni-rostock.de/IuK/gk/thm/devdef/.

Paternó, F. (2000). Model-based Design and Evaluation of Interactive Applications. Springer.

Szekely, P., Luo, P., Neches, R. (1993). Beyond Interface Builders: Model-Based Interface Tools, in Proc. INTERCHI'93 Human Factors in Computing Systems, ACM/IFIP, pp. 383-390.

Wilson, S., Johnson, P. (1996).Bridging the Generation Gap: From Work Tasks to User Interface Design, in Proc. of CADUI'96, Namur, Belgium, 5-7 June 96, pp. 77-94.

Accommodating Diverse Users in ETAG and ETAG-Based Design: task knowledge and presentation

Geert de Haan

Maastricht McLuhan Institute, Maastricht University
P.O. Box 616, 6200 MD Maastricht, The Netherlands
g.dehaan@mmi.unimaas.nl

Abstract

This paper discusses ETAG, a formal model for design representation and ETAG-based design, a method for user interface design. The paper discusses and exemplifies the two main facilities in ETAG to describe the differences and similarities in task-knowledge requirements between users. Using evidence from the Comris project (van der Velde, 1997), conclusions are drawn about ETAG-based design as a single-notation design method, and about the tension between ETAG as a model for design representation and as a representation of user competence knowledge.

1. Extended Task-Action Grammar

ETAG (Extended Task-Action Grammar; Tauber, 1988, 1990; de Haan, 2000a) is a formal language to represent user interfaces in terms of the knowledge that a perfectly knowing user would have (in a mental model) about performing tasks. To create a psychologically valid description of user interface for design purposes, ETAG stratifies user interface knowledge into a number of levels using existential logic and written down in a formal grammar. Stratification into levels intends to meet the existence of levels in human knowledge and to reflect the major decisions that occur during the design process. An ETAG representation consists of a canonical basis, a user virtual machine, a dictionary of basic tasks, and a section with production rules.

Comris (van der Velde, 1997) is an agent-based system to provide visitors of large conferences with context-sensitive information about nearby events and people, based on a personal interest profile, an active badge system, and information about the conference schedule In the Comris system, users carry a wearable computer, which stores a list of advisory messages from the agent system for textual and/or spoken presentation. By means of the wearable, users may, among others, play, present, delete and respond to messages, move to a previous or a next message, and adjust the volume and the minimal interest threshold.

The *canonical basis* of the ETAG representation of Comris contains general concepts such as object, position, state, attribute and event, to define the specific concepts that the system uses. The *type hierarchy* included the concepts that the user comes into contact with, such as messages and message list, and the concepts that the user may influence, such as the interest threshold that messages should exceed to be included in the message list. In addition, the type hierarchy specifies an agent concept as the external source of new messages, as well as interest and location concepts which are necessary to understand how the user's interests and whereabouts influence the advisory information in newly created messages. In the *type specification* each of the concepts in the type hierarchy is defined, and additional concepts and attributes, such as current_message and message_position are added as 'semantic sugar' to describe how the system works, as experienced by the user. Part of the type specification of Comris' UVM will read as follows:

```
type [message_list isa object]
        themes: message
        places: place_pos(x)(message_list)
        attributes: max_pos, current_pos
type [message isa object]
        attributes: message_text, next, previous
```

The specification states that a message_list provides space for messages, which each take up one place in the list, and that messages have a certain text, and a previous and next attribute which will be used later on, to describe the task of selecting another message to take the current_position.

The *event specification* is the part of the UVM describing the workings of the system as it virtually appears to work from the point of view of the users, which may be different from how it is actually build to work. Part of Comris' event specification will read as follows:

```
type [create_message isa event]
        parameters: relevance, threshold, message_list, message_text
        precondition: relevance > threshold
                create_on(message_list, max_pos, message)
                copy_to(text, message)
                current_pos = max_pos
                max_pos += 1
                play_message
```

The event specification describes how the system processes tasks from the user and other agents using a pseudo computer program notation. The create_message event reads as follows: when the relevance of the new message surpasses the user's interest threshold, it is added at the end of the user's message list. It then becomes the current message and an event is raised to play it (present it as speech and text).

The *dictionary of basic tasks* lists the tasks that the system provides to the user, and it links the command sequence, as described in the production rules, to the event specification. In the original proposal of ETAG (Tauber, 1988, 1990) it excluded system tasks and tasks which require user decisions during execution. Embedded tasks were introduced to describe system functions like those with clear visible effects, and so-called menu tasks were introduced to describe higher-level user tasks, such as making multiple selections from a list. The following task illustrates the dictionary of basic tasks:

```
type [create_message isa embedded task]
        effect: [create_message isa event][play_message isa task]
        remark: a system task to present a new message to the user
```

To complete the ETAG specification of the Comris system, the *perceptual interface* should be specified next. This consists of the user perceivable aspects of the system. It is not (yet) possible to specify the *presentation interface* in ETAG. which is concerned with the visual design. It is better to specify the user's knowledge about the visual parts of a user interface pictorially, and for design purposes, it is easier to use so-called interactive user interface builders. The specification of the *production rules* which describe the command procedures will be illustrated in section about design representation for diverse users/user task world presentation.

2. ETAG-Based user interface design

In ETAG-Based Design (de Haan; 1996; 2000a) user interface design is regarded as the incremental specification of the mental model of a perfectly knowing user. The design process is structured into a number of discrete steps, each covering a specific set of design decisions: task- and context analysis, task design or task synthesis, conceptual user interface design, and perceptual user interface design, which covers the presentation interface and the interaction language. ETAG-Based Design was inspired by the notion that user interface design is not so much concerned with creating software but rather with providing a particular view on a task domain. Given that the user's view of the task domain (the user's task world) and the system's view should correspond for optimal task performance, and that the two views are mediated by the knowledge that users (should) have about performing tasks, it seems reasonable to develop a method which represents user interface designs by means of modelling required user knowledge.

An important characteristic of ETAG-based design in comparison to other design methods, such as MUSE (Lim and Long, 1994), Paternò's Model-based approach (Paternò, 1999), or Usage Centered Design (Constantine and Lockwood, 1999), is that, as far as possible, it uses a single notation throughout the design process.

In ETAG-Based Design the ETAG notation is used to represent the analysis and design results. For this, it is necessary that the notation is flexibly adapted, beyond its original definition, to meet the specific purposes of all design stages. Originally, ETAG was intended only for user interface specification and not for representing the results of task analysis and task design. By altering the level of abstraction of the specification, the amount of detail, and the inclusion of special modelling concepts, the ETAG notation becomes useful for different purposes. For example, in modelling business procedures during task analysis, the representation is specified at a high level of abstraction without much detail, and special concepts are used to represent agency and ownership. In contrast, when the representation of the user interface is intended for conceptual design or for prototype generation, the representation includes more details at lower levels of abstraction and excludes special concepts.

Using a single notation throughout design facilitates the transitions from one design stage to the next. It also makes it easier to create tool support, which is particularly relevant for presenting the ETAG representation as a formal representation in a suitable way to all the stakeholders. A final advantage of using a single notation is that designers themselves are not required to learn and use a variety of different notations. In the approaches of Paternò (1999) and Constantine and Lockwood (1999), for example, designers must be able to deal with a variety of notations, in contrast to only one in ETAG-based design.

3. Design representation for diverse users

In ETAG-based design, differences between groups of users can be accommodated in two principally different ways, depending on whether users share common knowledge about the user's task world, and whether they share a common presentation of the user task world.

3.1 User task world knowledge

During task design, when the results of task analysis are synthesised into new task specifications, it may become clear that the design involves multiple users with rather different roles, tasks, and privileges in the work situation. In this case, different users have different knowledge requirements with respect to performing tasks and, consequently, it is necessary to create multiple ETAG representations of the UVM and the dictionary of basic tasks. Examples are, software to support business processes in which users have different responsibilities, and adaptive systems, which provide different task sets, depending on user characteristics, such as experience (Oppermann, 1994).

In the Comris example, conference administrators and ordinary users play a different role. Users and administrators use the same browser interface to provide the system with information and instructions, but administrators are also able to change the conference schedule or to create and distribute messages. In effect, the task dictionaries are very similar but the UVM and the type-specifications are considerably different.

The need to create more then version of the higher levels of the ETAG representation does not imply that separate design tracks are required or that the representations will be distinctly different. First, regardless of how many different types of users may be distinguished, they share the same work situation and need to have more knowledge in common than in separation. Secondly, at some point in the design process it may be practical to join the different representations, for example, to specify the implementation of the software, but, in principle, when users require different mental models, also different high-level ETAG representations are required.

3.2 User task world presentation

A different situation occurs when users need the same knowledge to perform their tasks but don't want to or cannot access the system in the same way. This situation occurs when different users or different work situations require that the user task world is presented in different ways, such as, for example, browsing the internet by means of mobile hardware, or using Braille or speech-output by blind users. In ETAG terms, the user virtual machine for browsing the web by means of a desktop PC or a web-enabled mobile phone are similar. In these cases, a single ETAG representation suffices but different presentation interfaces and production rules must be specified.

In the Comris system, users may choose between accessing advice messages by means of a small wearable, worn on the breast with a shoulder strap, and by means of an information kiosk, or rather, a concealed PC, which is also used to enter information and instructions to the agent system. With respect to browsing the user's message list, both devices provide the same information and similar functions: to start and stop playing a message, to start and stop displaying a message, to respond with Yes or No, to select the next or the previous message, to delete a message, and to increase or decrease the interest threshold

The Comris kiosk is supplied with a pointing device and a keyboard, whereas the wearable only has five buttons and a volume slider. The audio output is similar for both devices. For visual output, the kiosk is supplied with a PC screen, whereas the display area of the wearable device is limited to 2 lines of 16 character positions, also used to display labels for each of the buttons. Given the physical limitations of the wearable device, large part the dialogue and the presentation interface has to be laid-out in time whereas the kiosk interface can use a spatial layout.

The kiosk interface provides sufficient space to present the list of messages in textual form but the wearable device has to use either scrolling text or abbreviated versions of the messages. In a usability experiment (de Haan, 2000b)

both options were used: messages were presented by scrolling text when displaying a message, and presented in abbreviated form when selecting a next or previous message. The kiosk provides sufficient space to represent each function by a separate button. On the wearable there are more functions than there are buttons available, and alternative function-to-button mappings are required by means of dialogue steps or an additional mode function.

Four different interfaces were designed for the wearable device, each according to a general principle to map between functions and buttons, such that one interface used a control key and three others used a combination of modes and dialogue steps. The presentation and the dialogue interface are very different from those of the kiosk.

In ETAG, the user-system dialogue is represented in the production rules, which specify how basic tasks are translated into the physical actions to invoke them, along 4 levels of specification which describe, respectively: the command line syntax, the way of referring to commands (pointing, naming, etc.), the labelling of command elements, and the physical actions. The specifications of the wearable and the kiosk interface have in common that a single interaction style is used (pointing) and the physical actions are virtually identical (pressing buttons). Also, each button is labelled with the name of the command and, consequently, the lexical-level of production rules is superfluous. In effect, this means that all four levels of the production rules collapse into one level, describing the relation between a particular task and the associated action.

Specifying the kiosk interface is straightforward because there is no need for dialogue steps. Since, both, the list of textual messages and the list of function buttons fit one display screen, the presentation interface may be designed as a single web page and the dialogue interface only requires a single one production rules for all basic tasks:

T[task] ::= click_button[task]

Specifying the wearable interfaces is more complex because there is insufficient space for both messages and buttons. That messages are presented as scrolling text or in abbreviated form is relevant only to the presentation interface and not to the production rules because the presentation modes are associated with different functions. That functions are invoked by means of different dialogues is important to the design of the presentation interface because each state of the dialogue requires a description of the contents of the output window, the interpretation of each of the buttons, and the dynamic characteristics of the dialogue state. For example, upon invoking the function to present a message as scrolling text, the 'start' button may now operate as the 'stop' button, and after the message has finished scrolling, it may start again, or the dialogue may automatically change to a previous state or to some default starting point.

To design the presentation interface as a series of changes between window-definitions, a complete transition diagram of the dialogue is necessary. A rough sketch of the transition diagram results from using the entries in the dictionary of basic tasks as landmarks and from applying the function-to-button mapping principles. After this, the presentation design is a matter of filling in the details for each window and, whenever appropriate, adjusting the (e.g. automatic) transitions between dialogue states.

That functions are invoked by means of different dialogues may or may not be important to the specification of the production rules. For the purpose of designing the interface, the user-system dialogue should be completely specified, as described, from the dictionary of basic tasks, interaction principles, and some 'syntactic sugar'. However, to specify the interaction in terms of the user's competence knowledge, the production rule specification should not include elements of user control. The Comris wearable, as a highly interactive system in which the presentation of information by the system is interleaved with decision making by the user, shows the tension between ETAG as a flexible model for design representation and as a strict model of user competence knowledge.

When ETAG is applied very strictly and user control is not allowed within basic tasks then additional tasks should be added, such as "change_mode" or "select_cancel". For example, to specify that the user first has to select the proper mode, which may or may not already be the case, followed by selecting the intended task is simply treated as two different task invocations. In this case, the production rule specification of the wearable interface is similar to the specification of the kiosk interface. When ETAG is applied more loosely and some user control is allowed within basic tasks then composite production rules might be used. In this case, the example of selecting the proper mode and the intended task may be rewritten as a composite syntax-level production rule:

T[task] ::= click_button[task] ||
click_button[select_mode] + click_button[task]

With multiple modes, multiple dialogue steps, or both at the same time, as is the case in three of the Comris wearable prototypes, the composite rules become fairly complex. However, using the strict ETAG interpretation, the different states of the system should be represented by preconditions in the event specifications, and the UVM should become more complex. A decision between the two options regarding a strict or relaxed interpretation can be made on empirical and on theoretical grounds.

Empirically, a decision between production-rule and UVM complexity depends on whether users perceive a difference between functional and interaction complexity. The usability experiment reported in De Haan (2000b) tried to answer one side of the question, hypothesising that a user interface with many dialogue steps would hide conceptual information whereas an interface which avoids dialogue steps by using control key combinations would provide a better overview of the functionality of the system. The results show that the two prototypes at the extremes of the dimension are about equally effective as well as about equally preferred. Given that the prototypes were not build to test the hypothesis but rather to pick the best one, we tentatively interpret the result as an indication of a trade-off relation between dialogue consistency on the one hand, and system overview and dialogue guidance information on the other.

From a theoretical point of view, a decision for the strict interpretation allows that the perceptual interface influences how users interpret the workings of the system. For design purposes, this should be avoided because the conceptual design of the user interface is a different phase from the perceptual design, which should be independent of each other, as far as possible (de Haan, 2000a). For modelling purposes, it should also be avoided because dependencies between the conceptual and the presentation levels of a model make it more difficult to interpret and apply. As such, even though the implication is that ETAG would no longer be used as a pure model of user competence knowledge, the more relaxed interpretation of the model is to be preferred for both design and for modelling purposes.

4. Discussion

In comparison to other design methods ETAG-based design requires designers to learn and use only one notation and, as such, it is relatively easy to use and provides more opportunities to allocate design resources to other issues, such as user interface adaptivity and adaptability (Oppermann, 1994). With respect to the facilities which ETAG-based design provides to represent user interface designs to accommodate diverse users, ETAG is able to represent different knowledge requirements about task performance between users and it is able to represent different ways of accessing a system in terms of characteristics of hardware devices and of interaction styles.

There are a more strict and a more relaxed interpretation of the ETAG notation with respect to representing the structure of the user-system dialogue in different systems or design alternatives. For both, design as well as modelling purposes, the more relaxed interpretation must be preferred, even though, introducing performance elements into the model is at odds with the original specification of ETAG. However, to decide between the strict and the relaxed interpretation on empirical ground, further research is necessary about the organisation of the user's knowledge about interactive systems and how different knowledge structures influence task performance.

References
Constantine, L.L. and Lockwood, L.A.D (1999). Software for Use: A Practical Guide to the Models and Methods of Usage Centered design, Addison-Wesley.
de Haan, G. (1996). ETAG-Based Design: User Interface Design as User Mental Model Design. In P. Palanque & D. Benyon (Eds.), Critical Issues in User Interface Systems Engineering (pp. 81-92). Springer Verlag.
de Haan, G. (2000a). ETAG, a Formal Model of Competence Knowledge for User Interface Design. Doctoral Dissertation, Free University, Amsterdam (http://www.mmi.unimaas.nl/userswww/dehaan/thesis.html).
de Haan, G. (2000b). Interacting with a Personal Wearable Device. In P. Wright & S. Dekker & C.P. Warren (Eds.), Proceedings ECCE-10, HCI: Confronting Reality (pp. 2-15). August 21-23, Linköping, Sweden
Lim, K.Y., Long, J.B. (1994). The MUSE Method for Usability Engineering. Cambridge University Press.
Oppermann, R. (1994): Adaptively supported Adaptability. Int. Journal of Human-Computer Studies, 40, 455-472.
Paternò, F. (1999). Model-Based Design and Evaluation of Interactive Applications. Springer Verlag.
Tauber, M.J. (1988). On Mental Models and the User Interface. In G.C. van der Veer T.R.G. Green & J.M. Hoc & D.M. Murray (Eds.), Working with Computers: theory versus outcome (pp. 89-119). Academic Press.
Tauber, M.J. (1990). ETAG: Extended Task Action Grammar. In D. Diaper & D. Gilmore & G. Cockton & B. Shackel (Eds.), Proceedings INTERACT'90 (pp. 163-168). North-Holland.
van der Velde, W. (1997). Co-Habited Mixed Reality. In Proceedings of the Fifteenth International Joint Conference on Artificial Intelligence, Aichi, Japan, 23-29 August 1997.

A Visual Component Based Tool for Developing Embedded Application Software

Young-Ho Kim, Jin-Hyun Kim**, Ji-Young Kim***, Jee-In Kim*,
Jin-Young Choi**, Chang-Woo Pyo****

* Computer Science & Engineering, Konkuk University
** Computer Science, Korea University
*** Computer Engineering, Hongik University, Seoul, KOREA

Abstract

We propose a development tool for embedded application software. The main issues are: (1) *visual design*: The engineers can visually design the embedded application software by drawing digital logic circuit diagrams. (2) *design verification*: The engineers can verify the correctness of the design and remove errors at the early stage of the development process. (3) *code generation*: The C code generator automatically translates the visual design into computer programs running on the target system. (4) *simulation*: The generated C program is simulated so that the program can be checked before the target system hardware is ready.

1. Introduction

In general, an embedded system consists of hardware, a real-time operating system (RTOS), and embedded application software. A typical process of developing an embedded system is as follows: (1) The experienced engineers in the application fields such as electrical engineering design the target system. (2) The computer programmers understand their design and write computer programs in programming languages such as C. The software is usually developed under an independent computer system running Unix or Windows. There are number of development tools available such as a cross-compiler, a debugger, etc. (3) When the hardware of the target system is available, the application software can be run and tested on the target system.

We address some issues regarding to such a development process. (1) The engineers do not necessarily have programming skills. On the other hand, the computer programmers do not usually have the expertise of the application field. Therefore, it is not easy for the non-programming engineers and the programmers to communicate each other during the development process. (2) The programmers could be a major source of the system errors, because they manually translate the design into the computer programs. Even though the design is correct, the program could be wrong. (3) The debugging process of the system could be delayed, because the embedded software cannot be executed and tested until the hardware of the target system is available. Once we have the hardware, it is expensive to correct the errors and/or modify the system design. (4) The reliability of the software is not formally guaranteed, because any kind of testing or verification process is not explicitly defined during the development process.

2. Overview

We propose a visual component based tool for developing embedded application software used for power control systems. The electrical engineers use a graphical editor to develop the embedded software as if they design and test digital logic circuits. The engineers can verify the correctness of their design using verification tools. Then, the design is automatically translated into C programs. A simulator is used to examine the behaviors of the generated program.

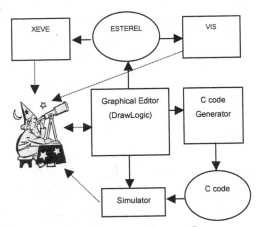

Figure 1. Software Development Process

The system development tool consists of the following tools:

(1) **a graphical editor** (called "DrawLogic" as shown in Figure 1) that helps the non-programming engineers in designing embedded application software. When the electrical engineers develop an embedded application software of a power control system, for instance, they draw the design of the embedded software using the standard graphical symbols such as "AND" gates, "OR" gates and lines in digital logic circuits.

(2) **formal verification tools** that are used to verify the correctness of the system design. Since the visual design is not "verifiable" yet, we must translate the design into a "verifiable" form. A formal language, ESTEREL[1], is used. We use two tools as shown in Figure 1. XEVE[2,3] is an ESTEREL verification tool and VIS[4] is used to verify the visual design using the model checking [5] method.

(3) **an automatic C code generator** that automatically translates the design to C programs. An optimization is also applied so that the generated C code can be efficiently executed on the target system.

(4) **Simulator** that simulates the behaviors of the C code. It aims to check if the generated C program works properly. The simulator graphically displays the behaviors in terms of its input and output values.

3. Related Works

There are number of tools for developing embedded application software: WindRiver's Tornado[6,7,8], Green Hill Software's MULTI[9], Integrated Systems' pRISM+[10], etc. Tornado supports real-time operating systems such as VxWorks. It provides shell, compiler, debugger, etc.

We adopt the basic concepts of the existing tools and extend them by applying the software engineering principles. The graphical editor is used to enhance the usability of the tool so that the non-programming engineers can develop the embedded software by drawing digital logic circuits. The formal methods are used to verify the correctness of the design. The design is automatically translated into C programs to minimize the human errors during the translation. The simulator is used to check if the generated C program properly works. The errors of the system could be removed at the early stage of the development process, even if its hardware is not available..

4. Visual Design

The graphical editor is called "DrawLogic". As an example, a half-adder is visually designed using the editor. The visual design is written in terms of digital logic circuit symbols as shown in Figure 2.

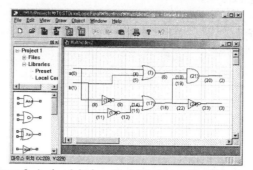

Figure 2. A visual design of a half-adder using DrawLogic

The circuit design is expressed in terms of **components (nodes)** and **edges** connecting them. The engineers design the software using icons of logic gates ("AND", "OR", and "NOT" gate symbols in Figure 2). The icons are selected from the "palette" of the editor. Then, the icons are put on the "canvas" of the editor. The icons represent the **primitive components** such as ports, gates, and the **complex components** made of other components. Each component design shows inputs, outputs, nodes, and edges. The nodes denote components used as parts of the component being defined. The nodes are connected by the edges to express the component connection.

The engineer can define a new component by combining existing components. They can assign an icon that represents the component. The engineers can use "predefined" components from libraries. A **library** contains a set of components. DrawLogic has many libraries and its **library manager** maintains them.

5. Design Verification

The system design should be verified, because its correctness is important and the errors can be inexpensively corrected at the design stage. For its formal verification, the design is translated into a formal language, ESTEREL, that can express a target system as a finite automaton. XEVE is an ESTEREL verification tool. VIS is used to perform the Model Checking method that is an automatic technique for verifying finite state systems.

Figure 3. Design verification using XEVE

First, we express the requirements of the target system using concurrent temporal logic (CTL) and use XEVE to verify if the design satisfies them. XEVE concurrently executes the requirements and the designed circuit. We examine if the designed circuit enters in any illegal state by observing whether the output signals are illegal. For instance, a requirement can be expressed in CTL as follows:

```
AG(!(osPSAB=1)->!(osOST=1))
```

The requirement stipulates that the output signal ("osOST") should never occur, unless the input signal ("osPSAB") occurs in the system. Figure 3 shows the verification process using XEVE. The result shows that if the "osPSAB" signal occurs in the system, then the output signal, "osOST", does not occur. So the requirement is satisfied.

Model checking has been successfully used to verify complex sequential circuit designs and communication protocol. In this method, the requirements are specified in CTL. Once all reachable states of the system are automatically generated, they are compared with the requirements written in CTL. VIS is a tool for model checking. To use VIS for the verification, the design must be specified with data paths. Then the values of the output signals are observed with the given input signals. The following is the results from VIS when the sample requirement is examined.

#MC: formula passed :
AG(!(osPSAB=1) -> !(osOST=1))

It means that if the value of osPSAB is not "1", the value of osOST cannot be "1" in the designed circuit. Thus the requirement is satisfied.

6. Automatic C Code Generation

The automatic C code generator translates the design into C programs. First, the design is translated into abstract syntax tree (AST). AST is transformed into graph data structure showing the connections among hardware components. An optimization procedure is applied to the graph. An edge simply connecting two ports would be translated into a copy statement, which copies a value from a variable to another. The copy statement contributes to nothing but degradation of the target system performance. The optimization eliminates the copying edges. As the effect of the optimization, unnecessary copying assignments from a variable to another disappear from the generated C code.

Following the optimization, sub-graphs, called **tiles**[11], are identified as units for the code generation. The tiles are discovered by comparing sub-graph with the graph patterns. There is a code template for each graph pattern. The code template shows how to produce a C code fragment matching the graph pattern. Every node should belong to one and only one tile. Each tile is translated into a sequence of assignment statements, which is attached to the tile.

The last phase of code generation visits the tiles in topological sorting order collecting the code fragments in the tiles. A proper function header and a return statement are added to the collected code sequences producing complete definition of functions in C.

The sample visual design of a half adder in Figure 2 is translated into its equivalent C code as follows:

```
<half_adder.h>
/* half_adder definition   */
typedef struct {
  unsigned int a : 1;
  unsigned int b : 1;
} half_adder_IN;
typedef struct {
  unsigned int S : 1;
  unsigned int C : 1;
} half_adder_OUT;
half_adder_OUT half_adder(half_adder_IN);

<half_adder.c>
#include "gate.h"
#include "half_adder.h"
half_adder_OUT half_adder(half_adder_IN arg) {
  OR2_OUT OR2_;
  NOT_OUT NOT_;
      …
  AND2_out_21 = AND2(AND2_in_21);
  result.S = AND2_out_21.out;
  return result;
}
```

7. Conclusions

A development tool of embedded application software is proposed. The methodologies of software engineering are applied to the development of embedded application software. They are visual design, design verification, automatic C code generation and simulation. We have implemented and integrated the graphical editor, the code generator, the simulator and the verification tools. A number of application software for embedded systems have been designed and their correctness have been verified using the tool. The corresponding C programs have been generated. The simulator graphically shows the behaviors of the generated programs.

It is expected that the tool would make the software development process more flexible. The reliability of the embedded software would be increased with the design verification. It follows that the productivity of the engineers would be enhanced.

Acknowledgements
This research is funded by Han-Woo Tech Co., Anyang, Korea.

References

1. Gerard Berry. *The Esterel v3 Language Primer Version 5.21.* Centre de Mathematiques Appliqueses, April 6, 1999.
2. Alain Girault and Geread Berry. *Circuit Generation for Verification of ESTEREL Programs.* Inria in France, 1997.
3. Amar Bouali. *XEVE: an Esterel Verification Environment.* Inria in France, 1997.
4. Adnan Aziz, Robert K.Rrayton, Gary D Hachtel, et.al. *VIS User's Manual.* University of California, Berkeley.
5. Edmund M. Clake, Orna Grumberg, and Doron A. Peled. *Model Checking.* MIT press, 1999.
6. WindRiver Systems. Tornado. http://www.wrs.com.
7. WindRiver Systems. Tornado API Guide. http://www.wrs.com
8. WindRiver Systems. Tornado user's Guide. http://www.wrs.com
9. Green Hills Software. MULTI. http://www.ghs.com.
10. Integrated Systems. pRISM+. http://www.isi.com.
11. A. Aho, M. Ganapathi, and S. Tjiang. Code generation using tree matching and dynamic programming, *ACM Transactions on Programming Languages and Systems,* 11(4), 491-516, 1989.

Managing Accessible User Interfaces of Multi-Vendor Components under the ULYSSES Framework for Interpersonal Communication Applications

Georgios Kouroupetroglou, Alexandros Pino, Constantinos Viglas

University of Athens, Department of Informatics and Telecommunications,
Panepistimioupolis, Ilissia, GR-15784, Athens, Greece, e-mail: koupe@di.uoa.gr

Abstract

In this paper we present a novel approach, under the ULYSSES framework, to cope with accessible user interface management issues that arouse during the design, installation and maintenance of independently developed prefabricated software parts derived from multiple vendors and each one having its own GUI. ULYSSES constitutes a software engineering framework that facilitates Component Based Development for the domain of Computer Mediated Interpersonal Communication systems. Technical guidelines are provided for both component developers and system integrators.

1. Introduction

In the emerging convergence of voice and data communication technology, Computer Mediated Interpersonal Communication (CMIC) (or its equivalent term *e-communication*) launches an important societal role for all citizens. In e-communication either voice or text is commonly used to achieve synchronous or asynchronous (e.g, messaging, mailing) communication between two or more individuals. In some cases an alternative symbolic communication system (such as DLISS, PCS and REBUS) can be also utilized (Von Tetzchner, 1992). Of special interest is the case where in a communication session the two partners apply different meaning representation chosen among voice, text and symbols. Interpersonal communication is traditionally referred in the context of assistive technology. Recently however, general solutions have been proposed allowing communication between able-bodied and the disabled (Kouroupetroglou et al 1997, Viglas, Stamatis & Kouroupetroglou, 1998).

Universal Access (UA) according to Stephanidis et al (1998) entails the development of systems that can be used effectively, efficiently, and enjoyably by all users. Accessibility refers to two things (Benyon, Crerar & Wilkinson, 2001): a) physical access to equipment (in sufficient quantity, in appropriate places, at convenient times), and b) the operational suitability of both hardware and software for any potential user (even effortful participation qualifies as minimal accessibility). Attainment of UA is crucial for the design and the implementation of interactive CMIC systems (Emiliani et al 1996, Kouroupetroglou, et al, 1995). Consequently it is imperative that we architect CMIC applications for maximum adaptability to changing requirements for accommodating both the users' differences as well as the user characteristics changing with time.

Typically a Component Based Development (CBD) platform prescribes requirements that must be satisfied by component interfaces, and it provides facilities that support communication and coordination among those components (Allen, Garlan & Ivers, 1998). Furthermore, CBD has better suitability for a Design for All in HCI, compared to development methods closer to the physical level of interaction (e.g. presentation-based, physical task based and demonstration based) (Stephanidis et al 2001).

In this paper we present a novel approach, under the ULYSSES framework, to cope with user interface management issues that arouse during the design, installation and maintenance of pre-fabricated components derived from multiple vendors. ULYSSES constitutes a software engineering technical environment that facilitates CBD development for the domain of CMIC (Pino & Kouroupetroglou, 2000). According to ULYSSES, each component can be developed independently, complying with specific technical guidelines and requirements and incorporates its own specific Graphical User Interface (GUI). Thus the system integrator has to efficiently arrange, resize, fine-place and fixate the various GUIs of different components in order to constitute a single GUI for a specific product.

2. The ULYSSES framework

Component Based Development (CBD) consists of a recent software engineering approach that promises to fulfill the emerging need to produce flexible, scalable, adaptable and customizable cost effective applications on time.

Components do not exist in isolation, but within an architecture that provides context across a technical domain or a business. First steps were taken already in constructing internet-based CBD infrastructures that allow teams around the world to collaborate on every phase in the life cycle of a global software product (Gao, et al 1999). Componentization impacts the entire life cycle of an application, including the activities of analysis, design, acquisition, build, assembly, deployment, adaptation and replacement. A CBD framework is a software engineering technical environment facilitating the construction and delivering of domain specific applications out of independently-developed prefabricated parts. CBD frameworks are becoming popular (Kobryn, 2000), (D'Souza & Wills, 1999).

ULYSSES is a framework we developed (Pino & Kouroupetroglou, 2000), which offers a CBD infrastructure in the domain of CMIC. It is based on existing standards, domain engineering, looking for commonalties and variability of possible products covering a wide range of user groups including the disabled. Proprietary software architectures for CMIC systems have been already proposed (Kouroupetroglou et al, 1996, Viglas et al 1998).

ULYSSES components are likely to be self sufficient and independent of each other. They are polymorphic entities of different scale developed with different software technologies by independent vendors. ULYSSES adopts a three-layer architecture scheme: component (or construction), services and application (or integration). Components have their own intrinsic architecture and offer services that form the service architecture. The services can be reused as high-level concepts and integrated into the overall application architecture purely by knowing about the interface. The sequencing and use of the services can be easily modified. Likewise the use of the services is independent of the implementation and can be replaced with a new component that offers new capabilities or uses new technology.

CMIC applications under ULYSSES may incorporate various kinds of components: a) with their own visible GUI, like an on-screen keyboard or a message editor, b) with a speech user interface (SUI), like the output of a text-to-speech system, c) with a combination of GUI/SUI, and d) without any particular UI, like a natural language syntactic parser. The COM+ model and its event services have been adopted in ULYSSES as the mechanism for connecting components.

For component developers ULYSSES provides an engineering-for-reuse environment with guidelines and tools to build software components, which can operate effectively and interact with each other transparently, without even being aware of each other's existence. This facilitates the convenient production of reusable components for a family of products rather than single monolithic systems. Therefore, components with different functionality, developed by independent manufacturers and with various programming languages can be deployed. Furthermore, ULYSSES grants a process of engineering-with-reuse for system integrators to take advantage of the reusable assets produced during engineering-for-reuse. Integration standards in ULYSSES are typically specified using a combination of informal and semi-formal documentation. On the informal side guidelines and high level descriptions of usage patterns, tips and examples are included. On the semi-formal side it provides a description of an application programmers' interface (API) that explains what kind of services are provided by the infrastructure. APIs are formal to the extent that they provide precise descriptions of those services-usually as a set of signatures, annotated with pre- and post- conditions. ULYSSES supports four types of components: 1) specific to a CMIC product or family line, 2) CMIC domain specific components, and 3) domain-independent components (reusable across domains) and 4) technical infrastructure components (standalone components). The last two cases are also referred as components of an open market.

3. User Interface Management under ULYSSES

An integrator under ULYSSES has to provide a single UI of the final CMIC system by combining several independently-developed prefabricated components, each one having its own UI. An additional constrain is that components have been compiled separately, and the integrator has not the opportunity to change their code. To relief the end-user from the burden of arranging and dealing with multiple UIs, ULYSSES enacted guidelines to accomplish the following general specifications: a) *All Graphical User Interface (GUI) elements belonging to the CMIC application should be configured and arranged by the integrator after composing the application from multiple components. b) The application integrator must be able to manage the sizes and positions of all windows and GUI elements selecting to give the look of a single-window or a multiple-window application.*

Component **developers** should implement the above specifications complying with the following guidelines:

Each GUI of a component must have two function modes: the normal "run" mode (Figure1.a) and the "configuration" mode (Figure 1.b).

- End users cannot be able to resize, move or reposition any of the communication aid windows.
- When needed, the fact that the application consists of multiple windows must be completely transparent to end-users.
- Visible windows during the "normal run" mode must not have borders, control boxes, title bars and standard minimize, maximize and close buttons. Windows lacking of these characteristics, give the impression of a single window or user interface when they are placed close to each other with zero distance between them.

Figure 1: Component's GUI in a) "run" and b) "configuration" mode

- End users must have access and be aware only of the normal run mode in which window sizes and positions are fixed.
- The appearance of these windows must be centrally controlled by the operating system settings for foreground and background colors and font sizes.
- The configuration mode must be available only to system integrators and during this mode the windows must have "Sizeable ToolWindow" border, title bar and "close" button.
- During configuration mode the integrator must be able to resize and move the windows, while all windows elements and controls are resized and moved accordingly.
- Configuration mode must be activated with a preassigned hotkey and can be protected with a password known to integrators only.
- Pressing the "close" button during the configuration-mode will cause the application to return to the normal run mode and all the settings for window sizes and positions to be saved in configuration files or in the system registry in order to be available in future component activations.

Careful design of all user interface elements and some extra code is needed during the implementation. Each programming language may need different effort for realizing the required characteristics for the user interface. For example, in Visual Basic extra code and calculations are needed to make all window controls and elements resizable relatively to the parent window size, while in Java this can be done automatically. In Visual C++ a form can change its border from fixed to resizable ToolWindow at run-time, while in Visual Basic there is need for double forms (one fixed with no border and another resizable with ToolWindow border).

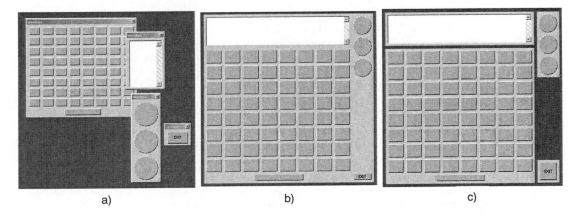

a)   b)   c)

Figure 2: Three phases of the UI during the integration process

In the UI initialization phase of the CMIC application during the integration process, system **integrators** must manage several separate UIs belonging to multiple software components. At this phase all components have been installed and the application is ready to run. During the first application start-up, all components operate in normal run mode and their UIs are shown on the integrator's screen using default size and positions according to the settings each developer used during the implementation of the component (Figure 2.a). There are specific steps that integrators must follow to set the application's final and unified user interface into an accessible and consistent format according to the available space on the screen and the number of visible windows:

1. Firstly, all windows must be set to window configuration mode. To do this the system integrator must give focus to the ULYSSES window and press a predefined hotkey. The ULYSSES window is a primitive GUI of the ULYSSES basic component. It consists of a simple window with only one button: "Exit". Pressing the hot key a message will be sent from the COM+ event system to all components with visible UIs causing them to enter the configuration mode.

2. All GUIs must be arranged on the screen in a way that can be visible and not overlapping or hiding each other.

3. The GUIs must be resized according to the user needs and the preferred layout on the screen. During this process the integrator must consider the consistence of font and control elements sizes between different components. Command buttons, for example, can have different sizes according to their functionality or the size of their parent window, but the integrator must ensure their homomorphy in order to avoid an odd and inconsistent appearance.

4. Positioning of the GUIs comes next. The communication aid integrator must place all visible UIs in suitable positions and in the right distances according to their functionality to achieve the final look preferred for the complete application. For example, if there are reasons for pointing out the difference between the various functional components of the application, this can be done by increasing the distances between their user interfaces (Figure 2.b). If there is need for the incorporated user interfaces to give the impression of a single unified user interface, they must be placed in a way that there is no visible space between them. At this point some fine placement and special attention may be needed to achieve with detail the final result. Keeping consistency and homomorphy in mind, the combination of all windows must present a robust user interface (Figure 2.c).

5. After the final GUI sizes and positions have been achieved, all settings must be saved and fixed. The integrator must press the "Close" button on the upper-right corner of each GUI in configuration mode. This will force all windows to turn to normal run mode and fix their sizes and positions, which will be "locked" for the end user. In all consequent executions of the CMIC application the multiple GUIs will retain their settings and only the integrator can reactivate the whole set up process.

Any specific accessibility option (such as scanning techniques) implemented either by the ULYSSES framework or by a dedicated component, according to the guidelines set by ULYSSES, will be inherited to the integrated application.

4. Conclusions

This paper has described an approach, under the ULYSSES CBD framework for the domain of CMIC, to cope with accessible user interface management issues that arouse during the design, installation and maintenance process.

As far as GUIs management is concerned, another known system is FRESCO, which is an OO user-interface system for developing windows-based applications supporting distributed component-based UIs (FRESCO, 1996). It includes, among others, a layout kit, which provides operations to manage the arrangements of GUI elements. The basic difference compared to the work described in this paper is that FRESCO is a CORBA based CBD environment for embedding and distributing graphic objects, while our approach is a COM+ framework facilitating the integration of independently developed prefabricated software parts derived from multiple vendors and each one having its own GUI.

Based on the guidelines and the tools of the ULYSSES framework, a number of user interfaces addressing both disabled and non-disabled users were built, under the AENEAS Project, on MS Windows 2000 platform, utilizing various accessibility options as well as different input/output devices and interaction elements. Multiple combinations of components implementing various functionalities and user interfaces for interpersonal communication applications only revealed the power and flexibility of the framework.

Acknowledgements

Part of the work reported in this paper was carried out within the framework of: a) the M-PIRO project (contract IST-1999-10982), funded by the IST Programme of the European Union and b) the AENEAS project (contract 98AMEA19), funded by the EPET II Programme of the Greek General Secretariat of Research and Technology.

References

Allen, R., Garlan, D., Ivers, J. (1998). Formal Modeling and Analysis of the HLA Component Integration Standard. *Software Engineering Notes*, 23(6), 70-79.

Benyon, D., Crerar, A., Wilkinson, S. (2001). Individual Differences and Inclusive Design. in Stephanides, C. (Ed) *User Interfaces for All*, (pp. 21-46). London: Lawrence Erlbaum Ass.

D'Souza, D., Wills, A.C. (1999). Objects, Components and Frameworks with UML: The Catalysis Approach. Addison-Wesley, Reading, MA.

Emiliani, P.-L., Ekberg, J., Kouroupetroglou, G., Petrie, H., & Stephanidis, C. (1996). Development Platform for Unified Access to Enabling Environments, in Klaus, J., Auff, E., Kresmer, W. & Zagler, W. (Eds) *Interdisciplinary Aspects on Computers Helping People with Special Needs*, Oldenbourg, Munhen and Wien, Proc. of ICCHP'96, July 17-19, Linz, pp. 69-75

FRESCO (1996) http://www.iuk.tu-harburg.de/Fresco/HomePage.html

Gao, J., Toyoshima, C.C., Leung, D. (1999). Engineering on the Internet for Global Software Production, ACM Computer, 32(5), 38-47.

Kobryn, C. (2000). Modeling Components and Frameworks with UML. Communications of ACM, 43(10), 31-38.

Kouroupetroglou, G., Viglas, C., Stamatis, C., Pentaris, F. (1997). Towards the Next Generation of Computer-based Interpersonal Communication Aids. in Advancement of Assistive Technology, Edit. G.Anogianakis, C. Buhler and M. Soede,. 3, IOS Press, p. 110-114, AAATE 97, Proceedings of the 4th European Conference for the Advancement of Assistive Technology, 29 Sep. - 2 Oct. 1997, Porto Carras

Kouroupetroglou, G., Viglas, C., Anagnostopoulos, A., Stamatis, C., Pentaris, F. (1996) A Novel Software Architecture for Computer-based Interpersonal Communication Aids. in Klaus, J., Auff, E., Kresmer, W. & Zagler, W. (Eds) *Interdisciplinary Aspects on Computers Helping People with Special Needs*, Oldenbourg, Munhen and Wien, Proc. of ICCHP'96, July 17-19, Linz, 715-720

Kouroupetroglou, G., Paramythis, A., Koumpis, A., Viglas, C., Anagnostopoulos, A., Frangouli, H. (1995). Design of Interpersonal Communication Systems Based on a Unified User Interface Platform and a Modular Architecture, Proc. TIDE Workshop on User Interface Design for Communication Systems, Brussels, July 7, 1995, 8-17

Pino, A., Kouroupetroglou, G. (2000). The ULYSSES Component Based Development Framework. Tech. Rep. 3.2. of AENEAS Project, Athens, Greece.

Stephanidis, C., Salvendy, G., Akoumnianakis, D., Bevan, N., Brewer, J., Emiliani, P.-L., Galetsas, A., Haataja. S., Iakovidis., I., Jacko, J., Jenkins, P., Karshmer, A., Korn, P., Marcus, A., Murphy, H., Stary, C., Vanderheiden, G., Weber, G., Ziegler, J. (1998). Toward an information society for all: An international R&D agenda. International Journal of Human-Computer Interaction, 10(2), 107-134.

Stephanidis, C., Kouroupetroglou, G. (1994). Human Machine Interface Technology and Interpersonal Communication Aids. Proc. of the Fifth COST 219 Conference: *Trends in Technologies for Disabled and Elderly People*, Tregastel, France: Gummerus Printing, 165-172.

Stephanidis, C., Akoumnianakis., D., Vernadakis, N., Emiliani, P-L., Vanderhheiden, G., Ekberg., J., Ziegler, J., Faehnrich, K.P., Galetsas, A., Haatajua, S., Iakovidis, I., Kamppainen, E., Jenkins, P., Korn, P. Maybury, M., Murphy, H., Ueda.H. (2001). Industrial Policy Issues. In Stephanidis, C. (Ed) *User Interfaces for All – concepts, methods and tools*, (pp. 589-608). London: Lawrence Erlbaum Associates.

Viglas, C., Stamatis, C., Kouroupetroglou, G. (1998) Remote Assistive Interpersonal Communication Exploiting Component Based Development, Proceedings of the XV IFIP World Computer Congress, 31 August - 4 Sept. 1998, Vienna – Budapest, Congress: *Computers and Assistive Technology*, ICCHP'98, pp. 487-496.

Von Tetzchner, S.(1992). Use of Graphic Communication Systems in Telecommunication. In Von Tetzchner, S. (Ed.). *Issues in Telecommunication and Disability*. CEC, DG XIII, Luxembourg, 280-288.

Mechanization of web design guidelines evaluation

Céline Mariage, Jean Vanderdonckt

Université catholique de Louvain – Institut d'Administration et de Gestion
Place des Doyens, 1 - B-1348 Louvain-la-Neuve, Belgium
{mariage,vanderdonckt}@qant.ucl.ac.be – http://www.qant.ucl.ac.be/membres/jv

Abstract
The systematic evaluation of Web design guidelines poses a series of challenges, including the ability to accurately interpret them and to transform them into a mechanized way that is machine processable. Systematic or automatic evaluation is particularly useful when the web site to evaluate across the designated guidelines is a large-scale, information-rich, potentially distributed web site (very-large web site). This paper attempts to analyze various ways to mechanize the evaluation of some design guidelines towards automated evaluation and identifies some shortcomings.

1. Introduction

Many guidelines have been published in document sources throughout the literature. These sources fall into five categories (Vanderdonckt, 1999):

1. *Design rules*: they comprise a set of functional and/or operational specifications that specify the design of a particular user interface. These specifications are presented in a form that requires no further interpretation, either from designers or from developers. Their straightforward format allows an immediate exploitation.
2. *Compilations of guidelines*: they comprise several prescriptions written for a wide range of user interfaces. Each prescription is presented as a statement, sometimes along with examples, with or without clarifying explanations and comments. Each prescription generally results from a human consensus between guideline users. This consensus is less relevant once a prescription is experimentally tested and verified. They can range from a small set of guidelines dedicated to a particular usability feature (e.g., interaction technique) to an extensive collection of guidelines covering a family of tasks and domains. For instance, guidelines for multimedia, interactive kiosks, accessibility (Constantine, 2001), for web site in general (Detweiler & Omanson, 1996). Some of these are validated by experimental results provided by user testing, laboratory experiences or others.
3. *Style guides*: they comprise a set of guidelines and/or functional or non-functional specifications aiming at consistency for a collection of distinct user interfaces. This collection can be based on an operating system, on a software editor, by a particular physical environment such as SUN (Levine, 1996), by a domain of human activity or by a corporate.
4. *Standards*: they comprise a set of functional and/or operational specifications intended to standardize design. Standards are promulgated by national or international organizations for standardization. They can be military, governmental, civil or industrial. There is no web dedicated standard we are aware of, but some important general standards are good candidates to tailor their guidelines to suit the exact issues faced by web sites, e.g., ISO 9241, HFES/ANSI 200.
5. *Ergonomic algorithms*: these aim to systematize one design aspect by building it in the same way, according to a same base of design rules. They appear as a software component that implements an algorithm rather than a paper procedure. Such algorithms are primarily intended to design a series of web pages that automatically respect some guideline (usually, design rules or style guides) by construction. In this way, they immediately enforce the address of guidelines without the need to take care of them. Of course, cascading style sheets or page templates can be resourceful, provided that their quality intrinsically mirrors the application of guidelines, which is not necessarily the case. For example, DON (Kim & Foley, 1993) is capable of automatically laying out widgets to minimize the screen space, while preserving label and fields alignments to fasten the reading path. We observed that we have not yet seen a true software providing designers with some assistance in designing usable pages, but they may come.

191

2. Some examples of automated testing of usability guidelines

2.1 Design rules

Let us imagine two design rules for the inclusion of a particular graphic: "Each personal home page should contain the company logo at the top left corner" and "Each personal home page should contain the photo of the person after the logo". The second can be formalized similarly. Capture-and-Replay testing tools represent some first attempt to automate compliance with design rules by performing four steps:

1. *Capturing*: the tool captures static properties of objects in page (e.g., color, size) as well as dynamic properties (e.g., activation, deactivation).
2. *Programming*: the properties are stored in a test script to be programmed with traditional algorithmic structures.
3. *Inserting checkpoints*: checkpoints regarding properties are inserted into the script.
4. *Replaying*: the test script is run on any page to be evaluated and any discrepancy with respect to reference properties and checkpoints is detected.

Several design rules can be tested this way, such as "Ok and Cancel are always located at the bottom or on the right, but never on top or on the left", "Submit and Reset push buttons in forms, if any, should be presented in this order". Fig. 1 reproduces a screen capture where the two above design rules have been captured and subsequently replayed with Watchfire's MacroBot (http://www.watchfire.com/products/macrobot). It basically consists of two test cases testing the presence of the graphics at their right location, plus the general composition. In this case, the two guidelines are not respected. Capture-and-Replay tools are dependent on the very specific parameters of a guideline. In the recording process, it recorded the index position of an image the user pointed to in the images array rather than more uniquely identifying information such as its source. Adding or removing a single image to the page would completely disrupt the test. Direct creation of JavaScript representations of guidelines instead of macro recording could help avoid some of this, but the definition of guidelines in this tool is highly dependent on a document's particular structure. A more generic way of expressing guidelines that depends, for example, on an object's resultant position on the screen rather than its position in the document structure could be expected.

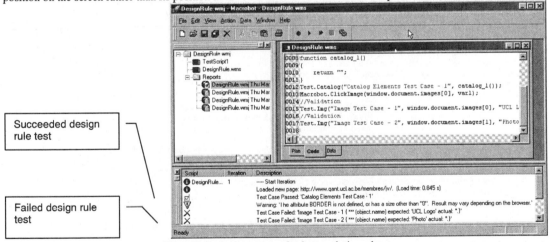

Figure 1. Replaying a test script for two design rules.

2.2 A guideline from a compilation

Let us consider now a guideline extracted from a guidelines compilation: "Select colors that will make your page easy to read by people with color blindness". One good test is to see if your page is readable in back and white. This guideline is a typical one that can be used in requirements engineering, in site specification, and in site evaluation. This guideline, as it is expressed, cannot be automated in a straightforward manner. However, if we refer to the research conducted by Murch as reported in (Scapin et al. 2000), a restriction of this guideline can be expressed as: the combination between the background color and the foreground color should belong to the best color combination or should not belong to the worst color combinations. This research contains figures that represent the best color combinations for thin lines and text. It can be read as follows for example: for thin lines and text displayed on a white background, the best color is blue in more or less 94% of cases, then black in 63% of cases, and finally red with 25% of cases (Murch's research is based on legibility tests). For thin lines and text displayed on a black background, the best color is white in 75% of cases followed by yellow in 63% of cases. The other lines can be interpreted similarly. For bold lines and panels displayed on a white background, the best color is black in more or less 69% of cases, then blue in 63% of cases, and finally red with 31% of cases. We consequently observe that usable color combinations also depend of the text style or area. While identifying bold text is easy thanks to the <B> bold text </B> tag, identifying bold lines and panels remains more challenging.

Now, having these experimental results in mind, we could more easily think how to automate the testing of this guideline. The different colors used in the HTML code should be identified for this purpose. Color is rendered on computer-based screens as an additive mixture of three primary colors: red, green, and blue, the intensity of which is indicated by a value ranging from 0 to 255. Thus, in the Red-Green-Blue (RGB) model, each color consists in a triple (red-intensity, green-intensity, blue-intensity). This triple is represented in HTML as a hexadecimal code of the type #XXYYZZ, where XX, YY, and ZZ represents the different intensities respectively. Therefore, these codes need to be transformed into decimal for comparison. When these values belong to the set {00,33,66,99,CC, FF}, the resulting colors are said to be principal as no other palette than the basic palette should be loaded to display the colors. The bgcolor and color HTML tags specify the foreground and the background text colors respectively. The resulting procedure for this automated testing is the following:

```
ListOfBestColors1  := {see figure 11}; /* Definition of best color combinations for thin lines and text */
ListOfBestColors2  := {see figure 12}; /* Definition of best color combinations for bold lines and panels */
ListOfWorstColors1 := {see figure 13}; /* Definition of worst color combinations for thin lines and text */
ListOfWorstColors2 := {see figure 14}; /* Definition of worst color combinations for bold lines and panels */
ListOfPages = {URL1, URL2,..., URLn} /* List of the URLs of the web pages to be tested */
for i=1 to # (ListOfPages)              /* for each web page to be tested */
  ListOfColorComb := DetectPresence ( ListOfPages(i), ColorCombinations )
    /* Detect the presence of multiple color combinations if any; for example, a same color background can be
       used, while the foreground color may be varying within a same page. Each element of the list is
       characterized by a triple (fgbolor, bgcolor, bold) */
  if ListOfColorComb <> ∅ then          /* if there is at least one color combination detected in the page */
    for k=1 to #(ListOfColorComb)
      CurrentComb.fgcolor := HEX2DEC (ListOfColorComb (k).fgcolor);
       /* Transform the hexadecimal code of the foreground color into decimal */
      CurrentComb.bgcolor := HEX2DEC (ListOfColorComb (k).bgcolor);
       /* Same for background color */
      CurrentComb.bold := ListOfColorComb (k).bold;
       /* Specifies if the bold tag has been detected at the same time */
      if CurrentComb.bold then
        if CurrentComb in ListOfWorstColors2 then /* if the current combination belongs to the worst list... */
          AddUsabPblm (ListOfProblems (ListOfPages (i)), ColorCombPblm, 1, CurrentComb)
            /* Add to the list of usability problems detected for the i-th page the problem identified by
               ColorCombPblm, which is a reserved constant indicating the type of problem, with the highest
               severity level (1) for the CurrentComb color combination */
        elseif CurrentComb not in ListOfBestColors2 then /* If the combination does not belong to any lists */
          AddUsabPblm (ListOfProblems (ListOfPages (i)), ColorCombPblm, 3, CurrentComb)
            /* Add to the list of usability problems detected for the i-th page the problem identified by
               ColorCombPblm, which is a reserved constant indicating the type of problem, with the moderate
               severity level (3) for the CurrentComb color combination */
        endif
      else
        if CurrentComb in ListOfWorstColors1 then /* if the current combination belongs to the worst list... */
          AddUsabPblm (ListOfProblems (ListOfPages (i)), ColorCombPblm, 1, CurrentComb)
        elseif CurrentComb not in ListOfBestColors1 then /* If the combination does not belong to any lists */
          AddUsabPblm (ListOfProblems (ListOfPages (i)), ColorCombPblm, 3, CurrentComb)
        endif
      endif
    endfor
  endif
endfor
```

This is a basic procedure which does not support the evaluation of hue, saturation and lightness: each composite color is reduced to its corresponding primary color, thus introducing a restriction of the initial guideline. Graphic backgrounds are not supported although the different colors used in GIF or JPEG files could be identified. But in this case, a spatial algorithm should take care of physical appearance of text on top of graphical background colors, which is far more complex. As we can see here, the cost of developing the complete procedure for automated testing of a single guideline can be very high. However, this implementation being done, the guideline can be tested rapidly.

2.3 A style guide guideline

Style guide guidelines are important to check adherence to some corporate "Look & Feel" guidelines. Style guide compliance increases usability whereas violation will normally decrease it. Let us consider the following style guide example: "All data fields must be accessible via a tabulation key in the same order. If fields are grouped together, then tab must access the fields within the group" (Levine, 1996). A procedure to automatically test this guideline would certainly exploit the appearance order of widgets in the HTML code and check whether this order is compatible (hence, the compatibility ergonomic criteria above) with the order recommended by the task. For this purpose, the test can only be achieved if such an order can be provided either manually (e.g., at run-time) or automatically (e.g., from a declarative task model).

2.4 A standard guideline

Let us consider the following guideline re-expressed from the ISO 9241-Part 12 standard: "Error messages should not use humor." Evaluating this guideline on web pages automatically is quite complex since identifying error messages on web pages is hazardous; testing if a text contains humor is even more hazardous as this is language-, context-, and user-dependent. To solve these problems, a possible restriction of the above guideline could be to submit the text of each web page generated by HTTP error codes (e.g., HTTP 404 File not found) to a detection of commonly used humorous words and expressions. In this case, the original guideline cannot be tested automatically, but its restriction is. The expressiveness and the completeness of the initial guideline have consequently been reduced in this transformation. Hence, it is important to warn the evaluator of this possible loss of expressiveness. It is indeed almost impossible to encompass all humorous words and expressions in a list to be tested for occurrence, even if this list is made user expandable. The resulting procedure for this automated testing is the following:

```
HumorWords = {blablabla, aha, hmm, ...}  /* To be enriched as experience is growing /
Humor Expressions = {joke1,joke2,...}
ListOfPages = {URL1, URL2,..., URLn} /* List of the URLs of the web pages to be tested */
for i=1 to # (ListOfPages)              /* for each web page to be tested */
  ListOfMsg := DetectPresence ( ListOfPages(i), ErrorMessages )
    /* Detects the presence of error messages produced by HTTP protocol and returns the results in a list */
    if ListOfMsg <> Ø then              /* if there is at least one error message detected in the page */
      for k=1 to #(ListOfMsg)
        if IsThereHumWords ( ListOfMsg(k), WordIndex, HumorWords) or
          /* Returns 'true' if a regular expression belonging to the list HumorExpressions is found in ListOfMsg(k) */
          IsThereHumExp ( ListOfMsg(k), WordIndex, HumorExpressions)
        /* Returns 'true' if a word belonging to the list HumorWords is found in ListOfMsg(k) */
        then AddUsabPblm (ListOfProblems (ListOfPages (i)), HumMsgPblm, WordIndex, ListOfMsg(k))
          /* Add to the list of usability problems detected for the i-th page the problem identified by
          HumMsgPblm, which is a reserved constant indicating the type of problem, at location WordIndex
          for the k-th message in the list */
      endif
    endfor
  endif
endfor
```

2.5 Guidelines in an ergonomic algorithm

Let us consider the following guideline re-expressed from an ergonomic algorithm (Kim & Foley, 1993): "Links should be consistent with related page titles." Two situations for testing this guideline may occur: either the link label is exactly the related page title or at least one difference exists. The automation is straightforward in the first situation, while impossible in the second: from a semantic point of view, it is impossible to prove that a link label is semantically consistent with the related page title. However, a similarity score used in Information Retrieval can be computed from the amount of shared words and shared ordering: if the two expressions share a significant amount of terms in a relatively similar ordering, then it can be assumed that a syntactical consistency is established. Again, the resulting guideline is a restriction to the syntactic domain. The resulting procedure for this automated testing is the following:

```
Threshold := constant;                    /* Definition of a certain threshold beyond which it is considered that
                                             two different expressions are not syntactically related with respect
                                             to the Similarity Factor */
ListOfPages = {URL1, URL2,..., URLn} /* List of the URLs of the web pages to be tested */
for i=1 to # (ListOfPages)                /* for each web page to be tested */
  ListOfLinks := DetectPresence ( ListOfPages(i), Links )
    /* Detects the presence of links in the i-th page and returns their names in ListOfLinks */
  if ListOfLinks <> ∅                     /* if there is at least one link detected in the page */
    then
    for j=1 to #(ListOfLinks)
      TargetTitle := DetectPresence ( ListOfLinks (j), Title )
                    /* Returns the <TITLE> tag of the related web page if any */
      if TargetTitle = ∅
      then /* The target title is missing: another potential usability problem can de detected here */
        AddUsabPblm (ListOfProblems (ListOfPages (i)), MisTitlePblm, null, null)
          /* Add to the list of usability problems detected for the i-th page the problem identified by
            MisTitlePblm, which is a reserved constant indicating the type of problem */
      else /* The target title exists */
        SimilarityFactor := ComputeSimilarity (Name(ListOfLinks(j)), TargetTitle)
          /* Compute the similarity factor used in information retrieval between the two expressions:
            the name of the link being currently used and the title of the related page */
        if SimilarityFactor <= Threshold  /* If this factor is not important enough, i.e. beyond a given threshold */
          then AddUsabPblm (ListOfProblems (ListOfPages (i)), InconsTitlePblm, null, Name(ListOfLinks(j)))
            /* Add to the list of usability problems detected for the i-th page the problem identified by
              InconsTitlePblm, which is a reserved constant indicating the type of problem, for the name of the
              j-th link in this page */
        endif
      endif
    endif
endfor
```

3. Conclusion

The provided examples can give a first idea on how complex the automated evaluation of web design guidelines can be. Beyond the facility of rapidly accessing the code of any web page through the Internet, it rapidly appears that either the evaluation dramatically becomes complex or should be reduced to a more simplistic, yet reduced, version. The static analysis of HTML code can certainly provide some support for automated evaluation of very-large web sites, but is clearly a hard to reach target.

References

Detweiler, M.C. and Omanson, R.C. (1996), *Ameritech Web Page User Interface Standards and Design Guidelines*, Ameritech Corp., Chicago. Accessible at
http://www.ameritech.com/corporate/testtown/library/ standard/web_guidelines/index.html

Kim, W.C. and J.D. Foley, (1993), *Providing High-Level Control and Expert Assistance in the User Interface Presentation Design*, in Proceedings of ACM Conf. INTERCHI'93, ACM Press, New York (pp.430-437).

Levine, R.(1996), *User Interface Design for Sun Microsystem's Internal Web*, Sun Microsystems Inc. Accessible at http://www.sun.com/styleguide

Scapin, D., Leulier, C., Vanderdonckt, J., Mariage, C., Bastien, Ch., Farenc, Ch., Palanque, Ph., Bastide, R. (2000), *A Framework for Organizing Web Usability Guidelines*, in Proc. of 6th Conf. on Human Factors and the Web. Accessible at http://www.tri.sbc.com/hfweb/scapin/Scapin.html

Stephanidis, C., (2001), User Interfaces for All. In Stephanidis C. (ed), *User Interfaces for All – Concepts, Methods, and Tools*, Mahwah, NJ:Lawrence Erlbaum Associates, Inc. (pp 3-17).

Vanderdonckt, J., (1999), *Development Milestones Towards a Tool for Working with Guidelines*. In Interacting with Computers. (Vol 12, no. 2. pp 81-118).

Remote access to public kiosk systems

Kai Richter

Zentrum für Graphische Datenverarbeitung e.V.
Rundeturmstraße 6
64285 Darmstadt
Germany
Kai.Richter@ZGDV.de

Abstract

Public kiosk systems (e.g. banking terminals or ticket machines) are designed for the average user. Interfaces of such machines generally represent something like the "lowest common denominator" between security, cost-efficiency and a most general user model. This implies the provider's silent agreement that people which do not fit into this scheme are excluded. This paper presents a remote access platform for mobile devices. A User Interface Management System (UIMS) has been implemented which provides entire separation of application logic and its representation by means of XML and Java technologies. Customers can use their own optimized input and output utilities to interact with a remote image of the terminal interface displayed on the mobile device. Specialized assistive technology for disabled persons have been integrated.

A fully operant implementation of the described platform has been presented at the conference "Mensch & Computer 2001" in Bad Honnef, Germany. The work is currently supported by the project EMBASSI, which is co-funded by the German Ministry of Education and Research (BMBΓ).

1. Introduction

Mobile computing provides users with remote access to resources that can be hosted on the other side of the globe. But still most of the technical systems within our reach have to be accessed through their appropriate interface. Some of the technically most elaborate machines in our everyday life are public kiosk systems like banking or vending machines. Most of them are simple to use by pressing a button or two, always assuming; (a) you know which buttons to press; and (b) you are able to press the buttons and they are not out of your reach or perception. Of course the developers of such machines are not to blame for this. They have to deal with so many problems as there are financial restrictions limiting the equipment of their machines to the bare necessary or security aspects forbidding components sensitive to manipulation or destruction, and finally there is a infinite number of different individuals to be served. The resulting compromise often excludes disabled persons from access, forces the novice back to the cashier or annoys the frequent user with cascades of dialogs.

Mobile devices with their increasing functionality in wireless communication can show a way out of this deadlock situation. By enabling portable devices to access common public kiosk systems the interaction can be relocated from the terminal to the individual client. This offers the opportunity to customize interface, input and output devices. For disabled persons this means that they can use their own familiar configuration of assistive devices allowing them to act independently of any aide.

2. Architecture

2.1 Overview

The project EMBASSI focuses on the development of assistance for the everyday user of technological devices. Therefore the goal of our platform as part of this project was not only to provide a secure architecture for remote access to public kiosks but to offer enhanced terminal functionality with additional scalable assistance for the user. First step in assistance was to establish an access-ability for everyone including normal and disabled users by providing and integrating innovative interaction tools.

196

Technically speaking this approach consists of three components including a common terminal software module, a communication platform of software agents and an interface player for the remote device. Due to this modularization existing terminals can be extended with minimal effort. Mobile clients log in to the communication platform through a wireless peer-to-peer connection allowing secure interaction. A generic description of the interface is transmitted to the client through an XML-dialect called User Interface Markup Language (UIML; Abrams, Phanouriou, Batongbacal, Williams, & Shuster, 1999). As resources on mobile devices can be limited all the client requires is a small module implementing an interface-player which interprets the interface descriptions, handles the connection and translates user input into commands to the terminal.

Unlike other approaches this solution realizes the strict separation of application logic and presentation not only logically but also physically, offering a secure platform for mobile e-business applications. By applying the XML-based interface description runtime manipulations and adoptions are possible, allowing optimal configuration and personalization for the individual user and access for the disabled.

The rest of this paper is organized as follows. The next section provides a detailed description of the system architecture. Assistance aspects of this architecture are illustrated in the subsequent part. Then the application of UIML as a device-independent description for user interfaces is discussed regarding the opportunities it offers for our approach. In a concluding discussion propositions for future elaboration are presented.

2.2 The remote access platform

The principal architecture is based on a software agent platform communicating over a TCP-layer by means of the Knowledge Query and Manipulation Language (KQML; Labrou & Finin, 1997) wrapping XML-based commands. A common terminal software is extended with a plug-in (see figure 1. where it is marked as a black box) which establishes a local TCP connection to the device manager agent. Information about terminal state (e.g. which is the present screen) and user inputs are exchanged over this connection. The plug-in works as translator between application specific commands and a text-based platform protocol. In this way not a single bit of application logic has to leave the terminal. The only terminal specific part is the translation between device manager and terminal by the plug-in.

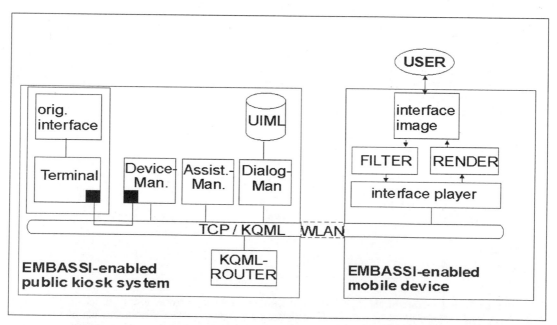

Figure 1. Architecture of the remote access platform. Interface description language is transmitted to the remote client running an interface player.

The software agent platform uses a KQML-router to address the single agents. After the terminal information is translated into the XML protocol the single agents' services are requested. The generic EMBASSI architecture specifies two additional agents. First, there is the assistance manager providing assistance-content like help texts or other resources not available on a standard terminal. Secondly, there is the dialog manager which administers the communication with the client. The dialog manager retrieves and prepares the requested interface descriptions and manages incoming user input validating it and sending it to the device manager.

Through the whole session the program flow stays under control of the terminal software. Only terminal responses call the next interface to be rendered or a user input to be confirmed. When the terminal signals the change of the current screen, the dialog manager retrieves the corresponding interface description and sends it to the client calling a *"NewScreen"* event. Further any change within an interface invokes a *"NewProperty"* event. User input is signaled by a number of different user events. To keep both sides synchronous every command has to be confirmed.

The interface player on the remote device consists of a software agent managing the communication with the terminal side. The player controls two components: a rendering engine which interprets the interface description and an input filter collecting the user's input. Both are platform specific depending on the configuration of the device. The rendering engine can be anything from a HTML browser or mobile phone to a special engine rendering UIML into a special toolkit. Presently the interface is rendered into a Java-Swing toolkit by means of the Harmonia UIML2 to Java Renderer™ (Harmonia, 2001).

User input is caught by the input filter reporting interactions to the player which then generates messages to the terminal. The interface description does not contain information about underlying logic. Thus the player interprets user input on a shallow level deciding only on the type of action and the elements involved but not on the procedure to be called. The response to user input can only be generated from the terminal's reaction.

If for instance the user presses a button an action event is communicated to the terminal supplemented with the identifier of the button. For ease identifiers are presently held synchronously but they could also be translated by the dialog manager into terminal-internal names.

The feedback loop necessary for confirmation of in- and output imposes a severe problem to the performance of this system. If every action has to be confirmed any button which is pressed invokes a call through the whole system. As the input is blocked for that time response latency becomes perceivable. We intend to solve this problem by defining portions of interaction (e.g. a transaction form) which are allowed to be validated locally through XML-schema validation.

2.3   Assistance

The assistance concept we employ here is based on three levels of assistance:

-   Assistance by access: as long as some users still are locked out from technical systems, any sophisticated wizard or alike must fail.
-   Assistance during interaction: by guiding the user, by providing adaptive help and interfaces the user can be supported all through the interaction process. The sequences of planning, execution and evaluation of action have to be taken into account.
-   Assistance by information: during acquisition of expertise and during expert use additional information can help the user to fulfill his task. This form of assistance has to be requested actively while the above assistance acts autonomously.

In the initial stage of the implementation process realization was focused upon assistance by access as this is the fundamental precondition for any further help. To provide access also for special user groups the integration of existing assistive technologies like switches for physically impaired has been taken into account in every stage of the project. Also new devices (e.g. an integrated sensory display) have been developed and were successfully applied to the platform.

Assistance during interaction should take into account the three basic stages of interaction as there are planning, action and evaluation (Norman, 1988). All the three stages profit from an optimal interface design. Depending on the target device complexity and presentation of an interface have to vary. UIML offers the possibility to statically define different presentation styles or structures within one description. In this first stage of the project two styles were defined, one which was a nearly original image of the original interface (see figure 2a; including pictured buttons etc.) and a second that contained only the absolutely necessary information (see figure 2b; textually based). Assistance by information can be accessed over the platform by requesting additional hypertext pages containing online help documents or additional information like terms and conditions.

Following the integral assistance concept described above both special and 'normal' user groups can be provided with the most appropriate solution. The platform not only allows remote access to mobile devices but also guarantees that this access is really applicable by every user group and thus a gain in accessibility and not limited to a one-to-one image to a remote device.

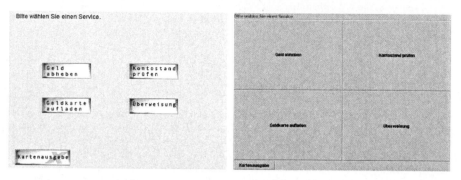

Figure 2a & b. The same interface rendered in two different styles which were statically coded into the same UIML description resource. The left picture shows the authentic image of the original interface. The right figure shows a text-only representation leaving out all additional information thus representing a more generic description.

2.4   Device-independent user interface description

Due to the multiplicity of portable devices any interface description based on a certain toolkit would be limiting the range of target platforms. What is needed is a generic description for interfaces which can be translated to any target representation.

Rather than describing a certain toolkit of widgets the User Interface Markup Language provides a means to specify an interface based on its dimensions. Those are defined as structure (e.g. what elements are contained in the interface and how are they combined), style (e.g. how are the elements displayed), content (e.g. textual contents of label and buttons) and logic (e.g. reference to local methods or RMI-calls). The interface elements which are used to fill this skeleton are not defined. Present rendering engines only support a certain set of elements depending on the specific renderer in use, namely Java AWT or Swing widgets, HTML or WML tags. Although targeting platform independence the absence of a generic set of interface elements allowing transition between the platforms is a fundamental problem of this approach. Further the fact that references to application methods are to be described within the interface description violates the separation of presentation and logic and constitutes a security problem.

Another approach recently introduced by the w3c HTML working group are the XForms (Dubinko et al., 2001) which represents an effort to specify a form-related set of interface elements on an abstract level in order to allow the free conversion between different target markup languages for usage in the web (Martin, 2000). Due to its special purpose XForms is too limited for the description of full application interfaces.

The Mozilla project (Mozilla.org, 1999) presented another example how to realize a platform-independent interface description with their Cross Platform Front End (XPFE) described in the XML-based User Interface Language (XUL) integrated in their browser. The toolkit developed here is designed for window based systems does not offer the flexibility required for the translation to other interface concepts.

Within the EMBASSI project a set of abstract interface elements is being developed which allows to use UIML as framework for a generic interface description which can be transported to any other toolkit or markup and which is not limited to graphical interface concepts but allows also translation to acoustic or haptic representations. For the design of this set of elements analysis of common interface elements goes beyond their presentation focusing on the functional aspect of an element as this is partly realized in XForms (Preim, 1999; Shneiderman, 1998).

3.   Summary and future work

The approach presented in this paper enables mobile devices to interact with public kiosk systems over a wireless connection. Terminals have to be equipped with a plug-in module that allows interaction with the remote client over a XML-based protocol offering a high level security as it is necessary for reliable e-business solutions. Remote clients only have to provide a lean interface player that feeds user input to the platform. Due to this fact the approach is applicable also for devices with limited resources. The interface player has to be implemented device dependently offering room for individual enhancements of in- and output devices.

By shifting the locus of interaction from terminal interface to individual mobile devices; (a) access becomes available for a broader range of users including people with limited means to interact; (b) customized interaction interfaces can be offered through runtime manipulation of interface descriptions allowing adoption to user needs and abilities; (c) user profiling allows recording of preferred action sequences and reduces user-input; and (d) additional assistive functions can be offered like online help, information bases, hotlines or offline forms that can be filled at home and automatically sent when passing the next terminal. As a first step the fully functional remote access platform presented here has been implemented and yet a variety of assistive technologies have been integrated on the client side, providing means for blind and physically handicapped. Psychological evaluation through all stages guarantees compliance with usability aspects.

Future development will concentrate on assistive functions served by the terminal. Present research addresses the design of a functionally oriented XML-based description of user interface elements. This set of cross-platform interface elements may then serve as a basis for dynamic XSLT transformations (Clark, 1999) of interface descriptions depending on user device and profile. This will offer the server to derive interfaces of different styles and structure from one resource. Also will there be the possibility to support different non-UIML rendering engines like HTML or WML browsers.

Further, offline interactions like form-filling will be allowed through modularization of interface parts into application-referred and information-referred parts. Information-referred parts (e.g. a shopping list received from a shopping terminal or a money transfer form from a banking terminal) will be downloadable once and applicable to any conform terminal application. Personalization and user profiling is another goal. The user shall have the option to record program paths frequently followed. He can save them in a user profile that serves as a basis for the automatic generation of application shortcuts allowing to skip irrelevant sequences and to access the desired dialog directly. This concept applies the possibilities XML offers to traverse and transform program structures represented in a DOM (Document Object Model; Le Hors et al., 2000) tree.

The approach presented in this article has its focus on the needs of special user groups as well as on usability aspects in general striking for the ambitious goal to offer a secure gateway to information and services for everybody.

References

Abrams, M., Phanouriou, C., Batongbacal, A. L., Williams, S. M., & Shuster, J. E. (1999). *UIML: Appliance-independent XML User Interface Language*. Paper presented at the 8th International World Wide Web Conference, Toronto.

Clark, J. (1999, 16 November 1999). *XSL Transformations (XSLT)* (Version 1.0). W3C. Available: http://www.w3.org/TR/xslt.html.

Dubinko, M., Dietl, J., Merrick, R., Raggett, D., Raman, T. V., & Welsh, L. B. (2001, 16 February 2001). *XForms 1.0 - W3C Working Draft*. W3C. Available: http://www.w3.org/TR/xforms/.

Harmonia, I. (2001). UIML2 to Java Renderer (Version 0.7a). Blacksburg VA: Harmonia, Inc.

Labrou, Y., & Finin, T. (1997). *A Proposal for a new KQML Specification*. Available: http://www.cs.umbc.edu/~jklabrou/publications/tr9703.ps.

Le Hors, A., Le Hégaret, P., Wood, L., Nicol, G., Robie, J., Champion, M., & Byrne, S. (2000, 13 November, 2000). *Document Object Model (DOM) Level 2 Core Specification* (Recomendation, version 1). W3C. Available: http://www.w3.org/TR/DOM-Level-2-Core/.

Martin, D. (2000, 23. August 2000). *Adapting Content for VoiceXML*. O'Reilly. Available: http://www.xml.com/pub/a/2000/08/23/didier/index.html.

Mozilla.org. (1999, October 1, 1999). *XPToolkit Project*. Mozilla.org. Available: http://www.mozilla.org/xpfe/.

Norman, D. A. (1988). *The Psychology of Everyday Things*. New York: Basic Books.

Preim, B. (1999). *Entwicklung interaktiver Systeme: Grundlagen, Fallbeispiele und innovative Anwendungsfelder*. Berlin: Springer.

Shneiderman, B. (1998). *Designing the User Interface: strategies for effective human-computer interaction*. (3 ed.). Reading, Massachusetts: Addison-Wesley.

The dialog tool set : a new way to create the dialog component

Guilaume Texier, Laurent Guittet, Patrick Girard

LISI/ENSMA
Téléport 2 – 1 Avenue Clément Ader - BP 40109
86961 Futuroscope Chasseneuil Cedex
Téléphone: (33) 05 49 49 80 67
Télécopie : (33) 05 49 49 80 64
texier, guittet, girard@ensma.fr

1. Introduction

To help graphical and interactive applications designers in their task, several architecture models have been proposed. They all recommend to separate the application in three components which are the presentation, the dialog component and the specific component. Beyond these architectures, several tools have been proposed to create these components. These tools can be divided in two great families: the presentation toolkits called toolkits in the following and the top down generators.

The toolkits are made of different classes that represent presentation elements called widgets. A crude control mechanism is offered by toolkits. It is a matter of the callbacks which are used to link the widgets to specific component actions, regarding the events received by widgets. This control mechanism can be easily used to create a small application, but when the application enlarges the encapsulated code in the callbacks becomes hard to write and to maintain. Meanwhile, toolkits are universally used, and the method to create an application which consists of using the callbacks to define the dialog component is very spread. Moreover toolkits permit to locally and dynamically modify the application because their widgets can be dynamically instantiated and taken into account by the application.

The top down generator and particularly the model based systems (Szekely, et al., 1993) use specific languages to describe the application and are able to generate the dialog component and the presentation component from this description. These tools permit to explicitly design the dialog component of the application which is necessary to write and to maintain the application when it becomes big (many actions provided by the dialog component) and/or when the dialog is structured (a structured dialog is a dialog mode where a task can use the result of an other task to be executed). Meanwhile, there are few or no top down generator that permit to dynamically and locally modify the dialog component. With most of them, the dialog component can be modified only by changing its description and then the application have to be rebuilt.

Our proposition is to use new tools as presentation widgets which permit to create explicitly the dialog component. This dialog specific toolkit is called the **dialog tool set**. It is composed of several classes which are reification of dialog elements. Thus, any people that understand how to create the presentation component using widgets could easily define a dialog component able to manage a complex dialog like structured dialog. So, these persons would be able to easily create more complex applications.

This tool set is completely independent from the presentation component. So, its behavior is completely disconnected from the manner the user provides data. Then, it can be used to easily create multi-modal applications allowing a universal access to the application functionality.

In the following section of this paper, the concept of presentation toolkit is discussed, and particularly the way to design the presentation using this kind of tools. The next section deals with the dialog tool set, the similitude and the differences between the dialog tool set and the toolkits are pointed out.

2. The presentation toolkits

A presentation toolkit (or toolkit) is a library of interaction objects called widgets. The presentation of an interactive application is created by instantiating widgets as windows, menus, buttons, etc. The widgets have two important roles in the presentation component :
1. They are in charge of transforming information that come from input devices (keyboard, mouse, microphone,…) into application usable data.
2. They offer services to manage output devices (screen, speakers, force feedback devices, …), permitting the system to show its state.

Several advantages are offered to the designers by toolkits. First, they permit to reuse interaction tools which have a higher abstraction level than the drivers provided by the input/output devices. Then, the implementation of the presentation component is really simplified. Second, the aspect and the behavior of the widgets are the same for all the application built with one toolkit. This make the application more usable regarding the familiarity of the application (Dix, et al., 1998).

2.1 Presentation definition

The aspect definition of interactive applications is entirely devoted to the toolkit regarding the Arch (Bass, et al., 1991) architecture model. To create the presentation component, several widgets are instanced. Then, at running time, the user interactions are transformed by a specific program (often called event handler) in event representing the user action on a widget. When an event is received by a widget, it has two behavior: an internal behavior and an external behavior.
The internal behavior represent a self reaction of a widget to an event. As an example, when the user push on the mouse button and when the mouse pointer is on a button, this last one sinks. Generally, the internal behavior can not be modified. Some widgets have only an internal behavior (menu buttons are only used to display menus).
The external behavior represent the application reaction to the user interactions. This behavior has to be explicitly described by the designer with callbacks. Callbacks are functions which are called when an event is send to a widget. They are different for each widgets and for each events. Thanks to this mechanism, the application can be directly controlled by the presentation component. Indeed, specific component actions can be called using callbacks. In this case, the toolkits insure two services which are a presentation service and permit an application control role.
Meanwhile, big problems can occur when the same callback has to perform a different work regarding the application state. As an example, a mouse click may has several meanings. It can be used to select an object or to compute a position for the system. Then, to manage this ambiguity, a lot of information have to be registered in the presentation component. In consequence, the code of the presentation component highly grows up when a new functionality is added to the application. Moreover, the experience has shown that real interfaces contain often hundreds of callbacks, which makes the code harder to modify and maintain (Myers et Rosson, 1992). That's why, the architecture models for interactive systems recommend to separate the presentation and the control.

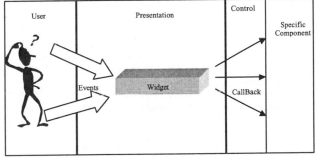

Figure 1: Interactive system control

The Figure 1 shows the manner toolkits control interactive systems using the event/widget/callback paradigm.

The way to design an application using a toolkit for presentation and control is very spread. Because widgets have several advantages.

- Firsts, they are very generic, they can be reused for any kind of applications.
- Second, they can be directly instanced using a standard programming language.
- And third, they can be dynamically instanced, so they allow the user to adapt the interface of its system to make it more usable (Dix, et al., 1998).

Meanwhile, they can not be used to control a structured and multi-object dialog. For this kind of dialog, the dialog component must be explicitly described (Pierra, 1995), and the callback mechanism can not be used because it represent an implicit dialog control that modelize only interactions but neither the application state nor the dialog state.

In the next section, the dialog tool set we have proposed is demonstrate. We focus on the ease and on the similarity of use of these tools and the presentation toolkits.

3. The dialog tool set

The dialog tool set is a set of gadgets that permit a designer to create a dialog component (respecting the ARCH (Bass, et al., 1991) architecture model and based on the element defined in H^d (Pierra, et al., 1995) by the same way he/she makes the presentation component using widgets.

3.1 The dialog tool set classes

This tool set is composed of four family of classes that represent different elements needed to describe a dialog component supporting structured tasks and multi-objects tasks.

3.1.1 The tokens

The Tokens are the information units used to capture, at various levels of abstraction the user input (e.g. number, position, circle, and line…) in the dialogue tool set.

3.1.2 The questionnaires

The questionnaires represent the system tasks in the dialog component. In fact, they are the signatures of the functions called by the dialog component to realize treatment on the specific domain component. So, they link the dialog component with the specific domain component. They are defined by the command used by the user to activate the task they represent, by the types of tokens used by the task to call functions of the specific domain component and by the types of the returned token when the task returns a result.

3.1.3 The diagets

The diagets organize the different questionnaires of the application in abstraction levels. All the questionnaires dedicated to the creation of geometric entity are brought together in the "creation" diaget. The "Calculator" diaget is shaped by the questionnaires dealing with arithmetic's operations. Diagets contain ATN (Augmented Transition Networks) (Green, 1986) that aggregate tokens to call a questionnaire implementation.

3.1.4 The monitor

The monitor manages all the diagets. It recovers the tokens coming from the presentation and sends them to all the diagets from the lower abstraction level to the higher abstraction level.

At running time, the monitor is in charge of recovering the tokens that come from the presentation component. Then, they are transmitted to each diaget regarding the order of the hierarchy. When, a diaget received a token that it waits, it changes of state (notion of consumption), when it has received all the tokens needed to call a questionnaire, it realize this action and then it transmit the resulting token to the monitor (notion of production). Then, the monitor transmit this token to the higher diaget in the hierarchy. The fact that the diagets do not know who produces the tokens they received and who consumes the tokens they produce insure the independence between the tasks. This independence permit to define a structured dialog.

It is important to notice that the tokens representing user input are created in the toolkit events. Meanwhile the dialog component does not know the manner the tokens are created. Thus, regardless the modality used to create a token, the dialog component behavior is the same. An example of multi-modal application has been created with this toolkit, the user could choose command with menu button or with a voice recognition system. The transition from a full WIMP application to the multi-modal application did not entail any modification of the dialog component.

3.2 Diaget vs widget

There are many similarities between the diagets and the widgets, but their respective goals are completely different. The widgets are use to create the presentation component, they are reifications of presentation concepts whereas the diaget are use to implement the dialog component, they are reifications of control concepts.

The first similarity concerns the types of elements composing these tools. To create the presentation component, two kinds of widgets are instantiated: the containers and the terminals (Fekete, 1996). The containers are only used to organize the terminal widgets in order to improve the presentation. The terminal widgets are used either to call an application functionality, either to display data, or both. By the same way, the dialog tool set has two kinds of control objects: the monitor and the diagets. The monitor is used to organize the diagets regarding the dialog semantics. The diagets are in charge of controlling the calls of the application functions.

Concerning the action calls, a similitude also exists between widgets and diagets. Widgets call actions (callbacks) when they receive some specific events. The called callback depends on the event received by the widget. Actions (the questionnaires) are called when the diaget receives a certain list of token. A similitude can be made between the widget behavior and the diaget behavior. In fact, in the dialog tool set, the diagets are like widgets, the token lists represent the events and finally, the callbacks are represented by the questionnaires (Figure 2). This similitude enlights that the proposed approach is the natural extension, for the dialog design of the way to create graphical interfaces with toolkits. Nevertheless, the proposed dialog tool set presents an essential advantage to control the application regarding standard toolkits: it permits an easy representation of the contextual aspect of user inputs because it is based on automata.

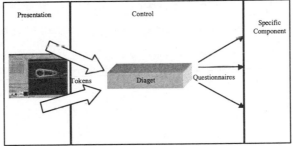

Figure 2: Diaget as widget

The last similar point between the diaget and the widget is the dynamic instanciation. Using the dialog tool set, the dialog component can be dynamically modified. Then, our tool set can be used to create an adaptive system where the user can define new functions using programming by demonstration techniques (Cypher, 1993), these functions are dynamically added to the dialog component by instantiating new questionnaires and new tokens. Anyway, this dialog tool set has been used in the TexAO (Texier et Guittet, 1999) system which is an adaptive CAD system.

4. Conclusion

Toolkits are the most used environments in the Human Computer Interaction field. They permit to reuse, rather easily, interaction tools which are friendly to create every kinds of applications. The callback mechanism appears to be adequate for the control of applications when the tasks are autonomous and slightly structured. Inversely, when the tasks are recursively decomposed into sub-tasks, the dialog language becomes strongly structured and contextual, the callback mechanism is completely inadequate. Then, the dialog component code is spread in callbacks, and the application becomes hard to maintain.

This paper presents a new kind of toolkit that permits the designer to explicitly create the dialog component. This toolkit, called the dialog toolset, is based on several classes that permit to define and to organize the system tasks in order to control their activation regarding to user inputs. The main characteristic of these tools is that they can be used exactly as the widgets. So, they keet the genericity and ease of use of the widget. The way to create the dialog component consists of creating diagets (diaget + gadget) (representation of the widgets in the dialog toolset) and to link them to the specific domain component functions using questionnaires (representation of the callbacks) and tokens (representation of the events). Then to create the dialog component, the designer retrieves the well known method consisting of linking objects to actions via callbacks regarding events.

Moreover, the dialog toolset is perfectly adapted to the management of multi-modal application. The independence between the dialog component and the presentation component permits to adapt existing WIMP application to over modalities without modification of the dialog component.

References

Bass, l., Pellegrino, R., Reed, S., Sheppard, S. et Szezur, M. (1991) *The Arch Model : Seeheim revisited*. In *Proceedings of User Interface Developper's Workshop* .

Cypher, A. (1993) *Watch What I Do: Programming by Demonstration*. The MIT Press, Cambridge, Massachusetts

Dix, A., Finlay, J., Abowd, G. et Beale, R. (1998) *Human-Computer Interaction*. Prentice Hall,

Fekete, J.-D. (1996) *Un modèle multicouche pour la construction d'applications graphiques interactives*. Doctorat d'Université (PhD Thesis) : Université Paris -Sud

Green, M.W. (1986) *A Survey of three Dialogue Models*.

Myers, B.A. et Rosson, M.B. (1992)*Survey on User Interface Programming*. In *Proceedings of SIGCHI'92* (May, Monteray, CA), pp. 192-202.

Pierra, G. (1995) *Towards a taxonomy for interactive graphics systems*. In *Proceedings of Eurographics Workshop on Design, Specification, Verification of Interactive Systems* (June 7-9, Bonas), Springer-Verlag, pp. 362-370.

Pierra, G., Girard, P. et Guittet, L. (1995) *Towards precise architecture models for computer graphics: the H4 architecture*. In *Proceedings of Eurographics Workshop on Design, Specification, Verification of Interactive Systems* Bonas), .

Szekely, P., Luo, P. et Neches, R. (1993) *Beyond Interface Builders: Model-Based Interface Tools*. In *Proceedings of InterCHI93* pp. 383-390.

Texier, G. et Guittet, L. (1999) *User Defined Objects are First Class Citizen*. In *Proceedings of Third Conference on Computer-Aided Design of User Interfaces (CADUI'99)* (21-23 October, Louvain-la-Neuve, Belgique), Kluwer Academic Publishers, pp. 231-244.

Towards a New Multi-Layer Approach for the Global Network architecture

Gianluca Vannuccini[a], Paolo Bussotti[a], Maria Chiara Pettenati[a], Franco Pirri[a], Dino Giuli[a]

[a]Electronics and Telecommunications Department, University of Florence
Via S. Marta, 3 50139 Firenze (IT)

Abstract

Proper exploitation of user subjectivity by all types of users is to be considered as a key requirement for the evolution from Internet to the Global Network. A new structure for the logical network architecture has thus been devised, as two integrated parts to propose a solution accounting individual subjectivity and to envisage compatibility with existent network functioning. One is the subjectivity engine, called the Subjectivity-Gene Imprinter (SGI); the other is the Actuative Stack (AS), conceived of as a conventional multi-layered architecture, suitably restructured and linked to the SGI. This paper focuses on the AS, in order to define its functionality and consolidate its technical feasibility.

1. Introduction

The present Internet is rapidly growing to become a global system of interconnectivity, a universal medium of communication, which, however, was initially conceived of as an informal and open access instrument. The natural evolution of this system tends towards the Global Network, a global interaction environment supporting direct or indirect relations among persons [1,2,3,4]. Given this tendency, many limiting factors have been highlighted over the last few years. A more symmetrical relationship between End Users and Providers has to be provided, and this symmetry can only be achieved if the End User is given some instruments to represent and portray his expressed will and his subjectivity in the Network. Some important pre-requisites for this symmetric relationship have been highlighted in [1,2,3,4,5], in which a Human Subjective Profile, and a collective intelligence together with knowledge have been defined, in order to represent, respectively, the User's subjectivity on and within the Network, as well as the social knowledge that comes from the relations among subjects. More specifically, the Subjectivizing Gene metaphor has been defined, as an imprinting factor which influences all the subjectivizing processes [2,3]. The subjectivizing driver, called the Subjectivizing Gene Imprinter (SGI) [6] correspondingly becomes the driver of the dynamic processes, which accounts for all the above factors, while transforming the classical Network operation as requested for the subjectivizing gene imprint exploitation. The SGI, instrumentally acting as subjectivity driver, can be embodied in the logical network architecture, envisaging it as re-conceived and re-structured to match the new requirements for the Global Network. The entire network is thus composed by the SGI, integrated with a multi-layered conventional network architecture. The latter, called the Actuative Stack, has to be re-defined to take into account the needs of the structural matching of personalized and socially relevant user requirements, to be associated with network relations and interactions. The integration between the SGI and the Actuative Stack, also implies a modification with respect to the classical multi-layered model at the application layers. In this paper, such a logic network architecture is introduced, while specifically focusing on the Actuative Stack, on its functions as well as on the integration or interoperation with currently available technologies which can validate its technical feasibility.

By using the proposed architecture, users might be provided with an efficient medium to balance their relationship with the Global Network, thus letting universal access be widespread in a symmetric way. The present paper reports on the specific results of current research within the research framework program, called *B.E.S.T* (**B**ridging **E**conomy & **S**ociety with **T**echnology) *beyond Internet* [1], while a general outline of such research is provided by a companion paper [3]. Focus on further specific and related subjects, thus completing the report on such research, is made in other companion papers [4,5,6].

2. Devising a new logical structure for the global network architecture

Exploitation of user subjectivity has to be considered as a key factor to overcome current inherent limits of the Internet, in order to proceed towards the Global Network development [1,2,3,7]. Based on the *Subjectivity Gene* metaphor, a re-conceptualization and re-structuring of the network logical architecture is required; this must, consequently, embody a cognitive processing engine, namely the Subjectivity Gene Imprinter (SGI) [2,7]. Such a new network structure could then be devised, while referring to the conceptual scheme reported in Figure 1.a.

According to this figure, the SGI engine which is introduced, acts as a user-gene imprinter of the process of network-user interaction and relations, in order to match it to the user needs while enabling user provision of the needed instrumental assistance. The SGI is instrumentally connected, and interacts with an Actuative Stack: the latter, which is the typical network multi-layer logical architecture, has been re-thought and structurally modified, also to embody the needed interaction with the SGI engine. Shared network-core resources are commonly made available for the technical and operational requirements of both the SGI and the Actuative Stack, according to the needs of each user and his/her socially valued relation through the network.

A sketch of the SGI functions is hereafter provided. The SGI exploits information contents of the user profile pertaining to the gene-imprinter (user), as well as of the Shared Signification Space, which is a representation of the Network Shared Knowledge illustrated in [6]. The latter pertains to the knowledge that can be shared in the network among users, as in a user community, or made available by network intermediaries. In the SGI engine, the active processing role is played by the Tutor Agent (TA). This is an intelligent software agent, which acts on behalf of the tutored user (the gene-imprinter). The TA coordinates the operation of the SGI, in particular, to produce the changes within the Actuative Stack, as needed for personalization of the network user interaction and relation processes. For example, in an object-oriented and polymorphic approach, personalization could be realized by "assistance modules", which are produced by the SGI and delivered to the Actuative Stack, where they are executed according to a standard interface. The SGI also operates the conceptual-based and inferential processes requested to match the user's needs and to provide him with the requested instrumental assistance. Social values to be attributed to the interaction and relation with other users, through the network, imply the TA interacts with the TAs of other users, even to allow negotiation of such values as well as of the associated modes of interaction and relation. At the same time, the TA coordinates the SGI processes requested for the needed interaction between the SGI and the Actuative Stack. Based on the above mentioned features, the TA also plays a key role in overcoming current relational asymmetry among Internet users. For further insights on the SGI, the reader is referred to the companion paper [6].

The Actuative Stack, as sketched in Figure 1.b, refers to a multi-layer architecture and completes the logical architecture we are considering for the Global Network, in order to match the user subjectivity requirements. In the next sections, more insight is provided about the Actuative Stack, which is the focus of this paper.

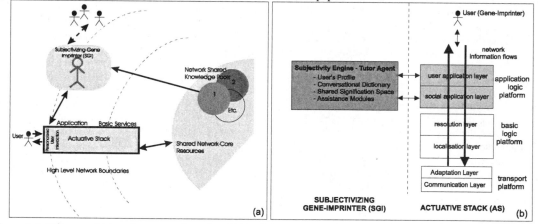

Figure 1: (a) The new Network structure based on Subjectivity-Gene Imprinter. (b) The Actuative Stack within a logical architecture for the Global Network.

3. The actuative stack

With reference to the scheme in Fig.1b, the Actuative Stack can be thought of as a classical OSI-like multi-layered logical architecture. In the field of Human Computer Interaction, a huge effort has been made to define special network architectures, which could reach the best level of interaction quality [7,8,9]. Most of them are indeed layered structures, in which, at each layer some main functions are defined, and some messages are exchanged with the neighboring levels. For our purposes, we consider it limiting to fully model the complex social and technical environment by means of a conventional multi-layer structure. Since multi-layered models presume a rigorous definition of functions and interface characteristics at each level, and since our work also deals with human relationships, indeed it does not appear possible, and it seems even troublesome, probably, to define all the complex links between human character, human subjective profile, shared knowledge, intelligent agents, and data-managing processes, through a strictly functional model. Therefore, the multi-layered conventional architecture, is not here applied to upper layers, which constitute the *application logic platform*. They still imply vertical interactions among vertically adjacent layers, but differing from the lower layers of the Actuative Stack, the application layers also imply horizontal and dynamical, though not independent, interactions with the SGI. The latter interaction supports the subjectivizing process, driven and imprinted by the SGI, in order to allow actualization of the network information flows from/to the user to/from the network, as needed to match user requirements, while assisting the user in his/her interaction. A conventional multi-layered architecture is adopted for the other layers, these structures are, thus, not dependent on the application, since only their operations are specified by the application platform. The lowest levels are grouped in the logic platform operating for information transport (*Transport Platform*). Two intermediate layers, the *Resolution* and *Localization layers*, are also included and grouped in the *basic logic platform*, its structure remaining independent of the application. In the following, we describe the main functionality of the Actuative Stack layers, also referring to some examples, as well as to possible exploitation of already available technologies, simply in order to consolidate technical feasibility.

3.1 The logic application platform

The subdivision of the logic application platform into two layers, the *User Application Layer* and the *Social Application Layer*, accounts respectively for:

- the operations which imply direct interfacing with the user, to be personalized to user application requirements, while providing utmost direct assistance for user interaction, both at the physical and cognitive levels;
- the operations which allow, for the desired application, to single out subjects of the implied social relations, as well as to specify the social values of such relations, also for negotiation of related features of the user interaction processes, as needed for definition of lower layer operations, while meeting associated personal requirements from the user; in the latter sense, the *Social Application Layer*, provides an additional instrumental direct assistance to the user.

Joint operation of the above application layers provides the final direct assistance to the user interaction, as enabled and optimized by means of the interaction with the SGI. For such a purpose, the SGI can also operate by dynamically delivering adequate *assistance modules*, as user knowledge-based effective solutions [6]. Each application layer is functionally described in more detail.

3.1.1 The user application layer

This layer deals with everything which directly relates to the User and, in particular, his/her direct interface with the Network. Here the User's requirements are definitively met, and the Subjectivizing Gene's personal features are implemented on data flows, and/or as requirements to be transferred to the lower layers.

For instance, if the user entered unconsciously into a critical Web site, where the privacy of the transaction could be violated, the User Application Layer would recognize this danger, and it would send a message to the SGI which, by consulting both the reliability of the Web site in the Network Shared Knowledge (NSK) and the user's personal opinion on non-secure transactions, would send to the AS the proper requirement to the Transport platform, in order to develop a suitable ciphering procedure.

On the User side, this layer also acts as a direct assistant. It recognizes as usual all the commands imposed by the User, either in a typing or voice mode. It can also actualize the help needed for natural language interaction. It could also operate for understanding the current emotional mood and activity of the User, whether he is busy, or he is driving, and so on, and it decides the best way to send him data from the Network.

Depending on this present user activity, user notified or instrumentally detected, this layer can make specific requests to the SGI and produce suitable requirements for the lower layers.

On the lower layer side, this level can act as the main monitor, since it can observe the lower levels status and eventually activate a call to the SGI in case the subjective profile, or the User's desire, have changed during his/her Network interaction, so as to call for new instrumental actions.

As shown in Fig.1b, this layer is horizontally linked to the SGI, as regards its subjectivizing processes, thus, acting as a network monitor, and as its network service device (from right to left), or as executing the support functions delivered by the SGI sub-system (from the left to the right), to be implemented in the Actuative Stack. Moreover, this layer vertically processes the data flows upwards, to the User, and downwards, towards the lower layers.

When the End User application requires, as typical, to communicate with some other entity in the Network, related directly or indirectly to other users, this layer can produce the suitable call to the lower Social Application Layer, sending it the relevant and appropriate parameters (metadata) for the activation of the call, also depending on the subjective profile of the User himself.

3.1.2 The social applications layer

At this layer, all the operations regarding the relational aspects of the End User, with other network subjects, are activated on the data flows. The SGI, also closely communicates with this layer, to select, for example, the suitable relational views [5], as originated by the End User's subjective profile, in order to correspondingly specify and start the applications involving other users. Since the SGI, depending on applications and involved users can provide different relational views, this layer can produce different requirements for the lower layers depending on the User's subjective needs. A special data flow coloring could be imprinted at this layer, on the basis of the subjective-relational value to be attributed to the data flow for the current interaction.

For example, if the relation involved simultaneously three users, two of which are employees and the latter is the office head, the SGI, by consulting the NSK and the users personal and relational profiles, could send to the Social Application

Layer a request to process three different communication links, inside the multicasting session, two of which in a colloquial, private, and informal way, and the latter with more formal and business-like features.

This layer will also act as a monitor for the SGI, for example requesting its activation if some requirements concerning the relational views are no longer met. It will produce direct specifications to the lower layers, through the Localization Layer, to which it will send the specifications of the particular network relations implicitly requested by the User

3.2 Basic logic platform

In the Basic Logic Platform, all the basic logical operations are executed, except for those which directly execute the transport of information, in order to build the basic platform, commonly supporting upper application layers services for a really personalized interaction.

From this level downwards, there is no longer a direct logical connection with the SGI, since the functions of the levels which are lower than the application platform, can be seen as automatic, once the upper platform layers have correctly embodied the SGI actions. Two main layers are singled out in this platform, associated with the main functions that still remain to be personalized in the Network.

3.2.1 Resolution layer

This layer includes the format conversion of the resources the user is exchanging with the Network. Depending on the User's request, the specific requirements produced at the upper layers also by the SGI assistance will determine the format conversion at this level. For instance, if the End User asks for some identification certificate from the Network (for some Civic Network Service), the upper layers will produce a query containing some metadata with the name, the birth date, and so on, and this layer will transfer the correct query to some remote database within the Network.

This layer can also acquire a greater importance when interacting End Users are using different devices and technologies to access the Global Network. This layer can also exploit the relational view from the upper Social Application Layer, to select the appropriate application view [5], and, correspondingly, determine the suitable format in which data coming from the upper layers must be converted.

An enabling technology at this layer could be the extended Markup Language (XML), which should, in this case, be specifically oriented to subjective user profiling (we could define a personal-XML).

3.2.2 The localization layer

Once the suitable format has been selected to send data between User terminals, the location of these terminals must be found over the Network. All the localization techniques already used in Internet, such as the URL and the DNS systems can be considered here. Nevertheless, our approach at this layer is again oriented to a real personalization: we can, therefore think about personalized addressing trees derived from the URL location address tree, where all the End User's resources are labeled depending on their use with respect to the user himself/herself. A special resource localization tree could be thought of not only for each user, but also for each service that he/she is using. As an example, a special resource tree can be built at the user's end containing all the providers with whom he is doing trading online, or a special tree could be used to address and easily locate all his work colleagues. These features are already implemented in many applications, such as e-mail address books, and browsing bookmarks. In our approach, a dynamic resource location is envisaged, depending on the End User's present attitude (i.e. his foreground subjective profile [5]), which could change the resources the user needs depending on his present activity, or his particular mood.

This layer will therefore receive, from the upper layers, information such those which can be derived from the relational views, the application views as subjective and relational requirements [5]. These data would then be processed at this layer and, using the above mentioned personalized location techniques, classical IP-like addresses will be produced and sent to the lower levels, where the typical IP location process will be performed to locate the real entity in the Network.

3.3 The transport platform

The layers described in previous paragraphs are conceptually situated over the Transport Layer, as it is intended in Networking literature. In the present model a Transport platform is then operating at the bottom for transport of information through the Network. In order to make the proposed model compatible with the current communication systems, the Transport platform as shown in Figure 1(b) has been splitted in two layers: i) the Adaptation Layer, that will provide up interoperability functions, to allow the proposed model to run over many different communication technologies, while also allowing match to the users subjectivity requirements transferred at the transport level; ii) the Communication Layer, which is intended to provide down the classical communication services. Considering the Internet model as the reference model for the integrated transport of information over the Global Network, we can assume an IP based architecture for the Adaptation and Communication layers and, then, correspondingly devise personalization services at the transport level.

3.3.1 The adaptation layer

One of the most promising IP Quality of Service techniques, the Differentiated Services Architecture, could be used at this level to personalize the flows which the End User exchanges with the Global Network. For example, we could figure out a "Privacy of Service" architecture, where DiffServ techniques would be used to mark data streams depending not only on usual transport quality features, but also on their subjective-social value as associated with the current network interaction, as coming from the upper. For instance, if a reserved communication is requested from the upper layers, this could determine, at this layer, a stream ciphering request for the underlying IP level for example by means of labeling (coloring) the data flows according to the subjective-social value. Coloring according to such values can also be exploited at this level in order to manage transmission and information routing in order to cope with network performance scalability requirements in case of network congestion or other failures: this could be for example the case for an emergency communication, that would specifically require a very high reliability quality of service level also in the case of network congestion. The Adaptation Layer will then read the data from the upper layers and it will actualize the requested personalization functions producing suitable commands for the classical communication layers, such as data flow coloration and stream ciphering in the IP communication environment.

3.3.2 The communication layer

This layer contains all the classical, already implemented communication protocol, services stacks, such as the IP-based protocols and services used in the Internet. We can include in this logical layer the actual OSI-like protocols from the TCP to the Physical layer. This logical Communication Layer is expected to actualize all the subjectivizing requests as processed at the higher layers and imprinted by the SGI engine.

4. Conclusions

A new logical network architecture has been introduced in order to cope with new key requirements posed by proper exploitation of user subjectivity. Such an architecture is composed of two parts: i) a cognitive part (the Subjectivity-Gene Imprinter - SGI), in which the operation of intelligent agents and exploitation of user subjectivity and collective knowledge, are asked to produce the specific requirements for the second part; ii) the Actuative multi-layered part (the Actuative Stack - AS). A really subjectivized End User's interaction with the Global Network is thus finally expected.

Focus has particularly been placed on the Actuative Stack, in order to outline its functions as well as its consolidation, especially as regards its technical feasibility. Much additional work is still needed for full specification of the AS, especially as regards its integration with the SGI. The proposed architecture gives raise to wide developments in research and applications on human-computer interaction. A real universal access can be devised while applying the concepts described above.

References

1. D.Giuli "Background and objectives of the B.E.S.T beyond Internet framework research program", Dept. of Electronics and Telecommunications, University of Florence, Internal Report DET-1-01, February 2001.
2. D. Giuli "Subjective information and the subjectivizing gene for a new structure of the instrumental relation and interaction process in the Global Network", Dept. of Electronics and Telecommunications, University of Florence, Internal Report DET 2-01, February 2001 (in Italian).
3. D. Giuli, "From Individual to Technology Towards the Global Network", HCI International 2001 9th International Conference on Human-Computer Interaction, August 5-10 2001 New Orleans Lousiana.
4. D. Calenda, D. Giuli, "Subjects, subjectivity and privacy in the Global Network", HCI International 2001 9th International Conference on Human-Computer Interaction, August 5-10 2001 New Orleans Lousiana.
5. M.C. Pettenati, D.Giuli "Human Subjectivity Profiling Factors for the Global Network", HCI International 2001 9th International Conference on Human-Computer Interaction, August 5-10 2001 New Orleans Lousiana.
6. P.Bussotti, G.Vannuccini, D.Calenda, F.Pirri, D. Giuli "Subjectivity and Cultural Conventions: the Role of the Tutoring Agent in the Global Network", HCI International 2001 9th International Conference on Human-Computer Interaction, August 5-10 2001 New Orleans Lousiana.
7. L. Breeland, "The rise of managed service offerings for contact centers", Alcatel Telecommunications Review, 2nd Quarter, 2000, pages 142 – 147.
8. P. Maglio, R. Barrett, "Intermediaries Personalize Information Streams", Communications of the ACM, August 2000 / Vol. 43, n°8, pages 96 – 101.
9. D. C. Schmidt, V. Kachroo, "Developing Next generation distributed applications with QoS enabled DPE Middleware", IEEE Comm. Magazine, Oct 2000.

Achieving universal usability through web services architectures

Charles Wiecha, Stephen Boies, Margaret Gaitatzes, Stephen Levy, Julie Macnaught, Paul Matchen, Scott Mcfaddin, David Mundel, Rich Thompson

IBM T.J. Watson Research Center, P.O. Box 704, Yorktown Heights, NY 10598 USA

Abstract

The goal of universal usability is to broaden access to information and communication technologies for a greater variety of users, interacting with a wider variety of devices, to perform an ever more valuable set of tasks. This paper starts by describing the emerging need to support use by a greater variety of the end-user population. We argue that new infrastructure for mass customization ("web services") is in fact emerging--not driven by but highly suited to the requirements of universal usability. We examine the capabilities of this emerging architecture to increase the variability of application presentation, navigation, and function. Next, we give examples of key variations in interface behavior important for universal usability that may be enabled through web services. Finally, we present a worked example of how we have used this approach to support a greatly increased level of interactivity and end-user assistance than is typical for web applications.

1. Introduction

Universal usability is important for both ethical and commercial reasons. Ethically, we seek to reduce the "digital divide" by ensuring effective—i.e. usable—access to computing technologies. Commercially, we seek future growth in the market for information services largely in finding ways to enable new types of users to perform new tasks any time and anywhere they like.

We seek to go beyond the common approach to universal usability wherein the requirements of selected populations are supported through special-purpose standards and technologies. Further, we see as impractical the construction of integrated applications that seek to support all types of users either through a "least common denominator" or perhaps worse through an "all possible functions" approach. Rather, the problem is to arrive at a new application technology that supports economic mass customization of information service functions and their interfaces.

1.1 The evolving application design problem

Application and solution developers now confront several "variety challenges" as they develop, implement, and evolve solutions that are to be universally usable: a variety of users, a variety of devices and channels, and a variety of roles and functions.

1.2 Logical decomposition, or factoring, of next generation web applications

The evolving web services architecture is creating the capacity for attacking these variety challenges. The computer science community has a long history of attempting to abstract user interface functionality from its look and feel. Web developers are beginning to apply some of these ideas using XML and XSL (see http://www.w3c.org) to separate style information from HTML using cascading style sheets. The trend in HTML is toward removing ever more style information and treating the XHTML document as abstract presentation markup. The user interface architecture then consists of content definition, abstract presentation, and concrete presentation produced by the application of one or more XSL style sheets to the abstract presentation.

The trend in next generation web architectures is to extend this practice and to factor the application design further in order to separate issues such as device, navigation style, geographic localization, and personal preferences into multiple transforms as shown in Figure 1. Each arrow in the figure depicts variations in the type of task, user, or device handled by applying an application transformation, typically implemented in the extensible style language, XSLT, of the web.

1.3 Physical decomposition of web applications

A second trend in the next generation web is the transition to a machine to machine architecture (see http://www.ibm.com/developer/webservices for tutorial information on web services). This approach provides access to web functions by programs through XML formatted remote method invocations, rather than by end-users through relatively unstructured pages.

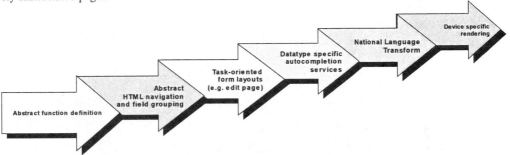

Figure 1: Logical decomposition of application design into an example series of sequential transforms. Data, presentation, navigation, and transaction specifications are all candidates for transformation.

A web services architecture leads to applications being developed through the composition of a series of services which either originate at content providers (so-called "origin" servers) or within the network (so-called "intermediary" servers). Intermediaries may add value through aggregation, augmenting or subsetting function, caching, and/or providing some or all elements of the user interface (Barrett & Maglio, 1999). Note that there is no requirement for intermediaries to be located on different machines across the network. Indeed, the benefits of physical decomposition of applications extend to co-located services and can be implemented with optimized connection paths avoiding the need for repeated serialization and messaging among the various components.

Together, the trends toward logical and physical decomposition of web applications lead to an architecture where any number of intermediate services may be inserted between content producers and consumers, each of which can be seen as implementing one or more transformations in the chain shown in Figure 1. These capabilities greatly reduce the cost of customizing web function and presentation for the requirements of populations too small to serve in the past.

2. Example variations in presentation, navigation, and collaboration

We believe that the evolving web services architectures outlined above significantly improve the prospects for realizing the goals of universal usability in mainstream practice. In this section we give examples of some of the dimensions along which applications may be varied which meet high-priority requirements of universal usability.

2.1 Supporting alternative and multiple interaction devices

The most obvious step toward supporting a variety of users is to deal with a greater diversity of devices for interaction. Many key aspects of human variation can be addressed through alternative devices without changing information content, modifying navigation, or transforming an individual interaction into a collaborative one. Screen readers and voice recognition are examples.

Historically, the greatest difficulty in effectively enabling these devices has been the need to communicate with them using overly concrete application specifications. In terms of the diagram in Figure 1, previous attempts to support alternative interaction devices have come at the end of the chain when interface decisions such as style, presentation, and navigation are already targeted to the wrong audience. Parsing pages filled with low level and irrelevant markup, and recovering the original abstract intent of the interface designer have been too difficult to achieve in scale and in practice.

Rather, the web services architecture allows for interface design to be factored into abstract intent and alternative chains of transforms to map that intent into realizations appropriate for given interaction platforms. Importantly, in the case of customized interaction devices, some of the final transforms should be seen as taking place either within or by the device itself. Thus we can connect directly to the underlying functional content and apply whatever transforms are appropriate to given case.

2.2 Expert/novice variations in navigation

A second technique helpful for first-time or occasional users in challenging interactions is to decrease the number and density of fields on a given screen. This approach trades off increased navigation among screens such as that in Figure 2 for the reduced complexity possible by working with one or a limited number of fields at a time. Those more expert with the application may prefer the opposite design tradeoff, shown in Figure 2 as well.

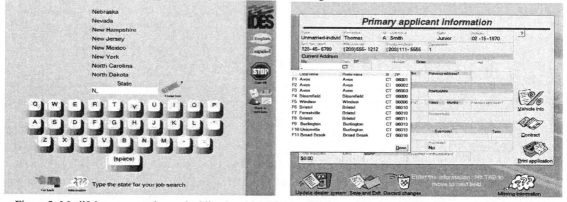

Figure 2: Modifying presentation and adding local intelligence for data validation. Novice screen used for interacting with a single data field at a time (left). Expert screen with high density of fields and pop-up data assistance (right).

We have seen, as well, that a given user may perform as an expert in some parts of an application, whereas in others (perhaps those used less often), they are novices. As a training device, therefore, it is extremely useful to be able to alternate on demand between the two modes.

2.3 Assisting users through collaborative transactions

We often think of universal usability in terms of adapting user interfaces to variations in individuals without thinking of expanding the social context of interaction to include additional participants as an aid to completing work successfully online. Children, for example, might be able or permitted to carry out various elements of a network-based service but not others. They might, for example, initiate the online purchase of book or CD but not be allowed to complete the purchase without parental approval. Typical web filters treat such interactions as either allowed or disallowed depending on user characteristics. Rather, we can build on the capabilities of intermediaries between content producers and end users to complete some tasks by individual interactions and others collaboratively. Either alternative can be invoked without the need for modifying the underlying structure of the back-end application.

The fundamental insight arising from experimenting with all of these techniques is that our challenge is to support a continuum of diversity, rather than simply its two endpoints. Our specific application experience strongly indicates that a combination of these techniques, including the example in the following section, is in fact required for the bulk of the population to be successful users.

3. A worked example: assisting complex data entry through successive transforms

The rapidly increasing number of interactions that can be carried out over self-service web sites promises to make users ever more self-sufficient in the many administrative tasks facing them day to day. Examples include on-line travel reservations, financial, insurance, or real-estate interactions, and healthcare advice and referrals. By their nature, most of these transactions require rich data about the people involved (applicants, citizens, students) their histories (work, educational, medical) their preferences, skills, and a variety of other numerical, enumerated, or semi-structured information. Rarely can a self-service channel be utilized without the need for the end-user to provide a rich and highly interconnected set of data as part of the interaction. How can we enable a wider range of people to benefit from the ever-increasing complexity of these self-service channels?

In traditional so-called "thick" client interfaces, implemented with conventional client-server tools, it is often possible to add some level of data validation given the ability to install extensive packages directly on the end-user's machine. The "thin" client (i.e. browser) of the web, where function usually is downloaded from a remote server, requires new techniques to distribute interaction intelligence dynamically out to the point of contact with the user. The amount of support required, whether it can be applied directly within the client or only remotely on a server, and the appropriate tradeoffs between help and performance, can not be known beforehand. These design decisions must take into consideration characteristics of the particular user, device, and network connection involved.

In the example presented in this section, both the logical and physical layering of the web services architecture described above combine to support variations in the appearance, timing, and actual data content of user interactions. Logically, the structure and scripting of the view will vary depending on the transformations applied. A single <input> statement in HTML, for example, will be expanded greatly with interaction techniques to provide feedback on partially entered data. Physically, an intermediary knowledgeable about the data domain of each field dynamically connects that field to a database elsewhere in the network that assist the user in completing its value to produce the interaction technique shown in Figure 3:

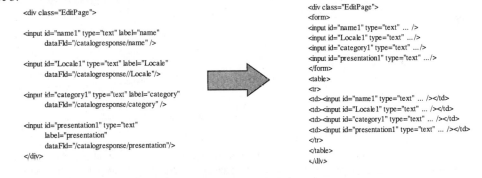

Figure 3: Transform from abstract field grouping to task-oriented edit-page form.

The XHTML form used to specify the desired inputs in this example is shown above. No styling or layout decisions are expressed in this first form. Elements such as <div>, <input>, or <button> are interpreted as not yet committed to a particular rendering as breaks, input fields, or concrete buttons (see www.uiml.org as well as Cover (2000) and Wiecha et al. (1990) for additional markup languages used for this purpose). As can been seen in the HTML code fragment, the <input> tags have not yet been refined in structure to build the label and input area structure that is repeated within each field on the page. This "leaf" level transform, controlling the local formatting in a single field, comes quite late in the chain and is independent of this first "panel" level transform.

3.1 Transforms between abstract presentations

The next transform in our example, corresponding to the "Task oriented form layout" arrow in Figure 1, takes as input the abstract form above, and further refines it to better support it for use in an editing task. In editing (as opposed to read-only browsing), we want the user to be able to select the products offered by navigating a selection list coupled together with an editing form in the same interaction. This design pattern is again captured in an XSLT style sheet, and is still independent of downstream look and feel considerations.

This transform to build an edit page recognizes the `class="EditPage"` attribute and copies the <div> and <input> fields of the original HTML into an editing form shown at the top of the figure above. A scrolling list is created by again taking the <input> elements from the original code fragment and placing them inside a <table> block. Only one set of row data elements is generated by the transform to describe the pattern of fields contained in one record of the table.

At run time, event handlers will replicate the HTML from the given record's pattern as many times as required for the desired number of rows to be displayed. Additional event handlers synchronize the currently selected record in the table with the elements in the editing form as the user interacts with the list. Together these two areas support the user's task of scrolling through a list of records with the contents of the currently selected record exposed in an editable form. Each field of the edit form may then have associated edit assists--such as auto-completion or local validation--added by the next transform discussed below.

3.2   Rendering and style transforms

Transforms closer to the right hand side of Figure 1 begin to address concrete issues of how abstract controls such as the <input> tags we have been dealing with so far will be rendered in the presentation. These are transforms that both alter the structure of the abstract presentation—typically by refining single elements such as the <input> into multiple ones to implement a particular visual design--and add style attributes such as font and color. A rendering transformation is the one most often thought of by application authors today who make use of automatic styling technologies such as XSLT. The style we desire for input fields places a left-justified label above the actual <input> entry field. In addition, event handlers are attached to the entry field to provide visual feedback with color changes as the user enters and leaves the field. Thus a typical transform result for a single field is as follows:

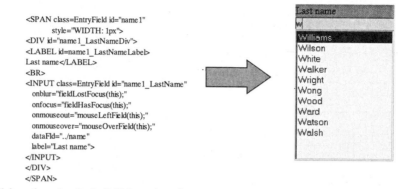

Figure 4: Elaboration of a single field from the edit-page of Figure 3 adding visual label and event handlers (left). Further transformation of the input element by adding dynamic autocompletion handlers and pull-down list (right).

Some devices may have the potential to support quite sophisticated and highly interactive displays. High-end browsers having support for dynamic HTML, local XML parsers, and XSLT transform tools are now available. Such platforms can be used to implement data-intensive user interfaces with functionality and performance at or above traditional, "thick-client" interfaces. When using such a platform, we apply an additional transform that uses information about the data types and valid ranges of common elements in an interface (e.g. names and addresses) to assist user interaction within those fields. When network bandwidth further allows, we couple this auto-completion logic with keystroke level validation at local servers. In this way we can achieve higher usability and lower error rates for complex interactions.

4.   Conclusions and directions for further work

While we have described our preliminary experience with assembling applications and presentations from multiple chaining transforms, much work remains to be done to understand the multiple dimensions along which it is (1) desirable and (2) feasible to factor user interface design. An important issue is to understand whether the approach to interface design through chaining multiple transforms is attractive to developers, designers, and managers from the standpoint of the programming model. A second issue is to understand whether a "fixed" set of layers can or should be identified as a standard approach to interface layering. Perhaps most important, we must understand in more detail what the performance implications are of this approach to interface design. How can sets of transforms be compiled to improve run-time performance while not losing the ability to respond to run-time circumstances in selecting the transforms that should be applied?

References

Barrett, R., Maglio, P. (1999). Intermediaries: An approach to manipulating information streams. *IBM Systems Journal 38* (4).

Cover, R. (2000). XUL, Extensible User Interface Language. From the world wide web: http://www.oasis-open.org/cover/xul.html

Wiecha, C., Bennett, W., Boies, S., Gould, J., Greene, S. (1990). ITS: A Tool for Rapidly Developing Interactive Applications. *Transactions On Information Systems 8*(3), pp.204-236.

MULTIMODAL, CONTINUOUS AND UBIQUITOUS INTERACTION

Ubiquitous access to community knowledge via multiple interfaces: Design and experiences

Alessandra Agostini °, *Giorgio De Michelis* °, *and Monica Divitini* ~

°DISCO - University of Milano Bicocca, Milano, Italy
~IDI – Norwegian Univ. of Science & Technology, Trondheim, Norway

Abstract
Campiello is a system supporting the exchange of information among communities living in art cities, with the goal of turning local inhabitants and tourists into active participants in the creation of knowledge. In order to provide accessibility to all the members of the communities, Campiello adopts multiple interfaces. The paper presents the experience with the use of Campiello in Venice with focus on how its interfaces support access to the underlying community knowledge.

1. Introduction

One of the challenges of systems addressing the needs of social communities is to "... ensure accessibility and usability of community-based information resources by all potential users ..." (Stephanidis et al., 1999). This is one of the key issues addressed in Campiello. Campiello (Campiello, 1997-2000; Agostini et al., 2000a) is a system supporting the dynamic exchange of information and experiences among communities of people living in art cities, and between these communities and foreign visitors. Its goal is to turn local inhabitants and tourists into active participants in the creation of local knowledge, enhancing at the same time their chance to comment on, critique, and make use of it. In this way Campiello aims at reinforcing the boundaries of local communities and opening them to visitors. In a system like Campiello the users' characteristics, preferences, and behaviors are particularly vast. Therefore we claim that the efficacy and effectiveness of its services must be coupled with the universal accessibility of the system. To reach this objective, in designing Campiello we pointed out six main requirements that should be met by the system. These requirements have a general applicability, but the relevance of each single one strongly depends on the typology of the system. Let us summarize the requirements:

1. The knowledge of the system must be locally accessible from the territory occupied by the communities; for instance, in the case of Campiello, the system must be as much as possible ubiquitously accessible within the city.
2. The knowledge of the system must be accessible to people who are physically distant, temporarily or not, from the communities. In other words, some of the services of the system, ideally all the services, should be remotely accessible.
3. The knowledge of the system must be accessible to all members of the communities, independently from their genre, age, cultural and educational background, and, specifically, whatever is their technical skill.
4. Different uses of the knowledge must be supported, in particular both active and passive uses. For instance, both creation of new knowledge as well as fruition of it must be allowed to users.
5. The access to the knowledge must be differentiated not only with respect to the specific aim of the user, but also depending on the user role, preferences, and behaviors. In particular, both the provided services and the filtering and visualization of the knowledge must be differentiated in terms of the above specificity.
6. The system must allow both individual and group access to knowledge since community practice includes both individual and group activities.

Due to the differences among these requirements, a single device for accessing the system cannot fulfill all needs or at least would poorly meet some of them. The above requirements imply in nuce diversity and multiplicity of interfaces, since they cannot be met simply providing a broad range of services on the same interface. Let us emphasize that, providing different interfaces, it is necessary to create a multi-dimensional continuity: conceptual continuity on perceiving the system as a unique entity; and, continuity on the use of different interfaces and on switching among them. In the next section we will briefly describe Campiello and its interfaces. Section 3 introduces the experimental adoption of the system in Venice. Section 4 presents the results of this experiment with respect to the usage and efficacy of the three interfaces. Section 5 concludes the paper and presents a reassessment of the requirements on the basis of the experiment.

2. Campiello and its interfaces: initial assumptions

In Campiello information is proposed to users according to their profile, their preferences, the history of their (inter-)actions with the system, and to their physical location. Information is filtered and recommended by agents that attempt to transform this large quantity of information into a qualitative selection of shared knowledge. Moreover, the system stimulates the reaction of users in the form of annotation: comments, ratings and further contributions can be attached to the 'core' piece of information (Agostini et al., 1998). The architecture of the system can be split into four levels. At the bottom level there are two interlinked repositories for the knowledge: a RDBMS server and a web server. Over the repositories there are the services constituting the Campiello core: mainly, the information filters (search and recommender (Glance et al., 1998;

218

Grasso et al., 1999)) and the map server for localizing each piece of information on a map. These services can communicate so to be used together. Over the core services, there are the display managers applying the algorithms specific of a class of interfaces, and the interfaces themselves. In Campiello, users can interact with the system through a variety of interfaces: large screen, paper, and PC. These interfaces have been selected considering their potentialities in supporting ubiquitous access to information and in assuring a high degree of accessibility and usability by the whole community. Other interfaces (e.g., mobile phones and palmtops) could be integrated in Campiello to enrich its capability to be used in any occasion by any person. Let us recall in the following the main characteristics of the three developed interfaces and how they meet, according to our initial assumptions, the above requirements for universal access to community knowledge.

The *PC interface* is a standard web-based interface to the knowledge base. With respect to the first requirement, since it does not require any installation at the client side, computers available in the territory (e.g., offices, schools, and homes) can be used for accessing Campiello obtaining a medium-low degree of accessibility from the spaces occupied by the communities. Second, the web-based interface has the advantage of providing remote access to users physically distant from the city. However, this interface requires the willingness of using a computer and therefore it is not usable by all members of the community. On the other hand, it allows both active and passive use of information. From the passive perspective, it allows easy surfing and searching. From the active perspective, it is still the most efficient way to insert, modify, and organize digital information. Thanks to the login mechanism, the identification of users is possible. This allows different accesses to information and to the offered services. Moreover, user's identification allows for the collection of preferences and behaviors of users and therefore the personalization of services. Lastly, the PC interface is more adapt for an individual use, even if it is not rare its use by small groups.

Figure 1: Two examples of the paper interface (NewsCards)

Thanks to a Xerox technology, DataGlyph (Johnson et al., 1993), users can access the system through paper. DataGlyph is similar to the barcode but it has a higher density of information and a higher robustness to errors; moreover, it can be printed by any printers and read through ordinary scanners and faxes. With DataGlyph you can, e.g., encode all the human-readable data in a flyer to become computer-readable or selectively launch programs depending on the 'state' of specific areas of a flyer (e.g., written or not). In Campiello, flyers including a DataGlyph are called *NewsCards* (Figure 1). Through NewsCards, users can ask for specified information simply by checking with an ordinary pen a box on the flyer. To get the answer, all users must do is to send the flyer in by fax to a specified number. Moreover, they can add their hand-written comments on a particular topic to share with other users. A Personal Identification sticker (PID-pad) is provided to a user when she registers with Campiello; it can be stuck on NewsCards so that the system can recognize the user as with the PC login.

Paper can be distributed at low cost all over the territory, creating a not intrusive form of ubiquity (Weiser, 1993). In addition, paper can be easily carried around so that people can collect a NewsCard in one place and process it later in another place. Paper can also be mailed and processed by fax allowing remote access to the system. However, in Campiello NewsCards have been designed for being distributed within Venice to *enrich a place with information* and they generally contain information related to the place in which users can find them. The paper allows access without requiring users any technical skill. Therefore, NewsCards make the system accessible by all members of the community. With respect to the passive and active use of knowledge, simplicity of use and easy distribution were expected to stimulate a more active role of people, for example in commenting and voting the knowledge. Thanks to the PID-pad, personalization of the services is possible. However, not all computer-available services are appropriate for being implemented on paper and some functionality needs to be redesigned. In particular, due to the higher answer time of paper with respect to computers, it seems inappropriate to implement functions requiring multiple steps of interaction between the user and the system. Paper has been designed for being used by single individuals even if circulation of flyers among people is a common practice.

Finally, users can access the system by means of large screens (denominated *CommunityWalls*, CW in short) where a selection of the information gets briefly visualized and where people, interacting with the screen, can request the complete visualization and/or a printout of specific articles. The items visualized on the CW change continuously. They are automatically selected, thanks to various parametric filtering algorithms, so to reflect the interests of the community, e.g. the most read or commented articles, the newest one, the ones related to the place where the CW is installed. The CW is intended to provide an overview of the '*most interesting*' information and, most importantly, it is designed for triggering conversations among its viewers. Though a wide distribution of CW is difficult, due to the high cost of hardware, we expected that even a limited number of installations could have a strong impact on the community in terms of visibility of the system. In addition, if social interaction takes place around the CW, then it is easier that the system becomes a part of the community life. All members of the community are able to read the information displayed on the CW (even the youngest illiterate kids are attracted by the continuously moving images of the articles, photos of the people who inserted comments, etc.). However, just a part of the community is able to interact with the screen for obtaining the full CW services. The CW is mainly devoted to support a passive fruition of the knowledge. Lastly, in the actual implementation, the CW is mainly intended for an undifferentiated access of people, and no personalization of the services is available. In fact, the key reason for developing the CW was exactly to support an undifferentiated group access to knowledge.

3. The experiment

To evaluate Campiello an intensive period of experimentation was set up (Agostini et al., 2000b). The period, from the end of May 2000 till beginning of July 2000, was selected so to encompass various types of community events. This allowed for the evaluation of the system in connection to special local events (e.g., the annual feast of the patron saint, the end of the school year) as well as to its day-to-day use. During this period both qualitative and quantitative data have been collected. Various representatives of the local communities participated actively in the project with the twofold role of influencing the design of the system and producing the content of the knowledge base. In particular, one school (P.F. Calvi), one bilingual magazine (Leo), one bilingual newspaper (VeneziaNews) and, one Museum filled in the majority of the Campiello content. The system has been accessible during the experiment, and still is, at http://www.campiello.org. For a higher availability of NewsCard, during the experiment one issue of the Leo magazine including a Campiello NewsCard has been distributed (Leo is distributed approximately in 15,000 copies). Complimentary copies of Leo have been distributed in the rooms of selected hotels. In those hotels, faxes were available to clients for processing NewsCards. Three Campiello areas have been set up: one at the Historical Naval Museum; one at Maria Ausiliatrice, a former church in a popular district; and one during the week-long annual feast of the San Pietro patron saint, in the bell tower of San Pietro church. All sites allowed access via PCs (2 for each site), large screens (1 for each site), and paper interface (15 different NewsCards were available).The areas have been selected in order to reach different types of potential users. The Naval Museum had been thought mainly for tourists, the other two areas mainly for locals, one for everyday conditions, and the other in connection with a community event. All three sites were located within the sestiere of Castello (Venetian neighborhoods are named "sesticri"). In fact, from the beginning of the project we choose this Venetian sestiere for its characteristics (e.g., number of inhabitants, number of tourists, associations, cultural and educational institutions). See (Agostini et al., 2000b) for more details.

4. Evaluation

One of the objectives of the experiment in Venice was to verify the validity and completeness of the six requirements for promoting accessibility by the whole community as well as the initial assumptions on how the selected interfaces meet these requirements. First of all, let us point out that multiple interfaces proved very useful in providing a wider access, because they allowed both to cover a wider territory and to involve people with different background, who would have not accessed the system through a computer. In general, people had no problems in perceiving the system as a whole. Accessibility of the knowledge from the territory of the community has proved to be an important requirement. The smooth integration with the territory is useful to foster casual encounters with the community as well as to assure that users come back to the system. In particular the diffusion in places familiar to the community has assured the access by people that would have never accessed the Campiello web site, especially old people and children. Many people came in contact with the system while strolling around or during the daily shopping. For some of them this casual encounter has turned into an almost daily habit. As expected the paper is particularly suited to assure the distribution on the territory at a relatively low cost. In fact it proved to be easy to distribute and easy to reproduce whenever needed. The assumption on the role of the CW was confirmed as well. The PC proved to be more useful than expected. The cost of PCs and the risk of vandalism make difficult to have a wide distribution of dedicated PCs in public spaces. However, as foreseen, PCs already in use in the community, in our case mainly in the schools and at homes, have become an access point at no cost. For instance, people with computers offered their availability to give access to others when specific tasks had to be performed, e.g. for inserting new content. We have however observed that a wide distribution on the territory is not sufficient for assuring the usage of the system. For example, we have experienced a considerable difference of interaction depending on the physical setting within which the system was made available, and the usage of the setting from people, e.g. during a feast or visiting a museum. The distribution of the system on the territory must be carefully planned because it has a key role in determining the success of the system within the community. The prominence of the location of the stands in the city; the organization of space within stands; the additional use of the place in which the stand is posed; etc. all resulted more relevant than foreseen. For example, during the experimentation the approach to the system changed quite dramatically in occasion of the S. Pietro Feast. The occasion allowed for a very easy going interaction both for locals and visitors, proving to be particularly appealing to the former who, feeling at home, were happy to find out what other people they knew had made to contribute to the system. When people simply stroll around the Feast area they seemed to be in the right situation for looking about and "lose their time" for something whose finality is not immediately apparent. Instead, in the Museum stand we have observed lack of time emerging as a major concern. Moreover, people at the museum were very focused in their expectation. They were mainly looking for information about the museum and they would have probably looked at the system if it were the "museum web site", but they were not always willing to act a shift in their focus.

The possibility to access the system remotely has not been used much. However tourists and locals alike ranked this possibility high. Though the PC interface was the one designed with this purpose, the possibility to submit NewsCard by fax proved to be as powerful, even if allowing limited interactions. The possibility to access the system remotely was appreciated especially in connection to the willingness to play an active role, e.g. tourists giving comments when back at home and locals inserting new content from home, when more time is available. The possibility of an active use of the knowledge proved particularly relevant, especially the insertion of new content and comments, rather than ratings. The active role that users can play is in fact essential for people to take possession of the system and make it a part of their life. During the month of experimentation we have seen a growing number of people coming to visit the system because attracted by the possibility to give their contribution, e.g. by advertising an existing association, for finding a wider audience to their work, for increase knowledge about an underrepresented community, and in general for disseminating and sharing knowledge. Independently by the motivations, all the people we have been contacted by were driven by the feeling of belonging to a community and the wish to give voice to this community. The PC interface proved to be the most effective for supporting people in playing an active role. Paper also proved useful, but to a much lower degree, for the insertion of comments, though some limitations have been detected in connection to the lack of flexibility of the writing area. However, we believe that paper could play a key role for the active participation of people if it were used to insert their paper material. In fact, most of the material that people wish to insert is available in paper format. Thanks to the paper interface, this insertion could be easily done, for example, by providing a specialized type of NewsCard that allows for the easy identification and classification of an item in Campiello, and then the insertion of scanned images and texts constituting the content of the item. The CW was intended only for passive use of the knowledge and this proved to be a limitation that can impact on its use. For example, some pupils of the Calvi School perceiving themselves as content providers almost completely neglected the CW because too passive. Regarding the personalized services, we experienced that, unfortunately, the recommending system —based on commonalties in rating articles on the same topic— never had enough start-up information. We believe that this happens for two reasons. First, most of the interactions are from non-identifiable users. In fact, due to time and to a less extent privacy concerns, people were often not registering, interacting with the system as guests. Second, the number of interactions with the systems, especially rating, needed for activating a recommendation is too high. We believe that other rules should be integrated with the existing ones; for instance, considering other kind of user actions (e.g., comments). Moreover, our experience observing the CW shows that the rules adopted for the selection of items in the CW could be positively integrated in the recommender. In fact, people appreciated very much the selections of information applied within the CW. Individual and group access to knowledge proved to be much more important than expected. In Campiello the CW was the only interface specifically designed for supporting the group fruition of knowledge and foster the socialization around the system. However, during the experiment we have observed that people tend to socialize around the system independently by the interface that they are using. These interactions are started, e.g., for helping people to understand the system, discussing/evaluating the content of the system, and adding personal knowledge that is not in the system. Campiello was observed as particularly effective in supporting socialization for locals, and in particular to ease the interaction between elderly and young. Interesting was the case of a man who, hating computers, categorically refused to use Campiello. However, he visited the Campiello area at Maria Ausiliatrice repeatedly and every time he got involved in long conversations with younger people (who were using the system) providing information on Venice. The discussions always started from the content of Campiello. Moreover, often students came with friends either to help them register or to show their work inserted in Campiello. Frequently, starting from browsing their work, they become interested in Campiello content authored by other people. In general, we have observed that young people tend to socialize more around the PC that around paper and CW, probably because of the richer interaction and active possibilities offered by the PC. The socialization around the system naturally leads to group interaction. Even if an interface is not specifically designed for group access, still we have experienced that people build practices around it to gain the group fruition of the knowledge. For example, we observed two friends that registered as a single user in order to work together, and groups of people talking about a NewsCard and then deciding together what to ask. During the experiment people with many different backgrounds, from young children to old people, accessed the system. Some of them, for various reasons, completely refused the use of a PC. It was therefore important for them to have different interfaces, at least for the first interaction. Paper was intended to be an easy interface for all. However, this assumption did not proved to be completely correct. Pen and paper is simple to use, their use is well founded in western culture and a high percentage of people is able to make a cross with a pen in a piece of paper. However, we discovered that it is quite difficult to understand what to do with a form other than for reading the information contained on it. People appreciated the available NewsCards, but they have not used them to interact with the system as much as we were expected. In general, we have observed that people have problems in understanding the interactive nature of NewsCards. Some people asked about the differences between a NewsCard and a printed web page, could not spot any difference despite they had spent some time looking at both of them. In general, comparing the experience done this year with the experiment of last year, we think that the less information is on a NewsCard the easier it is for users to understand its interactivity. Instead, most of the distributed NewsCards were taken as information sources worth to be collected per se, without triggering the users to get more information.

5. Discussion and Conclusions

To fully understand the lessons the experiment can give us about Campiello and its potential utility, the presented evaluation should be contextualized taking into account the limits of the prototype and of the experimentation itself. In short, the Campiello prototype implements only a part of the services needed by local communities and visitors of art cities

In particular, it has only three interfaces, while more of them are needed to assure universal access and ubiquity. Moreover, the knowledge stored in it is highly incomplete and can only be considered as indicative of what can be considered as the collective memory of the community. Finally, the limited number of Campiello stands, the unreliability of the network connecting the stands, and the weak performances of some software modules strongly constrained the user's interactions. If we take into account the above constraints, we can fully understand the causes behind the observations we have made and outline some general guidelines for systems supporting social interaction.

- The use of multiple interfaces is a good way to involve the whole target community. When possible different formats should be adopted so to suggest intended uses, even if there must be always space for unexpected usage. This is better understandable if we consider the richness of available formats of paper. With paper we can select a different format depending on the objective to be reached such as small-size business card in order to be more foldable and large-size posters to be well visible and readable at distance.
- The presentation across multiple interfaces and formats requires that information be recorded properly for being presented in all the interfaces. This requires designing the knowledge base taking into account all of them. In addition it is necessary to provide switching facilities among the interfaces preserving context.
- Social practice around the system is of paramount importance and should be analyzed and taken into account in the design. According to our experience, for example, the group access to the system is wider than expected. This implies that group access should be provided for all the interfaces.
- Personalization algorithms are important, but they must be carefully designed in accordance with user practice.
- It is important to have continuous accessibility from different places: this seems against the adoption of kiosks as the only access points.

However, technical issues are not enough to assure the success of the system: a well-designed system can fail; an easy interface does not assure users actually use it. Following are some more general aspects that have to be taken into considerations for integrating the system as much as possible into the life of the community.

- Accessibility to the system has to consider the context where the system is made available so to foster casual encounters and re-encounters with the community as well as to meet the expectation of potential users and their availability to use the system. In particular, it is not enough to consider visibility and readability but it is necessary to provide sufficient space around the interface to support social interaction and for users to "take possession" of the tool.
- Assistance at the access points is essential for assuring a wide use of the system, especially where the intended users have no familiarity with computers.
- The collaboration of *promoters* with firm roots in the user community must be looked for since the very beginning, even before the system is in place. In fact, these persons can increase the knowledge on and acceptance of the system in the community. It is however important to take into account also that people could reject the system not because of the system itself, but because they have a negative attitude toward the promoter.
- Early involvement of different potential users in the design of the system and its content is key to its success. This not only assures that the system meets user needs, but also assures allies within the community to promote the system.

Acknowledgement

The Campiello results are due to the efforts of all partners, so our thanks go to them all. In particular: the Grenoble Lab.-Xerox Research Centre Europe, developer of the paper and large screen interfaces; the Domus Academy Research Center (Milano, Italy) designing the interaction modes; the MUSIC Lab.-Technical Univ. of Crete (Greece) responsible for the knowledge base and the PC interface. Lastly, the authors thank Alan Munro for his precious work during the experiment.

References

Agostini, A., Grasso, A., Giannella, V., & Tinini, R. (1998): Memories and local communities: An experience in Venice. In *Proc. of the 7th Le Travail Humain Workshop "Designing Collective Memories"*, Paris, France.

Agostini A., Giannella V., Grasso A., Koch M., Snowdon D. (2000a): Reinforcing and Opening Communities Through Innovative Technologies. In M. Gurstein (Ed.), *Community Informatics: Enabling Communities with Information and Communications Technologies*, Idea Group Publishing, pp. 380-403.

Agostini A., Divitini M., Munro A. (2000b): *Campiello Deliverable 1.4 "Evaluation of the Campiello Project: Methodology, Experiment, and Results"*. Available on request at agostini@cootech.disco.unimib.it

Campiello (1997-2000): Esprit Long Term Research project #25572, PC interface available at: www.campiello.org

Glance, N., Arregui, D., & Dardenne, M. (1998): Knowledge Pump: Supporting the flow and use of Knowledge in Networked Organizations. In U. Borghoff & R. Pareschi (Eds.), *Information Technology for Knowledge Management*, Springer Verlag, Berlin.

Grasso, A., Koch, M., & Rancati, A. (1999): Augmenting Recommender Systems by Embedding Interfaces into Practices. In S. C. Hayne (Ed.), *Proc. of the International ACM SIGGROUP Conference on Supporting Group Work*, Phoenix, Arizona, pp. 267-275.

Johnson, W., Card, S. K., Jellinek, H., Klotz, L., & Rao R. (1993): Bridging the paper and electronic worlds: The paper user interface. In *Proceedings of INTERCHI'93*, ACM Press, April 1993.

Stephanidis, C. et al. (1999): Toward an Information Society for All: HCI challenges and R&D recommendations. *International Journal of Human-Computer Interaction*, Vol. 11(1), 1999, pp. 1-28.

Weiser, M. (1993): Some computer science problems in ubiquitous computing. *Communications of the ACM*, July 1993.

Audio augmenting physical navigation in art settings

Alessandro Andreadis, Giuliano Benelli, Alberto Bianchi

Dipartimento di Ingegneria dell'Informazione, Università degli Studi di Siena,
via Roma 56 - 53100 SIENA - ITALY

Abstract

This paper introduces new interaction paradigms for navigating physical spaces. The adoption of new technologies and interaction modalities allow people to navigate physical spaces augmented by related information spaces (Benelli et al., 1999). An audio interface results to be easily and widely accepted from any type of user. In fact, people are used to receive audio information while guided in art settings either by an expert or by pre-recorded audio guides. Hyper Interaction within Physical Space (HIPS) is a research project, partially funded by the European Commission (4th framework, Esprit program). The HIPS prototype is context sensitive on new types of inputs. The device may disappear and the implicit user action (movement), together with a suitable user modeling approach, is enough for the system to infer how to assemble highly context-related and dynamic information. The user can simply put the device in his pocket, listening to the audio.

1. Introduction

The system is composed by several modules, communicating with each other through appropriate software interfaces; the following functional requirements are covered: the detection of user position; the internal representation of the visitor behavior and personal interests; the dynamic builder of context-aware personalized presentation in audio modality, augmenting the immersion in a physical environment.

It is not reasonable to simplify the interaction by only limiting functionality. We cannot expect from the user an expert behavior. The information content may be far from being intuitive, even if it is context-related to the physical environment and to the exact position of the user. The user has to react to unknown situations, taking his decisions, without being conscious of what may be related to a selection criteria. Moreover, he may not have a clear understanding and interpretation of the objects around him. Audio results to be a general medium for delivering information and it does not require personal skills and attitudes towards new *Information Technology* (IT) devices. An audio interface results to be easily and widely accepted from any type of user. In fact, people are used to receive audio information while guided in art settings either by an expert or by pre-recorded (and thus static) audio guides. In both cases, communication cannot be synchronized with user movement. For example, for the expert it is not easy to match the different interests of each user belonging to a visiting group, while for an audio guide, the user is tied to follow a pre-defined itinerary (i.e., audio contents are already assembled and designed for fixed navigation paths).

The paper is based on the *Hyper Interaction within Physical Space* (HIPS) project, partially funded by the European Commission (4th framework, Esprit program). The HIPS prototype is context sensitive on new types of inputs. We call "input" to the HIPS system both an *explicit* selection on the user interface and an *implicit* action in the physical space (i.e., to turn around, to move faster/slower). The system keeps track of the already delivered information, so as to be able to provide new contents on the basis of a wide set of options.

Instead of simply basing its output on stereotypes, HIPS has a more dynamic imprint, which fits properly with an augmented reality situation (Marti et al., 1999).

The device may disappear and the implicit user action (movement), together with a suitable user modeling approach, is enough for the system to infer how to assemble highly context-related and dynamic information (Not et al, 1999). The user can simply put the device in his pocket, listening to the audio (Bianchi et al., 1999).

2. Technical issues

The user actual and previous position and orientation in indoor environments are respectively obtained through a coherent use of *InfraRed* (IR) transmitters and the adoption of an *Electronic Compass* (EC). A number of IR emitters are fixed to the rooms' walls and to the ceiling. The user handles a *Personal Digital Assistant* (PDA) (Benelli et al., 2000) that is connected to an IEEE 802.11 compliant wireless *Local Area Network* (LAN), and he is equipped with an IR receiver. A similar configuration could be adopted outdoor as well, especially in small historical centers like Siena (Italy), where the HIPS experimentation was carried out. In such a case, the cell coverage may be reduced up to 50 meters (see fig.1), by reducing the emitted power at the minimum.

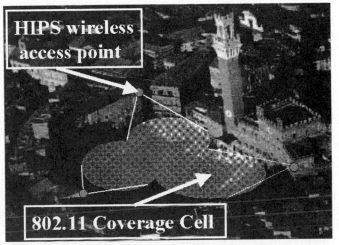

Figure 1: IEEE 802.11 HIPS test site coverage

A *Visiting Style Module* (VSM) estimates the degree of compatibility between the user's movement patterns and four movement categories defined in ethnographic studies on art settings (Veron et al., 1983). Hence the *Presentation Composer* builds a new presentation, picking out from an existing repository of short audio files (i.e., even a single word); the selection is based on the knowledge of the visitor's area/orientation, while information coming from the VSM is exploited to collect specific information formats (e.g., content length, reading style, deepenings) (Bianchi et al., 1999). The presentation is assembled according to the history of movement, of received contents and of interaction with the graphic interface.

The information is mainly audio, in order to let the user enjoy the artworks rather than having to interact with the tool. However, the visitor may decide to employ the visual channel, in order to listen to a more detailed description of an artwork or to orient himself through a map.

In this scenario, tracking the users position in the physical space is not only a matter of sensing: users move around in order to get information. In this sense, movement resembles a communicative act. As for all communicative actions, knowledge comes into play.

By developing the *User Interface* (UI) in Java (hoping in a quick achievement of a Java platform independent tool), the HIPS system relies upon a multi-platform user device. Having in mind third and fourth mobile phone generations with a reduced storage memory capacity and actual PDAs, the multimedia contents may also be kept server side and transmitted upon the 802.11 wireless link.

HIPS is based upon asymmetric connections, weighting differently up-link and down-link. The up-link is the channel used by the PDA to communicate to the Access Point, and the down-link is the opposite one. The up-link is used only for requests, explicit and/or implicit interactions of the user, while the down-link may carry the information payload. In fact not all the existing devices have enough memory capabilities to store the entire HIPS audio repository from which to play the audio files selected by the Presentation Composer. Our system implementation delivers to each PDA a differentiated amount of information (according to the user implicit and explicit interests). Users who are not very interested in detailed information require less network resources (i.e. bandwidth). In this way more collaborative users may take advantage of a higher user throughput in their own down-link. Moreover, the system has the following additional functionality, so as to save further network resources, to improve the QoS and the user satisfaction:

- it stops sending audio when it detects that the user has moved towards another object of interest (i.e., artwork);
- being aware of the already delivered contents, repetitions are avoided by dynamically assembling audio information
- audio length is adapted to the user movement style and to the average time he stays in front of objects

2.1 Architectural/integration issues

The architecture upon of the actual prototype is shown in figure 2. In indoor environments, a sensor position has been implemented by integrating IR with EC (while outdoor the Global Positioning System, GPS technology, could be used). The position recovery system has adopted the EC technology for orientation purposes and the IR for location purposes (Benelli et al., 2000). A JAVA based UI has been developed. Barriera is a C++ software module dedicated to the management of the data coming from the sensors and to their translation into a user position-orientation couplet. The position-orientation couplet is notified only when it changes, avoiding to waste time, HW/SW resources and wireless bandwidth.

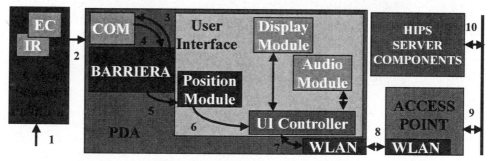

Figure 2: HIPS PDA side architecture

3. User movements and audio adaptation

According to the ethnographic studies of Veron & Lauvassier (1983), visitors of museums can be classified in four categories on the base of their movements (Marti et al., 1999).

The HIPS user (i.e., museum visitor) is free of moving without interacting or interfacing with a device, following all the stimuli coming from the physical environment, objects included. The user can move his hands freely, according to his common behavior in physical spaces, enjoying the exploration of the space and feeling supported and helped by an audio information related to what is attracting his attention.

The information received is presented according to the physical context and the history of the actual visit, having in mind the knowledge acquired, the visited places and the recorded explicit actions. This allows building the description, the conceptual fruition and the understanding of the art settings according to the user attitudes toward the physical exploration.

The information payload is strictly bound to the user behavior, avoiding to waste time and network resources with the transmission of audio files, which will be not listened to by the user. The audio is stopped automatically when the user moves towards another artwork, without obliging the user to listen to what is related to artworks he's no more looking at. At the same time the description can start again without loosing knowledge in the content, according to the new context, when the user comes back in front of a previous artwork. Boring repetitions are thus strictly avoided. Follow-ups and comparative explanations related to the already visited physical space are assembled reasonably according to the user profile. The couplet position-orientation is transmitted only when changes are detected and the adapted information payload is closely bound to the user physical navigation.

4. Evaluation

The user evaluation performed in HIPS, raised some interesting issues about the applicability and effectiveness of usability evaluation methodologies in the field of art, entertainment and leisure. Most of human activities in such contexts are not directly guided by a precise, aware objective defined ahead of time in the intentions of a person (Marti et al., 2000). The visit to a museum is an example of "non-goal oriented" activity since people can be pushed just by curiosity or pleasure, their behaviour is not predictable and their needs are "situated" (Suchman, 1987). Indeed when visiting a museum individuals generally do not anticipate alternative courses of action, or their consequences, until some courses of action are already under way. The visiting experience is a case in which individuals frequently "adjust" the way in which they interact with the environment, depending on many different and context-dependent factors (Marti et al., 2000). Visitors of the Museo Civico, who volunteered to try the system during their own visit to the museum and to be interviewed, evaluated HIPS prototype.

In general the first impression on the idea of HIPS is quite enthusiastic for many reasons:

- People felt very free in their movements during their visit
- Thanks to the dominant audio channel, users concentrated directly on the artwork they were looking at. Indeed many tourists say that they usually take a book during the visit, but they complain the fact to continually move their attention from the book to the artwork. People feel comfortable to use HIPS without interacting much with the PDA interface. Most of visitors who tried the system indeed used it only to listen to the audio comments.
- The User Graphical Interface was mainly used just in case of explicit user requests of additional and/or different information and it represented for them a suitable shortcut.
- The information is contextualised: HIPS follows people's movements and offers information related to each new context. All visitors appreciated this feature.

5. Conclusions

Several communication and position recovery technologies have been adopted to provide new types of inputs to the HIPS server and to allow the design of new approaches for the information retrieving.

Collaborative visitors of museums explore art settings in different manners according to their different level of knowledge of the artworks. If they already know the artworks or/and have already visited that museum they want to be free to move inside, perhaps standing in front of the artworks and listening for more detailed information. If they have no knowledge, they may want to learn as much as possible even reading or listening to very long explanations, and being also guided in fixed routes. It is out of doubt that in these cases their cultural enrichment does not follow their natural way of exploring the physical environment. Most of them seem not to suffer from these constraints, but is this the best way of visiting museums? Yes, of course, if there are no alternatives to this collaborative behavior. By using HIPS, the user feels explicitly the consciousness of receiving all the information details about the surrounding artworks, but coherently to his own style of visit, represented by his free movements in the museum physical space. This represents fundamental and innovative additional stimuli to the information fruition. Users don't loose time looking for signs when searching a famous artwork, consulting a printed guide, asking custodians around or hyper-navigating a fixed information space. It is enough to take the device out of their own pocket and ask the system where is that specific item. The system will support the user in the physical navigation to reach it and when the user gets close to it, the navigation can again be resumed accordingly.

References
Benelli, G., Bianchi, A., Marti, P., Not, E., Sennati, D. (1999). HIPS Hyper-Interaction within Physical Space. *IEEE ICMCS99*, Florence.
Benelli, G., Bianchi, A., Diligenti, M. (2000). A position aware information appliance. *EUSIPCO 2000*, Tampere.

Bianchi, A., Zancanaro, M. (1999). Tracking users' movements in an artistic physical space. *i3 Annual Conference*, Siena.

Marti, P., Rizzo, A., Petroni, L., Tozzi, G., Diligenti, M. (1999). Adapting the museum a no-intrusive usermodely approach. *UM99*, Banff.

Marti, P., Gabrielli, F., Bianchi A. (2000). Modelling non-goal oriented activities. *Building Tomorrow Today - I3 Third Annual Conference*, Jonkoping.

Not, E., and Zancanaro, M. (1999). Reusing information repositories for flexibly generating adaptive. *IEEE International Conference on Information, Intelligence and Systems*, Washington.

Suchman, L. (1987). Plans and Situated Actions. The problem of human machine communication, *Cambridge University Press*, Cambridge.

Veron, E., Levasseur, M. (1983). Ethnographie de l'exposition. Paris, *Bibliothèque publique d'Information*, Centre Georges Pompidou.

From Multimodality to Multimodalities: the need for independent models

Dominique Archambault[a,b], Dominique Burger[a]

[a]INSERM U483/INOVA - Université Pierre et Marie Curie
9, quai Saint Bernard - 75252 Paris cedex 05 - France

[b]Laboratoire d'Informatique du Havre, Université du Havre
BP 540 - 25, rue Philippe Lebon - 76058 Le Havre - France
Dominique.Archambault@snv.jussieu.fr, Dominique.Burger@snv.jussieu.fr

Abstract

This paper considers the general problem of media and modality conversion in human-computer interaction. It discusses a generic approach based on abstract models, independent of modalities. The particular case of interfaces for people with disabilities illustrates the discussion.

1. Introduction

Fifteen years ago, the development of personal computers, accompanied by the generalisation of networks - especially the Internet - and the apparition of Braille and speech devices, opened new perspectives for visually impaired people. At home, at work or at school, the use of electronic data should help social inclusion. Indeed, a priori, these data can be processed by a computer and used through different modalities; then, multimodal interfaces should be able to present them optimally according to the users' needs and specificities.

Electronic documents are usually planned to be used by one multimodal interface, implementing one static set of modalities, and not by several multimodal interfaces using various alternative modalities. The following discussion is illustrated by results from several researches we carried out during the last five years: Web accessibility [Archambault and Burger, 1999, Duchateau et al., 1999, Archambault et al., 2000b]; design of a specific Web browser [Burger and Hadjadj, 1999]; adaptation of workstations [Schwarz et al., 2000] with a view to enabling blind persons to work in an ordinary environment; pedagogical tools for blind pupils [Burger et al., 1996].

2. From Multimodality to Multimodalities

Human communication is performed through different channels corresponding to our five senses and to our means of expression (speech, gesture). Depending on the structure of the information transmitted, each channel corresponds to various modalities. For instance, Braille and tactile diagrams are two modalities corresponding to the tactile senses.

A multimodal interface is an interface which is able to use various modalities, and to provide users with various kinds of interaction, possibly through various channels of communication. For instance, standard user interfaces (Figure 1) involve a graphical screen, a mouse, a keyboard, and a loudspeaker. These devices correspond to specific modalities and types of interaction. As there is no or very little redundancy between them, it is called exclusive multimodality [Coutaz, 1991, Burger, 1994]. Usually the data which can be accessed through these interfaces are formatted to fit specifically to that scheme.

In more and more cases the need appears to access the same information using different types of devices, corresponding to alternate modalities; and this trend is increasing.

- Web Contents may be accessed using the reduced screen of a WAP mobile phone or PDA , or using the speech synthesis of an autoPC[TM]...

- CD-Rom, DVD-Rom or Web sites do not allow the same throughout of data. The same information can be presented as a high quality video on a DVD-Rom, and as a set of pictures or a very short and low quality video on a Web site.

Figure 1: Standard multimodal interface

- Visually impaired people can only access information via specific devices (corresponding to various modalities), namely: Braille displays, speech synthesis, tactile boards for children who cannot use a keyboard (because they are too young or because they have additional disabilities), or special devices like switches or move detectors. Adjustable screen displays are needed for numerous users: very large fonts for users with amblyopia, or very small fonts for users with a reduced field of vision; adapted contrast or colours for others; no animation for users who need some time to perceive still images. Thus each user may have use a specific multimodaj interface, composed with the devices that fits his needs (Figure 2).
- In addition, it seems interesting to develop cooperation between alternative modalities in order to compensate for the lack of vision [Hatwell, 19931: that is cooperative multirnodality. Cooperative multimodality is useful for developing adaptable interfaces [Emiliani and Stephanidis, 2000] which can exploit several alternative modalities according to the specificities of users, in order to, for instance, introduce redundancy (i.e. express the same information via several modalities simultaneously).
- Finally, especially in the case of young children, or children with multiple disabilities, it is important to design adaptive interfaces, that is interfaces which axe able to adapt themselves according to the user and the interaction context (repeated failure or success, action impossible to execute ...).

In each of these cases, information have to be adapted to fit the alternative device, that is information should be converted in order to fit the specific presentation rules which are associated with each modality. Therefore the data model should include all the elements which are necessary to correspond to the specific presentation rules of each modality.

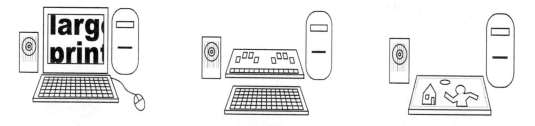

Figure 2: Three specific multimodal interfaces

3. Conversion

These conversions often request additional information that is difficult to find in the standaxd interface, because it is expressed graphically or through the layout of graphical objects in the window.

3.1 Reformulation

In order to optimise the data presentation, reformulation is often necessary. For instance in a Web page, the links have a special colour and are often underlined- the user can detect them easily. In order to facilitate the translation of such presentations into Braille, several reformulations are possible: put the link into brackets; or blinking the text of the link; using speech together with a special sound at the start of the utterance (using another audio modality) [Archarnbault et al., 2000].

3.2 Reorganisation

Even if the standard interface is purely textual, reorganisation of the data may be necessary to optimise the performance. In a workstation adaptation project, we noticed a remarkable improvement of the productivity of blind workers (from +25% to +60%, achieving performances similar to other workers); thanks to data reorganisation [Schwarz et al., 2000].

Web pages often include a large number of links grouped at the top of the page. The simple translation of such HTML documents into a sequential modality, like Braille or speech synthesis, will result in a drastic reduction of information accessibility, because of the very structure of these documents. This is not the case with database servers which make it possible to access raw data, then to structure its presentation and format it especially for each user. One may consider that the representation of information in a database is compatible with a data model independent of modalities, while it is not the case for HTML documents.

3.3 Special functionality fitting a specific modality

In some cases the specificity of a modality makes it necessary to provide special tools. This is the case with modalities which do not allow a global view of the document.

In the case of a web page, the user cannot know whether it is a long or a short page, and how many links it contains. The contextual global information available to a sighted user before he/she starts reading the page is not accessible to users of a non visual interface. A summary of the page content may compensate efficiently for this limitation, even if it is only based on a statistical analysis of this information.

Another case is when the information given by the context, for instance the layout of the screen, is necessary to understand the document. In some cases, specific help should be added. Once again, if the data model does not contain enough information, this help has to be designed specifically by an expert.

3.4 Data formatted exclusively for one communication channel

At the beginning, the use of the Internet seemed very promising for blind users because it was only textual. The fast development of graphical techniques made it more difficult, and the accessibility of the Web quickly became a problem. When the data model is linked to only one mode of communication, for example graphics, the only way to enable conversions into other modalities is to provide a textual alternative. Otherwise the conversion will cause an important drop in information accessibility. In the case of a Web server home page, the consequences may prove disastrous (inability to access the information available on the server or other related information sources).

To make HTML documents accessible, it was necessary to set up a large number of guidelines [WAI, 1998] - a sizeable work performed by the Web Accessibility Initiative. The main rule of the guidelines is that all information included in documents should be expressed in textual mode.

4. Toward new software models

Universal access to technologies means that user interfaces to these technologies become transparent. That is the interface should display information in the more appropriate way for each user. In other words the information should be converted to fit the modalities used by the user. To perform theses conversions, we can observe in the cases presented previously that:

- information should be stored in several redundant forms, to fit the presentation rules of every modality (for instance the alternative text of images in HTML documents)
- information should be stored in a well structured model that allows to change the structure of the display

For current projects, we need more and more robustness in adaptations, and more adaptativity. Therefore we have to design software models that fit these characteristics.

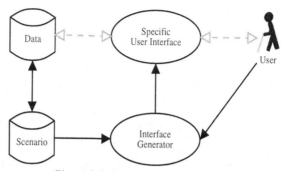

Figure 3: Independent software model

4.1   Computer games adaptation

The TIM project [Archambault et al., 2000a] concerns the adaptation and design of computer games for all visually impaired children (from 3-4 years old, including children with additional disabilities). The aim is to develop a tool allowing to produce adaptable and adaptative games. This implies the possibility for the game to change the modality according:
- to the devices used by the player
- to the player's competences in the course of interaction.

For instance, in the case of question-answer games, if the player is a young blind child, starting learning Braille, speech synthesis can be used for the questions, and the proposed answers displayed on the Braille display. Then if the child gives correct answers rapidly during a while, the questions will be displayed on the Braille display. Conversely when the child obviously has difficulties, the speech synthesiser can read the answers in the same time they axe displayed in Braille. For younger children, who can hardly understand synthesised speech the questions should be transmitted in audio format (with a recorded 'real' voice).

In fact, for young children, the adaptation implies larger conversions. A lot of games axe based upon the global vision of the layout and the visual memory. It is often not enough to simply present the game elements in an alternative way, but it is necessary to change the scenario of the interaction. Therefore the classical software models used until now to make adaptations of interfaces axe inappropriate.

The approach of the TIM project is to provide each user with a specific interface, instead of maintaining a standard interface and specific adaptations. The interface will be generated by an interface generator capable of building an adaptative interface for any specific user, from a set of default values corresponding to standard use, and in function of a scenario of interaction (Figure 3). Then the feedback from the user will allow the interface to adapt to his needs and competences.

4.2   Independent models

In most cases the conversion from a multimodal interface to another will cause an important drop in infor- mation transmission because data models are very close to the standard display, especially regarding the Web. In fact, technologies have developed much more rapidly than our conscience of their utilities. For most Web designers the HTML language is simply a language describing the aspect of Web pages, rather than the structure and the semantics of the corresponding documents.

The conversion of a standard interface to meet the requirements of a specific user or user community will be replaced by the generation of a multimodal interface adapted to these requirements. Thus we need the data model to be totally independent of the modalities, including, in function of the kind of application and the targetted group of users, some interaction scenarios and alternatives. Such models should allow the use of any modality (alone or in conjunction with other modalities), and any output or input device, including future devices which axe not developed yet.

References

Archambault, D. et al. (2000a). *TIM: Tactile Interactive Multimedia computer games for visually impaired children.* This project is funded by the European Commission - Information Society Technologies, ref. IST-2000-25298.

Archambault, D., Burger, D. (1999). *Better access to the WEB for blind and partially sighted people.* INSERM, Paris,
Le Havre. Second edition. [On-line]. Available in French, English, German and Spanish: http://mm.braillenet.jussieu.fr/accessibilite/livreblanc/.

Archambault, D., Duchateau, S., Burger, D. (2000b). Multisite: Build HTML documents according to the user needs. In Vollmar, R. & Wagner, R. (Eds), *Proc. ICCHP 2000 (International Conference on Computers Helping People with Special Needs)* (pp. 29-34), Universität Karlsruhe, Germany. Öbsterreichische Computer Gesellschaft.

Burger, D. (1994). Improved Access to Computers for the Visually Handicapped: New Prospects and Principles. *IEEE Transactions on Rehabilitation Engineering*, 2(3), 111-118.

Burger, D., Buhagiar, P., Ceserano, S., Sagot, J. (1996). Tactison : a multimedia tool for early learning. In Burger, D. (Ed.), *New technologies in the education of visually handicapped*, number 237 in Colloque Inserm (pp. 237-242), Paris: John Libbey Eurotext Ltd.

Burger, D., Hadjadj, D. (1999). Non Visual Surfing on the Internet: the BrailleSurf browser. In Bühler, C. & Knops, H. (Eds), *Assistive Technology on the Threshold of the New Millenium*, pp. 53-57, Amsterdam, The Netherlands: IOS Press.

Coutaz, J. (1991). Prospects in Software Design with Multi-modal Interactive Systems. In *Proceedings of SITEF International Symposium on Cognitive Interactions* (pp. 47-63), Toulouse, France.

Duchateau, S., Archambault, D., Burger, D. (1999). The accessibility of the World Wide Web for visually impaired people. In Bühler, C. and Knops, H. (Eds), *Assistive Technology on the Threshold of the New Millenium* (pp. 34-38), Amsterdam, The Netherlands: IOS Press.

Emiliani, P.L., Stephanidis, C. (2000). Web-Based Information Systems for People with Disabilities: the EC AVANTI Project. In *Proceedings of CSUN'2000 (15th Annual Conference "Technology and Persons with Disabilities")*, Los Angeles, California, USA.

Hatwell, Y. (1993). Images and non-visual spatial representations in the blind. In Burger, D. and Sperandio, J.-C. (Eds), *Non-visual Human-Computer Interactions*, number 228 in Colloque Inserm (pp. 13-35). Paris: John Libbey Eurotext Ltd.

Schwarz, E., Burger, D., Ferre-Blanchard, M. (2000). Revamping techniques for adapting workstations for the blind. In Vollmar, R. and Wagner, R. (Eds), *Proceedings of ICCHP 2000 (International Conference on Computers Helping People with Special Needs)* (pp. 649 - 654). Universität Karlsruhe, Germany. Österreichische Computer Gesellschaft.

WAI (1998). *Web Accessibility Initiative - Web Content Accessibility Guideline 1.0.* World Wide Web Consortium (W3C). [On-line]. Available: http://www.w3.org/WAI.

Computing in a Multimodal World

Chris Baber

Kodak/Royal Academy Educational Technology Research Group, School of Electronic & Electrical Engineering, The University of Birmingham, Edgbaston, Birmingham. B15 2TT. UK

Abstract

Mobile individual computers and communicators represent a generic class of computing device that can be carried or worn by people and can be used in conjunction with other activities. These devices are situated in a world that already presents information and requires activity using the full range of human sensory modalities. This means that rather than designing for two or more modalities on one product, the challenge is to design for situated multimodal computing. The scope of this challenge will be addressed in this paper, and prototypical applications will be presented. In particular, the paper reports a technique, based on critical path analysis, that can be used to model situated multimodal computing.

1. Introduction

The development of mobile technology has carried on apace over the past two decades. No longer are mobile telephones bulky objects connected to even bulkier battery packs (cf. the 'carphones' of the early 1980s), but are small, stylish, fashion accessories with respectable battery life and acceptable coverage. A typical business traveler today not only carries a mobile telephone, but also all manner of communications, information processing and data capture devices. Thus, our fictional business traveler might have two mobile telephones (one for Europe and one for the US) for contacting colleagues and customers, a WAP (wireless access protocol) telephone for checking web-based travel information, a PDA (personal digital assistant) for recording appointments and contact details, an MP3 player to listen to music (or watch movies), a digital camera for recording meetings or sight-seeing, and a laptop (or smaller) computer, perhaps with internet connectivity from the hotel room's telephone socket (or through one of the mobile telephones). The intention is for our business traveler to maintain contact with a digital world from anywhere in the physical world.

Efforts to develop 3G (3rd generation mobile communications) devices herald the rise of hybrid computers /communicators, e.g., Sagem have announced a combined PDA and mobile telephone, Samsung and Sony have both launched combined MP3 player with digital cameras, while Kodak's PalmPix allows PDAs to be transformed into digital cameras. A characteristic of 3G devices is their ability to take advantage of the increased bandwidth that future communications technologies will offer.

In addition to being mobile, this technology is designed for, and adapted by, individual users. I think the notion individuality (as opposed to merely 'personal') is a defining feature of future technologies. The devices are not simply used by a single person, but are adapted to specific users; this adaptation is performed by both the user (through changing settings, entering data etc.) and by the device (through agents that can modify the behaviour of the device based on characteristic user activity and features of the environment in which it is being used). Contemporary technology typically offers people unimodal access (this is not to deny that the very near future will see multimodal devices).

Of interest to our work is the fact that using such technology 'in the field' is, by definition, a multimodal activity. In this paper, the relationship between activity in the world and using Mobile Individual Computing / Communication (MIC) devices will be considered using the example of a wearable computer for paramedics. The application is characterized by a central concern in that the use of a computer is secondary to the main tasks being performed. It is proposed that this is, or is becoming, true of MIC, i.e., returning to our mobile telephone users – their intention is not to engage in use of the telephone per se (i.e., not to explore and play with the various functions offered by the telephone) but to engage in the activity of having a conversation while they are busy doing something else[1]. The person walking along a busy High Street while talking on their mobile telephone is performing multimodal activity. One might argue that the multimodal nature of such activity is not an issue for computer designers; after all, the designer ought not to be held responsible for the way in which people use the technology. However, given that MIC are designed with the specific intention of supporting mobile computing and communications, the interaction between use of the technology and a person's activity in the world is very important. Previous forms of portable technology (from books to watches to laptop PCs) relied on the user stopping one action and then focusing their attention on the technology, i.e., it is not easy to read a book while walking. This relationship changed some 25 years ago, e.g., one could listen to a Walkman while walking or jogging. This convergence of activities is the starting point for the discussion in this paper.

2. Being multimodal

One can take either a Human or Technological perspective on the term 'multimodal'. Although the two perspectives are not mutually exclusive, it is proposed that a Human perspective is couched in terms of the sensory and response modalities available to the person; whereas a Technological perspective is couched in terms of processing resources available to a computer. In both cases, one of the central concerns is the management of competing modalities, and the problem of ensuring that modalities are used efficiently. From a Human perspective, it could be argued that almost all human activity (outside the psychology laboratory) is multimodal; in HCI, the person's attention is directed at both the computer and the environment around them (if only to minimize distractions). The Human perspective on multimodal activity allows us to introduce concepts from cognitive psychology relating to the issues of attentional resource. It is clear that sometimes it is easy to do two things at once (walk and chew gum) and other times it is difficult, if not impossible (converse on two different topics simultaneously). One explanation for this is that attention has a limited pool of resource on which to draw.

The problem of limitation can be dealt with simply shedding tasks, or by changing strategy, or by redefining the activity. The ease with which activities can be shed or combined, or strategy modified is partly a function of expertise and partly a function of the activities. Thus, Wickens (1992) suggests that activities can compete for resource in terms of the modality in which stimuli are presented (e.g., visual or auditory or tactile), the state of processing (e.g., perception – central processing – response), the response employed (e.g., spoken or manual responses). More controversially, competition could arise as a function of the processing code employed, e.g., visuo-spatial versus verbal.

3. Acting in real and technical domains

Given that use of MIC occurs in parallel with 'real' world activity, there is an obvious question relating to the degree of overlap between activities in the real world and those in a 'technical' domain. In this paper, when I say technical domain I mean any domain created by or for technology; thus, the keypad and menus on a mobile telephone form a technical domain that is separate from the real world. The point is that users of MIC will need to switch attention between these different domains. Staying with the mobile telephone example, one could characterize a typical activity as shown in table 1. While this is a fictional description, there are several points to raise here. First, there is little obvious overlap between the real and technical domains (although the user stops walking to navigate the display). Second, the caller continues to walk while talking on the telephone – the mobile telephone serving to act a signal to other people that the caller is not talking to herself (people using hands-free headsets for mobile telephones continue to make some form of gesture to their face to signal that they are not talking to themselves). Third, while there is no task dependency in this description, the activities are interleaved in such a manner that the caller can perform the two tasks (walking and making a telephone call) as near simultaneously as possible.

Searching for a name in an address book could be demanding, both visually (with the need to read a small screen) and manually (with the need for several keypresses). The caller stops walking to perform these activities. In this

---

[1] Of course, the issue of whether it is socially acceptable to engage in a conversation without one's full attention on the conversation is interesting; if I ignore the person I am speaking to and stare into the distance for even a few seconds, then my behaviour might be deemed odd. Yet we might accept pauses and asides from our conversational partner if we know they are driving.

instance, the demands of one set of activities is high enough to need to shed other activities. Once a connection has been made and the telephone call can commence, the caller starts to walk again. Walking can now be performed in parallel with talking on the telephone. At one level, this example can be considered as a form of 'concurrent' multimodal activity, in that two distinct sets of activity are performed in parallel. However, it is apparent that there are points of overlap and dependency that link these into a coherent whole. In other words, this is not simply matter of making a telephone and walking as two parallel activities. For multimodal activity, a key issue relates to the interleaving of tasks. The interleaving might arise as a consequence of task-semantics, e.g., point at screen => speak command. Alternatively, the interleaving might arise from attention-switching, e.g., type SMS text=> look up to avoid bumping into someone. Taking the Human / Technical distinction raised earlier, it is proposed that the Technical perspective on Multimodality would see interleaving as the logical consequence of designing a specific task sequence to be performed on a specific product. On the other hand, a Human perspective would see interleaving as the consequence of the user developing, adapting and applying different strategies that are appropriate to contextual demands.

Table 1: Making a telephone call while walking

Domain	Decide to make call	Make connection	Make call	End call
Real: telephone call	Record shop which reminds you of X, and that you need to call him		Speak to X. Use left hand to emphasise points, smile and laugh.	Say 'Goodbye'
Real: navigate street	Walk past record shop.	Walk to shop window and stop	Walk along street, move aside for other pedestrians	Carry on walking
Technical: telephone call	Take telephone from pocket	Switch on telephone and wait for connection	Select 'Address book' icon and search using X. Press OK. Press 'telephone icon'. Hold telephone to mouth and ear using right hand.	Press 'red telephone icon'

4. Modelling multimodality

Previous work has demonstrated the possibility of using critical path analysis as a means of modeling multimodal dialogue (Baber and Mellor, 1997; 2001). In this work, activities are decomposed into constituent tasks and assigned to specific modalities. Figure 2 shows the application of this approach to the example described in table one. Each modality occupies a separate line (and is colour coded). Each unit-task is linked to subsequent tasks in terms of task or modality dependency. In other words, unit-tasks that follow a task sequence share task dependency, and unit-tasks that rely on the same modality share modality dependency. If the caller in this example did not stop, then there would be a Visual task associated with walking that would recursively interact with the unit-tasks associated with entering the telephone number. This recursive interaction would be likely to result in problems in both the walking and the entering telephone number tasks. Hence, in this example, stopping to enter the telephone number can be seen to be an optimal strategy for performing this task. Indeed, the optimality can be demonstrated by assigning numbers (to represent approximate times) to each of the tasks, and calculating the critical path for this model.

We have developed wearable computers for the emergency services. In one project, a wearable computer was developed for use by paramedics (Baber et al., 1999a; 1999b). Figure 1 shows a paramedic wearing the computer during user trials. The intention was to develop a device that could offer the user the opportunity to record information (through speech recognition) while performing another activity, i.e., treating a patient. The computer also offered the chance of displaying information to the wearer, e.g., displaying 'side-effects' (contra-indications) of specific drugs prior to administration. A CPA model allowed us to compare the possible effects of different design options on performance. Figure 3 shows a description of paramedic activity related to treating cardiac arrest. By introducing additional boxes to represent unit-tasks related to patient reporting and data collection, it is possible to determine the likely consequences of introducing computer-technology into this domain. For example, the central line of orange boxes describe manual activity that is currently being performed; introducing a manual data collection method (such a pencil and paper or computer keyboard) would disrupt this activity and shift the critical path onto data collection and away from patient care. On the other hand, the use of speech recognition was found to have relatively little affect on the critical path. Subsequent development of a prototype was based on speech recognition, and a user trial demonstrated that the use of speech recognition did not disrupt the manual activity (Baber et al., 1999). There was some change in performance with the introduction of the wearable computer; we employed a head-mounted display and the top (pink) line shows 'visual tasks', which we affected by the additional task of checking the display. We had anticipated that 'check display' tasks might cause other tasks to pause, and this is what was found during the trials.

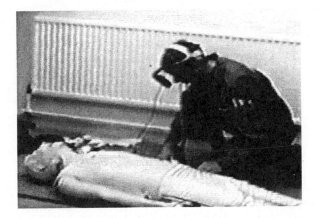

Figure 1: Paramedic, wearing computer, working with training doll

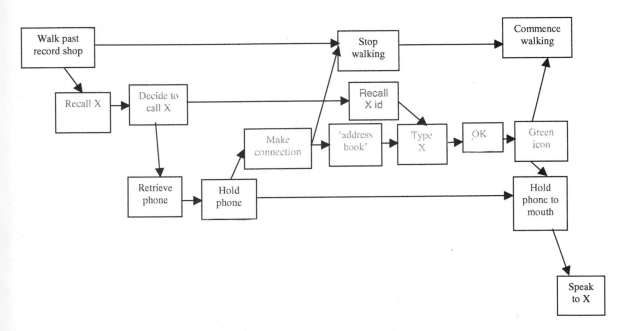

Figure 2: Part of Critical Path Analysis of Example in Table One

236

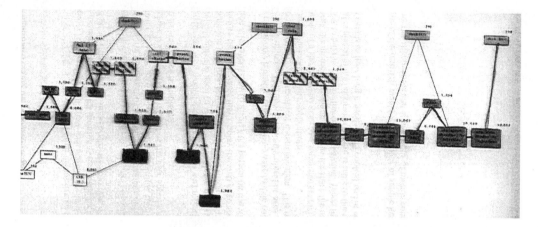

Figure 3: Critical Path Analysis of Original Activity

5. Discussion

In this paper, it has been proposed that the use of MIC is inherently multimodal. The need to combine activity in real and technical domains introduces problems related to attentional demand and allocation of resource. A simple technique, based on CPA, can be used to model the relationship between activity in real and technical domains. This technique requires consideration to be given to the possible impact of technical domain activity on real world activity.

References

Baber, C., Mellor, B. (1996). The effects of workload on speaking implications for the design of speech recognition systems. In S.A. Robertson (ed.), *Contemporary Ergonomics 1996* (pp. 513-517). London: Taylor and Francis.

Baber, C., & Mellor, B.A. (1997). Modelling dual-task performance time multimodal human-computer interaction using critical path analysis. In D. Harris (ed.), *Engineering Psychology and Cognitive Ergonomics volume II* (pp. 223-230). Aldershot: Ashgate.

Baber, C., Mellor, B.A. (2001). Modelling multimodal human-computer interaction using critical path analysis. *International Journal of Human Computer Studies.*

Baber, C., Haniff, D., Lindley, M. (1999). Designing the paramedic protocol and patient reporting computer, In D. Harris (ed.), *Engineering Psychology and Cognitive Ergonomics IV* (pp. 103-107). Aldershot: Ashgate.

Baber, C., Arvanitis, T.N., Haniff, D.J., Buckley, R. (1999). A wearable computer for paramedics: studies in model-based, user-centred and industrial design. In M.A. Sasse, & C. Johnson (eds.), *Interact'99* (pp. 126-132). Amsterdam: IOS Press.

Wickens, C.D. (1992). *Engineering Psychology and Human Performance*, New York: Harper Collins [2nd Edition].

Structureless, Intention-guided Web Sites:
Planning Based Adaptation

Matteo Baldoni, Cristina Baroglio, Viviana Patti

Dipartimento di Informatica – Università degli Studi di Torino
C.so Svizzera, 185 – I-10149 Torino (Italy)
Tel. +39 011 6706711, Fax. +39 011 751603,
E-mail: {baldoni,baroglio,patti}@di.unito.it
URL: http://www.di.unito.it/~alice

Abstract

The great variety of users of the services available on the internet raised the problem of finding flexible forms of presentation and interaction, which depend on the specific user's characteristics and needs. In this article we present an approach to adaptation, which exploits the reasoning capabilities of a rational agent. The key idea, which makes this approach orthogonal to personalization based on the "user model", is that the system adopts the user's intentions during the interaction. Then the system exploits its planning capabilities to dynamically generate a web site, adapted to the current user's needs. For the sake of understanding we also discuss a possible application of the approach to the construction of a virtual tutor.

1. Introduction

Many of the most advanced solutions on web site adaptation to the user, such as (Ardissono *and* Goy, 1999a; Marucci *and* Paternò, 2000; De Carolis, 1998), start from the assumption that adaptation should focus on the user's characteristics; in different ways, they all try to associate the user with a reference prototype, the "user model", to which the presentation is adapted. By doing so, such approaches catch important aspects connected to the personality and the general interests of the user but, in our opinion, they underestimate the role that the user's intentions and needs, which may change at every connection, have in order to achieve a real adaptation.

In this paper, we present an approach to web site adaptation which uses an agent logic programming language, called DyLOG (Baldoni *et al.*, 1998; Baldoni *et al.*, 2000a), for building cognitive agents that dynamically generate a site based on the user's goals and constraints. When a user connects to a site managed by one of our agents, (s)he does not access to a fixed graph of pages and links but (s)he interacts with a rational agent that, starting from a knowledge base specific to the site and from the requests of the user, builds an ad hoc presentation structure. We obtain run-time adaptation by exploiting the capability of a rational agent to reason about its actions and to make plans. So the problem to generate a web site adapted to user's needs is interpreted as a planning problem. In this sense, our web sites are *structureless*. Their structure corresponds to the plan built for pursuing the user's goal. Thus adaptation occurs at the navigation level rather than at the presentation level. Indeed, our focus is on the definition of the navigation possibilities available to the user and on determining which page to display based on the dynamics of the interaction.

This approach brings along various innovations w.r.t. what can be found in the literature. From a *human-machine interaction* perspective, the kind of adaptation that we propose is focused on the actual interests of the user for that specific connection and not, as it often happens in the user model approach, on his/her generic and cultural interests or past choices (past-oriented adaptation). From a web site design perspective, some advantage is expected for the *build-modify-update process*. In fact, while in order to modify a classical web site one has to change both the contents of the pages and the links among them, in our case the site does not exist as a structure, because the structure is built at the moment. All that we have is a data base containing the domain knowledge, whose maintenance is simple, and an agent program, that is not changed often.

Although our approach could recall some works on natural language cooperative dialogue systems, there are some differences. For instance, in (Bretier *and* Sadek, 1997) Sadek and Bretier proposed a logic of rational interaction for implementing the dialogue management components of a spoken dialogue system. This work is based, like DyLOG, on dynamic logic and reasoning capabilities on actions and intentions are exploited for planning dialogue acts. Our work, instead, is not focused on recognizing or inferring the user's intentions (which is also a very interesting task):

the user's needs are taken as explicit input by the software agent, which uses them to generate the goals that will drive its behavior. The novelty of our approach is that we exploit planning capabilities for building web site presentation plans guided by the user's goals rather than for dialogue act planning. The structure of the site is built by the system as a conditional plan, according to the initial user's needs and constraints. The execution of the plan by the system corresponds to the navigation of the site, where the user and the system cooperate for finding a solution to the user's problem.

We chose DyLOG as an agent programming language due to two of its main characteristics. The first is that, being based on a formal theory of actions, it can deal with reasoning about action effects in a dynamically changing environment and, as such, it *supports planning*. The second is that the logical characterization of the language is very close to the procedural one, and this allows to *reduce the gap* between theory and use.

In order to check the validity of the proposed approach, we have implemented a client-server agent system, named WLog. In this application the users connect to the structureless web site for asking for support in some task. As an example, we are currently implementing a virtual seller that helps clients to assemble a computer, which is good for their needs (Baldoni *et al.*, 2000b; Baldoni *et al.* 2001), but we will show in the next section that the same approach can be exploited in other, apparently very different, applications. The idea is that the user interacts with a software agent for solving a problem. Similarly to what happens in an interaction between humans, the solution is obtained thanks to a dialogue, in this case guided by the agent (see next section for a an example). Requests, information, and proposals are presented to the user in the form of web pages.

Technical information about the system can be found at http://www.di.unito.it/~alice.

2. An Example Application: the Virtual Tutor

In a parallel work, we are using WLog to tackle an e-commerce application: to build a "virtual seller" that helps users to assemble computers depending on which use that computer will have. To this aim the agent system adopts the user's goal in order to plan a solution. Another application in which a similar interaction is useful is the construction of "studiorum itinera" that depend: (1) on the competence a student wants to acquire, (2) on his/her current competence, (3) on the material available in the system's store and (4) on student's time constraints. Consider the following example ("Tutor" is the agent system):

Declaration of the user's main intention

Student: I wish to know something about programming languages.

Tutor (asking): Are you interested in an *advanced* or in an *introductory* course?

Student: I'm a beginner. I would like to start with an introductory course.

Definition of the learning goals

Tutor (thinking): Let me think, for an introductory course, first he needs a training module about the principles of programming, then some module of exercises to test the newly acquired knowledge. At this point he can continue by studying Object Oriented programming; afterwards, he will have to use an OO exercise module, and at last he will be ready to study some specific language. Let's list what he should learn...

Acquisition of the user's preferences

Tutor (asking): This is a sketch of the knowledge that you should acquire in order to reach your learning target, is there something that you already know?

Student: I already had a crash course in programming, so I would skip the part about the basic principles; I would start from OO theory.

Tutor(asking): Would you prefer a short course (maximum 20 hours), a medium course (maximum 40 hours) or time is not a problem?

Student: I'd rather choose a medium length course, I work part-time.

Planning

Tutor (thinking): I need to **plan** according to his actual competence, time constraints, and the availability of training modules...

Execution

Tutor (explaining): Well, I have many solutions that could satisfy you. For learning OO programming theory, you can choose between a 20 hours module by prof. Smith or a 10 hours module by prof. Black. Which one do you prefer?

Student: I prefer the 20 hours course by prof. Smith.

Tutor: Afterwards you can practice with the exercises contained in a 5 hours module by prof. Green; you could end your course by studying one of the following OO languages *[list]*. Which one do you prefer?

Student: I heard that my colleagues use java; I choose the java 10 hours course by prof. Giovannetti...
The student expresses his/her desire. The tutor starts a dialogue aimed at fixing both a set of subgoals and a set of constraints (such as the overall time the student can devote to the course); afterwards, it builds a conditional plan that will allow the student to grow knowledge about programming languages; the execution of the plan leads to show the web pages which correspond to the different actions in the conditional plan (see Sec. 4). The process is similar to the one presented in (Baldoni *et al.* 2001).

An agent system of this kind could be a valid interface for an on-line library of training modules (in the line of commercial systems, such as Competus Framework). Modules can be tutorials, CBT tools, links to web pages, etc. and have the most various origin. Each module has an associated description about the prerequisites necessary to use it and what it allows to learn. The agent system works on the meta-knowledge in order to build plans. Such a system would be extremely useful for all those persons who cannot attend regular classes: they could download the suggested modules and use them when and where they can.

Some examples of persons who could benefit of a similar system are students affected by serious illnesses (e.g. cancer-affected students often have to spend long periods of time at home or in a hospital), persons who want to increase their education but have a full-time job, persons in jail. Indeed, a system of this kind, with a reach enough module set, could serve different categories of people. Plans could therefore depend also on the category the user belongs to. In this way we would have a richer form of adaptation: on one hand, we would have a personalization according to the category (user model), while on the other we would have a more detailed adaptation to the specific user and situation (the study plan).

3. Modeling the Virtual Tutor as a Rational Agent in DyLOG

The language that we use for specifying our *virtual tutor* is DyLOG. It is based on a logical theory for reasoning about actions and change in a modal logic programming setting. It allows one to specify a rational agent's behavior by defining both a set of *simple actions* that the agent can perform, some of which are *sensing* and *suggesting* actions that allow to interact with the user, and a set of prolog-like *procedures* which build complex behaviors upon simple actions. Simple actions are defined in terms of preconditions and effects on a *state*, which consists in a set of fluent formulas representing the agent beliefs. Indirect effects of actions are expressed by means of *causal rules*. See (Baldoni *et al*, 2000a, 2001) for details on the language and on its use for agent programming. The behaviour of our *virtual tutor* is captured by a collection of procedures and it is driven by a set of goals. The top level procedure, *advise* is defined as follow:

    advise is *ask_student_competence*; *ask_available_time*;
        plan(combine_courses, total_time(H) ∧ student_time(T) ∧ (H ≤ T),Plan); *Plan*.

The virtual tutor starts by interacting with the student in order to find (and adopt) his/her learning goals, by asking if (s)he is interested in an *advanced* or in an *introductory* course and by checking if the user already has some of the competences which are the target of a standard program (*ask_student_competence*). Then it gets information about student's time constraints (*ask_available_time*). By using this information, the knowledge on the available training material and its expert competence about how to combine the courses, it starts to plan how to reach the goals of the program, predicting also future interactions with the student. Planning is needed to find *study itinera* by taking into account two interacting goals: the goal of combining courses into a program which satisfy the student learning needs and the goal to consider only programs affordable by the student's time constraints. Finally, the agent executes the conditional plan *P* resulted from the planning process.

The way the agent combines the courses into a program is specified by procedure *combine_courses* that, until the program is believed complete, tries to achieve the goal of getting a still missing training module.

    combine_courses is *program_complete?*.
    combine_courses is ¬*program_complete?*; *achieve_goal*; *combine_courses*.

Note that only when all of the goals to get the necessary training modules are fulfilled combining available courses, the main goal to have a program to propose to the student is reached and the program is considered complete. Until there is still a goal to fulfill, the program is considered not complete (this is expressed by the causal rule: ¬*program_complete* if *goal(X)*). Indeed, we assume the behaviour of a rational agent to be *driven by a set of goals*, which are represented as fluents having form *goal(F)*. The system detects the goals based on student's inputs and its expert competence about learning and courses combination. Initially the tutor does not have explicit goals, because no interaction with the student has been performed. The student's inputs are obtained after the first interaction phase (see *advise*) and they generate a set of goals that the agent has to achieve to compose a suitable program. In the language, we model this by describing the adoption of a goal as the indirect effect of requesting user's preferences.

240

For instance, the suggesting action *ask_program_level* (which is a simple action of *ask_student_preferences*) asks if the student is interested in an advanced or in an introductory course. This action has as indirect effect the generation of the goal to have a program leading the student to reach a given level of competence:

> *ask_program_level* **suggests** *requested(competence(Level))*.
> *goal(has(competence(Level)))* **if** *requested(competence(Level))*.

Let us suppose the student is interested in an introductory course (fluent *requested (competence(introductory))* is known by the agent), then, the causal rule above will generate the goal *goal(has(competence(introductory)))*. By means of an appropriate instantiation of the following causal rule, this main goal will generate a set of sub-goals for selecting the training modules that could allow the student to acquire the desired competence:

> *goal(user_knows(Subject))* **if** *goal(has(competence(Level)))* ∧ *part_of(competence(Level),Subject)*.

Roughly, this is the way in which the agent initializes the set of goals[1] that it will try to solve by means of planning, which is done as described in (Baldoni *et al.*, 2001). The extracted plan is one possible execution of procedure *combine_courses*. Indeed we take advantage of a specific instance of the *planning problem,* by looking for a possible execution of procedure *p* leading to a state where some condition *Fs* holds. In this case the search space for an action sequence leading to the goal is naturally constrained by the procedure definition.

Fig. 1 reports a conditional plan that could be obtained when the agent plan for composing an introductory program with medium length, i.e. maximum 40 hours. Each box represents a simple action, which will cause the creation of a web page. Those corresponding to sensing and suggesting actions allow an interaction with the user, and can be seen as queries for inputs whose values cannot be known at planning time. Each branch of the plan corresponds to one of the possible input values.

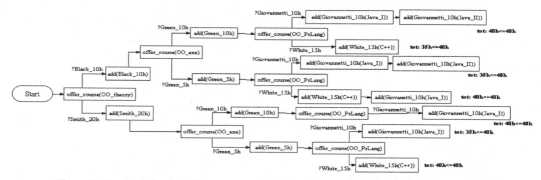

Figure 1. *Result of plan(combine_courses, total_time(H) ∧ student_time(40) ∧ H ≤ 40, Plan)*

4. WLOG: System Architecture

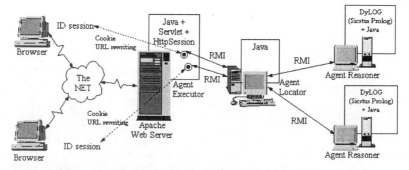

Figure 2: WLog multi-agent architecture

---

[1] Observe that agents use a *blind commitment strategy* to their goals

WLog is the system that we developed in order to verify the feasability of the intention-driven approach. The novelty respect to previous implementations is that it has a *multi-agent architecture*, which is sketched in Fig. 2. Briefly, the system consists of two kinds of agent: *reasoners* and *executors*. A DyLOG reasoner generates conditional presentation plans; once built the plan is executed. Actual execution it is the task of the executors, which are Java servlets embedded in an Apache web server. *Executing* a conditional plan implies following one of the paths; only the part of the site corresponding to this path will actually be built. The execution of the actions in the path consists in showing one or more web pages to the user; in particular, when a page that corresponds to a branching point is shown, a feedback from the user is required. Therefore, the execution of each action in the plan consists of the following steps: 1. the reasoner sends a message to the executor; 2. the executor produces a proper HTML page; 3. the page is shown to the user and a feedback is asked; 4. the executor sends the feedback to the reasoner; 5. the reasoner adds the new fact to the knowledge base and takes it into account for passing to the next step: sending to the executor the information about which page to show next. All unselected alternatives are forgotten. The connection between reasoners and executors has the form of message exchange in a distributed system; message exchange is FIPA-like (FIPA, 1997). Each agent is identified by its "location" which can be obtained by other agents from a facilitator ("Agent Locator" in Fig. 2). The interaction between the user and WLog starts with the declaration of the user's goal. The executor works as an interface between the reasoner and the user. It looks for a free reasoner; if one is available, from that moment till the end of the connection will be dedicated to serve that specific user.

5. Conclusion

In this paper, we have shown how to implement web applications, where a real-time adaptation is required, by exploiting planning capabilities of an agent logic language. Adaptation is obtained by dynamically generating presentations guided by the interaction with the user. The observation that user's goals may change at every connection, which motivates our proposal, seems to be particularly true for web-sites implementing *recommendation systems*. As representative of this category, we chose the virtual tutor case study. Of course, the planning task that we described is just one of the tasks that a real tutoring system should have. For instance, a real system should also be able to check the progress of each student and, if necessary, perform some replanning; all these problems are, however, out of the scope of the current example.

References

Ardissono, L. & Goy, A. (1999). Tailoring the interaction with users in electronic shops. In *Proc. of the 7th International Conference on User Modeling*, pages 35-44, Banff, Canada. Springer-Verlag.

Baldoni, M., Giordano, L., Martelli, A., and Patti, V. (1998). A Modal Programming Language for Representing Complex Actions. In the *Post-Conference Workshop DYNAMICS'98*, pages 1-15.

Baldoni, M., Giordano, L., Martelli, A., and Patti, V. (2000a). Modeling agents in a logic action language. In *Proc. of the Workshop on Rational Agents, FAPR'00*, London, September 2000. To appear.

Baldoni, M., Baroglio, C., Chiarotto, A., Martelli, A., and Patti, V. (2000b). Intention-guided Web Sites: A New Perspective on Adaptation. In Proc. of the 6th *ERCIM Workshop UI4All*, pages 68-82, Florence, Italy.

Baldoni, M., Baroglio, C., Chiarotto, A., and V. Patti (2001). Programming Goal-driven Web Sites using an Agent Logic Language. In *Proc. of the Third International Symposium on Practical Aspects of Declarative Languages*, volume 1990 of LNCS, pages 60-75, Las Vegas, Nevada, USA. Springer-Verlag.

Bretier, P. and Sadek, D. (1997). A rational agent as the kernel of a cooperative spoken dialogue system: implementing a logical theory of interaction. In *Proc. of ATAL-96*, LNAI 1193, pp. 189-204. Springer-Verlag.

FIPA (1997). Agent Communication Language. FIPA 97 Specification, Foundation for Intelligent Physical Agents, http://www.fipa.org.

De Carolis, B.N. (1998). Introducing reactivity in adaptive hypertext generation. In H. Prade, editor, *Proc. of the 13th European Conference on Artificial Intelligence, ECAI'98*, Brighton, UK. John Wiley & Sons.

Marucci, L. and Paternò, F. (2000). Designing an Adaptive Virtual Guide for Web Application. In Proc. of the 6th *ERCIM Workshop UI4All*, pages 57-67, Florence, Italy.

McTear, M. (1993). User modelling for adaptive computer systems: a survey on recent developments. In *Journal of Artificial Intelligence Review*, 7:157-184.

Navigating the voice web

Jennifer Balogh, Nicole LeDuc, Michael H. Cohen

Nuance Communications, 1380 Willow Road, Menlo Park, CA 94025

1. Introduction

The telephone drastically changed the way people communicate, and in recent years, the World Wide Web radically enhanced people's ability to share information and conduct business. Now is an exciting time since the power of the web and the phone are being combined to form a new means of communication: the Voice Web, which uses speech technology to integrate the telephone and Internet. The Voice Web is similar to the World Wide Web but instead of viewing content, the user talks to it. One benefit of the Voice Web is that it is available to anyone with a telephone, not just those with a computer. Also, callers who are mobile can obtain information from the Internet without extra equipment such as wireless modems or WAP phones.

While browsing the Voice Web is similar in purpose to browsing the World Wide Web, a number of issues must be reconsidered. Most important is the fact that the voice user interface (VUI) is not the same as the graphical user interface (GUI). There are no visual cues such as color that identify links; there are no Back buttons to click; there are no visual lists that tell callers what sites they can revisit. Therefore, new navigational techniques need to be used to allow callers to navigate the Voice Web.

This paper presents our findings of two important issues regarding navigation of the Voice Web: (1) how callers learn to break their traditional view of the telephone and apply the paradigm of the Internet to navigate to and from content and (2) how navigational features that allow movement between sites, in particular, audio equivalents of the Back button and hyperlinks, come into play in the Voice Web.

In order to investigate navigation of the Voice Web, we drew from our experience with a specific audio browser, Nuance Voyager[TM]. The design of Voyager, builds upon several navigational paradigms of the Internet and introduces them into the auditory domain of the telephone. The concept of an audio browser is similar to a web browser in that the browser is always available regardless of the content being visited. This continued presence of the browser introduces a leap in thinking about the telephone. Now instead of making separate phone calls to access information and talk to people, the browser allows callers access to a variety of information and people in a single call. Therefore, an important design feature of Voyager is continuous call control with the ability to navigate to and from content without hanging up the phone. Voyager also includes a mechanism to teach callers what content is available. Two other navigational features of Voyager focus on movement between content sites. One is the audio equivalent of a Back button (and Forward button), which allows callers to return to content sites previously visited; and the other is audio hyperlinks, or voice links, that enable users to jump from one content site to another. Also included in the Voyager design are bookmarks. When callers reach a particular site that they want to revisit later, they can ask the browser to bookmark the site. Then on subsequent visits, they can simply ask for the site by name instead of following a series of links to get there. However, this feature was not the focus of our studies since most content sites were easily accessible from the browser's main menu.

Using Voyager, we developed a system called V-Center in which callers could gain access to several voice sites. These included a corporate dialer, a personal dialer, voice mail, a room reservation system, a stock quote site, and several demos. To investigate callers' reaction to different navigational features and their ability to learn how to use them, we conducted usability studies, analyzed longitudinal usage data from internal callers, and ran laboratory experiments. We first discuss our findings of how callers learn to maintain continuous call control and get to and from content sites. Then we explore issues of navigational techniques that allow callers to move from place to place.

2. Learning How to Navigate the Voice Web

Since many concepts introduced by voice browsers are new to the telephone, especially with regard to navigation, teaching the caller how to take advantage of these features is critical to the success of the Voice Web. The two areas of interest are how to teach callers to change their traditional view of the telephone and adopt the paradigm of the Internet, and how to enable callers to learn what content they can browse. We address each of these in turn.

2.1 Learning How to Maintain Continuous Call Control

One of the most important features of Voyager is continuous call control - without this capability, the user could just as well make several phone calls to different information services already available, such as weather. One of the challenges for the caller is realizing that this continuous call control is available and remembering how to get back to the browser. This is particularly challenging when placing phone calls, which habitually end by hanging up the phone.

In order to lead callers through this conceptual shift in thinking about phone calls and navigation, we conducted a series of usability studies, specifically focused on a tutorial that teaches callers how to maintain continuous call control. In one of the first versions of the system, the tutorial was quite exhaustive in explaining the different features of the browser. One of the points made in the tutorial was that the caller should say the command "Voyager" to get back from a phone call. However, in the first usability session, only one out of the four participants remembered to say the command (one participant even commented that the system never explained how to get back). Because 75 percent of the participants failed to understand a fundamental capability of the system, the tutorial was revamped. Previous research has shown that interactive tutorials can successfully teach callers about spoken language systems (Kamm et al., 1998), so an interactive tutorial was chosen for the design of the second iteration. Instead of passively presenting information, the second tutorial had the caller interact with the system to make a sample phone call and return back to the system, thereby focusing on only core functionality. The tutorial was then assessed in a second usability study. As in the first study, participants were exposed to the tutorial and asked to place several phone calls. Nine out of the 11 participants who went through the tutorial, successfully returned from their first phone call by saying the command "Voyager" ($p < .05$).

The results show that navigation back to the browser, especially after making a phone call, is a conceptual leap for callers who are accustomed to hanging up the phone after completing a call. And a passive tutorial that explains how to navigate in this way is not effective in teaching callers this behavior. Instead an interactive tutorial that allows callers to experience navigation in this way is a better tool to help train users.

2.2 Learning How to Navigate to Content

The next issue we examined was how callers learn what content sites they are interested in and how to get to them. Our understanding of how callers move from the browser to content is based on the usage patterns of our system, V-Center. In this system, callers can request sites directly by name (analogous to typing in a URL), or they can listen to a list of available sites (similar to menus of popular categories provided by web portals).

The voice browser contained two important states relevant to the question of navigation: these were the Main Menu where callers could say the name of any content site available in the system, and the What's Available List, which read through all the offered sites - this list had to be explicitly requested by the callers. A two week period of actual system use was analyzed (6/1/00 to 6/14/00); during this time, there were 328 calls from 58 Nuance employees from all different divisions of the company. (Data from individuals working directly on the design or implementation of the system were not included in the analysis). Enrollment was open and on-going so a mixture of novice and expert users contributed to the data. Interactions between the caller and system reflected real usage (eg., callers were never given specific tasks to complete). All caller responses to the system during this time period were logged, recorded, and transcribed.

The results of the analysis show that the What's Available List was the most heavily used area of the system aside from the Main Menu and the state that announced transitions from the browser to the content site. Other areas used less frequently included enrollment, login, tutorial, voice authentication, bookmarks, personal profile, and help on specific topics. Thus, figuring out the names of content sites was a critical step in navigation. In fact, 39.6 percent of the callers accessed the What's Available List at least once during this time period (23/58 users). Across phone calls, however, the What's Available List was used only 8.5 percent of the time (28/328 calls), and usually only on the first and second calls into the system. This indicates that although a large number of callers asked for the list, they did so to become familiar with the content early on. After this learning took place, callers associated each content area with its name (URL) and took the more efficient path of requesting the sites directly.

This particular implementation of the system allowed callers to listen to the entire list, say the name of the item they wanted at any time, and also scroll through the list with commands such as "next" and "previous". Typically, the caller said the name of the content site while listening to the What's Available List (users did this 43 percent of the time, or 16/37 total uses of the list). Another common behavior was for callers to listen to the entire list before making a selection (even if the item they wanted appeared near the very beginning). This behavior was observed about 38 percent of the time (14/37 uses of the list). Only a few callers took advantage of the scrolling functionality. In general, most callers barged-in over the list when they heard the item they wanted - given this, the most efficient design was one in which frequently requested items were mentioned early in the list.

The findings reported here show that callers do learn the new navigational paradigms of the Voice Web. Mechanisms designed specifically to teach callers how to move around, for example interactive Tutorials and What's Available Lists, are effective ways to help callers learn how to navigate.

3. Navigating Between Content Sites

Two navigational features of the World Wide Web that allow callers to navigate from site to site are the Back button and hyperlinks. These features provide a powerful mechanism to browse content, and for this reason they were both adopted and integrated into the voice browser, Voyager. In the next section, we examine the audio equivalents of these two navigational features in more detail.

3.1   Going Backward and Forward

The "go back" command in Voyager draws from the analogy of the Back button of a web browser. By using this command, callers are able to return to content sites previously visited. The browser maintains a history of all the content sites visited during a session just as a web browser does. Similarly, the "go forward" command is associated with the concept of the Forward button.

To understand how callers took advantage of this navigational feature, we tracked real usage patterns of the internal V-Center system at Nuance. We examined both how often and in what context the "go back" and "go forward" commands were used. Out of 58 people, five of them used the "go back" command, (8.6 percent of users). Only one caller used the "go forward" command. All the callers used the "go back" command to go back one step (as opposed to multiple steps). Usually, the caller went to the corporate or personal dialers, made a phone call and then said "go back" to return to the dialer content site.

In the context of a small, closed system, the navigational feature was not widely used by callers. Most likely the limited uptake of this feature is related to the small number of content sites offered in the service compared to the vastness of the Internet. However, an interesting observation is that some users did take advantage of this feature, even though there were other means of getting to the same place. The callers could have just as easily said the name of the content site again, but chose to use navigational commands instead. What this suggests is that navigation to previously visited sites is a promising feature that will be realized once the Voice Web develops a critical mass; however, more research is needed.

3.2   Audio cues for Hyperlinks

As the Voice Web grows, many more connections *between* sites will start to appear, and the caller will eventually be able to jump from one piece of content to the next without having to visit the service's main menu. The analogous feature in the World Wide Web is hyperlinks. For the Voice Web, we refer to them as 'voice links'. Again, since the Voice Web is still in its early stages, there is not a large web of sites with interconnecting voice links yet, so in this section, we focus on cognitive issues rather than usage patterns. In particular, we examined what the best cue is to *identify* voice links. In the World Wide Web, hyperlinks are conveyed by color and underscored text. The challenge for the Voice Web is that the cues must be all auditory. Also, the acoustic signal is ephemeral, and therefore, the caller must not only detect the voice link, but must remember the word or phrase to repeat in order to follow it.

Research by Frankie James at Stanford University has shown that some cues such as relative tone, relative volume and pauses are not salient enough for listeners (James, 1998a, 1998b). Also, subjects become confused with marked text in the middle of a sentence indicated by a speaker change. The change is distracting, and therefore, the subjects focus on the speakers' voices instead of the content of the message. Too many sounds in the interface increases the cognitive load required to learn and remember the many sounds and map them onto a mental model of the system. Also sounds that are too long slow down the listener, and underlining a word or phrase with a sound (eg., playing a sound simultaneously with the word or phrase) results in user distraction.

Because of the many cognitive demands on the caller, we investigated other strategies not addressed in James' research, namely, filtered speech and sounds (or earcons). Two experiments were conducted: one comparing filtered speech against a bracketing strategy in which earcons occurred before and after the voice linked word or phrase; the other comparing the bracketing strategy against only one earcon at the beginning of the voice link. The method used to assess the best audio cue was the same for both experiments (aside from the actual audio cue used in the comparison). The goal was to collect subjective preferences for each voice link type while also measuring how the cue affected the subject's ability to remember the voice link while comprehending the sentences. Subjects listened to 24 sentences, each with a voice-linked phrase, for example, "If you're always on the go, [One-Minute Ready Meals] are the fastest way to put a healthy meal on the table without putting a hassle in your schedule." For the first 12 sentences, the subject listened to one type of cue and for the second 12, the other cue. Lists were counterbalanced so that half the subjects were exposed to one cue first, while the other half were exposed to the other cue first. After each sentence, the subject was asked either to repeat the voice link or to paraphrase the sentence. By asking the subject to paraphrase once in awhile, the subject was encouraged to attend to the sentence and try to comprehend it while retaining the voice link. If subjects were only asked to remember the voice link, they may have used subvocalization rehearsal strategies without attending to the content of the sentence (Baddeley, 1986). After all sentences had been read, the subject was asked which style he or she preferred. Both subjective and objective data were analyzed.

The first experiment compared two bracketing earcons with filtered speech. Six subjects participated in this study. Although there was no statistically significant difference between the number of voice links recalled as a function of the voice link cue, all the subjects preferred the bracketing earcons over filtered speech, ($p < .05$).

The second experiment compared bracketing sounds with only one earcon at the beginning of the voice link. Thirty-two subjects participated in this study. Unlike the experiment with filtered speech, there was no strong preference for one type of voice link cue over the other. Both were selected about 50 percent of the time. Also the overall number of voice links recalled was not significantly different between the two cues. Because these data revealed no clear answer as to the better strategy, we analyzed the error data *post hoc* to see if the types of errors were more benign for one cue compared to the other. The errors were sorted into five different categories, from the most to least severe types of errors: the caller had no recollection of what the voice link was (most severe); the caller said a word or phrase from the wrong part of the sentence; the caller substituted one word for another (for example "dinner" for "meal"); the caller omitted a word from the voice linked phrase; the caller knew the voice link, but waited too long to respond. Figure 1 shows the distribution of the error types associated with bracketing sounds (two earcons) vs. only one sound at the beginning (one earcon). Statistical analysis of the error type shows that there are indeed different types of errors associated with each voice link cue (χ^2 (4); p=0.03).

One sound at the beginning of the voice link results in more severe errors including not remembering the word or phrase (don't know) and saying words in the wrong part of the sentence (other part); two sounds bracketing the word or phrase result in more benign errors including substitutions and delays. Also, some subjects commented that one sound leaves too much room for ambiguity, since it is not always clear where the voice link ends. Therefore, if forced to choose between the two strategies, we recommend bracketing sounds.

Overall, when considering how callers will navigate from place to place using voice links, it seems that bracketing earcons are the best auditory cue. This strategy is either the most preferred by listeners or results in the most benign errors. In general, the content available on the Voice Web does not include many voice links because the caller can easily move around the finite number of content sites. But as the number of content sites on the Voice Web reaches a critical mass, voice links will become an important means of navigation, as it is on the World Wide Web. For this reason, understanding the cognitive issues surrounding voice links and audio cues is an important step toward enabling this functionality.

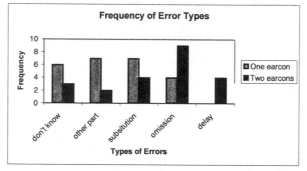

Figure 1. Distribution of Error Types for two different voice link strategies: one earcon versus two

4. Conclusion

Navigation of the Voice Web draws from many of the analogies of browsing the World Wide Web, for example, continued presence of the browser, menus that lead to interesting content, the Back button and hyperlinks. However in an all-auditory environment, with no buttons or colored text, new navigational techniques must be used to allow callers to take advantage of the Voice Web, and along with these new techniques are the human factors issues that accompany them.

First, callers must learn how to navigate. Something completely new to navigation on the phone, is how callers get back to the browser to maintain continuous call control. Data presented here suggest that an interactive tutorial is the best means of helping callers shift their thinking about phone calls and navigate back to the browser. In closed systems, such as V-Center, the list that provides callers with the system's available content is one of the most frequently visited areas of the system because it helps them learn what content is of interest. These findings support the notion that callers do learn new navigational paradigms to take advantage of content offered by the Voice Web. Navigational techniques that allow callers to move between sites are also of interest when brought into the auditory domain of the telephone. Interestingly, the navigational command "go back" is used by callers to return to content previously visited, even though these same sites could have been accessed by name. Although not a heavily used feature, our findings provide promising support for including a Back button analogy in the Voice Web. Finally, the best strategy for indicating audio voice links appears to be bracketing the word or phrase with sounds. This navigational features, like the "go back" command will become increasingly important as more sites become interconnected on the Voice Web.

Even though the Voice Web is still in its infancy, we are starting to make strides towards understanding how callers learn to navigate and how navigational analogies of the WWW can be applied to the auditory modality. By advancing our understanding of these issues and improving the voice user interface we help to mature the Voice Web and draw closer to the ultimate fusion of the web and telephone.

References

Baddeley, A.D. (1986). *Working memory.* New York: Cambridge University Press.
Kamm, C., Litman, D., Walker, M. (1998). From novice to expert: The effect of tutorials on user expertise with spoken dialogue systems. *Proceedings of International Conference Spoken Language Processing, ICSLP 1998.*
James, F. (1998a). Experimenting with audio interfaces. WEB Techniques. *Miller Freeman.* 3 (2), 55-58.
James, F. (1998b). Lessons from developing audio HTML interfaces. *Proceedings of ASSETS 98: ACM SIGCAPH's 3rd International Conference on Assistive Technologies*, 27-34.

Issues in the combination of visual and haptic interaction

Monica Bordegoni (1), Umberto Cugini (1), Piero Mussio (2)

(1) Dipartimento di Ingegneria Industriale, Universita' di Parma
(2) Dipartimento di Elettronica per l'Automazione, Universita' di Brescia

Abstract

This paper discusses possibilities and problems arising in the design of usable interactive systems when both visual and haptic interaction modes are available. Communicational and operational features of such a system are studied, and first set of problems that can arise for a typical end user are identified. Some experimental results are reported that give indications on how to face some of these problems integrating the two interaction modes.

1. Introduction

The successful use of haptic devices in several specific fields of application, and the increasing offer of haptic devices [1], suggest the possibility of generalizing their use to allow end users to realistically interact with the virtual scenes rendered by the computer. This paper explores how haptic devices can be integrated in an interactive system, taking into account the problems that can arise for a typical end user, and studying how to overcome those problems.

Problems arise in generalizing the use of haptic devices because of the novelty of the haptic technology and of the differences between computers mediated haptic interaction with virtual entities, and the natural haptic interaction with real entities, which users are familiar with.

Moreover, haptic tools seem to be per se difficult to use for the average users and some experiments suggest that haptic cues are not enough for the users to interpret correctly the virtual world they are facing [2]. These experiments as well as experience suggest also that haptic interaction can usefully complement visual messages.

The point of view assumed in this paper is that of designers of *usable* haptic-visual systems. A system is usable if it is easy to use and easy to learn [3]. The paper considers a necessary requirement for the usability of an interactive system the fact that it supports its users in reasoning with continuity on achievement of their tasks – learning or using the system – without being distracted by housekeeping or technological problems. To reach this goal, the designer must take into account communication and operational aspects of the interaction process.

To make these concepts precise, visual-haptic interaction is examined from the communicative and operational point of view within the frame of two models: the first, the Pictorial Computing Laboratory (PCL) model, focused on the communicational aspects of interaction [4], and the second, the Tacit Layered Reference Model [5] focused on the operational and continuity aspects of the process.

Haptic tools make available a stream of input–output events to the system, which appear to the user as a continuous signal. To decode and interpret these signals in order to assign them a meaning, users need to recognize event of interest in stream. The paper generalizes the concept of visual characteristic structures of the PCL model, defining characteristic structures to be sets of perceptible events that the user recognizes as a functional or perceptual units to which s/he can associate a meaning. Some sources of noise may force the users to distract from their goals, and break the continuity of the reasoning. The Tacit model helps in understanding this mechanism to suggest possible guidelines of design for the definition of usable visual-haptic interaction. The discussion is based on examples of haptic interaction developed using an existing device, the PHANToM haptic device [6].

2. Modeling the communication aspects

The PCL models WIMP HCI as a process based on the exchange of images between two participants, namely the human user and the computer [4]. Both human and computer interpret every event occurring during interaction with reference to the whole image appearing on the computer screen display. This image is formed by text, graphs, pictures, icons, etc.. and therefore represents a multimedia message, and materializes and conveys the meaning intended by the sender and must be interpreted – i.e. associated with a possible different meaning – by the receiver.

The images on the screen are produced (materialized) by the human or by the system through actions and computations. Human and computer communicate by materializing and interpreting a sequence of images at successive instants of time $t_1,....,t_n$. Humans interpret and materialize the images using human natural cognitive criteria, the computer using the criteria programmed in it. The images describe the state of the interaction, and their interpretation determines the next action of the human and the next computation of the computer.

During the interaction the computer assumes the double role of the tool used by the human to materialize his messages and of the second participant to the interaction. The human changes the image on the screen acting on the input devices of the computer. Therefore, the human uses the computer as the tool used to materialize the message. However, the computer also computes the reactions to the human's actions.

In this model, the I/O devices through which the human performs his/her actions, in particular the haptic ones – the mouse and the keyboard – appears as secondary channels of communication. Humans interpret the images on the screen by recognizing *characteristic structures* (css or structures for short), i.e. sets of image pixels that users recognize as functional or perceptual units. The cs recognition results into the association of a meaning with a structure. Humans express the meaning attributed to the cs by a verbal description. In an image a human may recognize several css: combining their meanings, humans derive the meaning of the whole image on the screen.

On the other hand, the computer associates graphical entities (set of pixels) on the screen with computational constructs. Each computational construct -here denoted by u- represents the meaning associated with a graphical entity on the screen, as intended by the system designer.

It is exactly this association that makes the computer able to interpret the captured user actions (such as clicking on a button), with respect to the image on the screen, possibly firing computational activities whose results are materialized on the screen, via creation, deletion, or modification of css. The association between a cs and the corresponding u is called a *characteristic pattern* and can be formalized introducing two functions intcp and matcp the first mapping the cs into u and the second u into the cs. Hence a *characteristic pattern*(cp) is specified by cp=<cs, u, <intcp, matcp>>.

The image I on the screen is a cs formed by a set of component css, each associated to its u. Hence the inage I can be associated to a description d, the set of all descriptions, associated to the component css. Two functions, int and mat, can be defined on the basis of the individual functions in the cps, and the relationship between the image i and its semantics d is summarized by a *visual sentence,* the triple: vs=<i,d,<int, mat>>. A set of vss is a Visual Language (VL).

Two semantics are always implicitly defined in any interaction: one internal to the computer, in which each image is associated with a computational meaning, as defined by the designer and implemented in the computer; and one proper to the user performing the task, depending on his/her role in the task, as well as on his/her culture, experience and skill. As observed in [7], users can achieve their tasks if they associate with each cs and with the whole image a meaning similar to the one associated by the computer, i.e. that is if an *adequate communication* is reached. From this point of view, the goal of a successful design is to bring the system semantics to reflect the user's one, so that the messages exchanged during the interaction are properly understood by the user and adequately managed by the computer.

3. Modeling the operational aspects

According to the Tacit model [5] and our experience, the physical events occurring in HCI are managed at different level of abstraction both by humans and computers. All events occur at physical level, resulting from user's actions performed using input devices, or resulting from system computations through output devices.. A human *input action* is defined as an action performed by a user through an input device connected to the computer. An *output action* is an action performed by a device connected to the computer on a user perceptive structure.

The computer detects the user's actions trough devices that convert user's actions into digital events. Those are characterized by variables having values in finite sets. For example, the keyboard translates the user's key press (action discrete in time) performed on one or more keys, into a binary code sent to the program (digital event). The mouse samples every movement performed by the user (continuous action) and computes the (x,y) coordinates of each position coded as finite binary words. Therefore, the action is translated into a sequence of events, and arranged according to time. Each event consists of sending the couple of digital coordinates to the program. Because the coordinate values belong to a finite set, the mouse movement performed on a plane surface in the real world corresponds to a movement performed on a discrete toroidal surface in the virtual world.

Computer performances can be modeled at different levels of abstraction, corresponding to different levels of architectural description available, e.g. instruction set architecture or high-level languages. However, given a program, i.e. a model expressed at a certain level of abstraction, its execution requires its translation into the lowest level, i.e. the machine language level [8].

Human performances can also be modeled as if perception, reasoning and interpretation and communicational processes occur at different levels of abstraction. However it seems that all these processes are all executed at the highest admissible level of abstraction suited for their necessity and skills, according to the environmental conditions in which they operate. Humans maintain the interaction at this level of abstraction adopting a frame of reference while performing an action and against which they monitor success or progress - a *prepotent identity* [9].

3.1 Events that causes the change of reasoning layers

Unexpected events or the lack of dexterity may oblige the users to renounce to reason at the level necessary to achieve their task, and force them to assume a different frame of reference. Therefore these events break the continuity of reasoning. For example, a skilled human needs to select a customer file using a command language. The skilled human looks at the screen and reasons at a high layer of abstraction (e.g., s/he realizes that s/he need to "select the customer file") and executes the required operations always reasoning at this *prepotent layer*, without focusing his/her attention to the lower layer articulation of high layers operations into simpler ones.

S/he creates the command "select" depending on the context, without thinking of the context itself and of the elementary operations needed to execute the command. If s/he is typing, s/he types the command "select" on the keyboard, without thinking explicitly to each key to be pressed; if s/he is using the mouse s/he moves it to the desired position and clicks on an icon denoting "selection", without being aware of the physical execution of the low level operations involved.

To obtain an equivalent result, a novice user has to focus his/her attention to this physical execution, looking on the keyboard for the key marked "s", then for the one marked "e", and so on. S/he has to explicitly design a strategy to execute a low level operation within the frame of high layer operation.

This same behavior can be observed for the skilled user, if an event impedes to maintain the action in terms of its *prepotent identity*. For example, if the user moves the mouse too quickly, the coordination with the cursor on the screen can be lost. The skilled user becomes disoriented so that the prepotent identity level changes in order to determine where the error is, and how to recover from it. In these cases, there is the necessity for a lower layer identity to become *prepotent*: the attention of the skilled user is switched to the lower layer, his/her capabilities are activated to translate the high layer operation into lower ones, the reasoning goes on at the new prepotent layer.

Forcing an analogy between human and machine models, the human can be modeled as if s/he has available several different machine languages wired in - prepotent identity frames - and with the capability of switching from one to another and trigger the required translators according to necessity. According to this view, for what concerns humans, the high layer of reasoning and communication is not virtual but corresponds to real processes performed until the current prepotent level can be maintained.

Ambiguous or equivocal situation may also cause the breaking of reasoning continuity. *Ambiguity* arises when one of the two communicants -the human or the computer- associates two different meanings with a message, and flips from one meaning to the other during a same reasoning process.

Fig. 1 shows a well-known case of ambiguity for the human. The computer generates an image based on well-defined geometrical model. The human is not able to interpret univocally the image and may flip between different interpretations: a cube with a hole, two boxes, and a box in a corner.

Equivoque arises when the two communicants associate two different meanings with a same message. Each communicant has no ambiguity in what s/he is doing, but different cultures or situational reasons determine different interpretation. Conflicts may arise between the two communicants, which break the continuity of reasoning. Equivoques are not treated here any more.

Figure 1. Images of three 3D models: each image can be interpreted as a carved box, two boxes or a box in a corner.

4. Haptic interaction

Several kinds of devices have been proposed for haptic interaction, from one-actuator (for example, the PHANToM device [6]) to multiple actuators (for example, the haptic display [10]). In any device, each actuator transmits a sampled, quantized signal, in terms of point applied forces reproducing the physical interaction with an element of a surface, having specific characteristics (roughness, softness, etc.). The human hand/finger receives this set of stimuli, and the person decodes all that as a local, tactile description of a 3-D surface.

In haptic interaction human and computer exchange:

a) a set of impulses which are applied by the computer through some device to the human's hand and/or fingers and;

b) a set of movements performed by the human through the hand and/or finger on an haptic device and mapped by the device into a time sequence of digital signals for the computer.

Due to current technological constraints, haptic devices are limited in resolution and dimensions. They send a high granularity set of signals that describe a limited part (or even a point) of the surface being explored. This limitation induces some aliasing effects that may determine ambiguous or equivocal situations.

250

In order to understand an environment, the human has to integrate in time the set of data that receives from the haptic device. The identification of haptic patterns is based on perceiving continuity or discontinuity in a set of signals to be interpreted, so as to form a mental model of the surfaces present in the environment. The frequency at which the haptic device transmits impulses to the human hand /fingers determines the human perception and influences interpretation of the set of signals. Similarly to the visual refresh rate, some studies have shown how continuous perception is achieved only if the haptic refresh rate is not below 1KHz [11]. If the refresh rate goes below this threshold, humans may feel some discontinuities that simulate non-existing patterns. Moreover, in present day haptic devices, other disorienting feature may exist, determined by the specific technology adopted.

Assuming haptic devices as primary communication devices changes the interaction scenario, and requires an enrichment of the PCL model. Still humans recognize characteristic structures in the stream of input events made generated by the system but events produced by the haptic devices are only local in space and generate *css* as stream of output events. The recognition of haptic *css* requires the human to integrate these events in this stream to recognize the subsequences that represent functional or perceptual meaningful units. The perceptual and cognitive mechanisms by which these haptic *css* are recognized are different. Their use poses the problem of defining what *haptic css* are, how the human and the computer relate these structures to their meaning, and how the meaning of a haptic *css* and a visual *css* are combined to derive the meaning of the whole situation.

To make the reasoning concrete, the exposition will be based on examples from some experiments performed using a PHANToM device.

4.1 Ambiguity in the haptic interaction

To use a PHANToM, a human wears a thimble that ends with a needle. The thimble is mechanically connected to a mechanism, controlled by the system. When the user touches a virtual entity through the thimble, the device reacts generating a point wise force simulating the entity reaction. The direction and intensity of the force depend on the movement performed by the user through the thimble. The human contacts the inside of the whole thimble and perceives the reaction through an area of the skin and therefore always perceives a bi-dimensional message. In this situation, the user cannot distinguish a point like event – the Phantom needle touching the point of a virtual needle - from a surface event – the Phantom needle touching a rigid surface. Humans can distinguish these events only dynamically, exploring in space a contact neighborhood and mentally integrating in time the continuity of a sequential signal. In other words, the patterns emerge from the integration of a set of messages in time. These features characterizing the PHANToM may contribute to disorient the human.

Consider for example, an interactive environment in which the human interacts with a system, which has created a wire, a surface and a body with edges in a 3-D virtual space. The virtual space is discrete for the system, but perceived and conceived as continuous by the human. Figure 2 shows the three virtual objects: a wire (a), an edge (b) and a line on a planar surface (c), created in the virtual space by the system.

If the user perceives a constant signal continuous in time, the corresponding mental model is that of a flat surface and if s/he perceives a discontinuity in the signal the corresponding mental model is that of an edge. If the human were able to follow a linear path when touching the three virtual entities, s/he would perceive an identical signal in the three cases, and possibly interpret the three situations as the exploration of a surface. This happens because the physical contact of the user's finger wearing the PAHNToM is mediated by the internal surface of the thimble. Therefore, the user always perceives the same set of signals in the three cases, and is not able to discriminate the three situations.\

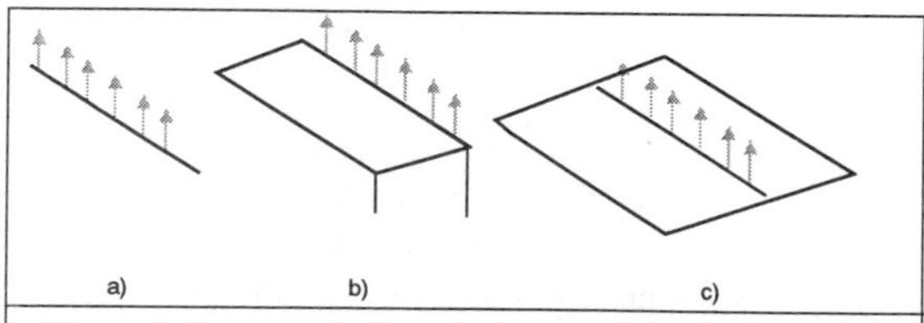

Figure 2. Representation of three virtual objects: a wire (a), an edge (b) and a line on a planar surface (c). The arrows represent the force reaction provided by the PHANToM device.

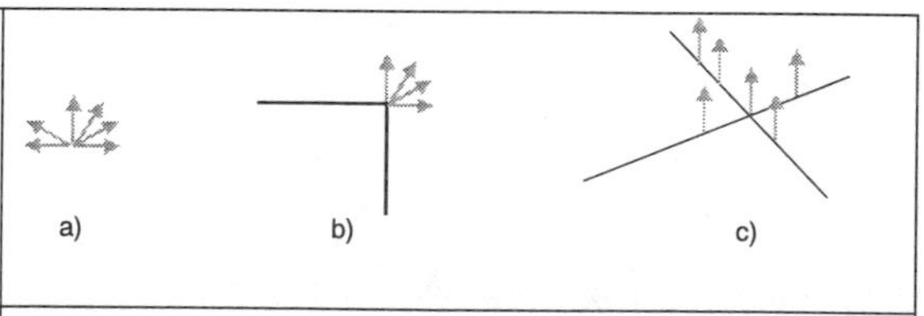

Figure 3. Exploration of the space neighborhood of a point in other directions other than the linear one, considering the objects of Figure 2.

In order to discriminate the single situations, the user should explore the space neighborhood in further directions other than the linear one. In this case the PHANToM device would provide consistent physical reaction, as shown in Figure 3.

The perception of continuity in spatial and temporal dimensions allows the user to perceive correctly and discriminate the three cases.

4.2 Combined visual-haptic interaction

To verify the power of haptic interaction we have run some experiments with twenty subjects [2]. The experiment is based on the 3D objects shown in Figure 1, and has been run in three steps. (Figure 4 shows another view of the models, in order to show the reader, what the objects are). The first time, the subjects have been shown the images of Figure 1, one after the other, and have been asked to say which object they think they recognize. The analysis of the answers shows an evident ambiguity in recognizing objects from their image.

In the second step, we have run the experiment providing haptic stimuli only of the 3D models. The result of this second experiment shows that none of the subjects was able to say which object they were touching, regardless of their experience and skills. These very poor results have several causes. The device by itself has a very low bandwidth of information considered both in absolute terms and when compared to the sense of touch in reality. Humans do not use only one finger during shape recognition and, even in that case, the number of receptors conveying haptic information is more than a thousand order of magnitude of what the PHANToM provides.

The third step has provided the subjects both visual and haptic stimuli for the three objects. This last experiment has shown better results. In fact, the coupling of two modalities providing redundant information on the scene to be explored and recognized increases the probability of a correct interpretation of the scene.

Figure 4. Objects of Figure 1 shown from another point of view.

5. Conclusion

A usable system allows its users to reason at the highest prepotent level, permitted by their culture and skill and by the current situation. Both inadequate communication and/or lack of dexterity force the user to lower the prepotent identity layer of reasoning. In multimodal system inadequate communication may be caused by ambiguity arising in the interpretation of events in one modality. Experiments suggest that in this case events occurring on the other channels can be used to resolve ambiguity.

However, to exploit this possibility some problems still open must be faced: in the case of visual-haptic interaction the definition of what haptic *css* are; how human and computer relate *css* structures to their meaning; how the meaning of a haptic *css* and a visual *css* are combined to derive the meaning of the whole situation.

References

[1] Cugini, U., Bordegoni, M., Rizzi, C., De Angelis, F., and Prati, M., (1999). Modelling and Haptic Interaction with non–rigid materials". *Eurographics'99 - State-of-the-Art Reports*, Milano, 1999.

[2] Faconti, G., Massink, M., Bordegoni, M., De Angelis, F. and Booth, S. (2000). Haptic Cues for Image Disambiguation. *Computer Graphics Forum. Blackwell, 19(3)*, pp. 169-178, 2000.

[3] Nielsen, J. (1993). Usability Engineering, Academic Press, 1993.

[4] Bottoni, P., Costabile, M.F., Levialdi, S., Mussio, P. (1996). A Visual Approach to HCI. *ACM SIGCHI Bulletin, Vol.28, No.3*, July 1996, pp.50-55, also available at the URL www.acm.org/sigchi/bulletin/1996.3/levialdi.html.

[5] Massink M. and Faconti G. (2000). *CHI2000 Workshop Report: Continuity in Human-Computer Interaction. SIGCHI Bulletin*, September 2000.

[6] SensAble Technologies, Inc., URL: www.sensable.com.

[7] Chang, S.K. and Mussio, P. (1996). Customized Visual Language Design. *International Conference on Software Engineering and Knowledge* Engineering - *SEKE'96*, pp. 553-562.

[8] Tanenbaun A.S. (1999). Structured Computer Organization. *Prentice Hal, London*, 1999.

[9] Buehener, M. (1999). Comments on the first draft of Characterizing Continuous Interaction in Systems for Information Access and Presentation. *TACIT Report*, 1999.

[10] Hirota, K. and M. Hirose, M. (1995). Providing Force Feedback in Virtual Environments. *Proc. IEEE Computer Graphics and Applications*, pp. 22-30, September 1995.

[11] Massie, T. (1998). Physical interaction: The Nuts and Bolts of Using Touch Interfaces, Physical Interaction: The Nuts and Bolts of Using Interfaces with Computer Graphics Applications, SIGGRAPH 98, Tutorial 1, 1998.

From User Notations to Accessible Interfaces through Visual Languages

Paolo Bottoni+*, Maria Francesca Costabile§*, Stefano Levialdi+*, Piero Mussio°*

+Dipartimento di Scienze dell'Informazione, Università "La Sapienza", Roma, Italy
§Dipartimento di Informatica, Università di Bari, Bari Italy
°Dipartimento di Elettronica per l'Automazione, Università di Brescia, Brescia, Italy
*Pictorial Computing Laboratory, Università "La Sapienza", Roma, Italy
[bottoni, levialdi]@dsi.uniroma1.it, costabile@di.uniba.it, mussio@bsing.ing.unibs.it

Abstract

This paper reports on the Pictorial Computing Laboratory methodology for designing interactive visual systems, stressing the need of system accessibility by end-users. The methodology is aimed at satisfying three principles: a) the user must always understand the effects of the system activity with respect to the execution of his/her task; b) the user must always be in control of the interactive computation, avoiding to get lost in the virtual space; c) the system has to trap users' errors and maintain itself viable, i.e. in a predictable set of states. The design of the visual system is accomplished through different steps, starting from the analysis of the notations end-users have developed in their working environments, and each step producing a new visual language that must be verified and validated.

1. Introduction

In the emerging information society technology, there is a need of universal design: it emphasises accessibility and high quality of interaction by an end-user population as broad as possible. Increasingly, end-users are people not expert in computer science, who use interactive computer systems to perform tasks of which they are responsible in a precise context and according to their working culture and experience. End–users have different needs and backgrounds, and operate in different contexts. As a consequence, our view of *universal design* does not imply that a single user interface is suitable for all users. Instead, as designers we put effort in proposing solutions tailored to the system users.

Since the early 80s, the community of researchers working in Visual Languages has devoted a lot of effort in designing environments that can facilitate the way people interact with computers. A visual approach is certainly not suitable for visually impaired people, but for many other people the use of visual formalisms has been very successful in several situations, such as working with a computer operating system or in the various phases of the process of building software that should satisfy the needs of different user populations. In particular, visual languages have been successfully exploited in programming environments for children and in areas of application such as laboratory instrumentation.

Brancheau and Brown define end-user computing as " ... the adoption and use of information technology by people outside the information system department, to develop software applications in support of organizational tasks" [1]. End-users (or users for short) are therefore experts in a specific domain, not necessarily experts in computer science, who use visual environments to perform their daily tasks. Users are responsible for the activities accomplished trough the system and for the produced results.

The member of the Pictorial Computing Laboratory (PCL) have gained experience in the design of interactive visual systems that provide appropriate environments for different types of experts. This experience has led to a methodology for designing interactive visual systems, which stresses the need of system accessibility by end-users. Such a methodology is described in this paper, after discussing our approach to the design of accessible systems.

2. Approach to the Design of Accessible Systems

The PCL approach to the design of end-user accessible interactive systems is motivated by three principles: a) the user must always understand the consequences of the system activity with respect to the execution of his/her task; b) the user must always be in control of the interactive computation, avoiding to get lost in the virtual space; c) the system has to trap users' errors and maintain itself viable, i.e. in a predictable set of states.

The PCL approach relies on a design methodology, presented in this paper, aimed at satisfying the above three principles on the basis of the following assumptions: a) *adequacy of human-computer communication* [2] and *system communicability* [3] are prerequisites to the user acceptance of the system; in our view they are important dimensions of the overall system usability; b) *users' notations* are the way in which users express their culture, therefore the development of usable systems has to capitalise on these notations [4]; c) *formalisation* is a prerequisite for verifying soundness, completeness and consistency of the progressively obtained results of the design process [5]; d) two *semantics* exist, one relative to the user and the other implemented within the system: system validation is an activity aimed at evaluating the match between these two semantics; e) a co-ordinated use of *validation* and *verification* activities is necessary to ensure both usability and viability of an interactive visual system. The system validation activity is primarily based on usability evaluation, stressing the adequacy and communicability dimensions. The verification activity is based on a formal theory for specification and design of interactive visual systems, derived from the theory of visual sentences, as developed in the last five years [6].

In the PCL approach, Human-Computer Interaction is modeled as a process in which the user and the computer communicate by materializing and interpreting a sequence of messages at successive instants of time, the human using his/her cognitive criteria, the computer using the programmed criteria [6]. The interactive visual system may be considered as a generator of messages from the designer to the system user [3].

Two semantics are always implicitly defined in any interaction; one internal to the system, in which each message is associated to a computational meaning, as defined by the designer and implemented in the system, and one proper to the user performing the task, depending on his/her role in the task, as well as on his/her culture, experience, etc. As observed in [2], the interactive achievement of the user task requires that a similar meaning be associated by the user and the system to each message, i.e. that an adequate communication be reached. In the case of two-dimensional direct manipulation, on which this paper focuses, the exchanged messages are the whole images represented on the screen display, formed by text, graphs, pictures, icons, etc.

Humans interpret such images by recognising *characteristic structures* (**cs**s or structures for short), i.e. sets of image pixels which they recognise as functional or perceptual units. The **cs** recognition results into the association of a meaning with a structure. Humans express the meaning attributed to the **cs** by a verbal description. Such identification of **cs**s is influenced by the (dis)similarity with graphical entities and constructs traditionally adopted in the user community.

On the other hand, the system associates graphical entities with computational constructs as well. It is exactly this association which makes the computer able to interpret the captured user actions (such as clicking on a button), with respect to the image on the screen, possibly firing computational activities whose results are materialised on the screen, via creation, deletion, or modification of **cs**s.

The problem of achieving adequate communication, therefore, requires that a precise correspondence be defined between the structures perceived by users and those foreseen by designers and implemented in the system. In general, this requires the formal definition of arrangements of pixels to be considered as **cs**s, and of the association between such structures and their perceived or intended meaning. To this end, we have introduced the notion of *characteristic pattern*.

3. Interaction Visual Language

A **cs** is a set of pixels, perceivable on the screen. The meaning associated with a **cs** is formalised as an *attributed symbol*, called u, consisting of a type name - the *symbol* - and a tuple of properties - the values of the attributes providing a description and interpretation of the **cs**. The association between a **cs** and its description u is expressed by two functions, intcp and matcp. The *interpretation function*, intcp, associates the **cs** with its description u. The *materialisation function*, matcp, associates the attributed symbol u with the **cs**. A *characteristic pattern* (**cp**) is the triple cp=<cs, u, <intcp, matcp>>.

In an image i, several **cp**s can be identified. Moreover, i as a whole can be associated with a symbol u_i synthesising the overall properties and meaning of the image i. In any case, the set d of all the attributed symbols appearing in the **cp**s constructed on i constitutes a description of the image i. Two functions, int and mat, can be defined on the basis of the individual functions in the **cp**s, and the relationship between the image i and its semantics is summarised by the triple: vs=<i,d,<int, mat>>. We call vs *visual sentence* (**vs**), i image component and d description component of vs.

From the user's point of view, each image is an electronic document. Performing a task with the support of an interactive visual system is equivalent to producing a set of electronic documents through the execution of a sequence of actions. In general, any interactive activity supported by the system can be reduced to the production of such a sequence of electronic documents. The users identify the system with the sequence of **vs**s, whose image part may be seen on the screen. Therefore, from the designer's point of view, the whole system is specified by defining the set of all admissible sequences that could be generated during an interaction process. This set constitutes a dynamic visual language, i.e. a set of **vss** structured by a preorder relation [7], called Interaction Visual Language (IVL). The design of the interactive visual system is reduced to the design of IVL. For this reason in the following we only address the IVL as the object of the design and implementation activities.

The **vss** in IVL are composed from a finite number of **cp** types. The set of **cp** types is described by representing each type by an exemplar **cp**. This finite number of exemplar **cp**s is called the alphabet K of IVL. A **vs** is composed by a finite number of instances of elements of K, suitably transformed and composed.

In our approach, these transformations and compositions are defined on K in finite terms [6]. Moreover, users communicate their intentions to the visual system via some actions, performed through gestures. The set of gestures is constrained by the technology of input tools and results to be a finite set of input signals to the system. The system interprets these gestures to determine the computation to be executed with respect to the coordinates of the objects on the screen and to the specific structures appearing on it. More precisely, the interpretation occurs with respect to the subset of pixels on the screen currently pointed at by the system, from which it derives the set of structures currently active and the computations that have to be performed. If the user executes a gesture that could result in an inconsistent state, the system has to trap this gesture, and execute an appropriate computation, for example the generation of a warning message or simply a null computation - in any case avoiding a system crash.

4. Visual Languages in Traditional Working Environments

In traditional working environments, humans executing a task (also called experts) inscribe their knowledge by organising graphs, images and texts into documents [4], with the aim of tracking their activities and products and communicating their knowledge to other humans. From the point of view of the PCL theory of visual sentences, words in the text and structures in graphs and pictures are all **css**, documents are image components of **vss**, where the descriptions of their meanings and correspondence relations are not necessarily made explicit, but are in any case present to the experts in their activity. A working procedure in a traditional environment is here modelled as document production and transformation, where a first document contains information about the initial state of the process and the succeeding documents represent the intermediate stages of the user activity to achieve a specific task, until the last one showing the final results.

The sequence of documents the experts produce during task execution is an observable result of their activity. This sequence of paper documents is analogous to the sequence of electronic documents observed when a user works with an interactive visual system. We exploit this analogy to match the user and the system semantics, by adopting the notation used by the experts in their traditional working environment as the kernel of the IVL definition. This is an important step in order to satisfy the three principles reported in Section 1, in particular to reach adequate human-computer communication. The design of an IVL based on user notation starts from the analysis, performed by the designer together with representatives of the users, of the process of production of a set of such documents made visible by the collaboration. Together with their accepted descriptions, these documents form what we call *observed Task Visual Language*, TVLo, while the analysis of the process enables one to establish relations among the documents.

In this view, TVLo is a finite set of interpreted documents constituting a sample of the documents which can be generated during execution of users tasks, which is the proper Task Visual Language. The design of an IVL requires the abstraction of a notation from TVLo, and the subsequent derivation of an intensional formal definition of TVL. However, when the performed activity is more abstract and/or their culture is oriented to formal reasoning, experts often develop a visual notation -and sometimes an intensional formal definition of the VL- through which they generate documents by a rigorous, even if not algorithmic, procedure. Typical examples are Petri Nets. Between these two extreme situations, several intermediate cases may be found when studying different expert communities. We observed at least three different types of situations:

1. Experts produce documents to specify some situations or products. TVLo is a set of documents, which do not describe the whole task, but account for some important moments and decisions along its achievement. Documents are produced without following any formal strategy and experts learn how to produce a document by observing the activities of other older experts [8].
2. Experts systematically transcribe each step of their activities, as occurs in laboratories where a logbook exists or in engineering activities. TVLo documents describe the strategies followed by experts to achieve their tasks. In some fields, guidelines and visual standardised glossaries for document production exist (e.g. mechanical drawings).
3. Experts systematically describe their activities, developing their documents according to a formal -in a mathematical sense- notation. Not only does a well established TVL exist -whose documents describe the

strategies followed by experts in performing their tasks- but also a notation is used, which defines the strategies followed by experts in developing their documents, as in the case of building Petri Nets [4].

Traditionally, experts perform their tasks without the support of a computer system. In order to design a system to help the users to enhance their performances, the designer collaborates with the users working in their environment, to identify the strategies by which users perform tasks and organise them into sub-tasks (*task analysis*). The designer also identifies the documents by which these strategies are outlined. Such documents are ordered according to the stage of the task in which they are produced. Together with their interpretations, the documents constitute the **vss** of the TVLo. Additionally, the designers can identify the ordered structure imposed on the production of **vss**, and derive a method to only allow the production of admissible sequences of **vss**.

5. The Methodology for IVL Design

Following a user-centred approach, in the design of IVL we start from the user and task analysis and from the observation of the produced documents (see Figure 1). The designer organises all the observations in a systematic way into an explicit ontology of the domain, an organisation that the user is not always able to reach. This organisation results into a visual alphabet and a set of rules expressed in the user language, namely TVLo (top box of Figure 1). The adequacy of such a glossary -and of the rules- with respect to the user significant constructs requires a first verification phase, to demonstrate that all the sentences in the TVLo can be produced and that no sentence violating the constraints required by the users can be created. Moreover, a first phase of validation is also necessary to check that the visual alphabet and the rules can be properly understood and managed by the user.

From the observation of the produced documents and of the users' activities, we abstract the definition of the Task Visual Language (TVL). To this end, we need to identify the visual alphabet and rules implicitly used by experts in building and transforming their documents. This process can go through the definition of several tentative generalisations of the TVLo before an agreement is reached. In particular, it is possible that the typical graphical constructs employed by the users are not sufficiently expressive or non-ambiguous to build a formal notation. Several steps are then required to reach a form adequate to be formalised, by eliminating the identified ambiguities and inconsistencies.

Once an agreement is achieved on what has to belong to the TVL, a formal definition of this language is needed (box 2 in Figure 1). In [9], details on this are provided. A semi-formal verification is possible by checking that every observed document can be produced with the use of the rewriting system that specifies the language, that incorrect documents cannot be composed, and that the legal sentences correspond to acceptable intermediate steps during document construction and transformation. The formal notation thus obtained must be evaluated with respect to the user semantics. In particular, it is necessary to verify that the users correctly understand the definition of the derived TVL and of its use.

The TVL thus defined is then augmented to take a communicational advantage from the interactive capabilities of the system (box 3 in Figure 1). Indeed, the sequences of TVL **vss** describe the development of experts' activity. TVL is an intermediate step in the design of IVL, which proceeds by enriching TVL with the definition of the **cps** through which the visual system communicates to the user the state of the interaction and provides access to its functionalities. Moreover, TVL itself can be augmented to exploit the possibilities of interaction and animation available in a computer system. The designer may augment TVL by introducing new **css** to make the **vss** in the language more compact and expressive. The designer can also use attention holders, feedback mechanisms, animation effects, detail exhibition, focus modification. For example, in order to provide feedback to the user action of pointing to a **cs** on the screen, such a **cs** is highlighted by changing its colour, thus generating a new **cs**. The notation must also be adapted -hence often changed and approximated- to the digital technologies by which the image on the screen is generated as well as to the fixed, limited dimensions of the screen.

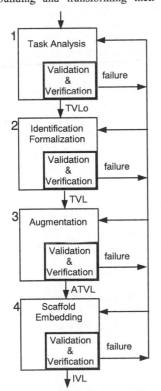

Figure 1. Steps in IVL design

The validation of the augmented TVL (ATVL) requires a partial implementation or mock-up of the system, so that the consistency of the dynamics can be checked with the ones expected by the user. In particular, the interaction dynamics can be simulated with mock-ups and simple prototypes to ensure that the behaviours of the **cps** are understandable by the user and that the user has access to all the needed details made implicit by the synthetic notations. The verification of ATVL requires that all the **vss** in the TVL remain accessible and obtainable through ATVL.

Finally, the Interaction Visual Language (IVL) is specified by defining how the **vss** of the ATVL can be integrated into visual sentences with additional support for the interaction, such as **cps** providing context for user orientation in the interaction space (see the concepts of frame and scaffold in [7]), or giving access to the system functionalities, tools to operate on the **vss** themselves, etc. The definition of the IVL also encompasses the constraints posed by the strategies of document construction identified during the Task Analysis. These may imply sequences of actions to be performed in the production of the document, checklists, levels of access to functionalities, etc.

As sketched in Figure 1, each step of the IVL design consists of: i) design and implementation of abstract or concrete tools; ii) verification of their soundness, completeness and consistency; iii) validation of their adequacy with respect to human-computer communication, using tools according to the technique chosen with respect to the users' semantics.

During this process, each step or sequence of steps may be iterated several times if the verification or the validation of the results of some steps in the procedure so requires. Each step in the IVL design requires the preliminary choice of tools necessary to execute the step and of a technique of assessment of the obtained results: new tools or new sets of data. In the tool choice and use, problems arise from the existence of the two semantics, the user and the program ones. Evaluation of the results must be performed with respect to both semantics and require verification and validation through experimental evaluation.

6. Conclusions

We have presented a methodology for designing interactive visual systems that stresses the need of system accessibility by end-users. It aims at satisfying three principles: a) the user must always understand the effects of the system activity with respect to the execution of his/her task; b) the user must always be in control of the interactive computation, avoiding to get lost in the virtual space; c) the system has to trap users' errors and maintain itself viable, i.e. in a predictable set of states. To satisfy these principles, the design of the system is broken into three main steps:
1. From task analysis the language TVL is derived that consists of all visual sentences required for the task.
2. A necessary augmentation of the TVL (ATVL) is derived to exploit the possibilities of interaction and animation available within a computerized system.
3. The interaction visual language (IVL) is obtained by specifying characteristic patterns which communicate to the user the state of the interaction process also providing access to the system functionalities.

The designer must provide a formal specification of all such languages [9], in a form suitable for usability validation and for consistency verification.

Acknowledgements
This work is supported by EC grant FAIRWIS n. IST-1999-12641, and by MURST 60% and Cofin 2000.

References
[1] Brancheau, J.C., Brown, C.V. (1993). The Management of End-User Computing: Status and Directions. *ACM Computing Surveys*, 25(4).
[2] Chang, S.K., Mussio, P. (1996). Customized Visual Language Design. Proc. of Eighth Int'lConference on Software Engineering and Knowledge Engineering, SEKE'96 (pp. 553-562).
[3] Prates, R., De Souza, C., Barboza, S. (2000). A Method for Evaluating the Communicability of User Interfaces. *Interactions*, 7 (1), 31-38.
[4] Latour, B. (1986). Visualization and Cognition: Thinking with Eyes and Hands. In: *Knowledge and Society: Studies in the Sociology of Culture Past and Present*, Vol. 6, pp. 1-40.
[5] Dix, A., Finlay, J., Abowd, G., Beale, R. (1998). *Human Computer Interaction*. London: Prentice Hall.
[6] Bottoni, P., Costabile, M.F., Mussio, P. (1999). Specification and Dialogue Control of Visual Interaction through Visual Rewriting Systems. *ACM TOPLAS*, Vol. 21, N. 6, pp. 1077-1136.
[7] Bottoni, P., Chang, S.K., Costabile, M.F., Levialdi, S., Mussio, P. (1998). On the Specification of Dynamic Visual Languages. *Proc. IEEE Symposium Visual Languages'98*, pp. 14-21.
[8] Bianchi, A., D'Enza, M., Matera, M., Betta, A. (1999). Designing Usable Visual Languages: the Case of Immune System Studies. *Proc. IEEE Symposium Visual Languages'99*, pp. 254-261.
[9] Bottoni, P., Costabile, M.F., Levialdi, S., Matera, M., Mussio, P. (2000). On the Specification of Dynamic Visual Languages. *Proc. IEEE Symposium Visual Languages 2000*, pp.145-152.

What Visual Programming Research Contributes to Universal Access*

Margaret M. Burnett[a]

[a]Department of Computer Science, Oregon State University, Corvallis, OR 97331

Abstract

There is a significant body of visual programming language research aimed at expanding the kinds of users who can succeed at programming, and research of this nature has close ties to issues of universal access. These ties are universal access issues as they arise in the domain of programming, and go in two directions. The first direction is that languages aiming to expand programming accessibility to new types of users can gain from emerging research from the universal access community. The second direction, and the subject of this paper, is that, because visual programming languages have made a number of inroads into a particular domain of accessibility not normally considered by universal access researchers—programming—they can contribute back to the universal access community ideas that may be useful in improving universal access in other domains with significant problem solving components. In this paper, we consider the following examples of such contributions: usability methods drawn from cognitive psychology research on human problem solving in programming, principles learned in exploring programming by new audiences, and supporting flexibility in both syntactic and semantic matters.

1. Introduction

Visual programming is programming in which more than one dimension is used to convey semantics (Burnett, 1999). Examples of such additional dimensions are the use of multidimensional objects, the use of spatial relationships, or the use of the time dimension to specify "before-after" semantic relationships. The overall goal of most visual programming languages (VPLs) is to strive for improvements in programming language design from the perspective of how well humans can use them (Blackwell, 1996). That is, VPLs aim to make programming accessible to new audiences and/or more accessible and usable by existing audiences. The role of visual and other techniques are that they are enabling mechanisms toward this goal of more accessible programming languages and environments.

A way VPL research intersects with universal access (UA) is in the goal of bringing *programming* power to audiences not served before. Thus, VPL work offers contributions that may be applicable to bringing UA to other types of problem-solving activities as well. Examples of ways VPL research can contribute to UA include mechanisms used by VPL researchers to measure usability related to problem-solving, mechanisms to support new audiences in this kind of problem-solving, and notions of flexibility not only at the syntactic level, but also at the semantic level. In this paper, we discuss each of these points.

2. Usability and Problem solving

Since the goals of VPLs have to do with improving humans' ability to program, research into cognitive issues relevant to programming and problem-solving has played an important role. In contrast to this, most usability evaluation mechanisms contributed by the larger HCI community have been more surface-level, and do not particularly focus on the problem-solving aspects of human-computer interaction. (A notable exception is the cognitive walkthrough (Polson et al., 1992).) The focus on problem-solving aspects is one that may also be important in UA work, since it too aims to help humans who are unaccustomed to computers use them in a variety of problem-solving tasks, from "programming" entertainment preferences to finding and navigating web pages.

The most influential of this body of work has been that of Thomas Green and his colleagues in their contribution of *cognitive dimensions* (Green & Petre, 1996), a set of terms describing a programming language's components as

* This work was supported in part by Hewlett-Packard and by the National Science Foundation under awards CCR-9806821 and ITR-0082265.

they relate to the cognitive psychology of programming and problem solving. For example, the cognitive dimension of "progressive evaluation" asks "Can a partially-complete program be executed to obtain feedback on 'How am I doing'?" This brings out the fact that, if impacts of a user's problem-solving activities are shown while they are being performed, the user is better able to judge the effectiveness of these activities. As another example, the cognitive dimension of "hidden dependencies" asks "Is every dependency overtly indicated in both directions?" This brings out the fact that settings or statements in one part of a program may affect other parts, and it saves users the burden of discovering and remembering these relationships if the system explicitly presents them in some way.

As these examples demonstrate, the emphasis of cognitive dimensions is on cognitive actions for problem solving. In applying these notions to various audiences, Yang et al. (1997) suggested the idea of measuring how well a VPL's appearance and interface match audience prerequisites. The idea of prerequisite measurements is different from other design-stage usability mechanisms in that it requires the designer to enumerate what intended users have been able to do in the past, rather than on predicting (perhaps too optimistically) what users will be able to do in the future.

To compute the audience prerequisite benchmarks, a designer compiles a list of all the prerequisites required of the intended audience. For example, is familiarity with popular web browsers required? Given these prerequisites, the designer then answers questions about every element in the VPL, of the form: "Does the <element> look like the <object/operation/composition mechanism> in the intended audience's prerequisite background?" Elements are semantic items of the VPL such as text, icons, buttons, connections among them, and so on, and object/operations/ composition mechanisms are items from the designer's list of prerequisites. The final score is the number of "yes" answers divided by the number of questions, but the real value is in the list of "no" answers. If an element does not look like a prerequisite, it could be difficult for the intended audience to understand and use effectively. Thus, these "no" answers point to potential problems.

3. Supporting new audiences

Some VPLs aim directly at supporting audiences without prior programming training. Concrete sample values and progressive evaluation have been used especially in this kind of VPL. These ideas have been motivated in part by a kind of "instant testing" goal, with the idea that if the user immediately sees the result of a program edit, he or she will spot programming bugs as quickly as they are made, and hence will be able to eradicate them right away. However, research in the VPL and office automation communities has shown, by empirically studying the spreadsheet paradigm, that this strategy is not enough for end users to find all their bugs. This is an example of the kind of approach in which designers omit support for powerful techniques (such as formalized testing) because the new audience does not have the training to perform them, at least in traditional ways.

However, an alternative strategy is to retain the powerful techniques, but by the system collaborating with the user in a non-intrusive way that rewards the user for each incremental contribution made. For example, to bring some of the benefits of applying formalized notions of testing to the informal, incremental, development world of spreadsheet-like VPLs, the "What You See Is What You Test" (WYSIWYT) methodology (Rothermel et al., 1998) encourages the user to "check off" any correct value noticed during the course of spreadsheet development, and uses this information to reason about testedness and to outline in color each portion of a program or spreadsheet along a continuum of red ("untested") to blue ("tested"). See Figure 1. Behind the scenes, determination of coloring is done using formal theories of test adequacy. In empirical studies, the methodology significantly reduced overconfidence about how tested spreadsheets were, as well as improving effectiveness and efficiency of testing (Rothermel et al., 2000).

1040EZ calculations:					
Presidential election?		yes			
1. Total wages		5132			
2. Taxable interest		297			
3. Adjusted gross		=C4+C5			
4. Parents?	yes	=IF(B7="yes",F7,5550)		Line E	1500
5. Taxable income		=C6-C7			

Figure 1. Mock-up of WYSIWYT in a spreadsheet's "formula view." The adjusted gross is blue-bordered, the next two cells are purple, and the arrows are blue (top) and red (bottom).

Another interesting line of work aimed directly at supporting new audiences has been the "natural programming" project of Pane and Myers (2000), who seek to empirically observe the fundamental principals of how end users think about queries and programming *before* designing a language to support these audiences. The participants in the empirical work done in this project have been children and adult end users, most without prior experience programming. The project has revealed a number of characteristics common in today's languages and search engines that are not particularly usable by these audiences. For example, the use of "AND" as a Boolean conjunction ("find all blue and green cars") is often misinterpreted by end users. Also, the use of parentheses for grouping is not effective for end users; many of the participants in the natural programming studies simply ignored them. Pane and Myers also found that participants were significantly more effective at querying when they constructed queries by filling out a table than when using a traditional textual syntax. This kind of information is critical for the design of VPLs aimed at new audiences, but is equally important for UA in other problem-solving domains, because it points the way to designing less troublesome interfaces for relatively untrained computer users to use in these domains.

4. Flexible Syntax, Flexible Semantics

Some VPL researchers have worked to support flexible syntaxes for a single VPL, with the idea that the user, not the language designer, chooses flexibly and opportunistically among multiple syntaxes (Erwig & Meyer, 1995; Grundy et al., 1995). In VPLs, these syntaxes are often grammar-based, being treated as multiple representations of a single language element. The UA community also supports multiple syntaxes, although they are typically implemented differently than in the VPL community.

While multiple input/output syntaxes provide some flexibility, these are still just different representations at a surface level, and do not address the fact that different users approach problem-solving in different ways. In programming language terms, the idea of supporting different approaches to problem solving suggests the need for multi-paradigm support.

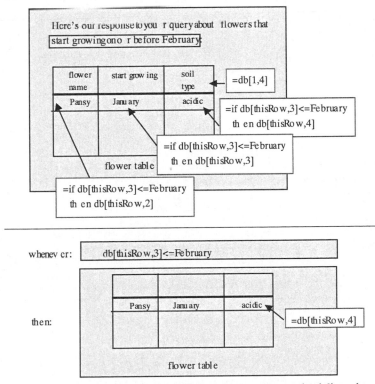

Figure 2. A design-stage mock-up of FAR. (Top) Defining web page content to be delivered to the consumer via formulas. (Bottom) Alternatively, defining the content using rules.

FAR ("Formulas And Rules") (Burnett & Chekka, 2000) is a new end-user VPL that uses this strategy, flexibly combining two paradigms—spreadsheet-like programming and rule-based programming—into a third paradigm: web page layout. The purpose of FAR is to allow end users to offer e-services. The design philosophy is to provide, to end-user service providers, programming devices that emphasize the "what" perspective of the user's goals (such as "I want to deliver a custom web page about a flower"), not the "how" perspective of the underlying events and communications required to fulfill the goals.

For example, using FAR, an end user with a local database of information, such as about certain types of flowers, could offer "flower encyclopedia" services as a cottage contribution to a larger digital library. The way such an end user uses FAR is to lay out a sample web page via direct manipulation and samples, specifying the semantics for replacing the samples via rules and spreadsheet-like formulas. The user specifies as part of the semantics the location of the necessary information about flowers, i.e., the location of her local database. In Figure 2, the user is referring to rows and columns in the database ("db": not shown in the figure) to fill in the blanks of the "flower table" that is to be included on the web page. The top portion shows these semantics as spreadsheet formulas, and the bottom shows the same semantics as rules. The user can view and edit either or both views opportunistically.

5. Conclusion

Visual programming researchers have been working to bring better accessibility to programming power to more users. This research shares common ground with UA work in domains with strong problem-solving components. This paper has discussed three examples of visual programming research that may be able to help with UA work in such domains. Not discussed in this paper are the contributions in the other direction, namely that UA research can bring to visual programming languages. Both subareas stand to benefit if the contributions of each are used as building blocks toward more progress in the other.

References

Blackwell, A. (1996). Metacognitive theories of visual programming: What do we think we are doing? *IEEE Symp. Visual Languages*, 240-246.

Burnett, M. (1999). Visual programming. In J. Webster (ed.) *Encyclopedia of Electrical and Electronics Engineering* (pp. 275-283). John Wiley & Sons.

Burnett, M., Chekka, S. (2000). FAR: An end-user WYSIWYG programming language for e-speak: Interim report. *Oregon State University, TR 00-60-10*.

Erwig, M., Meyer, B. (1995). Heterogeneous visual languages: Integrating visual and textual programming. *IEEE Symp. Visual Languages*, 318-325.

Green, T., Petre, M. (1996). Usability analysis of visual programming environments: a 'cognitive dimensions' framework. *J. Visual Languages and Computing* 7(2), 131-174.

Grundy, J., Hosking, J., Fenwich, S., Mugridge, W. (1995). Connecting the pieces. In M. Burnett et al. (eds.) *Visual Object-Oriented Programming* (pp. 229-252). Prentice Hall.

Pane, J., Myers, B. (2000). Tabular and textual methods for selecting objects from a group. *IEEE Symp. Visual Languages*, 157-164.

Polson, P., Lewis, C., Rieman, J., Wharton, C. (1992). Cognitive walkthroughs: A method for theory-based evaluation of user interfaces. *Intl. J. Man-Machine Studies*, 741-773.

Rothermel, G., Li, L., DuPuis, C., Burnett, M. (1998). What you see is what you test: A methodology for testing form-based visual programs, *Intl. Conf. Software Engineering* (pp. 198-207).

Rothermel, K., Cook, C., Burnett, M., Schonfeld, J., Green, T., Rothermel, G. (2000). WYSIWYT testing in the spreadsheet paradigm: An Empirical evaluation, *Intl. Conf. Software Engineering* (pp. 230-239).

Yang, S., Burnett, M., DeKoven, E., Zloof, M. (1997). Representation design benchmarks: A design-time aid for VPL navigable static representations. *J. Visual Languages and Computing* 8(5/6), 563-599.

Auditory Icons and Earcons: Categorical and Conceptual Multimodal Interaction

Myra P. Bussemakers, Ab de Haan

Nijmegen Institute for Cognition and Information
P.O. Box 9104, 6500 HE Nijmegen, The Netherlands

Abstract

Due to technological advances, our environment is more and more "enriched" by artificially generated multimodal information. There has not yet been much research on the influence of those co-occurring information streams. In this paper an approach for the investigation of this problem and some results on possible information processing characteristics is presented. It seems that in the integration of sound with visual displays it is mainly the categorical character that determines the speed of processing, while an explanation in terms of conceptual connotations should not be ruled out.

1. Introduction

The sensitivity of the human user and the increased capabilities and diversities of interaction systems allows for a substantial variety of modalities that can be applied when attempting to optimize the interaction between man and his environment. Information can for instance be presented through vision, touch, sound and in some cases even scent. A designer of interactive systems therefore has the difficult task of choosing the more optimal combination of information streams for supporting activities such as office work or shopping. The choice often is not merely trying to find the best solution from a performance standpoint, but the designer also may have to take more subjective appreciations into account to improve acceptance of the system. Sometimes it may be difficult to distinguish between usability issues and these more subjective opinions. In video games for example users report that they cannot perform as well without having the audio on (Edworthy, 1998). The effect may be caused by mere aesthetics, but it is also possible that there simply is more information available. Besides sound intuition and cultural conventions however, cognitive engineering as a scientific, investigative activity can provide basic theoretical principles to guide design.

2. An Empirical Perspective on Multi Modal Interaction

Investigating how to support certain activities can be approached in different ways. In this paper an experimental setup is discussed that is known to assist in the study of facilitation and interference effects between the visual and auditory modalities. Within the setup both the Stroop and related Stroop like paradigms such as the Simon effect can be manipulated. The Stroop paradigm (e.g. Stroop, 1935; MacLeod, 1991) looks at the integration of information within a modality at the level where feature information is coded and integrated. For instance, in a picture-sound integration task it investigates how the semantic nature of an auditory icon, a ringing bell or a meowing cat, influences the categorization of the picture of a dog. The Simon-effect (e.g. Simon, 1990) on the other hand is not supposed to be caused by the integration of relevant information streams but is seen as a facilitation or interference due to irrelevant features of the task situation. For instance, in the traditional Simon effect it is found that if a response to a stimulus on the left side of the screen needs to be made, the response is faster if the button that needs to be pressed is on the left side of the subject as well, compared to for instance the right side. The location information involved in this spatial Simon effect is irrelevant because it bears no weight in the semantic nature of the decision and the particular task that is being performed.

3. The Multi Modal Stroop Effect

In the project described here visual categorization was studied influenced by redundant sound information from the earlier mentioned perspective. This means that in a Stroop like task besides the visual display showing the to be categorized information, in some trials redundant sounds were presented. In such a task there are generally four types of conditions. The first condition consists of trials where the sound that is played is congruent with the visual information. This congruency can be on two levels. First of all the sound that is played can be the same as the picture that is shown. This is the case in the example of the picture of a cat that is accompanied by the sound of a cat. On a second level, the information can be congruent in the sense that the same category is represented in the sound and the picture, for instance when the sound of a dog barking is presented with the picture of the cat. The second type of condition is the incongruent condition, where the information presented through sound is incongruent with the visual information, for example when the picture of the cat is presented with the sound of another category in the experiment, for instance musical instruments, like the violin. Both types of information suggest an opposite response. In a neutral condition, the pictures are shown with a sound that is unrelated to the task, like the sound of water running. Finally, there are control trials, where the picture is presented with no sound. The auditory icons that were used in the experiments are supposed to have an influence on the primary task, i.e. pictorial categorization, because of their semantic nature. The results indicate that the picture of a cat with the sound of a cat leads to faster categorizations (see figure 1) (e.g. Bussemakers & de Haan, 2000). These experiments are similar to other Stroop like effects such as for instance found in picture word categorizations.

The examples described so far consisted of categorization in the context of redundant auditory icons such as sounds of concrete occurrences. The main research effort however was concerned with the investigation of categorization of information presented on visual displays accompanied by abstract sounds or earcons (e.g. Bussemakers & de Haan, 1998; Bussemakers, de Haan & Lemmens, 1999). These earcons have no relation to actual occurrences but may in an abstract way be used to provide information "about" ongoing processes. An example may be a tone of which the pitch is correlated to the speed of processing of a device. The earcons that were used were abstract, musical chords. The main distinction we investigated was the minor/major key distinction, as this seems to represent affective information differences. This distinction between major and minor key tones and especially its connotation has been the focus of research for hundreds of years. Historically there have been many studies that showed the existence or non-existence of a link between major tones and 'happy' and minor key tones and 'sad', and recently the work by Crowder (1984, 1985a, 1985b, 1991) has validated this relation. And in his reanalysis of earlier experiments he states that "There cannot be the remotest doubt that for both trained and untrained listeners, isolated chords, played in random order and out of all musical context, produce connotative judgments in accord with the conventional happy/sad dimension" (Crowder, 1984). For the current discussion is seems important how this connotation influences the perception of visual information. A multimodal Simon effect may shed some light on this issue.

4. The Multi Modal Simon Effect

Because the information in the earcons was not only redundant but also not semantically linked to the correct response the reaction to the accompanying sound can be treated as a semantic Simon effect. As mentioned earlier, in traditional studies looking at the Simon-effect the influence on the response is studied of an irrelevant distracter, like the location of a stimulus on the screen (e.g. Simon, 1990). Results show that although the irrelevant information should and could be ignored, there is facilitation in those trials where the location on the screen is congruent with the location of the response button. With a Simon-effect three issues are of importance. First of all, the relevant stimulus feature, i.e. that what the subject in the given task needs to respond to. In the experiments described here this was a pictorial display that had to be categorized. Subjects were presented with pictures of animals or non-animals. Secondly the irrelevant stimulus feature plays a critical part, since it is this feature that distracts the subject when responding. Major or minor key tones were used, that are not related to the pictures that were shown, but through their positive or negative connotation seem to be related to the response to the categorization, i.e. 'yes' or 'no'. Lastly the relevant response feature, meaning the response the subject needs to give to the relevant stimulus. In this case the response was an answer to the question whether the picture they see is of an animal, yes or no.

In these types of experiments, the irrelevant stimulus feature is related to the relevant response feature: the connotation of the earcon is related to the response to the category of the picture. However, the relevant stimulus feature (for example the category animal) is not related to the relevant response feature (the 'yes' response). At the same time the relevant stimulus feature is not related to the irrelevant stimulus feature (the category of the picture is not related to the major or minor earcon) (Kornblum, 1992; Simon, 1990). This experimental approach has advantages over traditional Stroop-like tasks, because in these types of experiments the targets not only are semantically related to the distracters, but also are related in other non-semantic dimensions (associative or perceptual for example). It is therefore difficult to assess whether the results in these experiments are a consequence of the automatic semantic processing or the non-semantic processing (e.g. LaHeij, 1988; Shelton & Martin, 1992; Williams, 1996).

5. Categorical and Conceptual Information Processing

These different types of sounds were used, both abstract (earcons) and concrete (auditory icons), as co-occurring with visual information that should be categorized in pre defined more or less abstract categories such as "fruit", "animal" and "musical instrument". Thus, while the primary task of the user was to make conceptual decisions with respect to visual displays that were accompanied by sounds that provided no strategic information with respect to the required decision, these sounds were supposed to influence the non-strategic aspects of the decision such as speed and accuracy.

The experiments suggest that with respect to the use of major or minor earcons and auditory icons two important temporal dimensions in the relation between visual and redundant auditory information are present. The first and main dimension is the difference between earcons and auditory icons. The second dimension is the difference within the conditions where auditory icons and earcons were present (for experimental results see figure 1).

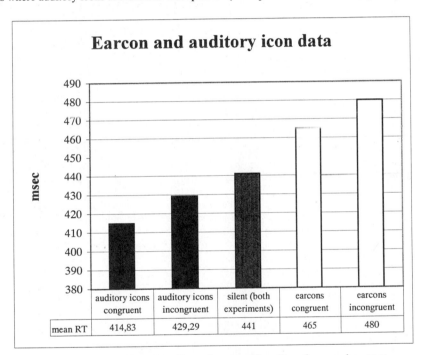

Figure 1: Combined results of experiments with auditory icons and earcons

Stroop and Simon effects can explain the first temporal dimension within the auditory icons and earcons. For earcons it is the semantic but for the task irrelevant connotation of incorrectness that slows down incongruent responses (yes responses combined with a minor key sound for example). In the case of auditory icons we have already shown that semantic distance can explain the findings. With respect to the second temporal dimension we seem to have identified a categorical effect, which resulted in a general decrease in reaction times for concrete auditory icons and an increase in reaction times for abstract earcons. This may be explained in terms of a Stroop effect. With respect to the Stroop effect it may be noted that most information in our memory is either categorical or conceptual[1] in nature. Categorical knowledge, which is perceptual in nature, enables us to generalize about what we

---

[1] This distinction is analogous to the previously mentioned where/what distinction, or direct perception versus representation (Neisser, 1994).

have learned about an object or an event to other similar objects or events. It enables us to adapt our behavior to our environment.

When we categorize an object, we are essentially assuming that this object is perceptually similar to other objects we know, contrary to the process of discrimination where we try to determine that an object is different (Shanks, 1997). For instance, the category "cat" consists of those objects that are perceptually similar to one another. This implies that the cat category also seems to include some other catlike animals (e.g. Mareschal, French & Quinn, 2000). However, without becoming to philosophical about this, sound can make a difference.

If the semantic distance between a visual display and an accompanying sound can cause the interference found in a Stroop task, one would expect that the reaction time to a picture of a cat-like creature accompanied by a bark of a dog would differ from a similar picture accompanied by a meow, as indeed is shown by our data (figure 1). The data also confirms the similar expectation that unrelated sounds like the sound of a violin even more interfered with the animal categorization. This semantic difference however cannot be used to explain the fact that accompanying auditory icons in general seem to lead to a facilitation, as reaction times of pictures accompanied by auditory icons are faster that the reaction times in the silent condition, while pictures accompanied by earcons lead to a slowing down. It thus seems that the categorical nature of the accompanying sound plays an important role in the integration of information.

As we already noticed, apart from categorical knowledge we humans also acquire and use conceptual knowledge such as the concepts animal, man and woman. Concepts are a representation of a class of objects. Where the usefulness of categorical knowledge may be limited, because it depends on the perceptual similarity between the object and previously encountered objects, conceptualization can assist in setting the boundaries of what belongs to the category and what does not by using principles or rules. Concepts are based on deeper and more abstract properties of objects (Goldstone, 1994). The general decrease in reaction times for auditory icons can then be explained as a consequence of reciprocal activation of both concrete auditory and visual stimuli dimensions. On the other hand the general increase in reaction time for the presence of abstract redundant auditory information suggest that here no reciprocal activation is present. The earcons, being abstract, are by definition not referring to some concrete object such as a violin or a dog, nor are they signaling processes that cause the sound. This suggests, in accordance with the above observations, that our memory is organized in different local regions of more or less abstract processing. This is in accordance with insights from the morphology of brain processes that are the substrate of our performance. With respect to the possible neural mechanisms involved it suggests for instance that the integration of concrete features of different modalities occurs relatively automatic, fast and within the more upstream synaptic regions (about the fourth synapse, see eg. Mesulan, 1997) of the posterior heteromodal association areas. The more abstract and contingent integration of heteromodal information occurs on the other hand more downstream and in prefrontal ventral areas and may take longer.
Another possible explanation of the general slowing down of the categorization in the earcon case may be that the integration is not semantic in nature but affective and due to the connotation of the abstract sounds. A final solution to this problem may be found in looking directly at the neural correlates during task performance. Regions of interest would then be hippocampus regions and prefrontal association areas. An activation of the latter areas would favor an explanation in terms of semantic information integration while relative activation of hippocampus regions would suggest an affective explanation.

6. Conclusions

When creating a universal interface, meaning that as many different types of people as possible need to be able to access the information, a strategy can be to use a multimedia environment with sound and graphics. Users would hopefully be able to work with the interface by integrating information from both modalities. However, if both visual information and auditory information need to be integrated, the effect of sound on a primarily visual task seems to be determined by whether the auditory information is categorical or conceptual in nature. When concrete, categorical information (auditory icons) is presented in a visual task, a Stroop-effect can be observed where the degree of overlap between the sound and the visual information is an indication of facilitation. The more the picture and the sound are similar, the higher the facilitation that occurs. With abstract, conceptual auditory information (earcons), a Simon-effect is found, where instead of facilitation, inhibition of the reaction times to the visual stimuli is seen if an earcon is presented as well. Here, the degree of inhibition is determined by the relation between the for the experiment irrelevant connotation of the sound and the response to the picture. If the response is incongruent with the connotation, the inhibition is greater than when the response is congruent with the connotation.

This research suggests that designers carefully need to consider the type of sounds they add to an interface, if for instance performance, i.e. speed of entry is critical.

References

Bussemakers, M.P., de Haan, A. (1998). Using earcons and icons in categorisation tasks to improve multimedia interfaces. In *Proceedings of ICAD '98* (Glasgow, UK) British Computer Society.

Bussemakers, M.P., de Haan, A., Lemmens, P.M.C. (1999). The effect of auditory accessory stimuli on picture categorisation; implications for interface design. In *Proceedings of HCII '99*, Munich, Germany (pp. 436-440).

Bussemakers, M.P., de Haan, A. (2000). When it sounds like a duck and it looks like a dog...Auditory icons vs. earcons in multimedia environments. In *Proceedings of ICAD 2000*, Atlanta, USA (pp. 184-189). International Community for Auditory Display.

Crowder, R.G. (1984). Perception of the major/minor distinction: I. Historical and theoretical foundations. *Psychomusicology, 4(1-2)*, 3-12.

Crowder, R.G. (1985a). Perception of the major/minor distinction: II. Experimental investigations. *Psychomusicology, 5(1-2)*, 3-24.

Crowder, R.G. (1985b). Perception of the major/minor distinction: III. Hedonic, musical, and affective discriminations. *Bulletin of the Psychonomic Society, 23(4)*, 314-316.

Crowder, R.G., Reznick, J.S., Rosenkrantz, S.L. (1991). Perception of the major/minor distinction: V. Preferences among infants. *Bulletin of the Psychonomic Society, 29(3)*, 187-188.

Edworthy, J. (1998). Does sound help us to work better with machines? A commentary on Rauterberg's paper 'About the importance of auditory alarms during the operation of a plans simulator'. *Interacting with Computers, 10*, 401-409.

Goldstone, R.L. (1994). The role of similarity in categorization: providing a groundwork. *Cognition, 52*, 125-157.

Kahneman, D. (1973). *Attention and effort.* New York: Prentice Hall.

Kornblum, S. (1992). Dimensional overlap and dimensional relevance in stimulus-response and stimulus-stimulus compatibility. In G.E. Stelmach & J. Requin (Eds.), *Tutorials in motor behavior*, Vol. 2, pp. 743-777. Amsterdam: North Holland.

LaHeij, W. (1988). Components of Stroop-like interference in picture naming. *Memory and Cognition, 16*, 400-410.

MacLeod, C.M. (1991). Half a century of research on the Stroop effect: An integrative view. *Psychological Bulletin, 109(2)*, 163-203.

Mareschal, D., French, R.M., Quinn, P.C. (2000). A connectionist account of asymmetric category learning in early infancy. *Developmental Psychology, 36(5)*, 635-345.

Mesulam, M.M. (1998). From sensation to cognition. *Brain, 121*, 1013-1052.

Neisser, U. (1994). Multiple systems: A new approach to cognitive theory. *European Journal of Cognitive Psychology, 6(3)*, 225-241.

Shanks, D.R. (1997). Representation of categories and concepts in memory. In M.A. Conway (Ed.), *Cognitive Models of Memory* (pp. 111-146). Hove: Psychology Press.

Shelton, J.R., & Martin, R.C. (1992). How semantic is semantic priming? *Journal of Experimental Psychology: Learning, Memory and Cognition, 18*, 1191-1210.

Simon, J.R. (1990). The effects of an irrelevant directional cue on human information processing. In R.W. Proctor & T.G. Reeve (Eds.), *Stimulus-response compatibility: An integrated perspective* (pp. 31-86). Amsterdam: North Holland.

Stroop, J.R. (1935). Studies of interference in serial verbal reactions. *Journal of Experimental Psychology, 18*, 643-662.

Williams, J.N. (1996). Is automatic priming semantic? *European Journal of Cognitive Psychology, 8*, 113-161.

Recommendations for the design of usable multimodal command languages

Noëlle Carbonell

LORIA, UMR 7503 (CNRS, INRIA, Universités de Nancy)
Campus Scientifique, BP 239, F54506 Vandœuvre-lès-Nancy Cedex, France
tel: +33 (0)3 83 59 20 32, fax: +33 (0)3 83 41 30 79, e-mail: Noelle.Carbonell@loria.fr

Abstract

We present and discuss the implications of the results of three empirical studies for the design of usable multimodal interaction languages based on the synergic use of controlled speech and gestures. In particular, we propose a method for designing such languages which represent appropriate substitutes for direct manipulation, in *all* contexts precluding the use of mouse and keyboard, and for *all* standard categories of users, especially the general public.

1. Introduction

The major characteristics of the coming world-wide Information Society may be foreseen through the key concepts which summarize the overall aims of present international research trends and projects in the area of human-computer interaction. These concepts are embodied in such phrases as: "disappearing computer", "pervasive or ubiquitous computing", "universal design and accessibility". They express the common will to implement computer accessibility for all users in any context of use (*e.g.*, wearable computers, mobile computing, smart home artefacts, virtual or augmented reality interactive environments, ...).

Multimodality, that is the simultaneous or alternate use of several modalities, appears as a promising short-term research direction for advancing the implementation of universal accessibility, thanks to the recent development of new interaction media and modalities.

Modality here refers to the use of a medium or channel of communication as a means to express and convey information (Coutaz & Caelen, 1991; Maybury, 1993); the recipient has to interpret the incoming signal in terms of abstract symbols, while the sender has to translate concepts (symbolic information) into physical events which are conveyed to the recipient through the appropriate medium. These processes involve the user's senses and motor skills and, symmetrically, the system input/output devices. According to the taxonomy presented in (Bernsen, 1994), several modalities or modes may be supported by the same medium (*cf.*, for instance, text and graphics as output modalities).

To characterize the various possible combinations of modalities, a taxonomy composed of four classes has been proposed (Coutaz & Caelen, 1991). As we focus here on issues relating to usage rather than implementation, we only need to consider two classes in this taxonomy: 'alternate multimodality' [1] which characterizes a multimodal sequence of unimodal messages, and 'synergic multimodality' which refers to multimodal messages or, in other words, to the simultaneous use of several modalities for formulating a single message.

The benefits that interface designers can draw from the implementation of multimodality are threefold (Carbonell, 1999). First, alternate multimodality can advance computer accessibility, in as far as it provides users with the means to choose among available modalities/media according to the specific usability constraints induced by certain contexts of use (Oviatt, 2000). Then, alternative media, modalities, and styles of multimodality are necessary for providing disabled users with computer access. Finally, synergic multimodality represents an indisputable improvement on unimodal interaction, as regards flexibility (*i.e.* expressive power), usability, and efficiency.

A significant advance in the implementation of universal computer access may be achieved now, thanks to the integration of speech into user interfaces either as an alternative or a supplementary modality.

The technology is available: the accuracy and robustness of present speech recognizers is sufficient to meet standard usability requirements. Integrating speech into user interfaces increases the diversity of available interaction media, as graphical user interfaces (GUI) have been so far the prevailing support of human-computer interaction (HCI). In addition, speech has strong assets as a means of expression: it is flexible and natural.

The design of speech-based user interfaces is also possible by reason of recent research advances concerning the design of appropriate strategies for promoting efficient cooperation between the processors in charge of speech recognition, natural language interpretation and dialogue control, respectively (*cf.*, for instance, Morin et al., 1992). Products

---

[1] This class results from the grouping of the 'alternate' and 'exclusive' multimodality classes in (Coutaz & Caelen, 1991).

offering generic software tools for developing oral dialogue interfaces are even available on the market (*cf.* Speechmania [2] among others).

Appropriate software architectures have also been proposed for integrating speech into multimodal user interfaces (*cf.* Nigay & Coutaz, 1993, namely).

However, issues concerning the usability of speech and natural language in unimodal or multimodal HCI environments have not yet been solved satisfactorily; further investigation is still needed.

The paper is a contribution to filling in this gap. It summarizes, for the benefit of speech interface designers, the main results of three empirical studies focused on the use of speech, or speech and gestures, in various simulated HCI environments. We first describe the features common to the three interaction situations which we chose to study. Then, we present the main results drawn from the analyses of the collected oral or multimodal protocols, and discuss their implications on the design of speech or multimodal user interfaces: after a brief discussion of the usability of spontaneous speech as a command or query language, especially vs controlled speech, we propose a few recommendations for the design of acceptable multimodal interaction languages. We describe as "*acceptable*" any artificial input language which is usable, that is easy to learn and to use, and which can be processed reliably by present recognition and interpretation software.

2. Overall description of our empirical studies on the usability of speech in HCI contexts

All three studies involved the same overall setup. Subjects had to interact, during three weekly sessions (of half an hour or so each), with an experimental user interface which could "understand" speech or multimodal inputs, and issue fluent verbal or multimodal (speech + graphics) outputs. The interface was simulated using the Wizard of Oz paradigm. The realistic simulation of the interface functionalities required two wizards who were assisted in their activities through software. Subjects, wizards and screens were videotaped. Analyses of the subjects' behaviors were carried out on written transcripts of the verbal exchanges, subjects' gestures, and system actions.

The objective of the first study (Amalberti et al., 1993) was to determine the possible influence of the dialogue context – human communication vs human-computer interaction – on the verbal expression and the cognitive behavior of the human partner. Two groups of subjects (six subjects per group) had to request information (specified in scenarios) from an operator in an information center; one group, the experimental group (*Gexp*) believed they were dialoguing with a "talking system"; as for the other group, or reference group (*Gref*), they believed they were exchanging with a human operator. Free speech was the only modality available to the subjects and to their interlocutor (human operator or simulated system). This study is referred to as *E1* in the remainder of the paper.

The second and third studies were focused on multimodal interaction. The two simulated user interfaces allowed the subjects to interact with a graphical application, using speech or/and 2D gestures (on a touch screen). The commands necessary for carrying out the prescribed design tasks could be expressed with equal facility, using either gestures or speech, or both speech and gestures, since the actions to perform amounted roughly to displacements (translations and rotations) of graphical icons on the screen. The outputs of the simulated interface were multimodal, the wizards having to associate appropriate pre-recorded oral messages with their actions on the graphical application. One group of eight subjects participated in each study; both groups performed the same set of tasks so that meaningful inter-group comparisons could be possible. In the second study (*E2*) spontaneous speech and gestures were allowed, while in the third study (*E3*) subjects had to comply with realistic expression constraints, that is restrictions (on the set of allowed multimodal utterances) which take account of the present limitations of speech and gesture recognition-interpretation systems. Such constraints are still needed at present (see paragraph 3), in order that the processing of speech inputs proves sufficiently accurate and robust for meeting standard usability requirements.

3. Is spontaneous speech a suitable HCI modality?

As an input modality spontaneous speech is far more difficult to implement than controlled speech. In so far as spontaneity encourages expressiveness, hence prosodic diversity, it enhances inter- and intra- speaker variability which is still a major obstacle to accurate recognition. In addition, the simultaneity between the thinking and formulation processes that spontaneity implies, is the source of many wording "errors" (*e.g.* hesitations, unfinished or incorrect sentences, repetitions, ...) which increase the complexity of the interpretation process. For instance, hesitations and syntactical errors were more frequent [3] during the first session of E2 than during the first session of E3 (Robbe,

---

[2] Speechmania is a software toolkit developed by Philips Speech Processing – Dialogue Systems, Aachen, Germany.

[3] The inter-group difference reaches statistical significance.

268

Carbonell & Dauchy, 1997). By reason of these specific difficulties, accurate automatic interpretation of spontaneous speech will probably be achieved in a more distant future than the accurate 'understanding' of controlled speech or text. Therefore, one can wonder whether it is advisable to implement spontaneous speech as an input modality right now. Besides, the usability of spontaneous speech in HCI environments is basically questionable. Various empirical evidence suggests that users are reluctant to use written of spoken natural language for controlling software or consulting databases.

In the eighties already, N. Borestein (1986) reported that inexperienced users of Unix who could use written natural language for consulting an online help database developed rapidly a tendency to express their requests in personalized telegraphic styles. Besides, subjects who participated in E2 complained during the final debriefings that, in the absence of guidance (*e.g.*, in the form of instances of allowed unimodal and multimodal commands), they had experienced some difficulty in interacting with the simulated interface using free speech and gestures.

These findings may be explained as follows. It has been observed that users, novice users included, focus their attention and efforts on achieving the tasks which motivate their use of a given software rather than on mastering the software operation; accordingly, they are likely to view the wording of natural language commands as a tiresome additional cognitive load, and to prefer using stereotyped or artificial interaction languages (such as direct manipulation or command languages); such languages encourage the development of cognitive routines for the generation of commands/requests, provided that they are simple and intuitive enough.

Other drawbacks to the implementation of spontaneous speech need to be taken into consideration by designers. If users are allowed to use unrestricted natural language (NL), they tend to assume implicitly that, for any action which lies within the scope of the application domain and which can be described by a verb, there exists a corresponding specific function in the software which can be activated using this verb. Thus, several E2 subjects formulated NL commands which referred to actions beyond the functionalities of current iconic and graphical interfaces, such as, for instance: *"Permute X and Y."*.

Therefore, to be successful, the implementation of unrestricted NL as a computer language should satisfy the following requirement: for each action-verb in the application domain of the software, there should exist a corresponding function in the software which should be capable of emulating the action expressed by the verb and could be activated by means of the verb. This requirement may imply the design of additional software functions, which might prove very demanding in the case of complex application domains.

Finally, empirical and experimental results in the eighties indicate that subjects spontaneously control their expression (rhythm, enunciation, vocabulary, and syntax) when 'talking' to a computer; see, for instance (Kennedy *et al.*, 1988), and the review in (Amalberti et al., 1993, paragraph 2.1). The findings stemming from the analysis of the verbal protocols collected during E1 contradict these results: we observed no significant statistical difference between the linguistic behaviors of the Gexp and Gref groups.

A plausible interpretation of this discrepancy may be inferred from the following observation. In the context of E1, the computer responses were simulated by a wizard who was instructed to control the variability of her prosody – otherwise she could speak naturally. In addition, the speech signal was minimally degraded through an equalizer [4] so as to achieve a realistic computer-like voice quality. Authors of previous studies on the other hand chose to simulate or to implement the technology available at the time [5], so that they generated computer speech which was very slow, monotonous, and whose intelligibility was rather poor. In all likelihood, participants in previous studies inferred from the poor quality of the system verbal expression that its speech recognition and NL understanding capabilities were also limited. So, in an effort to achieve cooperative communication, these subjects adapted their enunciation and language use to the supposedly low system 'understanding' capacity. On the other hand, E1 subjects were prompted by the verbal fluency of the system to rate its 'understanding' as excellent, hence to speak freely. Such cognitive adaptation processes are frequent in human-human verbal exchanges: humans tend to adapt their language use to the presumed knowledge and capabilities of their dialogue partner (see Amalberti at al., 1993, for references).

Designers should be aware of the possible influence of the system style of verbal expression on the users' mental representations of the system speech 'understanding' capacity. Generally speaking, it has long been argued that users develop 'system images' which tend to affect their interaction behaviors (Sutcliffe & Old, 1987). However, the knowledge necessary for making the most of this observed general tendency is not yet available. How to influence users' behaviors through system messages and responses is still an open issue; for instance, research is needed to determine precisely how each type of constraint which can be defined on the system enunciation or on the linguistic structure and vocabulary of the system messages affects the enunciation, syntax, and vocabulary of users.

---

[4] The transformation amounted to a change in the voice timbre.
[5] Many of them used 'vocoders' or similar devices.

4. Design of acceptable speech- and gesture-based multimodal user interfaces

The synergic combination of controlled speech with pointing/designation gestures represents an attractive form of input multimodality. Throughout the remainder of the paper, 'multimodality' (and 'multimodal') refers to this input mode, that is the synergic use of controlled spoken NL and designation gestures. The automatic interpretation of controlled gestural designations (of objects and locations in a 2D graphical presentation or in a virtual 3D space) is indeed easier than the 'understanding' of spatial linguistic references, which involves sophisticated inference mechanisms yet to be fully explained. In addition, accurate 'fusion' algorithms are now available (Nigay & Coutaz, 1993) for taking charge of the temporal and semantic integration of incoming speech and gestures. Then, multimodal commands can be processed by present software more reliably than equivalent speech utterances; recognition and interpretation rates are now sufficient for making it possible to consider launching on the market user interfaces which provide users in the general public with multimodal input facilities. Then, multimodality can be viewed as an appropriate substitute for direct manipulation in situations where the use of mouse and keyboard is inappropriate, provided that the expression constraints forced on users for ensuring accurate system 'understanding' are usable.

We propose a method for designing such acceptable multimodal command languages. In order to validate the efficiency of our method, we took the following steps. First we used it for designing an interaction language with the same expressive power as direct manipulation. This language is easy to interpret accurately, as its gestural component amounts to a few simple intuitive designation gestures, and its oral component is characterized by a vocabulary of about a hundred words, and a syntax which can be described by a simple context-free grammar (static branching factor 5.5, dynamic branching factor 2.6). Then, we assessed the usability of this language empirically. Comparisons between E2 and E3 first sessions (Robbe, Carbonell & Dauchy, 1997), together with a careful analysis of the evolution of E3 subjects' behaviors over the three sessions (Robbe, Carbonell & Dauchy, 2000), indicate that the expression constraints implemented in this language are easy to comply with; E3 subjects assimilated them rapidly without initial specific training, in the course of the first session; besides, neither the efficiency of their interactions nor their subjective satisfaction were noticeably affected by these constraints. We describe this design method in the remainder of the paragraph.

The method we propose can be used for designing both the oral and the gestural components of any specific multimodal command language. We shall focus exclusively on the application of this methodological framework to the design of acceptable oral interaction languages, as its use for defining acceptable sets of designation gestures is straightforward.

Our approach is based on the two following assumptions:

- H1: it has been observed that verbal exchanges between cooperating operators are limited to a restricted subset of natural language, the size of which varies according to the complexity of the collaborative activity.
- H2: users will easily comply with speech constraints, provided that:
    - A. the subset of natural language defined by these constraints includes all the necessary commands for interacting efficiently with the given software package;
    - B. ambiguity and synonymy are excluded from the structures and vocabulary of this subset.

An efficient approach for defining such subsets is to collect spontaneous speech interactions between potential users and the given software, using a simulated oral (or multimodal) user interface [6], and then to eliminate all ambiguous or synonymous words and structures from this subset.

The resulting oral command language (or component), which satisfies H1 and H2, should be easy to learn (*cf.* H2.A). In addition, multimodal interaction languages including oral components designed along these lines should prove usable, in-as-much as designation gestures can be easily and 'naturally' connected to speech utterances through deictic phrases such as "here/there" or "this/that" (used either as adjectives or pronouns). Besides, according to H1 and H2.B, present speech interpreters should be capable of processing any utterance belonging to such languages reliably and in real time, provided that the complexity of the application domain and tasks is not too high.

However, it should be noted that this approach is costly, since its implementation involves the collection and analysis of verbal or (multimodal) protocols in a simulated human-computer environment.

Our empirical results (*cf.* E3 especially) suggest other useful design recommendations which are listed hereafter.

- Designers of multimodal input languages should be aware that some users will not spontaneously resort to synergic multimodality. They need appropriate inducement, for instance in the form of a sample of allowed multimodal utterances. In the absence of an initial stage of training in the use of the language, such a set of instances is also necessary for prompting and helping users to comply with the constraints forced upon their spontaneous expression.
- In cases when reliable processing of the input multimodal language cannot be achieved continuously, designers should provide users with a robust alternative input modality (or form of multimodality).

---

[6] At least, the interpretation of the users' oral/multimodal inputs should be simulated by a human operator.

* Designers should be aware of the implications of the following finding. We observed two main styles of multimodal spontaneous expression (*cf.* E2) which denote two different 'system images' or mental representations of human-computer interaction (Carbonell & Mignot, 1994): communication with the system vs manipulation of graphical representations of application objects [7]. This finding provides designers of multimodal interaction languages with a way of inducing users to develop one representation or the other.

* Finally, we observed departures from standard NL syntax and semantics in the verbal protocols collected during E3 (Robbe, Carbonell & Valot, 1997). Although they are globally less frequent than in the context of E2 (Robbe, Carbonell, & Dauchy, 1997), designers should take account of such 'deviations or errors' in the design of acceptable multimodal languages.

Let us note that, although all these recommendations stem from empirical data on command languages, they are general enough to be applicable to the design of multimodal query languages for simple information domains.

5. Conclusion and future research directions

We have described a validated method for designing acceptable multimodal input languages. As these languages allow the synergic use of designation gestures and a subset of spoken NL, they will prove easy to learn through interaction, that is without specific initial training; in addition, their use is not likely to entail a noticeable increase in the cognitive workloads of users. Therefore, they may be viewed as attractive candidates for replacing direct manipulation in *all* contexts precluding the use of mouse and keyboard, and for *all* standard categories of users, especially the general public. It is still possible to increase the usability of such languages by helping users to master them through interaction. There are several ways for achieving this aim. One research direction is to improve the automatic 'understanding' of these languages. Another one is to explore and exploit the potential of alternative modalities and adaptive user interfaces. A third one would be to consider the design of systems capable of diagnosing expression 'errors' and correcting them; however this last research direction raises numerous difficult issues.

References

Amalberti, R., Carbonell, N., Falzon, P. (1993). User representations of computer systems in human-computer speech interaction. *International Journal of Man-Machine Studies*, 38, 547-566.

Bernsen, N.-O. (1994). Foundations of multimodal representations, a taxonomy of representational modalities. *Interacting with computers*, 6, 347-371.

Borenstein, N. (1986). Is English a natural language. In K.T. Hopper & I.A. Newman (Eds.), *Foundations for Human-Computer Communication*, Amsterdam: North Holland, pp. 58-72.

Carbonell, N. (1999). Multimodality: a primary requisite for achieving an information society for all. *Proceedings of HCI International'99*, London: Lawrence Erlbaum Associates, pp. 898-902.

Carbonell, N., Mignot, C. (1994). Natural multimodal HCI: experimental results on the use of spontaneous speech and hand gestures. *Proceedings of the 2nd ERCIM Workshop on Multimodal HCI*, pp. 97-112, Paris: ERCIM.

Coutaz, J., Caelen, J. (1991). A taxonomy for multimedia and multimodal user interfaces. *Proceedings of the 1st ERCIM Workshop on Multimodal HCI*, pp. 143-148, Lisbon: INESC.

Kennedy, A., Wilks, A., Elder, L., Murray, W.S. (1988). Dialogues with machines. *Cognition*, 30.

Maybury M.T. (Ed.) (1993). *Intelligent Multimedia Interfaces*. Memlo Park, (CA): AAAI/MIT Press.

Morin, P., Junqua, J.-C., Pierrel, J.-M. (1992). A flexible multimodal dialogue architecture independent of the application. *Proceedings of ICSLP'92*, Banff (C), pp. 939-942.

Nigay, L., Coutaz, J. (1993). A Design Space for Multimodal Systems: Concurrent Processing and Data Fusion. *Proceedings of INTERCHI'93*, New York (NY): ACM Press, pp. 172-178.

Oviatt, S.L. (2000). Multimodal system processing in mobile environments. *Proceedings of UIST'2000*, New York (NY): ACM Press, pp. 21-30.

Robbe, S., Carbonell, N., Dauchy, P. (1997). Constrained vs spontaneous speech and gestures for interacting with computers: a comparative empirical study. *Proceedings of INTERACT'97*, London: Chapman & Hall, pp. 445-452.

Robbe, S, Carbonell, N, Dauchy, P. (2000). Expression constraints in multimodal human-computer interaction. *Proceedings of IUI'2000*, New York (NY): ACM Press, pp. 225-229.

Robbe, S, Carbonell, N, Valot, C. (1997). Towards Usable Multimodal Command Languages: Definition and Ergonomic Assessment of Constraints on Users' Spontaneous Speech and Gestures. *Proceedings of EUROSPEECH'97*, Grenoble (F): ESCA, pp. 1655-1658.

Sutcliffe, A.G., Old, A.C. (1987). Do users know they have user models? Some experiences in the practice of user modelling. *Proceedings of INTERACT'87*, Amsterdam: North Holland, pp. 36-41.

---

[7] This second representation is based on the same metaphor as direct manipulation.

Virtual Tourist based on PeerRing – Communicating with people you have never met

Ola Carlvik, Ing-Marie Jonsson

Ericsson Research, 2100 Shattuck Avenue, Berkeley, CA 94704,
Dejima Inc, 160 West Santa Clara St, San Jose, CA 95113
ola@erilab.com, ingmarie@dejima.com

Abstract

We describe in this paper an application called "Virtual Tourist", that is implemented using PeerRing, a framework for applications based on three key technologies, location based services, self-presenting devices and user profiles, and peer-to-peer communication. Applications based on the PeerRing framework enable users to advertise their position, to find others, to find users in specific locations, and to make use of applications and devices on other users terminals and computers. Furthermore, applications can switch between being client and a server when being used.
Virtual Tourist enables people to become virtual tourists in remote places by engaging the other users of the application as guides. Virtual Tourist has a web interface, making it fairly platform independent, and can be accessed from different types of devices such as handheld computers and laptops. This interface allows users to setup profiles for preferences of services and for disclosure of personal data. User's of Virtual Tourist can locate and communicate with other users, and furthermore also locate and use open resources on other users devices.
In this paper we also present results collected during initial experiments of the applications, and the collected data reflect users preferences regarding communication methods, how to contact other users, choice of communication modality, and social aspects of communicating with people that are far away and that you have never met. Users show interest in using new ways of communicating, using cameras and microphones, and most importantly contacting and communicating with unknown people. They are also willing to let other users use peripherals on their devices, in the same manner as they in turn utilize peripherals on remote devices. These results indicate that the application group has interesting properties and can support and promote new types of communication. It is, however, important to focus future work on security and privacy issues, to ensure that the user always feels in control of the application and the data that is being disclosed. Furthermore, it is important to allow the user to turn off or refuse the use of functionality and peripherals by remote users at any time.
Virtual Tourist is an example of an application that enables people, that for various reasons cannot travel, to experience remote places. We emphasize on the ability to experience activities through the use of Virtual Tourist as one of the most important properties of the application. Being a tourist is just one activity that can be shared, and limited only by our imagination we see that activities such as, bicycling, mountain climbing, diving, and shopping are equivalent and have an equally exiting potential.

1. Introduction

Virtual Tourist is a PeerRing application (Jonsson & Carlvik, 2000) based on the combination of three technologies; Location Based Services (LBS) (Gabber & Wool, 1998), Self-Presenting Devices and User Profiles and Peer-to-Peer Communication (Bird et al 1995). With Virtual Tourist we exploit PeerRing to communicate with users that share similar interests and to initiate communication with users based on location.

Virtual Tourist was implemented with a web interface, depicted in Figure 1, thus enabling both platform and device independence. For the initial studies, the application was tested solely on devices with a camera, a microphone, and a speaker in addition to the conventional I/O devices. The middleware in PeerRing allows us to experiment with novel ways of sharing device resources such as cameras and microphones. This allows users to "see" through the eyes of the camera on a remote device, and to listen to sounds form a remote place. These mechanisms enable users to select which and how much information to disclose -

Figure 1:
Home page of Virtual Tourist

ranging from being virtually anonymous to full disclosure of personal information similar to the actions in NYNEX Portholes (Lee, Girgensohn, & Schlueter, 1997). The support of alternative interfaces and alternative interaction models is based on the use of web interfaces paired with mechanisms for self-presenting devices and user profiles. Furthermore, it also allows users access to a remote application even though they use devices without screens, or similarly, a format for presentation of data that excludes the presentation of images. For example, one could experience a city like Bangkok using a telephone, just by listening to the sounds from the street, the traffic, the vendors and the passers-by.

1.1   PeerRing Architecture

Location Based Services: These applications use a user's position to tailor information to local or regional offerings. The location can be determined using technologies like Global Positioning System (GPS), mobile 'phone base station tracking, or a combination of the two, (SnapTrack, 1998) since they are optimized for different, and complementing, environments.

Self-Presenting Devices: Both a user and the user's device present themselves via their profile when connecting to the network (Jonsson, 1999; Jonsson & Hodjat, 2001). These profiles contain the user preferences and the device capabilities and requirements respectively. Virtual Tourist requires information about the device, such as screen size, bandwidth capabilities, connected devices - speaker, microphone, camera, etc., when the application is started. Both the user and the device information are collected by a resource manager in the PeerRing network, and will be presented in response to queries about specific resources, locations or users.

Peer-to-Peer Communication: The application provides functionality that enables the device to act both as a client and a server for information and resources.

1.2   Related Work

The PeerRing architecture originally started as technology driven project, where three interesting technologies were combined into an application framework. After the initial prototype had been tested in a couple of informal studies, it became clear that the driving force of the architecture was not based on technological advances, but rather on the new types of communication offered by the system.

Studies and trends show that communication using text messages, chat or email, has become more popular than voice communication for the younger generation. The number of telephone calls per day being less than the number of emails and chat sessions (Noll, 1999). There is also data to support that audio conferencing outperforms face-to-face communication for information exchanges such as discussion of ideas and interviewing.

Results from studies of telework show that the physical distance induced by telework did not contribute to perceived psychological distance (Rathod & Miranda, 1999). Thus to some extent, neither the level of dependency of coworkers or the level of trust of coworkers was affected by physical distance. These results contradict other earlier studies of psychological distance and telework. In particular a study that showed that interpersonal contact is necessary for the development of trust (Macneil, 1980). Raising the question of the halo effect or a change in social response due to the use of new communication technology and impact of a well-developed technological infrastructure.

Other studies have indicated the need to learn both social and technological skills of use when introduced to advanced communication technology (Niederman & Beise, 1999). Interesting research questions raised by the authors included, "What influences people and groups to select mediated communication over face-to-face communication?" "To what extent do behaviors regarding conflict and conflict resolution vary between mediated communication and face-to-face communication?"

These results and questions together with data from our initial PeerRing prototype, lead to the implementation of Virtual Tourist as the first real PeerRing application. Virtual Tourist provides both communication using text, and communication using voice, allowing data to be collected for a comparison. The application relies to some extent on reciprocity, implemented as the idea that users of Virtual Tourist can be both guides and tourists. Users acting as tourists must initiate contact with users in desirable vacation spots, reciprocating the trust and reliance on physically distant and unknown people that others have shown towards them. With this issue Virtual Tourist add another dimension to the effect of physical distance, communication and collaboration with totally unknown people. Where the selection of communication partner is based on location, or preferred choice of communication method, or maybe on the resources open for use on the users access device.

2. Virtual tourist

Virtual Tourist enables a user to be a tourist and experience remote places without traveling. This is made possible by disclosing information about users in other locations, and by forwarding information from remote locations through the I/O peripherals of users' devices. The location information about a user, or rather a user's device, is presented as both text and graphics on the screen. In addition there is information about the remote user that corresponds to preferences and instructions in that user's profile - including how they prefer to be contacted, e.g. via email, chat or phone, and which resources that user allows others to use on their device. Both the user profile and the device information are communicated to Virtual Tourist when the user opens the web page with the user interface.

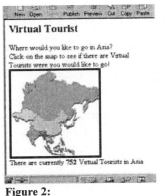

Figure 2:
Map of Asia in Virtual Tourist

Virtual Tourist will advertise users of the application that are willing to engage in communication, and also, most importantly, which resources are open for others to use on their access devices.

Consider this example; Joe connects to Virtual Tourist, transferring his preference to communicate via chat, and that his device has a color screen, a microphone and a speaker. He advertises that other users may use his microphone by asking. At the same time as this communication of information takes place, Joe is presented with a web page containing a map of the world – Figure 1. The web page also contains general information on how to use Virtual Tourist, what the application can do, and some data such as the current time and date and how many users are currently active in Virtual Tourist.

Joe would like to know more about Nepal, and especially the capital Katmandu. He clicks to select Asia, and is presented a new web page with a map of Asia and general information about the continent – Figure 2. Finally, Joe clicks on the location of the map where Nepal is and is presented a

map of the country – Figure 3. This page also contains more specific information about Nepal such as the capital of the country, the local time and date, number of citizens, politial system and climate.

In order to get more information about Katmandu, Joe selects Katmandu and is told that it is currently 7PM and that there are 11 Virtual Tourist users in Katmandu. Of these 11 users, Joe initiates a chat session with Mary after finding that she has advertised that she us willing to share the use of both a camera and microphone after being contacted via chat.

Virtual Tourist, will in this way, offer users like Joe the opportunity to experience both pictures and sounds from remote and exotic places, through the willingness of users such as Mary, to communicate and share resources.

Figure 3:
Map of Nepal in Virtual Tourist

3. Data from initial study of virtual tourist

The study we present here was conducted both face-to-face and using questionnaires over the Internet. The 10 subjects ranged in age from 14 to 35. The collected data show that this age group of users, which in essence spans two generations, are frequent users of chat applications and mobile phones. The users reported an understanding of the Virtual Tourist application and how it could be used to "travel without traveling". They felt that the application would provide new modalities such as sound and video, thereby enhancing communication. They also indicated that some emotions are hard to transfer using conventional technology, and that they feel the need of alternative ways to communicate. Virtual Tourist would provide the means to send a photograph, or a video stream, to convey the beauty of a landscape, rather than relying on a verbal description over the telephone.

Users that had experience with advanced communication, for example video telephony, voiced a concern about the quality of images in Virtual Tourist.

Most users also felt that some methods of communication are too intrusive, for example the lack of privacy due to always being reachable when carrying a cell phone. Data from this study indicates that a communication device that bears little resemblance to a traditional 'phone, would make these users feel more comfortable. The users indicated a strong need for staying in touch with their friends, but would prefer the notification of incoming communication, as well as the communication to be less intrusive than audio signals and voice.

Many of the users were comfortable with revealing information about themselves, for instance their location. They could understand how others would use the information, and they could see how they could make use the information to find other users. Most users voiced a concern about remote users privacy; they were reluctant to contact remote users in ways that would be considered intrusive, indicating a preference of text messaging over voice for initial contact. Most users also highlighted the importance of the ability to turn off functions and resources, such as location information, by dynamically changing the user and device profile.

Users prefer to have not one, but a number of different profiles, to be used for different situation such as work and leisure, and for different groups of people such as friends and colleagues. For example, the same user could reveal their location to remote users of Virtual Tourist, and at the same time hide the current location from their parents. Again, this indicates the fear of being under surveillance, at the same time as seeing benefits of revealing information for selected purposes and/or users.

Some users also raised the issue of trust, how can a user be sure that the information disclosed would not be misused.

The users that have tested Virtual Tourist have been mostly positive about the application, voicing some concerns relating to control and privacy, but mainly indicated that they would like to have access to the application.

Initial user responses to our first PeerRing application have been positive, but there are issues related to user experience that need to be further explored, such as volunteering location information and information about a user device, being monitored, and sharing resources either automatically or by request.

4. Conclusions and future work

Based on the data collected and processed from a small study of Virtual Tourist, some trends and research questions have surfaced. The trends are most pronounced in the area of communication method; text messages are preferred over voice communication, and chat is preferred over email. Voice communication is most often used to discuss the operation of the camera, and to clarify spatial instructions. Unlike today's touristing with the use of a tour guide/buddy that you sign up with at least for the day, VirtualTourist offers the opportunity to live vicariously with the ability to change guide/buddy as the interest/focus changes. However, for this to become acceptable we also have to find technical solutions for secure handling of the users privacy.

Locating and searching for information and other users based on profiles basically receive the same user responses as Napster (Napster, 1999) and iMesh (iMesh, 1999), i.e. reduced search space to locate users with similar interests. Information augmented with, or found by, position is desirable and highly liked by users. This is, however, not always true for disclosing information about the user's own position. Most users liked the idea of being allowed access to other user's I/O devices, and in the same manner as for position, skeptic when other user's used their I/O devices. This is an issue regarding control and reciprocity - a user must always be in control of the resources on his/her device, and a user is more likely to disclose information to people that disclose information about themselves.

Is there change in social response to the use of new communication technology and impact of a well-developed technological infrastructure, or is this just a result of the halo effect? There are clearly changes in communication patterns, such as the use of email and chat applications. These two applications have changed the way people communicate, engaging in chat sessions with unknown people, engaging in email discussions that rely on asynchronous instead of synchronous communication. Learning or adapting to a more terse style of communication - SMS messages, emails and chats all display relatively short sentences and follow their own social protocol that allow users to ask direct questions without being considered rude.

An interesting aspect of these computer mediated communication applications is how the lack of non-verbal cues and distance alters the social aspects of communication. Using chat, it is possible to get directly to the point without initial social communication expected in voice communication or face-to-face communication. It is also possible to learn to know a person really intimately from a distance, people disclose personal information in a setting where the physical non-verbal cues would hamper the communication, and the physical distance instead encourages disclosure. These are not new findings, but it is interesting to follow the trend, will users adapt to and use the possibilities of the new technology, or are we just seeing a halo effect? I.e. will traditional social communication resurface once this new technology has become ubiquitous?

We see Virtual Tourist as an example of an application that enables people, that for various reasons cannot travel, to experience remote places. The application raise many interesting research topics in the are of computer mediated communication, however, we would also like to emphasize on the ability to share the experience of activities as an important properties. Being a tourist is just one activity that can be shared, and limited only by our imagination we see that activities such as, bicycling, mountain climbing, diving, and shopping are equivalent and have an equally exiting potential.

References

Gabber, E., Wool, A. (1998) How to Prove Where You Are: Tracking the Location of Customer Equipment. *Proceedings of the 5th ACM conference on Computer and Communication Security,* 142-149.

iMesh (1999) provides access to Information on the Internet and communication with others with similar interests. [On-line]. Available: http://www.iMesh.com/.

Jonsson, I (1999) "The Decoupled Application Interaction Model", in *Proceedings of HCII '99* Munich.

Jonsson, I., Carlvik, C., (2000) "PeerRing - Combining Location Based Services with Self Presenting Devices and Peer-to-Peer Communication", in *Proceedings of HCII 2000,* Sunderland.

Jonsson, I., Hodjat, S. (2001). "Natural Interaction based on Host Agents, Profiles and Decoupled Applications", in *Proceedings of HCII 2001,* New Orleans.

Lee, A., Girgensohn, A., Schlueter, K., (1997), "NYNEX Portholes: Initial User Reactions and Redesign Implications", in *Proceedings of GROUP '97,* Phoenix.

Macneil, I., (1980), *"The New Social Contract: An Inquiry into Modern Contractual Relations".* Yale University Press, New Haven.

Niderman, F., Beise, C., (1999), " Defining the "Virtualness" of Groups, Teams and Meetings", in *Proceedings of SIGCPR ' 99,* New Orleans.

Napster, (1999), with a proprietary MusicShare technology. [On-line]. Available: http://www.napster.com/

ScourExchange (1998) media search engine and peer-to-peer file exchange. [On-line]. Available: http://www.scour.com/

Noll, M., (1999), "Does Data Traffic Exceed Voice Traffic?", *Communications of the ACM, June 1999, Vol. 42*

Rathod, M,, Miranda, S , (1999), " Telework and Psychological Distance: The Mediating Effects of Culture and Technology in Four Countries", in *Proceedings of SIGCPR '99,* New Orleans

SnapTrack (1995) Combines GPS with network based positioning. *An Introduction to SnapTrack Server-Aided GPS Technology.* [On-line]. Available: *http://www.SnapTrack.com/pdf/ion.pdf*

Gesture Query for the Sentient Map

Shi-Kuo Chang

Department of Computer Science
University of Pittsburgh, Pittsburgh, PA 15260 USA
Chang@cs.pitt.edu

Abstract

The sentient map is a new paradigm for visual information retrieval. It enables the user to view data as maps, so that gestures, more specifically c-gestures, can be used for the interaction between the user and the multimedia information system to improve user accessibility. Different c-gestures are then dynamically transformed into spatial/temporal queries, or σ-queries, for multimedia information sources and databases. We are currently working on the replacement of c-gestures by finger gestures to implement a versatile gesture-oriented sentient map.

1. Introduction

Maps are widely used to present spatial/temporal information to serve as a guide, or an index, so that the viewer of the map can obtain certain desired information. Often a map has embedded in it the creator's intended viewpoints and/or purposes.

A web page can also be regarded as a map, with the URLs as indexes to other web pages. In fact, any document can be regarded as a map in a multi-dimensional space. Moreover, with associated scripts, the maps can also be made active [Chang, 1995].

These two notions, that data can be viewed as maps and that maps can be made active, led us to propose a new paradigm for visual information retrieval - the **sentient map**. In practice, the sentient map is a gesture-enhanced multimodal interface between the user and the multimedia information system to improve user accessibility.

The natural way to interact with a sentient map is by means of gestures. In this paper we will use the word "gesture" in the general sense. That is, gestures can range from the traditional mouse clicks and keyboard strokes, to hand gestures, facial expressions, the body language, and even collective gestures of a group of people. More specifically, simple gestures called **c-gestures** consisting of mouse clicks and keyboard strokes are used to interact with a sentient map.

This paper is organized as follows. The sentient map and the three basic gestures are introduced in Section 2. Section 3 presents the underlying σ-query language, and Section 4 defines a simple gesture query language, the c-gesture language. A prototype sentient map system is described in Section 5. For empirical study, an e-learning environment called the Macro University serves as a test bed to evaluate this approach. An extension of the c-gesture query language is described in Section 6. In Section 7, we discuss the recognition of finger gestures to implement a versatile gesture-oriented sentient map.

2. The Sentient Map

The *sentient map* is a map that can sense the user's input gestures and react by retrieving and presenting the appropriate information. We use the term "map" here in the general sense. Geographical maps, directory pages, list of 3D models, web pages, documents, still images, video clips, etc. are all considered maps, as they all may serve as indexes and lead the user to more information. In practice, a sentient map is a gesture-enhanced interface for an information system.

A sentient map *m* has a *type*, a *profile*, a *visual appearance*, and a set of *teleactivities*.

m=(type, profile, v, IC)

A sentient map's *type* can be geographical map, directory page, web page, document and so on.

A sentient map's *profile* consists of attributes specified by its creator for this map type.

A sentient map's *visual appearance* is defined by a visual sentence *v*, which is created according to a visual grammar or a multidimensional grammar.

The *teleactivities* of a sentient map is defined by a collection of *index cells IC*. Each index cell *ic* may have a visual appearance, forming a *teleaction object*. These teleaction objects are usually overlaid on some background.

Ignoring the type and profile, a sentient map is basically a collection of teleaction objects including a background object.

A *composite sentient map* is defined recursively as a composition of several sentient maps.

As said before, in practice the sentient map is a novel user interface that combines the features of the map, the browser, and the index. Furthermore, the sentient map is accessed and manipulated using gestures.

In our approach we begin with the three basic gestures: (a) a *finger gesture* to **point** at an object on the sentient map, (b) a *moving finger gesture* to indicate a **time-line**, and (c) a *moving finger gesture* to draw a boundary for a **space** on the map. Any gesture query is composed of these three basic gestures for *point*, *time-line* and *space*.

These basic gestures are sufficient to express spatial/temporal queries [Chang, 1998]. For the user's convenience, the basic gestures can be augmented by other auxiliary gestures to add more constraints to the query, for example, indicating yes or no by clicking on a yes/no menu (c-gesture), making a circle/cross (finger gesture) or nodding/shaking the head (head gesture).

The three basic gestures are also not just restricted to point, time and space. As to be discussed in Section 6, the time-line gesture and space gesture can be replaced by other dimensions for maps whose types are document, web page, etc.

The three basic gestures can be represented by mouse clicks: a double-click to select an object, a click and a double-click at both end points to specify a time-line, and multiple clicks to specify an area. We will call gestures represented by mouse clicks and keystrokes **c-gestures** (see Section 4). Color can be used to highlight these basic c-gestures.

By combining the basic gestures with other auxiliary gestures the user can specify the spatial/temporal relationship between two objects on the sentient map. For example, *finger gestures* can be used to add/change certain attributes of objects on the map. *Head movement gestures* can be used to indicate yes (nodding head), no (shaking head), etc. *Motion-assisting speech gestures* can also be used to augment finger gestures, hand gestures and head gestures.

3. The σ-Query

We use the σ-query [Chang, 1998] as the intermediate representation in the transformation from gestures to queries. As described in previous sections, the user can utilize the three basic gestures to point at objects, draw time-lines, and indicate the boundary of search space. The combined gestures are then transformed into a σ-query.

An example of a σ-query is illustrated in Figure 1. The video source R consists of time slices of 2D frames. To extract three pre-determined time slices from the source R, the query in mathematical notation is: $\sigma_t (t_1, t_2, t_3) R$. The meaning of the σ−operator in the above query is to *select*, i.e. we want to select the three frames (time slices) along the time axis. The subscript t in σ_t indicates the selection of the time axis. In the SQL-like language ΣQL, a σ−query is expressed as:

```
SELECT t
CLUSTER t₁, t₂, t₃
FROM R
```

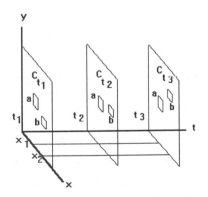

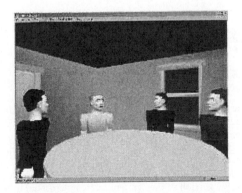

Figure 1: A video source R with three time slices

Figure 2. A virtual conference room [NetICE, 1999]

In the above a new keyword "CLUSTER" is introduced, so that the parameters for the σ–operator, such as t_1, t_2, t_3, can be listed. The word "CLUSTER" indicates that objects belonging to the same cluster must share some common characteristics, such as having the same time parameter value. A cluster may have a sub-structure specified in another (recursive) query. Clustering is a natural concept when dealing with spatial/temporal objects. The result of a σ–query is a string that describes the relationships among the clusters. This string is called a *cluster-string* [Chang, 1996].

The clause "CLUSTER *" means default clustering. For example, if the time axis is selected for a video source, the default clustering is to sample all time slices.

4. The C-Gesture Query Language

The c-gesture query language uses mouse clicks to represent gestures. The basic syntax is defined below:

```
<point> ::= <double-click>
<time-line> ::= <single-click> <double-click>
   <begin-time-value> <end-time-value>
<space> ::= <single-click> <single-click><single-click>
   <double-click>
```

Example 1: The user double-clicks on a video icon to **point** at it as the source. The user then single-clicks followed by a double-click to specify a *time-line*, and enters the values for a time interval such as [10, 34]. The σ-query *Query 1* is:

```
SELECT t
CLUSTER [10,34]
FROM video-source
```

The query result is the set of video frames in the time interval [10, 34] from the video source.

Example 2: The user double-clicks on the objects retrieved by *Query 1* to point at them as the source. The user then single-clicks twice followed by a double-click to specify a rectangular *space*. Since three points specify a triangle, the rectangle is defined as the triangle's minimum enclosing rectangle. For example, if the three points are (1,3), (6,2), (2,5), the rectangle is [(1,2), (6,5)]. The σ-query *Query 2* is:

```
SELECT x, y
CLUSTER [(1,2), (6,5)]
FROM   Query 1
```

The query result is the rectangles from the video frames in the time interval [10,34] from the video source.

Example 3: The user double-clicks on the objects retrieved by *Query 2* to point at them as the source. Using the extended c-gesture to be explained in Section 8, the user then single-clicks followed by a double-click followed by a click on an attribute name such as *type* in a pull-down menu to specify a *type-line*, and enters the value for a type such as `person'. The σ-query *Query 3* is:

```
SELECT type
CLUSTER `person'
FROM Query 2
```

The query result is the objects of type `person' in the rectangular areas from the video frames in the time interval [10,34] from the video source.

5. A Prototype Sentient Map System with Application to the Macro University

One application of the sentient map is the Macro University [MU, 2000], a consortium of universities and research groups interested in e-learning. It disseminates enabling technologies for e-learning, develops tools and contents for adaptive e-learning such as the cross-lingual courseware and provides a test bed for e-learning projects. Currently the Macro University has fourteen members that cooperate in research, development and e-learning courseware design.

For this e-learning environment, we implemented two versions of the sentient map user interface: a *PC version* implemented in Visual C++ and a *Web version* implemented in JavaScript.

The PC version of the sentient map implemented in Visual C++ is integrated into the NetICE environment for virtual teleconferencing developed by Dr. T. Chen at Carnegie Mellon University [NetICE, 1999]. Students and instructors can use this environment as a virtual classroom. Several users represented by their *avatars* are engaged in a discussion, as shown in Figure 2. When they need information they may turn their attention to the sentient map on the wall, and more information becomes available and is also visible to all the participants of the virtual teleconference.

The Web version of the sentient map implemented in JavaScript is independent of NetICE With a browser, a user can use c-gestures to query e-learning courses offered anywhere in the world by a member university of the Macro University Project. An example is illustrated in Figure 3, where the point c-gesture is used to select a query about instructor, subject or course title. A time line c-gesture or a space c-gesture, as shown in Figure 4, is used to formulate the query.

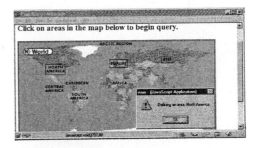

Figure 3. A point c-gesture is used to select a query   Figure 4. A space c-gesture is used to formulate a query

The two versions of the sentient map user interface are integrated so that the user can start in the virtual teleconference environment, and then go to the web to obtain more information from the e-learning universe.

6. Extended C-Gesture Query Language

We can expand the capability of the sentient map interface so that c-gesture queries are applicable to any document, any web page or any sensor. In order to do so, the c-gesture query language is extended. The extended syntax is defined below:

```
<attribute-line> ::= <attribute_name>
    <single-click> <double-click>
    <attribute_value_1> <attribute_value_2>

<attribute-space> ::= <attribute_name> <single-click>
    <single-click> <single-click> <double-click>
```

In addition to the *time-line* and *space* c-gestures, we introduce *attribute-line* and *attribute-space* c-gestures by prefixing the c-gestures with *attribute-names* which can be represented, for instance, by an icon or by color coding. For each document, web page or sensor, a number of characteristic attributes are defined so that the user can use attribute-line and attribute-space c-gestures to formulate a gesture query. The customization of c-gestures can be done either before a session or dynamically during a session. The user can then formulate c-gesture queries in a real space, or a virtual space, or a hybrid space composed of real space and virtual space.

7. Gesture Queries

A gesture query language is important for two reasons. First of all, gestures are trans-cultural so that people with different cultures and languages can use the same gesture query language. Secondly, hearing impaired people can also use gestures to communicate with the computer. We are developing a gesture query language so that people can browse the Macro University using gestures, and further investigate the design of a gesture-oriented sentient map as the standard interface for the Macro University.

280

Dynamic hand gestures can be reduced to a limited set of hand configurations, and a hand gesture can be defined as a sequence of hand postures with orientation and trajectory values [Bordegoni, 1994]. Indeed, a visual gesture language can be defined to generate hand-arm gestures, and the movement primitives have been identified to be: *pointing, straight-line, curve, ellipse, wave* and *zigzag* [Lebourque, 1999]. Luckily, these six movement primitives are more than sufficient for defining our gesture query language because we will mainly use finger gestures.

In our experimental setup, a video camera mounted on top of, or nearby, a PC will be used to capture a user's finger gestures for further processing. In our gesture language, the hand configurations are actually characterized by temporal pauses, i.e., a gesture consists of a sequence of point gestures, movement primitives and pauses (see Figure 6). These characteristics, we believe, will render the recognition of gesture queries feasible. Since gestures are inherently imprecise, we also will experiment with the effect of information presentation at different resolution on the recognition accuracy of the gesture query language, and the precision of the querying parameters.

Figure 5. A line gesture consists of a point gesture, pause, straight-line movement, another point gesture and another pause.

8. Discussion

To go beyond mouse clicks and key strokes as c-gestures, we need a system that understands more powerful human gestures. Real-time identification of finger gestures from a video source may be feasible when a certain degree of imprecision in determining spatial and temporal attribute values is acceptable. The use of contextual information in gesture understanding, and the inverse problem of interpreting context from gestures, will also be explored [Chever, 1998].

References

Bordegoni, M. (1994). An Environment for the Specification and Recognition of Dynamic Gestures. *Journal of Visual Languages and Computing*, Vol. 5, 205-225.

Chang, S.K. (1995). Towards a Theory of Active Index", Journal of Visual Languages and Computing, Vol. 6, No. 1, 101-118.

Chang, S.K., Jungert, E. (1996). Symbolic Projection for Image Information Retrieval and Spatial Reasoning. Academic Press.

Chang, S.K., Jungert, E. (1998). A Spatial Query Language for Multiple Data Sources Based Upon Sigma-Operator Sequences. *International Journal of Cooperative Information Systems*, Vol. 7, No. 2&3, 167-186.

Chever, A., et al. (1998). A Unified Framework for Constructing Multimodal Experiments and Applications. *CMC'98*, Tilburg, The Netherlands (pp. 63-69).

Lebourque, T., Gibet, S. (1999). High Level Specification and Control of Communication Gestures: the GESSYCA System. *Proceedings of Computer Animation'99*, Geneva.

MU (2000) www.cs.pitt.edu/~chang/cpdis/macro-u.html

NetICE (1999) http://amp.ece.cmu.edu/proj_NetICE.htm

Design for Accessibility: Meeting the "Section 508 Challenge"

Brant A. Cheikes

The MITRE Corporation, 202 Burlington Road M/S K302, Bedford MA 01730-1420 USA
bcheikes@mitre.org

Abstract

Recent regulations require agencies of the United States Government to provide accessible Electronic and Information Technology (E&IT) products to their employees who are individuals with disabilities, as well as to ensure that individuals with disabilities who are members of the general public have access to agency products comparable to that of individuals without disabilities. Responding to this challenge will demand new designs and the concerted effort of many individuals and organizations, including the developers and vendors of end-user applications, operating systems, and accessibility tools. The X.Org Consortium, steward of the X Window protocol standard, has chartered an Accessibility Task Force to develop and implement a plan to ensure that X-Windows meets or exceeds all applicable accessibility standards. This paper describes some new designs being considered, and identifies some of the opportunities and risks of enabling accessibility to X-Windows applications.

1. Introduction

This paper describes a software technology originally developed to facilitate intelligent computer-assisted instruction, discusses how this technology is being transferred to an industry consortium, and explains how this transfer promises to help the consortium address new Federal rules pertaining to Electronic and Information Technology (E&IT) accessibility. We identify areas in which the X Window System, a portable, network-transparent window management system for multitasking computer operating systems, will require new designs to enable X-based software applications to be made accessible to individuals with disabilities. We begin with an overview of the Federal rules in question, followed by a description of the consortium and MITRE's role in it.

1.1 Federal E&IT Accessibility Regulations

The Workforce Investment Act was signed into law in August of 1998. Included as part of this Act were the Rehabilitation Act Amendments of 1998. Section 508 of the Rehabilitation Act Amendments requires that when Federal agencies develop, procure, maintain, or use electronic and information technology, they must ensure that these technologies permit Federal employees with disabilities to have access to and use of information and data that is *comparable* to the access to and use of information and data by Federal employees who are not individuals with disabilities, unless an undue burden would be imposed on the agency (Access Board, 2000). Section 508 is far-reaching in that it also requires that individuals with disabilities *who are members of the general public* have access to information and services from Federal agencies that is comparable to that of individuals without disabilities. Section 508 placed responsibility for developing E&IT accessibility standards in the hands of the Architectural and Transportation Barriers Compliance Board (the "Access Board"), an "independent Federal agency devoted to accessibility for people with disabilities" (Access Board, 2000a). The Access Board published its final accessibility standards on December 21, 2000. These standards take effect for enforcement purposes on June 21, 2001, six months from the date of their publication in the Federal Register.

A crucial feature of Section 508 is that it both mandates Government compliance, and provides avenues for affected individuals to seek recourse through civil litigation. Consequently, Section 508 is a "big stick" that will motivate the E&IT industry, in concert with vendors of assistive technology, to provide accessible products. New designs will be required: existing E&IT products will need to be retrofitted with accessibility-enabling technologies, and future products will (we hope) be designed with accessibility requirements in mind from the outset. In the next section, we describe an industry consortium that is working to meet the "Section 508 challenge" for users of the X Window System.

1.2   The X.Org Consortium

X.Org is an industry consortium serving as the steward of the X-Windows protocol standard (the "X Standard"). X-Windows is a portable, network-transparent windowing system that has become the base technology for the implementation and management of graphical user interfaces (GUIs) for software applications running in multitasking computing environments such as Unix (Mansfield, 1993). X was originally developed at the Massachusetts Institute of Technology; early work was sponsored by the Digital Equipment Corporation and International Business Machines. Since the release of X-Windows Version 11 (X11) in 1987, the standard has been further developed by the MIT X Consortium, and most recently by X.Org, which has assumed stewardship of the technology. The current release of the X-Windows Sample Implementation is X11 Release 6.6, which can be downloaded from the X.Org website (X.Org, 2001).

All major vendors of Unix computing platforms are members of X.Org, and each of their proprietary windowing systems is derived from a release of the Sample Implementation maintained by X.Org. Other members include vendors of X-based products and end-user organizations with a stake in X-Windows technology.

Believing that enhancements to X-Windows would be needed in order for X-based products to comply with the Access Board's accessibility standards, X.Org chartered an Accessibility Task Force to guide the development of technical solutions that all members could share. MITRE, a non-profit operator of Federally-Funded Research and Development Centers, offered to lead this task force. Besides having a public-interest mission, MITRE has developed technology under its internal research and development program which, if contributed to X.Org, could help jump-start the consortium's response to the Section 508 challenge, and thereby provide benefits down the road to MITRE's U.S. Government sponsors. We describe that technology next.

2.   Software instrumentation for embedded training

Software applications play a critical role in many work environments. These applications may be general purpose, such as word-processing and spreadsheet tools, or tailored to specific business functions, such as air-traffic management and military command-and-control (C^2) systems. End-user training is critical if these applications are to be adopted and used effectively. In Cheikes *et al.* (1999) we advocate *embedded training* as one method of helping end users learn to operate complex information systems. Embedded training systems (ETSs) have been used throughout the military services, and the United States Army has mandated the use of embedded training techniques for all new systems it procures (Sherman, 2000).

MITRE research on embedded training has been investigating how intelligent computer-assisted instruction (ICAI) techniques can be used to train operators of C^2 systems, in particular, C^2 systems with so-called WIMP (windows, icons, menus, pointing device) GUIs such as those managed by X-Windows. Our focus is on *cognitive apprenticeship tutoring* (Collins *et al.,* 1989), where trainees acquire software-application skills by means of "learn by doing" exercises in which coaching support is progressively reduced.

Early on in our research we learned that adding ICAI-oriented embedded training to an existing C^2 application presented unique technical challenges. Chief among these was the need for the ETS to be able (1) to *observe* the trainee's actions on the target application and the application's response to those actions; (2) to *inspect* the target application's GUI at any time to determine its state, and (3) to *script* the target application, executing actions on behalf of the trainee. Unfortunately, we discovered that few software applications provide the facilities that would allow an ETS to observe, inspect, and script them.

To overcome this problem, we developed a technique which we call *software instrumentation*. We found that it is often possible to insert executable code into the *windowing environment* of a target software application transparently to the application itself, and without modifying it in any way. By monitoring events occurring within the window system and reporting them to interested clients, this code can provide the facilities needed to realize our vision of embedded training.

We have developed a prototype implementation of this concept for X-Windows applications, called WOSIT (Widget Observation, Scripting and Inspection Tool). WOSIT works by modifying the X-Windows standard toolkit (called the *Intrinsics*) used to build GUI widgets such as menus, buttons, text fields, etc., and by taking advantage of standard X-Windows facilities for event notification and inter-client communication. In essence, WOSIT installs "listening devices" that monitor widget activity within an application, and provides notification of significant events (e.g., button presses and changes to text fields) to its clients. WOSIT exposes an interface to its clients that allows them (a) to receive messages describing actions on the GUI of monitored applications, (b) to issue requests for information about active widgets, and (c) to generate widget events as if they were performed using the keyboard or mouse.

```
window(open, "TARGETS", application, 23.8);
radio(press, "New Project", "TARGETS", user, 42.1, "true");
button(press, "Ok", "TARGETS", agent, 46.2);
window(close, "TARGETS", application, 47.4);
window(open, "New Project", application, 47.4);
edit(change, "TextField1", "New Project", user, 370.3, "proj_3");
button(press, "Ok", "New Project", user, 372.0);
window(close, "New Project", application, 382.0);
```

Figure 1: Sample WOSIT Observations

Figure 1 illustrates a sequence of observations reported by WOSIT. Each message, a simple text string formatted to ease parsing by machine, describes the type of event that occurred (e.g., the opening of a window, or the pressing of a button), identifies the window involved (using the text in its title bar), the widget involved (by its text label or the contents of a nearby label widget), the agent of the event (the user, the application, or another agent via WOSIT's scripting capability), the time of the event (currently measured in seconds from the start of WOSIT observation), and any relevant modifiers.

WOSIT is in essence a form of "middleware" that was originally intended to mediate between an end-user application and an embedded training system. What we have since learned is that the observation, inspection, and scripting capabilities provided by WOSIT would also be useful to assistive devices such as screen readers and voice-navigation systems. Indeed, WOSIT's design both was inspired by and incorporates software from the (concluded) Mercator project at the Georgia Institute of Technology that focused on developing screen-reading technology for users of X applications (Mynatt & Edwards, 1992; cf. also Walker *et al.,* 1993). The Mercator team recognized the need for GUI-access facilities (including asynchronous notification of GUI events, and the ability to query widget state), and their design implemented some of those facilities in the windowing system, and the rest in the screen reader itself. WOSIT also places facilities in the windowing system, but goes further by encapsulating the rest of the processing (including the logic for tracking GUI states) in a general-purpose server application, rather than in a specialized assistive tool. This server application then provides observation, inspection and scripting services to its clients using a simple string-based message protocol.

WOSIT has been released to the public as open-source software, and is available for download from MITRE's website (WOSIT, 2000). In the next section, we discuss how WOSIT might help X.Org meet the Section 508 challenge.

3. Meeting the challenge

Section 1194.21 of the Access Board's E&IT accessibility standards defines twelve requirements for software applications and operating systems. These provisions are X.Org's initial concern, and are expected to drive new designs and extensions of the X standard. At least two of these provisions appear to be addressable by integrating WOSIT's functionality into X. Section 1194.21(c) states:

"[A] well-defined on-screen indication of the current focus shall be provided that moves among interactive interface elements as the input focus changes. The focus shall be programmatically exposed so that assistive technology can track focus and focus changes."

X-Windows currently complies with the first part of this requirement. WOSIT would enable it to comply with the second part. WOSIT could be the mechanism by which focus-related states and events are programmatically exposed. For this to work, however, an extension to X-Windows would be required that enables widgets to issue reports when they gain or lose input focus.

Section 1194.21(d) states:

> "[S]ufficient information about a user interface element including the identity, operation and state of the element shall be available to assistive technology. When an image represents a program element, the information conveyed by the image must also be available in text."

Compliance with this provision poses a more significant challenge. It begs the question of what constitutes "sufficient information" relative to the needs of assistive technology. Ultimately, this is a question that must be answered by the assistive-technology vendor community. In the interim, we believe that useful parallels can be drawn between ETSs and assistive technologies such as screen readers, voice-navigation and eye-tracking devices. In other words, perhaps what is sufficient for ETSs will also prove sufficient for assistive devices.

In developing WOSIT to support embedded training, we found that it needed (a) to provide unique, meaningful, human-readable names for each widget within a given display window, (b) to ensure that these names remain the same across each user session, and (c) to provide access to information about the *type* and *location* of each widget. We expect that assistive technologies will have similar requirements. For example, a screen reader should be able to describe the layout of a given application's GUI using the same terms each time that application is run. (Note that internally to X-Windows, widgets are assigned unique identifiers that change each time the application is run, and possibly each time a window is re-displayed.) WOSIT implements a widget-naming algorithm that is based on the structure of each GUI window, rather than the system-assigned identifiers. As long as the GUI structure does not change, WOSIT will assign each widget the same name every time the application is executed.

As another example, a screen reader may want to include the type of each widget in the referring expression generated for that widget, e.g., "the File menu", "the OK button", "the Box Size text field", "the right vertical scrollbar". WOSIT's inspection mechanism provides access to widget state information, including widget type, location, and value(s), if any. For example, the current contents of a text field can be retrieved, the state of a toggle button (*on* or *off*) can be determined, and the set of items selected in a list widget can be obtained.

Regarding information about widget operation, our work on WOSIT made clear that each widget type undergoes only a small number (typically one or two) of *event types* that are of interest to users. For example, action buttons can be *pressed*, toggle buttons can be *switched on* or *switched off*, menus can be *opened* and menu items *selected*, text fields can be *edited*. We also learned that actions on widgets by the end user, as well as *reactions* by the application, need to be observed and reported. We anticipate similar needs from assistive technologies; for example, a voice-navigation system may need to be able to confirm user actions (e.g., "You just selected 'Slow' in the 'Speeds' list") and describe system responses (e.g., "The 'File Open' window just appeared").

Although WOSIT's functionality appears likely to help X-Windows comply with the first part of 1194.21(d), the second part makes it clear that some cooperation on the part of application developers will also be necessary. For example, when a widget is labeled with an icon, WOSIT currently can only determine its type. Relying on this, a screen reader would only be able to describe a row of iconic action buttons as, e.g., "Button 1, Button 2, Button 3...", instead of as, "the New Document button, the Open Document button, the Save File button...". For this to change, new designs will be needed on two fronts. First, window management systems such as X-Windows will need to provide facilities that programmers can use to associate text labels with icons. Second, programmers will need to be informed about and encouraged to take advantage of those facilities. Section 508's "big stick" might help to provide the encouragement if groups like X.Org can provide the facilities.

4. Concluding remarks

The Access Board's E&IT accessibility standards are a clarion call to the E&IT industry and assistive-technology vendors to provide products that accommodate the needs of individuals with disabilities. This paper described a software technology called WOSIT that was originally developed to enable advanced embedded-training applications, but also appears promising as part of X.Org's response to the "Section 508 challenge." We have suggested that WOSIT may be useful as "middleware" that enables service-provider applications such as embedded training systems, screen readers and voice-navigation systems to query and track pertinent GUI states and state changes.

Our work on WOSIT, building on earlier work by the Mercator team, identified these facilities that need to be available in the window-management layer of the operating system:

- a *rendezvous* mechanism allowing service-provider applications to find and establish contact with end-user applications;

- an *event-notification* mechanism to receive asynchronous reports of significant changes occurring on the GUIs being monitored;
- a *query* mechanism allowing service-provider applications to inspect widget type, location, and value(s).

WOSIT extends these mechanisms by providing a reliable, structure-based widget naming scheme, and by exposing a simple, string-based message interface to its clients.

We expect that most, if not all, of the facilities needed by ETSs will also be needed by assistive devices such as screen readers and voice-navigation systems. If this proves to be true, it suggests that as window-management systems such as X-Windows are extended to become more supportive of the needs of individuals with disabilities, they will as a side effect become more hospitable to a potentially wide range of other performance-support systems and tools, of which ETSs are but one example.

Of course, many new technologies have attendant risks, and accessibility-support mechanisms are no exception. We believe, for example, that efforts to comply with provision 1194.2(d) of the Access Board's E&IT accessibility standards will lead E&IT vendors to build unprecedented forms of access into their software, operating systems and telecommunications products. Although this access will be intended for use by assistive technologies, there is reason to expect that developers of malicious software will attempt to exploit accessibility facilities. To avoid surreptitious observation of users (with or without disabilities), theft of confidential data, or wholesale subversion of end-user applications, developers of software accessibility services must take care to include information-security specialists in their design processes.

References

Access Board. (2000). *Electronic and Information Technology Accessibility Standards: Final Rule*. (2000). Federal Register, 65(246): 80500-80528. Online at http://www.access-board.gov/news/508-final.htm.

Access Board. (2000a). *About the Board*. Online at http://www.access-board.gov/indexes/aboutindex.htm.

Cheikes, B. A., Geier, M., Hyland, R., Linton, F., Riffe, A. S., Rodi, L. L., & Schaefer, H.-P. (1999). Embedded Training for Complex Information Systems. *International Journal of Artificial Intelligence in Education*, 10, 314-334.

Collins, A., Brown, J. S., & Newman, S. E. (1989). Cognitive apprenticeship: Teaching the crafts of reading, writing, and mathematics. In L.B. Resnick (editor), *Knowing, Learning, and Instruction: Essays in Honor of Robert Glaser*. Hillsdale, NJ: Lawrence Erlbaum Associates.

Mansfield, N. (1993). *The Joy of X: An Overview of the X Window System*. Harlow, England: Addison-Wesley.

Mynatt, E. D. & Edwards, W. K. (1992). *The Mercator Environment: A Nonvisual Interface to the X Window System*. Technical Report GIT-GVU-92-05, Georgia Institute of Technology.

Sherman, K. O. (2000). Intelligent Tutor for Today's Army. *Military Training Technology*, **5**(5). Online at http://www.mt2-kmi.com/5_5_Art6.cfm.

Walker, W. D., Novak, M. E., Tumblin, H. R., & Vanderheiden, G. C. (1993). Making the X Window System Accessible to People with Disabilities. *Proceedings of the 7th Annual X Technical Conference*, Boston, MA, January 8-20. Online at http://trace.wisc.edu/docs/x_win_disability/x_disabl.htm.

WOSIT. (2000). *WOSIT home page*. http://www.mitre.org/technology/wosit.

X.Org. (2001). *X.Org Home Page*. http://www.x.org.

A control centred approach to designing interaction with novel devices

Gavin J. Doherty[a], Tim M. Anderson[b], Michael D. Wilson[a], Giorgio Faconti[c]

[a]CLRC Rutherford Appleton Laboratory, Oxfordshire, U.K.
G.J.Doherty@rl.ac.uk, M.D.Wilson@rl.ac.uk
[b]Drake Music Project, U.K. - http://www.drakemusicproject.com -
timanderson@drakemusicproject.com
[c]CNR Istituto CNUCE, Pisa, Italy. G.Faconti@cnuce.cnr.it

Abstract

Modern information technology is becoming both increasingly ubiquitous and increasingly varied in the possible ways the user can interact with it. Accompanying this, there is a trend towards interfaces where the user is in constant interaction with the computer system, communicating with it by many different means, such as gestures, speech and haptics as well as discrete communication. This development requires new interface design approaches that allow for the analysis of the continuous as well as the discrete aspects of the interface, and that support reasoning about real-time issues. In the area of manual control, theories have been developed for continuous control of systems by human operators. In this paper we examine how we can apply manual control concepts in a qualitative fashion to the design and analysis of interactive systems. This involves a focus on control and feedback signals, transformations of these, and control characteristics of user, device and controlled process. While we make reference to the particularly challenging application area of performance control systems for disabled musicians, we believe that control issues of this nature will become increasingly common in interface design.

1. Introduction

In this paper, we look at the analysis of a class of novel and emerging interface technologies where interaction between system and user is, in some sense, continuous. In modern interfaces using techniques such as gesture recognition, speech recognition, animation and haptic feedback, the user is in constant and closely coupled interaction with the computer system over a period of time. The interaction is no longer based on a series of steps or discrete interactions, but the input provided by the user and/or the output provided by the computing system involve a continuous exchange of information at a relatively high resolution. Applications that use continuous interaction techniques can be found in virtual reality and teleconferencing, but also in less obvious areas such as ubiquitous computing (including active and intelligent environments), teleoperation, and alternative interface technologies for people with sensory and motor impairments. To help us with the analysis of such interfaces, we look to a branch of engineering psychology dealing with manual control.

2. Manual control

Manual control theory was originally developed by feedback control engineers modelling tasks such as tracking for anti-aircraft gunners. However, the theory is applicable to a wide range of tasks involving vigilance, tracking, stabilising etc. (for example driving a car or piloting an aircraft). The theory, particularly that branch developed from control theory, has been refined to a very high degree over the years. A general introduction can be found in (Salvendy, 1997). There is a large base of both predictive (McRuer 1980) and explanatory theory (Hess, 1985) based upon, and validated by, a wealth of experimental data. In the control theory approach, continuous mathematics is used to model human performance. The focus of the approach is on the perception and transformation of signals representing for example the actual and desired state of a process. Motor performance is viewed in terms of information transmission, with inaccuracy viewed as additive noise. A number of qualitative concepts from control theory can be used in describing human-machine interaction. This is shown by (Jagacinski, 1977) which looks at a number of these. The first of these is that of open and closed loop control. In open loop control, only the target signal is available to the user; thus there is no ability to account for noise or environmental interference. In closed loop control by contrast, both the target and output signal (fed back) are available, giving the user the opportunity to compensate for error. In musical performance for example, the target signal is the desired sequence of musical outputs; the fed-back signal is the current musical output. Another concept is that of positive and negative feedback. In a negative feedback system, the tendency is to minimise errors caused by disturbance; in positive feedback systems the tendency is to amplify disturbance (ie. they are unstable). While positive feedback may be useful in the design of experiments, the vast majority of manual control scenarios involve negative feedback systems. A third concerns gain and time delay, and is discussed below.

2.1 Gain and time delay

Consider a simple closed loop negative feedback system which we describe with two parameters, firstly a delay or latency t which is the time taken by the controlled element to react to it's input, and secondly the gain K which determines the rapidity of adjustment. If K is low, the system will respond very sluggishly moving only slowly towards the target signal. Conversely, if K is high, then the system is likely to overshoot, requiring adjustment in the opposite direction which itself may overshoot, leading to oscillation. The delay t can also contribute to this behaviour - a high delay makes oscillatory behaviour much more likely. Additionally, for most performance systems where music is directly output, t must be low (of the order of 20ms) to produced a perceived *immediate* output. If t is much higher, most musicians become unwilling to perform using the system.

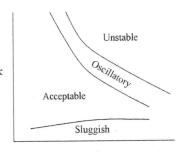

Figure 1: Effect of gain and time delay parameters on control (adapted from Jagacinski, 1977)

This is a particular type of real time control, but it is a particularly common one, useful wherever system delay is a performance shaping factor. For example, the above view was developed with reference to a small time scale (on the order of the delay between seeing a system output and carrying out the motor actions for an appropriate response), one can also consider it over longer time intervals. For example, consider an in-car navigation system where instead of gain, we have the frequency of decisions taken on which route to take and the time delay is that between a certain position being reached and appropriate instructions being displayed within the car. The parameter space of such an application should be very similar to that in figure 1 above.

3. Application: control systems for disabled musicians

In this section, we give some background on the particular application area which we focus on, before looking at the control and feedback signals present, possible transformations between these, and control characteristics of user, device and controlled process. The Drake Music Project has for many years been developing control systems to enable musicians with a wide range of physical disabilities to play music, solo or within a group (Anderson, 1997), adapting commercially available components, as well as developing its own "E-Scape" software. The latter can allow limited bandwidth control signals to be converted into complex or subtle music output, by allowing a performer to do some of the creative work "offline", ie. pre-compose and assemble musical material. This allows the number of input parameters and/or value ranges to be reduced during subsequent continuous interaction with the system, such as in a live performance (Anderson, 1999).

A user's input can be derived from a variety of actions, eg:

- movement in space, with position detected with up to 6 degrees of freedom (eg three x,y,z Cartesian coordinates, plus three orientations: azimuth, elevation, roll), eg by video, radio, ultrasonic, capacitive, or infrared sensors[gavin1]),
- interaction with physical devices, eg balls, joysticks, switches, pads, again with 1-6 degrees of freedom.

However, for disabled performers, it is most important to maintain flexibility of detection and filtering of actions: for example, many performers initially find their best musical results come from utilising only 1 degree, but then want to progress to more, with increasing input range and discrimination. User actions to operate music systems in current use can include:

1. Generating a series of related values, by:-
(i) gestural movement through an area in space. In typical music sensors, movement is detected in 1 direction, eg. radial movement within a hemispherical detection zone, or longitudinal distance along the axis of a conical zone.
(ii) 2D movement of finger, toe (or even nose) on mousepad, or trackball.
(iii) varying pressure on a squashable pad, eg. the "MIDIpad" device developed by Drake at the University of York, Department of Electronics.
(iv) 2D movement *and* pressure, eg. mousepad-like device "MIDIslate" (York).
2. Generating a trigger event, by movement *at a specified time*, by:-
(i) movement into or out of the detection zone (1.i. above)
(ii) movement (without contact) to a specific location in space,
(iii) movement to, and pressing of, a physical switch device (or a key or position on computer or concept keyboard)

These actions can then provide input signals to the E-Scape music engine, such as:

1.a. Apply continuously varying timbral parameters to music events already playing (or previously started by other signals). Such parameters can include:
  ➢ pitch "bending",
  ➢ low pass filter frequency (often used with some resonance to give a "filter sweep"),

> volume,
> stereo pan position,
> relative volume or pitch of sonic components ("vector synthesis")

1.b. Trigger events from a pre-composed set (usually related, eg. an arpeggio, or set of phrases which make up a piece). Points at specific locations within the detection zone are mapped to event triggers; a many to one mapping can enable low accuracy positioning to reliably trigger a smaller set of music events.

2.a. Trigger a pre-composed music event to start playing (a single note, a chord, short phrase, or larger music segment). Additional input parameters (eg. to set attack time, onset loudness etc.) also have their values pre-composed (ie. planned beforehand) and embedded in each event.

2.b. Trigger a music event, plus set parameter values (as in 1a). The range of input values needed can be reduced if desired, by again having pre-composed values embedded in the event, which the input value can then alter, to a greater or larger extent[gavin2].

2.c. Trigger events in turn from a pre-composed list, eg either the next or previous event.

Trigger events (2) can also be derived from *processing* continuous user movement (1). A good example in E-Scape is the creation of trigger events from analysis of user path. For example each reversal of motion direction could trigger the next event in a series. This can enable a performer to produce a natural "conducting" action, eg. by nodding head, or waving a leg.

The challenge facing the designer of this type of application is how to match the capabilities of the users, who have varying degrees of motor skill, via a control system, to the space of needed input parameters. Particularly interesting is the question of how to achieve acceptable performance in continuous real-time control, as might be the case in the context of a live musical performance.

3.1 Control characteristics of user, device and controlled process

A useful place to start is to examine how the capabilities of the user may vary, the kinds of input the system may require to achieve a task, and the possible mappings between the two. To match a control system to the *users abilities*, we require a characterisation of these abilities, since familiar results like Fitts' Law (see Mackenzie, for a review) may not hold for users with sensory and motor impairments, or where there are environmental constraints.

* What independent motor control capabilities does the user have available
* What range of movement (distance, angle, discrete values) can be produced
* What is the accuracy or precision of movement (avg. distance, angle or rate of error).
* The speed of the movement (m/sec, rad/sec, inputs/sec)

Embedded within the physical form of the *input device* there may be a number of transformations of the forces applied by the user, or their movement in free space detected by a sensing mechanism. For example a microswitched joystick transforms an angular input into a number of discrete possibilities; an ultrasonic beam (eg. EMS Soundbeam) has a number of discrete positions it can discriminate between, or can produce different output depending on the direction of approach to a position. The input device may also encapsulate a given *control dynamics*:

* the order of the control system (e.g. is it a distance, velocity (first order) or acceleration (second order) control). Higher order systems may allow us to produce a wider range of outputs, and quickly move between very different output values, but they are also more difficult to control, and more sensitive to feedback latencies.
* the input gain (what magnitude of change in output is produced by a given change in the input). In music applications, this can often be constrained, eg. a pitchbend of more than two semitones is musically inexpressive, and so the gain must be chosen with this limited range in mind.
* time delay (if there is feedback at input level). What is the delay between an input and some response to that input

An important distinction concerns the nature of the output of both the physical controller and the control system; whether it generates *event* or *continuous output*. An event output is generated at a particular point in time. With continuous output, some output is provided constantly. (This could be the "neutral" position of a bang-bang controller). The values of both event and continuous output channels can have either a discrete range or a continuous range. This characterisation applies to both the physical controller and the control system. The control system might in fact transform between these, eg. sampling will turn a continuous output channel to an event output channel. The point at which a sample is taken could also be assigned to a user controlled channel (eg. a single switch). Transformations can also be applied to the output values, for example quantisation to transform from a continuous range to a discrete range. In the context of the music application, the transformation of a continuous-valued output channel (eg lateral position of foot over a music keyboard) into one with discrete values is prone to errors. The number of discrete events depends on the accuracy of continuous output, and hence filtering of the signal produced is very important. Hence, it has been found to be useful to have "null" values within a continuous output channel, eg. where user cannot guarantee achieving the desired output value reliably enough. A good example of such processing in E-Scape is the dynamic re-mapping of music keyboard input - any zone of keyboard keys (eg foot positions) can map to a single music event (eg note), and other keys (eg the "black notes") can be mapped to zero output. Thus, if the user presses keys at the side of the desired white key, only that note will sound, and slipping onto the black keys makes no sound. Of course, the downside is that fewer output events can be directly controlled, but with careful preparation and/or splitting of control channels this is not a problem - see 3.2 below.

The *input parameters* to the system or controlled process could concern many different aspects of a given task. If the controlled process is to trigger phrases of music, then aspects would include which phrase to play, loudness, transposition, tempo, voice (instrument) used, etc. At the most basic level, we have:

- range - what possible values the input can take
- whether the inputs are continuous or discrete
- delay (if there is feedback at output level).

An additional concern where there are more variables to control than independent input channels is that some form of input moding must be implemented. For example, a music engine might accept a limited number of values from a user input to trigger musical events. When a particular value (eg. an end point) is received, the input is used to trigger events from a different set.

3.2   Design tradeoffs

We have from the above some requirements for information in order to design a system. There are a number of parameters we wish to control, for which we may identify a range of values, and the nature of the control required (discrete precisely timed inputs, rate of input needed, continuous control). For each user or class of user, there are the available motor control channels and the allocation of these channels to control parameters. As stated above, the control system which matches these two may itself transform the control signals. These three facets of the application are mutually constrained, and design is necessarily an iterative process. Inevitably, there is a tradeoff to be made of *expressiveness* against the *accuracy* required for real-time performance. This increased accuracy may be achieved by decreasing the resolution of the controlled process parameters and also by use of filtering. The main design goal here is to make *best use of available motor control*, in the case of our particular application area, giving the *greatest degree of musical expression*.

Where there is difficulty, there are two obvious design alternatives;

- Where the information required by a parameter is more than that on any available motor control channel, more than one channel can be assigned to a given parameter. An example of this would be to combine foot and knee position. The foot used to control a set of 10cm wide pads in a row on the floor, with a sideways range of 1m allowing 10 discrete pad values, and the knee rising to enter an ultrasonic beam. If the parameter to be controlled is pitch, then a number of combinations are possible, for example the foot selecting between 10 pitch values within a set, and the knee changing the set of pitches in use each time it is raised.
- Conversely more than one parameter may be assigned to a given control channel. An example of this would be use foot position to control chord type, root note and volume. Again with a set of foot pads, one might be used to toggle high or low volume, several could be used to pre-select (but not trigger) chord shapes (eg. minor triad, major 7th), with the remaining pads used to play a chord with a choice of root notes (eg. C,E,F,G,A) with the selected volume and shape.

Such compromises can still produce surprisingly good musical results if carefully designed, particularly when events consist of higher-level musical structures such as loops or phrases, which give "covering" delay while another input is selected. The designer should be aware that particularly for real-time tasks, the information transmission rates of the combined channels is likely to be less than the information transmission rates of the channels when used in isolation, and similarly for split channels. Where no acceptable assignment of motor control to process parameters can be found, the informational requirements of the controlled process must be decreased; in many cases preparation of pre-composed material can mitigate the loss of expressive control.

3.3 Design representations

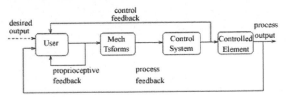

Figure 2 - Different levels of feedback within control loop

A simple box diagram and filter notation can be very useful for high level modelling, development of this might also be useful for resolving issues of software and hardware architecture. This can also encode issues such as event and continuous output channels, continuous and discrete values and so on. Having a structured design representation can help us to consider the likely effect of varying the dynamics of the controlled system (input gain, order), and system performance (feedback delay) or additional feedback earlier in the loop (figure 2). We can also anticipate the need for filtering (a tradeoff of expressiveness vs. accuracy). For example, consider the issue of feedback. We discussed above how feedback gain and latency parameters can affect real-time control performance. However, when discussing this, we

290

should remember that feedback can be present at several levels. At the lowest level, the user's own body provides proprioceptive feedback on position of limbs (although some disabled users have far less). Also, physical device characteristics may provide feedback at this level (eg. where there is a limited range of movement). Next, we may have feedback about the control inputs generated by the input device (eg. position of a head controlled switch). This is extremely important where there is significant latency in the controlled process, for example the position of an electric wheelchair (see Doherty and Massink, 1999 for more discussion of time issues), or where a delay is deliberately introduced (eg. where notes are placed in a queue before playing) and it becomes important to be able to tell exactly when (and whether) the input event was triggered. Finally we have the controlled process state (the music heard), which in many cases will be the most important source of feedback.

Consider for example the pitch control configuration described in section 3.2 above (see figure 3). The output channels

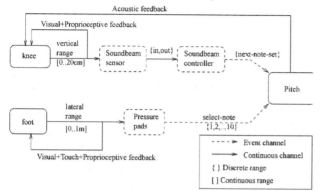

and the values carried can be concisely represented, allowing us to see at a glance the association between motor control channels and controlled process, and the transformations applied to both control and feedback signals. Such representations serve a dual purpose, helping us both to evaluate designs with respect to feedback and control issues (expressiveness), and to encode design alternatives in a fashion accessible to participants from different backgrounds in the design team. Representing design options in a structured and concise fashion helps us to reduce complex design issues to a form where we can hope to find answers in the human factors literature, or by means of straightforward reasoning.

Figure 3 - diagrammatic representation of a control configuration

4. Conclusions

The advent of a number of modern interaction technologies means techniques must be developed to allow designers to consider issues raised by continuous interaction between users and computer systems. In this paper, we have looked at the application of concepts from a branch of engineering psychology (particularly manual control theory) to the problem. This approach requires that special consideration is given to control and feedback signals, and transformations of these signals. We have investigated the application of this view to interactive system design by considering systems for music performance by disabled musicians. In this context, we have looked at characterisation of control aspects of user, device and controlled process, some of the tradeoffs involved in designing a control system, and simple graphical representations of control system configurations.

The needs of disabled musicians within the Drake Music Project for performing music provide a wide-ranging and demanding testbed for these ideas. In terms of future work we are interested in the guidance that manual control theory can give in the choice of control system. Qualitatively, we would like to develop some guidelines, driven by user and controlled process characteristics. A full control theoretic treatment of this application would be problematic; individual differences in sensory motor skills are very large, very few control scenarios are "pure" enough to facilitate an equational specification, and the benefits of the research effort would not justify the cost. However, with appropriate empirical data, we believe some simple quantitative estimations could be developed.

Acknowledgement

This work was supported by the TACIT network under the European Union TMR programme, contract ERB FMRX CT97 0133.

References

T.M. Anderson (1997). Making music with Computers. *Ability* 209:9-15.

T.M. Anderson (1999). Using Music Performance Software with Flexible Control Interfaces for Live Performance by Severely Disabled Musicians. *Proc. EuroMicro 25(Vol. 2):* 20-27.

G.J. Doherty and M. Massink (1999). Continuous Interaction and Human Control. *Proceedings of XVIII European Conference on Human Decision Making and Manual Control*, Group-D Publications.

R. A. Hess (1985). A model based theory for analyzing human control behaviour. *Advances in Man-Machine Systems Research*, 2:129-175.

R.J. Jagacinski (1977). A qualitative look at feedback control theory as a way of describing behaviour. *Human Factors*, 19:331-347.

G. McRuer (1980). Human dynamics in man-machine systems. *Automatica* 16(3):237-253.

G. Salvendy, editor (1997). *Handbook of Human Factors and Ergonomics*. Wiley Interscience, 2nd edition.

I. Scott MacKenzie (1992). Fitts' Law as a Research and Design Tool in Human-Computer Interaction, *Human Computer Interaction*, 7:91-139, Lawrence Erlbaum Associates, Inc.

Supporting Sign Language Users of Web-Based Applications: A Feasibility Study

Joanna Donkin, Cornelia Boldyreff, Liz Burd

Sarah Marshall

RISE, Department of Computer Science, University of Durham, UK

Council for the Advancement of Communication with Deaf People, Durham, UK

Abstract

This paper described the work performed jointly by the University of Durham and the Council for the Advancement of Communication with Deaf People (CACDP [1]) to migrate training and assessment courses for communication skills with deaf people to web based applications. Communication difficulties felt by deaf or hard of hearing members of the community lead to numerous problems including social isolation. Within the UK as a whole there is often desire or necessity to communicate with the deaf but there is a lack of experienced trained teachers to be able to train the hearing community to appropriate skilled levels where easy communication is possible. Furthermore, most of this training must be provided within people's leisure time so the training needs to be flexible and fun.
This project through the use of web based materials will seek to reduce this training need gap. The main objective of this paper is to promote discussion on the topic of using the Internet to promote awareness of the communication difficulties of the deaf. The approaches to be adopted will be described within this paper, as will some of the some of the preliminary results obtained for early software feasibility studies. Some early conclusions are finally drawn regarding the direction in which the joint project will take.

1. What is CACDP?

CACDP (The Council for the Advancement of Communication with Deaf People) is a registered charity that works to improve the quality and coverage of communication between deaf and hearing people. They currently offer qualifications in Deaf Awareness and British Sign Language (BSL), as well as provide publications and videos for sale.

2. Problems faced by Deaf People

There are approximately 8.7 million deaf and hard-of-hearing people in the UK, about 1 in 7 of the population. The problems they face when communicating with the hearing population are often not recognized, and even when they are, the hearing person will often not know how to communicate effectively. There are many differences in the ways deaf people communicate, as there are with the hearing population. Some deaf people can read English well and some cannot. Some deaf people prefer BSL but may have English as a second language, where others can lip-read to a level at which many hearing people would not realize they are deaf. Communication problems can lead to misunderstandings and other problems, as well as stress and anxiety and a feeling of discrimination or isolation. Effective communication is essential so that Deaf People can gain access to everyday activities without needing to focus on the communication rather than the content.

3. Aim of project

The aim of the project is to make the examinations already offered by CACDP, available to a wider range of people. Designing and implementing a web-based education and examination pack that will allow candidates greater access to the material in the Deaf Awareness or BSL courses will achieve this. Candidates will be able to access the information at times that suit them and from locations that suit them. They will be able to access as much or as little as they need or want at the time. It is hoped that the greater availability of the material will lead to an increase in the amount of people taking the courses and eventually an increase in the level of Deaf Awareness in the general population. This in turn would lead to less communication problems and less discrimination and isolation.

292

4. The Project Stages

The first stage of the project has involved the formation of a work plan based upon a two-year delivery phase. The first of the stages identified within this plan involves the conducting of a Market Survey. This survey will be carried out in order to determine the demand and interest in such a project from CACDP's current 'customers' i.e. those who are already taking the examinations and courses at the current time. It will attempt to gauge interest and any requirements they may have with regard to connections and hardware. Based on the results of the survey conducted a set of requirements will be derived. The resulting Requirements Report will serve as a means of formally determining the details of the requirements of CACDP. It will need to set out the exact specifications with regard to what the system needs to do and how it should do it.

The second stage of the project will involve the formation of a CACDP Products and Services Strategy Document. This document will define supporting development processes for identified range of CACDP applications, initially concentrating on the existing products to be migrated to the World Wide Web firmly based on an engineering approach to hypermedia development and the web. It will need to define the Web development processes and identify appropriate methods and tools to support those processes.

Also included in the Products and Strategies survey will be an investigation into the hardware requirements as CACDP have no existing web technology. This will be achieved through the proposed Specification of ITC Infrastructure Study. A major part of the project will be to determine what hardware support levels would be appropriate and resulting in a specification and strategy for the identification of the equipment required and then ultimately the installation of the equipment.

Once the hardware installations are in place then some of the interesting work based on that of the software requirements will commence. Specifically a Web Based Deaf Awareness training programme with integrated online examination is proposed. This programme teaches candidates to understand and appreciate the problems faced by deaf people, and equips the candidates with some basic communication skills. This programme is designed so that candidates can access the material in the course over the web, either as an addition to the traditional teacher lead course, or as a standalone set of lessons. It is proposed that candidates will even be able to sit their examination via the web. Thus the project affords greater flexibility to all students encouraging greater levels of participation but also open the course to those who may be otherwise unable to attend such as those with mobility difficulties. Some early development work has already been done in this area by CACDP.

Based on the experiences gained for the development of the Deaf Awareness Programme the engineering principles learned will then be applied and extended to support the more technical complex courses offered by CACDP. These include the training courses and examinations for BSL levels 1 and 2. Within these courses the complexities of sign language are covered and significant degrees of communication fluency are taught and examined. This has significant technological requirements necessitating video support. A detailed feasibility study and extensive trials will be used to ensure that the technology used is usable and feasible. Further technical complications are later introduced into the work with the formation of a web based Deafblind Awareness training programme with again an integrated online examination. This, since it is similar to the above training programme for Deaf Awareness, is to be used to evaluate if the proposed web engineering development processes are suitable. This use for constant process evaluation will ultimately be used to produce a web based development process improvement report. This report will identify a basis for continuous web development process improvement based on established software quality and process improvement standards. This review process will define benchmarks to ensure quality is maintained and ultimately improved within the company processes and procedures.

However, for CACDP the migration of their training packages will not be their only technological achievement to come from this work. Later stages of the project plan to extend e-commerce through the company. This will include a web based CACDP Directory and associated on-line database and services. The CACDP publishes a directory of Interpreters and Agencies. It is intended to make this available over the web as a searchable database. This would tie in with further advancements into the world of e-commerce, possibly to allow CACDP to charge money for the use of the Directory and to allow them to offer other products for sale e.g. publications and videos.

5. A Feasibility Study

Before the development work is to fully commence it has been decided to conduct a feasibility study. This feasibility study will be used to evaluate the provision of other existing products that offer similar functionality. This will allow the gathering of a base of knowledge on which to base the new system. The success or failure of the other similar products will help when determining the requirements of the new system. /

Although sign languages differ from country to country this study has not been restricted to BSL alone. Despite this, the results of the survey point to the fact that there are not many systems available that offer sign language training facilities. The study has identified that, of those that have been developed, they tend to fall into one of three categories. These are as follows:

1. **Dictionary based** - Some simply offer a dictionary function, that is, the user enters or, more often, selects a word from those on offer and this is signed to them by the use of a short video clip, or a succession of clips.
2. **Lesson based** - Other applications work using a lesson based approach where the user can progress through the language.
3. **Game based** - A third method makes use of games and tests to reinforce the vocabulary learning.

The feasibility study has analyzed examples of each of these basic types of systems and categorized its results based on these broad categories. In order to evaluate these systems a number of evaluation criteria have been devised. These criteria are based on the requirements of CACDP and that of web based non-vocation training courses in general. The criteria used within this evaluation are as follows:

1. **System usability characteristics** – this includes the presentation of the information on the computer screen, its layout and the general 'first impressions' felt by users.
2. **Language specifics** – this evaluation criterion investigates if the software is composed for a specific deaf communication language and what, if any, implications this support will have for other deaf communication languages.
3. **Pedagogical nature of the application** – including the training approaches used and whether this appropriately supports the learning process.
4. **Navigation capabilities** – this criterion evaluates if learners are able to appropriately navigate around the software, whether they become lost and disorientated and the overall effect this has on the teaching.
5. **Use of technology** – this aspect evaluates the types of technologies that are used and how effective each technology is in terms of supporting learning.
6. **Real time capabilities** – what speeds are demanded by the software and whether technology is able to deliver the required expectations and overall how this effect the learning process.
7. **Entertainment** – since many of the learners will be using these courses as non-vocational the entertainment factor is also consider an important aspect in the process of maintaining the motivation of the learner.

The CD-ROMs were all tested on the same Pentium III 650MHz laptop machine with a fast DVD Drive.

The results of the evaluation process using the above criteria are now discussed.
The dictionary based systems tend to simply offer functionality where the user can select a word from a list, e.g. *apple* and they are shown a short video clip which they have to copy in order to learn the word. One example of this type of system is the HandSpeak [2] Web Site. This site uses American Sign Language (ASL) as it is an American site, but the principles that can be applied remain the same. There are various problems with this type of site in that they tend not to offer a very wide range of vocabulary, although HandSpeak claims to offer 3,090 words, and they do not offer any help with grammar or syntax. Some of the sites in this category also offer the user the chance to type in a sentence and have the whole sentence signed back to them. This is a good idea in principle, but what happens in practice is the sentence is simply translated word-for-word and therefore is not any use for learning as BSL is not translated from English in this manner, but is a language with a grammar and rules. Therefore this is not the correct way to translate BSL. These sites are more useful for increasing vocabulary than for teaching ASL or BSL as languages.

A second problem with this method, and with all methods that use video clips, the video clips show the signing from the viewers point of view, and therefore this is very difficult to copy. On the other hand, the signing cannot be shown from the signer's point of view, as all they would see is the signers back.

There are various examples of this type of site. E.g. HandSpeak, or the ASL Dictionary Site [3]. The HandSpeak site is more effective as it uses real-life video clips, where the ASL Dictionary uses mainly line drawings with a few real-life video clips which are slow and jerky, and a description of the sign in words. However the HandSpeak site is confusing as the signed words are only shown in a small frame down the right hand side of the page, and not as the main feature.

A second method is the Lesson Based approach. This is a more traditional approach to learning and would allow the user to start with simple signing and progress as they learned. An example of this type of application is the "Sign Language For Everyone" CD-ROM [4]. Again, the language being taught is ASL as this is an American product. It uses a series of lessons made up from text, audio and video to teach ASL. The application itself is confusing as it has icons for controls where the designers have not made it clear what the icons actually do. The video clips used are fast and clear with a textual description of each sign. This is effective as a method of teaching. The program is slow to run from the CD though which can get frustrating and make the user lose track of the lesson.

A different approach is to teach the language using a series of games and make use of vocabulary in this way. This is termed the Game Based approach. An example of this way of teaching is used on the "Simple Signs" CD-ROM [5]. The user must select the correct answer to the question based on what he thinks the translation of the signed word is. This is a slightly different approach to the dictionary style as here the user is shown the signed word and asked the translation, whereas in the dictionary style they select the word and are shown the sign. The games approach should be more effective as the user must work out what the sign is, instead of being told what it shows. This approach is very useful when learning to read BSL, but does not have any formal sections for the user to attempt the signs themselves. Again, as with the dictionary approach, this method does not teach the user any of the rules or grammar of the language. It is a good method for teaching vocabulary as it will be fun and will reinforce the vocabulary being taught in a non-obvious manner.

6. Conclusion

The use of the evaluation criteria has afforded the opportunity to assess and compare the differing systems within our study. The findings of the study reveal that there are many good and interesting features of a good number of the systems, but in addition many of these have some drawbacks. The results of the feasibility study will be used as a means of composing a list of useful features and technological approaches for the requirements of the CACDP system.

Overall from conducting the feasibility study it has been identified that this type of system is not simple to design. There are complexities due to the need for cross-discipline knowledge from both Education and Computer Science, as well as a very good knowledge of the Deaf Awareness material. In order to be considered a success, such a system must be designed to teach rather than simply inform, and in addition the pedagogical approach adopted must serve to motivate students if it is to be successful as a non-vocation course.

One important feature of the CACDP system that is not inherent within the other systems evaluated is that of the technical issue of security. CACDP, in their proposed policy of open access to course materials (i.e. to allow disabled people to study from home) is yet to solve the problem of certification of the examination process to ensure the correct candidate sits the examination process. Additional work will now be conducted to ensure that the open policy principles are maintained together with the security of the qualification certification process.

The early work of this project is aimed at teaching communication skills to hearing students. This, it is hoped, through increasing public awareness and training will significantly assist the general communications problems that the deaf community faces. However, eventually it is anticipated that the technologies that are developed through this project will directly help the deaf community. Many of the deaf community prefer to communicate through sign language rather than the written word therefore the technologies devised within this project to train and educate could eventually be used as a direct communication mechanism to the deaf.

References

(1)   www.cacdp.demon.co.uk
(2)   www.handspeak.com
(3)   www.bconnex.net/~randys
(4)   Sign Language For Everyone CD-ROM, 2nd Ed., Higher Learning Systems
(5)   Simple Signs CD-ROM, Sign Communique

Human-Computer Protocols

David J. Duke[a], David A. Duce[b], Phil J. Barnard[c], Jon May[d]

[a]Department of Mathematical Sciences, University of Bath, Bath, BA2 7AY, UK
[b]School of Computing and Mathematical Sciences, Oxford Brookes University, Gipsey Lane Campus, Headington, Oxford, OX3 0BP, UK
[c]MRC Cognition and Brain Sciences Unit, 15 Chaucer Road, Cambridge, CB2 2EF, UK
[d]Department of Psychology, University of Sheffield, Western Bank, Sheffield, S10 2TP, UK

Abstract

We have recently developed an approach to modelling interaction that encompasses not just the device, but also aspects of a cognitive model. This integrated framework, called syndetic modelling, has been used to reason about the interplay between cognitive and computational resources deployed within an interaction. Here, a new view on this integrated framework is described. We consider the interaction between user and device as forming a hierarchy of protocols, covering different levels of abstraction over the information exchanged. As the protocol is layered, we can discuss interaction in terms of different levels of granularity, better accommodating the representation of new technologies such as vision tracking and speech which can be considered as "continuous" at some levels. We conclude by summarising classes of mathematical representation that can be utilised to represent and reason within such a model.

1. Introduction

Syndetic modelling is an innovative approach to understanding human-computer interaction through the development of models that provide an explicit link between the capabilities and resources of user and device. By linking user and device models, syndetic models [1] have allowed us to examine and reason about aspects of interaction within, for example, multimodal and gestural input, that are contingent on the resources and behaviour of the combined system. The long term objective of syndetic modelling is the development of a body of theory, and associated methods, that can be used as a theoretical framework for the description and analysis of interaction, and as the basis for rigorous design methods for advanced interfaces (e.g. continuous interaction methods such as speech and gesture, multimodal fusion and fission [2], distributed multimedia, and virtual and augmented reality [3]).

The key feature that distinguishes syndesis from other approaches to modelling interaction that make claims about user behaviour, for example [4] is the explicit use of a model of human information processing developed within applied psychology. In this way assumptions and properties of user behaviour within the conjoint system are grounded in empirical and theoretical results within cognitive science. To achieve this, we have formalised the structure and operating assumptions of such a model, specifically ICS (Interacting Cognitive Subsystems); the rationale for choosing ICS is set out in [1].

Our initial work with ICS involved reasoning about the stability of processing patterns [1]. While these issues are also relevant to continuous interaction, in this paper we are interested in the exchange of information between "bottom up" perceptual processing and "top-down", goal-directed activity. Interaction between different levels of representation is a key component of ICS. We want, however, to link this idea of levels of cognitive processing with the processes involved in interacting with an interface. The approach set out here borrows ideas from the layered architecture of computer networks [5] to organise a model of interaction. In this paper, we seek to motivate this approach, and then to discuss the implications of this view for the kind of modelling that we require to understand continuous interaction.

In describing the protocol-oriented view of interaction we will need to refer to elements of the ICS model. These are summarised in Section 2. Section 3 introduces the protocol-oriented view. The conclusions of the paper reflect on what may be some general lessons for "advanced" interfaces.

2. ICS

ICS is a cognitive model that describes human information processing in terms of a distributed architecture (see Figure 1) built from a collection of subsystems, and the mental representations and codes that are processed within these subsystems. Descriptions of ICS and its role in HCI, cognitive psychology (including clinical and experimental applications), and of course syndetic modelling can be found in the literature [6, 7]. All subsystems in the architecture have a common structure, consisting of:

- An *input array*, at which incoming data streams arrive.
- An *image record* representing an unbounded episodic store that holds a copy of all data received by that subsystem.
- A *copy process* that copies all incoming data from the input array to the image record.
- A set of *transformation processes* that convert data into certain other mental codes.

The representation in Figure 1 situates the subsystems within a data network over which mental codes are transferred. This "network" is a representational convenience that prevents the diagram becoming cluttered with the set of possible data paths. Within the brain, the "output" of a process is effectively co-located with the input array of the subsystem that receives itscodes. Nine distinct mental codes are posited, each associated with one of the subsystems:

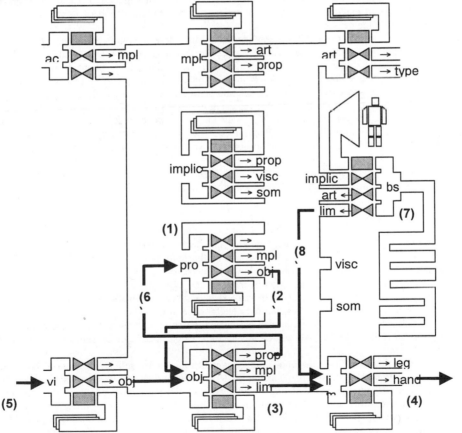

Figure 1. The ICS Architecture

Sensory Codes		Structural Codes	
vis	Visual	**obj**	Objects and shapes
ac	Acoustic	**mpl**	Morphonolexical
bs	Body state		
Semantic codes		Effector codes	
prop	Propositional	**art**	Articulation
implic	Implicational	**lim**	Limb control

Overall behaviour of the cognitive system is in part determined by the possible transformations and the demands of a particular task. Information in visual code (for example, derived from a display) cannot be translated directly into propositional code, but must be processed via the object system that addresses spatial structure. This flow of data between processes within a task determines an overall pattern of processing called a configuration.

More than one flow can be active in a configuration, for example the person writing this paper is attending to the content of the visual display, and in parallel generating effector code that results, ultimately, in the addition of text to the document. The ability of the architecture to carry out such processing tasks is subject to a number of key assumptions, for example:

- A process can augment an incomplete or unstable data stream by disengaging from its input array and instead operating on an extended representation produced at the proximal end of its subsystem's image record by the action of the copy process.

The representation in Figure 1 shows the architecture as configured for a task involving hand-eye coordination, for example finding an icon and moving the cursor (via the mouse) to that icon. The propositional subsystem (1) is processing information about the target and the current display contents through its image record, and using a process (written as :prop-obj:) to convert propositional information into an object-level representation. This is passed over the data network (2), and used to control the positioning of the hand through :obj-lim: (3) and :lim-hand: (4) transformations. However, both obj and lim are also receiving information from other systems. The users' view of the rendered scene arriving at the visual system (5) is translated into object code that gives a structural description of the scene [8], and from this a propositional representation of the scene is generated by :obj-prop: and passed to prop (6). In parallel with this 'primary' configuration, proprioceptive feedback from the hand is converted by the body-state system (7) into "lim" code (8) in a secondary configuration. If the mouse was 'sticky', it is likely that proprioceptive and visually derived feedback would become inconsistent, making it difficult for the :lim-hand: system to produce stable output, and eventually leading to a degradation in the overall configuration.

3. Layers and protocols

Interchange of information between a computer system and user is mediated by the interface of the hardware/software system and the perceptual/effector systems of the human. In many applications, the interface provides only a partial window onto the model maintained by the device, and the user's understanding of the state of interaction must be maintained through a combination of memory and exploration of the device. In other words, direct communication between the application and the user exists only via the user's perception of the rendered presentation, but the implication is that from this, the user ought to be able to construct an understanding of the model. Each stage in this process of understanding can be understood as an attempt to communicate a particular level of representation, an attempt which, like any communication mechanism, can be subject to the effects of distortion and error.

To understand the role of levels of representation and communication in modelling interaction, consider the following scenario, derived from the "Magic Board" exemplar [9] featured at the CHI'2000 workshop on Continuity in Interaction. In the scenario, the user is manipulating a combination of physical drawing and projected image, with the aid of a vision-based interface that can track the position and motion of the user's finger, for example to describe regions on the board. Performing a task such as selecting a region of interest involves communication both between the user and device (has the device registered the user's intentions?), and between the cognitive resources needed to perform this task. In the latter case, communication typically involves interaction between representations capturing actions or top-down goals, and a bottom-up stream of representations carrying information about the world.

One of the properties of interactions between subsystems or the user and their environment is that part of the "dialogue" can become proceduralised. Rather than requiring focal awareness and access to episodic memory, the task can be carried out using learnt patterns of behaviour. For example, most computer users are familiar with the use of the mouse to select a region of text or drawing, by first positioning a point at the start of the region, then depressing the mouse button and dragging the cursor across the region to be selected, and then releasing the button to complete the selection. When first undertaken, this task probably required deliberate action and attention, but over time it becomes routine – at a propositional level, we develop a "procedure" (or Common Task Record) for dealing with this. The states and transitions captured as part of this procedure are shown in the upper half of Figure 2.

Of course, as Figure 1 shows, performance of a procedure like this is not isolated at any one level of representation, but requires interaction with other levels of mental representation. At the moment, though, our concern is with the relation between the propositional level representation, and the states that the device goes through, which are shown in the lower half of the figure. Note that, in both cases, transitions between states are mediated by events in which both cognitive system and device participate.

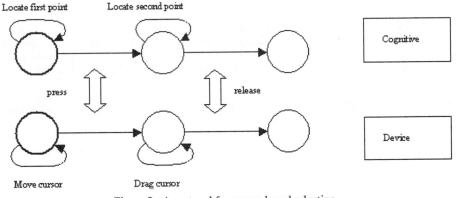

Figure 2. A protocol for mouse-based selection

In the Magic Board scenario, the situation is a little more complicated. The lower half of Figure 3 shows the corresponding device states needed to select a region. In order for the vision tracker to determine the location on the board to which the user is pointing, the user must leave their hand in the one position for a certain period of time (in the order of a few seconds). If the user stops, and then starts to move their hand before the delay period has passed, the device returns to tracking the position of the hand.

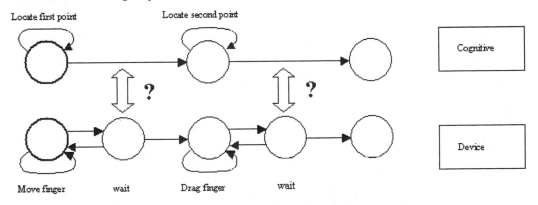

Figure 3. A protocol for Magic Board selection

Now, although this task can certainly be performed with practice, any task record that is formed for magic board selection is likely to co-exist with the task record for mouse-based selection. Both tasks support the same type of goal ("select a region"), and as a result, it is possible that the task record for mouse-based selection could be revived while interacting with the Magic Board. This combination is shown in Figure 3. The problem is that, at the propositional level, transition between "first point" and "second point" states is accomplished by an event. The corresponding transition in the device involves a new state, the exits from which are determined by the duration of the state. A user may well equate stopping motion of the hand with the "located point = press" event, and at a propositional level then begin the next phase of the task, locating the second point, without the device having made a similar transition.

4. Conclusion

We have sketched out a view of interaction that utilises levels of mental representation from a model of cognition. Interaction involves communication between levels of representation both within the cognitive systems, and between the cognitive systems and device or interface. Using this framework we described interaction with a novel vision-based interface, and argued that similaties between the interaction protocol for this device, and those for mouse-based selection, may lead to confusion.

The analysis presented here seems to point to a tension in the design of novel interaction methods. On the one hand, there are certain generic tasks that reappear across devices and interfaces, for example selecting objects or regions, moving objects, etc. This level of similarity, or metaphor, makes it possible for example for the user of one desktop UI to sit down in front of another and carry out simple tasks with little difficulty. However, if the interaction protocol is modified, then there is a danger of confusion between the two task records. In this case, it could be argued that new interaction protocols should be significantly different from those that have been used elsewhere. For better or worse, the desktop metaphor and environment is now part of working culture, and users are exposed to it from an early age. Old habits will die hard.

References

1. D.J. Duke, P.J. Barnard, D.A. Duce and J. May, *Syndetic Modelling*, Human Computer Interaction, 13(4): 337-393, Lawrence Erlbaum Associates, 1998
2. L. Nigay and J. Coutaz, *A Generic Platform for Addressing the Multimodal Challenge*, Proc. Of CHI'95, Addison Wesley, 1995.
3. S. Smith, D.J. Duke and M. Massink, *The Hybrid World of Virtual Environments*, Computer Graphics Forum, 18(3): 297-307, Blackwell, 1999.
4. T.G. Moher and V. Dirda, *Revising Mental Models to Accommodate Expectation Failures in Human-Computer Dialogues*, in P. Palanque and R. Bastide (Eds), Proc. DSV-IS-95, pp. 76-92, Springer, 1995.
5. A.Tannenbaum, *Computer Networks*, Prentice Hall, third edition, 1996.
6. P.J. Barnard and J. May, *Interactions with Advanced Graphical Interfaces and the Deployment of Latent Human Knowledge*, in F. Paterno' (Ed), Interactive Systems: Design, Specification and Verification, Springer, 1995.
7. J.D. Teasdale and P.J. Barnard, *Affect, Cognition and Change: Re-modelling Depressive Thought*, Lawrence Erlbaum Associates, 1993.
8. P.J. Barnard and J. May, *Representing Cognitive Activity in Complex Tasks*, Human Computer Interaction, *14*, 93-158, Lawrence Erlbaum Associates, 1999.
9. L. Watts, The Magic Board, an Augmented Reality Interactive Device Based on Computer Vision, Scenario description for CHI'2000 Workshop on Continuity,
    http://kazan.cnuce.cnr.it/TACIT/CHI2000/MagicBoard/MB.html

Continuous Interaction with Computers: Issues and Requirements

Giorgio Faconti, Mieke Massink

Istituto CNUCE, National Research Council, Via V.Alfieri 1, 56010 Ghezzano (PI), Italy

Abstract

This paper introduces continuous interaction as a requirement of interactive systems to support native human behavior. Subsequently it addresses a framework of modeling notations and tools to drive design decisions during the development of interactive systems.

1. Introduction

Interactive systems in the modern world are becoming both increasingly pervasive, and rich in the variety of tasks supported. In fact, one of the next major developments in computing will be the widespread embedding of computation, information storage, sensing and communication capabilities within everyday objects and appliances [1]. Interacting with such systems will involve multiple media and the coupling of distributed physical and digital artifacts, supporting a continuous flow of information. The concept of Continuous Interaction [2,3] was brought about by advances in technologies that will have a profound effect on the way we interact with computers and information, and ultimately the way we work and live. Continuous interaction differs from discrete interaction in the sense that the it takes place over a relatively longer period of time in which there is an ongoing relevant exchange of information between the user and the system at a relatively high rate, such as in vision based or haptic interfaces, that can not be modeled appropriately as a series of discrete events. This shift towards more continuous interaction between user and system is a direct consequence of two aspects of a new generation of interactive systems: *ubiquity* and *invisibility*. Since interaction devices will be spread in the surrounding environment, the distinction between *real world objects* and *digital artifacts* will become immaterial. Users will behave naturally as human beings without adopting a simplified behavior that characterizes a state of *technological awareness*. Consequently, there's a need for the systems to adapt to users, to be aware of their operating context, and to be able to take autonomous decisions to some extent.

2. Evolution of applications

Today miniaturization of hardware and increase in computing power makes it possible to put on one's pocket a relevant computing capability. Most every-day consumer goods incorporate microprocessors and support interaction: computer hardware becomes invisible and ubiquitous. In this context, sensing and effecting devices are developed that show an autonomous behavior and are able to cooperate to achieve a given goal (e.g. tele-passing facility in the highways). On the other end, *virtual* and *augmented reality environments* are being developed where common user tasks are supported by electronic devices (e.g. networks of domestic appliances on a *home LAN*). Although ubiquitous technologies were initially conceived as part of the infrastructure of work, these are now being extended to merge digital and physical worlds and the distinction between them becomes blurred. This process of merging is occurring at even the finest level, with the advent of *smart matter*. It incorporates small-scale sensors and effectors into physical materials, in such a way that the general behavior of the material may be actively altered and controlled by distributed techniques [21]. Those new environments are a great challenge for the future of Human Computer Interaction studies. More senses (vision, earring, touch) and more means of expression (gestures, facial expression, eye movement, speech) are involved in interaction from the human side with respect to traditional *WIMP* based applications.

3. Design complexity

As has become clear in the introduction, continuous interaction is closely related to the development of smart and virtual environments where the computing system is preferably invisible. The trend is that the system is adapting to the complexity of user's behavior rather than the user adapting himself to a simple system's interface. The

302

consequence is that the system needs to be able to deal with a larger part of the interaction. It must deal with the multiple and simultaneous way in which users and groups of users may interact with and via the system. This requires the coordination of many parallel, loosely coupled activities and processes, sensed by even so many different devices under tight real-time constraints in order to guarantee responsiveness of the system and a sense of *reality* to the user. The responsiveness needs to satisfy the constraints that make human perception and cognitive processes possible. All these aspects together lead to many design decisions that have to be taken into account during the development process. In order to address the behavioral issues as early in the design process, formal modeling techniques for distributed real-time and stochastic systems supported by powerful analysis tools could be considered. However, to address the numerous issues systematically, also a more general framework that can guide the modeling approach is needed.

4. A reference framework for continuous interaction

Structuring system design in hierarchical layers is a common approach to reduce design complexity in the software engineering practice. Reduction of the complexity can be obtained by defining for each layer what *services* are provided to the next layer. This way a designer can concentrate on the problems that are relevant to a particular layer while assuming that the lower layers are providing more basic, lower level means of interaction. He also can abstract from the higher level layer problems because he knows what are the means of interaction that the higher level requires. In the reference model presented in [3] and illustrated in fig. 1, we tentatively propose five different levels. A physical level where physical interaction takes place, followed by three levels of information processing: the perceptual, propositional and conceptual or task level. These are closely following the layers in well-known frameworks such as the Skill-Rule-Knowledge-model [4] and are not too far from the levels in Norman's Interaction model [5]. Finally we introduce a group level that explicitly deals with interaction problems related to coordination of tasks in the context of cooperative work.

5. Modelling notations and tools

The Reference Model on its own does not provide any specific technique or notation for the description of the behavioral aspects of interaction at the different layers of abstraction. It just provides a framework to guide the way in which a complex problem such as the design of continuous interfaces could be split into sub-problems at different levels of abstraction. It is well-known that in the software development process the requirements analysis and design phase are very important for the successful development of software. This holds even more so for the development of user interfaces. Formal modeling is increasingly recognized as a promising way in which behavioral and performance aspects of different designs can be assessed at an early stage of software development [6]. Given the particularities of continuous human computer interfaces, such as critical real-time dependency, distribution and parallelism, we discuss a number of formal modeling approaches that could provide appropriate concepts to deal with this form of interaction and the control issues this poses for the system.

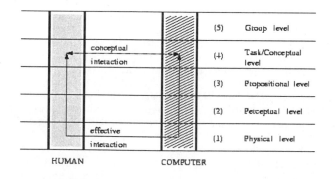

Figure 1: Reference Model for Continuous Interaction

Manual Control Theory

An approach that is used to analyze human and system behavior when operating in a tightly coupled loop is manual control [7]. The theory has been refined to a very high degree and there is now a very large base of predictive [8] and explanatory [9] theory validated by a wealth of experimental evidence. The general approach followed in manual control theory is to express the dynamics of combined human and controlled element behavior as a set of linear differential equations in the time domain. In order to obtain a solution this set of equations is transformed into a set of linear algebraic equations in the complex frequency domain by well-known mathematical techniques. In several models human related aspects of information processing are explicitly included in the model such as delays for the visual process, motor nerve latency and neuro-motor dynamics. Control theory can be linked to Fitts' Law [10] by viewing the pointing movement towards the target as a feedback control loop based on visual input and the limb as the controlled element.

Hybrid Modelling

Hybrid models allow to express both discrete and continuous aspects within the same formalism. Hybrid Automata [11] are seen as an extension of the purely control theoretic approach in manual control theory. It combines the description of continuous behavior as sets of algebraic equations with that of discrete behavior. In the hybrid automata approach behavior is described as sets of automata where each automaton consists of nodes and transitions between the nodes. The nodes are labeled by sets of equations that describe the continuous behavior when control is at that node. The transitions are labeled by guards on variables that define under which conditions one continuous behavior is replaced by another. The formalism is supported by software tools that provide reachability analysis of the state space and check properties of the system model formulated as temporal logic formulas. Several approaches to the application of hybrid modeling of continuous interaction have been examined. In [12] a case study is presented where the interaction between a pilot and an hydraulic subsystem of an aircraft is analyzed. In [13] a Hybrid Petri-Net model is discussed presenting two-handed navigation through a 3D architectural space in a VR-application. A hybrid automata approach to modeling a gesture language for 3D CAD applications is discussed in [14].

Stochastic approaches

In continuous interaction, the human variability in performance of activities is an issue. Stochastic modeling can provide qualitative information about the shape of the probability distribution of real-time related issues concerning the combined human and system performance. Estimates can be made of the effect of the combination of mostly log-normal distributions generated when the reaction time of humans is involved [15] and the system performance that is often characterized by variants of exponential distributions [16]. Stochastic modeling, simulation and analysis using stochastic automata is still a relatively new field. As such the expressivity of the specification languages based on the technology, the theories concerning analysis of specifications, and the incorporation of these into automated support are still at an early stage of development. We believe such techniques have an exciting potential for modeling performance in interactive systems, taking into account the abilities and limitations of both user and system. Such models allow us to generate a richer set of answers to design questions, which enables meaningful comparison of the results of our analysis to human factors data. Two preliminary case studies of the use of stochastic automata for the modeling of polling behavior and that of a pointing task can be found in [17,18].

The Unified Modelling Language

The Unified Modelling Language (UML) is a de facto industrial standard. The UML is semi-formal in the sense that its syntax and static semantics are defined formally, but its dynamic semantics are described only informally as are the relations that may exist between different diagrammatic notations. Several formal semantics have been proposed for UML Statechart Diagrams, that are meant for the modeling of behavioral aspects of systems. Currently, the UML provides only limited support for the modeling of continuous, real-time and stochastic behavior of systems. Nevertheless, it provides a rich set of notations that are widely adopted in industry and as such, they form an interesting starting point for further extensions of the notations for the modeling of complex interactive systems and intelligent environments. Preliminary work in this direction can be found in [17] where a stochastic model of a user checking the progress of a system is described using a stochastic extension of a subset of UML statechart diagrams [19] In [20] UML interaction diagrams are used for a high level specification of interaction in a VR application. It is shown how the interaction diagrams may be used to reflect the different levels of abstraction in the reference model in fig. 1.

6. Conclusions

Although continuous interaction is not a completely new phenomenon in user interfaces, it has become a much more prominent and critical issue in many future computing systems such as virtual reality applications and intelligent environments. To guarantee responsiveness of a system and to provide the user with a pleasant sense of (virtual) reality the designer is confronted with many new critical design decisions. The information needed to make these decisions require insight in the human cognitive aspects of processing continuous and often simultaneous information combined with the distributed and parallel operation of the system in which many non-homogeneous input and output devices are involved. This, in turn, requires modeling techniques in which these aspects can be expressed and analyzed before a complete system is implemented. It is well known that complex, distributed real-time systems like intelligent environments, are notoriously hard to debug after their implementation, and therefore need a design that is setup carefully and systematically supported by appropriate analysis tools.

References

1. D.A.Norman, The Invisible Computer, MIT Press, 1998.
2. The TACIT Research Network (http://kazan.cnuce.cnr.it/TACIT)
3. G.Faconti and M.Massink, Continuity in Human Computer Interaction, Workshop Report, CHI 2000, The Hague (http://www.acm.org/sigchi/bulletin/2000.4)
4. J.Rasmussen, What can be learned from human error reports?, In K.Duncan et al., eds, Changes in Working Life. John Wiley & Sons, 1980.
5. D.A.Norman, The Design of Everyday Things, Basic Books, New York, 1988.
6. I.Sommerville, Software Engineering, Fourth Edition, Addison-Wesley, 1992.
7. G.Salvendy (Ed.), Handbook of Human Factors and Ergonomics, Wiley, 1997.
8. D.McRuer, Human dynamics in man-machine systems, Automatica, 16(3):237-253, 1980.
9. R.A.Hess, A model-based theory for analyzing human control behaviour, Advances in Man-Machine Systems Research, 2:129-175, 1985.
10. P.M.Fitts, The Information Capacity of the Human Motor system in Controlling the Amplitude of Movement, Journal of Experimental Psychology, 47, pp. 381-391, 1954.
11. T.Henzinger, The Theory of Hybrid Automata, In: Proceedings of 11th Annual IEEE Symposium on Logic in Computer Science, pp. 278—292, 1996
12. G.Doherty, M.Massink, and G.Faconti, Using hybrid automata to support human factors analysis in a critical system, Journal of Formal Methods in System Design. Kluwer Academic Publishers}, 2000. To appear.
13. M.Massink, D.Duke, and S.Smith, Towards hybrid interface specification for virtual environments, In D.Duke and A.Puerta, editors, Design, Specification and Verification of Interactive Systems '99, pages 30--51. Springer Computer Science, 1999.
14. G.Doherty, G.Faconti, and M.Massink, Formal verification in the design of gestural interaction, To appear in Electronic Notes on Theoretical Computer Science, Elsevier.
15. A.D.Swain and H.E.Guttmann, Handbook of human reliability analysis with emphasis on nuclear power plant applications - Technical Report NRC FIN A 1188 NUREG/CR-1278 SAND80-0200, US Nuclear Regulatory Commission; Washington, D.C., 1983.
16. R.Jain, The art of computer systems performance analysis : techniques for experimental design, measurement, simulation, and modeling, Wiley, 1991.
17. G.Doherty, M.Massink, and G.Faconti, Stochastic modeling of interactive systems, In Int. Workshop on Towards A UML Profile For Interactive Systems, York, 2000 (http://math.uma.pt/tupis00/programme.html)
18. G.Doherty, M.Massink, and G.Faconti, Reasoning about interactive systems with stochastic models, TACIT Report, Under submission.
19. S.Gnesi, D.Latella, and M.Massink, A stochastic extension of a behavioural subset of UML statechart diagrams, In Fifth IEEE International High-Assurance Systems Engineering Symposium, pages 55--64. IEEE Computer Society Press, 2000.
20. L.Sastry, G.J.Doherty, M.D.Wilson, D.R.S.Boyd, Continuity of Interaction in VR Facilitated Groupwork, Submitted.
21. L.A.Watts, J.Coutaz, D.Thevenin, E.Dubois, M.Massink, G.Doherty, Environmental interactive systems: Principles of systematic digital-physical fusion, TACIT Report – TR0013, 2000.

Complexity measures in handwritten signature verification

Michael C. Fairhurst, Elina Kaplani, Richard M. Guest

Electronic Engineering Laboratory
University of Kent
Canterbury, Kent CT2 7NT, UK

Abstract

Biometric control of access to places, restricted information or personal data is an increasingly important consideration in many situations. The handwritten signature is a common biometric which has been in regular use, but with authentication decisions traditionally made by human visual inspection. This paper investigates several issues relating to signature complexity and authentication decisions, and points to the increasing need for an understanding of both human and machine-based perceptual mechanisms as biometric processing becomes more widespread.

1. Introduction

Facilitating universal access implies the need for an awareness of the wide range of characteristics of individuals, their similarities and differences, so that appropriate system design considerations can be invoked. System design to support universal access then seeks to define strategies and algorithms to minimise the effect of the observed differences with respect to their influence on the user and the way in which system/user interaction is achieved.

It is important and interesting to note, however, that there are also situations in which it is actually the differences between individuals which are important and which it is appropriate to seek to exploit. The most obvious such situation is when issues of security of information and control of access are of prime concern. Establishing a clear understanding of these differences can then allow the exploitation of *biometrics*, using individual characteristics, not as a means of extracting what is common within a population, but focusing on how individuals differ, to control the access to information based on a knowledge of the authenticity or otherwise of the recipient.

Biometrics is a diverse and increasingly important field, covering many modalities and types of individual characteristics (Nakamura, O., Mathur, S., & Milhami, T., 1991; Hrechak, A.K., & McHugh, J. A., 1990; Daugman, J., 1995; Ashbourne, J., 1994; George, M., 1995; Jain, A., Bolle, R., & Pankanti, S. (Eds.), 1999). The biometric modality with perhaps the longest history, and one which certainly enjoys the greatest degree of public acceptance, is the handwritten signature (Fairhurst, M.C., 1997). Thus, signature verification is an area which continues to attract research interest, and which is part of the complex pattern of interacting strands which impact on the theme of universal access.

There is another important aspect to consider, however. Checking and analysing handwritten signatures as a means of establishing or verifying identity is both a challenge for technology (i.e. algorithms for robust *automatic* signature verification are constantly sought) and for the powers of human perception, since there are many situations where signature checking by machine might be inappropriate or, at least at present, insufficiently reliable, for routine use. This is especially the case when the risk of forgery is high, or where acceptance of a non-authentic signature could have serious consequences. Furthermore, it can reasonably be claimed that a better understanding of human ability in analysing and authenticating signatures can lead indirectly to the specification of more accurate and perhaps more robust techniques which can be implemented automatically.

This paper reports on some important aspects of our work in the field of signature verification and, in particular, addresses some important issues relating to the human and machine identification of signature imitations/forgeries. As noted above, we can envisage the handwritten signature playing a key role in two related broad scenarios. First, human checking (by direct visual inspection) is still a common means of determining/confirming identity or authorising transactions, and the pervasiveness of this type of activity should not be underestimated. Secondly, there is an increasing need for *automated* (machine-based) verification of the handwritten signature (e.g. in electronic sign-on systems, or where improved objectivity and accuracy, such as in point-of-sale applications, is at a premium). It is this continuing need for both human and machine-based signature verification which provides the focus for the work reported, addressing issues concerning the form and structure of the signature in relation to the reliability with which genuine and fraudulent samples may be distinguished.

2. Signature characteristics and signature complexity

Signatures and signing styles can differ significantly, both within samples from the same signers, but self-evidently to a very large degree across a population of signers, and the susceptibility of a signature to false imitation is intuitively a function of the nature of the signature itself. This paper reports specifically on some experiments in human signature analysis.

Five writers were selected to use as reference or target signers, whose signatures were of varying length, number of strokes, and with differing degrees of embellishment in signing execution. The aim was to generate five target signature groups which could intuitively be expected to have a wide spread of "complexity" in visual appearance.

A group of 36 subjects were then presented with a sample from each of these groups and asked to assign a score to each sample (on a scale of 1 to 10) to indicate its perceived degree of complexity. Figure 1 shows the distribution of complexity estimates for each of the five target signatures. Even this simple data is revealing, since it demonstrates that, while at the extremes of the scale there is a modest spread in the perceived degree of complexity, with a relatively sharp "cut off", the intermediate signature samples appear to be much more difficult to assess and categorise quantitatively. This, in itself, represents an interesting problem which supports the notion of the desirability of some more objective or algorithmic means of analysing samples if complexity is to be considered for practical exploitation in increasing the robustness of signature checking procedures. More important, however, is to consider the implications of complexity in relation to the susceptibility to compromise of an individual signature model.

In order to investigate this aspect of signature perception, in a further experiment subjects were asked to view a range of signatures for each of the sample groupings, some of which were genuine samples and some of which were forgeries (generated in a separate experiment with a disjoint set of subjects who produced the imitations from a visual inspection of a genuine sample). In total each subject viewed 10 genuine and 10 forged samples from each of the five target groups. Each subject was asked simply to classify each sample as "genuine" or "forgery", in comparison with a genuine sample which was in view simultaneously, as would be the case, for example, in checking a signature against a "model" written on the back of a credit card.

Table 1 shows the results of this experiment. In order to give a clearer degree of distinguishability to the judgements made by participants, and the implications of these, the signature samples which were rated the least complex and the most complex were selected, together with the signature ranked in the middle of the overall ordering. These are designated sample A (least), sample B (middle) and sample C (most complex) in the results reported. The Table shows the average perceived complexity value, the total number of authentication errors made, and the split between false acceptance rate of forged signatures (FAR) and false rejection rate (FRR) of genuine samples.

An intuitive assessment of the relation between perceived complexity and the likelihood of errors in judging sample authenticity can lead to two (essentially opposing) hypotheses. On the one hand, it could be predicted that low complexity leads to ease of imitation and therefore potentially higher FAR. On the other hand, an alternative supposition might be that higher complexity makes imitation more difficult, and therefore more flexibility in assessment, leading to more errors in perceived authenticity, might be expected. Although the former hypothesis was the more popular among subjects tested, it is the latter hypothesis which is supported by the results of our study, as Table 1 makes clear.

Within the limited spread represented here, increasing complexity leads to a decrease in total errors in perceived authenticity. However, within this total, the FAR rises with increasing complexity (suggesting that observers find more complex signatures more difficult to judge, and show more tolerance with respect to the target signature) and the FRR falls (again, implying greater tolerance of variability in the more complex samples).

307

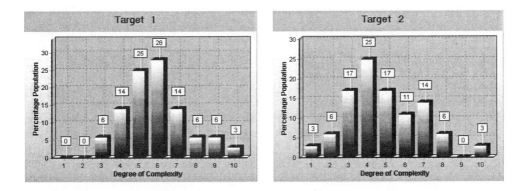

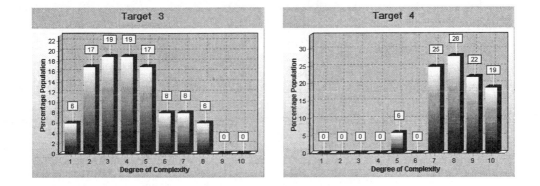

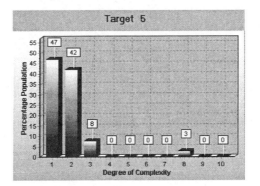

Figure 1: Distribution of perceived complexity scores for target signatures

308

Table 1: Error-rate performance in relation to complexity assessments

Signature Group	Avg. Perceived Complexity	Total error (%)	FAR (%)	FRR(%)
A	1.78	33.06	2.08	30.97
B	4.83	27.5	2.08	25.42
C	8.19	21.94	2.92	19.03

This is an interesting observation, especially when it is noted that the experimental conditions reproduce quite closely a typical practical scenario under which signatures are visually inspected. A more complex signature, potentially offering more challenges to the forger, is more likely to be susceptible to penetration than a more simply structured signature model.

3. Discussion and conclusion

It is known that measures of signature "complexity" can be defined and computed objectively and algorithmically, and the relations between some of these measures, the assessed similarity between test and target samples, and their likelihood of correct authentication have been explored, particularly from the point of view of "expert" analysts (Brault, J-J. & Plamondon, R., 1989, 1993). However, our study shows the nature of the problem to be confronted in routine situations where human signature checking is required. It also points to the value of devising an inspection protocol which could be adopted where human evaluation for authentication is necessary, where the procedures which may lead to greatest security and reliability may be seen as to some extent counter-intuitive. Strategies for human authentication may therefore benefit from a clearer understanding of the elements of complexity and, indeed, other features of a signature model which better reflect a justifiable confidence in judging authenticity.

Finally, there is an important issue raised here connected with a much more important question for the future, relating to the potential for the introduction of automated signature verification, since evaluations based on human perception and machine-based processing do not necessarily coincide, and the relations between the form and complexity of signature samples, human perception and interpretation, and algorithmic signature processing require considerably more investigation.

These distinctions may be very important in developing appropriate support for signature checking in different applications and different operational environments. This may especially be the case where automated systems and human intervention are likely to be integrated, a scenario which is likely to become increasingly common in practice. The results presented here therefore identify some important elements in a strategy to support improved human-machine interaction in relation to biometric identity checking for managing access control in practical situations, and which are therefore self-evidently of direct relevance in the broader context of universal access and its underlying themes.

References
Ashbourne, J. (1994). Practical implementation of biometrics based on hand geometry. *IEE Colloquium Digest 1994/100*, 5.1-5.6.
Brault, J-J., Plamondon, R. (1989). How to detect problematic signers for automatic signature verification. *ICCST, Zurich, Switzerland*, 127-132.
Brault, J-J., Plamondon, R. (1993). A complexity measure of handwritten curves: modeling of dynamic signature forgery. *IEEE Trans. Syst. Man Cybern.*, Vol. 23, No. 2, 400-413.
Daugman, J. (1995). High confidence recognition of persons by rapid video analysis of iris texture. *Proc. ECOS95, IEE Conf. Pub. 408*, 224-251.

Fairhurst, M.C. (1997). Signature verification revisited: promoting practical exploitation of biometric technology. *Electronics and Communication Engineering Journal*, 273-280.

George, M. (1995). A new strategy for low power, high discrimination voice biometrics. *Smart Card*, 130-134.

Hrechak, A.K., McHugh, J.A. (1990). Automated fingerprint recognition using structural matching. *Pattern Recognition*, 23, 893-904.

Jain, A., Bolle, R., Pankanti, S. (Eds.) (1999). Biometrics, personal identification in networked society. *Klewer Academic Publishers*.

Nakamura, O., Mathur, S., Milhami, T. (1991). Identification of human faces based on isodensity maps. *Pattern Recognition*, 24, 263-272.

Towards Emotive Captioning for Interactive Television

Deborah I. Fels, Lorelle Polano, Terry Harvey, Singh S. Degan, Charles H. Silverman

Ryerson Polytechnic University, 350 Victoria St., Toronto, Ont. M5B 2K3
dfels@acs.ryerson.ca

Abstract

The digital revolution has finally caught up to the television industry. As a result of this revolution and the increased available bandwidth accompanying it, access to television and television programming for people with disabilities is poised for significant improvements. The digital television industry is just beginning to develop and as a result there is an important window of opportunity to provide useful and cost-effective access solutions for digital television that can benefit many viewers. One salient and obvious focus of research is to examine opportunities for enhancements of closed-captioning. While the new digital standard for closed-captioning, EIA-708, provides for some limited enhancements such as different fonts and different colours, it does not address the need for capturing and transmitting all of the audio information in a broadcast with a different form (e.g., sound effects, music and emotion). In this paper, we present initial concepts and models that provide graphical and tactile expressions of non-speech audio information. The broadcast program, Il Menu, is used to exemplify and illustrate these concepts.

1. Introduction

Enhancing the productivity and communication of people with disabilities through assistive technologies is a very vibrant and active area of research. The new field of interactive television provides a forum for exploring multimodal I/O and interactive access to high-quality broadcast technologies.

Interactive television (iTV) can be described as an interactive system that combines traditional broadcast television with an Internet back channel connection (Krebs et al., 2000). By using line 21 of the vertical blanking interval (VBI) of the television signal, broadcasters can supply data, such as closed-captions, that accompany the broadcast.

Closed-captioning (Lewis, 2000) is one of the standard assistive technologies available for television. It gives viewers who are deaf or hard of hearing the option of displaying a text transcription of the spoken words in television programs. The text transcription may also be accompanied by one or two word text descriptors that provide speaker identification, accents, emotions, and other information such as music and background sounds. However, these additional descriptors are used sparingly and inconsistently (Harkin et al., 1995). As a result, the viewer may be left without a full appreciation for the affective state of the conversation (emotions such as happiness, sadness, sarcasm, etc.) and illocutionary force of conversation (urgency of the conversation) suggested by Dix at al. (1998) as critical in human conversation.

Mehrabian (1968) suggests that human-to-human communication consists of seven percent words, thirty-eight percent non-verbal linguistic modifiers (paralanguage), and fifty-five percent visual cues or gestures. While television viewers are not active participants in the human conversation of a television program, they are one of the intended receivers of this communication. It is through the interaction between humans on-screen, background sounds and music that much of a television program's semantics are conveyed.

The deaf or hard of hearing caption viewer must rely on visual-only cues such as body language and gesture combined with the words (which according to Mehrabian (1968), would only convey 62 percent of the communication) and short text descriptors of other information to understand a television program. Without continual access to conversation modifiers and access to the paralinguistic components of human conversation, people who are deaf and hard of hearing may mis-interpret or mis-understand the semantics of television, be unable to discuss the program with hearing viewing partners, and come away with an unsatisfactory experience.

We have begun to develop strategies to translate the emotion expressed through speech as well as non-verbal audio information (such as background noise and music) into alternative displays for WebTV. Graphical and tactile displays are used to convey this information. These strategies include using differently shaped speech bubbles, colour, font and overlay to identify a speaker and his/her emotions; using icons and symbols to represent background noise and music; and also using vibration and air pressure to convey emotions. We present our developments with example broadcast content containing enhanced captions preceded by a detailed discussion of closed captioning and interactive television technologies.

2. Closed-captioning

Existing analog television captioning technologies display closed-captions with a single font and size, a limited set of colors, and four text styles: normal, underline, italicized, or flashing where normal is all capitalized text (some current decoders allow for lower case text). Further, captions display an opaque, black background to enhance readability. Closed-captions are intended for a deaf or hard-of-hearing viewing audience.

To address the need for complete expression of audio information, captions should also deal with speaker identification, speaker emotions, background sounds, and music. Presently, this is often accomplished by adding text descriptors and various punctuations to a caption. These are usually limited to single or double words, such as "(Man screams)," or "(Woman laughs)", and so on.

While some auditory information may also be derived from visual cues, auditory information may carry its own information. It might enhance the viewer's understanding, or even be in collision with the visual information being represented. A speaker may cry, but the crying may subside or become louder over a period of time. This quality of crying would be obvious to the hearing viewer. The deaf viewer would have to be satisfied with the two word descriptors or summaries. The changing tone of a debate is another example where the hearing viewer has additional information about conscious and unconscious motive and intent of the speaker. Background sounds and background music are also pieces of information that convey the director's attempts to move, set or enhance the mood, or even prepare the viewer for changes. Hawkins (1995) reported that captioning does not provide sufficient access to the non-speech audio elements. As a result of this missing information, the deaf viewer may have an unequal grasp of the content. However, the extent of this inequality remains to be determined.

One of the current limitations in providing expansive descriptors of paralinguistic and background audio data is the amount of reading a viewer can manage at one time. The average television show is captioned at 141 words per minute (Jensema, 1998). This ranges from children's shows that are edited for readability and display anywhere from 60 to 90 words per minute, to live news programs, which present captioned content at up to 255 words per minute (Jensema et al., 1996). Although conversations between people can occur above 255 words per minute, 255 words per minute is the practical limit that captions can be displayed on analog television receivers.

Jensema (1998) reported that people are able to read captions on television comfortably at 145 words per minute, and for short period of times, tolerate faster speeds of up to 170 words per minute. The average of 145 words per minute does not allow the captioner to add much in the way of descriptor content, such as speaker identification, audience reaction, emotional delivery of the speaker, and other non-speech information. Providing this additional content using the current text-based display and style may well overtax the viewer's capacity to comprehend the increased text-based bandwidth.

A number of studies have demonstrated that closed-captioning also offers many crossover benefits for people who are hearing. For example, people with reading or learning disabilities, people who are learning a second language, and people who want to improve literacy levels including young children learning a language can benefit from closed-captions for television programming (Parks, 1994). Closed-captions may also contribute to bringing television into the digital age where users expect to have searchable access to all media. Closed-captioned content of a television program will enable the creation of a text database of all of the spoken dialog. Clearly, closed-captioning is a premier example of universal design in action.

The limitations of the analog television and closed captioning systems may soon be eliminated with the advent of digital television. However, current approaches to digital television closed captioning (DTCC) only offer a number of enhancements, such as multiple text fonts, sizes, and some additional modifiers to the text display. These may not be sufficient to convey the descriptor content in a way that is helpful to the deaf and hard of hearing viewer.

3. Interactive and Digital Television

The television industry is undergoing the most significant revolution since the origin of the NTSC standard in 1941. Television is without exception the last partner in the communications industry to embrace digital delivery of its transmission signal. The United States (US) and Canada have only recently begun the process of converting to digital transmission. In the US, the FCC has imposed a deadline requiring that all over-the-air (OTA) broadcasts be sent digitally by 2002 (FCC, 1998) with over one hundred stations already broadcasting digital signals. In contrast, the CRTC in Canada has chosen not to impose a timetable for digital transmission, preferring to allow the broadcast industry to set its own pace.

The Canadian Association of Broadcasters (1999) and the Forrester report (Forrester Research Inc., 2000) have predicted that interactive and digital television will replace traditional production processes, business models and distribution of television programming. As a result, television will be undergoing a revolutionary change over the next five years. Television standards are being re-written and revised in an attempt to keep up with changes in technologies and the demands of the industry and customers. Accessibility is an important and rich aspect of interactive television because access solutions can be greatly improved over existing systems (e.g., traditional captioning, described video and input devices).

Current television closed captioning in North America relies upon an industry-agreed standard, EIA 608. This standard was requested by the FCC as part of its mandate under the TV Decoder Circuitry Act of 1990 (TVDCA) and resulted in the pivotal law requiring television set manufacturers to install caption decoders in all television sets of 13" or greater beginning in 1993 (Robson & Hutchins, 1998). This system can display at most 255 words per minute, and does not support multiple fonts, text sizes, multiple text styles, expanded colors, let alone 3-D and animated text, graphical or other displays for captioning.

Since the mid-1990's, the broadcast industry worldwide has agreed to adopt a digital transport stream for digital television based on MPEG-2. Video carried by MPEG-2 is substantially more efficient than analog NTSC multiplexing for television signal carriage and as a result more bandwidth is available within the broadcast channel for future data services.

In preparation for digital television, the FCC recognized the need for upgraded caption standards as well. Stating that people who are deaf and hard of hearing should also benefit from the advances in digital broadcasting technology, the agency requested a new standard to reflect the advantages that the new technology held for closed captioning. The resultant work, known as EIA 708, was released in 1997. The FCC has recently endorsed a subset of EIA-708 as a requirement for digital broadcasts beginning in July 2002 (FCC, 2000).

3.1   EIA-708 – Captioning for digital television.

EIA-708 supports a data rate of 9600 bits per second of captioning data using an 8-bit word (twenty times the previous bandwidth in a highly optimized environment). The extra throughput offered by this increase in bandwidth will effectively remove many of the restrictions of EIA-608, such as single font, single size, restricted styles, limited colours, and limited options for caption background and positioning. However, since the standard was introduced, there have been few research initiatives investigating ways of incorporating, let alone expanding on, these enhancements. There are also few available captioned examples that demonstrate even basic EIA 708 enhancements for broadcast content. Today, the broadcast industry seems content to concentrate on providing a means of converting EIA 608 analog captions into digital signals.

Of the few examples of research and development initiatives to address the issues of enhanced captioning, the most extensive work has taken place at the National Center for Accessible Media (NCAM/WGBH) where a number of digital TV concepts have been tested (NCAM, 2000) with a variety of users. Much of this work has informed the EIA 708 committee. In addition, NCAM/WGBH has explored the multi-channel capabilities of DTV and has created an accessible prototype of the Arthur™ program that offers a method for selecting American Sign Language, multiple captioning services, described video, and even hyperlinks.

4.   The Enhanced Captioning Project

One of the main findings and a motivation for the passage of the U.S. Television Decoder Circuitry Act of 1990 was that deaf and hard of hearing people should have the fullest access to television in order to achieve equity. We live in a society that transmits much of its information auditorily. Without access to auditory information, individuals cannot fully participate or be full, contributing members.

Research in the non-verbal communication field over the last 50 years has shown that in order to understand the full meaning that a speaker conveys, information other than actual spoken word is necessary (Dix et al., 1998). This information, called paralanguage, consists of speech sounds other than words, along with style delivery. Without paralanguage, plus environmental sounds and music, the experience of the deaf person relying on captions to understand a television program will differ extensively from that of a hearing person.

The Enhanced Caption Project is developing a model of overlay closed captioning that includes a set of word, graphical, and haptics-based modifiers for speakers, environmental sounds, and music as well as speaker identification. We have created an example that demonstrates the concept of enhanced graphical captions for a 5-minute opera, Il Menu, a spoof on opera produced for television by MarbleMedia Inc. Graphical information replaces the extended word based descriptions currently used in closed-captions. By using a graphical basis for displaying other critical sound-based content, we intend to maintain reading rates at between 145 to 170 words per minute identified by Jensema (1998) as an optimal reading rate range.

Eight different emotions have been identified for Il Menu: happy; cheerful; somber; sad; irritated; frustrated; angry; and furious. Five modifiers, volume; rate; pitch; rhythm; and accent, have also been identified to allow these emotions to contain additional semantics such as emphasis. The emotions and modifiers along with speaker identification are expressed using comic book conventions such as the creation of speech bubbles, expanding and shrinking text display, color, and iconic and animated representation. Standard speech bubbles styles are paired with each of the emotions, and their modifiers.

Although in this particular example environment sounds have not been included, they remain an important consideration for future work. Environment sounds may be used to emphasize a point of view, provide the tone of a dramatic statement, or enhance a scene by conveying additional semantics. However, these types of expressions are similar to the set of emotions we have listed and thus we believe our model will also apply to environment sounds.

The Il Menu video lends itself to illustrating a number of challenges which conventional EIA 608, text-based captioning and even 708 digital captioning with its enhanced set of fonts and text styles do not address. Il Menu is a complex piece of work, which the viewer with full hearing is at once aware of, and takes for granted, the many nuances and drama conveyed by sound. For instance, the various sounds of the kitchen offer rhythmic tension and lightness. The music, which wafts through provides comic relief. To capture the background sounds, we report the various interferences via text descriptors and iconic representations. Speech bubbles are used to convey emotion and speaker identification.

A second more complicated example occurs as the speaker passes a lobster tank whose air pump disrupts the speaker's voice. In Figure 1, animated bubbles collide with the spoken text. A small, text-based description explains

that the words are being blurred by the disruption of the pump. By employing the animated graphics, the deaf viewer is able to have a real time experience closer to that of the hearing viewer. In this instance and others that follow in the Il Menu project, the real-time auditory experience is converted to a real time graphical modality.

Figure 1: Environment sound of an air pump interfering with speech. White text is the closed-caption equivalent.

WebTV, an interactive television platform, is used to display the graphical overlay captions for Il Menu. WebTV requires a special piece of hardware attached to a television set, a set-top box, to display the extra information available with interactive television. The WebTV box receives line 21-based triggers and calls the actual graphic related to those triggers via its PPP Internet connection. The received graphics are then displayed over the broadcast image at their required times.

5. Future work and Conclusion

We are currently in the process of developing tactile equivalents for the emotions expressed in Il Menu. These equivalences will allow people to "feel" the emotions in Il Menu. Audience reaction and environment sounds will be the initial components expressed through the tactile device by activating a series of air and vibration points on a modified footrest. Additional descriptor text and graphical representation may be needed to clarify the tactile experience.

Figure 2: Tactile Foot Device

WebTV has many limitations that restricted the possibilities we could explore. For example 3-D text enhancements and vector graphics cannot be displayed and thus they were not used. Instead we had to rely on bitmapped graphics that take much longer to transmit which reduces the quality and complexity of the graphics used. WebTV used a PPP only connection. Broadband connection is currently not supported at this time although future support for broadband connectivity is expected with new generations of interactive television technologies. Finally, WebTV accepts a subset of Javascript that further limits the speed and scope of the graphical displays, affecting such things as the timing of caption displays.

As a result of the limitations of existing interactive television technologies, we plan to implement a version of the enhanced captions in a QuickTime environment. QuickTime is an industry standard which supports multiple media tracks, wired sprites, animation, and text display and it is implemented in the MPEG-4 standard (Deck, 1998). While QuickTime is not one of the underlying technologies in MPEG-2 compression standard, it offers many of the features that we expect will be implemented in future MPEG-2 technologies.

Television is ubiquitous and an important part of the cultural fabric of many societies. Access to television content means that people who are Deaf, Hard of Hearing or Deafened can participate in this aspect of our culture. In addition, video is a powerful tool to present ideas, curriculum and entertainment. Research in enhanced captioning will provide new and innovative solutions for creating accessible video material for applications such as education, training, and entertainment.

Acknowledgements

We would like to acknowledge the Ontario Ministry of Energy, Science and Technology through the Communications and Information Technology Ontario. We also thank Haylea Systems Inc. and MarbleMedia Inc. for their support in this project.

References

Canadian Association of Broadcasters (1999). *FuturePlan*. http://www.cab-acr.ca/english/joint/futureplan.html.

Canadian Radio-television and Telecommunications Commission (1998, 13 June). *Industry Canada Notice Smbr-002-98*. Canada Gazette, DTV (Digital Television) Transition Allotment Plan. http://www.cab-acr.ca/english/television/submission/sub_sept1198.htm.

Centre for Applied Special Technologies (CAST). (1997). *Understanding science through captioning*. http://www.cast.org/initiatives/captioning.html.

Deck. S. (1998). *ISO taps QuickTime for MPEG-4 multimedia spec.* http://careers.computerworld.com/home/online9697.nsf/all/980211iso1CC3E.

Dix, A. Finlay, J., Abowd, G., Beale, R. (1998). *Human-computer interaction*. Toronto: Prentice Hall.

Federal Communications Commission (FCC). (1998). Digital Television Consumer Information. http://www.fcc.gov/Bureaus/Engineering_Technology/Factsheets/dtv9811.html.

Federal Communications Commission (FCC). (2000). *FCC adopts technical standards for display of closed captioning on digital television receivers*. http://www.fcc.gov/Bureaus/Mass_Media/News_Releases/2000/nrmm0031.html.

Forrester Research Inc. (2000). *The end of TV (As we know it)*. www.forrester.com.

Harkins, J.E., Korres, E., Beth R. Singer, B.R., Virvan, B.M. (1995). *Non-Speech Information in Captioned Video: A Consumer Opinion Study with Guidelines for the Captioning Industry*. Gallaudet Research Institute.

International Conference on Consumer Electronics (ICCE) Proceedings. http://www.robson.org/gary/writing/icce98.html.

Jensema, C.J. (1998). Viewer reaction to different television captioning speeds. *American Annals of the Deaf*. 318-324.

Jensema, C., McCann, R., Ramsey, S. (1996). *Closed-captioned television presentation speed and vocabulary*, 141(4), 284-292.

King, C.M. (1996). *CAP-Media Web Site*. http://www.cap-media.com.

Krebs, P., Hammerquist, J., Kindschi, C., (2000). *Building Interactive Entertainment and E-Commerce Content for Microsoft® TV*. Redmond: Microsoft.

Lewis, M.S.J. (2000). Television captioning: A vehicle for accessibility and literacy. *On-line proceedings of CSUN*. www.dinf/org/csun_99/session0057.html.

Mahrabian, A. (1968). Communication without words. *Psychology Today*. 2(4). 53-56.

National Center for Accessible Media (NCAM). (2000). www.wgbh.com/ncam.

Parks, C. (1994). Closed Captioned TV: A Resource for ESL Literacy Education. *ERIC Digest*. http://ericae.net/ericdb/ED372662.htm.

Robson, G.D., Hutchins, J. (1998). *The Advantages and Pitfalls of Closed Captioning for Advanced Television*.

Multimodal Access to Vector Graphics on the Web by Computer Users with Print Disabilities

John A. Gardner, Vladimir Bulatov

Science Access Project
Department of Physics, Oregon State University
Corvallis, OR 97331-6507
John.Gardner@orst.edu, bulatov@dots.physics.orst.edu
http://dots.physics.orst.edu

1. Introduction

One of the last "frontiers of universal accessibility" is access by people with print disabilities to graphics. The WWW recommendation for making web graphics accessible is to include an ALT tag and, for complex graphics, a longer description that can be opened and read by people who cannot see or understand the information presented in the graphic. A description is hardly adequate access to such important graphical information as charts, flow diagrams, graphically-displayed data, maps, and illustrative diagrams used in virtually all scientific educational and professional literature. Indeed such graphics are excellent illustrations of the proverb that "a picture is worth a thousand words".

Tactile copies of graphical information have historically been used to provide access by blind people to graphics when word descriptions were inadequate. Until fairly recently, tactile graphics were laboriously created by hand. Consequently there was very little tactile information available to blind people. A tactile graphics overview [1] was published several years ago by John Gardner describing both hand-creation and more modern methods for creating tactile graphics. The recent introduction of the Tiger Tactile Graphics Embosser [2], promise a vast improvement in access to tactile graphics. Tiger can print tactile copies of essentially anything that appears on a computer screen. Tactile images from computers can also be made using capsule papers as described in the overview linked above.

Some graphical information (e.g. color and shading of minerals, plants, and animals) is intrinsically visual and cannot easily be represented tactually. However most abstract information, represented by line and block graphics (e.g. maps, graphs, charts, and diagrams is generally representable, at least in principle, in a way that is quite accessible to blind people. Unfortunately, most abstract graphic information that is currently available in electronic form, including most World Wide Web graphics, are not really useful if just printed in tactile form. Images of typical text are not readable, and much fine detail is tactually unrecognizable. Typical figures contain a wealth of detail that is far less comprehensible to the fingers than to the eyes.

A technique pioneered by Dr. Donald Parkes [3] has proven to be very effective in providing blind readers access to complex graphical information. The blind reader uses a tactile graphic placed on a digitizing pad or some other device that can act as a computer mouse. It is attached to a computer that has a file of information about that figure and can run an application that gives access to this information file.

A user can feel the image and query the computer about objects. The object information can be spoken by a speech engine or be displayed on a computer braille display. A great deal of information can be provided if the computer brings up a text box and permits the user to browse the text in audio or braille before moving on to other objects. This technique permits blind readers to read fairly sophisticated and detailed tactile information that would be impossible to comprehend without the computer interaction. It also permits access by blind people who do not read braille.

We have previously explored several special methods [4] by which information can be included with web graphics that permit specialized browsers to provide access to blind people, but all required special links on web pages to the "special" accessible information.

In the past we have discussed the desireability of new technologies for making "smart figures" [5] with native ability to label and describe objects. We believe that such "hidden" information has enormous potential well beyond the narrow field of accessibility by people with print disabilities. However it has only become possible recently to use a mainstream graphics method that can make smart figures a reality. When Scalable Vector Graphics (SVG) [6] becomes a W3C recommendation, web authors have all the capabilities needed to make mainstream smart graphics that are fully accessible to blind people. These smart figures should also provide greatly improved accessibility for many sighted people with print disabilities, e.g. dyslexia.

2. SVG Overview

SVG is an XML application. Unlike current web graphics SVG figures are coded in marked-up plain text and read by the client computer as a text file. As its name implies it is scalable and can contain a wealth of fine detail even when enlarged hundreds of times. SVG figures are displayed by a SVG viewer program downloadable at no charge. The most popular is the Adobe viewer [7]. Although SVG is a very powerful language capable of use for dynamic display, we use SVG only as a tool for making smart static graphics.

SVG defines an image using a set of geometrical primitives, which are organized into a hierarchical tree of objects that can be arbitrarily complex. SVG is well suited for displaying technical drawings, maps, diagrams, and other abstract informational graphics. SVG is not directly usable for photographic images, which are best represented by bitmaps. However photographs may be embedded into SVG documents.

Every SVG element may have two special sub-elements <title> and <desc> (description). These elements are optional and are not displayed by the graphics viewer. Their role is purely informational.

3. The Accessible SVG Viewer

We have created an alternative SVG viewer that permits users to access the hierarchical element tree of SVG graphics and the title and description fields for those elements. This Accessible SVG Viewer will be demonstrated in the presentation at the meeting. When complete, it will be downloadable at no cost and is recommended for users with severe print disabilities. It uses a combination of modern hardware and software to permit blind or dyslexic users very flexible access to the visual image as well as the underlying descriptive information.

Hardware components are:
- The Logitech WingMan [8] force feedback mouse (based on the force feedback technology developed by Immersion, Inc [9]). This commercially available compact device similar in shape and functionality to a standard mouse can create haptic effects that feel like textures, edges of plateaus, ridges, grooves, etc. It can also pull the user's hand to selected positions.
- Either the Tiger Tactile Graphics Embosser or a capsule paper process to create a tactile image of the SVG graphic.
- A digitizing pad such as the Intuos Graphics tablet from Wacom [10] or Touch Pad from EloTouch [11].
- Win98 or Win2000 PC computer.

The Accessible SVG viewer software is written in java version 1.2 with the following libraries/tools:
- IBM XML parser [12],
- SVG toolkit from CSIRO, Australia [13],
- Microsoft Speech Engine [14],
- Java foundation classes (swing) from SUN [15].

The Accessible SVG Viewer is a self voicing program. We initially attempted to use a general adaptive technology such as the self voicing Java kit from IBM. However, the general technologies do not work properly for this viewer, because they are intended to work with text in forms or dialogs only and they are not useful for voicing text that doesn't appear on screen.

The Accessible SVG Viewer's screen has two areas: a graphics area and element text based area for browsing the SVG image's element tree (Fig.1). A user can browse the list of graphical objects as a hierarchical list or tree similar to the windows explorer display. Both tree view and list view are used for hierarchy browsing. We found, that list view is more suitable for navigation by a blind user.

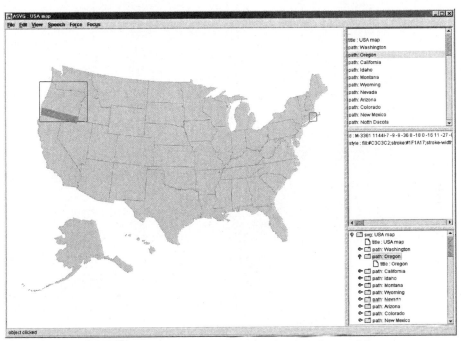

Figure 1: Overview of USA map with currently selected state of Oregon

4. Using the Accessible SVG Viewer

The Accessible SVG Viewer permits multi-modal access to the SVG graphics through visual, haptic, tactile, voice, and non-speech audio modes.

There are 3 modes for browsing the graphic image with the Wingman haptic mouse:

- feeling the whole picture,
- locating a selected object by having the mouse move to that object,
- feeling the outline of a selected object.

Haptic regime 1: feeling the whole picture
In this regime every object is represented as a small scratchy area in the center of the object. When the mouse pointer crosses an object's boundary, the object title is spoken. If the object has no title or description, a simple geometric description is automatically generated, e.g. circle, ellipse, rectangle, polygon, path, line.

Haptic regime 2: Locating a selected object
In this regime the WingMan mouse pulls the user's hand toward the center of a selected object. When the user selects another object from the SVG object list, the WingMan mouse pulls the user's hand toward the center of the other object.

Haptic regime 3: feeling the outline of a selected object
An object's outline is displayed haptically as a wall , which pushes the user's hand inside the polygon that approximates the shape of the object. A scratchy texture is felt inside the outline. This feedback tells the user that the mouse pointer is inside the selected object.

In our opinion, haptic access is of relatively limited usefulness for detecting subtle details, particularly of small objects. A zoom feature (Fig.2) permits users to magnify the image up to the size of the full screen for better haptic resolution.

318

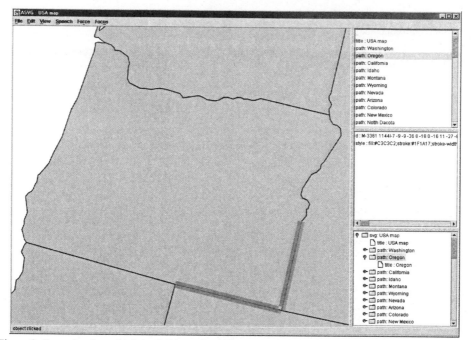

Figure 2: Zoom in view of selected object. Green bar in the bottom right corner of Oregon's boundary shows currently active haptic force field area

Access via tactile copy on a digitizing pad

A tactile copy of the whole image or any selected portion may be created by embossing on the Tiger or by using capsule paper. This tactile copy is placed on a digitizing pad and calibrated to the screen image by indicating two fiducial points. The digitizing pad is used to move the mouse pointer to an object. That object cand then be identified by a request to hear its label or description fielddisplayed in speech. The haptic mouse can follow the digitizing mouse movement if desired.

A user can locate a specific object on the tactile image by finding it in the object tree and asking the haptic mouse to go there. Alternatively, an audio guidance method allows a user to move around the tactile image until the object is located. A tactile image, supplemented by a haptic mouse, should provide excellent access for people who are blind or severely dyslexic.

5. Making SVG Graphics that are accessible

Graphics files in SVG format are currently exported by many popular graphics authoring programs including CorelDraw and Adobe Illustrator. These exported SVG documents typically have no <title> or <desc> elements. These documents are not really accessible, but even without object labels they contain far more information than bit map graphics. A visually impaired user can determine the overall structure of the image and the shapes of individual objects. Text elements in the document may be read as well.

However, <title> and <desc> elements can greatly increase the document accessibility. We have created a SVG editor that permits one to add title and description to elements if they do not have them or to edit them if they do exist. The modified document file may be saved in place of the original file. It is visually indistinguishable from the original.

An SVG document will be most informative if the elements have a correct hierarchy. Certain kinds of authoring tools, such as graphing software, flow diagram generators, chart creation tools etc. could certainly export SVG documents with good hierarchical structure and even with some label and description information. However good SVG object hierarchy is not automatically generated by every authoring program. For example, a graphics authoring program being used to create a US map cannot know whether islands are states or parts of nearby states unless the grouping can be indicated during authoring. Consequently it will occasionally be desirable to re-order the hierarchical structure of SVG documents to organize objects into proper groups. For example, a graphics program might export a US map with Staten Island as an object equivalent to a state or even as a sub-object of New Jersey. The authoring tool we are creating permits one to make it a sub-object of New York.

6. Conclusions

Static graphics in the form of SVG documents are far more accessible to visually-impaired users than bit map graphics. Many current graphics authoring applications permit one to export the document in SVG format, so it is already easy to author such graphics. A simple static SVG document can be made very accessible if objects in the document element structure are properly organized and are titled and labeled. We are creating a SVG editor that can be used to edit the structure, object labels, and object descriptions of existing SVG documents. Consequently, static SVG graphics can be easily created to be fully accessible or can be easily edited later to improve accessibility. Agencies legally obligated to make information universally accessible finally have a very easy way to make accessible graphics for electronic documents or web sites.

The Accessible SVG Viewer permits blind and dyslexic users to utilize the document structure and view the title and descriptions of document objects. The current Viewer is accessible only in audio, but braille access is easily achievable once a stable braille api (application programmer interface) is defined and supported by braille display manufacturers.

Acknowledgements

This research was supported in part by the National Science Foundation.

References

[1] John A. Gardner, Tactile Graphics, An Overview and Resource Guide
     http://dots.physics.orst.edu/tactile/tactile.html
[2] TIGER Advantage Tactile Graphics and Braille Embosser. http://www.viewplustech.com/products.html
[3] D. Parkes "Nomad": an audio-tactile tool for the acquisition, use and management of spatially distributed information by partially sighted and blind persons". Editors. A. F. Tatham and A. G. Dodds. Proceedings of the Second International Symposium on Maps and Graphics for Visually Handicapped People. King's College, University of London, April 20-22 1988, pp. 24-29
[4] John Gardner, Vladimir Bulatov. Accessing Maps, Diagrams, and similar Object-Oriented Graphics. http://dots.physics.orst.edu/graphics.html
[5] John A. Gardner, Hadi Bargirangin, Vladimir Bulatov, Hans Kowallick, and Randy Lundquist. *The problem of accessing non-textual information on the web. Proceedings of World Wide Web Conference. Santa Clara, 1996.*
     http://dots.physics.orst.edu/publications/www6.html
[6] W3C Scalable Vector Graphics. http://www.w3.org/Graphics/SVG/Overview.html
[7] SVG: Scaleble Vector Graphics. http://www.adobe.com/svg/main.html
[8] WingMan - force feedback mouse. http://www.logitech.com/
[9] Immersion, Inc. http://www.immersion.com
[10] Intuos Pad from Wacom Inc., http://www.wacom.com
[11] TouchPad. EloTouchInc. http://www.elotouch.com
[12] XML parser. http://www.alphaworks.ibm.com
[13] SVG toolkit, http://www.cmis.csiro.au/svg
[14] Microsoft Speech SDK http://www.microsoft.com/speech/
[15] Java foundation Classes http://www.javasoft.com

320

Enhanced Facial Feature Tracking of Spontaneous and Continuous Expressions

Amr Goneid, Rana Ayman El Kaliouby

Department of Computer Science, The American University in Cairo, Egypt
goneid@aucegypt.edu, ranak@aucegypt.edu

Abstract

The integration of multimedia technologies into mainstream computing have both raised the user's expectations of computer interfaces, and made possible the development of multi-modal emotionally intelligent systems. The true strength of facial expression recognition (FER) shows when seamlessly integrated into emotionally intelligent systems enabling applications to add facial expressions to traditional input modalities. FER systems must operate on natural human expressions in a real time environment. This paper presents efficient methodologies for tracking spontaneous facial expressions that are continuous in time. A model for natural facial expressions is developed, within which we propose three feature point tracking mechanisms that operate efficiently within that model; the adaptive facial feature tracking methodology developed to improve tracking results while decreasing the total number of computations, the predictive feature point tracking mechanism well-suited to large motions, and the non-exhaustive sum of squared differences (SSD) technique, an optimization on the SSD algorithm on which many tracking methodologies are based. Performance analysis carried out on different image sequences showed that adaptive and predictive mechanisms used in conjunction with the non-exhaustive version of the SSD yield up to a 1-10 increase in time efficiency, whilst improving tracking accuracy. These results help further the development of real-time natural facial expression recognition systems.

1. Introduction

The integration of multimedia technologies into mainstream computing have both raised the user's expectations of computer interfaces, and made possible the development of multi-modal emotionally intelligent systems that are able to process external emotional cues. Humans interact with their machines using facial expressions in the same way they interact with each other (Reeves and Nass, 1996), thus FER systems should be seamlessly integrated into emotionally intelligent systems by operating in a real time environment that supports natural human expressions.

Natural human facial expressions can be either spontaneous or deliberately expressed (Lisetti and Schiano, 2000), each proceeding in separate neural tracks. Whereas automated FER systems must take into account both scenarios of emotion expression, most of the attempts in the field have been confined to the recognition of deliberate expressions denoted by an image sequence expressed intentionally by a human actor (Tian, Kanade, and Cohn 2001). This bias towards deliberate expressions may be due to the fact that they are relatively easier to devise, experiment, and analyze when compared to spontaneous expressions. In addition, current research manipulates facial expressions as a discrete sequence of disjoint events, mastering the extraction and recognition of a single expression (Yacoob and Davis, 1994), (Lien, 1998), (Donato, Bartlett, Hager, Ekman, and Sejnowski, 1999), (Tian, Kanade and Cohn, 2001). Expressions are often likened to separate words in speech recognition (Picard, 1997), although studies of day-to-day computer usage patterns by Ehrich et al. (1998) show that expression boundaries (currently either manually separated or marked by a neutral frame) do not truly exist in spontaneous expressions (Lisetti and Schiano, 2000). Spontaneous expressions are perceived continuously rather than categorically, and unlike posed ones do not follow a pre-defined template responding instead to the user's emotional state. The disjoint model of facial expression oversimplifies the process of understanding human expressions, and rules out the possibility of doing any contextual processing.

Furthermore, we find that current facial feature tracking methodologies are fine-tuned to operate within models of disjoint facial expressions. Facial feature tracking methodologies frequently use the Sum of Squared Differences algorithm (SSD) (Lucas, 1984) to track facial features across image sequences (Lien, 1998). The SSD algorithm has the drawback of being exhaustive, largely affecting the number of computations performed.

Tracking methodologies also assign feature points constant search and feature window sizes although different feature points experience different motion magnitudes. The two parameters largely affect both efficiency and tracking accuracy, and while mechanisms try to find an optimal search window size, too small search windows can result in undersized search windows, and excessively large search window sizes can result in an overmatch.

Based on the above limitations of posed models of facial expression and tracking methodologies, we propose a model that more accurately captures natural human-machine interaction, and accordingly propose an enhanced facial feature tracking methodology that optimizes efficiency and tracking accuracy.

2. Natural model of facial expression

2.1 Model Characteristics

The model is built on spontaneous facial expressions because they are more accurate descriptors of human-computer interaction. Expressions occur continuously in time, contrary to the posed model of expressions where expressions occur disjointedly. While existing research has been concerned with defining expression boundaries (Otsuka and Ohya, 1996), we argue that facial expressions are a continuum in time that cannot be divided into disjoint expressions without losing valuable contextual information, and do not necessarily start or end with a neutral frame. In addition, an image sequence is continuous in time (neutral frames are not necessary to separate expressions) and can span any number of continuous expressions.

2.2 Model Parameters

We define the parameter set for a natural facial expression model λ as $\lambda = \{N, k, M, \{\Delta m(f_i)\}, \{\Delta m(n_i)\}, \{F_i\}, \{S_i\}\}$, where
- N is the number of frames in an image sequence.
- Frames are sampled at a constant rate k; no human intervention is required to separate expressions.
- M is the number of feature points tracked.
- $\{\Delta m(f_i)\}$ represents the motion magnitude for feature point f_i, across two consecutive frames, and is defined as the average velocity of the feature point between frames (t) and (t-1).
- $\{\Delta m(n_{t-1/t})\}$ represents the motion magnitude for every two consecutive frames t-1 and t in the sequence. For frames t-1 and t, motion is the average velocity of all the feature points between the two frames.
- $\{F_i\}$ defines the size of feature f_i for all feature points.
- $\{S_i\}$ defines the search window size for feature f_i, within which the point is tracked.

Accordingly, this paper proposes the adaptive facial feature tracking methodology, the predictive feature point tracking mechanism, and the non-exhaustive sum of squared differences technique as enhancements to current facial feature tracking methodologies, designed to operate within the natural model of facial expression.

3. Adaptive facial feature tracking

Finding the optimal search and feature window size is critical to the efficiency of the feature-tracking algorithm as this largely affects both the number of computations and the tracking accuracy of the results. Feature points in spontaneous expressions exhibit smaller motion magnitudes indicating that smaller search windows could be used to reduce the number of computations. Conversely, too small search window sizes can result in undersized windows whereas large search window sizes can in addition to increasing the number of computations, result in an overmatch. Undersized search windows occur when the new location of a feature is outside the bounds of the search window, and happens in two cases: the first is when the search window size is very close to the feature window size (i.e. the relative difference in size is small), and the second is in the case of large feature motion. An overmatch for feature point f_i at frame(t) on the other hand, occurs when the SSD algorithm returns a "best match" at frame(t+1) that has the same gray value distribution as the feature point but is very far off the feature point's new location. This occurs when the search window is too large relative to the feature point.

The natural model of facial expression however, has shown that feature points exhibit different motion patterns across frames in an image sequence. Based on this fact, the adaptive facial feature tracking methodology assigns different feature and search window sizes to feature points depending on their motion magnitude. The technique assigns the smallest possible search window sizes to begin with to all feature points because the spontaneous model of facial expressions assumes that expression sequences are slow to start with. Search window sizes are then adjusted as feature points start experiencing larger motions across consecutive frames. As shown in Figure 1 if the algorithm hits any of the search windows' edges in search for a best match, the algorithm expands the search window size in that direction and re-tracks.

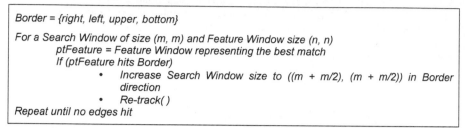

Figure 1: Adaptive Facial Feature Tracking Algorithm

4. Predictive feature point tracking

The predictive feature point-tracking algorithm uses the feature point acceleration to predict its displacement. The algorithm first calculates v_i, v_f the average velocities for the intervals [t-2, t-1] and [t-1, t]. These are used to compute the average acceleration $\mathbf{a}$. The feature point's displacement $\mathbf{d}$ (across frames t and t+1) is computed using the acceleration $\mathbf{a}$, and velocity v_f ($\mathbf{d} = v_f + \frac{1}{2}\,\mathbf{a}t^2$). Finally, the feature's location at frame (t) and displacement $\mathbf{d}$ is used to predict the new location at frame (t+1). At frame (t+1) the search window is centered on the feature point's predicted location, and the SSD algorithm begins its search.

5. Non-exhaustive sum of squared differences

The SSD algorithm determines the feature point's presumed position at frame (t) by computing the sum of squared differences between pixels in frame (t) and frame (t-1) over a specified region (search window), returning the minimum SSD as the best match.

The Non-Exhaustive Sum of Squared Differences we present here is an optimization on the sum of squared differences. Two optimizations are developed to reduce computations before the minimum SSD is found: the minimum crossovers and the thresholding technique. The former terminates SSD computations that have crossed the current SSD value, while the latter imposes a minimum threshold value on SSD values.

5.1 Minimum Crossovers

The minimum crossovers technique checks whether the computation in progress has exceeded the current minimum SSD value (i.e. is not a best match), in which case the current computation is terminated. Terminating unnecessary computations early reduces the total number of computations performed before the algorithm converges to a minimum value. Moreover, because the algorithm discards SSD values that have passed the current minimum SSD value, the optimization has absolutely no effect on tracking accuracy; it returns the same result set as that returned by an exhaustive SSD algorithm in all cases.

5.2 Threshold Values

One of the prominent characteristics of the SSD algorithm is that it magnifies mismatches, but returns a minimum value for close matches. Perfect matches (for which the SSD is 0) are recurrent across frames in which there is minimal motion of feature points. In consecutive frames where motion is not minimal the minimum SSD value is not by necessity zero.

If we define the minimum SSD value for feature point f_i (with gray value distribution of $I(f_i)$) to be SSD_{min}, the increase in SSD_{min} is proportional to the change in gray value distribution of f_i. Imposing a threshold level for SSD_{min} enables us to terminate the algorithm when the threshold value is encountered. This reduces the number of computations since it eliminates the need for exhaustive searches. Nevertheless if a threshold level above SSD_{min} is set, the algorithm returns imprecise locations for the tracked feature points.

6. Experimentation and results

Performance analysis has been conducted on 90 different image sequences of the Cohn-Kanade database for facial expression analysis (Kanade, Cohn, and Tian, 2000). Out of the total 33 action units (AUs), 15 were tracked (4 upper face AUs and 11 lower). Two testing criteria are used, efficiency (determined by the number of computations) and tracking accuracy (the tracked location is compared to the new location of the feature point, determined visually).

The results show that adaptive facial feature-tracking algorithm is particularly suited for spontaneous expressions that occur continuously in time as they are optimized for smaller search window sizes to start with, expanded on demand according to feature and frame motion factors. Figure 2 (left) shows the effect of the adaptive window approach on the number of computations and tracking accuracy (error coefficients) versus non-adaptive facial feature tracking given various window search sizes and a fixed feature window size of 10. The figure shows that efficiency increases for lower search window sizes, and larger search window sizes improve accuracy but efficiency increase decreases. This is natural because using large search window sizes is equivalent to adopting a non-adaptive approach - our reference for measuring the results – and hence approach equivalent performance results. The adaptive technique achieves over a 20% increase in efficiency over the non-adaptive version for equal accuracy results, and is capable of accommodating large feature point motion.

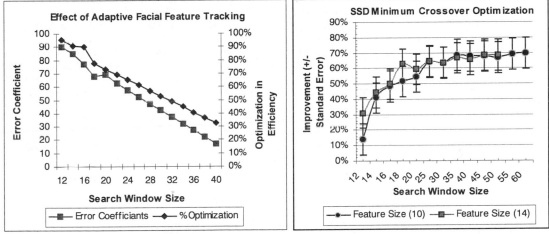

Figure 2 The effect of adaptive facial feature tracking on efficiency and accuracy (left)
and minimum crossover on efficiency (right)

The predictive feature point-tracking algorithm does not change search and feature window sizes and hence has no effect on efficiency, but is particularly suited to large motion. However, the algorithm has the potential problem of accumulating error largely because the locality of space is not preserved. When a wrong prediction is made about a feature points new location (either in terms of magnitude or direction), the probability of recovering that error prediction is lower than that produced by the adaptive algorithm and current methodologies.

Finally, Figure 2 (right) shows the percentage optimization obtained for difference search window sizes when using the non-exhaustive SSD relative to the exhaustive version using two different feature window sizes. As the search window size increases, the percentage improvement largely increases; the percentage improvement in number of computations for search window sizes over 18 (for both feature window sizes) is above 50% (sometimes as good as 70%), i.e. efficiency is more than doubled on any platform.

While the minimum crossover algorithm does not affect tracking accuracy in anyway, the threshold technique has the potential of degrading tracking accuracy. Thresholds can provide an average of 5% optimization in number of computations whilst maintaining 100% accuracy. The SSD optimization largely improved the efficiency of the algorithm and its effect on the number of computations can be largely seen and extrapolated to longer sequences.

7. Conclusions

In this paper we present the natural model of facial expression that operates on spontaneous expressions that are continuous in time. In contrast to posed models of facial expression where expressions are deliberate and disjoint, spontaneous expressions are the natural way with which humans interact with computers, and continuous expressions allow multimodal emotionally intelligent interfaces to fully utilize the input expressions without losing valuable contextual information. Inefficiencies inherent in current optical-based facial feature tracking methodologies, which are fine-tuned to operate within posed models of facial expression, are pinpointed.

An enhanced facial feature tracking methodology is presented to operate within a natural model of facial expression. The adaptive facial feature tracking methodology employs adaptive feature and search window sizes for different feature points across frames in an image sequence, based on feature point and frame motion parameters. The adaptive approach largely improves tracking accuracy and efficiency by growing the window size in the direction of the point movement. The predictive feature point-tracking algorithm predicts the new location of feature points based on their current acceleration magnitude and direction, and is particularly suited to large motion. The algorithm does not change the parameter values of the model (search/feature window size) hence produces the same number of computations as the optical-based tracking methodology, but has the problem of accumulating error; a problem tackled in future research. Another possible enhancement would be to use higher order predictors to better predict feature point displacements. The non-exhaustive optimization of the sum of squared differences algorithm improves the time efficiency of the algorithm while maintaining the same accuracy levels provided by the exhaustive version. Although the resolution for motion estimation used was 1 pixel, sub-pixel motion estimation should be considered in future work.

On the whole, the adaptive and predictive approach to feature point tracking, used in conjunction with the non-exhaustive version of the SSD has yielded a 1-10 increase in efficiency, whilst improving tracking accuracy. The results help further the development of real-time natural facial expression recognition systems and further research includes fine-tuning the predictive mechanism to solve the cumulative error problem, and doing more performance analysis on spontaneous expressions.

Acknowledgements
We would also like to express our gratitude to Dr. Jeffrey Cohn at the University of Pittsburgh who was kind enough to grant us access to the Cohn-Kanade Facial Expression Database, which we have used heavily throughout our research.

References
Donato, G., Bartlett, M.S., Hager, J., Ekman, P., Sejnowski, T. (October 1999). Classifying Facial Actions. *IEEE Transactions on Pattern Analysis and Machine Intelligence,* 21(10), 974-989.

Kanade, T., Cohn, J., Tian Y. (March 2000). Comprehensive Database for Facial Expression Analysis, In *Proceedings of International Conference on Face and Gesture Recognition.*

Lien, J. (1998). *Automatic recognition of Facial Expressions Using Hidden Markov Models and Estimation of Expression Intensity.* Doctoral dissertation, Technical report, CMU-RI-TR-98-31, Robotics Institute, Carnegie Mellon University.

Lisetti, C., Schiano, D. (2000). Automatic Facial Expression Interpretation: Where Human-Computer Interaction, Artificial Intelligence and Cognitive Science Intersect. *Pragmatics and Cognition (Special Issue on Facial Information Processing: A Multidisciplinary Perspective),* 8(1), 185-235.

Lucas, B. (1984). *Generalized Image Matching by the method of Differences.* Doctoral dissertation, Technical Report, CMU-CS-85-160, Carnegie Mellon University.

Otsuka, T., Ohya, J., (1996). Recognizing Multiple Persons' Facial Expressions Using HMM Based on Automatic Extraction of Significant Frames from Image Sequences. ATR Media Integration & Communications Research Laboratories, Seika-cho, Soraku-gun, Japan.

Picard, R. (1997). *Affective Computing.* Cambridge, MA: MIT Press.

Reeves, B., Nass, C. (1996). The Media Equation: How People Treat Computers, Television, and New Media Like Real People and Places. Cambridge, MA: Cambridge University Press.

Tian, Y., Kanade, T., Cohn, J. (2001). Recognizing action units for facial expression analysis. *IEEE Transaction on Pattern Analysis and Machine Intelligence,* 23(2).

Yacoob, Y., Davis, L. (1994). *Recognizing Facial Expressions by Spatio-Temporal Analysis. Technical Report CAR-TR-742.* Center for Automation Research, University of Maryland, College Park.

Bringing Gaze-based Interaction Back to Basics

John Paulin Hansen, Dan Witzner Hansen, Anders Sewerin Johansen

The IT University of Copenhagen, Glentevej 67, 2400 Copenhagen NV, Denmark[+]

Abstract

This paper argues for a joint development of an eye gaze-based, on-line communication aid running on a standard PC with a web-camera. Tracking software is to be provided as open source to allow for improvements and individual integrations with other aids. The interface design shall be defined by the achieved resolution of the tracking system. The design of a type-to-talk system with 12 large on-screen keys is described in the paper. In order for gaze tracking systems to become widely used, the strive for mouse-pointer precision should be replaced by a focus on the broad potentials of low-resolution gaze-based interactive systems.

1. Introduction

The ability to express oneself fast and effectively in a precise and extensive language is fundamental to the quality of life. Some people with special needs are not able to perform interpersonal communication fluently. Producing simple, short sentences may take several minutes, depending on their specific handicap and the communication equipment in use. Conversation tends to degrade towards a few standard phrases and "yes"/"no" responses to questions.

Input mode, input speed, learn-ability and mobility of the communication aid are often tradeoffs. For instance, a single-switch selection of characters or words in a typing system that scans stepwise through the rows and columns of a grid board, is slower than mouse-pointer selection on a full on-screen keyboard. It may however, be the only feasible input mode for the user. Morse code is a most efficient communication system, but the initial learning curve may be too steep to keep people motivated. Gaze-based interaction is fast and allows for dwell-time selection on standard graphical user interfaces, but most of the tracking equipments available are so sensitive to small changes in the head position or ambient light, that they can only be used in a totally controlled environment.

In addition, fatigue and medical prognoses have to be considered. Some people have to give up their preferred communication tool because of repetitive strain injury or because they eventually loose the motor control necessary for the input mode. In the end, the prevalence of a communication aid will depend on external factors like distribution form, system support and the overall system price.

The following scenario was made up by our development team to keep focus on usability:

> A person wakes up in a hospital after a traffic accident, paralyzed and unable to speak. In front of her is a mobile computer with a pen size web-camera. When she looks at a large moving button at the screen, the screen gradually turns blue. When she looks away from the button, it turns white again. She realizes that she is actually controlling the computer by her eye movements. She then fixates the moving button and holds it down for almost 30 seconds, until the screen turns all blue. The interface then changes to a board with 10 large letter buttons and a text field. In the text field, she reads the sentence: "By looking at the characters shown below, you can type a message. By looking at the dot button, your message will be spoken loud. Try to look at the dot button." She looks at the dot button, and a synthetic voice says: "By looking at the characters shown..." etc. When finished, the next message comes up: "Now try to spell your own message by looking at the letters you want to type". A nurse has been warned by the sound of the synthetic voice, and enters the room. "What has happened?" the woman types by moving her eyes around on the computer screen.

[+] This work is supported by the Danish Ministry of Information Technology and Research. Professor Kenji Itoh, Tokyo Institute of Technology, and Henning Dahl, The Danish Muscular Dystrophy Association, are also members of the research team. The team can be reached at gaze@it-c.dk.

The scenario is meant to keep three things in mind: The communication aid should be self-explanatory, the communication aid should have text-to-talk facilities, and calibration of the eye tracking should be embedded in the interaction (i.e. by moving the button).

Conversations with persons who have late-stage ALS (Lou Gehrig´s disease), neurologists, communication specialists, and caretakers gave us some clear indications on the priorities that should be given to system facilities. Synthetic and/or digitized voice output, fast word typing and e-mailing was consistently regarded as most important, while outdoor use and "control of all Windows commands" were regarded among the least important. High priorities where also given to robustness and forgiveness. Therefore it was decided that the first version of the system should be a stand-alone type-to-talk-device that can send and receive e-mails. This version is expected to be available for free download in a Danish, English and Japanese version by year 2003.

2. Eye tracking

Most people with special needs often have a superior control of eye movements compared to other muscles. In the 1980's, real-time eye-gaze tracking emigrated from military research to user interfaces for people with special needs (especially quadriplegics and ALS-patients). Chapman (1991) developed a system from which the user can select one of several controlling screens to typewrite, place a telephone call, control a TV-set, call upon the caretaker, play games etc. All selections are effectuated after the user has gazed at the selection for a latency period. Chapman (ibid.) reports that 75% of locked-in syndrome patients can significantly benefit from using their eye-gaze system.

Frey et al. (1990) implemented a similar system "ERICA" (Eye-gaze Response Interface Computer Aid). Six large on-screen keys were used instead of an entire keyboard due to limited resolution of the eye tracker. A character-prediction algorithm decreases eye-typewriting time by 25%.

In the past 10 years, several new eye tracking systems for people with special needs have been introduced. Some of the systems are almost totally unobtrusive and allow for (small) head movements. Most of the systems measure the reflection of light that is shone onto the eye. Typically, infrared (ir-) light is used in order to minimize distraction of the user and to avoid interference from other light sources. The precision achieved is usually 0.5 - 1.5 degree. This enables the user to control a standard graphical user interface, clicking by dwell-time activation or slow blinks. However, the strive for mouse-pointer precision through the use of ir-light introduces several drawbacks:

1. A large portion of the users is not able to get a sufficiently good calibration. This may be due to false reflections from glasses, occlusion by eyelids or eyelashes, interference with the ambient light, or low contrast between iris and the pupil. For instance, Schnipke & Todd (2000) were only able to get acceptable eye track data from 6 out of 16 subjects, using an advanced remote tracking system.
2. Systems that do not track head movements in addition to eye movements require that the user does not move, especially during calibration. While this might not be a problem for totally paralyzed people, it is very difficult for most people and impossible for people with involuntary, e.g. spastic, movements.
3. The use of ir-light sources and ir-cameras increases the overall system price considerably and requires installation by technical experts.
4. People's eyes tend to dry out due to the large amount of ir-light shone into them. This is very uncomfortable during continued use. "It fries my eyes", one of our subjects complained.
5. Present remote ir-tracking systems are only for stationary and indoor use, due to the low tolerance in the geometries of the tracking algorithms.

A non-ir system introduced recently by the company "Seeingmachines" is an exception to these constraints. The system uses two ordinary web-cameras to achieve a precision of 3 degrees. Tracking is done by visible light, and the system degrades gracefully in vibrating environments like automobiles. However, the expected market price (25,000 US$) is prohibitive to most people with special needs. Other systems have achieved similar precision with visible light (Pomerleau & Baluja,1993).

The algorithm developed by our team does real-time eye and head tracking in visual light. We are currently estimating a precision of approximately 4 degrees. Other developers may improve the tracking method when we release the system as open source by year 2003.

3. Design of a "large key" type-to-talk system

With an estimated tracking precision of 4 degrees, the interface will have to use larger-than-normal on-screen keys. The first version is designed on a 4 x 3 button grid. Further improvements of the tracking precision may consequently allow for smaller (and thus more) on-screen keys, or it may allow the system to be used in unstable environments (e.g. mounted on a wheelchair).

The limited amount of keys increases the number of strokes needed to type words. Recent developments in text entry methods for mobile phones and palm tops have introduced new ways to deal with the problem of a narrow input bandwidth. For example, an expert user of the so-called "T9 input method" on a 12-key mobile phone keypad can produce 33 words per minute (wpm) and 21 wpm using a multi-press method (1 word = 5 characters, including space) (Silfverberg et al., 2000).

Our system is a hybrid of several different methods. The basic version for novice users consists of a simple alphabetic typing system. Each character is entered by two strokes. After a few hours of use, a probabilistic character layout strategy (Bellman & MacKenzie, 1998) is introduced to the user. The 6 most likely next characters are shown on separate keys every time a character has been entered. In the advanced setting, introduced after several hours of use, one of the buttons displays the 6 most likely words predicted from the previous words in the sentence. When the user enters a word that is not already in the dictionary, it is added to the dictionary.

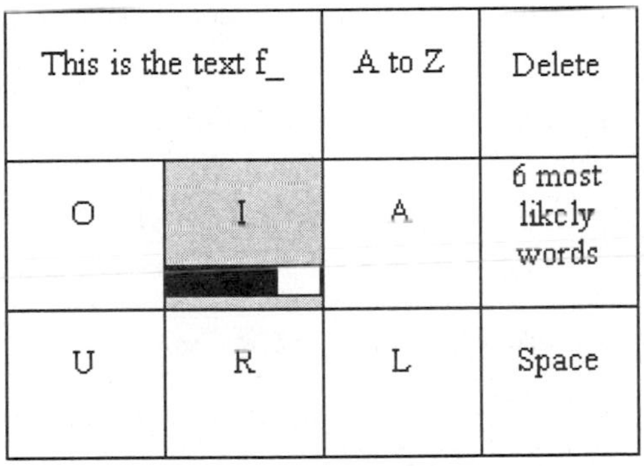

This is the text f_	A to Z	Delete
O	I	A	6 most likely words
U	R	L	Space

Fig. 1: Layout of gaze typing system

Keystrokes can be made by either clicking the mouse (or some other single switch) or by dwell time activation. A progress bar provides simple feedback to the user on the remaining time before a key is activated (c.f. the "I"- key in fig. 1).

Longitudinal testing of the system has yet to be done with real subjects in a daily life setting. Initial experiments with our prototype designs have been done using an ir-based tracking system. The experiments indicate that novices typically produce 4 words per minute. Once people get used to the probabilistic character layout and takes advantages of the "6 most likely words", performance increases up to 10 words per minute for novel sentences. Intensive use of the "6 most likely words" is possible for common expressions. In this case, we have observed input rates from 10 wpm to 20 wpm.

In order to increase the input speed of trivial expressions, we are currently working on refinements of the word predictor. To increase typing speed of novel sentences, we work on a number of minor improvements. We have estimated that a successive decrease of the dwell time from a start-up level at e.g. 750 ms to a working level of 500 ms (subtracting 1 ms each time a word is entered) will allow the user to type in one extra word per minute. If the user is able to click (using a mouse or a single switch), dwell-time activation can be avoided. This may increase typing speed with approximately one word per minute. Finally, cues like colour coding or increased font size may facilitate a more efficient visual processing of positions by peripheral vision.

4. Conclusion and future work

Our user research indicated that future communication aids should be on-line. The lack of physical mobility and the difficulties in using phones obviously increases the wish to be tele-present. This will allow users to communicate with family, friends and acquaintances and to join chat communities. In the future, protocols for short-range data exchange with external devices (like "Blue Tooth") may empower the user to control his wheelchair, open doors, operate elevators, turn on television, etc.

The development team has come up with the following user scenario to keep this perspective:

> Year 2010: A person is approaching you in his wheelchair. A mobile computer with a built-in camera is attached to his chair. The person is controlling his wheelchair by simply looking in the direction he wants to drive. Within talking distance, he introduces himself: "Excuse me, my name is John Smith. I am very thirsty. Would you mind assisting me for a second?" he asks, with a natural voice coming fluently out of the computers loudspeaker. He starts operating the nearby vending machine by looking at a graphical image of its buttons on his screen. Withdrawal is done automatically from his account. "Could you please pick up one of my straws? They are in the bag hanging on the back of my chair. Look at my screen and I will show you where they are." He starts a pre-recorded video sequence, and freezes an image of the opened bag. With a cursor following his gaze, he points at a pocket in the bag. You open the bag, find the straw in the pocket, and put it in the can and in his mouth. He obviously must have had the bag all filmed by his caretaker before leaving home this morning, you realize.

Gaze-based interaction has several application areas besides assistive technologies. It may supplement traditional manual mouse pointing (Zhai et al., 1999), it may replace the television remote control, it may be used for outdoor information kiosks, it may be used to operate head-up displays in vehicles, and may even be used in a Morse code manner to communicate with extremely small displays, using off-screen targets (Isokoski, 2000). In this sense, design of gazed-based interaction systems benefit "ordinary" users when designed for "extraordinary" users (Newell & Cairns, 1993).

In conclusion, gaze-based interaction holds potentials for becoming a standard input method if the noise sensitivity of the tracking systems decreases and their accessibility increases. Should this so happen, it may have a tremendous positive impact on the daily life of lots of people with special needs.

References

Chapman, J.E. (1991). Use of an eye-operated computer system in locked-in syndrome. *CSUN's Sixth Annual Int. Conference: Technology and Persons with Disabilities*, Los Angeles.

Frey, L.A., White, jr., K.P., Hutchinson, T.E. (1990). Eye-gaze word processing. *IEEE Transactions on Systems, Man, and Cybernetics, 20*(4), 944-950.

Schnipke, S.K., Todd, M.W. (2000). Trials and Tribulations of Using an Eye-tracking System. *Proc. ACM CHI '2000* (pp. 273-274).

Pomerleau, D., Baluja, S. (1993). Non-Intrusive Gaze Tracking Using Artificial Neural Networks. *AAAI Fall Symposium on Machine Learning in Computer Vision, Raleigh*, NC.

Silfverberg, M., MacKenzie, I.S., Korhonene P. (2000). Predicting Text Entry Speed on Mobile Phones. *Proc. ACM CHI '2000* (pp. 9-16).

Bellman T., MacKenzie, I.S. (1998). A Probabilistic Character Layout Strategy for Mobile Text Entry. *Proc. Graphics Interface '98* (pp. 168-176).

Zhai, S., Morimoto, C., Ihde, S. (1999). Manual and Gaze Input Cascaded (MAGIC) Pointing. *Proc. ACM CHI '99* (pp. 246-253).

Isokoski, P. (2000) Text Input Methods for Eye Trackers Using Off-Screen Targets. *Proceedings of Eye Tracking Research and Application Symposium 2000* (pp. 15-21).

Newell, A.F., Cairns, A.Y. (1993). Designing for Extraordinary Users. *Ergonomics in Design*, October 1993, pp. 10-16.

Word recognition for all: application to speech training

Marie-Christine Haton, Jean-Paul Haton

Université Henri-Poincaré, Nancy 1
LORIA BP 239, F-54506 Vandœuvre-lès-Nancy, France
+33 3 83 59 20 00
mchaton, jph@loria.fr

Abstract

This paper deals mainly with speech reeducation of deaf people at word level. A related domain is also concerned, i.e., foreign language teaching for hearing persons. The main problem in both cases is that speech signal is captured in an approximate way, so that important information can be lost. We present a set of lessons that have been developed and tested for the training of isolated words. These lessons extensively use automatic speech recognition algorithms based on global and analytical Hidden Markov Models (HMMs).

1. Introduction

This paper is concerned with speech training at the level of words, in the context of speech rehabilitation of verbal and written communication disorders. Deaf children are mainly concerned with this education level, since they perceive the oral language through the filter of their deafness. A hearing person perceives a foreign language in a similar way; language teaching is thus also an application field for this work [4]. In both cases, the speech signal is captured in an approximate manner, and some important information is not perceived. The work presented in this paper is the continuation of the SIRENE system that we have developed over a long period [3]. The basic idea is to use an automatic system to help a trainee. This has already been proposed in language teaching for hearing persons (see for instance [6]). Its implementation for hearing-impaired persons poses specific problems.

Enriching the vocabulary of a trainee does not only mean to orally present a new word. It also implies to introduce the underlying concept in various real situations [2]. Metalinguistic consciousness will then complete the speech education when the child reaches an acceptable maturity. That will lead him to play with words. For reaching this goal, we have developed a set of lessons that are described below. The validation phase of these lessons is also presented and discussed.

The paper is organized as follows. Our HMM-based speech recognition engine, called ESPERE, is first briefly presented. Then, we introduce the set of lessons that were developed for training at word level and we present the various aspects of the validation of these lessons with trainees. Further improvements and a conclusion terminate the paper.

2. The speech recognition system

ESPERE (Engine for SPEech REcognition) is a HMM-based toolbox for speech recognition that is composed of three processing stages: an acoustic front-end, a statistical training module, and a recognition engine [1].

2.1 The acoustic front-end

The parameterization is based on Mel frequency cepstral coefficients (MFCC). The user can customize this parameterization by choosing the size and shift of the analyzing window, the number of the triangular filters, the lower and upper frequency cut-offs of the filterbank, and the number of cepstral coefficients. Moreover, the user can remove C0, the first MFCC, or replace it by the energy computed as the log of the signal energy. Finally, the first and second derivative coefficients can be added. The acoustic front-end integrates a voice activity detector (VAD). All these facilities are particularly important in order to adapt the system to pathological voices.

2.2   The training module

The core of this module uses the Baum-Welch re-estimation algorithm with continuous densities. The user can define the topology of the HMMs, i.e., number of states and allowed transitions. The modeled unit can be a word, a phone or a triphone. These units are trained using either an isolated training (the boundaries of every unit are specified), or an embedded training (only the string of labels is specified). The number of probability density functions (pdf) per state is determined during the training phase: the number of mixture components is repeatedly increased until a desired value by a splitting mechanism. With regard to triphones, a decision tree is used in order to reduce the number of mixture pdfs and the states to be tied are then selected.

2.3   The recognition engine

The recognition engine implements a one-pass, time-synchronous algorithm. It requires the HMMs trained as described in the previous section, the lexicon of the application, and a grammar. The structure of the lexicon allows the user to give several pronunciations per word (including pronunciations related to a particular pathology). The grammar may be a word-pair or a bigram. In order to speed up the recognition, a pruning threshold can be applied. The implementation of ESPERE contains more than 20000 C++ lines and it runs on PC-Linux and PC-Windows.

3.   Lessons at the levels of syllables and words

3.1   Presentation

A prototype system using a word recognition module has been designed for training the pronunciation of words (or short sentences comprising a few syllables). This prototype is based on HMMs at the phonetic and co-articulation levels. HMMs are trained for signals sampled at 16 kHz, to keep information until about 8 kHz. Only the standard soundboard of the personal computer is necessary. For efficiency of programming and for being able to easily develop new lessons, exercises are automatically customized through the use of word containers. Such containers make it possible to select the set of words and the opposition between minimal pairs concerned by an exercise.

3.2   The lessons

Figure 1 summarizes the lessons developed at the word levels. Some details about the objectives of the lessons, their implementation, and their validation as training tools are presented in the next section.

Categories	Exercises
Syllables	CV, CVC syllables
Words	Word alone Words by minimal pairs Spelled word

Figure 1: Exercises implemented at the word level

The following sketch illustrates the lesson concerning single word training:
- At first, the menu allows to select the exercise.
- The word to be trained can be either chosen randomly by the system, or designated by the trainee. Let us assume that the word *château* (castle) is selected. Altered patterns of the word, together with phonetically close words constitute the vocabulary for the recognizer.
- The trainee is asked to pronounce the word.

- Let's suppose that he recognition system gives the highest probability to the word *chapeau* (hat). The place of articulation of the medial stop consonant is wrong, the rest of the word is judged as correct.
- The explanations about places of articulation of plosives are presented to the trainee, with HTML links to profile radiograms, to sound files, and to other articulation information helping to make the discrimination between close sounds.

3.3 Validation

The lessons at the word level have been experimented during a validation phase involving speech therapists (who had been involved in the specification and development of the system), interpreters for facilitating the communication with a deaf person, normally-hearing adults and phoneticians, normally-hearing children, young speech-impaired children, and a young hard-of-hearing boy.

The total duration of this validation was about six months. Figure 2 presents the conclusions drawn during this period and the improvements that have been brought to the lessons.

3.4 Conclusions

The set of exercises at the level of syllables, words, and short sentences are very promising for the therapists. Actually, in many cases, their patients present systematic articulatory defects that could be correct by intensive training.

For profoundly or severely deaf persons, it seems difficult to transpose the abilities acquired for the uttering of individual sounds to abilities to fluently pronounce an entire word. Intonation and rhythm must intervene for the intelligibility of speech. It is necessary to measure which are the most reasonable efforts to ask for. What is wished is intelligibility rather than quality.

Self-training may be envisaged for words already known by the trainee. Each time a word or a pair of words is new, supervised training is preferable. Face reading and residual hearing capability are catalysts for the trainee's performance.

Level of teaching and answers to the objectives are good. Much effort has still to be done to extend the acquired ability with words to everyday communication. Repetition and word chaining seems to be a good approach for that purpose.

As said before, no real-time feedback appears during the utterance. It can be disconcerting for persons who used to be attentive to this feedback in the previous exercises at sublexical levels. A particular attention must thus be brought to the way the conclusion of the exercise is presented. This remark holds for all the exercises using automatic recognition or labeling programs. In the future, it will be necessary to couple these programs with real-time displays, for example at the suprasegmental level, for keeping the attention and interest of the trainee. This coupling has the great advantage of favoring the acquisition of skills which are useful for intelligibility and mandatory for the quality of speech, like intonation, rhythm, and speech rate. This is still more important for foreign language learning by hearing persons.

Efforts have been done concerning illustrations and links with explanations. The orthographic forms of words have been coupled with images, LPC sequences, and dactylology. LPC (*langage parlé complété*, completed spoken language constitutes a gestual complement to face reading to raise lip-reading ambiguities. The linguistic structures of the spoken sentences are thus preserved. A simple example is given by [p], [b], and [m], which are labial doubles, but with different manners of articulation. Dactylology provides the hand-spelling of the word.

3.5 Further work

As stated in figure 2, additional displays would allow understanding why the system has recognized a word rather than another. For that purpose, it is necessary to memorize the scores obtained for the two words and to draw a conclusion from their values. A specific method must be developed to compute confidence measure from the outputs of the HMMs. This is a general issue in automatic speech recognition. The results could be displayed as illustrated in Figure 3 with two close words, i.e., [ʃapo] and [ʃato].

Exercises	Objectives/Lessons/What is the trainee asked for/ Observations and conclusions
Syllables and words	**Objective** The goal is to lead the trainee to spontaneously use the articulatory gesture previously acquired. Most of the exercises in this phase aim at practicing phones within opposite modes: oral/nasal, voiceless / voiced, constrictive / occlusive, and on the place of articulation. Indeed, a few sounds can raise a difficulty at the beginning or the end of a word. They can be correctly produced within the word with the help of the consonantal and vocalic environments.
Word alone	• Exercise: The trainee is asked to utter a word selected in a database, randomly or not. The evaluation takes into account the expected word and the nearest neighbor. • Conclusions: The exercise is useful and easy to understand.
Coupled words	• Exercise: Words are selected according to an opposition between minimal pairs. The evaluation takes into account the expected word and the second word of the pair. • Conclusions: The exercise is clear. In a first step, short words must be trained as significant syllables. Pictures and other illustrations would be useful and must be added systematically. The exercise should be interrupted if it appears to be too difficult. Additional displays would allow understanding why the system has recognized a word rather than another.
Spelled letters	• Exercise: The exercise aims at helping the trainee to master spelled letters for the situations where pronouncing a word is difficult. • Conclusions: The exercise interests very much professionals. However, it is difficult to successfully differentiate letters that differ only by a short burst at the beginning (/te/, /pe/, ...) or at the end (/El/, /Em/). The speech models used must be trained with normal voices but also with slightly pathological voices. It is also necessary to explain to the trainee that, in real-life, it is common to ask people to repeat when there is an ambiguity, or to use keywords like "b like Bravo" or "a like Alpha".

Figure 2: Evaluation of the exercises at the word level

Figure 3: Display of the proximity between the utterance (arrow) and reference words (at the extremities)

The evaluation could also be refined by coupling the speech recognition system with an analytical system that would bring additional information linked to the quality of speech.
Prosodic features are also to be introduced for words (intensity, intonation, duration, and rhythm...).

4. Conclusion

In most automatic speech education systems, focus is mainly oriented toward suprasegmental information or very local segmental information.

We have shown in this paper that the introduction of results obtained in automatic word recognition makes it possible to develop adapted lessons that improve the training of a vocabulary of words.

The quality of the obtained results may be improved through multiple training of speech models: discriminative training, training including common errors, training with slightly pathological voices, etc. The next step in the development of lessons will concern exercises using words and their semantic values: oral control of actions performed on the screen, and various word games.

Acknowledgments

Part of this work has been carried out in the framework of the ISAEUS project [5] funded by the European Community (TIDE program).

References

1. Fohr, D., Mella, O., Antoine, C. (2000). The automatic Speech recognition Engine ESPERE: Experiments on Telephone Speech, in *Proceedings of ICSLP'2000* (Beijing, China, October 2000) (pp.246-249).
2. Haton, M-C. (1998). Issues in Using Models for Self-Evaluation and Correction of Speech. In M. Ponting (Ed.), *Computational Models of Speech Pattern Processing*. Berlin: Springer Verlag.
3. Haton, M.-C., Haton, J-P. (1979). SIRENE, a System for Speech Training of Deaf People. In *Proceedings of ICASSP'79* (pp. 482-485), Washington DC (May).
4. Hiller, S. et al. (1993). SPELL: an Automated System for Computer-aided Pronunciation Teaching. In *Proceedings of Eurospeech'9* (pp. 1343-1346) Berlin, Germany.
5. ISAEUS (2000). *Speech Training for Deaf and Hearing-Impaired People*. TIDE project, Final Report.
6. Kawai, G., Hirose, K. (2000). Teaching the pronunciation of Japanese double-mora phonemes using speech recognition technology. *Speech Communication*, 30, n° 2-3, 131-143.

Modeling spontaneous speech events during recognition

Peter A. Heeman

Computer Science and Engineering, Oregon Graduate Institute of Science and Technology
20000 NW Walker Rd., Beaverton OR 97006, heeman@cse.ogi.edu

Abstract

In spontaneous speech, speakers segment their speech into intonational phrases, and make repairs to what they are saying. However, techniques for understanding spontaneous speech tend to treat these events as noise, in the same manner as they handle out-of-grammar constructions and misrecognitions. In our approach, we advocate that these events should be explicitly modeled. We modify the speech recognition process so that it not only models determines the words that the user is saying, but also models intonational phrasing and speech repairs. This not only improves speech recognition performance but also results in a much richer output from the recognizer, with speech repairs resolved and intonational phrase boundaries identified.

1. Introduction

To enable spoken dialogue systems to advance towards more collaborative interaction between humans and computers, we need to deal with language as it is actually spoken. In natural speech, speakers group their words into intonational phrases and make repairs to what they are saying. Consider the following speaker's turn from the Trains corpus [Heeman and Allen, 1995].

Example 1 (d93-13.3 utt63)
um it'll be there it'll get to Dansville at three a.m. and then you wanna do you take tho- want to take those back to Elmira so engine E two with three boxcars will be back in Elmira at six a.m. is that what you wanna do

From reading the word transcription, the reader should immediately notice the prevalence of *speech repairs*, where speakers go back and change or repeat something they just said. Fortunately for hearers, speech repairs tend to have a standard form. The *reparandum* is the stretch of speech that the speaker is replacing; it might end in the middle of a word, resulting in a word fragment. The end of the reparandum is called the *interruption point*. There can also be an editing term, consisting of fillers, such as 'uh' and 'um', or cue phrases, such as 'let's see', 'well', and 'okay'. This is then followed by the *alteration*, which is the replacement for the reparandum. Speech repairs are very prevalent in spontaneous speech. In the Trains corpus, 10% of all words are part of the editing term or reparandum of a speech repair, and 54% of all speaker turns with at least 10 words have at least one repair. To determine the speaker's intended message, speech repairs need to be *resolved*: they need to be *detected*, by finding their interruption point, and *corrected*, by determining the extent of their reparandum and editing term. In addition to making repairs, speakers also break their turn of speaking into intonational phrases, which are signaled through variations in the pitch contour, phoneme lengthening and pauses. Previous research has shown that intonational information can reduce syntactic ambiguity for humans [Beach, 1991] and for computer parsers [Ostendorf et al., 1993]. Although intonational phrases might not be the ideal unit for modeling interaction in dialogue, it is undoubtedly a major component of any definition. We refer to speech repairs and intonational phrasing as *spontaneous speech events*. We now reshow our earlier example annotated in terms of them. Reparanda are indicated in italic, with the alteration starting on a new line indented to start at the reparandum onset. Intonational phrase boundaries are marked with '%'.

Example 2 (d93-13.3 utt63)
um it'll be there
    it'll get to Dansville at three a.m. **%**
and then you wanna
       do you take tho-
          want to take those back to Elmira **%**
so engine E two with three boxcars will be back in Elmira at six a.m. **%**
is that what you wanna do **%**

Although the spontaneous speech events of speech repairs and intonational phrasing are much more common in human-human speech than in human-computer speech [Oviatt, 1995]. In fact, one line of research work involves investigating means of structuring the interaction with the user so as to reduce the complexity of the user's speech, both in terms of the number of disfluencies and the linguistic and intonational structure of the speech [Oviatt, 1995]. However, as time advances, we will want to build human-computer interfaces that can collaborate with us on difficult problem-solving tasks [Allen et al., 1995]. It will become imperative to allow both the human and computer to be able to freely contribute to the dialogue, rather than having them focus on the form their interaction [Price, 1997]. Thus, we will need to allow for the full richness of spontaneous speech, with its apparent imperfections, such as speech repairs, and complex intonational phrasing.

To deal with these spontaneous speech events, it has become popular to use robust parsing techniques. For understanding spontaneous speech, speech repairs are not the only phenomena that create problems; one also needs to deal with word misrecognitions and out-of-grammar constructions. All three of these problems tend to be lumped together and given to a robust parser. Ward [1991] used a robust semantic parser to look for sequences of words that matched grammar fragments associated with slots of case frames. The parser tries to fill as many slots as possible. If a slot is only partially filled, it is abandoned. If a slot is filled more than once, the latter value is taken [Young and Matessa, 1991]. Others have adopted a similar approach: Rose and Lavie [2001] describe a robust parser, which incorporates a skipping mechanism, with a feature unification grammar; and van Noord describes using a skipping mechanism in parsing word graphs.

Rather than view spontaneous speech events as noise in the input to a robust parser, we take a different approach. We advocate that speech repairs and intonational phrasing should be explicitly modeled. There are local cues, such as editing terms, word correspondences, and pauses, which give evidence for these events. Hence, we should be able to automatically identify intonational phrases and resolve speech repairs. By modeling these events, we will have a richer understanding of the speech. This will simplify later syntactic and semantic processing, since such processing can start from enriched output rather than trying to cope with the apparent ill-formedness that spontaneous speech events cause [Core and Schubert, 1999]. This should also make it easier for these processes to deal with the other problems of understanding spontaneous speech: namely misrecognitions and out-of-grammar constructions.

Speech repairs and intonational phrasing are intertwined with the speech recognition problem of predicting the next word given the previous context [Heeman and Allen, 1999]. Hence, our approach is to redefine the speech recognition problem so that it includes the resolution of speech repairs and identification of intonational phrases. We also include the tasks of part-of-speech (POS) tagging and discourse marker identification, since these tasks are also intertwined with resolving speech repairs and identifying intonational phrasing. Since all tasks are being resolved in the same model, we can account for the interactions between the tasks in a framework that can compare alternative hypotheses for the speaker's turn. Not only does this allow us to model the spontaneous speech events, but it also results in an improved language model. Furthermore, speech repairs and phrase boundaries have acoustic correlates, such as pauses between words. By resolving speech repairs and identifying intonational phrases during speech recognition, these acoustic cues, which otherwise would be treated as noise, can give evidence as to the occurrence of these events, and further improve speech recognition results.

In the rest of the paper, we first give a brief overview of speech recognition language modeling. We then present a simplified version of our model of speech repairs and intonational phrases. We then give the results of running our model on the Trains corpus. Finally, we present the conclusions and future work.

2. Speech recognition

The goal of a speech recognizer is to find the sequence of words $\hat{W}$ that is maximal given the acoustic signal A.

$$\hat{W} = \arg\max{}_W \Pr(W \mid A) = \arg\max{}_W \frac{\Pr(AW)\Pr(W)}{\Pr(A)} = \arg\max{}_W \Pr(A \mid W)\Pr(W)$$

The above shows the speech recognition problem rewritten as the product of two probability distributions that need to be estimated: the first is the *acoustic model* $\Pr(A \mid W)$ and the second is the *language model* $\Pr(W)$. The language model probability can be expressed as the product of the conditional probability of each word given the words that precede it. This is shown below, where we rewrite the sequence W explicitly as the sequence of N words.

$$\Pr(W_{1,N}) = \prod_{i=1}^{N} \Pr(W_i \mid W_{1,i-1})$$

To estimate the probability distribution in the above line, a training corpus is used to determine the relative frequencies. Due to sparseness of data, one must define *equivalence classes* amongst the contexts $W_{1,i-1}$, which can be done by limiting the context, or by using decision trees.

3. Augmenting the recognizer

The basic speech recognition model uses a simplistic language model, in which the probability of a word only takes into account the previous words. The occurrence of speech repairs and intonational boundaries, however, also affects which word will occur next. Consider the following example.

Example 3 (d93-3.2 utt 45)

which engine   are we   are we taking

           reparandum ip

After seeing the words "which engine are we," the probability of then seeing the word "are"' again would be very low. The word trigram `"are we are" rarely occurs. But, by hypothesizing a speech repair, its probability would be much higher. Furthermore, there are acoustic cues that can be used to give independent confirmation of the repair, thus further improving our ability to model these events.

To incorporate speech repair and intonational phrasing into our language model, we redefine the speech recognition problem so as to include extra variables that will be tagged for the occurrence of these events. Let I_i indicate whether word W_{i-1} ends an intonational phrase and let R_i indicate whether word W_{i-1} is the interruption point of a speech repair. The speech recognition problem can now be cast as finding the best interpretation for the repair and intonation variables along with the best word sequence.

$$\widehat{WR\widehat{I}} = \arg\max\nolimits_W \Pr(WRI \mid A) = \arg\max\nolimits_{WRI} \Pr(A \mid WRI)\Pr(WRI) \approx \arg\max\nolimits_{WRI} \Pr(A \mid W)\Pr(WRI)$$

The second probability distribution is our new language model, which can be rewritten as follows.

$$\Pr(W_{1,N} R_{1,N} I_{1,N}) = \prod_{i=1}^{N} \Pr(I_i \mid W_{1,i-1} R_{1,i-1} I_{1,i-1})\Pr(R_i \mid W_{1,i-1} R_{1,i-1} I_{1,i})\Pr(W_i \mid W_{1,i-1} R_{1,i} I_{1,i})$$

Our full model, which is described elsewhere [Heeman and Allen, 1999], includes five additional variables. One variable is used to model the POS tag for each word, which allows us to model shallow syntactic knowledge and how it correlates with speech repairs and intonational phrasing. Another variable is used to model editing terms, which sometimes accompany speech repairs. The other three variables are used to correct the speech repair, which is determining the extent of the reparandum. Modeling the reparandum helps in detecting speech repairs and improves speech recognition performance. This improvement is because there are often strong word correspondences between the reparandum and alteration, which gives evidence for the repair, and helps predict the words in the alteration. Furthermore, the alteration tends to be a fluent continuation of the words before the reparandum, giving us further evidence of the repair and of the words of the alteration. We use a variable toindicate the reparandum onset, and for each word of the alteration, a variable indicates which word of the reparandum it corresponds to, and another variable indicates the type of correspondence, whether it is a word match, same POS tag, or other. This gives us a total of eight variables for our language model of spontaneous speech.

4. Results

To test our model, we ran experiments on the Trains corpus using a six-fold cross validation procedure. We tested three versions three different language models [Heeman, 1999]. The first is a traditional word-based language model, the second incorporated modeling POS tags, and the third is our full model of speech repairs and intonational phrasing (and POS tags). The results are given in Table 1. We report the results in terms of perplexity and word error rate. *Perplexity* is a measure of how well the probability distributions of the language model predict the data in

a test set $w_{1,N}$, and is calculated as 2^H, where
$$H = -\frac{1}{N}\sum_{i=1}^{N}\log_2 \hat{\Pr}(w_i \mid w_{1,i-1})$$
. Lower perplexity values indicate improvements in predicting the test data. *Word error rate* measures the percentage of mistakes that a speech recognizer makes using the language model. We used a large vocabulary continuous speech recognizer that has completed in the Broadcast News task developed at OGI.

Table 1: Impact on Speech Recognition

	Perplexity	WER
Word – Based Model	24.8	26.0
POS – Based Model	22.6	24.9
Full Model	21.3	24.6

We see that by incorporating the additional modeling, we are able to improve the speech recognizer's performance by 5.4%, as measured with word error rate. This model only makes limited use of acoustical cues, namely pauses between words. By using richer acoustic cues of speech repairs and intonational phrases, we should be able to further improve the results (cf. [Stolcke et al., 1999]).

We have also run tests to determine how well we can identify intonational phrase boundaries and speech repairs [Heeman and Allen, 1999]. We again used a six-fold cross validation procedure. We report results in terms of *recall* and *precision*. The recall rate is the number of times that the algorithm correctly identifies an event over the total number of times that it actually occurred. The precision rate is the number of times the algorithm correctly identifies it over the total number of times it identifies it.

For intonational phrase boundaries, we distinguish between those that occur within a speaker's turn from those that occur at the end. This is because our model uses end-of-turn information as part of its input, and since almost all turns end with an intonational phrase boundary, it easily learns this regularity. As for the within turn boundaries, the model achieves a recall rate of 71.8% with a precision of 70.8%.

For detecting speech repairs, we achieved a recall rate of 76.8% with a precision of 86.7%. Although our model classifies repairs into three different types (fresh start, modification, and abridged), we count a repair as correct as long as its interruption point was identified, without regard to whether its type was correctly determined. Furthermore, when multiple repairs have contiguous reparanda, we count all repairs involved (of the hand-annotations) as correct as long as the combined reparandum is correctly identified. For correcting speech repairs, we achieved a recall rate of 65.9% and a precision of 74.3%. A repair is counted as correctly corrected if it was identified and the extent of the reparandum was correctly determined.

5. Conclusion and future work

Previous work in spoken language interfaces has regarded speech repairs and intonational phrasing as noise in the dialogue, which needs to be skipped over and not understood (e.g. [Ward, 1991]). Others have investigated means of structuring the user's interactions so as to reduce the complexity of the user's speech, thus making it easier to understand [Oviatt, 1995]. However, as time advances, we will want to build human-computer interfaces that can collaborate with users on difficult problem-solving tasks [Allen et al., 1995]. It will become imperative to allow both the human and computer to be able to freely contribute to the dialogue, using the full richness of language, with its apparent imperfections, such as speech repairs and complex intonational phrasing.

In this paper, we reported on our work in which we model these spontaneous speech events. We redefined the speech recognition task so as to also include identifying intonational phrases and resolving speech repairs. This allows us to better account for the words involved in a speaker's turn, as evidenced by the improved word recognition rates. It also allows us to return a more meaningful analysis of the speaker's turn, with speech repairs and intonational phrases identified. This richer output should make it easier for later processing to deal with spontaneous speech. Hence, interfaces of the future will be able to better deal with the occurrence of spontaneous speech events. In fact, human-computer interfaces might when a user makes a speech repair, this gives evidence that the speaker might be uncertain about what he is saying. The human-computer interface should then modify its response appropriately, for instance by double checking the information.

Much work remains to be done. One area of research that we are pursuing is incorporating higher level syntactic and semantic processing. This would not only allow us to give a much richer output from the model, but it would also allow us to account for interactions between this higher level knowledge and modeling speakers' utterances,

338

especially in detecting the ill-formedness that often occur with speech repairs. It would also help in finding richer correspondences between the reparandum and alteration, such as between the noun phrase and pronoun in the following example.

Example 4 (d93-14.3 utt27)

the engine can take as many ‖ um ‖ it can take up to three
    reparandum   ip *et* *alteration*

A second area of research is to incorporate better acoustical cues. We currently only exploit inter word pauses. This is a rich source of information for detecting (and distinguishing between) intonational phrases and interruption points of speech repairs. It would also help in determining the reparandum onset [Nakatani and Hirschberg, 1994], which tend to occur at intonational boundaries. Acoustic modeling is also need to identify word fragments, which accompany many speech repairs. As we further exploit acoustic cues, this will improve our ability to detect speech repairs and intonational phrases, which will have the added benefit of improving our speech recognition results.

Acknowledgments

This work, and the results it builds on, has been supported by funding from NSERC Canada, the NSF under grant IRI-9623665, DARPA - Rome Laboratory under research contract F30602-95-1-0025, ONR/DARPA under grant N00014-92-J-1512, ONR under grant N0014-95-1-1088, ATR Interpreting Telecommunications Laboratory, CNET France Télécom, and the Intel Research Council.

References

Allen, J., Schubert, L., Ferguson, G., Heeman, P., Hwang, C. H., Kato, T., Light, M., Martin, N., Miller, B., Poesio, M., and Traum, D. R. (1995). The Trains project: A case study in building a conversational planning agent. *Journal of Experimental and Theoretical AI*, 7:7-48.

Beach, C. M. (1991). The interpretation of prosodic patterns at points of syntactic structure ambiguity: Evidence for cue trading relations. *Journal of Memory and Language*, 30(6):644-663.

Core, M. and Schubert, L. (1999). A syntactic framework for speech repairs and other disruptions. In *Proceedings of the 36th Annual Meeting of the Association for Computational Linguistics*, College Park, Maryland.

Heeman, P. A. (1999). Modeling speech repairs and intonational phrasing to improve speech recognition. In *Automatic Speech Recognition and Understanding Workshop*, Keystone Colorado.

Heeman, P. A. and Allen, J. F. (1995). The Trains spoken dialog corpus. CD-ROM, Linguistics Data Consortium.

Heeman, P. A. and Allen, J. F. (1999). Speech repairs, intonational phrases and discourse markers: Modeling speakers' utterances in spoken dialog. *Computational Linguistics*, 25(4):527--572.

Nakatani, C. H. and Hirschberg, J. (1994). A corpus-based study of repair cues in spontaneous speech. *Journal of the Acoustical Society of America*, 95(3):1603-1616.

Ostendorf, M., Wightman, C., and Veilleux, N. (1993).Parse scoring with prosodic information: an analysis / synthesis approach. *Computer Speech and Language*, 7(2):193--210.

Oviatt, S. (1995). Predicting spoken disfluencies during human-computer interaction. *Computer Speech and Language*, 9:19-35.

Price, P. (1997). Spoken language understanding. In Cole, R., Mariani, J., Uszkoreit, H., Zaenen, A., and Zue, V., editors, *Survey of the State of the Art in Human Language Technology*. Cambridge University Press.

Rosé, C. P. and Lavie, A. (2001). Balancing robustness and efficiency in unification-augmented context-free parsers for large practical applications. In *Robustness in Language and Speech Technology*. Kluwer Academic Publishers.

Stolcke, A., Shriberg, E., Hakkani-Tür, D., and Tür, G. (1999). Modeling the prosody of hidden events for improved word recognition. In *Proceedings of the 6th European Conference on Speech Communication and Technology*.

van Noord, G. (2001). Robust parsing of word graphs. In *Robustness in Language and Speech Technology*. Kluwer Academic Publishers.

Ward, W. (1991). Understanding spontaneous speech: The Phoenix system. In *Proceedings of the International Conference on Audio, Speech and Signal Processing (ICASSP)*, pages 365-367.

Young, S. R. and Matessa, M. (1991). Using pragmatic and semantic knowledge to correct parsing of spoken language utterances. In *Proceedings of the 2nd European Conference on Speech Communication and Technology*, pages 223-227, Genova, Italy.

Visual languages to bridge the gap between software developers and their clients

Bertrand Ibrahim

Computer Science Department, University of Geneva,
rue du Général Dufour 24, CH-1211 Geneva 4, Switzerland
E-mail: Bertrand.Ibrahim@cui.unige.ch

Abstract

Software engineering has mostly focused on providing software developers with methodologies and tools that help them develop more reliable software, more easily, more in time and within budget. However, because of their complexity, these tools and methodologies are invariably inaccessible to those people for whom the software is being developed, i.e. the clients. Using visual languages for software design can help compensate for this deficiency and bridge the conceptual gap between developers and users. Visual languages can indeed allow the clients to participate in the detailed design of the software they are going to use, while preserving the potential for using sophisticated CASE tools and environments for software development, for instance through automatic code generation techniques. The visual formalisms thus work as a communication medium between the various actors in the development process, including the clients who generally don't have any background in programming. We have successfully used this approach in the development of large computer aided learning (CAL) programs. We used a graph-based visual formalism to allow educators and psychologists to specify the detailed behavior of the CAL programs they wanted to build. These specifications were then given to programmers who were in charge of implementing them. The visual formalism was simple enough to be quickly mastered by teachers with no background in programming, but detailed enough to allow for partial automatic code generation and for the programmers to implement what couldn't be generated automatically from the specification. This article presents our experience with this approach, describing the pivotal role of the visual formalism to help people with very different backgrounds collaborate in the development of courseware.

1. Introduction

The benefits of good software engineering practice have long been recognized. Indeed, the use of strict development methodologies have been a key factor to the success of many large projects in institutions like the American Department of Defense (DoD) or the National Aeronautics and Space Administration (NASA). However, all too often, software engineering techniques are viewed strictly as methods by which computer scientists can improve the quality, ease and timeliness of the software development process as well as the robustness of the resulting software products. Many such development strategies involve the clients, i.e. the people for whom the software is developed, only in the very early stages of the development, to define informal requirements or specifications, usually in natural language. Software engineering techniques are then used to develop formal specifications, produce the detailed design and help the computer scientists in later stages of the development. These techniques are, however, too complex to allow the clients to follow the development process and ensure that what is produced really corresponds to what they need. It is only when the first prototype becomes available that the clients can determine whether their needs have been correctly understood and adequately reflected in the formal specifications and the actual implementation. This delay can be very detrimental, both in terms of additional costs due to additional implementation efforts, and in terms of delay in delivering the final product.

To prove that a piece of software is correct cannot just be reduced to proving that the code produced matches the formal specifications. Above all, one must ensure that the specifications and the code match the needs of the clients. Correctness should thus be viewed not only in terms of theorem proving, but most of all in terms of adequacy to the clients' needs.

Furthermore, the later in the development process an error is discovered, the more difficult and costly it is to correct it. It is therefore of paramount importance that the clients be allowed to participate as far down in the development process as possible to ensure that the software developed really corresponds to what the clients want. There is therefore a communication need that can be best mediated through the computer. One way to achieve this is to use CASE tools and semiformal visual languages in the detailed design phase. The visual language specification can then be used all along the development process as a human-readable description of the software artifacts produced further down along the development process chain. Appropriate tools can show synchronized views of the visual specification and of the software artifacts, produced by automatic code generators or hand-coded by programmers, to help work on these artifacts. The visual specification language can thus be viewed as a communication medium between the clients/designers and the developers/computer scientists.

2. Visual languages

Various visual language paradigms exist, which have their own individual application domains. For instance, for a long time now, state-transition diagrams have been used to design user interfaces {Parnas, 1969; Wasserman, 1985}. More recent design and modelling techniques such as UML {OMG, 1999} arw making heavy use of visual formalisms (class and object diagrams, use case diagrams, sequence diagrams, collaboration diagrams, statechart diagrams, activity diagrams, implementation diagrams). In this latter case (UML), most of the diagrammatic formalisms are meant for software developers to make their task easier and more manageable, and each notation is used at a different stage of the development or to illustrate a certain aspect of the software. Even if they are mostly used by developers, use case and activity diagrams can nevertheless probably be understood by people without any programming background and could therefore be viewed as a form of universal communication medium.

For domain-specific applications, it is often possible to have a reasonably simple visual specification formalism that is understandable and usable by the people for whom the software is developed. Such a notation might even already exist, as a design documentation tool, in which case the visual specification formalism should embody this existing notation as much as possible. The specifications produced with this notation can then be used to automatically generate executable code, or be directly executed by an interpreter. To illustrate this, we will describe, in the following sections, the "IDEAL" development environment, which allows teachers and pedagogues to participate in the detailed design of computer-aided learning software through a control-flow-based visual formalism.

2.1 A visual language for the specification of CAL software

Computer aided learning, and more specifically highly interactive tutorial learning, is an application domain that is more control-driven than data-driven. A control-flow representation is therefore more appropriate than a data-flow one. In our case, the representation we use is based on directed graphs, with various node types corresponding to various types of elementary actions the CAL program should perform, and with directed edges expressing the relative order in which the nodes should be executed. For a more in-depth description of the visual formalism, the reader can refer to {Ibrahim, 1999}.

Thanks to its visual nature and to the fact that it includes natural language components, the formalism is fairly easy to learn and to use. Usually, teachers and pedagogues who participate in our design sessions only need a short introduction to the formalism before they can start using it for the detailed design of CAL material. From our own experience, less than an hour is necessary to get the designers, the teachers and pedagogues, acquainted with our notation and to allow them to become active participants, even if they are not familiar with any programming activity. At the beginning, it is however more effective if a person already familiar with the formalism and the methodology participates in the process.

The overall development process is illustrated in Figure 1. It shows the major steps involved in the whole life- cycle of a CAL program. The first phase, initial planning, consists in finding funding for the development and in putting together a team of domain experts (teachers and pedagogues) for the pedagogical design. The second phase, pedagogical design, can actually be broken down into two sub-phases: overall design and detail design. Overall design consists in brainstorming sessions to decide exactly what subject matter will be covered. It results in an outline of the educational material that will be produced, as well as a subdivision of this material into units, as well as general indications regarding major learning difficulties with each unit. The result of the overall design will serve as a guideline for the detail design.

Detail design is usually done by multiple teams, each working on separate units. For each unit, one or more "scripts" will be produced to describe in detail the behavior of the unit, including all the interactions with the learner. These scripts give all the details of what is to happen, without necessarily going into how it is to be implemented. Implementation details should not get in the way of the line of thought of the designers.

341

Following the design phase, the design teams review each other's scripts, suggesting possible improvements to other teams or considering improvements to their own work based on what others have done.

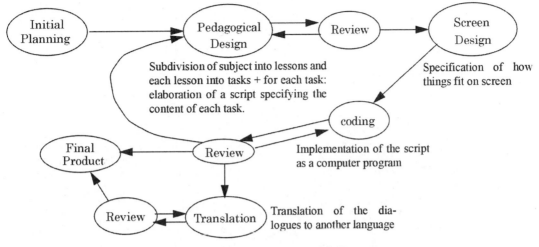

Figure 1: Development life-cycle of CAL programs

Once the pedagogical design is completed, graphic designers can work on the screen design of the various scripts, adding details about screen layout and creating the graphics that will improve the visual appearance of the material. After the screen design phase, the implementation phase follows, during which programmers write the code necessary to run the scripts. Once the code is tested, the resulting programs are reviewed by the design teams who might consider revising their pedagogical design based on the impression they get from running the actual implementation. A formative evaluation can then be undertaken, by testing the material with students. This can in turn lead to a revision of the pedagogical design.

Once the tests are satisfactory, the programs can be used on a larger scale and possible translations of the dialogues can be prepared to make the CAL software available to a wider audience. The translations must of course be reviewed within the context of the CAL software, to check the adequacy of the choice of terms in the translations.

Based on this development life-cycle, we created a CASE environment centered around the visual specification formalism, as illustrated in Figure 2. During the pedagogical design sessions, the designers (teachers and pedagogues) use a "script" editor to create the visual specification of the CAL programs that need to be developed. This specification is then used by an automatic code generator to produce a finite state machine in a general-purpose target programming language (a few commonly used programming languages are already supported). Since some parts of the specification are in natural language, the code generator cannot produce all the code needed. The programmers then use the synchronous editor to view both the visual specification and the code at the same time. The synchronous editor allows them to see the exact location in the specification, as they move in the code, making it easier for them to understand what the code should do. Once the code is completed, it can be run on student machines. If the students' machines are networked with the teachers' workstations, the latter can view, with a supervision tool, directly on the visual specification what the students are doing. If the software is to be used by students from various linguistic origins, the dialogues and the textual answer analysis criteria need to be translated to other languages. This can be done by professional translators using the translation editor. This multi-window tool presents, in one window, the content of the message nodes of the specification to the translator, one message at a time. In another window, the translator can type in the translation of the message.

In the system, a script is internally represented as a set of at least two files, one for the graph structure (graph file), and one or more files for the actual content of all the nodes that deal with user input and output (message file, one for each language supported). This separation makes it easier to translate the dialogues of the CAL programs into many different languages, as the message file contains all the text that will appear on the user's screen and all the answer analysis criteria used to handle textual user input. Once the original message file is translated to other languages, the

342

graph file can be used in combination with any of the available message files. The visual specification, through these two files, is used in most steps of the development life-cycle.

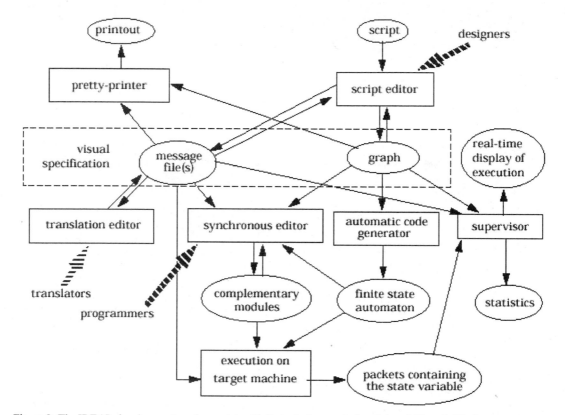

Figure 2: The IDEAL development environment, with its tools (rectangles) and data (ellipses). The human actors are also shown: designers, programmers and translators.

3.   Computer-mediated communication

Many people have to be involved in the development of large-scale computer-aided learning software. The various actors cannot necessarily meet with each other all the time, assuming that it is ever at all possible. Indeed, very often, the designers create the scripts during brainstorming sessions, over a short period of time, and it is only weeks, if not months later, that the programmers can come up with a first prototype. It is therefore very important that all the protagonists have a powerful and versatile communication medium. Natural language is often too ambiguous and imprecise to be used for a detailed specification. On the other hand, formal specification languages are too complex to be understandable and usable by people with no computer science background.

We have found that a combination of a visual notation and short natural language descriptions were a good compromise between the two above mentioned extremes. The visual, formal, part of the notation is not as intimidating for non-computer scientists as other formal notations, and the natural language component helps in keeping the overall notation reasonably simple. Of course, this can be complemented with programming by example elements whenever an easy-to-use direct manipulation interface is available. For instance, the insertion of multimedia sequences can be done by using interactive tools such as audio and video editors to define the exact content that will be provided at run- time.

A closer look at Figure 2 shows that the visual specification plays a central role in the overall development process. The internal representation of the specification is used in one way or another by most of the tools: the automatic code generator uses it directly to produce code in the form of a finite state automaton, the synchronous editor uses it to display synchronized views of the specification and the complementary code that the programmers need to produce, the remote supervision tool uses it to display the current state of execution of the final program running on a target machine. It is worth noting that the visual specification is thus used as formal input to the code generator, as a specification for the programmers who have to implement the parts that could not be generated automatically, as human-readable documentation to help the programmers who do the maintenance better understand the code, and as a visual map of the software for the teachers who want to follow what the learners are doing.

In addition, a version control mechanism embedded in the synchronous editor makes it easy to locate changes in the visual specification since the last running version of the code was created. This allows the programmers to quickly locate the parts of the code that need to be changed to bring it up to date with the latest version of the specification.

A similar version control mechanism is embedded in the translation editor to make it easy to locate changes in the text of the dialogs since the last translation was made. The translators first indicate what language is the original one, and what translation they want to work on. Then, they only need to check what has changed on the specific messages concerned with the changes. In different windows, they are shown the content of a message node in the original language as it was before the change, and after it. The translators are also shown the translation, if one already existed and they can change it as needed, based on the changes made in the original language, or they can create a new translation if none existed before. They can thus step through the changes and quickly bring a translation up to date.

In typical projects we have worked on with this methodology, the visual specifications included many thousands of nodes and edges, usually broken down into modules of two to three hundred nodes. Each design team was composed of two to four teachers or pedagogues, and a few design teams worked in parallel. The automatic code generator would typically produce fifty percent of the overall code.

4.   Conclusion

In this paper, we have described the use of visual languages as a means to bridge the gap between the people who build software and those for whom the software is built. Through an example we have shown that a visual specification formalism could allow the "clients" to participate in the detail design of their software and that this visual specification acts as a computer-mediated communication means between the designers and the implementors. Indeed, for people without a background in computer science, a formal textual specification language is too intimidating. On the other hand, a visual specification language combined with natural language can be much more accessible while still allowing specifications written with it to be detailed and precise enough to be useful all along the development chain.

References

Ibrahim, B. (1999). Semiformal Visual Languages, Visual Programming at a Higher Level of Abstraction. In Proceedings of the Fifth Conference of the ISAS (Information Systems Analysis and Synthesis) / The Third Conference of the SCI (Systemics, Cybernetics and Informatics), Orlando, Florida, USA, July 31-August 4 1999, ISBN: 980-07- 5916-6, pp 157-164.

Parnas, D.L. (1969). On the Use of Transition Diagrams in the Design of a User Interface for an Interactive Computer System. In *Proceedings of the 24th National ACM Conference* (pp. 379-385).

Wasserman, A.I. (1985). Extending State Transition Diagrams for the Specification of Human-Computer Interaction. *IEEE Transactions on Software Engineering*, Vol SE-11 No 8, Aug.1985, pp 699-713.

OMG Unified Modeling Language v. 1.3 specification, Rational (June 1999). [On-line]. Available: http://www.rational.com/media/uml/post.pdf

Natural Interaction based on Host Agents, Profiles and Decoupling Modalities from Interaction Devices

Ing-Marie Jonsson, Siamak Hodjat

Dejima Inc, 160 West Santa Clara Street, San Jose, CA 95113, USA
ingmarie@dejima.com, siamak@dejima.com

Abstract

Information is neither accessible nor inaccessible per se - it is the form in which it is presented that makes it one way or the other. Based on this assumption our main focus is to provide access to computer driven systems based on their users' needs, preferences and abilities – a personalized approach we term "natural interaction".

In this paper we introduce two new techniques; decoupling devices from interaction modalities, and dynamically reconfigurable host agents. We discuss how these techniques simplify the realization of natural interaction interfaces. This is implemented using two technologies; Adaptive Agent Oriented Software Architecture, (AAOSA), and the Total Access System, (TAS).

Decoupling interaction devices from application control modalities enables us to manage a dynamic configuration of interaction methods, adding, replacing, and removing interaction devices with minimal effort.

Host agents that can be updated using rules and descriptions provide a mechanism for dynamic reconfiguration of software applications and functions.

We also show that we achieve a higher level of personalized interaction by adding user profiles, device profiles, and descriptions of the applications user interface capabilities. These descriptions and profiles are automatically combined to provide users with a more natural interaction experience - offering an improvement in personalized interaction methodology.

1. Introduction

Users are often presented information in forms that are not accessible. Therefore, we propose a framework that provides a more natural interaction with information technology and appliances. This framework is based on two technologies, Adaptive Agent Oriented Software Architecture, AAOSA, (Hodjat & Amamiya, 2000) and the Total Access System, TAS (Scott, 1991). AAOSA provides a middleware and an engineering methodology where complex software is partitioned into a community of simpler, collaborating, message-driven components known as agents. TAS, originally developed to enable disabled users access to computers, provides a base for decoupling applications from interaction modalities (Jonsson, 1999). Using these two combined technologies appropriate input and output modalities can be selected to provide a personalized natural interaction with information technology equipment.

Unlike a traditional User Interface Management System (UIMS), our approach does not require a redesign of either the application or the current User Interface (UI). Our framework supports different levels of personalized interaction, from keyboard and mouse replacement to more complex interaction models outside the Windows-Icons-Menus-Pointer (WIMP) paradigm (Hinckley, 1996; Morrison, 1998.

We add profiles with user preferences, device descriptions, and descriptions of the applications together with software that handles profile matching and negotiation to implement higher levels of personalization - here called natural interaction.

1.1 Adaptive Agent Oriented Software Architecture

Agent Oriented Software Engineering (AOSE) is an approach based on active, cooperative, and persistent software components, i.e. agents (Huhns, 2000). AOSE can significantly enhance the ability to model, design and build complex (distributed) software systems by producing robust and reusable software (Jennings, 1999). Adaptive Agent Oriented Software Architecture (AAOSA) is an extension to AOSE emphasizing the agent's adaptability.

Applications built using AAOSA are designed by dividing the problem into sub-domains, represented by a network of AAOSA agents. An interpretation policy describes the responsibility of each individual agent. This policy defines how an agent should interpret its input, and how to coordinate activities from underlying agents. These two properties make AAOSA highly adaptable since adding or removing a sub-domain only affects the interpretation policies of the coordinator of that sub-domain

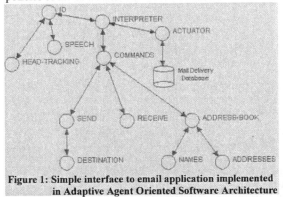

Figure 1: Simple interface to email application implemented in Adaptive Agent Oriented Software Architecture

In Figure 1 a network of AAOSA agents represents a simple email application, where speech and head tracking are used as input modalities. In our example, the IO agent is responsible for combining the inputs from the SPEECH and HEAD-TRACKING agents. The IO agent delegates the interpretation of this multimodal input to the INTERPRETER agent. The INTERPRETER will further delegate the processing of the input to its underlying component agents, ultimately deriving a command based on the user's input and the system state. This interpreted command is finally forwarded to the ACTUATOR (in this case the Mail Delivery Database) for execution. For example, the user might command "new letter to Bill". The IO agent hands the recognized phrase text to the

INTERPRETER. The INTERPRETER hands this to the other interface sub-agents for the e-mail system, which return the commands necessary to drive the ACTUATOR to create a new outgoing e-mail with the correct address for Bill.

1.2 Total Access System

Conventional computer systems are typically directly connected to their input and output devices. Reliance on devices such as a keyboard, mouse and display restricts access to only those users who are able to use these devices. The Total Access System, (TAS) is designed to enable alternative input and output modalities to be used with any

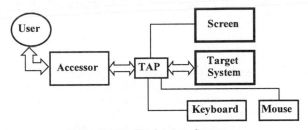

Figure 2: The Total Access System

computer based system. Inserting a device between the computer system and its traditional input and output ports essentially does this.

This device, called a Total Access Port, (TAP), can monitor, redirect and emulate input and output devices.

Different input or output technologies such as head trackers and speech technology may drive the emulation. This enables a user to choose input and output devices that match their personal preferences and collect them into a personal

access device, an accessor in TAS terminology. The TAS, even though designed for computers, will work for any system in which input and output devices can be decoupled from the application (Jonsson, 1999). Multiple TAPs can be connected through a local area network, (Scott et al., 1999) making it possible for an arbitrary collection of input and output devices to operate with an arbitrary collection of target systems. It also enables a user to utilize a single set of accessors for working with multiple target systems. This benefits users that invest time and effort into customizing and learning to use accessors such as speech recognizers and handwriting recognizers.

2. Profiles for natural interaction

Combining AAOSA and TAS we get a framework that supports users using their own accessors to interact with a variety of computer based systems and appliances. The framework implements plug-and-play mechanisms, resource management and a communication protocol to be used by all components on the network. Each user and each device connected to the network must identify themselves and preferably provide a profile of their abilities and requirements to the networks resource manager. The descriptive profiles are independent of the network, and it is the responsibility of users and the owners or manufacturers of devices to provide and maintain these profiles.

A resource manager uses the profiles to present information about each connected resource and to help users of the system to locate devices and applications (Gensereth et al. 1994; Jacobs & Shea, 1996). We achieve automatic configuration of the interface and the interaction by using the resource manager services to combine the user profile with information about the functionality of the application and the accessor (Titmuss et al, 1996).

2.1 Profiles - formats and management

The self-descriptive profiles for users, applications and devices all have the same format. A typical profile contains at least two parts:

> **An Identifier** – a unique identifier, analogous to an Internet Protocol (IP) address, for a networked resource.
> **A Description** – the features, functionality, methods and modes of operation and preferences of the resources.

The profile is described in Extensible Markup Language (XML) (Bray, 1998), and while a profile always contains an identifier and a description in a fixed format, the content and format of auxiliary information will vary according to the resource being described.

Although some adaptation can be automated (Adomavicius, &Tuzhilin, 1999), we acknowledge the effort needed to define, update and maintain profiles, consequently we envision three levels of access offered:

> **Full access**; The entire application or service is made accessible,
> **Partial functionality access**; Some key services are made accessible, but not all, and
> **Base access**; This is a closed service, but even so, our system still allows interaction alternatives such as speech and head-tracking to replace or supplement mouse and keyboard functions.

This is the property that we emphasize as the main difference when compared to other approaches and solutions. It should be noted that applications that offer full access would not solve all acknowledged access problems. The original UI design and its profile are the limiting factor in how well the system can be configured.

When computer based devices connect to the network, they present themselves by sending the identity and description part of their profile to the resource manager. Other parts of the profile such as personal information, databases and speech files are kept private. The resource manager stores the public profiles and can hence match requests for help and services from connected devices. Typical requests are to locate applications or to locate input and output devices on behalf of a user. Requests are either issued explicitly by the user of device or implicitly based on preferences or requirements contained in device or user profiles.

3. Host Agents

Natural interaction reflecting a user's preference regarding interaction model requires us to implement dynamic mechanisms. To support this we introduce host agents, agents that can dynamically update their interpretation policies based on the user's preferences and the device and application profiles. These host agents are used throughout our framework, located in all places where dynamic configuration is needed, including the TAP and the accessor. The host agents are implemented as a new type of AAOSA agents and can be freely combined with the original AAOSA agents as part of an application network.

We use the host agents to implement the default functionality and behavior of a component until a user preference requests a modified behavior. The modified behavior, i.e. the agent's interpretation policy, is forwarded to the host agent. The host agent will install and use the new interpretation policy for the duration of the interaction session. When the user disconnects, the host agent will resume its default behavior by reloading the original interpretation policy. In our example, depicted in Figure 1, we see that the IO agent, the INTERPRETER agent, and the ACTUATOR are typical AAOSA host agents. A new interpretation policy for the IO agent would, for example, change the multimodal behavior of the headtracker and the speech recognizer. A second language could be added by replacing the email network, for instance with a version updated with Spanish. The behavior of the ACTUATOR agent would be modified by a user's request to change the profile of the email application. We can also implement the mechanisms for plug-and-play of input devices by modifying the IO agent based on the input device profiles.

The reconfiguration of a host agent would be based on the communication between the host agent and other agents in the system. These are typically agents that represent other components in the network such as the resource manager, user profiles, input and output devices, and applications. In a typical scenario, the resource manager will require that all connected components identify themselves. The resource manager will then act as a directory service, a profile matcher and a negotiator to setup the communication paths necessary to configure the interaction model. Finally, the host agents would be activated and modified according to the user's preferences.

4. Decoupling interaction modalities from interaction devices

To further simplify the implementation of natural interaction, we propose a change in the relation between an interaction modality and a device. In general, some interaction modalities will have a one-to-one relationship with their devices, such as position information and pointing devices. Other interaction modalities, such as voice and gestures, will be based on information from multiple devices. Interaction modalities that are tightly coupled with their devices makes it easy to replace one device with another device from the same category, such as replacing a mouse with a head-tracker or a laser pointer. Interchanging multimodal input schemes, such as gestures and tactile interaction, is much more complex since these modalities are interpreted through a combination of devices such as cameras, sensors, and microphones. Replacement often means that a number of devices and software modules must be added to the system.

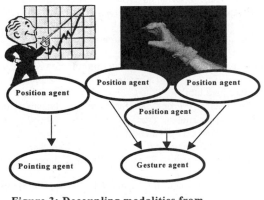

Figure 3: Decoupling modalities from Interaction devices

To simplify managing and replacing interaction modalities we propose a model that decouples modalities from interaction devices. In this model, each interaction modality is a low-level description of the task it performs. A complex interaction modality is described as the combination of low-level output from simple dedicated devices. This is regarded as a low level description of the complex modality. For example, a pointing device generates a stream of X-Y screen positions, and a gesture recognizer generates a stream of coded body positions and movements containing multiple X-Y-(Z) positions.

Host agents are used to implement the decoupling of modalities and devices. They generate the low-level output stream of the modality and communicate this to other agents in the system. Simple modality agents interpret the event stream from a device and generate an output stream, complex modality agents can receive input from multiple agents and produce an output stream based on the interpretation and combination of these input streams.

Agents can be located in any where in the system, on a networked computer, in an I/O device, or in an accessor. Changing a host agent's interpretation policies can alter the functionality of input modalities. For example, a user or an application might change the interpretation policies of a mouse, by having it behave as a one-button mouse, a two-button mouse, a three-button mouse, or even act as the return key.

5. Conclusion

There are systems that implement personalized user interaction using various methods, OAA with a task based delegation model that offers alternative interaction methods to applications (Cheyer, 1997). Most often these systems are found in the disability domain, (Law & Vanderheiden, 1998). Our solution differs from this by providing basic interaction alternatives without configuration and additional software. However, to implement natural interaction, we need to add software in the form of profiles based on descriptions of capabilities and preferences, and software that provides profile management.

In this paper we introduce the AAOSA agent network as a design methodology to decouple the interaction model from the application. Such a decoupling provides a highly configurable platform. Furthermore, we introduce the AAOSA host agents that make it possible to dynamically reconfigure the behavior of a system based on new information about user preferences, device profiles and application profiles.

We also define a model that separates interaction modalities from their devices. We use host agents as the method of implementation, interpreting information from input and output devices at a low level, and generating low-level output streams that are communicated to other agents in the system. This enables new interaction modalities to be relatively easily defined by combining multiple low-level streams.

Using these techniques we have simplified the implementation of natural interaction making it possible to approach and immediately interact with an application. All of the information required to automatically configure the user's interaction is in the user's profile, and will be forwarded and installed upon contact.

References

Adomavicius, G., Tuzhilin, A., (1999), "User profiling in personalization applications through rule discovery and validation". In *Proceedings of the 5th ACM SIGKDD 1999.*

Bray, T., (1998), Extensible Markup Language, W3C Recommendation. http://www.w3.org/TR/1998/REC-xml-19980210

Cheyer, A., Moran, D., Julia, L., Martin, D., (1997), "The Open Agent Architecture and its Multimodal Interface". in *Proceedings of IUI'97,* Orlando.

Genesereth, M. R., Singh, N. P., Syed, M. A., (1994), A Distributed and Anonymous Knowledge Sharing Approach to Software Interoperation. http://logic.stanford.edu/sharing/papers/fgcs.ps

Hinckley, K. (1996), "Haptic Issues for Virtual Manipulation", *PhD Thesis in CS, University of Virginia.* http://www.research.microsoft.com/Users/kenh/thesis/front.htm

Hodjat, B., Amamiya, M., (2000), "Introducing the Adaptive Agent Oriented Software Archicture and Its Application in Natural Language User Interfaces", in *Proceedings of 1st international workshop on AOSE,* Limerick

Huhns, Michael N., (2000), "Interaction-Oriented Programming", in *Proceedings of 1st international workshop on AOSE,* Limerick

Jacobs, N., Shea, R., (1996), The Role of Java in InfoSleuth: Agent-based Exploitation of Heterogeneous Information Resources, *IntraNet96 Java Developers Conference,* http://www.mcc.com/projects/infosleuth/publications/intranet-java.html.

Jennings, Nicholas R., (1999), "Agent-Oriented Software Engineering" Proc. 12th Int Conf on Industrial and Engineering Applications of AI, Cairo, Egypt, 4-10 [Also appearing in Proc. 9th European Workshop on Modelling Autonomous Agents in a Multi-Agent World (MAAMAW-99), Valencia, Spain

Jonsson, I. (1999), The Decopled Application Interaction Model, D.A.I.M, *Proceedings of HCII'99,* Munchen.

Law, C., Vanderheiden, G., (1998) , "EZ Access Strategies for Cross-Disability Access to Kiosks, Telephones and VCRs", Presented at *CSUN 1998,* Los Angeles

Morrison, A., (1998) A Specification Paradigm for Design and Implementation of Non-WIMP User Interfaces. Proceedings of CHI'98, Los Angeles.

Scott, N. (1991) The Universal Access System, American Voice Input/Output Society Conference, Atlanta.

Scott, N., Jackson, J., Jonsson, I., (1999) The TASCLOUD: a Networked Total Access System, CSUN'99 , Los Angeles.

SUN Microsystems, The Jini Whitepapers and specifications, Waldo, J., (1998), http://java.sun.com/products/jini/specs/

Titmuss, R., Winter, C.S., Crabtree, B. (1996), Agents, mobility and multi-media information, *Proceedings of First International Conference on Practical Application of Intelligent Agents and Multi-Agent Technology,* London.

Voice Windows : A Script-Based Platform in Windows for Visually Impaired

Tuneyoshi Kamae[a], Hirohiko Honda[b], Takayuki Watanabe[c], Tomio Koide[d], Tohru Kurihara[e]

[a] Stanford Linear Accelerator Center, Stanford University,
MS 43A, P.O. Box 4349, Stanford, CA 94309, USA
[b] Dept. of System and Communication Eng, Shonan Institute of Technology,
[c] Department of Information Science, Shonan Institute of Technology,
[d] Create System Co Ltd, Tachikawa, Japan,
[e] Department of information sciences(for the blind), Tsukuba College of Technology

Abstract

Voice Windows (VW) is a platform on Windows that facilitates script-based barrier-free programming for users including the visually impaired. It is an extension of Windows Scripting Host (WSH) of Microsoft. VW adds to WSH a set of user interfaces with built-in voice guidance and large-format Graphic User Interface (GUI). These new interface can be accessed as objects or called as processes from VBScript codes on WSH. VW comes with a voice-assisting editor that suggests objects, methods and properties to users, similar to Visual Basic Editor of Microsoft does. Through use of Visual Basic for Application (VBA) supported in WSH, users can easily write barrier-free programs with voice guidance and large format GUI. To enhance joy of programming, VW supplies objects to use animation such as Microsoft's Agent and Hitachi's TVML. Present work aims to assist the visually impaired in programming for office work; to assist programmers in writing applications with voice-guidance; and to encourage the visually impaired students to program on computers.

1. Introduction

Personal computers have been used by the visually impaired, as by the sighted, to produce documents, navigate through web pages, and, read and write emails. In fact visually impaired programmers developed sophisticated application programs such as Emacspeak by T. V. Raman [1] and VDM100 by Masao Saito [2]. The two programs run on text-based platforms, Emacs and MS-DOS, respectively. As GUI-based Windows systems became available, however, many MS-DOS-based programs ceased to operate. Awareness to assist the visually impaired has risen and many new programs are available now. However, they fall short of assisting programming for practical office work. Barrier-free computer interface is now becoming the norm in the public service sector. Programmers' work will be much simplified, should they be supplied with a standard set of user interfaces with needed functionality. The present work aims at supplying a platform on which the visually impaired can write sophisticated programs for their office work, and the sighted as well as the visually impaired can write barrier-free applications with voice-guidance on Microsoft Windows [3,4]. We have added a third goal, accessibility to animation programs, hoping to invite students to the joy of computer programming. Included in VW is a dedicated editor (VWEditor) that holds all objects, methods, and properties in a database and advises users in coding through voice-guidance [5,6]. Voice Windows runs on the Japanese version of Windows98 and requires Toshiba's Japanese Text-to-Speech (TTS) and Microsoft's English TTS packaged with Windows98. An audio driver conforming to the Windows Driver Model (WDM) is also required for some applications. To run Microsoft's Agent [7] and TVML [8], users have to install respective programs.

2. Windows Scripting Host, VBScript and Voice Windows

The Windows Scripting Host (WSH) is an Object Orientated (OO) platform to run script-based Web languages outside of the Web browser. VBScript, JScript, and Perl run on WSH, but only VBScript is supported in the current version of Voice Windows. MS-DOS Shell scripts are also supported in WSH and thus in VW.

VBScript includes a wide variety of objects, methods, and properties to manipulate files and access networks. WSH has its own system related objects, methods, and properties. VW supplies additional I/O objects furnished with elaborate voice-guidance. Objects supplied by WSH, VBScript and VW often perform similar function and users are left to choose according to their need. For example, to transfer a file by FTP, users can specify the host address, the path and the file name as properties of an Internet FTP object supplied by WSH or call a dialogue box with voice-guidance supplied by VW. Similarly users can open files by specifying its path and name speechlessly or invoke a VW dialogue to search for a file assisted by voice guidance.

Users can also control the Text-to-Speech (TTS) functionality in various ways. When VW objects are called, a standard Japanese and English TTS are built in. Users can set up a TTS in WSH to speak whatever sentences in their VBScript program. In either case, users can specify the TTS engine, the voice character, and the voice speed.

All VW programs are written in Visual Basic 6 and most are made accessible as objects, methods, and properties. They can also be run through the MS-DOS shell. Microsoft Agent has its own objects, methods, and properties. TVML is an independent package running on its own platform, but has been adopted to VW through a set of dedicated script interpreter. Windows runs processes on multiple threads and users often face difficulty in controlling timing among processes. VW supplies a program to control the timing but leaves problems when multiple processes are launched from WSH. At present we see no easy way to solve this problem.

3. Programming on Voice Windows

Voice Windows supplies utility programs of the following categories:
   A) I/O programs with voice-guidance for data input and message output.
   B) Common Dialog Boxes with voice-guidance for files and printers.
   C) Process control programs with voice-guidance for launching a new process and waiting for a process to end.
   D) Internet programs with voice-guidance for email and FTP.
   E) The dedicated editor (VW Editor) with voice-guidance for VW programming.
   F) Programs with voice-guidance to make a TVML presentation.

All programs can be called through MS-DOS's shell. A), B), C), and D) are also accessible as objects, methods, and properties from VBScript programs.

A typical user program is shown in Table 1: it consists of WSH scripts for various setups and a main program. In the program, two ways to use VW I/O programs are shown. One is by calling an executable through WSH shell. The other is to use as an object, set properties, and call methods. WaitDoEvents.exe towards the end of the program holds execution of the VBScript commands after the line for 5 seconds.

4. Use of Animation Programs

One aim of this work is to bring joy of computer programming to the visually impaired and to introduce barrier-free programming to high school and college   computer classes. Animation with synthetic voice is included as an optional with a hope to attract younger generation to the programming. To facilitate otherwise complex animation sequence control, various short moves are available to WSH/VBScript. Users can arrange these moves in a sequence and insert speech texts for moves with speech.

MS-Agent supported by Microsoft offers various characters and their moves. Many animation creators offer, worldwide, other MS-Agent characters for free download [8]. They can be incorporated as objects and controlled via WSH and VBScript as shown in Table 1. Software to create new characters and new moves is also available in market.

TVML is software developed by Hitachi Kokusai Electronics and Japanese Public Broadcasting Corporation (NHK) to produce television shows using animation characters. TVML allows coordinated moves (eg. dialogue) between two characters, scenery settings, and camera motions. Short scenes and acts of TVML have been provided to users that can be combined to make a show, a presentation, or a dialogue [9].

5. Editor for the program development

Voice Windows (or WSH) operates on scripts written in relatively simple rules and vocabulary. We developed an editor (VWEditor) similar to Microsoft's VB Studio for programming on the Voice Windows platform with proper voice guidance for the visually impaired.

VWEditor holds database of objects, methods, and properties accessible to users and presents possible choices by voice as users type in program codes. The Editor also has a standard set function with voice guidance as in typical screen editors and screen readers. Users can also assign shortcut keys to frequently used functions. Anticipating use of users' favorite screen reader concurrent to VWEditor, a function is added to VWEditor to mute its voice guidance.

Table 1: A WSH/VBScript sample program on Voice Windows

```
'Definition of paths.
Const VWHomePath = "C:\VWHome"
Const VWExePath = "C:\VWHome\VWExe"
'Setup to use WSH Shell.
Dim WshShell
Set WshShell = WScript.CreateObject("WScript.Shell")

'Setup to use Toshiba TTS from VBScript.
Dim TOSTTSX1
Set TOSTTSX1 = WScript.CreateObject("TOSTTSX.TOSTTSXCtrl.1")
TOSTTSX1.Speed = 20                 'Set the voice speed of Toshiba's TTS

'Setup to use characters in Microsoft Agent.
Dim Agent1, Genie
Set Agent1 = CreateObject("Agent.Control.1")
Agent1.Connected = TrueAgent1.Characters.Load _
              "Genie","C:\WINDOWS\MSAGENT\CHARS\Genie.acs"
Set Genie = Agent1.Characters("Genie")    'Define a new name
Genie.LanguageID = &H0409                 'Let Genie speak English
Genie.Get "State", "Showing, Speaking"    'Load animation sequences

'Setup for VW objects.
Dim ObjMsg
Set ObjMsg=WScript.CreateObject("VWWSHMsgBox.clsVWMsg")

'Sample use of a VW object, VW message box
ObjMsg.Message="Good morning friends."
ObjMsg.VoiceSpeed=m          'Medium speed (1.5 times normail) chozen.
ObjMsg.VoiceChar=f           'Female voice chozen.
ObjMsg.TimeOut=3             'Timeout set.
ObjMsg.VWMsgBox             'Message box displayed and spoken on screen.

'Setup to use VWMsgBox through WSH shell.
Dim Msg1
Msg1="Soon you will see Genie on the screen."
WshShell.Run VWExePath & "\VWMsgBox.exe ;" & Msg1 & ";m;m;00;4",1,True

'Genie appears and speaks.
Genie.MoveTo 150,150
Genie.Show
Genie.Speak "Hello, Friends!"

'Wait until Genie's speech ends.
WshShell.Run VWExePath & "\WaitDoEvents.exe ;5;0", 1, True

'Hide Genie from the screen and unload the objects.
Genie.Hide
set Genie = Nothing
```

6. Summary

Voice Windows (VW) is a platform based on Microsoft's Windows Scripting Hosts that supports barrier-free programming in MS's VBScript, for the visually impaired as well as for the sighted. It adds a new set of voice-guided interfaces for standard I/O operations, controlling other processes, and running animation sequences. VW has its own editor (VWEditor) which facilitate VW programming. The platform is aimed at the visually impaired students and the programmers unfamiliar with barrier-free voice-assisted programming.

The present Windowes98 version of VW is still in developmental stage. Functionality of WSH has been expanded in Windows2000, where VW will become even more powerful allowing many standard VBScript subroutines be supplied as an XML library. We hope to migrate to Windows2000 so that accumulation of user-supplied programs makes VW programming even easier. In addition, other languages WSH supports (JScript and Perl) can also be implemented in Voice Windows.

Acknowledgements

This work has been supported by Internet Technical Research Committee (ITRC) and by the Mirai Kaitaku project of Japan Society for Promotion of Science (JSPS-RFTF97R16301). We gratefully acknowledge their financial aids.

References

1. Raman, T.V., *Audio System for Technical Readings*, PhD Thesis, Cornell University (1994); *Audio User Interface – Toward the Speaking Computer*, Kluwer Academic Publishers (1997)
2. Saito, M. (1995). *Development a software for visually impaired, Information processing*, Vol. 36, No.12. 1116, (in Japanese).
3. Kamae, T. (1999). Use of Internet Based on a Voice-Based Platform for the Visually Disabled. In *Proceedings of 1999 Annual Meeting of Information Processing Society of Japan* (in Japanese).
4. Kamae, T., Watanabe, T., Honda, H., Koide, T., Kurihara, T. (1999). Introduction to Voice Windows. *Presentation in the 5th ITRC Symposium* (in Japanese).
5. Honda, H., Kamae, T., Watanabe, T. Koide, T., Uno, S, Kurihara, T. (2000). *Voiced Scripting Host – Development of a script-based Win98 System for the visually disabled*--, Tech. Report of IEICE SP2000, Vol. 99 No. 2, 53 (in Japanese).
6. Honda, H., Kamae, T., Watanabe, T., Koide, T., Uno, S., Kurihara, T. (2000). *Utilization of Windows applications By Voced Scripting Host System, Tech. Report of IEICE SP2000*, Vol. 100 No. 256, 23 (in Japanese).
7. See the following Web site for Microsoft Agent: msdn.microsoft.com/workshop/imedia/agent/ and www.msagentring.org/
8. See the following Web site for TVML: www.strl.nhk.or.jp/TVML/English/Esitemap.html

Cognitive considerations in the design of multimodal input systems

Simeon Keates[a], P. John Clarkson[a], Peter Robinson[b]

[a]Engineering Design Centre, Department of Engineering, University of Cambridge, Trumpington Street, Cambridge CB2 1PZ. United Kingdom
[b]Computer Laboratory, University of Cambridge, New Museums Site, Pembroke Street, Cambridge CB2 3QG. United Kingdom

Abstract

Whilst it is generally viewed that offering a higher bandwidth for interaction between a motion-impaired user and a computer will improve the quality of the interaction, this paper offers evidence that simply providing a wider bandwidth is not enough. Unless there is a consistent approach to mapping the system to behave as the user expects, then the potential for confusing the user is great. This is particularly true for motion-impaired users. This paper will show that such input systems must not become too cognitively demanding for users and hence urges that only additional modes that truly augment the input system be used. This discussion shall draw on the results from user modeling and cursor control experiments and a prototype multimodal input system using gesture input.

1. Introduction

Users with a number of different physical impairment conditions have the same desire to use computers as able-bodied people [1], but cannot cope with most current computer access systems [2]. Such conditions include Cerebral Palsy, Muscular Dystrophy, spinal injuries or disorder, arthritis and strokes. Frequent symptoms include tremor, spasm, poor co-ordination, restricted movement, and reduced muscle strength. Any computer input system intended for use by people with varying physical capabilities and designed around one method of input is unlikely to be flexible enough to cope with the diverse needs and demands of the users satisfactorily. This is not to say that it might not suffice, but for extended computer usage something more flexible and with a broader bandwidth may be required. This idea is supported by evidence that suggests increasing the degrees-of-freedom of input devices, such as incorporating finger flexion, can improve interaction rates [3].

Extending this principle to include more degrees-of-freedom through multiple input channels, implies that this should also yield improved information transfer rates. When designing a multimodal input system, the immediate concerns of designers will usually be focused on the technical aspects of the system, how to allocate I/O resources and so on. When the user is considered, the temptation is to regard physical load alone as the most important factor. However, by examining the results from three case studies assessing how motion-impaired users interact with computers, this paper will show that cognitive loads are as, or arguably even more, important.

The first case study shall address how motion-impaired users interact with computers through the calibration of a straightforward user model. The second case study will examine attempts to offer assistance to cursor control tasks and the third will investigate the effects of cognitive loading on a multimodal gesture input prototype. The case studies are individually comprehensive studies that cannot be described in completeness within the limits of this paper. Consequently, the discussions of the methods and results analysis are kept necessarily brief. References are included for the more detailed analyses.

2. Case Study 1: Modeling the motion-impaired user

User models are almost always calibrated exclusively on able-bodied subjects. One of the most straightforward user models is the *Model Human Processor* [4]. This model segments the interaction into three functional steps: the time to perceive an event; the time to process the information and decide a response; and the time to perform the response. Total response times to stimuli can thus be described by:

$$Total\ time = x\tau_p + y\tau_c + z\tau_m \qquad (1)$$

where x, y and z are integers and τ_p, τ_c and τ_m correspond to the times for single occurrences of the perceptual, cognitive and motor functions. User trials were performed to calibrate the user model with four able-bodied and six motion-impaired users. The results of the calibrations are shown in Table 1, along with the values obtained by the developers of the model, Card et al. Of particular interest are the motor function times, which were derived from the times to press and release a key. The theory predicts that these actions should be automatic and correspond to τ_m only.

Table 1: The calibrated Model Human Processor times averaged across the user groups.

User group	τ_p (ms)	τ_c (ms)	τ_m (ms)
Card et al.	100	70	70
Able-bodied	80	90	70
Motion-impaired	100	110	210

A cursory analysis of the results indicates that motion-impaired users have similar perceptual and cognitive performance to able-bodied users and that the source of the delays in their interaction is purely due to slow motor function times. However, examination of the distribution of τ_m times for the users showed a different explanation. Figure 1 shows the key-down times obtained for one of the motion-impaired users. The largest peak, the second one, corresponded to the average value of τ_m observed. However, the first peak is very similar to the times obtained for the able-bodied users. Further analysis showed that the differences between the peaks corresponded to τ_c and indicated that the cause of the apparently slow motor function times was extra cognitive steps being inserted into actions that were presumed to be automatic.

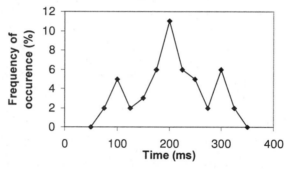

Figure 1: The key down times for one of the motion-impaired users

The extra cognitive loading is thought to originate from the extra demands required to overcome the impairments and produce sufficiently accurate motion. The effect of those cognitive steps is to slow down and degrade the interaction. Consequently, the hypothesis is that the rate of interaction of motion-impaired users using multimodal input systems can be adversely affected by cognitive overloading, more so than for able-bodied users. The aim of this paper is to examine whether there is any validity to this hypothesis.

3. Case Study 2: Cursor control

The existing keyboard/mouse/monitor paradigm relies principally on visual feedback, sometimes supported by sound. It is known that motion-impaired users often have difficulty with point-and-click tasks, so the aim of this case study was to investigate whether different methods of assistance could help make the task easier [5]. The methods of assistance chosen could be classified as approaches to make it easier to either: (i) move the cursor towards the target; or (ii) recognize when the cursor was over the target. Both approaches were attempts to make the task cognitively easier and should have improved the times to completion.

Four motion-impaired users participated in a series of trials using a Logitech force feedback mouse. The users exhibited a range of symptoms, including spasm and constant tremor, arising from medical conditions such as Cerebral Palsy and Friedrich's Ataxia. The experiment was performed with the following types of assistance to support the user:

- *None* - no additional feedback offered, i.e. normal mouse operation
- *Pointer trails* - cursor trails from the MS Windows accessibility options
- *Color* - the target changes color once the cursor is over it
- *Gravity well* - a force feedback 'gravity well' effect added to the target
- *Vibration* - the mouse vibrates once positioned over the target
- *All* - all of the above assistance was combined

The gravity well effect for the force feedback is best described by imagining that there is a circle around the target that corresponds to the extent of the gravity field. Entering that outer circle causes the cursor to become subject to the gravity and it is attracted by a spring force towards the center. The average times obtained across all the users are shown in Figure 2.

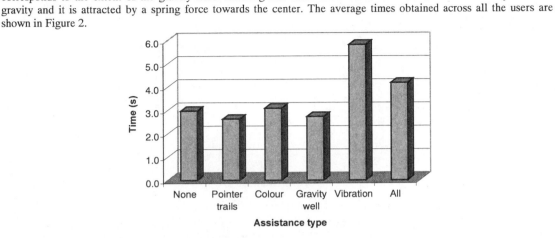

Figure 2: The average time per target for the different types of user assistance.

It can be seen from Figure 2 that the use of a gravity well around a target and pointer trails both appear to improve the time taken to complete the task by approximately 10-20%. Both types of assistance aid the guidance of the cursor. It can be argued that the gravity well provides physical motor assistance to the user and that this is responsible for the time improvement, rather than the cognitive assistance. However, this argument does not apply to the use of pointer trails, which showed a similar improvement. The addition of vibration and color were retrograde steps, with vibration almost doubling the time to perform the task. The vibration results mirror those obtained from other research with able-bodied users [6]. All of the users expressed displeasure at the vibrating sensation when performing this task and the color change of the target was distracting. In both cases, the negative cognitive effects of the assistance outweighed the anticipated benefits from knowing that the cursor was on the target. The negative effects of the presence of these forms of assistance in 'All' also outweighed the beneficial effects of the pointer trails and gravity well.

4. Case Study 3: Gesture-based multimodal input

A prototype system was developed for investigation of the interaction involved in multimodal input [7]. For the purposes of the prototype system, two gestural input channels: the head and the hand, and an in-house gesture recognition system, Jester [7] were used. A Polhemus was used for the head gesture recognition, as the users had previously experienced using this input mode and device in an earlier set of trials [8]. An analogue joystick was used for the hand gestures, being similar in nature to a wheelchair joystick.

Rather than just record recognition rate, a scoring system was developed to reflect the presence of non-recognition and misrecognition. Correct recognition earned +1 point, non-recognition scored 0 points, and an incorrectly recognized gesture received a -1 score. These scores were then scaled to a maximum score of 100 to eliminate dependence on the number of gestures recorded per test. This gave a complete potential score range of -100 to +100. By definition, any score below 0 would be entirely unacceptable, because the user would spend all the time trying to correct errors, and could never produce any useful input beyond the simplest of tasks.

The gesture alphabet was chosen to be easy to learn. It consisted of 4 directional gestures, UP, DOWN, LEFT and RIGHT, and 2 oscillatory gestures, YES (UP-DOWN-UP) and NO (LEFT-RIGHT-LEFT). Two experiments were conducted. In both cases, the users were directed to produce the gestures by verbal and on-screen prompts. The gesture names were used as prompts to minimize the cognitive load on the users and a visual cue providing symbolic representation of the gestures was drawn and positioned at the bottom of he screen as an additional aid.

The first experiment involved measuring the effect of alphabet size on recognition rate for an individual input mode. User trials were conducted with 8 motion-impaired users and one hour per user every alternate week. Figure 3 shows the comparative scores obtained for Single Mode input with an alphabet of three gestures, YES, UP and DOWN, and those for the 6 gesture alphabet. It can be seen from Figure 3 that the 3 gesture alphabet consistently outscored the 6 gesture one. The three extra gestures were defined to be similar in nature to the original three, so should not have required an additional cognitive effort to produce. A brief study also showed that they were also more reliable than the original three gestures when used as a 3 gesture alphabet. Consequently the only explanation for the discrepancy between the two alphabets seen in Figure 3 is the effect of having to remember, decide between and cognitively process the six gestures.

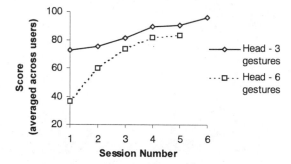

Figure 3: Single Mode head gesture input scores averaged across the users

The second experiment combined the inputs into a multimodal input system. Tests were completed with each mode individually (*Single*) and combined with both inputs being either identical gestures (*Duplicated*) or different ones (*Different*). Exposure to the tests was staged in the order Single, Duplicated and Different so as to coincide with the increasing cognitive load. For the multimodal tasks, Duplicated and Different, users were encouraged to produce both gestures simultaneously. However, this proved too difficult for Different and all users produced sequential gestures instead, typically the head and then the hand gesture. Figure 4 shows the scores obtained, averaged across all the users.

Under BM Different input, the users eventually ceased to look at the screen prompts, choosing instead to rely solely on the aural prompt from the computer operator reading the required gestures aloud. They later reported that this was because it was easier to remember the instructions when spoken. This enabled the users to recall the entire instruction without having to continually refer to the information source and has implications for future multimodal systems if the users become so overloaded by the interaction process that they cannot produce inputs without constant double-checking of the prompt.

The scores generated by the multimodal input test for the 6 gesture vocabulary are shown in Figure 4. The highest scores overall came from the Single Mode head input, and show a typical learning curve as predicted by the Power Law of Practice [4]. That Single Mode head scored more highly than Single Mode hand is probably because the users were already familiar with head gestures from the first gesture experiment, but that effect was virtually canceled out by the 5th session, when the users were becoming equally familiar with both.

Both the Single Mode inputs scored more highly than their direct counterparts in the (Both Modes) Duplicated tests.

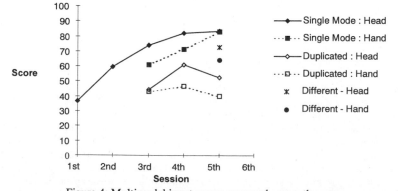

Figure 4: Multimodal input scores averaged across the users

This appears to confirm the observations from the first experiment that increasing the users cognitive load has an adverse effect on the interaction rate. In this the extra cognitive load arises from remembering the task, learning a new process and controlling a wider range of muscle groups simultaneously, which had increased from the original Single Mode test.

The (Both Modes) Different trial was similar to the Single Mode ones, no more than one gesture was being produced at any moment in time, but the cognitive load was increased, because the user was responding to more complicated instructions. It can be seen from Figure 4 that the Different scores were lower for both input modes when compared to the Single Mode score from the same session. This is the fairest comparison, because the users were significantly more familiar with generating gestures by this stage, so comparison to earlier sessions would not give an accurate reflection.

Analyzing the scores, whilst giving a valuable insight into the relative reliability of the particular mode combinations, does not fully describe the data transfer rates achieved as no account is made of the relative alphabet sizes involved. To rectify this, the bit rates of useful information transfer were calculated from the following equation from information science theory [10]:

$$Rate = \frac{\log_2(Vocabulary\_size) * \frac{Score}{100}}{Time\_taken} \tag{2}$$

For each of the Single modes and for both modes Duplicated, the alphabet size was 6 gestures, whilst for Different it was 36 gestures (6*6). Using equation (2) to calculate the data bit rates achieved, the Single Mode head and hand inputs achieved peak rates of 0.72 and 0.77 bits/second respectively. This compares to the 0.65 bits/second obtained for Duplicated and 0.56 bits/second for Different. Thus the highest data transfer rates were achieved with the cognitively easiest task, despite having the equal smallest alphabet size.

5. Conclusions

It was seen in the first case study that motion-impaired users add extra cognitive load to actions that would usually be assumed to be automatic. This implies that extra effort needs to be placed on minimizing the cognitive load on the users to prevent the need for even more extra cognitive steps and slowing the interaction down further. The second case study showed an attempt to achieve that by offering different kinds of assistance to cursor control activities. However, while some of the assistance types were helpful, some were definitely unhelpful. This shows that adding extra feedback is not in itself a solution, unless the feedback is genuinely helpful. The final case study looked at the effectiveness of a multimodal input system. The results of that case study showed that, as expected, the cognitive demands on the user had an adverse effect on the quality of the interaction. This implies that simply adding more input channels does not necessarily improve the interaction. Therefore, as with the assistive feedback from the second case study, it is necessary to add only genuinely useful channels, otherwise the users can become overloaded and less able to complete their intended tasks.

References

1. Busby, G. (1997) Technology for the disabled and why it matters to you. *IEE Colloquium Digest Computers in the service of mankind: Helping the disabled*, 1/1-1/7.
2. Edwards, A.D.N. (1995) Computers and people with disabilities. In A.D.N. (ed.), Edwards *Extra-ordinary Human-Computer Interaction* (pp. 19-44). CUP, Cambridge.
3. Zhai, S., Milgram, P., Buxton, W. (1996). The Influence of Muscle Groups on Performance of Multiple-Degree-of-Freedom Input. *Proceedings CHI 96* (pp. 308-315).
4. Card, S.K., Moran, T.P., Newell, A. (1983). *The Psychology of Human-Computer Interaction*. Lawrence Erlbaum Associates, Inc.
5. Keates, S., Langdon, P., Clarkson, P.J., Robinson, P. (2000) Investigating the use of force feedback for motion-impaired users. *Proceedings 6th ERCIM UI4ALL Workshop* (pp. 207-212).
6. Oakley, I., McGee, M.R., Brewster, S., Gray, P. (2000). Putting the feel in 'Look and Feel'. *Proceedings CHI 2000* (pp. 415-422).
7. Keates, S., Robinson, P. (1999). Gestures and multimodal input. *Behaviour and Information Technology*, Taylor and Francis Ltd. January-February, 1999, 18(1), 36-44.
8. Perricos, C., Jackson, R.D. (1994). A head gesture recognition system for computer access. *Proceedings RESNA 94 Conference* (pp. 92-94).
9. Keates, S., Potter, R., Perricos, C., Robinson, P. (1997). Gesture Recognition - Research and Clinical Perspectives. *Proceedings RESNA 97* (pp. 333-335).
10. Welsh, D. (1998). *Codes and Cryptography*. Oxford University Press.

User interface and usability for phone based operation of remittance service in Automatic Teller Machine (ATM)

Iwao Kobayashi[a], Akihiro Iwazaki[b], Katsuhiro Sasaki[c]

[a]Faculty of Software and Information Science, Iwate Prefectural University,
152-52 Sugo, Takizawa, Iwate 020-0193, Japan

[b]Product Design Department, Design Center, Fujitsu, Ltd.,
4-1-1 Kamiodanaka, Nakahara-ku, Kawasaki, Kanagawa, 211-8588, Japan

[c]Miyagi Prefectural School for the Blind
6-5-1 Kamisugi, Aoba-ku, Sendai, Miyagi 980-0011, Japan

Abstract

In this paper we designed and evaluated the user interface of remittance operation of Automatic Teller Machine (ATM) for the blind users using a phone-based handset. The experiment, conducted by 13 normal subjects without any visual information except 12 keys of handset presented on a screen, showed that our new model was more effective in shortening the operation time than that of our previous model. However, it still took longer time compared to normal subjects with visual UI and both visual and auditory UI as in the previous experiment. In discussion we described the design of the model and future problems.

1. Introduction

Automatic Teller Machine (ATM) is set everywhere in the world and we are using it in our daily lives. Although it is convenient for people who can use it well, it has been shown that people with disabilities, especially blind people, have difficulties in its uses (Gill, 1997) and some researchers have examined the adequate ATM design for them (Vanderheiden & Law, 1998; Center for Accessible Environments, 1995).

In Japan, specially designed ATMs for blind users to access bank services have some limits in their function (Kobayashi & Sasaki, 1996). In particular, remittance service is not accessible for the blind though it is frequently used in the country (Japan Electronic Industry Development Association, 2000). In our previous research, we designed the system model for the remittance operation by the blind users, based on the operations by voice instruction and/or the feedback using a phone-based handset (Kobayashi, Iwazaki & Sasaki, 2000a). Then we implemented the prototype system and evaluated it by an experiment. The experiment clarified that such system is workable for the blind. On the other hand, mean time of operation was statistically longer than that of users without vision impairment, using visual interface and both visual and auditory interface of ATM. As a future problem, the design improvement was needed to reduce the time of total operation.

The purpose of the paper is to examine the design improvement to reduce the time of operation by the phone based ATM interface.

2. System model

2.1 Device

We used a handset on which fixed numbers of push buttons were built-in like a phone as a device to manipulate the ATM by blind users. They input the number of buttons to operate the ATM following the guidance announced via the phone such as a conventional phone-based user interface (UI) (Halstead-Nussloch, 1989).

Moreover, the buttons are used for input of Japanese character that is necessary during the operation of remittance. Although it has not been applied to ATMs for the blind yet because of the complexity of the input process of Japanese characters, recently Japanese character input using telephone keys has become wide spread in Japan (Kobayashi, Iwazaki & Sasaki, 2000b). So we used the input method in our system.

2.2 Operation

The operation for remittance, paying by cash, consists of 12 handling tasks. In our previous model, each task is a series of acts of guidance, selecting the key, feedback, and select for confirmation. Key assignment is decided as follow. Number keys are used to input the number or Japanese characters. Sharp key is to decide the key input or return. Asterisk key is to repeat the guidance or sub-guidance for beginners.

We improved the design guideline of remittance operation by phone based UI from the following points of view that were requested by subjects in the authors' previous research (Kobayashi et al., 2000a).

1) We simplified the verbose guidances.
2) Users can skip the guidance whenever they would like to do and go to the next step.
3) In every task, we skip the confirmation of contents of each task, and users confirm all contents as the last task instead.

3. Experiment

3.1 Subjects

The Subjects were 13 men who have neither vision impairments nor other physical impairments. Their ages were between 18 and 32 years old (mean age = 22.07). All subjects are the same subjects in our previous experiment. They have enough knowledge of the input method of the system and operation of ATM.

3.2 Experimental system

We implemented the system according to the above design guideline for evaluation of our new system model by normal subjects without visual information of operation except 12 keys of handset presented on a screen. The system was implemented on a personal computer based on Mac OS9 as an operating system.

The function of ATM was realized using macromedia Director 6J. Subjects used a mouse to operate 12 keys on the screen that corresponds with the same 12 keys of telephone. Guidance and sub-guidance were provided via a speaker.

3.3 Procedure

The subjects attended to the experiment one at a time. Before trials, an experimenter gave them the information which were necessary for remittance operation: names of a remitter and a remittee, the telephone number of the remitter, the bank and branch names, bank account number, and the amount of money to be remitted.

We took two trials in an experiment. In the first trial, subjects practiced how to use the system, step by step with the experimenter's explanation if necessary. After the trial, they started the second trial. In the trial, experimenter instructed the subjects that they should do the remittance operation by themselves as quick as possible. The experimenter measured the time of subjects' operations.

3.4 Analysis

After the experiment, the experimenter analyzed the mean and standard deviation (SD) of 13 subjects' time of operations and compared with the results of previous phone-based UI and with those of normal subjects with visual UI and both visual and auditory UI. We used SPSS 10.0J for Windows on a PC for the analysis.

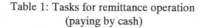

Table 1: Tasks for remittance operation
(paying by cash)

1. Selection of "remittance"
2. Selection of "payment cash"
3. Input of the remittee's name
4. Input of the name of the bank
5. Input of the name of the branch
6. Selection of "ordinary/current account"
7. Input of the bank account number
8. Input of remitter's name
9. Input of remitter's telephone number
10. Input of the amount of payment
11. Confirmation of input
12. Payment

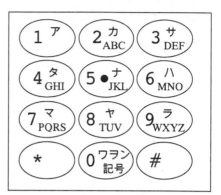

Figure 1: Key layout

3.5 Results

Mean time of operation was 403.0 sec., and SD was 29.7 sec. In order to compare the result with that of our previous experiment, in Figure 2 we illustrated results of both experiments.

We analyzed the data of the four conditions by Kruskal-Wallis test, which is a non-parametric test. This is because of the small number of subjects. The test yielded the significant effect on the 4 conditions ($p < .05$). Bonferroni-type multiple comparison tests, compared 6 times among all condition groups, showed that the operation time of our new model was significantly shorter than the operation time of the previous phone-based UI by 4 blind subjects ($p < .05$). However, it was significantly longer than that of normal subject with visual UI and both visual and auditory UI in the previous experiment ($p < .05$).

4. DISCUSSION

Statistically the time of operation was significantly shorter than that of the previous phone-based UI, but still longer than that of normal subject with visual UI and both visual and auditory UI. It should be added from Figure 2 that the minimum of our new model was smaller than the maximum of the visual and auditory UI in the previous experiment, and that no subjects took longer times in the experiment of our new model than the minimum of the previous phone based UI operated 4 blind subjects. While all blind subjects in our previous experiment had requested for improvement to reduce the time of operation, there was no such demand regarding the length of operation time in the new model of the experiment. These results indicate that the new model of the phone-based UI was effective in shortening the operation time than that of our previous model. The main reason for reduction in time seems to be skipping to the next step in the middle of guidance. Actually, all subjects began to operate next task before the end of the guidance for the present task.

For some reason, remittance service had not been accessible for the blind even by specially designed ATMs, though it is frequently used in Japan as we described in Chapter 1. We think that this is because adequate user interface for remittance service had not been examined well, for the simple reason that remittance needs a longer and more complex operation than other operations such as withdrawal and deposit. But our experiment showed that the user could operate the ATM for the remittance service without any trouble taking the mean time of 403.0 sec. by our new model. We think the function of the model should be included in ATMs, and "FACT-V" has been constructed by Fujitsu Ltd.. This is the only product including the function (Kobayashi et al., 2000b). However, it took still longer time of operation by normal subjects with the model than that with visual UI and with both visual and auditory UI in the previous research. It is necessary to improve the model to short much time of operation. We also need to examine the use of our new model by the blind subjects in the next research.

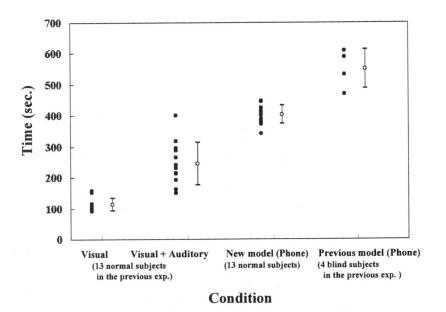

Figure 2. Time of operation

On the other hand, we need and would like to examine other banking methods and accessing of them by people with disabilities. For example, e-banking and the access by a mobile phone and/or a personal terminal has been the focus of attention and it is expected that the method will spread widely in future. Although some researcher have examined the UI of the mobile phone for people with disabilities or elderly persons (Gjoderum, Hypponen, Nordby, Ruud, Ekberg, & Martin, 1999; Ekberg, 2000; Gill, 2000), most studies have described only possibilities of the use of mobile phone or hand-held terminals as public access terminals.

Whereas the technology of mobile phone has been rapidly increasing in recent years such as i-mode in Japan, some researches showed the negative opinion on the use of mobile phone as public terminals from the viewpoint of UI (Reid, 2000; Nielsen, 2001). Therefore, it is also necessary to research on the other banking methods to provide the independent life of the people with disabilities and to avoid the expected digital divide.

5. Conclusion

In this paper we designed and evaluated the phone-based UI for blind users to operate remittance services in Automatic Teller Machine (ATM). Wc improved our system model of the previous research and implemented the system. Our experiment showed that the new model of phone-based UI was effective in shortening the operation time than that of our previous model. However, it took still longer time compared to normal subjects with visual UI and both visual and auditory UI in the previous research. In future we would like to examine the system for the blind subjects and to improve the design guideline much better.

References

Center for Accessible Environments. (1995, September 13). *Proposal for UK guidelines for improving access to ATMs and similar equipment.* Retrieved November 25, 2000, from the World Wide Web: http://trace.wisc.edu/world/ kiosks/cae/caeguide.doc

Ekberg, J. (2000, July 7). *Final Public Report: Telematics Applications Programme (Disabled and Elderly).* Helsinki: Retrieved January 15, 2001, from the World Wide Web: http://www.stakes.fi/include/final-report.htm

Gjoderum, J., Hypponen, H., Nordby, K., Ruud, S. E., Ekberg, J., & Martin, M. (1999, October). *Guidelines-Booklet on Mobile Phones.* Retrieved January 5, 2001, from the World Wide Web: http://www.stakes.fi/cost219/mobiletelephone.htm

Gill, J. M. (1997, May). *Access Prohibited? Information for designers of public access terminals.* London: Retrieved January 20, 2001, from the World Wide Web: http://www.tiresias.org/pats/index.htm

Gill, J. M. (2000, June). *Mobile telephony: Will future developments be accessible to visually impaired users?* Retrieved January 20, 2001, from the World Wide Web: http://www.tiresias.org/reports/potsdam.htm

Halstead-Nussloch, R. (1989). The design of phone-based interface for consumers. *Proceedings of the CHI'89,* Austin TX, USA, 347-352.

Japan Electronic Industry Development Association. (2000). *A Research on the Future Trend of Terminals in Finance.* Tokyo: Japan Electronic Industry Development Association. (in Japanese)

Kobayashi, I., & Sasaki, K. (1996, October). *The use of bank and ATM/CD of the Blind.* Paper presented at the 12th Symposium on Human Interface, Yokohama, Japan.

Kobayashi, I., Iwazaki, A., & Sasaki, K. (2000a). An inclusive design of remittance services for the blind users' operation of automatic teller machines (ATMs). *Proceedings of the Conference on Universal Usability,* USA, 153-154.

Kobayashi, I., Iwazaki, A., & Sasaki, K. (2000b). FACT-V: universal access and quality of interaction for automatic teller machine (ATM). *Proceedings of the 6th ERCIM Workshop "User Interface for All",* Italy, 360-361.

Nielsen, J. (2001, January 7). Mobile phones: Europe's next minitel? *Alertbox.* Retrieved January 30, 2001, from the World Wide Web: http://www.useit.com/alertbox/20010107.html

Reid, T. R. (2000, September 15). WAP, Europe's wireless dud? *The Washington Post,* pp. E01.

Vanderheiden, C. G, & Law, C. M. (1998). EZ access strategies for cross-disability access to kiosks, telephones, and VCRs. In I. P. Porrero, & E. Ballabio (Eds.), *Improving the Quality of Life for the European Citizen* (pp. xxxi-xl). Netherlands: IOS Press.

Sign Language Digitization and Animation

Tomohiro Kuroda, Yoshito Tabata, Mikako Murakami, Yoshitsugu Manabe, Kunihiro Chihara

Graduate School of Information Science, Nara Institute of Science and Technology
8916-5, Takayama, Ikoma, Nara, 630-0101, JAPAN

Abstract

In order to be accessible, computer systems should let users interacting in users' mother tongue. Therefore, computer systems must handle sign language to make them accessible for the hearing impaired and the Deaf. To handle sign language, computer system must have function to obtain and to display signs. This paper presents sign language digitization and animation techniques.

1. Introduction

Foregoing researches on HCI emphasize importance to make computer systems available for natural communication in user's mother tongue. Therefore, many researches have been working on natural speech recognition and generation. And several products are available now in the market.
The mother tongue of the Deaf is sign language. Therefore, computer systems must handle sign language to make them accessible for the hearing impaired and the Deaf.
To handle sign language, computer system must have function to obtain and to display signs.
This paper presents sign language digitization and animation techniques.

2. Sign language digitization

As sign language is given by gestures and facial actions, computer systems need to utilize motion capture or recognition techniques to digitize signs. Foregoing and ongoing researches on sign digitization can be classified into two types; video-based approach and sensor-based approach.

Video-based approach utilizes many foregoing researches on image-analysis and recognition. Vogler et. al. [1] realizes to recognize American Sign Language by a single camera mounted on user's head using HMM. Video-based systems can be a user-friendly input interface as it is available without any attachments to put on users. However, to arrange image sensors to avoid occlusions is almost impossible when target motions are so complicated as sign languages. Moreover, a previous research of the authors [2] shows that video-rate is insufficient for sign language.

Sensor-based approach utilizes motion-capture systems. Motion capture system can provide detailed motion in high frame-rate. However, foregoing motion capture systems have two problems to use as sign language input interface. One is that too many sensors interfere with users' free motion. Another is that foregoing systems cannot obtain sufficient motions to distinguish given signs.

2.1 Three-Sensor Motion Capture System

A motion capture system with a few sensor sets makes a computer system available in daily lives. The authors developed three-sensor motion-capture system [3]. Figure 1 shows the prototype.
The system obtains whole upper body motion using three sensor units; two gloves and a headset. The gloves obtain hand posture and 6D trackers attached on the wrists and top of the head obtain positions and orientations of wrists and heads.

The system estimates whole upper body motion from these obtained information depending on skeletal constraints of human body. The system utilizes Bezier curve, which holds C^2 connectivity, as an approximation of the backbone. Additionally, the system introduces an assumption that human shoulder tends to stay its default position. These assumptions enable the system to estimate whole upper body motion with a set of linear equations.

Figure 1: Motion capture result

2.2   Glove Type Input Device

Most important and complicated motion of signs is hand posture. Foregoing hand posture sensors such as or CyberGlove [4] cannot obtain sufficient information to distinguish signs. Research on sign notation codes such as HamNoSys [5] clears appropriate sensor arrangement to make glove available with the least number of contact and bend sensors. The developed prototype (Fig. 2) guarantees the appropriateness of the proposed sensor arrangements.

Figure 2: Prototype of glove type interface

3. Sign Animation

To be accessible for the hearing impaired and the Deaf, computer systems must display legible signs. Most of foregoing computer systems utilizes avatar-based CG animation to show signs, called "sign animation". This chapter presents two methods to make sign animation legible.

3.1   Finger Character Visualization

A sign animation should present whole upper-body motions for signs and small hand motions for finger spellings clearly at the same time. Most of avatars for sign animation are in the shape of standard human body. However, finger character given by the standard size avatar is illegible, because the size of the avatar's hand is few percent of whole screen.

Foregoing research using eye-tracker [6] clears that sign experts like the Deaf looks on face of speaker in most cases and watches strong hand while speaker speaks signs in which hand posture takes important role, such as finger characters. Therefore, the authors proposed a method that expands the avatar's strong hand while avatar spells as shown in Fig 3 [7]. The method increases legibility of sign animation without any visual mismatch.

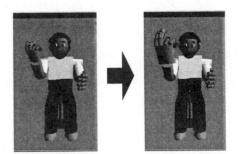

Figure 3: Legible visualization of finger character

3.2 Visualization of modifier expressions

Generated sign animation must be natural and grammatically correct. Although grammar of sign languages is decisively different from audible languages, basic grammatical structure is still unclear. Therefore, foregoing sign animations are generated via word-for-word translation. The authors tried to introduce several grammatical rules into sign animation. Preliminary research through books and interview clears that native signers express adjuncts by varying range, speed and frequency of modified words. The authors examined modifier expressions using the motion capture system mentioned above, and abstracted the decisive parameters of several words [8]. Experiment using computer generated sign animation varying abstracted parameters (Fig. 4) clears that the generated animation conveys modifier expressions and looks natural.

"Heavy"   "Very heavy"   "Extremely heavy"

Figure 4: Automatic Generation of Modifier Expressions

4. Applications

The authors developed several applications applying the developed techniques to testify effectiveness of the proposed techniques.

4.1 S-TEL: sign language telephone

The authors developed a prototype of sign language telephone named S-TEL integrating the developed motion capture system and the sign animation engine [1].
The prototype system is evaluated over UDP/IP satellite communication connecting two cities in Japan. The prototype developed on Win95 based PC with Pentium100MHz transferred 26.1 frames per second over 14 Kbps communication channel, and the users could understand 75% of spoken signs at first glance. Treatment of facial

expression may increase the legibility of given signs. The result guarantees an effective communication for sign conversation via a lossy channel.

4.2 Finger Character Learning System

The authors developed a finger character self-learning system based on the developed glove type input interface [9]. The comparison between users who learned finger characters via textbook and users who learned them via prototype self-learning system shows that the effectiveness of the developed interface and system.

5. Conclusion

This paper presents several methods for sign language digitization and animation. Precise digitization of signs is indispensable to make computer systems to recognize and handles given signs. To be accessible, computer system must be available on any kinds of natural communication. These proposed methods may increase accessibility of computer systems for the hearing impaired and the Deaf.

Acknowledgement
This research is partly supported by Telecommunication Advancement Organization International Communication Foundation, and CREST of JST.

References
[1]   C. Vogler, D. Metaxas, "A Linguistic-based Approach to American Sign Language Recognition", Conference on Gestures: Meaning and Use (2000), In Print.

[2]   T. Kuroda, K. Sato, K. Chihara, "S-TEL: An Avatar Based Sign Language Telecommunication System", International Journal of Virtual Reality 3.4 (1998): 21-27.

[3]   T. Kuroda, K. Sato, K. Chihara, "Reconstruction of Signer's Action in a VR Telecommunication System for Sign Language", International Conference on Virtual Systems and Multimedia, (1996): 429-432.

[4]   Virtual Technologies: CyberGlove User's Manual, (1994)

[5]   HamNoSys: http://www.sign-lang.uni-hamburg.de/Projects/HamNoSys.html

[6]   S. Kamei, Y. Nagashima, N. Seki, "Fundamental Study of Sight Line in Dialogue of the Sign", Human Interface News & Report, 12 (1997): 51-54, Japanese.

[7]   T. Kuroda, K. Chihara: "A Research on Visualization of Finger Spelling in Sign Animation", International Conference on Virtual Systems and Multimedia (1999): 46-52.

[8]   M. Murakami, T. Kuroda, Y. Manabe, K. Chihara, "A study on Sign Animation Including Modifier Representations", Technical Report of IEICE, WIT99-47 (2000): 137-140, Japanese.

[9]   Y. Tabata, T. Kuroda, Y. Manabe, K. Chihara, "Finger Character Learning System with Visual Feedback", International Conference on Disability, Virtual Reality and Associated Technologies (2000): 219-224.

Eye-scan patterns of Chinese when searching full screen menus

Wing Chung Lau, Ravindra S. Goonetilleke*, Heloisa M. Shih

Human Performance Laboratory
Department of Industrial Engineering & Engineering Management
Hong Kong University of Science and Technology
Clear Water Bay, Hong Kong

Abstract

With the rapid globalization of Information Technology, it is important not only to evaluate performance, but also to understand the underlying causes of performance differences among populations when using computer interfaces. In this study, the search strategy of three different groups of users (Mainland Chinese, Hong Kong Chinese and Non-native Chinese readers) was assessed. A full screen search task was used to understand the search strategies and their relationship to cultural and linguistic aspects. The stimulus materials were high or low complexity Chinese words having one of three layouts: Row, Column or Uniform separation. The search patterns were recorded using an eye tracking system. To categorize these patterns, a new measure called HV-ratio was developed. Analysis of variance on HV-ratio showed that the search patterns between Hong Kong Chinese and Mainland Chinese are significantly different. The HK Chinese used predominantly horizontal search patterns in all three layouts. In contrast, Mainland Chinese changed their search pattern depending on the screen layout. HK Chinese used a more systematic horizontal search pattern. These strategies explain the overall performance findings in relation to search time.

1. Introduction

As China enters the information technology era, it becomes important to understand Chinese users in an effort to produce software that is efficient and usable. It is known that primitives such as culture and language play an important role in interfaces made for people. In this study, we attempted to understand the impact of the Chinese language on visual search patterns.

Despite the fact that Chinese are one of the largest ethnic groups in the world, little is known about their visual search performance and search patterns. Most past studies have focused on visual search performance of Western populations even though differing language backgrounds may have differing scanning patterns (Nielson, 1990). The English language reader has an effective visual field or perceptual span that is asymmetric, extending about 4 characters to the left of the letter being fixated and about 15 characters to the right (Rayner et al., 1981). Pollatsek et al (1981) has shown language dependencies in relation to this asymmetry. The Chinese language is somewhat different as it can have two different orientations (a horizontal or 'Z' type orientation, starting in the top left corner of the paper and a vertical or inverted 'N' type of orientation, with text starting in the top right corner of the page) and the Chinese people are accustomed to both these orientations. A search field that fits the visual lobe efficiently is known to facilitate visual search performance. Hence, this presents us with an important parameter that needs investigation in relation to the visual search process of Chinese users. We hypothesized that the cultural difference could possibly interact with different types of layouts and have an impact on search performance.

In one of our previous publications (Lau et al, 2000), we presented the effect of culture when searching full screen menus and showed that Hong Kong Chinese, Mainland Chinese and non-native Chinese readers had significant differences in terms of preference and search time. In general, Hong Kong Chinese had significantly better performance on the top horizontal area of the search field, while Mainland Chinese had better search performance on the top left side of the screen. Non-native Chinese readers had no "preference" on any particular area.

Differences in search performance however did not provide a complete picture of the cultural effects on search strategy. In this study, we report the eye tracking data results in an attempt to quantify search strategies and understand the differences in search performance among the three "cultural" groups.

2. Methodology

2.1 Experimental task

The objective of the task was to find a Chinese word (target word) in a full screen search field. This was a two-step process; the target word was shown on the screen first (target word screen). After the subject "studied" this word, the search screen was shown. The search screen had the target word among many other words. The target word screen had the same layout as the search screen and had one target word laid out all over the screen. The reason for filling the screen with the target word was to prevent any starting position bias when the target word screen changed to the search screen. The experimental time was about 1 hour.

2.2 Experimental design

The independent variables of the experiment were population (Hong Kong Chinese, Mainland Chinese, and Non-native Chinese), Layout (Row, Column and Uniform), Word-complexity (High and Low). The experiment was a 3 (Population) × 3 (Layout) × 2 (Word-complexity) × 10 (Trial) design. Each participant completed a total of 60 trials (10 trials each for the 3 Layouts and 2 Word-complexities). The experiment was "blocked" by the 6 conditions (three Layouts and two Word complexities) with each block having 10 trials. The sequences were balanced among participants by using the six conditions to form a Latin square-like design.

The difference between the three layouts was the separation between row and columns. The format of the "Row" layout was similar to the horizontal writing format ("Z" type) of Chinese text, while the Column layout was similar to the vertical writing format ("N" type). The third layout was called "Uniform separation" where the horizontal and vertical separations between words were the same. In all these three layouts, the numbers of rows and columns were divisible by three thereby giving nine macro areas as shown in Figure 1.

1	2	3
4	5	6
7	8	9

Figure 1. The nine areas in the search field.

The Chinese words used in the experiment were selected based on their word complexity. Two word complexities (High and Low) were used in this experiment. 'Low complexity' words were those that had 10 to 12 strokes. 'High complexity' words were those that had 16-18 strokes. According to the Chinese dictionary (Hu-pei tzu shu chu pan she, 1989), low complexity words comprise 15-34 percentiles while High complexity words correspond to 60-81 percentiles of the total word distribution. Only Chinese words with a left-right format (for example 講 for "Talk") were selected. Those Chinese words with a top-down format (for example 葉 for "leaf") were not selected as Chinese words with a left-right format are generally much clearer on a computer display. The size of the Chinese words were 9mm*9mm, equivalent to a visual angle of 1 degree.

2.3 Participants

In this experiment, three groups (Hong Kong Chinese, Mainland Chinese, and Non-native Chinese readers) of 6 participants (a total of eighteen) were tested. Their age range was 13-36 years. All participants recognized all the Chinese words.

2.4 Materials and equipment

The experiment was programmed using Visual Basic 6.0 and run on a Pentium 200MHz computer in the Microsoft Chinese Windows 98 environment. The subject responses were acquired through a touch screen monitor and two external push buttons. The Applied Science Laboratories (ASL) 5000 eye tracking system was used to record the subject's eye movement.

2.5   Procedure

Prior to the actual experiment, each participant was given three practice trials. In all trials, the target word screen was shown first. After the participants had memorized the target word, they were asked to press the external "Continue" button to proceed to the search screen. When the search screen was shown, the participant had to find the target word as fast as possible. The time limit for each trial was set to 90 seconds. Once the participants found the target word, they were asked to "touch" that word on the touch screen monitor. If they were unable to find the target word or thought that the word did not exist on the screen, they were asked to press the external "Give up" button to end the trial.

3.   Results and analysis

The performance results of this experiment were reported in Lau et al (2000). This paper focuses on the eye tracking data. Researchers have categorized search into two main types - systematic search patterns and random search patterns. In this experiment, the systematic search patterns were most frequent (about 90%) and the random search patterns were less seen (about 10%).

In the context of our experiment, systematic search patterns are those performed row by row or column by column, until the target is identified. In this experiment, participants usually started a search at specific starting points (usually in the top left or top right corner) and then searched that row or column. After finishing that particular row or column, participants searched the next row or column respectively.

Eye scan patterns are generally not quantified and are very subjective. As a result, a new measure was developed to quantify the search patterns. The quantified variable is denoted as the HV-ratio. The direction of a saccade can be identified based on the coordinates of two consecutive fixations. To determine the HV-ratio, each saccade was divided into a vertical movement and a horizontal movement (Figure 2). The sum of all the absolute values of the vertical movements within a trial corresponds to the total vertical movement. The sum of all the absolute values of the horizontal movements within a trial corresponds to the total horizontal movement. The total vertical movement was then normalized with respect to the height of the search field and the total horizontal movement was normalized by the width of the search field. The ratio of these two quantities was defined as the HV-ratio.

$$\text{HV-ratio for a trial} = \frac{\Sigma \text{ (Horizontal movement i)/ (width of the search field)}}{\Sigma \text{ (Vertical movement i)/ (height of the search field)}}$$

If the scan pattern of a participant in a particular trial was dominantly horizontal (that is a search performed row by row), the HV-ratio would be large. If the scan pattern was dominantly vertical (column by column search), the HV-ratio would be low.

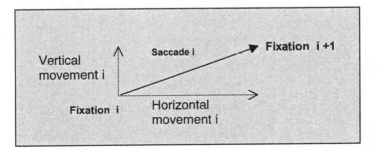

Figure 2. Components of the HV-ratio calculation with respect to the computer screen.

A 4-way (Population × Layout × Word complexity × Target position) ANOVA on HV-ratio showed a significant effect for Population (F (2, 709) = 118.93, p<0.0001). The *post-hoc* Student Newman Keuls (SNK) analysis showed Hong Kong Chinese had a significantly higher HV-ratio than both Mainland Chinese and Non-native Chinese readers. There were no differences between Mainland Chinese and Non-native Chinese readers. Layout was also significant (F (2, 709) = 4.75, p<0.009). The Row layout had a higher HV-ratio than both Column and Uniform separation layouts. The interaction between Population and Layout was also significant (F (4, 709) = 5.49, p<0.0002). Figure 4 shows the interaction plot between Population and Layout.

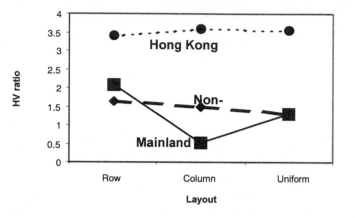

Due to the significant interaction, a simple effects ANOVA was performed for each Population group (Table 1). It was found that Layout was not significant for the HK Chinese and Non-native Chinese reader groups. However, the Layout was significant for the Mainland Chinese group (F (2, 243) = 48.21, p<0.0001), with each of the three layouts being significantly different from each other. For the Mainland Chinese group, the Row layout had the highest HV-ratio. On the other hand, the Column layout had the lowest HV-ratio, and the Uniform separation had a value between those of the Row and Column layouts.

Table 1. Simple effect ANOVA of HV-ratio for each Population group.

Population groups	F value (df) (probability)	SNK grouping of Layout* HV-ratio increases from left to right
HK Chinese (H)	F (2, 257) = 0.09 (0.91)	R E C (3.42) (3.53) (3.57)+
Mainland Chinese (M)	F (2, 243) = 48.42 (0.0001)	C E R (0.5) (1.29) (2.07)
Non-native Chinese readers (N)	F (2, 209) = 0.88 (0.42)	E C R (1.29) (1.47) (1.64)

+ The mean values are in parenthesis ( ).
*Levels sharing the same underline indicate no significant difference.

4. Discussion and conclusions

HK Chinese had relatively high HV-ratios in all three layouts implying that they mainly used a horizontal search pattern independent of layout (Table 1). The Non-native Chinese readers also had no significant difference in HV-ratio among the three layouts. For this group of subjects, the HV- ratio is around 1, implying that the Non-native Chinese reader group, in general, had no preference towards a horizontal or vertical search pattern in all three layouts. Mainland Chinese on the other hand seem to change their search patterns in different layouts. For the Row layout, Mainland Chinese used more horizontal search and in the Column layout, they used more vertical search (HV-ratio < 1 has been regarded as a vertically dominant search pattern), and for the Uniform Separation layout, Mainland Chinese have a HV-ratio that is in-between the value of other two layouts.

The differences in strategies between the two Chinese groups may be due to many different factors. Some of which are:

1. Differences in reading materials: Nowadays, in Hong Kong, the most common Chinese reading materials are newspapers and magazines printed having a vertical format. In Mainland China on the other hand, most of the reading materials (including newspaper and books) are printed in a horizontal format even though there is other reading material (e.g. fiction) printed with a vertical orientation. The reading materials in Mainland China are a mix of both orientations and as a result Mainland Chinese adopt a pattern depending on the stimulus material.

2. Another possibility may be related to the differences in computer usage. Hong Kong Chinese subjects have immense experience with the use of computers. The software packages that are frequently used in Hong Kong include Microsoft Office and Windows based software, which generally have a horizontal structure in their menu layouts. The icons in these packages are square in shape and have a pictorial format, a characteristic similar to Chinese characters. Therefore the search pattern of HK Chinese on the Chinese full screen menu may be influenced by the format of icon menus in computer software. The Mainland Chinese subjects had less experience with computers before they came to Hong Kong and as a result, they may have chosen a search pattern based on the reading pattern of a given Layout.

Thus, it may be concluded that the search patterns between Hong Kong Chinese and Mainland Chinese are significantly different. The HK Chinese used predominantly horizontal search patterns in all three layouts. In contrast, Mainland Chinese change their search pattern depending on the screen layout. HK Chinese used a systematic (about 90%) horizontal search pattern. This caused the HV-ratio of HK Chinese to be significantly higher than that of Mainland Chinese. The changing patterns of Mainland Chinese in the three layouts led to a significant interaction between Population groups and Layout (Figure 3). These results explain the findings of our previous study (Lau et al., 2000) that Hong Kong Chinese had better performance when the target was in the top horizontal area of the search field.

For interfaces similar to the ones we used in this experiment, it may be appropriate to consider horizontal layouts as Hong Kong Chinese have a horizontal scan pattern, and the Mainland Chinese tend to show flexibility in their search patterns depending on the layout. The results of this study have important implications for designers. Even though different population "groups" are familiar with similar characteristics (in this case, Chinese words), their performance on any given interface can be different depending on the underlying mechanisms that govern their usage. Understanding the underlying differences between "similar" groups or cultures and catering to such differences is the key to producing optimized interfaces.

Acknowledgments

The work was supported with funding from a HKUST Direct Allocation Grant (DAG99/00.EG14).

References

Han yu ta tzu tien pien chi wei yuan hui 漢語大字典編輯委員會 (1989). *Han yu ta tzu tien* 漢語大字典, Hu-pei tzu shu chu pan she 湖北辭書出版社, Wu-han 武漢。

Lau, W.C., Shih, H. M., and Goonetilleke, R. S. (2000). Effect of cultural background when searching Chinese menus. In *Proceedings of the 4th APCHI / 6th SEAES Conference 2000 (Ed. K. Y. Lim)*, (pp. 237-243), Amsterdam: Elsevier.

Nielsen, J. (1990). Usability testing for international interfaces. In J.Nielsen (Ed.), *Designing user interfaces for international use* (pp. 39-44) Amsterdam: Elsevier.

Pollatsek. A., Bolozky, S., Well, A.D., and Rayner, K. (1981). Asymmetries in the perceptual span for Israeli readers. *Brain and Language*, *14*, 174-180.

Rayner, K., Inhoff, A. W., Morrison, R. E., Slowwiaczek, M. L., and Bertera, J.B. (1981). Masking of foveal and parafoveal vision during eye fixations in reading. *Journal of Experiment Psychology: Human Perception and Performance*, 7, 167-179.

Measuring the performance of speech applications:
a user-centered approach

Nicole Leduc, Melissa Dougherty, Vytas Ankaitis

Nuance Communications, 1380 Willow Road, Menlo Park, CA 94025

Abstract

Well-designed speech recognition systems have the power to revolutionize the way people interact with computers in networked applications. This technology allows us to move beyond "Press or say One" type interactions which are widely available over the phone, to interfaces that allow a caller increasing latitude in saying what they'd like to do, when asked a naturally-phrased question. In the last five years, speech recognition systems like these have been implemented in hundreds of companies, replacing human operators and touch-tone systems.

In this paper, we address the issue of whether callers to speech recognition systems are satisfied with this way of accessing information and conducting transactions over the phone, and investigate which characteristics of a speech interface are most desirable to callers. We present findings summarized from our 1999 and 2000 Performance Benchmarking Surveys of deployed speech recognition systems and observations from usability studies conducted over the last two years that provide qualitative support for these survey findings. Finally, we discuss how this user research and the resulting insights have directly impacted our design considerations for speech-enabled services, especially in terms of conversational prompting, consistency, and the importance of information structure.

1. Introduction

In the last five years, speech recognition systems in the network have progressed from a handful of innovative companies, such as Charles Schwab (brokerage, transactions) and American Airlines (information), to hundreds of deployed systems in 25 languages around the world. More and more companies are replacing older touch-tone interactive response systems with speech recognition systems and as a more natural and friendly way of reaching out to their customers and potentially the billions of phone subscribers internationally. The types of speech applications that are most common are systems that allow users to access general information or information related to their own accounts, reservations, and orders. In addition, the emergence of voice dialing, through telecommunications companies, and voice web applications, through voice portals like TellMe and BeVocal, that allow consumers to access the Internet through their cellular phone, support mobile users who don't always have access to their PCs.

Given the millions of callers to speech recognition systems in the United States, there are surprisingly few publicly available studies that measure user satisfaction with these systems, or verify through usability studies what are the design elements and considerations that make the caller successful in using a speech system. Those that exist are usually company confidential. The information we present in this paper will therefore focus first on the annual quantitative Performance Benchmarking Survey that has been conducted by Nuance for the past two years to measure caller satisfaction with deployed speech recognition systems.

2. Performance benchmarking survey

Nuance conducts an annual quantitative survey designed to provide insight into consumer satisfaction, attitudes, and usage of V-Commerce, Voice Web, Call Center, and other deployed speech recognition systems. This study fulfills two objectives. It provides statistics on the impact of speech-enabled systems on callers to these systems, giving us the insight necessary to answer questions about the acceptance of voice interfaces. Second, it provides a baseline against which companies can compare their own applications, and gauge the strengths and weaknesses of these applications, comparing information from year to year.

The Performance Benchmarking Survey was conducted in 1999 and in 2000, with more than 500 users of six deployed speech recognition systems, using different voice-driven applications each year. The speech applications were in a range of vertical markets (travel, financial, communications and other business-to-consumer applications) and enabled the caller to perform a variety of tasks (seeking information, connecting phone calls, checking status, ordering products, accessing information, etc.). The survey was conducted as an over-the-phone interview by an independent research agency, Evans Research, Inc. and analyzed by Nuance's Voice Interface Services group. Users were recruited through lists of recent callers to the speech systems, provided by the companies deploying the systems. A primary goal of the survey was to measure user satisfaction with deployed speech recognition systems. The survey also included demographic, attitudinal and usage–related questions. All the systems surveyed were in the United States.

2.1 User profile

The users interviewed in the Performance Benchmarking Surveys were actual callers to the "live" systems included in the survey. They were not necessarily representative of speech users overall in the United States, but they do provide an important initial measure for real-world caller reactions to speech applications.

The 1999 Performance Benchmarking Survey suggested that the more classic early adopters of technology were using speech. The 1999 users were higher-income individuals (median income of $82,000), primarily male (79 percent), younger, (61 percent aged 30-49) and users of multiple technologies (personal computers, cell phones, handheld devices, etc.). Seventy-three percent of the speech users had also used the web site of the company offering the speech system, with respondents under 50 even more likely to have used the company web site. In the 2000 Performance Benchmarking Survey there appears to be a shift towards a more balanced user sample, including: more female users (35 percent, up from 21 percent in 1999), more users over 60 years old (29 percent up from 18 percent in 1999) and more users who do not access the Internet (28 percent, up from 18 percent in 1999). These figures suggest that the speech recognition is becoming a more mass-market option. This trend will be further verified in the 2001 Performance Benchmarking Survey.

2.2 General user satisfaction with voice-driven systems

In Table 1, data from the 1999 and 2000 Performance Benchmarking surveys are presented. Eighty-seven percent of callers in the 2000 survey were somewhat to completely satisfied with the voice–driven applications they used, an increase of 4 percent over the 1999 survey. Twelve percent expressed dissatisfaction with the system in the 2000 survey, down from 17 percent in the 1999 survey.

Table 1 User Satisfaction with Voice-Driven Systems

SATISFACTION	1999 SURVEY	2000 SURVEY
Completely Satisfied	22%	27%
Very Satisfied	34%	35%
Somewhat Satisfied	27%	25%
TOTAL SATISFIED	83%	87%*
TOTAL DISSATISFIED	17%	12%*

*Note: Total of 99% reflects a rounding error.

While the number of mobile/car-based users included in the survey was small (about 15 percent), these individuals were the most satisfied with the speech recognition system they used. In the 2000 Performance Benchmarking Survey, 94 percent of those that use a voice-driven service in the car and 92 percent who use speech recognition on a mobile phone were satisfied with the experience. This suggests that there may be a latent demand for better ways to obtain information while mobile which is being satisfied in part by access to voice-driven solutions like voice portals, voice-activated dialing and voice activated call centers. Recent research (Salvucci, 2001) indicates that voice interfaces may be better suited to multitasking while driving. In addition, anecdotal evidence from usability testing and user debriefing interviewsindicate that users, seeking a safer way to use mobile phones in the car, are looking for a hands-free alternative to touch-tone input. In the 2000 survey, satisfaction with the speech system was measured across demographic groups. Satisfaction did not vary appreciably across demographic groups, or by technology usage. Voice–enabled systems achieved high levels of satisfaction regardless of age, gender, income, or technology used, suggesting that speech systems are appropriate as a mass-market technology. When asked why they are satisfied with speech recognition, users cited its speed, efficiency, and ease of use, similar findings in both the 2000 and 1999 surveys. About a quarter of the surveyed users (24 percent) said they were satisfied because of the performance and reliability of the system.

2.3 Specific attitudes towards speech

Both the 1999 and the 2000 Performance Benchmarking Survey asked users to rate how well various attitudinal statements matched their feelings about the voice-enabled service they used. Our analysis suggests that there is a range of attitudinal statements that tend to be rated highly by those who are most satisfied with the speech recognition system they used. Looking at Table 2, there were increases in most of the key attributes from 1999 to 2000. Speech recognition was more likely to be rated as *comfortable* and *fast to use*. It was more likely to be preferred to touch-tone. And, interestingly, it was seen as an indication that the company cares about giving callers what they need. One reason for these increases may be that the voice interface is better matched to caller interests and usage as experience in designing voice interfaces increases.

Table 2 Specific Attitudes Towards Speech

Statement	1999 Survey Ratings 4 + 5*	2000 Survey Ratings 4 + 5*
"I like the freedom to use the system whenever, wherever I want"	85%	78%
"This is a system I feel comfortable with"	69%	75%
"I like speaking my responses better than pushing buttons"	63%	68%
"This system is really fast to use"	52%	68%
"The fact that the system exists tells me that the company cares about giving customers and employees what they want and need"	54%	65%
"This system understands what I say"	58%	64%
"I feel in control when I'm using the system"	58%	62%
"I can always get what I want from the system"	53%	60%

* Ratings were on a 5 point scale where 5 is "fits the way I feel perfectly and 1 is "does not fit the way I feel"

2.4 Comparison to users previous method

Before users started accessing a speech system, they acquired the information they were seeking, or accomplished their transactions through other sources: Through talking to a human being, accessing a web site, using a touch-tone application, reading newspapers, etc. How does a voice-driven system compare to these previous methods? Seventy-five percent of the users in the 1999 survey said that a speech recognition system was the same or better than the method they were using before. In the 2000 survey this number was up to 82 percent, with 38 percent indicating that speech was a "great improvement" versus the method used before, up from 27 percent in the 1999 survey.

In the 2000 survey, a more detailed analysis was done to compare the users' previous method to the speech system they were currently using. The results are reported in Table 3. Fifty-five percent talked to a human being prior to using their speech recognition service. When asked why they preferred the speech system, users had a range of responses including 24 hours a day/seven days a week access to information and transactions, not having to wait on hold, not being concerned that live agents are entering the incorrect data or that they are taking up too much of their time if they check the same information several times a day. Forty percent of the users surveyed had used touch-tone applications prior to using a voice-enabled service. When asked why they preferred using a voice-enabled service, they tended to reply that speech recognition systems are easier and faster to use. We believe this is due to not having to associate numbers on the keypad with the different options offered (Boyce, 2000), and the shorter length of the system prompts used in speech systems. Users perceive speech recognition as faster, because menuing proceeds more quickly. Also the speech system can be used hands free, without visual attention. Internet users tend to mention mobility-related benefits as the reason for their preference.

Table 3 Comparison to Users Previous Method

How does it compare to previous method – Talk to a Human	2000 Survey
Great Improvement	38%
Somewhat better/Same	46%
Somewhat worse/Much worse	16%
How does it compare to previous method – Use DTMF System	
Great Improvement	34%
Somewhat better/Same	46%
Somewhat worse/Much worse	20%
How does it compare to previous method – Use the Web	
Great Improvement	29%
Somewhat better/Same	55%
Somewhat worse/Much worse	16%

Overall, we have seen from the Performance Benchmarking survey results that speech recognition systems are very promising. In summary, most callers to these systems, regardless of age, gender or Internet experience, are satisfaction using this form of communication, even when compared to the methods they used previously. Using findings from the surveys, how can we design even better systems? One way is to incorporate the findings and lessons learned from usability testing.

3. Key usability issues and design considerations

Nuance consistently explores and evaluates the latest dialog techniques, including speech user interfaces, through usability research to support the development of our own products and on behalf of our customers. In looking at the aggregate result from this research over the last two years, we have identified a range of usability issues that impact the ability of users to complete a task, and diminish caller satisfaction. In this section, we will discuss three of the issues we consider most important: Information structure, consistency, and conversational interface, and present related design considerations.

3.1 Information Structure

Usage of a less than optimal information structure requires additional user effort: Users of speech applications all have pre-existing ways of accomplishing the tasks handled by the speech system. Users bring this pre-existing frame of reference to the speech application. If the speech application does not map well onto these expectations, the mismatch causes user confusion, error and frustration. Relating this back to the attitudes towards speech presented in Table 2, we feel that a poorly designed information structure does not help the user to "Always get what I want from the system" or "Feel in control". In usability studies, we have identified many causes of this mismatch. Some of the most important are:

- Using the company's internal structure as a basis for designing the call flow, rather than considering the user's mental model. For example, forcing consumers who are interested in information about products to ask for departments or product categories. Many callers are not necessarily aware of the company's terminology for those departments or product categories, so their product need cannot be easily matched to a system offering. This mismatch is not easily resolved, as a speech system provides a linear flow of information, and it is not always easy for users to get a holistic view of all the possibilities that are available to them.
- Frequently asked for items buried low in the menu structure. In this situation, callers frequently have to search for their desired item in the overall menu structure. This results in longer interactions with a potential for increased user error, frustration and diminished call completion.
- Too much complexity/too many choices for the user to retain at any given level. Based on their existing frame of reference, users expect to entertain a set of viable options at various stages in their interaction. These options should all relate to a particular stage in their decision process, and should use terminology that is familiar and useful. Options that are either (1) ambiguous, or not familiar or (2) not viable choices at a particular stage can negatively impact the caller's efficiency in determining their next step—by wasting time and making inappropriate choices.

There is a benefit in conducting pre-design work with target market users, including task analysis, before determining a high-level call flow design for a speech system. This step will help the designer understand the user's previous experience, set priorities for design, determine the overriding structure of the speech system and which options should be available at each step in the dialog flow. The more frequently requested items should be available early in the interaction, either by providing: (1) an ability to request these items quickly at the main menu; or (2) shortcuts that allow those who frequently use these options to have access to them without burdening other users.

3.2 Consistency

Users tend to assume that the techniques and methods used in previous interactions, or in one specific part of an application they are using, will be useful in another. Specifically, they assume that the interaction will be:

- Consistent with previous usage: As mentioned earlier, our research suggests that users tend to apply their existing frame of reference to the functional organization of the system. Users' experience with a web site, corresponding software or familiar interaction with an operator or touchtone system will form the basis for their expectations of the speech system. As shown in section 2.4, callers to speech systems obtained their information from various sources before using a voice-enabled system.
- Internally consistent: Users seem to assume that similar tasks will be fulfilled in a similar manner using identical terms throughout the speech application. Areas that fulfill similar goals but function differently are often mis-used initially.

Consistency has long been a general rule for human-human interactions (Fiske and Taylor, 1991). We believe that it is important in the design of speech interfaces.

Designers should be mindful of users pre-existing frame of reference. This can be accomplished by maintaining consistency in the information structure and terminology used between the speech system and the other systems the user interacts with—either currently or previously. Designers should also strive for internal consistency in the following areas: (1) Availability of help and other universally available commands; (2) Providing a consistent navigation paradigm through: navigational universal commands, for example a word which will always take the user to the main menu; (3) Terminology, and (4) Task process for similar tasks.

3.3 Conversational interface

Because speech recognition is a new technology for most users, many callers will experience some learning curve. When callers are not told they are entering a speech system, we have observed difficulties with usage, such as inappropriate speech. This lack of familiarity with a speech system can be minimized by using the familiar constructs of conversation, in combination with good dialog design practices. Users have been trained to conduct a conversation since birth, but many have never used a speech system. Allowing callers to talk to the system in a more natural way is closer to how humans interact. We believe that certain attitudes towards speech, as reported in Table 3, such as "I like speaking my responses better than pushing buttons" and "This is a system I feel comfortable with" are related to providing the conversational cues that the caller is expecting. When this is absent, or social errors are present in the dialog, (Nass and Gong, 2000) we have observed users doing the following:

- Not speaking, or speaking too much or inappropriately-leading to poor recognition and inefficient use of the system. These responses suggest that the caller is having trouble parsing the information provided by the system.
- Losing their sense of where they are. This suggests that prompts are not maintaining the expected context for the interaction—Similar observations have been made in past research regarding the need for common ground in conversation (Clark and Schaefer, 1989 and Clark and Brennan, 1991).
- Mentioning that systems which are less conversational/more mechanical are "not helpful" or "not listening" in debriefing interviews.

Speech systems that include the following are likely to be most effective for callers: (1) A welcome message, or a musical tone, that tells callers that they are entering a speech system; (2) An optional tutorial for new users giving basic information about how to use the system and providing valuable context for the caller; (3) A conversational dialog design and prompts that provide a familiar interaction with the caller through speech techniques like landmarking, which provide linguistic cues that a user has reached a new area of the application or new functionality; (4) a comprehensive support system (including error recovery and help) that helps the user get back on track by giving the caller information on what to do next, or helping them figure out what they are doing wrong—just as a person would in the same situation.

4. Conclusions

Nuance research has shown that over the phone speech recognition systems have the ability to satisfy mass-market callers. The availability of both market and usability research has helped us more fully understand user performance in a speech system, in terms of both perceived success and actual performance in the interaction—laying the groundwork for a complete, and multi-dimensional understanding of successful speech systems. Key attributes which satisfied individuals associate with speech systems include: "fast", "easy to use", "comfortable", "better than pushing buttons", "in control" and the ability to "use whenever and wherever I want". In order to create speech systems that are satisfying to users, designers must support the caller in a sequential, non-visual user experience by recognizing the importance of information structure, consistency and established conversational paradigms.

References
Boyce, S. (2000). Natural Spoken Dialogue Systems for Telephony Applications. In *Communications of the ACM 43*, 9, 29-34.

Clark H.H., Schaefer, E.F. Contributing to discourse. In *Cognitive Science 13* (1989), 259-294

Clark H.H., Brennan, S.A. Grounding in communication. In L.B. Resnick, J.M. Levine and S.D. Teasley (eds.), *Perspectives on Socially Shared Cognition* (1991), 127-149

Fiske, S.T., Taylor, S.E. (1991). *Social Cognition*. New York: McGraw-Hill

Nass, C., Gong, L. (2000). Speech Interfaces from an Evolutionary Perspective. In *Communications of the ACM 43*, 9, 36-43

Salvucci, D. (2001). Predicting the Effects of In-Car Interfaces on Driver Behavior using a Cognitive Architecture. In *Proceedings of CHI 2001*, 3(1), 120-127

Físchlár on a PDA: Handheld user interface design to a video indexing, browsing and playback system

Hyowon Lee, Alan F. Smeaton, Noel Murphy, Noel O'Connor, Seán Marlow

Centre for Digital Video Processing
Dublin City University, Glasnevin, Dublin 9, Ireland

Abstract

The Físchlár digital video system is a web-based system for recording, analysis, browsing and playback of TV programmes which currently has about 350 users. Although the user interface to the system is designed for desktop PCs with a large screen and a mouse, we are developing versions to allow the use of mobile devices to access the system to record and browse the video content. In this paper, the design of a PDA user interface to video content browsing is considered. We use a design framework we have developed previously to be able to specify various video browsing interface styles thus making it possible to design for all potential users and their various environments. We can then apply this to the particulars of the PDA's small, touch-sensitive screen and the mobile environment where it will be used. The resultant video browsing interfaces have highly interactive interfaces yet are simple, which requires relatively less visual attention and focusing, and can be comfortably used in a mobile situation to browse the available video contents. To date we have developed and tested such interfaces on a Revo PDA, and are in the process of developing others.

1. Introduction

Físchlár is a web-based digital video system currently used by about 350 users within the campus environment in Dublin City University. The system allows the user to record broadcast TV programmes, and allows browsing and playback of the recorded programmes on a web browser. The user can easily browse 8 terrestrial TV channels' schedules for today and tomorrow, and by simply clicking on a programme sets the recording. The system then encodes the programme in MPEG-1 format when the broadcast time comes. The encoded programme is then subjected to automatic camera shot and scene boundary detection to extract representative keyframes, which are used for the user's browsing and playback interface. The interface allows the user to select one of the several browsing methods we have developed to see the keyframes and clicking on any of the keyframes will pop up a new window which starts streamed playback of the video from the clicked keyframe onwards. The system is capable of streaming to about 150 users concurrently. The system is a testbed for our technology development, wherein any implemented techniques such as various shot/scene boundary detection, integration with a programme recommender system (Smyth & Cotter, 2000), mobile application for video browsing and playback, and various user interface ideas are easily plugged in to the system and the outcomes are visibly demonstrated to our current user base. Users of the system are an important element of our work, as they provide new ideas from their own, real usage context. The Físchlár system is further described in O'Connor, et al.(2001).

In the recognition of the diversity of users' preferences and task contexts, we have developed a design framework for video browsing interfaces which allows us to come up with many different styles of browsing interface, and we have implemented 8 different browsers suitable for a desktop environment for the user to choose and use. As we are presently working on the use of mobile devices to access the Físchlár system's video contents, porting the system's browsing/playback feature to mobile devices has become an important concern for us. In this paper we apply the developed browsing interface design framework to a small, handheld PDA (Personal Digital Assistant) and demonstrate the resultant example browsing interfaces for the PDA. Section 2 briefly explains the rationale for developing such a design framework, and summarises the actual framework for keyframe-based browsing interfaces and how it can be used to design a specific browser. In section 3 we apply general interface design concerns for mobile devices to the design framework and demonstrate two example designs suitable for a PDA. Section 4 concludes with some future directions in video browsing interface design.

2. Design framework for video browsing interface

One of the main problems in designing a user interface for a novel system such as a digital video browser is the absence of usage information and lack of new ideas, making the conventional process of the initial design and subsequent iterative refinement through usage monitoring difficult. User interface design is a highly creative process (Shneiderman, 1998) requiring intuition and an artistic sense from the designer, and also some past design experience or mimicking from other systems' design (Sanderson, 1997) which should be understood as part of the design process. The designed interface usually goes through user testing and is iteratively refined. However, a single "optimised" interface developed this way cannot satisfy everybody, because people come to the system with different aptitudes, attitudes, preferences and task contexts. Furthermore, the current trend in technology shows diversification of devices for a single underlying system sharing the same data, such as email software accessed from an office desktop PC, a PDA or a mobile phone. This results in the need for designing different user interfaces for different devices suitable for different users. To address these problems, there have been efforts to streamline and turn the fuzzy, unpredictable and ill-defined interface design concept into a more structured and formalised process, exemplified by "design space analysis" (MacLean, Young, & Moran, 1989) and further adapted in various forms such as Stary (2000). In this approach, roughly the following steps are followed:

(i)   analyse and identify important elements and alternatives in designing an interface, resulting in an exhaustive sets of possible design options, or design space;
(ii)   consider the particular environment where the interface in concern is to be used, and
(iii) select a suitable set of options from the design space

This way, designing an interface becomes less of an intuitive, artistic task and more of a concrete, clear and simple decision-making process where the designer can come up with many different interfaces by selecting different combinations of options suitable for the target usage. A crucial part in this approach is the initial construction of the design space and the selection of the right set of options for the target usage. In designing video keyframe browsing interfaces for the Físchlár system, we have constructed a design space by identifying 3 important facets (or dimensions), and the possible options (or values) for each dimension.

2.1   Layeredness

The amount of keyframes extracted from a video programme crucially affects the user's browsing. A very large number of keyframes would provide detailed browsing, but become unsuitable for quick browsing. Provision of only a small subset of keyframes (for example, 10 to 30 keyframes extracted from one programme) could be a useful way of providing an overview of the programme content. The *Layeredness* dimension is concerned with the different detail or granularity of the keyframe set and the transition between different granularity, and has to be considered in designing a keyframe-based browsing interface. Some of the typical options for this dimension are:
Single layer - provides only a single set of keyframes, whether very detailed or selective;
Multiple layer without navigational link - provides more than one set of keyframes in a browser, thus the user can select the granularity s/he wants in the browsing;
Multiple layer with navigational link - provides more than one set of keyframes, and the user can jump between different sets of keyframes while maintaining the current point of browsing.

2.2   Temporal Orientation

Video is a time-based medium and the keyframes extracted from video are an ordered set of images by time. Thus an important concern is *what kind of time information*, if at all, should be provided to the user when browsing the keyframe set. Some of the typical options for this dimension are:
No time information - provides no explicit time information regarding each keyframe;
Absolute time - provides exact time information in numeric form (for example, timestamping each keyframe such as "15 minutes 30 seconds into the video");
Relative time - shows the time of the current browsing point in relation to the whole length of the video (for example, a timeline bar indicating the current viewing point).

2.3   Spatial vs. Temporal Presentation

There are two distinctive ways of presenting keyframes on the screen, and the designer has to decide which one should be adopted for the interface in concern:

Spatial presentation - displays many miniaturised keyframes side by side, allowing quick spatial browsing;
Temporal presentation - displays keyframes one by one.

2.4   Specifying a Browsing Interface

The three dimensions and their typical options introduced above form a design space where the designer selects one (or more than one) option from each dimension. Because each design option represents a distinctive design decision in a dimension, different combinations of options result in distinctive browsing interfaces. This makes it possible to design all conceivable and different browsing interfaces. For example, selecting the following options

> *Single layer* from Layeredness dimension
> *Relative time* from Temporal orientation dimension
> *Spatial presentation* from Spatial vs. Temporal presentation dimension

results in a specific and distinctive browsing interface. What options to select should be based on the target usage context. The target users and the kind of devices they will be using roughly indicate which options are preferable and which are not. For example, on a very small, low-resolution screen such as a mobile phone, spatial presentation is not suitable because each keyframe content would be unrecognisably small if we try to further miniaturise keyframes and display many of them on one small screen.

In designing the browsing interfaces to the Físchlár system on a PDA, general guidelines and common sense can be used in this selection process. The following section considers these general ideas suggested for mobile device interface design, to be applied to the design framework to select suitable options, and two example interfaces emerged as a result.

3. Video Browser for PDA

User interface design for mobile devices is getting more attention nowadays, as more and more computational power and higher wireless bandwidth are becoming available for these devices. However, the usability and interface design of today's mobile devices has not been studied enough and most elements are simple replicas of desktop interface elements. The problem is that a good design for a desktop application with a large monitor and a pointing device might not be suitable for a small, handheld device such as PDA or mobile phone. Well-established graphical user interfaces with a direct manipulation style will enhance the usability greatly for a desktop environment where the user generally keeps on looking at the screen holding the mouse, ready for reacting to any visual feedback from the system (Kristoffersen & Ljungberg, 1999). This is far from the truth in the mobile environment (in a bus, on the street, etc.) where the user may be unable to keep on focusing on the screen and small visual details can easily be overlooked. However, a successful interface style on one environment can be completely unsuitable for another environment. While there should be more studies in developing a new interaction paradigm for mobile interfaces (Marcus, Ferrante, Kinnunen, Kuutti, & Sparre, 1998), in this paper we concentrated with designing interfaces based on the framework described in previous section, for a particular PDA called Revo from Psion. Revo was chosen within our project administrative resources and not necessarily ideal for developing fully-working versions. The interfaces designed were not connected to the Físchlár system via a high bandwidth wireless network to show live online information stored on the server, and the limited memory size (8 MB) and display rate of the device meant that a realistic evaluation involving user testing was not possible. However, for the purpose of demonstrating our design directions and ideas, this initial phase mock-up demonstration was sufficient.

Figure 1. Single layer / Relative time / Temporal presentation

The Revo has a landscape-oriented screen of 480 by 160 pixels with 16 shades of grey. The screen is touch-sensitive and is usually used with either a finger or a stylus that comes with the device. The first obvious design decision is that the low resolution and small screen of Revo makes spatial presentation unsuitable. Figure 1 shows one of our designed interfaces. On the right side of the screen, the list of recorded and processed TV programmes is displayed with a scroll bar, which

380

the user can use with the right thumb while holding the device with the same hand. On selecting one of the programmes, the first keyframe extracted from that programme is displayed on the left side of the screen. Below the keyframe there are two buttons (previous/next) for the user to flip through keyframes one by one, using the left thumb while also holding the device with the left hand. As the user flips through the keyframes by repeated tapping on the buttons, a timeline bar beside the buttons indicates relative time showing the current point of browsing in relation to the whole programme. Automatically flipping through keyframes (true temporal presentation) rather than using the buttons would be possible, but because it should not force the user to keep on concentrating on the screen, having the user controlling the keyframe flipping was considered better. Requiring a high degree of interaction (repeated tapping on the previous and next buttons) should be okay when the number of the interaction objects (buttons) is only two and these are always under the user's thumb. Note that it is possible to grab the device and interact with it with only one hand at either stage of interaction (either browsing to select a TV programme or flipping through a programme content), freeing the other hand for something else (holding a bag, opening a door, etc.). When both hands are available, the user can simply grab the device firmly with both hands, and use the right thumb for scrolling and selecting a TV programme, and the left thumb for flipping through the selected programmes keyframe content.

Figure 2. Multiple layer with navigational link / Absolute & Relative time / Temporal presentation

Another designed interface for Revo is in Figure 2 below. In this interface, browsing the keyframes of a single programme is considered, with multiple layers of keyframes available. With the two buttons on the right side (up/down buttons), the user can jump between 6 different layers, with the layer indicator beside the buttons showing currently selected layer. The top layer has very selectively-chosen 10 keyframes providing an overview of the whole programme; the bottom layer has the full camera shot-level amount of keyframes depending on the programme (usually 300-700 keyframes); other 3 middle layers provide in-between details of keyframes to browse. With the two buttons on the left side (previous/next buttons), the user can flip through the keyframes in the currently selected layer, and the current temporal position in the programme is indicated with the timeline bar above the buttons. The layers have navigational links between them, meaning that when the user jumps up or down a layer, the current point of browsing is maintained (i.e., when the user taps on the up/down button on the right, the timeline bar on the left will still be indicating the same position). Thus, the user can easily jump up a layer to move the current position more quickly to right or left, and when an interesting position is reached, jump down a layer to browse more detailed keyframes in that area. Note that all the widgets on the interface are on the sides of the screen and the keyframe displayed is in the centre. This browser is mainly meant to be used with both hands grabbing the device and continuously tapping buttons as if playing a pocket video game console, the user continuously moving up and down the layers while flipping through keyframes left and right in a highly interactive manner.

In both designed interfaces, the user has full control over the displayed information on the screen while the widgets to trigger any system feedback (i.e., buttons) are very obvious and always in easy reach. This style makes it acceptable and natural for the user to casually take attention away from the screen while stopping interacting with the system with thumbs, and a few seconds later focusing back on the screen, no more different from casually playing a pocket video game while sitting in a busy bus on the way home. Both interfaces were designed in such a way that the user need not pay careful visual attention or point at a small area in the middle of the screen, unlike the majority of desktop application interfaces (as well as the PDA interfaces which adopted the desktop interaction style) require. A rather simpler and blunt interaction, but still efficiently allows highly interactive and pleasant browsing of video content in different details. Although more sophisticated applications with more functionality would inevitably require finer-level pointing on the screen with the stylus, the overall styles used in the above example interfaces would be a good direction and guideline in designing other applications on a mobile device.

The design framework used in this design provided building blocks and the way to consider different elements in designing the interfaces, while general ideas on designing interfaces for the mobile environment provided selecting particular elements from the framework as well as dictating the overall interaction style of resultant designs. Revo provides as an option a small "tick" sound as aural feedback whenever a screen is touched, thus the user knowing that his/her action was sent to the system. However, mapping the "virtual buttons" on the above interfaces to physical buttons on the device would enhance the tactile feedback for the user (Myers, Lie, & Yang, 2000), which was not possible to implement with Revo for technical reasons. With the technical limitations of the Revo, informal discussion with a small number of test users has been conducted during and after the implementation mainly to get rid of minor usability problems (for example, the initial implementation of the first interface had smaller button sizes than the current version, which was picked up by a few test users), often with positive comments on the overall idea. As an alternative to using Revo, we are presently developing a mobile PDA interface to Físchlár for the Compaq iPAQ which has more RAM (32 MB as opposed to 8 MB), a more detailed screen (4,096 colours as opposed to 16 greys), a faster processor (206 MHz as opposed to 36 MHz) and has wireless LAN connectivity.

4. Conclusion

In this paper, designing a keyframe browsing interface to a PDA was considered with a specially constructed design framework for video browsing interfaces as a base. The commercial and research community are more and more aware of the importance of recognising people's individual differences and personal preferences, as evidenced in the idea of "personalisation" both as a live research area and appearing frequently in popular technology products and services. In the user interface design field, attempts to cater for the diversity makes it difficult to have a single user interface for a system which supports everybody's needs. Furthermore, the diversification of different devices for very different environments makes it impossible to stick to a single interface to support these different environments. Identifying all possible interface elements and specifying an interface from this list can be a good starting step for heading towards realising universal access which supports potentially all users and their circumstances. Already research is underway in the area of "unified interface design" (Akoumianakis, Savidis, & Stephanidis, 2000), which considers developing methodologies and tools for an intelligent system which can eventually automatically identify each individual user's preferences and needs at the time of use, and assemble suitable interface elements to provide this to the user dynamically, thus eventually being able to support everyone's needs on an individual basis.

Mere technological progress does not guarantee a wide acceptance of usage of that technology in the end product. Numerous failures in usability are found in small, handheld devices because the same interface paradigm for the so-far dominant desktop systems were used without further elaborate consideration. It is thus important to consider in depth the context of the use of the particular interface in concern.

References
Akoumianakis, D., Savidis, A., & Stephanidis, C. (2000). Encapsulating intelligent interactive behaviour in unified user interface artefacts. *Interacting with Computers, 12*, 383-408.
Kristoffersen, S., & Ljungberg, F. (1999). "Making place" to make IT work: empirical explorations of HCI for mobile CSCW. *Proceedings of the ACM Conference on Supporting Group Work (GROUP '99)*, Phoenix, AZ, 276-285.
MacLean, A., Young, R., & Moran, T. (1989). Design rationale: the argument behind the artifact. *Proceedings of the ACM Conference on Wings for the Mind (CHI '89)*, Austin, TX, 247-252.
Marcus, A., Ferrante, J., Kinnunen, T., Kuutti, K., & Sparre, E. (1998). Baby faces: user-interface design for small displays. *Proceedings of the ACM Conference on Human Factors in Computing Systems (CHI '98)*, Los Angeles, 96-97.
Myers, B., Lie, K., & Yang, B-C. (2000). Two-handed input using a PDA and a mouse. *Proceedings of the ACM Conference on Human Factors in Computing Systems (CHI 2000)*, The Hague, Netherlands, 41-48.
O'Connor, N., Marlow, S., Murphy, N., Smeaton, A., Browne, P., Deasy, S., Lee, H., & Mc Donald, K. (2001). Físchlár: An on-line system for indexing and browsing of broadcast television content. To appear in the *26th International Conference on Acoustics, Speech, and Signal Processing (ICASSP 2001)*, Salt Lake City, UT.
Sanderson, D. (1997). Technology design and mimicry. *Proceedings of the ACM Symposium on Designing Interactive Systems: Processes, Practices, Methods, and Techniques (DIS '97)*, Amsterdam, 311-313.
Shneiderman, B. (1998). *Designing the user interface: Strategies for effective human-computer interaction (3rd ed.)*. Reading, MA: Addison Wesley Longman, p.89.
Smyth, B., & Cotter, P. (2000). A personalized television listings service. *Communications of the ACM, 43*(8), 107-111.
Stary, C. (2000). A structured contextual approach to design for all. *Proceedings of the 6th ERCIM Workshop "User Interfaces for All"*, Florence, Italy, 83-97.

Universal Multimedia Information Access

Mark T. Maybury

Information Technology Division
The MITRE Corporation
202 Burlington Road
Bedford, MA 01730, USA
maybury@mitre.org
http://www.mitre.org/resources/centers/it

Abstract

Efficient, effective and intuitive access to multimedia information is essential for business, education, government and leisure. Unfortunately, interface design typically does not account for users with disabilities, estimated at 40 million in America alone. Given broad societal needs, our community has a social responsibility to provide universal designs that ensure efficient and effective access for all to heterogeneous and increasingly growing repositories of global information. This article describes information access functions, discusses associated grand challenges, and then outlines potential benefits of technologies that promise to increase overall accessibility and success of interaction with multimedia.

1. Multimodal information processing

With the explosion of the web, increasing amounts and kinds of multimedia and multimodal information is available worldwide. Following Maybury and Wahlster (1998), by *media* we mean the carrier of information such as text, graphics, audio, or video. Broadly, we include any necessary physical interactive device (e.g. keyboard, mouse, microphone, speaker, screen). In contrast, by *mode* or *modality* we refer to the human senses (more generally agent senses) employed to process incoming information, e.g., vision, audition, and haptics.

Table 1 outlines a range of functions that users need to perform when manipulating media. These include retrieving it from storage, extracting elements from it, translating from one media to another (e.g., text to graphics, text to speech), asking (possibly multimedia) questions about it, generating multimedia presentations, and summarizing media content. Each of these areas has associated grand challenges and promise of benefit for universal access. In the remainder of this paper we articulate the nature of these challenges and illustrate potential benefits if they are solved. We exemplify some of these by illustrating a content-based multimedia news understanding system (Maybury et al. 1997; Maybury 2000a) that incorporates several of the discussed multimedia access functions.

Multimedia Retrieval. Users need mechanisms to retrieve artifacts such as text, audio and video. Analysis of the content of these artifacts can enable more direct access to underlying materials. Users with memory or motor skill impairment may benefit as methods that retrieve the most relevant materials can offer fewer media objects to the user.

Multimedia Extraction. The ability to identify and extract media objects from sources enables direct access to subelements of media. This also supports the automated population of databases as well as the resequencing of media into new artifacts. Extracted objects might include named entities and events from natural language text, people and faces from pictures, and animate and inanimate objects from video. As we articulate in a concrete example below, this can enable summary forms or resequencing of material tailored to user needs.

Table 1: Information Access Challenges and Benefits

Multimedia Access Function	Grand Challenges	Example(s) of Universal Benefits
Retrieval	Content-based retrieval of text, imagery, audio, video.	Direct access to media easing navigational burden. Reduction of search time.
Extraction	Segmentation, object and event identification, and extraction from multimedia sources (text, audio, video).	Direct access to media elements, including specific types that may be user preferred. Reuse of media elements enabling user tailored selection or presentations.
Translation	Verbalization of graphics, Visualization of text.	Cross media/mode information access, especially important for modality (e.g., visual, auditory) impaired users.
Question Answering	Question analysis, response discovery and generation from heterogeneous sources.	Overcome time, memory or attention limitations required to sift through many returned web pages following search.
Multimedia Query	Spoken query processing, visual query analysis (e.g., sketching), Mixed media query (e.g., text and graphics)	Enable users to select form of input for query (text, speech, graphical) if motor impaired.
Presentation Generation	Media selection/allocation, coordination and realization. Animated life-like characters.	Sensitivity to specific physical, perceptual, or cognitive challenges. Agents engaging/motivating to younger and/or less experienced users.
Summarization	Multimedia content selection, condensation, and presentation. Cross media summarization.	Facilitates access to large amounts of material for users with memory and attention limitations.

Multimedia Translation. Research has explored the conversion of elements in one media to another, for example, visualizing natural language text descriptions as graphics or verbalizing graphical depictions as text. This is an extension of more popular "screen readers" which synthesize auditory version of written text, and as such would support access, for example, of blind users to a spoken language translation of graphics or to deaf hearers to a visual form of music (e.g., an animated score).

Multimedia Question Answering. Question answering promises to replace today's paradigm of keyword queries returning web pages with a natural language question which is interpreted and then satisfied with a specific response culled from structured, semi-structured and unstructured multimedia sources (Maybury 2000b). This kind of capability promises to overcome the memory and/or attention limitations required to sift through many returned web pages following search.

Multimedia Query. Query formulation has classically been confined to keyword selection or expansion, limiting some modality disadvantaged users. In contrast, multimedia query provides users with both alternative media with which to formulate their question as well as cross media query (e.g., searching for text using graphics, searching for graphics using text). Query analysis includes language analysis, sketch analysis, and even sound analysis (e.g., humming a few notes into a microphone to match and retrieve audio samples).

Multimedia Presentation Generation. Presentation generation includes content selection, allocation (to media), coordination, and layout of media. While dynamic generation of presentations requires (some level of) understanding of source material, it enables the creation of user tailored presentations that utilize the media, language, and format required by the end user based on a model of their perceptual, psychological or cognitive state. On-the-fly design also requires the allocation of content across media that can be sensitive to the individual user needs. For example, approximately 8% of western males have some degree of color blindness (1/30,000 can perceive no colors), with their biggest challenge distinguishing red and green or blue and yellow, so these combinations could be avoided during graphical design. Information about the mypoia (near sighted) or hyperopia (far sighted) of the user could further constrain design choices. Simple and inexpensive magnifying glasses can assist in this, however, the user and system have no control of the content and form of the underlying material. Just as judicious mixes of media have been shown to enhance the accessibility of information (Merlino and Maybury 1999), mixed media displays, such as generated Braille tactile displays combined with speech, promise disadvantaged users synergistic access, in this example the speed of spoken language combined with spatial context of a physical layout. More recently, researchers have begun to explore more engaging interfaces that promise not only perceptual and cognitive benefits but also social and/or emotional ones. For example, animated interface agents are being explored to coach less experienced users (André, Rist, and Müller, 2000), to motivate and engage users (e.g., those that might be introverted or disinterested) or even to support individuals in psychological states such as depression (Johnson and Rickel, 1998).

Multimedia Summary. Summarization may entail extraction and translation from individual or groups of media artifacts, condensation of this material, and then generation of a compact form. This summarization can occur within and across media. Summaries promise to help memory and attention limited users, for example, by enabling users to interact with summaries as opposed to the entire source document.

2. Personalcasting

In order to explore how some of the functions might be realized in a concrete and ubiquitous application, we consider an every day task, watching the news. We explore how interaction can be optimized to individual user requirements by considering the Broadcast News Navigator (BNN) (Maybury et al. 1997; Boykin and Merlino 2000). The web-based BNN gives the user the ability to browse, query (using free text or named entities), and view stories or their multimedia summaries.

As detailed in the above referenced publications, BNN creates summaries by first using a range of audio, imagery and text cues to discover the start and finish of news programs and individual stories. For example, BNN detects story changes using evidence combined from multiple linguistic cues (e.g., "In our top story tonight …"), linguistic patterns (e.g., <reporter-name> <news-organization> "reporting from" <location>), audio cues (e.g., silence between stories), and visual cues (e.g., cuts from reporters in the field back to anchors in the studio). BNN then extracts media elements from these segmented stories, such as the most frequently occurring names or keyframes from the middle of reporter shots following anchor shots on the same story. Finally, BNN generates several types of news summary presentations from these segmented stories and associated media elements.

For example, Figure 1a is a display of all stories about Florida on multiple North American broadcasts from October 24, 2000 to January 22, 2001 (only CNN Morning Headline programs are visible). Figure 1a illustrates a "story skim", in which each retrieved story is represented by an extracted key frame and the three most frequently occurring names of people, places or things in the story, called "named entities". If the user clicks on the key frame from 9 November with "Palm Beach County" as the first name, a "story detail" view is generated (shown in Figure 1b) which displays multiple media elements extracted from the source. This includes the story (closed caption) text transcript, a one line summary, all named entities mentioned in the story (including time and money expressions), a list of subjects or topics addressed by the story, and a link to the full original (audio and) video of a story.

An empirical user study (Merlino and Maybury 1999) revealed that users can perform information retrieval and extraction tasks on these mixed media displays of the news as well as on individual media elements (e.g., only text, keyframe, one line summary, audio, or video). However, users retrieve information more effectively (measuring a weighted combination of precision and recall) using the mixed media presentations show in Figures 1a and 1b. Furthermore, on a scale of 1 (dislike) to 10 (like), users report the highest satisfaction level (8.2) for these mixed media displays.

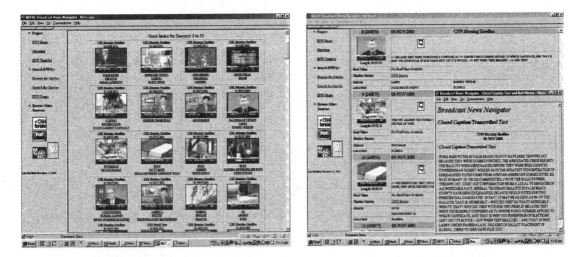

Figure 1a. BNN "Story Skim"   Figure 1b. BNN "Story Detail"

Not all users, however, may have the luxury of perceiving all media. By extracting individual media elements from the original source we are not constrained to provide only the original video source. In fact, in contrast to the traditional mixed media broadcast, we can present either single media (e.g., only text or images) extracts or generate a mix of media tailored to user needs or desires. Moreover, by stating particular queries (either keywords or named entities), the user can retrieve a custom set of news stories, thus automatically creating a personalized news program, a "personalcast".

Finally, just as news programs utilize editors, reporters and anchors to organize and present material, so too we might expect animated digital agents (Johnson and Rickel 1998, André et al. 1999) to take on organization and delivery roles in future presentations. One promise is that animated interface agents have been shown to increase user motivation by enhancing user engagement, although task performance improvements have not yet been empirically proven.

3.   Implications for universal access

In order to enable access to the broadest possible community of users, systems should be both adaptable, i.e., modifiable by the user, as well as adaptive, that is observe user behavior and learn to optimally meet the user's needs. Even where we don't have detailed user models (that can be expensive, errorful, and invasive to collect), we might be able to anticipate key user needs. For example, the retinas of users over 60 years old let in only about one third of the light allowed in by 20 year olds. Users over 40 may suffer from presbyopia or decline in "accommodation", the ability of the eye to change its focus. To make matters worse, (in the UK) 35% of the visually disabled population have a hearing deficit (Bruce et al. 1991). Limited physical dexterity can further exacerbate these conditions. The intelligent multimedia interface can respond with automated designs that not only select the optimal presentation media, but also incorporate clear contrast, higher illumination, and more accessible controls.

One of the challenges will be the effective blending of inexpensive devices such as magnifiers or screen readers, with much more sophisticated systems such as animated interface agents. Another challenge will be designing interfaces that are sensitive not only to the physical challenges faced by the user, but equally the psychological and social ones (e.g., mediating intra/extraverted personalities and/or individual mood).

Applications that promises on-demand, personalized access to multimedia information on a broad range of computing platforms (e.g. kiosk, mobile phone, PDA) offer new opportunities for challenged users. Synergistic processing of speech, language and image/gesture promise both enhanced understanding of artifacts such as web, radio, and television sources (Maybury 2000a) and enhanced interaction at the interface. Coupled with user and discourse modeling, new services such as delivery of intelligent instruction, custom help, and individually tailored personalcasts become possible.

References

André, E., Rist, T., Müller, J. (1999). Employing AI Methods to control the behavior of animated interface agents. *Applied Artificial Intelligence*. 13, 415-448.

Boykin, S., Merlino, M. (2000). A. Machine Learning of Event Segmentation for News on Demand. *Communications of the ACM*, 43 (2), 35-41.

Bruce, I., McKennell, A., Walker, E. (1991). *Blind and Partially Sighted Adults in Britain: The RNIB Survey*. HMSO, ISBN 0 11 701479 6. See also http://www.tiresias.org/reports/linkopin.htm.

Gill, J. (2000). *Which Button? Designing user interfaces for people with visual impairments*. Royal National Institute for the Blind (RNIB). http://www.tiresias.org/controls.

Johnson, W. L., Rickel, J. (1998). Steve: An animated pedagogical agent for procedural training in virtual environments. *SIGART Bulletin*, 8, 16-21.

Maybury, M., Merlino, A., Morey, D. (1997). Broadcast News Navigation using Story Segments. In *ACM International Multimedia Conference*, Seattle, WA (pp. 381-391), November 8-14.

Maybury, M. (2000a). News on demand: Introduction. *Communications of the ACM*, 43(2), 32-34.

Maybury. M. (2000b). Keynote. Ask Questions, Get Answers. In *First International Conference on Adaptive Hypertext '00*, Trento, Italy August 29, 2000. [On-line]. Available: http://AH2000.itc.it/Maybury-AH2000.pdf

Merlino, A., Maybury, M. (1999). An Empirical Study of the Optimal Presentation of Multimedia Summaries of Broadcast News. In Mani, I. & Maybury, M. (eds.), *Automated Text Summarization* (pp. 391-401). Cambridge, MA: MIT Press.

Stephanidis, C. (Ed.). (2001). *User Interfaces for All - Concepts, Methods, and Tools*. Mahwah, NJ: Lawrence Erlbaum Associates (ISBN 0-8058-2967-9, 760 pages).

Present Stage and Issues of Sign Linguistic Engineering

Yuji Nagashima[a], Kazuyuki Kanda[b]

[a]Kogakuin University , 2665-1 Nakanano-machi, Hachiouji-shi, Tokyo, 192-0015, Japan

[b]Chukyo University, 101-2 Yagotohoncho, Showa-ku, Nagoya-shi, Aichi, 466-8666, Japan

Abstract

Sign language is one of the communication means for the hearing disabled. It is a visual language and the articulatory components of a sign are presented in three-dimensional space. Moreover the so-called non-manual signals including facial expressions and body postures are attached to it. They are linguistically characteristic and are different from those in vocal languages. Social interest for sign language has been increasing in Japan and Sign Linguistic Informatics is attracting their attention. On the other hand, the descriptional system for the articulation of signs is not established yet. The history of sign language analysis by linguists and engineers is too short and researches have not been fully advanced. In this paper, we analyze linguistic characteristics of sign language to verify our notational system. The newly proposed notational system for JSL utterances is the kernel subject of our project SiLE.

1. Introduction

We have some 350 thousand deaf population in Japan and ten times as much as the whole hearing impaired. Sign language is their communication means which has a peculiar linguistic system of three dimensional visual presentation with mouthing, eyebrow movement, eye movement and other facial expressions, moreover head movement, shoulder movement and body postures as well as manual movements. However linguistic analysis for sign language is incomplete, so that IT researches of sign recognition, sign generation, computer dictionary, machine translation, educational system, etc are underdevelopment through the former analysis of native intuition with those data a priori. This reports firstly the linguistic characteristics of sign language, focusing on the roles of sight line, secondly notational systems, thirdly introducing machine translation, and the lastly general view and scope of researches in Japan.

2. Linguistic characteristic

2.1 Variety of sign language

We have three types of sign language, Japanese Sign Language (or Traditional SL), Signed Japanese (SimCom) and Pidgin Sign Japanese. JSL is that of the Deaf and it has a different lexicon and grammar from Japanese language. SJ is a manual representation of Japanese according to Japanese grammar. PSJ is a mixture of JSL and SJ.

2.2 Characteristics

Sign language is articulated in the three dimensional space with manual movement and non-manual signals (NMS). On the contrary in the vocal language, articulational components are linearly presented and combined on a single dimension. According to this characteristic, sign language has the following{1-2}.

1. Simultaneity and non-linearity: multiple articulators move in a space at a moment so that phonemes and morphemes are represented simultaneously.
2. Iconicity: meaning is combined to form of a sign, such as CLs.
3. Spatial function: 1st person and 2nd person are presupposed on a sight line. Direction and start/end point represent subject-object function or active-passive voice.
4. Pronominal replacement: specific handshape or location preserves its meaning through the sentence. It functions as pronoun or relatives.
5. Predication: nominal arguments are internalized in many of the verbs. Verbs inflect adverbially or by case, numbers and other syntactical elements.
6. NMS: supra-segmental elements called NMS function as prosody in lexical, phrasal or clausal level.

2.3   Sight line

According to the Eye Mark Recorder analysis of the sight lines of native signers in a dialogue situation, the followings are found.

1. Receiver's eye

   The receiver always watches the sender's face focusing eyes and lips still recognizing movements around the breast. Large and rapid signings seem to be recognized by peripheral gaze. The rate of center focus is shown in Table 1.

2. Sender's eye

   The sender recognizes the specific sight lines: now-on-coding sight when he is thinking something, sight to the hands and sight to pointing to personal location or pronominal space.

Table 1 : Ratio of gaze point of subjects(hearing impaired)

subject	face	hand	others
A	95%	0%	5%
B	87%	9%	4%

3. Notational system

There is no IPA(International Phonetic Alphabets) for sign languages. Neither Stokoe's System nor Friedman's System in USA nor HamNoSys in Germany is universal{3-5}. Thus we proposed our notational system good for computer dictionary or for network use.

3.1   Hierachic NVG morphemic notational model

This model reflects the linguistic analysis and the sight line analysis. We suppose it is good for phono-morphemic description of sign language{6}.

1. N-morph: it is often represented as a handshape. When a sign has a movement, the target location is described and when the terminal handshape is different from the initial one, those informations are done.

2. V-morph: it is subcategorized to Movement morph, Path Movement morph and static Deictic morph.

3. G-morph: grammatical morph includes facial expressions, nodding, mouthing and other NMS, considering those syntactic factors are often realized in word formation in sign language.

4. S-morph: sight light morph seems to be functionally independent to other NMS.

We have resulted from describing JSL words with the NVG model. It was confirmed that some sign structures have the similar morphemic structure, but only their values of parameter are different. Then we introduce the meta morph as a more abstract and higher level concept as compared with a morph. As a result, we are able to describe the morphemic structure with less elements of information by introducing the morph structure having higher abstraction as compared with the morphs. Example of the concepts of meta morphs are "Movement", "Convergence", "Divergence", "Tense"etc.

Table 2 shows a sample of a hierarchic NVG morphemic model. Description example of the NVG model are shown below.

```
<Description Examples>   JSL = {asu} {Hokkaido} {hikouki}
                Japanese sentence = asu Hokkaido ni ikimasu
                English equivalent  = I'll go to Hokkaido by airplane tomorrow
                NVG model = [N(#CL:H1)V(m#tense:future,t)]S(n)+
                        [[NN(H12)]V(locuse)]S(n)+
                        N(#CL:H45)V(m#movement,per1 - per3)]S(Ns)
```

Table.2 Hierarchic NVG Morphemic Notational Model

morph	variable	parameter	value	value for S Morph
V Morph	Kinetic Morph	Kinetic Morph	target	target
		Locus Morph	locus	trace
	Pointing Morph	Pointing Morpheme : #PT		target
		Attribute Morpheme : #AT	target	target
		Present Morpheme : #PR		#PTi or (j)
N Morph	Initial Hand Shape	Classifier : #CL	position	#PT$_i$
		Specific Hand Shape	local movement : lm-i	
	Final Hand Shape	Classifier : #CL	target	
		Specific Hand Shape		
G Morph	NMS(except for Sight Line)	parts of body, nodding, shoulder, mouthing etc		
	AU number of FACS			
S Morph	Sight Line Morph	value to N Morph	N, Ns, Nw, n	
		target for V Morph	target	
		target for remembrance sight line	r$_{(j)}$	
		target for pronoun space	#PTi or (j)	

Aspect	^ *variable*
morpheme	# *variable*
Meta morph	m# *variable*
allomorph	[*morph*]am
dominate relation	[*N Morph*](strong hand) or w(weak hand)
range of S Morph appearance	[*NV descriptions*] S(value)
range of G Morph appearance	[*NVGS descriptions*] G(value)
preservation of hand shape	& [*N Morph*]
#PTi	i = personal pronoun space for 1st, 2nd, 3rd, s, w
#PT$_{(j)}$	j = pronoun space for a,b,c, etc.
lm-i	i=counting, wiggling

3.2 sIGNDEX

In 1996 we established a working group for labeling signs in Sign Linguistic Engineering Association. Their results are opened to public as sIGNDEX V.1 (Lexis) in 1998 by CD-ROM of 515 signs, and sIGNDEX V.2 (Utterances) forthcoming{7-9}. Their guidelines are as follows.

　　1) Not describing phonetic contrasts.
　　2) Linear description.
　　3) All the symbol is on PC keyboard.
　　4) Network use (consistency with SGML, etc)
　　5) The initial letter is small and the rest are capitalized to differentiate sign label with explanation.
　　6) Deductive approach starting from description.
In sIGNDEX V.2 the following two conditions are added.

　　7. Collection of pure JSL utterances.
　　8. Description of NMS's

4. Recognition and generation of sign language

Our final goal is to develop a automatic mutual translation system for Japanese and JSL. Figure 1 shows a concept structure of the bi-directional translation system now being under construction.

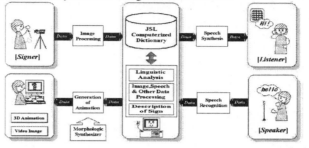

Figure 1. Concept structure of bi-directional translation system between JSL and Japanese

4.1 Input
Considering the characteristics of JSL cited above, we attempted Cyber Glove™ which is good for collecting and processing data, and video recognition which removes the signers obsession and enables their free movement.

4.2 Recognition
Segmentation of transition between the two signs is inevitable. The segmented movements of the signs are recognized by pattern matching method, DP (dynamic programming) or HMM (hidden Markov model). NMS are the problem.

4.3 Generation
In sign generation by rule synthesis, each angle of the finger joints at the starting, changing and ending point and spatial location of the hands are symbolized and registered. The inverse kinetics generates its natural movement. It economizes data volume and is adoptable to sign inflections. The movement of the animation is rather mechanical however. Sign generation by motion capture is expected to move naturally. It needs time to register signs and is not flexible to inflections.
Figure 2 shows the flow of the process of animation generation by NVG model and sIGNDEX{10}.

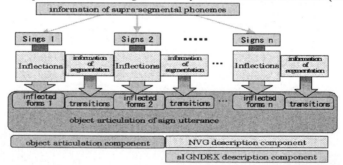

Figure 2. Flow of generation process by NVG and sIGNDEX

4.4 Present stage of researchers
Many organizations and laboratories have developed their own models in various ways. In our system data input of a signing is made by the video camera and recognized by HMM processing of the two dimensional traces. We adopted sIGNDEX and NVG Morphemic Model to describe the data.
Hitachi{11}, Matsushita{12-13}, and other laboratories developed their own systems but they are not cited here because of a short space.

5. Computer dictionary

The computer dictionary is a kernel research in application of sign linguistic researches. It will contain the following functions.

1. Easy search for a sign by the Japanese label.
2. Lexical information of compounding, constructional homonymity and polysemy
3. Optional observation of a sign from any direction.
4. Sample sentences and usages.
5. Etymology
6. Search for a sigh by the strings of sign phonemes and NMS's.

The newer systems add animation generation and networking to the above{14-15}.

6. Conclusions

This reported linguistic characteristics and description of sign language, introducing NMG Morphemic Model and sIGNDEX with their practical application to them. It briefly reported the recent research of translation system.

Researches for vocal language are highly developed. Phonological, morphological, syntactic and semantic construction is rather declared, however, that in sign language is under development now. In recognition process, words and situation are limited and inflections, assimilation are neglected.

But those practical applications are expected to be quickly expanded and to be advanced in the new century. However those technologies should not be ad hoc without taking the linguistic characteristics of sign language for account. Hearing impairment is invisible so that researches for them are less advanced than in those in other handicaps. But the base of technology for them has been growing and they are to be integrated to go forward soon.

References

1. Kanda, K. (1991). Lectures on Sign Linguistics, Fukumura Press (in Japanese).
2. Nagashima, Y. (1999). Present State and Issues of SiLE, Technical Report of IEICE, 99, 449, 57-64 (In Japanese).
3. Stokoe, W.C. et al. (1965). A Dictionary of American Sign Language on Linguistic Principles, Gallaudet College Press.
4. Friedman, L.A. (1977). On the Other Hand, Academic Press.
5. Siegmund, P., et al. (1989). HamNoSys Version 2.0, Signum Press.
6. Nagashima, Y., Kamei, S., Sugiyam, Y. (1996). Sign Language Animation System Morpheme Information Driven Type, Technical Report of IEICE, 96, 604, 73-78 (In Japanese).
7. Kanda, K., Nagashima, Y., Ichikawa, A. (1996). Signdex -- The Data Base of Vocabulary in Signdex --, Technical Report of IEICE, 96, 604, 53-58 (In Japanese).
8. Kanda, K. et al. (2000). Signdex V.1 & V.2 The Computerized Dictionary of Japanese Sign Language, Proceedings of TISLR, 195.
9. Kanda, K. et al. (2000, in print). Notation System and Statistical Analysis of NMS in JSL, Proceedings of Gesture Workshop.
10. Fukuda, Y., Nagashima, Y. (2001). Synthesize of the Sign Language Animation using sGIGNDEX and NVG Morphological Notation Model, Technical Report of IEICE, WIT00-4, 39-44 (In Japanese).
11. Sagawa, H. et al. (1998). Method to Describe and Recognize Sign Language Based on Gesture Components Represented by Symbols and Numerical value, Knowledge Based Systems, 10, 5, 287-294.
12. Lu, S. et al. (1997). Towards a Dialogue System Based on Recognition and Synthesis of Japanese Sign Language, Proceedings of Gesture Workshop, 259-271.
13. Matsuo, H. et al. (2000). The Japanese Sign Language using General Movement Traits and Localized Characteristic in Images, Proceedings of SCI2000, 5, 1, 189-195.
14. Kamei, S. Nagashima, Y. (1997). Development of Computerized Dictionary for the WWW, Proceedings of WWDU, 77.
15. Kuwako, H., Horiuchi, Y., Kanda, K., Ichikawa, A. (2000). The JSL Electronic Dictionary on the WWW, Correspondences on Human Interface, 2, 5, 11-14 (In Japanese).

Usage of a stenographic typewriter for the hearing impaired at NRCD in Japan

Tsuyoshi Nakayama[a], Noriyuki Tejima[b], Shigeru Yamauchi[a]

[a]Research Institute, The National Rehabilitation Center for the Disabled,
1, Namiki 4-chome, Tokorozawa City, Saitama Prefecture 359-8555, Japan*

[b]Department of Robotics on Faculty of Science and Engineering at Ritsumeikan University,
1-1-1 Noji-Higashi, Kusatsu, Shiga 525-8577

Abstract

In 1987, utilizing stenographic technique an assistive device, called *Stenopcon*, was developed at the National Rehabilitation Center for the Disabled in Japan. It can convert voices into Japanese sentences and display them with only a few seconds delay for people with hearing impairments. Since then, it has been used at many meetings and conferences relevant to hearing impairments, including international ones and weekly meetings for the hearing impaired at the training center of the NRCD. The problems, future views of the devices and recent studies with respect to the device are also discussed in this paper.

1. Introduction

People with hearing impairments often find difficulties in acquiring information from speeches. Sign language is a very powerful and important method of communicating. However, it is not easy to master sign language especially for people who were deafened or became hard of hearing in their adulthood. In fact the rate of the hearing impaired who master sign language is not so high in Japan. And so another method of communication is sometimes required.

From 1987, an assistive device has been studied and developed at the National Rehabilitation Center for the Disabled in Japan, which helps people with hearing impairments. The device was converted from a stenographic typewriter, called Soku-type, which many years ago was remodeled in order to suit the Japanese language. The device uses a system, which is now called the Stenopcon system. The Stenopcon system can convert speeches to Japanese sentences and display them quickly. Reading the sentences the hearing impaired can understand what speakers are saying. In this paper, how the device is used at the NRCD and the latest studies related to the device, such as, a study of displaying Japanese sentences for the hearing impaired, are reported.

2. The Stenopcon system

The Stenopcon system consists of a special keyboard, a personal computer, particular software and a display unit, such as a large TV monitor and a video projector. This system requires two operators, a stenographer and an editing operator. A diagram of the Stenopcon system is shown in Figure 1. A stenographer types on a Stenopcon keyboard according to what speakers are saying. The information from the keyboard is sent to a personal computer via RS-232C. Particular software translates the electric signals to Japanese sentences and displays them on a monitor immediately. An operator edits the Japanese sentences and corrects them if there are any errors. After editing, these sentences are displayed on a large TV monitor or screen via a video projector or an overhead projector. A photograph of a Stenopcon keyboard is shown in Figure 2 and another photograph of the Stenopcon system being used at the NRCD is shown in Figure 3.

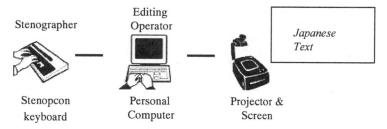

Figure 1. Stenopcon system diagram

Figure 2. Stenopcon keyboard

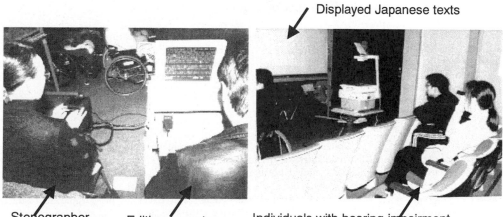

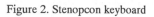

Figure 3. The Stenopcon system at a certain meeting at the NRCD

3. Characteristic feature of the stenopcon system

The Stenopcon system has several good points. When other methods such as displaying a summarization of spoken words by computers or handwriting are used, the audience might misunderstand what speakers really want to present because not all of a speech is displayed, instead only a summary is displayed. And speakers sometimes might complain about the summaries because they are not always summarized properly. On the other hand, the Stenopcon system can follow any speech even if its speed is very fast thanks to the stenographic technique. Since it displays texts after errors are corrected, there are few errors in the displayed sentences. Stenographers can type on the Stenopcon keyboard without any special training. And it can create verbatim records automatically while recording displayed texts on a floppy desk.

3.1 An evaluation of the Stenopcon system by a questionnaire to an audience with hearing impairments

A questionnaire was given to people with hearing impairments after they attended a certain meeting in which the Stenopcon system was used. Most of them felt that the Stenopcon system was helpful to understand the speeches. Some respondents explained the reasons why this system was helpful. 44 out of 81 people, that is, about 53% of the respondents answered that they can understand the speeches from the displayed texts even when they missed watching sign language interpreters or listening to them through their hearing aids. In this case, the Stenopcon system worked as an assistant of other communicating methods with the hearing impaired. And 30 respondents, about 37%, answered that they could understand shades of meaning in the speeches owing to the Stenopcon system. This is because it can display the whole words of the speeches. 28 respondents, about 35%, answered that the displayed texts were very easy to read. 24 respondents, about 30%, answered that it can present much more information of the speeches.

On the other hand the Stenopcon system has several weak points. According to the result of the questionnaire 44 respondents, about 53%, mentioned that there were not enough Chinese characters in the displayed texts and because of it they felt difficulties in reading them to some extent. The proportion of Chinese characters from all of the characters in ordinary Japanese sentences is estimated at about 40%. Only about 10% of texts displayed by the Stenopcon system contain Chinese characters. Of course, the rate can increase or decrease depending upon the contents of the speeches. Problems with respect to the Stenopcon system including this are discussed in the next chapter.

4. Problems surrounding the Stenopcon system

The Stenopcon system has been used at many meetings and conferences relevant to hearing impairments including a weekly meeting for the hearing impaired at the training center of the National Rehabilitation Center for the Disabled in Japan. However, the Stenopcon system hasn't become so popular so much in Japan. On the other hand, in the USA, real time captioning services represented by Rapidtext(R) have become popular recently {1}. In this chapter, several problems with respect to the Stenopcon system are described.

4.1 Problem of written words in Japanese

Written words in Japanese are quite different from English. In Japanese sentences apart from the Roman alphabet, three other kinds of characters such as kanji, katakana and hiragana are used. Both katakana and hiragana consist of about 50 phonetic letters. On the other hand kanji characters are ideograms and the number of them in general use is over 3000. As mentioned above, special software on a personal computer translates stenographic codes into written words in the Stenopcon system. Unfortunately stenographic codes can't cover all these Japanese characters. Without kanji characters, it takes much time to read written Japanese texts because the Japanese language has too many homonymous words. Although an automatic conversion algorithm has been developed and used practically, it is still necessary to select the proper word out of some words nominated by the algorithm. And so it is much more difficult to realize real time captioning in the Japanese language. Therefore the Stenopcon system requires an editing operator in addition to a stenographer.

4.2   Problems with stenographers and stenographic typewriters

Most stenographers work at courts in order to create verbatim records both in Japan and the USA. However, the Supreme Court of Japan decided to adopt tape recording in stead of stenography as the recording method several years ago. In Japan a national training center for stenographers was established in the 1950's and about 30 stenographers had graduated from the center every year. Because of the decision of the Supreme Court of Japan this training center was closed recently. Now the number of stenographers including the retired in Japan is estimated to be 1,500. It is said that this is much less than the number of those in the USA. The difference of these two numbers will increase more. Speaking of the Japanese steno typewriters, the Soku-type, it is difficult for the Japanese manufacture to continue producing them. In fact it is nearly impossible to purchase a new Soku-type now in Japan. The conditions surrounding the Stenopcon system have become severe recently.

4.3   Problem of fatigue of the audience in reading texts

When the Stenopcon system is practiced audiences with hearing impairments continue to read Japanese texts displayed on a screen or a large TV monitor. Normally texts are displayed over the whole area of the screen and about 160 characters are displayed at the same time. A New line of texts appears depending upon the stenographer's typewriting and the oldest line on the screen disappears. The speed of displaying texts depends upon the speed of speaking words and is sometimes very rapid. And so audiences with hearing impairments sometimes feel a little fatigue while reading texts. Some people from the audience mentioned that they experienced a slight eye fatigue when the meeting lasted long. In order to lessen the fatigue of readers easily read texts are required as much possible. However, an optimum method of displaying Japanese texts on a screen by the Stenopcon system has not been established.

5.   Current study and future works

As mentioned above the Stenopcon system has been developed and used practically. In this chapter current study and future works relevant to the Stenopcon system are described.

5.1   An alternative method of the Stenopcon system

In consideration of the problem with stenographers and stenographic typewriters, alternative measures of the Stenopcon system are studied. For instance, a continuous speech recognition system was tried. Because the accuracy of recognition was not good enough it could not be used in practice. However, the technology in continuous speech recognition has advanced remarkably in this decade and TV closed captions for broadcasting news started last year in Japan {2}. Because this system may be too expensive and not on the market it can't be still used as an alternative of the Stenopcon system now. When a continuous speech recognition software or hardware is available reasonably and has enough abilities to be an alternative of the Stenopcon system, we will test it again.

5.2   Readability of texts on a large screen

In order to lessen the fatigue of audience with hearing impairments while reading texts, the readability of texts on large screens was studied {3}. The size of the screen was about 1200mm wide and 780mm high. And the resolution of the screen was SVGA. At the first stage, the relation of reading speeds and character-viewing angles or viewing distances was studied. The relation of the most readable character-viewing angles estimated from reading speeds of six subjects and viewing distance is shown in Figure 4. It is clear that the most readable character-viewing angle decreases gradually as the viewing distance increases. In this way the relationship of readability of texts on a large screen and other factors such as contrast and luminance which can influence it will be studied.

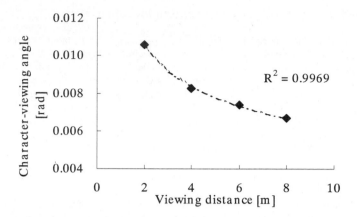

Figure 4. Relation of viewing distance and the most readable character-viewing angles estimated from reading speeds.

6. Conclusion

An assistive device called Stenopcon system is described which can convert spoken words to written ones with only a little delay for people with hearing impairments. In order to evaluate this system a questionnaire was given to an audience with hearing impairments. From the result of the questionnaire, many respondents thought that this system is useful for them to understand speeches well. And the problems and future views of the device and recent studies with respect to the device are also discussed in this paper.

References
1. Linke-Ellis, N. (1999). *Program & Proceeding of TAO Workshop*, 31.
2. Ando, A. (1999). *Program & Proceeding of TAO Workshop*, 121 (in Japanese).
3. Nakayama, T., Tejima, T. (2000). *Jpn J. Ergonomics,* Vol. 36, No. 2, 81 (in Japanese).

Affective Pedagogical Agents and User Persuasion

Chioma Okonkwo, Julita Vassileva

Computer Science Department, University of Saskatchewan,
1C101 Engineering Bldg., 57 Campus Drive, Saskatoon, S7N 5A9, Canada
email: {cno156, jiv} @cs.usask.ca

Abstract

The use of animated pedagogical agents with emotional capabilities in an interactive learning environment has been found to have a positive impact on learners. Aristotle contended that three elements; emotion, logic, and character are crucial for successful persuasion, i.e. in winning others over to one's way of thinking. We have designed a pedagogical agent that acts in an interactive learning environment, using Ortony's Cognitive Structure of Emotion model and McCrae and John's five-factor model of personality. We investigated the persuasive impact of this emotional pedagogical agent on a group of learners. The results show that while not contributing any significant performance gain in learning, the incorporation of emotion changes the way students perceive the learning process, and makes it more engaging. We also found out that there are some gender-based and individual differences in the user perception of an emotional agent, which need to be taken into account when designing a more adaptive and "intelligent" emotional pedagogical agents.

1. Introduction

The use of personified agents in the interface has been a controversial issue in both the HCI and AI community. However, studies have continued to show that using personas significantly improves users' satisfaction with the system (Mulken, Andre and Muller 1998; Koda and Maes 1996). The explanation for this is that these interfaces create a two-way face-to-face communication that enriches the feeling of a personal and social interaction. Researchers and interface designers are striving to enhance the believability of animated interfaces by equipping these personas with qualities that would enable them to communicate in a more human-like fashion. Traits such as distinct personality and expressing affect in communication are believed to increase believability thereby significantly enhancing the user's total experience (Huang 1999; Bates 1994). Emotional agents are believed to have advantages in the context of learning environments. It has been argued by a number of researchers that personification of the environment (e.g. by introducing an animated character/persona) positively affects the student's perception of their learning experience and integrating emotional traits into these personas would result in more effective and motivating instruction (Elliott, Rickel and Lester 1997; Rickel and Johnson 1997). The belief is that these agents have the ability to change students' perception of learning from something that is dull and boring into a fun and engaging activity. Ultimately the student that enjoys a learning environment will spend more time there, which is likely to increase learning (Elliott, Rickel and Lester 1997). So far no empirical study has been carried out to determine how much impact integrating an emotional model into a pedagogical agent would have on students' learning experience. We have conducted an empirical study to investigate the impact of integrating personified pedagogical agents with emotional model, personality traits and affective reasoning on the learning experience of students. Our goal is to find answers to the following questions:

- Does the integration of an emotional persona have any impact on user's learning experience and performance?
- If there is an impact, how significant is it?
- What individual user factors influence the impact of an emotional persona on the users (e.g. sex, knowledge-level, previous experience with educational software and with animated characters)?

The paper is structured as follows: First we describe the pedagogical agent and the learning environment. Next we describe our experimental methodology. Then we present the experimental results, and finally in the discussion section we try to answer the research questions we have outlined above.

2. The Learning Environment and the Affective Pedagogical Agent

We developed a simple courseware application, which takes the student through an introductory lesson on the structure of a C++ program. A simulated programming environment allows students to practice coding skills and

398

make intermediate quizzes. At the end of the training session users are presented with a set of questions testing their knowledge and understanding of the materials presented. The application is designed using Visual Basic, while the materials for the training are delivered using audio output (using pre-recorded female voice). The entire training takes about 30 – 45mins depending on the user's speed.

For the study, we designed "Smiley" - an emotional pedagogical agent that delivers the training. Smiley exhibits verbal and facial expressions in response to user actions and progress during the learning process. The character is designed to appear as a caring pedagogical agent that is concerned about the student's progress and performance throughout the training. Smiley can display six major emotional states: sad, surprised, angry, happy, pleased and neutral (see Figure 1).

Figure 1: Smiley's facial expression for the six major emotional states

We chose to keep the agent's facial expressions simple by reducing the level of abstraction. Some studies (Bartneck, 2001) show that human beings tend to find the emotional expressions of a cartoon-like face more distinct and easily recognizable, in comparison with a real human face. An affective reasoning engine based on the student's action, the agent's preceding emotional state and the agent's overall goal controls Smiley's facial expression and emotional state. Smiley's goal is to motivate the student by convincing him/her that it really cares about his/her performance. To achieve this, Smiley emits a sign of pride and joy whenever the student's performance on a quiz or test item is positive and exhibits an expression of worry and concern (sometimes even disappointment and annoyance) when the student's progress is less than desirable.

3. Methodology

The students' were exposed to two versions of the application in a controlled learning environment; one with the emotional engine switched off and the other with the engine switched on. The experiment ran on high-end Pentium PCs with color monitors. The experiment and data collection phase lasted two days, with each student going through the training and tests, after which they answered a questionnaire to evaluate the students' perception of their learning experience.

3.1 Participants, Experimental Setting and Procedure

The participants 12 individuals, 6 male and 6 female, with an average age of 25 years, were enrolled as first year computer science students at the University of Saskatchewan with no previous experience in programming with C++. In order to encourage participation, the students were offered a free entry ticket to a movie. In order to ensure uniformity and to account for all systematic differences in the groups, participants were randomly assigned to the two test groups. There were an equal number of males and females in each of the groups.

Each participant was given an overview of what was expected of him/her and what to expect during the experiment. Care was taken not to give out information that may bias the user's perception of the animated character. The participants were assured that they were not under evaluation and that they could quit the experiment at any point. The participants were then provided with consent forms assuring anonymity and confidentiality. The participants were then asked to fill out a pre-questionnaire that was used to gather information about their background; sex, age, computer skills, experience in C++, e with software training programs and with animated characters. Figure 2 shows the experimental design.

After the information gathering phase, each student proceeded through the training program. Headphones were provided for each participant to avoid any distractions. Each participant completed the training and tests. The tests were aimed at assessing the knowledge acquired during the lesson. Users actions and responses were tracked and recorded as they went through the entire experiment.

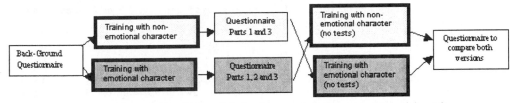

Figure 2: Experimental design (arrows show the groups of participants)

PART ONE

1. Did you find the training entertaining?
2. Was the training difficult?
3. How easy/difficult were the tests?
4. Did you find the persona sympathetic?
5. Did you find the persona distracting?
6. Did the persona help you concentrate on the training?
7. Did the persona motivate/encourage you to further pay attention to the training?
8. Would you choose to have a persona present in your future trainings?

PART TWO

9. Where you able to distinguish between the varying emotional responses of the persona?
10. Did you feel that the persona behavior / responses were appropriate for the situations?
11. Whenever you made a mistake, was the persona concerned?
12. Whenever you made a mistake, was the persona irritated?
13. Did you find the emotions displayed by the agent convincing?
14. Did you feel a need to perform well on the test because you didn't want to disappoint the persona?

PART THREE

15. What did you like or dislike about the training?
16. How would you summarize your experience?

Figure 3: Assessment Questionnaire 1

On completion of the training module, participants were then asked to complete a questionnaire aimed at eliciting responses to be used to assess their perception of the learning experience. The questionnaire consisted of 3 parts (Fig. 3). The first part contained questions which all participants where required to complete. This part contained questions that concerned the general difficulty of the training material/tests, how entertaining the experience was and whether or not the persona helped the participant concentrate on the training. The second part consisted of 6 questions which where specific to participants who had undergone the training with the emotional engine switched on. This part was concerned with the possible impact of the emotional responses. The third part of the questionnaire consisted of open-ended questions, which asked all the participants to comment on how they felt about the training. The questions on part one and two were answered on a five-point scale, users were also asked to comment on why they selected a particular option. Each participant then went through the training again under the alternate persona condition. Those participants who first used the persona with the emotional engine set to "off" where asked to undergo the training with the engine switched "on" and vice-versa. No test was administered during this second round of the experiment (since otherwise it would have been unclear if the students have gained knowledge during this or the previous session). On completion they where given a final questionnaire (Fig. 4).

1 Did you notice any difference between the two versions of the training?
2. If yes which of the two versions do you prefer and why?

Figure 4: Assessment Questionnaire 2

4. Observations

To aid in the analysis, we defined the following metrics. The independent variable is the persona, with emotional engine switched on and off, while the dependent variables are the users' performance on the test and their subjective responses on the questionnaires. The persona variable was manipulated between participants and the results were analyzed using t-test with an α level of 0.05. Our null hypothesis was that there was no difference between the two groups; i.e. whether the emotional engine was switched to on or off, performance was the same. Also for each of the individual responses in the questionnaire, we state a null hypothesis suggesting that there is no difference between the two groups for each of the effects of interest.

4.1 Analysis

To determine the impact of the persona on the performance of participants, the test scores for the participants were analyzed using a two-tailed t-test. Each correct answer in the test was awarded a score of 12.5, leading to a maximum score of 100. The analysis revealed that there was no real effect of emotional persona on student's performance as there was no significant difference between the means of the emotional and non-emotional condition ($t (10)= -0.24$; $p=0.82$). See Table 1 below for a summary of the test scores.

Table 1: Summary of participant's test scores

Partici-pant	Non-Emotional	Emotional
1	37.5	62.5
2	75	87.5
3	87.5	62.5
4	75	62.5
5	50	75
6	75	62.5

Table 2: Mean and t-values for questions in Part 1

Questions	Mean Emotional	Mean Non-emotional	t-value
1	4.17	3.83	+0.95
2	3.67	3.67	0.00
3	3.67	3.33	-0.85
4	3.67	1.83	-2.36
5	4.83	4.17	+1.32
6	3.00	1.50	-3.00
7	1.83	3.67	-4.04
8	3.50	3.33	-0.25

Table 3: Mean scores for questions in Part 2

Q	Mean
9	4.8
10	4.8
11	4.8
12	2.3
13	4.8
14	3.7

For the subjective assessment, the data from the first part of the questionnaire were subjected to t-tests. For all questions besides question 4, 6 and 7, the analysis showed no significant effects. For question 4 (see Figure 3), a significant effect was discovered ($t (10) = -2.36$; $p = 0.04$). For question 6, also significant impact was also discovered ($t (10) = -3$; $p = 0.01$). For question 7, the data also showed a significant effect ($t (10) = -4.04$; $p = 0.00$). Questions 1, 2, and 3 asked about the degree of entertainment and the difficulty of the training/test. All participants found the training to be entertaining. Participants from both groups did not find the persona to be distracting, however participants under the non-emotional said they didn't pay much attention to the persona. On question 8, 9 participants commented that they would like to have an agent present during their training, while 3 participants where indifferent. Table 2 shows the mean and t-values obtained for the questions in part 1.

For the questions in part 2 of the questionnaire, which were related to the impact of the personality and emotion displayed by the pedagogical agent on users' perception, the following results were obtained; For questions 9, 10, 11 and 13, five participants gave a value of 5 points, and 1 participant gave 4 points; thus indicating that they felt that the persona was very concerned about their progress, its emotional responses were adequate and convincing, and that they were able to clearly distinguish between the agent's responses. On question 12, there were mixed answers. These users who felt that the persona was irritated commented that this was the case only when they continually failed answer a question correctly. Interestingly, on question 14 (on whether they felt a need to perform well on the test in order not to disappoint the persona), three participants, all female, gave a value of 5 points indicating that they felt a need to do well so as not the disappoint the persona. From the other three male participants, one gave 1 point while the other two awarded gave 3 points. So the male participants didn't seem to feel pressured to perform better in order to please the persona. This suggests that females are influenced differently from males by the persona. Table 3 shows the mean values obtained for the questions in part 2. When asked to choose between having between the two versions of the persona all participants voted that they would rather have the persona with the emotional engine switched on. Most participants felt that the emotional version made the training more interesting and motivational and prevents them from quitting the application altogether.

5. Discussion and Conclusions

The results showed that the participants found the persona motivating and all of them said they enjoyed the training with the emotional persona more. Participants who worked with the emotional persona commented that they could tell what their progress on the training was by simply looking at the facial expression displayed by the persona or noting the joy or disappointment it displayed. This observation supports the speculated impact of affective pedagogical agents discussed in literature (Elliott, Rickel and Lester, 1997). On completion of the experiment, a number of participants under the emotional condition commented that the they were surprised that they were not bored and were actually able to go through the whole training twice, as they have never viewed going through a training as a fun experience. Participants under the emotional condition believed that the persona acted as a motivational factor as they all rated the questions on concentration and encouragement very highly. The persuasive impact that the emotional agent had on these participants' perception of their learning experience was clearly evident.

There also seems to a difference in the way the female participants viewed the persona when compared to the male participants. While the women felt the need to do well so as to elicit a smile from the agent, the men felt no such compulsion but merely used it as a tool to determine their progress on the training. However, all the participants under the emotional condition felt that the persona was truly concerned about their progress.

However, our findings showed that an emotional pedagogical agent had no impact on the learning performance of users. This is in agreement with findings of other authors. For example, (Mulken, Andre and Muller, 1998) show that the presence of a persona has no significant impact on the users' understanding when technical explanations were being presented. This finding is also consistent with the comments made by participants; 80% of whom stated that they paid little attention to the agent since they were busy trying to grasp the material and paid more attention to the audio output and the visual reading material. Though most of the students reported that they did not find the persona distracting, there is still a possibility that the persona while entertaining, had some negative impact on the student's concentration during the training.

One possible explanation is that our study didn't account for the user's personality and preferences. A good tutor takes the student preferences and behavior into account. Incorporating user-modeling capability in the agent will help to generate an adaptive affective pedagogical agent that is more suited for a particular student. Further work is required to see if incorporating user modeling of the affective preferences of the user as well adapting to the user's emotional response will help to influence learning.

Finally, it would be interesting to investigate the effect of a longer-term exposure to an emotional persona on the user. If, as suggested by our results, the persona makes learning more entertaining and fun, the students may spend more time in the environment and some of them may actually strive to perform better in order to please the persona, which will eventually tie up in better performance results. However, a long-term experiment has to account for the effect of novelty that an animated persona introduces in the interface. It is possible that the motivating effect of the agent in our study was due only to the entertainment effect of the first encounter. Defining the threshold of the user's "saturation" to the point of getting bored and annoyed with the agent is an important task for future work, not only for animated personas in learning environments, but in general.

References

Bartneck C. (2001). Affective Expressions of Machines. *Proceedings CHI2001 Conference*, Seattle.

Bates J. (1994). The role of emotions in believable agents. *Communication of the ACM*, 37(7): 122-125.

Elliott C., Rickel, J., Lester, J.C. (1997). Integrating affective computing into animated tutoring agents. In *Proceedings IJCAI Workshop on Animated Interface Agents: Making Them Intelligent*, Nagoya, Japan (pp. 113-121).

Huang H.-Y.(1999). The persuasion, memory and social presence effects of believable agents in human-agent communication. In *Proceedings of the Third International Cognitive Technology Conference, CT'99*, San Francisco/Silicon Valley.

Koda T., Maes, P. (1996). Agents with faces: The effect of personification. In *Proceedings 5th IEEE International Workshop on Robot and Human Communication (RO-MAN'96)*, Tsukuba, Japan.

Lester, J.C., Converse, S., Kahler, S., Barlow, T., Stone, B., Bhogal, R. (1997). The Persona Effect: affective impact of animated pedagogical agents. In *Proceedings CHI'97 (Conference on Human Factors in Computing Systems)*, Atlanta GA (pp. 359-366). ACM Press.

McCrae R.R., John O.P. (1992). An introduction to the five-factor model and its applications. *Special Issue: The five-factor model: Issues and applications. Journal of Personality*, 60: 175-215.

Ortony, A., Clore, G., Collins, A. (1986). *The Cognitive Structure of Emotions*. Cambridge, MA: Cambridge University Press.

Rickel J., Johnson L. (1997). Integrating pedagogical capabilities in a virtual environment agent. In *Proceedings First International Conference on Autonomous Agents*.

Rist T., Andre, E., Muller, J. (1997). Adding Animated Presentation Agents to the Interface. In J. Moore, E. Edmonds, & A. Puerta, (Eds), *Proceedings International Conference on Intelligent User Interfaces*, Orlando, Florida (pp. 79-86).

van Mulken S., Andre, E., Muller, J. (1998). The Persona Effect: How Substantial Is It?. In H. Johnson, L. Nigay, & C. Roast (Eds), *People and Computers XIII: Proceedings HCI '98*. London: Springer-Verlag.

Speech recognition for data entry by individuals with spinal cord injuries

Kweso Oseitutu[1], Jinjuan Feng[1], Andrew Sears[1], Claire-Marie Karat[2]

[1]Laboratory for Interactive Systems Design, Information Systems Department, UMBC, 1000 Hilltop Circle, Baltimore, MD 21250
[2]IBM TJ Watson Research Center, 30 Sawmill River Road, Hawthorne, NY 10532

Abstract

Speech recognition is an important technology that is becoming increasingly effective for dictation-oriented activities. While speech recognition can be a convenient alternative for traditional computer users, it can be a powerful tool for individuals with physical disabilities that limit their ability to use a keyboard and mouse. In this article, we report on a study that highlights the differences between the usage patterns of traditional users and users with high-level spinal cord injuries. We discuss the implications of these differences.

1. Introduction

Speech recognition (SR) is an increasingly important technology. Effective SR systems can result in fundamental changes in the lives of some individuals with disabilities that limit their ability to physically manipulate their environment. For individuals with physical disabilities but no speech impairments, speech activated environmental control systems can increase independence while SR-based dictation systems provide an additional method of communicating with others. As a result, SR can be an important technology for individuals with a variety of physical disabilities including spinal cord injuries, severe repetitive stress injuries, upper extremity amputations, arthrogryposis, arthritis, and muscular dystrophy.

In this article we report on an experiment designed to provide insights into the efficacy of SR-based dictation systems when used by individuals with physical disabilities. More specifically, we focus on these issues in the context of individuals with high-level spinal cord injuries. Our goal was to determine whether or not traditional computer users and individuals with spinal cord injuries employ the same process as they interact with this technology.

2. Related research

While many researchers have investigated the use of SR, many have focused on small-vocabulary SR systems and relatively simple tasks. For example, Ainsworth (1988) examined the optimal string length for digit input and more recently Ainsworth and Pratt (1992) explored several error-correction strategies for use in a phone dialing scenario. Both of these studies used isolated-word SR systems, short input strings (i.e., 1-14 digits), and small vocabularies (i.e., 11 or 14 words). Noyes and Frankish (1994) explored the efficacy of auditory or visual feedback after every word or sequence of six words while Baber and Hone (1993) found that the optimal error correction strategy depends on the accuracy of the system and the length of the input strings. Karat, Halverson, Karat, and Horn (1999) compared large-vocabulary SR software to the traditional keyboard and mouse for text entry when used by traditional computer users with no physical disabilities. In addition to productivity, they also examined correction episodes and the strategies used. They noticed with experience that the number of correction episodes decreased (from 11.3 to 8.8) and more importantly the number of steps per episode also decreased (from 7.3 to 3.5) as participants gained experience. Their investigation of strategies used to correct errors indicated that users often fixate on a single error correction strategy, frequently redictate words that are initially recognized incorrectly, and spend significant time correcting cascading errors. While each of these studies was designed to provide insights that could improve dictation-oriented speech recognition application, none included participants with physical disabilities.

3. Research objectives and hypotheses

The purpose of the current study is to provide the first empirical evaluation of current SR technology when used by individuals with physical disabilities to generate nontrivial quantities of text. Currently, the focus is on individuals with high-level spinal cord injuries. By investigating any differences that may exist between traditional computer users and those with high-level spinal cord injuries, we will create a foundation for the future research and development aimed at enhancing the usability of SR for individuals with physical disabilities.

4. Method

4.1 Subjects

Fourteen individuals were recruited to participate in this study. Seven participants had no documented physical impairments that would hinder their ability to use the keyboard or mouse. The remaining seven participants had spinal cord injuries (SCI) at or above C6 with American Spinal Cord Injury Association (ASIA) scores of A or B. As a result, our SCI users had limited or no use of their hands. All participants were recruited from the Maryland area. Participants received payment of $30US as compensation for their time. To address the additional time and effort required by the SCI participants, they were offered an additional $20US in compensation.

Participation was restricted to individuals who did not have any uncorrected visual impairments or documented cognitive, hearing, or speech impairments. All participants had prior experience using a commercial speech recognition software product for dictation-oriented activities. Thirteen participants were male, the participants' average age was 34.4 (stdev: 11.7), and participants averaged 15.6 years of computing experience (stdev: 6.8).

4.2 Apparatus

Participants utilized a Gateway Solo Pro 9300 laptop computer with a 600MHz Pentium III processor, 128M of memory, and a VXI Parrot 10-3 microphone. Participants interacted with a custom speech recognition application, TkTalk 1.0, that uses IBM's ViaVoice Millennium Edition speech recognition engine. TkTalk allows users to dictate text and to edit that text using a full range of speech-activated editing capabilities. By developing new SR software we were able to provide the core editing commands that were expected by individuals with previous experience using virtually any commercial SR product, including IBM's ViaVoice and Dragon's Naturally Speaking software. The room contained multiple cameras, a scan converter, and microphones that enabled audio/video recording of all activity by the participants. In addition, the TkTalk software was instrumented to provide a detailed record of all speech-based activity.

4.3 Tasks

Each participant completed four tasks that were presented on separate pieces of paper. Two transcription tasks required users to enter a predefined paragraph of text (67 and 71 words respectively). Two composition tasks required users to respond to several questions posed in a hypothetical email (3 and 4 questions respectively). Responses for composition tasks could be as brief or lengthy as the participant desired.

4.4 Procedure

After reading and signing the consent form, participants were guided through the standard enrollment process utilized by the Millennium edition of ViaVoice. Each participant completed one training passage, ensuring that the recognition engine had an equivalent amount of information about each participant's speech patterns. Participants were guided through 30 minutes of training and practice using two sample tasks. At the end of the training session, the experimenter left the participant alone to complete the tasks.

Participants were given four tasks, one at a time, to ensure that the tasks were completed in a predefined order. A typing stand was available to hold the current task. After completion of each task, participants responded to questions about the ease of use of the software and whether or not they were satisfied with the amount of time required to complete the task. After completing all four tasks, participants completed a final questionnaire regarding their satisfaction with the software in the context of transcribing predefined text and composing new text, feelings about locating and correcting recognition errors, comparisons with their normal method of completing these types of tasks, recommendations for changes to the software, the tasks for which they would use SR, and how important SR was as a method of interacting with computers. They also provided a variety of demographic information at the conclusion of the study.

At the end of the session, the experimenter answered any questions, thanked participants for their time, and provided the appropriate payment for the session. Sessions lasted approximately 1.5 hours, including several required breaks.

5. Results

No significant differences were detected as to the productivity of the two groups of users, but SCI users provided more positive satisfaction ratings (Sears, Karat, Oseitutu, Karimullah & Feng, 2001). To better understand the process participants employed when interacting with the SR software, we analyzed the number of dictation episodes; how long users delayed correcting recognition errors; how likely users were to use the "scratch that" command to correct recognition errors; and the percentage of time allocated to dictation, navigation commands, and non-navigation commands.

Table 1: Means (and standard deviations) for the number of dictation episodes.

		Transcription		Composition	
		T1	T2	C1	C2
Group	Traditional Users	4.00 (1.16)	4.29 (1.38)	3.86 (1.57)	4.57 (1.40)
	SCI Users	7.57 (6.16)	10.57 (13.51)	11.29 (12.75)	7.43 (3.55)

5.1 Dictation episodes

The number of dictation episodes measures how willing a participant is to interrupt dictation to correct errors. Means and standard deviations for the number of dictation episodes are reported in Table 1. The number of dictation episodes was analyzed using a one-way ANCOVA with repeated measures for type of task (transcription vs. composition) and task number (e.g., first transcript vs. second transcription). This analysis identified a significant effect due to group (F(1,10)=5.12, p<0.05), but there was no significant effects due to type of task or task number. Neither importance nor experience had a significant effect on the data entry rates and there were no significant interactions. Overall, there were 9.2 dictation episodes per task for SCI users and only 4.2 episodes per task for traditional users.

5.2 Amount of deferment

The amount of deferment refers to the number of words a user dictates after a recognition error before they correct the error. The amount of deferment is directly related to how often the user interrupts their dictation to correct errors. If errors are corrected as they occur, there is no deferment. At the other extreme, a user may dictate the entire document and then correct all errors. An independent sample t-test on the average deferment for each participant indicates that there is a significant difference between the two groups (t(12)=2.8, p<0.02). Traditional users delayed corrections longer (mean=7.6, stdev=6.2) than SCI users (mean=2.2, stdev=2.0).

5.3 Use of "scratch that"

Use of the "scratch that" command provides another indication that users are carefully watching the output of the speech recognition software and that they are willing to interrupt their dictation to correct errors as they occur. We counted the number of correction episodes for each group. We also counted the number of these episodes where "scratch that" was the first command issued. These episodes highlight situations where users corrected errors as soon as they occurred. A CHI-Squared analysis revealed that the SCI users were significantly more likely to correct errors as they occurred using the "scratch that" command ($X^2(1) = 176.0$, p<0.001). For SCI users, 60% of the correction episodes involved using the "scratch that" command to correct errors as they occurred as compared to less than 1% for traditional computer users.

5.4   Time allocation

Figure 1 illustrates how time was spent by the two groups of users. The time allocated to dictation was analyzed using a one-way ANCOVA with repeated measures for type of task and task number. This analysis identified a significant effect for group ($F(1,10)=6.76$, $p<0.05$), but did not identify any significant effects due to type of task or task number. Neither importance nor experience had a significant effect and there were no significant interactions.

An equivalent analysis of the time spent issuing navigation commands also identified a significant effect due to group ($F(1,10)=7.77$, $p<0.02$), but there was no significant effect due to type of task or task. Neither importance nor experience had a significant effect and there were no significant interactions. Finally, the time spent issuing non-navigation commands was also analyzed using a one-way ANCOVA with repeated measures for type of task and task number. This analysis did not identify any significant effects due to group, type of task, or task. Neither importance nor experience had a significant effect and there were no significant interactions.

6.   Conclusions

While performance did not differ, traditional users were significantly more negative about their experience after completing their tasks. Given no differences in productivity, the more negative satisfaction ratings provided by the traditional users may attract attention from both researchers and developers. As a result, future studies could continue to focus on usability for traditional computer users with the assumption that changes that make the system more effective for traditional users will also benefit disabled users.

The two groups of users included in our study employ different strategies when interacting with SR software. Traditional users spent approximately 40% of their time navigating from one location to another within the document. This suggests that more efficient navigation should result in substantial benefits for traditional users. It also suggests that techniques that make locating incorrectly recognized words while reviewing dictation results may benefit this group of users. In contrast, SCI users interrupt their dictation more frequently, navigate shorter distances, and as a result they spend less of their time issuing navigation commands. As a result, SCI users will benefit less from improved navigation. These users may benefit more by improving commands that are used when you interrupt dictation to correct errors, like "scratch that."

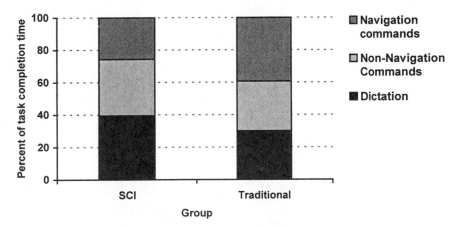

Figure 1: Allocation of time to dictation, non-navigational commands, and navigation commands for both groups of users.

Most ongoing research and development of dictation-oriented SR applications currently focuses on traditional users. While traditional users appear to be more difficult to satisfy, the underlying processes adopted by SCI users are fundamentally different. Therefore, our results demonstrate that a new approach, that more effectively integrates disabled users, is required to make SR systems universally accessible.

References

Ainsworth, W. A. (1988). Optimization of string length for spoken digit input with error correction. *International Journal of Man-Machine Studies, 28*, 573-581.

Ainsworth, W. A. and Pratt, S. R. (1992). Feedback Strategies for Error Correction in Speech Recognition Systems. *International Journal of Man-Machine Studies, 36*, 833-842.

Baber, C., and Hone, K. (1993). Modeling error recovery and repair in automatic speech recognition. *International Journal of Man-Machine Studies, 39*, 495-515.

Karat, C-M., Halverson, C., Karat, J. and Horn, D. (1999). Patterns of Entry and Correction in Large Vocabulary Continuous Speech Recognition Systems. *Proc. of CHI 99*, 568-575.

Noyes, J. M. and Frankish, C.R., (1994). Errors and error correction in automatic speech recognition systems, *Ergonomics, 37*, 1943-1957.

Sears, A., Karat, C-M, Oseitutu, K., Karimullah, A., J. Feng. Productivity, satisfaction, and interaction strategies of individuals with spinal cord injuries and traditional users interacting with speech recognition software. Universal Access in the Information Society, 1, 1-12.

Designing Robust Multimodal Systems for Diverse Users and Environments

Sharon Oviatt

Computer Science Department, Oregon Graduate Institute of Science & Technology
20000 N.W. Walker Road, Beaverton, Oregon, 97006 USA[*]
oviatt@cse.ogi.edu; www.cse.ogi.edu/CHCC/

Abstract

Multimodal interfaces are being developed that permit our highly skilled and coordinated communicative behavior to control system interactions in a more transparent and flexible interface experience than ever before. The presence of modality choice per se is an important feature and design issue for multimodal interfaces. As applications become more complex, a single modality does not permit varied users to interact effectively across different tasks and environments (Oviatt et al., 2000). While individual input modalities are well suited in some situations, they may be less ideal or even inappropriate in others. However, a flexible multimodal interface offers people the freedom to use a combination of modalities, or to switch to a better-suited modality, depending on the specifics of their abilities and preference, the task at hand, or the usage conditions.

1. Accessibility for diverse users and contexts

Among the most important reasons for developing multimodal interfaces is their potential to greatly expand the accessibility of computing for diverse and non-specialist users, and to promote new forms of computing not previously available (Oviatt, 1999; Oviatt, 2000). Since there are large individual differences in ability and preference to use different modes of communication, a multimodal interface permits the user to exercise selection and control over how they interact with the computer (for examples, see Fell et al., 1994; Karshmer & Blattner, 1998). For example, a visually impaired user may prefer speech input, as may a manually impaired user with a repetitive stress injury or her arm in a cast. In contrast, a user with a hearing impairment, strong accent, or a cold may prefer pen input. Well before the keyboard is a practiced input device, a young preschooler could use either speech or pen-based drawing to control an educational application. In this respect, multimodal interfaces have the potential to accommodate a broader range of users than traditional graphical user interfaces (GUIs) and unimodal interfaces— including users of different ages, skill levels, native language status, cognitive styles, sensory impairments, and other temporary or permanent handicaps or illnesses. A flexible multimodal interface also permits alternation of individual input modes in a manner that can be critical for preventing overuse and physical damage to any single modality, especially during extended periods of computer use. Just as the forearms can be damaged by repetitive stress when using a keyboard and mouse, the vocal cords also can be strained and eventually damaged by prolonged use of a speech system (Markinson, 1993).

Multimodal systems also can expand the usage contexts in which computing is viable— for example, in natural field settings and while users are mobile. In a physical sense, multimodal interfaces that include input modes well suited for mobility (e.g., speech, pen) free us to cross the artificial borders representing on- and off-the-desktop computing. Their support for switching among modes permits users the flexibility that is needed to accommodate the continually changing conditions of mobile use. For example, since speech and pen input are complementary along many dimensions, their combination provides broad utility across varied usage contexts. A person may use hands-free speech input for voice dialing a car cell phone, but switch to pen input to avoid speaking private information during a public transaction. There is a sense in which mobility can induce a state of "temporary disability," such that a person is unable to use a particular input mode for some period of time. For example, a user carrying a child may

---

[*] This research was supported by Grant No. IRI-9530666 from the National Science Foundation, Special Extension for Creativity (SEC) Grant No. IIS-9530666 from the National Science Foundation, Contracts DABT63-95-C-007 and N66001-99-D-8503 from DARPA's Information Technology and Information Systems offices, Grant No. N00014-99-1-0377 from ONR, and by grants, gifts, and equipment donations from Boeing, Intel, Microsoft, Motorola, and SAIC.

be temporarily unable to use pen or touch input at a public information kiosk, although speech is unaffected. Within a multimodal architecture, adaptive weighting of the input modes during environmental change can further enhance and stabilize the system's overall performance.

2. Robustness and performance stability

Another major reason for developing multimodal systems is to improve the performance stability and robustness of recognition-based systems (Adjoudani & Benoit, 1995; Oviatt, 1999 & 2000; Oviatt et al., 2000; Tomlinson et al., 1996). From a usability standpoint, multimodal interfaces provide an opportunity for users to exercise their natural intelligence about when and how to use input modes effectively. For example, people will avoid using an input mode that they believe is error-prone for certain lexical content. When a recognition error does occur, they typically will switch input modes. This usually resolves the error, since the confusion matrices are different for lexical content involving the two different modes being recognized.

A well-designed multimodal architecture also can support the *mutual disambiguation* of two input signals. For example, if a user says "ditches" but the speech recognizer confirms the singular "ditch" as its best guess, then parallel recognition of several graphic marks in pen input can result in recovery of the correct spoken plural interpretation. This kind of *architectural pull-up* can result in more accurate and stable system performance. It also can produce a relatively greater performance advantage precisely for those users and usage contexts in which unimodal systems typically fail— for example, with accented speech or in noisy field environments for which a speech system is error prone. In such cases, a multimodal architecture can reduce or even close the recognition rate gap between these "challenging" cases and less difficult ones (Oviatt, 1999 & 2000).

To reap these error-handling advantages fully, the unique semantic complementarities of a given pair of input modes needs to be identified and exploited. For example, in the speech and lip literature, natural feature-level complementarities have been identified between visemes and phonemes for vowel articulation, with vowel rounding better conveyed visually, and vowel height and backness better revealed auditorally (Robert-Ribes et al., 1998). In speech and pen input literature, the main complementarity involves visual-spatial semantic content (Oviatt, 1996; Oviatt et al., 1997). Whereas visual-spatial information is uniquely and clearly indicated via pen input, the strong descriptive capabilities of speech are better suited for specifying temporal and other non-spatial information. To further optimize robustness, a multimodal system must be designed so that the two input modes (e.g., speech and pen) provide parallel or duplicate functionality, so that users can accomplish their goals using either mode.

3. Multimodal system performance for accented speakers

At the Center for Human Computer Communication (CHCC) at the Oregon Graduate Institute, we recently conducted performance evaluations with the QuickSet multimodal system (Cohen et al., 1997) to investigate whether a multimodal architecture could be designed to support mutual disambiguation of incoming signals, yielding higher recognition rates than a traditional spoken language system. We also wanted to explore whether larger performance improvements could be obtained for accented speakers (compared with native ones), and whether an alternate input mode (pen input) could be used to disambiguate and stabilize speech input when processed within a multimodal architecture.

In this first study, the participants were eight native speakers of English, and eight accented speakers representing different European, Asian and African languages. Everyone communicated 100 commands multimodally using QuickSet while they completed simulation exercises involving community flood and fire management (see Oviatt (1999) for procedural details). A record of users' speech and pen input, along with the multimodal system's performance, was recorded during 2,000 multimodal commands.

The results of this study confirmed that the QuickSet multimodal architecture supports significant levels of mutual disambiguation (Oviatt, 1999), with one in eight user commands recognized correctly due to mutual disambiguation. Overall, a 41% reduction was revealed in the total error rate for spoken language processed within a multimodal architecture, compared with spoken language processing as a stand-alone. These results indicate that a multimodal system can be designed to function in a substantially more robust and stable manner than unimodal recognition-based technology.

While the spoken language recognition rate was much poorer for accented speakers (–9.5%), as would be expected, their gesture recognition rate averaged slightly but significantly better (+3.4%). Although mutual disambiguation was present for both groups, the rate of mutual disambiguation was significantly higher for accented speakers (+15%) than native speakers of English (+8.5%) — by a substantial 76%. As a result, the final multimodal recognition rate for accented speakers no longer differed significantly from the performance of native speakers. The main factor responsible for *closing this performance gap* between groups was the higher rate of mutual disambiguation for accented speakers, for whom two-thirds of all signal pull-ups involved retrieving poorly ranked speech input.

4. Multimodal system performance during mobile use

In a second study, we were interested in determining whether the multimodal performance improvements obtained with accented speakers were specific to that population, or whether they would generalize to challenging usage contexts. In this case, we investigated whether a multimodal architecture could support significant levels of mutual disambiguation, as well as higher rates of mutual disambiguation during mobile use in a noisy field environment, compared with quiet stationary system use.

Twenty-two native English speakers interacted multimodally using the QuickSet speech/pen system on a hand-held PC. They completed 100 commands using a similar procedure to the first study. However, in this study each volunteer completed 50 commands while working alone in a quiet room that averaged 42 decibels (e.g., "stationary" condition), and another 50 while walking through a moderately noisy public cafeteria that ranged 40-60 decibels (e.g., "mobile" condition). Testing also was conducted with two opposite microphones, including a high-end Andrea close-talking noise-canceling microphone, and a low-end built-in microphone that lacked noise-cancellation (see Oviatt (2000) for procedural details). In total, data was analyzed on over 2,600 multimodal commands.

In this mobile study, 19-35% reductions in the total error rate (for noise-canceling versus built-in microphones, respectively) were observed when speech was processed within the multimodal architecture. Once again, this substantial improvement in robustness was a direct result of the disambiguation between signals that can occur in a multimodal architecture. The spoken language recognition rate was significantly degraded while users were mobile in the noisy setting (–10%), although their gesture recognition rates did not decline during mobility. The mutual disambiguation rate also averaged significantly higher when users were mobile (+16%), compared with when the same users were stationary (+9.5%). Depending on which microphone was used, this mutual disambiguation rate ranged from 50-100% higher during mobile system use. Since mutual disambiguation was elevated while mobile, a significant *narrowing of the gap* occurred between the mobile and stationary recognition rates (to –8.0%) during multimodal processing, compared with spoken language processing alone.

In summary, although speech recognition as a stand-alone performed poorly for accented speakers and in mobile environments, a multimodal speech/pen architecture was demonstrated to decrease failures in spoken language processing by 19-41%. This performance improvement mainly occurred because of the mutual disambiguation between input signals that is supported within a unification-based multimodal architecture— which occurs at higher levels for challenging user groups and usage environments. That is, one very interesting property of multimodal architectures is that they can reduce or in some cases eliminate the performance gap for precisely those users and contexts in which a traditional spoken language system typically would fail.

5. Conclusion and future directions

Multimodal systems are capable of expanding the accessibility of computing for diverse and non-specialist users, and in field and mobile environments. Multimodal systems also can be designed that perform more robustly and with greater stability than unimodal recognition technologies. To further improve both the performance and accessibility of next-generation multimodal systems, many key research topics still remain to be addressed. Among these topics are the development of new natural language and dialogue processing techniques, hybrid symbolic/statistical architectures, and techniques for adaptive multimodal processing.

References

Adjoudani, A., Benoit, C. (1995). Audio-visual speech recognition compared across two architectures. In *Proceedings of the Eurospeech Conference*, Madrid, Spain, (2) 1563-1566.

Cohen, P.R., Johnston, M., McGee, D., Oviatt, S., Pittman, J., Smith, I., Chen L., Clow, J. (1997). Quickset: Multimodal interaction for distributed applications. In *Proceedings of the Fifth ACM International Multimedia Conference* (pp. 31-40). New York: ACM Press.

Fell, H., Delta, H., Peterson, R., Ferrier, L., Mooraj Z., Valleau, M. (1994). Using the baby babble-blanket for infants with motor problems. In *Proceedings of the Conference on Assistive Technologies (ASSETS'94)*, Marina del Rey, California (pp. 77-84). [On-line]. Available: http://www.acm.org/sigcaph/assets/assets98/assets98index.html.

Karshmer A.I., Blattner M. (organizers) (1998). *Proceedings of the Third International ACM Proceedings of the Conference on Assistive Technologies (ASSETS'98),* Marina del Rey, CA. [On-line]. Available: http://www.acm.org/sigcaph/assets/assets98/assets98index.html.

Markinson, R. (1993). *Personal communication.* University of California at San Francisco Medical School.

Oviatt, S.L. (1996). Multimodal interfaces for dynamic interactive maps. In *Proceedings of Conference on Human Factors in Computing Systems: CHI '96* (pp. 95-102). New York: ACM Press.

Oviatt, S.L., DeAngeli, A., Kuhn, K. (1997). Integration and synchronization of input modes during multimodal human-computer interaction. In *Proceedings of Conference on Human Factors in Computing Systems: CHI '97* (pp. 415-422). New York, N.Y.: ACM Press.

Oviatt, S.L. (1999). Mutual disambiguation of recognition errors in a multimodal architecture. In *Proceedings of the Conference on Human Factors in Computing Systems (CHI'99)* (pp. 576-583). New York: ACM Press.

Oviatt, S.L. (2000). Multimodal system processing in mobile environments. In *Proceedings of the Thirteenth Annual ACM Symposium on User Interface Software Technology (UIST'2000)* (pp. 21-30). New York: ACM Press.

Oviatt, S.L., Cohen, P.R., Wu, L., Vergo, J., Duncan, E., Suhm, B., Bers, J., Holzman, T., Winograd, T., Landay, J., Larson J., Ferro, D. (2000). Designing the user interface for multimodal speech and gesture applications: State-of-the-art systems and research directions, *Human Computer Interaction*, vol. 15,26 3-322. (to be reprinted in J. Carroll (ed.) Human-Computer Interaction in the New Millennium, Addison-Wesley Press: Boston, to appear in 2001).

Robert-Ribes, J., Schwartz, J-L., Lallouache, T., Escudier, P. (1998). Complementarity and synergy in bimodal speech: Auditory, visual, and audio-visual identification of French oral vowels in noise. *Journal of the Acoustical Society of America,* 103 (6), 3677-3689.

Tomlinson, J., Russell, M.J., Brooke, N.M. (1996). Integrating audio and visual information to provide highly robust speech recognition. In *Proceedings of the IEEE ICASSP* (pp. 821-824).

Continuity through User Interface Adaptation: a perspective on Universal Access

Alexandros Paramythis[1], Frank Leidermann[1], Harald Weber[1], Constantine Stephanidis[1, 2]

[1] Institute of Computer Science
Foundation for Research and Technology – Hellas
Science and Technology Park of Crete
GR-71110 Heraklion, Crete, Greece

[2] Department of Computer Science, University of Crete, Greece

Abstract

This paper proposes *user interface self-adaptation* as an effective approach towards meeting the HCI challenges posed by Universal Access with respect to the quality criterion of continuity. These challenges arise through the different dimensions of diversity, namely diversity in users, contexts of use, and platforms and devices.

1. Universal access

Universal Access in the context of Human-Computer Interaction refers to the conscious and systematic effort to practically apply principles, methods and tools of universal design in order to develop high quality user interfaces, accessible and usable by a diverse user population with different abilities, skills, requirements and preferences, in a variety of contexts of use, and through a variety of different technologies [1].

Consequently, Universal Access poses various requirements on user interfaces. These requirements can be expressed as quality criteria for human-computer interaction, describing key characteristics of universally accessible user interfaces. Exemplary criteria are acceptability, accessibility, learnability, or ease of use. This paper advocates the introduction of the criterion of *Continuous Interaction* (or, in short, *Continuity*) as established by the *TACIT* network [2], as a further such prerequisite towards Universal Access. The intention is to identify those aspects of continuity, which correspond to challenges posed by Universal Access, and which are not covered by other quality criteria in the literature. This paper views Continuity and Universal Access as related, but distinct concepts. Specifically, while the two concepts have a number of commonalities under the perspective of HCI design (e.g., supporting different contexts of use), each of them additionally focuses on unique areas without overlap (for example, Universal Access with respect to inter-personal or cultural diversity, and Continuity with respect to synchronization of different output modalities [3], or interaction with shared input devices [4]).

This paper proposes *user interface self-adaptation (UI Adaptation)* [8] as an effective approach towards Universal Access in HCI, with respect to the quality criterion of Continuity. Section 2 points out exemplary relationships between Continuity and Universal Access, while section 3 presents the "role" of user interface adaptation as one attempt to achieve continuous and universally accessible systems.

2. Continuity in the context of universal access

The concept of Continuous Interaction [3] was brought about by advances in technologies, their wide dissemination in all aspects of life, and in the way people experience them. Currently, there exists no final definition of continuity [5]. Nevertheless, the notion of Continuity focuses on human-computer interaction *over a certain period of time*. Of particular interest is the degree of "continuity" or "stability" of interaction within this period. These depend both on the duration of such a period, and, more generally, on the level of observation ranging from the physical level to more abstract levels [6]. The latter distinction between levels of interaction refers to the basic assumption in user

412

interfaces design, that on a certain level, interaction tasks cannot be appropriately split into isolated "discrete" subtasks [3].

The challenges that Universal Access poses on human-computer interaction, can be analyzed across different dimensions of diversity, relating to (intra- and inter-individual) user characteristics, contexts of use, and the interaction devices and platforms employed (see Figure 1). Changes along any of these dimensions have the potential to cause discontinuities in interaction. While there exist approaches that seek to address single dimensions of diversity, so as to support Continuous Interaction along these dimensions, the vision is to provide a continuous interaction experience within the complete space of diversity.

In order to further highlight the relationship between Continuity and Universal Access, this section describes some examples of potential discontinuous interactions, while the next section focuses on potential solutions to these discontinuities through UI Adaptation.

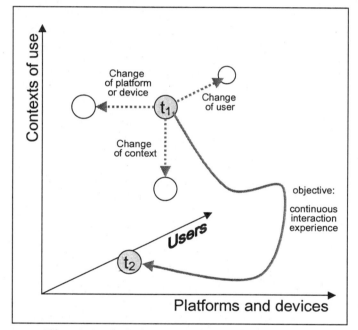

Figure 1: Continuity in the framework of Universal Access

As a first example, discontinuities can be observed across *different interaction sessions*. A typical user group that might face discontinuities while interacting with a complex software system is that of novice users, who are not aware initially of all system features. It can be observed that frequent use of the system results in increasing awareness about a system's full functionality, which should be taken into account by user interfaces striving to ensure continuity in interaction. Although one could argue that two different sessions should be treated as separate interactions, the attempt to construct interfaces that are continuously usable also implies a perspective on longer periods of time. Therefore, to increase the usability of a system and to facilitate continuity, the system needs to be aware of the user's intra-individual changes across interaction episodes, and should reflect this appropriately in the design of the user-system interaction.

The same source of intra-individual changes might also generate discontinuities within a *single* interaction session. A user might not be able to memorize a menu or dialogue structure (e.g., in the case of occasional system users). However, with appropriate (e.g., timely, efficient, non-obtrusive) system help, the user might learn this task during a single session, such that the system does not need to provide extensive help on this issue any longer. In a different case, a user's attention might decrease due to tiredness or distractions from the environment. Therefore, to avoid discontinuity, the system needs to reflect such intra-session changes (e.g., the increase of user errors) through appropriate support measures.

Another source of discontinuities might arise from changes in the physical and organizational context in which the system is located, as well as from varying goals / tasks on the part of the user [7]. It is noteworthy that the relation between contextual changes, as described below, and continuity refers not just to the notion of continuity in time, but also to continuity in space [3].

Variations in the physical environment may also have an influence on the interaction. Some examples that highlight the necessity for systems that are able to recognize certain states of the physical surrounding to facilitate continuous interaction, include: operating a device while being in a vehicle (taking, for example, into account the presence of acceleration, vibration); using audio output in noisy environments; or perceiving visual outputs in an environment with high illumination (e.g., sunlight).

Furthermore, varying user goals or tasks may necessitate changes regarding the priority of usability objectives. For instance, while a user of a tourist information system may expect a learnable and easy to use interface while browsing through history-related information, the same user might prefer an efficient system in case of an urgent need for healthcare related information. In the case of business software, changing goals / tasks are usually caused within dynamic organizational contexts, thus introducing other potential sources for discontinuity.

Finally, diversity in technological platforms or devices has to be addressed. For instance, "switching" between several devices (like PC or PDA) while using one software system might cause discontinuities, due to, for example, different screen sizes or interaction metaphors. Moreover, the input or output modalities may need to change, e.g., when the user switches from a desktop PC to a device installed in a car. In this case, the user becomes contextually blind (with respect to visual system output) while driving, and the use of hands to operate the device is usually restricted. In consequence, to provide continuous interaction with a certain system while performing a primary task, the modalities of interaction need to adapt to the specific context of use.

3. Addressing continuity through adaptation

The potential discontinuities pointed out in the above examples imply usability problems caused by the diversity of users, usage contexts, or platforms and devices, which have not been addressed during the interface design process. One approach to anticipating and preventing such discontinuities is the attempt to design interfaces that are capable of automatic self-adaptation [8] to a wide variety of user- and context- related requirements. Having the focus on continuous interaction, it is of special interest that these requirements usually change over time, either between, or during interaction sessions. This distinction corresponds to two types of self-adaptation, *adaptability* and *adaptivity* [16].

Adaptability refers to self-adaptation, which is based on knowledge available to the system prior to the initiation of interaction, while adaptivity is based on knowledge that is acquired and / or maintained by the system during interactive sessions (e.g., through monitoring techniques), and which leads to adaptations that take place during interaction [9]. For the purpose of a timely prevention of discontinuities, monitoring mechanisms are required that are capable of analyzing the "path" an interaction follows within the space of diversity (see Figure 1), and of initiating appropriate adaptations in interaction.

Similar to the fact that continuity can be observed at different levels of abstraction, self-adaptation can relate to three levels of interaction at which it is realized, namely, the semantic, syntactic and lexical level [10].

At the *lexical* level, automatic modifications of the *presentation attributes* of interactive elements, e.g., font size or speech volume, can cater for changing user- or context- related requirements, as well as the support for, and conditional (de)activation of, *multiple interaction modalities* (such as rendering information content in speech, rather than visual, or activation of interface scanning to support switch-based interaction).

Examples of related adaptations on the *syntactical* level include *adaptive error prevention and correction* [11], *adaptive awareness prompting* in order to raise the users' awareness to unused or under-utilized features or functionalities [12], and *adaptation to support alternative task structures* [13].

At the *semantic* level, the diversity of requirements can be addressed by the interactive metaphors (e.g., desktop, book, rooms), which are used to embody different functional properties of the system [14].

The above exemplary adaptations were derived from the AVANTI browser, a multi-modal, adaptable, and adaptive Web browser, which is capable of self-adaptation at all three levels of human-computer interaction [15]. The important point regarding these examples is that, although not explicitly designed for the purpose of preventing discontinuities, adaptations such as the ones presented have the potential to contribute towards Continuous Interaction.

4. Conclusions

Continuity has been discussed in this paper as a quality criterion of universally accessible user interfaces. In the framework of Universal Access, some aspects of continuity become essential system characteristics, required to provide users with a "stable" interaction over a certain period of time. Potential sources of discontinuities, some of which have been presented through examples, arise from different dimensions of diversity with respect to the users, context of use and the platforms and devices employed. Although not explicitly designed for the purpose of preventing discontinuities, examples for successful solutions towards this objective taken from the AVANTI Web browser led to the proposal that *user interface self-adaptation* can serve as an effective approach towards meeting the HCI challenges posed by Universal Access with respect to the quality criterion of continuity.

References

1. Stephanidis, C. (Ed.) (2001). *User Interfaces for All - concepts, methods and tools*. Mahwah, NJ: Lawrence Erlbaum Associates. (ISBN 0-8058-2967-9, 760 pages)
2. *Theory and Applications of Continuous Interaction Techniques*: http://kazan.cnuce.cnr.it/TACIT
3. Massink, M., Doherty, G., Faconti, G. (2000). Continuity in Human Computer Interfaces. *Position paper, CHI 2000*, The Hague (http://kazan.cnuce.cnr.it/TACIT/CHI2000/contrib.html)
4. Watts, LA.: *The Magic Board, an Augmented Reality Interactive Device Based on Computer Vision*, (http://kazan.cnuce.cnr.it/TACIT/CHI2000/MagicBoard/MB.html)
5. Watts, L.A., et al.: *The TACIT Glossary, Technical Report TACIT-TR010* (http://kazan.cnuce.cnr.it/TACIT/TACITweb/DOCUMENTS/TechnicalReports/TACIT-TR010.html)
6. Faconti, G., Massink, M. (2000). Continuity in Human Computer Interaction. *Workshop Report, CHI 2000*, The Hague (http://www.acm.org/sigchi/bulletin/2000.4)
7. ISO 9241, Part 11: Guidance on Usability
8. Dieterich, H., et al. (1993). State of the Art in Adaptive User Interfaces. In: M. Schneider-Hufschmidt, T. Kuehme, U. Malinowski (Eds). *Adaptive User Interfaces* (Amsterdam 1993).
9. Stephanidis, C., Paramythis, A., Akoumianakis, D., Sfyrakis, M. (1998). Self-Adapting Web-based Systems: Towards Universal Accessibility. In C. Stephanidis, A. Waern (Eds.), *Proceedings of the 4th ERCIM Workshop "User Interfaces for All"*, Stockholm.
10. Hoppe, H., Tauber, M., Ziegler, J. (1986). *A Survey of Models and Formal Description Methods in HCI with Example Applications*. ESPRIT Project 385 HUFIT, report B.3.2.a., 1986
11. Browne, D., Norman, M., Riches, D. (1990). Why Build Adaptive Systems?, In: D. Browne, P. Totterdell, M. Norman (Eds.), *Adaptive User Interfaces* (pp. 15-57).
12. Kühme, T., Malinowski, U., Foley, J.D. (1993). *Adaptive Prompting*. Technical report GIT-GVU-93-05, Georgia Institute of Technology.
13. Paetau, M. (1994). Configurative Technology: Adaptation of Social Systems Dynamism. In R. Oppermann (Ed.): *Adaptive User Support: Ergonomic Design of Manually and Automatically Adaptable Software* (pp. 194-234). Hillsdale, NJ.
14. Stephanidis, C., Akoumianakis, D. (1998). Multiple Metaphor Environments: Issues for effective interaction design. In *8th ERCIM - DELOS Workshop on User Interfaces for Digital Libraries*, Stockholm.
15. Stephanidis, C., Paramythis, A., Sfyrakis M., Savidis A. (2001). A Case Study in Unified User Interface Development: The AVANTI Web Browser. In C. Stephanidis (ed.) *User Interfaces for All – Concepts, Methods and Tools* (pp. 525-568). Mahwah, NJ: Lawrence Erlbaum Associates (ISBN 0-8058-2967-9, 760 pages).

16.   Stephanidis, C., Paramythis, A., Sfyrakis, M., Stergiou, A., Maou, N., Leventis, A., Paparoulis, G., Karagiannidis, C. (1998). Adaptable and Adaptive User Interfaces for Disabled Users in AVANTI Project. In S. Trigila, A. Mullery, M. Campolargo, H. Vanderstraeten & M. Mampaey (Eds.), *Intelligence in Services and Networks: Technology for Ubiquitous Telecommunications Services - Proceedings of the 5th International Conference on Intelligence in Services and Networks (IS&N '98)*, Antwerp, Belgium (pp. 153-166). Berlin: Springer, Lecture Notes in Computer Science, 1430.

Speech technology for universal access in interactive systems?

Régis Privat[1], Nadine Vigouroux[1], Caroline Bousquet[1], Philippe Truillet[1,2] , Bernard Oriola[1]

[1] IRIT, UMR CNRS
118, Route de Narbonne
F-31062 Toulouse Cedex - France
{privat,vigourou,bousquet,oriola}@irit.fr

[2] CENA DGAC - Division PII
7, Av. Edouard-Belin BP 4005
31055 Toulouse Cedex - France
truillet@cena.fr

Abstract

The problem discussed in this paper concerns how spoken interaction research can give a solution for the universal access for all. This paper will firstly report on the state of the art of speech technology for older people. Secondly, it will describe the case study of the selection of a French dictation system for a fundamental evaluation by elderly people. This session must be considered as a case study. We will place our approach with respect to NIST (http://www.nist.gov/speech/) and EAGLES [Gibbon 1997] evaluation contexts. Finally, we will give preliminary results mainly on the effect of the people age.

1. Introduction

Speech is natural, efficient and flexible for human-human communication. Is the use of Speech Recognition (SR) technology still science fiction, or is it an opportunity to obtain a universal access for ubiquitous human-computer interactions?

Research in speech technology has been underway for decades and a great deal of advances have been made in reducing the word error rate and in allowing natural spontaneous speech. On the other hand, recent studies demonstrate that machine performance is still quite far from human performance across a wide variety of conditions: vocabulary size, environment noise, bandwidth, familiarity with speech technology, age, social and cultural characteristics of the speaker, speech disorders, etc. [Lippmann 1997] reported that for switchboard tasks the machine performance involved 43% of word errors vs. 4% for human performance.

Speech research is in progress to avoid specific adaptation to application domains: [Padmanabhan 2000] proposes to develop a generic speech recognition system that can deal with linguistic as well as acoustic problems occurring in different domains/tasks.

Speech technologies[1] as they mature represent great opportunities in the way people work with others or/and with information through Interactive Voice Systems (IVS). The providers of continuous speech technologies are trying to propose speaker-independent and large-vocabulary SR systems [WWW 2001]. This is one of the challenges of IVS like Web-Galaxy [Lau 1997] and the ARISE project [Baggia 1999] for use by the general public.

For users with speech or language disabilities or normal elderly people, the challenge could be *not focusing on the technology but on the needs of the individual user in regards to the technology accuracy.*

From the many challenges to overcome in developing human-like performance, we believe that spontaneous speech in different age groups is one of the most broadest and most serious. As reported by various studies of human-human communication, the main problem is that there are variations in spontaneous speech (faltering, false starts, ungrammatical sentences, emotional speech such as laughter, slowing down, etc.) which are attenuated in read speech. However, similar variations are observed on ageing (http://www.pitt.edu/~hennon/speech.htm).

The objective of this work is to start investigating the spoken interaction as a solution for the concept of the **universal access for all** [Emiliani 2000]. Speech technologies can be evaluated by means of the EAGLES/ISO methodology which consists in assessing software quality characteristics (**accuracy[2]**, recoverability, **usability[3]** and changeability) [Canelli 2000] and quality in use (effectiveness, satisfaction).

---

[1] There are a lot of speech recognition products for Dictation (IBM, Dragon Systems, Lernout & Hauspie, Philips, etc.). There are also companies such as Nuance, SpeechWorks, AT&T Bells Labs, Philips, IBM, etc. developing IVS packages for use over the telephone or Internet.
[2] ISO definition of accuracy : "the capability of the software product to provide the right or agreed results or effects".
[3] ISO definition of usability: "the capability of the software product to be understood, learned, used and attractive to the user, when used under specified conditions".

2. Use of voice dictation systems and elderly people

Few works have studied the use of the speech technology for handicapped and/or elderly people. [Kambeyanda 1996] reported problems associated with the use of isolated speech recognition products: difficulties to maintain constant pitch, volume and inflection while dictating word by word.

The elderly people have a definite need for IVS to access information. The first study was conducted by [Wilpon 1996] and reported that the error rates of recognition increase for speakers over 70 years old. [Yato 1999] described an experience conducted on 469 elderly Japanese people (age range 59 to 85). The IBM Via Voice system was running on 600 place name utterances. It also seems that the average error increases with the age of the speaker.

Another interesting experiment was conducted by [Anderson 1999a,b]. He studied the effect of "user training", that is to say the ability of a speaker to improve the accuracy rate of the system by modifying his way of speaking. The results showed small changes, but some users discovered on their own solutions —speaking slowly or loudly— to decrease their word error rate. But, as they said, the study did not take into account speaker fatigue which could, with the long time the experiment took, interfere with the accuracy.

Generally, Speech Recognition Dictation Systems (SRDS) are used to input text by "standard[4]" speakers. The IRIT's research goals for the use of speech technologies by "all" are:

a) to identify the characteristics of SRDS that are needed to be considered for universal use by "all", mainly during the training phase (lexicon personalization, adaptation of the language model, decreasing of the training duration) and for different elocution mode to avoid problems caused by the improper use of SRDS;

b) to study if accuracy problems are independent of the technology or the user;

c) to understand/identify the influence of age, social class, etc., on spoken communicating functions;

d) finally, to propose a methodology for SRDS configuration to reduce both the training time and the cognitive loading charge —fatigue, stress— for elderly people, and to specify intelligent assistant in IVS according to the user's needs.

3. The case study

3.1 The objectives

The challenge is: can the SRDS be used by the elderly to input text or/and to interact through an information service? The standard evaluation methodology would have involved conducting a series of experiments to evaluate the main SR systems[5] for dictation by two age classes of speaker: standard and elderly persons (elderly means more than 60 years old). As recommended by NIST (National Institute of Standard Technology, http://www.nist.gov/speech/) and EAGLES [Gibbon 1997], ASR evaluation needs large speaker numbers and huge corpora sizes to have reliable results.

The aim of this paper is not to evaluate the accuracy of the SRDS (commercial tests already give these results) but to try to determine the effect of some specific use conditions on accuracy. To avoid long processing times for evaluation of several SRDS, we decided to conduct a case study to determine for the French language which SRDS gives the best results for our purpose. This study was performed with a limited number of speakers. It consists of measuring/evaluating the facilities of adaptation (acoustic model, language model, lexicon), the compatibility with the Speech Application Programming Interface (SAPI version 4, http://www.microsoft.speech) for their integration in an interactive application, the accuracy robustness with regards to the speaker-dependent model but also the "extra-use" with non normal elocution (spontaneous way, elderly persons, speech disorders...).

3.2 Method

3.2.1 Definition of study factors

The three main commercial SRDS[4] for French were selected for this case study. Two factors were retained: *the training effect and the elocution mode.*

These SRDS are speaker-dependent: this is why it is necessary, before the recognition phase, to proceed to the adaptation of the speaker acoustic model. Two states of training models were defined: 1) the implicit acoustic model given by the SRDS just smoothed by the signal-noise ratio (SNR) process applying in the use context (environment and speaker) named **reference model**; 2) the acoustic speaker-dependent model, named **SDM**.

To measure the effect of the elocution mode on the accuracy, three different modes of elocution were defined: text reading with (**EM1**) and without pronouncing the punctuation markers (**EM2**) and spontaneous speech mode (**EM3**) as naturally as possible. These three modes were suggested by both the results of [Kambeyanda 1996], who reported that SRDS accuracy is dependent on the elocution, and by the IRIT's hypothesis: it is reasonable to use SRDS for spoken interactions with a computer application. So, each speaker has to undergo the training phase in front of a computer (the texts were presented on the screen of the computer) for the three modes.

---

[4] Standards means male or female speakers of 25-60 years old.
[5] L&H Voice Xpress Professional V4; IBM ViaVoice Pro, Millenium Edition, release 7; Dragon Naturally Speaking Prefered V4.

418

For this case study, only two computer science students did these trial series but various tests were extensively conducted with three different elocution modes and three speaker-dependent acoustic models. We decided to test all the speaker-dependent acoustic models for each SRDS with the two users in order to determine the real effects of the training phase: firstly, to evaluate the benefits or not of the training phase on the accuracy (second column of each SRDS sub table in Table 1); secondly to evaluate the accuracy deviation of SRDS trained for one speaker and tested with another (third column). This value could be interpreted as an indicator of the system ability to recognize different voices without SNR smoothing and speaker training. This is due to the way we want to use SRDS, since vocal input for IVS has to be independent of the speaker.

For each SRDS, 18 tests were run. Each test took about fifty minutes for the training phase (adjustment, elocution phases and the automatic smoothing by the SRDS). For the purpose of this paper, the three systems will remain undifferentiated.

3.2.2 Apparatus

Tests were run on a portable Compaq Armada M-700, PII366 with 128 Mo of RAM, with Microsoft Windows95. The training and test voice corpora were recorded at a sampling rate of 22KHz through an Andrea Technology ANC-600 headphone. They were transmitted to the SRDS by a cable directly plugged in to the microphone entry of the sound card. As speech variability must not interfere with the results, the test corpora were recorded in order to compare the recognition for a single input.

The test corpora is a phonetic calibrated text [Cadilhac 1997] of 196 words.

3.2.3 Accuracy

Each SRDS output is compared to the orthographic reference transcription by means of an alignment tool. The accuracy measurement is based on the ratio between the number of words truly/wrongly transcribed and the number of words in the text. The results are expressed as the proportion of words correctly recognized.

3.3 Results and discussion

Table 1: Some of the case study results

| | | SRDS 1 | | | SRDS 2 | | | SRDS 3 | | |
| | | Speaker | | Other | Speaker | | Other | Speaker | | Other |
		Reference model	SDM	SDM	Reference model	SDM	SDM	Reference model	SDM	SDM
	EM1	76.28%	3.83%	-10.71%	84.44%	-1.53%	-5.61%	82.14%	-2.30%	-14.03%
Spk 1	EM2	78.32%	2.04%	-14.80%	79.59%	-0.77%	-3.83%	81.38%	-4.59%	-14.29%
	EM3	68.11%	4.85%	-14.29%	74.49%	-3.06%	-7.40%	79.34%	-10.20%	-22.96%
	EM1	61.48%	3.57%	14.03%	76.02%	5.61%	-9.18%	63.78%	12.76%	9.44%
Spk 2	EM2	61.99%	2.30%	14.29%	71.68%	-3.06%	-19.90%	58.93%	15.82%	12.24%
	EM3	42.86%	8.93%	20.41%	68.62%	-4.85%	-19.64%	52.55%	8.67%	11.73%

The results (Table 1) are given in terms of word recognized rate for the reference model. The gap of recognition rate due to the training phase is then given for the two speakers (SDM for the speaker performing the training phase).

Two interesting points can be noticed: the first is the quite comparable results for the two first elocution modes, especially with the SRDS 1 (average increase of 0.13% for the elocution mode without punctuation) and the SRDS 3 (average decrease of 2.25%); the second is that the results obtained seem to confirm that the SR performance is dependent of the speaking mode (an average decrease between 10 and 13% from EM1 to EM3 can be noticed), and this with all the SRDS.

4. Towards an analytic methodology of SRDS evaluation

Our purpose is to build an analytic methodology for the evaluation of SRDS in order to determine whether they can be integrated in interactive systems or not, as a solution for enabling speech input. The case study confirms the needs on the SRDS selected to:

a) investigate the use of speech technologies by "all" in how important it is to have tools to adapt/personalize linguistic knowledge (lexicon personalization taking into account phonological variants of pronunciation, adaptation of the language model used by the SRDS);

b) study the effect of people age on the elocution mode, speech rate, sound energy and phonological transcription;

c) analyze if accuracy problems are independent of either the technology or the user.

4.1 Methodology

Three factors were retained for this study: *the age effect, the elocution mode and the training model length* of the SRDS chosen. For the two first factors, the same principle as the one described in 3.2.1 is applied. Concerning our efforts to reduce the training time length, we plan to build a set of acoustic models for each speaker by changing the training text: we want to evaluate the re-use of the short texts required for smoothing the implicit model of the SRDS. This manner aims to reduce the effort, the stress, the fatigue caused by the long training time. As for the case study, the three elocution modes will be used (EM1, EM2, EM3).

4.1.1 Participants

The study deals with two speaker's classes: students in computer science (20-30 years old) and elderly persons (elderly means more than 60 years old). For each population, fifteen speakers have been selected.

4.1.2 Apparatus

We decided to use a more powerful workstation, to avoid any material limitation for the SRDS. The computer is a PIII933 with 396Mo of Ram and a Sound Blaster Live Player sound card. The operating system is Microsoft Windows2000 Professional. All the corpora (training, tests) are recorded on a DAT with a sampling rate of 48KHz using an Andrea Technology ANC-600 headphone.

4.1.3 Procedure and study factors

Our aim is to study the age, the training time reduction and the elocution mode effects on the recognition results in terms of accuracy (word-recognized percentage), but also to identify some extra-linguistic phenomena which could occur during spoken human-computer dialogs. The same principles as in 3.2.3 have been applied to compute the accuracy. A satisfaction questionnaire will be added to estimate the quality in use and the cognitive load.

4.1.4 Preliminary results and discussion

Global results show that successful uses of SRDS require some considerations: the training phase, consisting in two preliminary steps (SNR adjustment, benchmarks of audio devices), and the reading of a text is primordial. In this paper, we decided to focus our discussion on the speech rate analysis. This will be done partially on four of the thirty speakers. The speech rate (in words per second) is computed from the number of words in the texts divided by the duration of the recording of each corpus.

Results analysis of the word rate for all elocution modes: As shown in Table 2, the standard deviation varies between 0.304 and 0.512. This last value can be explained by the speaker behavior. This deviation is due to the strategy taken by the speaker 3 during the EM3: this speaker has a word rate very different according to the elocution mode EM1 (+ 67 words per minute in a spontaneous way).

Results analysis for reading modes (Training phase + EM1 + EM2): For all speakers, the standard deviation varies between 0.31 and 0.34. For all speakers, the average is 2.57 words per second (around 154 words per minute). This rate is very close to the rate announced by [Price 1999] who reported that a user can dictate 160 words per minute with the Dragon System.

For the speaker 1 and 2 (20-30 years old), the word rate per second increases (+ 22 words per minute). For the speaker 3 and 4 (more than 60 years old), the word rate decreases (-22 words per minute).

These preliminary results seem to attest the effect of the age on accuracy. They need to be confirmed on the whole set of speakers.

Table 2: Speech rate given in words per second.

	Speaker 1	Speaker 2	Speaker 3	Speaker 4
Training phase	2.86	3.05	2.16	2.23
Test text (EM1)	2.62	2.72	1.85	2.34
Test text (EM2)	2.93	3.27	2.36	2.39
Test Text (EM3)	2.84	2.18	2.97	2.23
Average	2.846	2.986	2.181	2.245
Standard deviation	0.304	0.390	0.512	0.302
Average - EM3	2.846	3.041	2.154	2.245
Standard deviation - EM3	0.340	0.310	0.328	0.332

These first results point out the more important deviation observed for the spontaneous way mode (EM3). It means that the guidelines to pronounce the text in the EM3 condition needs to be more refined for the next stage. At the present time, two ways are explored: 1) one way is based on explanations and examples; 2) another way is to define dialogue scripts to try to obtain more natural corpora.

Correlation analyses are being conduced on the SRDS accuracy and the word rate per minute.

These first results will provide knowledge about the behavior of speakers according to their age and face to the speech context of interaction (reading or spontaneous speaking).

5. Conclusion

The preliminary results of both the case study for selecting a SDRS for French language and the evaluation of the system chosen for standard (versus elderly people) show that it is necessary to carry out of the experimental protocol for standard and elderly speakers. It seems that the age of people have an effect on the speech rate. A number of opened questions have not yet been addressed, including the following:

a. User's characteristics need to be taking into account such as general cognitive level (fatigue, stress), speech disorders (difficulties to pronounce syllables, unknown words), syntactic structures, self adaptation ability to realize the reading and talking exercise. The first diagnostic analyses seem to show that the word error rate depends on the quality of the speaker's voice, such as the fluency, the prosodic stress and the respect or not of punctuation for a reading task.

b. It is necessary to design the lexicon and language model for both the faltered pronunciation and the extra-linguistic words.

Important effort of experiment and analyses need to be pursued to improve the universal access by means of speech interaction for elderly persons. The final goal is to give guidelines to implement/run these technologies for elderly persons.

Acknowledgments

We wish to thank Région Midi-Pyrénées (France) for supporting this work, and Mr. Winterton for his help in correcting this paper.

References

[Anderson 1999a] Anderson S., Liberman N., Gillick L., Foster S., Hama S., The effects of speaker training on ASR accuracy, in Proceedings of EUROSPEECH'99, Budapest, Hungary, CD-ROM.

[Anderson 1999b] Anderson S., Liberman N., Bernstein E., Foster S., Cate E., Levin B., Recognition of Elderly Speech and Speech Driven Document Retrieval, IEEE International Conference on Acoustics, Speech and Signal Processing, Phoenix, AZ, March 1999.

[Baggia 1999] Baggia P., kellner A,. Pérennou G, Popovici C, Sturm J,. Wessel F, "Language Modeling and Spoken Dialogue Systems - the ARISE experience", in Eurospeech'99, Budapest, Hongrie, 5-9 septembre 1999, Vol. 4, pp.1767-1770.

[Cadilhac 1997] Cadilhac Cl., Des structures textuelles à leur traitement : compréhension et mémorisation d'un récit par déments de type Alzheimer et sujets normaux âgés, Thèse d'Université, Toulouse II, Décembre 1997

[Canelli 2000] Canelli M., Grasso D., King M., Methods and Metric for the Evaluation of Dictation Systems : A Case Study, in Second International Conference on Language Resources and Evaluation, Proceedings Volume III, 31 M1y-2 June 2000, pp. 1325-1331.

[Emiliani 2000] Emiliani P.L.; Stephanidis C., From Adaptations to User Interfaces for All, 6 TH ERCIM Wporkshop, User Interfaces for All, Florence, Italy, 25-26 october 2000, pp.313-323.

[Gibbon 1997] Gibbon D., Moore R., Winski R., Handbook of Standards and Resources for Spoken Language Systems, Editors D. Gibbon, R. Moore, R. Winski, Berlin, New-York 1997.

[Kambeyanda 1996] Kambeyanda D., Cronk S., Singera L., Potential Problems associated with use of speech recognition products, RESNA'96 Proceedings, pp. 119-122.

[Lau 1997] Lau R., Flammia G., Pao C., Zue V., "WebGALAXY: Integrating Spoken Language and Hypertext Navigation" Appears in: Proceedings of Eurospeech '97, Rhodes, Greece, pp.883-886, September, 1997. From the World Wide Web: http://www.sls.lcs.mit.edu/raylau/publications.html#WebGalEuro

[Lippmann 1997] Lippmann R., Speech recognition, by Machines and Humans, Speech Communication, Vol. 22, N° 1, 1997.

[Padmanabhan 2000] Padmanabhan M., Picheny M., "Towards super-human speech recognition", in Automatic Speech Recognition, Challenges for the New Millenium, September 18-20 September, 2000, Paris, France, pp. 189-94.

[Yato 1999] Yato F., Inoue N., Hashimoto K, A study of Speech recognition for the Elderly, in Proceedings of EUROSPEECH'99, Budapest, Hungary, CD-ROM.

[Wilpon 1996] Wilpon J.G., Jacobsen C.N., "A study of speech Recognition for Children and the Elderly", Proc. of ICASSP96, pp.349-352.

[WWW 2001] http://www-4.ibm.com/software/speech/   http://www.dragonsys.com/   http://www.lhsl.com/ http://www.speech.philips.com/ud/get/Pages/psp_home.htm   http://www.nuance.com/

Adaptation of information delivery to support task-level continuity

Thomas Rist   *Stuart Booth*

DFKI Saarbrücken, Germany   Universita' degli Studi di Parma
email: rist@dfki.de   email: stuartandrewbooth@yahoo.com

Abstract

This contribution firstly introduces the notion of task-level continuity by means of the TACIT reference model for continuous human interaction. Secondly, it addresses the relationship between computer-based information delivery and task-level continuity in the context of an intelligent driver support system.

1. Introduction

In applications like car navigation or intelligent driver assistance, computers are increasingly used to actively assist users in task performance. An important form of assistance is the "just-in-time" provision of information on what to do next, and under certain circumstances, how to do it. Information delivery can take advantage of multiple presentation modalities and media, such as verbal utterances, acoustic markers, static and dynamic graphical displays. Content, form, and timing of the information delivery should be chosen so that the user can perform his/her task without disruption or delays. To achieve a high degree of what we call task-level continuity, the assistance system must carefully monitor progression of the task performance by the user. Therefore, it must be able to adapt information delivery to the particular needs of the user, and as much as possible contribute to the overall well-being of the user. In a car navigation scenario, for instance, the status of the task performance can be derived from sensed positioning information. In contrast, decision making on questions, such as which information to convey next, and how to present it to the user requires the consultation of a user model. Ideally, such a model covers all aspects of potential relevance for successful task performance. Among other aspects, the user model must comprise a representation of the task as it is understood by the user. By comparing this representation with a computer-internal representation of a validated task model, the assistance system can make decisions on what to present. Concerning the issue of how to present selected information units, however, the assistance system must also take into account the user's currently available perceptual resources, assuming that some of them are already bound by the user's engagement in the task performance.

2. Task-level continuity

The term task-level continuity was coined in the context of an initiative carried out by the TACIT network [1, 2] toward the development of a reference model for human computer interfaces that is able to deal with continuous interaction. At the current state of development, the reference model comprises a layered schema that allows description of interaction phenomena from different points of view and at different degrees of granularity. As illustrated in Fig. 1, the scheme foresees five different layers for each entity (i.e., human(s), and system(s)) involved in an interaction. From bottom to top the layers are labelled *(i) Physical/Physiological, (ii) Perceptual/Information-Theoretical, (iii) Algorithmic/Individual-Cognitive, (iv) Individual Goal and Task Level,* and *(v) Group Goal and Task Level.* When describing a concrete interaction scenario, however, not all layers are necessarily of interest. For instance, in a one-user-one-system scenario, the fifth layer may be neglected, or when describing the interaction with a light-weight computer-based tool, the system may not have individual goals nor a representation of the task for which it is used. In this case, an instantiation of the layers four and five may be omitted when describing the tool. Observable interactions take place at the *Physical/Physiological Layer* only. From the point of the user, perhaps the most interesting question concerns the system's contribution to a successful and efficient task performance/ completion (layers iv and v). The notion of continuity at the task layer means that a task can be performed without delays and interruptions.

422

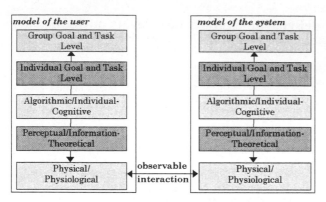

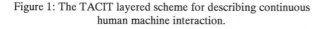

The purpose of the layered scheme is to provide a frame that allows to investigate in a systematic way the relationship between a continuous task performance and the required physical interaction, taking into account all processes located within the intermediate layers. In many scenarios, task-level continuity in fact requires continuous interaction at the *Physical/Physiological Layer.*

Driving a car provides a good example for this. In the next sections we investigate the driving scenario further with a special focus on new upcoming driver support systems.

Figure 1: The TACIT layered scheme for describing continuous human machine interaction.

3. Information Management Support in a Car Driving Scenario

Despite various technical improvements that enhanced both safety and comfort, driving a car remains still a complex task – partly due to the continuous increase in traffic density, but also because of the larger number of information, communication, and entertainment services becoming available in the interior. Some of these new services, such as navigation systems, are directly related to the driving task since they aim at providing further driver assistance. In contrast, other services, such as playing a radio program or a CD, making phone calls, or querying internet-based information sources, serve complementary purposes and may even distract the driver. Such observations have stimulated the development of support systems that assist drivers in managing the information flood and eventually make driving more safe and comfortable. To specify the functionality of such support systems it seems practical to distinguish between passive and active information offers, and also between the perceptual channels through which the driver can access the information offers.

Active versus passive information sources available to the driver: The term active information offer comprises all kinds of information services, communication and entertainment services which are directed towards the driver, and which are under the control of the envisaged driver support system. Examples are scales and displays in the dashboard that offer information on the status of the driving situation (e.g., speed, amount of navigated kilometers, outdoor temperature, etc.) and the status of the car (e.g., remaining fuel, temperature of the engine, air pressure of the tires etc.), navigation systems providing route information, traffic services that inform about traffic congestion, radio programs, private music programs from the CD player, and so forth. In view of the total information offer available to the driver, the term "passive information offer" refers to the complement of the set of active information offers. For instance, by looking through the front shield or into the driving mirror, or by noticing traffic noise of close-by vehicles the driver usually acquires a great portion of important information from the out-side environment (cf. Fig.2). Though several attempts have been made to sufficiently sense the environment for the purpose of building autonomous vehicles, for the time being, we still assume that sensing the environment will remain to a large extent a task of the driver, mainly because drivers still like to keep the control, but also because the support systems will only have partial access to the outdoor environment.

Multiple perceptual channels: The second distinction considers the driver as an information recipient and refers to the way in which the driver interacts with an information source. Most of the information will be perceived visually and acoustically. However, even though modern cars are equipped with lots of servo motors and electro-mechanic parts and hence maneuvers like steering, accelerating, breaking etc. could be accomplished with almost no physical engagement, direct force-feedback intentionally remains another important source of information for the driver, e.g., when turning the steering wheel or pushing the pedals.

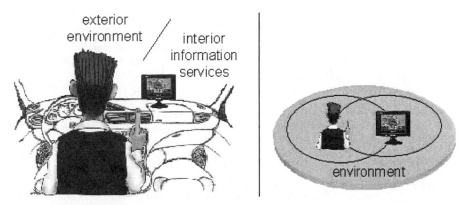

Figure 2: Left: The driver acquires information from the exterior environment (passive offers) and from on-board information, communication and entertainment services (active offers). Right: the circles indicate that parts of the environment are mutually/exclusively accessible by the driver and the support system.

Driver model: In order to make appropriate decisions on which information to offer, and when and how to present it, a driver model is required that allows to reason about the driver's:

-   most urgent information needs with regard to the driving task;
-   information needs relating to subordinate tasks/activities;
-   available cognitive resources to acquire new information units;
-   available cognitive and physical resources to follow instructions and suggestions.

As with any model, one has to choose an appropriate level of granularity to capture the most relevant aspects. For instance, a driving task can be roughly modeled by a sequence of subsequent state transitions of the form $s_i -a_i->$ s_{i+1} whereby s_i and s_{i+1} denote particular driving situations and a_i stands for a particular action or intervention by the driver performed in the situation s_i with the goal to achieve the driving situation s_{i+1}. Assuming further a start state s_0 and a final state s_n in which the driver has safely and efficiently arrived at her/his destination, the act of driving can be seen as a means to accomplish the transition from s_0 to s_n under the constraints that safety is maximized and deviations and delays are avoided as much as possible. A driving situation in turn stands for a complex set of observations (e.g., made by various sensors) comprising external conditions, such as the roadway and weather conditions, range of vision, volume of traffic, the distance to the car in front, the remaining driving distance to destination and so forth. In addition, there are observations concerning the car, such as current speed, remaining gas, etc. Depending on the observations made in a driving situation s_i a smart driver support system will encounter the need to suggest to the driver a certain course of action a_i, e.g., to reduce speed in order to increase the distance to the next car. In some cases, more complex maneuvers may be required in order to get closer to the final state s_n. For instance, if a road gets blocked due to an accident, the driver may have to turn back in order to follow the route newly calculated by the car navigation system.

Since the driver support system has to coordinate all available active information offers, it should have also a model of how additional information, communication and entertainment services can be used by drivers in a meaningful way. For instance, if a driver asks for the display of a her/his favorite song and at the same time, a traffic notice arrives, the support system should be able to evaluate the urgency of the notice and if possible postpone it's announcement so that the song can be played without interruption. In other cases, information may be buffered so that it can be replayed again after an interruption. This is for example the case if the car navigation systems need to urge a maneuver and therefore interrupts the news speaker in the middle of the sentence. If buffering is available, the news can be properly played but with a delay compared to the original broadcast.

424

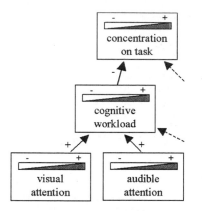

Figure 3: Fragment of a model
showing some interdependencies
between cognitive variables

The most complex and difficult task concerns the modeling of the driver's resources and capacities to perceive information, as well as the modeling of the potential impact of new information on the driver's mental state. Given the overwhelming complexity of human cognition, a model can only capture a few fragments, ideally those most relevant for the task under consideration. Inspired by work in the area of AI-based user modeling [3], one can try to model the driver's cognitive workload and remaining resources by means of a network of interrelated variables. For the purpose of illustration, Fig. 3 shows a fragment of such a model assuming that the driver's ability to concentrate on the driving task depends on the overall cognitive workload which in turn, among other factors, depends on the degree to which the driver's visual and audible attention are bound to information acquisition tasks. Arrows annotated with a plus symbol indicate proportional interdependencies whereas a minus indicates versed proportionality.

4.   Monitoring the driver's concentration on the driving task

One of the main tasks of the support system is to coordinate the active information flow to the driver in order to avoid information overload and associated distraction in crucial driving situations. In practice this means that the support system has to carefully filter and, under certain circumstances, even to withhold information so that the total amount does not exceed maximum processing capacity. However, when monitoring the driver's concentration on the driving task, one has to consider a somewhat contrary phenomenon as well - some driving situations actually result in *under-stimulation*. For example, when driving under constant conditions for an extended period of time, such as on a quiet and boring motorway, drivers may lose concentration. Without novel environmental stimulation, mental habituation tends to occur due to lack of demand on processing resources. In other words, the mind tends to wander from the driving task to other, more stimulating ends, such as daydreaming or thinking about what to have for lunch. Furthermore, attention to the driving task may suffer due to a reduction in overall level of arousal. The driver may, therefore, become lethargic and, in the worst instance, actually fall asleep. The result may be catastrophic. For example, steering may be become erratic and the vehicle may wander from its lane. Attention may only be regained once a warning has been provided by another driver or when the 'rumble strip' at the edge of the road is crossed. Ultimately, a crash may be the inevitable end.

A solution might, however, be possible. By monitoring driving conditions and driver inputs it may be feasible to produce a probabilistic model of attention to the driving task. Several variables could, therefore, be assessed, such as road speed, steering input, and velocity change, in order to identify potential problem situations. Given this data, it may then be possible for information to be presented to the driver in a manner designed to redirect their mental resources appropriately. Evidence from experimental psychology is available providing an indication of how this might be achieved [4]. For example, participants may be presented with a monotonous stream of identical, consecutive stimuli in which infrequent target stimuli occasionally occur. They are given the task of responding, as quickly as possible, to *targets* only, by making a simple key-press. This situation can be compared to driving on a boring motorway – the participant need do nothing until a target appears and, as a result, loses concentration. When reaction times are assessed, it becomes evident that attention to task begins to wander - key-presses gradually become slower. The interesting finding here, however, is that, when irrelevant stimuli are presented that differ from the stimulus sequence but require no response, reaction times to subsequent targets are much faster. In other words, the presentation of novel information results in a re-orientation to task.

In a similar manner it may be possible to employ an analogous approach in re-directing a driver's attention back to the driving task. Information may, therefore, be provided, such as a verbal comment on the current state of driving conditions, an estimate of the remaining travel time, tourist information on the region, or even a direct warning not to fall asleep. More specifically, the variable "driver's_concentration_on_driving_task" may, for the purpose of illustration, be modeled roughly as a probabilistic function of 'time since last salient task-related stimulus'. The probability of attention being directed to driving will reduce over time, starting from a maximum value at the time when the stimulus is provided (cf. Fig. 4 left-hand diagram). A threshold can then be applied beneath which the concentration value must not be permitted to fall. That is, whenever the value falls below the threshold, the support system triggers the production of another salient task-related stimulus. The expected run of the resulting curve is illustrated in the right-hand diagram of Fig. 4

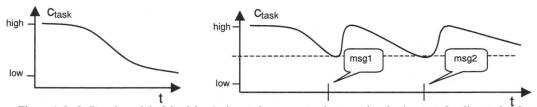

Figure 4: Left: Rough model of the driver's decreasing concentration over time in absence of salient task-related stimuli. Right: Expected periodical recovering of the concentration value due to interspersed task-related messages and comments. Note that the diagrams are meant to illustrate the approach- they may not be taken literally as they are not based on empirical findings.

5. Conclusions

This contribution introduces a future driving-scenario in which the driver receives assistance in managing information sources through an intelligent driver support system. In order to maintain continuity at the task level – foremost in driving, but also in other tasks in which a driver may engage during driving - such a system must have a clear understanding of the task and the driver's capability to cope with multiple information streams. The TACIT layered scheme for describing continuous interaction has been adopted as a starting point for the required research. However, further work is required, especially in view of the modeling of a driver's cognitive resources.

Acknowledgment
Many thanks to Gerd Herzog and Ralph Schäfer for fruitful discussions on the topic.

References
[1] TACIT homepage: http://kazan.cnuce.cnr.it/TACIT/
[2] Faconti, G. and Massink, M.: *Continuity in Human Computer Interaction*. Workshop Report, CHI 2000, The Hague. Available under: http://www.acm.org/sigchi/bulletin/2000.4
[3] Jameson, A., Schäfer, R., Weis, T., Berthold, A., & Weyrath, T. Making Systems Sensitive to the User's Changing Resource Limitations. *Knowledge-Based Systems, 12*, 413-425. 1999
[4] May, J., Barnard P., and Booth, S. (Title and details of report to be provided)

Universal Access Design for Information Kiosks

Hirohiko Sagawa[a], Haru Ando[b], Masaru Takeuchi[a] and Nobuo Hataoka[b]

[a] Multi-modal Functions Hitachi Laboratory, RWCP,
1-280, Higashi-koigakubo, Kokubunji-shi, Tokyo 185-8601, Japan

[b] Hitachi Central Research Laboratory,
1-280, Higashi-koigakubo, Kokubunji-shi, Tokyo 185-8601, Japan

Abstract

A multi-access interface is described that provides several types of user interfaces and input/output systems. These user interfaces allow the handicapped or the aged to easily access information systems. We especially focused on an information kiosk as an application of the multi-access interface. And an information kiosk with a sign-language interface was designed for hearing impaired people. Sign language is a usual method of communication for hearing-impaired people, and we have developed a sign-language recognition system to support the communication between hearing-impaired people and hearing people. Hearing-impaired people can easily operate this information kiosk by using sign-language gestures. This kiosk was tested in a government office and accepted favorably by its users.

1. Introduction

We are developing a multi-access interface that allows the handicapped or the aged easy access to information systems such as PCs and information terminals. The concept of a multi-access interface is similar to that of universal design, that is, the creation of products and environments completely accessible to all people. These products and environments should not need adaptation or specialized design [1]. Systems following the universal design concept usually have only a few types of interfaces designed to be accessible to many users. On the other hand, a multi-access interface provides several types of user interfaces and input/output systems that can be adjusted to individual user characteristics. Therefore, a multi-access interface is regarded as an expansion of barrier-free design.

One application of a multi-access interface is an information kiosk. An information kiosk is an information terminal that provides various kinds of public and corporate services. However, an ordinary information kiosk is not well designed for use by the handicapped or the aged. A multi-access interface is necessary to give them easy access to information kiosks. Figure 1 shows an information kiosk with a multi-access interface. Users can access information in this kiosk by using an input/output means adjusted to their ability.

We supposed that the main users of this information kiosk would be the aged, the visually impaired and the hearing impaired. Therefore, we provided a user interface using speech recognition and synthesis systems for the aged and the visually impaired and another using sign-language recognition and generation systems for the hearing impaired. Our group is also developing methods for recognizing and generating sign-language gestures to automatically translate Japanese sign language (JSL) into spoken Japanese and vice versa [2, 3, 4]. These methods can be an additional user interface for the hearing impaired.

We designed an information kiosk with a sign-language recognition system [5]. In this paper, we will describe this information kiosk. We tested this kiosk in a government office and surveyed users' impressions of it. We will also report the test results in this paper,.

2. Information kiosk with a JSL recognition system

2.1 JSL recognition system

The JSL recognition system works as shown in Figure 2. Signed sentences are comprised of hand gestures, nods, glances, and facial expressions [6]. However, our system recognizes only hand gestures. Users employ a glove-based input device, such as the CyberGlove[1], and a magnetic sensor device, such as the FASTRAK[2] to input hand gestures into a computer.

The word-recognition step identifies each signed word represented by an input gesture. The words are identified by combining gesture primitives: hand shape, palm direction, linear motion, circular motion, etc. Each primitive consists of a gesture-primitive type, the part of the hand that represents the primitive, and gesture-primitive attributes, such as the start position and motion direction. During the recognition process, the gesture primitives are identified from the input gesture, then the signed word is recognized base on the time and spatial relationship between the gesture primitives [2].

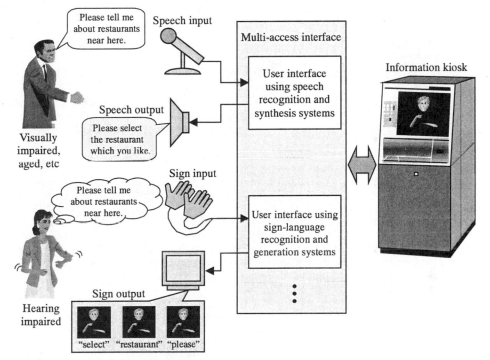

Figure 1. Information kiosk with a multi-access interface

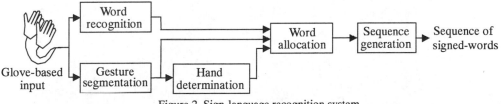

Figure 2. Sign-language recognition system

The gesture-segmentation step detects the borders of the signed words and divides the input gesture into several segments. The hand-determination step determines whether the gesture in each segment is one- or two-handed [3]. The word-allocation step analyzes the relationship between the recognized signed words and segments, then assigns the words to the segments. Finally, the sequence-generation step combines the recognized signed words and generates a word sequence.

---

[1] CyberGlove is a trademark of Virtual Technologies.
[2] FASTRAK is a trademark of Polhemus Inc.

428

2.2 Information kiosk

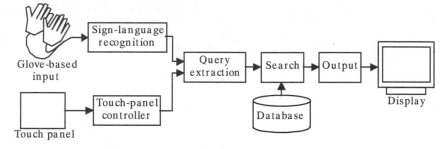

Figure 3. Structure of information kiosk with JSL recognition

The structure of the information kiosk is shown in Figure 3 [5]. All the components except sign-language recognition are the same as in an ordinary information kiosk. The hearing impaired can search for information by using JSL. JSL gestures are input into the system by wearing a glove-based input device, and the responses appear on the screen as JSL animation and as text.

The system can accept two types of signed gestures.

(1) Words written on buttons on the kiosk's console

A user can input signed gestures that represent words written on kiosk buttons. Thus, the user can operate the system by using JSL rather than by touching the buttons.

(2) Sentences in JSL

A user can ask for desired information by expressing sentences in JSL. To enable this, we stored templates of signed sentences in the system. The JSL-recognition system searches for and extracts the appropriate signed sentence from those stored in the templates. For example, if the user wants to know about restaurants in Nagasaki City, he/she can input signed gestures representing the following sentences.

(a) Please tell me about restaurants in Nagasaki City.
(b) I want to know about restaurants in Nagasaki City.
(c) Are there restaurants in Nagasaki City?

Here, the words "restaurants" and "Nagasaki City" can be replaced by other words.

The required information is extracted from the database based on keywords represented in the recognized sentence.

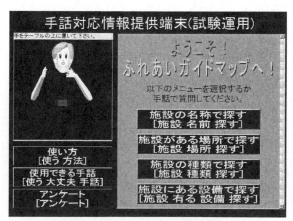

Figure 4. Example screen image

Figure 5. System appearance

3. Test run

The information kiosk was tested for three months in the Isahaya City Office in Nagasaki Prefecture. During the test run, visitors to the city office were given free access to the kiosk system.

This kiosk provided information on public and corporate services in Nagasaki, Sasebo, and Isahaya Cities in Nagasaki Prefecture. The hearing impaired could use JSL gestures to search for information about these areas. Figure 4 shows an example of the screen image, and Figure 5 shows the appearance of the system. The JSL recognition system in the kiosk can recognize 268 JSL words and 52 kinds of JSL sentences.

To ascertain whether this information kiosk with its JSL recognition system is useful to the hearing impaired, we surveyed the users impressions. The questions were as follows.

(1) Is such a kiosk system necessary?
(2) How difficult is it to input JSL gestures while wearing the glove-based input device?
(3) Where do you need such a kiosk system?

We collected the impressions of 32 users (9 hearing-impaired people and 23 hearing people). The answers to the questions are summarized in Tables 1, 2 and 3.

Table 1. Answers to question 1

Answer	Hearing-Impaired	Hearing	Total
Necessary	2	4	6
Somewhat necessary	5	11	16
Not necessary	0	1	1
I don't know	1	2	3
No answer	1	5	6

Table 2. Answers to question 2

Answer	Hearing-Impaired	Hearing	*Total*
No problem	4	1	5
A little tight But no problem	3	6	9
A little hard	2	3	5
Very hard	0	10	10
Impossible	0	1	1
No answer	0	2	2

Table 3. Answers to question 3

Answer	Hearing-Impaired	Hearing	*Total*
Tourist resort	2	3	5
Government office	6	12	18
Police station	3	10	13
Railroad station	3	4	7
Travel agency	0	4	4
Museum	0	2	2
Hospital	3	10	13
Art museum	1	3	4
Aquarium	1	4	5
Amusement park	0	3	3
Library	2	7	9
Department store	3	8	11
Bank	4	10	14

430

4. Discussion

As shown in Table 1, 22 of the users acknowledged the value of this kind of system. This result indicates that this method of using JSL gestures is significant not only for the support of communications between hearing-impaired people and hearing people, but also in that it allows hearing-impaired people to interact with information equipment. However, many users did not regard such a kiosk system as indispensable. This is because most young hearing-impaired people can read sentences in Japanese.

The answers to question 2 from hearing-impaired people were clearly different from those of hearing people. Most hearing-impaired people accepted the glove-based device, however, many hearing people rejected it. Even so, we believe the kiosk system with a glove-based input device will be accepted because hearing-impaired people are intended to be the main users of the system.

Table 3 shows that many users need this system in the places they often visit in daily life such as government offices, hospitals, banks, and department stores. Putting a practical kiosk system in these places is difficult because highly accurate sign-language recognition is necessary. However, we must develop a practical kiosk system for such places.

5. Conclusions

We developed an information kiosk with a sign-language recognition system as an example of an information system with multi-access interfaces and tested it in a government office. The system was received favorably by most users. We are planning to develop a more practical information kiosk for hearing-impaired people by improving this kiosk.

We are also planning to install a speech-recognition and synthesis systems in this kiosk as an interface for the visually impaired and the aged. To devise a multi-access interface, user interfaces based on speech and sign language are not enough. In the future, we will take other user interfaces into account for use in our information kiosk.

Acknowledgments
We thank the staff of Jumon-kai, a social welfare corporation in Isahaya city , and the Isahaya city office who supported the information kiosk's test run.

References
1. Connell, B.R., Jones, M., Mace, R., Mueller, J., Mullick, A.. Ostroff, E., Sanford, J., Steinfeld, E., Story, M., Vanderheiden, G. (1997). *Principles of Universal Design*. http://www.design.ncsu.edu/cud/univ_design/princ_ overview.htm.
2. Sagawa, H., Takeuchi, M., Ohki, M. (1998). Methods to Describe and Recognize Sign Language Based on Gesture Components Represented by Symbols and Numerical Values, Knowledge-Based Systems, Vol. 10, No. 5, 287-294.
3. Sagawa, H., Ando, H., Koizumi, A., Iwamura, K., Takeuchi, M. (2000). Sign Language Recognition and Its Applications, Proceedings of 2000 RWC Symposium, 143-146.
4. Sakiyama, T., Oohira, E., Sagawa, H., Ooki, M., Ikeda, H. (1996). A Generation Method for Real-Time Sign Language Animation, Transactions of IEICE, Vol. J79-D-II, No.2, 182-190 [in Japanese].
5. Sagawa, H., Takeuchi, M. (2000) .Development of an Information Kiosk with a Sign Language Recognition System, CUU 2000 Conference Proceedings, 149-150.
6. Kanda, K. (1994). Lecture on Sign Language Study, Fukumura Publisher [in Japanese].

The ViSiCAST[1] Project:
Translation into Sign Language and Generation of Sign Language by Virtual Humans (Avatars) in Television, WWW and Face-to-Face Transactions

Rolf Schulmeister

Universität Hamburg, Institut für Deutsche Gebärdensprache

Abstract

The ViSiCAST Project aims to facilitate and enhance the access of deaf people to spoken or written information by providing a sign language translation in television, in the WWW, and in face-to-face transactions by means of avatars (virtual humans) which are able to generate sign language sequences in these media.

1. Aims of the ViSiCAST project

The aim of the ViSiCAST project (Virtual Signing, Capture, Animation, Storage and Transmission; http://www.visicast.co.uk) is to facilitate the access of deaf people to spoken or written information in television, internet, and in face-to-face transactions. The project develops photorealistic avatars (virtual humans), able to generate sign language on the screen from spoken language or from text. The project is a contribution to improve the situation of the Deaf in the information society and their participation in the social and political area.

Why are we interested in this topic? Before describing the project in more detail, I would like to answer this question. There are several aspects to consider:
*   the communicative situation of the Deaf and their sign language
*   social barriers for the Deaf in the information society
*   scientific and technological progress in the fields of sign language recognition, generation, and translation.

1.1 The communicative situation of the deaf and their sign language

Deaf people have developed their own fascinating means of communication or symbolic system: sign language. What is even more fascinating is that deaf people communicating in sign language are no longer handicapped. They only have some problems learning the oral language and the written language of hearing people. I am often asked, why the Deaf cannot read and write as well as hearing people. The reason is the lack of auditory feedback: Since the deaf person cannot directly receive the sound of the hearing partner and never receives an immediate feedback for his own oral language production, there is no natural way leading to a better command over this language. Verbal and written language is learned as an artificial language, just as if we would try to learn to speak Latin as native language today.

In comparison, sign language is learned as a vivid means of communication, owning the rich linguistic repertoire of a natural language. Only in this language do deaf people have a chance to fully participate in social affairs, leading to the consequence, that participation by way of interaction with the hearing world is only possible by using interpreters.

---

1 The ViSiCAST Project is funded within the 5th Framework in the Information Societies Technology (IST)-Programme of the European Commission since 1.1.2000 for three years. Partners in the project are the Independent Television Commission (ITC) in Winchester, UK, the Institut für Rundfunktechnik in München, Televirtual in Norwich, UK, the University of East Anglia (UEA, School of Information Systems) in Norwich, the Institut National des Télécommunications (INT) in Evry, France, the Instituut voor Doven (IvD) in Sint-Michielsgestel, Netherlands, the UK Post Office, the Royal National Institute for Deaf People (RNID), UK, and the Institut für Deutsche Gebärdensprache, University of Hamburg.

1.2   The socio-political impairment of the deaf in the information society

The participation of the Deaf in the information society thus is heavily restricted. News and other information features in broadcasting or television, but also in newspapers and magazines do not grant the Deaf true participation. Following a resolution of the European Parliament in some European states the national sign language has been recognized as official language or as a legitimate language for education, recently in Greece.

Television companies have for a long time opposed translation into sign language and have instead insisted on providing subtitles. Subtitles, however, do not provide an advantage for the Deaf, because they reduce the amount of information transmitted and because they heavily rely on the reading capabilities of the Deaf which are indeed poor. The fact that the main news of the two public television channels in Germany have been interpreted for the last few years on a separate special channel Phoenix seems to provide an excuse for not offering more interpreter services. The percentage of interpreting in television is rather low. In Great Britain the parliament has ordered the terrestrial digital television companies to raise the percentage of interpreted hours. This has not been embraced with enthusiasm, since an additional window or signal with a sign language interpreter raises the amount of data being transmitted, which otherwise could be profitably sold for other purposes. Raising the percentage of interpreting thus means to look for a method of transmission which uses less capacity. In that case it might be possible that the television companies are willing to raise the amount of sign language interpreting.

In a similar way the actual development stage of the Internet poses an additional barrier against the participation of the Deaf in the information society. Since the world of the Internet actually is predominantly text-based, the majority of the Deaf is excluded from the wealth of information although being highly motivated to make use of new media. Presentations of web-based information resources in sign language are rare, only offered in special communities and expert circles. And even if there was more net-based information in sign language, then probably the capacity and speed of transmission today would not be sufficient to allow for convenient communication in sign language.

1.3   Progress in the generation, transmission and translation of sign language

The videophone today has reached a technical standard that enables two deaf people to have convenient communication in sign language at a reasonable price. But unfortunately the videophone does not allow communication between Deaf people and people who do not know sign language. To make this possible, there must be either a permanent interpreter relay service installed or there must be automatic machine translation from sign language to oral language and vice versa between sender and receiver. This idea has been in our minds and research during the last decade [Schulmeister 1992; Schulmeister 1994a; Schulmeister 1994b], but only now with the ViSiCAST project are we able to make a contribution to solve this problem.

The technology of set-top boxes attached to a digital television set allows broadcasters to mix signals and thus enable the user to decide, if he/she wants to blend the interpreter into a picture or not. ViSiCAST will make a contribution to this problem: to reduce the amount of data being transferred in low bandwidth, thus enabling the receiver, to overlay an avatar over the TV image, producing the subtitles in sign language. The transfer of signs will be compliant with standards for bodies and faces in MPEG-4 and MPEG-7.

The plug-in technology for web browsers makes it possible, to call up additional functions, e.g. a translator, translating text into a foreign language. ViSiCAST will develop a plug-in, calling up an avatar, which will translate selected text in sign language.

Our avatars are three-dimensional figures with a photorealistic look and feel which are able to sign realistically. Sign language will be notated in our notation script language HamNoSys and turned into XML (in our project called SiGML — Signing Gesture Markup Language) and then sent to the avatar.

2. The idea of the ViSiCAST project

It is not the aim of ViSiCAST, to substitute human interpreters by avatars. Besides, regarding the present state of machine translation, that would be an unrealistic and unsocial aim.

But it is the aim of ViSiCAST, to deliver translation **there and then** where translation is needed and no interpreter is available or where translation is needed independent of time and location, e.g.

- for emergency messages in TV, when an interpreter is not available,
- in the World Wide Web, where one cannot foresee, what kind of information is requested by a deaf person,
- in face-to-face transactions, which occur irregularly and too seldom to engage an interpreter, e.g. at the counters of the Post Office or banks and other public services.

The heart of the ViSiCAST project is the development of a machine-readable system for the description of sign language. The system is based on the HamNoSys-Notation for sign languages, developed at the Institut für Deutsche Gebärdensprache [Prillwitz et al 1989; Prillwitz & Zienert 1990] with added features for mimics, mouth patterns and syntax. ViSiCAST develops translation technology on top of this description language. The system aims at

1. Transmitting sign language in television
2. Generating sign language in real time
3. Being compliant with MPEG/DVB and XML

Sign languages have special morphological, phonological and syntactic features. Phonological components of sign language are handshape, hand position, location in space, and movement. Signs and mimics are synchronously executed meaningful grammatical functions. Several signs are frequently incorporated in one single sign and are not executed sequentially as in oral languages. ViSiCAST in its first phase aims at an interactive semi-automatic translation in two stages: The text to be translated will be notated in Discourse Representation Structure (DRS; Kamp and Reyle 1993), then translated into HamNoSys notation and then turned into SiGML. ViSiCAST will produce tools for this semi-automatic translation. The result will be transferred to an animation machine, the avatar. ViSiCAST uses avatars with high resolution working in real time.

3. Applications for the ViSiCAST technology

ViSiCAST has selected three application fields as testbeds for this technology:

- The translation and transmission of signs in television.
- The translation of weather reports in the Internet.
- The translation of transactions at counters of the UK Post.

3.1 Translation in television

Interpreting in TV is rare (less than 1%). The reason is not the fees for interpreters, but prejudices of the program directors and the limits of transmission capacity in terrestrial digital TV. In order to raise the amount of interpreting, a method must be chosen, to mix the interpreter signal with the TV image intentionally by the user and to reduce the amount of transferred data substantially. Two methods are to be developed in ViSiCAST:

- During the first stage of the project a human interpreter wears a data suit, and his data are transmitted to a set-top box on the receiver's side. The set-top box contains the avatar that transforms the data into a human figure. The datastream transmitted in VBI or MPEG-2 (MPEG-4)-format is thus substantially reduced [Mozelle & Preteux 1998]. This method only modifies the transmission type. It is unintelligent, and needs no translation. Subject of research are data compression, methods of encoding and decoding, and quality of the image.
- A second method works with the same set-top box, but develops a system, which is able to translate text, e.g. subtitles, into sign language. The system is intended for situations in which no interpreters are available, e.g. short emergency messages.

434

3.2 Translation in the Internet

Deaf people not only have problems to hear or articulate oral language or read it from the lips, but having no direct feedback from oral interaction they also have enormous difficulties learning to read an to write. In spite of visual interfaces they encounter great difficulties in the predominantly text-based internet world. The aim of ViSiCAST is to offer sign language translation in the Internet. A "Viewer" will be developed as a plug-in for WWW-Browsers which generates an avatar, converting the SiGML sequences into sign language. This technology may later be also used for tutorial applications for hearing people, enabling the user to enter text and see the result in sign language. The application in ViSiCAST will be weather reports. We intend to set up a web-site producing daily weather forecasts in Dutch sign language.

3.3 Translation in Face-to-Face Transactions

ViSiCAST also aims at face-to-face communication. An installation is currently being evaluated by the UK Post Office [Pezeshkpour et al 1999]. When a deaf person comes to the counter, the clerk can speak his or her answers and questions into a microphone. On a PC speech is translated into text, and the text is translated by an avatar into sign language. The system is at the moment still semiautomatic. Most signs are contained as captured data in a database and retrieved by a lookup method. But the avatar can incorporate variable data for numbers with nouns for time, money etc. Two enhancements are projected: to improve the voice recognition and make the language input more flexible, and to base the translation method on the technology developed by ViSiCAST. The present Post Office system, called TESSA, has received the BCS IT Award in 2000 (www.bcs.org.uk/awards/itawards/press/index.html) and (www.thetimes.co.uk/article/0,,30343,00.html).

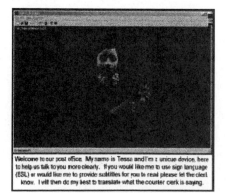

Welcome to our post office. My name is Tessa and I'm a unique device, here to help us talk to you more clearly. If you would like me to use sign language (BSL) or would like me to provide subtitles for you to read please let the clerk know. I will then do my best to translate what the counter clerk is saying.

435

References

Kamp, H./Reyle, U.: From discourse to logic: introduction to modeltheoretic semantics of natural language, formal logic and discourse representation theory. Dordrecht et al.: Kluwer 1993.

Prillwitz, S. et al (1989): Hamburg Notation System for Sign Languages — An Introductory Guide. -In: International Studies on Sign Language and the Communication of the Deaf, Vol. 5. Institute of German Sign Language and Communication of the Deaf: University of Hamburg 1989.

Prillwitz, S./Zienert, H. (1990): Hamburger Notation System for Sign Language: Development of a sign writing computer application. In: Prillwitz, S./Vollhaber, T. (eds): Current Trends in European Sign Language Research. Proceedings of the 3rd European Congress on Sign Language Research. Hamburg July 26-29, 1989. (International Studies on Sign Language and the Communication of the Deaf; 9) Hamburg: Signum (1990) - S. 355-380

Schulmeister, R. (1992): Generierung und Erkennung der Gebärdensprache. -In: ISI 92. Proceedings des 3. Internationalen Symposiums für Informationswissenschaft, Reden zur Eröffnung (= Bericht 22), Universitätsverlag Konstanz 1993 (ISSN 0942-2625).

Schulmeister, R. (1994a): Computer Assistance in Learning Sign Language. In: Brunnstein, K./Raubold, E. (eds.): Applications and Impacts. Information Processing '94. IFIP-Transactions A-52. Volume II. 1994. North-Holland, pp. 702-707.

Schulmeister, R. (1994b): Evaluation des Bildtelefons für Gehörlose. In: Das Zeichen 28 (1994), S. 204 - 216.

Mozelle, G./Preteux, F. (1998): „Tele-sign: A compression framework for sign language distant communication", Proceedings SPIE Conference on Mathematical Modeling and Estimation Techniques in Computer Vision, San Diego, CA, Vol. 3457, July 1998.

Pezeshkpour, F./Marshall, I./Eliott, R./Bangham, A.J. (1999): Development of a legible deaf-signing virtual human. In Proc. IEEE Conf. Multi-Media, Florence, 1999.

Approach-and-Use Technology

Neil G. Scott

The Archimedes Project, CSLI, Stanford University
210 Panama Street, Stanford CA 94305, USA
ngscott@arch.stanford.edu

Abstract

The Archimedes Project has developed a universal access strategy called the Total Access System (TAS) that enables individuals with disabilities to access any type of information appliance. Traditional access strategies require modifications to the computer that is being accessed often with the result of making the system unusable by anyone other than the disabled user and, in many cases, making the disabled person totally dependent on a single machine. The widely ranging needs of disabled people generally make it impractical for manufacturers to make their equipment accessible to everyone. The TAS overcomes these problems by separating the access issues from the application that is being accessed. It achieves this through the use of a small device called the Total Access Port (TAP) that emulates all of the functions performed by the keyboard, mouse and screen of the information appliance without disturbing the operation of the real input and output devices. The disabled operator uses a personal "accessor" to control the TAP. Personal accessors provide each person with an interface that is precisely matched to individual needs, abilities and preferences. A standardized communication protocol enables any accessor to work with any TAP thereby enabling users to mix and match different input and output modalities without needing special software in the target machines. Multimodal combinations such as speech recognition and head tracking, for instance, can provide significant performance improvements over the standard keyboard and mouse. In this paper we describe strategies for using networks of distributed agents in conjunction with the TAS to automatically link multiple input and output resources. In addition to managing the discovery of new accessors and TAPs and the flow of information between all of the devices that make up the multimodal interface, the distributed agents are also used to perform natural language processing that disambiguates all of the input information generated by the user.

1. Introduction

The term digital convergence keeps cropping up when people try to rationalize what is happening with Electronic Information Technology (EIT). Supposedly, digital electronics will enable a person to use a single device to satisfy all of their information processing requirements. This concept is at odds, however, with what you see when you visit an electronics superstore or search on the web for EIT devices. In each situation, you are bombarded with options for devices that perform all manner of tasks and, in most cases the devices employ interface strategies and protocols that are unique to each manufacturer.

While the concept of digital convergence is frequently touted as the next big step in the evolution of electronic devices such as computers and Audio Visual equipment, what we are seeing is really a digital divergence. Manufacturers are making things that are different in an attempt to differentiate their products from their competitors and to lock users into their particular style or brand of device. The components that make up a smart house, for example, will only work properly if each of those components and the underlying networks are designed around a common philosophy and protocol. While a few companies are striving to standardize the infrastructure that will be used to build smart environments, most are building proprietary solutions that are incompatible with products from their competitors. I recently met a person in Japan who lives in a smart house that was make by Panasonic. He said the good thing is that everything in the house is made by Panasonic and is therefore very compatible and reliable. He also said, that the bad thing is that everything in the house is made by Panasonic and he has limited choice about what new appliances he can include. Archimedes researchers are working to bridge the differences between manufacturers by developing a consistent, personal user interface that works equally well with any of the different product lines. In other words, we are creating convergence at the user interface without restricting the functional choices available to the user.

2. Issues in the design of smart environments

Around the world, there are many research and design groups exploring concepts for how people will interact with objects in smart environments such as smart houses, smart offices and smart cars. Most of the current interaction strategies are based on extensions to existing AV style remote controls or on hand-held tablets with some variant of the GUI interface that is commonly used on office computers. There are also a few groups working on the concept of using highly instrumented environments in which arrays of microphones and cameras track the movements and actions of the people who inhabit the smart space. This "big brother is watching" paradigm is at the core of futuristic designs in which the environment observes what is happening, interprets the observations and acts accordingly. There is a major problem with this approach, however, in that we are still a very long way from having the ability to remotely track and interpret the intentions and actions of one person in a reliable way, let alone a group of people. This is further complicated when we must allow for everyone as a unique individual with different intentions, levels of understanding, physical abilities, cognitive abilities, skills, age, acceptance and trust of the technology and so on. Taken together, these factors are an enigma for any software that must make instant decisions about what the person wants to do, intends to do, or is in the process of doing.

Quite apart from the problems associated with the people in the smart environment, there are also many problems with the technologies that are used to observe the people. In spite of more than two decades research and development, technologies such as speech recognition, eye tracking and gesture recognition still don't work well enough for general purpose applications, particularly when more than one person is being observed.

There are three major requirements to be considered when we think of establishing a smart environment:

1. How the smart environment will connect to the information infrastructure
2. How appropriate information will be collected from both the smart environment and the information infrastructure, identified, and processed, and
3. How the smart environment will connect to and interact with individuals within the smart environment.

In working out how to meet these requirements, there are two basic questions that must be asked: where do we place the smarts; and how smart will every piece need to be?

3. The ARCHIMEDES project approach

The Archimedes Project has been studying a similar set of requirements and questions for more than a decade in its search for ways to enable individuals with disabilities to interact fully with EIT such as computers, (Scott, 1991), ATMs and home appliances. Our overall conclusion is that each object in a smart environment must be smart enough to perform its assigned function perfectly, and to be able to communicate with other objects in the smart environment whenever necessary. For example, the function of my alarm clock is to wake me up at a particular time, the function of my coffee maker is brew a cup of coffee, and the function of the scheduler in my PDA is to remind me when it is time for me to do something. So, before I go to bed, I check the time of my first appointment in the morning, set the alarm clock to wake me up in time and set the dial on my coffee maker to have a cup of coffee ready when the alarm goes off in the morning. In a smart environment, all of these actions could happen without any action from me apart from initially entering the appointments in my PDA, and even that might be done automatically as the result of an email request from my secretary. But the question still remains, "where are the smarts?"

> One approach is to provide each device with a built in scheduler that receives updates at the same time as the PDA. This is not a good idea, however, because we are making the individual devices more complex and therefore more expensive than they need to be and then filling them with information they don't need to know. The alarm clock only needs to know the time of the first appointment for the day, the coffee maker only needs to know the time the alarm clock is going to operate so that it can compute when to start making the coffee.

> Another approach is to integrate the coffee maker and alarm in a single product but this isn't satisfactory because I always take my alarm clock with me when I travel and it isn't practical to take the coffee machine as well.

The approach we have adopted at the Archimedes Project is to treat each object as an expert that knows how to perform a specified task and how to delegate any subtasks it can't handle (Norman, 1998). Hence, a coffee maker makes great coffee but it obtains time information from another source such as the alarm clock or scheduler. While it was originally developed meet the needs of disabled people, the Archimedes strategy can be generalized to provide a practical framework for interactions between people and objects in any smart environment.

438

3.1 Using the Total Access System in smart environments

Traditional methods for accommodating individuals with special needs are based on changing each item within the living or working environment to suit the particular needs of the user. This process is labor intensive and time consuming and falls far short of the goal of providing universal access. In contrast, the approach developed by the Archimedes Project requires no changes to the information infrastructure and can be deployed very quickly. Rather than modifying the target computer, the Archimedes Total Access System (TAS) provides individuals with a personal information appliance, called an accessor, that provides alternative ways to perform all keyboard, mouse and/or monitor functions of any target system.

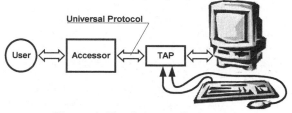

Figure 1. Total Access System

As depicted in Figure 1, the TAS consists of three components; (i) a Total Access Port (TAP) that emulates the operation of the keyboard, mouse and screen of the target computer, (ii) a personal accessor that implements whatever interfaces are needed by the disabled individual, and (iii) a universal communications protocol that connects any accessor to any TAP.

An important part of the TAS concept is that individual accessors and TAPs do not require prior knowledge of each other to be able to communicate. This contrasts with conventional personal computers that must have a matching driver program installed before they can interact with a new peripheral device. Maintaining the correct drivers is a major logistical problem when adding and removing peripheral devices. The TAS protocol gets around this problem by including a self-description within each accessor and TAP that other devices use to learn how to interact with it (Jonsson, 1999). The TAS has two levels of communication, a basic level that supports all of the normal functions provided by a standard keyboard and mouse, and an advanced level that supports the special functions contained in the self-description. When equipped with a personal accessor, a disabled individual is able to interact with any EIT device using whatever access strategies they know and prefer. Once a person crosses the IT threshold with their personal accessor disguised as a standard keyboard, mouse and/or monitor, disabled individuals can transparently use the full capabilities of the target systems without penalty, and with whatever performance augmentation they require to be competitive.

The long-term goal of the Archimedes Project is to create an infrastructure (Scott, Jackson & Jonsson, 1999) that enables any person to walk or wheel up to any target device and gain full access using whatever interface, or combination of interfaces, best matches his or her individual needs, abilities and preferences. This is a large undertaking that involves many different areas of research. For example: (i) psychological studies are necessary to determine how individuals with different disabilities perceive information, and to understand which types of interface work best in various settings, (ii) linguistic research is necessary to determine how to create natural language interfaces, and to understand how to augment the language and or communication abilities of communicatively impaired users, (iii) innovative human interface strategies must be researched and developed and (iv) many hardware and software engineering challenges must be overcome to provide a stable platform that will deliver results derived from the other areas of research.

4. Current status of TAS for smart environments

The Archimedes Project has adopted a divide-and-conquer strategy to reach its goal. The TAS provides the core technology for implementing practical solutions and incorporating emerging technologies wherever appropriate. TAPs have been developed for PC, Sun Mac, SGI and HP workstations, the LonWorks industrial control network, and the X10 home control network. Other TAPs are being developed for specific targets such as household appliances and TV game consoles. Over the past five years, TAS has been installed at key sites around the country including Sun, SGI, Lockheed Martin, Bank of America, National Security Agency and Boeing. The response from users has been very positive and negotiations are underway for licensing the technology to companies in the U.S., Europe, and Japan.

The commercially available version of the TAP connects a single accessor to a workstation. A four-way splitter is also available that allows a single accessor to operate with up to four target systems (Synapse Adaptive, 1997).

4.1   Using agent technology to achieve distribution, plug&play, and natural interaction

Ongoing research is focused on creating a truly universal access strategy in which any number of accessors and TAPs can be automatically connected in whatever configuration may be necessary to perform a desired operation. Our strategy for achieving universal access is to incorporate distributed agent software in each accessor and TAP. These agents provide the necessary intelligence for identifying and communicating with whatever devices are required to perform a desired interaction. Several agent technologies have been evaluated and we have settled on Distributed Agent Technology from Dejima Corporation (Hodjat & Amamiya, 2000). Dejima's has developed an extremely flexible approach for designing, implementing and testing networks of intelligent agents. The agents that make up a particular network can be located in a single device or distributed among many different devices. Dejima's agent networks follow a claim-delegation model in which the agent or agents designed to perform a particular task will identify and operate on input data directed towards the task. Each agent will also delegate all input data to all connected agents, this will ensure that agents can be kept small and dedicated to a simple task and that data will reach all agents in a network. Agent networks can be configured as arbitrary graphs, and communication between the agents by message passing on a one-to-one or one-to-many basis.

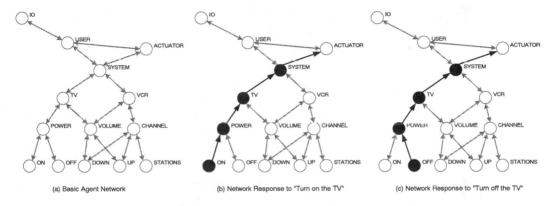

(a) Basic Agent Network   (b) Network Response to "Turn on the TV"   (c) Network Response to "Turn off the TV"

Figure 2. Dejima Distributed Agent Network Controlling a TV and VCR.

One of the strengths of Dejima's distributed agent network is that the intelligence required to perform any particular task is contained within the agents that perform that task. This allows new capabilities to be added merely by adding new agents, and redundant or obsolete capabilities to be removed merely by removing the appropriate agents. Thus, capabilities can be added or removed without disturbing the rest of the system. This property is particularly relevant for the TAS because of the need to add or remove accessors on the fly. For example, an accessible ATM machine will have a selection of built-in access capabilities for common types of disability and a TAP that allows people with special needs to connect their own personal accessor. When a new accessor is connected to the ATM, the agent networks contained in the accessor and the ATM TAP are temporarily integrated into one larger network in which each device has full access to the capabilities of the other. Any new capabilities such as speech recognition or eye tracking that are provided by the accessor are fully integrated into the ATM. When the accessor is removed, the ATM no longer sees the accessor network and therefore the additional capabilities disappear and the ATM reverts back to its normal capabilities (Jonsson & Hodjat, 2001).

In addition to providing the capability for automatically adding and removing accessors to a TAS system, the Dejima Agent Technology also provides natural interaction processing capabilities to disambiguate user inputs.

5. Future work

The TAS is an evolving strategy that saves individuals from the need to configure and learn new interfaces whenever they begin using a new application. Future work is aimed at streamlining the interfaces between the user and their personal accessor and the interface between the TAP and the system to which it is connected. We plan on automating the configuration of TAPs either by recognizing the self-descriptions some manufacturers are beginning to include in their products or by downloading configuration data from the Web. Accessors will be improved by giving them the ability to behave more like humans. Whereas most of today's computer-human interfaces are limited to digital on-off switches (keyboards and mouse buttons) or very basic two-dimensional pointing, future interfaces will utilize multiple channels of simultaneous analog information. Fuzzy processing and neural networks will be used to analyze data derived from many different biological sources. We plan to develop accessors that can detect subtle relationships between the stimuli provided by computer generated outputs and the reactions or intentions of the user. Body Media (Teller, 2000) has developed wearable body monitors that transmit comprehensive physiological data that will be combined with speech and video-derived gesture and tracking data to dramatically change the way we interact with computers.

We explored the use of bioelectric signals (EEG, EOG, and EMG) several years ago but did not pursue it because the results were too unreliable. The interesting EEG signals are severely masked by the EMG and EOG signals and are very difficult to separate with current analog and digital circuitry. Recent developments in neural networks, however, are making it possible to identify and separate the EEG components. (Suppes, Han, & Lu, 1997; Suppes, Han, & Lu, 1998).

6. Conclusions

As Internet applications extend beyond the office and telecommunications arenas we are beginning to see fragmentation among the standards adopted by vendors as they strive to carve out their own territory by introducing proprietary networks for home, automobile, industrial and personal environments. Competing standards are beginning to force users to either stick with one vendor's products or learn to use several different systems. The TAS, combined with Dejima's Agent Technology makes the human computer smart. This not only simplifies the user interface to a particular device, it also provides a strategy for enabling individuals to mix and match information appliances from different vendors in ways that satisfy their own needs, abilities and preferences. The advantages for the individual, therefore, are that they can become proficient with a single interface that can be used with any information appliance. The advantages for the vendor are that they can provide the same information appliance to people with different needs and abilities without needing to customize the user interface.

References

Hodjat, B., Amamiya, M., (2000). Introducing the Adaptive Agent Oriented Software Architecture and Its Application in Natural Language User Interfaces. In *Proceedings of 1st workshop on AOSE,* Limerick

Jonsson, I. (1999). The Decoupled Application Interaction Model, D.A.I.M. In *Proceedings of HCII'99*, Munchen.

Jonsson I., Hodjat, S., (2001). Natural Interaction based on Host Agents, Profiles and Decoupling Modalities from Interaction Devices. In *Proceedings of UAHCI 2001*, New Orleans (current volume)

Norman, D. A. (1990). *The design of everyday things.* New York: Doubleday.

Scott, N. (1991). The Universal Access System. In *American Voice Input/Output Society Conference*, Atlanta.

Scott, N., Jackson, J., Jonsson, I., (1999) The TASCLOUD: a Networked Total Access System. In *Proceedings of CSUN'99* , Los Angeles.

Suppes, P., Han, B., Lu, Z.-L., (1998). Brain-wave recognition of sentences. In *Proceedings of the National Academy of Sciences, 95, 15861-15866.*

Suppes, P., Han, B., Lu, Z.-L., (1997). Brain-wave recognition of words. *Proceedings of the National Academy of Sciences, 94, 14965-14969.*

SynapseAdaptive (1997). http://www.synapseadaptive.com/

Teller, A., (2000). Wearable Body Monitors. In *Proceedings of TTI/Vanguard Conference, "Where Technology and People Meet"*.

Analysis of haptic data for sign language recognition[1]

Cyrus Shahabi, Leila Kaghazian, Soham Mehta, Amol Ghoting,
Gautam Shanbhag, Margaret McLaughlin

Integrated Media Systems Center and Computer Science Department
University of Southern California, Los Angeles, CA 90089-0781
{shahabi, kaghazia, srm, ghoting, shanbhag, mmclaugh}@usc.edu

Abstract

For the past two years we have been addressing the challenges involved in managing the data generated within immersive environments. We together with many other researchers have addressed the management of obvious data types such as image, video, audio and text. However, we identified a set of less familiar data types, collectively termed as *immersidata*, that are specific to immersive environments. In this paper, we focus our attention on analysis of a kind of immersidata known as *haptic* data. We propose to analyze the haptic data acquired from CyberGlove to recognize different static hand signs automatically. The ultimate objective is to understand how to model and store haptic data in a database, for similar types of applications. We propose several techniques to analyze subtle changes in hand signs and words (a series of signs). We show that our techniques can recognize the most important features to distinguish between two letters and several preliminary experiments demonstrate more than 84.66% accuracy in sign recognition for a 10-sign vocabulary.

1. Introduction

Recently, several research efforts have been directed towards immersive environments. Such environments can facilitate the virtual interaction between people, objects, places and databases. Data types such as image, audio, video and text are an integral part of immersive environments and many in the past have addressed their management. However, we identified a set of less familiar data types, collectively termed as *immersidata* [SBE99], that are specific to immersive environments. *Haptic* data-type is a kind of immersidata. CyberGlove is used as a haptic user interface and it consists of several sensory devices that generate data at a continuous rate. The acquired data can be stored, queried and analyzed for several applications. In this paper, we focus our attention on the analysis of haptic data with the objective of modeling these data in a database. An application may need haptic data stored and modeled at different levels of abstraction. For now, we consider three levels of abstraction. First, in [SBK 01], we made our first attempt to model haptic data conceptualized as time series data sets, at the lowest level of abstraction. Second, in this paper, we move a level up from our previous work in using raw haptic data, by trying to understand the *semantics* of hand actions, and we employ several learning techniques to develop this understanding. The application that we focus on is *limited vocabulary American Sign Language recognition* that involves the translation of American Sign Language (ASL) to spoken words. Finally, for the third level of abstraction, there exists a class of applications that need to analyze *pre-processed* data as opposed to analyzing raw haptic data. We intend to study this final level of abstraction as part of our future work. We investigate three different analysis techniques for the automatic recognition of signs and evaluated their accuracy over a 10 sign vocabulary. We use C4.5 Decision Tree, Bayesian Classifier and Neural Networks for the recognition of static signs. Our experiments show that Bayesian Classifier can classify haptic data with an average error of 15.34%. Bayesian Classifier appears to be the fastest classification technique providing the best classification accuracy for our experiments. Our research is distinct and novel in the following three aspects.

To begin with, we are distinct with respect to the framework we have used for our research and experiments. All our analysis and experimentations were performed on raw haptic data without any kind of pre-processing. In addition to a novel framework, we have taken a new approach for modeling haptic data, which is based upon learning techniques such as Decision Trees, Bayesian Classifier and Neural Networks. The comparison of these three techniques within the same environment and experimental setup is also novel and unique. Finally, the ultimate objective of our research is to model and store haptic data at different levels of abstractions.

The remainder of this paper is organized as follows. Section 2 discusses how we acquire haptic data using CyberGlove. Section 3 explains the three learning techniques that we use for sign recognition. The results of our experiments in

This research has been funded in part by NSF grants EEC-9529152 (IMSC ERC) and ITR-0082826, NASA/JPL contract nr. 961518, DARPA and USAF under agreement nr. F30602-99-1-0524, and unrestricted cash/equipment gifts from NCR, 3M, Intel and SUN.

comparing the three analysis techniques have been reported in Section 4. Section 5 covers various other research efforts in the area of sign language recognition. Finally, Section 6 concludes this paper and provides pointers to our future research plans.

2. Data Acquisition

The development of haptic devices is in a very early stage. We have focused our research and experiments on the CyberGrasp exoskeletal interface and accompanying CyberGlove from Virtual technologies. It consists of 33 sensors as described in [SBK 01]. We developed a multi-threaded double buffering technique to sample and record data asynchronously. We used 10 alphabets (A to I and L) from the American Sign Language (ASL) for our experiments. We term each of these 10 alphabets a *sign*. The 22 sensor values (excluding sensors 23 to 33 in [SBK 01]) are recorded in a log file for each sign made by a *subject,* termed as a *session.* Each session log file contains thousands of rows of sensor values sampled at some frequency, which depends on the sampling technique used. We denote each such row as a *snapshot.* We thus have thousands of snapshots for each session.

2.1 Sampling Techniques

In order to record several snapshots for each static sign, made within a session, we need to sample the values of sensors for each subject making a sign. Thus sampling the sensors at a rate, which would lead to lower storage space requirements and better accuracy is central to the task of data acquisition for any haptic device. We designed and implemented the following sampling techniques for our experiments (see [SBK 01] for more details).

First, Fixed Sampling was used to record the session. This technique is wasteful since it records data for each sensor at each possible opportunity regardless of the sensor type or the semantics of the session. Next, Group Sampling was used to record the session. We can isolate a sampling rate for each group and acquire data at different rates, based upon the group membership for each sensor. The advantage of this technique is its improvement over the fixed sampling technique by further reducing storage space and transmission requirements, while maintaining accuracy. Finally, Adaptive Sampling was used to sample the sensors. This is a dynamic form of sampling where we try to find an optimum rate r_{ij} for each sensor i during a given window j of the session. The benefit of this technique is that the sampling rate changes with the nature of the sessions. This makes the approach more efficient and robust as compared to the fixed or group sampling approach. In [SBK 01], we provide details on these sampling techniques and the various tradeoffs among factors like bandwidth, storage and computational complexity.

3. Classification methods

In this paper we explore three different classification techniques, and evaluate the accuracy of each technique to detect 10 different hand signs. The employed techniques are C4.5 Decision Tree, Bayesian Classifier and Neural Networks. Each classification technique is implemented in two different stages, the training phase and the recognition phase.

3.1 C4.5 Decision Tree

Tree induction methods are considered to be supervised classification methods, which generate decision trees derived from a particular data set. C4.5 uses the concept of information gain to make a tree of classificatory decisions with respect to a previously chosen target classification [Quinlan 93]. The output of the system is available as a symbolic rule base. The cases, described by any mixture of nominal and numeric properties, are scrutinized for patterns that allow the classes to be reliably discriminated. These patterns are then expressed as models, in the form of decision trees or sets of *if-then* rules which can be used to classify new cases, with an emphasis on making the models understandable as well as accurate [Quinlan 93]. For real world databases the decision trees become huge and are always difficult to understand and interpret. In general, it is often possible to prune a decision tree to obtain a simpler and more accurate tree.

We employed C4.5 Decision Tree because it provides a model to build a sign recognition language. In addition, decision trees in general and C4.5 in specific provide results as a set of understandable and interpretable rules. Finally, C4.5 has been used as a benchmark in several other works in machine learning, artificial intelligence and data mining literature. C4.5 complexity is $O(nt)$ where t is the number of tree nodes and the number of tree nodes often grows as $O(n)$ where n is the number of sessions. The complexity for non-numeric data would be $O(n^2)$, for numeric $O(n^2 logn)$, and for mixed-type data, somewhere in between.

3.2 Bayesian Classification

Bayesian Classifier is a fast-supervised classification technique. Bayesian Classifier is suitable for large-scale prediction and classification tasks on complex and incomplete datasets. Naïve Bayesian Classification performs well if the values of the attributes for the sessions are independent. Although this assumption is almost always violated in practice, recent work [DP96] has shown that naïve Bayesian learning is remarkably effective in practice and difficult to improve upon systematically. We have decided to use the naive Bayesian Classifier in our application, for the following reasons. First, it is efficient for both the training phase and the recognition phase. Second, its training time is linear in the number of examples and its recognition time is independent of the number of examples. Finally, It provides relatively fine-grained probability estimates that can be used to classify the new session [Elk97]. The computational complexity of Bayesian Classification is fairly low as compared to other classification techniques. Consider a session with f attributes, each with v values. Then with naive Bayesian classifier with e sessions, the training time is $O(ef)$ and hence independent of v.

3.3 Neural Network

We use Neural Networks for the recognition of static signs with a limited vocabulary. Supervised learning is being used for the classification. A Multi Layer Perceptron is a feed-forward network with one or more layers of nodes between the input and output layers of nodes. These additional layers contain hidden nodes that are not directly connected to both the input and the output nodes. The capabilities of the multi-layer perceptrons stem from the non-linearity used in these nodes. The number of nodes in the hidden layer must be large enough to form a decision region that is as complex as required by a given problem. A Multi Layer Perceptron (MLP) is trained using the Supervised-learning rule. The most commonly used algorithm for such training is the Error-back-propagation-algorithm. A three-layer perceptron can form arbitrarily complex decision regions. Hence, usually, most problems can be solved by 3-layer (1 hidden layer) perceptrons.

3.3.1 Implementation of Neural Network Classification over Static Signs:

We used 22 nodes for the input layer, in addition to one threshold value node for the next layer. The hidden layer required 10 nodes. Output layer required 10 nodes, each corresponding to one sign. With each of the 23 inputs (22 haptic glove values + 1 threshold) connected to each of the 11 hidden layer neurons (10 neurons + 1 threshold), and again each of these hidden layer neurons being connected to each of the 10 output neurons, the total number of weights in the network is equal to $(23 \times 10)+(11 \times 10) = 340$ weights.

For our experiments, we establish the cardinality of the training set, to achieve a good generalization as propounded in [VC71], approximately 10 times more, to cross the "*VCDim*" threshold. Hence, we have training data set of 10 subjects, each making 10 signs, and for each session we have 40 snapshots, resulting in 4000 sets of sensor values. We train the network for 500 epochs. We generate pseudo-random weights, the range of which is −1.0 to +1.0. We strived to make the neural network learn on raw haptic data so that it learns to handle noisy data. This can be useful when we try to use the classifier in real-time immersive applications. Work done by [WH99] yields very good results on the training set, but the ability of this approach to be generalized needs to be ascertained. Our approach provides a good promise for an overall generalization.

4. Performance evaluation

4.1 Experimental Setup

We conducted several preliminary experiments to evaluate each classification method. Fifteen subjects were asked to generate the following signs: *a, b, c, d, e, f, g, h, i* and *l*, and data has been stored in a database. The signs *j and k* are complicated and taking the novice subjects into consideration, the signs were skipped for simplicity reasons. To evaluate each algorithm we used the cross validation technique. We split the data into 3 sets, trained the system using two of the sets and conducted the tests using the third set. We implemented the test procedure in a round robin fashion (shuffling the training and test sets) and computed the average error.

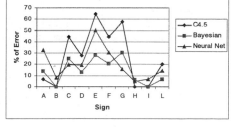

4.1.1 Storage of the Input

Neural Network is trained using 4000 snapshots, as described earlier. These data are retrieved from 100 session log files (10 subjects, 10 signs each, 40 different snapshots). For our experiments on static signs, we analyzed the recorded log files stored in a database and extracted the snapshot that has the sensor values consistent over a substantial period of time. Classification algorithms can then be developed using incremental learning.

4.2 Results

Figure 1: Sign recognition error

Figures 1and 2 compare the average recognition error for each sign using the three different classification techniques. The naive Bayesian Classifier has the highest average accuracy with 50 training examples: 84.66% (with standard deviation sd = 2.94).

Sign	A	B	C	D	E	F	G	H	I	L
C4.5	**6.6**	**0.0**	44.4	27.7	64.4	44.4	57.7	**0.0**	**0.0**	20.0
Bayesian	14	**0.0**	25	**13.06**	**28.4**	**20.83**	30.52	6.4	**0.0**	**6.4**
Neural Net	32.8	7.87	**19.6**	19.5	50.04	30.55	**15.73**	4.9	6.67	14.24

Figure 2: Sign recognition error comparison

	Error	Standard Derivation
C4.5	22	8
Bayesian	15.34	2.94
Neural Net	20.18	7.92

Figure 3: Overall classification error

4.3 Analysis

The Bayesian Classifier shows a very efficient and accurate result as compared to other classification techniques. The results of our experiments illustrate that C4.5 Decision Tree is not suited to the task of sign recognition. Both Neural Networks and C4.5 have a large amount of variation in their performance. However, most often, C4.5 results are more interpretable and understandable. In contrast, the Neural Network architecture and procedure are not interpretable, and it is similar to a black box in which case, we only have access to input and output. Our experiments indicate that all of the classifiers relatively performed quite well on signs 'b', 'h', 'i' and 'l'. Inspecting the signs, it occurs that it was intuitive for subjects to perform these signs most consciously. Considering all the signs as points in 22-dimension hyperspace, and computing the Euclidian distance among them, we realized that on the average, these four signs are quite apart from the rest of the signs, which justifies our observation. On the other hand, letter 'e' was quite close in distance to all the other signs and hence all classifiers were confused one way or the other with the recognition of letter 'e'.

The performance variation of individual classifiers over the signs can be traced back to the performance characteristics of each classifier. A neural network, inherently tries to draw crisp distinguishing boundaries between groups of signs in the 22-dimensional hyperspace. Hence, it distinguished all the signs made when the hand is in the horizontal position (i.e., 'c', 'g' and 'h') quite well. Note that although C4.5 was the best classifier for letter 'h', it had the minimum recognition error among the other letters with the neural net. With C4.5 and Bayesian Classifiers, the main assumption is that all features in a given space are independent. In general any strong dependency increases the level of error for both methods, while a low degree of dependency among features might be negligible. Further, since C4.5 produces decisions based on a set of 'if-then' rules, it tends to be relatively rigid, resulting in a high standard deviation as well as a high overall error. Since Bayesian classifier decides based on probability distribution of the input samples, it tends to perform quite well overall despite intuitive variations in performance of signs by different subjects. We illustrate that even with a small pool of snapshots, a fast learner such as Naïve Bayesian Classifier and an appropriate I/O design we can achieve an acceptable performance.

5. Related work

Various research groups worldwide have been investigating the problem of sign recognition. We are aware of two main approaches. Machine-Vision based approaches analyze the video and image data of a hand in motion. The Haptic based approaches analyze the haptic data from a glove. These efforts have resulted in the development of devices such as CyberGlove. Due to lack of space, we refer the interested readers to [WH99] for a good survey on vision based sign recognition methods. Using gloves and haptic data, Fels et. al [FH95] employ a VPL Glove to carry out sign recognition. Sandberg [San97] provides an extensive coverage and employs a combination of Radial Basis Function Network and Bayesian Classifier to classify a hybrid vocabulary of static and dynamic hand signs. One more variant exists [MT91] using Recurrent Neural Networks to classify Japanese sign language. Hidden Markov Models are popular here too, which is reflected in [NW96] and [LY96]. The work of Lee et. al in [LY96] is particularly relevant because it presents an application for learning of signs through Hidden Markov Models taking the data input from CyberGlove. Other neural network algorithms- Radial Basis Function Network, Orthogonal Least Squares and Self – Organizing maps have also been tested on various kinds of data-glove inputs, in [LIN98] and [IM99].

Our work is distinguished from all of the above-mentioned works, because we provide a complete system including I/O unit, data acquisition module, database structure and classification methods for American Sign Language recognition. All our analysis is carried out on raw haptic data. We are the first to use Decision Tree for the analysis of haptic data. We are also the first to use Bayesian Classifier for *raw* (i.e. no pre-processing) haptic data analysis. Taking our framework into consideration, we are also the first to use and compare Back Propagation Neural Networks with Bayesian Classifier and C4.5 Decision Tree for the recognition of static signs.

6. Conclusion and future work

In this paper, we analyzed three different classification techniques for sign language recognition. We showed that Decision Tree, Bayesian Classifier and Neural Networks could be used for American Sign Language recognition. Bayesian Classifier proved to be the fastest classification technique amongst the three classification techniques. It also proved to have the best classification accuracy for static sign recognition. We carried out several preliminary experiments and the results of our experiments suggest that Bayesian Classifier can be used to develop a real time sign language recognition system. However, more work needs to be carried out in order to establish the validity of our results, which are very encouraging in the early stages of experimentation.

We intend to extend our work in several ways. First, we intend to investigate Time Delay Neural Networks and Evolving Fuzzy Neural Networks for the recognition of *dynamic* signs. It would be interesting to compare the effectiveness of these two techniques. Second, we want to utilize the lessons learned in our analysis of haptic data to model haptic data in a database. Our analysis would determine what data we need to store at which level of abstraction for a given application. Third, we would like to analyze haptic data at the third level of abstraction, which requires us to analyze pre-processed haptic data. Finally, we propose to use shape recognition techniques for the recognition of dynamic signs based upon a fixed sign language vocabulary. Using a shape to represent the dynamic part of a sign would let us view the dynamic sign in a time independent manner.

References

[DP 96] P. Domingos, M. Pazzani.Beyond Independence: Conditions for the Optimality of the Simple Bayesian Classifier Proceedings of the Thirteenth International Conference on Machine Learning (pp. 105-112), 1996. Bari, Italy: Morgan Kaufmann.

[ELK 97] Elkan, C. (1997), Boosting and Naive Bayesian learning, In proceeding of KDD -97, New Port beach, CA.

[FH95] S. Fels and G. Hinton. Glove-talkii: An adaptive gesture-to-format interface. In Proceedings of CHI95 Human Factors in Computing Systems, 1995.

[IM99] Ishikawa, M.; Matsumura, H, Recognition of a hand-gesture based on self-organization using a DataGlove. Neural Information Processing, 1999. Proceedings. ICONIP '99. 6th International Conference on , Volume: 2 , 1999 Page(s): 739 -745 vol.2

[LIN98] Daw-Tung Lin, Spatio-temporal hand gesture recognition using neural networks Neural Networks Proceedings, 1998. IEEE World Congress on Computational Intelligence. The 1998 IEEE International Joint Conference on , Volume: 3 , 1998 Page(s): 1794 -1798 vol.3

[LY96] C. Lee and X. Yangsheng. Online interactive learning of gestures for human/robot interfaces. In Proceedings of IEEE International Conference on Robotics and Automation, pages 2982-2987, 1996.

[MT91] K. Murakami and H. Taguchi. Gesture recognition using recurrent neural networks. In Proceedings of CHI91 Human Factors in Computing Systems, 1991.

[NW96] Y. Nam and K. Wohn. Recognition of space-time hand-gestures using hidden markov model. In Proceedings of ACM Symposium on Virtual Reality Software and Technology, pages 51-58, 1996.

[Quinlan 93] J.R. Quinlan, C4.5: Programs for Machine Learning, Morgan Kaufmann, 1993.

[San97] A. Sandberg. Gesture recognition using neural networks, 1997.

[SBE99] C. Shahabi, G. Barish, B. Ellenberger, N. Jiang, M. Kolahdouzan, S. Nam and R. Zimmermann, Immersidata Management: Challenges in Management of Data Generated within an Immersive Environment, In proceedings of the Fifth International Workshop on Multimedia Information Systems, 1999

[SBK 01] C. Shahabi and M. R. Kolahdouzan and G. Barish and R. Zimmermann and D. Yao and K. Fu and L. Zhang, Alternative Techniques for the Efficient Acquisition of Haptic Data, to appear in Sigmetrics'2001.

[VC71] Vapnik, V. N. and Chervonenkis, On the uniform convergence of relative frequencies of events to their probabilities. Theory of Probability and its Applications, 16:264--280. 1971

[WH99] Y. Wu and T. Huang. Vision-based gesture recognition: A Review. In Proceedings of the International Gesture Recognition Workshop, pages 103-115, 1999. [RJ93] Rabiner, L. and Juang, B.-H. (1993). Fundamentals of Speech Recognition, Prentice Hall, Englewood Cliffs, NJ.

Bilingual Emacspeak Platform
-- A universal speech interface with GNU Emacs --

Takayuki Watanabe[a,A], Koichi Inoue[A], Mitsugu Sakamoto[A], Masanori Kiriake[A], Hirohiko Honda[b], Takuya Nishimoto[c], Tuneyoshi Kamae[d]

[a] Department of Information Science, Shonan Institute of Technology,
1-1-25 Tsujido-NishiKaigan, Fujisawa, Kanagawa, 251-8511, JAPAN
[b] Dept. of System and Communication Eng, Shonan Institute of Technology,
[c] Dept. of Electronics and Information Science, Kyoto Institute of Technology,
[d] Stanford Linear Accelerator Center, Stanford University,
[A] ARGV (Accessibility Research Group for the Visually-Impaired)

Abstract

The Bilingual Emacspeak Platform is the system that aims to extend T.V. Raman's Emacspeak to bilingual, Japanese and English, and to two operating systems, Linux and Windows, to meet the requirements of Japanese visually impaired (low-vision and blind) users who want to use Emacs. Emacspeak is a self-voicing GNU Emacs editor that has a concept of Auditory User Interface to provide information in the best way for the sense of hearing. The current system is developed by both sighted and visually impaired Japanese users. The system, consisting of two bilingual software speech servers for two operating systems and a bilingual extension of Emacspeak, is distributed as an open source program. The current system is still under development but the alpha version was released at http://www.argv.org/bep/. By using the current system, Japanese visually impaired users can use Emacs at any place either under Windows and Linux without any external devices. They can read and write Emails, develop programs, and browse WWW pages outside their home and office.

1. Introduction

Personal computers in 1980s used a character to transfer data to and from a user. Screen-readers at that time could easily obtain the text information that should be outputted to a user. In 1990s the advent of Microsoft Windows, which uses a point-and-click user interface, popularized personal computers. At the same time Internet became popular and these computers and Internet produced a so-called the information age. It was not until Windows 95 that screen-readers for Windows were sold in Japan. Graphical User Interface is easy to use for sighted users, whereas it is hard to use for visually impaired. There are some Japanese visually impaired users who still use DOS when they do not need to use Windows applications. Unix is another high-capability OS but there have been no Japanese screen-readers until 2001, which obliges a user to have to login the system from a remote terminal that has a speech output.

In late 1990s T.V. Raman developed a new speech system Emacspeak, a self-voicing GNU Emacs. Emacs is not only an editor but also a platform of Emacs Lisp and there are many useful applications written in Emacs Lisp. Emacspeak adds an audio feedback to these applications in a manner best suited for the sense of hearing, i.e. as shown in Figure 1, it has AUI (Auditory User Interface) (Raman, 1994; Raman, 1997). For example, Emacspeak speaks a calendar using calendar's two-dimensional information. It uses various voices to effectively display various kinds of information such as comment strings, key words, and quoted strings of source code.

In 1999, we, the sighted and the visually impaired who love Emacs, launched the Bilingual Emacspeak Project (Watanabe, 2000a; Watanabe, 2000b) that would extend Emacspeak into Japanese and English and run under Windows and Linux. The aim of the current system is to assist Japanese visually impaired graduate students, researchers, and programmers to use computers at high levels, which is not offered by existing screen-readers. The current system is bilingual because Japanese users need to read English as well as Japanese at workplace and Internet. It runs under Windows because there are many users who use Windows. It also runs under Linux because

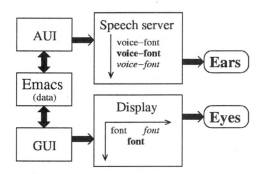

Figure 1. Emacspeak uses AUI to display the information to the sense of hearing. It uses various voice-fonts to express various kinds of information.

the original Emacspeak works only under UNIX and UNIX could be an alternative accessible OS for the handicapped.

Unlike Windows whose GUI cannot be isolated from OS itself, UNIX has a text-based console and most of the operations and applications can be carried out under this console, which makes it easy to add a speech feedback to UNIX. Some UNIX families and many applications are open source; therefore, it might be easy to add assistive functions to UNIX.

2. Bilingual Emacspeak

2.1 Structure

As shown in Figure 2, the current system consists of 2 parts, an Emacs Lisp program package that acts as AUI and a software speech server. Emacs Lisp program represents the bilingual version of Emacspeak. The speech server, which accepts commands of DECtalk format and text in Japanese Shift_JIS character encoding scheme, controls English and Japanese text-to-speech (TTS) engines. Windows' speech server, which is a modification of G. Bishop's one, uses Microsoft's American English TTS engine and a commercial Japanese TTS engine that are conformable to Microsoft Speech API 4 (Watanabe, 2000a). Linux's speech server uses a Japanese TTS engine developed by us from a commercial software development kit (Watanabe, 2000b). It speaks English with a Japanese TTS engine using a Japanese-English dictionary made by J. Ishikawa.

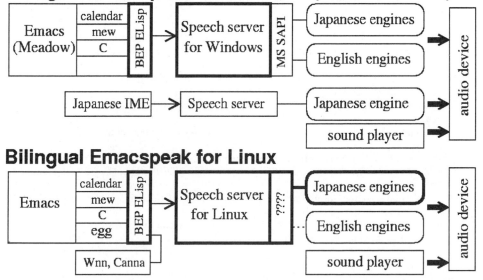

Figure 2. The structure of Bilingual Emacspeak. Bold boxes are modules developed by the current project. Meadow is a Windows' porting of Emacs 20.x, Mew is a Japanese Email package, and Wnn and Canna is a Kana-Kanji translation server.

The current version identifies a language in a speech server and automatically uses an appropriate engine according to the language. Japanese TTS engines are used if there is a leading byte of two-byte characters in the text to be spoken. This language identification will be carried out in an Emacs Lisp level in the near future to make a robust and flexible discrimination of the language in multilingual information.

Sound mixing is carried out in a WDM sound driver under Windows, while it is carried out through the 'esound' (Enlightened Sound Daemon) under Linux, which enables a simultaneous output of voices and sounds of auditory icons.

Audio feedback of Kana-Kanji translation in an Input Method Editor is treated differently for two operating systems. Windows version uses a Windows' screen-reader and does not produce any audio feedback inside Emacs, while Linux version gives an audio feedback in an Emacs Lisp level of `egg' package.

2.2 Auditory representations of Japanese characters

Japanese language has Kanji (a Chinese character), Hiragana, Katakana, and other multi-byte characters. As Kanji is homonym, there are multiple Kanji characters that correspond to one pronunciation. There also are same Hiragana and Katakana that have exactly the same pronunciation. Thus a user cannot identify these characters by hearing the pronunciation and needs additional information. Bilingual Emacspeak is to use two mechanisms to give this information to the user. One is an explanatory reading of Kanji, which is the same function as ordinary Japanese screen-readers have. The current system has 3 fields for one Kanji. They are very brief explanation used for reading a character according to a cursor movement, short explanation used for explaining a Kanji at Hiragana (Kana)-Kanji translation of a Japanese input method, and long explanation used for other cases. The dictionary of this explanatory reading is based on the one used in J. Ishikawa's screen-reader. The other mechanism is audio formatting, or use of various voice-fonts. For example, Hiragana and Katakana will be read with different voices or different pitch of the same voice.

2.3 Distribution of the current system and its merit

The current system is still under development but some members of the project use the system daily. We decided to distribute it as an alpha version because the current version already has some unique features that are not provided by existing Japanese screen-readers as follows.

1. Users can read English and Japanese with native engines. Japanese graduate students, programmers, and researchers have much opportunity to read English information; however, there has been no speech system that meets this requirement.
2. Users can use Emacs in the same manner regardless of the operating system. They can read Email with the `mew' package not paying attention to the difference of OS.
3. Users can use Email and other applications inside Emacs world, which is very useful if they know Emacs.
4. Voice-fonts and auditory icons of Emacspeak make it easy to recognize information through the sense of hearing.
5. Japanese students who study a programming language, C for example, can use the system to edit, build, debug, and execute the program with the appropriate speech feedback and expressive sounds of auditory icons.
6. W3, a Web browser for Emacs, can use aural style sheets of CSS2 (Cascading Style Sheets, level 2). Users can browse web pages with various voice-fonts.
7. Japanese Linux users can use Linux computers without using an external speech synthesizer. They can use a notebook computer at any place.

3. Concluding remarks

3.1 Problems and prospects

There are several problems in the current system.

1. Lack of free Japanese TTS engines. TTS engines are basic components of a speech system; therefore, a free engine must be available to prevail a free speech system.
2. The speech server for Linux is not bilingual. There is no standard speech interface for Linux; therefore, it is difficult to handle different engines at the same time. It is strongly desirable that a free multilingual (American, European, Asian, etc) TTS engines with a unified interface would be available for any operating systems.
3. Localization of Emacspeak is still in progress.
4. Emacs is not an application that can be used by anyone but an application that requires some skills and efforts to a user; therefore, enough documents and supports would be required to use the current system especially for Windows users.
5. No Braille output. There are many users who need Braille display in some cases such as programming.

3.2 Multilingual audio desktop

Raman developed Emacspeak by adding a speech feedback to Emacs. We extend it into bilingual and to multi-platforms. Emacs itself can treat multilingual texts at the same time; therefore, it would be nice to extend Bilingual Emacspeak into multilingual one to meet the requirements of international users, especially of Asian users.

3.3 Linux accessibility

Linux is a free operating system that can be used as an alternative OS to Windows. Linux becomes popular for general users and there are many distributions and applications that use GUI, which might make Linux hard to be accessed by the visually impaired users. Thus, we must claim the importance of accessibility and universal design to the Linux community.

4. Summary

Bilingual Emacspeak provides a new accessibility for Japanese visually impaired users who want to use Emacs. They can use Emacs' high capabilities that are already used by sighted users. Bilingual Emacspeak is an extension of Emacspeak to bilingual and multi-platforms. It adds some functions that are necessary for Japanese. Bilingual Emacspeak is distributed as open source except commercial TTS engines. In this regard, we need free or open-source Japanese TTS engines. Development is still in progress but an alpha version was released at http://www.argv.org/bep/. Please contact bep-contact@argv.org if you are interested in this system.

References
Raman, T.V. (1994). *Audio system for technical readings.* Ph. D thesis, Cornell University.
Raman, T.V. (1997). *Auditory user interfaces -- toward the speaking computer --*. Kluwer Academic Publishers.
Watanabe, T., Inoue, K., Sakamoto, M., Honda, H., & Kamae, T. (2000a). Bilingual Emacspeak for Windows -- a self-speaking bilingual Emacs -- *Tech. Report of IEICE SP2000*, 100(256), 29-36, (in Japanese).
Watanabe, T., Inoue, K., Sakamoto, M., Kiriake, M., Shirafuji, H., Honda, H., Nishimoto, T., & Kamae. T. (2000b). Bilingual Emacspeak and accessibility of Linux for the visually impaired *Proceedings of Linux Conference 2000* Fall, 246-255, (in Japanese).

Configuring Social Agents

Charlotte Wiberg, Mikael Wiberg

Department of informatics & Center for digital business, Umeå University
901 87 Umeå, Sweden
[colsson, mwiberg] @informatik.umu.se

Abstract

Social agents have recently been more frequently used in the user interface. However, so far not many studies have been conducted on what impact such interfaces have on users behavior. This paper discusses this and reports on empirical findings, which focus on impact of social agents on user behavior. We talk of *social agents* as interfaces that act autonomously but are related to the actions of the user. However, to really figure out what social impact these interfaces have on humans, we discuss what characteristics of social agents that should be possible to configure, in order to establish, maintain and develop a fruitful relation with the user. In order to do so, we needed to explore the impact for real users. The exploration of the impact of social agents such as BonzyBuddy the Parrot and Bob, the Paper-clip guy, was done empirically through observations and interviews with users. Based on empirical data collected in the study, a user-agent interaction model was constructed. The model illustrates three dimensions for configuration of social interfaces. Given the interaction model the two agents investigate are discussed followed by a discussion on what implications these observations has for design of social agents. Having identified the need for self-examining and self-adapting social agents and related problems we then conclude the paper and points at some future work.

1. Introduction

Social agents are applications that act both autonomously and on behalf of the user. They are becoming more and more common (Jensen, 2000). The purpose of social agents is twofold. Social agents should help users perform meaningful tasks. At the same time a social agent should be joyful and pleasurable for the user to interact with, a situation not always compatible with traditional approaches in HCI (e.g. Jordan, 2000; Olsson, 2000; Picard, 1998). Since there is an autonomous part of these agents the relation between help and fun can turn into a paradoxical situation if the relation between the user and agent is not configured properly. To grasp what kind of configuration is needed to avoid this paradox empirical research is needed on how social agents influence users and what users expect from their electronic pals. This paper deals with the question of what kind of impact social agents have on users and how that knowledge can be fruitfully reframed into a model of how social agents should be configured to enable the establishment, maintenance and development of a sound relation to users. The rest of this paper is structured as follows: First, the concepts of social agents is outlined followed by a brief description of two applications, BonziBuddy and Bob. After that an empirical study of the use of BonziBudy and Bob and their impact on users is described, followed by a summary of the main findings concerning the use of the social agents. Finally, a social agent interaction model is outlined, based on the empirical findings. Our intention with the developed model is that it could provide designers of social agents with a tool to think about how to allow users to configure their relation to their social agents. Future work includes the design and evaluation of a social agent that is built according to the model. Hopefully, that work can give more insights on fruitful ways to design social agents that can be configured by the user in a manner so that a fruitful relation can be established, maintained and developed.

2. Social agents

In the following discussion the term social agents refers to applications that are focused on the experience and enjoyment of the user. Instead of supporting a specific task related to work, social applications focus on the satisfaction of the user. This type of applications is mostly related to, what we call, *interactive intelligent agents*. The term *agent* is very broad. A more specific definition of the concept of *software agent* is given as follows:

> *"[Agents are] ..semi-intelligent computer programs which assist a user with the overload of information and the complexity of the online world." (Schneiderman & Maes, 1997)*

The difference between interface agents and other type of applications is mainly their ability to act autonomously on behalf of the user in different situations (Dehn & van Mulken, 2000). Also, intelligence is something that relates to some of them (all of them, according to Dehn & van Mulken, 2000). Intelligence, according to Dehn & van Mulken, is related to their ability to perform tasks delegated to them in a context- and user-dependent way.

Mainly, what we put in this category are agents, that are (1) intelligent, in that they are aware of the user's activities, and (2)interactive in the sense that users are aware of their existence on the screen. The last point excludes, for instance, agents that work as background processes, without user's knowledge, as for instance the feature in Word 2000 that keeps track of your menu choices and changes the menus from this.

Examples of social agents are *Bob, the Paper-clip guy*, his 'Mac-Cousin' *Max*, and *BonziBuddy the Parrot*, as well as other interactive assistants and pets. The examples of this category are numerous and also, the number of them are growing on the web. The purpose of these types of applications seems twofold; (1) to help and assist in the performance of the users activity, and (2) to do this in a social and more or less, enjoyable way. By this we mean that the 'feature' acts in such a way that, for instance the help should be helpful *and* fun. These two social agents are further described below:

BonziBuddy - the Parrot

BonziBuddy the Parrot is an application easily downloaded from the web. This is an example of a social agent that tries to combine different types of input and output. BonziBuddy uses gestures and pet-like moves, as well as sound to call upon your attention in different situations. The pet metaphor is obvious and can help you to surf on the web, tell you jokes, read your e-mail to you and do all kind of 'tricks'. He can keep the user occupied for hours if the user so chooses. Also, there are more features added all the time and more songs, jokes and so on to buy from the software company that has the rights of *BonziBuddy*.

There is an uncertainty about how autonomous *BonziBuddy* is. Information in the application itself tells you that the parrot for instance is able to learn patterns in your web surfing, and predict how you would rank choices of web sites. All this based on your earlier surfing patterns. However, whether this is the case or not is questionable. The parrot gives a quite personal impression though, due to the fact that it repeatedly calls you by name, as you have printed it into the application.

Bob - 'the paper-clip guy' and Max - his Mac-cousin

This agent is wide spread and known all over the world as a person-like paper-clip moving on screen when one uses *MS-Word*. It is a part of the help system in the application. It is not as noisy and messy as the parrot described above, but there are some similarities between the two. The similarities are that they are both person like and that they are animated. *Bob*, however, is more connected to the specific activities in a single application though, while the parrot is more of an social agent not specifically connected to a specific application.

The appearing social agent is the default choice when downloading the application. However, the feature is easy to take away and there is also some ways in which the user can customize the social agent. In general, this agent is more anonymous on the screen compared to the above-mentioned parrot.

3. Empirical study

The study lasted for 3 months, and it was conducted by use of participant observations and semi-structured interviews. The number of users in the study was all together ten. The main purpose of the observations was to investigate the user-agent relation. The purpose of the interviews was to get a more genuine and longitudinal view of the whole process - from downloading to finally end of usage of the agent. Another strong reason to use interviews is to collect the rate of the subjective satisfaction, and then interviews could be a proper choice (Nielsen, 1993, p. 209 ff.). The users in the study could be divided into two groups; (1) users that voluntarily downloaded the application, and (2) users that got the application when buying or downloading other types of applications or visiting web sites with attached 'social' features. In the study, both types were represented. Our intention of using both types of users was initially to find two types of sets of aspects. However, so far, the indications from studies show that there are mostly the same results coming from the two groups. The spectrum of findings are, of course, wide, however there are some significant findings to extract. These are related to a timeline of usage.

452

The timeline could be divided into three stages. (1)*The phase of acquaintance*, when the social application either were downloaded or followed another application. (2) *Maturing phase* where the user tried and elaborated with the social application, and finally (3) *End phase* which occurred in those cases when the second phase had resulted successfully, i.e. the user had accepted the existence and functionality of the companion.

4. Results

Below we outline three identified phases of the user-agent relation (i.e. The phase of acquaintance, the Maturing phase, and finally the End phase).

Findings related to the timeline are first surprise. The users report on moments when they giggled surprised when the feature did something. On the question of how to measure the social agent in this phase, autonomy came up. This aspect was very important in the first phase.

> *"Why I started to use it at the first place? Because it was cool and that it surprised me! I was laughing at the screen and I had found myself a new friend. I felt as he was a completely autonomous feature. It was really fun!"*

Second, findings from the second phase show that elaboration and number of embedded features were of importance.

> *"I wanted to test what BonzieBuddy could do. They said that he had build in intelligence in what I searched for on the web. I tried to find out if and how that worked "*

Finally, in the last phase, if this occurred at all, aspects as easily used features for shutting of, quieting down and putting the application in the back of the screen were of importance.

> *"I still use the feature [in this case a virtual pet]. I tend to give him less attention nowadays though... However, one good feature is his sleeping mode, in which he relaxes if I do not give him any feedback. Also, the sounds in the application is now turned off"*

In the cases where the last phase did not occur, situations of disappointment of the application in one way or the other were often reported.

> *" I stopped to use the paper-clip guy when it couldn't understand my questions that I typed into the field. The commercials said that I could ask questions as I talked. That did not work so I found no use of him. He was just in my way."*

5. A user-agent interaction model

Based on our collected empirical data, an agent interaction model was constructed. The model illustrates the relations between a user and his/her agent. Hopefully the model can serve as a basis for discussing possible configurations of agents. Given the three identified phases above a social interface needs to be configured on at least three different dimensions to establish, maintain and develop a fruitful relation with the user. These three dimensions include a *distribution* dimension, an *access* dimension, as well as a *filtering* dimension. As stated by the quote above from phase 1 *distribution* was important (e.g. the agent should reach out and surprise the user, suggest things, play a song, etc and be very entertainment. In the next phase however, the entertainment requirements where exchanged for features more related to the task that the user tried to perform. In this phase the user needed *access* to certain information and be able to quickly find relevant web pages etc. As a consequence the interface does not need to be much entertaining in this phase. However, ease of use is very critical. Finally, as phase three illustrates it is important that the agent can adopt of the way that the user wants to interact with the computer (e.g. through some kind of defined *filters* for the interaction), otherwise it will just be time consuming and not used at all. The figure below illustrates how these three dimensions can be plotted in the user-agent relation.

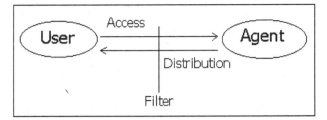

Figure 1: The interaction model based on the empirical findings.

Related to the two agents described we, based on the model above, assumes that *BonzyBuddy* could be much more used if filtering of different kind could be implemented. As some users find the social agents loud speaking and even disturbing in some situations they could gain from customization of sound in a transparent way. *BonziBuddy* also provides some features concerning configuration of the distribution level by allowing the user to install add-ons to the agent. However, how the user is able to access the agent is not configurable. Bob, on the other hand is much more low on the agency scale and is not as critical as the parrot. However, in order to give the paper-clip more status and intelligence, the filtering might be the most critical of the three aspects. If the user should trust the paper-clip to be of more help than it is today and carry specific information about the user, filters for how the agent access information about the user must be in the control of the user.

6. Implications for design of social agents

As the phases identified suggest a user-agent relation grows over time. Given the three phases outlined and the dimensions presented in the interaction model we therefore suggest that social agents should be able to analyze their own interaction history with their user and then self-adapt their way of interacting with the user accordingly. For instance, when the user-agent relation is in phase 1 the agent could be very active. Then, as the phases pass along the agent could transform to be more task-oriented. However, this suggestion also opens up questions concerning if and how the user then should be able to trigger the agent to be regressive and go back one or several phases (i.e. the question of levels of autonomy of agents). One way of doing that, of course could be to let the user define a current state in the relation as a mode and then be able to shift back to that mode whenever appropriate.

7. Conclusions and future work

We have identified three phases of the user-agent relation. Based on these phases we have constructed a user-agent interaction model. Given the developed model and the phases of the user-agent relation, a social interface needs to be able to analyze its own interaction history with the user and then self-adapt their way of interacting with the user accordingly. As stated before, this opens up new questions concerning levels of autonomy of social agents. Future work includes design and evaluation of a social agent that provides the user with interaction configuration support according to the three dimensions outlined in the developed model. Hopefully, this will give more insights into how social agents should be designed, evaluated and made possible for configuration by the end user.

References
Dehn, D. M., van Mulken, S., (2000) The impact of animated interface agents: a review of empirical research. Int. Journal of Human Computer Studies. Vol. 52, (pp.1-22).
Jensen, J.F. (2000) Trends in Interactive content & Services. In proceedings of WebNet2000, San Antonio Texas (pp. 281-286).
Jordan, P. (2000) Designing Pleasurable Products: An Introduction to the New Human Factors.
Nielsen, J. (1993). Usability Engineering. Academic Press.
Olsson, C. (2000) To Measure or Not to Measure: Why Web Usability Is Different From Traditional Usability. In proceedings of WebNet2000, San Antonio Texas (pp. 425-430).
Picard, W. R. (1998) Affective computing. The MIT Press. Boston, Massachusetts.
Schneiderman, B., Maes, P. (1997). Direct Manipulation vs. Interface Agents. In Interactions of the ACM (pp. 42-61). Nov/Dec 1997.

Supporting older adults at the interface

Mary Zajicek

The Speech Project, School of Computing and Mathematical Sciences, Oxford Brookes University, Headington Campus, Oxford OX3 OBP, UK
Tel: +44 1865 483683, Fax: +44 1865 483666, Email: mzajicek@brookes.ac.uk

Abstract

In an information society for all, everyone must have equal access to information in order to function effectively. Those who are unable to access vital information, which is increasingly distributed via the Internet, will become marginalised within society. Older people using computers, and hence Internet browsers, for the first time are faced with new ways of thinking and have little experience to draw from. The faculties that decrease with age such as memory, sight and strategy building are precisely those that are required for successful computer use. We require new forms of interface design to support these users. A voice help interface to support memory loss on BrookesTalk, a Web browser for visually impaired users, is described together with results from experiments. These experiments show that voice help can get older adults to use the Web in ways they could not without the voice tool. Most important, experiments also show that shorter voice help message length is important for older adults but not for younger ones.

1. Introduction

This paper is concerned with interface design for older adults with age related memory and sight loss lacking in confidence to use the Internet for the first time. It introduces a new approach to interface design to compensate for these factors, and reports experimentation carried out by the author with new forms of interaction for this group.

Older adults find computers difficult to use (Zajicek & Hall, 2000). The ageing process affects an individual's ability to function successfully with the standard graphical user interface. In fact the facilities, which are required for this kind of interaction, are the very ones that deteriorate most markedly with age. Deteriorating visual acuity makes the interface difficult or impossible to see. Memory impairment reduces the ability to build conceptual models of the working of the interface since this activity relies on remembering sequences of actions and reasoning about them. In addition the ability to navigate successfully and build strategies, deteriorates with age (Wilkniss, 1997) an aspect, which is especially important when moving from page to page in the World Wide Web environment. Manual dexterity is also affected (Walker, 1996) making mouse use difficult and user's confidence in tackling new situations diminishes (Zajicek & Arnold, 1999) promoting a reluctance to tackle new tasks.

It is generally agreed that older people are still able to learn but knowledge of the effects of age associated memory impairment indicates the need for a different type of interaction, which uses aspects of cognition that are less likely to be impaired.

This paper describes studies carried out with a specially modified Web browser for the blind, with voice help, which speaks out instructions to the user as they interact. This paper also examines the effect of personal support in promoting confidence to use the software and getting elderly visually impaired users up and running with the World Wide Web. It also reports experimentation to determine optimum amounts of information that can be absorbed from voice help by older adults. The issue of message length in the voice help is also discussed, since it impacts the levels of functionality in a system and functionality progression for older adults.

2. Difficulties experienced by older first time users

A Web browser for blind and visually impaired users, developed at Oxford Brookes University, called BrookesTalk (Zajicek, Powell and Reeves, 1998) was used as platform to evaluate new interface designs. Difficulties experienced by elderly visually impaired users were first brought to light during evaluation trials of BrookesTalk. The system was distributed free to over 200 blind and visually impaired users and evaluated (Zajicek & Arnold, 1999). There was no control group of sighted users as the issue of Web use for the targeted user group involves a range of different factors. Browser uptake by elderly visually impaired first time users was disappointing, 82% were unable to get up and running on the Web.

Analysis of their interaction showed that they were unable to build useful conceptual models of the functionality of BrookesTalk and of the workings of the Web. The users found it difficult to remember sequences of actions they had previously performed. Impaired memory seriously interferes with exploratory activity that involves remembering many combinations of actions and outcomes. They had no confidence in making the decisions needed for the construction of conceptual models and became confused and frustrated. For example some users were unsure as to the functionality of a link. Sighted users can see the link, how it relates to other text on the page and follow it to reinforce their concepts, easily returning to their original position.

Interviews with older visually impaired adults showed they did not understand the way a computer application works. Some 'borrowed' the model of a video recorder and expected one press of a button to make everything 'happen'. They were afraid that they would 'break' the software if they did something wrong. The concept of dialogue, and learning to use a language at the interface through trial-and-error was new. They did not understand the relationship between the function keys and the functions the keys represented. Furthermore, they did not understand the concept of mapping the task in hand onto the appropriate sequence of functions, and function keys, to achieve a goal.

Incomplete conceptual models of the Web, as distinct from the browser, also form a major impediment, where they cannot use the many visual clues available to help sighted users find information on the Web (Zajicek et al, 1998).

3. The voice help interface

At the interaction level difficulties experienced by older adults can be attributed to two interrelated factors that interfere with conceptual model development; age associated memory impairment and visual impairment. Both these factors reduce the user's ability to benefit from visual clues and context. To accommodate memory loss and visual impairment, a speaking front end was built onto BrookesTalk, to support the user in their construction of conceptual models by 'talking' them through their interaction.

For each possible state of BrookesTalk spoken output is provided. The user is told where they are in the interaction and which actions are possible at this point. Further details are also available to describe the consequences of each action. After listening to the message the user can choose an option, press the appropriate function key and then receive a message describing the new state of the system. As an example, the spoken output for those who have just started up BrookesTalk would be:

> *'Welcome to BrookesTalk your speaking web browser. There is currently no page loaded. Would you like to:*
> *Enter the URL of a page, press F1*
> *Start an Internet search, press F2*
> *Change the settings of the browser, press F7*
> *Hear more details about options available to you, press F3*
> *Repeat the options, press return'*

This and other messages reinforce the users' knowledge of the state of the system and explain what they can do next. The development of conceptual models should be supported through repetition and the user will no longer need to rely on memory. The user can function with no conceptual model of the system, by using it like a telephone dialogue system, where you simply answer questions. The aim of voice help was to familiarize the user with the steps needed to achieve Web interaction goals, eventually the spoken instructions would become superfluous and the user would 'know' which function key to press for the required result.

The system was used in a pilot study (Zajicek and Hall, 2000) with eight users drawn from our large-scale evaluation, who had previously been unable to get up and running with standard BrookesTalk. The aim was to determine the effects of personal support in developing confidence and whether people who used voice help BrookesTalk could move on to standard BrookesTalk.

It was found that personal support is very important elderly for visual impaired users using a computer application for the first time. In our small sample users with personal support, in the form of a helper who would answer yes or no to questions, were more likely to get up and running with voice help BrookesTalk as a step towards using standard BrookesTalk, than the users not given personal support. Our results also showed that people who had been unable to use standard BrookesTalk could use the voice help BrookesTalk and then move on to standard BrookesTalk.

4. Reducing memory overload

It was clear from observation of users struggling to recall voice help messages that wherever possible, messages should be simple and short since older adults cannot absorb or remember large amounts of information. Conceptual grouping of functions determines voice help message length. BrookesTalk is operated using twelve function keys, which were divided into two conceptual groups. One for functions involved with page retrieval, and the other for functions concerned with different views of the page once it was loaded.

Grouping functions provides for a smaller set of function keys to choose from at any one time and therefore also fewer and shorter messages with less to remember. However, conceptual groups of options rely on the user understanding the concepts behind the groupings. First time Internet users cannot be assumed to know where the function, and the function key, they want to invoke may be found. This is furthermore not required of sighted users making selections on a Graphical User Interface since they can see the options available at any time.

A smaller number of options in a message make selection easier. However presenting a smaller number of options means that commands must be grouped in a deeper menu structure, which is more difficult to conceptualize.

The functionality of BrookesTalk was reduced (Zajicek, 2000), in order to reduce both the message length and the number of conceptual groups. The advantage gained by multi functionality, where several different views are possible, is lost when the user cannot even load a page, and cannot visualize what the different views mean.

5. Information retention and different message lengths

Thus dialogues, for older adults, should be designed with messages of the shortest possible length that allow enough functionality for successful interaction. Those designing interfaces for this user group should be aware of how much advantage is gained by shortening messages. Experiments were carried out to determine whether older subjects remember more when the communicated information was presented using shorter messages.

Subjects were older adults, seventy years of age and older, attending day centres run by 'Age Concern' in Oxfordshire. The subjects showed normal age related sensory impairment, none had used a computer before and subsequently, had little or no idea of what the World Wide Web was.

The subjects were allocated to two groups of comparable memory levels, one was given shorter messages to listen to, and another was given longer messages. The first group, Group 1, listened to the two short messages below which represented output from the version of BrookesTalk with reduced functionality.

> *This is the main menu.*
> *There is currently no page loaded.*
> *To enter an Internet address, press F1.*
> *To perform an Internet search, press F2.*
> *To repeat these instructions press R.*

You have a page loaded.
To read the current page, press the right arrow key.
To follow the links on this page, press F4.
Go back to the main menu, press M.
To repeat these instructions, press R.

The second group, Group 2, listened to the longer messages in the original voice help system. Both groups were asked questions about the information in their messages.

The percentage of correct answers to the questions was computed for each subject. Two different percentages were calculated for Group 2 namely; the percentage of all correct answers, and the percentage of correct answers to those questions also asked of Group 1. The resulting percentages for Group 1 and Group 2 were tested for normality and a one-tailed t-test was carried out at 5% level of significance, on the mean values to determine if the differences were significant.

The average of correct answers to questions for Group 1 was 54.81%, this was significantly higher than the average of correct answers to all questions for Group 2 (26.67%). If we instead only consider the questions that were common to both Group 1 and Group 2, we see that the average of correct answers for Group 1 (54.81%) is still significantly higher than the average of correct answers for Group 2 (31.1%).

These results show that users are able to answer significantly more questions correctly when messages are shorter, meaning that more information is being retained from shorter messages.

The experiments described above were performed with a control group of young adults, which showed that shorter messages are not significantly helpful for younger users.

6. Conclusion

The results indicate that personal off-line support is important for elderly visual impaired users, who are using a computer application for the first time. Furthermore, the results also indicate that voice help is useful where memory impairment precludes the building of strategies and experimental learning at the interface. The combined effect of these two factors appears to be significant although the level of their individual effects is unclear at the moment.

We have found that retention of spoken messages works differently for older adults. While younger adults are able to accommodate differences of length of output message, older adults are not. Message-length is therefore an important factor in interface design for older adults with visual impairment, making the difference between a usable and unusable interface. Other research into interfaces for older adults, points to the need for low functionality systems, with the possibility of adding extra facilities when a few simple ones are mastered. For example Czaja et al in (Czaja, 1990) found that older adults were happy to add extra facilities once they had mastered a reduced functionality email system.

We have at this point identified that off-line personal support, message length, and level of functionality are important factors that affect the usability of systems for older adults.
Many other issues concerning speech output in interfaces for older adults need to be resolved. Work is currently underway to investigate the use of speech output in conjunction with text, where users can select to use either simultaneous speech and text output or only one of them. More research is needed in the are of human-computer interaction for this user group if they are not to be excluded from the information age.

References

Czaja, S., Clarke, C., Weber, R., & Nachbar, D., (1990): Computer communication among older adults, Proc. The Human Factors and Ergonomics Society 33rd Annual Meeting , pp 146 – 148.

Walker, N., Millians, J., & Worden, A., (1996): Mouse accelerations and performance of older computer users', Proc The Human Factors and Ergonomics Society 40rd Annual Meeting, pp 151 – 154.

Wilkniss, S. M., Jones, M,. Korel, D., Gold, P., & Manning, C.,(1997): Age-related differences in an ecologically based study of route learning. *Psychology and Ageing* 12(2) 372-375.

Zajicek, M., & Arnold, A., (1999): The 'Technology Push' and The User Tailored Information Environment, 5th European Research Consortium for Informatics and Mathematics. Workshop on 'User interfaces for all', pp 5-12.

Zajicek, M., (2000): 'The construction of speech output to support elderly visually impaired users starting to use the Internet', ICSLP 2000, Vol. I pp 150 – 153.

Zajicek, M., & Hall, S., (2000): 'Solutions for elderly visually impaired people using the Internet', HCI 2000, pp 299-307.

Zajicek, M., Powell, C., & Reeves, C., (1998): A Web Navigation Tool for the Blind, In Proc. 3rd ACM/SIGAPH on Assistive Technologies pp 204-2.

PART 5

USER DIVERSITY AND USER PARTICIPATION

Evaluation of user customisation in e-mail and mobile telephones

Margit Biemans, Jan Gerrit Schuurman, Janine Swaak

Telematica Instituut
P.O. box 589, 7500 AN Enschede, The Netherlands
http://www.telin.nl - {mbiemans, schuurman, swaak}@telin.nl

Abstract

Nowadays, systems have customisable designs in order to meet very dynamic and diversified requirements. We refer to customisation as user initiated adaptation of a software system. Users can fit generic applications to their personal preferences, to specific tasks, to changed work situations, or to broader contexts of use. In this way, customisation is a means to ensure accessibility by all users to information and to satisfy experiences in the use of systems. We evaluated two customisable applications, i.e., mobile telephones and e-mail. The emphasis in the evaluation was on *how* do users customise at this moment, and can customisation *patterns* be identified?

1. Introduction

Nowadays, systems have adaptable designs in order to meet very dynamic and diversified requirements. Adaptation can be 'performed' by the system, by users, or by a combination of both. In this paper we evaluated how end-users adapted systems to their personal preferences, to specific tasks, to changed work situations, or to broader contexts of use. In this paper, we call user-initiated adaptation of a software system *customisation*. This is in line with comparable studies as performed by Mackay (1991) and Page et al. (1996).

Customisation of generic applications is getting more and more important, for example to ensure accessibility by all users to community wide information and communication sources, as well as to satisfy experiences in the use of systems that carry out a broad range of social activities (Stephanidis et al., 1999). At this moment, customisability is a hype. All systems should be customisable to better serve the user, which leads to situations in which designers do not exactly know what the user wants to have customisable. They provide the users with all possible aspects and consequently the users can not see the wood for the trees. This may lead to a lack of usability in customisable systems. We evaluated the use of two customisable systems in order to identify user customisation. The aim is to identify customisation patterns, which can be useful to define profiles. Profiles are predefined sets of customised aspects for systems. Of course, over the years some more research has been done on customisation. However, none of these studies explicitly focused upon the usability of customisation. Moreover, both the software and the users have changed a lot since the eighties. A small summary of the results of these studies is described as it provides a starting point for our own evaluation of customisation behaviour.

Rosson (1984) studied customisation of a text editor. Her main findings were that both programming experience and the amount of experience with text-editors are good predictors for the amount of customisation. Mackay (1991) studied customisation behaviour in a Unix environment. In general, she found that '*data support the idea that users 'satisfice' rather than optimise'*. The main reason for this is that people can not calculate on forehand the efficiency of customisation. She categorised reasons for customisation in external events, social pressure, software changes and internal factors. The most cited triggers for customisation are repeated patterns, getting annoyed and retrofitting. The far most cited barriers for customisation are lack of time and too hard to modify. Oppermann and Simm (1994) also studied customisation in a business environment. They looked at applications like word-processing, spreadsheets, database and drawing applications. They distinguish two kinds of adaptations to a system, i.e., adaptations of functionality (e.g., Macros) and interface adaptations (e.g., Menu bars). Users rather customise interface adaptations than adaptations of functionality. The main reasons for customisation are making work easier and saving time, although they found that curiosity and fun were also mentioned as reasons for customising. In 1996, Page et al. studied customisation changes of word-processors. They found that 92% of word processing users customise their software. They think the reasons for this are related to both work needs and ease of customisation. Amount of use of the application is a good predictor for the amount of customisation and 'simple to perform' customisation features are also more used. They categorised 5 types of software customisation, i.e., general preferences (changed by 86%), functionality (changed by 79% in total), access to functionality (71%), access to interface tools (25%) and visual appearance (15%).

In general, the problem of customisation can be studied in two ways: observation based and theory based. Our study has an exploratory character and is therefore observation based. It will give input for further research (both observation and theory based) in order to develop guidelines for customisable designs.

The main issues of the literature review that gave input for our own evaluation are:

- The research is conducted in a business environment and with business applications. As customisation is a personal matter, we like to evaluate customisation behaviour in at least one 'personal system', i.e., a system that users perceive to be owned by them.
- Since the eighties, software has become user-friendlier (less experience is necessary), does this also count for customisation?
- The research is more oriented towards *what* they do instead of *how* they perceive it and *how they were triggered* doing it.
- Only one application at the time is studied. Comparing two (or more) applications customised by the same person might provide insights into 'steady' behaviour, more independent of the application.
- The research was mostly conducted on applications with a specific content, e.g., word editors. It those situations, it is difficult to abstract from the content. Therefore we will choose communication media.

2. Our view on the customisation process

Mørch (1997) makes a distinction between open and closed customisation. Open customisation, (or tailoring as he names it) can be regarded as continuous design, as it allows users to make deeper changes to the system by giving them the same kind of flexibility as the original designer had.

Closed customisation does not allow users 'under the hood of the application' (Mørch, 1997). Closed customisation can be accomplished by switching parameters in predefined settings or by modifying the appearance of user-interface objects. In this paper, we concentrate on closed customisation, i.e., predefined settings and user-interface appearances. We call the options to customise 'fixed design aspects' or *customisation constructs* of an application.

For the construction of our questionnaire, we take the customisation process as a starting point. This process can be considered in several ways, e.g., psychological oriented and system oriented. For our explorative study in this paper, we chose the system view. This is depicted in Figure 1. We consider it to be a cyclic process of both using and customising the system. While using the system, a user might get a need to change an aspect. Based upon (fixed) design aspects (or customisation constructs) of the system, the user gets the possibility to change it. After having changed that aspect, the user starts using the customised system, until he gets another need to customise and so the process continues.

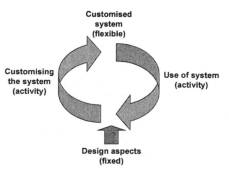

Figure 1: Customisation model for the construction of our questionnaire. The possible customisation aspects (=design aspects) are provided. Users were asked for information about their activities in relation to these aspects.

3. Method

We choose to evaluate mobile-telephones and e-mail, because they both have clear customisation possibilities, wide spread use, distinct use groups (e.g., business and private users) and at least one of them is a 'typical owner' application (mobile telephone). In order to be able to identify customisation patterns, we chose to evaluate three use groups, i.e., consultants, students and a mixed group of people, referred to as 'ordinary people'. In order to compare the data of mobile telephones with e-mail (for example to identify the influence of the context of use), only users of *both* applications were asked to complete the questionnaire.

The selection of customisation constructs (or fixed design aspects) was guided by the customisation categories of Page et al. (1996). We choose customisation constructs related to general preferences, functionality and access to functionality (the three most customised categories). Four customisation constructs for mobile telephones were selected, i.e., type of sound, volume, phonebook and quick access and three customisation constructs for e-mail, i.e., signature, address book and filters. These selected constructs differ in customisation complexity. In relation to Page et al. (1996), type of sound, volume and signature are regarded as general preferences (GP), while phonebook and address book are regarded as access to functionality (AtF), and quick access and filters as functionality (F). For each of the seven customisation constructs, the same kinds of questions were constructed in relation to the '*who, when, why, how* and *usability*' of customisation. The outline of the questionnaire is depicted in Figure 2.

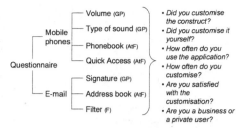

Figure 2: Outline of the questionnaire. Depicted are the applications with corresponding constructs. For each construct the same kinds of questions are asked.

4. Results

This study has two objectives, *how do users customise at this moment* and *can customisation patterns be identified*. The results in relation to the first objective are described in the first sections about the *subjects*, the *why*, *how* and *usability* of customisation. These results are of descriptive nature. The results in relation to the second objective are described in the *who* and *when* customisation sections. These results are derived using a Chi-Square analysis (Kruskal-Wallis).

Subject description : 56 users (32 male and 24 female) completed the questionnaire. These subjects are divided into three distinct use groups, i.e., students (N=30), consultants (N=16) and a more mixed group of people, called 'ordinary people' (N=10). Their age ranged from 14-52 years old, on average 22.5, with two peaks (15-16 (students) and 29-30). The subjects also differed in, education, amount of use and goals of use of the application (business and private users). 54 subjects (96%) had customised at least 2 constructs of the 7 mentioned in the questionnaire. For the specific applications, 3 mobile telephone constructs and 1.3 e-mail constructs were customised. On average, they customised 4.3 constructs out of 7.

Why customising? : This aspect is evaluated by looking at the opposite; 'why was a construct NOT customised'? 20% of the constructs were not customised because the user was not aware of the possibility to do so. Users reported that when they had been aware of the possibility, 29% more would have been customised (19% mobiles and 33% e-mail). Of the subjects who had been aware of the customisable construct, 3% of the subjects indicate that they had not customised because they did not know how to do it. Another group of subjects indicates that they do not see a profit for customising a specific construct (mobiles 24% and e-mail 21%). Other subjects report very diverse reasons, e.g., no time, no need to do so, etc.

How customising? : 71% of the constructs were customised by a trial and error process; the customisation constructs of the application were explored by the user. 23 % of items were customised using the manual or help function. 63% of the respondents used more than one means to customise, i.e., mostly both trial and error and the manual/help function. 86% of the subjects was satisfied with the customisable options. Reasons for dissatisfaction were mainly too many or too few options.

Usability of customising?: In relation to the usability of performing customisation, users reported this as rather easy to do. To be more precisely, on average 1.4 on a scale of 1 (easy) till 4 (difficult). 86% of the subjects is satisfied with the customisable options. Reasons for dissatisfaction are mainly too many or too few options.
The following two sections aim at identifying customisation patterns. The results of the Chi-Square analysis are described in the sections *who* and *when* customising. The analysis took place on two levels: firstly on a general level for the whole application (all customisation constructs together), secondly on a detailed level (each customisation construct separately). This section ends with a table providing an overview of all the significant results of the Chi-Square analysis.

Who customises? : More than 90% of the customisation in mobile telephones is performed by the user himself, while in e-mail only 69%. In more detail, other people customise general preferences (mobiles 2%, e-mail 24%), access to functionality (mobiles 10%, e-mail 36%), and functionality (mobiles 38% en e-mail 56%). User characteristics as male/female and programming experience had no significant relation with customisation. The

464

subjects are divided in different groups, based on different use characteristics. First, they are divided into business, private and combined users. Of course, most subjects in the business use group are consultants. There was no significant relation found for both applications in general. When looking into more detail in the specific customisation constructs, we see that business users customise less the phonebook of their mobile telephone (Chi square=6.111, df=2, p=0.047). For e-mail, there was no such significant relation found. Second, we divided the subjects in the type of use of the application. We called the groups sending messages, receiving messages and combined use. There was no significant relation found for both applications in general. When looking into more detail in the specific customisation constructs, we see that the group of 'sending' users of e-mail, had customised more filters (Chi square=7.857, df=2, p=0.02). Third, the subjects are divided into students, consultants and 'ordinary people'. Only in e-mail significant relations are found; 'ordinary people' customise more (Chi square=10.144, df=2, p=0.006). When looking into more detail in the specific customisation constructs, we see that they customise more filters (Chi square=13.946, df=2, p=0.001).

Table 1: Results of the Chi-square analysis on who customised and when did they customise.

	E-mail	Mobile telephones
Who customises?		
• Business or private use	No significant results	Business users customised their phonebook less (p<0.05).
• Type of use (sending, receiving messages or a combination)	Users, who sent more messages than they received customised more filters (p<0.05)	No significant results
• User groups (students, consultants, ordinary people)	Ordinary people customised more in general (p<0.05). In detail, they customised more filters (p<0.05).	No significant results
When customising?		
• Timeframe of use of the application	No significant results	When someone used to have another mobile telephone before, the more customisation has been done (p<0.05). Especially, the Quick access functionality (p<0.05).
• Amount of use of the application	In general, the more the application was used, the more it was customised (p<0.05). In detail, more use of the application, more customisation of the signature (p<0.05), and the less use of the application, the less the address book was customised (p<0.05).	No significant results

When customising? : This aspect is evaluated by the amount of use of the application and the timeframe in which the application was used. For e-mail, there was a significant relation in the amount of use. The more the e-mail application was used, the more it was customised (Chi square=7.835, df=2, p=0.02). When regarding this into more detail, we see that both the construct signature and address book have a significant relation. For signature, the more the e-mail application was used, the more a signature was customised (Chi square=8.590, df=2, p=0.014). While for the address book, the less the e-mail application was used, the less the address book was customised (Chi square=5.052, df=2, p=0.080). For mobile telephones, there was no such significant relation found. For the timeframe of use, a significant relation was found for mobile telephones. When a subject used to have another mobile telephone, the more customisation on his present mobile telephone was done (Chi square=3.465, df=1, p=0.063). Looking into more detail, we see that especially quick access functionality is customised more (Chi square=10.385, df=1, p=0.001).

5. Discussion/Conclusion

We performed an exploratory study on user customisation of mobile telephones and e-mail. The emphasis was on *how* do users customise at this moment, and which customisation *patterns* can be identified? We asked users of both applications to complete a questionnaire. The questionnaire concentrated on '*who, why, when, how* and *usability*' of 7 customisation constructs. The results that almost all the users customise their e-mail program and/or mobile telephone in some way. We think this very high number is due to both the ease of use of customisation and the 'ownership or personal envision' of the application. Most people perform the customisation themselves. When other people assist them, it mostly regards more complex issues like customising functionality. Awareness of customisation possibilities is found to be an issue. Almost one third of the subjects said they would have customised specific aspects when they had been aware. Identification of customisation patterns for the design of profiles was difficult. We mainly succeeded on a very detailed level. Therefore we suggest exploring this further, for example by looking into a more detailed level, on more content related applications or by other research methods.

Our first objective was to identify how users customise at this moment. We matched the results of our study with comparable studies (see literature review in section 1). We found that users do customise (96% of the subjects). This is comparable to Page et al. (1996) who found that 92% customised. On average, our subjects customised more than half of the constructs mentioned in our questionnaire. Both Page et al. (1996) and Rosson (1984) found a positive relation between the amount of use and the amount of customisation. We only found this relation for e-mail, not for mobile telephones. Possibly because the average amount of customised constructs of mobile telephones was very high, compared to that of e-mail (74% versus 44%).

In relation to this, we found that the user himself mostly performed the customisation, even more on mobile telephones than on e-mail applications. This is consistent with our idea that mobile telephones are perceived as typical personal or owner applications. The sets of constructs customised by other people than the user himself also support our idea of complexity of customisation, i.e., general preferences (volume, signal and signature) are more easy to customise than functionality aspects (filters and quick access). These findings can be compared to the results of both Oppermann & Sim (1994) and Page et al. (1996).

We also found that subjects who previously had another mobile telephone customised more than other subjects do. In some way, it needs to be studied whether this is comparable to Mackay's finding that a major trigger for users to customise is retrofitting. On the other hand, the specific construct, which is customised more, is the quick access functionality. This is the most complex construct of mobile telephones, so maybe only after a long period of use, users perceive the benefits or even become aware of the possibility to customise this construct.

When looking at why subjects did not customise, a major finding was that they were not aware of the existence and possibility to customise, especially in e-mail. As they had been aware almost one third of the subjects indicated that they would customise it in the future. This result indicates a need for future design of customisable systems: make customisable options and possibilities more transparent to users. When looking at the results that are specifically related to usability, than we can see that the two applications studied are rather usable.

Our second objective was to identify customisation patterns, based upon use aspects as type, task and goal of use. On the generic level, we only found three relations, i.e., ordinary people customise more in e-mail, the more use of e-mail, the more customisation and when someone used to have another mobile telephone before, he customised more. Based upon literature, we expected to find a difference between business and private users, as Page et al. (1996) discussed that the amount of customisation depends upon work needs. Possibly, the difference between these findings can be related to the applications evaluated, i.e., word-editors versus communication media. On a more detailed level (insight the customisation constructs), we found 6 relations of very specific aspects.

Based upon these results we can not identify generic use aspects as type, task and goal of use to guide the design of profiles. However, we found some more results on a detailed level, so maybe a specific task level or likely more content related aspects should be used to design profiles.

6. Future work

The results of this evaluation indicate that generic customisation profiles are difficult to relate to generic use aspects, or tasks. Future research will concentrate on a more detailed task level, and on more content related aspects. Another aspect for further research is the lack of transparency in customisable systems. The results of this study indicate that some users not only want to know how things can be customised, but also what can be customised (and what its effect is). The ongoing research presented in this paper is part of the GigaMobile and GigaCSCW project, both projects that are embedded in GigaPort, the Dutch next generation internet initiative. For more information see: http://www.gigaport.nl

References

Mackay, W.E. (1991). Triggers and Barriers to Customizing Software. In *Proceedings of the conference on Computer Human Interaction* (pp. 153-160). New York: ACM Press.

Mørch, A.I. (1997). Three levels of end-user tailoring: Customisation, Integration and Extension. In, Kyng, M. & L. Mathiassen (Eds), *Computers and design in context* (pp. 51-76). Cambridge, MA: The MIT Press.

Oppermann, R., Simm, H. (1994). Adaptability: user-initiated individualisation. In R. Opperman, (Ed.). *Adaptive user support, Ergonomic design of manually and automatically adaptable software* (pp. 14-66). Hillsdale: LEA.

Page, S.R., Johnsgard, T.J., Albert, U. Allen, C.D. (1996). User customization of a Word processor. In *Proceedings of the conference on Computer Human Interaction,* (pp. 340-346). New York: ACM Press.

Rosson, M.B. (1984). The effects of experience on learning, using and evaluating a text-editor. *Human Factors, 26,* 723-728.

Stephanidis, C. (Ed.), Salvendy, G., Akoumianakis, D., Arnold, A., Bevan, N., Dardailler, D., Emiliani, P.L., Iakovidis, I., Jenkins, P., Karshmer, A., Korn, P., Marcus, A., Murphy, H., Oppermann, C., Stary, C., Tamura, H., Tscheligi, M., Ueda, H., Weber, G., & Ziegler, J. (1999). Toward an Information Society for All: HCI challenges and R&D recommendations. *International Journal of Human-Computer Interaction*, 11(1), 1-28.

Design and the Cultural Significance in International Communication

Red Keith Bradley

Department of Communication, Slippery Rock University
Slippery Rock, PA

Abstract

In a five-year period, between 1993 and 1998, the number of Internet sites grew 200 fold (U.S. Department of Commerce, 1998). Accordingly, there are now an estimated 304 million global Internet users, an increase of 78 percent from 1999 (NUA, 2000). Another appreciable trend is who is using the Internet and from where. Recently released research found that the United States and Canada now account for less than 50 percent of total user growth (NUA, 2000). This is a trend that most researchers believe will continue during the next decade. This belief is also strongly supported by research conducted by the World Economic Forum. The WEF Task Force on the Global Digital Divide Initiative issued a report to the Kyushu-Okinawa G-8 Summit in June disclosing research and suggestions for action. The WEF report noted that "Market-oriented policy reforms, local entrepreneurial efforts and the support of the international community have combined to significantly increase the deployment and usability of telecommunications, Internet and related technologies in many countries across Asia, Latin America, Africa and the Middle East" (WEF, 2000). It is apparent that design plays a significant role in effective commerce. The Italian designer Lidia Guibert Ferrara noted that, "The goal of cross-cultural design is to respect the subtleties of one's audience, an effort that requires constant questioning of both the designer's personal beliefs *and* the troubling misconceptions that might be lurking in the assignment" (Steiner, 1995). The major concerns addressed here may seem common place. In practice, however, designers and usability experts continue to overlook basic communication considerations as noted by example throughout this paper.

1. Connectivity: Technical Issues

Most large corporate Internet site development involves project teams who work together to create visual messages that inform, interact or to sway visitors to purchase commodities. In many cases, however, elements of visual entertainment are infused with textual messages that are packaged without consideration of the technical issues, cultural practices, religious heritage or symbolic interpretation of that content. It appears, at times, that the audience is secondary to the "one size fits all" design mentality of internationalization. In other words, content and localization issues are overshadowed by elements of technological feats of programming and visual elements that may be unviewable in developing countries. Perceived design objectives may be lost as users move to sites that are user-friendly, localized and attentive to the needs of the marketplace. The company that is savvy to the idiosyncrasies of localization can attract customers regardless of monetary backing and without a sales force – a 24 hour, seven day a week global storefront. Approximately 90 percent of users go online to search for information of some type, including news, product information, weather, statistics, and travel, because the information is current and accessible anytime (Maddox, 1997). With a projected number of users doubling every 100 days, the potential market for those who invest in localization projects grows astronomically (U.S. Department of Commerce, 1998).

Each layer in the technical design process has the potential of generating a set of problems. Technical considerations like the use of plug-ins, large images, video clips, and dominant colour selections are just a few of the potential land mines designers face. The problem is compounded by hardware, bandwidth and connection speed. To illustrate the point, downloading a 3.5-minute video sampler at 28.8 Kbps will take 48 minutes or longer if the connection speed is slower (FCC, 1998). Connection speeds internationally are problematic when coupled with complex site designs. In Spain and Italy, half the online users are limited to 33.6 Kbps connections or slower (Fineberg, 2000). In Germany and Denmark, only 60 percent of those online have connection speeds of 56Kbps or greater. John Roth, chief executive officer of Nortel Networks, in his address to the Telecom '99 audience, noted that Nortel researchers found that 2.5 billion hours were spent waiting for pages to download in 1998 (Associated Press, 1999). As the data attest, designing technically complex sites may have an adverse effect on delivering content, even in developed nations, if downloading time exceeds user patience.

Moreover, the number of turns (hops between routers requiring an out-and-back trip to create a page) is increasing. In 1995, an average page required 22 turns, today that number has doubled resulting in slower downloading and increased user waiting (Wilson, 1999). To illustrate this point ArchiText a localization service company based in the United States advertises its localization services globally. The company homepage, however, is heavily dependent on large graphics and an animated cube that notes the company's qualities. Downloading the homepage takes one-minute using a T-1 connection translating into 20-minute wait for those using a 56.6Kbps modem. This problem is annoying in situations where older technology, coupled with excessive connection costs, may serve as a deterrent in effective eCommerce.

2. Not everyone in the world speaks English

Globalization and localization have recently begun to take on added significance as business begin to tap into the rich online resources and vast audience potential. For a modest investment, a company can join the online world and become an international participant. The fundamental obstacle for these companies, however, is that more than 50 percent of the 147 million Internet users are non-English speakers (Dubie, 2000). Furthermore, estimates note that within three years, 60 percent of the one billion estimated global Internet users will be non-English speakers, according to International Data Corporation (Dubie, 2000). English as the predominant Web language continues to erode, even though data suggest that 85 percent of the current Websites are based in the United States (Howell, 2000). One in five non-English users are fluent in Japanese. Spanish (15.4%), German (14%), French (9.9%) and Chinese (9.9%) follow Japanese in commonly used Internet languages (Lawrence, 2000). As developing countries begin to expand their reach into cyberspace, accommodations in localization will need to be made on both ends of the line. The essential understanding of one's audience and how to clearly communicate with that audience will determine the effectiveness of online commerce outside one's own borders.

Are visitors more likely to stay and peruse Websites if the site is in their native language? Research conducted by Alison Toon, the worldwide localization program manager for Hewlett-Packard's IT Resource Center, found that less than 5 percent of its Korean customers would be happy if English were used instead of the local language (Dubie, 2000).

Research conducted by Pro Active International of Amsterdam also found that Europeans preferred the use of their native language on Websites. In both France and Spain, 80 percent of respondents acknowledged knowing enough English to navigate a Website, but preferred those sites in the French or Spanish language. Similar results were found in Scandinavia, although the number dropped to 60 percent of online users (Arlen, 2000).

Another critical design problem in translating text into various languages is text swell – the difference between the width of text between various languages. Accommodation in design and layout needs to be considered during the design process allowing for this occurrence. Typically, German translations require 40% more space that English and while browsers can accommodate some swell depending on monitor resolution, it is much easier to design for the user's lowest common denominator. Contrary to English languages, Arabic and Hebrew run right to left, while Japanese and Chinese character sets run top to bottom, creating a different set of design problems.

3. What is the key to internationalization?

What are the implications of doing business globally? Most acknowledge that language, religious heritage, date and time formats, iconography, writing system (Latin, Semitic or Asian), cultural subtleties, scripting accommodations and character formatting (two-digit formatting known as double byte character sets or DBCSs required for Asian characters versus the one-digit scripting for English and European languages) have the greatest implications when designing for a culturally diverse audience. Likewise, pricing, payment structures, systematic currency exchanges, shipping costs, exchange policies and customer support may further complicate transactions with customers on the other side of the globe.

While a few developed countries have unlimited Internet access time at a flat rate, many users are charged on an hourly rate – a rate that sometime exceeds the average hourly salary of its residents. The monthly Web charge in Argentina is $50 (all monetary figures measured in USD) or 17 times the average hourly salary of its residents. In Kenya, the monthly fee is $100 (email-only service is available for $10 per month), in India the cost is $82 monthly and in Armenia an astounding $121 per month for Internet connection (Petrazzini 1999). It is easy to understand the international implications of the designer's work when placed in the context of cost in developing countries.

It is also likely that the majority of designers have little experience designing in Arabic scripts that are bi-directional (right to left, except for numbers and foreign-language words that run from left to right. Far Eastern languages require twice the space of English for each letter because of DBCSs. Localizing a site requires special attention to these design considerations.

4. Designing for the global audience

Most designers have little or no training in addressing localization issues, although numerous articles and special issues focusing on globalization and localization have recently appeared. The majority of these articles have, however, centered on language and very little attention has been paid to the larger issues of globalization. The lack of formal design training (focusing on designing for global audiences), special programming needs required for various linguistic groups and content interpretation in cross-cultural applications allow design firms to bypass a true globalization effort, thereby affecting content and possibly message reception. The majority of these issues are either overlooked or ignored in an effort to visually enhance a site with state-of-the-art feats.

"The desire to become a cool site is often detrimental to good design. Certainly, boring or confusing sites will not attract many users, but the use of advanced design elements simply for the sake of adding more stuff to the page will discourage users from repeat visits to a site," said usability expert Jakob Nielsen to the 1998 gathering of the Association of Computer Machinery Special Interest Group on Computer-Human Interaction (ACM/SIGCHI). With the rapid growth of eCommerce, the need to effectively communicate globally challenges designers to adhere to the principles of simplicity rather than those of technical complexity. The classical role of designer is becoming that of information architect/information designer by encompassing the tools of design aesthetics with user interfaces and navigation tools that promote, rather than deter, usability. Neilsen's statement has added significance when applied in the global context.

The Federal Express site for Brasil demonstrates a business-to-business localization attempt that may fall short of corporate strategy. Gains made by the conversion of navigation aids into the national language, the placement of the flag in the design and inclusion of local contact numbers may be offset by the representation of an Asian carrier. Moreover, visitors to the site have a one-in-four chance of seeing this particular carrier and a one-in-three chance of seeing a woman.

The United Arab Emirates site is localized using the national language, local charter set, and reading preference while using the same carrier image as noted above, although it has been reversed to follow the text flow. Bthe cultural problem here, however, is the use of not only the Asian carrier, but that of a woman with exposed arms – a cultural taboo in the Arab region.

Both Federal Express examples adequately illustrate the problem of site development for international audiences. To be fair, however, Federal Express maintains 216 separate sites one of the largest online localization efforts.

5. Theory versus practice

Two other noteworthy examples using the Federal Express site illustrate the complexity of global design. The site for Hong Kong, ignoring any associated political aspects, uses double-digit coding rendering the text unreadable in the United States. This problem is potentially resolved by the use of Unicode and its 65,536 character combinations, although it is not necessarily a catchall fix for the problem. The Unicode site itself also has its own problems in translating "About our services" page into Arabic. The translation error occurred employing Internet Explorer v5.0 in the International translation setting. While formatted for the geographical area (right-to-left) the Arabic words did not display correctly.

Three other considerations worth noting in undertaking international translations:

- Spelling varies between countries that share similar dialects. In the United States, "color" is different from its counterparts in the United Kingdom and Canada where it is spelled "colour." Other words include "tire" versus "tyre" and "localization" versus "localisation."

- Variations in colloquialisms are also important to consider during localization. The use of elevator versus lift or apartment versus flat is equally important in accommodating your audience. Attention to detail leaves the user with the impression that time and care was taken during the process. This process should also take into account special accent marks used in various languages.
- If computerized translation services are used, they should be manually checked for accuracy. While the use of the service certainly speeds the translation process, the service is not error free. For example, the translation of *browse* from English to French will return *Regardez* or watch, whereas in German it will return *Blättern Sie* or page (www.translations.com).

6. Conclusion

Many aspects of information design have been presented in this chapter. It is a holistic review of writings available from numerous sources. The author acknowledges that while an intensive search of literature took place there is an immense likelihood that additional material exists.

There are roughly 5,000 languages and dialects used globally. Out of this group, about 100 are used in business and technical communication. It has, however, been adequately addressed that translation is only a portion of the overall process. Issues related to text, colour selection, cultural preferences and technical issues also play a pivotal role in the process.

Other considerations that were not included, but worth observation:

- Values associated with direction. Power and righteousness are associated with the right hand in Western cultures, whereas in Chinese traditions honor dwells in the left hand while destruction is on the right (Cooper 1978);
- Carefully consider the use of icons for communicating concepts. The mailbox, standard in the United States, does not carry the same interpretation in other parts of the world;
- Attention should be paid to national differences in graphical representations. A power plug in the United States has different physical features to those in Switzerland, the United Kingdom and other parts of Europe;
- Limit the use of alphabetic characters that have little representation or comprehension by cultures that utilize a different alphabetic system;
- When picturing people, be sensitive to the customs and practices of other cultures. Nudity in one country may be acceptable, but not in others. In some Asian cultures, bare feet or the sole of the foot send an entirely different message than in the United States, where the practice is common. Showing women with bare arms is prohibited in Islamic countries, where only the eyes and hands are shown;
- Avoid hand gestures to illustrate a point. They may not be interpreted the same way internationally. Hands are used to perform a task, although care should be used to ensure the proper, or acceptable ethnic representation appears on your site; and

There are numerous considerations in developing a site for internationalization or localization. While the suggestions noted here give a brief checklist of considerations, it is advisable to utilize the services of local usability experts when preparing sites for an audience outside ones own borders. Nor is the author suggesting that all sites need to be translated. At a cost of 40-60 cents per word (Horton 1993), translating every page and word could become cost prohibitive and unnecessary.

It is noteworthy to suggest that the segmentation between the "information haves" and "information have-nots" will continue until the digital divide is eliminated. Singapore, with a population of 3.7 million, has a thousand times as many Internet hosts as the 60 poorest countries that account for more that 3 billion people. Iceland, with a population of 250,000, has 20 times the number of Internet hosts as 100 of the world's poorest countries (Petrazzini 1999). The World Economic Forum suggested in their report that "the digital opportunity will not be realized fully, and indeed could be squandered, unless developing country governments take decisive and enlightened action" (WEF 2000). Without assistance and persistence of the "information haves", however, it is likely that this opportunity may pass.

The author closes this discourse with the following design considerations:

1. Content should account for 80 percent of the usable window space acquainting the user with as much information as possible about your site. This, however, may vary on the homepage to allow for navigation aids and corporate identifiers and/or text accommodations;

2. The strict use of "websafe" colours is no longer a significant obstacle for designers. It is estimated that the majority of users can now view thousands of colours on their monitors opposed to the restrictive websafe palette of 216 for Windows machines and 256 for Macintosh;

3. Page size should be kept to 34 kilobytes for optimum downloading by modem users. Proven Edge's research suggests that users will bailout (leave the site) if pages exceed 40 kilobytes at the rate of 25-30 percent. This is in sharp contrast to a bailout rate of 7-10 percent where pages are 34 kilobytes or less (Nielsen 1999);

4. Let your users know who you are as soon as they enter your site. While this suggestion appears elementary, there are instances where this practice was not undertaken and users are left wondering where they have landed.

5. Leave some white space for the user – it allows the designer to control the flow of information. It also allows the user some space to absorb the content without overwhelming the user with wall-to-wall graphics, text and animations;

6. The standard hyperlink colour is blue and most usability experts suggest its continued use. Most online users have become accustomed to its meaning and can instantly recognize its intended purpose;

7. When internationalizing, a staging page with language choices should be created to allow users the choice of language. The use of international flags is at best can be confusing – i.e. which flag would you use for English – the United Kingdom, United States or Canada? Most usability experts suggest that language choices should be spelled out including the use of special markings, spellings and symbols.

Finally, indigenous users should test pages being localized before they are made "public." Performing this task will ensure that both the content and translations are correct and culturally acceptable.

References

Arlen, G. (2000). *Internet Imperialism? Mais Oui!* [On-line] Available: http://www.washtech.com

Associated Press reported by NUA Internet Surveys (1999). *2.5 Billion hours Spent Waiting in 1998.* [On-line] Available: http://www.nua.ie/surveys/?f=VS&art_id=905355331&rel=true

Cooper, J.C. (1978). *An Illustrated Encyclopedia of Traditional Symbols.* London: Thames and Hudson

Dubie, D. (2000). Going Global: Break The Language Barrier To Get Your Message Across. *Publish*: 66.

Fineberg, S. (2000). *Euro-style Research is On the Way* [online] Available http://www.channelseven.com/adinsight/surveys_research/2000features/surv_20000201.shtml

Horton, W. (1993). The Almost Universal Language: Graphics for International Documents. *Technical Communications.*

Howell, D. (2000). Is Net's Growth Slowing? Depends How You Look. *Investor Business Daily*: A6.

Kuehl, C. & Virzi, A.M. (1999). *StarMedia's Serenade of Latin America.* Internet World: 36

Lawrence, S. (2000). *Behind the Numbers: The Mystery of B-to-B Forecasts Revealed* [On-line] Available: http://www.thestandard.com/research/metrics/display/0,2799,11300,00.html

Maddox, K. (1997). Information still killer app on the Internet. *Advertising Age.* [On-line] Available http://adage.com/interactive/articles/19971006/article7.html

Petrazzini, B., & Kitbati, M. (1999). The Internet in Developing Nations. *Communications of the ACM*: 31

Steiner, H. & Hass, K. (1995). *Cross-Cultural Design: Communicating in the Global Marketplace.* London: Thames and Hudson Ltd.

U.S. Department of Commerce, & Netcraftt (1998). *The Emerging Digital Economy.* [online] Available: http://www.doc.gov/ecommerce/emerging.htm

U.S. Department of Commerce (2000). Digital Economy 2000. [On-line] Available: http://www.doc.gov/ecommerce/emerging.htm.

Wilson, T. (1999). Complex web Apps Will Degrade Response Times. *Technology News.* [On-line] Available: http://www.techweb.com/wire/story/TWB19991013S0019

World Economic Forum (2000). *From the Global Digital Divide to the Global Digital Opportunity.*

Using user profiles to customise the user interface

Marcelino Cabrera, Miquel Gea, Juan Carlos Torres

Dpto. de Lenguajes y Sistemas Informáticos- Universidad de Granada
Avda. Andalucía s/n. 18071 – Granada. Spain
e-mail: <mcabrera,jctorres,mgea>@ugr.es
Web site: http://giig.ugr.es

Abstract

New technological improvements comprise a continuous challenge to society, and can change our very conception of the world around us. We are living in an *information society* that relies on computer-based communication within the *global village* concept. Nevertheless, the ambitious challenges and opportunities offered are not, in fact, accessible to user communities with any kind of disability, because no attention is paid to user adaptation with respect to the disabled. This paper presents the structure of a tool capable of producing user-adaptable interfaces obtained from a single definition of the application interface.

1. Introduction

Technological advances continually provide society with new challenges and thus change our very conception of the world around us. New concepts such as *global village, Internet, electronic mail* or *wireless telephony* are used everywhere in a natural way; these media allow us to characterise communication in the twenty first century, which has been called the *information society*. In this context, Computer Science allows us to develop complex applications for industry, communication, education, commerce, etc.; in other words, computer programs that are focused on trained users with experience and skills in the domain in question.

Due to the growing expansion of Information Technology at all levels of society, there is a need for the development of applications that require very little user skill. The *information society* embraces a large number of new users with scant knowledge of computers. Such users, therefore, need applications that are easy to learn and to use.

In this scenario, furthermore, a large number of people suffering some kind of disability have serious problems in benefiting from the advantages offered by Information Technology. This is due to the fact that interfaces are currently limited to 'typical' users. Moreover, the experience level of each user varies depending on previous knowledge and needs.

This situation demands an adjustment/adaptation of application interfaces to user needs. At present, user adaptation is often limited to little more than minor features such as the sensitivity of mouse movement, adaptability for left-handed users, colour coding, shortcut keys, etc.

Current desktop environments allow us to carry out modifications that enhance several features of user interfaces according to the individual preferences and capabilities but do not, however, reflect sufficient adaptability to cater for a wider population with special cognitive or motor capabilities.

Research activities are classified by how users' needs are addressed during the design process.

1. **Usability**
   Usability is a software feature oriented towards the ease of use of a program. This feature may be obtained by using a design rationale based on empirical studies that take into account factors such as predictability, consistency and familiarity. Obviously, usability is a desirable property in all software programs, regardless of the user's perceptual/motor abilities. Several guidelines have been proposed to include these features in the design phase (Stephanidis & Akoumianakis, 1999).

2. **Accessibility.**
   Accessibility is an application property that considers the use of the program by a wider community. This approach focuses on the input/output suitability for a range of human motor abilities. In this context, disabled users and their requirements (mobility, sensorial perception, etc.) are the focus of the design process. Accessibility is oriented towards user groups with similar circumstances. An illustrative example of this

feature is the creation of tools to improve the web browser (Emiliani & Stephanidis, 2000; Baldoni et al., 2000).

3. **Adaptation**

User adaptation is a property that is related to improving the user's performance. This goal is achieved by optimising the surrounding software environment to the user's cognitive skills. Novice users can easily find out how to use the tools while expert users obtain better results when they improve their knowledge by working within the system. Thus, adaptation is oriented to each individual. In this approach, several tools have been developed to create adaptable user interfaces (Martins, 1996; Cooper et al., 2000; Koutsabasis et al., 2000).

This classification is not complete, and our goal is to clarify the application scope of current research according to the target users and the current design phase.

2. Goals

In this scenario, our proposal aims to create a user interface that is adapted to a wider community, covering individuals with differences in culture, cognitive skills and abilities. These differences concerning the learning process, data acquisition and manipulation should be incorporated into the system to be developed, and therefore the human factor should be taken into account during software development. Users may encounter various problems in the use of a computer program, related in some way to the user's capabilities; such issues, therefore, should be considered in the design process.

The above goal requires us to define several *interface designs* according to the characteristics of all potential users of our application. This approach, though, would increase the complexity and development cost of such an application. The problem could be solved if we had tools that would allow us to reflect these user characteristics as profiles in the design phase. In other words, for a single application, tailored user interfaces would be developed depending on the users' capabilities.

Our objective in this study is to develop a user interface methodology and related tools which allow us to generate user interfaces that are adapted to the individual user, thus saving time and effort on the development of applications. This approach could be achieved by special purpose tools adapted to groups with special needs, or on the other hand, as part of a standard User Interface Design tool. In the second case, the advantage is that the experience and flexibility of these tools may be applied to any user independently of capabilities. Furthermore, these tools, also known as MBUID (Model Based User Interface Design) incorporate methodologies (task analysis, verification, rapid prototyping, etc.) that are also necessary for developing interfaces for any kind of user (Szekely et al., 1993; Bodart et al., 1995; Schlungbaum & Elwert 1995; Lonczewski & Schriber, 1996; Puerta & Eisenstein, 1998).

Our methodology is oriented towards embracing the special needs of a user by means of unique User Interface Design tools, allowing the generation of several interface designs adapted to different types of users in an automatic way, starting from a single application definition, and by introducing user requirements as a profile: each user profile would contain a description of the individual's characteristics, preferences and needs.

This work is based on an extension of GUIDE (Cabreara et al., 1999) developed by our group. This tool works with a formal specification of the application, in which its functionality is described, permitting interactive definition and modification of the visual features (layout) of the interface, and generating an executable prototype of the application. Formal methods give us enough abstraction to cover user needs and design requirements without unnecessary implementation details. It is also interesting to compare different design solutions for a single formal definition, and to analyse the best strategy.

3. Application Architecture

GUIDE is oriented towards traditional event-based windows systems; these have a hierarchical structure which allows us to store elements consisting of an interface starting from the window or main container. Inside this, we can find other interface elements or containers with new elements.

The data structure reflects the hierarchy of the interface. Two types of elements can be defined, the branch (container element) and the leaf (interface element), which store the principal features of the layout design. GUIDE was originally oriented towards fitting the visual layout of the interface (fonts, colour, size, etc.) on a classical windows system.

This paper presents an extension of this tool to embrace user needs. Depending on the type of user and individual characteristics, the user interface can be adjusted, thus changing the appearance of the interaction mechanism. In this way, user inputs and computer outputs are modifiable and customisable. These adjustments can also be seen as a modification of the application's functionality.

To enable this type of adjustment, the user is considered part of the system development process, in which the user's features, style of interaction and preferences are described. This goal is achieved by including a "*user profile*" which is used to choose the best interface components for each user. These profiles include information about the user's input and output devices, colour preferences, and also other more subjective information such as level of experience in the use of devices and/or accessibility requirements.

The choice of components, features and interaction protocol to be incorporated into the user interface is made through a series of rules which decide among the most suitable elements for each user. For instance, if we have a colour-blind user we will choose those elements that allow us to select a combination of colours specific for this person, avoiding colour ranges that could cause confusion. The functional description of the application includes specific parameters which allow us to decide which type of object in the structure is the most appropriate for the implementation of each function.

To generate prototypes, we have chosen the Java language, because its portability onto various operating systems and platforms, including its suitability for Internet. Moreover, our data structure is based on an object hierarchy, which is useful for a hierarchical description of user interface components based on functionality and user adaptation.

4. Functional description of GUIDE+

This section focuses on the generation process of user interfaces in GUIDE+ with user profiles (Figure n 1).

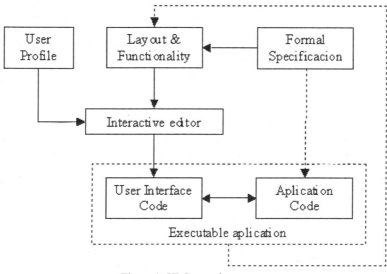

Figure 1: UI Generation process

The initial point is the definition of application functionality, including the interactive components (this definition can be generated from a formal specification). This definition does not contain information about particular interactive components, and only reflects its functionality and the relationship with other high level components or within the non-interactive application. This information is supplied to the interactive tool by means of a text file.

Once we have described the application, the features of each potential user are introduced, using the profile editor. These features determine the user interface adjustment including the choice of appropriate interaction style, component properties and layout.

For instance, suppose that we create a profile for a person with sight problems and limited experience in computer use. The following user interface component may be associated with this profile: a button capable of emitting an audible message when the cursor is over it. The related action is a temporary increase in the size of buttons to show their associated text; moreover, the colours of all the applications are adjusted in order to generate a greater contrast.

In the case of user characteristics that affect functionality, such as application accessibility, the functional description of the application must incorporate qualifiers to allow the range assigned to the user to be matched with a subset of the application functionality. According to the user's level of experience, it is also possible to generate appropriate interfaces. For instance, for a novel user, the interface may contain lists (JList) with all the possible options always visible, from which the user can choose, instead of a ComboBox (JComboBox), which shows only the default option, making it necessary to click to see the remaining options.

When the interface components have been chosen from the user profile description, the designer considers the user interface layout. This is designed by using an interactive tool, which allows the distribution of elements within the interface as well as setting all the features that are not predetermined by the user profile.

Finally, the automatic translator tool is used to generate the interface code. By using this structure, if we change the user profile the code is automatically changed to match the new adjustment, while other component properties remain unaffected.

The structure associated with the component definition is an object hierarchy with a container element (Branch) and final elements (Leafs). The branch is an abstract definition of an interactive component while each leaf defines special purpose configurations adaptable for a range of users. In the first stage of the interface definition, the elements are defined as classic windows environment elements (Button, Label, Textfield, Panel), but once the user profile has been defined these elements are replaced by suitable components, suited to the user's needs.

To implement this replacement, the elements are chosen starting from a parallel structure which allows the definition of new elements as an extension of already existing ones. Each element must have assigned a specific functionality depending on the level of experience of the user and his level of access to the application.

As an example of this structure consider a possible extension of a button.

As the element structure is object-orientated, every element can be extended, with its basic characteristics being inherited and the necessary components added for its specialisation For instance, when defining a standard GButton, its associated action, position, size, predetermined background and foreground colours and text font, are specified. If we need a button for a person with limited sight, a BigButton can be defined, using inheritance, for which we can activate the focus event to increase the size of the font and the contrast between background and foreground.

5. Conclusions and future work

This paper presents the extension of GUIDE to support user needs by using user profiles. The generation of adaptable user interfaces is obtained from a single formal definition of the application interface. The tool is currently being used in the development of software tools for a virtual campus for disabled people.

We are at present working on the implementation of the tool to support user profiles. Future work will be oriented towards evaluating the tool and to applying reengineering to an existing application in order to modify the interface.

Acknowledgements

This research, specified as the User Interface of GIRT, an object-oriented rendering system, is funded by the Comisión Interministerial para la Ciencia y la Tecnología (CICYT), grant number TIC98-0973-C03-01.

References

Baldoni, M., Baroglio, C., Chiarotto, A., Martelli, A., Patti, V. (2000). Intention-guided web sites: a new perspective on Adaptation. In *Proceedings of the 6th ERCIM Workshop "User Interfaces For All"*. Italy. [On-line]. Available: http://ui4all.ics.forth.gr/UI4ALL-2000/files/Long_papers/Baldoni.pdf

Bodart, F., Hennebert, A.M., Leheureux, J.M., Provot, I., Sacré, B., Vanderdonckt, J. (1995). Towards a Systematic Building of Software Architectures: the TRIDENT Methodological Guide. In P. Palanque, R. Bastide (eds.), *Design, Specification and Verification of Interactive Systems DSV-IS'95*. Wien, Springer Verlag.

Cabrera, M., Torres, J.C., Gea, M. (1999). Towards User Interfaces Prototyping from Algebraic Specification. In D.J. Duke & A. Puerta (eds.), *Design, Specification and Verification of Interactive System '99* (pp: 67-83). Wien: Springer Verlag.

Cooper, M., Santacruz Valencia, L. P., Donnelly, A., Sergeant, P. (2000). User interface approaches for accessibility in complex World-Wide-Web applications- an example approach from the PEARL project. In *Proceedings of the 6th ERCIM Workshop "User Interfaces For All"*. Italy. [On-line]. Available: http://ui4all.ics.forth.gr/UI4ALL-2000/files/Position_Papers/Cooper.pdf

Emiliani, P.L., Stephanidis, C. (2000). From Adaptations to User Interfaces for All. In *Proceedings of the 6th ERCIM Workshop "User Interfaces For All"*. Italy. [On-line]. Available: http://ui4all.ics.forth.gr/UI4ALL-2000/files/Position_Papers/Emiliani.pdf

Koutsabasis, P., Darzentas, J.S., Spyrou, T., Lambrinoudakis, K., Darzentas, J. (2000). Aiding Designers to Design for All: The IRIS Approach. In *Proceedings of the 6th ERCIM Workshop "User Interfaces For All"*. Italy. [On-line]. Available: http://ui4all.ics.forth.gr/UI4ALL-2000/files/Posters/Koutsabasis.pdf

Lonczewski F., Schriber S. (1996). The Fuse-System: an integrated User Interface Environment. In J. Vanderdonck (Ed.), *Proceedings of the 2nd International Workshop on Computer-Aided Design of User Interfaces CADUI'96*, Namur, Belgium (pp. 39-56). Presses Universitaires de Namur.

Martins, F. (1996). Semi-Automatic Design and Prototyping of Adaptive User Interfaces. In *Proceedings of the 2nd ERCIM Workshop "User Interfaces for All"*, Czech Republic. [On-line]. Available: http://ui4all.ics.forth.gr/UI4ALL-96/martins.pdf

Puerta, A., Eisenstein, J. (1998). Interactively Mapping Task Models to Interfaces in MOBI-D. In *Proceedings of the 5th Eurographics Workshop on Design, Specification and Verification of Interactive Systems (DSV-IS'98)*, Abingdon, United Kingdom.

Schlungbaum, E., Elwert, T. (1995). Modelling and Generation of Graphical User Interface in the TADEUS Approach. In P. Palanque, R. Bastide (eds.) *Design, Specification, and Verification of Interactive Systems* (pp. 193-208). Wien: Springer Verlag.

Stephanidis, C., Akoumianakis, D. (1999). Accessibility guidelines and scope of formative HCI design input: Contrasting two perspectives. In *Proceedings of the 5th ERCIM Workshop "User Interfaces for All"*, Germany. [On-line]. Available: http://ui4all.ics.forth.gr/UI4ALL-99/Stephanidis_1.pdf

Szekely, P.; Luo, P.; Neches, R. (1993). Beyond Interface Builders: Model-Based Interface Tools. In *Proceedings of InterCHI'93* (pp 383-390).

Applying User Dichotomies to the Design of Web User Interfaces

Lynne Dunckley[a], Matthew Dunckley[b]

[a] The Open University, Walton Hall, Milton Keynes, MK7 6AA, UK *
[b] United Business Media International, CityReach, 5 Greenwich View Place,
London, E149NN, UK

Abstract

This paper describes how Web technologies can be combined in a simple way to provide user interfaces that can be adapted by a process of differentiation based on a generic or stem interface to meet the needs of different user groups. The user groups are characterized by means of dichotomies in order to simplify the design space. User dichotomies are end-points of various dimensions on which users differ. In addition to addressing user issues such as age, visual and motor abilities this approach can be applied to the development of multicultural interfaces.

1. Introduction

At present the pervasiveness of the Internet has transformed user interface style so that a Web browser style is predominant. Even many traditional application areas that are unrelated to the Web adopt this style. For Web developers this gives both a challenge and an opportunity to provide interfaces that can adapt dynamically to fit the needs and requirements of different user groups. Web developers are only too aware that the content of their pages may be displayed on a large variety of devices. For example users with different visual and motor impairments may use non-standard devices and browsers as assistive technology. The designer can no longer dictate the 'look & feel' the user will experience and has to yield control to some extent to the user. In addition the user's environment may no longer be known to the developer in any detail.

1.1 The impact of Internet technologies

Traditionally user interfaces have been designed to provide a static product intended to accommodate the perceived traits of the majority of future users. This approach is difficult to follow or justify with Web interfaces where there are many variations in the computer and browser environment. Web technology enables the provision of interfaces across cultural and geographical boundaries. However this should not mean that every user regardless of their cultural, cognitive and physical abilities and personnel preference should be faced will 'one fits all' interface. By taking the opportunity to separate the information content from its presentation, the developer can adapt the presentation in many different ways, one of which can be to take into account the specific abilities as well as the cultural characteristics and expectations of the users. Style sheets and XML are one way forward that can provide relatively manageable and low cost adaptation in such a way as to provide a partnership between the user and the developer. Style sheets can be embedded in a Web page or linked style sheets can be referenced by a Web page to insert a style definition. This can result in improved performance, for example more efficient downloading and reduced maintenance, since a whole Web site can be organized through a single style sheet to provide a consistent 'look & feel'.

There are a number of different style sheet facilities. JavaScript is a scripting language that enables programming commands to be embedded into Web pages supporting high levels of interaction, triggered by actions performed by users as they read the page. Client side JavaScript is interpreted within the user's browser, embedded in HTML and executed by the browser. Server side JavaScript commands can also be embedded within the HTML within the same page but they are processed on the server so that the results of the processing is displayed on the user's browser. Alternatively the style sheet could be based on Cascading Style Sheets(CSS) so that blocks of HTML can have their own style margins, fonts etc. Once the style has been defined it can be applied automatically. In the future XSL (Extensible Style Language) will offer a significant alternative through a combination with XML alone or by transforming XML documents into HTML. In addition XML tools need to be developed to support developers. At present only a few proprietary tools appear to address these issues.

1.2   The need for adaptation

Users may have needs and attitudes that are significantly different from the norm because of age, physical or cognitive disabilities. In addition users from different cultural backgrounds differ in objective factors (e.g. gender, age, ethnic background, mother-tongue) and subjective cultural factors (e.g. values, beliefs and rituals) and have different preferences (Dunckley et al., 1999).

Accessibility is an important issue that designers cannot ignore. However designing for a diverse user groups is complex and difficult as it involves linking user characteristics with design artefacts such as objects, images and metaphors. Akoumianakis et al. (2000) discuss the problems of diverse user requirements from the interaction designers viewpoint, advocating the development of adaptive user interfaces through the identification of plausible alternatives, describing the building of multiple metaphor environments. Multiple metaphor environments are needed because of the diversity of users, diversity of contexts and the diversity of platforms.

Much of the diversity arises from the users' physical abilities and cultural backgrounds. Rowan et al. (2000) have summarised the problems of designing Web sites with effective disability access. Current accessibility guidelines require developers to fully understand the requirements of each guideline, the reasoning behind that guideline and the steps that have to be taken to meet that guideline. They conclude that no single evaluation method exists which will highlight every accessibility issue in a simple form for designers. Therefore, without a clear framework, the choice of interaction objects and metaphors is particularly difficult for Web designers.

The situation is actually simpler for international Web interfaces. Hofstede's work is particularly relevant because of its focus on human interaction where cultural differences clearly matter. He carried out a study of 116,000 IBM employees distributed through 72 countries using 20 languages in 1968 and 1972. He conceptualised culture as 'programming of the mind', meaning that certain reactions were more likely in certain cultures than in others as a result of differences between basic values of the members of different cultures. Hofstede proposed that all cultures could be defined through three dimensions:
- power distance (PD) - the degree of emotional dependence between boss and subordinate
- collectivism versus individualism (IC) - integration into cohesive groups versus being expected to look after him/her self
- femininity-masculinity - which could be interpreted as toughness versus tenderness.
- In addition he recognised that, for Western cultures there was another important dimension:
- uncertainty avoidance - the extent to which members feel threatened by uncertain or unknown situations;

and for Eastern cultures
- long-term Confucian orientation - which represented a philosophy of life that was prepared to sacrifice short-term results for long-term gain.

Using this framework Marcus (2000) evaluated Web sites in terms of Hofstede's dimensions. He suggests that high PD cultures such as Malaysia generate Web pages with strong axial symmetry, include images that feature prestigious buildings and include symbols such as official seals. Low PD cultures in contrast will develop Web pages with asymmetric layouts that emphasise the role of people using images that suggest the same status for different genders, ages and ethnicity. Similarly individualist cultures generate sites with images of individual success, materialism and 'what's new' versus collectivist cultures which emphasise relationships.

Consequently, specialists in software internationalisation have realized that simply translating text and changing the format of dates and numbers is not enough. Localization is the name given to the process of adapting user interfaces to the culture of the end-user. In the past this has usually been a permanent transformation of a software product into a localised product with changed features, particularly the interface. A number of different localization approaches have been considered, for example:
- full localization through the development of new versions of the software, evaluated using appropriate methods for the target culture,
- developing an adaptive interface that anticipates the user's preferences,
- developing an intelligent interface, where the appearance of the interface can be changed by the interface itself in response to the user's interactions with it. The system will usually have a component that receives signals or events from the interface as the interaction proceeds. (Keeble & Macredie, 2000).

It is not possible to transfer the responsibility for localization to the user. Research has also shown that very few users customize interfaces, even though the latest browsers provide such facilities for the user. For example both IE5 and Mozilla provide the facility for the user to specify their own style sheet as well as more basic options to change font size, colours and modify the display of graphics and animations.

2. Differentiation approach

We are suggesting differentiation as a concept that could be particularly helpful for the design of Web interfaces where the abilities and cultures of the remote users will be relatively unknown. Localizing an interface developed in one culture for another is resource intensive. Instead, this approach would take an undifferentiated Web interface and modify it to provide interfaces that were more acceptable to users from different cultures but not fully localized. The concept of differentiation is borrowed from cell biology. Cells in tissues specialize in structure and function through a process known as differentiation. A cell is not born in an information free environment but in a universe of signals sent out by its fellows. In general cells can be divided into stem cells which divide repeatedly and end-stage cells which cannot divide e.g. red blood cells. Stem cells are an important group of cells that can divide and reproduce but they are undifferentiated and contain all the genetic information that is present in the DNA. However they can divide to create other cell lines that can become differentiated by signals from the environment so that the differentiated cells can adopt a very different appearance and provide different functions. The complexity of the social life of a cell depends on its environment and the range of signals it is capable of responding.

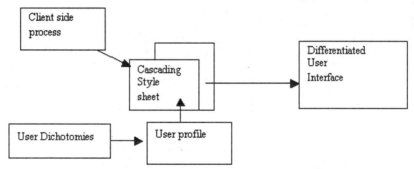

Figure 1: Architecture/Framework for differentiated site

In an analogous way the stem interface, controlled by a master style sheet would be adapted by receiving signals from the user to present a differentiated interface provided by cascading style sheets. Many differentiated interfaces can then evolve from the original stem interface. One of these could be an internationalised version. The communication of the user's signals is a key question. What will be the source of the user's signals and how will they be communicated. This could be through the site collecting user metadata which is used to select a style imposed by the designer. This data collection relies on the users either being repeat visitors with a history of interaction or being recognized as similar to another user. Another possibility is to provide a series of thumbnail images of possible interface displays that the user selects from. Alternatively the user could be asked key questions to identify those cultural factors that they are particularly sensitive to. We need to know which of these approaches will be acceptable to users. What we are suggesting falls short of full adaptability. Related to adaptability is proactivity which is the ability of an application to interact with the user on its own initiative. Applications can be passive i.e. provide information only in response to user's requests or be proactive – send information to users on the occurrence of certain events (Fraternali, 2000).

This interface differentiation can be achieved in practice by using an architecture /framework, such as that illustrated in Figure 1, that allows the interface to be differentiated for different users. In this paper we describe a practical approach by using user dichotomies to simplify the design space. User dichotomies are end-points of various dimensions on which users differ. In addition to addressing user issues such as age, visual and motor abilities this approach can be applied to the development of multicultural interfaces. The method we have been working on exploits features of XML and Style Sheets. Instead of using a single style sheet to define a centralised design that ensures consistency and uniformity across the Web site style sheets can merge a user's style sheet with a generic style sheet to provide a user with an adapted interface presentation. For example font, colour and the position of objects on the Web page can be altered. A layering approach to design can facilitate this, creating layers of objects when the page is loaded, testing for browser functionality and assigning different functions.

3. XML Documents

The main application at the moment of XML tends to be to send information from the server to the client in a format that the browser understands. However XML when combined with style sheets can be used to display format replacing HTML completely. A middle ground is to combine this with HTML in an improved structure.

Replacing HTML with XML will result in smaller and more efficient Web pages because the documents will have a better structure. XML tags can be used by search engines since they encapsulate metadata about the content of the document allowing content based searches. Agents will be effective in terms of performance because they will only need to locate and search specific elements within the XML documents processing and bandwidth requirements will be reduced by the use of XML. Connections between agents and search engines will be made easier by making it possible to exchange data a users agent will be able to search data selected by a search engine in an intelligent way. XML and relational databases can be combined since XML can be used to describe the content and the structure to exchange information between databases from different suppliers and models. XML can send information to the clients that they can understand interpret and use locally. Templates are also very useful since it is possible to create a template which is saved in the database. Templates can be included in other templates and access global variables. For example a single Web site can be differentiated in five different ways. Keywords attached to the end of the URL for the main site, result in the display of the differentiated sites. Users interests and preferences can be stored so that a user is redirected to the appropriate customised site.

3.1   User Dichotomies

This differentiation requires the identification of a 'stem' style and a small number of differentiated styles. In this paper we describe a practical approach based on user dichotomies to simplify the design space.

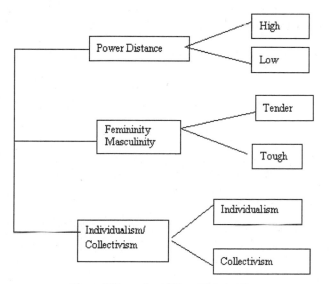

Figure 2 Examples of User Dichotomies

The process of design differentiation is assisted by completing the analysis shown in Table 1 where the user factors are related to possible design factors. The user factor 'Age' is simplified into the dichotomies 'older' and 'younger'. The design factors that are seen as sensitive to that user factor are specified for each dichotomy. For example in Table 1 font, colour scheme and spacing are identified as sensitive to 'Age'. Walker, Millians and Wooden (1996) found there is a critical minimum target size below which older adults cannot effectively use a mouse. Dark text on a light background (Murch, 1987) is more efficient for older adults. Highlighting important information is important to support the perceptual organisation of older adults (Czaja, 1997).

Table 1 User Factor Design Factor Mapping

User Factor	Dichotomy	Design Factor	Requirement
Age	Older	Font	14pt san serif e.g. Arial
		Colour scheme	Dark type on light background
		Spacing	Large buttons in space

4. Conclusion

The method we are proposing provides a systematic way of approaching the design of Web interfaces for diverse user groups without falling into the pitfalls of either 'one size fits all' or adopting crude stereotypes. By linking user dichotomies directly to design factors that are relevant to Web interfaces it is possible to consider users in the terms of a number of dimensions and to check that their needs are being addressed in the design. This approach does not lead to a fully adapted interface which may, however, become achievable as technologies improve. However it does enable the needs of many user groups to be met in a practical and cost effective manner.

References

Akoumianakis, D., Savidis, A., Stephanidis, C. (2000). Encapsulating intelligent interactive behaviour in unified user interface artefacts. *Interacting with Computers*. 12, 383-408.

Dunckley L, Hall, P.A.V., Smith. (1999). Software International Architecture and User Interface Design. In Prabhu, G and del Galdo E. M. (Eds.) *Proceedings of First International Workshop on Internationalization of Products and Systems,* Rochester USA

Czaja S.J. (1997). Computer technology and older adults. In M.E. Helander, T.K. Landauer & P.Pabhu (Eds.), *Handbook of human-computer interaction* (2nd. Ed., pp.797-812). Elsevier.

Fraternali, P. (2000). Development of Data-Intensive Web Applications. *ACM Computer Surveys*, 31(3), 230-263.

Keeble, R.J., Macredie, R.D. (2000). Assistant agents for the world wide wed intelligent interface design challenges. *Interacting with Computers*, 12, 337-381

Marcus, A., Gould, E.W. (2000). Crosscurrents: cultural dimensions and global Web user-interface design. *Interactions*, 7(4), 32-46.

Murch, G.M. (1987). Colour graphics – blessing or ballyhoo?. In R.Baecker & W.A.S. Buxton (Eds.) *Readings in human computer interaction: A multidisciplinary approach* (pp.333-341). Morgan Kaufmann.

Rowan, M., Gregor, P., Sloan, D., Booth, P. (2000). Evaluating web resources for disability access. In *Proceedings of the 4th international ACM conference on Assistive Technologies*, Arlington, VA USA (pp. 80-84).

Walker, N., Millians, J., Worden, A. (1996). Mouse accelerations and performance of older users. In *Proceedings of the Human Factors and Ergonomics Society 40th Annual meeting* (pp. 151-154).

Identifying Critical Interaction Scenarios for Innovative User Modeling

V. Katie Emery[1], Julie A. Jacko[1], Thitima Kongnakorn[1], Vipat Kuruchittham[1], Steven Landry[1], George McLeland Nickles III[1], Andrew Sears[2], Justin Whittle[1]

[1]School of Industrial & Systems Engineering
Georgia Institute of Technology
765 Ferst Drive, Atlanta, GA 30332-0205

[2] Information Systems Department, UMBC
1000 Hilltop Circle
Baltimore, MD

Abstract

Usability testing typically focuses on methodology and metrics, while the specific interactions being tested are chosen in an ad hoc way. This paper demonstrates a framework for organizing interaction scenarios for graphical user interfaces (GUI). The framework is an adaptation of the two-dimensional abstraction hierarchy introduced by Rasmussen [1] in which an interaction consists of a purpose, functionality, and form. Interactions for a GUI are organized into four main categories, with numerous subtasks. The four main categories determined are 1) object manipulation, 2) content manipulation, 3) view manipulation, and 4) information presentation. The general framework can guide evaluators in choosing key interaction scenarios for GUI applications across a diverse array of user capabilities.

1. Introduction

Many tools exist to facilitate the evaluation of computer interfaces (e.g., cognitive walkthroughs, heuristic evaluations, user testing, etc.). Ideally these methods would investigate all possible interaction scenarios that might occur in a particular application between the user and the computer. However, this is not realistically feasible. As a result, evaluators often employ only small subsets of interactions or selected representative interactions. To our knowledge no research to date has provided a systematic method to determine what interactions should be investigated, or how interactions could be compared to other interactions within the same interface, or with interactions in other interfaces. A hierarchical framework that organizes potential interaction scenarios would provide support for choosing interactions to be tested, and allow comparison of interactions within and between applications.

One difficulty in creating such a framework is that an interface abstracts the user from the task they are trying to accomplish. The interface can be viewed as a "cognitive agent" on which the user must act to accomplish the desired goals [2]. The user's ultimate goal (e.g., producing a document, creating a chart, coordinating data, etc.) is not directly performed, but instead is composed of a sequence of low-level interactions with the interface. In this regard the "usability" of an interface is a combination of low-level actions (e.g. moving a mouse, pressing a key combination, etc.), the sequence of low-level interactions required to accomplish a goal, and the ability of the interface to satisfy the user's goal. These different aspects of the task reflect both physical and cognitive behavior from the user. Additionally, interactions can be initiated by the interface (e.g. providing feedback, alerting the user to system status, etc.), further complicating efforts to create a general framework of interactions.

For a high level goal, such as inserting a chart into a document, there are a number of lower level objectives that must be accomplished, typically by performing actions on objects through the functionality of the interface. This structure, characterized by purpose at the higher levels, functional aspects in the middle levels, and form at the lower levels, can be represented as an abstraction hierarchy. This type of hierarchy, applied to interactions in a graphical user interface (GUI) can elicit key interaction scenarios. In the succeeding sections, a description of the general framework is followed by an example of how it may be applied.

2. Abstraction hierarchy

Rasmussen's abstraction hierarchy [3] serves as the foundation for the framework and possesses two dimensions, as shown in Figure 1. On the horizontal axis is decomposition, where moving to the right can be considered "zooming in" on components of the system. On the vertical axis is the level of abstraction, which breaks down the task, from goals or purposes at the highest level down through the physical form of the interface at the other. Goals propagate down through the hierarchy, eventually affecting the physical functioning of the system, while the physical form can produce effects that propagate up the hierarchy, and can affect the purposes, either by confounding them, or creating new purposes.

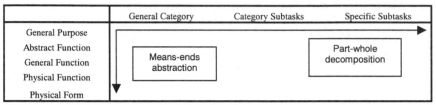

Figure 1: Abstraction hierarchy

It should be noted that this use of the abstraction hierarchy differs from its use by Rasmussen in ecological interface design (EID). In EID, the hierarchy is used to describe the process that is being controlled through the display, so that the display reflects the underlying functionality of the process. For its use in eliciting interaction scenarios, the process (e.g. creating a document, manipulating data) is not being shown. Instead, the abstraction diagrams the functionality of the interface.

The value of this structure is that, in segmenting an interaction into several levels of abstraction, the interaction can be compared with other interactions at each level. For example, within a completed framework, one could compare different physical actions that can accomplish the same purpose, or compare the ability of someone to accomplish the same purpose across different interfaces. The hierarchy also clearly indicates the tangible connections between purpose, function, and form.

Applying the abstraction hierarchy to interaction scenarios, the high level General Purpose is given by the particular goal of the user, and is not dependent upon the application. Each succeeding level answers the "how" that accomplishes the preceding level, in increasing detail. The Abstract Function is the high-level description of the method by which the General Purpose will be accomplished. The next level, General Function, describes how the Abstract Function is accomplished in terms of the functioning of the general class of applications in question. The Physical Function level describes how, in regards to the functionality of the particular interface, the General Function is accomplished. Finally, the Physical Form describes the items necessary to accomplish the Physical Function.

2.1 The general framework for graphical user interfaces

For a particular class of applications, such as graphical user interfaces (GUI), a portion of the framework can be completed, easing implementation of the framework for particular interfaces. In this regard a literature search for interaction scenarios was conducted for GUI research. The interactions described in the literature were catalogued and categorized into General Functions. These General Functions were then classified into the type of manipulation that was being accomplished, resulting in the level of Abstract Function. The results of the literature search and the resulting categorizations are shown in Table 1. The four Abstract Functions are 1) object manipulation, 2) content manipulation, 3) view manipulation, and 4) information presentation. Note that the level of General Purpose, and Physical Function and Form are removed, as these relate to specific objectives in the case of the former, and in specific interfaces for the latter. Short descriptions of the categories, including examples of objects to which they apply, are also given.

3. Application to specific interfaces

The first few hierarchies for specific interactions involved in the opening of a document in Microsoft Word 2000 is shown in Table 2. For clarity and brevity, the Physical Form appears in parentheses on the same level as Physical Function. In this example the advantages of the hierarchical form can be seen. Lower level activities can be identified and measured (e.g., mouse movements, keyboard movements, etc.). The structure of the interaction shown by the various hierarchies and how the hierarchies are combined in many cases reflect the navigation required, both within the general function, and for the general purpose overall.

Each interaction at each level must be executed for the General Purpose to be accomplished successfully. The complexity of the interaction, and the eventual success of the interaction, is dependent not only on the underlying low-level interactions (e.g., ctrl+o, select "file" menu, etc), but also on the structure of the high level interactions and the feedback provided by the interface.

Table 1: Hierarchy Framework and References

	General category	General subtasks	Specific subtasks	References
Abstract function	Object manipulation			
General function		Open/Close (application, file, window ...)		(Miah & Alty, 2000; Wixon, Williges, & Coleman, 1985)
		Add/Delete (file, application, graphic, ...)		(Bordegoni, 1994)
		Change attributes (color, text properties, ...)		(Wixon et al., 1985)
		Change position		(Bordegoni, 1994; Bowman, Johnson, & Hodges, 1999; Foley, Wallace, & Chan, 1984; Miah & Alty, 2000)
		Change orientation		(Bordegoni, 1994; Bowman et al., 1999; Foley et al., 1984; Gallimore & Brown, 1993)
		Change size		(Miah & Alty, 2000)
		Change membership (objects, files, ...)		(Bowman et al., 1999)
		Activate/Deactivate (window, object, ...)		(Bordegoni, 1994; Bowman et al., 1999; Foley et al., 1984; North & Shneiderman, 1997; Perlman, Green, & Wogalter, 1992)
		Export (print, send to audio channel, etc.)		(Shneiderman, 1998)

	General category	General subtasks	Specific subtasks	References
Abstract function	Content manipulation			
General function		Text entry		(Foley et al., 1984; Kishino & Hayashi, 1995; Perlman et al., 1992; Thimbleby, 1983; Wixon et al., 1985)
		Object creation		(Bordegoni, 1994; Foley et al., 1984)
		Selection from a set of continuous values		(Ackerman & Cianciolo, 1999)
		Selection from a discrete array		(Ackerman & Cianciolo, 1999; Buxton, 1988; Instance & Howarth, 1993; Perlman et al., 1992)
		Import (scan, tactile interface, sound, ...)		(Brewster, Raty, & Kortekangas, 1996)

	General category	General subtasks	Specific subtasks	References
Abstract function	View manipulation			
General function		Scrolling in 2,3 dimensions		(Bederson & Meyer, 1998; Darken & Sibert, 1996; Kaptelinin, 1995; North & Shneiderman, 1997; Swierenga, 1990)
		Paging		(Paap, Noel, McDonald, & Roske-Hofstrand, 1987)
		Zooming		(Beard & John Q. Walker, 1990; Bederson & Meyer, 1998; Kommers, 1991; North & Shneiderman, 1997)
		Changing perspective		(Arsenault & Ware, 2000; Darken & Sibert, 1996)
		Channel control (i.e. direct input to particular instance, channel, etc.)		(Arnold, A. G., & Roe, R. A. 1989; Goldstein, J., & Roth, S. F. 1994)

	General category	General subtasks	Specific subtasks	References
Abstract function	Information presentation			
General function		Alerting (any of the senses)		(Ware, Bonner, Knight, & Cater, 1992; Wiener & Curry, 1980)
		Feedback (any of the senses)		(Arsenault & Ware, 2000; Bowman et al., 1999; Dennerlein, Martin, & Hasser, 2000)
		Attention directing (any of the senses)		(Ware et al., 1992)
		Unrequested information (balloon help, ...)		(Freeman, 1994)
		Requested information (document, image...)		(Jacobson, Fusani, & Yan, 1993; Wixon et al., 1985)

3.1 Applications to User Needs

The failure of any key interaction in the task sequence, either due to the human or machine side of the system can potentially result in a failure to achieve the system goal(s). It is therefore important for the evaluator to consider the user in the development of the hierarchy. The evaluator needs to discover what assumptions each interaction assumes about the user's skills and abilities, and about the context of use. Those assumptions that are incorrect will result in an inability for the user to achieve her or his desired goal using that interface.

Table 2: Completed hierarchy for two tasks

	General category	General subtasks	Specific subtasks
General purpose	Open a file		
Abstract function	Object manipulation		
General function		Open a new document	
Physical function (Physical form)			Ctrl+o (keyboard) Select "file" menu, select "open" (mouse) "Open" button on toolbar (mouse)

	General category	General subtasks	Specific subtasks
General purpose	Open a file		
Abstract function	Information presentation		
General function		Visual feedback	
Physical function (Physical form)			Blank file opens (monitor)

	General category	General subtasks	Specific subtasks
General purpose	Open a file		
Abstract function	Content manipulation		
General function		Selection from a discrete array	
Physical function (Physical form)			Select file to open (mouse) Move to file to open (keyboard)

4. Conclusions

The hierarchy presented in this paper provides practitioners and researchers with a way in which to organize interaction scenarios to achieve more systematic interface evaluations. In addition, this organization scheme, which is grounded in Rasmussen's hierarchy, enables researchers to more quickly identify and attend to interaction scenarios that may pose particular challenges for people with limited capabilities due to perceptual, physical and/or cognitive impairments. By organizing and categorizing interaction scenarios in this manner, researchers and practitioners are better equipped to engage in accurate modeling of human performance for computer-based tasks. Accurate user modeling is particularly challenging for users with reduced capabilities due to perceptual, physical, and/or cognitive impairments because, to-date, researchers have not comprehensively characterized the influence of these types of impairments on computer-based task performance. The organization scheme introduced in this paper will facilitate modeling of this type by providing an organization scheme of interaction scenarios that should be investigated across diverse user groups. This paper serves as a launching point as the development of this scheme is ongoing.

References

1. Rasmussen, J., Pejtersen, A.M., Goodstein, L.P. (1994). *Cognitive Systems Engineering.* New York: John Wiley & Sons, Inc.
2. Norman, D.A. (1991). Cognitive artifacts. In J.M. Carroll (Ed.), *Designing Interaction: Psychology at the Human-Computer Interface.* New York: Cambridge University Press.
3. Rasmussen et al., 1994.
4. Miah, T., Alty, J.L. (2000). Vanishing Windows Technique for Adaptive Window Management. *Interacting with Computers, 12*(4), 337-355.
5. Wixon, D.R., Williges, R.C., Coleman, W.D. (1985). Collecting Detailed User Evaluations of Software Interfaces. In the *Proceedings of the Human Factors Society 29th Annual Meeting.*

6. Bordegoni, M. (1994). Parallel Use of Hand Gestures and Force-Input Device for Interacting with 3D and Virtual Reality Environments. *International Journal of Human-Computer Interaction, 6*(4), 391-413.

7. Bowman, D.A., Johnson, D.B., Hodges, L.F. (1999). Testbed evaluation of virtual environment interaction techniques. In the *Proceedings of the ACM symposium on Virtual Reality Software and Technology.*

8. Foley, J.D., Wallace, V.L., Chan, P. (1984). The Human Factors of Computer Graphics Interaction Techniques. *IEEE Computer Graphics and Applications, 4*(11), 13-48.

9. Gallimore, J.J., Brown, M.E. (1993). Visualization of 3-D Computer-Aided Design Objects Articles. *International Journal of Human-Computer Interaction, 5*(4), 361-382.

10. North, C. Shneiderman, B. (1997). *A Taxonomy of Multiple Window Coordinations*, [On-line]. Available: ftp://ftp.cs.umd.edu/pub/hcil/Reports-Abstracts-Bibliography/3854HTML/3854.html.

11. Perlman, G., Green, G.K., Wogalter, M.S. (1992). A Comparison of Direct-Manipulation, Selection, and Data-Entry Techniques for Reordering Fields in a Table. In the *Proceedings of the Human Factors Society.*

12. Shneiderman, B. (1998). *Interaction Devices, Designing the User Interface: Strategies for Effective Human-Computer Interaction* (3rd ed., pp. 305-349). Reading, MA: Addison-Wesley.

13. Kishino, S., Hayashi, Y. (1995). A Method for Estimating Code Key-In Times of College Students. *International Journal of Human-Computer Interaction, 7*(2), 123-134.

14. Thimbleby, H. (1983). Guidelines for 'Manipulative' Text Editing. *Behaviour and Information Technology, 2*(2), 127-161.

15. Ackerman, P., & Cianciolo, A. T. (1999). Psychomotor abilities via touch-panel testing: Measurement innovations, construct, and criterion validity. *Human Performance, 12*(3-4), 231-273.

16. Buxton, W. (1988). The Natural Language of Interaction: A Perspective on Non-Verbal Dialogues. *INFOR: Canadian Journal of Operations Research and Information Processing, 26*(4), 428-438.

17. Instance, H., & Howarth, P. (1993). *Performance Modeling and Investigation of User Performance using an Eyetracker as an input device.* Paper presented at the the Polish-English Meeting on Information Systems, Bialystok, Poland.

18. Brewster, S. A., Raty, V.-P., & Kortekangas, A. (1996). *Enhancing Scanning Input with Non-Speech Sounds.* Paper presented at the Second Annual ACM Conference on Assistive Technologies.

19. Bederson, B., & Meyer, J. (1998). Implementing a zooming user interface: Experience building Pad++. *Software - Practice and Experience, 28*(10), 1101-1135.

20. Darken, R. P., & Sibert, J. L. (1996). Navigating Large Virtual Spaces. *International Journal of Human-Computer Interaction, 8*(1), 49-71.

21. Kaptelinin, V. (1995). *A Comparison of Four Navigation Techniques in a 2D Browsing Task.* Paper presented at the Proceedings of ACM CHI'95 Conference on Human Factors in Computing Systems.

22. Swierenga, S. J. (1990). *Menuing and Scrolling as Alternative Information Access Techniques.* Paper presented at the Proceedings of the Human Factors Society 34th Annual Meeting.

23. Paap, K. R., Noel, R. W., McDonald, J. E., & Roske-Hofstrand, R. J. (1987). *Optimal Organizations Guided by Cognitive Networks and Verified by Eyemovement Analyses.* Paper presented at the Proceedings of IFIP INTERACT'87: Human-Computer Interaction.

24. Beard, D. V., & John Q. Walker, I. (1990). Navigational Techniques to Improve the Display of Large Two-Dimensional Spaces: The Visual Presentation of Information. *Behaviour and Information Technology, 9*(6), 451-466.

25. Kommers, P. A. M. (1991). *Virtual Structures in Hypermedia Resources Congress II: Design and Implementation of Interactive Systems.* Paper presented at the Proceedings of the Fourth International Conference on Human-Computer Interaction.

26. Arsenault, R., & Ware, C. (2000). *Eye-hand co-ordination with force feedback.* Paper presented at the Proceedings of the ACM Conference on Human Factors and Computing Systems, New York.

27. Arnold, A. G., & Roe, R. A. (1989). Action Facilitation; A Theoretical Concept and its Use in User Interface Design. Paper presented at the Proceedings of the Third International Conference on Human-Computer Interaction.

28. Goldstein, J., & Roth, S. F. (1994). Using Aggregation and Dynamic Queries for Exploring Large Data Sets. Paper presented at the Proceedings of ACM CHI'94 Conference on Human Factors in Computing Systems.

29. Ware, C., Bonner, J., Knight, W., & Cater, R. (1992). Moving Icons as a Human Interrupt. *International Journal of Human-Computer Interaction, 4*(4), 341-348.

30. Wiener, E. L., & Curry, R. E. (1980). Flight-Deck Automation: Promises and Problems. *Ergonomics, 23*(10), 995-1011.

31. Dennerlein, J. T., Martin, D. B., & Hasser, C. (2000). *Force-feedback improves performance for steering and combined steering-targeting tasks.* Paper presented at the Proceedings of the ACM Conference on Human Factors and Computing Systems, New York.

32. Freeman, D. (1994). *Object Help for GUIs.* Paper presented at the ACM Twelfth International Conference on Systems Documentation.

33. Jacobson, T. L., Fusani, D. S., & Yan, W. (1993). Q-Analysis of User-Database Interaction. *International Journal of Man-Machine Studies, 38*(5), 787-803.

Ambiguity Problems in Human-Computer Interaction

Ivan Kopeček[1], Karel Pala[1], Markéta Straňáková-Lopatková[2]

[1]Faculty of Informatics, Masaryk University, Botanická 68a, 602 00 Brno, Czech Republic
E-mail: kopecek@fi.muni.cz, pala@fi.muni.cz
[2]Faculty of Mathematics and Physics, Charles University, Prague, Czech Republic.
E-mail: stranak@ufal.mff.cuni.cz

Abstract

The paper deals with an algebraic approach to the problem of solving ambiguities in dialogues. A method based on formal models of dialogue strategies is introduced and briefly illustrated in a connection with the developed dialogue system that supports blind and visually impaired programmers in creating the source code of computer programs.

1. Introduction

Solving ambiguity problems in human-computer interaction, especially in spoken dialogue systems, is one of the most important tasks related to practical implementation of dialogue systems. The ambiguity can be related to syntax, semantics and pragmatics of language and it can also arise as a consequence of our incapability to detect or to understand some relevant features (e.g. prosodic (Kopecek, 2000a) or emotional attributes (Kopecek 2000b)).

Current research concerning dialogue systems covers a wide range of approaches, e.g. speech act theory, language modelling, theory of common knowledge, lexical semantics and pragmatics, conversational implicature theory, dynamic semantics and dynamic epistemic logic, conversation theory, user modelling, issues related to emotions and prosody, agents theory and many others. On the other hand, practical applications like solving ambiguity problems in real dialogue systems demand solutions of many tasks of different types, often having very interdisciplinary features. In this situation, there is a natural need to use general models that can be applied to a wide variety of related problems, both theoretical and practical. Dialogue automata that can model different aspects of dialogues in relation with ambiguity problems offer a general framework for this task. (Kopecek 2000b).

Dialogue automata use simple algebraic structures relating internal states of a participant of a dialogue (agent) with the attributes of dialogue utterances by means of transition and output functions. A powerful feature of the dialogue automata approach lies in the possibility to determine unknown attribute values in the context of the dialogue communication. It can be used for finding unknown or unsure parameters and values in the dialogue context.

Appropriate choice of attributes of dialogue automata model enables us to model different instances of the dialogue communication. For example, dialogue automata may relate emotions (that are described by the attributes of the internal states) with prosody (which is described by the attributes of the dialogue utterances) and other parameters that are entering the model. Another choice of the basic attributes allows us to analyse semantic ambiguity in dialogue, etc. In order to determine the structure of the dialogue automaton, we need to have sufficient data describing the internal states and dialogue utterances. To this aim we can use user inquiring, deriving the emotional state from facial expressions, using measurements, etc.

In what follows, we explain basic ideas of this approach, discuss a general ambiguity problem and in addition we briefly illustrate this idea with respect to some applications.

2. Preliminaries, basic notions

In the following text we use standard terms and notation of algebraic theory of formal languages and automata. If M is a set, then M^* will denote the free monoid over the set M, i.e. the set of all strings consisting of the elements of the set M (including the empty string). An alphabet is a finite nonempty set. If U and V are sets, $U \times V$ denotes the Cartesian product of U and V.

Let us recall the notion of *information system* (in the sense of Pawlak, see e.g. (Pawlak, 1981). Let U, T, V be nonempty sets and let f be a mapping of the set $U \times T$ into V. Then the ordered quadruple $S = (U, T, V, f)$ is said to be an information system. The elements of U are called objects, the elements of T attributes, and the elements of V

values of attributes. Two objects u, v are called interchangeable if $f(u, t) = f(v, t)$ for all t from the set T. In what follows, we assume that no objects are interchangeable. Usually, it is assumed that the sets U, T, V are finite. We do not suppose this and use the notion of information system in this slightly generalized sense.

Further, we will give a brief overview of some basic related notions (Kopecek, 1999, 2000b).

If $U = (U, T, V, f)$ is an information system, and $u \in U^*$, then u will be called *dialogue*. This definition can be interpreted as follows: Let $u = (u_1, u_2, ..., u_n)$ be a dialogue. Then the odd indexed elements of u correspond to the utterances of the first participant of the dialogue and the even indexed elements correspond to the utterances of the second participant of the dialogue. Empty string corresponds to the empty dialogue.

Let $U = (U, T, V, f)$ be an information system and # a special symbol not belonging to U. Then a function $s : \{U^* \cup \{\#\}\} \to \{U \cup \{\#\}\}$ is called *dialogue strategy*. The set of all dialogue strategies related to U will be denoted by $Stg(U)$.

Suppose s is a dialogue strategy, $u \neq$ # and $s(u) = v$. Then the interpretation is as follows: based on the present dialogue u, the dialogue strategy s determines the continuation of the dialogue, which is uv, (concatenation of u and v) provided $v \neq \#$, or terminates the dialogue, if $v = \#$.

Suppose now that $u = \#$ and $s(u) = v$. Then the interpretation is as follows: the dialogue strategy s determines the beginning element of the dialogue, which is v, provided $v \neq \#$, or terminates the dialogue, if $v = \#$ (resulting in the empty dialogue).

Let s_1, $s_2 \in Stg(U)$. We shall say that the ordered pair of strategies (s_1, s_2) *generates* the dialogue $u = (u_1, u_2, ..., u_n) \in U^*$ (or that the dialogue u is generated by (s_1, s_2)), if $s_1(\#) = u_1$, $s_2(u_1) = u_2$, ..., $s_1(u_n) = \#$ provided n is even or $s_2(u_n) = \#$ provided n is odd. We will use the notation $u = [s_1, s_2]$.

A function E from U^* into the set of real numbers will be called *evaluation function*. The set of all evaluation functions related to U will be denoted by $E(U)$. The evaluation function is a subjective evaluation of the dialogue by the user. Let S_1, $S_2 \subset Stg(U)$ and E_1, $E_2 \subset E(U)$. Then the ordered quadruple $M = (S_1, S_2, E_1, E_2)$ can be considered as a dialogue communication. *Goal of the dialogue* for the first (second) participant of the dialogue is to minimize the function E_1 (E_2), i.e. to choose such a strategy $s_1 \in S_1$ ($s_2 \in S_2$) that $E_1([s_1, s_2]) = min$ ($E_2([s_1, s_2]) = min$). For instance, the participant could minimize the number of turns that is needed to obtain a desired information, or minimize his or her lost in the dialogue. Based on the previous notions we can characterize dialogues as: cooperative, if $E_1 = E_2$; *non-cooperative*, if $E_1 \neq E_2$; *zero-sum*, if $E_1 = -E_2$. (This formal approach enables a direct application of the game theoretical approach, which seems to be promising especially if combined with formal models in social and behavioural sciences (see e.g. (Greenberg, 1997)).

3. Dialogue automata

Dialogue automata suitably generalize the notion of Mealy-type automata (see e.g. (Geczeg & Peak, 1972)) and can be seen as a very simple user model of the participant of dialogue, which includes pragmatics. States of the automaton correspond to the internal states of the user, and input and output symbols correspond to the dialogue utterances. Transition functions then represent the scheme of the user behaviour.

Let $S = (S, A_S, V_S, f_S)$ and $X = (X, A_X, V_X, f_X)$ be information systems. We will assume that the sets of attributes f_S and f_X are finite. *Dialogue automaton* is an ordered quadruple $A = (S, X, \lambda, \delta)$, where $\delta : S \times X \to S$ and $\lambda : S \times X \to X$ are transition function and output function, respectively. Sets S and X are the *sets of states* and *dialogue utterances*, respectively. Being in the state $a \in A$, the automaton detects the actual input symbol (dialogue utterance of the other participant of the dialogue), changes its state according to the function δ and outputs a dialogue utterance according to the function λ. A dialogue automaton for which the set of the states and the set of the dialogue utterances are finite is said to be *finite*.

Provided that we have selected a state to be the initial state and that the special symbol # belongs to X, a dialogue automaton represents a dialogue strategy.

Let A_1 and A_2 be dialogue automata. For the sake of simplicity, by A_1 (A_2) we will understand also the set of all corresponding strategies (i.e. strategies defined by the corresponding automata and a chosen initial state). In accordance with this convention, we denote by $[A_1, A_2]$ the set of all dialogues generated by A_1 and A_2 (for all possible initial states).

In reality, one of the most crucial problem in constructing such models lies in the fact that we do not know the internal structure of the used models – we have to consider them as black boxes and deduce or investigate their structure from their behaviour.

It can be shown that: given a (finite) set of dialogues D, we can construct (finite) dialogue automata A_1 and A_2 such that $D \subseteq [A_1, A_2]$. Moreover, minimizing their structures we obtain the corresponding canonical forms (Geczeg & Peak, 1972).

In (Kopecek, 2000b) an illustrative example is presented, in which the model is applied to dialogue communication between two gamblers playing poker, involving emotional and prosodic attributes. In this example, the following set A of attributes of the states has been chosen: A_S = {$AS1$ - my_cards, $AS2$ - bet_in_pot, $AS3$ – self_confidence, $AS4$ - counter_player_cards, $AS5$ - plan}, whereby (for instance) the attribute $AS1$ has taken the values {*bad, medium, good*}, the attribute AS2 has taken the values {*high, medium, low*}, the attribute AS5 has taken the values {*careful, risky, bluffing*}, etc. We omit more details and discussion here, just mentioning that even the finite model shows quite powerful generality.

In many situations, we can represent the values of attributes of internal states and dialogue utterances in a numerical form. Real numbers, which express a measure of the intensity of an attribute, are a typical instance. In such cases, we may obtain models, for which the number of internal states as well as the number of dialogue utterances is infinite. Then, we can assume that the sets of attributes are indexed and that the values of attributes are real numbers. Hence, both internal states and dialogue utterance can be considered as real vectors. This leads us to the following problem:

Given a finite set of dialogues D and two finite sets of attributes (one set of attributes related to the states and other set of attributes related to dialogue utterances), find (infinite) dialogue automata A_1 and A_2 such that $D \subseteq [A_1, A_2]$, and the internal states of the automata coincide with the observed states for all the dialogues from the set D.

To obtain an approximate solution of this problem, we suppose that the functions $\delta : S \times X \to S$ and $\lambda : S \times X \to X$ are sufficiently smooth. Let us remark that this assumption is quite acceptable – it expresses that a small change of the attribute will also imply a small change of the function and that this relation is "smooth". Accepting this assumption, we can – at least locally – approximate the functions linearly. Hence, we will suppose that the functions $\delta_i : S \times X \to s_i$ and $\lambda_j : S \times X \to x_j$ are bilinear forms for all $I \in (1,...,m)$, $j \in (1,...,n)$ ($s = (s_1, ..., s_m)$, $x = (x_1, ..., x_n)$, $s \in S$, $x \in X$). Under this assumptions, the functions δ_i and λ_j can be represented by matrices D_i and L_j in the form: $\delta_i (s, x) = s D_i x$, $\lambda_j (s, x) = s L_j x$. Thus, if we have sufficient data (i.e. a set of dialogues for which the corresponding utterances are classified by means of the attributes and the set of corresponding states, also classified by means of attributes), we can compute all the matrices D_i and L_j, generally from an overdetermined system of linear equations by means of Least Square Method (see, e.g. (Demmel, 1999)).

4. Ambiguity problem

A powerful feature of the dialogue automata approach lies in the possibility to determine unknown attribute values in the context of the dialogue communication. It can be used for determining the unknown or unsure parameters and values in the dialogue context. To illustrate this briefly, let us consider a very simple example of prepositional group ambiguity (Pg-ambiguity) that typically appears in the cooperative information retrieval dialogue systems (the example comes from a Czech dialogue corpus):

Chci jet rychlikem do Brna.
[(I) want – to go – (by) express (train) – to – Brno.]

and in Armenia an astounding $121 per month for Internet connection (Petrazzini 1999). It is easy to understand the international implications of the designer's work when placed in the context of cost in developing countries.

It is also likely that the majority of designers have little experience designing in Arabic scripts that are bi-directional (right to left, except for numbers and foreign-language words that run from left to right. Far Eastern languages require twice the space of English for each letter because of DBCSs. Localizing a site requires special attention to these design considerations.

4. Designing for the global audience

Most designers have little or no training in addressing localization issues, although numerous articles and special issues focusing on globalization and localization have recently appeared. The majority of these articles have, however, centered on language and very little attention has been paid to the larger issues of globalization. The lack of formal design training (focusing on designing for global audiences), special programming needs required for various linguistic groups and content interpretation in cross-cultural applications allow design firms to bypass a true globalization effort, thereby affecting content and possibly message reception. The majority of these issues are either overlooked or ignored in an effort to visually enhance a site with state-of-the-art feats.

"The desire to become a cool site is often detrimental to good design. Certainly, boring or confusing sites will not attract many users, but the use of advanced design elements simply for the sake of adding more stuff to the page will discourage users from repeat visits to a site," said usability expert Jakob Nielsen to the 1998 gathering of the Association of Computer Machinery Special Interest Group on Computer-Human Interaction (ACM/SIGCHI). With the rapid growth of eCommerce, the need to effectively communicate globally challenges designers to adhere to the principles of simplicity rather than those of technical complexity. The classical role of designer is becoming that of information architect/Information designer by encompassing the tools of design aesthetics with user interfaces and navigation tools that promote, rather than deter, usability. Neilsen's statement has added significance when applied in the global context.

The Federal Express site for Brasil demonstrates a business-to-business localization attempt that may fall short of corporate strategy. Gains made by the conversion of navigation aids into the national language, the placement of the flag in the design and inclusion of local contact numbers may be offset by the representation of an Asian carrier. Moreover, visitors to the site have a one-in-four chance of seeing this particular carrier and a one-in-three chance of seeing a woman.

The United Arab Emirates site is localized using the national language, local charter set, and reading preference while using the same carrier image as noted above, although it has been reversed to follow the text flow. Bthe cultural problem here, however, is the use of not only the Asian carrier, but that of a woman with exposed arms – a cultural taboo in the Arab region.

Both Federal Express examples adequately illustrate the problem of site development for international audiences. To be fair, however, Federal Express maintains 216 separate sites one of the largest online localization efforts.

5. Theory versus practice

Two other noteworthy examples using the Federal Express site illustrate the complexity of global design. The site for Hong Kong, ignoring any associated political aspects, uses double-digit coding rendering the text unreadable in the United States. This problem is potentially resolved by the use of Unicode and its 65,536 character combinations, although it is not necessarily a catchall fix for the problem. The Unicode site itself also has its own problems in translating "About our services" page into Arabic. The translation error occurred employing Internet Explorer v5.0 in the International translation setting. While formatted for the geographical area (right-to-left) the Arabic words did not display correctly.

Three other considerations worth noting in undertaking international translations:
- Spelling varies between countries that share similar dialects. In the United States, "color" is different from its counterparts in the United Kingdom and Canada where it is spelled "colour." Other words include "tire" versus "tyre" and "localization" versus "localisation."

Second basic strategy is the strategy with initiative of the user. This strategy is complementary to the strategy with initiative of the system and allows the user to be more efficient in the situations when the strategy with initiative of the system shows drawbacks. Roughly speaking, using this strategy the user freely formulates key words characterizing the meant construction (statement) and the system guesses the construction. Clearly, ambiguities can often appear in these situations.

In contrast to the above-mentioned illustration of PP-attachment ambiguity, let us mention a very simple example of lexical ambiguity. Suppose that the user wants to create a cycle statement and gives the system the keyword "cycle". Does the user mean the general form of the cycle statement or does he or she mean the standard form (which is more simple to define)? This ambiguity can be resolved if we know whether the user is an experienced user or novice. Our reasoning can be as follows. The experienced user would add the keyword "standard" and moreover, the experienced user will probably use more often the general form of the cycle statement. This situation can also be influenced if the user has used the keyword "cycle" in a short history before the actual time point. It can be supposed, that the user remember the history and that he or she would use the keyword in accordance with the previous usage. All such aspects can be effectively and concisely incorporated in the dialogue automata user model.

5. Conclusions and future work

In the paper we have presented a discussion of the ambiguity in dialogues and a possible method that could help to solve some ambiguity problems in the context of the dialogue system. Further development of the discussed method involves studying the problem from linguistic point of view as well as further developing of the theory of dialogue automata and its applications to dialogue ambiguities.

Acknowledgement

The research has been partially supported by Grant Agency of the Czech Republic under the Grant 201/99/1248 and by Czech Ministry of Education under the Grant LI200027.

References

Batusek, R., Kopecek, I. (1999). "User Interfaces for Visually Impaired People". Proceedings of the 5th ERCIM Workshop on User Interfaces for All, Dagstuhl, pp. 167 – 173.

Demmel, J.W. (1999). *Applied Numerical Linear Algebra*. SIAM Books.

Gecseg, F., Peak, I. (1972). *Algebraic Theory of Automata*. Akademiai Kiado, Budapest .

Greenberg, J. (1997). "Situation Approach to Cooperation". in Hart. S, Mass-Colell, A. (Eds.): *Cooperation: Game-Theoretic Approaches*. NATO ASI Series, Springer Verlag.

Kopecek, I. (1999). "Modelling of the Information Retrieval Dialogue Systems". Proceedings of the Workshop on Text, Speech and Dialogue - TSD'99, LNAI 1692, Springer-Verlag, pp. 302-307.

Kopecek, I. (2000a). "Active and Passive Strategies in Dialogue Program Generation". Proceedings of the Workshop on Text, Speech and Dialogue – TSD 2000, LNAI 1902, Springer-Verlag, pp. 427-432.

Kopecek, I. (2000b). "Emotions and Prosody in Dialogues: An Algebraic Approach Based on User Modelling". Proceedings of the ISCA Workshop on Speech and Emotions. Belfast, pp. 184-189.

Osborne, M.J., Rubinstein, A. (1999). *A Course in Game Theory*. MIT Press, London.

Pawlak, Z. (1981). "Information Systems, Theoretical Foundations". *Information Systems 6* (1981), pp. 205-218.

Sadek, D.M. (1992). "A Study of the Logic of Intention". Proceedings of the 3rd Conference on Principles of Knowledge Representation and Reasoning, Cambridge, Massachusetts, pp. 462-473.

Stranakova, M. (1999). "Selected Types of Pg-ambiguity". *Prague Bull. of Math. Linguistics 72*, pp. 29-58.

Involving Chinese users in analyzing the effects of languages and modalities on computer icons

Sri H. Kurniawan[1], Ravindra S. Goonetilleke[2], Heloisa M. Shih[2]

[1]Institute of Gerontology and Dept. of Industrial and Manufacturing Engineering, Wayne State University, 226 Knapp Bldg, 87 E. Ferry St., Detroit, MI 48202 USA. af7804@wayne.edu.
[2]Dept. of Industrial Engineering and Engineering Management, Hong Kong University of Science and Technology, Clear Water Bay, Kowloon, Hong Kong. ravindra@ust.hk.

Abstract

The present study investigated the effect of modality and language on icons' appropriateness and meaningfulness for Chinese users. Based on the findings of the previous studies of computer icons, three hypothesis were developed: H1: Bimodal icons rate the highest; H2: Icons with Chinese characters rate higher than icons with English words; and H3: Pictorial icons rate higher than verbal icons. Fifty Hong Kong Chinese daily computer users participated in the experiments. The results suggested that H1 and H2 are partially supported and H3 is not supported. A significant interaction between the language and modality was observed. The best icon group for Chinese users is the bimodal Chinese group.

1. Introduction

The use of icons to represent interactive objects in computer systems has become a common phenomenon in human-computer interface design. Various studies testing icon preference, appropriateness and meaningfulness have resulted from this trend (e.g. Lodding, 1983; Stephanidis & Akoumianakis, 1997; Choong & Salvendy, 1997; Tudor, 1994). Parallel to this line of studies, various studies on the effects of icon modality have also been conducted (e.g. Strijland, 1993; Young & Wogalter, 1988; Selcon, Taylor & Shadrake, 1992; Guastello, Traut & Korienek, 1989). Different studies adopted different concept of what modality represents. While some research studied modality as an avenue of sensation, such as auditory, visual, etc (Brewster et al, 1996; Brown et al, 1989), others defined modality as the form of representation, such as pictorial and verbal (Guastello et al, 1989; Bernsen, 1994), which is also the definition adopted in the present study.

Findings on different aspects of icons were reported in various studies. Guastello, et al (1989) found that (1) Mixed modality icons were rated as distinctively more meaningful than their alternatives. (2) Ratings were occasionally bolstered by population stereotypes acquired through experience. (3) Long abbreviations are preferable to short ones. (4) It is possible to construct pictograms that are more meaningful than the industry's standards, and (5) Verbal icons are sometimes preferred over pictorial icons when mixed modes are not available. The author also stated that "ratings [of meaningfulness] were occasionally bolstered by population stereotypes..." (p. 99).

In an experiment investigating the use of warning/caution icons and verbal warning messages, the combination of icons and verbal warning was proved to improve response latencies (Selcon et al, 1992). Young and Wogalter (1988) performed an experiment to determine whether the salience of warning messages would improve the memory of warnings in proceduralized instructions by giving subjects verbal warning messages, either accompanied by meaningfully-related icons or without the icons, and found that content recall and semantic learning was significantly better for subjects who received the combination of verbal warning and icons.

Studies have suggested that the results of a particular study on icons can not be generalized across different user populations (e.g. Guastello, 1989) In other words, there might be a need to investigate icons in a specific population of prospective users. Recently some researches investigated the effect of using native language verbal representation on icon-related tasks or icon characteristics (e.g. Choong & Salvendy, 1997, 1998; Sacher, 1998) to investigate whether the results obtained for English-speaking population yield the same findings for non-English-speaking population. In line with Guastello et al's suggestion, non-English-speaking user population showed different preference and task performance when using icons with local characters.

Choong and Salvendy (1997) conducted a study on the effect of icon modality and language on the task performance of Chinese and American user groups. They found that while bimodal icons were superior for both groups, Chinese users performed better on pictorial icons when bimodal icons were not available. The term bimodal in their study refers to pictorial icon complemented by labels. Pictorial icons complemented by Chinese labels were tested on Chinese users (similarly, English labels were tested on American users). The present study is based on Choong and Salvendy's study, focusing on Chinese users and extending the number of tested functions.

In the current study, different combinations of modalities (verbal, pictorial, and bimodal) and languages (English and Chinese) were used to create some of the tested icons that would be compared to icons currently displayed in commercial software. Most commercial icons are pictorial icons (e. g. icons on the "Toolbar" in Microsoft Office™ products). Bimodal icons in the current study were defined as icons where some parts of the function were represented by a (Chinese or English) word and the other parts by non-word objects. The example of tested function and icons is shown in Figure 1.

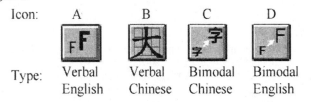

Icon: A B C D

Type: Verbal / English   Verbal / Chinese   Bimodal / Chinese   Bimodal / English

Note: The Chinese character in Icon B means "big" and in Icon C means "font"
Figure 1: Example of icons for the function "Increase Font Size" used in the experiments

Based on the findings of the previous studies, three hypothesis were developed:
H1: Bimodal icons rate highest in appropriateness and meaning
H2: Icons with Chinese characters rate higher than icons with English words in appropriateness and meaning.
H3: Pictorial icons rate higher than verbal icons in appropriateness and meaning.

2. Methods

Fifty (26 male and 24 female) Hong Kong Chinese, all daily computer users, participated in the experiments. Twenty-six of the participants were familiar with Chinese software, while the other 24 were not. The participants included office workers and students from various educational backgrounds and occupations. The ages ranged from 18-40 years (mean = 25.3, std. dev. = 5.5 years).

A paper-based questionnaire was used to test the hypotheses. All of the instructions, questions and ratings were displayed in both English and Chinese languages. Typical time to fill in the whole set was half an hour. The subjects were compensated with HKD 30 (USD 3.86) at the end of the experiment. The sample of the questionnaire is depicted in Figure 2.

作用: 改變字型

Function: **Change Font-type**

你認為上圖能否適當地表達上述的作用?
Is the icon **appropriate** for the function?

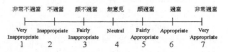

非常不適當	不適當	頗不適當	無意見	頗適當	適當	非常適當
Very Inappropriate	Inappropriate	Fairly Inappropriate	Neutral	Fairly Appropriate	Appropriate	Very Appropriate
1	2	3	4	5	6	7

Figure 2: Example of the paper-based questionnaire used in the experiments

The following variables were analyzed in the present study:

Appropriateness. Subjects' opinions of icons' appropriateness were assessed with 7 discrete scale bipolar semantic ratings from "Very Inappropriate" (rated as 1) to "Very Appropriate" (rated as 7).

Meaning. The questionnaire was presented with the same format as appropriateness. The differences were the replacements of the question "Is the icon appropriate for the function?" with "How well does the icon represent the meaning of the function?" The ratings ranged from "Very badly" (rated as 1) to "Very Well" (rated as 7).

Language. Language variable is coded in SPSS into three values, '0' for icons with no word component, '1' for icons with English word, and '2' for icons with Chinese word in them.

Picture. Picture variable is also coded in SPSS into two values, '0' for icons with no picture (only contains word) and '1' for icons with picture in them. Therefore, for example, a bimodal English icon will have a value of '1' for the variable Language and a value of '2' for the variable Picture.

3. Results and Discussions

The purpose of this study was to investigate the effects of modality and language on icon's appropriateness and meaning for Chinese users. Analysis focused on: (1) examining the effect of bimodality on icons' appropriateness and meaning, (2) comparing the effect of the use of Chinese characters and English words on icons' appropriateness and meaning, and (3) comparing the appropriateness and meaning of pictorial and verbal icons.

Across all of the tested icons, the bivariate correlation between appropriateness and meaning is 0.78, suggesting that appropriateness and meaning are strongly related, in conformance with the suggestion of ISO DIS 9186 (1989) that meaning is one of the strongest measures of appropriateness.

The descriptive statistics of the measured variables are shown in Table 1-2.

Table 1: Descriptive statistics of icons' appropriateness and meaning: Language effects

Language	Cases	Mean(St Dev)	
		Appropriateness	Meaning
0 (No word)	1075	4.44(1.75)	4.47(1.71)
1 (English)	250	5.15(1.72)	5.18(1.71)
2 (Chinese)	375	5.52(1.60)	5.51(1.55)
All	1700	4.78(1.77)	4.80(1.74)

Table 1 showed that icons with Chinese words rate higher than icons with English words or pictorial icons in both appropriateness and meaning. The Bonferroni post-hoc analysis showed that while the difference between Pictorial icons and English icons or Chinese icons is significant with $p \leq 0.05$, the difference between English and Chinese icons is not significant (p=0.023). In other words, icons with words (English or Chinese) are considered more meaningful and appropriate than icons with no words (just picture). The result is further verified by Table 2. Hence, Hyporthesis 2 is only partially supported.

Table 2: Descriptive statistics of icons' appropriateness and meaning: Picture effects

Picture	Cases	Mean(StDev)	
		Appropriateness	Meaning
0 (W/o picture)	400	5.46(1.56)	5.46(1.54)
1 (With picture)	1300	4.57(1.78)	4.60(1.74)
All	1700	4.78(1.77)	4.80(1.74)

The fact that icons with English words and Chinese words are not significantly different is intriguing, considering that in icons with Chinese words, because of the logogram[1] nature of Chinese words, the words in the icon represent the meaning of the function precisely (see Figure 1 for example). On the other hand, in icons with English words, in order to fit the word in the icon, the word was abbreviated (e.g. 'CLR' for 'CLEAR'), which might lead to misinterpretation and/or require previous exposure to the abbreviation in order to understand it. Therefore, icons with Chinese words in theory should be considered to be more appropriate and meaningful. One of the possible

---

[1] Logogram means "the actual representation of the object it described" (Gittins, 1986, p. 520)

494

reasons is that the bilinguality of our Hong Kong Chinese subjects, which treated English and Chinese words to have the same quality of information.

Contrary to the finding of Choong and Salvendy's study (1997), Table 2 showed that verbal icons are better than pictorial icons in appropriateness and meaning. Hence, Hypothesis 3 is not supported. Studies showed that symbols have lower degree of representation than words (Guastello, et al, 1989). Therefore, icons representing their functions with words are considered to have higher meaningfulness and appropriateness than pictorial icons.

The general factorial ANOVA analysis showed significant effect of Language and Picture on Appropriateness and Meaning ($p \leq 0.001$). In both measures of Appropriateness and Meaning, the interaction between Language and Picture is significant with $p < 0.001$ as shown in Figure 3.

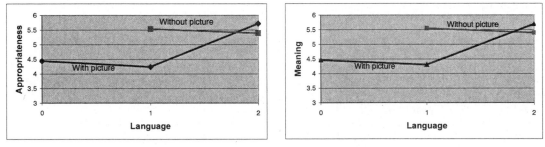

Figure 3: Interaction Plot of Appropriateness and Meaning

The interaction plots in Figure 3 showed that, while in icons with Chinese words complementing the word with picture produced higher ratings, the case is reversed in icons with English words. Visual observation revealed that the highest rating in both appropriateness and meaning is for icons with Chinese words complimented with pictures (in short, bimodal Chinese icons). However, visual observation also showed that the rating of bimodal Chinese icons is very close to icons with just English word. Therefore, Hypothesis 1 is only partially supported.

Tudor (1994) mentioned that "icons and symbols in general need to be empirically evaluated among all groups of prospective users" (p. 62). Based on the result of the analysis, the implication for the icon design is that for Chinese users, the best icon in terms of appropriateness and meaning is bimodal Chinese icon (although the ratings of verbal Chinese icons or verbal English icons are not significantly different). This is an interesting finding considering that although many software have been translated into Chinese, the icons used were not translated or adapted to Chinese users.

4. Conclusions and Further Research

The results of the study showed that using Chinese characters in icons improves subjective ratings of appropriateness and meaningfulness for Chinese users. Designers of computer interfaces who plan to create programs in Chinese should consider using icons incorporating Chinese words in addition to the common practice of translating the menu and help items into Chinese.

More generally, the study also showed that it will be fruitful to perform usability testing using the prospective users of a human-computer interface. This study and various previous studies had revealed that there is a different perception between Chinese users and English users or, in general, different user populations. Understanding who the prospective users are might contribute to the success of the designed interface.

There are some limitations of the study. The present study was done with bilingual subjects which consider English and Chinese words to have the same degree of representation in their daily life. It would be interesting to test the same sets of icons to Chinese subjects from countries where English is not as extensively used as daily language as it is in Hong Kong.

Even though in this study there is an indication that the quality of information might affect subjective ratings of icon, the quality of information was not quantified nor obtained from subject's data. Rather, the result was analytical. Further research focusing on quantifying the quality of information might bring a useful contribution to this area.

Further research should be directed toward finding out why certain types of modalities and the use of Chinese characters were considered more appropriate and meaningful by Chinese users. Finally, this study is a preliminary study that would hopefully trigger similar studies in the future, considering the numbers of Chinese computer users

This sentence has two different meanings:
1. *I want to go to Brno by an express.* (= the express train may go elsewhere via Brno)
2. *I want to go (somewhere) by an express to Brno.* (= it is the express to Brno, I can go elsewhere)

In the dialogue context, the ambiguity can be resolved by the following diagnostic questions:
1. Where do you want to go?
2. How do you want to go (there)?

The ambiguity of prepositional group can be detected by NLP module incorporated into the dialogue system. Two types of Pg-ambiguity are distinguished – the ambiguity of syntactic relations (structural ambiguity) and the ambiguity originated in ambiguous morphological information. Analysis by reduction allows to apply linguistically based criteria for the detection of Pg-ambiguity (Stranakova, 1999). Such criteria include (i) general purely syntactic rules (e.g. rules based on surface word order of a sentence) as well as (ii) the rules exploiting the characteristics of concrete lexems (the valency information of verbs, nouns and adjectives is crucial, optionally also rules based on semantic features can be used).

In the example, structural ambiguity of the Pg *do Brna* [to Brno] is identified. Both types of criteria are used: (i) the ambiguous word order pattern is met in the sentence; (ii) the valency requirement for Direction (obligatory in this case) of the verb *jet* [go] is fulfilled by the Pg. Thus two syntactic structures are created – the Pg modifies either the verb *jet* [go] (valency modifier) or the noun *rychlik* [express (train)] (free modifier).

Analysing the sentence, we get just the presence and the type of ambiguity, but not the meaning relevant in the given context. In our model, the type of ambiguity can be described as an attribute, whereby the solution of the ambiguity corresponds to the attribute value that can be in our example *1* (first meaning) or *2* (second meaning). (The attribute value *0* is reserved for sentences with no Pg-ambiguity detected.) This value is ambiguous and has to be determined in the dialogue context. (Let us mention that even this example is substantially simplified).

Let us formulate the ambiguity problem more generally. Let A_1 and A_2 be dialogue automata and $X = (X, A_X, V_X, f_X)$ be the corresponding information system describing dialogue utterances. Let $X' = (X, A_X, V'_X, f_X)$ be an information system that differs from X only in the set of attribute values, to which a new element '?' is added. This new element will denote unknown values of attributes.

Let $u = (u_1, u_2, ..., u_n)$ be a dialogue whose elements belong to X'. Let $y \in X'$, $z \in X$. Let us denote y ~ z if the attribute values of both objects differ only at the positions where the value '?' is presented. Now, our ambiguity problem lies in determining whether there is an (unique) object $x = (x_1, x_2, ..., x_n)$ satisfying $x \in [A_1, A_2]$ and $x_1 \sim u_1, ..., x_n \sim u_n$ such that $x \in [A_1, A_2]$.

The application of the considered model of dialogue strategies in real dialogue system can be described by the following scheme:
1) Is the actual user known to the dialogue system? In the positive case assign the corresponding dialogue system to the user and go to (3), else go to (2)
2) Analyse the behaviour of the user to obtain the corresponding model (i.e. the transition and output function of the corresponding dialogue automaton), assign the corresponding dialogue system to the user and remember the assignment.
3) Use the model for elimination of the ambiguities (and for possible other purposes as well).
4) Evaluate the effectiveness of the disambiguation procedure to ensure a feedback and possibly correct the user model.

Let us present a simple illustration of this approach in relation to the developed dialogue system for generation of the program source code (Kopecek, 2000a). The system is primarily oriented to the blind and visually impaired programmers and the purpose of this system is to help the user in creating the source code by automatic eliminating most of the possible syntactic errors and by using interactive context sensitive dialogue help. The system uses two basic dialogue strategies. First basic strategy is the strategy with the initiative of the system. This strategy is based on inquiring the user and building up the structure of the created source code according to the information given by the user. This strategy ensures eliminating the syntactic errors and gives the user the opportunity to be led by the system; nevertheless, it can be ineffective in some situations or bothering for experienced users.

Involving seniors in designing information architecture for the web

Sri H. Kurniawan, Panayiotis Zaphiris, R. Darin Ellis

Institute of Gerontology and Dept. of Industrial and Manufacturing Engineering
Wayne State University, 226 Knapp Bldg, 87 E. Ferry St., Detroit, MI 48202 USA.
af7804@wayne.edu, p.zaphiris@wayne.edu, rdellis@wayne.edu.

Abstract

The present study utilized the card sorting technique and cluster analysis to define the best information architecture of Web health information for seniors. Sixteen seniors participated in the card sorting, twenty in the category identification and thirteen in category labeling experiments. Seniors participated in the study tend to group the tested items conceptually at the higher level but tend to group the items based on common words found in the titles at the lower level of the hierarchy. The study also found that user grouping produced more heterogeneous structure than the Dmoz set-up information architecture. Category labels suggested by seniors were observed to be less formal than the Dmoz category labels.

1. Introduction

1.1 Health Information and Aging

U.S. Census Bureau (2000) projected that by the year 2030 people aged 65 and above will represent 20% of US population. Although older adults are still underrepresented as computer users, more and more are beginning to incorporate the Web as part of their lives, currently representing 13% of the online users (Cury, 2001). One of the main uses of computers for seniors is to look for health information (White et al., 1999; Cochrane, 1999).

This significant increase in the older computer user population has led to various studies investigating the age effect in utilizing the Web. Some findings suggested that older adults have some disadvantages in fully utilizing the Internet as an information source. That is, older people have more trouble finding information in a Web site than younger people (Mead et al., 1997). However, little effort has been placed to ensure that health information on the Internet is structured to help older computer users to find the desired information easily and efficiently.

In addition to age-related problems of searching for information on the Internet, seniors often face various problems when searching specifically for health information. Research reviewing medical and health related web sites listed the following problems: although those sites are attractive, they only function as "yellow pages" (Cochrane, 1999); it is difficult to find specific information and it contains content of varying quality (Hersh, 1999); there is a potential for serious abuse and conflict of interest because of the profit acquired from selling an advertiser's products (Bloom & Iannacone, 1999). For most topics, inability to find the proper information (or to find the information in a timely manner) might not bring severe consequences. However, the case is different in health and medical areas. Hence, it is crucial to make certain that online health and medical information is structured in a way that would enable users to find the information easily and efficiently.

This study investigated ways to better structure and information architecture of Web health resources and to be more intuitive to seniors. Previous studies showed that unintuitive web link labels might lead users to a wrong path or caused disorientation, which render users unable to find the information (Oliver & Oliver, 1996).

1.2 Card Sorting Technique

An important step in organizing the content of a web site is to place the information on the web site according to how individuals typically view information (Bernard, 2000b). A well-constructed taxonomy will help end users to locate the desired information quickly and accurately and a badly constructed taxonomy might contribute to a waste of time and effort without obtaining the desired information (Karneva et al., 1997).

The organizational structure of a Web site can have a profound effect on its ease of use. An ideal structure would allow users to navigate efficiently through the site, while a less-than-ideal structure might make users unable to achieve their goals in information search. For example, many corporate Internet sites design their web site's structure based on the internal structure of their companies, no matter how conceptually unintuitive the structure is. Unfortunately, most visitors are unfamiliar with the actual structure of the company, and are unlikely to find this kind of site easy to navigate as these sites do not match the user's mental model of the structure of the company.

To make certain that the information architecture design matches the way users view information, usability testing needs to be conducted. Usability testing should follow the philosophy that the successful design of a web interface is very dependent on the typical user's mental model of the structure of a web site (Bernard, 2000a). Thus, in web site design it is essential to ensure the design of a meaningful structure of information organization, as a meaningful information organization will promote efficient navigation (Shneiderman, 1992). The optimal information structure should fit the user's mental model (Lisle, Dong & Isensee, 1998).

One of the ways to understand user's perceived relationship of various components of the web sites using a formative design support technique is by utilizing the card sorting technique. In the card sorting exercise, participants are presented with randomly ordered cards representing pages of a Web site, and they are asked to group the cards based on their perceived fit (Martin, 1999).

Card sorting is considered as one of the best usability methods for investigating users' mental model of an information space (ZDNet Developer, 1999). The resulting tree structures can be used as a base for organizing the site and for identifying meaningful patterns in the resulting hierarchy which are indicative of general underlying cognitive processes or user mental models. These patterns can then be generalized to form principles and guidelines for organizing Web content (Karneva et al., 1997).

There are two ways to analyze collected data from card sorting experiments: by "eyeballing" the card's grouping trend (Martin, 1999), which is tedious for large number of users; or by utilizing the cluster analysis technique. Cluster analysis of card-sorting data is a promising method for understanding and summarizing multiple participants' input to the organization of Web site pages. Cluster analysis quantifies card-sorting data by calculating the strengths of the perceived relationships between pairs of cards, based on how often the members of each possible pair appear in a common group (Martin, 1999). The degree of the relationship between any two cards is represented by their similarity score. The output can be displayed in the form of tree diagrams in which the relationship between any two groups of cards is represented graphically by the distance between the origin and the branching of the lines leading to the two groups.

2. Methodology

2.1 Participants

The experiment was designed specifically to include a representative pool of the prospective users of the tested health information web sites. Sixteen seniors aged 55 and above (mean age = 68.69, SD = 5.97 years) participated in the card sorting experiment (experiment 1). A different pool of twenty seniors (mean age = 70.41, SD = 13.41 years) participated in the category identification follow-up experiment (experiment 2). Another new pool of thirteen seniors (mean age = 70.69, SD = 8.18 years) participated in the category labeling experiment (experiment 3).

All participants live independently in the community (non-institutionalized) and have no visual and cognitive impairment and functional illiteracy. All participants have at least 13 years of formal education.

2.2 Stimulus Material

The pages used for the whole card sorting experiments were sixty four (64) leaf items taken from the "Health: Aging" hierarchy of Dmoz (**http://www.dmoz.org**) web site from four main categories: Geriatrics, Diabetes and Alzheimer, Life Cycle and Life Expectancy. The example of the original web site structure is pictured in Appendix A-1. Items were carefully selected to be of interest to our participant's age population and of similar title complexity.

2.3 Apparatus

3'x5' Index Cards with the Web link names and short descriptions of the content of Web pages were used in the Card Sorting test. USort and EZCalc software by IBM[TM] were used for the cluster analysis of the card sorting data. Paper and pencil questionnaires were used for category identification and category labeling experiments.

498

2.4 Procedure

2.4.1 Experiment 1: The card sorting
The link title and a short description of the web page of each of the 64 items was printed on a separate index card. An example of those index cards is depicted in Figure 1.

Alzheimer's Outreach
Information for caregivers of Alzheimer's patients and caregivers. Includes message board, poetry and numerous places to relax.

Figure 1: An example of Index Card used in this experiment

Each participant was given one set of sixteen (16) randomly ordered index cards from one of the four Dmoz main categories. The participant was tested in an individual session to make sure that the grouping was based on individual observation rather than group observation. The participants were asked to sort the cards into logical groupings based on the following instructions (ZDNet Developer, 1999):

- Please sort these cards into piles such that things that you think go together are in the same pile.
- You can have as many or as few piles as you like.
- The piles do not need to contain the same number of cards: some piles may be very big and others may have one or two cards if you don't think they are sufficiently similar to anything else.
- You can change your mind and move cards around and merge or split piles as you go.

When they were comfortable with their final sorting arrangement, they were asked to record their card groupings on paper. To aid in understanding the underlying concept of how they group the information, participants were also asked to write down group names and descriptions of why they grouped the items that way. Each participant was then asked to repeat the experiment by using a different grouping strategy.

Because it was predicted that seniors might be confused and worried when finding an item that they couldn't group with any of their groups, two options were given: to place the item to any of the existing categories or to list the item as a separate group with the item's name as the group name (some help was provided by asking the participants to use the "thinking out loud" method while creating the categories). In cases where the participants felt that some items could fit into more than one group, they were allowed to list that item in one or more categories.

Next, a cluster analysis was conducted using EZCalc across all participants' card groupings to produce final hierarchical structures. The final hierarchical information architecture can be found in the Figures 2-5 of http://agrino.org/pzaphiri/Papers/HCII2001.

2.4.2 Experiment 2: The category identification
In this follow-up experiment, participants were given pages containing items from the same main category of Dmoz. Each of the pages contained items that were suggested to belong to the same group by the participants of Experiment 1. Each participants in Experiment 2 was then asked to write down a suggested label for each of the groups. In general, across the twenty participants, 3-5 names were proposed for each group.

2.4.3 Experiment 3: The category labeling
The last experiment, the category labeling, involves presenting users with the category labels suggested by the participants of Experiment 2 and the items that belong to that category. Each participant was asked to rank the suggested labels based on their fit to the group (lower number means higher fit). The number was then added up and the label with the smaller sum was the chosen label for the group.

3. Results and discussion

The new hierarchical structure (see Appendix A-2) revealed that, through the use of user feedback, the information structure has transformed from a homogeneous (four items per branch) design of Dmoz to a heterogeneous hierarchy (ranging from two to six items per branch). The users' mental model is more in agreement with a heterogeneous information architecture than a homogeneous one.

Another interesting observation is that seniors participated in our experiment tended to group items conceptually at the higher level of the information structure (e.g. by putting items related to Organizations or Diabetes in one group) but tended to be influenced by common words found in link name titles (e.g. "Longevity" or "Anti-aging") when grouping items at lower level of the hierarchy. In contrary to the commercial grouping of Dmoz where the information was often grouped based on geographic location (e.g. Research Institutions in USA versus Research

Institutions abroad), our senior participants tend to group items based on their functionality or service provided (e.g. Institutions about Aging, research centers about diabetes).

From the category labeling experiment, new names for the proposed categories where obtained. Category labels suggested by seniors were observed to be less formal than the category labels designed by web site experts. Our hypothesis is that these category labels match the user perception better than the information architecture proposed by Dmoz designers. The validity of this hypothesis needs to be tested by a formal usability experiment.

4. Conclusions

This study applied a series of user-centered design exercises to build a senior-oriented information architecture for health-related information on the Web. The results of the study showed that involving prospective users in the design can capture users' underlying perceptions of different components of the information architecture, including the structure and the labels of the hierarchy. The resulting information architecture is expected to be more user-friendly as we believe it is a closer fit to the user's mental model. The study suggests that web designers should accommodate the needs of users to ensure that their products would appeal to the end-users. More generally, with the Internet being more integrated in various aspects of life, it is necessary to accommodate users with different characteristics (e.g. people with disabilities or older users) when designing the online information architecture.

References
Bernard, M (2000a). Constructing User-Centered Websites: The Early Design Phases of Small to Medium Sites. *Usability News* [On-Line]. Available: http://156.26.16.3/newsurl/usabilitynews/2W/webdesign.htm.

Bernard, M (2000b). Constructing User-Centered Websites: Design Implications for Content Organization. *Usability News* [On-Line]. Available: http://wsupsy.psy.twsu.edu/surl/usabilitynews/2S/webdesign.htm.

Bloom, B.S. and Iannacone, R.C. (1999). Internet availability of prescription pharmaceuticals to the public. *Annual Intern Medicines, 131,* 830-833.

Cochrane, J.D. (1999). Healthcare @ the speed of thought. *Integrated Healthcare Reports, May 1st,* 16-17.

Cury, J.O. (2001). *Young at heart, and online: The over-50 crowd finds empowerment, community and new friends on the Web* [On-Line]. Available: http://www.buyingarizona.com/senior_source.html.

Elliott, B. and Elliott, G. (2000). High volume medical websites. *Delaware Medical Journal, 72*(1), 21-29.

Hersh, W. (1999). "A world of knowledge at your fingertips": the promise, reality, and future directions of on-line information retrieval. *Academic Medicines, 74,* 240-243.

Kanerva, A., Keeker, K., Risden, K., Schuh, E. & Czerwinski, M. (1997). Web Usability at Microsoft Corporation. In J. Ratner, E. Grosse and C. Forsythe (Eds.), *Human Factors for World Wide Web Development.* New York, NY: Lawrence Erlbaum.

Lisle, L., Dong, J., & Isensee, S. (1998). Case Study of Development of an Ease of Use Web Site. *Proceedings of the 4th Conference on Human Factors and the Web* [On-Line]. Basking Ridge, NJ: HFES. Available: http://zing.ncsl.nist.gov/hfweb/att4/proceedings/lisle/

Martin, S. (1999). Cluster Analysis for Web Site Organization. *Internetworking 2(3)* [On-Line]. Available: http://www.internettg.org/newsletter/dec99/cluster_analysis.html.

Murray, G. & Constanzo, T. (1999). *Usability and the Web: An Overview* [On-Line]. Available: http://www.nlc-bnc.ca/pubs/netnotes/notes61.htm.

Nielsens, J. & Sano, D. (1994). SunWeb: User interface design for Sun Microsystem's internal web. *Proceedings of the 2nd World Wide Web Conference '94: Mosaic and the Web* [On-Line]. Available: http://www.ncsa.uiuc.edu/SDG/IT94/ Proceedings/HCI/nielsen/sunweb.html.

Oliver, R. & Oliver, H. (1996). Information access and retrieval with hypermedia information systems. *British Journal of Educational Technology 27,* 33-44.

Shneiderman, B. (1992). Designing the User Interface: Strategies for Effective Human-Computer Interaction, 2nd edition. Reading MA: Addison-Wesley.

U.S. Census Bureau. (2000). *Projections of the Total Resident Population by 5-Year Age Groups, and Sex with Special Age Categories: Middle Series, 2025 to 2045* [On-Line]. Available: http://www.census.gov/population/projections/nation/summary/np-t3-f.txt

White, H., McConnell, E., Clipp, E., Bynum, L., Teague, C., Navas, L., Craven, S., & Halbrecht, H. (1999). Surfing the net in later life: A review of the literature and pilot study of computer use and quality of life. *The Journal of Applied Gerontology 18,* 358-378.

ZDNet Developer. (1999). *Card Sorting* [On-Line]. Available: http://www.zdnet.com/filters/printerfriendly/0,6061,2253113-84,00.html.

500

Appendix A-1: An example of web site structure from Dmoz of the main category: Geriatrics & Anti-aging

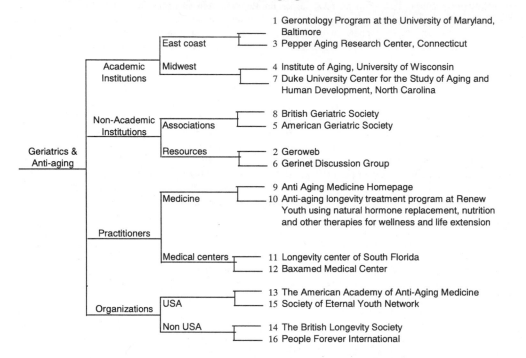

Appendix A-2: An example of the result from the card sorting experiment from Geriatric & Anti-aging

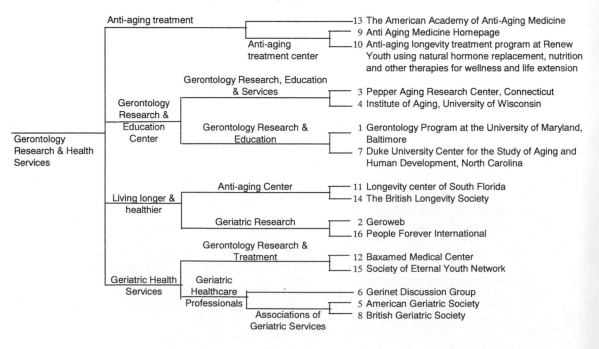

Engineering multilingual internet commerce

Aarno Lehtola, Jarno Tenni, Kuldar Taveter, Tuula Käpylä, Paula Silvonen, Kristina Jaaranen

VTT Information Technology
P.O.Box 1201, FIN-02044 VTT, FINLAND
email: aarno.lehtola@vtt.fi

Abstract

In the paper we describe our three tools that address the WWW multilinguality, namely Webtran, CONE and Unifier software. Webtran is a machine translation software that has been specifically designed for fully automatic translation of domain specific texts in online services. It is in daily use translating product descriptions in a mail-order company in Finland. CONE is a conceptual network software that includes an editing tool for defining domain ontologies embedding linguistic information and a user interface tool enabling multilingual navigation to information. It has been used to provide cross-lingual access to consumer protection legislation of some countries in European Union (EU). Unifier software helps in correcting erroneous or inexact text inputs of end-user customers in online services. For instance, it can be embedded to correct and homogenise address information.

1. Introduction

B2B electronic commerce has been estimated by Gartner Group to grow at over 40 % a year for the following five years (Tieke 2000). Even more enthusiastic growth figures have been presented. For instance, Boston Consulting Group has forecast that Internet commerce now triples in each forthcoming year. This growth takes place in tandem with the liberation of the world trade and the formation or expansion of common markets. While the trade transactions more and more frequently cross country and language borders, there is an extensive need for technical solutions to make the e-commerce services multilingual. VTT responds to these challenges by developing tools for building multilingual, localized services on the internet. In the following we review some central requirements that a multilingual e-commerce service should fulfil.

High quality of user interface localisation will be crucial in building positive image of an e-commerce service. Especially in the consumer markets, there will be many users that are illiterate in any other language than their native language. That is why multilinguality must be implemented in a consistent and comprehensive manner. McKinsey & Company has found in a survey that from all visitors of e-commerce sites (over 1.8 million visitors considered in the survey) only 7 % became customers and only 1,3 % became repeat customers. According to a survey by Jupiter Communications 39 % of consumers are more likely to shop online at merchants with whom they have had an offline experience (Ramsey 2000). There are plenty of inexperienced computer users among the clientele, so the user interface must be as simple as possible to use by a customer. It is crucial for an e-commerce service to get a high-quality image starting from the first access by a new customer.

Adaptation to the circumstances of the target market area is important. A seller may have some external reasons, why the product repertoire is not the same in every country. E.g. the seller may have rights to sell in only some countries, some items may have luxury taxes etc. At latest, when a customer has selected some product, there must be shown: price in local currency, methods of payment, methods and costs of delivery, delivery time and other terms of delivery. As far as the transactions cross over borders of the EU also taxation/customs needs to be considered. There are issues that depend on local legislation, such as returning and refunding of the goods. The multilingually provided information must be reliable to avoid legal disputes. The used language adaptation techniques must guarantee accuracy.

E-commerce services should be quick enough to prevent the customer from changing the channel. In a survey by Service Metrics Inc. (U.S.) it was found that the average download time for e-commerce sites is between six and seven seconds. It was also recognized that most customers will get annoyed if forced to wait more than eight seconds for a downloading of pages. There are several surveys on the percentage of aborted trading sessions, i.e. "abandoned shopping carts". For instance, both Andersen Consulting and Forrester Research have resulted with 25 % of all sessions, eMarketer with 31 %, Visa with 43 % and Greenfield Online with even 67 % (Ramsey 2000). Quick response times are of essence for customer satisfaction.

Additional requirements include support of a wide range of terminals, high security requirements, good integration to legacy systems, portability and homogeneity of the solution, and scalability and expandability.

502

Figure 1. E-commerce processes in company-to-customer interface

Figure 1 outlines the processes, that are prevalent in company-to-customer interfaces. In the boxes are presented the technical solutions that provide multilinguality. Cross-lingual information retrieval (IR) may involve correcting and unifying input search terms, ontology based translation of the query and finally translating the search results. Multilingual forms adapt to locale by their graphical appearance (layout, colours, icons, text orientation and direction, sounds etc.), translated fixed texts and processed form data. The last one involves, e.g., correction and unification of user inputs, text generation and machine translation. Multilingual helpdesk could be based on a database of centralised questions and answers to be maintained in one pivot language, and cross-lingual IR. Automatic cross-lingual cataloguing finds for a new product the suitable categories from locale specific catalogues. These examples involve functions that can be implemented with our software tools Webtran, CONE and Unifier.

2. Automatic translation of in-company language

Webtran software (Jaaranen 2000, Lehtola 1999a, Lehtola 1999b) is a tool for authoring and automatic translation of domain specific texts that are written in a human sublanguage characterised by selected domain, vocabulary and sentence structure, such as technical documentation: manuals and product descriptions, and formal reports: e.g. weather forecasts, medicine effect descriptions and epicrises. Webtran is currently in production use in the mail-order company Ellos Postimyynti Oy in Finland for automatic translation of the Ellos sales catalogues from Swedish into Finnish. Automatic translation shortens the time-to-market of products and improves the overall competitiveness of the company.

Figure 2 below illustrates an arrangement with Webtran translating fully automatically texts of a repository. The figure subsumes that the original texts have been authored and edited in a pivot language with controlled conformance to the in-company language model and its constraints, and the texts therefore include a minimised amount of ambiguities. The in-company language model is extended and maintained using the Webtran language modelling tool. The translation engine can automatically and accurately translate approved texts to multiple target languages. Major savings are obtained as post-editing of the translations can be avoided.

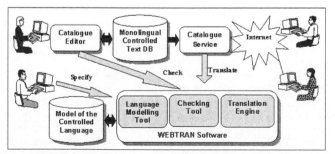

Figure 2. Embedding fully automatic translation into an online service

3. Cross-lingual information retrieval using ontologies

CONE software for conceptual network modelling and reasoning (Taveter 1999) is a tool for supporting cross-lingual information retrieval, building of market specific product models (ontologies), and data mining for e-commerce. CONE enables the modelling of concepts and interconceptual relationships at different logical levels. The models also include linguistic knowledge: the ALE-rules (Augmented Lexical Entries) (Lehtola 1999b) attached to the concepts and relations conduct the recognition of different ontological constructions that correspond to certain expressions in natural language, and the generation of textual expressions on the basis of ontologies. In other words, the use of linguistic information combined with ontologies enables the interaction between the user and the online service provider in human language, thus making virtual shopping experience more natural. Use of ontological approach also enables a more efficient way for the customer to make searches and to find relevant products in online product catalogues than with a normal keyword–based search.

As Figure 3 reflects, linguistic information in form of terms in different languages is expressed as concept properties denoted by the language abbreviations in parenthesis. Each term is further characterized by a fuzzy value, expressing the degree of exactness with which the term confirms to the concept. For instance, the utterance "*Two-piece jogging outfit with elastics at the leg cuffs*" might be a part of the user dialog in an e-shop for clothing products. Based on matching the user utterance against the terms related to the concepts of the ontology, the concepts *two-piece* (product model), *jogging outfit* (product type), *elastic*, *leg*, and *cuff* (product parts) are recognized. Relations between the concepts are also recognized by using the ALE's shown beside the relations in Figure 3.

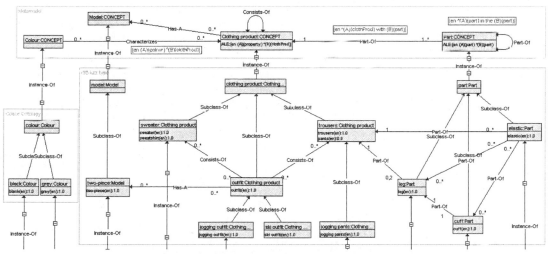

Figure 3. An excerpt from an ontology covering clothing products

In addition to identifying product articles based on indistinct user utterances, ontological approach of the described kind also enables online upselling (offering a bit more expensive and better product article) and cross-selling (offering additional product articles based on the user's selection). Our approach also helps in finding suitable products to be bundled for offering and selling together. CONE has also been used for ontology-based translation of query terms in a legislative domain. Here, the conceptual approach is preferred over traditional thesaurus because translation and use of legal terms depends on the differences between law systems which can be described by ontologies.

4. Correcting and unifying input texts

Unifier software helps in correcting and unifying erroneous or inexact text inputs of online customers in e.g. search engines or order and registration forms. It also facilitates the management of customer information, e.g., in order processing and in data mining. The software also applies to correction of inexact database material collected, for

504

example, by optical character recognition (OCR) devices. Databases often contain many variations of the product names, only slightly modified. It is easier to utilise this information, if the used terminology is analogous; hence the products are easier to be categorised according to their names, e.g. for auditing purposes.

Unifier checks and corrects errors both on word and phrase level. Phrase level error detection is integrated with Webtran language modelling tool described above. It is used to guide the user in word and sentence selections to ensure that the generated text confirms to the definition of a domain specific language model. Word-level errors are detected with dictionary lookup. The dictionary structure can also be used as a thesaurus, which is helpful in applications where it is important to homogenise the terminology. For example, different variations of product names and features can be declared as synonyms. This improves the use of search engines, as different variations produce the same search result.

Unifier can make use of domain knowledge, which enables better accuracy and performance. The contextual information can be used for dictionary partitioning: in order forms the possible values for a field can be declared in a special dictionary. Different dictionaries can be attached to each other: e.g., when checking addresses, the postal codes can be attached to post office names to ensure their mutual accordance.

Unifier ranks the correction candidates according to keyboard neighbourhood. Different keyboard structures can easily be added to the system. Unifier adapts to user's misspelling habits, which is done by collecting user history statistics. The history is used in ranking of the correction methods.

Unifier has been piloted in address correction, which can be embedded in e-commerce order forms. The following Figure 4 shows the process of postal address verification and correction in online services. The process starts when a user enters postal address in online form (e.g. order form, registration form) and clicks on "Submit"-button. Unifier first verifies the postal place (phase 1) and finds out that there is a typo in it and thus returns the only correction suggestion resembling the user entry. At the next phase (phase 2),, Unifier verifies the postal code. As several postal codes match the postal place they all are returned. At the phase 3, the street name is verified. Again, there are multiple correction suggestions, of which 5 first are returned (as 5 is here the maximum number of return values). At the phase 4, street names are combined with 'postal code-postal place' -tuples. At this time, impossible combinations are discarded. The two satisfactory entries are sorted, first using information of typical error types found from user history and then according to the keyboard. The correct address entries are shown to the user in probability order. The user selects the preferred one, and the selection is then used to update the user history.

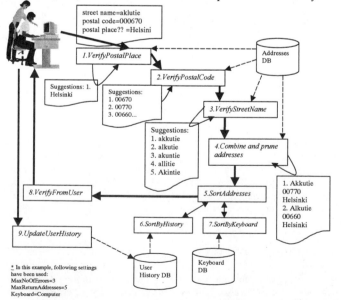

Figure 4: Practical arrangement for correcting postal addresses in online services

5. Conclusions

Webtran, CONE and Unifier are language independent. Tests have been carried out with English, French, Swedish, Estonian and Finnish. In the future, the number of languages will be increased. Currently, VTT is participating in the EU-project Mkbeem (Multilingual Knowledge Based European Electronic Market place, 2000-2002) where a mediation system is developed to enable the access to online products and services in the native languages of customers (Léger 2000). Mkbeem explores new ways of combining human language processing with ontologies for realising three very basic use scenarios of a multilingual online shop: multilingual cataloguing, multilingual information retrieval, and multilingual trading. In MKBEEM, the semantics of the products, i.e., their properties and internal relationships, are desribed in domain ontologies whereas the the content provider ontologies describe product categories of a product catalogue (Gómez-Pérez 2001). These semantics are used to cope with human language ambiguities in order to provide efficient and accurate human language processing. As a proof of concept, the project is implementing a mediation service which supports three European languages: French, English and Finnish.

References

Gómez-Pérez A., Corcho García O., Fernández López M., Lehtola A., Taveter K. , Sorva J., Käpylä T., Tourmani F., Soualmia L., Barboux C. , Castro E., Sallantin J., Arbant G., Bonnaric A. (2001). Requirements, Choice of a Knowledge Representation and Tools. 126 p. http://mkbeem.elibel.tm.fr/

Jaaranen, K., Lehtola, A., Tenni, J., & Bounsaythip, C. (2000). Webtran tools for in-company language support. *Language Technologies for Dynamic Business in the Age of the Media.* Köln, 23 - 25 Nov. 2000. Vereinigung Sprache und Wirtschaft. Köln, pp. 145 - 155

Léger, A., Michel, G., Barrett, P., Gitton, S., Gomez-Pere, A., Lehtola, A.., Mokkila, K., Rodrigez, S., Sallantin, J., Varvarigou, T., & Vinesse, J. (2000). Ontology domain modeling suppor for multi-lingual services in E-Commerce: MKBEEM. *ECAI'00 Workshop on Applications of Ontologies and Problem-Solving Methods.* Berlin, Berlin, 4 p.

Lehtola, A., Tenni, J., Bounsaythip, C., & Jaaranen K. (1999a). Controlled Languages as the Basis for Multilingual Catalogues on the WWW. Jean-Yves Roger, Brian Stanford-Smith and Paul T. Kidd (Eds.). *Business and Work in the Information Society: New Technologies and Applications.* IOS-Press, Amsterdam, pp. 207-213.

Lehtola, A., Tenni, J., Bounsaythip, C., & Jaaranen, K. (1999b). WEBTRAN: A Controlled Language Machine Translation System for Building Multilingual Services on Internet. In: *Proceedings of Machine Translation Summit VII `99 (MT Summit 99),* Singapore, pp. 487 - 495.

Ramsey, G. (2000 March). Online sales surged over the holidays. *Business 2.0 Magazine,* pp. 433-434.

Tieke (2000). Edisty Journal, No. 4, *Finnish Information Technology Development Centre.*

Taveter, K., Lehtola, A., Jaaranen, K., Sorva, J., Bounsaythip, C (1999). Ontology-Based Query Translation For Legislative Information Retrieval. In: *Proceedings of the 5th ERCIM Workshop On User Interfaces For All (UI4All),* Dagstuhl, Germany, 1999, pp. 47-58.

Involving users with learning disabilities in virtual environment design

Helen Neale[1], Sue Cobb[2], John Wilson[3]

Virtual Reality Applications Research Team (VIRART)
University of Nottingham, University Park, Nottingham NG7 2RD UK
[1]helen.neale@nottingham.ac.uk, [2]sue.cobb@nottingham.ac.uk, [3]john.wilson@nottingham.ac.uk

Abstract

The authors have developed and evaluated a number of virtual environments (VEs) to teach life skills to people with learning disabilities. We have found that involving representative users and professionals in the design and development process is invaluable in ensuring that the end product is relevant, usable and acceptable. However, the process of user-centred development is not straightforward. This paper describes the process and activities that we have carried out in order to involve end users in product design. The paper also highlights some of the major difficulties that we have faced and lessons learned in attempting to carry out these activities.

1. Introduction

The Virtual Reality Applications Research Team (VIRART), at the University of Nottingham have been developing Virtual Environments (VEs) for people with learning disabilities since 1991. A selection of experiential learning environments, used to situate life skills practise in an interactive 3-D representation of the real world, have been designed and evaluated with the involvement of representative users (Brown, Neale, Cobb, & Reynolds, 1999; Cobb, Neale, & Reynolds, 1998; Neale, Brown, Cobb, & Wilson, 1999). Users have influenced VE design at many stages of the development process. Section 2 describes the methods and activities undertaken to involve users in the process, detailing some of the obstacles encountered during this. The two main roles of a product designer are to understand the requirements of the product and to develop it (Preece et al., 1994). User-centred design aims to involve users in the development process as much as possible. Some approaches to user-centred design, such as co-operative or participatory design, involve users as active collaborators or partners in the design process (Bodker, Ehn, Sjogren, & Sundblad, 2000). As well as ensuring that the technical system meets user and task requirements, this type of approach also ensures that the system will fit into the context of use, taking into account social and organisational factors. The process that we describe in this paper contains some of the sub-activities common to these types of approaches. Chappell (2000) reports that recognition that people with learning disabilities are capable of insight into and analysis of their experiences is relatively recent, as is their role in participatory research. In order for research to focus on issues of real importance to people with disabilities they must be involved in the research process. Minkes et al (1995) state that the end result of such collaborations will result in: *"...research which more clearly benefits the people whom it is supposed to help."* (P. 97). End-user populations with disabilities may be difficult to design for. Their specific needs may not be well understood by the developer and may be very different from the needs of other mainstream populations. Although there are guidelines for the reduction of cognitive loads for computer use for people with cognitive disabilities (Cress & Goltz, 1989; Scadden & Vanderheiden, 1988), these are quite general, and different task objectives and user groups will require different designs. As Newell (1995) points out, the implementation of any such guidelines for computer and interface design will be difficult because of their general nature and the subjective interpretation of such guidelines. It is also not clear that such guidelines are wholly appropriate for VE design and so it has been imperative that we involve users in the development process.

Virtual Environments present the user with 3D graphical representations of real or imaginary worlds. Within VEs the user may navigate around and interact with objects and characters. The features specific to VEs mean that they have been reported to be more difficult to design than standard 2D interfaces (Herndon, Van Dam, & Gleicher, 1994). The VE interface may need to facilitate a number of tasks that are not found in 2D interfaces. The first is 3D navigation, this includes identifying your position, searching for, locating and approaching (Kaur, Maiden, & Sutcliffe, 1999). Another task is recognising that an object is interactive and knowing how to interact with it. There are also a number of tasks that the user may perform in parallel (for example, moving forwards, looking for and object and interacting with another object). This combination of new task types, a large 3D design space and a lack of existing VE design rules or guidelines (Hix et al., 1999; Wilson, Eastgate, & D'Cruz, In Press) means that it is important to involve users in the design process in order to shape the VE. How to actually go about involving users with disabilities in the development process is not straightforward. Due to the specific requirements of a learning disabled population, some of the methods most commonly used in HCI evaluation could not be used or have had to be adapted in order to be applicable. Iterative involvement of representative users has been critical to the design process, however, care must be taken to ensure that we truly are accessing opinions of users and not just the views of the professionals that work with them. The methods used to facilitate user involvement in the design process need to be appropriate for both the user group and the technology worked with. Bodker et al (2000) emphasise the importance of seeing the conditions of the participatory design process from the point of view of the users as well as the researchers.

2. The development process

This paper presents a design process that involves users at many different stages of the VEs life cycle see figure 1. The approach taken has attempted to involve users with learning disabilities initially in selecting what content should be included in this environment (Brown, Kerr, & Bayon, 1998; Meakin et al., 1998) and later in the review of the design of the interface (Cobb et al., 1998; Neale, Cobb, & Wilson, 2000). The techniques that have been used include brainstorming and storyboarding, VE review by users and professionals from learning disabilities services, observation of VE use, user interviews and questionnaires. Each method plays a different role in the design process and yields different information from the users about their views and behaviour.

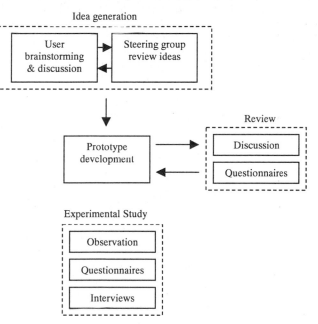

Figure 1 User-centred development process

2.1 Idea Generation

Brainstorming and storyboarding techniques have been used with groups of representative users to specify VE content. These sessions have usually taken part early on in the process at the VE specification stage. Usually a facilitator would chair the discussion and between 5 and 15 representative users (and their advocates) would attend. As well as verbally discussing the ideas, they would be represented as pictures on a large flip chart for the whole group to see. During early sessions the users defined scenarios within which they would like to learn daily living skills (e.g. shop, cinema, hospital); later sessions concentrated on defining specific tasks that the users would like included in these scenarios (e.g. in the shop, use a shopping list). Selection and prioritisation of these scenarios and tasks occurred at steering group meetings, where professionals from special needs schools and services as well as VE developers reviewed storyboards representing user ideas. Ideas were prioritised and refined to ensure they met practical and educational goals. Learning objectives were assigned to each VE, and scenarios within the VE were structured to facilitate them. From this idea generation and design specification stage more complex storyboards were produced for the start of VE development.

2.2   Prototype Review

Representative users and professionals from learning disabilities services reviewed VE prototypes at different stages of the VE life cycle. Early prototypes were reviewed through demonstration and use followed by direct discussion with the VE developer. This provided initial, direct feedback about acceptability from users and usefulness in terms of its application from experts. Later prototypes were reviewed in a more formal manner, where tasks within the VE had been set and questions directed at particular aspects of the environments. This allowed a more in depth review to take place that focused on specific aspects of the VEs as directed by the instructions given. However, this meant that the feedback mechanism was less direct, as questionnaire answers had to be analysed before any changes could be made the VE.

2.3   Experimental Study

Representative users and their support workers participated in experimental or quasi-experimental studies towards the end of the development cycle. Users were set tasks to carry out within the VE and their support workers were asked to help them with their task as needed by the individual. The interactions were video-recorded for analysis.

Observation of interaction was used to measure the number and types of tasks completed by the user, the level of support provided by the support worker for each task and emotional reactions to VE use. The dialogue between the participants and any other relevant behaviour was also recorded. This method enabled a detailed examination of the interaction of the participants with both each other and the VE. Analysis of behavioural data alongside data from the VE using a multiple activity chart (Neale et al., 1999) enabled the identification of a number of usability issues.

The researchers worked with learning disabilities professionals in order to adapt questionnaires so they were suitable to be used with a learning-disabled user group. Questionnaires were used to enquire about general likes and dislikes, specific problems experienced as well as attitudes towards using the VE in general. There were limitations to the types of data that could be collected by questionnaire use. Support workers co-participating in VE use completed interviews and questionnaires about their opinions on user behaviour. They were also questioned on their opinions of the VE as a teaching tool and their ideas for changes and improvements to aid usability and teaching relevance. Usability reports were used to present the results of the experimental study to the VE development team. Each report focused on a task or task component and displayed a picture of the VE alongside observational and subjective data relating to any usability problems experienced when carrying out this task. A sub-section of the steering group reviewed these reports and used them as a basis for deciding revisions to be made to the VEs and how these revisions should be prioritised.

3.   Outcomes

The involvement of representative users in the design process mainly influenced the design of the user interface. A summary of usability problems identified and refined is reported in (Brown et al., 1999). Changes that were made to the VEs as a result of user involvement in the design included the way in which 3D objects were highlighted to indicate they were interactive, and the redesign of confined spaces such as corridors within the VE to facilitate easier navigation. The role of learning disabilities professionals focused on the design of VE content; they specified which life skills to be included, how tasks should be presented to the user, and specified the language and symbols to be used for communication.

4.   Discussion

The researchers found that it was important for idea generation brainstorming and discussion sessions to be facilitated by a professional, who was skilled in accessing the ideas of people with learning disabilities. The use of picture-based storyboards for the representation of ideas was also found to be beneficial in the improving the users understanding of the discussion.

During the steering group review of ideas it was often difficult to get all parties to agree to the same priorities for development and strategies for learning. The professionals that attended were from different backgrounds, including education, social and residential services. It had to be kept in mind that one of the goals of the final product was that it was to be suitable for as wide a spectrum of users as possible for the learning of general daily living skills.

Informal prototype review with representative users and professionals allowed the VE developers to directly observe and talk over any problems of VE suitability with the end-users. This direct feedback mechanism meant that this process was quick and effective at highlighting any major problems. This method is therefore best used in the early stages of development. During the later stages of development, a more structured review session allowed the developers to target reviewers responses to particular features of the programme by using questionnaires. This method is similar to the predictive inspection methods carried out by HCI experts, but the involvement of professionals that work with learning disabled clients has been necessary in order to take into account the abilities and training requirements of a learning-disabled user group.

Observation methods were heavily relied upon for evaluation of VEs with this user group. This method allows the collection of a great range of data, from the number of attempts each user took to carry out a task to qualitative descriptions of how the task was carried out. Observation data was particularly important in this project, where self-report data has been limited. However, this data can be time consuming to analyse.

In this type of research it is difficult to access users true opinions. None of the standard questionnaires used to assess the VE experience were suitable for use with a learning-disabled user group and so the researchers had to adapt and create new ones. They were often difficult to implement, as the users were generally more interested in carrying on with using the VE than listening to the researcher and answering questions. Even though the questionnaires had been adapted for the user group some of the users still found the questions difficult to understand; their support workers sometimes had to re-iterate the question or prompt answers. Many of the users also found it difficult to criticise the VE, and when asked what they didn't like about the programme they would often answer "nothing."

We found it to be important to access the opinions of the support workers that had worked with the user throughout the experimental programme. They held important knowledge about how the VE would be used for skills teaching in practise. Some of the support workers questioned commented that it would be beneficial to have a copy of the virtual shopping list (that can be created in the virtual supermarket) to print out and take to the real supermarket when virtual training is interspersed with training in the real world.

It was crucial for members of the VE development team to attend many of these user-centred design activities, as they could then better understand the rationale for decisions made about content and style of the VE. This also allowed them to explain to users and professionals the opportunities and limitations of the VE software and programming resources.

5. Conclusions and further work

Allowing users to be involved in many stages of VE design and taking multiple evaluation measures helps to ensure that end-user requirements are taken into consideration. It is important that we access and take into account the views of the users and not just of their support workers or advocates. When involving users with learning disabilities in this process we face extra challenges and the methods used should be carefully chosen, and if necessary adapted for the particular user group. We have presented a selection of activities and methods that we have used to inform the VE development team of user opinions, interactions and difficulties and it is important to recognise when we should use each of these methods in order to optimise the design process.

In general, the professional groups have had the most influence over the content of the VE. This includes defining what the VE is used to teach, and which strategies are used to teach them, for example, deciding which areas of shopping skills are key to this experience, making a list, matching the list with real items, and then paying for them. This is based on a combination of their teaching experience; their knowledge of skills that are difficult to teach in the real world (due to constraints of time and/or money) or require repeated practise. For example, although the user group identified a number of areas within which they would like to learn life skills, the steering group prioritised the final 4 areas that were developed, based on their own judgement of which contained the most critical life skills.

Although the user may influence the content to some degree, through brainstorming and storyboarding sessions, they have less overall control of the final decisions made about this. The users did however directly influence the usability of the VE; how messages are communicated, metaphors used, the look and feel of the VE. For example, to make navigation tasks simple, door frames and corridors were widened; to ensure that users understood messages from the system, simple text, Makaton symbols and verbal messages were used; to identify interactive objects to the user the objects were highlighted red.

It is thought that improvements could be made to the questionnaires answered by the people with learning disabilities. In order to address the distraction they face when concentrating on the questions a computer-based interactive questionnaire could be used. Questions could more heavily rely on pictures and have verbal instructions so they had a greater understanding of what is required of them.

The process described in section 2 is being constantly adapted for the development of new products with differing groups of end-users. The authors are currently developing VEs for social skills practise for individuals with Autistic Spectrum Disorders (Parsons et al., 2000). Many different activities are being carried out under the 'idea generation' phase of the cycle, including brainstorming activities with groups of teachers, and 3D low technology modelling of scenarios.

The products using the process shown in figure 1 could be improved with greater numbers of iterations of user involvement followed by technology development. This cycle is usually dependent on the resources of the project.

Acknowledgements

This work described in this paper was carried out under funding by the UK National Lottery Charities Board. We would like to acknowledge the contributions of the other members of VIRART, the project consortium and the user group in supporting this work.

References

Bodker, S., Ehn, P., Sjogren, D., & Sundblad, Y. (2000). Co-operative Design - perspectives on 20 years with the Scandinavian Design Model'. Paper presented at the *NordiCHI*, Stockholm.

Brown, D. J., Kerr, S. J., & Bayon, V. (1998). The development of the virtual city: a user centred approach. Paper presented at the *2nd European Conference on Disability, Virtual Reality and Associated Technologies*, Skovde, Sweden.

Brown, D. J., Neale, H. R., Cobb, S. V. G., & Reynolds, H. (1999). Development and evaluation of the virtual city.*International Journal of Virtual Reality, 4*(1), 28-41.

Chappell, A. L. (2000). Emergence of participatory methodology in learning difficulty research: understanding the context.*British Journal of Learning Disabilities, 28*, 38-43.

Cobb, S. V., Neale, H. R., & Reynolds, H. (1998). Evaluation of virtual learning environments. Paper presented at the *2nd European Conference on Disability, Virtual Reality and Associated Technologies*, Skovde, Sweden.

Cress, C. J., & Goltz, C. C. (1989). Cognitive factors affecting accessibility of computers and electronic devices, *Resna 12th Annual Conference* (pp. 25-26). New Orleans, Louisiana.

Herndon, K. P., Van Dam, A., & Gleicher, M. (1994). The challenges of 3D interaction: a CHI '94 workshop. *SIGCHI Bulletin, 26*(4), 36-43.

Hix, D., Swan II, E., Gabbard, J. L., McGee, M., Durbin, J., & King, T. (1999). User-centered design and evaluation of a real time battlefield visualisation virtual environment. Paper presented at the *Proceedings IEEE Virtual Reality '99*.

Kaur, K., Maiden, N., & Sutcliffe, A. (1999). Interacting with virtual environments: an evaluation of a model of interaction.*Interacting with Computers, 11*, 403-426.

Meakin, L., Wilkins, L., Gent, C., Brown, S., Moreledge, D., Gretton, C., Carlisle, M., McClean, C., Scott, J., Constance, J., & Mallett, A. (1998). User group involvement in the development of a virtual city. Paper presented at the *2nd European Conference on Disability, Virtual Reality and Associated Technologies*, Skovde, Sweden.

Minkes, J., Townsley, R., Weston, C., & Williams, C. (1995). Having a voice: Involving people with learning difficulties in research.*British Journal of Learning Disabilities, 23*, 94-97.

Neale, H. R., Brown, D. J., Cobb, S. V. G., & Wilson, J. R. (1999). Structured evaluation of Virtual Environments for special needs education.*Presence: teleoperators and virtual environments, 8*(3), 264-282.

Neale, H. R., Cobb, S. V., & Wilson, J. R. (2000). Designing Virtual Learning Environments for People with Learning Disabilities: usability issues. Paper presented at the *ICDVRAT, The International Conference of Disability, Virtual Reality and Associated Technologies*, Sardinia.

Newell, A. D. N. (1995). Computers and people with disabilities. In A. D. N. Edwards (Ed.), *Extra-Ordinary Human-Computer Interaction* (pp. 19-43). Cambridge: Cambridge University Press.

Parsons, S., Beardon, L., Neale, H. R., Reynard, G., Eastgate, R., Wilson, J. R., Cobb, S. V., Benford, S., Mitchell, P., & Hopkins, E. (2000). Development of social skills amongst adults with Asperger's Syndrome using virtual environments. Paper presented at the *International Conference on Disability, Virtual Reality and Associated Technologies*, Sardinia.

Preece, J., Rogers, Y., Sharp, H., Benyon, D., Holland, S., & Carey, T. (1994). *Human-Computer Interaction*. Harlow, England: Addison Wesley.

Scadden, L. A., & Vanderheiden, G. C. (1988). *Considerations in the design of computers and operating systems to increase their accessibility to persons with disabilities* (Document of the Industry/government Cooperative Initiative on Computer Accessibility): Trace-Center, University of Wisconsin-Madison.

Wilson, J. R., Eastgate, R., & D'Cruz, M. (In Press). Structured Development of Virtual Environments. In K. Stanney (Ed.), *Virtual Environment Handbook*.: Lawrence Erlbaum.

Some Impacts of International Web Access and Agent Technology on the Evolution of User-Centered Design: A Study in Universal Access

Theresa A. O'Connell

14322 Old Marlborough Pike, Upper Marlboro, MD, USA
toconnell@acm.org

Abstract

User-centered design is evolving to accommodate the impacts of internationalization and new technologies such as intelligent agents. Because Web site users are potentially scattered around the globe, traditional user-centered design practices are evolving. Under the pressure of Internet time, design proceeds with little direct knowledge of user characteristics. An investigation into serving the goals of user-centered design faced the challenges of introducing intelligent agent technology over the Web to an international user group. Projecting a trustworthy agent personality to this diverse user set became a primary design goal. Commonly held usability principles were applied to user needs encapsulated in a synthesized culture. While this method is not based on traditional understanding of users and their needs, it centers on making a site's offerings available to its targeted users across the globe.

1. Introduction

User-centered design (UCD) is entering a new phase. This evolution is both heightened and hindered by the Web, where universal access can mean access to users all over the globe. Simply put, UCD has encountered internationalization. Internationalized UCD faces new dimensions of fulfilling the expectations of a diverse, remote, often ill-defined user group. The goal of international user-centered design is to grant universal access by providing a common ground which at best meets and at a minimum does not offend the cultural and linguistic expectations of targeted users. At the same time, Web access to new technologies, such as personal autonomous intelligent agents, exhibits the potential of radically impacting human-computer interaction, and, therefore, UCD.

This paper explores impacts on UCD of both these forces, internationalization and agents. A pilot study initiated ongoing efforts to design a Web site to deliver agent technology to a projected disparate, multi-national, multi-linguistic, multi-cultural, novice user group (O'Connell, 2000). Under the pressure of Internet time, design proceeded with little direct knowledge of user characteristics. For the most part, it was based on knowledge of the multi-disciplinary principles behind usability engineering applied to delivering the site's offerings. Yet a successful and satisfactory user experience persisted as design's primary responsibility; design still had to be user-centered.

2. Approach

Traditional user interface design proceeds from a simple maxim: know your users. Because UCD focuses on or involves real users to increase the likelihood of user success and satisfaction, UCD has included an understanding of user's cultural and linguistic requirements. Today, this understanding is increasingly difficult to achieve when designing an international Web experience. How can designers grant international access without interacting with users all over the globe, looking over their shoulders, engaging them in think-aloud protocols, talking to them about design ideas? The crux of the question is: How can design be user-centered when users only reveal themselves after the fact?

A pilot study investigated how to render mobile phone account services over the Web, with the help of personal autonomous intelligent agents, first through standard browsers and later through a WAP phone. Projected user locales spanned Europe, North America, North Africa and Australia. Due to time constraints and the diverse user base, the traditional stress on knowing the user gave way to a stress on the site's context of use. This approach defined user characteristics in terms of the site's functionality. Design started with identifying tasks a user has to perform to maintain a pre-paid mobile phone account. Usability engineers investigated the impacts of internationalization and agent technology on this task set and then applied an understanding of usability principles.

The agent had the ability to autonomously and automatically take on tasks commonly associated with pre-paid mobile phone accounts. These included building and modifying a user profile, recharging the balance, finding a new best plan and displaying account activity. The site scenario added two new tasks: displaying a log of agent activity and logging out.

Instead of focusing on the unknown users, the design team started with what was known about the site and then progressed to what could reasonably be postulated about its users. An ancillary goal was to prepare for later localization while doing the least harm now to the international user experience.

2.1 Site user as a cultural group

Time constraints ruled out standard UCD practices such as interviewing or observing targeted users, or studying their cultural characteristics and differences. The site offerings were considered to be users' needs. Designers compiled these needs as characteristics of targeted users and grouped them into a synthetic cultural group called "Site Users."

Site Users were considered to be a cultural group based on an adaptation of criteria described by Yeo (1996) and Hoft (1996). Designers expected that shared characteristics would lead Site Users to react to the site's functionality in the same way. These characteristics included knowledge of the Web and of mobile phone services. Site Users had domain experience in all tasks except displaying a log of agent activity. It was posited that users knew little about agents. The model also included those few cultural and linguistic characteristics known to occur across targeted locales. While this approach offered a pragmatic and rather simple beginning, internationalization and the introduction of agent technology proved to be complicating factors.

2.2 Relying on usability engineering principles

Usability engineering principles are guidelines drawn from all the disciplines that converge in usability engineering. By addressing the attributes that characterize usability (Helander, Landauer & Prabhu, 1997, Nielsen, 1999, & Shneiderman, 1997), they guide designers in promoting user success and satisfaction. Characteristics such as domain knowledge and lack of agent experience were expected to impact on Site Users' success and satisfaction. Mapping known usability principles to such characteristics served the goal of quickly creating an international design that lowers the risk of impeding or offending users. Table 1. partially lists these mappings.

Table 1. The characteristics of Site Users map to usability principles

Site Users' Characteristics	Usability Principles
Based on previous experience with computers, have a strong need to feel in control of interaction	User control Informativeness
Novice to agent technology	Ease of use Learnability
Most likely distrustful of agent technology	Trustworthiness
Prefer to work in their own language	Speak the user's language
Familiar with pre-paid mobile phone accounts	Build on domain knowledge

These principles intertwine. Users need to feel in control of interaction with a system. This system must provide all information needed for a successful and satisfactory user experience. An easy to use system leads a user through tasks without unnecessary impediments. It facilitates the user experience. A learnable system helps a user understand how to access and execute its functionality quickly and comfortably. A trustworthy system sets a user at ease by demonstrating that it will follow a user's directions and act in a user's best interest. Speaking the user's language includes using familiar domain vocabulary as well as translation into a user's native tongue when possible. Building on domain knowledge incorporates metaphors and language familiar to the user to minimize learnability and enhance ease of use. These principles introduced usability attributes integral to the success of the Web site.

3. Focusing on Trust

Looking at the attributes of a personal autonomous intelligent agent through the lens of Site User experiences disclosed that the integral usability attribute that design had to instill in the users was trust. For example, in order for agents to recharge mobile phone account balances, Site Users must trust them to a degree where they permit agents to debit funds from their bank accounts. Trust is closely linked to the users' expectation of control over a system. Based on prior computing experience, Site Users need to feel in control of interaction (Chaudhury, Mallick & Rao, 2001). Only if Site Users agree to relinquish some of their accustomed control will they be able to receive the benefits autonomous agents offer. However, Site Users are unlikely to give up control to the agent unless they trust it. An untrusted agent is impeded from succeeding on its user's behalf. For example, by the very nature of the technology, the behavior of a personal autonomous intelligent agent adapts to accommodate the personal preferences and perhaps even the culture of its user. To realize this benefit, the user has to trust the agent with personal information and with the authority to perform tasks over a period of time while building its knowledge about the user.

Just as any culture is assimilated through learning, so the Site Users' synthetic culture can expand to include learning that an agent is trustworthy. An integral attribute that impacts on a successful user experience with an agent is the agent's personality (Bates, 1994, Reilly, 1995, Walker, Cahn & Whittaker, 1997). If the agent projects a personality, Site Users will tend to interact with it in the same way that they interact with other people (Nass, Youngme, Fogg, Reeves & Dryer, 2000). It became clear that the agent not only had to project a personality, but that this personality had to influence Site Users to trust the agent. Concurrently, while the agent needed to engage its user, its personality had to be as culturally neutral as possible to avoid offending its user. To make the agent as generically trustworthy as possible, usability engineers looked at the agent's attributes and its capabilities for interacting with users in light of the characteristics of Site Users. To project the agent representation's personality in a way that built Site Users' confidence in the agent's intentions and competence, design strategies addressed the look and feel of the site and linguistic issues.

3.1 Look and Feel

Site Users form a dynamic cultural group whose behaviors are evolving in reaction to the new experience of intelligent agent technology. Because an agent acts on behalf of its user, one potential of agent technology is to replace a significant portion of human-computer interaction with agent-to-agent and agent-to-system interaction. Next, human-to-agent interaction will decrease as trust and delegation increase. This will pave the way for a simplified interface and decrease the localization workload. The requirement arose for minimalist universal access design criteria to accommodate these gradual changes in user behavior. Reducing elements in the interface also reduced opportunities for unintentional offensiveness. Design strategies addressed the choice of a simple agent representation, its range of movement and the color scheme of the Web site. Goals included building trust and avoiding cultural offensiveness.

In order to win Site Users' trust, it was decided to represent the agent onscreen. As users become more trusting and used to the agent, it is expected that this representation will no longer be necessary. In addition, over a phone, constantly showing an agent is problematic.

To avoid offensive cultural implications of animals and human body parts such as pointing hands and backs of feet, (Fernandes, 1995) the representation chosen was a cell phone. This choice benefited from drawing on a metaphor within the experience of Site Users. Few humanistic details are needed to depict character (Loyall, 1997, Nass, Steuer & Tauber, 1994). The agent has no arms or eyes. Simplified human behaviors and characteristics were enough to suggest trustworthiness. Curved lines and a slightly cartoonish look were known to be at best engaging and at least acceptable to most Site Users. The agent's onscreen environment and the representation use a blue palette. Light blue is a color known to signify trust in most of the targeted areas.

Users don't have expectations for movement in inanimate objects (Reilly, 1995). The agent's range of movement is limited to moving toward and away from the user. Blinking lights while the agent works stress diligence, and thereby connote trustworthiness. Toward these ends, the agent curves forward when listening to the user.

514

Figure 1. The agent's look is slightly cartoonish and takes on curved lines.

From their prior computing experience, Site Users will need to feel that they control interaction. The representation offers a degree of user control. The user always controls viewing the phone in two states, open and closed. The open state offers a menu for the user to set the agent to work.

Users from different cultural groups require different Web page designs (Marcus & Gould, 2000). Since it was impossible to address this in the first phase of the study, the strategy was to avoid offending by minimizing detail on the Web page. A help button is a question mark, known to Site Users as an icon for help. Forward and backward buttons draw on the familiar metaphor used in Web browsers and give the user control over the dialog's progression. No other within-site navigation is offered other than through the phone agent representation's menu.

3.2   Linguistic Issues

Users access agent technology through dialog. Linguistic style is a powerful tool to communicate the agent's personality (Walker et al., 1997). To further convey personality, the agent always speaks directly to the user in the first person (Walker et al., 1997). The dialog attempts to do no harm by avoiding cultural pitfalls such as humor. To avoid offending, the dialog projects as little emotion as possible. Because courtesy is valued across most of the targeted user locales, the agent is polite; it says "please" and asks permission.

A trustworthy personality became a primary goal of the dialog. The dialog attempts to express trustworthiness by stressing personal qualities such as diligence, competence and beneficial intent. It stresses the agent's abilities with statements such as "I can" or "I will." It demonstrates that it promotes the user's best interests with statements such as, "My only job is to act on your behalf." It promises to perform "only as you instruct me" and tells the user, "I'll protect your anonymity." The dialog builds on the Site Users' knowledge of pre-paid phone accounts by drawing on domain vocabulary. The agent responds immediately to all user input. This increases its own believability (Loyall, 1997) and thus helps to build trust.

The international nature of Web access added design criteria to the exercise of dialog composition. It is a characteristic of Site Users that they prefer to use their own languages. While there is a base set of concepts that the agent will share with all users, it will have to communicate in different languages. Localization will include dialog translation. Every effort was made to constrain the dialog to a simple vocabulary and simple grammatical constructs. This does more than facilitate translation. Because it reduces ambiguity, such a controlled language approach has been shown to also foster human-computer interaction (Lehtola, Bounsaythip & Tenni, 1998).

The dialog bore the burden of introducing novices to agent technology, a task taken on when the agent introduces itself at logon and explains its capabilities. By leading users through the process of building a user profile, the agent reduces Site Users' need to learn the system. At the same time, the dialog makes it easy for the user to provide information that empowers the agent to work on its user's behalf. The dialog serves informativeness when it expands domain knowledge by explaining concepts such as "peak hours." In the dialog, the agent's immediate responses to user input also afford system informativeness.

The dialog enhances ease of use. With dialog, the agent leads its user through the site's functionality, building trust and gradually eliminating the need for many screen elements. Widgets that redundantly offer what the agent can do for the user will disappear as users become more comfortable with dialog interaction and as they increasingly trust their agents to perform tasks for them.

4. Conclusions

Accommodating characteristics common to members of the synthetic cultural group, Site Users, accomplished the first steps toward an internationalized design. At the same time, it laid the groundwork for later localization as more is learned about real users. The tactic of pre-defining users based on a site's offerings may appear to turn UCD inside out, but the focus remains the user. The difference is the substitution of a largely hypothetical user for the real user. This approach is simply a new front-end to a tried and true system. It is a step to be bypassed as soon as the information that informs localization is available. Its validation can only be known in light of subsequently applied traditional UCD practices during localization. The pilot system that resulted from this study is now undergoing its first localization. This exercise will shed light on whether the approach described herein serves the localized user group and facilitates localization efforts.

In this study, introducing intelligent agent technology to a remote international audience moved user-centered design a step away from actual users, but did not lose sight of the fact that good design focuses on users. If designers are to grant access to users all over the globe, they must accommodate such impacts and address them to define the next steps in the evolution of UCD.

References
Bates, J. (1994). The role of emotion in believable agents. Communications of the ACM, 37, 122-5.
Chaudhury, A., Mallick, D.N., & Rao, H.R. (2001). Web channels in e-commerce. Communications of the ACM, 44, 99-104.
Fernandes, T. (1995). Global interface design. Boston: AP Professional.
Helander, M. Landauer T., Prabhu P. (Eds.). (1997). Handbook of human-computer interaction. Amsterdam: North Holland.
Hoft, N. (1996). Developing a cultural model. In E.M. del Galdo & J. Nielsen (Eds.), International user interfaces (pp. 41-73). New York: Wiley Computer.
Lehtola, A., Bounsaythip, C., & Tenni, J. (1998). Controlled language technology in multilingual user interfaces. In C. Stephanidis & A. Waern (Eds.), Proceedings of the 4th ERCIM Workshop on User Interfaces for All (pp. 73-78). Stockholm. Retrieved from the World Wide Web: http://ui4all.ics.forth.gr/UI4ALL-98/proceedings.html
Loyall, A. (1997). Some requirements and approaches for natural language in a believable agent. In R. Trappl & P. Petta (Eds.), Creating personalities for synthetic actors (pp. 113-119). Berlin: Springer-Verlag.
Marcus, A., & Gould, E.W. (2000). Crosscurrents: Cultural dimensions and global Web user-interface design. Interactions, 7, 33-46.
Nass, C., Steuer, J., & Tauber, E. (1994). Computers are social actors. In B. Addelson, S. Dumais & J. Olson (Eds.), Proceedings of ACM SIG CHI Annual Meeting (pp. 72-78). Boston: Addison-Wesley.
Nass, C., Youngme, M., Fogg, B.J., Reeves, B., & Dryer, D.C. (2000). Can computer personalities be human personalities? International Journal of Human-Computer Studies, 43, 223-229.
Nielsen, J. (1999). Designing web usability: The practice of simplicity. Indianapolis: New Riders.
O'Connell, T.A. (2000). A simplistic approach to internationalization: Design considerations for an autonomous intelligent agent. In P.L. Emiliani & C. Stephanidis (Eds.), Proceedings of the 6th ERCIM Workshop, "User Interfaces for All" (pp. 15-27). Florence: Consiglio Nazionale delle Ricerche and Istituto di Ricerca sulle Onde Elettromagnetiche.
Reilly, W. (1995). The art of creating emotional and robust interactive characters. In Working Notes of the AAAI Spring Symposium on Interactive Story Systems. Palo Alto, CA: Stanford University. Retrieved from the World Wide Web: http://www.cs.cmu.edu/afs/cs/user/wsr/Web/research/oz-papers.html
Shneiderman, B. (1997). Designing the user interface: strategies for effective human-computer interaction. Reading, MA: Addison-Wesley.
Walker, M., Cahn, J., & Whittaker S. (1997). Improvising linguistic style: Social and affective bases of agent personality. In Proceedings of the First International Conference on Autonomous Agents (pp. 96-105). Marina del Ray, CA: Association for Computing Machinery.
Yeo, A. (1996). World-wide CHI: Cultural user interfaces, a silver lining in cultural diversity. SIGCH Bulletin, 28. Retrieved from the World Wide Web: http://www.acm.org/sigchi/bulletin/1996.3/international.html

Design and Implementation of Universal End-User Commands, Interfaces and Interactions

Basawaraj Patil, Klaus Maetzel, Erich. J. Neuhold

German National Research Center for Information Technology (GMD-IPSI)
Dolivostrasse 15, D-64293, Darmstadt, Germany.
E-mail: {patil, maetzel, neuhold}@darmstadt.gmd.de

Abstract

The power of the Internet in its universality and accessibility is regardless of any disabilities or limitations of the end users. The Internet holds promise of access of information to all people, but offers a few opportunities and limits the access to large section of global users – native end-users, those who lack English skills and prefer to interact in their native language-like command languages. Current command languages and their interpreters are available in English and are not flexible enough to incorporate the requirements of universal, native end-users. Hence, native end-users encounter serious barriers at conceptual, logical, semantic and syntactic level of interactions. We propose a human-centric design methodology of commands languages and their interpreters, which are more flexible and adaptable to the requirements of the native end-users. The design methodology is embodied in EQUAL, a system of design and implementation of native commands languages and applications in a few representative end-users languages such as German, Italian, Cyrillic and Chinese.

1. Introduction

The power of the Internet in its universality and usability is regardless of any disabilities or limitations of the end-user populations. The Internet holds promise of access to information to all people, but offer a few opportunities and limits the access to a target population, *native end-users* – those, who lack English knowledge and prefer to interact in their native language-like command languages. The Web Accessibility Initiatives (WAI, 2000) and Universal Accessibility Protocol (UAP, 2000) have contributed to the development of universal access and usage of the Internet. But, these initiatives mainly focus on accessibility issues concerned with disabled people or disabling conditions. However, there are important gaps or *"digital divide"* among large section of the native end-users, who have not received the full benefits of Internet technologies. Due to rapid globalization and Internet explosion, native end-user issues may become serious because the vast majority of the world population who do not and will not in the foreseeable future speak English will be excluded from the system. People lacking computer skills may become a serious individual and social problem (Madon, 2000) and in the worst case may lead to - Internet Apartheid (Shneiderman, 2000). Lack of native end-user interfaces, linguistic barriers, complex social and cultural factors will continue to hinder the universal access and effective participation in Internet technologies. A wide range of native end-user requirements and related issues must be included in the universal design of user interfaces, command languages and should provide equal opportunities to all the people to interact and use Internet technologies.

2. Native end-users: a profile

The research studies on needs and requirements of native end-users are scarce and poorly understood. We initiate a few basic studies on native end-user requirements of textual languages and give broad perspectives on native end-user issues of command languages and applications. Within the scope of our research work concerned with universal design and universal accessibility of command languages with a long term research goal of *"Information Technology for All (IT4ALL)"*, we define and focus our studies to a target population called *"native end-users"*. Native end-users may be defined as end-users with one or more characteristics such as: (a) They prefer their native language-like commands, and operations in human-computer interactions. (b) They have minimal or no English knowledge. (c) Their medium of education and instruction is in their native language and socio-cultural environment. (d) They may or may not posses computing skills. (e) They would like to acquire information manipulation, programming and computing skills. Native end-users referred to here include students (e.g., high schools, colleges and universities), adults with general educational background. The majority of command languages and applications are based on English semantics and syntax and *"do not speak the languages"* of the native end-users. Hence, native end-users, especially, CJK (Chinese, Japanese and Korean), Asian, non-English speaking European end-users encounter serious barriers at conceptual, logical, semantic and syntactic level of human-computer interactions, commands and operations.

2.1   Native end-user command languages and interfaces: universal usability issues

The design for diversity and universal usability of command languages and applications is difficult and a challenging task. The universal design targets the broadest possible range of user capabilities and limitations. The principles of universal design explore to understand what factors contribute to the universal accessibility and usability and how to accommodate complex and wide spectrum of end-user population requirements. Universal usability means removing barriers that prevent native end-users from using new Internet technologies.

Important barrier-free end-user technologies are conversational interfaces, natural language interfaces and dialogues, and visual languages. One promising approach to the *"universal access"* especially for the technologically naïve users is conversational interfaces. The complexity of spoken language, limitations of speech recognition and understanding systems will hinder the effective adoption of these technologies. Natural language interfaces and dialogue systems are useful in restricted application domains. Again, the complexity of natural language grammar and ambiguous language constructs is a serious obstacle to the design of native, natural language end-user interfaces and dialogues. One of the effective end-user interface technologies are visual and graphical languages such as WIMP (Windows, Icons, Menus, a Pointer) interfaces and Direct Manipulation (DM). In graphical and visual languages, effective approaches are to support native end-users in native versions of software and GUI components by exploiting internationalization, localization and language engineering (Galdo & Nielsen, 1996). The process of localization begins with a set of original English command terms and makes a series of decisions on the appropriate conversion, translation or replacement for these terms within the native language and in turn may lead to the serious *"second language problems"* (Kukuska, 2000). These graphical and visual languages are useful to native end-users, but have limitations such as they lack expressive power, do not scale well and useful only in simple applications (Green & Petre, 1992).

We consider universal command languages and applications as another effective strategy for the native end-users. Surprisingly, there has been little effort to *"internationalize"* command languages, interpreters and applications. The majority of command languages, interpreters and applications are based and derived from English. The syntax and semantics of current command languages are not compatible and comprehensible to the native end-users. Current command languages are not flexible to support the needs, requirements and preferences of native end-users. Even in the era of visual and graphical languages, we believe that the textual languages have their own unique HCI role in specific situations (e.g., low bandwidth communication link, visually impaired users etc.) or applications (e.g., mobile, embedded etc.). Command languages can be extended with conversational interfaces for phonetic-based, textual input methods (e.g., Pinyin for Chinese, Indian languages, etc.). Command languages provide a sound foundation for learning advanced computing techniques, programming skills and automatic processing. Apart from that, the majority of end-user programming languages and XML technologies are textual in nature.

3.   EQAUL command languages and interfaces: basic principles

We have developed a research prototype called EQUAL (Easy, Quick, Universal, Accessable and Leverage) system used for the design, implementation and evaluation of native end-user textual languages. An in-house preliminary survey was conducted with native end-users and experts from various discipline and nationalities. The preliminary survey mainly focused usability and accessibility issues on native end-user command languages and interpreters. The important observations are (a) Command languages that are syntactically and semantically compatible to the native end-users are useful and enhance accessibility, learnability, comprehensibility and ease of operation. (b) Special global initiatives and research studies are necessary. (c) There is a critical need of universal design methodologies and human-centric languages, interpreters and tools. (d) Standardize native end-user command languages.

Command languages are simple textual languages that are used extensively in operating systems and command-based applications. Extensive work (Shneiderman, 2000; Moran, 1981) has been explored in the design of command languages, syntax and semantics of command languages and empirical studies. Human learning, problem solving, and memory are greatly facilitated by meaningful structure. If the command languages are well designed, users can recognize the structure and easily encode it in their semantic knowledge. Meaningful structures are beneficial for task concepts, computer concepts and syntactic details of command languages. As a central feature of many forms of communicative dialogue, names and naming have been extensively examined in a number of core disciplines such as linguistics, philosophy and empirical psychology (Black & Moran, 1982; Carroll, 1978, 1982). Both formal and informal observational analyses repeatedly identified names, naming and structural contexts as an important practical problem (Rosenberg, 1982). Several researchers (Rosenberg, 1982) tried to produce clear demonstrations that the way commands were named exerted an influence on user learning and performance. It is observed (Ledgard et al., 1980; Landauer et al., 1983) that natural language-like constructs and commands are effective in the design of end-user interactions and systems. We extend the above concepts and provide the end-users with native equivalent commands in their language that are natural and compatible to the needs and requirements of the native end-users – *Support naturalness*. We exploit the basic principle of usability engineering and support command names and its components in native end-user languages – *Speak end-user's language*. The EQUAL command language features are semantically adaptable to the requirements of the native end-users and match well with the task characteristics – *Support closeness of mapping*. Since, the medium of education and instruction is in native languages – *Leverage native knowledge and skills.*

EQUAL provides a unique opportunity to native end-users and can interact, command and operate in their native language–like constructs (Table 1). In this paper, we discuss our design principles to command languages, command interpreters and applications, but the basic principles are general and extensible to the design and implementation of textual languages and applications (e.g., spreadsheets, scripting, query, and programming languages).

4. EQUAL design methodology

The EQUAL design methodology is a universal, human-centric concepts designed to accommodate native end-users with diverse requirements and preferences. Important native end-users' syntactic components and semantic concepts (Table 2) are clearly separated and stored separately in a database. In general, verb-object (e.g., European users etc.) or object-verb (e.g., Asian, Indian users etc.) or mixed formats are used for command specification. The language designers can specify these design constraints and use them effectively to generate command interpreters in native end-users languages. The EQUAL interpreter is flexible to map different syntactic components of native end-user languages onto the formal semantics of command languages and generate interpreters for different native end-user languages. Language engineering skills and special knowledge is necessary to deal with complex morphological and linguistic constraints in CJK (Chinese, Japanese and Korean), Indian and Asian languages.

4.1 EQUAL design environment

The EQUAL design environment and important functional components (Figure 1) are:

- **Command Database:** EQAUL command language syntax and semantic features, vocabularies are based on the in-house survey results and recommendations from the HCI experts. The command specifications and vocabularies, linguistic and cultural data play an important role in native end-user interactions and are stored in the database and may be enriched by the domain experts. New command languages, linguistic and cultural data can be extended and integrated with the EQUAL system modules. In the present version, the multilingual error messages, warnings are hand-coded and stored in the external database.

- **Command Interpretation and Generation:** EQUAL command grammar, command interpreter and generator form the core module of the EQUAL system. It mainly consists of a syntactic analyzer (Watt & Brown, 2000) used to identify the lexical category of the input commands and its parameters. In general, the semantics of a command does not depend on the syntax of the command, hence, we exploit this feature to map different syntactic components of native end-users languages onto the formal semantics of a command. The complexity of the translator or interpreter depends upon the design constraints, task characteristics and system functionality.

Table (1). EQUAL command samples in native end-user languages

English	German	Greek	Chinese	Cyrillic
DIR	Anzeigen	Εμφάνιση	显示	ЮђюсNэрιрђ к
OPEN <file>	Offnen <datei>	Άνοιγμα <αρχείο>	打开<文件>	ЮђэNеђђк <ерιιсr>
PRINT <file>	Drucken <datei>	Εκτώπωση <αρχείο>	打印<文件>	Яхïрђк <ерιιсr>
CLOSE <file>	Schließen <datei>	Κλείσιμο <αρχείο>	关闭<文件>	ЧрaNеђђк <ерιιсr>

Table (2). EQUAL command language design constraints

Command Order	Positional or Non-positional
Command Format	Verb-Object or Object-Verb or mixed format
Command Name Set	Names, Number of commands
Command Abbreviation Rules	Standards, Truncation, Deletion of Vowels or Consonants
Command Sequence	Key, Key Strokes, Function Keys etc
Command Morphology	Morphological Constrains
Command Linguistics	Linguistic Constrains
Command Delimiters	Blank Space, Colon and Semi, Comma etc
Command Arguments	Objects, Names, Flags, Switches etc
Command Environment	Environment Variables

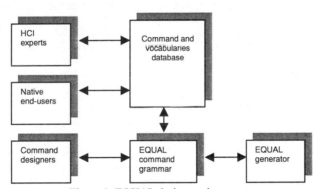

Figure 1: EQUAL design environment

5. EQUAL system implementation

The EQAUL design principles and methodology are demonstrated by generating native end-users command languages and interpreters for DOS or FTP applications. The major components of the system are based on Unicode and Java technologies. The system may be deployed as a standalone application running on a local machine as an interpreter/shell or server side component of client-server architecture. During interaction and dialogue, commands are encoded in Unicode and communicated to the interpreter or server for the execution. The command shell or server component can interpret the commands and execute them. The results, errors and warnings are displayed in the native end-user languages in a separate window.

5.1 EQUAL commands and examples

A sample EQUAL DOS (e.g., DIR, PRINT <file-1>) and FTP (e.g., PUT <file-1>, GET <file-2>) commands are shown (Table 1) in a few representative native language-like commands. The native end-users can command, interact and initiate a few DOS or FTP commands. The semantics of FTP commands are more complex than DOS commands. The EQUAL commands may be separated from the arguments by a blank or other delimiter, and the arguments may have blanks or delimiters between them. Note that there is no concept of delimiters in Chinese language, but, we recommend and use comma ',' as a delimiter. All EQUAL commands end with a carriage return.

The errors, warnings and results are shown in a separate window in native languages. The system is flexible to incorporate native end-user macro commands.

6. Conclusions

We have discussed the human-centric, universal design principles of command languages, interpreters and applications in a few representative native end-user languages. EQUAL system provides a unique opportunity to native end-users to interact, command and operate simple command-based applications in their native language-like information structures. We also discussed a flexible method of generation of interpreters of native end-users command languages. EQUAL system may also be used as a scaffolding system for teaching and learning command languages and computer concepts for native end-users. The EQUAL design principles and methodology may also be used in the generation of international user interfaces, GUI components, textual languages and native end-users interfaces to command-based legacy applications. The usability analysis and critical evaluation are planned for the future work.

References

Black, J., Moran, T. (1982). Learning and remembering command names *in Proceedings of Human Factors in Computer Science* (Gaithersburg), ACM, New York.

Carroll, J.M. (1978). Names and naming: An Interdisciplinary Review. *IBM Research Report* RC 7370.

Carroll, J.M. (1982). Learning, using and designing command paradigms. *Human Learning*, 1,31-62.

Galdo del, E.M., Nielsen, Jakob. (1996), *International User Interfaces.* New York: John Wiley & Sons, Inc.

Green, T.R.G., Petre, M. (1992). When visual programs are harder to read than textual programs, *Human-Computer Interaction: Tasks and Organisation, Proceedings of ECCE-6 (6th European Conferenceon Cognative Ergonomics)*, van der Veer, Tauber, M., Bagnaroa, S. and Ichikawa, T. (Eds.), Rome, Italy.

Kukuska-Hulme, A. (2000). Communication with users: Insights from second language acquisition. *Interacting with Computers*, 12, 587-599.

Landauer, T.K., Glotti, K.M., Hartwell, .S. (1983). Natural command names and initial learning: A study of text editing terms. *Communications of the ACM,* 26, 495-503.

Ledgard, H., Whitehead, J.A., Singer, A., Seymour, W. (1980). The natural language of interactive systems, *Communications of the ACM,* Vol. 23(10), 556-563.

Madon, S. (2000). The Internet and socio-economic development: Exploring the interaction, *Information Technology & People*, Vol.13(2), 85-101.

Moran, T.P. (1981).The command langauge grammar: A representation of the user interface of computer systems. *International Journal of Man-Machine Studies,* 15, 3-50.

Rosenberg, J. (1982). Evaluating the suggestiveness of command names. *Behavior and Information Technology*, No. 1, 370 – 400.

Shneiderman, B. (1998). Designing the User Interfaces: Stratgies for effective human computer Interactions, 3rd Edition. Amstrdam: Addison-Wesley.

Shneiderman, B. (2000). Universal usability. *Communications of the ACM*, Vol. 43(5), 85-91.

UAP (2000). Universal Accessibility Protocol. http://www.uniac.com/reqts.html

WAP (2000). Web Accessibility Initiatives. http://www.w3.org/WAI/

Watt, D. A., Brown, D.F. (2000). *Programming language processors in Java.* New York: Prentice Hall.

Designing for users with color-vision deficiency: effective color combinations

Pamela Savage-Knepshield

Lucent Technologies/Bell Labs
101 Crawfords Corner Road, Room 1K-534, Holmdel, NJ 07733

Abstract

The primary goal of this research program was to design effective color palettes to indicate four alarm severities for users with color-vision (C-V) deficiency. Previously, a single color palette was available for all users regardless of their ability to discriminate colors. This color palette used red to indicate a critical alarm, yellow for a major alarm, light gray for a minor alarm, and gray with a small green box in the lower right corner of the alarm bell icon to indicate no alarm. Research has shown that some users, who have C-V deficiency, have difficulty perceiving or lack the ability to perceive some colors (e.g., red, cyan, and green). Our current color palette included two of these colors. Therefore, a need existed for additional color palettes to optimize the user experience for those with C-V deficiency and promote universal access for network-monitoring engineers. Based on this study's results, six color palettes are recommended as options for C-V deficient users. Furthermore, the quantitative and qualitative data collected during this study stress the importance of using redundant visual coding for indicating relationships between objects.

1. Background

1.1 Color-vision deficiency

C-V deficiency occurs when people have problems in identifying various colors and shades, which can vary from having slight difficulty in distinguishing among different shades of the same color to the rare inability to distinguish any color at all. Approximately nine to twelve percent of males and less than one percent of females have some difficulty with color vision (Hoffman, 1999). C-V deficiency is usually hereditary, however, some problems occur with the normal aging of the eye's lens, retinal or optic nerve disease, and as a side effect of certain medications.

1.2 Using color in interface design

Colors connote importance, specify categories, draw attention, emphasize information, indicate relationships between objects, reflect a mood, and establish an element of style. There are numerous research findings that are related to the general use of color in interface design, the use of color in emergency situations, and the use of color by process control engineers. Several of these recommendations that are relevant follow:

- In the US, red is associated with stop/hot/danger/on, green with go/safe, yellow with caution, white with cold, and blue with off (Brown, Earnshaw, Jern, & Vince, 1995).
- As a professional group, process control engineers associate green with nominal/safe, yellow with caution, and red with danger (Thorell & Smith 1990).
- Effective colors for process control applications to show life support status include using: blue/green/white for OK, yellow/gold for CAUTION, and red (flashing option) for EMERGENCY status according to Smith (as cited in Brown, Earnshaw, Jern, & Vince, 1995).

1.3 Designing with color for the color-vision deficient

There is a growing body of literature that relates to the use of color in interface design when the target user population has C-V deficiency. A brief sample follows:
- Lighten the light colors and darken the dark colors (Arditi, 1999).
- Use blue, yellow, white, and black if color must be used (www.cimmerii.demon.co.uk).
- Do not use: cyan with gray, yellow with light green, green with brown, and red with black (Thorell & Smith, 1990).

2. Method

2.1 Subjects

Sixteen Lucent Technologies employees volunteered to participate in this study. The five normal-vision participants ranged in age from 25 to 61 and included two females and three males. All C-V deficient participants were male between the ages of 35 and 61. Seven self-reported red-green deficiency, 2 blue deficiency, and 2 claimed difficulty with most colors.

2.2 Test Stimuli

Based on recommendations in the literature, eight color palettes were created. Each palette used a different combination of the following seven colors (RGB values for each color are provided in parentheses): red (FF0000), yellow (FFFF00), light gray (CCCCCC), green (00CC00), orange (FF6600), black (000000), and blue (0066FF). The following eight color combinations, which are presented in order of alarm severity (critical/major/minor/no alarm), resulted:

- red/yellow/light gray/green
- black/dark gray/light gray/white
- black/yellow/blue/white
- black/yellow
- light gray/white
- black/blue/light gray/white
- red/orange/yellow/green
- black/yellow/white/green
- red/yellow/light gray/white.

As shown in Figure 1, alarm indicators were either iconic representations or color filled blocks.

Figure 1. Alarm severity representations:
alarm with hammer and color filled block

The experimental task was designed to simulate a network management task in that subjects learned the pairing between colors and alarm severities for a given palette, then they were presented each color multiple times and selected the alarm state that matched the color each time the color was presented. Therefore, it tested the memorability of the color to severity matching, as well as color discrimination among alarm states.

3. Procedure

Participants completed a brief questionnaire to collect demographic information. Next, they were given as much time as needed to familiarize themselves with a palette's four alarm severities and associated indicators. Then, participants were shown an indicator and responded by indicating which alarm severity it represented. After making 16 judgments for a palette (4 judgments for each of 4 alarm severities), participants rated how difficult they found it to use that particular color-coding scheme to identify alarm severities. Participants were given the opportunity to take a break half way through the study; however, none chose to do so. After completing this process for all 16 color palettes, they rank ordered their preference for the color palettes. All participants were tested individually and required approximately 45 minutes to complete the study.

Four trials of each alarm severity were presented in random order for each color palette. Color palettes were presented in random order, with either all eight block palettes or all eight icon palettes presented first. Presentation of the block/icon order was also random across participants.

4. Results and discussion

4.1 Color-vision deficient vs. color-vision normal performance

Participants rated the ease of use for each color palette using a 5-point rating scale, which ranged from very difficult to very easy. Accuracy was used to measure the percentage of correct judgments made for a color palette (i.e., identifying the correct alarm severity for a specific alarm indicator). Response time was used to measure the time in seconds that subjects took in order to make their judgments. Table 1 contains the means and standard deviation for the ease of use rating, accuracy, and response time measures for both C-V deficient and normal participants.

Table 1: Mean rating, accuracy, and response time for each C-V population

	C-V Deficient		C-V Normal	
	Mean	SD	Mean	SD
Rating	3.90[*]	1.28	3.42[*]	1.22
Accuracy (% correct)	94.66[*]	11.35	98.34[*]	4.35
Response Time (in secs)	1.52	0.49	1.62	0.56

$p < .01$.

The rating and response time results are not surprising when it is considered that the color palettes were created for optimal use by specific vision-color deficient populations. Apparently, when C-V deficient participants encountered a palette with alarms that were not easily distinguishable, they quickly "guessed" which alarm it was. They did not find this problematic because they are aware that they are C-V deficient and did not expect to make these distinctions accurately.

The C-V normal participants were less satisfied with the ease of use of the color palettes because they did not follow the typical color conventions of using a red indicator for critical alarms and a green indicator for no alarm. However, since the C-V normal participants have good color-vision, they were less prone to error than were the C-V deficient participants. For example, the red-green blind participants tended to have more difficulty with those palettes that were not optimally designed for their use. These participants expressed extreme dissatisfaction with palettes that contained red, orange, and green.

4.2 Color Block Palettes vs. Icon Palettes

One-half of the color palettes were created as alarm icons, while the other half were simply square blocks consisting of a black border filled with a contrasting color. The same color-coding scheme was used for both types of palettes. The blocks were designed because there was concern that the intricate design of the alarm icon may add further complexity to the alarm severity identification task. Results demonstrated, however, that there were no significant performance or preference differences between the icon and block palettes.

4.3 Color Palette Performance and Preference

The palette that is currently in use for network-monitoring (red = critical, yellow = major, white = minor, and a gray alarm with a green box in the lower right corner of the icon = no alarm) resulted in the best overall performance and was the most preferred palette by C-V deficient participants in this study. The next most preferred palettes consisted of Red-Yellow-Light Gray-White and Black-Yellow-Blue-White color-coding schemes. The least preferred palettes were those that contained grayscale shades and Red-Orange or Yellow-Green color combinations.

The only significant differences found by a Bonferroni test between the color palettes for the four dependent measures were for the rating and ranking measures. In general, most palettes were rated significantly easier to use and more preferred than the Red-Orange-Yellow-Green color combination palettes, with the exception of the grayscale palette.

Correlation coefficients were computed among mean accuracy, response time, ease of use rating, and ranking measures. A Bonferroni approach was used to control for Type I error across the multiple correlation comparisons. All measures were significantly correlated with each other. Accuracy was negatively correlated with response time. The lower a palette's accuracy rate was for identifying alarm severities, the more time was spent making the judgments. Accuracy was positively correlated with both rating and ranking. The higher a palette's accuracy rate was for identifying alarm severities, the higher its ease of use rating and the higher it was ranked for overall preference.

Table 2: Correlation data for mean accuracy, response time, rating, and ranking measures

	Accuracy	RT	Rating	Ranking
Accuracy	—	-0.583[*]	0.802[**]	0.667[**]
Response Time	-0.583[*]	—	-0.541[*]	-0.622[*]
Rating	0.802[**]	-0.541[*]	—	0.850[**]
Ranking	0.667[**]	-0.622[*]	0.850[**]	—

* $p < .05$, ** $p < .01$

5. Conclusions And Design Recommendations

Based on the quantitative and qualitative data collected during this study it is recommended that color should not be the sole discriminator when indicating relationships between objects. During debriefing several subjects explained that one reason they preferred the currently used palette over all the others was that its no alarm indicator contained a small green square in the lower right section of the icon. This redundant coding allowed them to immediately dismiss the no alarm severity as one of the four severity choices when making their judgment. Textual labels would also benefit users in discriminating among the alarms.

The following six palettes are recommended as options for C-V deficient users:
- Modified Red/Yellow/Light Gray/Green Icon Palette in which the critical alarm contains a triangle with an exclamation point in the upper right corner of the alarm and the no alarm has the "hammer" removed from the icon.
- Modified Black-Yellow-Light Gray-White Icon Palette in which the light gray alarm is texture-filled (e.g., crosshatch or diagonal line).
- Black-Yellow-Blue-White Icon Palette.
- Modified Black-Yellow-Light Gray-White Block Palette in which the light gray block is texture-filled (e.g., crosshatch or diagonal line).
- Black-Yellow-Blue-White Block Palette.
- Modified Black-Dark Gray-Light Gray-White Block Palette in which the dark gray block is crosshatch texture-filled and the light gray block is diagonal line texture-filled.

References

Arditi, A. (1999). *Effective color contrast: designing for people with partial sight and color deficiencies.* [On-line]. Available: www.lighthouse.org/color_contrast.htm.

Brown, J. R., Earnshaw, R., Jern, M., & Vince, J. (1995). *Visualization: using computer graphics to explore data and present information.* New York: John Wiley & Sons.

Hoffman, P. (1999). *Accommodating color blindness.* [On-line]. Available: www.stc.org/pics/usability/newsletter/9910-color-blindness.html.

Thorell, L. G., & Smith, W. J. (1990). *Using computer color effectively: an illustrated reference.* Prentice Hall: Englewood Cliffs, NJ.

The ICS model: simultaneous support to stand–alone and cooperative work

Christian Sifaqui

Zentrum für Graphische Datenverarbeitung e.V.
Rundeturmstr. 6, D–64283 Darmstadt, Germany

Abstract

This paper presents a novel model to shape objects with dynamic attributes and values. This model is the base for a new software architecture for managing object systems based on conversations. The objects can be located in the personal PC or can be distributed. The different instantiations of the model allows a gradual transition between stand–alone and cooperative use, achieving this way a transparent access to distributed resources.

1. Introduction

Searching and finding files in a personal computer and the diverse modalities that each person uses for it is a subject that already has a couple of years of existence (Barreau & Nardi, 1995; Nardi & Barreau, 1997). This subject has expanded now to the Internet, which on its part floods the same user with irrelevant pages as results of his/her searching.

While on the one hand there is the problem of the local searching for the previously filed files, on the other hand there exists the problem of searching for the relevant information in the Internet. If the user manages to solve the search problem, he/she now must solve where to store that obtained information, and for that he must use a hierarchical file system (HFS) or bookmarks, returning to the starting point.

The other aspect to integrate, is the search and storage of the information generated in a conversation (mail, chat, for example) and the resources that could have been referenced during the conversation. For this a model is needed that incorporates the knowledge generated at personal level and the generated one at social level.

2. Related Work

Some of the representative work for the landscape of solutions for the problem described on section 1 is presented now.

The Brain (Natrificial Software Technologies, 1998) is a first attempt to achieve a transparent interface to the desktop and to the Web, where by linking desktops objects, like files, e-mails and web-pages, a graphical graph with the relevant information can be built. The MIT Semantic File System (Gifford, Jouvelot, Sheldon, & O'Toole, 1991) provides associative access to a file system via virtual directories. Using native directory commands, virtual directory names are interpreted as associative queries. The results of a query are computed via an automatically indexed set of attributes. It provides a virtual organization documents through search. KnowSpace (Rawlins, 1999) is a data administrator of all the user's data that can come from the Web or the desktop. This data is automatically catalogued, and processed for later search. Placeless Documents (Dourish et al., 2000) is a document management system that, by using active attributes on documents, improves the interaction and search.

The first system shows an interesting interface, that is based on a more cognitive aspect, the last three present a similar model in their conception, that is adding attributes to existing objects. All the approaches described above have developed solutions for more or less specific problems. Our aim now is to consider this entire topic on a more comprehensive, more abstract level. For that we have designed the ICS model which defines a generally valid personal and social view to information and knowledge.

3. The ICS model

Very briefly, the autopoiesis theory (Maturana & Varela, 1984) states that the human being is a autopoietical system that has a complex nervous system, that is immersed in a social system and that uses language. Within the language it located itself as observer making distinctions. A distinction (to bring forth simple or composited unities) is a cognition act made by the observer and it only happens in the domain of structural coupling in

526

which he or she operates. The distinctions made by the observers not are necessarily coincident. This way an **observer** by means of **interactions** can distinguish **simple** or **composite unities,** that are characterized to have an organization and a structure.

Based on this terms we can sketch a model as shown in figure 1.

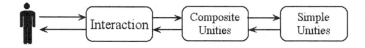

Figure 1: Model

This model is divided into three layers:
* Simple unity: is brought forth in an operation of distinction that constitutes it as a whole by specifying its properties as a collection of dimensions of interactions in the medium in which it is distinguished. We can represent a simple unity by the tuple $U_S = (A_S, V_S)$ where A_S is the set of dynamic attributes that do not make the unity loose its ownership to a specific class and V_S is the set of dynamic values assigned to A_S.
* Composite unity: is a simple unity in a metadomain with respect to the domain in which its components are distinguished because it results as such from an operation of composition as a simple unity. We can represent a composite unity by the tuple $U_C = (A_C, V_C)$ where A_C is the set of dynamic U_S and V_C is the set of dynamic values assigned to A_C.
* Interaction: refers to the actions that the observer can make with his/her distinguished unities, for example, searching for some unity, alteration in the state of a unity by means of copy or replication generating a history, set operations: union, intersection and substraction; navigation by the unities.

3.1   Stand–alone

This model can have several instantiations, depending on the number of observers that interact with it. 0 represents a stand–alone use and 1 represents a shared one. The figure 1 shows the instantiation 000 where an observer interacts in a stand–alone form with the model.

3.2   Cooperative use

The social systems emerge from or are constituted by the interactivity of their participants (Hejl, 1984; Maturana & Varela, 1984). In the shared use different combinations from the model arise, so we have the simplest case as shown in figure 2:

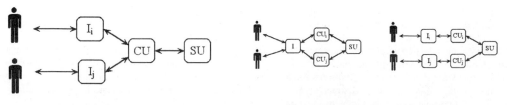

Figure 2: Instantiation 100   Figure 3: Instantiations 010 and 110

In this instantiation two (or more) observers use the same simple and composite unities, but the interaction is different, an example could be that an observer uses a desktop PC and another observer uses a Palm, or that the observers use different metaphors to access the distinctions (compare figures 5 and 6).
The instance 100 proved to be the most suited to model workflows, because the unities represent unequivocal tasks, that must be carried out without incompatibilities by all those involved in the social system.

The emphasis in the modes 010 and 110 lies in the discrepancy of the complex unities among the observers, a representative example would be the discrepancy that two people can have when navigating through the same Web–pages and storing them in personal bookmarks. Each person will be generating different composite unities. We must remember that the act of distinguishing only depends on the observer and his/her ontogeny, and because there are many observers, the distinctions made by everyone not necessarily must be the same ones. We will denominate multiverse the existence of heterogeneous groups of distinctions.

According to Maturana a structural coupling exists among the observers with the purpose of generating a domain of consensual coordinations of actions or distinctions. This process operates in the language and it allows the observer to operate in a domain of descriptions with other observers generating consensual objects.

To transfer this process so natural for us automatically to the computer has been extremely complicated. McCarthy (McCarthy, 1993) proposes as central idea to handle context like a formal object, allowing to establish a proposition that is true in a context: $\text{ist}(\kappa, \varphi)$ where φ is a proposition and κ a context. This way, the axioms in a limited context could be expanded transcending their original limitations.

Two context can be differ in, at least, three forms:

* It can have different vocabularies
* It can have the same vocabulary, but describe different situations
* It can have the same vocabulary, but treat it differently

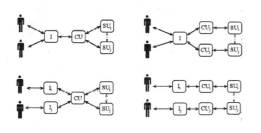

Figure 4: Instantiations 001, 011, 101 and 111

Ontologies allow us to solve the difference between different contexts. Ontologies provide a shared and common understanding of a domain. Ontologies are used to define the relationships between concepts. They record agreed sets of relationships that are relevant to a particular community. Nowadays, in order to solve the problem of the different vocabularies, diverse languages have been specified that deal with the ontology problem, this way we have, SHOE, CKML, KQML/KIF and Topics Maps among others. The proposed model outlines also different instantiations for the emergence of the multiverses (see figure 4).

Context arise on cases 001 y 101, and ontology on all this four instantiations.

4. Applications of the model

Two prototypes have been developed, BACán for the stand–alone use and Pulento for the cooperative one.

4.1 Stand–alone implementation: BACán

BACán is the first functional prototype that implements the 000 instantiation of the ICS model. BACán has been developed in Java and Java 3D.

Figure 5: Visualizing with K–Planes

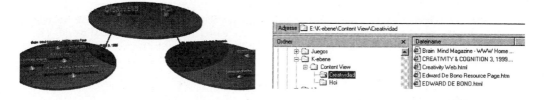

Figure 6: Interaction as Hierarchical File System

The layer I allows a hierarchical graphical visualization of the composite and simple unities. The composite unities exhibit hierarchical characteristics that agree with McNamara (McNamara, Hardy, & Hirtle, 1989), who says that the human being when forming knowledge organizes and presents hierarchical structures, even when these are not formally described, but only empirically observed. We choose to use the K–Planes (Sifaqui, 1999) (see figure 5) as an interaction and visual representation metaphor of the composite unities.

Another possible interaction form is the use of the HFS concept, to which every user is familiarized (see figure 6).

4.2 Cooperative implementation: Pulento

Once having BACán running and tested, we extended it to support cooperative use, sustaining the exchange of unities and allowing direct group communication. Pulento instantiates the cases 100, 010 and 110 of the ICS model. We choose this approximation, because according to Maturana social systems are realized primarily in linguistic domains. We have in addition that the participants may operate in multiple social systems, although within each one they function as if engaged in a distinct domain of interactions, that is, every observer may have different roles within one social domain.

Pulento has support to different roles (or supports distinct domain of interactions), the observer can choose between different variations of the possible interactions styles, among others:

- Variation on the number of participants: dual, only 2 observers are involved on the interaction (Whittaker, Swanson, Kucan, & Sidner, 1997); or multiparty, where more than 2 observers are involved (Stefik et al., 1987).
- Variation on the extent of the interaction: one–shot, where the interaction takes place only one time and is not required to register the activities (Stefik et al., 1987); or repeated, where the interaction takes place in an indefinite time period or with successive interactions separated by non definable time periods (Whittaker et al., 1997).
- Variation in the formality: formal, where a correct spelling, presentation, header and footer is necessary (Stefik et al., 1987); or informal (Erickson & Kellogg, 2000).
- Variation in the extension of the interaction: extensive, where the text generally involves long paragraphs (Stefik et al., 1987); or brief, where the interaction is in not more than a couple of lines (Whittaker et al., 1997).
- Interaction type: asynchronous, with the well-known examples of e–mail, news; or synchronous, like IRC or babble (Erickson & Kellogg, 2000).
- Interest in the preferences of the community: without interest, the opinions of the social groups do not have interference; or with interest, where the opinion of the group is relevant, news with priorities (Rose & Bornstein, 1998), AIR (Belew, 1989), Tapestry (Goldberg, Nichols, Oki, & Terry, 1992).
- Variation in the intervention: active, where each member participates actively in the interaction; or passive, with lurking as fundamental example
- Variation in the propagation of the message: one–to–one, like e–mail; or one–to–many, manifested in the mailing lists; or many–to–many, with examples like news and chat.
- Variation of the communication basis: the unity is the basis and the communication is build on it, for example, a document is used as conversation basis (Churchill, Trevor, Bly, Nelson, & Cubranic, 2000); or the conversation is the basis itself and the unity is build on it (Whittaker et al., 1997), for example, during an interaction, a document is sent as attachment.
- Variation on pervasiveness: invasive, the communication pop-ups on the other client machine; no-pervading, where the client is not aware of the communication until he/she looks at it.

This way each observer can interact with other observers in the way that agrees more to his/her interests and needs.

For Pulento we choose to implement the group communication with the abstract approach (Erickson & Kellogg, 2000) that represent the social information so that it does not have correlation with the physical world. This interaction is basically textual. The disadvantage of the text is reflected clearly in the hermeneutics, but in our case the aspect that the text arises as a observed unity for the members of the social system is interesting for us, that is to say, that the text itself becomes an own distinction.

On Pulento every observer works with the same simple unities repository, thus having the same vocabulary and the same basic distinctions.

Take for example, that observer A wants to do a search about HCI and Creativity, he feels that he has a good collections of Creativity Web–page links but only several links about HCI, than he chooses to do a communication one-shot, informal, asynchronous, one-to-many with all the members, by saying that he is looking for HCI links. Some observers may reply with their personal links by giving a composite unity that matches this criterion. Then observer A can do the intersection on his Creativity links and all the links he received and thus generate a new composite unity that matches the given attributes. In this way a scenario is possible, because the instantiation 010 or 110 have the same simple unities, that is, the same set of attributes A_S for all the observers.

5. Conclusions and future work

The phenomenological aspects of the autopoiesis theory were applied to model the way in which the human beings can interact with and through computer systems. A model has been developed that can be used as foundation for the development of personal systems in accordance with the different human ways of thought, for example, the variety of contexts when using information and knowledge sources and variations on the interactions styles. The implementation of the prototypes BACán and Pulento was made as a "proof of concept", that has allowed us to experiment with the model's key ideas. The next step is the implementation of a prototype that realizes the instantiations with shared multiverses.

References

Barreau, D., & Nardi, B. (1995, July). Finding and Reminding: File Organization from the Desktop. *SIGCHI Bulletin, 27*, 39-43.

Belew, R. (1989). *Adaptive Information Retrieval: Using a Connectionist Representation to Retrieve and Learn About Documents.* Paper presented at the SIGIR-89.

Churchill, E., Trevor, J., Bly, S., Nelson, L., & Cubranic, D. (2000, May 2000). *Anchored Conversations: Chatting in the Context of a Document.* Paper presented at the Proceedings of the ACM Conference on Human Factors in Computing Systems, CHI'2000, The Hague, Amsterdam.

Dourish, P., Edwards, W., LaMarca, A., Lamping, J., Petersen, K., Salisbury, M., Teng, D., & Thompton, J. (2000). Extending documents managements systems with user-specific active propertics. *ACM Transactions on Informations Systems, 18*(2), 140-170.

Erickson, T., & Kellogg, W. (2000). Social Translucence: An Approach to Designing Systems that Support Social Processes. *ACM Transactions on Computer-Human Interaction, 7*(1), 59-83.

Gifford, D., Jouvelot, P., Sheldon, M., & O'Toole, J. (1991, October 1991). *Semantic file systems.* Paper presented at the 13th ACM Symposium on Operating Systems Principles.

Goldberg, D., Nichols, D., Oki, B., & Terry, D. (1992). Using collaborative filtering to weave an information tapestry. *Communications of the ACM, 35*(12), 61-70.

Hejl, P. (1984). Towards a Theory of Social Systems: Self-Organization and Self-Maintenance, Self-Reference and Syn-Reference. In H. Ulrich & G. J. B. Probst (Eds.), *Self-Organization and Management of Social Systems: Insights. Promises, Doubts, and Questions* . Berlin: Springer Verlag.

Maturana, H., & Varela, F. (1984). *El árbol del conocimiento*: OEA.

McCarthy, J. (1993). *Notes of formalizing context.* Paper presented at the Proceeding of the Thirteenth International Joint Conference on Artificial Intelligence.

McNamara, T., Hardy, J., & Hirtle, S. (1989). Subjective Hierarchies in Spatial Memory. *Journal of Experimental Psychology: Learning, Memory and Cognition, 15*(2), 211-217.

Nardi, B., & Barreau, D. (1997, January). "Finding and Reminding" Revisited: Appropiate Metaphors for File Organization at the Desktop. *SIGCHI Bulletin, 29*.

Natrificial Software Technologies. (1998). The Brain: Information Management Software.

Rawlins, G. (1999). KnowSpace.

Rose, D., & Bornstein, J. (1998). Information rendezvous. *Interacting with Computers, 10*, 213-224.

Sifaqui, C. (1999). Structuring User Interfaces with a Meta-model of Mental Models. *Computer & Graphics, 23*(3), 323-330.

Stefik, M., Forster, G., Bobrow, D., Kahn, K., Lanning, S., & Suchman, L. (1987). Beyond the chalkboard: Computer support for collaboration and problem solving in meetings. *Communications of the ACM, 30*(1), 32-47.

Whittaker, S., Swanson, J., Kucan, J., & Sidner, C. (1997). TeleNotes: Managing Lightweight Interactions in the Desktop. *ACM Transactions on Computer-Human Interaction, 4*(2), 137-168.

Long term change in computer attitude, computer usage and skill transfer in elderly users of touch screen interface

Hiroyuki Umemuro

Department of Industrial Engineering and Management, Tokyo Institute of Technology, 2-12-1 O-okayama, Meguro-ku, Tokyo 152-8552 Japan

Abstract

Relations between computer attitude, computer use, and skill transfer of elderly Japanese computer users were investigated for six months. Eight males and eight females aged sixty to seventy-six participated. Each participant was provided one touch-screen-based computer specialized for e-mail handling in his or her home for six months. Participants' usage of e-mails, usage of mouse and/or keyboard as well as their computer attitude were investigated with a questionnaire. The results showed that while the activity level of e-mail usage did not change significantly over six months, the computer attitude of less active users showed a negative shift during third and fourth months, whereas the attitude of more active users stayed at the same level or became more positive. Computer attitude in the first month did not differ significantly between more and less active users and thus it did not predict the computer usage. Skill transfer from touch screen operation to mouse operation typically occurred in one or two month while skill transfer to keyboard operation took longer.

1. Introduction

Information technology including the Internet has been introduced not only into workplaces but also into our daily life very rapidly and widely. Now everyone has more opportunities than ever before to take advantage of the information society and make their life rich and efficient. The advantage of an information society can be especially beneficial for aged people in various ways (Czaja, 1996; Czaja, 1997), as they tend to stay in their homes for greater period of time and need means to interact with society despite sometimes having restrictions on their ability to travel. Thus efforts to enable easy access of elderly populations to information technology is of increasing importance.

Although elderly people are generally willing to use computers (Rogers & Fisk, 2000), they are considered to have some difficulties in adapting to the new technology. The factors that have been believed as major sources of difficulty are natural declines in physical abilities (e.g., motor control, see Smith et al., 1999) and cognitive abilities (e.g., memory), and resistance to new technology of elderly people. Careful investigations in various abilities and performance of elderly computer users revealed, however, that elderly people have the ability to make use of computers when supported by appropriate interface design and/or training (Czaja & Lee, 2001). For example, redesign of the mouse-driven pointing interface with careful consideration of age-related changes in motor-control characteristics may compensate for elderly users' performance (for a review, see Rogers & Fisk, 2000). A direct pointing interface using touch screen technology can also be an alternative solution (Karat et al., 1986). Czaja and Sharit (1998) recently showed elderly people, even with no or very limited experience with computers, can perform computer tasks such as data entry after a short period of training, and that their performance is not qualitatively different from that of younger people. The research also suggested that age-related, quantitative differences in performance are largely due to differences in fundamental skills, such as visiomotor, rather than to higher-order cognitive abilities.

Although elderly people have enough potential ability to acquire computer skills, older users' resistance to new technology is reducing many of their opportunities to become acquainted with computers. Ellis and Allaire (1999) showed that computer anxiety is a major factor in older users' intent to use computers. On the other hand, computer experience (Czaja & Shair, 1998b) and computer knowledge (Ellis & Allaire, 1999) can moderate such resistance and give elderly users a more positive attitude to computers. Thus it is expected that once such resistance can be eliminated, and elderly people have more positive attitudes, then they are more likely to use computers and acquire computer skills. Thus efforts to lower elderly users' resistance toward computers through careful design of interfaces and training methods are essential to promote their use of computers.

For elderly users, the effect of such efforts may appear in a long time period, rather than immediately. As computer attitude can be formed in relation with computer experiences, and elderly users require longer time to acquire knowledge and skills of computers (Czaja & Sharit, 1998a), computer attitude in elderly users may change slowly. As a consequence, change in computer attitude needs to be assessed over a longer period and in relation with computer usage. However, most previous studies seem to have assumed computer attitude as an invariant and stable characteristic of people, and have tried to examine the relationship between computer attitude and performance at a moment in time, without investigating their dynamic changes.

The objective of this research was to investigate the relations among computer attitude, computer use, and skill transfer of elderly Japanese computer users. E-mail communication was chosen because elderly people are supposed to be able to do that task well and they also may find it meaningful and valuable (Czaja, Guerrier, Nair, & Landauer, 1993). A personal computer specialized for e-mail communication, with a touch screen technology, was developed and provided to participants for six months, and participants' computer attitude, usage, and skill transfer were investigated during the period.

2. Method

2.1 Participants

Participants were sixteen adults between sixty and seventy-six years old (mean = 66.1yr), voluntarily enrolled in this research. Of these, eight were males and eight were females. All participants had no or limited experience using computers either in their workplaces or homes.

2.2 Apparatus

A touch-screen-based computer specialized for e-mail communication was designed for elderly Japanese users. The western typewriter ("qwerty") keyboard is thought to be one of the major sources of elderly Japanese users' anxiety about computers as most of them have very little experience in general with typewriters. Handling Japanese characters with typewriter keyboards requires complicated mental processes, including converting Japanese characters into their alphabetic expressions. To avoid this problem, the interface used in this research was designed with touch screen technology to enable direct input of Japanese characters as well as alphabetic and numeric characters.

The designed interface was implemented on Apple Computer's iMac personal computer with a 15-inch CRT equipped with an additional touch screen by Elo Touch Systems. The interface was based on Qualcomm's Eudora 4.2J e-mail client and NTT PC Communication's Easy Prolog 2.0 software keyboard package.

2.3 Procedure

Each participant was provided one computer in his or her home and could use it any time. Their computer use, computer attitude, and usage of alternative input devices were investigated with a questionnaire every month for a six-month period.

Participants were first interviewed about their demographic profiles and experiences with computers, then given training on how to use their computer for two hours. At the end of the training, all participants were able to send and receive e-mail without the instructors' help.

After the training session, computers were set up in participants' home with a free Internet connection. A conventional mouse and a keyboard were also left unconnected in their home as alternative input devices, with brief printed instructions for connection and usage. Every month thereafter, participants received a questionnaire asking about their computer usage, computer attitude, and skill transfer to the alternative input devices during the latest month. Participants were required to answer the questionnaire and mail it to experimenter.

2.4 Measurement

Participants' computer usage was measured with two indices: frequency of computer use and the number of people with whom e-mail was exchanged. Frequency of use was asked with a multiple-choice response set like "everyday", "once in two or three days", "once in a week", and so on.

Participants' attitude to computers was measured using 21 Likert-style items representing statements of general attitudes toward computers based on the Computer Attitude Scale, originally proposed by Loyd and Gressard (1984) and revised by Bandalos and Benson (1990). The original scale by Loyd and Gressard had 29 items. Bandalos and Benson revised this into a 23-item scale and showed that these items could be categorized into three subscales: liking, confidence, and achievement. The present study employed the Bendals and Benson scale with items that were not applicable removed or modified. Participants responded to each item with a number between 1 to 5, corresponding to strongly disagree and strongly agree, respectively.

Skill transfer to alternative input devices was assessed with questions about whether the participant had used the mouse and/or the keyboard during the latest month.

3. RESULTS

3.1 Computer usage

Participants could be divided into two groups according to their frequency of computer use. One group of nine participants, three males and six females, was relatively active, using e-mail more than once a week on average, while the other group of seven participants, five males and two females, was less active, using e-mail only a few times a month or less. This pattern of usage was stable through the experiment. No significant change in the frequency of use or the number of e-mail correspondents was observed during the six-month period.

3.2 Computer attitudes

Figure 1 shows the averages of three computer attitude subscales (liking, confidence and achievement), calculated across participants in each group, for each month of the experiment. In contrast to computer usage, computer attitude changed over the course of experiment. In the first month, there was no significant difference between the two groups on any of the three subscales. Subsequent performance is complicated in its details, but measures were little changed in months 2 and 3, and remained stable for the more active group through month 4 through 6. For the less active group, however, values for months 4 through 6 show a more negative computer attitude. Especially the "liking" subscale showed a significant decrease, while the reduction in "confidence" subscale was not significant.

3.3 Skill transfer

There were skill transfers from the touch screen operation to both of the two conventional input devices. The total number of participants who used mouse or keyboard each month is shown in Figure 2. In the first month, seven out of the sixteen participants already tried the mouse and the number of participants increased to nine by the end of the second month. By the sixth month, seven out of nine active users successfully adapted to the mouse operation. On the other hand, only three participants tried the keyboard in the first month and it was not until the fourth and fifth months that four more participants tried it.

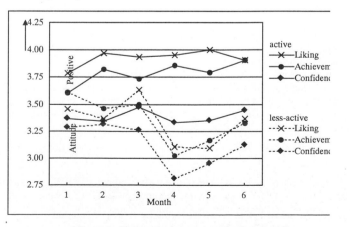

Figure 1. Change in computer attitude subscales.

4. Discussion

The objective of this research was to investigate how computer usage and attitude may change and how skill transfer may occur in elderly computer users over time. Although it has been assumed in many studies that users' computer attitude is rather consistent, results of this study indicate that it may change over a several month period when users are exposed to technology, and that the changes are related to actual computer usage. Because the attitude in the first month does not predict the computer usage, the results imply that computer usage and attitude of elderly users need to be assessed over a longer time period, and alternative interventions used for the different groups that are then able to be identified.

Another implication from the results can be derived in terms of the time difference between changes in attitude and usage. It has been believed that computer attitude, either positive or negative, affects activity in computer usage. However, the results of this research show that computer usage may not change significantly over time while computer attitude changes over a period of several months. It implies that the relationship between them is more interactive, and exposure to technology may also work to form attitude in either positive or negative ways. As discussed in introduction, computer experience is considered to affect computer attitude, while attitude is considered to be a major factor to determine actual computer usage. Since actual computer usage can be additional computer experience, computer usage and attitude are assumed to form a causal loop when considered in a long time period.

In terms of skill transfer, the result shows that the skill of the touch screen operation can transfer to the mouse easily, while the transfer to the keyboard seems to be more difficult and takes longer. This result suggests that once elderly users get accustomed to new technology with an easy-to-use interface, the skill may transfer to conventional interfaces also in a term of several months, depending on the nature of the alternative interfaces. This implies that offering elderly users choices among alternative interfaces is beneficial to support them getting accustomed to computer usage. If elderly users can choose "easy" interfaces first and then change to more "efficient" interfaces anytime when they feel confident, their resistance to the computer should be minimum and adaptation to computer technology can be done very smoothly. Touch screen technology is considered to be an important element to implement such "easy" introductory interfaces.

This research investigated the changes in computer use, attitude and skill transfer for the relatively long term of six months. It is still possible, however, that the results here are capturing only the first phase and changes may continue for a year or more. Thus, in order to understand the dynamic nature of computer usage, attitude and skill transfers, further research for a longer time period should be pursued.

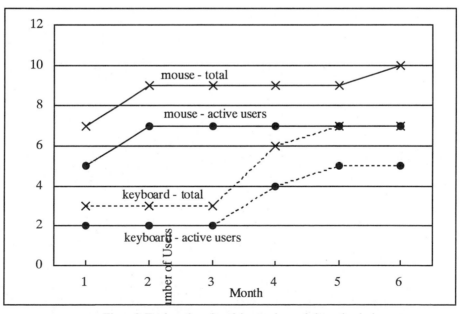

Figure 2. Total number of participants who used alternative devices.

References

Bandalos, D., & Benson, J. (1990). Testing the factor structure invariance of a computer attitude scale over two grouping conditions. *Educational and Psychological Measurement, 50*, 49-60.

Czaja, S. J. (1996). Aging and the acquisition of computer skills. In W. A. Rogers, A. D. Fisk, & N. Walker (Eds.), *Aging and skilled performance: Advances in theory and applications* (pp. 201-220). Mahwah, NJ: Lawrence Erlbaum Associates.

Czaja, S. J. (1997). Using technologies to aid the performance of home tasks. In A. D. Fisk & W. A. Rogers (Eds.), *Handbook of human factors and the older adults* (pp. 311-334). San Diego, CA: Academic Press.

Czaja, S. J., Guerrier, J. H., Nair, S. N., & Laudauer, T. (1993). Computer communication as an aid to independence for older adults. *Behavior and Information Technology, 12*, 197-207.

Czaja, S. J. & Lee, C. C. (2001). The Internet and older adults: Design challenges and opportunities. In N. Charness, D. C. Parks, & B. A. Sabel (Eds.), *Communication, technology and aging: Opportunities and challenges for the future* (pp. 60-80). New York, NY: Springer.

Czaja, S. J., & Sharit, J. (1998a). Ability-performance relationship as a function of age and task experience for a data entry task. *Journal of Experimental Psychology: Applied, 4*(4), 332-351.

Czaja, S. J., & Sharit, J. (1998b). Age differences in attitudes toward computers. *Journal of Gerontology: Psychological Sciences, 53B*(5), P329-P340.

Ellis, R. D., & Allaire, J. C. (1999). Modeling computer interest in older adults: The role of age, education, computer knowledge, and computer anxiety. *Human Factors, 41*(3), 345-355.

Karat, J., McDonald, J. E., & Anderson, M. (1986). A comparison of menu selection techniques: touch panel, mouse and keyboard. *International Journal of Man-Machine Studies, 25*, 73-88.

Loyd, B. H. & Gressard, C. (1984). The effects of sex, age, and computer experience on computer attitudes. *AEDS Journal, 18*(2), 67-77.

Rogers, W. A., & Fisk, A. D. (2000). Human Factors, Applied Cognition, and Aging. In F. I. M. Craik & T. A. Salthouse (Eds.), *The handbook of aging and cognition* (2nd ed., pp.559-591). Mahwah, NJ: Lawrence Erlbaum Associates.

Smith, M. W., Sharit, J., & Czaja, S. J. (1999). Aging, moter control, and the performance of computer mouse tasks. *Human Factors*, 41(3), 389-396.

Localisation and linguistic anomalies

George R. S. Weir[1], Giorgos Lepouras[2]

[1] Department of Computer Science, University of Strathclyde, Glasgow G1 1XH, UK
[2] Department of Informatics, University of Athens, Athens 157 71, Greece

Abstract

Interactive systems may seek to accommodate users whose first language is not English. Usually, this entails a focus on translation and related features of localisation. While such motivation is worthy, the results are often less than ideal. In raising awareness of the shortcomings of localisation, we hope to improve the prospects for successful second-language support. To this end, the present paper describes three varieties of linguistic irregularity that we have encountered in localised systems and suggests that these anomalies are direct results of localisation. This underlines the need for better end-user guidance in managing local language resources and supports our view that complementary local resources may hold the key to second language user support.

1. Introduction

In the quest for wider computer access, many authors have stressed the need to accommodate users whose first language is not English. This is often expressed as recommendations for internationalisation while others focus on the desirability of software localisation (e.g., Uren et al, 1993, O'Donnell 1994, Belge, 1995). Internationalisation and localisation can be understood as inter-related aspects of 'globalisation', or the move toward provision of globally accessible interactive systems. Internationalisation is the process of developing applications that can easily be converted to operate in different cultural or linguistic environments (Russo & Boor, 1996). The cultural environment extends beyond the local language to aspects such as beliefs, customs and ethics of a society. The internationalisation process ensures that culture-specific characteristics of an application are isolated, thereby enabling easy localisation.

Localisation is the process of converting applications to operate in a specific cultural environment. Occasionally, software is localised without first being internationalised, though, if internationalisation has been employed during the development of the application, localisation time can be reduced.

The localisation process reasonably consists of two main stages. The first step is translation of language resources to reflect the local language. In this stage, all language resources are translated to the local language, including features such as menus, commands, help texts, etc. Occasionally, the translation process employs a local software company or a company with expertise in translation. However, the outcome of the translation process is not always as anticipated (Lepouras & Weir, 1999).

The second step in localisation adjustms software to local cultural habits. Here, the application is adapted to reflect local customs. If the application is already internationalised, this stage may be unnecessary, since changes, such as special sorting algorithms, may have already been met during the development stage. For symbolic features such as the currency or the comma delimiter, the application may simply inherit configuration from the operating system.

While generally endorsing the importance of such linguistic and cultural accommodations, our purpose in this paper is to highlight several concerns in the use of localisation. Through problem illustration in the context of Windows-based applications, we stress the need to understand the technical mechanisms that afford software localisation. Without adequate grasp of the operational alternatives and their possible impact, there is serious risk that localisation may lead to linguistic anomaly and sub-optimal user interaction. We conclude with a set of recommendations that aim to minimise the threat of anomalies when developing or working within a localised setting.

2. Linguistic anomalies

The illustrations below are drawn from the context of Microsoft Windows™ and its applications. In this setting, several localisation features are available and are in common use wherever English is not the native user language. Such features vary from local font support within the operating system through localised (extensively translated) applications. This fact, coupled with the prevalence of the Microsoft environment, affords a credible setting in which to reveal the occurrence of linguistic anomalies. We have identified three varieties of anomaly: instances arising within mixed linguistic environments, terminological problems, and peculiarities resulting from extreme localisation.

2.1   Mixed environments

Perhaps the most apparent problem with localised applications arises in mixed environments. International users rarely enjoy a working environment that is fully localised. An environment where every application including the operating system speaks the user's native language is uncommon outside the English-speaking world. More often, the user's environment comprises a mixture of English and local language. This mixed language scenario may take two forms: between applications or within the same application. In either case, the dual language requirement impairs the uniformity and consistency of the user's working environment. This contravenes an explicit objective of most windowing environments (e.g., see Microsoft, 1992, Apple, 1992, Sun, 1999) and users confronted with such inconsistency face an additional burden of comprehension.

2.1.1   Mixed language between applications

In a mixed environment of localised and non-localised applications, the user has to learn both the original and the local language terminology to be able to interact with the applications installed. In such a context the putative advantage of using localised applications may become an extra overhead for the user. Imagine a user faced with a localised word-processing application and non-localised drawing application. The inevitable overlap of functionality requires that the user manage two contexts of interaction. In this case, two different applications share common menus, commands and underlying functions yet the user is faced with two sets of descriptors.

2.1.2   Mixed language in an application

Recently, another type of linguistic anomaly has emerged. This is characterised by the appearance of mixed language (English and local) within a single application. In such cases, the application maintains some resources in English and other resources in a local language. The result is often a peculiar mix of English and local language, with no obvious rationale for the mix. This is evident in Figure 1 wherein the Help for MS-Word shows its first tab with a Greek label (Contents), while the remaining labels, and all other displayed text, appears in English.

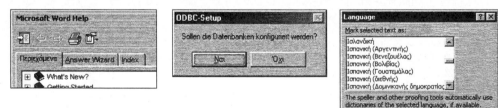

Figure 1: Mixed Greek and English in a single application.   Figure 2: Mixed English, German and Greek in a single application.   Figure 3: Further Mixed English and Greek in a single application.

A second illustration of mixed language proves that such linguistic anomaly is not exclusive to English and the local language. Figure 2 shows the setup procedure of an application developed by a German software company. The title of the dialog box is in English, the text of the dialogue is in German and the button labels are in Greek!
This section's penultimate example comes from a non-localised word-processor. Figure 3 illustrates a dialogue box with a choice of dictionaries for use in a spell checker. Ironically, all dialogue text is in English, except for the choice of dictionary language. All of the languages appearing in this list box are named and listed alphabetically in Greek.
Our final example of intra-application language mixing is even more bizarre. Figure 4 shows an initial screen (a) entirely in English. For no obvious reason, the next screen in the interaction (b) replaces the English title with a Greek language equivalent.

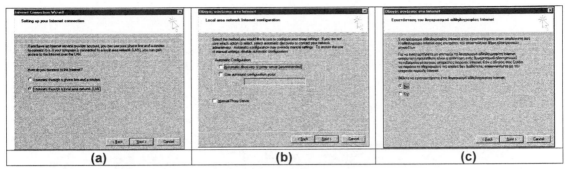

Figure 4: Mixed language within a single application
(Greek and English within Microsoft's Internet Connection Wizard)

Pressing the *Next* button in screen (a) takes us to screen (b). The linguistic strangeness continues at the third step in the dialogue. Pressing *Next* from screen (b) takes us to a third (more localised) display in which only the button labels remain in English (screen c).

With the possibility of such, apparently, arbitrary mixed language dialogues there is good reason to expect user confusion. One must take care however, to avoid the assumption that increased localisation will minimise user difficulties. There is strong evidence to suggest that full localisation without co-ordination between software houses leads to inconsistency across applications. Examples of such terminological linguistic anomalies are described below.

2.2   Terminological issues

Language problems are often caused by terminology. Whenever English language software is translated to a local language decisions are taken on mapping from English terms to local terms. Inevitably, some measure of arbitrariness is attached to this procedure. In consequence, some aspects of localised software may appear stranger to the local audience than the English (foreign language) original. This goes some way toward explaining why many users when faced with a choice between a localised (fully translated) application and an English-language original, express a preference for the latter (Lepouras & Weir, 2000).

Inconsistent localisation is evident between applications. For instance, localised Greek word processors show a measure of arbitrariness in choice of Greek terms. In a Greek version of Microsoft Word the term 'header' was replaced with 'κεφαλίδα' (derived from the Greek root for 'head'). A Greek version of Lotus Ami Pro replaced 'header' with 'υπέρτιτλος' (meaning 'supertitle'). Finally, a Greek version of WordPerfect has 'header' replaced with 'τίτλος σελίδας' (literally, 'page title'). A further selection of such terminological variety is given in Table 1 (cf. Lepouras & Weir, 1999).

Table 1.: Example variations in localised terminology

English Term	Word 6.0	Lotus Ami Pro	WordPerfect
footer	υποσέλιδο	υπότιτλος	υποσέλιδο
bullet	κουκκίδα	σύμβολο	σφαίρα
undo	αναίρεση	ακύρωση	ακύρωση
view	προβολή	όψη	θέα

Anomalies in terminology take forms other than variations in vocabulary. There are instances of neologism, transliteration (e.g., 'zoom' becomes 'ζουμ', 'style' becomes 'στυλ') and retention of original terms (e.g., 'Small Caps', 'textart' and 'kerning' are used verbatim in the localised versions of Greek word processors).

2.3   Extreme localisation

Our final category of linguistic anomaly derives from applications that are 'over-localised'. A classic example arises where a document was originally written in a localised word processor, and subsequently opened for editing in the original version of the same application. In this instance (Figure 5), the localised version had saved some style information (the paragraph numbering scheme) in Greek. As a result, when transferred to the original (English) version of the word processor, errors were reported of an *'unknown switch argument'*. Viewing the underlying 'Field Codes' revealed that the Greek version had saved the numbering style as 'Αραβικά' instead of 'Arabic', yet the corresponding code for page numbering (i.e., 'PAGE') was not also translated from English.

Figure 5: Extreme localisation of field codes

2.3.1   Criteria for localisation

There are varying degrees of localisation. Yet there are no obvious criteria for guiding the appropriate level of localisation. Evidence from our study of localised word processors shows that original menu shortcuts (such as Ctrl-C for 'Copy') are consistently retained rather than changed to accord with localised menu commands. The assumptions underlying this decision are obscure but may presume that retaining shortcut consistency across localised versions is beneficial to local users. Even this strategy can lead to anomalous results. Shortcut keys are often mnemonics for the English command names, e.g., Ctrl-N for 'New, Ctrl-O for 'Open' and Ctrl-S for 'Save' and Ctrl-P for 'Print'. When mapped to a localised Greek version of the 'File' menu, these mnemonics are inappropriate yet this set of shortcut keys from the English context are retained. This contrasts with the other set of application shortcut keys. Alt-F (mnemonic in English for 'File') invokes the File menu in the English version yet Alt-A is the equivalent Greek shortcut (mnemonic for 'Αρχείο'). In other words, one set of shortcuts is not localised (remains largely as English mnemonics) while another set is converted to localised mnemonics. Such examples illustrate a fundamental tension within localisation efforts, viz., the need to change interface characteristics whilst attempting to maintain consistency.

3.   Problem causes

Inevitably, one must wonder about the causes for many of the linguistic anomalies described above. We must stress that no steps were taken to create artificially any of the reported examples. Our feeling is that many of the more peculiar anomalies result from vagaries of the localisation process. For instance, localised applications may invoke some non-localised language resources from the operating system. Similarly, non-localised applications may invoke localised language resources from the operating system. In either case, the result is likely to be linguistically anomalous. Extreme localisation, such as translating application codes and other invisible elements seldom accessed by users, are a potential risk. Reasonably, any term with operational effect (such as a parameter) should work in each language.

Some anomalies arise from conflicts between different versions of the same application, as in the transition from localised to non-localised word processor. Installed language resources may be compromised following installation of unrelated applications. Shared libraries are common in a Windows environment and not always evident to the end-user. In cases where users are asked to confirm the replacement of shared files there is an implied risk in either response and no information on the likely impact on common resources. The shared library approach is fundamental to the Windows architecture. For this reason, such linguistic anomalies persist through new releases of the operating system. Figure 6 shows a now familiar syndrome in the context of Windows 2000.

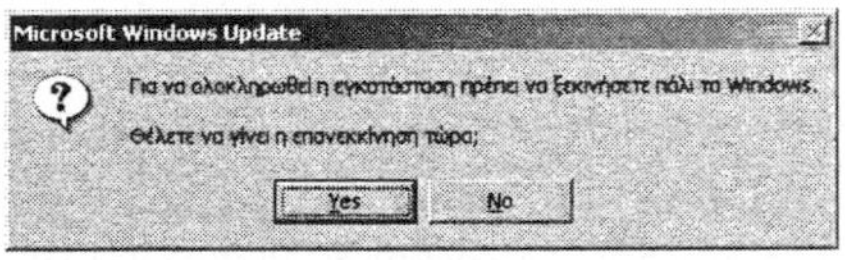

Figure 6: Linguistic anomaly in Windows 2000

4. Conclusions

Attempts to accommodate non-native speakers of English by translating or localising computer applications often give rise to unexpected difficulties. We have characterised three varieties of linguistic anomaly that may arise in such contexts. Although many of our examples are drawn from the Greek context, we are confident of similar experiences in other localised language communities. Notably, Pemberton (Pemberton, 2000) expresses similar concerns about localisation in a Dutch setting.

Many of the reported problems arise directly from localisation efforts, for instance, through conflicts in available language resources. Such anomalies can only occur because the language specific resources have been isolated from other functions of the application. We have focussed our descriptions on the user perspective. Evidently, some of the recounted anomalies arise through user efforts to deploy localisation resources within an environment that is often less than transparent. This suggests lessons both for end-users and for software designers.

End users who seek to deploy localisation facilities must be conscious of the coarse level of control over language resources. Additionally, users must appreciate how easily they may disturb the balance of localised resources within the operating software. The lesson to software designers is to facilitate easier and less tenuous language resource management. Responsibility falls on the designer to protect the end-user from anomalous effects. Ultimately, the range of potential difficulties, combined with the opacity of any applicable criteria is further reason for considering alternatives to localisation, e.g., through complementary support in the user's native language (cf. Weir et al, 1996, Lepouras & Weir, 1999, Lepouras, 2000).

References

Apple, Macintosh Human Interface Guidelines, Addison-Wesley, Mass., 1992

Belge M., 'The Next Step in Software Internationalization', Interactions, ACM, 1995, Vol. 2. No 1, pp. 21-25.

del Gardo E., 'Internationalization and Translation: Some Guidelines for the Design on Human Computer Interfaces', in J. Nielsen (Ed.) Designing User Interfaces for International Use. Elsevier, New York, 1990, 1-10.

Lepouras G., User-Computer Interaction: The methodology of supplementary support in the service of different cultural communities, PhD Thesis (in Greek), Department of Informatics, University of Athens, 2000.

Lepouras G. & Weir G. R. S., 'Its not Greek to me: a cross language study of three word processors', SIGCHI Bulletin, ACM, 1999, Vol. 31 No. 2, pp. 17-24.

Lepouras G & Weir G. R. S., 'Mind your language: A study of language preference in Greek users', paper submitted to International Journal of Human-Computer Studies.

Microsoft, The Windows Interface: An Application Design Guide, Microsoft Press, Washington, 1992.

O'Donnell S. M., Programming for the World: A guide to Internationalization, PTR Prentice Hall, 1994.

Pemberton S., 'Reflections: so much for WYSIWYG', Interactions, ACM, 2000, Vol. 7, No. 5, pp. 60-61.

Russon, P. & Boor, S., 'How fluent is your interface? Designing for international users', Proceedings of InterCHI'93, ACM Press, pp. 342-347, 1993.

Sun, Java look and feel design guidelines, Sun Microsystems, California, 1999.

Uren E, Howard R, Preinotti T., Software Internationalization and Localization: An Introduction. Van Nostrand Reinhold, 1993.

Weir G.R.S., Lepouras G. & Sakellaridis U., 'Second-Language Help for Windows Applications', In M.A. Sasse, R.J. Cunnigham, and R.L. Winder (Eds.), People and Computers XI, Proceedings of HCI '96 Conference, Springer, 1996, 129-138.

Age Differences and the Depth - Breadth Tradeoff in Hierarchical Online Information Systems

Panayiotis Zaphiris

Institute of Gerontology and Industrial & Manufacturing Eng.
Wayne State University, Detroit, MI 48202
+1 313 577 297 - p.zaphiris@wayne.edu

Abstract

This paper examines previous research on the topic of depth versus breath in hierarchical menu structures, and explains why searching for information on the world wide web follows a similar model. It also proposes age related enhancements to available mathematical models, design guidelines where possible and future research topics in the area.

1. Introduction

The World Wide Web (WWW) is exponentially increasing, both in terms of user population, as well as the amount of available information. In the beginning, people tended to concentrate more on aesthetics; how to make pages "look nice" by overloading them with images, java applets and other gadgets. Currently we are seeing a shift towards web usability, where the topic of information architecture is gaining momentum. People are using the WWW more and more in searching for specific information (eg. medical, travel) than for random browsing. Taking into account the user needs for fast and accurate access to information, big commercial search engines are evolving into a hierarchical collection of links and data (eg. Yahoo, Lycos, Altavista, Google). It is not surprising that all of the popular search engines now include hierarchical indexes/organizations of "important" links.

Both the exponential increase of use and information available on the web makes it necessary for information system designers to allow for a more diverse user population. Research has shown that older people, the fastest growing segment of the population (Charness and Bosman, 1990), are willing and able to learn to use computers, but that they have more difficulty than younger adults in doing so (Czaja, 1996). More and more older adults today use the WWW for information retrieval (especially health related), communication with family members and for keeping up with daily news. There is little existing research on the older adults' Web use, particularly on their ability to navigate complex sites (Mead, 1997). But it is expected that older adults' limitations in working memory and motor skills effects their performance when browsing the world wide web.

This paper examines previous research on the topic of depth versus breath in hierarchical menu structures, and explain why searching for information on the world wide web follows a similar model.

2. Depth versus Breath in Menu Selection

The topic of menu selection and especially depth versus breath tradeoffs has been extensively examined, both empirically and analytically. Menu panels usually consist of a list of options. These options may consist of words or icons. The word or icon, conveys some information about the consequences of selecting that option. Sometimes the options are elaborated with verbal descriptors. When one of the options is selected and executed a system action occurs that usually results in a visual change on the system. The total set of options is usually distributed over many different menu panels. This allows the system to prompt the user with options that are unlikely or illegal.

Web indexes are organized in a similar structure. Links (very often 2-3 words, sometimes elaborated with verbal descriptors) are arranged in various levels of homepages. These links convey information about the page (with information or further sub-categories) that will be displayed if that specific link is selected. It has been shown experimentally (Zaphiris (2000), Larson, K. & Czerwinski, M. (1998)) that hierarchical menu design experiments can be replicated when applied to hierarchies of web links.

Hierarchical decomposition of the user's selection of action is often necessary to facilitate fast and accurate completion of search tasks, especially when there is insufficient screen space to display all possible courses of action to the user. Hierarchical structures also help the novice user who lacks sufficient memory capacity to learn and recall all of the commands necessary to execute the desired actions.

On the other hand, the navigation problem (i.e. getting lost or using an inefficient pathway to the goal) becomes more and more treacherous as the depth of the hierarchy increases. Research has shown (Snowberry, Parkinson, and Sisson (1983)) that error rates increased from 4.0% to 34.0% as depth increased from a single level to six levels.

The challenge, therefore, is to enable the user to select the desired course of action using a clear, well defined sequence of steps to complete a given task (Wallace, Anderson, & Shneiderman, 1987).

2.1 Empirical Results

The trade-off between menu depth and breadth is considered by some researchers as the most important aspects that must be considered in the design of hierarchical menu systems (Jacko & Salvendy, 1996). Miller (1981) found that short-term memory is a limitation of the increased depth of the hierarchy. His experiment examined four structures (64^1, 2^6, 4^3, and 8^2) with a fixed number of target items (64). As depth increased so did response time to select the desired item.

Snowberry, Parkinson & Sisson (1983) replicated Miller's study by examining the same structures but this time including an initial screening session during which subjects took memory span and visual scanning tests. They found that instead of memory span, visual scanning was predictive of performance, especially in the deepest hierarchies.

Kiger (1984) extended Miller's research by doing an experiment that provided users with five modes of varying menu designs of 64 end nodes (2^6, 4^3, 8^2, and 16x4, 4x16). Performance and preference data were collected. The results of the experiment showed that the time and number of errors increased with the depth of the menu structure. The 4x16 structure had the fastest response times and the fewest errors. The participants ranked the menus with least depth as the most favorable (The 8^2 structure was favored).

An experiment by Jacko and Salvendy (1996) tested six structures (2^2, 2^3, 2^6, 8^2, 8^3, and 8^6) for reaction time, error rates, and subjective preference. They demonstrated that as depth of a computerized, hierarchical menu increased, perceived complexity of the menu increased significantly.

Wallace, Anderson and Shneiderman (1987) confirmed that broader, shallower trees (4x3 versus 2x6) produced superior performance, and showed that, when users were stressed, they made 96 percent more errors and took 16 percent longer. The stress stimulus was simply an instruction to work quickly ("It is imperative that you finish the task just as quickly as possible"). The control group received mild instructions to avoid rushing ("Take your time; there is no rush").

Zaphiris (2000) replicated Kiger's (1984) structures but this time on the WWW using hyperlinks. Overall, the results were in agreement with those of Kiger (1984). They found that of the structures tested (2^6, 4^3, 8^2, and 16x4, 4x16), the 8^2 structure was the fastest to search.

Larson & Czerwinski (1998) carried out an experiment using 512 bottom level nodes arranged in three different structures (8x8x8, 32x16, 16x32). Subjects on average completed search tasks faster in the 16x32 hierarchy, second fastest in the 32x16 hierarchy, and slowest in the 8x8x8 hierarchy. Also, on average, subjects tended to be lost least often in the 16x32 hierarchy.

2.2 Mathematical Modeling

Regarding the depth vs. breadth tradeoff in hierarchical information structures, researchers initially provided qualitative recommendations rather than theoretical or empirical predictions (Shneiderman, 1980; Norman, 1990). Starting in the mid-80's a stream of quantitative modeling in this area emerged.

Lee and MacGregor (1985), broke down the search time in hierarchical menu retrieval into two factors, the human factors and the machine factors. The human factors include search strategy, the strategy employed by a user in searching through the alternatives on an index page; reading speed, the rate at which users read or scan the alternatives; and key-press time, the time required to press the appropriate key(s) and/or make the necessary mouse move to select an alternative. With respect to scanning, people typically employ one of two basic strategies for searching through a list of alternatives: exhaustive search and self-terminating search (Norman, 1990).

Hierarchical menu structures of n items obey an inverse relationship between breadth b and the depth d:

$$d = \frac{\ln n}{\ln b} \qquad (1)$$

The total search time through the index, ST, is the product of the number of menus accessed and the average access time per menu:

$$ST = d(E(I)t + k + c) \qquad (2)$$

where $E(I)$ is the expected number of items examined by a user on one menu frame before making a decision, t is the time to process one option, k is human response time and c is computer response. For exhaustive search the number of alternatives per index page that minimizes search time can be computed using

$$b(\ln b - 1) = (k + c)/t \qquad (3)$$

Assuming random sequencing of the alternatives, a self-terminating search would require reading on average one-half of them before encountering the appropriated one. Thus,

$$ST = \frac{((b+1)t/2 + k + c)}{(\ln b)} \ln n \qquad (4)$$

Taking the derivative of the above equation and setting it to zero, it can be shown that the optimum b assuming a self-terminating search is given by:

$$b(\ln b - 1) = 1 + 2\frac{k+c}{t} \qquad (5)$$

3. Age Related Differences

In a study by Mead (1997) age related differences and training on WWW navigation strategies were examined. They state that older and younger adults who report low levels of computer experience were more likely to employ high visual momentum navigation strategies when searching a hierarchical library database than were younger adults who reported high levels of computer experience. Low computer experience participants were more likely to move up to higher levels in the hierarchy (zooming out) than were participants with high computer experience. They also observed that novice searchers are more likely to "get lost" in a hierarchical structure than were more experienced searchers. In an experiment they conducted, where they asked their participants to complete nine search tasks on the WWW, they showed that seniors were significantly less like to complete all tasks than were younger adults. However seniors were as likely to complete the five tasks with short optimal path length (two moves or fewer) as younger adults. Seniors on the other hand, were significantly less likely to complete the tasks with long optimal path length (3 moves or more) than were younger adults. Error results showed seniors may experience considerable difficulty finding information on the WWW when path length is long. Finally seniors were reported to adopt less efficient search strategies and had more problems remembering which pages they had visited and what was on those pages than did younger adults.

Apart from the study mentioned above, there does not seem to be any other previous research on the effect of age in searching hierarchies of links and there has not been any previous research on the topic of depth versus breath in menu selection applications, as applied for people of age. In our analysis we will use the age sensitive components of the human processor model (Card, 1984) proposed by Charness and Bosman (1990). Since previous mathematical models take into account reading speed and motor speed these factors would need to be adjusted for aging in our new proposed mathematical model.

4. Analysis

In this paper a sensitivity analysis, taking into consideration age related differences, of the linear model (Lee & MacGregor, 1985) is presented.

Two age related sensitivity parameters were defined

1. a_1 - represents an age related parameter for human processing time, estimated using perceptual processor cycle time (McFerland, 1958) as

$$Y_i = 104.16 + 1.05age_i \qquad (6)$$

2. a_2 - represents the age related parameter for human motor response time, estimated using the extended Fitts' Law (Fitts, 1954)

$$MT = IM \log_2(D/S + 0.5) \qquad (7)$$

where MT represents movement time, IM an age related variable (in ms/bit); IM = 60.68 + 1.68(age), D the distance of movement from start to target center and S the size of the target.

The following modified linear model is proposed

For exhaustive search

$$TEA_{b,d} = d(b * a_1 * t + a_2 * k + c) \qquad (8)$$

For self-terminating search

$$TSA_{b,d} = d\left[\frac{(b+1) * a_1 * t}{2} + a_2 * k + c\right] \qquad (9)$$

Where TEA and TSA represent the total search time through the index for exhaustive and self-terminating searches respectively. Contour plots of various combinations of breadth and depth were plotted for two age groups (25 and 70 years of age).

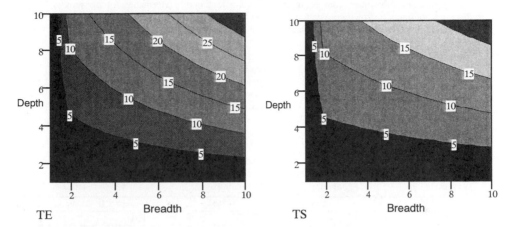

Figure 1: Contour plots for exhaustive and self-terminating search times for young (age = 25) users at t=0.25, k=0.5, c=0.5

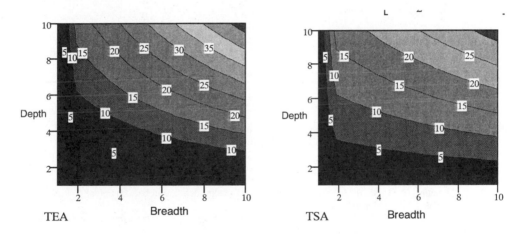

Figure 2: Contour plots for exhaustive and self-terminating search times for senior (age = 70) users at t=0.25, k=0.5, c=0.5

Finally an optimization analysis (using Newtons Method) was carried out in order to obtain optimum value for breadth that will minimize the total search time. Table 1 shows the results of this analysis.

5. Conclusion

The contour plots of the various search completion times show that seniors are expected to be slower, than their younger counterparts, in completing search tasks in hierarchical information structures associated with the decline in perceptual and motor abilities as we age.

Seniors also tend to have limitations in spatial abilities, which is a key issue in menu/link hierarchies. It is predicted that older users will tend to get more easily "lost" in very deep and very broad hierarchies thus resulting in a lower performance (the a_2 factor in the proposed model).

Also seniors are expected to have limitations in motor abilities. Since menu/link driven hierarchical user-interfaces are mainly mouse driven, it is predicted that older users will have a harder time selecting the appropriate choice among the ones presented to them (the a_1 factor in the proposed model).

All these factors combined result in slower completion times for seniors both for exhaustive and self-terminating designs.

Finally, only minor differences of optimum breadth were found for the two age groups (optimum breadth ranges in value from 3 to 18 depending on expertise.) It should be pointed out that even though the mathematical model predicts that performance of both age groups results in optimum performance for the same values of breadth this does not imply that their completion times will be equal. Also it is expected that designs with larger than optimum breadth will result in high age related differences in performance.

5.1 Suggestions to Practitioners and Researchers

Overall, difficult tasks, over-crowded interfaces, very deep hierarchies on slow computer networks, result in big age related differences in performance. Shallow hierarchies designed with optimum breadth will result in optimum performance with smaller age related differences among users.

Further research is needed on the topic. Experimental data needs to be collected and the proposed model tested against those data. Also experience and skill related parameters need to be calculated and incorporated into the proposed model.

Table 1: Optimization results for optimum value of breadth that minimizes total search time (where results represent the optimum number of items per screen E = Exhaustive, S = Self terminating, EA = Exhaustive aging, SA = Self terminating aging.)

k	t	c=0.5				c=1.0			
		E	EA	S	SA	E	EA	S	SA
0.5	0.25	6	6	8	8	7	6	10	10
0.5	.5	4	4	6	6	5	5	7	7
0.5	1	4	4	5	5	4	4	6	6
0.5	2	3	3	4	4	3	3	5	5
1	.25	7	7	10	10	8	8	12	12
1	.5	5	5	7	7	6	6	8	8
1	1	4	4	6	6	4	4	6	6
1	2	3	3	5	5	4	4	5	5

References

Card, S. (1984). *Control of Language Processes.* Hillsdale, New Jersey: Lawrence Erlbaum Associates.

Charness, N., & Bosman, E. (1990). *Handbook of the Psychology of Aging.* (3rd ed.). Academic Press.

Craja, S. (1996). *Aging and Skilled Performance.* Mahwah, NJ: Lawrence Erlbaum Associates.

Fitts, P.M. (1954). The information capacity of the human motor system in controlling the amplitude of movement. Journal of Experimental Psychology, 47, 381-391.

Jacko, J., & Salvendy, G. (1996). Hierarchical menu design: Breadth, depth, and task complexity. *Perceptual and Motor Skills, 82,* 1187-1201.

Kiger, J. I. (1984). The depth/breadth tradeoff in the design of menu-driven interfaces. *International Journal of Man-Machine Studies, 20,* 201-213.

Larson, K. & Czerwinski, M. (1998). Page design: Implications of memory, structure and scent for information retrieval. *Proceedings of CHI '98 Human Factors in Computing Systems* (pp. 25-32). ACM Press.

Lee, E., McGregor, J. (1985). Minimizing User Search Time in Menu Retrieval Systems. *Human Factors, 27* (2), 157-162.

McFerland, R. A., Warren, A. B. & Karis, C. (1958). Alterations in critical flicker frequency as a function of age and light: dark ratio. *Journal of Experimental Psychology, 56,* 529-538.

Mead, S., Spaulding, A., Sit, A., Meyer, & B., Walker, N. (1997). Effects of age and training on world wide web navigation strategies. *Proc. Human Factors Society, Fourty-First Annual Meeting* (pp. 152-156).

Miller, D. P. (1981). The depth/breadth tradeoff in hierarchical computer menus. *Proceedings of the Human Factors Society* (pp. 296-200).

Norman, K. (1990). The Psychology of Menu Selection: Designing Cognitive Control of the Human/Computer Interface. Norwood, NJ.: Ablex Publishing Corporation.

Shneiderman, B., (1980). Software Psychology: Human Factors in Computer and Information Systems. Cambridge, Mass.: Winthrop.

Snowberry, K., Parkinson, S. R., & Sisson, N. (1983). Computer display menus. *Ergonomics, 26,*(7), 699-712.

Wallace, D., Anderson, N. & Shneiderman, B. (1987). Time stress effects on two menu selection systems. *Proceedings of the Human Factors Society, Thirty-First Annual Meeting* (pp. 727-731).

Zaphiris, P. (2000). Depth vs Breadth in the Arrangement of Web Links. *Proc. Human Factors Society, Fourty-Fourth Annual Meeting.*

PART 6

HUMAN FACTORS, ERGONOMICS, GUIDELINES AND STANDARDS

W3C User Agent Accessibility Guidelines

Jon Gunderson

Chair, W3C User Agent Working Group
Division of Rehabilitation – Education Services,
College of Applied Life Studies
University of Illinois at Urbana/Champaign
1207 S.Oak Street, Champaign, IL 61820
E-mail: jongund@uiuc.edu

Abstract

The W3C User Agent Accessibility Guidelines (Jacobs, Gunderson & Hansen, 2001) are being developed as part of the W3C Web Accessibility Initiative (http://www.w3.org/WAI). UAAG is designed to provide information and standards on how to design technologies that render Web resources to be more accessible to people with disabilities. Accessibility specifications include direct accessibility through the built-in features of the standard user interface, and in-direct accessibility through compatibility with assistive technologies which provide alternative user interfaces required by some disabilities. UAAG is being developed through the consensus-based process of the World Wide Web Consortium (W3C, http://www.w3.org), which includes participation from developers, researchers, service providers, disability organizations, and individuals with disabilities. The guidelines use priorities for different types of accessibility features to assist developers in understanding the most important issues they need to address for improving the accessibility of their product. Developers can make conformance claims to specify that their product has satisfied a particular set of accessibility requirements defined in the conformance section of the guidelines. Guideline conformance criteria helps consumers understand the accessibility features they can expect from a user agent technology making a conformance claim.

1. Introduction

In the original design of the World Wide Web, Tim Berner's Lee envisioned a means for people to share and exchange information using a wide variety of technologies. The heart of this new technology to share information was the use of open standards to support interoperability between computing and software technologies. Open standards to support interoperability is still the primary objective of the W3C. One of the ways in which the W3C supports access by all people is to make sure that each new W3C recommendation is reviewed for disability access issues and that the recommendation includes support for the electronic curb cuts and ramps needed by people with disabilities. The W3C Web Accessibility Initiative is the primary group within the W3C to review recommendations for accessibility issues. In addition to reviewing recommendations, WAI is also developing recommendations to help developers understand how to design their resources, authoring tools and rendering technologies to be more accessible to people with disabilities. The User Agent Accessibility Guidelines (UAAG; Jacobs, Gunderson & Hansen, 2001a) is one of the guidelines being developed by WAI. UAAG helps developers of technologies that render web resources to understand how to make their technologies more accessible.

Most of the features needed by people with disabilities are not that much different than the needs of all users, especially as technologies for rendering Web content expand to mobile devices and people use web technologies in environments that preclude the use of some types of web content. For example most hand held PDA devices do not have full size keyboards or high resolution graphical displays, but people still are using them to surf Web content. This situation is similar to someone with a physical impairment that cannot use a standard computer keyboard or a person with a visual impairment that cannot see all of the information on a standard graphical monitor at the same time. In public places or noisy industrial environments people may not be able to hear audio information, so that access to text descriptions and captioning information is important for people to have a text alternative to the information encoded in the audio format. Therefore most of the accessibility requirements for people with disabilities are the same requirements that make the web accessible to everyone.

548

The UAAG guidelines promote the following features:
- Access to all content
- User control over the types of Web content rendered
- User control over the style used to render Web content
- Enhanced navigation of Web content
- Help the user to remain oriented
- Reconfiguration of input controls
- Documentation and increased the visibility of accessibility options and features
- Compatibility with assistive technology for alternative user interfaces
- Fully support W3C recommendations

2. User Control Over Content and Style

One of the main issues for Web accessibility is the ability of user agent to adjust the rendering of information to the capabilities, needs and preferences of the user. Adjustments include allowing the user to choose which types of web resources they would like to have rendered and the styling of information used to render content. Early web browsers provided users with almost complete control over the types and style of information they render. As the web became more popular, web browsers offered new capabilities to allow authors more control over how a web resource is rendered and content authors embraced these new capabilities. These capabilities shifted the perception of authors and the general public to think that the rendering of Web resources should not be under the user's control, but under control of the author. Many people are unaware of the current capabilities of most graphical browsers to allow control the font sizes and colors used to render Web resources and therefore they do not even think to look for the configuration options or user interface commands to make rendering adjustments. This type of awareness is even more essential for people with disabilities who often cannot access Web resources in the styles specified by the author.

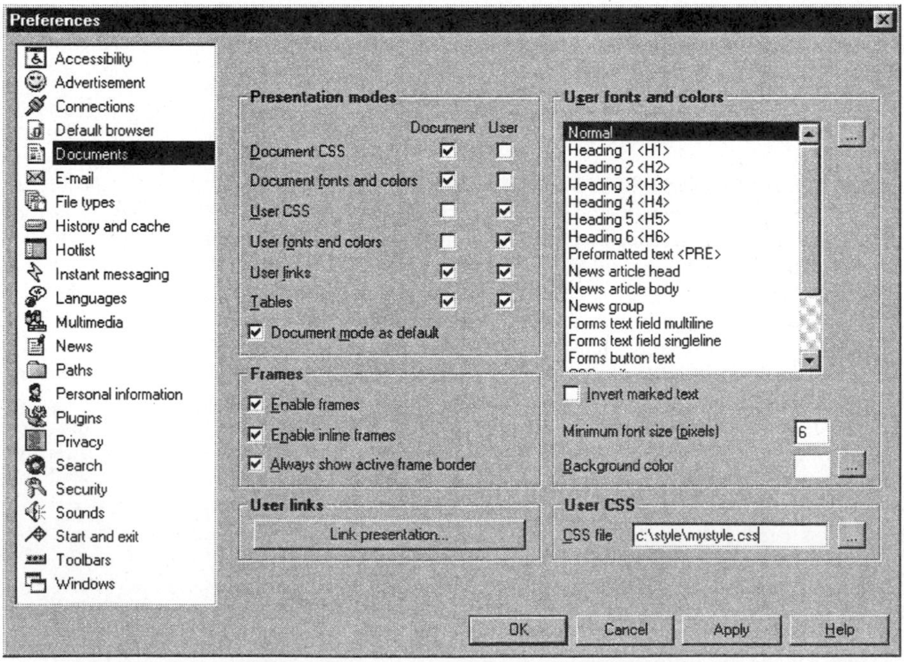

Figure 1: Opera 5.0 Preferences Control Panel

Many of the content styling requirements needed by people with disabilities can be satisfied by fully implementing and using W3C recommendations. For example the use of Cascading Style Sheet (Lie & Bos, 1999) technology can be used to control font characteristics and colors used to render text. CSS technology is designed to provide a powerful means to separate the structure of content from the style used for rendering. CSS also is designed to help manage the styling of an entire web site and the change of one style sheet can immediately affect the style of any page referencing the style sheet. One of the key provisions of the CSS recommendation is the concept of author and user style sheets. CSS requires that users be allowed to configure the user agent to ignore author supplied style information and use their own style information. Users should be able to change the default rendering style or link to their own externally referenced style sheet. Figure 1 shows the preference panel for the Opera (http://www.opera.com) browser which does support author and user style sheets. The control panel allows the user to configure a document and user rendering modes that allow the user to ignore author styling information and change the default style sheet of the browser.

3. Navigation and Orientation

One of the key characteristics of hypertext technology is the ability of users to move between Web links based on the user's own needs and task requirements. Usability studies of Web surfing have found that people tend to visually skim graphically rendered text for key words and phrases, and use the available links to explore information in unpredictable paths based on their own needs and interests. User agents need to make sure that users with disabilities have the same capability to easily navigate the active elements of a document and to review the structure of the document efficiently. One of the key provisions is the ability to navigate between groups of links and form controls. UAAG requires that users be able to use the keyboard to navigate to and select links in a Web resource; and to move the focus between form controls and to set the state of the controls. Navigation between individual links and form controls is only part of the requirements. Users also need to be able to navigate between other types of structural markup, like headers, groups of links and items in a list, to help users easily find major topics and important information in a Web resource. Current technologies like IBM Home Page Reader (http://www.ibm.com/sns) and Opera (http://www.opera.com) allow users to navigate element by element or to the headers (elements marked as H1-H6) in a resource.

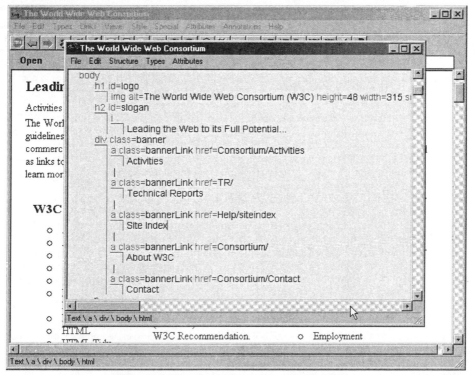

Figure 2: Amaya structured view of a document

Other types of browsers like Amaya (http://www.w3.org/Amaya/) provide a tree view of the document and allow the user to navigate through the document tree to explore content (see Figure 2). If the author uses HTML markup correctly the tree view can be a powerful means for the user to explore the contents of a Web resource.

Users need to be able to control changes to the information rendered in the document window for graphical user agents and the creation of new document windows to remain oriented. Authors can include information that will automatically generate new windows or reload the current page with updated or new web resources without explicit user requests. For some types of disabilities unexpected changes can be very disorienting. For example if a blind user is using a screen reader and a new window automatically is opened by the browser due to scripting, the blind user may not be notified of the new window opening and changes the focus changes to the window. When the blind user tries to use the back button to go the previous document and they find that the feature doesn't work, since the new window does not inherit the previous browsing history, resulting in user frustration and disorientation. Users need to be able to configure the user agent to prompt the user before any new document window is opened. The prompting includes giving the user a choice between:

1. Not to open the new window
2. Open the resource in a new window
3. Open the resource in the current browsing window.

Giving the user a choice allows the user to build their own internal map of the resources they have opened and remain oriented.

4. Assistive Technology Compatibility

One of the corner stones of UAAG is the requirement for compatibility with assistive technologies. Assistive technologies provide alternative user interfaces for people who cannot use the built-in user interface. For example, most popular user agent technologies support a graphical rendering of Web resources. Users who are blind cannot use the graphical rendering. Through assistive technology, called a screen reader, users who are blind can have a synthetic speech or refreshable Braille view of the graphical user interface and the web resources it is rendering. Another example is a user with a physical impairment that limits them to the use of a few switches. The switches can be configured to produce Morse Code and the codes can be used to simulate the keys on the standard keyboard. In this way the user can access the functionalities of the user agent by simulating user agent keyboard commands.

Alternative user interfaces need information about the standard user interface controls and the Web content being rendered for them to provide an alternative rendering or input for the user. UAAG requires that the W3C Document Object Model (Apparao et. al., 1998), which is a parsed tree representation of the Web resource, be exported for use by assistive technologies. This is very important since it provides assistive technologies with direct access to a structured representation of the Web resource. This allows the alternative rendering to be optimized for speech or Braille rendering mediums. For example, current screen reading technology often relies on intercepting screen drawing APIs to learn what text is being rendered on the computer screen. At this point in the rendering the information on what type of element the text originated from is lost. Assistive technologies relying on this technique cannot provide information on whether the text was part of header, paragraph, table cell or any other HTML structure. By using the DOM to communicate Web resource information, the assistive technology has access to the same information as the graphical rendering to provide contextual and orientation information in the alternative user interface. Other user interface controls, like menus, dialog boxes, status lines and other components are required to use accessibility APIs or the standard input and output APIs that already support the accessibility APIs. By supporting standard input APIs alternative input assisitve technologies can be used to control the user agent.

5. Conforming to the Guidelines

The conformance section of the guidelines is designed to indicate that a user agent complies with the requirements of the document and therefore is likely to be accessible to people with disabilities. There are three levels of conformance a developer can claim to the guidelines: single A, double A and triple A. These are the same levels as defined for the W3C Web Content Accessibility Guidelines (Chisholm, Vanderheiden & Jacobs, 1999) and W3C Authoring Tools Accessibility Guidelines (Treviranus, McCathieNevile, Jacobs and Richards, 2000). The single, double and triple A levels of conformance correspond to satisfying all the priority 1, priority 2 and priority 3 level checkpoints. Priority 1 checkpoints are defined as features that are needed for accessibility or it will be impossible for some users with disabilities to use the product. Priority 2 checkpoint features are need or it will be difficult for some users to access content, and priority 3 checkpoints are features that will make it easier for users with disabilities to access content. As each checkpoint was developed the group needed to reach consensus on the priority of the checkpoint.

The second issue is minimum requirements. During the spring of 2000 the working group submitted the guidelines as a proposed recommendation (the last stage before the document becomes a W3C recommendation) and many developers commented on the document. One of the main comments was how does a developer know that they have satisfied the requirements of a particular checkpoint. Many of the checkpoints identified problems that had a wide range of potential solutions and these were outlined in the associated techniques document (Jacobs, Gunderson & Hansen, 2001b) and developers were unsure of what they had to do to satisfy the checkpoint. To address the minimum requirement issue the group chose to clarify the minimum requirements for each checkpoint identified by the group as having vague requirements for conformance. The group spent most of the summer and fall of 2000 defining the minimum requirements. These are being reviewed in a second last call during the fall and winter of 2000.

The third component for conformance was applicability. Some of the checkpoints dealt with technologies or features that a particular user agent did not support. In this case the developer making a conformance claim must document which of checkpoints they felt did not apply to their product. To assist developers and consumers in simplifying these claims, the working group choose to make up predetermined sub groups of checkpoints for particular media types like text, graphical, video, audio and speech. In this case developers can use a label instead of specific checkpoints to indicate the groups of checkpoints that apply to their product. The labels will make it easier for consumers to understand what types of media the user agent is claiming conformance for accessibility.

6. More Information

The user agent working groups activities are all public. The current working drafts of the guidelines and techniques documents, minutes from meetings, working group participant information and the e-mail list archives can all be accessed through the working group's home page at: http://www.w3.org/wai/ua.

Acknowledgments
I would like to acknowledge the support of the user agent working group members, Ian Jacobs (W3C staff contact and editor), other W3C WAI staff, W3C member companies, disability organizations, and the United States and other governments who support the W3C Web Accessibility Initiative for their support in developing and implementing the UAAG.

References
Apparao, V., Byrne, S., Champion, M., Isaacs, S., Jacobs, I., Le Hors, A., Nicol, G., Robie, J., Sutor, R., Wilson, C., Wood, L. (1998). *Document Object Model (DOM) Level 1 Specification*. Cambridge, MA: World Wide Web Consortium. [On-line]. Available: http://www.w3.org/TR/REC-DOM-Level-1/

Chisholm, W., Vanderheiden, G.V., Jacobs, I. (1999). *Web Content Accessibility Guidelines 1.0*. Cambridge, MA: World Wide Web Consortium. http://www.w3.org/TR/WCAG10/

Jacobs, I. (2001). *World Wide Web Consortium Process Document*. Cambridge, MA: World Wide Web Consortium. [On-line]. Available: http://www.w3.org/Consortium/Process-20010208/

Jacobs, I., Gunderson, J., Hansen, E. (2001a). *User Agent Accessibility Guidelines Working Group*. Cambridge, MA: World Wide Web Consortium. [On-line]. Available: http://www.w3.org/WAI/UA/

Jacobs, I., Gunderson, J., Hansen, E. (2001b). *Techniques for User Agent Accessibility Guidelines 1.0*. Cambridge, MA: World Wide Web Consortium. [On-line]. Available: http://www.w3.org/WAI/UA/

Lie, H. W., Bos, B. (1999). *Cascading Style Sheets (CSS1) Level 1 Specification*. Cambridge, MA: World Wide Web Consortium. [On-line]. Available: http://www.w3.org/style

Treviranus, J., McCathieNevile, C., Jacobs, I., Richards J. (Eds.) (2000). *Authoring Tool Accessibility Guidelines 1.0*. Cambridge, MA: World Wide Web Consortium. [On-line]. Available: http://www.w3.org/TR/2000/REC-ATAG10-20000203/

MAGUS: Modelling Access with GIS in Urban Systems

Linda Beale, Hugh Matthews, Phil Picton, David Briggs*.

University College Northampton, The Graduate School, Nene Centre for Research, Park
Campus, Northampton, NN2 7AH, U.K. Tel: 01604 735500 Fax: 01604 791114
Email: linda.beale@northampton.ac.uk
*Department of Epidemiology and Public Health, Imperial College of Science, Technology and
Medicine, Norfolk Place, London W2 1PG, U.K.

Abstract

The project aims to develop, test and apply a Geographical Information System (GIS) for modelling access for
wheelchair users in urban areas. The principal focus is to provide a GIS system that will work as a decision support
and mapping tool for urban planners. In addition, the system has been designed to encompass future developments,
including use as a guidance device to help wheelchair users to assess and select optimal routes through the urban
environment. In devising the system for use by both planners and wheelchair users the system must be designed for
a diverse end user population and facilitate both those accustomed to personal computers and those new to their use.
Throughout its development the system is being evaluated by both wheelchair users and urban planners through
collation of user feedback at workshops and field validation, within the study area of Northampton, U.K.

1. Background

The project aims to develop, test and apply a Geographical Information System (GIS) for modelling access for
wheelchair users in urban areas. The study employs and integrates two contrasting methodological approaches:
social science techniques, aimed at investigating and quantifying the factors that need to be considered in modelling
access in the urban environment by wheelchair users; and computer technologies (including GIS and AI) to translate
this information into an operational route assessment and modelling package. This tool is to be applied and tested
using volunteers from both wheelchair users and urban planners.

Initially, 400 wheelchair users were contacted from across Northamptonshire (a county in the East Midlands, U.K.)
and invited to complete a postal questionnaire. To explore emergent themes and issues three focus groups of 6-8
participants were convened. In addition, all wheelchair users were taken into an urban environment for on-site
observations. Aggregation of these results provided an overview of the those features that most impede, allowing a
further insight into how the configuration of built environments can often disadvantage those who are mobility
impaired. The second stage of the project involved using such data and precepts to define the GIS. Increasingly the
role that GIS has to offer within both urban design (Singh, 1996; Dodge & Jiang, 97) and personal guidance system
development (Golledge *et al.*, 91; Golledge *et al.*, 98) is being recognised. Using GIS the individual routes or access
surfaces can be derived and the results output in both statistical description and map form within the GIS. Network
analysis provides the capability for route modelling and route finding along networks such as pavements and other
similar urban surfaces. The user may select a route in either of two modes; an optimal 'from-to' route, or all
wheelchair accessible routes' outwards from a specified location. The system determines the optimum route for the
user based upon artificial intelligence (AI) techniques that rely on the accumulative weighting or 'impedance' of
encountered urban barriers, and key input information, such as wheelchair type and personal preferences (e.g.
avoiding slopes > 4°).The assigned impedance values have been determined using a combination of the gathered
qualitative data, field measurements and visual images. All this information is available on screen throughout use of
the system, with these visualisation tools acting upon intelligence amplification (IA) by using human perceptual
skills to make the link back to the qualitative and visual data.

2. Interface Design

The system is built entirely within ESRI™ ArcView® GIS, using AVENUE™ programming language and Dialog Designer™ for the development and customisation of the interface. ArcView® works on Microsoft® Windows™, Apple® Macintosh®, and UNIX® whilst Dialog Designer™ is also a cross-platform development environment.

2.1 Rationale

A GIS is a powerful set of tools for collecting, storing, retrieving at will, transforming, and displaying spatial data from a real world for a particular set of purposes (Burrough, 1986). A GIS offers the capabilities for creation of an operational route assessment and modelling package. It is important that such GIS capabilities can be transferred to a wide range of users with a broad spectrum of skills. Most users will not have used a GIS, many will be unaccustomed to computing, and a number will be elderly or have mild mobility impairments. A key feature of the ArcView GIS is the easy-to-use interface. However, GIS software is unlike the personal computing software such as word processing packages that many people are accustomed to. Additionally, few will be comfortable with the analytical operations necessary to make decisions with maps (DeMers, 1997). Automating many of the required processes and providing interaction through a user-friendly interface provides the opportunity to offer some GIS capabilities to non-GIS experts. The user interface is one of the most important parts of any program as it determines its ease of use. A powerful program with a poorly designed user interface has little value (Apple® Corporation, 1996). The graphical user interface (GUI) should allow the user to make as few choices as necessary to accomplish the program task whilst the GIS programmer controls the complex interactions such as data manipulation, spatial processing and visual outputs (Wagnitz, 1999).

The most common 'golden rule' of user interface design is "know the user", including familiarity with the task, personal abilities, and likely response to the system (Witten et al., 1993). The users of the MAGUS system will be diverse both in computer literacy and skills and, so the system must be designed to be as inclusive as possible. In designing a system for modelling access for wheelchair users in urban areas, the knowledge and requirements of wheelchair users must be incorporated to inform urban planners. Ultimately, both wheelchair users and urban planners will be users of the system, but clearly knowledge of the subject lies predominately with one user group. Two important aspects for good human-computer interaction are accessibility and reliability (Greenburg, 1993). Accessibility must be designed to permit a diverse end user population ranging from I.T. experts to complete novices, all from various backgrounds and experiences. For those with computing skills the user interface should reflect a 'commonality' as consistency across interfaces reduces user learning time, facilitates user progress, and greatly reduces errors. The system must be stable and include error-checking routines to ensure reliability and ease of use. Programming design should, however, in all possible situations, follow the rule of 'prevention rather than cure'. Preventing, rather than permitting, erroneous input can negate user frustration. Further consideration should be focussed upon attaining a balance between the capabilities that a GIS offers and the interaction required from the user. Limiting the amount of actual user input would simplify use of the system, accelerate problem solving within the system and reduce error introduction.

2.2 The User-Interface

The user-interface is a menu interface that allows any choice available at a given point, within the program, to be visible to the user. Selection is made primarily through this 'main menu' which is permanently visible on the display screen providing a fixed location for finding commands or navigating through the system. All choices are made using "buttons" with each displaying a simple textual description of the option. The key commands for inclusion were minimised, permitting a number of advantages:
1. prevents accidental selection of neighbouring choices;
2. button descriptions are clearly visible;
3. attains a simple system appearance.

It has been found that elderly users are often afraid that erroneous input will cause damage to the software (Zajicek, 2000) - a matter of considerable importance given that a substantial proportion of wheelchair users are aged over 60 years. It was important, therefore, that interaction with the interfaces could be accomplished without fear of 'clicking' a control by accident. Furthermore, such spacing facilitates interaction by those with mild mobility impairments again, by reducing the need for very precise cursor placing. The spacing of the command buttons also enabled the textual description to be clearly read further enabling many elderly users. The reduction of buttons on the main menu simplified the appearance of the system and required that users digest less information. This minimalism was hoped to increase confidence and 'useability', whilst reducing demands upon those with memory impairments.

To incorporate all required commands a number of further pop-up menus were used. The use of pop-up menus allowed the programmer to navigate the user through the system. This was advantageous for the user as it provided ease of stepping through the system, whilst for the developer it reduced the need for error trapping codes. The use of numerous smaller menus reduced the number of decisions and interactions required at each stage therefore giving the system a simpler, 'easier to use feel'. However, such simplicity can only be maintained if the number of pop-up menus is kept to a minimum. GUIs with multiple menus and tools have been found to cause difficulties for infrequent users (Greenburg, 1993). Navigation through the system is further controlled by using progressive disclosure and only permitting interaction at appropriate points (i.e. controls only become visible as required). In controlling user interaction the system appears less intimidating to those less confident users as well as minimising the opportunity for usage error. Furthermore, in leading the user through the system, interaction does not detract the user from the system application and in no way compromises the usability of the system for those who are more confident users.

Further to this main menu, a tool bar menu is located across the top of the system. This menu is always visible, replicating the functions available on the main menu (figure 1). This tool bar was incorporated in an attempt to maintain a consistency and, therefore, familiarity to those accustomed to using personal computers, with most GUIs adopting windows, icons, and pop-up menus. Simplicity was a key feature within the design of the 'main menu'. Although this led to a minimisation of functions, the tool bar provided an opportunity to add functionality for those more confident users. GIS features such as 'pan', 'measure', and 'query the spatial database', are all available via 'clickable icons'. The inclusion of these tools was not deemed to confuse the more 'nervous' user as observations with some elderly computer novices revealed a general reluctance to interact with unfamiliar controls and a willingness to ignore the unfamiliar.

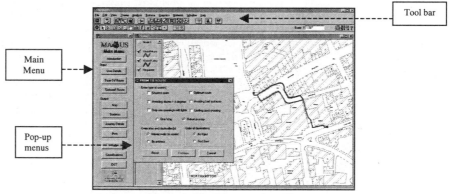

Figure 1: The Menus and tool bar

Through the design process the demands of I.T. novices, the elderly and those with mild mobility impairments were considered. It was felt that user input and required keyboard strokes must be minimised. To achieve this balance, several simplifications have been made to the route modelling system. The type and size of wheelchair will have a direct affect upon the accessibility of urban areas. Furthermore the physical ability of the wheelchair user or 'assistant', where the wheelchair is being pushed, will have an absolute effect upon possible routes around an urban area (table 1).

Many of these questions may be unknown to the user or are very personal questions. It must also be remembered that many of the users of the MAGUS system will be urban planners of whom few, if any, will be wheelchair users, so this information must be minimised to ensure that the system is equally accessible to non-wheelchair users. These factors were reduced to two salient features including type of wheelchair and an all-encompassing factor termed 'fitness' - this incorporated the wheelchair type e.g. lightweight frame, and the fitness level of the user/assistant/carer (figure 2).

Abbreviating the questions simplifies the interface, minimises user input, reduces the awkwardness of asking personal information, and allows non-wheelchair users to select an option without difficulty. Further input is required to identify route destinations. These locations can be entered either interactively (requires mouse action) or by entering the address (keystroke input). A user must select a point that lies along a route - a large buffer has been created around each route so that point location need not be too precise. Ease of entering a point is furthered by the ability to zoom in an out of the area, whilst the point symbol retains a fixed size. Allowing the addresses of destinations to be entered with keystroke input enables certain shops, for example, to be located without knowledge from the user. A minimum of one field e.g. street name must be entered to locate points.

Table 1. Individual factors affecting route selection

Wheelchair type	Wheelchair dimensions	Personal details	
Manual	Width	Weight of chair	Ability of user
	Length	Weight of occupant	Strength of user
	Height		
Assisted	Width	Weight of chair	Strength of assistant/carer
	Length	Weight of occupant	Ability of user
	Height		Strength of user
			Does the wchair user assist
Powered	Width	Weight of chair	Battery(s) power
	Length	Weight of occupant	
	Height		

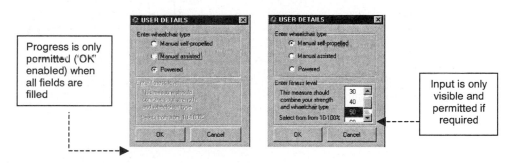

Progress is only permitted ('OK' enabled) when all fields are filled

Input is only visible and permitted if required

Figure 2: User input defining wheelchair type and 'fitness'

2.3   User feedback

Following selective trialing changes have been made to screen displays. Initially any determined routes remained on screen but could be made visible or not by clicking upon the desired route on the table of contents (figure 3). This operation was only determined by users with GIS knowledge, whilst the remainder of users found the accumulation of routes visually confusing and frustrating. All users reported that after any route has been selected, requesting further route selection

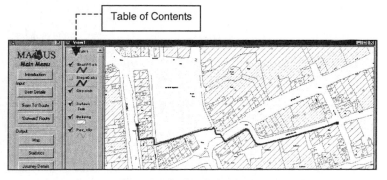

Table of Contents

Figure 3. Viewing derived routes

should invoke a pop-up menu offering a choice to delete or save the previously determined routes. This was felt important to reduce confusion whilst increasing satisfaction that a task had been successfully completed.

The initial pilot also highlighted the need to report the selected parameters for route determination. The use of pop-up menus led the users through selection choices (e.g. wheelchair type and route preferences) with ease. In a number of cases, however, the selections made were forgotten and retrieval of this information was not possible, inevitably reducing usability for urban planners investigating wheelchair access. Further alterations were required in relation to the input of destinations within route selection. The user was required to select the choice of user input at a previous menu to that which permitted interaction with the route. This meant that selection of input type was fixed for each route-finding operation rather than for each point.

Most feedback however, related to the use of geographic space and representation of the area which caused confusion for those unaccustomed to spatial data. Although the pedestrian routes around the town were derived features, representing the pavement centre lines, it was found easier to determine locations and plan routes when these pedestrian routes were visible. Further evaluation will be conducted using repeated testing of the prototype system using parallel observation. These trials are to be conducted with volunteers of wheelchair users and urban planners.

3. Conclusions

Throughout system development users have informed the needs of the system. The very nature of the GIS system ensured that many wheelchair users were involved throughout all stages of development. Further trials will ensure that users continue to inform development ensuring that interaction with the system does detract the user from the purpose of the system. The MAGUS system manages, analyses, and manipulates spatially referenced information in a problem solving synthesis making it by definition a GIS model (Fisher & Lindenberg, 1989), however, the operational route assessment and modelling package does not require a GIS expert to attain useful or meaningful results.

Acknowledgements
This research is supported by the EPSRC (Engineering and Physical Science Research Council) as part of the EQUAL (Extending Quality Life) initiative.

References
Apple Corporation (1996). *Macintosh Human Interface Guidelines*, Retrieved January 5th, 2001. [On-line]. Available: http://devworld.apple.com/techpubs/mac/HIGuidelines/HIGuidelines-2.html.

Burrough (1986). *Principles of Geographical Information Systems for Land Resources Assessment*, Oxford University Press.

DeMers M.N. (1997). *Fundamentals of Geographic Information Systems*, John Wiley & Sons, Inc.

Dodge M., Jiang B., (1997). Geographical information systems for urban design: providing new tools and digital data for urban designers, *Learning Spaces? Conference*, De Montfort University, Leicester, November.

Fisher P.F.and Lindenberg R.E. (1989). On distinctions among cartography, remote-sensing, and geographic information-systems, *Photogrammetric Engineering and Remote Sensing*, vol. 55: no. 10, pp. 1431-1434.

Gollege R., Loomis J.M., Klatzky R.L., Flury A., Yang X.O. (1991). Designing a personal guidance system to aid navigation without sight: progress on the GIS component, *International Journal Of Geographical Information Systems*, vol. 5, no.4, pp. 373-395.

Gollege R., Klatzky R.L., Loomis J.M., Speigle J., Tietz J. (1998). A GIS for a GPS based personal guidance system, *International Journal Of Geographical Information Systems*, vol. 12, no.7, pp. 727-749.

Greenburg (1993). *The computer user as toolsmith*, Cambridge University Press, New York.

Singh R.R. (1996). Exploiting GIS for sketch planning, *ESRI User Conference*, Palm Springs, California.

Wagnitz R.D. (1999). Communicating with Users: The "Art" of Program Composition and Teaching Techniques in a Multidisciplinary User Environment, *ESRI User Conference*, San Diego, July 26-30.

Witten, I.H., Greenberg S. (1993). User Interfaces, *Encyclopedia of Computer Science,* Eds. Ralston A., Reilly E.D., pp 1411-1414, Van Nonstrand Reinhold, New York.

Zajicek, M. (2000). Interface Support for Elderly People with Impaired Sight or Memory. In P.L. Emiliani, & C. Stephanidis (Ed.) Proceedings of the 6th *ERCIM Workshop: User Interfaces For All, Special Theme: Information Society for All*, Florence, Italy.

A proposed standard for consumer product usability

Nigel Bevan

Serco Usability Services, 22 Hand Court,
London, WV1V 6JF, UK
nbevan@usability.serco.com

Roland Schoeffel

Designafairs GmbH
D-81316 Munich, Germany
roland.schoeffel@designafairs.com

Abstract

Purchase decisions for everyday products are frequently made on the basis of matching a list of the product's functions to the user's needs. However, for the functions to be useful, the user needs to be able to operate them successfully. If the functions are not easy to operate, many users will find it difficult or impossible to achieve the main goals of use of the product. Unfortunately it is often not possible for a purchaser to judge at the time of purchase whether or not the product is easy to operate. Information about the ease of operation of a product would therefore be of great value to potential purchasers.

A standard is currently under development to provide guidance on how to take account of the needs of the widest possible range of potential users, and to define a standard usability test method and a statement describing the usability of products. Once agreed, this standard would offer a base line to communicate a product's ease of operation.

Several issues remain to be resolved. To what extent can the special needs of the disabled be included? How many users should be tested to obtain a reliable indication of usability? What criteria should be used for products to pass the test? How widely would such a standard be adopted by industry?

1. Why aren't consumer products usable?

Usability is still an optional feature in consumer products. For example, there have long been complaints about the usability of video recorders (Thimbleby, 1991), and more recently mobile phones (Helyar, 2000). Almost all consumer organisations and magazine product reviews rate usability as an important factor. So why do manufacturers not give usability a higher priority?

Unfortunately people have to purchase many products without the opportunity to try using them. Demonstrations in shops are limited. For example, in major UK high street electrical stores it is not possible for customers to try using a video recorder prior to purchase. The only information available is a partial list of features, the appearance of the equipment, and the personal opinion of a shop assistant.

Keinonen (1997) found that even after users spent an average of 20 minutes examining wrist heart rate monitors, the expected usability of the monitors had little impact on their preference.

In these circumstances there is little incentive for a manufacturer to worry about whether consumers can use their products. Indeed, the early adopters of innovative technology are more concerned with the excitement and status of owning and using an innovative device, than with whether it is actually useful.

But for users of mature technology, lack of usability is a time-consuming frustration. Thimbelby (1991) argues that manufacturers have a moral responsibility to supply usable products. In practice, commercial pressures mean that the priority is to make a profit while meeting legal obligations. Some manufacturers, e.g. the German household appliances manufacturer BSH, have actively decided to develop best possible usability, and also to reach the elderly market (Lohrum, 2000). But manufacturers that seek to obtain a reputation for quality do not always see usability as a necessary component (Bevan, 2000a).

2.   Consumer product usability: solutions

There are three factors that could contribute to the design of more usable consumer products: legislation for accessibility, usability labelling, and appreciation of the benefits of user centred design.

2.1   Accessibility

One group of consumers that have particular difficulties operating consumer products is those with physical or mental disabilities. As a small percentage of the market, they have not been a priority for manufacturers. With increasing concerns that this sector of society is being excluded from use of new technology, several countries have passed legislation requiring goods and services to be accessible to as wide a range of the public as possible (UK, 1995, USA, 1998, Portugal, 1999). Although legislation for accessibility is primarily focussed on physical capabilities, to be accessible a product must also have cognitive accessibility, one of the main prerequisites for usability.

One approach to accessibility is to provide specialised assistive technology to enable users with physical disabilities to make use of existing equipment. Another approach is Universal Design (also known as Design for All) in which products have the built in capability to support a wider range of users including those with functional deterioration.

ISO DTS 16071: Guidance on accessibility for human-computer interfaces (derived from ANSI HFS 200), provides guidelines and recommendations for the design of systems and software that will enable users with disabilities greater accessibility to computer systems (with or without assistive technology). Part 1 of ISO 20282 (described in Section 3 below) explains how to take account of the diverse range of human capabilities when designing products for a universal user population.

2.2   Usability labelling

There is currently a market failure: consumers want more usable products but have no information on usability when they make a purchase decision. One possible solution is to use a standard usability test method and a label describing the usability of products. Consumers would then be able to consider usability in conjunction with other features. This is one objective of ISO 20282-2 described in Section 4 below.

2.3   Benefits of user centred design

In manufacturing organisations that have usability groups, the extent to which usability techniques are used is often left to the discretion of individual project managers (Wiklund, 1994). Incorporating usability in the design process is perceived as adding extra cost and delaying product release.
Yet there is compelling evidence for the cost benefits of user centred design (Bias and Mayhew, 1994): development time can be reduced, the productivity of users improved, and the documentation and maintenance costs reduced.

Development: Usability engineering can reduce the time and cost of development efforts through early definition of user goals and usability objectives, and by identification and resolution of usability issues. Keil and Carmel (1995).
Use: Users benefit from:
*   *Increased effectiveness*. Avoiding inconsistencies, ambiguities or other interface design faults will increase effectiveness by reducing user error.
*   *Increased efficiency*. A system incorporating a user interface designed to meet the needs of the task will allow the user to be more productive.
*   *Improved satisfaction*: User acceptance is particularly important for applications like web sites where usage is discretionary.
Documentation and Support: A well-designed system designed with a focus on the end-user will reinforce learning, thus reducing the need for documentation and support services .

There is an international standard for users centred design (ISO 13407), and according to IBM (1999) "It makes business effective. It makes business efficient. It makes business sense".

In order to produce usable products, manufacturers need cost-effective methods to incorporate user centred design into the development process. This is an ongoing research area with recent practical solutions (Bevan, 2000b).

3.   ISO 20282 Part 1: Universal user profiles

The objective of ISO 20282-1 is to provide guidance on the design of products for the widest possible range of users. It is envisaged that products could be designed with three potential objectives:
*   *universal:* usable by the widest possible range of users without use of assistive technology
*   *accessible:* usable by a wider range of users with assistive technology
*   *skilled*: only usable by users with special skills or training

The standard is primarily concerned with the design of universal products that are easy to operate, especially without instructions, and without previous experience or training.

Guidance is given on how to take account of the following factors in universal design, while questions regarding accessible design still remain open.

Strength and Biomechanical Abilities: Can the product be used by those with limited physical abilities (particularly the young, elderly and those with disabilities)? The characteristics of the young and elderly are well-understood, but to what extent can people in wheelchairs, people without arms or legs, or people with shaking limbs from cerebral diseases be included in everyday product design?

Handedness: Can the product be used by a left-handed person? 4% of the Western population are left handed, and following universal design principles, everyday products should be designed for use by either hand.

Body Dimensions: Can the product be used by smaller and larger people, and those in wheelchairs? The range in height worldwide is from more than 2 metres to less than 1 metre. Wheelchair users are not only at a low height, but are also have a restricted arm reach.

Visual Abilities: Can the product be used by the blind and people with limited vision? 6% of males have a colour deficiency, and visual acuity and the near point deteriorate with age. Colour should always be complemented by another form of coding (e.g. position). Blind people can also use many products that have multimodal coding, e.g. with additional acoustical information for feedback.

Auditory Abilities: Can the product be used by the deaf or people with hearing aids? With increasing age the upper frequencies of hearing are lost.

Cognitive Abilities: Can the product be used by those with limited cognitive abilities (the young, elderly and those with disabilities)? Specific professional knowledge cannot be assumed. It is unclear how far accessive technology could compensate for mental diseases like Alzheimer, dementia or schizophrenia.

Language and literacy: To what extent can the intended users understand the language used on the product and the instructions? Even within a single country many different languages may be spoken, and there are still regions of the world where 59% are illiterate.

Culture: Which cultures will use the product: there are potential differences between cultures in the conventional meaning of colours, direction and pictograms.

Age: In universal design for everyday products, it is particularly important to take account of the needs of the elderly who account for an increasing part of the world population. The elderly have reduced sensory and physical capabilities, but often have good cognitive function.

Gender: As a principle of universal design, products should be usable by both genders.

4.   ISO 20282 Part 2: Usability test method

At the time of writing (January 2001) the test method is still under discussion. The current intention is to specify a user test procedure that can be used to make reliable statements about usability for the universal user population. The test method is expected to be a refinement of established procedures for measuring usability, as currently described in:
*   ISO 9241-11 Guidance on usability (ISO, 1997)
*   ISO DTR 9126-4 Quality in use metrics (ISO/IEC, 2001)
*   Common Industry Format for usability test reports (NIST, 1999)

560

The test method would require that a representative sample of users carry out the major intended tasks with the product in a realistic context of use, without any assistance (except use of instructions if required). The issues discussed below remain to be resolved.

4.1 How many users should be tested?

How many users should be tested to obtain a reliable indication of usability? For testing of professional products, 8-10 people from each distinct user group are regarded as adequate (Macleod et al, 1997). But even this is expensive, and how can a reliable assessment be made for a product that is intended to be used by the general public in several different countries? Is it better to test one representative group or to divide the population into several selected homogeneous groups?

Statistical tests (analysis of variance) can be used to determine whether results are statistically significant. But what level of probability is acceptable? A disadvantage of conventional testing is that it is difficult to determine in advance how many users should be tested: for a well-defined homogeneous group 10 users may be sufficient to obtain reliable results, while for a heterogeneous group many more will be required. One possibility is to use a sequential testing method (Brigham, 1989) in which statistical tests after each user determine whether additional users should be tested.

How can reliable statements be made about usability for people with special needs? The needs are so diverse that it is not realistic to expect that 8 users with every possible type of disability can be tested. Current thinking in this area is that some form of testing with simulated disabilities may be more realistic (Law et al, 2000).

4.2 What are the criteria for usability?

Usability is defined in ISO 9241-11 as:
> The extent to which a product can be used by specified users to achieve specified goals with effectiveness, efficiency and satisfaction in a specified context of use.

This implies testing:
- Effectiveness: what percentage of users can complete the task?
- Efficiency: how long do they take?
- Satisfaction: are they satisfied with the process?

For simple consumer products (such as a light switch) the major issue may only be whether or not the user can complete the task. For this type of product, the suggested approach is:
> A product is easy to operate if xx% of users in all tested user groups (or xx% of users in each main intended user group) can successfully achieve the main goals of use of the product.

While for a more complex product all the usability criteria may be relevant:
> A product is usable if xx% of users in all tested user groups (or xx% of users in each main intended user group) can successfully achieve the main goals of use of the product in the times stated for each goal, and the mean satisfaction score is greater than the criterion specified

Which criteria are appropriate for which products, and what should be the values of xx% in each case? These issues are currently under discussion.

5. Can this all be done?

One objective of the standard is to allow products of universal design to carry a label indicating how usable the product is. This may be difficult to achieve, and even if it is achieved, it is not clear how widely such a standard would be adopted. If it can be achieved, purchase decisions should be easier in the future, especially for the elderly and the disabled, but also for everybody else caring about usability.

The standard is being developed by ISO TC159/SC1/WG4. If you would like to contribute to the development of the standard, or to comment on current drafts, please contact your national standards body, or one of the authors. Roland Schoeffel is working group convenor and editor of ISO 20282 Part 1, and Nigel Bevan is editor of part 2.

References

Bevan, N. (2000a). Quality in use for all. In C. Stephanidis (ed), *User Interfaces for All* - concepts, methods and tools (pp. 353-368). Mahwah, NJ: Lawrence Erlbaum Associates.

Bevan, N. (2000b) *Cost effective user centred design.* http://www.usability.serco.com/trump

Bias, G., & Mayhew, D. *(Eds)* (1994). *Cost-Justifying Usability.* Academic Press.

Brigham, F. (1989). Statistical methods for testing the conformance of products to user performance standards. *Behaviour and Information Technology,* 8, 279-283.

Helyar, V (2000). *Usability issues and user perceptions of a 1st generation WAP service.* http://www.usability.serco.com/research/research.htm

IBM (1999). *Cost Justifying Ease of Use* http://www-3.ibm.com/ibm/easy/eou_ext.nsf/Publish/23

ISO/IEC PDTR 9126-4 (2001). *Software product quality - Part 4: Quality in use metrics.*

ISO 9241-11 (1998). *Ergonomic requirements for office work with visual display terminals (VDT)s - Part 11 Guidance on usability.*

ISO 13407 (1999). *User centred design process for interactive systems.*

ISO WD 20282 (2001). *Product usability.*

Keil, M., & Carmel, E. (1995). Customer developer links in software-development. *Communications of the ACM,* 38(5), pp33 - 44.

Keinonen, T (1997). Expected usability and product preference. In *Proceedings of DIS'97.*

Law, C., Barnicle, K., Henry, S.L. (2000). Usability screening techniques: evaluating for a wider range of environments, circumstances and abilities. In *Proceedings of UPA 2000,* Asheville, NC, USA. law@trace.wisc.edu

Lohrum, M. (2000). Relationship Marketing Success Through Investments in Products – The Case of BSH. In: Thurau, T.H. & Hansen, U. (Eds) *Relationship Marketing.* Springer, Munich.

Macleod M., Bowden R., Bevan N., Curson I. (1997). The MUSiC Performance Measurement method. *Behaviour and Information Technology,* 16, 279-293.

NIST (1999). *Common Industry Format for usability test reports.* http://www.nist.gov/iusr.

Portugal (1999). *Resolution of the Council of Ministers concerning the accessibility of public administration web sites for citizens with special needs.* http://www.acessibilidade.net/petition/government_resolution.html

Thimbleby, H. (1991). Can anyone work the video? *New Scientist,* 129 (1757), 48-51.

UK (1995). *Disability Discrimination Act , Part III, 1995.* Stationery Office Books.

USA (1998). *Rehabilitation Act Amendments of 1998, Section 508.*

Wiklund M.E. (ed) (1994). *Usability in practice.* AP Professional, Cambridge, MA, USA.

Architectural design in virtual media - "Holohouse Technique"

Jerzy Charytonowicz   *Krzysztof Sztajkowski*

Wroclaw University of Technology   College of Art and Design
Department of Architecture   Str. Pomorska 163/165, 90-236 Lodz, Poland
Str. B. Prusa 53, 50-317 Wroclaw, Poland   Tel:(0048-42) 2141251,
Tel/fax:(0048-71) 3212448   fax: (0048-42) 6399798
georgech@novell.arch.pwr.wroc.pl   krzysztofSzt@poczta.onet.pl

Abstract

"Holohouse Technique" is an idea of a new technology - oriented on architectonic design based on reflecting of the designer's vision in the computer's virtual reality as an electronic model of the building, expressed by means of virtual and holographic models. The difference between the holographic and the virtual models is in the change of the imaging format, and the consequences derived from that

A brand new attitude towards technical documentation enables not only simultaneous preparation of a technical documentation along with a design process, but also careful observation and design correction with the aid of the holographic system . The idea of the " Holohouse Technique " opens a way to making a building process more automatic by means of so called "characteristic points in the holographic model ". For the changes to come, however, some crucial improvements are necessary in construction of building objects. A major change will happen to staking out process, as there will be no need to mark outside foundation lines, the "laser position" coordinates will be enough for a holographic image that will show where the parts of the building are located . Application of the virtual and holographic reality in architectonic design not only opens new possibilities, but sets new standards to the jobs of design and construction. On one hand, the "Holohouse Technique" is a continuation to processing of a design as a spherical record through a virtual and holographic model. It also sets new standards in technical documentation as in a spatial virtual and holographic model record. On the other hand, it reaches for very basic rules of the architectonic design dating back to thousands years ago, this is, to preparing a model of a building object.

1. Introduction

"Holohouse Technique" is an idea of a new technology - oriented on architectonic design based on reflecting of the designer's vision in the computer's virtual reality as an electronic model of the building, expressed by means of virtual and holographic models. The difference between the holographic and the virtual models is in the change of the imaging format, and the consequences derived from that. This document contains a concept of changes of the leading idea on architectonic design with the aid of the Virtual Reality Technology, and of a number of virtual model representations of a building with the aid of a holographic picture that, depending on the representation level, lead to quite different design modifications :

- A change of the concept of an electronic design to a three- dimensional imaging of a space treated as an electronic model of a building, expressed as virtual and holographic models.
- A change of a traditional form of technical documentation of a design worked on in an architectonic studio to multi- dimensional and colorful, instead of black-and-white, two- dimensional traditional documentation (even the model alone is already a design, drawings- not).
- A change of a construction design imaging at a building site, not as a two- dimensional documentation, but as a spatial technical documentation, i.e. an electronic model of a building visualized by a holographic picture.
- A major breakthrough in the form of traditional technical documentation.

A brand new attitude towards technical documentation enables not only simultaneous preparation of a technical documentation along with a design process, but also careful observation and design correction with the aid of the holographic system . The "Holohouse technique" is aimed at embodiment of a quest of many architects for a technology that enables combination of an architectonic vision with a technical design of the object under construction. The new designing system may be useful regarding the whole object, both for technology design and interior design, which will allow to avoid the existing discrepancy between technical and art criteria.

2. Architectonic design methods

Even in ancient Egypt, the architects of those times were not only preparing "technical documentation" of their buildings, but were also reaching for a model. Such statement was announced by prof. Kazimierz Michałowski, a recognized specialist in ancient Egypt, with the example of the Great Pyramid of Cheops: "...completion of such structure could not have been based merely on plans and drawings, but in the meantime the models had to be used, as well. The three small pyramids at the southern side of the Great One are nothing else than just the models (at a scale of 1: 5) imaging the three phases of changes introduced during the construction of the Great Pyramid ... "[1] .

In our times, the traditional process of design is based on creation of a chain of two- dimensional drawings connected by their meaning with each other, combining together a technical design of a building object. In some cases during a design process, there are prepared three- dimensional drawings, or even models of the building object, used as completion of the general vision of a design and visualization of a prospective building. Along with invention of computers as modern tools of data operation, there occurred a natural tendency to making technical documentation with their aid. In the beginning, the process of technical drawings preparation was similar to that without computers, the only difference was visualization in a new environment- i.e. CAD programs.

The next step in architectonic design is preparation of virtual models and using them as the grounds for both technical documentation and visualization of building objects. It seems hat using a virtual reality for making technical documentation was a major breakthrough in architectonic design and a right way of using possibilities a computer offers as a logical algorithm machine. However, using a computer for mere visualization is a complete misunderstanding. A computer is not just an electronic drawing board, should not be only used for drawing lines or filling a needed area with colors. It is more of a data processing and their visualization machine; otherwise a designer comes back to the times before that programming revolution, to the outdated technical documentation criteria and the traditional form of representation.

3. Virtual technologies in design

Progress in computer techniques provokes both new attitudes and new standards in the field of processing, transmission and storage of information. Newer and newer interfaces, or places of functional form representation, make it possible to communicate with a computer by means of pictograms as an interactive place for questions and answers. Contemporary Computer Aided Design offer possibilities of observation and visualization of space through an interactive change of the Point of View and the Point of Observation in a real time, i.e. the Virtual Reality (VR).

In numerous programs, the Virtual Reality is a base of architectonic design. A typical example could be 'Archicad' by Graphisoft. The company describes the product as follows:

" ... Working with Archicad means actually building a virtual model, being a stock of all integrated data about the building, allowing creation of a technical documentation, sections and facades, balance of material, as well as creative idea presentation with photo-like perspectives, animation and virtual reality made in the Quick Time VR technique (Macintosh) and Real VR (Windows). The last feature makes it possible for a user to twist and to animate a model in a real time ... " [2] .

Unfortunately, the design in a virtual reality acts only in a three- dimensional setting of a computer, but the result is still seen as two- dimensional on a screen. A step forward is a research and a concept of applying three- dimensional views for industrial purpose done in Fraunhofer Institut Arbeitswirstschaft und Organisation under the leadership of prof. Hans Jörg Bullinger. The possibilities of applying 3D pictures for design and visualization of a prototype in industrial conditions were shown in the article " Virtual Prototyping- State of the Art in Product Design " [3] .

Very much attention is paid to a full use of spatial representation in a natural scale as to a 3D prototype for production documentation. The advantages to a 3D design as an electronic model of an object were described as:

"... The Virtual Reality software Boule supports engineers in the evaluation and quality assurance of virtual prototypes. Models constructed with Computer Aided Design (CAD) can be examined and documented using three-dimensional Virtual Reality visualization in natural scale. Supporting functions like a virtual cutting plane, allow the inner part of the components to be analyzed. The reliability of the process for evaluating virtual prototypes makes expensive foam models redundant and shortens the design cycle . ... " **[4]** .

A 3D design seems to be a today's routine in numerous architectural offices, not only research centers. However, the visualization of a 3D prototype in 3D pictures still remains to take some time comparing to the today's design practice. The IBM Company is working on such project in the angle of a new computer interface, which was named 'IBM Dream Space'. The IBM specialist claim that a 2D interface being visualized on a screen is only an introduction to a new kind of display- a holographic display reacting to motion, gestures and voice commands to a computer. A research done in making a computer interface more compatible with human natural reactions lead to introduction of a new type and manner of human- computer communication with the aid of spatial pictures, body language and human voice.   " ... Humans discover and understand their world through visual and conversational interactions. Computers (information/communication systems in general) can be designed and built to allow humans to interact in natural ways, using the common skills of speaking, gesturing, glancing, moving around, reaching out. Our DreamSpace allows users to collaborate in a shared space. The system "hears" users' voice commands and "sees" their gestures and body positions. Interactions are natural, more like human-to-human interactions. The "computer" understands the user, and -- just as important -- other users understand. Users are free to focus on virtual objects and information and understanding and thinking, with minimal constraints or distractions by "the computer", which is present only as wall-sized 3D images and sounds (but no keyboard, mouse, wires, wands, etc.). As shown in the schematic below, this intuitive human-like interaction is made possible by emerging interface technologies:
- voice input: user-independent, continuous speech IBM ViaVoice(tm)
- vision input of gesture and body: camera and machine vision algorithm;
- wall-sized stereoscopic "3D" display;
- high-bandwidth networks.

The DreamSpace (also called "Visualization Space") is a networked workspace where the computing system adapts to the human to optimize ease of use, enjoyment, and the organization and understanding of information. The DreamSpace paradigm of computing is ideal for many   applications ... " **[5]** .

Such assumptions may be recognized as an " ergonomic rule of a user- friendly computer creation ", this is accessible to everyone, regardless to a user's special field knowledge. Design based on spatial pictures should revolutionize not only the manner of visualization alone, but improve the design as such as well, thanks to features of human natural reactions and the natural way of communication.

Today's research in new forms of design is not limited to a mere 3D design, but to making it more automatic and to a full use of logic algorithms of a computer. The research team of the Kure National Institute of Technology introduced a new spatial design technique at the 6th International Symposium in Hague, called Virtual Kansei Engineering Applied to House Designing. The Kansei Method is can be applied in making a design of the whole building, called a spatial building- the " V HOUSE ". The method is a development of a design process using new techniques of visualization through the Virtual Reality. It underlines the ties between making a design process more automatic and application of the new tools as the computer and the virtual reality. " ... The stage of V-HOUSE starts from house exterior appearance and the consumer represents his or her kansei words concerning exterior appearance . The system moves the inference engine referring to the databases and rules , and decides a candidate of house exterior appearance . If the consumer does not like the outcome and wishes to change it slightly , the system changes a part of the design using " Detail changer " . If the result is fit to the consumer's feeling and image , the consumer can move in the CG and walk through it in order to check all parts . Secondly , the next stage of " entrance " will start . the consumer inputs his or her kansei about entrance and if every thing is satisfied , the consumer walks through the CG of the entrance again . This procedure is continued to the final section of bathroom . ...

The main system has the interference engine and house design system . Kansei Engineering has the role of calculation of each part design , and Virtual World System serves the consumer to have the virtual experience in the virtual world concerning the kansei design " **[6]** .

The methods mentioned before display information and knowledge leading to a conclusion that architectonic design with the aid of a computer is not only a design as such, but new automatic possibilities, new perception and also new possibilities in human- computer communication. Given such context, one can hardly agree on the traditional form of record and design of building objects. The new approach to the problem is the idea of the " Holohouse Technique ".

4. New architectonic design technique- " Holohouse Technique "

So far, the majority of the design concepts with the aid of a computer reflect more or less the traditional " drawing board " rules of architectonic design processing. Even when a part of a design is made with a virtual model, there exist some assumptions about the discrepancy between technical design processing and visualization of a building object. There are, of course, some methods of processing of the architectonic design as a virtual model, being a base for both a technical design and visualization. Most often, however, virtual models are carried out solely for object visualization, while all the technical documentation is made in the traditional " drawing board " approach, even if with the aid of a computer. As mentioned before, many engineers have been "cheating " the software and its capability for many years, working in the virtual reality and, having prepared a virtual 3D model, transferring to particular 2D projections to accommodate the design to the technical record standards. It is obvious that using a computer for traditional technical documentation does not take advantage of its true abilities, and treating it as an

" electronic drawing board " is a total misunderstanding of its idea and the computer techno- revolution initiated long ago. Surely, there are some professional programs for virtual reality visualization, as 3D Studio MAX or programs for graphic pictures processing, but there is still a lack of professional- level programs combining technical documentation processing and visualization at logic level, not pure representation. The formerly mentioned problem of using a computer for preparation of traditional documentation should be regarded as a transition period, still to be improved as for the prospective benefits of the idea of a computer, as well as the demands of architects already working in the virtual reality.

" Holohouse Technique " is the idea of a new technology aimed at architectonic design as an architectonic vision imaging thanks to the computer virtual reality: an electronic model of a building displayed with the aid of the virtual and holographic model that is both the documentation and the technical prototype of the building object. Such technology contains all of the so-far technical documentation requirements, this is: of architecture, construction, branch designs, as well as enables presentation of the interior design along with its equipment. The " Holohouse Technique " is the idea that greatly changes designing as such, and technical documentation that is simultaneously a visualization and, partly, a computer display or interface.

The following changes in perception while space processing (i.e. building objects design) are expected on introduction of the new tool:
- Architectonic design will be a processing of a 3D space, represented by a holographic model of a building.
- Architectonic design is a physical construction of a building object in a natural scale computer virtual reality. It means everything is exactly the same as in reality at the building site. The only difference is that it is actually in the computer virtual reality and in the holographic picture of the CAD program interface.
- The building documentation is an electronic model (in a given scale) as an electronic prototype of the building object.
- Correction of the design vision in three dimensions is done with the designer's sight, hand/palm gestures and the holographic model in a given scale.
- The virtual and holographic design is the spatial technical documentation of the constructed building, is also its visualization and a base for optimization of the building process.

The idea of the " Holohouse Technique " opens a way to making a building process more automatic by means of so called "characteristic points in the holographic model ". For the changes to come, however, some crucial improvements are necessary in construction of building objects. A major change will happen to staking out process, as there will be no need to mark outside foundation lines, the laser position coordinates will be enough for a holographic image that will show where the parts of the building are located.

The construction of the building object will be carried out accordingly to the holographic mist model, without any further calculations or measurements, judging solely by one's own sight that will tell the location of the item, its size, color, structure. The construction or installation descriptions may be displayed by e.g. touching with a hand in the so-called characteristic point of the holographic picture. The use of such technology will change working conditions at the building site completely, as well as accuracy of construction accordingly to its design. The use of characteristic points in a holographic image along with changes in the construction- the way of assembling of building objects, will allow making the process almost completely automatic, as in the case of car assembly lines.

5. Conclusion

Application of the virtual and holographic reality in architectonic design not only opens new possibilities, but sets new standards to the jobs of design and construction. On one hand, the "Holohouse Technique" is a continuation to processing of a design as a spherical record through a virtual and holographic model. It also sets new standards in technical documentation as in a spatial virtual and holographic model record. On the other hand, it reaches for very basic rules of the architectonic design dating back to thousands years ago, this is, to preparing a model of a building object. Both in the technology and the design spheres, the " Holohouse Technique " will use all major advantages of human perception (e.g. human sight), as well as of the computer abilities (e.g. the ability of processing logic algorithms). Is such technology going to revolutionize building sites of today and to make constructing of buildings more automatic? Is it really going to replace human builders with robots? It is actually a sort of a rhetorical question. It surely is. The only real question here is when- it is just the matter of time. The " Holohouse Technique " will be helpful in a design teaching and training process in addition to doing well in practice. It may also find its place as a bridge between architectonic technology and art in modeling abstract spheres, which is very important to both. The idea of spherical design, spatial documentation and holographic imaging at a building site is already here.
It is the " Holohouse Technique ".

References

Michałowski K. (1986) . „ Nie tylko piramidy... Sztuka dawnego Egiptu " Warszawa , Wiedza Powszechna ,136.
Graphisoft Retrieved March 02 , 2000 ,from the World Wide Web http://www.archicad.pl/archicad.html
Bullinger H., Breining R., Bauer W., Virtual Prototyping – State of the Art in Product Design Fraunhofer Institute for Industrial Engineering (Fraunhofer IAO) Retrieved July 22 , 2000 , from the World Wide Web http://www.vr.iao.fhg.de
Fraunhofer Institute for Industrial Engineering Retrieved February 02, 2000, from the World Wide Web http://www.vr.iao.fhg.de/boule/technologydesign.html
Lucente M. „DreamSpace : Natural interaction" Retrieved January 22 , 2000 , from the World Wide Web http://www.research.ibm.com/natural/dreamspace/
Nagamashi M., Sakata Y., Imamoto Y. (1998). Proceedings of Sixth International Symposium on Human Factors in Organizational Design and Managment in Hague,The Netherlands , Elsevier Science Ltd., 404 .

Communication Evaluation in Multimedia: Metrics and Methodology

Francisco V. Cipolla-Ficarra

Human-Computer Interaction Lab.
F&F Multimedia Communic@tions Corp.
Cas. Pos. 60 - 24100 Bergamo, ITALY
Email: f_ficarra@libero.it

Abstract

We introduce a set of metrics and methodology for the analysis in order to increase the communication quality in multimedia/hypermedia system. Thus we selected a multimedia application primarily because of its unique ability to use and merge different media types and narrative devices as appropriate for achieving more than one communication objective. Our goal contained three distinct objectives. The first one, to introduce a new vision of the content evaluation for multimedia system: communication. The second, to present the heuristic evaluation of the transparency of meaning in multimedia systems. The third, to present the principal results of the intersection between Metrics for the Communication Evaluation in Multimedia (MECEM) and a Methodology for Heuristic Evaluation in Multimedia (MEHEM).

1. Introduction

During the 1990s, hundreds of usability-oriented methods were developed in all corners of the world, with the aim of increasing usability. Today, we need a set of the metrics for communication evaluation in multimedia "off-line" and "on-line".

The user interface is understood to be the component that mediates between the user and the application program, translating the user's demands into an acceptable size for application and the answers of the program into a language which the user can understand. That is, it is a "mechanism" through which a dialogue between the system and the user is established.

In our case, the aforesaid "mechanism" consists of four categories: content, presentation, dynamism and structure (Cipolla-Ficarra, 1997b). By presentation is understood the visual part of the multimedia systems and how their elements are distributed on the frame; the content is the hypertextual information that is stored in the hyperbase; the structure is the organisation of the information and the access to it; and dynamism is the navigation of the user through the different nodes. Furthermore, communication in multimedia is the intersection between content, presentation, dynamism, structure and user's.

2. Interdisciplinary evaluation

Most concepts about methods and techniques, which are at present applied to the assessment of the systems, derive mainly from the human-computer interface (Raskin, 2000), software engineering (Pressman, 2000; Fenton & Lawrence, 1997) and usability (Nielsen & Mack, 1994). From present research, the first theoretical framework about the studied subject has been created using concepts stemming from software engineering, the human-computer interface and the design models for hypermedia/multimedia systems. Besides, several of these concepts have been adapted because of the interdisciplinarity existing at the moment of carrying out the assessment of a multimedia system.

Another interesting field for the present theoretical framework are the design models for multimedia/hypermedia systems. There are several design models, which can be classified according to their use in the following way: hypermedia, hypertext, multimedia documents and database for multimedia systems.

The suggested method belongs to the category of usability assessment techniques, which is named in a simplified way inspection. This inspection is located more precisely in the subcategory heuristic inspection (Nielsen & Mack, 1994). Heuristic inspection does not require final users because one works with experienced evaluators. This kind of inspection is based on the definition of a set of usability attributes named in a simple way heuristic. It is through these heuristics that usability can be fragmented for analysis.

The main existing models in the multimedia/hypermedia context are AHM (Amsterdam Hypermedia Model); Dexter; HDM (Hypermedia Design Model); OOHDM (Object Oriented Hypermedia Design Model), etc. A group of primitives has been determined which serve to detect the first failings in the multimedia system (Cipolla-Ficarra, 1999).

3. The heuristic evaluator for multimedia systems

Together with the suggested methodology, the profile of a new kind of professional to carry out the assessment has also been added. This professional is called heuristic evaluator for multimedia/systems (Cipolla-Ficarra, 1997a; 1998). With the word heuristic one refers to the method used for the assessment of usability. The heuristic method is made up of several rules and techniques among which is observation. Among the several kinds of observation, direct and structured observation is the main technique in use. It is a direct observation because between the interface and the heuristic evaluator at the moment of carrying out the assessment, no other person or technological resource is involved for the measurement. It is a structured observation because there is a set of stages to be followed inside a procedure, which have been previously defined by the evaluator. This technique has proved the reliability of the obtained results and has cut down costs. The costs cut is due to the fact that it does not require:

- A laboratory to carry out the evaluation.
- Equipment and maintenance of specialized devices.
- A group of evaluators or a sample of users who interact with the system at the moment of the evaluation.
- Previous training of the users and eventual aids of the evaluators.

Furthermore, the evaluation can be carried out from the different departments which participate in the creation of the multimedia system, such as design, programming, implementation, etc., thus immediately communicating the discovered failings.

The profession of evaluator or heuristic analyzer has been included in the worldwide spread multimedia systems. It is a new profession which requires specialists with knowledge and experiences in the field of the multimedia/hypemedia systems. It is located on the borderline between formal sciences and factic sciences (Bunge, 1974). The assessment work of this professional can be carried out in an isolated way or in a team. Teamwork allows verfication of the results obtained in previous evaluations with other experts of the evaluation sector. The advantage of the method is that every expert can work on a common share of the systems, but also on different aspects, and of a comparable kind, such as an access and navigation analysis in the guided connections of the entities.

4. Metrics

A set of heuristic attributes which concern the quality of the multimedia systems has been adapted and generated. These are the first quality attributes which have been used in four design categories which are interrelated among themselves in a bidirectional way. During the research these attributes have been constantly modified in order to reach a better definition and relationship between them.

Quality heuristic attributes have been divided into measurement factors which have been named "metrics" (Fenton & Lawrence, 1997). In these metrics descriptive statistics have been used to quantify several aspects of heuristic evaluationn such as the total of the system components or the total of detected failings. The results of descriptive statistics have allowed graphic representation of components and failings detected in the system in order to simplify the understanding process for those involved in the evaluation, such as designers, computer makers, clients, final users, etc. Graphics have also been included for the system components to be evaluated, thereby speeding up not only the evaluation task, but also knowing the scope of the component to be examined inside the system and whether it is in the whole structure or only in a part of it (Cipolla-Ficarra, 1999).

These are metrics where a special set of primitives and descriptive statistics are involved: universality, inference, empathy, panchronics, competence, control of fruition, isomorphism, motivation, naturalness of metaphor, phatic function, transparency of meaning, accessibility, self-evidence, richness, consistency, predictability, orientation and reuse. The interested reader can refer to the bibliography in (Cipolla-Ficarra, 1999; 1997a; 2000) for a complete description.

The set of the presented metrics has several advantages, among which are stressed:

- Carrying out the evaluation of the whole system or part of it establishing two modes of evaluation: partial and total. In the partial mode one works with a meaningful sample of the system to be evaluated. The total mode encloses all the primitives or constituents of the system. In both modes the reliability of the results is similar.
- Checking the results obtained in the partial evaluationn through a total evaluation.
- Changing the order of the stages inside the evaluation procedure.
- Detecting the scope of the failings in the system structure, that is, whether it concerns the whole system or just some areas.
- Adjusting or changing the metrics in view of the presence or absence of the components existing in the system (i.e., guided tours, links, index links, nodes, and frames).
- Previously determining the dimension of the system to be evaluated. This makes it possible to work on a budget on the likely cost of the evaluation.

With the results of all these metrics it is possible to graphically illustrate all the detected failings. In this way, in the design stage of the system, all the participants have a visual representation which assembles the failings of the system, thereby making it possible to determine responsibilities, as well as the necessary actions to solve them in an easy and fast way.

A metrics defined as belonging to the binary presence kind has been suggested. These metrics have been carried out through the table for heuristic evaluation aimed at the design of a multimedia/hypermedia system (Cipolla-Ficarra, 1997a). The table that has been created with the present research is the first in the multimedia sector and among its main qualities it can be mentioned that it has allowed the detection of the first usability mistakes and has helped to establish a close dimension of the system to be analyzed, through the totals registered in some components. Therefore, the table has made possible the quick realization of a previous budget on the evaluation cost. This table has constantly been modified. Besides, the components in the table have been classified by design categories, which has facilitated results that are more detailed. These kinds of results in the design stage have highlighted the cause for component failure in the making of the multimedia system.

5. Methodology and metrics: Intersection

The work made in the investigation has focused on the creation of a method and metrics for the communications evaluation in a multimedia system. Below are summaries of the intersection between MEHEM and MECEM:

- The methodology (MEHEM) is compatible with other evaluation techniques. The results reached with the suggested methodology can be verified through the use of other techniques or methods, which require a usability laboratory such as in empirical evaluation. In this case, the empirical evaluator can establish, in a precise way in a laboratory, a set of tasks to be accomplished by the users. These tasks are based on the results obtained in a previous evaluation of the system with the suggested methodology.
- A method that involves a two-stage procedure has been established which is to be combined with other evaluation techniques. Firstly, the heuristic evaluator detects the mistakes or errors of the system. With the errors a new series of tasks or actions to be developed by a group of users is elaborated. In the second stage a new evaluation of the system is carried out, where all users participate and the existence of the mistakes detected in the first stage is verified. The purpose of employing the users is to control the results achieved in the first stage.

- The precision of the evaluation is reached in two stages. The first through the knowledge the evaluator has of the failings that have been detected in the system. The second through the verification of the failings in the laboratory by a group of users by carrying out a set of tasks in face of the evaluated system. The failings in the empirical evaluation are registered in the form of a list. The advantage of having a previous list of the potential failings in the system makes the quality of the empirical evaluation results be more efficient at a lower cost.
- It has been determined through the analysis of multimedia systems the existence of many failings in the communication of global speed commercial systems (Cipolla-Ficarra, 1998). The research universe has focused on multimedia systems. These systems were chosen at random and classified according to their use and content. The advantage of the classification has been shown at the moment of coming to conclusions from the mistakes in relation to the potential users (children, youngsters, adults) and of the purposes in the use of the systems (teaching, entertainment, consulting, information, etc.) (Reeves & Nass, 1996). Results interpretation in terms of the application goals and the user's requirements, with the purpose of identifying what is objectively positive or negative for the real users.
- The functions of the differents nodes have been identified, such as the hyperbase nodes and the nodes of access to the structure, and the distribution of information on the screen has been examined. A new primitive named "main screen" has been introduced for interface analysis in the multimedia systems.

6. Example: Transparency of meaning

Analyses the usage mainly of terms (also, images and sounds related to the words) of the interface that do not cause ambiguities between contents and expression. The elements that make up the interface are all correlated, so accomplishing a stressing or meaning-reinforcement function. CD-ROM "Renoir" (Istituto Geografico De Agostini, 2001) is an excellente example (see Annex 1).

6.1 Establish

The coherence between image and text; the function of reinforcement or explanation of the elements that accompany the metaphors; the correlation of dynamic elements that accompany the text; the concord of the titles with the pre-established order of the textual nodes divided into "sections".

6.2 Set of the hypermedia primitives for the evaluation

Evaluation extension and desing categories are content (C), presentation (P), structure (S) and dynamism (D).

Primitives	Evaluation Extension	Design Categories
Collection	In-the-large	S
Element	In-the-small	S – C – P
Guided tour	In-the-large	D
Node	In-the-small	S
Frame	In-the-small	S – C – D
Node Type	In-the-small	S
Entities	In-the-small	S

6.3 Procedure

- To choose the first frames of the different entities and examine whether the different elements of the images (dynamic and static) and the texts (statics) on screen maintain a relationship between meaning and significance. For example, a sign. One of the essential correlates of the sign as defined by Saussure. For him, a sign is a correlation between a meaning and significance (Nöth, 1995).
- To establish the total metaphors in the application and the reinforcement functions of the application itself, with dynamic and static elements at the beginning of the entities (to check the first fifteen nodes).
- To choose at random five guided links and check whether the audio or the locution of the different nodes correlate (similar to the karaoke system). In an entity, to check fifteen nodes and evaluate whether the contents of the text are consistent with the title and with the order of the titles at the beginning of the node.

7. Conclusions

The intersection between MECEM and MEHEM offers a useful tool to facilitate the utilization of information technology for the design and development of multimedia systems for the support of interactive communication. This paper has introduced some elementary metrics for evaluation of the communication needed to facilitate the implementation of different types of multimedia contents. Metrics can represent a tool that, when properly used, enhances management control over the development process and product quality. These metrics are very important because the access of the multimedia information today is more or less global. Evidently, we need distinghuish between two heuristic evaluation multimedia system: communication, or high-level, evaluation and usability, or low-level, evaluation. Usability is not a synonym of the communication.

References
Bunge, M. (1974). *Treatise on Basic Philosophy: Semantics*. Dordrecht: Reidel.
Cipolla-Ficarra, F. (1997a). Evaluation of Multimedia Components. *Proc. IEEE Multimedia Systems '97*, Ottawa (pp. 557-564).
Cipolla-Ficarra, F. (1997b). Method and Techniques for the Evaluation of Multimedia Applications. *Human-Computer International (HCI '97)*, San Francisco: Elsevier.
Cipolla-Ficarra, F. (1998). Explorative Navigation for Multimedia: Results of a Heuristic Evaluation. *Proceed. International Conference on Information Systems Analysis and Synthesis - ISAS '98*, Orlando (pp. 140-147).
Cipolla-Ficarra, F. (1999). MEHEM: A Methodology for Heuristic Evaluation in Multimedia". Proceed. *Sixth International Conference on Distributed Multimedia Systems (DMS'99), IFIP*. Aizu: Elsevier.
Cipolla-Ficarra, F. (2000). MECEM: Metrics for the Communications Evaluation in Multimedia. *International Conference on Information Systems Analysis and Synthesis – ISAS'00*, Orlando (pp. 23-31).
Fenton, N., Lawrence, S. (1997). *Software Metrics: A Rigorous & Practical Approach*. Cambridge: ITP.
Istituto Geografico De Agostini (2001). Galleria d'arte: Renoir CD-ROM (Novara).
Nielsen, J., Mack, R. (1994). *Usability Inspection Methods*. New York: Willey.
Nöth, W. (1995). *Handbook of Semiotics*. Indiana: Indiana Universitary Press.
Pressman, R. (2000). *Software Engineering, A Practitioner's Approach*. New York: McGraw-Hill.
Raskin, J. (2000). *The Humane Interface*. New York: Addison Wesley.
Reeves, B., Nass, C. (1996). *The Media Equation: How People Treat Computers, Television and New Media Like Real People and Places*. Cambridge: CSLI Publications.

Annex 1: Transparency of meaning

This is an excellent model for transparency of meaning.
For example, on the frame principal of CD-ROM "Renoir" we can observe the coherence between image and text.

An International System for Quality Labelling from a User Perspective

Kjell Fransson[1], Jan Rudling[2]

1. Vikbyvägen 9, SE-181 43 Lidingö, Sweden. bok.fransson@telia.com
2. TCO Development, SE-114 94 Stockholm, Sweden. jan.rudling@tco.se

Abstract

TCO'99 is an internationally successful environmental and quality system for the certification of office equipment, in particular modern IT equipment. The system requirements are based on user experience and knowledge, research results and expert opinion, and are adaptable to keep pace with rapid progress in development. The requirements are intended to provide the maximum number of users, despite widely differing employment circumstances, the capability of participating in working life.

1. Background

As far back as the early 1980s The Swedish Confederation of Professional Employees (TCO) foresaw that personal computers would become the principal tools for the 1.3 million members of its associated unions. At the same time, the first disturbing reports began to appear of a possible link between magnetic fields, pregnancy problems and foetal injury. Later, terms covering such aspects as electromagnetic hypersensitivity associated with the work in front of computer displays, computer stress and physical strain injuries began to appear. Members wanted to know more about health risks that could arise from the new technology, and above all wanted the answer to the question "What could be done to minimise the risk?"
At that time, there were no reliable methods available to measure how good a display should be, nor to check how much risk was associated with the new technology. There is still very little concrete scientific knowledge about the possible risks. Instead, TCO, together with user representatives, experts and scientists, formulated a set of demands directly from a user perspective – and this time TCO chose to put these demands squarely to the market. Thus began a completely new way for trade union organisations to apply influence. Hitherto, a very time-consuming method had been used whereby the authorities, even in the best cases, took several years to devise and issue regulations to prevent operations that could pose harm to workers' health. TCO assumed that ambitious manufacturers would appreciate close co-operation to make good products – and it was thus important for them to have access to the fund of knowledge and experience acquired by a large user group. At the same time, users wanted good quality products that would maintain their professional competence.

2. The TCO quality and environmental labelling system

TCO'92 was the first international quality labelling scheme to appear on the market. Introduced in 1992, it contained emission and energy requirements for office computer displays. Since none of these displays were being manufactured in Sweden, the requirements and labelling system had to be of an international character. The labelling scheme developed into TCO'95, which added such areas as ergonomics and ecology. Keyboards and computer system units were new types of product that could be certified. The latest version of the scheme is TCO'99, which also includes the most stringent requirements so far for this type of product.

Figure 1. The three labels for TCO'92, TCO'95 and TCO'99

There has been a subtle change in emphasis between TCO'92 and TCO'99. TCO'92 was intended as a method to protect users from difficulties and injuries, whilst the requirements embodied in TCO99 are now development-driven quality requirements for the products in this branch. These quality requirements have in turn become a challenge to be met by only the most quality-seeking international companies. Today all the world's leading display manufacturers, with a few exceptions, have entered into certification contracts with TCO Development.

TCO'99 is a revision of TCO'95, insofar as the requirements continue to cover CRT displays, flat screens, system units, portable computers and keyboards as well as alternative keyboard designs. The environmental areas are unchanged, but a large number of additional restrictions are added. The demands have been further tightened up in TCO'99. Advances in technology, working life and knowledge development have necessitated the new requirements.

Because the quality requirements are very high, development has led to differences in the demands on CRT displays and their flat screen equivalents, and these in turn are not the same as those for portable computers. The requirements concerning alternative designs of keyboard in TCO'99 can be tested using methods that include practical user tests, which are different from those carried out on traditionally designed keyboards.

Today, TCO labelling is a global standard. No other labelling scheme has had such a powerful effect on information technology, and the Swedish requirements now lead the world. There is a strong increase in international interest. The reason for this is that more and more companies and their employees are becoming aware that high productivity and the quality of working life are setting high demands on such tools as displays and keyboards. There is also an increasing consciousness that the local and global environments are threatened. New technology often has only a limited life-span, and in just a few years a large proportion of today's computer products will be scrapped. The choice of environmentally-labelled products lightens the load for both people and Nature.

The possession of certification means that the manufacturer guarantees that every example of the certified product model will meet the requirements specified in the relevant TCO certification scheme. For its part, TCO Development undertakes to check random samples bought on the open market, to ensure that the manufacturers are still meeting the demands. This responsibility applies to the entire international market.

3. Office equipment, furniture and mobile telephones

- The current TCO'99 certification scheme include the possibility of certifying printers, faxes and copiers.
- A new system for mobile telephones is at present (February 2001) out on international review.
- Various types of office furniture are being investigated, as are different kinds of data entry systems, such as mice and trackballs.

4. Experience gained from working life as compared with research knowledge

The knowledge and experience garnered from working with IT tools by various professional groups represented by the TCO association is very wide, and often covers a long time spent in the job. This knowledge accumulated by users is in fact greater than the scientific knowledge base in this area. Above all, the time factor is important in a branch where developments are proceeding so rapidly. User experience is built up over a long time, and expressed in the work place long before signals arrive in the research field that there is an area that needs to be analysed and structured. There must also be sufficient interest for a researcher to have the motivation to spend a lot of time on a particular project, enough to solve the problem. It may take years from a problem arising in the work place until researchers can begin to provide preliminary results, and then even more time passes before appropriate measures can be introduced. A typical time frame is 5 to 10 years between the creation of a problem and the discovery of a solution. One example of this is work with a keyboard or a mouse, where everyone is aware of the consequences but researchers have yet to clearly define what dangers are inherent for users.

The TCO certification requirements are based on the experience and knowledge that the members have gained through their daily work. Hence the requirements that are embodied in the TCO schemes are even well ahead of standardisation. Obviously, if proven experience and accumulated knowledge from working life conflict with scientific knowledge, due consideration is applied in the certification process.

Technical developments today are mainly influenced by the powerful development of techniques and technology for handling information. From using computers to speed calculations, via word and text processing, layout and graphics, databases, computer assistance in complex systems and up to today's offerings of digitised music and films, this has been a steadily widening development that now affects everyone in their working life. The professional becomes more productive and effective in his or her job, while computers pose new human challenges which bring stress and problems in many areas.

At the same time the reverse is true, in that all jobs are affected by this development and everyone who works must use their IT equipment efficiently. In other words, IT equipment must be adapted to the users who are working. This applies across the spectrum, to users with different tasks, education, ethic backgrounds, gender, and even those who may have temporary or permanent handicaps of various degrees.

5. The requirements are for everyone in working life

The central aim of TCO certification is to provide all TCO members with better tools to use in their professions as office workers. This means that the majority of members should regard IT equipment characteristics as improving. The value of a task has much to do with such intangible aspects as professional pride, a sense of community and the possibility of developing one's own competence while doing a better job. Having a job – a meaningful task– gives an individual an identity and is in the best case a channel for creativity, commitment and participation.

In Sweden and Europe we have statistics to show that about 30 % of all workers will, at some time or another, suffer one or more temporary or permanent functional difficulties. It is therefore vital that tools are designed to permit their use by as many people as possible. The design of IT equipment must most definitely not be a reason for excluding certain people from working life.

The most common strain injury in Sweden at present as the result of long and intensive work in front of a computer display is pain in the upper spine and neck, shoulders and arms, along with itching or irritated eyes. The chief group of sufferers consists of secretaries, clerical workers, data input staff, computer specialists and administrators in public authorities. More recently this has extended to include, because of major reorganisation of their work, customs officers, taxation clerks and social service workers. The pace of work for all these people has increased considerably during the last 5 years. Affected areas also include women in financial jobs, along with public administrations, defence, civil authorities, etc.

One may also say that IT equipment that is designed with functionally handicapped people in mind will probably benefit most users in general. However, when the hindrance is too great, or the task too specific, TCO switches over from its certification system requirements to providing good advice to individuals who themselves may introduce or strive to obtain suitable conditions at work.

6. Main requirements in the TCO quality system

The most important requirements for the various product groups are:

Ergonomics: Visual ergonomic characteristics of displays, for example even luminance, good contrast, minimal reflections from the housing, colour fidelity, and the provision of a high refresh rate. In system units the principal regulation concerns noise level. Traditional keyboards have requirements in accordance with classical standards. Alternative designs of keyboard can be tested by typical users with the aim of reducing ergonomic strain on such users.

Emissions: TCO has traditionally included under the heading of emissions the electric and magnetic fields generated by electrical equipment. With the addition of mobile telephones, this concept has been extended to electromagnetic fields (at radio frequencies). In all cases, the aim is to reduce user exposure to as low levels as possible and to prevent different items of office equipment from affecting each other. Chemical emissions in the work environment have so far been kept down by the prohibition of certain chemicals that are known to cause problems, such as brominated flame retardants in plastics.

Energy: Here there are requirements for energy saving functions, which are always related to short restart times in order to reduce stress at work and also to ensure that the energy saving mode will actually be used and not regarded as a nuisance and switched out.

Ecology: Prohibitions on using solvents in production and on heavy metal additives in detail parts. Since 1995, brominated and chlorinated flame retardants have been forbidden in plastics. The capability of recycling of equipment, and the requirement for companies to implement approved environmental practices.

These requirements enable the user base of everyday IT equipment to be broadened. People who have allergies, poor vision and the elderly are given the chance to use it. Good visual ergonomic quality in a display permits increased range of movement at the work place, countering strain problems.

7. The future and conclusions

The TCO-labelling is a quality and environmental labelling of office equipments used in working life. Children, young people and pensioners are also users of these equipments and have their own specific requirements but TCO can not act in their places as their experiences and needs are not mapped and known by TCO. However these groups can derive advantage from the high quality requirements needed in working life and presented in TCO-labelling.

The requirement to provide greater possibilities for temporarily and permanently disabled users is more developed in respect of certification of printers, faxes and copiers. The goal of allowing everyone to take part in working life in accordance with their capabilities has gained more ground in the future certification systems for mobile telephones, office furniture and data entry devices.

The future development can be followed at TCOs web page: www.tcodevelopment.com

Learning Universal Access Guidelines by Solving HCI Problems

Elizabeth Furtado, Vasco Furtado

Universidade de Fortaleza
Washington Soares, 1321 - Fortaleza - CE - 60455770 – Brasil
{Elizabet, Vasco}@feq.unifor.br

Abstract

This work addresses the problem of learning guidelines to both (i) design user interfaces and (ii) evaluates the quality of interfaces, for universal access systems. We have developed a collaborative environment for helping the designer of user interfaces in solving problems of accessibility, usability and acceptability. The designer accesses the environment as a tool for assistance, aiming at solving difficulties he encounters when the user is not satisfied with the system with which the latter is interacting. During the problem solving process, the environment provides the designer with assistance by showing graphic scenarios, texts and by retrieving similar past solution cases. Moreover, the designer has a collaborative support that allows him to interact, in a synchronous or asynchronous way, with other designers.

1. Introduction

Many research works have studied ways to create and evaluate user interfaces which facilitate and promote universal access to information systems. These works contribute to define tools (Grammenos, Akoumianakis & Stephanidis, 2000), methods (Furtado, 1999) and guidelines (Bastien & Scapin, 1993) (Bodart & Vanderdonckt, 1993), which should be considered in Human Computer Interaction (HCI) design. Guidelines can provide designers with some assistance by helping them to design user interfaces (UI) which guarantee a minimal threshold of usability (Vanderdonckt, 1999a) and are more adapted and tailored to contextual needs (i.e., user population needs, physical environment constraints, and task needs). However, the integration of guidelines into the development of usable systems has shown to pose problems for the selection, validity and applicability (Vanderdonckt, 1999b). This work addresses the problem of learning guidelines to design UI for universal access systems.

We describe a collaborative learning environment of universal access guidelines for helping the designer in solving accessibility, usability and acceptability problems of interactive systems. This environment, called (Collaboration and Adaptation of Didactics - CADI) (Furtado, Vieira, Lincoln & Maia, 2000), makes it possible for the designer to reflect by himself and with others, the applicability of guidelines in a practical way. CADI is based on educational theories, which claim that learning must be a collaborative, constructive process (Shank, 1994) through problem solving in order to develop the users' capacity for reflection (Dewey, 1959) through practice (Schön, 1987).

In CADI, designers should be able to learn guidelines and solve problems through the collaborative creating and/or visualization of interactive examples and scenarios. The interactive examples expressed through interactive objects of a UI show compliance or violation cases in information presentation and dialog design. A scenario is a prototype situation representing HCI problems or solutions. A scenario about a usability problem could be, for instance, a picture illustrating a user entering data and, accidentally, selecting a function, which closes the data input window. Mainly, scenarios are used to express human factors (e.g., users' characteristics, position and expression, tasks and context). These factors are crucial for knowing which guidelines to use and why to use them. For example, in noisy environments such as shopping malls during holiday time, we must create a UI, that works well for people who cannot hear well or not at all. Thus, the urgent need for modeling and graphically recovering scenarios has led us to develop a scenario editor. In addition, the need for supporting the collaborative work has led us to develop a multimedia application for the World Wide Web.

The further sections of this paper will present: an overview of guidelines for user interfaces for all, the organization and representation of problems of HCI in scenarios; the CADI description; the guideline learning method, and the final conclusions.

2. Guidelines for User Interfaces for All

In the globalization era, universal interfaces (usable by a diverse population with different abilities, skills, requirements, and preferences) play a vital role in improving the communication between a user and any interactive software accessible anytime and anywhere (Stephanidis, 20001). The integration of guidelines into design or the evaluation of universal interfaces is one of the possible ways to improve accessibility, usability and acceptability of a software.

According to Benyon, Crerar & Wilkinson (2001), the accessibility refers to: i) accessibility to equipment (in sufficient quantity, in appropriate places, at convenient times), and ii) the operational suitability of both hardware and software for any potential user.

The usability refers to the quality of the interaction in terms of number of errors made, for example. Many problems with the usability of interfaces comes from the fact that they don't fit into user's needs. This fact is partly because, many human factors are not taken into account in the development methods of interactive systems. User-centered design methods (Nielsen, 1993) and tools which work in a proactive way, define a collection of guidelines for involving the users in the development of the interactive system. However, these methods do not specify how designers can treat different user groups whose requirements are not known a priori. Actually, its difficult for designers to define a profile to any kind of user which the system could have. It follows, in some cases, that a evaluation process of the system's UI is still necessary. The acceptability refers to fitness for purpose in the context of use, and also covers attitudinal factors, such as personal preferences about metaphors that contribute to users "taking to" an artifact, or not. Marcus (2001) describes the general guidelines for globalization of use of metaphors in the following: its necessary to check for hidden miscommunication and misunderstanding and determine the optimal minimum number of concepts and primary images to meet target user needs.

In our work, the accessibility, usability and acceptability problems caused by violation guidelines in information presentation and dialog design are organized in accordance with the concepts of use cases and scenarios.

3. Scenarios and Use Cases on HCI

Scenario represents a situation experienced by a user when performing a specific operation with a specific goal in mind. In order to create a scenario, we must specify all the variables (such as the context of use as class room/ head office/ laboratory/ home/ bank; the state of context of use, as cleanliness/lighting/ambient noise, and so on) of the situations occurring in the system. Each scenario is documented with text describing its content. To increase the effectiveness of representation and recovery of scenarios, its necessary to associate them to general situations which a user interacts with an interactive system, such as "entering data" and "visualizing data". These general situations are associated with the notion of use case, described in the Unified Modeling Language [UML]. Scenarios are therefore variations (instances) of a use case type happening on HCI.

The use cases are organized in a diagram. A use case may be split up into several specific use cases existing in a relationship of generalization, where a specific use case is a kind of a general use case. A use case may be linked to others from anywhere in the hierarchy. In this situation, there is the relationship *use*, which means that if A uses B then B can happen during the realization of A. Thus, the use case hierarchy described is not a strict one. In some level of the hierarchy, problem cases can be found. A problem case may be represented by a scenario, called a problem scenario. A problem scenario about usability of the use case "visualizing data" could be a picture showing a user visualizing data but without realizing what the most relevant information on the screen is. In relation to problem use cases, there is one or several analysis cases. An analysis cases is a possible cause of the problem case associated to and can be a problem case of another user case. For the problem case described previously, if the most important information is already in the top left corner of a window, three analysis cases can be presented: (i) the user is tired; (ii) the window is not easily visible; (iii) the user has visual impairments. An analysis cases can also be expressed by a scenario and may have one, several or no solution at all, expressed by a text about guidelines, for instance.

4. Description of CADI

Figure 1 depicts the architecture of CADI including five main components:

- The assistance module, which coordinates the recovering of assistance (scenarios, texts or interactive examples) which must be presented to the designer.
- The TELE module, which allows the cooperation and communication over the web by chat and the sharing of any application, such as an interface editor.
- The scenario editor, with which the designer can appreciate already existing scenarios and build the new ones.
- The user interface (UI), which is used by the designer to interact with the environment.
- The internet servers, which maintain the database and the knowledge base. The database contains texts and interactive examples about guidelines. The knowledge base contains the frequently encountered HCI problems.

The cooperation between the main components of the CADI is the following: the designer requests the analysis of a problem case[1]. The UI component, during this analysis, gives the designer some assistance. The designer is assisted by interactive examples, texts and/or scenarios, which are recovered from the servers by the assistance module. If the designer wants to communicate with others designers whom have passed by the similar problem cases and creates interface or scenario together as a possible solution for the identified problem, the UI component calls TELE module. The designer solution is stored in his local database and validated by the HCI experts before being stored in server.

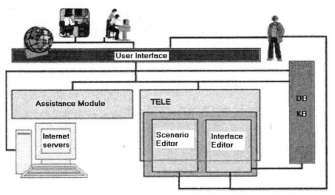

Figure 1 – CADI architecture

5. Learning Guidelines in CADI

Using CADI the designer must at first identify his problem. When identifying the problem, the designer chooses a use case which defines the context of the problem and subsequently the problem itself by selecting a problem case. Then the designer gives some general information about the user and context of use. At this moment the designer can create a scenario problem or see scenario problems already existents. When using a scenario, and through the thoughts and images associated with it, the designer can reflect better about the aspects of his problem. Once the problem case has been accurately defined, the assistance module starts to help the designer to solve his problem through the analysis cases retrieved. It should be pointed out that the recovery of the analysis cases is not based

---

[1] Problem cases which the designer must analyze can be suggested by CADI or defined by the designer. The first situation happens when the designer accesses the CADI to follow a course about HCI design. The second situation happens, when the designer has to evaluate the user interface in order to solve the problems the user encounters when he is not satisfied with his computational environment. In this paper, we will describe only the second situation.

solely on the choice of use cases, problem case and answers of the designer. Analysis cases are also retrieved by taking into account the relationship use between the use cases.

In order to illustrate the concepts presented we will analyze a situation considering the generic diagram of use cases of an user interacting in a computational environment (see Figure 2).This is a situation where the user doesn't know how to buy a product using an e-commerce system. The designer identifies this situation by selecting the use cases following: Interacting on the WEB, Navigating and Paying/buying, then by selecting a problem case related with orientation problems.

After the designer gives some general information to the system about the user misunderstanding in relation to the tasks to be executed to buy the products selected, such as, the user knows what to do and what he wants but he does not understand how to do. Through these information, the assistance module infers the problem is not exactly an orientation problem. Then this module investigates two possible analysis cases: *bad interpretation of metaphors* and *misunderstanding of feedback*, which were retrieved considering the relationships *use* of the use case *Paying/buying*. The use case *utilizing metaphors* holds general knowledge about metaphors (e.g., their ambiguity, concepts and images associated to) and also organizes problem scenarios of using of metaphors, which expectations were violated or anomalies were encountered. Depending on the designer´s preferences, the designer can see some problem scenarios, create another one and/or collaborate with someone, to help him to specify the general information described above associated to his problem. Though these specifications, the assistance module suggests to the designer that the user has a problem with the utilization of metaphors, because his experience with the trash icon induces him to doubt the use of the shopping basket icon. The user considers this icon ambiguous because, based on his experience, the action of picking the selected object up and putting it into an icon, means discarding something he doesn't want. In order to help the designer to solve this new problem identified, the assistance module suggests to him some texts and examples of guidelines about the choice of metaphors.

In this example we could show that the designer may initially choose a problem case thinking he knows the problem well and, after answering questions about related use cases, the assistance module may determine that the problem in question is not the problem selected, but a consequence of a problem related to the one selected. Such may happen because of the cross-contextual retrieving of cases which in its turn is due to the not strictly hierarchical nature of the case organization.

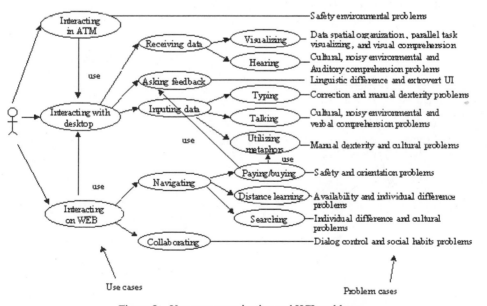

Figure 2 – Use case organization and HCI problems
A uses B means the problems of A can be caused by problems of B. ATM means Automatic Teller Machine.
Extrovert UI contains more words, a sense of humor, more sounds and a quicker pace of screen change compared with the introvert UI (Benyon, Crerar & Wilkinson, 2001).

580

6. Conclusion

The environment described here is considered essential for transporting the designer into a more sophisticated universe where he can explore all kinds of multimedia facilities for the solution of design problems about accessibility, usability and acceptability. The designer reflection about these problems, the possible consequences of his solution and the collaborative support of assistance help him learn universal access guidelines.

References
Bastien J.M.C., Scapin, D.L., (1993). *Ergonomic criteria for the evaluation of user interfaces*. INRIA, No.156.
Benyon, D., Crerar A., Wilkinson, S., (2001). Individual Differences and Inclusive Design. In Stephanidis C. (ed), *User Interfaces for All – Concepts, Methods, and Tools* (pp 21-46). Mahwah, NJ: Lawrence Erlbaum Associates, Inc. (ISBN 0-8058-2967-9)
Bodart, F., Vanderdonckt, J., (1993). Expressing Guidelines into an Ergonomical Style-Guide for highly Interactive Applications. In S. Ashlundn, K. Mullet, A. Henderson, EL Hollnagel, T. White (eds.). *Proceedings of InterCHI'93,* (pp. 35-36).
Dewey, J. (1959). *Como Pensamos*. São Paulo: Companhia Editora Nacional.
Furtado, E. (1997). *Mise en oeuvre d'une méthode de conception d'interfaces adaptatives pour des systèmes de supervision à partir des Spécifications Conceptuelles*. Thèse de doctorat. França.
Furtado, E. (1999). An Approach to improve design and usability of user interfaces for Supervision systems by using Human factors. In *Proceedings of HCI' 99*, Germany. (Vol 1, pp. 993-997).
Furtado, E., Vieira F., Lincoln F., Maia M. (2000). Auxílio à Solução de Problemas no Processo de Ensino Através de Cenários e do Contador de Estórias. In *Proceedings of SBIE'2000*, Brasil.
Grammenos, D., Akoumianakis D., Stephanidis C., (2000). Sherlock: A tool towards computer-aided usability inspection. In J. Vanderdonckt & C. Farenc (eds.), *Proceedings of TFWWG' 2000*, Biarritz, França (pp. 87-97). London: Springer-Verlag.
Marcus, A. (2001). International and Intercultural User Interfaces. In Stephanidis C. (ed), *User Interfaces for All – Concepts, Methods, and Tools* (pp 47-63). Mahwah, NJ: Lawrence Erlbaum Associates, Inc. (ISBN 0-8058-2967-9)
Nielsen, J. (1993). *Usability engineering*. Boston: Academic Press.
Schank, R., Kass, A.E, Riesbeck, C. (1994). *Inside Case-based Explanation*. Lawrence Erlbaum Associates, Institute for the learning Sciences.
Schön, D. (1987). *Educating the Reflective Practitioner*. New York: Jossey-Bass.
Sebillotte, S. (1994). From users' task knowledge to high-level interface specification. *International journal of human-computer interaction*, Vol 6(1), pp.1-15.
Stephanidis, C. (2001). User Interfaces for All. In Stephanidis C. (ed), *User Interfaces for All – Concepts, Methods, and Tools* . (pp 3-17). Mahwah, NJ: Lawrence Erlbaum Associates, Inc. (ISBN 0-8058-2967-9)
Vanderdonckt, J. (1999a). Automated Generation of an On-Line Guidelines Repository. In *Proceedings of HCI' 99*, Germany (Vol 1, pp. 998-1002).
Vanderdonckt, J. (1999b). Development Milestones Towards a Tool for Working with Guidelines. *Interacting with Computers*, Vol 12(2) pp 81-118.

Accessibility through standardization

Jan Gulliksen[a,b], Clemens Lutsch[c], Susan Harker[d],

[a] Department of HCI, Information Technology, Uppsala University, Uppsala, Sweden
[b] Center for user-oriented IT-design (CID), Royal Institute of Technology, Stockholm, Sweden
[c] Icon Medialab AG, Büro München, Germany
[d] Department of Human Sciences, Loughborough University, United Kingdom

1. Introduction

Accessibility and universal access have been gaining increasing attention recently due to the increasing recognition of the need to promote equal opportunities for all users of interactive systems. Growing dependence on information technology based products and services, both at work and at home, and for leisure and travel, means that ever increasing numbers of the population are users of the technology. Thus if the technology raises barriers for some users they will face potential exclusion from activities which have become routine within society. This has also been recognized by the International Organization for Standardization (ISO). The member bodies of ISO Technical Committee 159, SC4 (Ergonomics of Human-System Interaction) have adopted a work item on software accessibility that has resulted in the development of a Technical Specification, ISO TS 16071 Ergonomics of human-system interaction – guidance on software accessibility [1]. ISO standards are adopted according to the following three guiding principles; consensus, industry-wide and voluntary. ISO standards do therefore not represent the general truth, or state of the art, within the field, but rather a general agreement between the international member bodies participating in ISO on the topic in question.

2. Defining Accessibility

The purpose of ISO TS 16071 is to provide guidance to developers on designing human-computer interfaces that can be used with as high a level of accessibility as possible. Designing human-computer interactions to increase accessibility promotes increased effectiveness, efficiency, and satisfaction for people who have a wide variety of capabilities and preferences. Accessibility is therefore, according to ISOs definition of it, strongly related to the concept of usability as defined in ISO 9241 Part 11 – Guidance on usability [2]. However, it places the focus upon the specification of criteria which include the great potential diversity of user characteristics and needs. ISO TS 16071 defines **accessibility** as:

The usability of a product, service, environment or facility by people with the widest range of capabilities.

Usability is defined as "the extent to which a product can be used by specified users, to achieve specified goals, with effectiveness, efficiency and satisfaction, in a specified context of use" [2]. Through its relation to the definition of usability accessibility becomes a measurable entity, and subsequently developers could acquire the goal of increasing the level of accessibility of the products they develop rather than assessing if a product is accessible or not. ISO TS 16071 do not present methods or procedures for actually specifying metrics to identify specific levels of accessibility, just as ISO 9241 part 11 does not provide usability metrics. But, by promoting such a definition we hope to encourage research and development of methods for measurable accessibility.

ISO TS 16071 provides specific guidelines to take account of software attributes to increase accessibility by ensuring that both platforms and applications offer support for users with the widest possible range of capabilities. It does not specifically address the design of assistive technologies but strives to make sure that hooks should be provided for using assistive technologies whenever appropriate. In addition to the guidelines supplied by ISO TS 16071 the most important methodological approaches to increase the accessibility of a given human-computer interface are:

- The use of human centered design principles (ISO 13407 - Human centered design process for interactive systems) [3]

- Task-oriented design of user interfaces.
- Customization
- Individualized user instruction and training.

3. Addressed capabilities

The ISO TS 16071 is based mainly on the prevalent knowledge of individuals with sensory and/or motor impairments in a work context. However, accessibility is an attribute that affects a large variety of capabilities and preferences of human beings in a range of different settings. These different capabilities may be the result of age, disease, and/or disabilities. Therefore, ISO TS 16071 addresses accessibility for a widely defined group of users including:

- people with physical, sensory and cognitive impairments by birth or acquired during life
- elderly people who can benefit from new IT products and services but experience reduced physical, sensory and cognitive abilities
- people with temporary disabilities such as a person with a broken arm or someone who forgot their glasses
- people who are experiencing difficulties in certain situations, such as a person who works in a noisy environment or has both hands occupied by other work

Unfortunately a standard of this sort cannot address the quality of an impairment, for example, regarding its medical and/or legal implications. The purpose can never be to standardize impairments by describing them, defining limitations and borderlines and to identify where "regular" capabilities end and where the impaired capabilities begin. The same goes for the temporal aspects of a disability.

Therefore, in order to address the problem of incorporating the widest range of capabilities, it occurs sufficient and even most effective to work with paradigms. As long as a community agrees on characteristics of an impairment or disability, we can use this concept for our work. Therefore, this concept becomes a paradigm for our standard. A standard has to engulf established paradigms to address the widest range of capabilities. A consequence of this concept is that accessibility becomes the major strategy regarding all topics dealing with the use of a system, i.e. usability.

To become usable to a developer the accessibility guidelines should also be connected to and incorporated in a design process to provide designers with a decision strategy explaining what to address when. Initially it should be sufficient to link accessibility to the human-centered design processes (such as ISO 13407), But, given the increased addressing of the area, specialized cases of user centered design for increased accessibility must be developed. Issues arise that are needed to address the circumstances of accessibility during processes (assessment, design process, project management, etc.). The work on subsequent international standards developed based on this technical specification dealing with accessibility of human-system interaction will have to cover this area as well.

4. Rationale and benefits

In order to determine the level of accessibility that has been achieved, it is necessary to measure the performance and satisfaction of users working with a product or interacting with an environment. Measurement of accessibility is particularly important in view of the complexity of the interactions with the user, the goals, the task characteristics and the other elements of the context of use. A product, system, environment or facility can have significantly different levels of accessibility when used in different contexts.

Planning for accessibility as an integral part of the design and development process involves the systematic identification of requirements for accessibility including accessibility measurements and verification criteria within the context of use. These provide design targets that can be the basis for verification of the resulting outcomes of each iteration in the human-centered development process.

The approach adopted in ISO Technical Specification 16071 has benefits, which include [4]:

- The framework can be used to identify the aspects of accessibility and the components of the context of use to be taken into account when specifying, designing or evaluating the accessibility of a product.
- The performance (effectiveness and efficiency) and satisfaction of the users can be used to measure the extent to which a product, system, environment or facility is accessible in a specific context.
- Measures of the performance and satisfaction of the users can provide a basis for the comparison of the relative accessibility of products with different technical characteristics, which are used in the same context.
- The accessibility planned for a product can be defined, documented and verified (e.g. as part of a quality plan).

5. Discussion

The concepts of Universal Access and Design for all have gained increasing attention lately, due to the release of several publications in the area and the launching of this very conference on Universal Access in Human Computer Interaction. Universal access refers to the conscious and systematic effort to proactively apply principles, methods and tools of universal design, in order to develop Information Society Technologies which are accessible and usable by all citizens, including the very young and the elderly and people with different types of disabilities, thus avoiding the need for a posteriori adaptations or specialized design [5]. International standards play an important role in providing agreed sets of formal criteria that apply across different countries and which support the development of products and systems that will meet common needs wherever they are developed and wherever they are used. The development of a standard for software accessibility poses new challenges of which we do not have much experience, and the relevant body of knowledge continues to grow. It was therefore decided to develop the document as a Technical Specification, which is a normative precursor to the development of a full standard. Experience in the application of the TS and the increasing knowledge base will then be used to refine and extend the full standard.

As an international standard it has the broadest possible audience that can work to make the guidelines provided in this specification serve the purpose of universal access for all types of users, either through procurement, legislation, or quality certification, or, by influencing the way in which systems, services and products are designed to make them accessible to the widest possible audience.

For example addressing the needs of the growing number of elderly people (25 % of the population in Europe will be aged 60 or above in the year 2020 [6]), is an important goal that has not so far been subject to international standardization. Therefore the goal, when working to turn this technical specification into a standard must be to incorporate all existing knowledge on accessibility guidelines that has so far not been incorporated into ISO standards into this document, to benefit the widest possible range of users. This requires some modifications to the procedures that organizations, such as ISO, adopt, e.g. by referencing unpublished documents, such as web sites. The WAI guidelines for accessibility are a particularly important example of this issue [7].

One area that needs further elaboration to fit into such a procedure is the processes with which we develop accessible products. It is not obvious that the human centered design principles described in [3] could easily be applied to solve accessibility problems, even though the concept of accessibility is related to usability. Therefore the standards work need to review and analyze not only criteria for design of accessible user interfaces, but the processes with which accessible products should be developed.

Work on international standards for accessibility is important, not only to provide tools for the developers of systems, but for the discussion and marketing of the issues that in itself probably has the highest effect in increasing the collective awareness of universal access of human-computer interfaces.

By defining the character of Accessibility as a view of usability, it becomes a essential understanding of all aspects of access and usage of systems, environments, information etc. The recommendations included in the TS are aimed at system design issues that are supportive for every kind of user (impaired/disabled, aged, young, unimpaired). Universal access is an ideal quality of every system, environment etc. The TS provides an elaborate tool to communicate the model of universal access into system design. By addressing users with the widest range of capabilities the paradigm of Accessibility will be relevant for every design process.

Acknowledgements

The authors wish to acknowledge the entire subgroup on accessibility from ISO Technical Committee 159/Sub Committee 4/Working Group 5, and especially the editor of ISO TS 16071, John Steger. The input provided by Human Factors and Ergonomics Society in terms of their outline of the accessibility part of ANSI 200 is a significant contribution to the making of the Technical Specification.

References

1.   International Organization for Standardization (2000) *ISO 16071 Ergonomics of human-system interaction – guidance on software accessibility.* Technical Specification. (Switzerland: International Organization for Standardization)
2.   International Organization for Standardization (1998) *ISO 9241 Ergonomic requirements for office work with visual display terminals (VDTs), Part 11 Guidance on Usability.* International Standard. (Switzerland: International Organization for Standardization)
3.   International Organization for Standardization (1999) *ISO 13407 Human centered design process for interactive systems.* International standard. (Switzerland: International Organization for Standardization)
4.   Gulliksen, J., Harker, S., & Steger, J. (2001) The ISO Approach to the development of Ergonomics Standards for Accessibility. In Colette Nicole and Julio Abascal (Eds.) *Inclusive design guidelines for HCI.* Taylor & Francis Ltd.
5.   Stephandis, C. (2000) International Journal of Universal Access in the Information Society.
6.   Thorén, C. (ed.) (1998). Nordic Guidelines for Computer Accessibility. 2nd ed., (Vällingby: Nordic Cooperation on Disability).
7.   World Wide Web Consortium (W3C), (1999) WAI - Web content accessibility guidelines 1.0, May 5, 1999.

Analytical Evaluation of Interactive Systems regarding the Ease of Use

Nico Hamacher[1], Jörg Marrenbach[2]

Department of Technical Computer Science
Aachen University of Technology, Germany
Ahornstr. 55, D-52074 Aachen
Tel.: +49-241-44026105 - Fax: +49-241-8888308
[1]EMail: hamacher@techinfo.rwth-aachen.de
[2]EMail: marrenbach@techinfo.rwth-aachen.de

Abstract

This paper describes a new tool for the generation and analysis of normative user models based on the GOMS theory for the evaluation of interactive systems and the analysis of the usability. Furthermore, a comparison of user models and generated action protocols is facilitated. Additionally, design alternatives can be compared. The results of the analysis are visualised in various ways.
The introduced tool supports the system engineer in a considerable way to evaluate interactive systems and produces suitable analysis data as a base for decisions while the systems are developed.

1. Introduction

Today, usability engineering of interactive systems is paying increasing attention. In general usability is determined by testing prototypes. The disadvantage of this approach is that these tests can be performed only in late stages of the development process. An early analysis of usability would be a significant advantage in regard of saving time and resources. In this paper the tool TREVIS (Tool for Rapid Evaluation of Interactive Systems) is presented. TREVIS enables the design engineer to model the behaviour of an user while interacting with a device and derive usability measures from this simulation

2. Usability Evaluation

To analyse the usability of interactive systems an empirical evaluation is commonly used. This type of evaluation requires a prototype and a couple of qualified testing subjects. Mostly, this procedure is very expensive and time-consuming. Furthermore, this evaluation is feasible only in late stages of the development process when a prototype is available (fig. 1), so that the results of this evaluation often cannot be used for a redesign in a sufficient way. Even though if this empirical evaluation cannot be omitted, because of the generation of plenty of useful information for improving the usability, an earlier evaluation would be very helpful for the design and specification of the system.

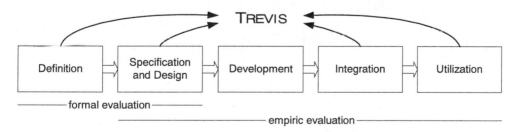

Figure 1: The different phases of a development process.

586

To analyse the usability in an early stage of the development process a formal evaluation is needed. One method of formal evaluation is developed by Card, Moran and Newell for modelling interactions between an user and an interactive system. It is called GOMS, which is an abbreviation for the components of the model: Goals, Operators, Methods and Selection rules. Over the years it has been shown that GOMS is able to sufficiently describe the interactions of an operator. Its simple and plain structure makes this method easy to understand particularly for development engineers which do not have a psychological background.

A GOMS model is also called user model. So far some GOMS variants have been introduced. The dialect with the most extensive analysis results is NGOMSL (Natural GOMS Language), first introduced by Kieras in 1988. An analysis based on NGOMSL generates qualitative as well as quantitative predictions, like execution and learning time. The execution time describes the time to reach the goal whereas the learning time specifies how much time an operator needs to learn the whole task.
Although the use of GOMS models is not very complicated, it is very tedious to build these models manually. Hence, a tool is needed which integrates the GOMS theory into the development process and supports efficiently creation and analysis of user models.

3. System Architecture

For the evaluation of usability three criteria are considered essentially. According to ISO 9241 part 11 these are *effectiveness*, *efficiency* and *satisfaction*. User performance is measured by the extend to which the intended goals or subgoals of use are archived (*effectiveness*) and the resources such as time, money or mental effort that have to be expended to archive the intended goals (*efficiency*). *Satisfaction* is measured by the extend to which the user finds the use of the system acceptable. Regarding these criteria the tool TREVIS (**T**ool for **R**apid **E**valuation of **I**nteractive **S**ystems) is described that supports the synthesis and analysis of user models based on NGOMSL.
TREVIS includes four main modules: the user model editor, the device model converter, the handbook generator, and the analysis module. The system architecture is depicted in figure 2.

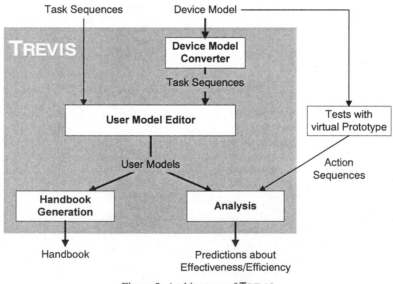

Figure 2: Architecture of TREVIS

Based on the task sequences as one result of the requirement analysis the user models can be created manually in the user model editor. The tool supports this process e.g. by offering a graphical editor and a library for reusing components. Moreover, user models can be stored in projects, where the project represents the interactive system and the user models describe the tasks which have to be done.

In contrast to other GOMS-editors which only allow to edit the user models manually, TREVIS is able to generate user models semiautomatically. A device model contains details about the inner work of the device. Using the device model converter the task sequences can be generated semiautomatically. The user models can also be created from these sequences.

In the analysis module the following four different analysis methods are included, which depend on the development phase, in which TREVIS will be used in:

- The user model analysis generates qualitative as well as quantitative predictions, like execution and learning time (as already described with NGOMSL).
- A comparison between different user models is implemented in the design analysis module, which can be used as a basis for design decisions. Although this comparison presentation is a helpful functionality, no other tool includes it.
- In the action sequence analysis, action sequences resulting from testing a prototype can be imported and analysed. A grouping of different sequences is possible, e.g. to perform an analysis of significance. With this feature, TREVIS is also applicable in late stages of a development process as depicted in figure 1.
- A fourth method analyses these action sequences in comparison with the user models. This analysis shows the differences between the actions specified in the user models and the activities, the users performed while interacting with a prototype. Based on this analysis, predictions about the effectiveness and efficiency can be made.

Figure 3 shows the user interface of the tool with the project management area on the left and the editing area on the right side. In the middle the different user models and action logs are shown.

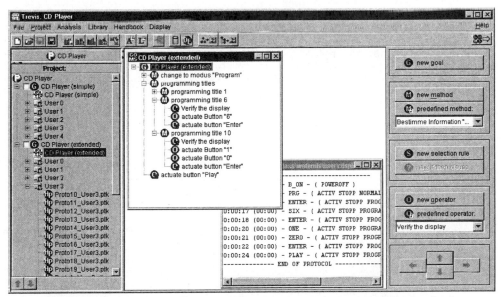

Figure 3: The user interface of TREVIS with an exemplary project, user model and action log.

With the possibility to analyse both user models and action sequences, TREVIS is the first tool, which can be utilised in early phases as well as in late phases of the development process. Beside this, a comparison of user models with action sequences has never been implemented before.

The user models contain the complete description of the procedural knowledge, which the user has to know in order to perform tasks using the device. Hence, Elkerton suggests to build a handbook by following and describing the tasks, a user has to do. Regarding this proposition, a handbook based on the user models can be created by the handbook generator.

4. Application and Evaluation Results

A CD player prototype was created to evaluate the tool features. The user interface of the two players differs primarily in the number of buttons. Prototype 1 (fig. 4, left) implements a simple CD player only with the most important buttons. The second prototype (fig. 4, right) includes additionally a number field to select the tracks in a direct way.

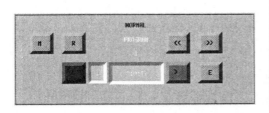 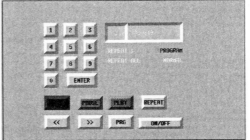

Figure 4: User interface of a simple CD player (left: prototype 1) and
extended CD player (right: prototype 2)

Therefore, the user models for two different types of CD players were modelled. The modelled task is "to programm a couple of songs and to start this program".

The user models were analysed by the user model analysis and the design analysis. Further on, several people interacted with the prototypes and action logs were recorded, which were also analysed. Finally, for each CD player a comparison between the corresponding user model and the action logs was carried out. Two of the resulting diagrams are depicted in figure 5. The execution time calculated from the user models is shown as a line where the crosses specifying the execution time of the action logs. It can be seen, that most of the empirical times only vary up to 10% (grey region). Wrong actions, done by the users, were counted and diagrammed as columns.

5. Summary

In this paper the tool TREVIS is introduced, which enables the development engineer to evaluate interactive systems during the whole development process. Thereto it offers a method for the analytical evaluation by using GOMS models. With these user models interactions between an operating user and an interactive system can be described. A manual editing is implemented as well as a semiautomatical generation from formal specifications. Furthermore action sequences resulting from testing a prototype can be imported. With these data various analysis can be performed. Beside the user model analysis, a comparison between different user models and a comparison between user models and associated action sequences.
For the evaluation of the tool CD player prototypes were created and an exemplary testing is performed.
The results indicate, that TREVIS is a useful tool to support the development engineer during the development process.

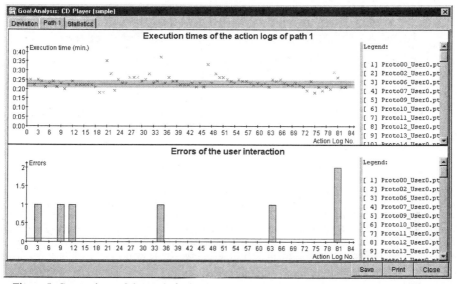

Figure 5: Comparison of the analytical calculated and rised execution time (prototype 1)

References

Card, S., Moran, T., Newell, A. (1983). *The psychology of human computer interaction.* Lawrence Erlbaum.

Elkerton, J. (1988). Online Aiding for Human-Computer-Interfaces. In: M. Helander (ed.), *Handbook of human-computer interaction* (pp. 345-362).Amsterdam: North Holland.

Hamacher, N. (2000). *Entwicklung und Implementierung eines Werkzeugs zur Bewertung interaktiver Systeme basierend auf normativen Benutzermodellen.* Department of Technical Computer Science, Diploma Theses, Aachen University of Technology.

ISO 9241-10 (1996). *Ergonomic requirements for office work with visual display terminals - Dialogue principles.* International Organisation for Standardisation, Genf.

ISO 9241-11 (1998). *Ergonomic requirements for office work with visual display terminals - Guidance on usability.* International Organization for Standardisation, Genf.

Kieras, D. (1988). Towards a practical GOMS model technology for user interface design. In M. Helander (ed.), *Handbook of human-computer interaction* (pp. 135-157). Amsterdam: North Holland.

Leuker S., Marrenbach J. (1998). Integrating User Models into Device Prototype Design. In *Proceedings of the 24th Annual Conference of the IEEE Industrial Electronics Society IECON, Volume 4/4* (pp. 2532-2534), Aachen.

Marrenbach, J. (1999). Rapid Development and Evaluation of Interactive Systems. In *Proceedings of the 5th ERCIM Workshop User Interfaces for All , Volume Report 74* (pp. 81-86), Dagstuhl.

Marrenbach J. (1999). Konzept eines Werkzeugs zur formalen und empirischen Evaluierung der Gebrauchstauglichkeit interaktiver Endgeräte. In K.-P Gärtner (Eds.): *Ergonomische Gestaltungswerkzeuge in der Fahrzeug- und Prozeßführung* (pp. 217-230), Stuttgart

Considerations on ergonomic work places for blind and visually impaired persons in Call Centers

Martin Jung

University of Applied Sciences Giessen-Friedberg
Department of Mathematics, Computer Science and Natural Sciences
Wiesenstrasse 14; 35390 Giessen; Germany
martin.jung@mni.fh-giessen.de

Abstract

For average users of video display terminals without vision impairments there are European regulations on ergonomic aspects (EN 29241-1, RL 89/391/EWG, RL 90/270/EWG). Since some of them can not be applied to persons with visual impairments, we will investigate different scenarios for work places of visually impaired or blind persons in Call Centers. Our suggestions for the setting up of ergonomic work places will be presented.

1. Introduction

Living in the Information Society, the nature of employment has been subject to rapid changes. Employers are expecting an adaptable and highly skilled workforce. While there is danger of impaired persons being excluded from the developments of the Information Society, there are some new opportunities, too. [1]

For blind and visually impaired persons there has always been a tradition of jobs related with telephony. One of the new job areas is associated with Call Center work. Given appropriate access to the required infrastructure, blind and visually impaired persons can work in a Call Center environment [2].

1.1 Statistical overview of the situation of visually impaired and blind persons in Germany

All provided numbers are based on estimations of Germany's umbrella organization of self-help groups (Deutscher Blinden- und Sehbehindertenverband e. V., German Federation of Blind and Visually Impaired People, Bonn [3]), since there are only a few official data available from the German government.
These estimations mostly rely on the statistics of some of the Bundeslaender (states) and an investigation of the situation of blind and visually impaired persons, carried out in 1998 by the German polling institute *infas*, Bonn [4].

- In Germany there are about 155,000 blind (i. e. residual vision less than 2%) persons and about 550,000 persons with vision impairments (residual vision between 2% and 30%).
- Only 30% of these people between the age of 18 and 60 are employed.
- Blind and visually impaired people are mostly employed in a few working areas.

More than 50% are working in only three working areas (masseurs/physiotherapists, telephone operators, office workers/administration). Due to technical development, there are less jobs offered for telephone operators.

The job profiles of the traditional working areas are changing; many of them have already changed. This requires development of new job areas and updating the old ones. Workforce has to be flexible and lifelong learning is a major issue for the future [2].

2. Ergonomic work places in Call Centers for blind and visually impaired workers

2.1   General considerations

Most of the physical conditions of a work place, such as humidity or temperature, do not have to be adapted to persons with special needs. General regulations for work places with video terminals exist and some publications dealing with ergonomic aspects of work places in Call Centers are available [5]. However, there are virtually no publications on the ergonomics of Call Center work places with respect to blind and/or visually impaired persons.

Thinking of a person with low vision, one would easily expect some problems. Regarding lighting, most persons with residual vision require more light to read printed material, but there is a considerable amount of persons irritated by glare in higher lighted environments.

Information on lighting with respect to vision impairment is not easy to be found [6]. For each case the personal requirements have to be analyzed, depending on the impairment and the tasks to be performed in daily work.

2.2   Different users – different needs

First we have to identify and categorize our user types.

Partially sighted people can see more than blind people but less than the visually handicapped. People are partially sighted when, by simplification, their vision is 5 percent or 1/20 in the better eye and with correction. An individual is considered visually handicapped when his or her vision, in the better eye and with correction glasses is no better than 10 percent or 1/10 [3].

Since these categories are not useful for our purposes, we will distinguish between persons with sufficient residual vision to optically read printed materials and use large print software on computers and frequent users of Braille displays and speech synthesizers. This is a more practically oriented approach.
Like every simplification, this does not represent all of our users. Of course, there are persons using Braille and large print software simultaneously or alternating on their computers.

2.2.1   Example of a work place of a large print user

A work place should include a computer equipped with an appropriate (mostly larger) monitor, a large print software to magnify the screen contents, several optical magnifiers and/or a video magnifying aid (CCTV). If lots of printed documents have to be handled, a scanner with optical character recognition software, a personal printer for (magnified) printouts and a headset is necessary for the work place.

Because of the additional devices for adapted access (CCTV, scanner, larger monitor) taking some space, additional desks are needed.
Chairs and desks should be chosen carefully. Not all of them give a visually impaired person the freedom to sit in a position very close (i. e. a few inches) to the monitor. Special monitor stands can fix this problem.

For a non-impaired person radiation and the flicker effect of monitors due to insufficient refresh rates are not critical with state-of-the-art technique. Considering a user sitting only a few inches from the monitor, this poses some serious problems. As radiation is proportional $1/r^2$, with r being the viewing distance, radiation can be estimated to be about 100 times higher than with a usual viewing distance. The flickering effect increases as you approach your monitor and, since the peripheral parts of the retina are involved, which are more susceptible to flicker, a more than linear increase in total can be expected.

Vertical frequencies higher than 75 Hz are sufficient for average users, but I have been reported of visually impaired persons with severe eye irritation (running tears) caused by flickering at vertical frequencies up to 110 Hz of cathode ray tube monitors. For some of these extremely sensitive users a tft-display (thin film transistor flat panels, backlit, without crt) is superior. Due to the different technique there is less eye-strain, even at comparably low refresh rates.

This type of display also has very low radiation and, being a flat panel display, is easy to place close to a visually impaired user.
Some of the visually impaired computer users prefer inverted color schemes and most of these individuals do not notice flickering.

Independent switchable light sources to accommodate the user completes our suggested work place. According to the personal needs, ceiling lights and a flexible lamp (e.g. an architect lamp) at the work place have to be installed. The users have different light preferences. Some prefer lights with a very high spot-like intensity, while others prefer lamps with larger equally lighted areas. If colors have to be discriminated, the color of light has to be taken into account.

Fig. 1: Example of a work place of a Braille user (with residual vision)

2.2.2 Example of a work place of a Braille user

A working place for a Braille user has to include a computer, a Braille display, speech output, a scanner for printed material and a tactile printer for Braille print (figure 1). An additional acoustically isolated printer housing is highly recommended, since these printers produce noise during printing.
Usually a standard monitor is included, at least for service purposes and additional desks for the aforementioned devices will be mandatory.

Using Braille does not necessary mean someone has no vision at all. The lighting of the work place should be sufficient to give some orientation (300 Lux, according to DIN 5035-8), but is not as important as for large print users (see above).

Computer Braille readers usually place their keyboard on the Braille display. On desks complying with DIN 45 49 this raises the keyboard by about two inches. From an ergonomic point of view this increases the risk of injuries (like RSI-syndrome) or tiring. Special furniture can solve this problem.

If a speech output device is used exclusively for adapted access, measures should be taken to protect the work place from the background noise of the other work places and vice versa. In Germany the additional costs of the technical aids for disabled workers (Braille display: $ 10,000, all technical aids can sum up to $ 50,000) are sponsored by governmental money, so Braille displays are standard for Braille users.

3. Call Centers – How will they change in future?

New economy leads to new forms of work, alternating tele-work is one of them. In terms of Call Centers, there is a trend towards outsourcing, but on the other hand people have realized the value of in-house Call Centers.

Even with the latest developments in telecommunication and information technology, there are some benefits of in-house Call Centers. For example information will always be slightly delayed for external facilities, not to forget interpersonal contacts of co-working persons busy in the different stages of support. These human factors cannot be measured easily, but do contribute to the overall result of the working process.

A large part of Call Center work is considered to be support. For first line support and time critical questions in-house Call Centers will be superior to external solutions.

Companies in the field of the new economy have to react very fast to changes of the market. Changes within Call Centers can be done easier in an in-house Call Center. Since money is always an issue, the combination of in-house and external Call Centers seems to be a promising approach.

Today there are already some examples, backing up this approach:

The company behind Germanys leading LINUX distribution, SuSe (Nürnberg), offers software support in many ways. From general topics, like installation, to the most advanced problems, everything is covered. They run an in-house Call Center for support, as well as some tele-working external employees.

Two of these external workers are blind. They do not live in the vicinity of the company. During their working hours support requests are routed via ISDN to their homes. At home they have computers equipped with adapted access, i. e. screen reader software, Braille displays, speech synthesizers and the necessary telecommunication devices (ISDN modems, headsets), of course.

They received an initial training at the plant in Nürnberg and now have been working for a few months successfully. To make this clear: it is not a matter of big numbers, only a few of SuSe's staff is blind or visually impaired. But it is a positive sign for some other companies and the blind community as well.

Call Centers in general will change towards Communication Centers. One has to deal with the integration of services and information management as main objectives for the future. The telephony centered work will be just another part of the work of Communication Centers, value added services have to be set up to keep the pace of the market.

4. Conclusion

Call Centers are an interesting work place for many reasons. They offer new job opportunities and require new skills from the future workforce. Blind and visually impaired persons can participate in this development. The ergonomic requirements of blind and visually impaired persons are different in comparison with "average" video display terminal users. Some of the regulations simply do not fit and have to be adapted. Analyzing the special needs of the handicapped and the job profiles (i.e. tasks to be performed) is essential for the design and the set up of ergonomic work places.

There are no general rules to be applied. Every handicapped person has to be treated as a special case. The furniture, like desks and chairs, the lighting of the room and the work place, the hardware and the software for adapted access are the main factors to be considered. Measures to prevent unwanted light from causing glare or reflections on the screens and additional acoustical shielding for noise reduction might be necessary. Optimizing these factors leads to a less tiring work place and increased productivity.

594

References

[1] IBM European Seminar, (1996, June 17 and 18), Social Exclusion, Technology and the Learning Society, *Briefing paper on technology applications for training and job placement*, Salzburg, Austria

[2] Miesenberger, K., Stöger, B., (2000) *Access to and usage of Call Centre Applications for blind and Visually Handicapped People,* Computers Helping People with Special Needs, ICCHP 2000, Oldenbourg, Wien, München, pp. 465-472

[3] German Federation of Blind and Visually Impaired People, retrieved on the web January 8, 2001, http://home.t-online.de/home/dbsv_/engl/e_facts.htm

[4] infas - Institut für angewandte Sozialwissenschaft GmbH, (1998, June 2), *Die Beschäftigungssituation von Blinden - Untersuchung im Auftrag des Landschaftsverbands Rheinland*, Bonn, Germany

[5] Schubert, P., (1999), Erfolgsfaktor Einrichtung im Callcenter: Praxisorientierte Marktstudie von Peter Schubert und Guido Eisfeller, Deutsche Verlags-Anstalt, Stuttgart, Germany, pp. 65-72

[6] Buser, F., (1993), *SZB Kurs für Low Vision Trainer und Trainerinnen: Grundkenntnisse der Basis-Optik*, St. Gallen, Switzerland

[7] Degenhardt, S., Kalina, U., Rytlewski, D., (1996), *Der Einsatz des Computers bei blinden und sehbehinderten Schülern*, Lebenswelten und Behinderung Bd. 5, Hamburger Buchwerkstatt, Hamburg, Germany, pp. 176-193

The effect of human factors in flexibility management - an international survey

Berman Kayis, Sami Kara, Shaun O' Kane, Andres Dingwall

School of Mechanical and Manufacturing Engineering
The University of New South Wales, Sydney, NSW 2052 Australia
b.kayis@unsw.edu.au

Abstract

Although flexibility of manufacturing systems has received much attention especially during the last decade, confusion exists in flexibility related issues and in managing manufacturing flexibility. Especially the role of human factors in clarifying and unifying several perspectives or views towards better implementation of flexibility management need further research to be carried out. The work presented in this article attempts to summarise the literature based on five groups; namely economic, organisational, strategic, operational and horizontally extended flexibilities and the role and importance of human factors in strategy setting, implementation and performance of manufacturers. A world-wide web based survey was designed and responses from Swedish and Australian manufacturers were received and analysed to explore different attitudes of manufacturers to flexibility management.

1. Background

Although theories of flexibility can be found in literature for many decades, most of them were published over the last decade (Sethi and Sethi, 1990, Beach et al, 2000). The introduction of automated production and Flexible Manufacturing Systems (FMS) escalated interest in this field. With the popularisation of business strategy in management literature, also flexibility, as a strategic competitiveness, has increasingly received attention in operations management literature. The tendency among the papers published over the last decade has been to extend the view of flexibility from the operations of the work flow, to a tactical or strategic level. however, although the material on the subject can be considered as vast and articulated, focus and standpoints sometimes differ considerably between different authors, due to their multi-dimensional nature of flexibility. Thus, the literature may be grouped so that each group and interactions with other groups could be carried out.

Among the five groups identified as economic, organisational, strategic, operational and horizontally extended, the role of human factors has been found to be very dominant. A summary of literature is as follows.

1.1   Economic Perspectives

The early interests in flexibility were in purely economical terms. Considering flexibility in production, a number of authors cite Stigler (1939) as the first author to note the impact of the slope of the production cost curve on the (volume) flexibility. Other economic aspects of flexibility was presented by Mandelbaum (1978), who observes the possibility of monitoring the economic consequences of either state or action flexibility and Son and Park (1978) who propose four different flexibility types with related measures in terms of cost. Pyoun and Choi (1994) also attempts a quantification of flexibility value, and preludes by citing Swamidass and Waller (1990), who recognised that, despite a diversify of approaches of financial justification models, flexibility has not yet been proven to have strategic benefits.

1.2   Organisational Perspectives

Sethi and Sethi (1990) cite some of the early definitions of organisational flexibility as the ability to change without severe disorganisation. A much referred to article on organisational flexibility is the one by Burns and Stalker (1961), who differentiate between and *organic* structure and a *mechanistic* structure of the organisation. Mintzberg (1979), with the concept of *adhocracy*, has also been referred to as one of the more influential contributions (De Toni and Tonchia, 1998).

The organisational side of flexibility also includes the flexibility of the human resources within an organisation. The pioneering work on labour flexibility was done by the Institute of Manpower Studies (UK), and a report of their findings is found in Atkinson (1985). Gupta and Cawthon (1996) concentrates on "organizational and personnel implications of flexible manufacturing for small/medium sized enterprises (SMEs)", in which they cover organisational culture and responsibilities of management, suitable organisational design, worker empowerment and team formation etc. A differentiation between organisational and management, i.e. software, flexibility and manufacturing technology, i.e. hardware, flexibility, is made by Carlsson (1989). He later also presented a comparison of the use of software flexibility between Swedish and American companies (Carlsson, 1992). The concept of software and hardware flexibility is also used by Hyun and Ahn (1992). They present a framework unifying system, environment-associated, and decision-hierarchical oriented aspects of flexibility, with much emphasis on the importance of organisational hierarchy and dynamic aspects of flexibility (including managerial implications etc.).

1.3 Strategic Perspectives

Hayes and Pisano (1994) characterise the *strategic approach* to flexibility as an emerging approach, which considers the ability of a firm to change its competitive priorities. Originating in business-strategy literature, Buffa (1984) and Wheelwright (1984) include flexibility as one of the major dimensions of manufacturing strategy (the other dimensions are quality, cost and reliability). Swamidass and Newell (1987) also approaches flexibility as a part of a manufacturing strategy, and found that even though flexibility is highly influenced by environmental uncertainties, these uncertainties have received little attention.

1.4 Operational Perspectives

A larger portion of literature on flexibility is a manufacturing context is concentrated on flexibility of the manufacturing system, which corresponds to an *operational approach*. Much of the early work concentrated on explaining the benefits and usefulness of manufacturing flexibility. Nagarur (1992) focuses on the flexibility at the machine level of the FMS, and presents a measure of 'producibility' said to reflect both flexibility and reliability at component and system level. Other studies with focus on the basic flexibilities include Tsubone and Horikawa (1999), who present a comparison between machine and routing flexibility in various shop environments. Zelenovich (1982) is the first author to recognise the need to define the limits of the production system to differentiate between internal and external needs for flexibility. He also advocates flexibility as a condition for effective production systems, and concludes that humans are the single most important part of a (flexible) production system. A common feature of literature on flexibility is the development of taxonomies of flexibility types. A detailed analysis and discussion of taxonomies can be reviewed in Kayis et al (2001). Shewchuk and Moodie (1999) presents a (rather complex) classification scheme. The result is a somewhat useful tool for comparing different flexibilities, but the classification fails due to the ambiguity of the different definitions. They later present a set of generic flexibility measures with the prospect of future practitioners to derive appropriate, situation based flexibility types.

Slack (1987) presents a multi-dimensional approach to the measurement of flexibility, with time, range and cost as the three dimensions. He also later presents a taxonomy of flexibilities based on findings of managerial attitudes, which was collaborated by Correa (1994). Correa's (1994) work expands many of Slack's theories on flexibility, and by combining literature and fieldwork in the form of extensive interviews of manufacturing managers.

Another recent review of manufacturing flexibility is presented by De Toni and Tonchia (1998). Their structured review takes a schematic approach to the different aspects of manufacturing flexibility, and includes the first classification of different flexibility definitions.

1.5 Horizontally Extended Perspectives

There are also a number of authors who take a *horizontally expanded approach*, using a larger frame of the manufacturing process, including the supply chain. Chen et al (1992) points out the lack of guidance for integration of flexibility in the marketing/manufacturing interface. In their paper they concentrate on market focused flexibility, and includes aspects of product life cycle, buyer concentration, and unexpected competitors. Aggarwal (1997) have presented articles on flexibility from an overall perspective, covering many important aspects.

Viswanadham and Raghavan (1997) present a comprehensive paper, which considers the different aspects of flexibility in the manufacturing enterprise, including the supply chain and the order-to-delivery process. An interesting fact of Narasimhan and Das' work is that they come from the theoretical background of agile manufacturing. Agile manufacturing is closely related to flexible manufacturing and much of the literature is of relevance to flexible manufacturing. The concept of agile manufacturing was coined in 1991 by a group at the Iacocca Institute (Nagel and Dove, 1991) and a number of articles on the subject have been published since then. Definitions of operational agility have been given by Kegg (1996) and Kuhmar and Motwani (1996), among others, and for a comprehensive definition, Youssef (1992) presents a model of agile manufacturing and defines a framework for agility. Yusuf et al (1999) explores the variety of definitions in a comprehensive review of literature on agile manufacturing and a reflection on its use as competitive advantage. The benefits of human factor considerations at each stage of manufacturing are also discussed in detail.

Later Towill (1997) discusses process integration and recognises a movement towards the 'seamless supply chain' and a supply chain for agile and customised manufacturing is collaborated in Christopher and Towill (2000). Naylor et al (1999) also looks at agility in the total supply chain and use the term leagility as an interface of lean and agile manufacturing, where human plays a crucial role.

2. Scope

As Narasimhan and Das (2000) suggest, further research to investigate how manufacturing firms compete through flexibility management would help extending the concept of flexibility from theory to better and more manufacturing applications. Thus, the aim of the research undertaken at UNSW, was to provide some evidence to certain theories concerning manufacturing firms' views and use of manufacturing flexibility and its management. Accordingly, several hypothesis were set and tested in Australian and Swedish manufacturing firms to:

- Investigate whether any relationship exists between manufacturers whom place emphasis on manufacturing flexibility and supply chain management.
- Determine the major difference(s) (if any) between Australian and Swedish managers' attitudes towards flexibility.
- Examine how companies of different categories e.g. size, industry, market share etc. may focus on flexibility.
- Investigate the different emphasis on flexibility between top-level and mid-level managers.
- Investigate the role of human factors in flexibility management.

In this paper, emphasis is only given to the last objective.

3. Methodology

Fifteen Swedish and five Australian manufacturers among two hundred, responded to our survey which was designed to be a world-wide web based survey. A total of 20 top-level managers, 13 mid-level managers and 10 engineers/operators responded.

3.1 Survey Design

Three questionnaires were prepared aiming to receive as much feedback as possible, from top and middle management as well as production operators and engineers. It was geared towards the examination of the consistency in views and opinions at different levels of management based on the manufacturing strategy of the firm. Two versions of the questionnaires, English and Swedish, helped facilitate the responses from Sweden.

Extensive work on designing the Internet web site was conducted and material was published. The web site was designed in Hyper Text Mark Up Language (HTML) with the aid of Microsoft Front Page. Through the design of the survey, the responses of the survey were automatically sent to several e-mail addresses in an electronic format.

The structure of the questionnaire and the number of questions in each category (in parenthesis) are as follows

1. Top Level Management Questionnaire
   - Description of the firm (e.g size, ownership, market share, annual gross sale, etc (7)
   - Manufacturing strategy (12)
   - Manufacturing decision-making structure (2)
   - Human resource management, team formation and training (15)
   - Sources of problems (14)
2. Mid Level Management Questionnaire
   - Manufacturing set up (1)
   - Manufacturing communication and decision-making structure (2)
   - Supplier strategies (6)
   - Tools, techniques and technologies used (12)
   - Human resource management, team formation and training (10)
   - Distributor strategies (6)
   - Sources of problems (7)
   - Appropriate levels of flexibility (33)
3. Engineers/Production Operators Questionnaire
   - Communication (6)
   - Human resource management, team formation and training (12)
   - Sources of problems (7)

All the responses are measured on a 5 to 10 point Likert scale.

598

4. Findings

Amongst several hypothesis tested, the following are within the context of the paper:

H1:   Companies with a lateral communication structure show more organisational flexibility compared to companies with a vertical communication structure.

H2:   Companies which give emphasis on human resource management tend to utilise several tools, methods and techniques.

Both Australian and Swedish managers stated that communication structure of their companies are getting less centralised over the past five years. However, the degree of vertical communication is still higher in Australia (Table 1). But it is worth noting a larger step towards de-centralisation of the communication structure over the past 5 years. The cut-off score was set to be 8 where 1 corresponds to a formal and vertical communication structure and 10 to an informal and lateral communication structure. Two groups of respondents were then compared on the basis of questions relating to the organisational flexibility. These are:

1.   Craft flexibility (level of worker skills)
2.   Functional flexibility (ability of changing workers' tasks)
3.   Numerical flexibility (ability to alter the number of workers)
4.   Financial flexibility (ability to financially support craft, functional and numerical flexibility)

The above questions may refer to company's infra-structural flexibilities regarding the labour flexibility. The group emphasised having vertical communication scored an average of 5.8 whereas the group with lateral communication scored 8.0. Companies with a lateral communication structure show more organisational flexibility compared to vertical communication structured companies.

It was interesting to note the consistency of the responses from top and middle management. The companies which give emphasis to human resources issues proved to utilise total quality management, Kanban, Just-in-Time, Concurrent Engineering, and group technology. The importance of team formation and cross-functional team were also noted. The high rating of tools like total quality management, Just-in-Time and Kanban among Australian manufacturers may be due to increased awareness towards these concepts during the last decade unlike Swedish manufacturers who have implemented them for over a decade (Table 1).

Empowerment is a crucial issue in human resource management and flexibility management. After grouping the respondents into two groups, based on the cut-off point of 5, on a Likert scale of 7 (1: Low, 7: High), an interesting pattern was noted. The group with high empowerment found worker communication a minor problem. This result is consistent with the responses from managers towards perceived problems in the company. Thus, managers working with empowered and autonomous workers do not experience workforce communication as a problem.

Organisational flexibilities seem to vary between Swedish and Australian manufacturers. Although craft, functional and financial flexibilities were observed among Swedish respondents which is a consistent result with multi-skilled, more empowered workforce, numerical flexibility failed. The major reason would be the differences in industrial regulations and restrictions for lay-offs. Therefore the variations of the size of the workforce may be harder in Sweden than in Australia.

References

Aggarwal S. (1997), "Flexibility Management: The Ultimate Strategy", *Industrial Management*, Vol. 39, No. 1 Jan-Feb, 5 pages.

Atkinson J. (1985), "Flexibility: Planning for an Uncertain Future", *Manpower Policy and Practice*, Vol. 1.

Beach R., Muhlemann A.P., Price D.H.R., Paterson A., Sharp J.A. (2000), "A Review of Manufacturing Flexibility", *European Journal of Operational Research*, Vol. 122, Issue 1, pp. 41-57.

Buffa, E.S. (1984), "Meeting the Competitive Challenge: Manufacturing Strategy of US Companies", Irvin, Homewood, IL.

Burns T., Stalker G.H. (1961), "The Management of Innovation", Tavistock Publications, London.

Carlsson B. (1989), "Flexibility and the Theory of the Firm", *International Journal of Industrial Organisation*, Vol. 7, pp. 179-203.

Chen I.J., Calantone R.J., Chung C.H. (1992), "The Marketing-Manufacturing Interface and Manufacturing Flexibility", *OMEGA International Journal of Management Science*, Vol. 20, No. 4, pp. 431-443.

Christopher M., Towill, D.R. (2000), "Supply Chain Migration from Lean and Functional to Agile and Customised", *Supply Chain Management: An International Journal*, Vol. 5, Issue 4.

Gupta M., Cawthon G. (1996), "Managerial Implications of Flexible Manufacturing for Small/Medium-Sized Enterprises", *Technovation*, Vol. 16, No. 2, February, pp. 77-83.

Hayes R.H., Pisano G.P. (1994), "Beyond World Class: The New Manufacturing Strategy", *Harvard Business Review*, January/February, pp. 77-86.

Hyun J.H., Ahn B.H. (1992), "A Unifying Framework for Manufacturing Flexibility", *Manufacturing Review*, Vol. 5, No. 4, pp. 251-259.

Kayis B., Kara S. (in print, 2001), "The Role of Human Factors in Flexibility Management, *Human Factors and Ergonomics in Manufacturing*.

Kegg R.L. (1996), "Flexible Manufacturing Trends and Data Base Requirements", *Proceedings of the 5th International Conference on Production Engineering*, ISPE, Tokyo.

Kumar A., Motwani J.A. (1995), "Methodology for Assessing Time-Based Competitive Advantage of Manufacturing Firms", *International Journal of Operations and Production Management*, Vol. 15, Issue 2, pp. 36-53.

Mandelbaum M. (1978), "Flexibility in Decision Making: An Exploration and Unification", *PhD Thesis, Department of Industrial Engineering*, University of Toronto, Toronto, Canada.

Mintzberg H. (1979), "The Structure of Organisation", Prentice-Hall, Englewood Cliffs, NJ.

Nagarur N. (1992), "Some Performance Measures of Flexible Manufacturing Systems", *International Journal of Production Research*, Vol. 30, No. 4, pp. 799-809.

Nagel R., Dove R. (1991) (Eds), "21st Century Manufacturing Enterprise Strategy", Iacocca Institute, Lehigh University, Bethlehem, PA.

Narasimhan R., Das A. (1999), "Manufacturing Agility and Supply Chain Management Practices", *Production and Inventory Management Journal*, Falls Church, Vol. 40, No. 1, First Quarter, pp. 4-10.

Narasimhan R., Das A. (2000), "An Empirical Examination of Sourcing's Role in Developing Manufacturing Flexibilities", *International Production Research*, Vol. 38, No. 4, pp. 875-893.

Naylor J.B., Naim M.M., Berry D. (1999), "Leagility: Integrating the Lean and Agile Manufacturing Paradigms in the Total Supply Chain", *International Journal of Production Economics*, Vol. 62, No. 1, pp. 107-118.

Pyoun Y.S., Choi B.K. (1994), "Quantifying the Flexibility Value in Automated Manufacturing Systems", *Journal of Manufacturing Systems*, Vol. 13, pp. 108-118.

Sethi A.K., Sethi S.P. (1990), "Flexibility in Manufacturing: A Survey", *International Journal of Flexible Manufacturing Systems*, Vol. 2, pp. 289-328.

Shewchuk, J.P., Moodie C.L. (1998), "Definition and Classification of Manufacturing Flexibility Types and Measures", *The International Journal of Flexible Manufacturing Systems*, Kluwer Academic Publishers, Vol. 10, pp. 325-349.

Slack N. (1987), "The Flexibility of Manufacturing Systems", *International Journal of Production Management*, Vol. 7, No. 4, pp. 34-45.

Son Y.K., Park C.S. (1987), "Economic Measure of Productivity, Quality and Flexibility in Advanced Manufacturing Systems", *Journal of Manufacturing Systems*, Vol. 6, No. 3, pp. 193-207.

Stigler G.J. (1939), "Production and Distribution in the Short Run", *Journal of Political Economy*, Vol. 47, No. 3, pp. 305-327.

Swamidass P.M., Newell W.T. (1987), "Manufacturing Strategy, Environmental Uncertainty and Performance: A Path Analytic Model", *Management Science*, Vol. 33, No. 4, pp. 509-524.

Tonchia S. (1995), "Production Performance Measurement Systems-Theory and Empirical Evidence on Structure, Characteristics and Indicators", PhD Thesis, University of Padua.

Towill, D.R. (1996), "The Seamless Supply Chain - The Predators Advantage", *International Journal of the Technology of Management*, Vol. 13, No. 1, pp. 37-56.

Tsubone H., Horikawa M. (1999), "A Comparison Between Machine Flexibility and Routing Flexibility", *International Journal of Flexible Manufacturing Systems*, Vol. 11, pp. 83-101.

Upton D.M. (1994), "The Management of Manufacturing Flexibility", *California Management Review*, Vol. 36, No. 2, Winter, pp. 72-89.

Upton D.M. (1995), "What Really Makes Factories Flexible?" *Harvard Business Review*, July/Aug, pp. 74-78.

Upton D.M., Fliedner G., Vokurka R. (1997), "Agility: Competitive Weapon of the 1990s and Beyond?" *Production and Inventory Management Journal*, Vol. 38, No. 3, pp. 19-24.

Viswanadham N, Raghavan N.R.S. (1997), "Flexibility in Manufacturing Enterprises", *Sadhana Academy Proceedings in Engineering Sciences*, Vol. 22, No. 2, pp. 135-163.

Wheelwright S.C. (1984), "Manufacturing Strategy: Defining the Missing Link", *Strategic Management Journal*, Vol. 5, No. 1, pp. 77-91.

Youssef A.M. (1992), "Agile Manufacturing: A Necessary Condition for Competing in Global Markets", *Industrial Engineering*, Vol. 24, No. 12, pp. 18-20.

Zelonovic D.M. (1982), "Flexibility - A Condition for Effective Production Systems", *International Journal of Production Research*, Vol. 20, No. 3, pp. 319-339.

Ergonomic Design of Call Centers

Dieter Lorenz

University of Applied Sciences
Wiesenstr. 14, D-35390 Giessen, Germany

Abstract

Many different aspects of stress in call centers have to be considered. Within the framework of a large scientific project financed by the "German Federal Institute for Occupational Safety and Health" (Bundesanstalt für Arbeitsschutz und Arbeitsmedizin), all important aspects of the work design in call centers have been investigated. This paper refers to the design of workplaces, their position in rooms and the effects of lighting.

Stress related to the design of workplaces effects the muscular-skeletal system. As employees in call centers have to work with telephone and VDU approximately 85% of their time at the company, the lack of movement is one of the major problems. Theoretical and practical solutions will be reported with relevance to improving movement at the workplace.
The position of the workplace and its position in relation to the other workplaces in a room effects the level of disturbance by glare and reflection on the screen (caused by lighting and windows) as well as visual disturbance and noise. With the example of two typical call center layouts (radial placement and position in line) the advantages and disadvantages of the solutions are discussed.

The quality of light in a call center is very important for the visual work that has to be done with VDU's as well as for well-being at work. The advantages and disadvantages of different technical solutions are discussed in this paper and the author makes proposals for the best solutions in practice.

1. Stress due to inadequate Workplace Design

This predominantly effects the muscular-skeletal system. As workers in call centers spend an average of 85% of their working life on the telephone at the workplace, lack of movement represents an area of a particularly problematic nature. With the exception of the hand and the arm for using the mouse and the keyboard, the muscular-skeletal system remains primarily static. To vary the burden on particular parts of the body, workplaces should be installed for work in a sitting or a standing position, allowing the call center workers to choose one of these two positions while on the phone. Appropriate training courses can teach the agents to adjust the height at which they work, whether in a sitting or standing position, as well as their chair to the correct position relative to their own height. The work surface has to be quickly and easily adjustable so as to guarantee that the function is actually used. In addition to this it is crucial that that agents have sufficient opportunity to move around and that they do not sit down during breaks; the installation of tables for standing only has proved a successful solution to this problem.

The so-called radial placement in the relation of workplaces to each other is still in use (cf. Figure 1). This form of layout, which is often seen as both supportive to communication flow as well as more sophisticated from an aesthetic point of view, nevertheless shows serious disadvantages. The radial formation of several workplaces in relation to each other does not fulfil the demands of concrete findings in the industrial sciences(nor does it fulfil the regulations of the European laws governing work at a VDU). For example, radial placement leads to a line of vision which puts extreme pressure on workers' eyes due to low density of light on the monitor and high light density at the window behind the monitor.

Figure 1: Radial placement of workplaces in a call center

Even partitions cannot improve these conditions. On the other hand, the window can be reflected in the monitor screen, which also causes undue pressure on the eyes. The accessibility of the cable conduits in the middle column of the radial design is also dramatically decreased. Above all it is the acoustic and visual disturbances by other colleagues which are of prime importance. The agents can be said to talk directly into each others' ears and can be distracted by the body movements of their colleagues. These problems can be avoided by positioning the workplaces in a line in combination with acoustically suitable partitions(cf. Figure 2). Figure 3 illustrates a direct comparison of the advantages and disadvantages of radial placement and linear positioning of workplaces in call centers.

Figure 2. Linear positioning of workplaces in call centers

2. Comparison of Call Center Furnishings

	"Radial" Form	*"Linear" Form*
• utilisation of surface area	high	low
• size of work surface	small	variable
• technical connections	point	finger
• glare/reflection	present	avoidable
• lighting	problematic	unproblematic
• visual contact	very positive	positive
• communication	very good	good
• visual distractions	frequent	negligible
• social control	high	adaptable
• visual contact with central information	not possible	possible

Figure 3: Advantages and disadvantages of radial and lineal positioning in call centers

3. Stress due to inadequate Lighting Design

The correct lighting design of a call center workplace with both daylight and artificial light can decrease unnecessary stress on the agent. The monitors must be installed parallel to the windows to avoid light density differences within the field of vision. Light density variations of not more than 1:3 are recommended by the relevant literature in the field of industrial science. Should direct lighting be used, a mirror light louvre technique is recommendable, which avoids any direct glare when looking upwards at the ceiling as well as preventing any reflections on the monitor screen by means of the limited angle of reflection for the light. In the author's experience, so-called dual or double component lighting has proved particularly successful. In this system the ceiling is lit evenly and glare-free by means of a wide-beam reflector, while the second component of the light source directs light at the work surface in an equally glare-free manner. If such lighting systems are additionally individually adjustable, then this leads to a high level of acceptance and an economically efficient illumination of the call center.

4. Stress due to inadequate Acoustic Design

An important part of the demands made on call center agents is the acoustic disturbances due to the phone-calls made by colleagues as well as the background noise innate in large and group offices. Stress due to noise is dealt with in a separate paper.

5. An Example of a well-designed Call Center

An example of well-designed call center layout and fittings is depicted in figure 4.

The large office of the firm Witt in Weiden shows a sophisticated design in both form and colour. The illumination by means of standard lamps for indirect/direct illumination lights the ceiling and makes the room appear lighter and airier on the one hand and, on the other, allows for individually adjustable light at the workplace. The workplaces with free-form desks (160 x 80/90), which can be easily adjusted by the workers for working in a sitting or a standing position, can be individually arranged. The workplaces were equipped with noise attenuating partitions to improve the acoustics. The placement of the workplaces in the room itself is generous and thus contributes further to the amelioration of the acoustics in it and to the well-being of the workers. All workplaces fulfil legal requirements with regard to concrete knowledge in the industrial sciences.

Figure 4: Call center Witt in Weiden with individual lighting of room and workplace, with individually adjustable desks for working in a sitting or a standing position. The workplaces were equipped with noise attenuating partitions to improve the acoustics. The placement of the workplaces in the room allows all employees a fine look through the windows (outside awareness).

Call Centre –Vocational Integration and New Software Applications for Blind and Visually Handicapped People

Klaus Miesenberger[1], Erdmuthe Meyer zu Bexten[2]

[1] University of Linz, i^3s^3; Altenbergerstr. 69, A-4040 Linz; phone: +43 732 2468 -9232
k.miesenberger@aib.uni-linz.ac.at
[2] Fachhochschule Gießen-Friedberg, BLIZ; Wiesenstraße 14, D-35390 Gießen;
phone: +49 641 309 23 69
erdmuthe.meyer-zu-bexten@mni.fh-giessen.de

Abstract

Telephone communication always has been an important field for vocational integration of blind and visually handicapped people. This field has witnessed significant changes the last years. Work with telephones got integrated into the much broader field of Customer Relationship Management (CRM) and into specialised units – call centres. Skills in handling standard and special computer applications are key qualifications today.

This paper reports on research done in computer supported telephone communication by blind and visually handicapped people. After an introduction to CRM and IT used a special focus will be given to the accessibility and usability of CRM systems for blind and visually handicapped people.

1. Customer-Relationship-Management (CRM) – a new field of work

In competitive markets and complex organisations external and internal contact management according to the needs and demands of partners becomes a key factor for success. To tell a client phrases like "hold the line", "call later", "please wait", "not now, please", "hold on, I am searching for ..." or "call our other office" will be interpreted as an invitation to search for a new partner. In addition increasing communication tends to information and contact overflow what may cause a waste of time of highly qualified staff.

These two often contradicting trends ask for specialisation, professionalism, support, automation and delegation. First areas like marketing, acquisition of clients and complaint management started to ask for professional Customer-Relationship-Management (CRM) and call centre services. Later on CRM and call centres became used to manage all internal and external communication in a professional way. Call centres could be defined locally, economically and organisationally as specialised communication units. Call operating, surveys, hotlines, ordering, booking, technical support, complaint management, information services, emergency calls, marketing, acquisition ... are examples of call centre activities. "*Inbound*" call centres receiving and handling incoming contacts and "*outbound*" call centres establishing contacts actively can be found. Call centre activities are organised as "*in-house*" services or outsourced to "*external*" service providers.

CRM is a fast growing market. More than 40% of all companies already use call centre services. CRM computer systems, e.g. in Germany, have reached a market volume of one billion DM in 2000, four times more than 4 years ago. Almost any company feels the pressure to invest in this field. The volume of expenditure for direct marketing reached a level of 32 billion DM. About 1500 call centres have been established till the end of 1999 employing more than 120.000 employees. In the USA, the biggest and most advanced call centre market, already in 1998 3% of the whole working population - 1.55 million people – have been employed in call centre related activities in almost 70.000 companies; 2 million employees are estimated for 2002. (Arndt & Jahnke,1999; CCA, 1998)

2. CRM and call centre technology – state of the art

Over decades telephone work has not changed much. Modern information technology is revolutionising this field. Today telephone work more and more gets integrated into CRM activities and is combined with other communication and management tasks. IT develops telephones further to multimedia PC workstations. Call centre and CRM applications are based on a marriage of telephone and computer technology. Three different levels of this integration can be seen:

- The first level is related to the handling of data: *"Front office systems"* provide facilities to find, enter and search the data needed during a call. *"Back office systems"* help to collect, monitor, analyse, control, finalise and present the data collected. Functionalities for automated document production and interfaces to standard office products support this process. *Forecasting/scheduling* helps agents, those people doing the communication, in planning and managing calls according to different persons, projects and communication histories.

  First *standard office products* were used for these purposes. Later on more sophisticated *individual software*, tailored database applications, was introduced. Today the fast growing market asks for systems, which could be costumed to almost any CRM tasks. (CCA, 1998)

- The second level is related to an efficient management of calls: *Computer Telephone Integration (CTI)* has become the key technology offering functionalities like *Automatic Call Distribution (ACD)* to route incoming calls to an agent based on skill associations. ACD systems also include so-called Agent Screening Functions like monitoring, statistics and accounting for agents or managers. Detailed and accurate information on agents, calls, groups, call flows, etc. is made available for supported or (semi-) automated staff planning and allocation. Besides controlling this supports activities like call load balancing, counselling, and training planning for agents.

  Another useful feature of CTI is the "*screen pop*" bringing information about the caller and the communication history to the screen of an agent when a call comes in or she/he dials a number. The caller is either recognised by the number he/she is calling from *(Dialled Number Information System – DNIS)*, by the number he called *(Calling Line Identification – CLI)* or by an identification number. Referring data are presented before establishing the connection. This allows a personalised and "to the point" communication taking the history of contacts into account. If the agent changes or another agent with different expertise should be involved all data are transferred in the same way before establishing the connection. This helps to avoid restating details.

  Interactive Voice Response (IVR) systems which ask for entering digits via touchtone key pad of a phone and/or via speech recognition are used for automatic information provision via recorded or synthetic speech. (Irish, 2000)

- The third level is related to the usage of multimedia and networking facilities: CTI allows building *"Web-enabled call centres"* (Irish, 2000). Beside telephone, snail mail and fax new means like E-mail, WWW, and chat are expected to be used for communication. These possibilities change the expectations and demands concerning quality and effectiveness of communication. The most efficient combination of media for a certain task should be employed. Internet telephony using Internet protocol (IP) or voice over IP (VoIP) will support this multimedia usage in the future. This relatively new concept offers features like a) asking questions via e-mails and routing or answering them automatically using keywords, b) asking for a call from an agent by activating a link on the WWW-page, c) building up chat rooms. Surveys for example can be done by sending a mail or pointing a person to a WWW form. After that an agent calls the candidate in order to motivate him to fill in and to assist him. Pictures, videos, and other media can be used to inform, support and motivate the candidate.

  Remote Agent packages support the integration of agents working from home or from any other location. The considerations for that could be many and amongst others like those concerning reduced costs, agreeable working conditions also the integration of people with disabilities.

3. CRM systems and accessibility problems

Modern information technology, as mentioned above, is revolutionising telephony and thereby also a traditional field of work of blind and visually handicapped people. Telephone operator and other jobs related to telephone communication have been important areas for the vocational integration of blind and visually handicapped people. The traditional telephone and professional applications linked to it can be seen as accessible tools and therefore possible areas of vocational integration.

Call centres and CRM are often seen as synonyms for outsourcing, increasing efficiency and thereby for reduction of cost by reducing labour force. Their image is very much related with professionalism empowered by complex IT applications. The often weak legal situation of employees in call centres has to be taken in consideration, too. Last but not least a tendency to bring "difficult" workers out of the office and place their jobs anywhere else by using remote agent systems may tend to exclusion. Such arguments are sometimes (mis) used to adapt the working environment to the needs of people with disabilities by avoiding their integration into call centres.

This at a first glance does not give a very optimistic perspective for the integration of blind and visually handicapped people into call centre activities. Nevertheless the increasing flexibility and adaptability of the working environment concerning time, location and organisation powered by IT makes call centres attractive for integration. New forms of jobs, companies and co-operatives (integration into call centres, special call centres, self-employment, home/distance work ...) could be tested. (Miesenberger & Stöger, 2000)

The technology driven changes in telephony may be seen, from a more conservative point of view, as a danger to loose a job because of accessibility problems or problems in learning and adapting to new tasks. From a more forward pushing point of view these changes in traditional telephony and the fast developing new fields of CRM have to be seen as a challenge to conquer new and more interesting jobs.

Taking up these new chances very much depends on the accessibility and usability of IT used in call centres. During the "BlindTrain" project Man-Machine Interfaces (MMI) and the usability for call centre work of standard applications, special call centre applications and CTI systems were tested. In the following important results are reported. More details can be found in (Miesenberger & Stöger, 2000):

- Standard office products, especially when used as "back office" applications, have an acceptable level of accessibility and usability. Over the last years screen readers, speech output as well as screen enlargement systems made these systems usable for blind and visually handicapped people. Training is available for the target group. In any case excellent skills in using these applications for preparing, performing and evaluating call centre work are seen as prerequisites for successful integration into this area of work.
- Nevertheless there is a need for special training and adaptations when standard office products are used as "front office" systems. Effective front office work using telephone and PC simultaneously is very time critical. Blind and visually handicapped people need more time for orientation and navigation. In addition to that blind Braille readers use their fingers both for typing and reading what causes time lags. The usage of speech output is also limited due to the usage of the audio channel in the conversation. Developing a "mind map" of the interface and knowing hotkeys for direct access to interface components have to be trained intensively.
- Special CRM applications are used to give better and faster access to data and for entering data at "front office" work. These systems structure data according to levels like companies, people, projects and actions as well as they provide navigable relations between them. This generally supports orientation, but still blind and visually handicapped users have to go hierarchically through long lists of data in one level without having access to the visual map the screen offers to sighted people.
- In addition in these special applications data elements are most of the times not ordered in the way an agent needs them in changing situations of communication. Searching a list of ten or more elements in a sequence during conversation lasts too long. Without developing a mental map of the interface or knowing hotkeys for direct access to data fields an efficient use during telephone communication is almost impossible.
- Special systems often do not support keyboard navigation, which is essential for blind and partially sighted users. Some fields, buttons and other elements cannot be reached or activated by hotkeys what makes navigation slow and cumbersome. Keys to navigate within a data record on the screen (e.g. Tab or Shift-Tab), between the different levels (e.g. Page-Down, Page-Up), between different data records in a level (e.g. function or hotkeys) are often not supported. Keyboard support in individual software is often incomplete, not intuitive and not definable.
- Further on screen readers very often have difficulties in getting access to buttons in individually developed software. Many buttons do not follow established standards of graphical user interfaces (GUI). Bitmaps without textual labels are used. This denies the screen reader the possibility to recognise them as buttons and to detect their function. This forces to manually assign textual labels to them, which is difficult and cannot be carried out by a blind person independently.
- Screen readers also have problems with synchronising to some data fields. Normally, a screen reader can identify focused data fields automatically. Fields for example that are arranged in a multi-column array may

cause such problems. To make these fields accessible, special scripts in the screen reader's built-in scripting languages need to be written.

- CTI offers possibilities to route only those calls to blind and visually handicapped agents, which they are able to manage. The screen pop function of call related information helps blind and visually handicapped people to prepare for conversation. Beside that the impact of CTI on the MMI for blind and visually handicapped people at front office work is rather low. The screening function and statistics are very useful for preparing special training and counselling for the target group.
- The integration of Internet, of services like E-mail, WWW and chat expands possible areas of work. The work of a single person might be restricted to one kind of contact, but in the long run skills in handling these tools and integrating them into the working field are necessary. The trend to multimedia in communication asks for skills in handling and combining these media. Otherwise agents with disabilities are in danger of being employed only for very low-level tasks. There is an increased need that the target group gets used to these tools.
- The introduction of standard CRM systems has to be seen as a major step forward for blind and visually handicapped people. MMI standards help blind and visually handicapped people in making systems accessible, in increasing usability and also flexibility to change between different tasks, jobs and call centres. They can relay on skills they acquire once and which they can use over a longer period of time in changing technical and organisational environments. Especially these trends, driven by general efforts to increase accessibility and usability of software, let us draw a much more optimistic picture about the vocational integration of blind and visually handicapped people.

Research performed during a Call Centre training course for blind and visually handicapped people in Austria (Miesenberger & Stöger, 2000) showed that most of these software tools, as long as they are based on modern GUI concepts, could be made accessible. There is of course a need for optimisation concerning an effective use of front office applications, which are very time critical for agents.

4. Training needs

The analysis above shows that the integration into training programmes or setting up special trainings will be most important for using the chances of call centre work for the target group. Due to the standardisation of MMI concepts there seems to be evidence that blind and visually handicapped people, when profound basic skills in handling a PC are available, can work successfully in call centres. They can also follow trainings independently which are most of the time organised in-house.

Although screen reader macro programming can solve accessibility problems the usability especially of front office applications is still to low for blind and visually handicapped people. This asks for intensive training because blind and visually handicapped users have to develop a "mind-map" instead of using the visual structure on the screen for fast navigation. This intensive training to get to know the interface and the hotkeys "by heart" still seems to be the only possibility to reach the level of speed necessary. This is in some respect in accordance with the style of working of blind and visually handicapped people and if the necessary training is made available they are able to manage the task. In any case more sophisticated methods to increase usability and speed should be worked out. Some first trials were made by using screen reader's built in macro programming language (Miesenberger & Stöger, 2000).

Market surveys report that basic IT and business knowledge are expected as a prerequisite to get a job in a call centre. Call centres provide trainings in using their software applications, in communication, organisation and especially in handling the contents that should be communicated. More than 80 % of these trainings are organised in-house what might cause problems to blind and visually handicapped people when trainers are not able to take their needs into account. The materials evaluated show that more than 13 days a year are used for training. Additionally social skills, frustration tolerance, creativity in communication and the will for continued learning are expected from applicants for call centre work. (ECMC, 1998)

It is also expected that special training programmes at different levels to prepare people for call centre work will be worked out in the near future. A special focus should be made on accessibility of these courses. The orientation of courses for blind and visually handicapped people towards accepted standard certificates like the European

Computer Driving Licence is recommended. This could raise the awareness of employers for the skills of blind and visually handicapped employees and thereby foster their chances of being hired. (Miesenberger & Stöger, 2000)

In accordance to other research related to the vocational integration of disabled people these studies showed that the biggest problems are related to prejudices and lack of information and awareness at employers site. Interviews and discussions with experts confronted with the idea of bringing blind and visually handicapped people into a call centre showed that after an initial refusal of the idea of integration employers begin to understand the possibilities and chances.

The evaluation also showed that the different tasks of call centre work are well defined and structured. From very basic to extremely complex ones can be found. This offers well-defined levels of work for integration as well as for a step-by-step approach in preparing blind and visually handicapped people for call centre work. Telephone communication trainers agreed that especially the ability of describing very precisely in verbal form when other media and forms of contact can't be used give blind and visually handicapped people special skills for these jobs. (Arndt & Jahnke, 1999; Miesenberger & Stöger, 2000)

One must not forget the fact that not more 20% of the qualification needed for call centres is IT oriented. (CCA, 1998) The importance of human-human communication, especially verbal telephone communication will increase because of the quality human contact offers in the growing multimedia jungle. Most important are communication, organisational and management skills as well as social skills to be successful in call centres. Although call centres might seem to be technology driven we must not forget the need for these skills in the long run.

5. Conclusion

The growing market of CRM and call centre activities should make us optimistic concerning the possibilities to find new fields for the vocational integration for blind and visually handicapped people. There are still a lot of problems in the accessibility and usability of applications used, especially when front office work and the time critical handling of software applications in parallel to telephone conversation is concerned. The tendencies towards standardisation of these applications let us estimate better accessibility and usability in the future.

These positive trends must not make us neglect that profound training programmes for blind and visually handicapped people are not available today. Courses concentrate on very specific aspects of call centre activities, which are low-level agent tasks most of the time. Without any perspective to get a broad training call centres might turn out to be a dead end of low-level jobs for blind and visually handicapped people. Research and development efforts are still needed to find new methods to increase the speed of the target group at front office work.

Last but not least one has to be careful concerning IT powered "remote integration". This seems to solve a lot of barriers at a first glance. But we must not forget that handicap is first of all a social construct defined by a lack of direct social interaction. Such systems are always in danger of establishing exclusion.

Acknowledgement
This research was done in the frame of the BlindTrain project supported by the European Social Fund, the Austrian Ministry for Labour and Social Affairs, Palm and Apriori Company.

References
Arndt, C. & Jahnke, J. (1999). Call Center; Jobchance für Menschen mit körperlicher Behinderung. ECMC, Marl
Call Center Akademie (CCA) NWR (1998). Qualifizierungsmaßnahmen für Call Center Agents im internationalen Vergleich, from the World Wide Web: http://www.callcenterakademie.de/06/f_ergebnisse/start.html
Irish, C. (2000). Web-enabled Call Centres: BT Technology Journal, Vol. 18, No 2, pp. 65-71
Miesenberger, K. & Stöger, B. (2000). BlindTrain - Call Center Ausbildung für blinde und sehbehinderte Personen; Evaluierung des Einsatzes von Informationstechnologie. Universität Linz, 2000
ECMC – Europäisches Zentrum für Medienkompetenz (1998). Ergebnisse der GfK – Marktstudie zur Entwicklung der Call Center-Branche in Deutschland, from the World Wide Web: http://www.callcenterakademie.de/pdf/gfkpraes.pdf

The assessment of the software quality by users

Jerzy Olszewski, Katarzyna Lis

University of Economics Department of Labour and Social Policy
61-875 Poznań, Al. Niepodległości 10
Poland
olszewsk@novc1.ae.poznan.pl

Abstract

The purpose of the paper is presentation the method of the assessment of the software quality by users and the results of researching works in several enterprises. The paper consist of four parts: introduction, two chapters and conclusions. The software quality is defined in to the introduction. This definition is contain in norm ISO 8402: the totality of features and characteristics of product or service that bear on its ability to satisfy stated or implied needs. Very important in this definition is fact that use of the simplest method of the assessment of software quality. The second part of this article presents the software quality method. This method consist in alternative assessment and selection of features which are connected with software abilities to satisfy needs and expectation of users. The principally features are: functionality, reliability, usability, efficiency, maintainability, portability, security and easy to use. The third part of this article includes the results of research and its analysis. With special attention to users qualifications. The last part of the article contains conclusions of the assessment of the software quality by users.

1. Introduction

The modern tendency of the society's development all over the world is focused on information. It's communicative role will increase together with the demand of global exchange of information. The extensive use of new information techniques is directly connected with the exchange of information. The computer: hardware and software has become the indispensable tool of any work. The computer user nowadays associates the quality of a computer with the respective make of the computer equipment. The quality of a computer is the notion one understands However, software quality is the notion which has different definitions. Therefor, it is essential to define the notion of software quality. Traditional approaches to quality put emphasis on meeting the specified requirements which are primarily functional but today it is so not much. Current approach to the software quality is defined in norm ISO 8402: the totality of features and characteristics of product or service that bear on its ability to satisfy stated or implied needs [2]. The attempts have been made to broaden the perception of quality, for examples in ISO 9126 which categories quality from a user perspective as functionality, reliability, usability, efficiency, maintainability and portability. The ISO 9126 definitions acknowledge that the objective is to meet user needs. But ISO 8402 makes it clear that quality is determined by the presence or absence of the attributes, with the implication that these are specific attributes which can be designed into the product [1]. It means that the measure of quality is the degree of satisfying the user's needs.

$$Q = \frac{\text{factual state}}{\text{expectations}}$$

If:

- $Q < 1$ - the degree of the satisfaction of the user's needs is imperfect, which leads to the decline of the position of the product on the market .If this happens the product requires modification.
- $Q = 1$ - the degree of the satisfaction of the user is ideal.
- $Q > 1$ - in this case the features of the product exceed the expectations of the user. It is connected with additional production costs and a higher price which a computer user is not likely to afford.

2. The method of assessment of software quality

In the method of assessment of software quality presented in this paper the definition of software quality ISO 8402 has been used In the method criteria of software quality, which are connected with the software ability to satisfy the user's needs and expectations, have been chosen These criteria include security, easy to use and the criteria from ISO 9126 (Fig. 1).

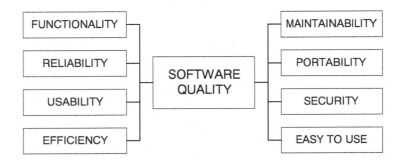

Fig. 1. Software quality characteristics.

In ISO 9126 software quality may be roughly evaluated by the following characteristics [3]:
– function - a set of attributes that bear on the existence of a set of functions and their specified properties that satisfy stated or implied needs,
– reliability - a set of attributes that bear on the capability of software to maintain its level of performance under stated conditions for a stated period of time,
– usability - a set of attributes that bear on the effort needed for use, and on the individual assessment of such use, by stated or implied set of users,
– efficiency - a set of attributes that bear on the relationship between the level of performance of the software and the amount of resources used, under stated conditions,
– maintainability - a set of attributes that bear on the effort needed to make specified modifications,
– portability - a set of attributes that bear on the ability of software to be transferred from one environment to another.

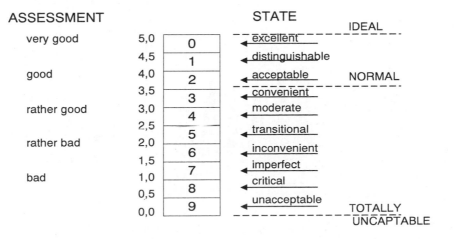

Fig. 2. The universal scale of relative states.

610

The users assess each of the criterion directly in the scale from 0 to 5. Where 0 means the lack of answer, 1 means bad, 2 means rather bad, 3 - rather good, 4 - good, 5 - very good. This is a universal scale of relative state thanks to which it is possible to estimate the software quality (Fig. 2). The assessment of the user automatically reflects the software quality and indicates the criteria influencing the worsening of software quality.

3. The results of the research

The research has been carried out among the people chosen at random. But the condition was that those people worked in companies located in Poznań. The total population was 219 people. Among these there were 164 (74,9%) computer users who did not work as computer scientist (101 - 61,6% women and 63 - 38,4% men) and 55 (25,1%) computer scientists (17 - 30,9% women and 38 - 69,1% men). All together there were 118 (53,9%) women and 101(46,1%) men. In the population of computer users there were more women while in the population of computer scientists there were more men.

The level of education of the population was 58% with higher education, 40,1% without higher education and 1,9% with vocational education only.

56,6% of the respondents when evaluating software quality assessed software functionality as good while 21,5% as rather good and only 15,5% as very good. Computer scientists consider the functionality of software as good.

47,0% of the respondents assessed the reliability as rather good. The same opinion is shared both by the computer users and computer scientists.

The usability was most valuable criteria assessed be the respondents because 54,3% considers it as good and 26,5% as very good. The efficiency of software was assessed as good (51,8%) and rather good (23,4%).

The maintainability of software was evaluated by the computer users as good (34,2%) and rather good (22,4%) However, computer scientists evaluated the same criterion as only good.

The portability of software which is understood as a set of attributes that bear on the ability of software to be transferred from one environment to another, has been assessed by 29,2% of the respondents as good but 24,2% did not respond to the very question because it was too difficult for them.

43,6% of the computer users and 40,9% of computer scientists assessed the security of data as good.

The easy to use criterion was assessed by the computer users as rather good (14,0%), good (47,6%) and as very good (31,1%). In the case of computer scientists the same easy to use criterion was assessed as rather good (30,9%), good (41,8%) and very good (16,4%).

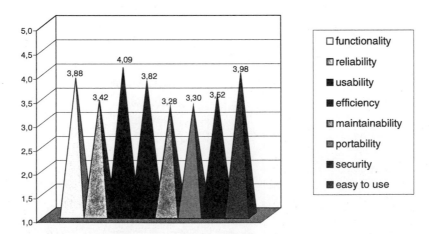

Fig. 3. The assessment of the criteria of software quality.

On the basis of the respective assessment of the criteria of software quality one can establish the state of software quality as follows: distinguishable state, acceptable state and convenient state (Fig. 3). The distinguishable state has the criterion of usability, the acceptable state has the criteria of functionality, efficiency and easy to use. The remaining criteria are in the convenient state.

4. Conclusion

The method just presented enables the measurement of the software quality. It compares the factual state of the software with the expected state. The quality state factors are the criteria defined as: functionality, reliability, usability, efficiency, maintainability, portability, security, easy to do. The results of this research have been obtained thanks to the use of the universal scale of relative states. Thus, it was possible to establish the normal state of the software quality assessed by users the most highly assessed criterion was the usability.

References
[1]   Bevan N. (1999). Quality in use: Meeting user needs for quality, The Journal of Systems and Software 49, pp. 89-96.
[2]   ISO 8402: 1994, Quality management and quality assurance - Vocabulary.
[3]   ISO 9126: 1991, Software product evaluation - Quality characteristics and guidelines for their use.

612

Towards Appropriate Design Tools for Inclusive Kitchens

Ken. J. Parker, Xie Hongyan

School of Design and Environment
National University of Singapore
bdgkjp@nus.edu.sg   Hongyanamtb@hotmail.com

Abstract

Computer aided design and drafting (CADD) methods are becoming popular in many areas. These tools make it possible to design, analyze and evaluate from the concept design through to final production. Many software packages used in architecture are optimized for drafting and presentation (in two and three dimensions). These packages provide little or no information concerning the end users' abilities, activities, vision, clearance and interaction between users and physical environment.

On the other hand, software packages are available for ergonomic analyses that are very powerful and often used to provide detailed man-models based on an anthropometrical database and the exact movements necessary for designing complex environments (such as vehicle cabs and airplane cockpits). In addition, these software packages are expensive, complicated and used on powerful workstation computers. It is suggested that a balance should be made between the two extremes, towards sculpting tools for architects and designers, that can conduct some basic ergonomic analysis as well as provide / organize related codes and design guidelines.

This paper discusses the information that should be involved in computer aided design tools for designing universal kitchens. Anthropometrics, code inspection and evaluation methods are discussed; this includes how a tool should be used to evaluate aspects such as accessibility, clearance and reach, taking into consideration a wide spectrum of users, including elderly persons and persons with disabilities.

A case study was conducted, evaluating several compact kitchens, which shows that CADD can support the understanding of a person's body capability, their functional reach, and the application of Codes. Visible simulation allows easier planning, design evaluation and manipulation. Suggestions are also given for future improvements of design tools for Universal Kitchen design.

Kitchens were chosen because they are the next most important personal space, after the toilet/washroom, for the continuance of the activities of daily living and personal independence within dwellings. The design of kitchens appears to be evolutionary rather than based on sound ergonomics and anthropometric data; this data should comprise the ergonomic operator needs of the equipment and the anthropometric abilities of users. This data should not be of an "average" person, but rather a compilation of the "worst case" scenarios of a range of users to ensure that a very wide spectrum of users can be accommodated. This includes those of short or tall stature, elderly persons, person with physical and sensory disabilities and persons using assistive devices (wheelchairs, etc.). Assistive devices (fixed or personal) can, in some instances, help bridge the gap between what the built environment provides and what a user needs. Singapore's public, high-rise, housing has been on a downsizing trend and the resulting smaller kitchens impact upon kitchen usability and efficiency for a wide spectrum of users with their varied personal ability sets.

Better design tools and exemplars (in the form of computer parametrics, design templates, Codes or guides, etc.) will provide better information for designers that will result in environments that are more inclusive. It is imperative that minimum criteria for more than "the majority" are established and satisfied.

1. Introduction

For both elderly people and persons with disabilities, whose mobility is limited and more vulnerable both physically and psychologically, the living environment, especially the domestic interior environment should be supportive, meeting their various and specific needs. Therefore, it is important that domestic environments be designed to be safe, barrier-free, supportive and inclusive.

For many years, Computer Aided Design and Drafting (CADD) methods have been very effective in the area of Architecture. There are many software packages available for both drafting and presentation (both in two and three dimensions). However, "... *as most CAD/CAM systems provide little or no information concerning the needs of the end user, there is considerable danger that design decisions may only consider the engineering, styling, legislative, and financial constraints for the product, with the ergonomics issues comparatively unaccessed until the design is completed*" (Porter et al., 1999). The prospective users' needs concerning their abilities, activities, vision, clearance and interaction between users and physical environment should be considered at all stages, from the preliminary design to final production.

Also, software packages are available for special ergonomic analyses that are very powerful and often used to provide detailed man-models based on anthropometrical databases. The man-models can simulate the exact movements necessary for designing complex environments (such as vehicle cabs and airplane cockpits). In addition, most of these software packages are expensive, complicated and used on powerful workstation computers (Porter et al., 1999). It is suggested that a balance should be made between the two extremes, towards sculpting tools for architects and designers, that can conduct some basic ergonomic analysis as well as provide / organize related codes and design guidelines.

2. Inclusive kitchen design

A successful design for elderly people and persons with disabilities (PWD) is based on the logical planning process and an understanding of the abilities, wants and needs of elderly people and PWD.

Recent, and on-going, research at the National University of Singapore has concentrated on design /adaptation /evaluation of domestic kitchens in public flats in Singapore, constructed by the Housing Department Board (HDB).

Kitchens were chosen because they are the next most important personal space, after the toilet/washroom, for the continuance of the activities of daily living and personal independence within dwellings. Kitchens are also multi-use and space-constrained. Kitchens are social spaces and are often used by two persons collaboratively. This is very different to the definite parameters and use patterns of automobiles and aircrafts; and the limited solo activities of washrooms / toilets.

The design of kitchens appears to be evolutionary rather than based on sound ergonomics and anthropometric data; this data should comprise the ergonomic operator needs of the equipment and the anthropometric abilities of users. This data should not be of an "average" person, but rather a compilation of the "worst case" scenarios of a range of users to ensure that a very wide spectrum of users can be accommodated. This includes those of short or tall stature, elderly persons, persons with physical and sensory disabilities and persons using assistive devices (wheelchairs, etc.). Assistive devices (fixed or personal) can, in some instances, help bridge the gap between what the built environment provides and what a user needs.

Undoubtedly, the application of CADD can increase the efficiency of designs. In order to know what is needed and required in an appropriate designing tool, a preliminary case study was conducted on Universal Kitchen design. The software used was 3D Studio Max 2.0 with the Character Studio plug-in. The planning tool prototype introduced by Eriksson et al. was adopted (Eriksson et al., 1995). In this prototype, the planning work was divided into four areas: Collecting Data, Modeling, Manipulation and Analysis, and Presentation.

3. Study : Inclusive kitchen

3.1 Collecting data

Two data sets were compiled; the first regarded the anthropometry of elderly persons and wheelchair users. Pilot surveys showed that there is a significant difference in anthropometric dimensions between western and asian elderly females. The average height is more than 15cm lower than their British counterpart and upper finger tip reach is about 22cm lower. For wheelchair users, the reachable range is a little less than recommendations given in western guidelines. The other data set was of the physical environment; five kitchens were selected and the main barriers for wheelchair users were identified. Dimensions of the kitchens and installed equipment were recorded (Hongyan & Parker, 2000).

3.2 Modeling

Modeling sometimes can be difficult and time consuming. Providing a model library can reduce the time by picking suitable models and putting them together. The models should be reshapable by changing the parameters. Another important feature of the model is hierarchy. Hierarchy is the logical and functional relationships between parts in the model. For example, doors and drawers are made as child objects attached to a cabinet as a father object. Thus

614

the software operator can move a cabinet as a whole or just open a door or pull out a drawer. Presently, existing models which can be purchased often lack this feature. The ideal software package for Universal Design should have a modeling system to make hierarchical models or the capability to adapt existing non-hierarchical models.

3.2.1 Man modeling

"There are five basic functionality of man modeling system: a. 3D modeling of people of the selected sex, age, nationality and occupational groups; b. knowledge base of comfort angles for the major joints of the body; c. ability to model the proposed workstation in 3D, together with the simulation of range of adjustment to be incorporated into the design; d. ability to assess the kinematic interaction between the models of people and the workstation; e. ability to make iterative modifications to the design to achieve optimum compromises" (Porter et al., 1999).

In Character Studio, a man-model was combined with a "biped" (bone system) and a mesh (flesh). Modification of the biped and mesh make it possible to build plump, muscular and lean man-models, but not very accurately. In an ideal man-model making system, man-models should be built and modified based on anthropometric dimensions such standing height, elbow height, arm length, etc. Thus the published anthropometric data can be used directly in man-model making.

It is important that man-models can have set limitations in joint range and body part, to simulate hemiplegia, paraplegia and the abilities of persons with other physical disabilities. But most man-model making software do not have this function, or are too complicated or time consuming for practical use by designers and architects. Therefore man-model making should become more realistic and user-friendly.

3.2.2 Motion library

Setting motion to a man-model is time consuming and complicated. A motion library could be set up to make this work step simpler for the software operators. In the planning tool prototype introduced by Eriksson et al. (Eriksson et al., 1995), the motion library was not mentioned. This conception came from the feature in Character Studio software where any motion file (*.bip) can be used on any figure (*.fig).

Previous research has studied the activities of daily living (ADL) and instrumental activities of daily living (IADL) of sixty older adults (Clark et al., 1990). This work used four different approaches to analyze the data: global task descriptions, task component profiles, task-specific profiles and action profiles. The motions can be defined through 3 aspects: 'Body Posture' (bend, lean reach, standard work, high reach, etc.), 'Action' (lift/lower, push/pull, fold, shake, etc.), and 'Hand Function' both of left and right (finger, pull, palm, etc.). The structure of the motion library should be built according to these four approaches and three aspects. Essential, basic, motion files can be pre-set, which will tend to simplify the work of software users.

3.3 Manipulation and analysis

At present, there are many software packages for Architects. Most of these packages are powerful for drawing and presenting, but few of them provide analysis functions. For a Universal Kitchen Design, analyses are necessary and very helpful. These analyses include: detecting code application, the moving space analysis of the users (a user in a wheelchair or with a walker, etc), work routine and routes analysis, storage and area analysis, and vision analysis.

3.3.1 Detecting code application

The software should allow checking a building model using a human model to determine if the building meets codes. The operator can serve as a 'virtual building inspector' or simulate a 'virtual user'.

"The virtual inspector and/or the simulated visitor who analyze the digital building may (1) discover and record new findings about accessibility issues, (2) include detection of discrepancies within the existing building code, and/or (3) add new applications of the building code which have not yet been documented" (Haley & Bermudez, 1999).

3.3.2 Moving space analysis of the users

In a Universal Kitchen, space must be provided for wheelchair users and users with other assistive devices (walkers, crutches) to maneuver close enough to cabinets and appliances to reach the knobs and controls of the devices. Since different approach methods and directions need different sizes of space, a database of the various maneuver spaces is important.

These maneuver spaces can be attached to the man-models by selecting the approach method. The ranges of the spaces can be shown on, and hidden from, the computer screen. The wheelchair's turning space (both circle and "T" turning) and knee spaces can be shown too.

3.3.3   Work routine and routes analysis

Most studies show that the most frequent movements occur between sink, working surface and cooker. Work-triangle, as a means of assessment in America, can be used to check the rational arrangement of working centers. In Germany, a method called *"Fadenstudie"* (line-studies) has often been used (Grandjean, 1973). In this method, all the paths flowed in the procedure of a certain job are recorded and drawn as lines in the kitchen ground plan. If a path is marked more often, that means this path is more important on the ground plan. Some typical tasks such as noodle-making or soup making are suggested for testing kitchen layouts.

Sometimes a Universal Kitchen may be used by more than two persons for different tasks – one is a wheelchair user (cooking) and the other is a standing user (doing laundry). The analysis on work routine and routes can show how the two users interfere and interact with each other and how much space they need.

3.3.4   Reach, storage and area analysis

Considering a Universal Kitchen which is used by a wheelchair user, the countertop frontage should be longer than that in a normal kitchen. The reason is that most necessary kitchen tools and equipments must be stored within the reach range of the wheelchair users, which is less than for the ambulant. Another reason is because the knee space provided under the sink and cooker, and low cabinets, serve to reduce the usable storage capacity of the kitchen (Figures1 & 2).

Aided measuring tools provided in software for the purpose of measuring area, length and calculating the storage in reachable range can be quite useful for analysis and evaluation during the design process.

3.3.5   Vision analysis

For a child, adult of short stature or wheelchair user, the viewpoints are lower than for others. Vision analysis can show what a user can really see; for example, if he/she can see inside the cooking pot, view the shelf contents or read warning signage.

3.4   Presentation

The important aspect of the output 'presentation' is not only the photo-like rendered pictures and animations. These vivid materials can stimulate discussion between designers, users and people of other areas: occupational therapists, kitchen fitment workers, etc. The analysis results, such as figures, charts or tables, are very important too for project evaluation and better, more inclusive, design decisions.

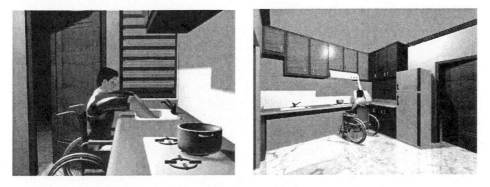

Figures 1 & 2: Height analysis of sink and reach at cooker hood.

4. Conclusion

Present kitchen design tools for inclusive kitchens are inadequate; the properties of suitable design tools have been identified. Better design tools and exemplars (in the form of computer parametrics, design templates, Codes or guides, etc.) will provide better information for designers that will result in environments that are more inclusive. Appropriate design tools for inclusive kitchen design, and other important spaces and tasks, are needed for interior designers and architects. These aided design tools should have better model-making systems and analysis functions. A well-built database (both of anthropometric and maneuver spaces), a model and a motion library are essential components of designer-friendly software for inclusive spaces.

References

1. Porter, J. M., Case, K. & Freer, M. T., (1999), Computer-aided Design and Human Models, in The Occupational Ergonomics Handbook, edited by Waldemar Karwowski and William S. Marras., USA.
2. Eriksson, J., Johansson, G. I. & Akselsson K. R., (1995), A Planning Tool Prototype for Environment Adaptations, in IEEE Transactions on Rehabilitation Engineering, Vol. 3, No. 3. September, 1995. New York.
3. Clark, M. C., Czaja, S. J., & Weber, R. A., (1990), Older Adults and Daily Living Task Profiles, in Human Factors, 1990, 32(5), 537-549. USA.
4. Hongyan, Xie & Parker, K.J., (2000), The Universal Design of Asian Kitchens, in Campaign 2000 : Breaking Barriers to Promote the Social Integration of Disabled Persons, 11-15 December 2000, Bangkok, Thailand.
5. Haley, Margaret & Bermudez, Julio, (1999), Evaluating Accessibility through Computer-Aided Design, in Enabling Environments: Measuring the Impact of Environment on Disability and Rehabilitation, edited by Edward Steinfeld and G. Scott Danford, USA.
6. Grandjean, Etienne, (1973), Ergonomics of the Home, Taylor & Francis Ltd, London.

Somatic symptoms in computer terminal workers

Ewa Salomon, Anna Janocha, Ludmila Borodulin-Nadzieja

Department of Physiology Medical University of Wroclaw, Chalubinskiego 10, 50-368 Wroclaw, Poland

Abstract

The examinations were carried in the two groups of 100 VDT operators and 100 traditional office workers (non-VDT). Both groups were subjected to an examination twice a day, before and after five hours work. The anterior segment of the eye was examined both before and after work. Additionally, after work, after dilating of the pupil, lens and vitreous were tested in a slit lamp and the fundus of the eye. A questionnaire was completed by all subjects to study the effect of a five hours work on the manifestation of subjective symptoms. The objective study of the eye in the VDT operators group proved that they suffered from the chronic inflammmation of the protective apparatus of the eye more often than the control group did. Moreover the subjective symptoms reported by the VDT operators were not only more varied but also appeared in much greater frequency than the control group.

1. Introduction

Widespread computerization in office work may produce new types of job-related disorders that involve the eyes. In particular, in performing their work, operators stay in constant contact with a screen monitor, the source of many physical factors that may affect their health. The aim of the study undertaken was to examine the condition of the eye in a VDT operators group. At the same time subjective symptoms were also investigated.

According to published data, the frequency of different complaints of subjective symptoms in VDT operators varies to a great extent from a few to several dozen percent and is generally greater than those found in non VDT workers.Taking into account the manifestation of subjective symptoms, different relationships have been examined.

There have been studies of the correlation between the frequency of symptoms and following factors : age, service duration, time, the frequency of breaks, optical features of the monitors and lighting (Smith et al. 1981, Shimai , Iwasaki & Takahashiet ,1986, Ong & Phoon, 1987, Shimai, Iwasaki & Suzuki, 1988, Rechichi & Scullica, 1990, Tarumi, Nagami & Kadowaki, 1990, Bergqvist et al. 1995). However, the authors did not take into account the factors influencing subjective complaints, such as the general health - status of a patient, eye disease, errors of refraction and functional visual system disturbances. Some authors observed that the VDT operators with refraction errors showed a greater tendency to report subjective symptoms (Luberto, Gobba & Broglia, 1989).

The aim of this study was to evaluate subjective complaints in group without refraction errors and functional disturbances in the visual system, after 5- hour work with display units.

2. Material

A 200 subjects took part in the study , 100 of them were the VDT operators (40 male and 60 female). A 5-hour task of the dialogue type was performed by these operators. A control group consisting of 100 persons (35 male and 65 female) performed traditional office work. Both groups were of similar age and service duration.

The average age of the VDT operators group was $41,5 \pm 5,9$ years (29 - 55 years) the average service duration was $18,6 \pm 6,5$ years (8 - 30 years) and the average experience with a monitor was $9,7 \pm 1,2$ years. In the control group, the average age was $42, 5 +- 7,6$ (29 - 55 years) and average service duration was $20,1 \pm 7,9$ years (8- 35 years).

According to age the two groups under examination were divided into three subgroups:
under 40 years (I), between 40-44 years (II), above 44 years (III).

The VDT operators performed their work using monochromatic monitors with a graphic card "Hercules". A „ light on dark „ mode of presentation was used on the VDT screen (green and amber characters against dark background). No protective filters were used. The distance between the eyes of VDT operators and the monitor was about 60 cm and the magnitude of presented characters was about 4,5 mm. Monitors were situated on the ordinary desks with their sides facing the windows. It was not possible to regulate the height of different elements used during VDT operations, such as monitors, documents and keyboards nor different parts of chairs used on work places.
Thus, from an ergonomics view the workplace conditions of this work were inappropriate.

The general lighting in the laboratory was natural, in case of need supported by fluorescent light. Intensity of lighting was 500 lx. No local lighting was used.

The noise varied from 50 to 55 dB . The operators did not have typical rest-breaks but rested when they felt fatigued. The work conditions of the control group (persons performing the traditional office work) were close to those of the VDT operators.

3. Methods

3.1   Objective study

Both groups were subjected to an examination twice a day, before and after work. The anterior segment of the eye was examined both before and after work. Additionally , after work , after dilating of the pupil with 1% Bistropamide lens and vitreous were tested in a slit lamp and the fundus of the eye.

3.2   Subjective study

A questionnaire containing items such as age and subjective symptoms concerned with "visual" and "general" status was completed by both groups under examination. The Chi- square test was employed for the evaluation of statistical signification of the difference in subjective symptoms reported by both groups under study.

4. Results

4.1   Objective measures

The objective study of the eye in the operators group revealed changes characteristic of chronic conjunctivitis in 24 persons. Blepharitis was recognized in 8 cases. The examinations carried out after work disclosed irritation symptoms i.e. the rubefaction of conjunctiva in 19 persons. Moreover, the tests revealed delicate, small, individual opacifications of the posterior capsule and cortex of the lens in 2 persons. They were young men of 40 and 41 who had no significant pathological record. No occurrence of cataract among their family members was noted in the interview, either. As far as the control group is concerned, except for 5 cases of conjunctivitis chronic, no changes in the organ of sight were observed.

4.2   Subjective measures

Table 1 presents the percentage of "general" and "visual" symptoms in VDT operators group against the control group. The subjective symptoms reported by the VDT operators were not only more varied but also appeared in much greater frequency than in the control group.
The most frequently reported subjective symptoms concerned the "visual" system in the VDT operators and were: visual fatigue (80 %), blurred vision (50 %) and burning sensation (29%). At the same time, in the control group, visual fatigue appeared in 48%, blurred vision in 16% and burning sensation in 12% of subjects.
Statistical analysis revealed significant differences between the VDT operators and control group in visual fatigue ($p < 0,05$), blurred vision ($p < 0,01$) and burning sensation ($p < 0,05$).
Analyzing complaints concerning general status, similar tendencies were observed ."General" symptoms of the VDT operators have been more varied, and what is more, the rates of complaints were relatively higher than those in the

control group.The most frequent symptoms in both analyzed groups were headaches, backaches, feeling of general fatigue and hyperexcitability.

The statistical analysis conducted by the use of chi-square test revealed significant differences for headaches (p < 0,05), for backaches (p < 0,05), for hyperexcitability (p < 0,05) and for fatigue (p < 0,01). It was noted that the VDT operators group showed other symptoms not occurring in the control group e.g. : gastrointestinal upset, wrist and fingers pain and facial erythemas. Since both groups under study consisted of persons of different age, we have analyzed the relation between age and the frequency of subjective symptoms.

Table 1. Percentage of "general"and "visual"symptoms in operator group against control group
* p< 0,05 , ** p< 0,01.

SYMPTOMS	OPERATORS	CONTROL
GENERAL		
Fatigue	27% **	12%
Headaches	48% *	36%
Backaches	58% *	32%
Gastrointestinal upset	11%	-
Hyperexcitability	37% *	28%
Wrist and fingers pain	9%	-
Cutaneus rash	2%	-
Facial erythemas	3%	-
VISUAL		
Ocular fatigue	80% *	48%
Burning sensation	29% *	12%
Itch	13%	-
Lacrimation	14,5%	4%
Blurred vision	50% **	16%
Scotomas	13%	4%

The rate of the visual symptoms by age of VDT operators and control group is presented in table 2.

Table 2. Rate of " visual " symptoms by age of VDT operators and control group.

SYMPTOMS	OPERATORS			CONTROL		
	AGE GROUPS					
	I	II	III	I	II	III
Ocular fatigue	84%	83%	63%	43%	70%	25%
Burning sensation	31%	22%	35%	-	20%	12%
Itch	21%	11%	11%	-	-	-
Lacrimation	10%	15%	17%	14%	-	-
Scotoma	10%	19%	-	14%	-	-
Blurred vision	52%	49%	52%	20%	28%	-

In both groups no relation between age and frequency of subjective visual symptoms was noted. Similar tendencies were found in the rates of general symptoms (table 3).

Statistical evaluation showed no significant differences between age and frequency of subjective symptoms in either of the groups under examination.

Table 3. Rate of "general" symptoms by age of VDT operators and control group.

SYMPTOMS	OPERATORS			CONTROL		
	AGE GROUPS					
	I	II	III	I	II	III
Fatigue	21%	31%	29%	14%	10%	12%
Headaches	57%	49%	41%	57%	30%	25%
Backaches	74%	49%	52%	43%	40%	12%
Gastrointestinal Upset	16%	4%	17%	-	-	-
Hyperexcitability	32%	42%	35%	57%	20%	12%

5. Discussion

In order to eliminate the influence of refraction errors upon the inflammation of the eye-protective apparatus eye irritation and fatigue, persons with emmetropia were classified as the subjects under study.

The objective study of the eye in the VDT operators group suggested that they suffered from the chronic inflammation of the protective apparatus of the eye more often than the control group did.

Chronic conjunctivitis was recognized in 24% operators, 8% operators suffered from chronic marginal blepharitis . The examinations carried out after work revealed bloodshot eyes in 19% operators.

Chronic inflammations of the protective apparatus of the eye and eye irritation in the VDT operators can be due to faulty lighting at the place of work, especially due to the following factors : glare, flicker, reflections, and dry air around the monitor. It was observed in many ophtalmological studies that VDT operators suffer from the inflammation of the protective apparatus of the eye more often. However, other eye diseases, like glaucoma, iris inflammation , maculopathy were not recognized as more frequent in VDT operators (De Groot & Kamphius, 1983, Cullen, Cherniack & Rosenstock, 1990). The examination with a slit lamp after dilating of the pupils with 1% Bistropamide revealed in both eyes of 2 persons (the operators were 40 and 41 old men) delicate, individual opacifications in the cortex and the posterior capsule of the lens. The opacitications were situated on the circuit and did not reduce visual acuity. Both operators had insignificant pathological record. The occurrence of cataracta in their families was not mentioned in the interview.

Zarret (1984), noted several cases of cataract in young VDT operators, claiming that the reason lies in the ultraviolet radiation emitted by the monitor screen.

A potential cataractogenic factor at the VDT operator's place of work is electromagnetic radiation: ionizing, ultraviolet-A, infra-red and microwave radiation. The field intensity of these radiations is considerably lower than those causing cataract (Boos et al. 1985, Knave et al. 1985). However , some of the epidemiological studies reveal that VDT operators suffer from cataract more often than controls (Boos et al. 1985).

The biological activity of the fields with complex spectral picture has not been thoroughly researched yet. However the number of the reports on their adverse effects is increasing .

Because of the physical dangers at the VDT operator's place of work, organization of the work space plays an important role in improving working conditions. A room with one monitor is relatively easy to arrange. When there are more monitors to install, the situation becomes more complicated. Then the parameters of the individual screens interfere with each other and the field intensity depends on the location of the monitors. If a location is faulty, not taking into account the resolution of the electromagnetic field around the individual monitors, the operators may find themselves in the area of field interference. Therefore it is necessary to pay special attention to the arrangement of the monitors in the room. Knowing the exact resolution of the field strength around the individual monitor, the optimum arrangement of many monitors can be chosen.

Working with display units mainly involves vision - controlled performance. This may be the reason for strain of the visual system . It was estimated that during operations with display units, 50% of all the work load represents "visual work". Therefore it is not supprizing that the subjective symptoms complained by the VDT operators were not only more varied , but also more frequent than in the control group (table 1).

The analysis of the frequency of complaints concerning vision suggested no dependency on age (table 2), similar results were obtained by Shimai, Iwasaki & Takahashi,1986.However, these authors claimed that the frequency of subjective symptoms reported by VDT operators depended mainly on their service time. The complaint rates of those workers with two or three year experience seemed to be relatively higher than those with shorter VDT experience. In the experiment reported here, the VDT operators were of similar service time, and there was not estimate of the the influence of this item on complaint frequency .

It can also be noted that musculoskeletal symptoms, headache and emotional problems are pre-eminent among general complaints (table 1). As in the case of visual symptoms, the frequency of the reported general symptoms was independent of age (table 3).

The present results and the literature data indicate that the work with display units may be the source of much more varied subjective symptoms than traditional office work (Luberto, Gobba & Broglia, 1989, Ong & Phoon, 1987, Smith, Cohen & Słammarjohn, 1981, Rechichi & Scullica, 1990, Shimai, Iwasaki & Suzuki 1988, Tarumi, Nagami & Kadowaki, 1990).

The organization of the work place of a VDT operator should take into account ergonomics parameters. Here "external factors" influencing development of fatigue process may be of great significance(Marumoto et al. 1997, Punnett & Bergqvist, 1999, Sotoyama et al. 1996, Villanueva et al. 1996) Among these factors lighting conditions and the visual properties of monitor screens play the main role.

References

Boos, S.R.,Calisendorff, B.M., Knave ,B.G., Nyman, K.G., Voos, M. (1985). Work with video display terminals among office employees III. Ophtalmological factors, *Scandinavian Journal of Work, Environment and Health, 11, 475 - 481.*

Bergqvist, U., Wolgast E.,Nilsson B., Voss M. (1995). Musculoskeletal disorders among display terminal workers: individual, ergonomic, and work organizational factors. *Ergonomics, 38:4, 763-76.*

Cullen, M.R., Cherniack, M.G., Rosenstock, L. (1990). Occupational Medicine (Second of Two Parts). *The New England Journal of Medicine , 322, 675-683.*

De Grott, J.P., Kamphius, A. (1983). Eyestrain in VDU users : Physical correlates and long-term effects. *Human Factors, 25, 409 - 413.*

Knave, B.G., Wibom, R.J., Bergqvist, U.O., Carlson, L.L., Levin, M.J., Nylen, P.R. (1985). Work with video display terminals among office employees II. Physical exposure factors. *Scandinavian Journal of Work, Environment and Health ,11 ,6, 467 - 474.*

Luberto, F., Gobba, F., Broglia, A. (1989). Temporary myopia and subjective symptoms in video display terminal operators. *La Medicina del Lavaro, 80 , 155-163.*

Marumoto, T., Sotoyama, M., Villanueva, M.B., Jonai, H., Yamada, H., Kanai, A., Saito, S. (1997). Correlation analysis between visual aciuty and sitting postural parameters of young students. *Nippon Ganka Gakkai Zasshi, 101:5, 393-9.*

Mbaye, I., Fall, M.C., Sagnon, A., Sow, M.L. (1998). Survey of pathology associated with the use of video display terminals. *Dakar Med. 43:1, 37-40.*

Ong ,C., Phoon ,W .(1987). Influence of age on VDT work. *Annuals of the Academy of Medicine of Singapore, 16 , 42-45.*

Punnett, L., Bergqvist, U. (1999). Musculoskeletal disorders in visual display unit work: gender and work demands. *Occupational Medicine 14:1, 113-24, iv.*

Rechichi, C., Scullica, L .(1990). Asthenopia and monitor characteristics. *Journal Francais D Ophtalmologie, 13 , 456-460.*

Shimai, S., Iwasaki, S., Takahashi, M. (1986). Subjective symptoms in VDT workers complaint rate and years of service. *Sangyo Igaku, 28, 87-95.*

Shimai, S., Iwasaki ,S., Suzuki, H. (1988). Survey on subjective symptoms in office workers using visual display terminals. *Fukushima Journal of Medical Science, 34, 45 -54.*

Smith, M.J, Cohen, G.F, Słammarjohn,L.W., Hoop, A. (1981). An investigation on health complaints and stress in video display operations. *Human Factors , 23 , 387-400.*

Sotoyama, M., Jonai, H., Saito, S., Villanueva, M.B. (1996). Analysis of ocular area for comfortable VDT workstation layout. *Ergonomics, 39:6, 877-84.*

Tarumi ,K., Nagami, M., Kadowaki, I. (1990). An inguiry into the factors affecting the complaints of subjective symptoms in VDT operators. *Sangyo Igaku , 32, 77-88.*

Villanueva, M.B., Sotoyama, M., Jonai, H., Takeuchi, Y., Saito, S. (1996). Adjustments of posture and vieving parameters of the eye to changes in the screen height of the eye to changes in the sreen height of the visual display terminal. *Ergonomics, 39:7, 933-45.*

Zarret, M. (1984). Cataracts and visual display units. In Pearce, B.G., (ed) Health hazards of VDTs ? Chichester, Wiley, 47 - 54.

Accessibility Guidelines:
Current status and future prospects in standardization

Constantine Stephanidis[1, 2], Demosthenes Akoumianakis[1], Anthony Savidis[1]

[1]Institute of Computer Science (ICS)
Foundation for Research and Technology-Hellas (FORTH)
Science and Technology Park of Crete, Heraklion, Crete, GR-71110, Greece
Email: cs@ics.forth.gr

[2]Department of Computer Science, University of Crete

Abstract

The accessibility of interactive computer-based products and services has long been an issue of concern to the Assistive Technology (AT) and Human-Computer Interaction (HCI) communities. Several efforts have aimed to document the consolidated wisdom in the form of general guidelines and examples of best practice. Despite their sound human factors content, these guidelines require substantial interpretation by designers, before they can generate practically useful and context-specific recommendations. The accumulated accessibility wisdom is now being consolidated into a draft Technical Specification (16071) by ISO 159 / WG 5 / SC 4. In this paper, we review this work and analyse its content and scope. Then, we advance a proposal for new activities, which could be accommodated in the technical specification (or as another part of the standard) in the course of on-going standardisation work.

1. Introduction

For several years, the available knowledge on user interface design for people with disabilities consisted primarily of anecdotal experience (Casali, 1995). With the exception of a few systematic studies (e.g., HFES/ANSI, 1997), the amount of readily available information on design practice and experience was far less than needed. By implication, accessible user interface design was largely based on intuition and on the competence of individual user interface designers. This state of affairs has been brought about partly due to some distinct peculiarities of the assistive technology sector (e.g., low technology adoption rate, lack of a solid science base), but also due to the apparent limitation of mainstream technology to impact and influence accessibility practices (Vernardakis, Stephanidis and Akoumianakis, 1997).

Of particular interest to this paper is the limited input of popular HCI design strands to the study of accessibility. For the purposes of our argumentation, we will briefly review popular HCI design traditions and examine their input to the study of accessibility in recent years. One such popular strand is that of cognitive / engineering modelling, which, over the years, has become the most prominent expression of social science input into the field of HCI. Despite this, the accessibility community has not only dismissed cognitive models, such as GOMS (Card, et al., 1983), but has even strongly criticised their application. One study, by Horstman and Levine (1990), concentrated upon a word prediction task in an augmentative communication system. Their work came under fire by Newell and colleagues, which has resulted in a lively exchange of opinions (Newell et al., 1992; Horstman and Levine 1992), regarding the suitability of cognitive user modelling in augmentative and alternative communication. In general, the models produced by cognitive science are designed to capture the generaliti﹏ of a population and are poor at addressing individual differences, or small groups of specialised users. This is due to the need for collecting statistically valid data to justify the models. On the other hand, accessibility work stresses the importance of individual differences and seeks to develop models and tools for accommodating such differences throughout a product's life cycle.

Another popular engineering discipline with major influence on HCI design is Human Factors evaluation. In HCI, Human Factors evaluation has delivered a wide range of general and platform specific design guidelines, as well as a rigorous scientific approach to system's evaluation. The available collections of guidelines cover a variety of topics and application domains. As with cognitive models, human factors evaluation has had limited impact on work on accessibility by disabled users. This is evident from the genuine lack of experimental evidence to characterise prevalent accessibility solutions. Despite this, in recent years there have been efforts to document the accumulated experiences on accessible design in the form of general, application-specific or user-oriented guidelines.

Accessibility guidelines are typically documented on paper, and reflect previous experience and best practice available for designing accessible interactive software. At present, there are several sources of guidelines for computer accessibility by disabled users (for a review see Nicolle and Abascal, 2001), several WWW sites containing relevant documents[1] as well as a collection of universal design principles (Story, 1998). These documents contain good design principles, design criteria, or design rules which have been found useful or applicable in specific application domains. Examples of such applications can be found in Web browsers (for example on the AVANTI browser see Stephanidis et al., 2001), public access terminals, point-of-sale equipment and cellular phones. Despite the sound human factors input propagated through accessibility guidelines, a number of problems impede the use of such guidelines (Akoumianakis and Stephanidis, 1999). Given the above, this paper offers an outline of a proposal which seeks to complement existing efforts in the area of accessibility, while, at the same time, it aims to broaden the covered application domains so as to include new accessibility issues. The latter, when tackled, would strengthen inclusiveness of the emerging Information Society and would respond to the recognised need for "umbrella" standards falling under the principles of universal access.

2. Extending the standardization agenda: A proposal

Today the accessibility challenge should be re-framed and articulated in a generic manner, to account not only for diversity in human abilities, but also diversity in the technological environment and the emerging contexts of use. This new frame of reference offers a broader insight to the study of accessibility in the emerging Information Society. At the same time, it also entails a thorough understanding of diversity in the user community, including disabled and elderly, in the technological environment (i.e., platforms, access terminals), and in the contexts of use (i.e., desktop, mobile, nomadic), as well as generic approaches to address diversity. To this end, there is a compelling need to expand the standardisation agenda to include several new items, which are progressively becoming crucial in the conduct of HCI design. Our proposal synthesises the above and comprises three critical elements which are not currently addressed by the current standardisation work. These are briefly described below.

2.1 The need for process guidance

Process guidance should identify, detail and elaborate on the process-oriented quality targets involved in designing for universal access. In the present context, process guidance on universal access should either exemplify ISO 13407 to resolve some of the problems identified above, or extend human-centred development by means of introducing more detailed design protocols. An important issue to be addressed in such a process guide is that of understanding the global execution context of tasks. To understand the needed shift in design perspective, let us contrast this principle (i.e., understanding the global execution context of tasks) against the more traditional focus on human centred design, namely understanding the user. In fact, understanding the user is a necessary but not sufficient condition to facilitate universal access. As a necessary condition, it needs to be treated appropriately and indeed, this the core thematic topic of human-centred design. Nevertheless, it can only offer a partial insight, which is sub-optimal when initial conditions or assumptions change, as, for instance, the context of use or the terminal used by the same user to access the same functionality. Understanding the global execution context of tasks reinforces the view that design should tackle as equally important quality targets variations in the technological and usage context. This entails that designers engage in processes to anticipate potential break downs, envision problems and solutions, and provide for incremental updates in their designs. In a nutshell, what designers need to be provided with is a design protocol for the "management of change" in HCI. To this end, it is appropriate to review macro-methods and instruments, which either directly facilitate such an objective, or provide the grounds for it. For example, the iterative nature of human-centred development offers a cost-effective basis for managing design changes. Additionally, suitable micro-methods are needed to facilitate specific objectives within the context of an iterative lifecycle. An example of the latter is the unified design method (Savidis et al., 2001).

---

[1] See for example http://www.w3.org/WAI/

2.2 Reactive versus proactive engineering

Either as part of the process guide or as a complement to it, review of engineering insights may be also useful. Two engineering perspectives on acessibility can be indentified (Stephanidis and Emiliani, 1999). The first roots to the reactive protocol, whereby adaptations are introduced a posteriori into a product or service to provide an alternative access system to be used by specific user categories. The alternative is to employ practices leading towards generic solutions to the problem of accessibility, and therefore closer to universal access. It is important to note the trade-offs between the two perspectives. The former is rather easier as it results in immediate gains. However, its short-term benefits are quickly outweighed by software updates, versioning, the rate of technological change, but also the short lifecycles of present day products. In contrast to this, the latter perspective entails proactive engineering practices to alleviate rather than remedy accessibility problems. This may require a "heavier" initial investment in design and usability engineering, but, on the other hand, it facilitates longer-term solutions, flexibility, quality in use and ease of maintenance.

2.3 Tool requirements for user interface development

In previous work, we have elaborated on the technical challenges involved in the implementation of a truly proactive engineering stragegies for user interface development (Stephanidis, 2001). In this section, we reflect on such work and summarise key outcomes regarding required and recommended quality attributes to be observed by tools for user interfrace development. Such tool requirements are needed in order to provide the grounds for constructing universally accessible user interfaces. These requirements can be summarised as follows:
- importing interaction resources offered by different interaction platforms (platform integration);
- augmenting the originally supported interaction techniques with new ones, suitable for specific users and contexts of use (platform augmentation);
- specifying interactive behaviours through abstract interaction elements relieved from platform-specific properties (platform abstraction);
- exposing and making use of information produced by external software tools (orthogonality).

In the following sections, we review each one of those requirements and discuss required and recommended features that development tools should possess to meet them.

2.3.1 Integration

Platform integration entails the capability to import any interaction platform that may be required for the development of interactive applications, so that all interaction elements of the imported platform can be directly accounted for by the same interaction building techniques. Platform integration is necessary in cases where the interaction elements originally supported by a particular interaction platform do not suffice to provide support for a particular type of interaction (e.g., non-visual). In such cases, it is important for the development tools to be able to utilise interaction elements from alternative sources (e.g., external object libraries) offering the interaction facilities required. It is important to note that, usually, interaction building blocks and re-usable interface components which are provided from different software vendors do not follow interoperability guidelines, thus introducing several impediments to platform integration. The required and recommended properties for a development tool to support integration are summarised in Table 1 (Savidis and Stephanidis, 2001).

Table 1: Integration: Required and recommended features of user interface development tools

Required	Recommended
• ability to link / mix code at the software library level; • documented hooks provided by platform developers, in order to allow mixing at the source code level.	• single implementation model made available for all integrated platforms, irrespective of the style of interaction supported; • Ability to modify aspects of the programming interface (i.e., the programmable view) of each imported platform; • Inter-operability of the integrated platforms.

2.3.2 Augmentation

Platform augmentation refers to the process through which additional interaction techniques are injected within the original collection of interaction elements of a particular platform. The rationale for platform augmentation arises from the fact that it is sometimes necessary to provide extended interaction facilities, beyond the original collection, which could be useful in specific contexts of use (e.g., voice-control of windowing application, scanning). In augmented development platforms, newly introduced interaction techniques should become an integral part of original toolkit elements, while old applications re-compiled with the augmented toolkit version automatically inherit the extra interaction features. The required and recommended features for a development tool to support augmentation are depicted in Table 2 (Savidis and Stephanidis, 2001).

Table 2: Augmentation: Required and recommended features of UI development tools

Required	Recommended
• ability to link / mix code at the software library level;	• single implementation model made available for all integrated platforms, irrespective of the style of interaction supported;
• documented hooks provided by platform developers, in order to allow mixing at the source code level.	• ability to modify aspects of the programming interface (i.e., the programmable view) of each imported platform;

2.3.3 Abstraction

Platform abstraction refers to the capability of a platform to specify interactive behaviours by means of abstract interaction objects. Platform abstraction is necessitated due to the fact that different interaction platforms offer different programming interfaces and calling conventions, thus complicating the development of an interface that makes use of several such platforms. For platform abstraction to be possible (see Table 3 for required and recommended features), there should be a well-defined protocol for mapping abstract interaction objects to concrete interaction elements as supported by a target platform (Savidis and Stephanidis, 2001).

Table 3: Abstraction: Required and recommended features of UI development tools

Required	Recommended
• pre-defined collection of abstract interaction object classes;	• facilities to define new abstract interaction object classes;
• for each abstract object class, a pre-defined mapping scheme to various alternative physical object classes;	• methods to define alternative schemes for mapping abstract object classes to physical object classes;
• alternative physical instances for each abstract object class;	• methods to define run-time relationships between an abstract instance and its various concurrent physical instances;
• activation of more than one physical instance for a particular abstract object class.	• methods to enable direct programming access, through the abstract object instance, to all associated physical instances.

2.3.4 Orthogonality

Orthogonality refers to the ability of a platform's run-time libraries to expose and make use of information produced by external software tools. When a user interacts with a particular application, there are issues that relate to the specific contexts of use and which can only be determined during the interactive session. In such cases, the interaction platform should provide the means to expose and receive information relevant to the context of use. Typically, what is required is well-established communications protocols and some extensions in the Application Programming Interface (API) of the interaction platform. Recommended feature is to provide support for inter-operability with other (external) software components. Table 4 summarises the required and recommended features for orthogonality (Savidis and Stephanidis, 2001).

Table 4: Orthogonality: Required and recommended features of UI development tools

Required	Recommended
• well-established communications protocols	• inter-operability with other (external) software components
• API extensions of the platform.	

3. Conclusions

This paper has presented the motivation and rationale for new standardisation activities in the area of universal accessibility. In contrast with other efforts, which build standards on the notions of "understanding the user" (e.g. ISO 13407), or that of a "universal user" (e.g. ISO 20282), our proposal emphasises the need for standards which promote and facilitate an understanding of the global execution context of task. This implies a process-oriented view emphasising the need for a generic universal design guide focusing on managing change and diversity as related to human abilities, technological platforms and contexts of use. It is expected that such an effort will be complementary to on-going activities (i.e. ISO TS 16071) and future ones. The primary target of such a process guide is to provide designers with a structured instrument which promotes and facilitates mechanisms through which designers can explore alternatives to obtain an understanding of the global execution context of a task, and devise accessibility solutions.

References

Akoumianakis, D., Stephanidis, C. (1999). Propagating experience-based accessibility guidelines to user interface development. *Ergonomics*, 42 (10), 1283-1310.

Card, S.K., Moran, P.T., Newell, A. (1983). *The psychology of Human Computer Interaction*. Hillsdale, N.J.: LEA.

Casali, S.P. (1995). A physical skills-based strategy for choosing an appropriate interface method. In A.D.N. Edwards (Ed.), *Extra-ordinary Human Computer Interaction: Interfaces for people with disabilities* (pp. 315-342). Cambridge University Press.

HFES/ANSI 200 (1999). *Draft HFES/ANSI 200 Standard, Section 5: Accessibility*. Santa Monica, USA, Human Factors and Ergonomics Society.

Horstman H, Levine S. (1992). Modelling AAC user performance: response to Newell, Arnott and Walter. *Augmentative and Alternative Communication*, 8, 231-241.

Horstman H, Levine S. (1990). Modelling of user performance with computer access and augmentative communication systems for handicapped people. *Augmentative and Alternative Communication*, 6, 231-241.

Newell A.F., Arnott J.L., Walter A. (1992). Further comment on the validity of user modelling in AAC. *Augmentative and Alternative Communication*, 8, 252-253.

Newell A.F., Arnott J.L., Walter A. (1992). On the validity of user modelling in AAC: comments on Horstman and Levine (1990). *Augmentative and Alternative Communication*, 8, 89-92.

Nicolle, C., Abascal, J. (2001). *Inclusive Design Guidelines for HCI*. NY: Taylor & Francis.

Savidis, A., Akoumianakis, D., Stephanidis, D. (2001). The Unified User Interface Design method. In C. Stephanidis (Ed.), *User Interfaces for All: concepts, methods and tools* (pp. 417-440). Mahwah, NJ: Lawrence Erlbaum Associates (ISBN 0-8058-2967-9, 760 pages).

Savidis, A., Stephanidis, C. (2001): Development requirements for implementing unified user interfaces, In C. Stephanidis (Ed), *User Interfaces for All: concepts, methods and tools* (pp. 441-469). Mahwah, NJ: Lawrence Erlbaum Associates (ISBN 0-8058-2967-9, 760 pages).

Stephanidis, C. (Ed) (2001). *User Interfaces for All: concepts, methods and tools*. Mahwah, NJ: Lawrence Erlbaum Associates (ISBN 0-8058-2967-9, 760 pages).

Stephanidis, C., Emiliani, P-L. (1999) Connecting to the information society: a European perspective. *Technology and Disability*, 10 (1), 21-44.

Stephanidis, C., Paramythis, A., Sfyrakis, M., Savidis, A. (2001). A Case Study in Unified User Interface Development: The AVANTI Web Browser. In C. Stephanidis (Ed), *User Interfaces for All: concepts, methods and tools* (pp. 525-570). Mahwah, NJ: Lawrence Erlbaum Associates (ISBN 0-8058-2967-9, 760 pages).

Story, M.F. (1998). Maximising Usability: The Principles of Universal Design. *Assistive Technology*, 10 (1), 4-12.

Vernardakis, N., Stephanidis, C., Akoumianakis, D. (1997). Transferring technology toward the European Assistive Technology Industry: Mechanisms and Implications. *Assistive Technology*, 9 (1), 34-36.

Noise in call centers

Charlotte A. Sust

ABoVe Ltd.
PO BOX 100 106, D-35331 Giessen, Germany
Kerkrader Str. 9, D-35394 Giessen, Germany

Abstract

The Federal Institute for Occupational Safety and Health" ("Bundesanstalt für Arbeitsschutz und Arbeitsmedizin") financed a scientific project to examine – within a wide range of different questions - the effects of noise in call centers. In a first phase about 1000 callcenters in Germany were contacted. The call center managers were asked to fill in a questionnaire about the working conditions (workplace, time, organisation, environment etc.). After analyzing these data, the second phase started with a detailed analysis of several callcenters. That means, beside analyzing the workplaces and collecting data about the environment (mainly lighting, climate conditions, noise) at least 15 agents in every callcenter were interviewed about their working conditions. Mainly they had to describe their work and to evaluate their working conditions. The interviews usually lasted about 45 to 60 minutes. In addition to the interview every participant had to fill up a questionnaire.

This contribution deals mainly with the agents' perception and evaluation of the acoustic environment, which is described with objective data (different noise measures). The acoustical environments includes machinery (air condition, personal computers etc.) as well as social noise (people going around, talking, laughing etc.). While fulfilling their main task – talking to their customers – employees in callcenters produce sound, which is "noise" to their colleagues, because of its disturbing quality. Talking as a sound has a high content of information and meaning.

In order to reduce the diversion of attention agents take different measures to comply with the negative effects of noise. Usually they have no other possibilities to cope with the noise other than to concentrate harder, to focus their attention by shutting out the noise (press their headsets to their ears f.e.). Especially in "rush hours" this is stress, which can reduce performance (more errors) and in the long term could result in impairment of health.

To reduce the noise levels different possibilities are discussed: f.e. sound screens, headset devices, office layouts. Best practice solutions will be shown.

1. Introduction

At first sight call center agents just seem to be a specialised office job. But there are some important differences concerning stress conditions. This contribution deals with one of the main causes of stress, that is noise.

But what is noise in call centers. Usually we think about noise in the context of very loud sound and hearing loss and the like. Actually we rarely find any of these loud noises in call centers. Still the agents complain especially about noise (see picture 1).

So I want to present a broader definition of noise which we use in Germany. We define noise as unwanted sound (see Guski, 1987, p.9):

> ➢ unwanted because it can cause hearing loss,
> ➢ unwanted because it can cause illness (f.e. cardiac-circulatory or gastro-intestinal diseases),
> ➢ unwanted because it can divert attention, reduce concentration, disturbs communication and regeneration.

That means this definition exceeds the technical definition: "noise" is not exclusively reduced to high levels of sound. It takes into account the evaluation of those who are troubled by noise. The disturbing quality results from its content of information – that is the number of changes over time – and/or its meaningfulness. In general we can say

628

the higher the content of information or meaningfulness the higher is the probabilty of stress reactions which can have effects on performance as well as in the long term on health.

2. Noise in call centers – some examples

Taking a closer look to what agents in call centers complain about, we find beside the general acoustic condition – background noises from PCs, air condition and the like – the main source of disturbing noise are "talking neighbours" (see picture 1).

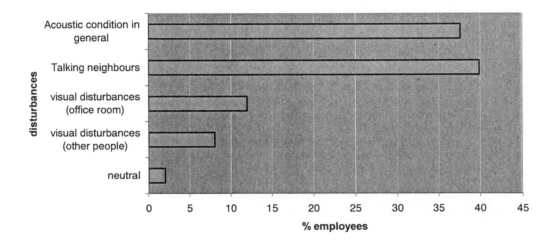

Picture 1: Sources of disturbances for employees in call centers

In our project, which was financed by the Federal Institute for Occupational Safety and Health, in Dortmund, Germany, we examined several call centers more closely , not only from the acoustic viewpoint but also with concern to other environmental conditions like lighting and climate as well as to organizational aspects like work flow, working time, tasks, qualification, training etc. These call centers were representatives of different types of call centers.

We examined the acoustical conditions that is we took acoustic measures ($L_{Aeq,}$) for different work places and we asked the employees about their perception and evaluation of the acoustic condition.
Two callcenters are shown which present the positive and negative examples how to comply with the problems of disturbances caused through meaningful noise.

The negative example: its call center for travelling information, altogether about 120 employees. We interviewed 20 of them. Acoustic measurements show noise levels of about 60 dB(A) over the day which is easy to understand if you see how employees are "packed" in their office room (see picture 2). They don't have any attenuating screens.

Result is a high disturbing quality, employees have to concentrate very much to fulfill their task. Additionally they receive about 10 to 20 % calls from mobile phones. That reduces speech intelligibility on both sides (calling customer and speaking employee) and forces them to talk even louder with the effect of a still higher noise level (up to 65 dB(A) speech level).

The results of the objective measurements are reflected in the answers of the interviews. 80 % of the employees complain about the noise: 90 % of these complain about talking, talking of neighbours, laughter, crying babies (they are situated on a gallery over a hall with public traffic).

In reaction to this they try to ignore the noises – most often in vane -, they press their headset against their ears (which results in louder speaking – Lombard effect), or to concentrate more intensely. After a working day most of them feel testy or exhausted or complain about headache and the like.

Picture 2: "Packed" work places in a call center for travelling information

Looking at the positive example – a call center of an insurance company - a lot more room is to be seen (picture 3).

Picture 3: Work places in a call center of an insurance company

Between the work places and in front of the employees are sound absorbing screens, which reduces the noise level remarkably (picture 4).

Picture 4: Sound-absorbing screens between work places

Measurements show a noise level of around 50 dB(A), which is fairly quiet. Still people interviewed complain about the noise, again sources named are talks, laughter, talking neighbours, but also about noises which occur rather rarely over the day or the week, just like the fire alarm or the cleaning woman with the vaccuum cleaner. If one wonders why people still complain about the noisy conditions, one has to take into account, that these employees have to fulfill much more difficult task as those mentioned before. They have to consult customers concerning insurances while the tasks of the employees of the "packed" call center mainly consists in giving information concerning travelling conditions etc..

3. Noise in call centers – what can be done?

If you are interested to reduce the noise induced stress in call centers, there are several options

- ➢ Give people room – noise level is divided in half by every doubling of distance
- ➢ Install sound absorbing floor, ceilings and walls, which reduces reverberation time.
- ➢ Install sound absorbing screens between workplaces, which reduces speech level
- ➢ Care for good quality headsets, so employees are well to understand and don't need to talk louder as necessary or to repeat their information to the customer

Even if one doesn't think it necessary to better the working conditions for the employees one should take into account the reaction of the customer calling. Imagine somebody who has f.e. a computer problem, then calls the hotline and is confronted with a noisy background which makes it difficult for him to follow the employees advices. If he is already confronted with that computer problem, he should not have an additional problem caused by noise, for which he cannot made responsible. There are solutions: You can reduce the noise for the customer. He might ask himself why the agent is not as attentive as he should be.

References:

Guski, R. (1987). Lärm – Wirkungen unerwünschter Geräusche (Noise - Effects of Unwanted sounds). Bern Stuttgart Toronto: Hans Huber

Sust, C.A. & Lazarus, H. (in prep.), Büroarbeit und Geräusche (Office work and noise). Bremerhaven: Verlag Neue Wissenschaft

Emotion and stress in Call Centers

Andreas Utsch

ABoVe Ltd.
PO BOX 100 106, D-35331 Giessen, Germany
Kerkrader Str. 9, D-35394 Giessen, Germany

Abstract

Working in call centers is not only demanding because of noise, inadequate workplaces, software and/or hardware problems but also because the care for customers could become in itself a stress factor. This is especially true if the agent is confronted with customers who give way to their frustration and/or aggressive behaviour. This might be caused if either service level of the call center is low (which means long periods of waiting or busy lines) or if there are problems with products or services of the company itself. So call center employees should be able to communicate properly with aggressive or frustrated customers – which means to keep polite and friendly even if customers are insulting, try to solve the customer's problems even if it means to solve the company's problems, to leave the customer satisfied even if there is hardly a chance to. If the call center agent is not able then sooner or later he or she will suffer from "burn-out"-symptoms.

The Federal Institute for Occupational Safety and Health" ("Bundesanstalt für Arbeitsschutz und Arbeitsmedizin") financed a scientific project to also address the problem of "emotion and stress" in call centers. In a detailed analysis of several call center agents answered questions concerning emotion and stress. These interviews lasted nearly about one hour each. Additionally the participants also received a questionnaire with a lot of questions to different topics, among them they had to rate their stress in general and especially during work.

1. Introduction

Stress is not a new phenomenom in psychological research. A lot of studies demonstrate a significant relationship between stressors and strains (cf. Kahn & Byosiere, 1992). Recent psychological stress research takes a new look at this area, because job demands are changing and with changing demands stressors change too (Morris & Feldmann, 1996). Especially in the area of service, new jobs are created. These new jobs involve emotion work. A typical new job in the service industry is the call center agent. This study focuses on emotion work as a source of stress for call center agents.

2. Emotion Work

The definitions of emotion work vary. Morris & Feldmann (1996) describe emotion work as "the effort, planning and control needed to express organisationally desired emotion during interpersonal transaction" (p.987). Baily (1996) defines emotion work as a "work role requirements concerning the display of appropriate emotions to create an impression as desired by an employer" (p.2). Both definitions describe emotion work as conscious and intentional action demanded by an organisation.

2.1 Emotion Work and Stress

Work is not necessarily a stressor. Work can produce well-being as well as strain. Similarly, emotion work should not lead necessarily to strain. Work overload, uncertainty, or obstacles are reasons why work can become a stressor. Emotion work becomes a stressor when people are forced to show emotions they wouldn't display otherwise. In this case emotional dissonance is a stressor because dissonance is an obstacle to do the job properly. Therefore, emotional dissonance is a mediator between emotion work and strain or well being. For example, showing positive emotions should only, lead to strain if they are not in line with person's true emotions. This means one has to show positive emotions although one does not feel that way and expierence emtional dissonance. Also, the frequency of emotional dissonance depends on the frequency of interactions.

➤ H1 The frequency of interactions is positively related to emotional dissonance
➤ H2 Emotional dissonance is negatively related to well being.
➤ H3 Emotional dissonance is a mediator between emotion work and well being.

2.2 Stress Resources

From psychological stress research, we know four types of stress resources: Internal resources (for example qualification), external social resources (for example social support), external task related resources (for example control) and external conditions (for example participation) (Karasek, 1979; Zapf, Dormann, & Frese, 1996). For emotional work, social resources and task-related resource are most important. Both of these resources can buffer the relationship between emotional dissonance and strain.

➤ H4 Social resources moderate the relationship between emotional dissonance and strain.
➤ H5 Task resources moderate the relationship between emotional dissonance and strain.

3. Method

The Federal Institute for Occupational Safety and Health" ("Bundesanstalt für Arbeitsschutz und Arbeitsmedizin") financed a scientific project to also address the problem of "emotion and stress" in call centers. In a detailed analysis of several callcenters agents answered questions concerning emotion and stress. These interviews lasted nearly about one hour each.

Participants were 60 call center agents (preliminary sample size) of four German call centers. Emotion work was measured by the FEWS (Frankfurt Emotion Work Scales (Zapf, Vogt, Seifert, Mertini, & Isic, 1999)). The FEWS includes the scales: frequency of emotion, showing sympathy, and emotional dissonance. Social resources were measured with the scale social support (Caplan, Cobb, French, van Harrison, & Pinneau, 1975). Task related resources were measured with the work control scale by Semmer (1984). Well being and strain were measured with three different scales: Job satisfaction (Warr, Cook, & Wall, 1979), emotional exhaustion (measured by the Maslach Burnout Inventory- German version of (Büssing & Perrar, 1992)) and psychosomatic complaints (Mohr, 1986).

4. Results

Emotional dissonance is positively related to the frequency of showing emotions and of showing sympathy. H1 was confirmed. H2 was also confirmed. There is a positive relationship between emotional dissonance and emotional exhaustion, psychosomatic complaints, and job satisfaction. (see Table 1).

The frequency of showing emotions and showing sympathy was not related to emotional exhaustion, psychosomatic complaints and job satisfaction. Therefore the prerequisite to test for mediation was not fulfilled. H3 has to be rejected. H4 and H5 hypothesized moderator effects of social and task related resources. These hypotheses have also to be rejected. The interaction terms do not increase the R^2.

Table 1: Means, standard deviations and correlation

	(1)	(2)	(3)	(4)	(5)	(6)	(7)	(8)
(1) Frequency of emotions								
(2) showing sympathy	.474**							
(3) work control	.193	-.112						
(4) social support	.072	.099	.199					
(5) emotional dissonance	.429**	.370**	-.328**	-.283*				
(6) psychosomatic complaints	.198	.234	-.260*	-.040	.321*			
(7) emotional exhaustion	.056	-.027	-.198	-.140	.361**	.605**		
(8) Job satisfaction	.061	.048	-.570**	-.299*	.421**	.437**	.554**	1.000

Note: N=60, * p<.05; ** p<.01

5. Discussion

The major result of the study is that emotional dissonance has a strong relationship with well being. Emotion work itself is not a key factor for well being. Additionally, the study has not found a resource to buffer the relationship between emotional dissonance and strain. New research should look in to ways on emotional work. How could we reduce emotional dissonance and are there other moderators between emotional dissonance and strain.

References
Baily, J. (1996). *Service agents, emotional labor and costs to overall customer service.* Paper presented at the Annual Conference of the American Society for Industrial and Occupational Psychology, San Diego.
Büssing, A., & Perrar, K.-M. (1992). Die Messung von Burnout. Untersuchung einer deutschen Fassung des Maslach Burnout Inventory (MBI-D) [The measurement of burnout: Studies on a German version of the Maslach Burnout Inventory]. *Diagnostica, 38*, 328-353.
Caplan, R. D., Cobb, S., French, J. R. P., van Harrison, R., & Pinneau, S. R. (1975). *Job demands and work health .* Washington: National Institute of Occupational Safty and Health.
Kahn, R. l., & Byosiere, P. ((1992)). Stress in Organizations. In M. D. Dunette & L. M. Hough (Eds.), *Handbook of industrial and organizational psychology* (2 ed., Vol. 3, pp. 571-650). Palo Alto, CA: Consulting Psychologists Ress.
Karasek, R. A. (1979). Job demands, job decision latitude, and mental strain: implications for the job redisign. *Administrative Science Quarterly, 24*, 285-308.
Mohr, G. (1986). *Die Erfassung psychischer Befindensbeeinträchtigungen bei Arbeitern [The measurement of psychological dysfunction of workers].* Frankfurt a.M.: Peter Lang.
Morris, J. A., & Feldmann, D. C. (1996). The dimensions, antecedents, and consequences of emotional labor. *Academy of Management Journal, 21*, 989-1010.
Semmer, N. (1984). *Stressbezogene Tätigkeitsanalyse: Psychologische Untersuchungen zur Analyse von Stress am Arbeitsplatz.* Weinheim: Beltz.
Warr, P. B., Cook, J. D., & Wall, T. D. (1979). Scales for the measurement of some work attitudes and aspects of psychological well-being. *Journal of Occupational Psychology, 52*, 129-148.
Zapf, D., Dormann, C., & Frese, M. (1996). Longitudinal studies in organizational stress research: A review of the literature with reference to methodological issues. *Journal of Occupational Health Psychology, 1*(2), 145-169.
Zapf, D., Vogt, C., Seifert, C., Mertini, H., & Isic, A. (1999). Emotion Work as a Source of Stress: The Concept and the Development of an Instrument. *European Journal of Work and Organizational Psychology, 8*(3), 371-400.

Development of Generic Accessibility/Ability Usability Design Guidelines for Electronic and Information Technology Products

Gregg C. Vanderheiden

Trace R&D Center
University of Wisconsin – Madison
Madison, WI 53719

Abstract

In designing products it is useful to have a set of design guidelines or standards which can guide one's efforts. This is particularly true when trying to design a product for users for whom they do not have complete expertise. These guidelines or standards can take a wide variety of forms. They can be general principles such as the principles for Universal Design which provide broad concepts not specific to any particular type of product; design standards such as those issued by the Access Board which, for the most part, focus on what should be accomplished, and checklists or design standards that provide specific, testable criterion for design but are limited to particular technologies and even particular form factors. Each of the different types of guidelines or standards (which the above are just three examples) have different roles. For example, principles are useful in teaching the underlying concepts but do not give much specific guidance. Regulations may specify what is to be achieved but, for the most part, are intentionally designed to not specify how it should be achieved. Design standards and checklists can be very specific as to how accessibility should be implemented but often lack generalizability. What is appropriate and effective on a ticket machine in an airport may not be appropriate, practical, or even possible on a pocket sized or wristwatch sized device. Having the variety of different guidelines, standards and checklists can be confusing to people just coming into the field and trying to understand what makes a product accessible or usable. This paper is part of an ongoing effort to create a set of guidelines for electronic and information technologies which captures both the requirements and the underlying strategies for meeting the requirements in a simple and compact form and comments are solicited.

1. Background

At the present time, there is a growing number of design guidelines to address the needs of people with disabilities or those operating under constrained conditions. There are web guidelines, software guidelines, consumer product guidelines, telecommunication guidelines, electronic product guidelines, etc. Some of these guidelines resemble each other and others have entirely different bases. we continue to evolve and develop new technologies, it will be difficult to find or develop a set a guidelines to match each technology as it comes out.

In an attempt to identify the basic principles underlying universal design, a group of individuals from multiple institutions were brought together by the University of North Carolina's Center for Universal Design and developed a set of universal design principles Connell, B. R., & et al. (1997)

These principles are:
1. Equitable Use
2. Flexibility in Use
3. Simple and Intuitive
4. Perceptible Information
5. Tolerance for Error
6. Low Physical Effort
7. Size and Space for Approach and Use

Each of these is then broken down into some basic guidelines. From there different sets of specific guidelines can be designed for different fields.

636

Basic Guidelines and Strategies for Access to Electronic Products and Documents		
Basic Access Guideline	**Why**	**How – General**
Make **all information** (including status & labels for all keys & controls) **perceivable** - Without vision - With low vision and no hearing - With little or no tactile sensitivity - Without hearing - With impaired hearing - Without reading (due to low vision, learning disability, illiteracy, cognition or other) - Without color perception - Without causing seizure - From different heights - NOTE: other aspects of cognition covered below	Information which is presented in a form that is only perceivable with a single sense (e.g., only vision or only hearing) is not accessible to people without that sense. **Note**: This includes situations where some of the information is only presented in one form (e.g. visual) and other information is only presented in another (e.g. auditory). **In Addition:** Information which cannot be presented in different modalities would not be accessible to those using mobile technologies e.g. - Visual only information would not be usable by people using an auditory interface while driving a car. - Auditory only information would not be useable by people in noisy environment	**FOR INFORMATION:** Make all information available either in a) **presentation independent form** (e.g., electronic text) that can be presented (rendered) in any sensory form (e.g. visual-print, auditory-speech, tactile-braille) **OR** b) **sensory parallel form** where redundant and complete forms of the information are provided for different sensory modalities (synchronized). (e.g., a captioned and described movie – including e-text of both), **FOR PRODUCTS:** Provide a mechanism for presenting all information (including labels)in visual, enlarged visual, auditory, enhanced auditory (louder and if possible better signal to noise ratio) and, (where possible), tactile form. **NOTE:** - this includes any information (semantics or structure) that is presented via **text formatting or layout.**

Provide **at least one mode (or set of different modes)** for **all product features** that **is operable**:
-    Without pointing
-    Without vision
-    Without requirement to respond quickly
-    Without fine motor movement
-    Without simultaneous action
-    Without speech
-    Without requiring presence or use of particular biological parts (touch, fingerprint, iris, etc.)
	Interfaces which are input device or technique specific cannot be operated by individuals who cannot use that technique (e.g., a person who is blind cannot point to a target in an image map; some people cannot use pointing devices accurately).
	
	In Addition:
	Technique specific interfaces may not be accessible to users of mobile devices. For example people using voice to navigate may not be able to "point".
		Provide at least one mode (set of modes) where...
		a) all functions of the product are controllable via tactilely discernable controls and both visual and voice output is provided for any displayed information required for operation including labels,
		  AND
		b) there are no timeouts for input or displayed information, OR allow user to freeze timer or set it to long time (5 times default or range), OR offer extended time to user and allow 10 seconds to respond to offer.
		  AND
		c) all functions of the product operable with:
		- No requirement for simultaneous activations, twisting motions, fine motor control, biological contact, user speech, or pointing motions
		  AND
		d) If biological techniques are used for security, have at least two alternatives with one preferably a non-biological alternative unless biological based security is required.

Facilitate Orientation and Navigation • Without sight • Without pointing ability • Without fine motor control • Without prior understanding of the content • Without the ability to hear • Without good memory	Many individuals will have trouble using a product (even with alternate access techniques) if the **layout / organization** of the information or product **is too difficult to understand.** Many individuals will not be able to operate products, such as workstations, with sufficient efficiency to hold a competitive job if navigation is not efficient.	a) Make overall organization understandable (e.g. provide overview, table of contents, site maps, description of layout of device, etc.). b) Don't mislead/confuse. (Be consistent in use of icons or metaphors. Don't ignore or misuse conventions.) c) Allow users to jump over blocks of undesired information (e.g., repetitive info – or jump by sections if large document) , especially if reading via sound or other serial presentation means. d) Make actions reversible or request confirmation. e) Consider having different navigation models for novice vs expert users
Facilitate Understanding of Content • Without skill in the language used on the product (poor language skills or it is a second language for them • Without good concentration, processing • Without good memory • Without background or experience with the topic	People with cognitive or language difficulties [or inexperienced users] may not be able to use devices or products with complex language.	a) Use the simplest, easiest to understand language and structure/format as is appropriate for the material/site/situation. b) Using graphics to supplement or provide alternate presentations of information c) If phrases from a different language (than the rest of the page) are used in a document either identify the language used (to allow translation) or provide a translation to the document language.
Provide Compatibility with Assistive Technologies commonly used by people • With low vision • Without vision • Who are hard of hearing • Who are deaf. • Without physical reach & manipulation • Who have cognitive or language dis.	In many cases, a person coming up to a product will have assistive technologies with them. If the person cannot use the product directly, it is important that the product be designed to allow them to use their assistive technology to access the product. [This also applies to users of mobile devices people with glasses, gloves or other extensions to themselves.]	a) Do not interfere with use of assistive technologies - Personal aids (e.g. hearing aids) - System based technologies (e.g. OS features) b) Support standard connection points for - audio amplification devices - alternate input and output devices (or software) c) Provide at least one mode where all functions of the product are controllable via human understandable input via an external port or via network connection.

The effort reported here takes a slightly different approach. It is an attempt to come up with a simple underlying set of guidelines which can be expressed concisely and could be used to guide the design and development of electronic and information and telecommunication products and systems. In developing this set of guidelines and strategies, one of the chief goals was to try to keep the list as short and succinct as possible and yet to cover the major dimensions of cross disability issues and strategies.

2. Overview of "Guidelines and Strategies for E & IT"

Figure 1 provides a one page summary of the "Principles and Strategies" for access to electronic and information technologies. The figure is composed of three columns. The first column provides the basic access guidelines or objectives. These basically take the form of requirements. They are user centered rather than being device centered and structured to give the designer a better idea for what is being sought. For example, rather than saying that a product should be usable across disabilities (which leaves a designer wondering what the different disabilities are and what the implications would be) the requirements take a form which lists the major dimensions that "cross disability access" would involve.

These are not necessarily comprehensive. That is they do not necessarily list all of the many types, variations and degrees of disability. They do, however, provide a good overview of the basic dimensions. The attempt is to strike the balance between no specifics (e.g. "people with disabilities" or "people with vision, hearing, physical and cognitive disabilities") and a full description of the various types, degrees and combinations of disabilities which could total more than one hundred.

Column two provides a short description of why it is important or essential for products to be accessible along each dimension. It also provides a brief description of how the access strategies would benefit people who do not have disabilities.

The third column provides the basic strategies for addressing each access requirement. The goal was to capture the essential strategies and to show that a relatively limited number of strategies can, if carefully implemented together, address the wide range of disabilities and their requirements.

2.1   Make all information perceivable

The first principle is that all information which is presented must be perceivable. This includes the information on any displays, indicators (e.g. LEDs, buzzers, etc.), as well as any printed information on the product, the labels on all of the keys, controls, etc. It includes the information whether it is presented visually, auditorially or via shape or vibration. Because people may be missing any particular sense it is important that all information be available across sensory modalities. There are two basic strategies for doing this.

One is to make sure that the information is in a form which is *sensory modality independent*. For example, electronic text is not tied to any particular mode of presentation; it be can presented visually (small or large), auditorially, and even tactilely. It is also machine readable which opens possibilities for translation and changing the language level (although this may result in some loss of resolution or detail).

Where this is not possible, information could be provided in a *sensory parallel form*. The most common type of sensory parallel form would be movies with captions and audio description. Information which is presented visually is also described auditorially so that people who cannot see can hear what is happening visually on screen. Information provided via the soundtrack is also provided in a parallel fashion via visual captions which describe not only the dialogue but the sounds which would be not be obvious from the visual presentation. In electronic forms the information could also be made available in e-text which could be rendered in tactile form as well, as noted above.

If the information is available in these forms a product can give the user the option of presenting the information in the form which best fits their situation. Individuals who are blind can ask for the information to be presented auditorially and individuals who are deaf can ask for all of the information to be presented visually. Similarly, people without disabilities can benefit from the availability of parallel forms of information. A spouse who wants to finish watching a movie after their mate wants to go to sleep can turn on captions and finish watching the movie silently. People whose eyes are busy could also follow action in an all auditory fashion.

2.2 Provide a least one mode for operating all product features (that is flexible)

The objective here is to provide the same flexibility with operation of the product that is provided for viewing information from the product. It is important to note that this does not mean avoiding popular and effective interface techniques such as pointing. However some people, either because they cannot see (and therefore cannot point to non-tactile objects) or because of physical disabilities, would not be able to point. Thus an alternate way of achieving this same effect should also be provided. On a computer this may mean that activities can be accomplished either by using the mouse or by using the keyboard. The key is to eliminate not a mode of control or operation (and thus make the product less usable by some) but to provide alternate modes of operation making it easier to use by all. Making a product accessible to people with a wide range of limitations thus becomes a matter of providing a carefully selected set of methods for operating the device, one o f which would allow a user to access each of the functions of the product even if they are experiencing constraints.

The primary strategy for doing this is to ensure that there are modes which are operable without vision, without timing constraints, without requiring fine motor movements, simultaneous activities, speech or body contact. Again these strategies not only provide access by people with disabilities but also are important to providing access in mobile computing. Strategies such as ensuring keyboard access also facilitates access via speech and via artificial intelligent agents.

2.3 Facilitate orientation and navigation

Although the first two principles by themselves give people basic physical and sensory access to the information and operation, it is also important to provide information to facilitate the user's orientation and the navigation within the information or product. This is important for several reason.

First, individuals without sensory disabilities but who do not have high cognitive function may find it too difficult to figure out how to operate the product. Second individuals who have sensory disabilities with average or above average intelligence may still have difficulty using a product non-visually where the product as originally designed and conceived to be operated by people with sight. With such products many of the orientation and navigation cues may be visual, making it easy to operate the product in that mode but making it cognitively confusing and difficult to operate non-visually. It is therefore important that the orientation and navigation information be usable by people with different types of disabilities.

A third reason for facilitating orientation and navigation has to do with efficiency. If a product is designed to be very easy to use with a mouse (less than one second per choice) but very hard to navigate using the keyboard (five to ten seconds and ten to fifteen tab strokes between choices) then it will not be possible for a person with a disability to be able to operate a product *efficiently* enough to study or work competitively. If the task is seldom required and enough time is always available then this may not be a problem. However, on workstations for example, the inability to work efficiently (e.g. four or five times slower) may mean the inability to get or keep a job.

There are a wide range of strategies for facilitating orientation and navigation. These are extensions of general human factors principles. The key is to apply these same principles, remembering that the user may not be able to see, point, use fine motor control or hear, and that the user may have lower memory or cognitive processing levels than is generally assumed.

2.4 Facilitate understanding of content

This principle is similar to the orientation and navigation principle. The information may be physically accessible or perceivable by an individual but may be presented in a form which is above the ability of the individual to effectively understand it. This may be due to a number of factors including cognitive level or the fact that the information being presented is not being presented in an individual's primary language. (It should be remembered that American sign language is a completely different language than English.)

The strategies used here would again be an extension of the usability principles for the general population. These include using the simplest and most straightforward language and structure as is appropriate for the material, using graphics to supplement the text or present the information in an alternate form, and minimizing the use of foreign language phrases or providing an explanation of them.

2.5 Provide compatibility with assistive technologies commonly used by people with disabilities

No matter how much one tries to design accessibly directly into a product, there will always be individuals for whom it is commercially impossible to make a product directly usable. For example, individuals who are deaf-blind may require a dynamic braille display. However, it is not practical to build one into each product. Similarly, individuals who are completely paralyzed and who get around using a sip and puff controlled wheelchair would find it difficult to operate a touch screen information kiosk - even with special accessibility considerations. They physically are not able to reach out and touch the product at all and may be not able to speak.

For those users where it is not possible to build in access (or efficient enough access) it is important that products be compatible with assistive technologies. This includes making sure that the controls can be operated with prosthetics and mouth sticks (e.g. no requirement of biological contact) as discussed above, not interfering with assistive technologies (e.g. electromagnetic interference with hearing aids or products which override access features in operating systems) and providing mechanisms for connecting alternate input interfaces. This latter item will be considerably facilitated by the completion of the current effort underway to create an industry standard for alternate interface interconnection..

3. Summary

The task of providing access to standard products for individuals with a complete range of types, degrees and combinations of disabilities can seem a daunting task. Although designing truly effective products would require that the designers take the time to master the needs of the population and the known strategies for addressing disabilities, it is useful to provide a summary of the major issues and strategies in a concise fashion in order to allow people pursuing this area to better get their arms around the problem. It is also highly beneficial to designers if general strategies and approaches can be identified which can work together and be flexible enough to address the wide variety of users without requiring an inordinate number of different interface strategies on a product. This document is an attempt to begin the process of creating such a summary document. There are also efforts underway such as the EZ Access project at the University of Wisconsin and the reference design program which are aimed at creating flexible cross disability accessible designs using a minimum of additional access strategies and which emphasize increasing usability for consumers without disabilities as well. (Vanderheiden, G. C., & Law, C. 2000)

Acknowledgements
This work is funded by the National Institute on Disability and Rehabilitation Research of the Department of Education under Grants #H133E980008 and #H133E990006. The opinions contained in this publication are those of the grantee and do not necessarily reflect those of the Department of Education.

References
Connell, B.R., et al. (1997). The principles of universal design: Version 2.0. Raleigh, NC: The Center for Universal Design.

Vanderheiden, G.C., Law, C. (2000). Cross-disability access to widely varying electronic product types using a simple interface set. In Proceedings of the XIVth Triennial Congress of the International Ergonomics Association and 44th Annual Meeting of the Human Factors and Ergonomics Society, 4, 156. Santa Monica: Human Factors and Ergonomics Society.

Vanderheiden, G.C. (2000). Fundamental principles and priority setting for universal usability. Proceedings of the ACM - Universal Usability Conference, Washington, DC. New York, NY: ACM Press, pp. 32-38.

Can Standards and Guidelines Promote Universal Access?

Jürgen Ziegler

Fraunhofer IAO
Nobelstr. 12, D-70569 Stuttgart, Germany
e-mail: juergen.ziegler@iao.fhg.de

Abstract

This paper reviews international standardization activities in the area of user interface design, usability and accessibility and discusses their impact on the concept of universal access. Different types and levels of current standards and options for future standardization are presented.

1. Introduction

The concept of universal accessibility aims at removing barriers for using and utilizing information and communication technologies that may stem from a wide range of factors such as personal, social or cultural characteristics. The design of systems and, specifically, of user interfaces should address those impediments and make information technology accessible to the widest possible range of users in the intended target groups (cf. Stephanidis 2001).

In designing for universal accessibility, different factors need to be taken into account, that determine the relevant requirements and constraints influencing the ability to use IT systems . The following areas typically have considerable implications for the accessibility or uptake of information technology:
- Usability of the system as determined by the effectiveness, efficiency and satisfaction by which users can achieve their goals (ISO 9241-11). Low usability can significantly deteriorate accessibility due, for instance, to insufficient suitability to the task, high cognitive complexity, or inefficient dialogs
- Personal characteristics such education, individual differences or cognitive or physical disabilities. Such factors may generate the need for alternative interaction modes, adaptation of content or functionality of the system, or special assistive techniques.
- Cultural and social factors such as language or level of income. If culturally determined preferences or conventions are violated in a system design (e.g. with respect to colors or dialog styles) acceptance of the system may be significantly lowered. Similarly, as current studies of Internet use show, the social background is an important determinant of technology access and use.

These areas define a variety of potential constraints that may impede universal access. Standardization, although not being sufficient to remove all these constraints, can contribute to lowering some of the barriers and especially to forming a consensus as to what the goals and approaches in designing, deploying and operating systems should be.

2. Types of standards and levels of guidance

The provision of standards and guidelines plays an important role in improving universal access as they can establish consensus on design goals, harmonize design solutions, or to a certain extent even influence the development of technologies. In order to discuss the potential influence of standards it is useful to distinguish different types of standards and levels at which they can be allocated.

User interface-related standards can usually be distinguished in product and process standards. Product-related standards describe properties of interactive systems which must be fulfilled if conformance with the standard is claimed (see e.g. Dzida 1997). On the other hand, process-related standards describe the different activities and their sequence involved in the design, development, evaluation or introduction and use of a system. ISO 13407, for instance, defines a human-centered design process involving activities such as task analysis or usability testing.

In order to determine in which way standards can be applied in the design of a system, it is useful to distinguish three levels of definitions which differ with respect to their degree of detail (see IBM 1998):

- Principles describe commonly agreed usability goals which should be achieved when designing and building interactive systems. Examples of such principles are "suitability for the task" or "self-descriptiveness" as described in ISO 9241-10.
- Guidelines are at a middle level of concreteness, stating recommended or required characteristics of a system or process, which are still independent from the concrete design but close enough in order to derive those concrete characteristics.
- Concrete specifications represent the most detailed level of standardisation. Specifications relate to directly observable or measurable properties of a system or a process. Industrial user interface styleguides typically provide concrete specifications of well-defined user interface components such as buttons, drop-down lists or windows.

These levels are associated with different degrees of concreteness and detailedness of the guidance provided. It is often argued that high-level principles and guidelines are less useful as they do not enable designers and developers to translate them into actual system designs without additional knowledge (see e.g. Blanchard 2000). Although some short-term projects may indeed benefit less from general principles if the designers involved are not sufficiently experienced, this statement neglects the fact that more abstract guidance can be instrumental in interpreting and understanding concrete guidelines and is also usually more stable over time. In standards like ISO 9241 or 14915, high-level principles and guidelines are usually accompanied by concrete examples, that provide instantiations of the abstract guideline in a concrete design context.

Principles can also provide a basis for reasoning about potential design trade-offs. An example of such a potential trade-off can be found in the high-level design principles given in ISO 9241-10 and ISO/DIS 14915-1. "Suitability for the task" as specified in ISO 9241-10 is a dialog design principle that can be considered as one of the cornerstones of ergonomic system design. While this principle stipulates that interfaces should be designed with respect to the user's tasks and goals, new applications of IT such as the World Wide Web have often the purpose of conveying information to a recipient which draws attention the goals of the provider of that information. In fact, user goals and goals of the information provider may often not be identical as one can see, for instance, in some e-commerce applications. In ISO 14915-1, the principle "suitability for the communication goal" is introduced as a complement to suitability for the task in order to make this distinction evident and to support designers in explicitly considering potentially conflicting goals in their design.

While the example discussed tries to point out the usefulness of abstract guidance for informing the design process itself, generally agreed (for ISO standards at an international level) design guidelines can also form a basis for regulatory or legislative activities such as the European VDU directive. Moreover, standards can play the role of a reference framework against which concrete specifications, for instance, developed in in-house styleguides can be checked.

3. Standards related to universal access

With respect to the areas relevant to universal access described in the introduction, there is a range of standards pertaining to an improved accessibility of systems. These standards are either promoted by single industries or industry groups or "official" standards developed by accredited national or international standardization bodies such as ISO or IEC. ISO TC159/SC4 is developing a comprehensive set of user interface related standards which comprise both interface design and process oriented guidance. Figure 1 gives an overview of the standards developed in this context.

In the following, we will discuss a selection of the most important user interface-related standards and their contribution to the issue of universal access. We will focus here on international standardization, a more complete overview comprising also industry standards and styleguides can be found in (Ziegler 1999).

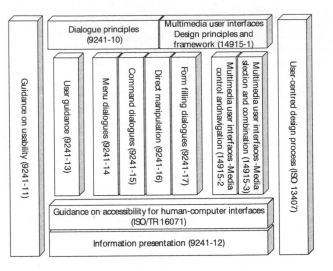

Figure 1: Overview of ISO standards related to usability and accessibility

3.1 ISO 9241 parts 1-17: Ergonomic requirements for office work with visual display terminals (VDTs)

ISO 9241 is the most comprehensive series of standards addressing both hardware and workstation aspects (parts 1-10) and software user interface aspects (parts 10-17). While the hardware parts address areas potentially relevant for accessibility (such as keyboard design) they do not specifically deal with such issues.

The software parts provide essential guidelines for the usability of a system, providing general dialogue design principles (part 10) as well as guidelines related to specific dialogue techniques such as menus or direct manipulation (parts 14-17). Information presentation and user guidance are covered in parts 12 and 14 respectively.

Although none of the 9241 parts is dealing explicitly with accessibility issues, the principles and guidelines described in those standards form a basis for ensuring the usability of software user interfaces. If we see usability as part of a broader concept of accessibility, this standard provides a baseline for taking into account the capabilities and constraints of an average user population without considering specific disabilities.

3.2 ISO 14915: Software-ergonomics for multimedia user interfaces

ISO 14915 is a relatively recent activity within ISO which has produced 3 parts so far. It extends the software parts of ISO 9241 with guidance specifically related to multimedia user interfaces.

Part 1: Design principles and framework: Part 1 introduces the main concepts, describes design principles particularly relevant for multimedia interfaces and sets up a framework of issues pertaining to the design of multimedia systems. Complementary to the principles of ISO 9241-10, four additional principles are introduced which are pertinent to multimedia design comprising suitability for the communication goal, suitability for exploration, suitability for engagement and suitability for perception and understanding.

Part 2: Multimedia control and navigation: This part provides guidance on navigating in hypermedia structures and on controlling the different media. An important concept used in this standard is the mapping between content structure and navigation structure. The standard describes different types of navigation structures (linear, tree, network), recommendations when to select a particular structure and navigation functions appropriate for the structure selected. Navigation aids such as bookmarks, overviews or history mechanisms are introduced and described which support the user's navigation in larger structures.

General guidelines for designing controls are provided addressing issues like minimum control sets or grouping of controls according to their type are provided. More specifically, controls for dynamic media are specified and guidelines concerning their design are provided.

Part 3: Media selection and combination: This part focuses on the selection, combination and integration of different media. While the standard does not deal with the design "within" a single medium (e.g. how to produce a video for a particular instructional purpose), it provides guidance on selecting a medium for a specific type of information and integrating different media to convey different aspects and perspectives of the content.

In addition to providing general guidelines for selecting media and media combinations on the basis of different content types and communication goals, particular attention is paid to the transition points between different media which can guide the user's attention when following a presentation thread through different media components.

In the 14915 series of standards, part 3 has the most direct relation with accessibility issues, as the selection and combination of media may immediately affect users with sensory or cognitive impairments. A general guideline is given that the selection of media shoud take the users' characteristics into account (with an example of blind users). The selection of media is based on the underlying type of information to be conveyed. This selection is not a strict one-to-one mapping but indicates preferred selections and alternative options. Although disabilities are not specifically addressed, these guidelines allow a designer to consider alternative media for specific user groups. Several guidelines refer to information overload, compatibility with the users' existing knowledge and perceptual or semantic conflicts that may be relevant for users with perceptual or cognitive impairments.

3.3 ISO/TR 16071 "Ergonomics of human system interaction – Guidance on accessibility for human-computer interfaces".

This new technical report is focusing on accessibility for people with special needs or disabilities. Its main emphasis is on enabling persons with sensory and/or motor impairments to use human-computer interfaces while taking a broad view on the range of capabilities and preferences of users that may affect accessibility including age, disease or temporary injuries as potential reasons for impairments. In this report, the concept of accessibility is seen as being closely related to the notion of usability with a rather continuous transition between those two concepts.

The technical report provides guidance on user interface design issues related to disabilities as well as on required technical capabilities of the system (operating system or application) to connect and interact successfully with assistive technologies such as speech synthesizers or Braille input and output devices. An example of a design related guideline is, for instance, the recommendation to "enable sequential entry of multiple keystrokes" in order to allow users who cannot press several keys at the same time due to e.g. motor impairments to enter complex key commands. This guideline is defining a specific interaction method suitable for particular user groups.

On the other hand, the report contains guidelines concerning enabling functions built into the operating system or the application such as "make event notification available to assistive technologies". While this function does not specifiy any specific interaction technique per se, it constitutes a prerequisite for connecting assistive technologies and for designing effective interactions with them.

ISO/TR 16071 is based on work done by HFES and ANSI. It has passed the ISO voting and (at the time of this writing) is being edited for publication in the near future. It is planned to further develop this technical report as an international standard.

3.4 Industry Initiatives and Guidelines

The World Wide Web Consortium (W3C) has established the Web Accessibility Initiative (WAI) in order to promote and support accessibility for Web technologies and applications. Due to the importance and ubiquity of web applications, the design of accessible web sites is an important measure for addressing a large number of disabled users and for accommodating a range of individual differences and preferences. Three sets of guidelines are currently under development and available as working drafts (W3C 2001). The documents provide information mainly at the level of guidelines and detailed specific conventions.

The "User Agent Accessibility Guidelines" provides twelve important guidelines with checkpoints for making client software like HTML browsers or media players accessible. "Web Content Accessibility Guidelines" describe a variety of recommendations for designing accessible web content, containing, for instance, advice on designing multimedia web pages and on structuring the content in order to facilitate navigation with alternative input techniques. Finally, the document "Techniques for Authoring Tool Accessibility Guidelines" gives guidance on making the authoring of web content accessible.

Initiatives driven by individual companies such Microsoft's Active Accessibility (http://microsoft.com/ enable/msaa/default.htm) and Sun's Java Accessibility Program (http://java.sun.com/ products/jfc/jaccess-1.2.2/doc/) provide guidelines and concrete technical solutions for implementing accessibility features on the respective technical platforms.

4. Discussion and conclusions

The range of standards presented shows, that both the breadth and the depth of the potential impact on accessibility varies considerably. User interface and usability standards attempt to maximize the efficiency and effectiveness of use for the largest possible "standard" user population in a particular task domain. "Enabler" standards like parts of ISO/TR 16071 or some industrial initiatives define technical features required for making systems adaptable to user groups with special needs without detailing how the resulting interface should be designed. General guidelines for designing interfaces for frequent disabilities such as perceptual or motor impairments are also given in ISO/TR 16071 as well as, for instance, in the W3C WAI guidelines. Detailed interface designs for specific disabilities, however, are less amenable to standardization due to large variety of impairments and the specificity of the resulting requirements which may often lead to individual adaptations.

The contribution of the types of standards discussed above might be grouped in different levels: the level of operability (can users operate the system at all?), usability in terms of whether they can use it efficiently and effectively, and the level of personal satisfaction which may additionally require adaptation to personal preferences, working styles, attitudes etc.

Cultural and social aspects, on the other hand, are hardly addressed in current standardization activities. Although, for instance, ISO/IEC JTC1/SC35 has started a work item on cultural and linguistic adaptability, there doesn't seem to be a coherent body of knowledge that could be translated into a comprehensive standards document. Social aspects and resulting constraints on accessibility are inherently difficult to address in standardization as they are critically dependant on basic societal processes such as legislation, regulatory activities, employer-union negotiations or market forces. Although it will remain difficult – and may not be desirable – for standardization to address these areas directly, it can play an important supportive role in building consensus, defining goals and principles and providing guidance that could influence decision processes at a more general level.

References
Bergman, E. (1997). The role of accessibility in HCI standards. In: *Proc. 7th Int. Conf. on Human-Computer Interaction* (San Francisco Aug. 24-29, 1997)/ ed. G. Salvendy; M.J. Smith; R.J. Koubek, 441-444.
Blanchard, H. (2000). When should standardization take place? Part 1: Implementation and specificity. *SIGCHI Bulletin*, Nov./Dec. 2000.
Dzida, W. (1997). International User-Interface Standardization. In: *The Computer Science and Engineering Handbook* / Ed. Tucker, Allen B. Jr.. Boca Raton, Fl.: CRC Press, 1474-1493.
ISO DIS 13407 (1998). *Human-centered design process for interactive systems.*
ISO DIS 14915-1 (2001). *Software ergonomics for multimedia user interfaces: Introduction and framework.*
ISO CD 14915-2 (2001). *Software ergonomics for multimedia user interfaces: Media control and navigation.*
ISO DIS 14915-3 (2001). *Software ergonomics for multimedia user interfaces: Media selection and integration.*
ISO 9241-10 (1996). *Ergonomic requirements for office work with visual display terminals: Dialogue principles.*
ISO DTR (2000). *Ergonomics of human system interaction – Guidance on accessibility for human-computer interfaces.*
Stephanidis, C. (Ed.) (2001). *User Interfaces for All - Concepts, Methods and Tools.* Mahwah, N.J.: Lawrence Erlbaum Associates.
W3C (2001*). Web Accessibility Initiative.* http://www.w3.org/WAI/
Ziegler, J. (1999). Standards for multimedia user interfaces - opportunities and issues. In *Human-Computer Interaction - Communication, Cooperation and Application Design, Proceedings 8th Int. Conf. on Human-Computer Interaction*, Vol. 2 (Munich, Germany Aug. 22-26, 1999). Mahwah, N. J.: Lawrence Erlbaum Associates, pp. 858-862.

Usability Evaluation of Universal User Interfaces with the Computer-aided Evaluation Tool PROKUS

Gert Zülch, Sascha Stowasser

ifab-Institute of Human and Industrial Engineering, University of Karlsruhe,
Kaiserstrasse 12, D-76128 Karlsruhe, Germany
Email: ifab@mach.uni-karlsruhe.de

Abstract

The rapid dissemination of computer systems and accessible information systems (e.g. internet) is still unbroken. On the individual level, human-computer interfaces and dialogues change many people's lives: examples are the daily work at visual display terminals, digital lectures on the web and home working software tools. A few years ago, it was unusual to assess and evaluate such software or interfaces. Nowadays, researchers and practitioners are recognising the problems which grow up when usability evaluations are neglected. Usability evaluations can be carried out in different phases at the software lifecycle. It requires appropriate procedures with various techniques and methods to request and scope these different occasions.

For the purpose of systematic usability evaluation, the ifab-Institute of Human and Industrial Engineering of the University of Karlsruhe (Germany) has developed an evaluation tool called PROKUS (Programmsystem zur kommunikationsergonomischen Untersuchung rechnerunterstützter Verfahren). This evaluation tool can be used for designing and carrying out usability evaluations in a reproducing and numerical process adapted to different occasions of the software lifecycle. The following paper will describe the evaluation process with PROKUS. Moreover, it will awake special attention to the scientific background of the development and the realisation of the evaluation tool PROKUS. A recently completed evaluation study on universal human-computer interfaces with PROKUS will complete the paper. For illustrating the procedure of PROKUS two web browsers exemplarily tested to what extent it fulfils the required standards of ISO 9241-14, 16 and 17.

1. Necessity and occasions to evaluate the usability of human-computer interfaces

Due to the increasing use of data processing in all fields in human life (e.g. work, education, leisure) the attention given to a universal user-friendly software design is rising. The quality and acceptance of an interactive human-computer system is highly dependant upon the ergonomic design of the user interface. In the past, ergonomic aspects were often neglected when developing software and information systems. Thus, many users complain about the troubles in learning to use a software product or its insufficient functionality and complexity of interactions. Once the user becomes dissatisfied with a software product, a dialogue system or an accessible interface, this dissatisfaction dominates the functionality and the productivity of the system. A few years ago, it was unusual to assess and evaluate such software or interfaces. Nowadays, researchers and practitioners are recognising the problems which surface when usability evaluations are neglected. Furthermore, the evaluation of accessible interfaces (e.g. web pages) is becoming more and more important. New standards for multimedia interfaces, web pages or further advanced information and data technologies (virtual reality etc.) have to be developed and have to be transferred into the design of these modern human-computer systems.

The analysis and evaluation of software products or accessible interfaces, which should examine the adequate transfer of user-friendly standards to the design and development of products, can be realised at several occasions in the software lifecycle (figure 1). The first occasion is during the development process of the product by choosing an adequate human-computer interface. Quality tests assess mainly the functionality of the software. Conformity tests compare the product with the requirements of the standards and evaluate the user-friendliness of interfaces and dialogues. Another occasion, performed by the customer or the future user, is helping to select a suitable product by examining existing products on the market and comparing these products (comparison test). In this product selection process, conformity tests are made to assess the adherence to user-friendly standards. When the software has already been selected, several usability tests with additional goals and needs can be carried out in order to support an eventual redesign phase.

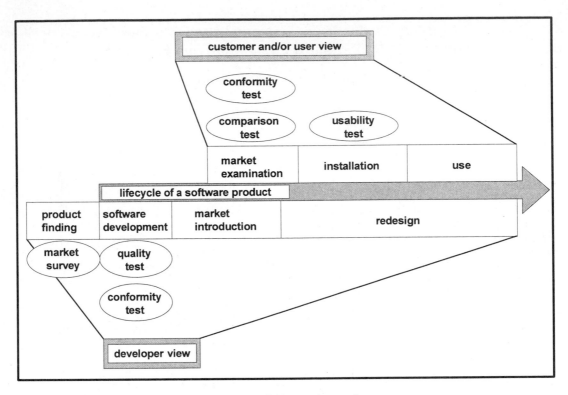

Figure 1. Occasions to evaluate software
(following Zülch & Stowasser, 2000, p. 2/17; Zülch, Englisch & Grundel, 1993, p. 268)

2. The computer-aided evaluation tool PROKUS

PROKUS (Programmsystem zur kommunikationsergonomischen Untersuchung rechnerunterstützter Verfahren), which has been developed at the ifab-Institute of Human and Industrial Engineering of the University of Karlsruhe, is a computer-system for the design of evaluation procedures and the performance of usability evaluations according to different evaluation situations. This tool can be classified as an evaluation method with guidelines and is applicable for systematic market examinations, conformity tests, quality tests and comparison tests of various kinds of software. This evaluation tool can be used to design and perform usability evaluations in a reproduction and numerical process adapted to different occasions of the software lifecycle.

2.1 System Elements of PROKUS

PROKUS is based on a catalogue with questions which are to be filled-in by an expert during the evaluation procedure. A central element of PROKUS is the exerciser database which consists of different series of investigations (figure 2). Each series of investigations contains several examinations which consist of a number of questions. Each question is described using the elements "criterion", "component", "task", "method", "importance" and "rating scale". The element "criterion" of the question represents the focused usability criterion. This criterion can be exemplarily derived from the seven general ergonomic principles (e.g. conformity with user expectations) which are described in ISO 9241-10.

The element "component" represents essential characteristics of the software or the interface which is to be evaluated with the respective question. One approach to arrange the components is the IFIP model for user interfaces developed by a working group of the German Informatics Community (GI) (IFIP stands for International Federation of Information Processing; Dzida, 1987, p. 339) . As an example, the components of human-computer systems include user, task and computer components with input/output, dialogue and tool components. The element "task" describes the function or purpose of the software or the interface according to the evaluated characteristic of the human-computer system (e.g. the button F7 is consequently used to save the data by the user). The element "method" represents the various test and evaluation methods with which the expert can evaluate the software and measure the required data in order to answer the question. ISO 9241-14 recommends the application of the evaluation methods "measurement", "observation", "expert judgement", "documental evidence" and "user testing". Regarding the special occasion of the evaluation, the use of a special method may be required. Evaluation with these methods means, that the evaluator tests an existing system and measures the required data or he derives these data from existing documents. The element "importance" presents the possibility of defining a ranking of the questions (e.g. class of question "1" means a "very important" question). The element "rating scale" represents the answering field for the actual question. Depending on the type of questions, PROKUS offers answering fields for three different scales, namely the nominal, ordinal and interval scales. An exact description of the evaluation system and the evaluation process with PROKUS can be read in Zülch and Stowasser (2000, p. 5/17).

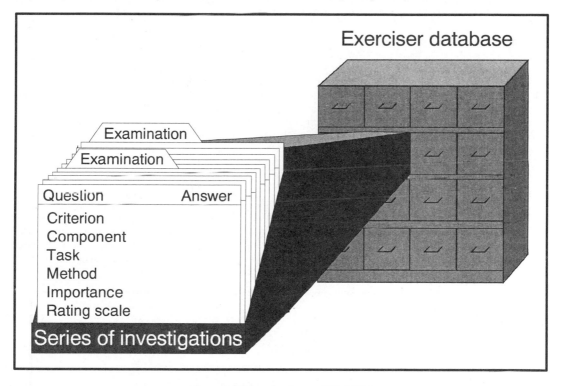

Figure 2: The evaluation tool PROKUS

2.2 Benefits in evaluating with PROKUS

Many procedures with different techniques and methods already exist for the evaluation of human-computer work systems and each evaluation refers to an adaptation to a special situation. For this purpose, PROKUS has been developed to design evaluation procedures according to different evaluation environments, e.g. resources or levels of detail, and to carry out different processes in usability evaluations. The complexity of an evaluation procedure designed with help from PROKUS can range from expert reviews and acceptance tests up to usability tests in laboratories.

A continual task in the conception of evaluation systems is the combination of the evaluation system with a system that stores different design guidelines, standards or other aspects concerning the evaluating matter. The evaluation tool PROKUS supports evaluators, designers and users on one hand with practicable and testable guidelines for human-computer interface design and on the other hand with methodological procedures for collecting evaluation and applying questions. These questions can be selected and combined in such a way which is desired by the evaluator (of course a combination of questions of different standards, guidelines etc. is possible). For combining several questions PROKUS offers different sorting functions: e.g. choosing questions corresponding to just one evaluation method, questions for evaluating specific aspects (for instance only questions which can be applied to testing the dialogue component).

3. Application of PROKUS for web browser usability studies

The pursuit of internet connections has lead millions of users to navigate through the variety of web pages, join mailing lists, visit chat rooms, and fill newsgroups with useful information. Many results from other user-interface topics (e.g. menu selection, direct manipulation, screen design) can be applied to web-site design. However controlled experimental studies are necessary to evaluate the universal human-oriented internet performance (Shneiderman, 1998, p. 581). A recently completed evaluation study of universal human-computer interfaces web browsers carried out with PROKUS has demonstrated an application of this evaluation tool.

3.1 The evaluation process

Two common web browsers (called anonym as Browser A and Browser B) have been tested with an extensive series of usability evaluations. Both tests were performed independently: they were not carried out simultaneously. For each conformance test a series of investigations had to be defined (compare chapter 2.1). This series of investigations contains several examinations for a conformity testing in correspondence with ISO 9241 (Parts 14, 16, 17). These examinations were again arranged by a series of questions which are stored in the exerciser database of PROKUS. The evaluation was performed with a total of 288 questions (63 criteria, 24 scales, 21 methods, 13 tasks). The user group consisted of 5 persons (3 students, 2 non-academic persons).

3.2 General results of the evaluation

The PROKUS interpretation system includes various possibilities to aggregate and interpret the evaluation data. One possible result of the conformity test is expressed as the level of compliance (*LCO*) with chosen standards, which is represented by the number of fulfilled questions divided by the number of tested questions as a percentage. Figure 3 summarises selected *LCO*s considering both web browsers. For example, both systems provide nearly excellent navigation options (like "forward arrows", "history-lists"). Another criterion is the structure of the menus: e.g. grouping options in a universal user-compatible way.

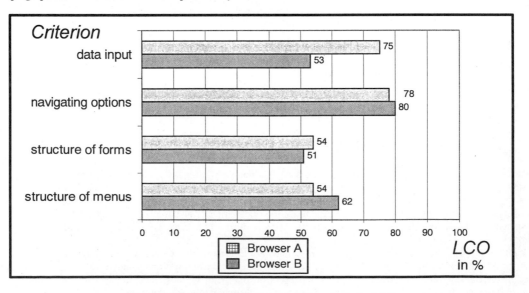

Figure 3: Level of compliance concerning selected criteria

The PROKUS interpretation system includes various possibilities to aggregate and interpret the evaluation data. One possible result of the conformity test is expressed as the level of compliance (*LCO*) with chosen standards, which is represented by the number of fulfilled questions divided by the number of tested questions as a percentage. Figure 3 summarises selected *LCO*s considering both web browsers. For example, both systems provide nearly excellent navigation options (like "forward arrows", "history-lists"). Another criterion is the structure of the menus: e.g. grouping options in a universal user-compatible way.

As a general result of the evaluation study, the overall level of compliance for Browser A is 69 %. Compared with this, Browser B measures 75 %. However, neither browser interface has a complete conformance with the considered standards. The results of the conformity test carried out with PROKUS shows that both interfaces need corrections in order to meet the required user-friendliness for universal application.

References

Dzida, W. (1987). On Tools and Interfaces. In: M. Frese, E. Ulich, W. Dzida (Eds.), *Psychological Issues of Human Computer Interfaces in the Work Place* (pp. 339-355). Amsterdam: North-Holland.

ISO 9241-10 (1996). *Ergonomics requirements for office work with display terminals (VDTs)*. Part 10: Dialogue Principles.

ISO 9241-14 (1997). *Ergonomics requirements for office work with display terminals (VDTs)*. Part 14: Menu dialogues.

ISO 9241-16 (1999). *Ergonomics requirements for office work with display terminals (VDTs)*. Part 16: Direct manipulation dialogues.

ISO 9241-17 (1998). *Ergonomics requirements for office work with display terminals (VDTs)*. Part 17: Form filling dialogues.

Shneiderman, B. (1998). *Designing the User Interface* (3rd ed.). Reading, MA: Addison-Wesley.

Zülch, G., Englisch, J., & Grundel, C. (1993). Beurteilung der Benutzungsfreundlichkeit von Programmsystemen. *Fortschrittliche Betriebsführung und Industrial Engineering*, 42 (5).

Zülch, G., & Stowasser, S. (2000). Usability Evaluation of User Interfaces with the Computer-aided Evaluation Tool PROKUS. *MMI-Interaktiv*, 2 (3).

PART 7

ACCESS TO INFORMATION

Smart interfaces supporting physical navigation while experiencing a heedful audio information space

Alessandro Andreadis, Alberto Bianchi

Dipartimento di Ingegneria dell'Informazione, Università degli Studi di Siena, via Roma 56 - 53100 Siena - Italy

Abstract

This article describes the problematic related to the physical navigation of areas of relevant cultural interest (e.g. museums) while making use of mobile devices. The expected solution is to provide the user with a heedful, highly contextualized and dynamic audio information space, which does not divert the user attention from the real experienced world. Position recovery techniques are embedded in the system in a seamless way. The user is not required to specify (interrupting the information fruition) where he is and what he is looking at. In this way the gap between the information space and the physical space is reduced to the minimum. The user has appropriate feedback and full control over the system behavior, thanks also to a graphic *User Interface* (UI) that is here carefully described. Implicit/explicit user interactions are taken into account and on-line/off-line ways of use of the instrument are differentiated accordingly and made accessible in real-time, in order to avoid hw/sw unpleasant switches.

1. Introduction

Mobile users strongly need to be supported in their everyday working and leisure activities while on the move. Tourists and businessmen as well need many different services and information, while moving in an "unknown" environment, and often they have not enough reference points or supports. Information is often lacking; static and generic paper guides or tourist offices are usually inadequate to offer real-time reaction to user requests. Existing electronic guides are generally limited to information about cultural heritage of a specific area or an entire city and use different technologies depending on the site.

The position recovery techniques, integrated with suitable context-aware servers (based on position, user profile, movements and user interactions history) represent a real innovative support to mobile customers, a support not well exploited up to now by existing systems. The *Hyper-Interaction within Physical Spaces* (HIPS) project (EU Esprit Research Program) demonstrated how helpful is for tourism the user immersion in a highly contextualised audio information space and how this choice increases the user satisfaction (Benelli, et al. 1999). Navigating museums represents an expression of high interest in audio explanations, not diverting the user from appreciating the museum artworks.

2. Information need for physical navigation

What is limiting the development of such smart interfaces? Users must cope with different applications, devices and wireless technologies for each different purpose; in most cases, the knowledge that the system has about the user is mostly fragmented and not consistent. Different mobile services and applications often adopt divergent user profiling approaches, thus making it difficult for them to interoperate. There is also lack of scalability and dynamic adaptation. Last, but not least, most of the available systems are a replication of a desktop *Information Technology* (IT) approach on mobile devices. Smart interfaces are designed for multipurpose devices, which take into account innovative interaction paradigms oriented towards mobile users. The metaphor to be followed is the replication of common human-to-human interaction modality in really mobile contexts, which are mostly device independent. The main issues that have been addressed by the HIPS UI are:
- to avoid unpleasant and useless information payloads for the user,
- to find suitable and consistent short-cuts to timeless information/services searches,
- to eliminate hard switches on the user terminal in order to get access to different usage modalities,
- to ease the mobile user while accomplishing his tasks in the physical environment,

- to better know the user on the basis of his preferences, knowledge, interest, time/space constraints and expectations,
- to provide a multipurpose and intuitive UI that can be widely accepted.

By using a mobile device (e.g., a PDA), HIPS users are immersed in a heedful audio information space, which is aware of their behavior (Bianchi et al., 1999). The system reaction is highly contextualised in respect to the user proximity to objects of interest. We believe that the audio interface is probably the best solution helping to satisfy all the previous listed points. A further support in graphic mode is needed to help the user not to loose his bearing (e.g., maps) and to answer to unexpected needs and new wills (e.g., menu choices, search facilities). In fact a completely audio interface can adapt itself only with respect to the "implicit" user behavior represented by his movement. A smart interface has been designed and implemented, partly having in mind a third generation (e.g., UMTS) mobile telephone and partly thinking to a more interactive game-boy like device. In this way we managed to simplify the HIPS controls to allow a natural and simple interaction.

3. UI design and implementation

Starting from the user-centred design approach, we conceived a prototype development process based on the well-known "design-evaluation-redesign" cycle. We developed a first UI in the first release of the HIPS prototype, which has been deeply evaluated according to:
1. Expert evaluations and preliminary user assessments
2. Identification of functionalities to be re-designed and functions to be added
3. Technical features re-design and parallel prototyping
4. Iterative refinements and specifications assessment for the implementation team

The re-design process leads us to substantial changes in the visual and auditory interface. Even the PDA hardware platform was redesigned to allow a more natural and simple interaction. Such a device is not available on the market and we simulated also the entire hardware platform (including hardware buttons). The addressed issues are the following:
- Need of clear audio control mechanisms.
- Need of a simplified interaction modality to interact with the system.
- Need of a clear feedback about the status of the system (e.g. loading of a presentation, positioning, wireless connection,..).

Our UI prototype design consists of these main features:
A Screen, 3:4 ratio, 240x320 resolution with (see fig.1):
- **An image Area**, where images, maps and other graphic information are shown
- **An information Area,** where different information can be displayed according to the status of the system: if the system is localised through wireless connection, the "Presentation Pane" (fig. 2) is displayed automatically.
- **A contextualised commands bar** with two labels (e.g., stop and more, see fig.1), each one corresponding to the related contextualised button described below.
- **A control bar** composed of:
- **A central "Game-boy like" control**, that consists of *up* and *down* buttons to easily scroll the list of items shown on the screen, and *right* and *left* buttons to select the right contextualised command (e.g. Stop/More, Play/Pause).
- **Two Lateral Buttons:** the "Menu" button for accessing other services, the "Hotspot" for creating a physical bookmark associated to the retrieved information, which can be shared with other users.

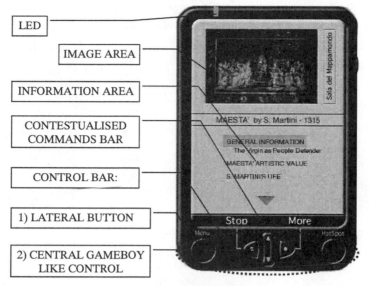

Figure 1: The UI components

3.1 Interaction modalities

The following modalities have been conceived, according to the way the user interacts with the UI:
- The *on-line* modality allows dynamic system reactions to user *explicit* and *implicit* actions on the UI. An *explicit* action is undertaken when the user requests a different style of reading, skips a running presentation, selects a different follow-up among those available in the list, makes a generic request for more information or pushes for further information, by clicking in a detailed area of the artwork represented in the image area of the display. An *implicit* action is undertaken with different user movement styles related to the specific physical space: going in another direction, staying a long time in front of an artwork, moving just in the middle of the room without getting closer to artworks, moving faster or lower, visiting sequentially all the artworks or jumping from an artwork to another one on a different wall (Veron et al., 1983). The information is adapted according to both *explicit* and *implicit* interaction modalities at the same time with an ad hoc synchronization and giving a higher priority to *explicit* interaction.
- The *off-line* modality allows accessing other functionality not so strictly related with the user actual *'audio-guide'* situation. The main difference is that in the off-line modality the dynamic narrative behavior of the system is to be stopped to accomplish other user tasks.

3.2 The "LED feedback"

As in some cellular phones, on the top of the palmtop a led shows the status of the system using different color states. If the user device is localised, the led appears as a stable green light. During the information loading, the led appears as a blinking green light. If the device is not localised, the led appears as a stable red light while in the *Presentation Pane* modality. The led is red also in *off-line* modality.

656

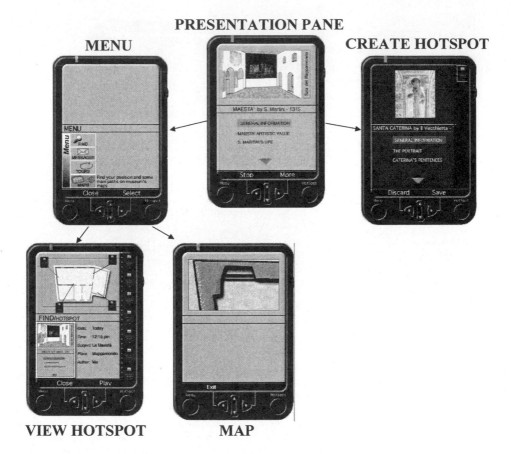

Figure 2: The different UI on-line and off-line modalities

3.3 The Hotspot modality

The *Hotspot* idea is that of taking a snapshot of the current moment of the visit and delivering it to another user if desired, after having added personal notes and messages. This implies for the system to record all the data delivered in a specific episode of the visit (text, audio description, images etc). The Hotspot creation is accessed in real-time by pressing the right lateral button labeled *Hotspot*. During the Hotspot creation it can be decided whether to share it with other users (e.g. belonging to the same group or not). When creating a Hotspot, the narrative is not stopped by default, even if the user may still decide to pause the actual presentation acting on the control bar. The view Hotspot is accessed differently through the Menu. The selection (play) of a Hotspot can be made with different criteria: choosing the author, the artwork, the space.

3.4 The map

The *Map* idea is that of browsing other physical spaces (rooms, floors) with different zooming levels. In fact the *on-line* navigated physical space does not need a map modality, being this functionality embedded in the 3-D images displayed on the image area of the *Presentation Pane*. The Map modality may support the "find" (artwork, room, art theme, friend) functionality, the "take me there" direction indication and so on. Other *off-line* modalities designed according to HIPS concepts can be implemented and differently related to the user actual position, to the community the user belongs to, to his profile and to the history of his visit.

This mobile device can offer many other features, contributing to provide contextualised and personalised information. It should be clear that users can decide at any time to get further support from the graphic modality other than the provided audio, extracting the device from their pocket and interacting with it through the described tools. In this way, contents can be chosen dynamically, so as to diminish/augment received information or direct the system focus towards other objects. Even if the system automatically provides them with an audio output (Not et al., 1999), mostly based on *implicit* interaction represented by the user movements, the user always has the full control of the interaction modality given by the graphical UI.

The user evaluation performed in HIPS, raised some interesting issues about the applicability and effectiveness of usability evaluation methodologies in the field of art, entertainment and leisure. The first impression on the idea of Hips is quite enthusiastic for many reasons:

- People felt very free in their movements during their visit.
- The heedful audio channel helped users to concentrate directly on the artworks.
- The User Graphycal Interface was mainly used just in case of explicit user requests of additional and/or different information and represented for them a suitable shortcut.
- The information is contextualised: Hips follows people's movements and offers information related to each new context. All visitors appreciated this feature.

4. Conclusions

HIPS has developed a handheld device through which the visitor of a museum can contextually obtain appropriate narratives regarding his physical environment. Descriptions of the artworks in the visual focus of the visitor are provided. These descriptions are tailored to both the characteristics of the visitor (e.g., movement style, interests, etc.) and also to the discourse history: what they have previously seen and heard in the tour.

The system continually tracks the user path throughout the physical space (Benelli et al., 2000), mapping this path onto the informational space: the body of history, technical details, anecdotes, etc., which lie behind each artwork. At the same time the device, thank to the graphical UI, is ready to take notes of the current visit, to communicate with other users, to browse off-line the existing exhibits and/or to move in a different museum area. The graphical UI has been carefully designed to provide an appropriate visible feedback to the user who feels this need while navigating an artistic physical space and contemporarily experiencing a heedful audio information space. The graphical UI also gives to the user a full control over the information assembled by the system, which has to carefully consider the user *explicit* interactions.

References

Benelli, G., Bianchi, A., Diligenti, M. (2000). A position aware information appliance. *EUSIPCO 2000*, Tampere.

Benelli, G., Bianchi, A., Marti, P., Not, E., Sennati, D.(1999). HIPS Hyper-Interaction within Physical Space. *IEEE ICMCS99*, Florence.

Bianchi, A., Zancanaro, M.(1999). Tracking users' movements in an artistic physical space. *i3 Annual Conference*, Siena.

Not, E., Zancanaro, M. (1999). Reusing information repositories for flexibly generating adaptive. *IEEE International Conference on Information, Intelligence and Systems*, Washington.

Veron, E., Levasseur, M. (1983). Ethnographie de l'exposition. Paris, *Bibliothèque publique d'Information*, Centre Georges Pompidou.

OFFICE 21 –
Inventing an Interactive Creativity Landscape

Wilhelm Bauer[1], Udo-Ernst Haner[2] and Alexander Rieck[1]

[1]Fraunhofer Institute for Industrial Engineering, Stuttgart, and
[2]Institute for Human Factors and Technology Management, University of Stuttgart,
Nobelstrasse 12, D-70569 Stuttgart, Germany

Abstract

Future information technology will allow mobile use of real time simulations and visualizations. If one is able to carry this kind of powerful and highly networked technology almost anywhere, why will there still be a need for an office (building) in the future? The answer is knowledge and creativity.
In the OFFICE 21 Interactive Creativity Landscape currently an office installation is being developed which targets the users' creativity and knowledge deployment. The landscape consists of three different zones. Each serves as support for a special phase in the creativity process. A sustainable increase of creativity and intelligence is only possible with a personalized, but simultaneously always changing environment. In this lab it is possible to demonstrate all needs for training a knowledge worker's brain.
This article presents the key ingredients of the OFFICE 21 Interactive Creativity Landscape.

1. Introduction

In a dynamic, competitive environment organizations are required to innovate in order to improve or just maintain their positions. Innovations, in turn, require the co-existence of a set of favorable factors which relate to ability, possibility, and feasibility. Ability subsumes all factors which influence people and their innovation attempts, like knowledge, imagination, motivation, creativity, and the absence of psychological barriers. Possibility refers to the supportive nature of organizational and other environmental factors. To this category belong issues like corporate structure, leadership, reward schemes, and working environment. Feasibility includes technological as well as economical factors.
Innovation is a process where the first stages are mainly characterized by creativity and sometimes by constructive chaos, whereas towards the end efficiency and order are increasing. Here, the focus will lay on the beginning of the process, on the supporting environment for innovation seeds. Since creativity is a dominant ingredient one has to ask, how a creativity supporting (office) environment can look like. In this article an answer is given by jointly considering human behavior, working environment and technological support.

2. Creativity as a Process

2.1 Creativity

Creativity, etymologically originating from the Latin word "creare", refers to the process of creating something new. It describes the ability of thinking by oneself or in a team in new and unusual patterns and thereby generating new ideas and new solutions. However, these generated ideas and solutions are themselves regarded as "creative" only if two conditions are present:
  a) The result of the creative process is new - in the sense of being completely different from the usual and the ordinary - and, in fact, often in contrast to the likely or the expected.
  b) The result is useful, meaning that it has some value for somebody besides the creator, especially to experts in the particular field.
Creative action therefore aims at generating new problem solving patterns for previously defined issues. This requires the combination of superficially not related facts, which have their origin in individual knowledge, imagination, or information.

2.2   Creativity Process

The process of creativity itself has four phases, which have to be distinguished:

I)   *preparation*: cognitive aspects of identifying situations where an improvement might be suitable and definition of the focus for change;

II) *incubation*: approach to the issue via conscious and subconscious processes;

III) *illumination*: spontaneous generation of possible clues to the issue; and

IV) *verification*: intellectual clarification, whether the illumination output respects the initially stated requirements, and formulation of a concept.

These phases have different requirements. In the preparation phase tools and methods are needed to structure the issue at hand, information has to be easily accessible via different channels, communication has to be facilitated. The incubation and illumination phases require the support of subconscious idea generation, fostering the concentration on intra-personal processes, possibly the opportunity to disguise the problem, and the freedom to cross perceived borders. The verification phase, in turn, requires visualization and discussion facilitation, support of co-operative work, feasibility analysis, and documentation. Differently stated, in the preparation and verification phase systematic and logical (i.e. convergent) thinking are dominant, whereas the incubation and illumination phase heavily rely on intuitive and lateral proceeding (see Fig.1). These different thinking styles are supported by different states of mind (or mental levels) which in turn are supported by different spatial zones. These relationships will be discussed in the following sections.

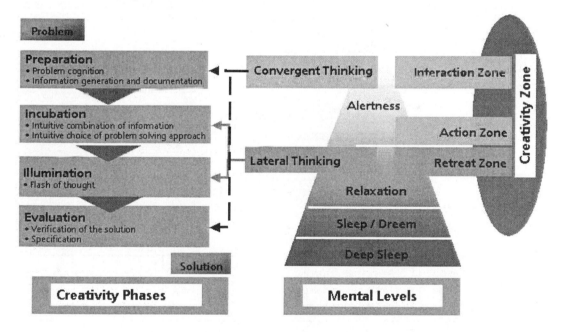

Figure 1: Relationship between Creativity Process, Thinking Styles and Creativity Zones

3.   Thinking Styles and Knowledge Workers

Convergent thinking requires a state of mental alertness (see Fig.1) since the tasks at hand demand predominantly concentration and focus. Measures supporting this thinking style therefore have to aim at increasing the state of alertness and thereby sharpening the ability to concentrate. External stimuli, e.g. air exchange, light regulation and physical action, have impact on the degree of alertness. Mental activity is also stimulated by the degree of information access and its usage, which again requires communication opportunities. So besides the environment also its infrastructure determines how well convergent thinking can be undertaken.

Lateral thinking on the other hand is fostered in the mode of mental relaxation. Supporting measures might include detaching the person from its environment by not supporting visual or acoustic information transmission. But physical detachment does not yet guarantee mental detachment, which has to be additionally supported by the opportunity to escape the issues at hand. Spatial solutions like a comfortable retreat room support the necessary focus change.

These requirements of the thinking styles have to be considered in designing creativity facilitating environments, especially within office environments for knowledge workers.

Knowledge workers often have the opportunity to use increasingly powerful information technology, which is very efficient in displaying heavy-load simulations and visualizations. Simultaneously, the mobility of this technology and thereby the mobility of the knowledge workers is rapidly growing. So, if one is able to carry this kind of powerful and highly networked technology almost anywhere, why caring for an office environment in the future? The answer is knowledge and creativity.

The know-how of a company is usually interchanged by personal contacts and non directional communication. Communication includes senses like smell and touch as well as short face acts. Therefore technology has to become more sensual and allow communication "by accident".

In the OFFICE 21 Interactive Creativity Landscape currently an office installation is being developed which targets the users' creativity and knowledge deployment. A long-term, sustainable increase of creativity and intelligence is only possible with a personalized, but always changing environment. In this demonstration center it will be possible to perfectly simulate all aspects needed for training a knowledge worker's brain.

So the office of the future becomes a place for meeting and communication. This real place has to be linked with the virtual mobile world through real time and multi-sensorial technology. From this perspective the office space becomes the interface to the integrated technology (see Fig. 2).

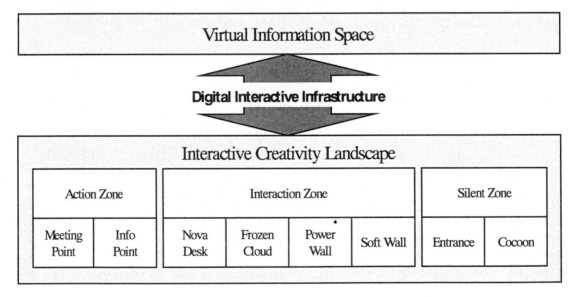

Figure 2: Components of the Interactive Creativity Landscape and
its Connection to the Virtual Information Space

4. The Interactive Creativity Landscape

Recognizing and considering the different needs during a creativity process, the Interactive Creativity Landscape incorporates the spatial support for this process. Therefore the landscape can be interpreted as an ensemble of different work spaces in a specially created office environment: an action zone, an interaction zone and a retreat zone. These will be described briefly in the following paragraphs.

4.1 Action Zone

The purpose of the action zone (marked with the letter A in Fig. 3) is to ease the information retrieval in the preparation phase of the creativity process. Within this zone a wide variety of information sources is available, anything from internet-connection to television and books and journals – basically covering all media types. This allows the user to get fast to know the state of the art in the particular area of interest, which might give impulses for the solution which is searched for. Additionally to being an information gathering point, the action zone is also the place where informal, face-to-face communication can take place.

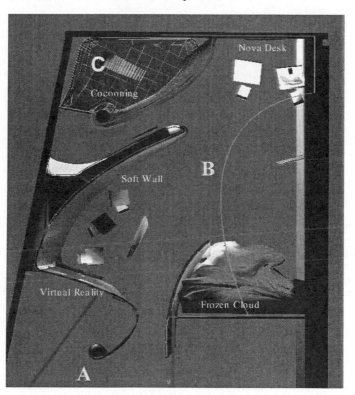

Figure 3: The Interactive Creativity Landscape

4.2 Interaction Zone

The Interaction Zone (marked with the letter B in Fig. 3) represents the largest part of the landscape. It facilitates the interaction of teams. A variety of technologies is available for enhancing the joint elaboration and evaluation of solution attempts. Included are visualization and simulation equipment as well as moderation and presentation tools. The main elements of the Interaction Zone are Nova Desk, Frozen Cloud, Power Wall and Soft Wall. Nova Desk is an example of so-called smart furniture, which serves as integrated computer and interaction device. Like an "intelligent" space wall Nova Desk replaces monitors and keyboards through projections. The Power Wall allows immersive projection and therefore intuitive real-time interaction with the virtual world. It additionally serves as a tool for large-scale multi-media conferencing. The Frozen Cloud is a device which facilitates physical stimulation. It is a sofa-like chill out landscape where spontaneous communication is facilitated by unusual forms and surface structures as well as the availability of special creativity tools. The Soft Wall is the device for producing and representing analog information. With its white board like-surface it is meant for large-scale manual activity like writing or drawing. The size intends to stimulate the movement of the whole body which is important for enhancing brain activity. Different flexible multifunctional walls will allow to vary the size of the rooms to accommodate different team structures.

4.3 Retreat Zone

The Retreat Zone is an individually adjustable – from the rest of the landscape separated – room for personal self-finding processes. Here the lateral, subconscious search for solutions during the incubation and the illumination phase is supported. One can even go that far to state that the available relaxing tools facilitate special creativity forcing processes.

The Retreat Zone consists of two major elements: the entrance area and the cocooning place. The entrance is equipped with an automatic environment adjustment system. It allows the individual configuration of the cocoon based on speech and gestures. Also, at the entrance a light shower system is integrated which stimulates certain, individually predefined human feelings. Both light and cocoon configuration will adjust automatically as soon as a particular user is identified. At the beginning a card based login station will serve as recognition device. It will be later replaced by a sensor system based on sound and/or visual recognition.

The cocoon itself is a small amorphously built room. The person inside is totally surrounded by translucent fabrics. Different technologies are provided to support the creativity task. The inspiration infrastructure allows sense stimulation within the wellness area. It will allow the exposing of the individual to environmental stimuli. As example may serve the change of the atmosphere by producing smells of nature or amusement parks. Also, colors and lights, video projection and digital walls are individually adjustable. Computer interaction, in the sense of creative output documentation, will be possible via touch screens or voice typing.

5. Conclusions

With increasing technology performance, the knowledge worker no longer needs to go everyday to a cubicle-type office to sit there in front of a computer or make a phone call. Rather, his intention will be to occasionally seize the interaction and communication opportunities – the social infrastructure – and thereby perform tasks which he might not be able to perform at other places. One of this particular tasks relates to creativity.The OFFICE 21 Interactive Creativity Landscape documents a shift from pure infrastructure provision (e.g. secretary, computer network, etc.) to a support of face-to-face communication, wellness and creativity. Thereby it is affecting the exterior and the interior architecture of the office in general since it merges digital media and the real haptic environment. With its different zones the landscape offers a variety of creativity enhancing instruments and systems. The built-in technology is almost invisible. Sensors and displays are hidden in the environment but permanently stand by for use. They become effective through the fostering of personal wellbeing, specific stimulation of individual senses, atmospheric changes and nevertheless a personalized environment. The architectonic concept provides an inspiring office atmosphere and an interesting to investigate office environment, for both the permanent user and the visitor.

References
Ceserani, J. & Greatwood, P. (1996). *Innovation & Creativity; Getting Ideas – Developing Solutions – Winning Commitment*. Great Britain: Clays Ltd.
Frey, H. (1999). *E-Mail: Revolution im Unternehmen -Wie sich Motivation Kommunikation und Innovationsgeist der Mitarbeiter wandeln*. Neuwied, Kriftel: Hermann Luchterhand.
Haffelder, G. (1998). Lernen optimieren – Lernstörungen verhindern, 'Ein-Blick' in die Gehirnforschung. In Geist & Gehirn, *CO-med*, Heft 10/98.
Haubl, R. (1998). Mit Sinn und Verstand - Einführung in die Umweltästhetik. In Günther, A., Haubl, R., Meyer, P., Stengel, M., Wüstner, K. (Eds.) *Sozialwissenschaftliche Ökologie: eine Einführung*. Berlin, Heidelberg, New York: Springer.
Koffka, B.M. (1996). *Die kreative Arbeitsumwelt, Ein Beitrag zur Ökologischen Psychologie und Architektur-psychologie*. Inaugural-Dissertation in der Philosophischen Fakultät I der Friedrich-Alexander-Universität Erlangen-Nürnberg.
Uebele, H. (1992). Kreativität und Kreativitätstechniken. In Gauglar, E. & Weber, W. (Eds.) *Handwörterbuch des Personalwesens*; Sp. 1165-1178. Stuttgart: J. B. Metzlersche Verlagsbuchhandlung und Carl Ernst Poeschel.

Web Accessibility for Seniors

Vicki L. Hanson[1], John T. Richards[1], Peter G. Fairweather[1], Frances Brown[2], Susan Crayne[1],
Sam Detweiler[2], Richard Schwerdtfeger[2], Beth Tibbitts[1]

[1]IBM Research, T. J. Watson Research Center, 30 Saw Mill River Rd, Hawthorne, NY 10532

[2]IBM Accessibility Center, 11400 Burnet Road, Austin, Texas 78758

Absract

Despite age-related disabilities that make web access difficult, seniors are flocking to the Internet. In this paper we describe a research project that examines the reformatting of web pages to make them accessible to seniors experiencing some age-related visual disabilities.

1. Introduction

The senior population worldwide is increasing rapidly, with the ratio of older to younger persons changing dramatically. Interestingly, seniors also constitute the fastest growing demographic group of Internet users (*Older Users Take to the Internet in Droves*, 2000; *Adults Over 50 Tune in to Technology*, 2000), and spend more time on the Internet than other demographic groups. One U.S. estimate puts their average online time at 8.3 hours per week (*Senior Citizens to Embrace the Web*, 2000). This compares to an average online time of 7.8 hours per week for college students. The reasons for their Internet usage have been attributed largely to factors such as increased amount of leisure time, a desire to communicate (especially with grandchildren), and web facilities that allow for comparison shopping and financial services.

There are, however, age-related disabilities that hinder seniors' access to the web. While we may not typically think of seniors as persons with disabilities, it is clear that a variety of physical and cognitive changes accompany aging (McNeil, 1997). For example, some degree of vision impairment is nearly universal for seniors. As we age, the eye goes through changes of pupil and lens structure resulting in reduced light and contrast perception, reduced color discrimination, and reduced acuity (Arditi, 1991). This is the result of normal aging, with older adults experiencing a world that is dimmer, has less vivid colors, and is somewhat blurred when compared with the visual experience of young adults. Some seniors will experience greater visual problems, such as macular degeneration, that result in more severe impairments, sometimes to the extent of "legal blindness". Such severe disability, while associated with aging, is not characteristic of the majority of seniors.

For most seniors, simply reformatting web pages could make a significant difference in their ability to access web content. We are currently investigating a method of transforming web pages such that seniors may access reformatted pages through any standard browser. Our system affords individual configuration for each user, even on machines that are shared with others (a characteristic, for example, of senior centers and libraries). The interface required for seniors to be able to perform this configuration easily is a key element of our design.

2. Background

Our project is not the first to attempt to make web content more accessible to users with vision disabilities. Importantly, the W3C not only has devised specific tags to make the web more accessible to all users (e.g., the ALT tag that provides for a textual description of images), but also has made more general recommendations for web accessibility (Dardailler, Brewer, & Jacobs, 2001). There are a number of places to which web authors can turn to learn how to make these accessibility enhancements, such as web reference books (e.g., Niederst, 1999; Paciello, 2000) and online resources that detail accessibility guidelines, including checklists (http://www.w3.org/WAI/, http://www-3.ibm.com/able/accessweb, http://www.rit.edu/~easi/webkit.htm). In addition, researchers in Japan are investigating a way to simplify web page layouts so as to alleviate problems encountered when using screen readers (Asakawa & Tagaki, 2000; Tagaki & Asakawa, 2000). Their approach relies on marking up pages with the appropriate annotations for page simplification.

As another approach, a few web sites have assumed the burden of creating web pages with font and layouts specifically designed for persons with vision impairments. Examples of this approach are sites such as EASI (http://www.rit.edu/~easi/), Betsie by BBC Education (http://www.bbc.co.uk/education/betsie), and the SETI-search (http://www.seti-search.com).

There are other means for making the web accessible for users with visual difficulties that do not require action on the part of web authors, but rather shift the burden to *users* to be savvy enough about computers to set their system in a manner optimal for them. For example, the two major web browsers, Netscape and Internet Explorer, both have options for setting font size, color, and style and for setting background color.

We differ in our approach in several major respects. The first is that we seek to provide users with accessible web pages irrespective of whether web authors have followed W3C guidelines, provided annotations for their content, or designed the site with low vision users in mind. In addition, we do not require that individual users have knowledge of their operating systems or browser settings to accomplish this.

3. Approach

Our goal in this research is to devise a system that requires no specialized hardware or software that must be installed by seniors. Our design point is that through any standard browser a user should be able to access web pages reformatted in a manner most readable for them. The system also must afford individual configuration for each user. The system employs an intermediary server to reformat web pages. Using a browser set to go through this HTTP proxy server, seniors access web pages the same as they would normally. The difference is that the pages served to them pass through this proxy and are reformatted according to their individual preferences. We use Cascading Style Sheets (CSS) to apply style modifications (e.g. fonts, colors, inter-letter spacing) to the entire page. We also provide for changes to the document HTML to modify such components as image size, removal of conflicting color information, and removal of backgrounds.

4. Interface

A challenge in implementing these text and image transformations for individual users was how to make these transformations selectable by the users. The changes needed to be selectable individually or in combination, such that one user might only want to take advantage of the font enlargement capability, while another might choose to both enlarge the font and change the foreground and background colors. The amount of enlargement and the color combination must be at the discretion of the individual user. The interface, additionally, had to be designed with the knowledge that many seniors have limited experience with computers.

We realized at the outset that we could not simply ask seniors to directly specify their preferences (e.g., 18 pt font, removal of background images). We decided to obtain information about user preferences by presenting users with a prototypical web page to which different transformations could be immediately applied. This page is representative of most web content and has both text and graphics. On the page, a small set of large, easily seen buttons cycle the page content through a range of preference settings for both text and graphics. Based on recommendations in the literature for making text legible for users with vision disabilities (Arditi, 1999) and our interviews with instructors at senior centers, we identified an initial set of critical variables that would impact text and image legibility for seniors. For text these manipulations included size, inter-letter spacing, color, and background. For images, these manipulations included size and sharpness. Our initial interface designs focused on simplicity and allowing seniors to make changes to web pages as quickly as possible.

To illustrate the settings process, let's consider the setting of font size. Our prototypical web page is initially displayed in 12-point Times New Roman (a typical browser default). Each click redisplays the text in the next larger size. In all cases, our design goal was to make the results of applying a transformation immediately evident. For simplicity, repeated clicks of the size button cycle through size options. The user can try different settings along any dimension in any order until they are satisfied. When the user has selected the set of transformations most appropriate for them, they simply click a "Use Settings" button to send the complete configuration to the intermediary server. This configuration is associated with the user name and applied to all pages subsequently loaded by the browser. Moreover, once set for a user, these preferences will be available to the individual from any other machine. Of course, a user's configuration may need to be changed at a later date or for a specific web page. To this end, we provide the ability to change their configuration from any web page.

Privacy was a key issue in our design due to the fact that personal health-related information can be inferred from the settings chosen by a user (Bennett, 1998). To meet this requirement, all sessions are secured by userid and password. A current design issue is the development of a logon screen that is truly accessible. This is critical since logon occurs before the system can load the user's individual preferences.

Our testing at local senior centers has provided us with valuable feedback about our design concepts[i]. The general interface proved easy to use, but we have gone through several improvement cycles based on user experience. While our initial interface designs focused on simplicity, as with all complex systems it became apparent that users desired more choices as they became familiar with the system. We are currently in the process of designing advanced options. For example, while our initial set of color selections should satisfy most users, it is clear that not everyone will find these choices optimal. Recent research with dyslexics provides an interesting illustration of individual differences in color preferences. This work found dyslexics' preferred combination for foreground and background to be the low contrast combination of brown on muddy green (Gregor and Newell, 2000).

5. Future Directions

The focus of this initial phase of our research has been on users with vision impairments. While our preliminary tests with seniors have validated our approach, we are aware that vision impairments are not the only disability experienced by seniors. Other age-related disabilities come in a variety of forms such as reduction of manual dexterity, hearing, and cognitive ability. Therefore, we are working to expand the types of features we can provide to seniors to promote their access to the web. We are now turning our attention to manual dexterity problems resulting, for example, from tremors or arthritis. These problems make it difficult for seniors to accurately use the keyboard and mouse due to difficulty in typing, double clicking, and pointing. We are investigating ways to make these input devices more usable for seniors (Trewin, 2000).

References

Adults Over 50 Tune in to Technology (2000). *NUA Internet Surveys*. [On-line]. Avalable at: http://www.nua.ie/surveys/

Arditi, A. (1991). A simple classification for functional aspects of age-related vision loss. In *Proceedings of the Symposium on Aging and Sensory Loss,* Mount Eliza, Victoria, Australia (pp. 64-80). National Centre for Ageing and Sensory Loss.

Arditi, A. (1999). *Making text legible: Designing for people with partial sight.* New York, Arlene R. Gordon Research Institute, Lighthouse International.

Asakawa, C., Takagi H. (2000). Annotation-based transcoding for nonvisual web access. *Proceedings of the Fourth International ACM Conference on Assistive Technologies, ASSETS 2000* (pp. 172-179). New York, NY: ACM.

Bennett, C.J. (1998). Convergence revisited: Toward a global policy for the protection of personal data? In P.E. Agre & M. Rotenberg (Eds.), *Technology and privacy: The new Landscape* (pp 99 – 123). Cambridge, MA: The MIT Press.

Dardailler, D., Brewer, J., Jacobs, I. (2001). Making the web accessible. In C. Stephanidis (Ed.), *User interfaces for all: Concepts, methods, and tools* (pp 571- 588). Mahwah, NJ: Lawrence Erlbaum Associates.

Gregor, P., Newell, A.F. (2000). An empirical investigation of ways in which some of the problems encountered by some dyslexics may be alleviated using computer techniques. In *Proceedings of the Fourth International ACM Conference on Assistive Technologies, ASSETS 2000* (pp. 85-91). New York, NY: ACM.

McNeil, John M. (1997). *Americans with Disabilities: 1994-95*. Washington, CD. Bureau of Census, U.S. Department of Commerce. [On-line]. Available: http://www.census.gov/prod/3/97pubs/p70-61.pdf

Niederst, J. (1999). Web design in a nutshell: A quick desktop reference. Cambridge, MA: O'Reilly.

Older Users Take to the Internet in Droves (2000). *NUA Internet Surveys*. [On-line]. Available: http://www.nua.ie/surveys/

Paciello, M.G. (2000). *Web accessibility for people with disabilities.* Lawrence, KS: CMP Books.

Senior Citizens to Embrace the Web (2000). *NUA Internet Surveys*. [On-line]. Available: http://www.nua.ie/surveys/

666

Takagi H., Asakawa, C. (2000). Transcoding proxy for nonvisual web access. In *Proceedings of the Fourth International ACM Conference on Assistive Technologies, ASSETS 2000* (pp. 164-171). New York, NY: ACM.

Trewin, S. (2000). Configuration agents, control and privacy. In *Proceedings of the Conference on Universal Usability* (pp. 9 – 16). New York, NY: ACM.

---

[i] We wish to thank Ann Wrixon, her team at SeniorNet, and the participants at SeniorNet Learning Centers for their cooperation in this research project.

Working Memory Capacity and Universal Access to the World Wide Web: Towards a User Adaptive Web Navigation Support System

David Heathcote*, Jennifer Jerrams-Smith**

* School of Design, Engineering & Computing, Bournemouth University
Bournemouth BH12 5BB, UK

**University of Portsmouth, Department of Information Systems
Portsmouth PO1 3HE, UK

Abstract

This work promotes Universal Access to the WWW by extending the currently existing user-model of the CAIN system. CAIN provides adaptive navigation support for the Web. It uses route-finding heuristics, which consult its model of the Web and a model of the individual user, to select a sequence of the most suitable Web pages to display to the individual user. Its adaptivity is therefore dependent on its Web model and its user models. Currently the model of the individual user is limited to domain-dependent user knowledge. However, we are now ready to add further domain-independent factors to CAIN's user model. One such factor is a measure of working memory capacity known as working memory span (Turner & Engle, 1989). One example of this extension to CAIN will be an ability to guide users to domain-specific Web pages whose content imposes a working memory load consistent with the user's working memory capacity. We intend to rate text in Web pages on the basis of their syntactic and semantic complexity, so that a rating of page comprehensibility can be added to CAIN's Web model. We intend to automate this process and to incorporate additional measures of working memory load. CAIN's route finding heuristics will be modified to take into account the new user model attribute (working memory span) and the new text comprehensibility index for Web pages.

1. Introduction

The World Wide Web (WWW) appears to provide an immense educational resource to be used in both teaching and research. However, in practice, its usefulness as a pedagogical aid is limited by its disorganised nature. The Computer Aided Information Navigation project (CAIN) aims to provide adaptive navigation support as a way of increasing the Web's value as a pedagogical tool (Lamas, Jerrams-Smith, Heathcote & Gouveia, 2000). Navigation support is accomplished by a basic route-finding heuristic which selects context specific Web model items, sorts them using associated qualitative ratings, and presents them to the user, one at the time, based on the attributes of the user's model (Lamas, Jerrams-Smith and Gouveia, 1999). Previous empirical work demonstrated that when students were required to use the Web to complete learning tasks, CAIN produced superior search and comprehension performance compared to conventional tools (see Lamas, Jerrams-Smith, Heathcote & Gouveia, 2000). In addition, the facilitatory effect of CAIN was found to interact with the user variables of domain knowledge and Web expertise suggesting that cognitive load influences performance in Web based learning tasks. This has important implications for the development of a fully adaptive Web navigation support system. Such a system would need to incorporate within its user model information relating to the cognitive capacities of individual users. This would be necessary in order to guide a user to sites whose content imposed a cognitive load consistent with the user's information processing capacity. However, before such a variable can be added to the CAIN user model, a valid a reliable measure of user cognitive capacity must be found.

2. Working Memory

One of the most important determinants of individual variation in cognitive capacity is working memory span (WMS) (see Miyake & Shah, 1999; Turner & Engle, 1989). Working memory refers to the system responsible for the temporary storage and concurrent processing of information (see Baddeley, 1986; 1997). Working memory is central to human cognition and plays a major role in comprehension, learning, reasoning, problem solving, reading and human-computer interaction (see Miyake & Shah, 1999). In terms of its ability to predict and explain a wide range of experimental findings, Baddeley's model of working memory is unrivalled in contemporary cognitive psychology (Heathcote, 1994). The working memory system has a tripartite structure consisting of a supervisory 'central executive' and two slave systems: the 'phonological loop' and the 'visuo-spatial sketch pad'. The central executive 'manages' the contents of working memory by executing a variety of control processes. Executive control processes include: updating the contents of working memory, monitoring and updating goals and sub-goals, monitoring and correcting errors, initiating rehearsal, inhibiting irrelevant information, retrieval of information from long term memory, switching attention between different elements of a task and co-ordinating concurrent activities. The central executive also co-ordinates the activity of the phonological loop and the central executive. The phonological loop is a speech-based processor consisting of a passive storage device, the 'phonological store', coupled to an active subvocal rehearsal mechanism known as the 'articulatory loop' (Baddeley, 1997). It is responsible for the short-term retention of material coded in a phonological format. The visuo-spatial sketch pad (VSSP) retains information coded in a visuo-spatial form. In recent formulations the VSSP has been decomposed into two functionally separate components: the 'visual cache' which provides a passive visual store and an active spatial 'inner scribe' which provides a rehearsal mechanism. Thus the organisation of the VSSP is similar to that of the phonological loop.

3. Individual Differences in Working Memory Capacity

Given the centrality of working memory in human information processing, we would expect individual differences in working memory capacity (WMC) to manifest themselves in the performance of a range of information processing tasks. Indeed, Engle et al. (1999) identify a number of published empirical studies which demonstrate a relationship between working memory capacity and performance in reading comprehension, speech comprehension, spelling, spatial navigation, learning vocabulary, note-taking, writing, reasoning and complex learning. Thus performance in these and related tasks can be partly predicted by working memory capacity. The measure of individual working memory capacity (WMC) is known as 'working memory span' (WMS) and several tests of working memory span have been devised (e.g. Daneman & Carpenter, 1980; Turner & Engle, 1989). All such tests require participants to engage in concurrent processing and storage. For example, Daneman & Carpenter's (1980) span test requires subjects to read lists of sentences. In addition to processing the sentences for meaning, the subjects are also required to recall the last word in each sentence (the storage load). Turner & Engle (1989) developed a span test in which subjects are required to store words while processing arithmetic problems. In both tests the subject's working memory span is taken to be the number of words correctly recalled. Thus, unlike I.Q., working memory span can be measured relatively quickly and unobtrusively. Indeed, for most Web users, the automated assessment of working memory span would be a rapid and straightforward process.

4. Working Memory Span as a Web Usability Factor

Clearly, many of the tasks found to be dependent on working memory span are likely to be present in Web based learning assignments. For example, most Web based learning tasks will involve text comprehension. Therefore the use of the Web as a pedagogical tool may be facilitated by matching the user's working memory span to the working memory load imposed by the Web page. In order to provide an accurate match, a reliable and quantifiable measure of Web page load is required. One such measure may be provided by objective indices of the comprehension complexity of passages of text. For example, the Gunning Fog Index (Gunning and Kallan, 1994) is a widely-used measure of readability. We intend to incorporate a similar measure into CAIN's Web model in order to guide users to suitable sites.

We have a developed a Complexity Index that is easily automated. It is a modified version of the Gunning Fog Index. Text is analysed by: 1) determine average sentence length; 2) determine the percentage of words of three or more syllables; and 3) add together. The following text was analysed, using our Fog Index:

> "A method was devised which enabled the development ofa taxonomy of error types so that errors could be classified. Subsequent investigation indicated the misconceptions (inaccuracies in the mental model) which were associated with some of the errors, and hence the remediation requirements."

The result was as follows :

Total word count = 43; Number of sentences = 2; Average sentence length = 21.5;
Number of words of three or more syllables = 12; Percentage of words of three or more syllables = 12/43 = 27.9%; Fog Index = 49.4

We intend to extend our Complexity Index by including additional factors relating to syntactic complexity. It will therefore include the percentage of sentences that include verbs that are in the passive voice (50% in the above example). It will also include the percentage of sentences that contain a negative.

The Fog Index for the Web model will therefore provide a numerical value for each Web page. This will provide a relative rating for Web page text comprehensibility. Web pages that are part of CAIN's Web model can therefore be compared. It is likely that more than one page will carry the required information for the user. If the user has low WMS then an appropriate the page with the lowest Fog Index would be presented by CAIN. We intend to investigate performance of non-expert users on comprehensibility tests when using pages that we have rated as having a low Fog Index. We would expect to find that although low Fog Index pages should produce better comprehension for all users, low WMC users will derive greater benefit from them than high WMC users. However, there may be an interaction between domain expertise and the user's ability to comprehend text (Doell, D.F. n.d.). It may be that greater domain expertise tends to negate the effect of low WMS. Therefore we also intend to investigate the interaction between expertise and WMC.

CAIN uses route-finding heuristics, which consult its model of the Web and a model of the individual user, to select a sequence of the most suitable Web pages to display to the individual user. Its adaptivity is therefore dependent on its Web model and its user models. CAIN's route finding heuristics will be modified to take into account the new user model attribute (working memory span) and the new text Complexity Index for Web pages.

References

Baddeley, A.D. (1986). Working Memory. Oxford, UK: Oxford University Press.

Baddeley, A.D. (1997). Human Memory: Theory and Practice. Sussex, UK: Psychology Press. Hove.

Daneman, M., Carpenter, P.A. (1980). Individual differences in working memory and reading. *Journal of Verbal Learning and Verbal Behaviour*, 19, 450-466.

Doell, D.F. (n.d.) *Gunning Fog Index*. Retrieved March 9, 2001 from the World Wide Web: http://www.pima.edu/~ddoell/tw/gfiex.html

Engle, R. et al. (1999). Individual differences in working memory capacity and what they tell us about controlled attention, general fluid intelligence and functions of the prefrontal cortex. In Miyake & Shah (Eds.), *Models of Working Memory*. Cambridge, UK: Cambridge University Press.

Gunning, R., Kallan, R.A. (1994). *How to Take the Fog Out of Business Writing*. The Dartnell Corporation.

Heathcote, D. (1994). The role of visuo-spatial working memory in the mental addition of multi-digit addends. *Current Psychology of Cognition*, 13, 207-245.

Lamas, D., Jerrams-Smith, J., Gouveia F.R. (1999). Adaptive Navigation on the WWW: an individualised and contextualised rough guide metaphor approach. In *Proceedings of 8th International Conference on Human-Computer Interaction (HCI International'99)*, Fraunhofer Gesellschaft, Munich, Germany, August 1999.

Lamas,D., Jerrams-Smith, J., Heathcote,D., Gouveia, F.R. (2000). Using directed world wide web navigation guidance: an empirical investigation. In *Proceedings of World Conference on Educational Multimedia (EdMedia 2000)*, Association for the Advancement of Computers in Education, AACE, June 2000

Miyake, A., Shah, P. (1999). Models of Working Memory: Mechanisms of Active Maintenance and Executive Control. Cambridge, UK: Cambridge University Press.

Turner, M.L., Engle, R.W. (1989). Is working memory capacity task dependent? *Journal of Memory and Language*, 28, 127-154.

Some HCI Challenges of the Virtual Exhibition

John Hiller   *Terry Postero*

University of New South Wales   Buffalo State College

Abstract

Organisation of images of art objects to produce *virtual exhibitions* is a development that soon will be with us. The selection of facets for the multi-media descriptions of the objects, the presentation of these objects, their sequencing, all present HCI (human-computer interaction) challenges.

The paper illustrates the importance of HCI issues in developing the *virtual exhibition*. It notes these exhibitions are not the equivalent of exhibitions in 'real art museums' but they can provide a related interactive experience that is fulfilling.
The discussion is based on the triad < *person, message, display* >.

1. Introduction

Fine art museums are digitising their collections; entrepreneurs from the computer industry are purchasing photographic series; national galleries are producing CDROMs of their works; e-commerce companies are being formed by major museums. The electronic or virtual gallery is coming.

The term *virtual gallery* as used here embraces collections of digital images and sets of multimedia objects that include these images. Descriptions of the items can be rich. For example, the multimedia object for a painting can include : its provenance, a bio of the artist, the circumstances of its commissioning, medium and technique, critical reviews, relationships to other works, in addition to an image.

Before moving to the specifics of the virtual gallery it helps to reflect on the typical visit to a gallery or museum exhibition.

Think for a moment about reviews of (real) gallery exhibitions that appear in our broadsheets. These refer to more than particular objects. Aspects of the artist's life, motivating influences, side preoccupations – all get a mention. In 1500 words or so the critic weaves a fabric from these strands. The display items contribute only part of the picture. Exhibition catalogues do something similar, and they can include observations from curators, art historians, and professionals from other disciplines. As we read these reviews and catalogues, particularly after visiting an exhibition, we may be humbled by the extent of their view in comparison with our narrower perceptions. Of course we can also become bored or find the descriptions pretentious and we may stop reading. There is an issue of the *appropriate breadth* for any discussion. Likewise there is the point about *commitment*. Further, it is important not to force over-faceted discussions on the reader. In virtual gallery terms, the viewer needs to be in control.

This paper is about journeying through virtual galleries; in essence it is about *virtual exhibitions*. It is about structuring the decisions needed from a visitor so that a destination is reached and the trek is worthwhile. It is about displays that border the highways and the possibilities of side-trips. It is about ensuring the viewer does not get lost while exploring these byways.

The interaction with a set of art objects can be based on the triad

< the person, the message, the display >.

The body of the paper treats each in turn.

2. The Person

The challenge in design of the navigation system is to set up appropriate facet (or mode) selection. First, we have to identify the facets. Looking at the conventional exhibition visit can give us some clues, not because we are consciously aiming for a similar experience, but rather because we expect many 'virtual patrons' will be comfortable if we provide familiar crutches. As Riemer reminds us '[a] common feature of most technological innovations is that their early products mimic those of the technologies they seek to displace' [Reimer 2001]. What does the conventional gallery visit tell us about the virtual gallery?

Recall the limited options in the conventional exhibition visit. A visitor to a virtual exhibition (or virtual gallery) accustomed to this simplicity may expect few selections and be prepared to accept standard options. This has the apparent advantage of allowing cognitive resources to remain focussed on the aesthetics. As the virtual experience becomes more common, the acceptable and required complexities of the interface will increase – indeed the cognoscente have such a requirement now. A range of sophistication is available in the interface options – the design issue is to decide how to offer it.

If I go to a museum that has a special exhibition my options can include :
> going to the exhibition
> viewing a video that may talk about the reasons for the exhibition, refer to its support, describe the assembly of the material
> getting an audioguide
> joining a tour
> visiting the exhibits in the 'standard order' or following an idiosyncratic order
> getting a catalogue.

My choices are simple and amount to selection of facets. Contrast this with the possibilities in a virtual exhibition where each facet may have multi-level options. Greater discrimination is now possible. The design issue is provision of an *acceptable level* of discrimination rather than offering a *range of complexities.*

With the real gallery my choices largely amount to selection / non-selection of the facets and possibly to the order of handling them. There is limited variety for any facet and my decisions amount to taking or not taking each of them *as is.* For example, in the guided tour the patron has little opportunity of nominating a theme of interest – the guide has a well-rehearsed patter and will not deviate much from this. Contrast such a limit with the possibilities of a virtual exhibition. There the *virtual patron* has options within each of the modes. From a decision point of view we are working in a < *mode, option* > space rather than in < *mode* > space. Richness is on offer; indeed richness may be expected in a computer-provided system. However, we must ask whether this richness adds to the experience. The threshold for acceptability in a virtual gallery differs from that in a real gallery. Can we achieve acceptability without giving full detail of the options? This is very important at the start of a visit to a virtual exhibition.

When I come to an exhibition I bring with me a set of previous experiences. What I see will be melded with my initial views. I also come with an expectation. It makes sense to find out a little about me and my goals. It also makes sense to use adaption in the navigation path. Note this is unusual in virtual galleries.

An example illustrates the point. Paterno and his colleagues have developed options for the (virtual) Marble Museum at Carrara. An initial question enquires whether the person is a tourist visiting the area, is an art student, or is an expert. Tourists need basic information, expressed clearly. They may want to organise a visit to the town and need to know their options. They are intolerant of unnecessary detail. Students have some knowledge of the application domain and they are likely to have specific projects. They need facts and formal discussions. They also need to be able to be challenged (Peters 2000). Finally, the experts want full access to all of the information available. Members of this group need minimal support to formulate their requests and they should be given this flexibility. There is a requirement for links to external references.

The initial question does not delineate my < *goal, skill* > pair but it makes a start. Noting that the virtual museum has high complexity Paterno argues for autonomous agents in the navigation. He sees the interface as adaptive with many decisions being made for the viewer. Of course it is easy to advocate such an approach but realising it is difficult due to the complexity of art objects and the search spaces they inhabit. A strength of the approach is provision of simple choices without requiring precision in definitions. A partial span of search space fits in with the cognitive style of the non-expert and echoes the simplicity of choice of the conventional museum. The initial choice may not be final; the interface should support changes in classification of the user during the interaction.

Classification of the user is not the only requirement for deciding on the navigation. Works may allude to another culture / time. Part of their meaning is lost if the viewer does not have the requisite background. However, the feature of interest need not depend on this knowledge or the viewer may be knowledgeable. Selection of issues is involved. Further, as noted in the *Introduction* there is an appropriate breadth in a presentation. Paterno sees the techniques for controlling this breadth as including : direct guidance, adaptive ordering, hiding and annotation – we tend to remember the last but not the other three. A dialogue with the data gives clues for the control. Note the word 'dialogue'. The viewer of a virtual exhibition has a solitary experience. Stratagems to give a feeling of a two-way interaction are important (Paolini 2000, Milekic 2000).

3.   The Message

The organisation of art objects into exhibitions involves themes, ordering, supplementing with commentary or references to comparable works. Shedroff refers to interactive design as story telling. A memorable experience comes from cohesion in encounters with individual items. The extra modes available to the virtual exhibition improve the chances of achieving this cohesion.

Integration of new information is important – integration means associating with what exists. The starting point may not be the same for different users; likewise cognitive styles can differ. As exhibition curators intend integration, it is to focus on the themes *they propose*. A rich description can help to clarify – if it does not confuse because of cognitive style. The way art professionals reason is not the way of novices, and the clues they plant may be missed. We need to note also the potential audience for a virtual exhibition is broader than those frequenting museums.

One group of art objects can support several themes. The organisation involves physical adjacency (in real galleries) or sequencing (in virtual galleries), selection and augmentation. In the art museum there is a limitation to one arrangement (see Falk 1992 and Serota 1996) and to selection from the objects that are available. In the virtual exhibition the curator for a particular theme can select from what is local and augment this with links to other collections[#]. In a similar way the one item can occur in multiple adjacencies and be used to illustrate several points. Sequencing for a particular theme involves both individual items and groupings of items. Such grouping is straightforward electronically.

The 'real exhibition' has items and labels. The latter are limited – Sweeney notes '…. A museum is not the right place for the transmission of the written word. A museum is not a book' (quoted in Falk 1992). The appeal of the conventional art exhibition is to those who are *visual* rather than to those who are *verbal*. However, the virtual exhibition can appeal to both groups. The visual need not dominate. The issue to ponder is whether one style of presentation can suffice. Can one leave it to the viewer to select from a range of options or is it appropriate to prompt? Falk comments : '[v]isitors look at a collection of objects and wonder: "What is it?" … "What is it used for?" … Rarely do they wonder "Why is this painting a landmark in abstract art?". Most visitors … deal with exhibits on a concrete, rather than at an abstract level.' (Falk 1992, p77). Having an item in a virtual exhibition can ease answering the concrete questions. The curator needs to decide whether abstract issues should be raised. In part this is about 'chunking'. The easier it is to become preoccupied with detail, the less likely it is to appreciate the wider view. In effect availability of detail may make it less likely that the theme message will get through.

[#]   Most discussions of virtual galleries refer to material owned by one gallery. The number of themes is limited but complications associated with copyright are avoided. These installations address only the viewer facet of the *limitation of place* (Perlin 1988). It is expected that developments will link images of works owned by several galleries to allow exhibitions that to be curated by staff who may not come from any of the source galleries.

There are some subtleties in the message. We need to distinguish the *message sent* from the *message received*. Expert users who use the virtual gallery to assemble information are generating their own messages. Some museum exhibits make their themes clear and have a physical organisation that continues to draw them to attention. For example, the exhibition *The Model Wife* at the Art Institute of Chicago in early 2001 had a plaque in the centre of the room that posed a number of queries. As the viewer moved around the exhibition it was necessary to continue to pass in front of this. Essentially one was forced to trip over these questions. The room organisation encouraged 'thinking big'. Virtual exhibitions have a propensity to encourage 'thinking small'.

The virtual exhibition gives control to the viewer. If extensive detail is available, the intended messages can remain unfound. However, this need not mean the journey is unrewarding. The requirement on themes is to be able to follow them if there is a continuing wish to do so. Curators need an incentive to set up *tours* of art object space that encourage visits. If this is not done we will merely have on-line art encyclopaedias, or the equivalents of the CD-ROMs now put out by the national galleries, or advertisements for current exhibitions. Of course developing a significant virtual exhibition will be expensive – how will these funds be provided? Is this someone's job or will there be charges for access? Do we look for virtual add-ons to the blockbusters provided by national galleries?

4. The Display

Questions to ask include :

> What it is possible to display.
> What it is desirable to display.
> What pattern of simultaneous viewing is desirable.
> How movement between pages should be supported.
> What requirements are for response time.

Limits on the display can be important for paintings, objects d'art. The conventional monitor with its 1024 * 768 pixel map is fine for thumbnails but may not be adequate for larger views. One example is the representation of pointillist work, eg artists like Seurat. There is much interest in the way that the dots of strong colour are assimilated by the eye. Uniform sampling of the image to produce the pixel map smears the colours and the effect is lost. Another difficulty with sampling is experienced with hatching in work by Van Goth. A second challenge comes up in colour. This has two sources : limits on the bits (eg 8-bit colour) and uncalibrated digitisers or monitors. Both provide substantial (and possibly significant) colour shifts. It is important to be aware that the display is limiting and not to expect too much. Recall that the viewer of a virtual exhibition is offered an experience *related to* what occurs in the real exhibition – not an *equivalent experience*.

Organisation of a display is a design challenge. In part this amounts to decision on the facets to provide directly, and those to handle with hot buttons. Both cognitive style issues and personal characteristics are important. An example of the latter is the confusion felt by a visual person when using hot buttons. The thread of a relationship to the base image may be lost after selecting hot buttons for artist detail and the like. It has been suggested that the image be included as a background or that thumbnails for superior levels of the hierarchy be included on all displays.

Along a similar line the requirement for backtracking is not as simple as usual. The familiar approach of providing the last 6 or 8 URLs is fine for searching but limiting when reviewing works that are essentially at the same level in a hierarchy. In a virtual exhibition hierarchies are important. These associations need to be available to allow a re-review of items on the basis of a hierarchic level or keyword equivalence. Displaying these and facilitating such a selection basis are the challenges. The familiar chronological order has limited (but non-zero) uses.

Since one experiences a virtual exhibition at one's desk, and there are significant applications for study or research, it makes sense to support note-taking, keywording, and indexing. Allowing a few lines of comment to appear at the bottom of a screen loses little real estate and may give a good basis for a hierarchic review. In short, the design involves facilitating user comment as well as selection. Having the user type comments is less than desirable – the technology needs to facilitate the interaction (Milekic 2000).

674

Finally in these brief comments on displays note the scope for an adaptive selection of what is displayed (Paterno 2000). Again decisions may be made for the viewer rather than by the viewer.

5. Conclusions

Much is possible with virtual exhibitions and rich experiences are on offer. Typical virtual exhibition journeys will be solitary. It will not be possible to converse with family members or friends or to get clarification from docents. The layout of information on screens has to offer ways of filling these 'human roles'.

The viewers will have a wide range of initial knowledge. Their expected outcomes and hopes for a virtual tour will be many and various. One could subject them to a set of defining questions, both initially and during the course of their journey. This is unlikely to be acceptable, either at a personal level or from the viewpoint of diverting cognitive resources away from the aesthetic. HCI professionals are one group who can provide a solution.

Some in the art world see the Internet in terms of art – *'What I am interested in is how the Internet can be a medium about art, rather than just information'* (Kasner 1998). The Internet through virtual exhibitions can open up art both to those whose primary interest is art and to others who focus information. A mix of facets is possible for different viewers.

References

Davis, D. (2000). The Virtual Museum, Imperfect But Promising. *The New York Times*, Arts & Leisure Section, 24th September, pp 3,33.
Dierkling, L.D, Falk, J.H. (1988). Audience and Accessibility. In S. Thomas & A. Mintz (eds), *The Virtual and the Real: Media in the Museum* (Chapter 4, pp. 57-70). Washington DC: American Association of Museums.
Falk, J.H., Dierking, L.D. (1992). *The Museum Experience*. Whalesback Books, Ann Arbor (205p).
Kasner, J. (1998). A Site for More Eyes. *ARTNews,* November (p141).
Milekic, S. (2000). Designing Digital Environments for Art Education/Exploration. *Journal of American Society for Information Science,* 51(1):49-56.
O'Doherty, B. (1986). Inside the White Cube: The Ideology of the Gallery Space. Santa Monica: The Lapis Press (91p).
Paolini, P., Barbieri, et al, (2000). Visiting a Museum Together: How to Share a Visit to a Virtual World. *Journal of the American Society for Information Science,* 51, 33-38.
Paterno, F., Mancini, C. (2000). Effective Levels of Adaptation to Different Types of Users in Interactive Museum Systems. *Journal of American Society for Information Science,* 51, 5-13.
Perkin, R.R (1998). Media, Art Museums, and Distant Audiences. In S. Thomas & A. Mintz (eds), *The Virtual and the Real: Media in the Museum* (Chapter 5, pp. 73-87). Washington DC: American Association of Museums.
Peters, O. (2000). Digital Learning Environments: Some Possibilities and Opportunities. *International Review of Research in Open and Distance Learning,* 1(number 1):1-10.
Riemer, A. (2001). Reading Between the Dots. *Spectrum, Sydney Morning Herald,* March 3rd.
Serota, N. (1996). Experience or Interpretation: The Dilemma of Museums of Modern Art. New York: Thames and Hudson (63p).
Shedroff, N. (1995). Information Interactive Design: A Unified Field Theory of Design. In R. Jacobson (ed), *Information Design*. Cambridge, Mass: MIT Press.

Presenting data as similarity clusters instead of lists -
Data from local politics as an example

Mauri Kaipainen, Timo Koskenniemi, Antti Kerminen, Antti Raike, Antti Ellonen

Media Laboratory, University of Art and Design Helsinki
Soft Computing Interfaces Group (SCIG), mauri.kaipainen@uiah.fi

Abstract

Similarity-clustering achieved by means of self-organizing map (SOM) is suggested as an alternative or complementary for conventional list-based manner of organizing large data characterized by significant patterns of multiple criteria. A sketch for a general-purpose browser interface is introduced as a tool to visualize and manage data using the SOM as the starting point. A data corpus of city council candidates' political opinions is used as an example.

1. Motivation and background

Immense reservoirs of information should already be open to everyone with a computer connected to the Internet. However, only a fraction of this potential is effectively in the layman's use. The usefulness of the information found depends critically on not only on the search methods but also on the tools of managing the found information. The standard way of presenting complex information is a standard list format allowing data to be ordered and viewed only one dimension at a time, such as alphabetical order or measure of goodness. This is handy when the information can be filtered down to a relatively small number. However, sometimes it is necessary to view large numbers of items to figure out holistic patterns. Examples of such cases include polls of qualitative opinions, complex system states consisting of multiple technical measurements, or combinations of economical indices. Conventional list presentations fail to describe such data optimally, because significant patterns tend to be of multi-dimensional character. From the users' viewpoint, it is often time-consuming and requires a relatively high level of academic skill to find coherent patterns in complicated data, because the significant information has to be collected from several lists and integrated in a relatively complex cognitive process. The aim of our report of a work in progress is to demonstrate an alternative to the conventional way of presenting data.

The cognitive load in such multi-criteria matching tasks can be reduced by presenting the user with data pre-ordered into similarity clusters, which take full advantage of internal correlations within the data. In our implementation, such order is automatically generated by the self-organizing map (SOM), introduced by Kohonen (1982), more concisely described e.g. in Kohonen (1990).[1] It is an artificial neural network based on a grid of abstracted neural-like units. It accepts lists of descriptive features, i.e. feature vectors consisting of numeric values of primitive feature descriptions, as its input. The output of the SOM for each input vector is a location on a two-dimensional map, surrounded by a distributed field of graded responses. The particularly useful feature of the map is that inputs that are similar, defined in terms of proximity of their descriptive vectors in the multidimensional input space, tend to be mapped close to each other on the map.

From the point of view of matching users' cognitive capabilities, it is worthwhile to point out that the fields of neural activity on the cortex respond to perceptual or behavioral stimulus patterns in a manner similar the SOM, i.e., with the point of maximal response surrounded by graded degrees of activity with a tendency to weaken as a function of distance. In such a mapping that can be legitimately called natural, relatively similar stimuli correspond to relatively similar activity patterns on the cortex. Tonotopies (Hood 1977), somatotopies (Merzenich 1988; Wall 1988) and spatial representations (Olton 1977) are well-known examples of such cortical mappings. They suggest that categories, groups and clusters emerge in the brain in some process of neurodynamical self-organization, whether of not it is comparable to the SOM algorithm.

---

[1] For an explication of the algorithm, see http://www.cis.hut.fi/research/som-research/som.shtml

2. The SOM Browser

Our contribution is to suggest both a computational solution combined with an interface design concept, in order to increase accessibility and readability for the future use of such services. Practically, our interface, a further development of (Kerminen, Raike & Kaipainen 2000), lets the user browse the map and retrieve information in an intuitive similarity-based manner which we claim is native to human cognition. The interface, implemented as a JAVA applet, consists of 1) *the map panel*, 2) *the questionnaire panel* and 3) the *data panel*. In addition, there is 4) a separate *display window* for the data that is retrieved.

3. Example corpus and demonstration

As an example from politics, it is often difficult for a layman to make sense of the candidates' agendas and in particular to compare them with each other, because there are too many issues to take into account. In the last local elections of Finland[2], all candidates for the city council of Helsinki (we exclude other communes of the country, originally included in the data) were asked to fill a form to indicate their stances to a list of specific political claims by choosing a degree of agreement with each. An Internet site[3] was established to facilitate the comparison of the user's opinion with those of the candidates. The user filled in his/her opinion using a standard radio button interface (Figure 1). The output of the service was a list of best matching candidates (Figure 2), resolved using a simple point-counting method.

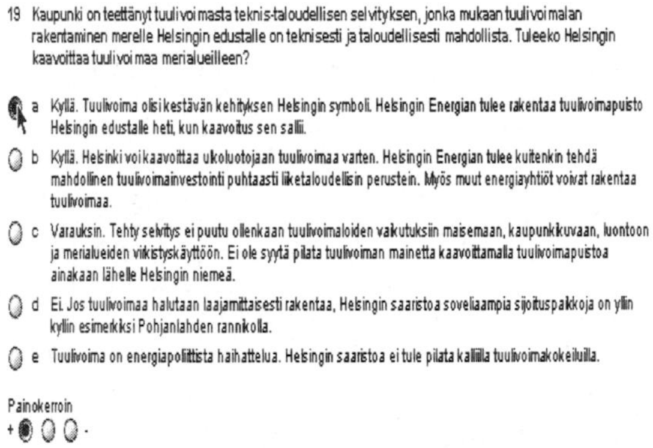
19 Kaupunki on teettänyt tuulivoimasta teknis-taloudellisen selvityksen, jonka mukaan tuulivoimalan rakentaminen merelle Helsingin edustalle on teknisesti ja taloudellisesti mahdollista. Tuleeko Helsingin kaavoittaa tuulivoimaa merialueilleen?

a Kyllä. Tuulivoima olisi kestävän kehityksen Helsingin symboli. Helsingin Energian tulee rakentaa tuulivoimapuisto Helsingin edustalle heti, kun kaavoitus sen sallii.

b Kyllä. Helsinki voi kaavoittaa ulkoluotojaan tuulivoimaa varten. Helsingin Energian tulee kuitenkin tehdä mahdollinen tuulivoimainvestointi puhtaasti liiketaloudellisin perustein. Myös muut energiayhtiöt voivat rakentaa tuulivoimaa.

c Varauksin. Tehty selvitys ei puutu ollenkaan tuulivoimaloiden vaikutuksiin maisemaan, kaupunkikuvaan, luontoon ja merialueiden virkistyskäyttöön. Ei ole syytä pilata tuulivoiman mainetta kaavoittamalla tuulivoimapuistoa ainakaan lähelle Helsingin niemeä.

d Ei. Jos tuulivoimaa halutaan laajamittaisesti rakentaa, Helsingin saaristoa soveliaampia sijoituspaikkoja on yllin kyllin esimerkiksi Pohjanlahden rannikolla.

e Tuulivoima on energiapoliittista haihattelua. Helsingin saaristoa ei tule pilata kalliilla tuulivoimakokeiluilla.

Painokerroin

\+ -

Figure 1. The users of the original web site defined their opinions with respect to specific issues of local politics using a radio button interface. This example is about whether wind power should be built on islands of Helsinki archipelago. (Original web site.)

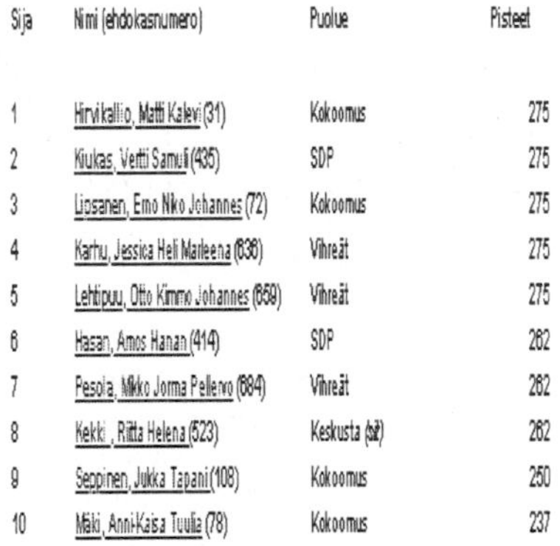

Sija	Nimi (ehdokasnumero)	Puolue	Pisteet
1	Hirvikallio, Matti Kalevi (31)	Kokoomus	275
2	Kiukas, Vertti Samuli (435)	SDP	275
3	Lipsanen, Erno Niko Johannes (72)	Kokoomus	275
4	Karhu, Jessica Heli Marleena (636)	Vihreät	275
5	Lehtipuu, Otto Kimmo Johannes (659)	Vihreät	275
6	Hasan, Amos Hanan (414)	SDP	262
7	Pesola, Mikko Jorma Pellervo (684)	Vihreät	262
8	Kekki, Riitta Helena (523)	Keskusta (sit)	262
9	Seppinen, Jukka Tapani (108)	Kokoomus	250
10	Mäki, Anni-Kaisa Tuulia (78)	Kokoomus	237

Figure 2. The original service output, a list of 10 candidates in the order of matching with the voter's opinion given above. (Original web site.)

Before the election date the service was accessed more than 70 000 times for or all 15 communes included and more than 33 000 times for the Helsinki city. The data consisted of a questionnary collected from city council candidates from Helsinki area in year 2000. Out of total 903 candidates 565 answered to some or all of the 20 questions. Answers to questions were multiple selections of three or four choices and possibility to give weight to the topic as 1-3. The data was converted for the self-organizing map as follows: The answers were either in nominal or ordinal scale, according to question. In the latter case, each choice was mapped to a individual component and marked with either 1 or 0. For each of weights and answers in ordinal scale, only one component was allocated and the numerical value of choice or weight was used after normalizing it to fit in range between 0 and 1. The procedure above resulted in 565 vectors each with 58 dimensions.

---

[2] In Finland citizens choose their local council candidate from a relatively large group of candidates. In the last local elections, several media companies built their case-specific search and evaluating engines to help citizens to find out most suitable candidates. The first application was done by Finnish Broadcasting Company in 1996 for the European Parliament elections.

[3] http://www.vaalikone.net/kysymyslomake.jsp?Kunta=HEL

A SOM for the data was calculated using rectangular topology and Gaussian neighborhood. The dimensions of the map were 16 times 20. During the teaching iteration the map was exposed to 26000 steps of data altogether.

4. Demonstrations

How can I find my candidate? The user can set his/her own preferences using the questionnaire panel, which is responded by the best matching node of the map to be indicated. (Figure 3).

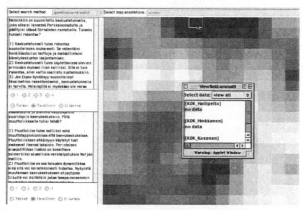

Figure 3. The user's preferences selected in the questionnaire panel. The best fitting node of the map is indicated, and the information corresponding to the candidates mapped to that area is displayed in a separate window.

Show all candidates that are for, e.g. improving communal services in circumstances of positive economical growth. The questionnaire panel lets the user choose a single question or a set of questions to be highlighted in the map panel. Graded degrees of a dedicated color's darkness indicate the degree of agreement of the candidates with the questions, in terms of generalized prototypes. An example is depicted in Figure 4.

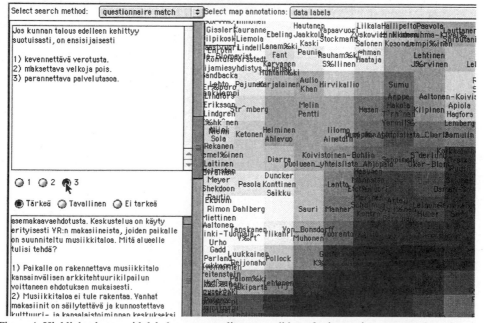

Figure 4. Highlighted map with labels corresponding to candidates for improving communal services in circumstances of positive economical growth. Dark color corresponds to strong agreement.

How can I find out facts about these candidate?. The user can take the advantage of the visual indications by selecting items on the map for further viewing, either by clicking a point or dragging over a rectangular area. *The display panel* (on the right) allows the user to choose from among the available information associated with the selected items, e.g., fields of the database. The chosen information will then be shown in the *display window*, as shown in Figure 5.

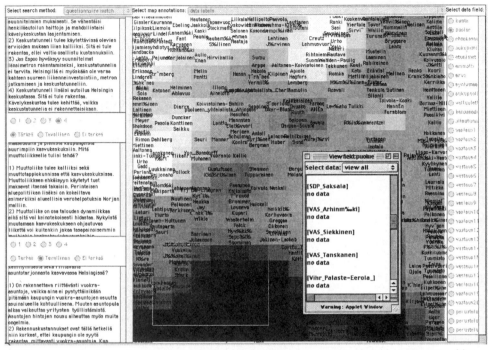

Figure 5. A map area is chosen by mouse-dragging, corresponding to the candidates with strongest position against the planned tunnel under downtown Helsinki. The candidates and associated information are listed in the display window.

5. Aspects of accessibility

The SOM browser facilitates reading complicated information for ordinary people, in particular for those not familiar with scientific presentations. Moreover, people with a degree of dyslexia may find this kind of visualized data presentation more accessible than strictly text based representations. From the point of view of the visually impaired, similarity grouping is a way to focus on topologically laid-out clusters controllable by mouse-hand movements. Combined by voice synthesis, such a presentation makes it easier to focus to the desired topic and renders it needless to read out loud extensive lists, typical to hierarchical representations search engine outputs.

6. Conclusions

We have introduced an approach and a sketch for a browser interface for information in which significant patterns are defined by multiple dimensions. Our intention is to make such information more accessible and graspable for all users. The idea is to organize large numbers of data items by content similarity using the self-organizing map paradigm, and to present the map so achieved with a set of easy-to-use online visualization and navigation tools allowing the users explore the material starting from their individual interests. These online tools give a possibility to explore data from the users' individual points of view in terms of highlighting map areas, responding immediately to their choices of descriptive feature dimensions. Such a hands-on approach to data is most likely to support grasping and learning more effectively than fixed presentations.

The approach is particularly well suited for internet or database searches with a large number of arguments. For example, it can be applied in conjunction with the WEBSOM technology (Honkela et al. 1996 etc.) for analysing massive numbers of free-format text documents, such as archives of patent applications and popular newsgroup articles.

We demonstrated the browser interface with data collected from the candidates for Helsinki city council, each describing his/her stance toward a list of specific political issues. With this choice we intended to suggest that in the future accessible knowledge management designs may have large socio-political value. We have also hinted at the advantages of the knowledge presentation method for special groups such as the dyslectics and the visually impaired. This work is an example of soft computation used as a method of knowledge management and visualization. It can be considered to be "softer" than conventional solutions in the sense that it serves the information to the user in a preclustered but not a priori indexed manner, yet offering tools to visualize and manage the information in a holistic manner.

Acknowledgements

We thank Helsingin Sanomat[4] for collaboration and access to the data.

References

Honkela, T.; Kaski, S.; Lagus, K.; Kohonen, T. (1996). *Newsgroup Exploration with WEBSOM Method and Browsing Interface*, Helsinki University of Technology, Lab. of Computer and Information Science. Report A 32.

Hood, J. (1977). *Psychological and Psychological Aspects of Hearing*, -Critchley, M.; Henson, R. 1980. Music and the brain. London: Heinemannss.

Kaipainen, M.; Karhu, P. (2000). Bringing Knowing-When and Knowing-What Together. Periodically Tuned Categorization and Category-Based Timing Modeled with the Recurrent Oscillatory Self-Organizing Map (ROSOM), Minds and Machines 10. 203-229, 2000.

Kerminen, A.; Raike, A.; Kaipainen, M. (2000). *Self-organizing map browser for database retrieval*, -Emiliani, P.L.; Stephanidis, C. (2000). User Interfaces for All. Special Theme: Information Soociety for All, Consiglio Nazionale delle Ricerche, Istituto di Ricerca sulle Onde Eöettromagnetiche "Nello Carrara", Firenze. 6th ERCIM Workshop, Florence, Italy 25-26- October 2000.

Kohonen, T. (1982). Self-organized formation of topologically correct feature maps, Biological Cybernetics 43:59-69.

Kohonen, T. (1990). *The Self-Organizing Map*, Proceedings of the IEEE, Vol. 78, No.9, Sept. 1990.

Merzenich, M. M.; Recanzone, G.; Jenkins, W. M.; Allard, T. T.; Nudo, R. J. (1988). *Cortical Representational Plasticity*, -Rakic, P.; Singer, W. (Eds.) 1988. Neurobiology of Neocortex. John Wiley & Sons Limited.

Olton, D. S. (1977). *Spatial memory*, Scientific American.

Wall, J. T. (1988). Variable organization in cortical maps of the skin as an indication of the lifelong adaptive capabilities of circuits in the mammalian brain, Trends in Neuroscience, Vol. 11, No. 12.

---

[4] The data was collected by Helsingin Sanomat, the biggest daily newspaper in Scandinavia with a circulation of 4469 729. The internet service, provided for 15 communes, was used more than 70,000 times before the election date (33 000 for Helsinki).

Electronic Kiosk Provision of Public Information:
Toward Understanding and Quantifying Facilitators and Barriers of Use

Kathy Keeling, Linda Macaulay*, Denise Fowler, Peter McGoldrick,
Konstantina Vassilopoulou *

School of Management and *Department of Computation; University of Manchester
Institute of Science and Technology; P.O. Box 88; Manchester; U.K.
+44 161 2757095 - kathy.keeling@umist.ac.uk

Abstract

Electronic information kiosks are seen as providing 'user friendly' access to public information and services. Research so far has shown that there is a range of issues apart from inadequate hardware and software maintenance that contribute to user acceptance. This paper presents an organizing framework based in the theories of the Diffusion of Innovations (Rogers, 1995) and the Technology Acceptance Model (1989) to aid understanding of the factors emerging from the literature, reports of kiosk implementation and qualitative research on two commercial kiosk trials. The model is tested using data from quantitative research from the latter. The results indicate that the model is valid, accounting for nearly 53% and 41% respectively of the variation in future intentions. Both studies, using different systems, in different circumstances, demonstrate the direct effects of accessibility and relative advantage on intentions to use electronic kiosks again. Further, the addition of affect, compatibility and normative influence are valid and aid understanding in environments where the use of a technological option must compete with other means of task achievement. The results highlight the necessity of ensuring the compatibility between the tasks the user might wish to perform and a) the content and other features of the interface design and b) the location of the kiosk, e.g., the need for privacy. Even in the presence of a perceived need, motivation may be depressed by social factors. This combination of effects indicates that the provision of public information access through electronic kiosks must consider the total experience of the user and the social situation in which use might be embedded if they are to understand fully the potential barriers to use. Together, the constructs of the extended Technology Acceptance Model can help designers to understand the relative importance of factors helping or hindering adoption of electronic information provision. Keywords: Kiosks, barriers, understanding users, contexts, models.

1. Introduction

Electronic kiosks are seen as providing 'user friendly' access to electronic information for those who might otherwise be subject to new forms of social exclusion. Touchscreen operation offers the advantages of lowering the entry barriers for those with little familiarity with computers. Commercial sites can be run alongside non-commercial, offering banking, sales, couponing and promotions with government and public utility information and services. However, potential users have freedom of choice between electronic means and other ways of carrying out information gathering or other tasks in their daily lives. Thus, electronic kiosks have to attract the user in a competitive environment under diverse situations and conditions. In addition, interaction times are likely to be short leaving little scope for training.

This paper considers the adoption of electronic kiosks for information provision by the general public and presents a model that helps organize the issues affecting uptake of electronic kiosk use by the general public. This model, based in theory, quantifies the interplay between the factors and their relative importance as facilitators and barriers to use and thus provides recommendations for implementation. The paper also presents the results of two studies of usage of commercial kiosks for information access where the model is applied. Commercial concerns bring the benefits of marketing experience, large design budgets and in-store locations. Given the interest in increasing information access and political participation by 'going where the people are', the results from studies of commercial kiosks should have valuable lessons for the location, design and marketing of 'not for profit' access kiosks.

The literature on public information user-interface design (e.g., Maguire, 1998) suggests that, in addition to addressing technical concerns, factors such as a) physical location; b) motivation for use c) user perceptions of self-efficacy c) user needs and d) the need for encouraging use should be considered. A review of commercial access points also underlines the importance of these features. In sum, there is a need to ensure that the ease of use and content of kiosks fulfills user needs and engages them in such a way that they would wish to use the kiosk again.

2. Theoretical framework

The theoretical framework for quantifying the effects of the facilitators and barriers to uptake is based on the Diffusion of Innovations (DOI) literature (e.g., Rogers, 1995) and the Technology Acceptance Model (TAM) (e.g., Davies, Bagozzi and Warshaw, 1989). The five key acceptance components identified in the DOI literature are Trialability, Observability, Compatibility, Relative Advantage and Complexity. The TAM explains system acceptance as a function of user perceptions of Usefulness and Ease Of Use. The Relative Advantage dimension of the DOI can be associated with Usefulness; Complexity with Ease Of Use. To incorporate all the factors that the Diffusion of Innovations encompasses, additions to the basic TAM model are necessary. The addition of aspects of Compatibility is indicated from the literature (Al-Gahtani and King, 1999; Green, 1998; Keeling, Fowler and McGoldrick, 1998). The addition of a further aspect, affective reaction, is also evident in the literature (e.g., Igbaria, Schiffman and Wieckowski, 1994). Figure 1 gives a schematic diagram of the extended model with the additions as indicated by the literature and the results of the qualitative data (see study one). Additions to the TAM are represented in the dashed boxes, affect is included as a component of Usefulness. This model gives a framework allowing the organization and estimation of the importance of factors affecting electronic kiosk uptake. The results of two studies using qualitative and quantitative data to assess the position of these constructs with the framework are discussed below.

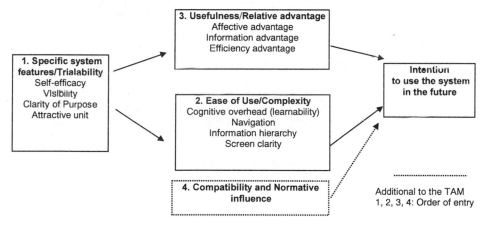

Figure 1. Incorporation of Diffusion of Innovation constructs within the TAM framework

3. Study one

The first study involved a prototype of a kiosk application for a cinema foyer. Both qualitative and quantitative research were undertaken to obtain a comprehensive understanding of potential user beliefs and motives underlying the use of a cinema touchscreen kiosk. Following qualitative research with 15 respondents, 79 respondents conducted a short ticket buying task on the Cinema Touch Screen system and completed a questionnaire immediately afterwards. The questionnaire was structured to reflect the factors identified by Rogers (1995) and discussed above as significant in the diffusion of innovations. The majority of items were derived from the analysis of the qualitative phase, however, some were generated from scales and recommendations in the literature. The trail use of the kiosk system did not take place in an actual cinema therefore questions relating to physical features, visibility or physical situation of the kiosk were not appropriate. However, trialability was represented by four items relating to respondents' beliefs about their general abilities to use touchscreens. Ease of Use was represented by two items concerning overall evaluations of perception of 'cognitive cost', adapted from the ease of use scale of the TAM. Also included were items regarding reducing mental load based on navigation and ease of finding information (Ebersole, 1997). To accommodate the diverse visual abilities of public users, two items asking for judgments of the text and picture clarity on the screens were also included. Affect, information quality and efficiency were

682

distinguished as potential contributors to the relative advantage of a kiosk system. Topics relating to these three factors identified from the results of the qualitative interviews were: the friendliness and impersonality of the system; whether the system provided sufficient information and allowed an easy comparison between films; and the effort and time needed to use the system. Three aspects of compatibility were represented in the questionnaire: compatibility with the purpose of the task itself; compatibility of use with the situation of use and compatibility with the social context of use.

Table 1. Results for cinema kiosk data (* indicates significance at 0.05 and beyond)

	1	2	3	4
Trialability				
Self-efficacy	.268*	.403*	.263	.087
Ease of Use (Complexity)				
Cognitive overhead		-.227	-.271	-.120
Navigation		.182	.095	.082
Information hierarchy		.174	.043	.079
Screen clarity		.009	.196	.225*
Relative advantage				
Efficiency advantage			.378*	.210*
Information advantage			.050	.006
Affective advantage			.397**	.375*
Compatibility and Normative Influence				
Fit how choose a film				.096
If I were alone				-.274*
With a group				.146
If someone watching				-.264*
R Square	.072	.151	.428	.607
Adjusted R Square	.060	.093	.363	.528

3.1 Results

The sample of respondents was drawn from undergraduate and postgraduate management degree students. Thus, the majority was in the 18-24 age group; a potential target market for the use of the touch screen. However, the results from this analysis must be interpreted bearing in mind that these respondents may have a greater familiarity with computing than the general public. Reaction to the prototype was good, the majority of respondents (87%) reported they were likely to use the system again. To assess the relative contribution of each of the constructs in the model to potential uptake, the variables were entered in a hierarchical regression with order of entry as shown in figure 1. The dependent variable was 'intention to use the system again'. The results are given in table 1. The results show that the constructs in the model account for nearly 53% of the variation in intentions to use the system again. This is a good result for questionnaires of this kind.

At entry, self-efficacy has a significant effect on intention, the higher/lower the self-efficacy, the higher/lower the intention to use the system in the future. Perceived complexity of the system also contributed a significant effect to intention: higher/lower evaluations of ease of use lead to higher/lower intentions. However, there is no significant individual contribution discernible. Notable is the interactive effect between self-efficacy and complexity, high/low self-efficacy gives rise to higher/lower evaluations of ease of use. Notwithstanding, the combined effects only account for around 9% of the variance in the intention to use the system again.

Most important for these users are efficiency and affective advantage, as both self-efficacy and complexity are mediated by the constructs of Relative Advantage. Saving time and effort, whilst having an enjoyable experience contribute an additional 27% of the explanation of future intentions. Interesting to note is the increase in the importance of screen clarity when Relative Advantage is considered. Indeed, screen clarity continues to increase in importance when Compatibility factors are added. The most important Compatibility factors are 'if I were alone' and 'if I thought someone was watching me'; together adding around 16% to the explanation of intention.

4. Study two

The second study involved a fully operational kiosk application that provided information about prices and system requirements for software and users were able to watch short multimedia demonstrations. The kiosks were situated in three stores of two retail chains, one selling general electrical goods, the other selling computer hardware, software and peripherals. The locations of the three stores were chosen to introduce geographic and demographic differences to the sample. Again, both qualitative and quantitative research was undertaken and the results of the qualitative phase used to generate wordings for the quantitative phase where appropriate. The quantitative data collection took place over a week and was spread throughout the opening hours of the stores.

The questionnaire was constructed in a similar manner to the first study. However, being conducted in real-world situations, there were constraints on questionnaire length to maintain respondent interest and motivation. The first study established the validity of self-efficacy in the prospective use of public kiosks; that construct was not used in this study. The results also indicated that representing ease of use by three indicator items concerning ease of navigation, clarity of on-screen instructions and clarity of the text on screen was sufficient.

These constraints allowed the inclusion of items to cover issues that were not addressed in the first study. The first study could not assess the impact of visibility, accessibility or kiosk housing design, therefore these were represented by four items with wordings arising from the qualitative phase. Two items were also included about the clarity of purpose of the kiosk. Topics relating to two usefulness factors of information quality and efficiency were identified from the results of the qualitative interviews. Normative influence was represented by two items: whether other people's opinions about a piece of software were important to them and whether they asked friends and family for information on software before purchase. Items were included concerning compatibility with the purpose of the task.

4.1 Results

Data was collected from users and non-users. Users were approached after using the system, non-users were approached after having walked past the unit. (They were asked initial questions about their non-use of the kiosk and then asked to give it a short trial.) Of the 208 respondents, 72% indicated that they would use the system in the future. New variables were computed as per study one. There were no significant differences for ratings on any of the constructs by gender, age or the store that data was collected in. As in study one, the constructs were entered in a hierarchical regression in the order shown in figure 1. The results are given in table 2. These show that the constructs explain 41% of the variation in Intention to use the system again. The findings from previous trials and the qualitative studies are supported; at entry the clarity of purpose of the unit and the attractiveness of the unit itself have a significant effect on Intention. However, this accounts for only just over 8% of the variation. With the significant entry of ease of use, the attractiveness of the unit still displays a significant effect. Together these constructs explain around 12% of variance in Intention. The effects of ease of use and attractiveness of the unit are mediated by the relative advantage of the system as shown by the drop in the size of the coefficient. Each of the three facets of advantage is significant, adding a further 27% to the explanation of Intention. Thus, the model shows the trade-off between ease of use and attractiveness with the perceived advantages of the system. Compatibility with the task is important in this model, normative influence less so. Together, they add around just 2% to the R square.

Table 2. Results for software kiosk data (* indicates significance at 0.05 and beyond)

	1	2	3	4
Trialability				
Visibility and accessibility of kiosk	-.042	-.118	-.185*	-.200*
Clarity of purpose	.146*	.115	.109	.094
Attractive unit	.251*	.271*	.124	.126*
Ease of use		.223*	.043	.066
Relative advantage				
Efficiency advantage			.294*	.186*
Information advantage			.239*	.165*
Affective advantage			.198*	.188*
Compatibility and Normative influence				
Compatibility with task				.202*
Friends and family opinion				.116*
Others' opinions important				-.036
R Square	.097	.140	.412	.442
Adjusted R Square	.084	.124	.392	.414

5. Discussion

Public access information providers wishing to maximize uptake of electronic delivery channels are faced by a combination of factors, not least is that users have freedom of choice between electronic and other ways of carrying out tasks in their daily lives. Thus, electronic kiosks have to attract the user in a competitive environment under diverse situations and conditions. In addition, interaction times are likely to be short, leaving little scope for training. This paper has examined the factors that emerged from the literature and the results of trials of public access kiosks as facilitators and barriers to use. The examination has taken place in the context of an organizing framework based

684

in the theories of the Diffusion of Innovations and the Technology Acceptance Model. The constructs in the model account for reasonable amounts of the variation in intentions to use the system again. It is especially significant that both studies have demonstrated the effects of accessibility, relative advantage, affect and normative influence in environments where the technological option must compete with other means of task achievement. What this combination of direct effects indicates is that those wishing to provide public information and services access through electronic kiosks must consider the total experience of the user and the social situation in which use of a kiosk might be embedded if they are to understand fully the potential barriers to use. One barrier is location and design of the kiosk housing. Not only must locations be chosen carefully, to be visible whilst affording some degree of privacy, but the design of the kiosk housing must not be intimidating. Another barrier is low self belief in the ability to use technological products: this is an additional reason to ensure the 'friendliness' of the kiosk housing. However, direct experience or training increase self-efficacy (Martocchio, 1994). Direct experience can be tackled in three ways, which have reinforcing benefits if used together. Firstly, enticing and engaging the passer-by to try the system out of curiosity by providing clear signage of the purpose of the kiosk and off-line usage instructions. Secondly, prior marketing of the presence and purpose of the kiosk and, thirdly, using personnel to promote use. However, trialability factors only account for about 6% and 9% of future intentions respectively in the two studies. Through the mediation of trialabilty factors by relative advantage factors, models also show the trade-offs between concerns about ability when the advantages of using the system become apparent to the user. Therefore, the main thrust of resources should be on marketing the advantages to the user. This does mean that design and content requirements must be assessed in terms of value and benefit to the end user. Furthermore, the results highlight the necessity for understanding the compatibility between the tasks the user might wish to perform and a) the content and other features of the interface design and b) the location of the kiosk, e.g., the need for privacy. Even in the presence of a perceived need, motivation may be depressed by social factors. In study 1, the most important compatibility factors are 'if I were alone' and 'if I thought someone was watching me'. Together these added around 16% to the explanation of intention. In another study of Internet access provision, through a kiosk in a supermarket, the authors found that concerns about the lack of privacy afforded by the positioning and bright screen prevented the use of a banking option. Potential users defaulted to the darker screen with smaller print of the external ATM, even in the rain! Together, the constructs of the extended Technology Acceptance Model can help designers to understand the relative importance of factors helping or hindering adoption of electronic information provision. Nevertheless, the conclusion that must be drawn is that, for maximum effectiveness, prospective providers must undertake comprehensive research of user needs and requirements from the user point of view. This is unlikely to be achieved without consulting potential users at all stages of planning and design.

Acknowledgments
This work was carried out with the support of the Engineering and Physical Sciences Research Council, UK. We thank all respondents, companies and store staff involved in the research and reviewers.

References
Al-Gahtani, S.S., King, M. (1999). Attitudes, satisfaction and usage: Factors contributing to each in the acceptance of information technology. *Behaviour & Information Technology, 18(4)*, 277-297.

Davis, F.D., Bagozzi, R.P., Warshaw, P.R. (1989). User acceptance of computer technology: A comparison of two theoretical models. *Management Science 35(8)* 982-1003.

Ebersole, S. (1997). Cognitive issues in the design and deployment of interactive hypermedia: implications for authoring www sites. *Interpersonal Computing and Technology, 5(1-2)*, 19-36.

Green, C.W., (1998). Normative influence on the acceptance of information technology: Measurement and effects. *Small Group Research, 29(1)* 85-123.

Igbaria, M., Schiffman, S.J., Wieckowski, T.J. (1994). The respective roles of perceived usefulness and perceived fun in the acceptance of microcomputer technology. *Behaviour & Information Technology. 13(6)* 349-361.

Keeling, K. A., Fowler, D., McGoldrick, P. J. (1998). Electronic Banking Customers: Determinants of Adoption and Retention. *In Proceedings of 27th European Marketing Academy Annual Conference*, Stockholm, 535-554.

Maguire, M. C. (1998). A review of user-interface design guidelines for public information kiosk systems available at http://www.lboro.ac.uk/research/husat/

Martocchio, J. J. (1994). Effects of conceptions of ability on anxiety, self-efficacy and learning in training. *Journal of Applied Psychology, 79(6)* 819-825.

Rogers, E. M. (1995) *Diffusion of Innovations* 4th Edition The Free Press New York.

Functional modeling of process systems for interface design

Qiao Liu[*], *Keiichi Nakata, Kazuo Furuta*

[*]Kondo Lab, Dept. of Quantum Eng. & System Sci., School of Engineering,
The University of Tokyo, 113-0033, Tokyo, Japan. Email: liu@sk.q.t.u-tokyo.ac.jp

Abstract

The using of piping and instrumentation diagram in industry has resulted in innumerable accidents because they are not enough to fulfill the needs of operators. Therefore, a systematic way that supports designing better interfaces has to be considered. In this paper, firstly, a functional modeling method that is used to define the contents of a process system that should be displayed in the interface is introduced. By capturing the analogies among different physical systems, a new set of functional primitives, which are governed by different equations, is proposed. These equations define the constraints of these functional primitives when they are used to describe a process system. Moreover, an example is given and its functional model is illustrated. Based on the functional model, we can define the necessary contents that should be displayed in the interface.

1. Introduction

In recent years, complex systems have been increasingly applied in industry. The common way to present information of these systems is to use the piping and instrumentation diagram. There are several problems with such kind of interfaces:

(1) The goal-relevant information is not explicitly considered because the presentation is based on the piping and instrumentation diagrams. Thus, the functions and goals are not conveyed.

(2) Single-sensor single-indicator approach is generally used, so the operators have to derive the overall status by integrating information about the plant from individual indicators. This places a heavy burden on operators because they have to store these relations in their memory.

(3) Display formats are designed without explicitly considering operators' cognitive characteristics.

Thus, it results in some problems for operators to manage a system correctly. Therefore, a systematic way to support designing interfaces has to be considered. In the new design framework, two questions must be considered. First, a modeling method, which can define the informational content that must be displayed in the interface, is needed. Second, how to communicate this information to the operator efficiently must be considered. This will provide a basis for determining the visual form that the information should take. The relations between operator, information content and visual form are illustrated in Figure 1. The focus of this article is on the first question. As shown in Figure 1, when we want to use a modeling method to define the content of the system, the cognitive characteristics of operators must be taken into consideration. We propose the adoption of functional modeling method to specify the information content. In this paper, firstly, the concepts of goal, function, behavior and structure are introduced. Then, we present a means-ends based functional hierarchy. By capturing the analogies between different physical systems, such as hydraulic, mechanical, electrical, thermal, etc., we propose a new set of functional primitives, which is used to describe the functional model of a process system. Finally, an example — a central heating system, is given and its functional model is illustrated.

2. The fundamental concepts: goal, function, behavior, and structure

Generally, there are four concepts that are used when modeling a process system. The four types of knowledge can be defined and distinguished from each other as follows:

(1) A goal represents the purpose of a system assigned by its designer.

(2) A function represents the explanation of how the roles that the designer intended a component or a sub-system should have in the achievement of the goals of the system of which it is a part. It serves as a bridge between behavior and goal.

(3) Behavior describes how a component or a sub-system works to realize its function. It is generally governed by physical laws.

(4)   Structure describes the physical arrangement of components.

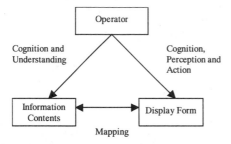

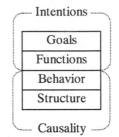

Figure 1: Relations between operator, information content and display form

Figure 2: The relationships between goal, function, behavior and structure (Lind, 1994)

The relations between goal, function, behavior and structure can be shown in Figure 2 (Lind, 1994). While the pair of structure and behavior describes the causality of the system, the pair of goal and function captures the intentional aspect of the system. The two aspects are connected to different scientific domains, the natural sciences (causes) and the social sciences (intentions). Here the concept of function plays an important role to bring the two domains together. On one hand, it is a part of the intentional knowledge of a system, which is related to the goal of the system. On the other hand, it is the interpretation of the behavior realized by the physical structure of the system. Therefore, the concept of function must be considered in engineering design.

3. Functional modeling method

3.1 Functional hierarchy

As shown in Figure 1, the modeling method should not only describe the contents of a system correctly, but also should be in accordance with the cognitive features of the operators. It is argued that goal-relevant information must be considered in the modeling method (Vicente et al, 1990). Thus, the means-ends relation is used for functional modeling. The functional hierarchy based on means-ends relation is illustrated in Figure 3. In this hierarchy, the system's goals are described on the top level, the functions of sub-systems are described in the middle levels, and the functions of components are described on the lowest level. Lower level functions explain "how" their next higher level functions are achieved. And higher level functions explain "why" their next lower level functions are needed. The means-ends relation provides the possibility for operators to solve complex problems. By moving up to the higher levels, there is a selective loss of detail, and thus problems can be made manageable. By looking downward to lower levels, the higher level function can be described in detail by the sub-functions, which are connected with it. In conclusion, the constrains posed by the higher level functions on lower functions in this structure show a way for an operator to navigate through the various levels and concentrate on the most relevant part of the system. In this manner, it provides a basis for interface design.

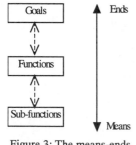

Figure 3: The means-ends functional hierarchy

The relations between goal, function, behavior and structure are shown in Figure 4. The structural hierarchy is based on whole-part decomposition. On the top level, the whole system is described, and on the lowest level the components are shown. The relation between function and structure is shown by the arrow from the structural hierarchy to the corresponding function in the functional hierarchy. The arrow represents the concept of behavior, which serves as the medium between structural and functional knowledge. It should be stressed that the mapping from a sub-system/component to its function is not one to one. One sub-system/component may realize more than one function. And conversely, one function may be realized by more than one sub-system/components.

3.2   Definition of a set of functional primitives

3.2.1   Motivations

Among the approaches in functional modeling, there is a type of representation called flow-based representation. Flow-based representation is based on the concept of flow (and sometimes effort). In this type of representation, function is separated from goal and treated as a relation between the input and output of energy, mass and information. A set of functional primitives is defined, and functions of all components/sub-systems are expressed in terms of these primitives. It is interesting to note that several flow-based modeling methods define similar sets of primitives (Chittaro et al, 1993; Karnopp et al, 1990; Lind, 1994). Moreover, by consulting with the domain expert, it has been shown that experts often refer to a limited number of abstract processes involving substances (such as transporting, storing, transforming, etc.). These words can be mapped onto generic arrangements of functions. It seems that adopting a functional modeling language will not result in acceptance problems by the domain expert.

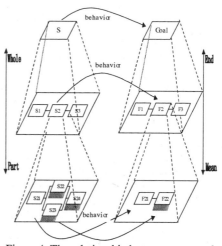

Figure 4: The relationship between structural hierarchy and functional hierarchy

In the following part, we will propose a set of functional primitives based on the work of MFM (Lind, 1994), Multimodeling (Chittaro et al, 1993) and bond graph theory (Karnopp et al, 1990).

3.2.2   A new set of functional primitives

In our approach, the concept of functional primitive is defined as a basic representation unit of function; it is an interpretation of a system/component's behavior, which is described by physical equations. All the functional primitives defined in MFM, Multimodeling and bond graph theory are listed in Table 1. The corresponding behavior expressions (generalized physical equations) are also shown in the first column of Table 1.

Table 1: Functional primitives defined in different modeling methods

Mathematical Formulation	Bond Graph	Multilevel Flow Modeling	Functional Role Model	New set of primitives
$e = E(t)$	Effort source		e-generator	generate
$f = F(t)$	Flow source	Energy/Mass source	f-generator	
$C \cdot de/dt = f$	Capacity	Energy/Mass store	q-reservoir	Store
$L \cdot df/dt = e$	Inertia		p-reservoir	
$f_{out} = f_{in}, \Delta e = R \cdot f$	Resistance	Energy/Mass transport	f-conduit	Transport
$e_{out} = e_{in}, \Delta f = R \cdot e$			e-conduit	
$e_{out} = k \cdot e_{in}, f_{out} = f_{in} / k$	Transformer			Transform
$e_{out} = k \cdot f_{in}, f_{out} = e_{in} / k$	Gyrator			
$\sum f = 0$	Flow distributor	Energy/Mass balance		Balance
$\sum e = 0$	Effort distributor			
$f = 0$		Energy/Mass barrier	f-barrier	Barrier
$e = 0$			e-barrier	
$f = F(t)$		Energy/Mass sink		Sink
$e = E(t)$				

Although the three sets of primitives seem sufficient to describe different physical systems, there is still a possibility to expand them. By analyzing the three sets of primitives in Table 1, a new set of functional primitives are proposed, which is shown in the fifth column of Table 1. These primitives are explanations of the generalized equations shown in the first column. At this point, it should be noted that all the generalized equations would be meaningful when a specified physical system is under study, such as an electrical system, and so on.

3.3 Functional representation of system/sub-systems/components using functional primitives

As shown above, we have shown seven terms in the new set of functional primitives. However, in different physical domains, domain operators and experts might use different terms they prefer to explain the behavior described by the same generalized equation. The different terms that are explanations of the same generalized equation must belong to the same class. Thus, the seven primitives can be seen as a representative of seven classes of primitives. Since all these terms are domain dependent and cognition relevant, it is difficult to list all these terms. Table 2 shows part of these terms.

Table 2: Some examples of the seven classes of functional primitives

Classes of functional primitives	Some terms
Generate	Create, form, make, produce, induce, provide, cause
Store	Deposit, stock, fill, reserve, reposit, conserve
Transport	Carry, move, transfer, deliver, shift, pass, transmit
Transform	Change, convert, transmute
Balance	Maintain, keep, preserve, hold, conserve, retain, remain
Barrier	Prevent, suppress, forbid, avoid
Sink	Discharge, release, assimilate, consume, empty

In addition, when modeling process system, control information needs to be considered. Control actions can be performed either by operators or by automatic control systems. The control algorithm can be either very simple (e.g., PID control) or very complex (e.g., neural control or adaptive control). In our functional modeling method, we use the primitive "control" to represent control information. The terms related to control information are shown in Table 3.

It should be noted that some terminologies could belong to different classes. For example, the term "maintain" can belong to either the class "balance" or the class "control" based on different contexts.

For functional modeling, a common syntactic structure is defined to describe functions of a system. It is shown as:

{functional primitive}{physical variable}{sub-system/component}{context}

The structure is not unique and the parts "physical variable", "system/component" and "context" are optional. One example of this structure is:

{Keep} {the water level}{of the empty tank} {between 1.1 meters and 1.2 meters}

3.4 An example — the application of the functional modeling method

The central heating system is shown in Figure 5, and its functional model is shown in Figure 6. Higher-level functions describe "why " the functions are needed, and next lower functions describe "how" the higher-level functions are achieved. For example, on the top level, the goal of the system - "keep the temperature of the room within limits" is described; on the second level, the functions – "generate heat", "transfer heat" and "release heat" describe how the goal can be achieved.

Based on the functional model, we can define the contents for interface design. For example, the information about the system goal must be included in the interface, i.e., the actual room temperature should be displayed. Thus, the operator can judge directly whether the system's goal is achieved or not. Then, the information about how much heat is produced by the boiler, how much heat can be transferred to the heat exchanger and the heat exchange ratio should be also displayed, etc. Thus, by moving from the top level to the lowest level, all the information that is necessary for the user can be defined for interface design.

It should be noted that control information is also included in the model. For simplicity, suppose that all control actions are done by operators manually. For example, when the goal is not achieved, the operator will have to take a series of procedures to make it realized. When the operator finds that the fuel in the boiler is not burning sufficiently, he/she will adjust the amount of fuel or oxygen provided to the boiler; when the water level is under the limit, he/she will open the feed water valve to provide more water to the system, and so on. Such control information should also be visualized in interface display.

Table 3: Some primitives for modeling
control information

functional primitive	Some terms
Control	keep, adjust, maintain, provide, regulate, change

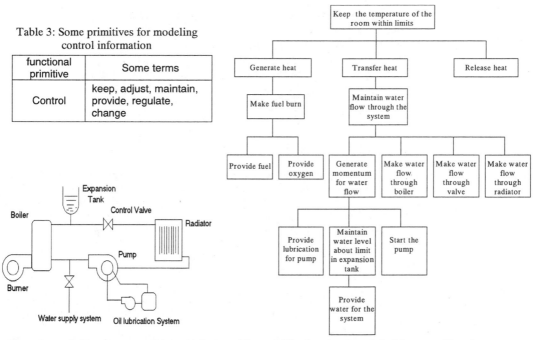

Figure 5: a central heating system (Lind, 1994)　　　Figure 6: The functional model of the central heating system

4. Conclusions

By examining some problems of current industrial interface design, a systematic way to display design is emphasized. The systematic approach concentrates on two questions. One is display content, the other is display form. The former is described in this paper. Firstly, the means-ends based functional modeling method is proposed as a solution to define the display content for interface design. Then, the functional hierarchy is introduced and the relationship between functional hierarchy and structural hierarchy is described. By capturing analogies among different physical systems, a new set of functional primitives is proposed, which serves as the component in the functional model. Finally, the functional model of a central heating system is presented. Based on the functional model, how to extract the information content for display design is illustrated. For further work, we will consider how to visualize these information contents and their relations by examining the principles of EID (Vicente et al, 1992) and configurality (Bennett et al, 1992).

References

Bennett, K.B., Flach, J.M. (1992), Graphical displays: implications for divided attention, focused attention, and problem solving, Human Factors, 34(5), 513-533.

Chittaro, L., Guita, G., Tasso,C., Toppano, E. (1993), Functional and teleological knowledge in the multimode ling approach for reasoning about physical systems: a case study in diagnosis, IEEE Transactions on Systems, Man and Cybernetics 23, 1718-1751.

Karnopp, D.C., Margolis, D.L., Rosenberg, R.C. (1990), System Dynamics: a Unified Approach, John Wiley & Sons, Inc.

Lind, M. (1994), Modeling goals and functions of complex industrial plants, Applied Artificial Intelligence, 8, 259-284.

Vicente, K.J., Rasmussen, J.(1990), The ecology of human-machine systems II: mediating "direct perception" in complex work domains, Ecological Psychology 2(3), 207-249.

Vicente, K.J., Rasmussen, J. (1992), Ecological interface design: theoretical foundation, IEEE Trantractions on Systems, Man, and Cybernetics 22(4), 589-606.

690

Web page adaptation for Universal Access

Juan F. Lopez, Pedro Szekely

University of Southern California - Information Sciences Institute
4676 Admiralty Way, Marina del Rey, California 90292-6695, US
juan@isi.edu, szekely@isi.edu

Abstract

Universal Access to the Information Society requires the adaptation of the web content and information to multiple devices and users. This paper presents a generic system that allows users to view any web page on any device. The system automatically gets a web page, transforms it and adapts it to the features of the device and the preferences of the user. The objective of the system is to allow users access the same information anywhere, anytime, and on any device. The approach used is to restructure web pages (typically designed for desktops), so that they can be adequately viewed in small portable devices.

1. Introduction

The exploding growth of the Internet and the Information Society is creating new challenges in UI design, because of the universal access requirements for a variety of users and devices. Users expect similar capabilities and accessibility on a small screen device and on a big screen desktop.

It is estimated that by 2003 more than 50% of the Internet access will be through non-PC devices, such as Internet appliances, PDA, web enabled cellular phones and other various handheld devices. All these new portable devices have very different capabilities and features (e.g., display size and resolution, bandwidth, processing power, input/output). Maintaining different versions of a web site for each one of these new devices is labor intensive and impractical.

This paper presents a generic system that allows users to view any web page on any device. The system gets a web page, transforms it and adapts it to the features of the device and the preferences of the user. The objective of the system is to allow users access the same information anywhere, anytime, and on any device.

2. Example

The following example clarifies what we want to achieve with this system. The left side of the figure shows the input of the system: a page from Amazon books website seen with a regular PC web browser. The right side shows the desired output (several adapted pages) when the user sees the Amazon website with a Palm VII.

Both systems present and make accessible the same information to the user, adapted to the capabilities of the device. Some of the adaptations and transformations illustrated in this example are:

- Each one of the generated adapted pages contains parts of the original HTML code of the page, but adapted to the capabilities of the device (e. g., Palm VII only supports a subset of the HTML 3.2 specifications). Some of the tags or attributes are added, removed or replaced to accommodate the limitations or constraints (e. g., Palm VII does not support JavaScript, nested tables, frames, cookies or Java) [1].
- Graphics and other multimedia elements are adapted: Some graphics are scaled, or the number of colors is reduced (Palm VII supports only 2-bit depth per pixel). Other graphics are replaced by their Alt attribute, a link, or simply removed (e.g., decoration graphics).
- The top of each Palm page contains a navigation menu to enable the user to navigate through the different sections. The system also generates index pages (like the one in the screen of the Palm) with links to the different sections (generated pages). These indexes list the available information to the user. All the HTML code for the navigation and indexes is added by the system (since the navigation did not exit in the original page). Navigation mechanisms allow the user to navigate through the information (the generated adapted pages).

691

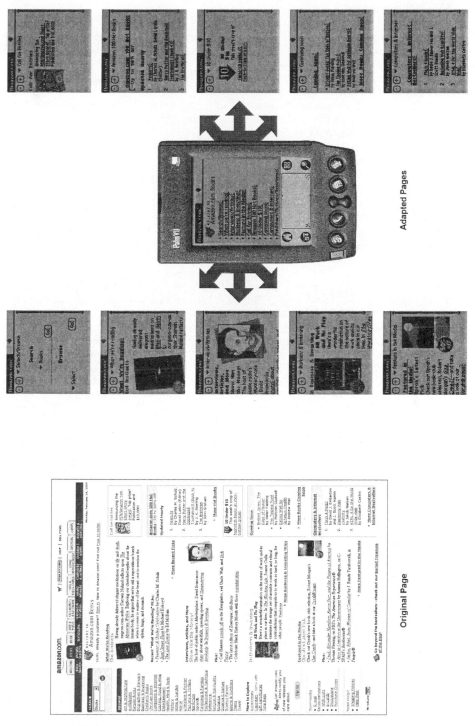

Adapted Pages

Original Page

Figure 1: Example

3. The Variety problem

The main problem to implement a system that allows universal access to the information on the web, as the one described above, is what can be called *'the variety problem'*: The system should be generic enough to manage very different classes of information, devices and users:

- Web pages may be very different in content, layout, structure, features, style, languages, standards, etc. There is a very broad variety of pages in the web: *variety of information*.
- New web enabled devices are continually appearing in the market (e.g., Internet appliances, PDA, cellular phones, handhelds) with very different capabilities and features (e.g., display size and resolution, bandwidth, processing power, input/output): *variety of devices*.
- Different kinds of users with different backgrounds, expertise, roles, skills, capabilities, navigation styles, localization, etc. try to access the same information on the web: *variety of users*.

4. Approach

The main elements to handle this variety problem are:

- Clean up of the original, desktop-specific HTML [2] and conversion to XHTML [3]: This makes easier to parse and work with the documents, since it is well-formed XML [4]. Transformations can be implemented using XSLT [5]. This also provides a transition path for existing HTML content. The system also accepts content in the new XML/XSL standard.
- Analysis of the document layout and structure to generate an ontology of the document: This includes layout analysis and generalization, main content detection, section classification, and multimedia elements classification (e.g., ornamental vs. information images)
- Document ontology: The document is parsed and analyzed to generate a tree with object classes and relations. Having a well designed and generic enough ontology enables the system to classify each node accurately, and establish the relations with other nodes. A generic ontology makes possible to accommodate very different classes of web pages (information).
- Classification and generation independence: Classification and Generation are independent processes, cleanly separated by the ontology. The system can use different classifiers or generators (automatic or manual) depending on the needs. Publishers can have control on the adaptation by including the ontology into the pages.
- Transformation based: The original document is adapted by a set of transformations. Transformations can be applied sequentially to the original page. Transformations can be reused for different devices or user profiles. New transformations can be easily added to the system.
- Device description and user description: By defining a set of characteristics or capabilities the system can adapt to a great variety of devices and users. Characteristics/capabilities can be added later to expand the system.
- Navigation strategy and navigation mechanisms: Enable the user to navigate through the different pages. Code for navigation may be added to each adapted page to enable the user to navigate through the different pages. The system may also generate index pages with links to the different generated pages.

5. System architecture

Our approach is to restructure and transform web pages (typically designed for desktops), so that they can be adequately viewed in small portable devices (automatic adaptation of desktop content for small devices). In a small display all the information cannot be shown at the same time. Normally one desktop page will be displayed as several smaller pages. It is essential to decide what to display and when to display it. Another problem is how the user can navigate between the different elements of information displayed in the page.

The new standards (XML, XSLT, and XHTML) enable us to separate information (content) from presentation (style) and to apply different transformations to the original documents. The system restructures web pages by applying a set of transformations/adaptations to the original document.

The system acts as a proxy, getting the real pages, classifying the contents, transforming the pages into a set of adapted pages, and sending the adapted pages to the client.

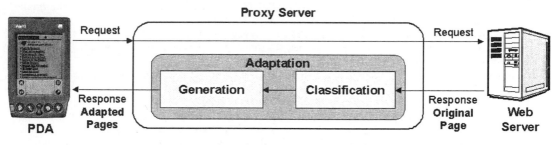

Figure 2: System Architecture

Two main components can be differentiated in the system:
- *Original Web Page Classification:* Classifies the different elements of the original page, divides the page into sections, classifies nodes, etc. Classification may be done manually or by an automatic classifier.
- *Adapted Web Page Generation:* Generates the web pages adapted to the device and user preferences, using the classification and the device and user description. The adapted pages are generated by a set of transformations.

5.1 Original Web Page Classification

This component of the system gets the original web page and classifies the different elements of the original page. Classification may be done manually by the publisher (special tags embedded in the HTML) or automatically by a software classifier.

The classification divides the page into different sections. The classification depends on the document structure, layout, discontinuities, fonts, styles, etc. Several types of relations (e.g., spatial, discourse, typographical, contextual, structural, semantic) can be used to segment the document.

Two basic steps are completed in this phase:
- Original page acquisition: The input for the system may be HTML, XHTML or XML/XSLT. If the page is HTML then the system cleans up the HTML code and converts it to well formed XHTML.
- Document parsing and classification to generate the ontology of the page: The ontology is a tree that represents the different sections, elements or nodes of the document, and the relations between them. In a typical web page different elements can be found: e.g., main content, navigation elements, images. To segment the document the system parses the page and identifies and classifies the different elements in the web page, and the relations between them. The ontology represents an *abstract model* of the web page.

5.2 Adapted Web Page Generation

This component of the system generates web pages adapted to the device and user preferences from the classification, the device and user description. To generate the adapted web pages the system does several transformations or adaptations, according to the device capabilities and user description:
- Navigation Strategy Generation: Depending on the ontology of the web page, the device features, and the user characteristics, the system constructs a navigation strategy. Each device is characterized by a description of their capabilities: screen resolution, number of color or gray-levels, bandwidth, processing power, standards supported (e.g., HTML, WML, XML, CGI support), multimedia capabilities, input/output, etc. Likewise the user characteristics can be described: expertise level, skills, preferences, etc.
- The navigation strategy represents how the information is presented to the user, which information can be removed, and how the user can navigate between the different elements in the page. The navigation strategy optimizes how the information is transmitted and displayed to the device-user, considering characteristics of the document, devices features and limitations, and user knowledge and preferences.
- Adapted Web Pages Generation: Using the navigation strategy the system generates a new page or set of pages with the necessary navigation elements. The new set of pages is a *specific view* of the ontology according to the navigation strategy. One or several sections in the ontology are transformed into one adapted page. The title of each section is used as an anchor for the navigation between different sections/pages

To create the new pages the system generates a set of XSLT transformations that are applied to the original document. These transformations include:

- HTML Code Adaptation: Each one of the generated adapted pages contains parts of the original HTML code of the page, but adapted to the capabilities of the device. Some of the tags or attributes are added, removed or replaced to accommodate the device limitations and constraints.
- Layout Adaptation: The layout and distribution of the different elements in the page is modified to fit the device screen. Each section can be represented with the full content or just the title of that section as an anchor link to that section. Each section can be displayed with different levels of detail.
- Images and Multimedia Adaptation: Some graphics are scaled, or the number of colors is reduced according to the device capabilities and user preferences. Other graphics are replaced by their Alt attribute, a link, a simplified version, or just removed (e.g., decoration graphics).
- Navigation mechanisms: Code for navigation is added to each adapted page to enable the user to navigate through the different pages. Also the system may generate index pages with links to the different generated pages.

6. Conclusion

In summary, the system classifies the web page according to the ontology of the document, and together with the device and user description it decides a navigation strategy. Using the selected navigation strategy the system generates and applies a series of transformations that converts the original document in a new set of web pages adapted to the device and the user.

The main features and benefits of this system are:

- Adaptive and universal access to the information on the web (browser, device and user independence). Access to the information anywhere, anytime, on any device.
- Acceptance of XML/XSL, XHTML or HTML as input (to have a transition path for existing web content).
- Transformation based: Transformations can be applied sequentially to the original page, and can be reused for different devices. New transformations can be easily added to the system.
- Ontology based: Classification and Generation are independent processes, and cleanly separated by the ontology. The system can use different classifiers or generators (automatic or manual) depending on the needs. Publishers can have control on the adaptation by including the ontology into the web pages.

References

[1]   Palm, Inc. Web Clipping Developer's Guide. 3 November 1999. [On-line]. Available:
        http://www.palmos.com/dev/tech/webclipping/
[2]   World Wide Web Consortium. HTML 4.01 Specification. [On-line]. Available:
        W3C Recommendation 24 December 1999. http://www.w3.org/TR/html4/
[3]   World Wide Web Consortium. XHTML™ 1.0: The Extensible HyperText Markup Language. A Reformulation of HTML 4 in XML 1.0. W3C Recommendation 26 January 2000. [On-line]. Available:http://www.w3.org/TR/xhtml1/
[4]   World Wide Web Consortium. Extensible Markup Language (XML) 1.0 (Second Edition). W3C Recommendation 6 October 2000. [On-line]. Available: http://www.w3.org/TR/REC-xml
[5]   World Wide Web Consortium. XSL Transformations (XSLT) Markup Language (XML) 1.0. W3C Recommendation 16 November 1999. [On-line]. Available: http://www.w3.org/TR/REC-xml

Towards Generating Textual Summaries of Graphs

Kathleen F. McCoy, M. Sandra Carberry, Tom Roper   *Nancy Green*

Dept. of Computer and Information Sciences
University of Delaware
Newark, DE 19716, USA
mccoy@cis.udel.edu, carberry@cis.udel.edu,
roper@cis.udel.edu

Dept. of Mathematical Sciences
University of North Carolina at
Greensboro
Greensboro, NC 27402, USA
green@uncg.edu

Abstract

For people who use text-based web browsers, graphs, diagrams, and pictures are inaccessible. Yet, such diagrams are quite prominent in documents commonly found on the web. In this work we describe a method for rendering important aspects of a line graph as plain text. Our method relies heavily on research in natural language generation and is motivated by an analysis of human-written descriptions of line graphs. In this paper we concentrate on determining which aspects of the line graph to include in a summary.

1. Introduction

The amount of information available electronically has increased dramatically over the past decade. The challenge is to develop techniques for providing effective access and utilization of such information so that all individuals can benefit from these resources and so that the information is readily available when needed. Unfortunately, many knowledge sources are provided in a single format; thus they are not accessible to individuals who cannot assimilate information in that format. For example, individuals with visual impairments have limited access to data stored as graphs; this prevents them from fully utilizing available information resources. In addition, even for individuals without disabilities, such graphical information is readily accessible only in environments with high-bandwidth transmission and viewing facilities. For example, recent developments have led to cellular telephones for accessing the Web. The use of such devices highlights the need for tools that convey information in other modalities.

In this work we attempt to make the major information content of some kinds of graphs available in textual form. We are not interested in creating a description of the graph itself, rather we wish to produce a summary of the major points the graphic is intended to convey in an understandable and well-structured text. To illustrate our approach, consider the case of a line graph. In choosing what is important to include in a summary, we attempt to take into account the presentational goals that led the author to produce a line graph (as opposed to another form of representation) in the first place. In addition, we consider what aspects of line graphs are "typically" included in descriptions of such graphs, and look at organizational methods found in naturally occurring line-graph summaries.

In this paper we first outline some related research. This includes work on both providing alternative modes of delivering graphical information, and on natural language generation research that is relevant to this problem. Next we lay out our proposed architecture for summarizing a graph. We give some detail of two components of this architecture: the intention recognition component and the component responsible for choosing aspects of the graph to include in the summary.

2. Related research

There has been some work on providing alternative modes of presentation of graphical/diagrammatic information for people who have visual impairments. For example, [Zhang & Krishnamoorthy, 1994] describe a package designed for people with visual impairments to study graph theory. Input to their system is an adjacency matrix for a graph. The system generates a description of the graph tailored to the particular purpose of graph theory applications. Because of this, items of importance can be predefined. In our work we wish to take in graphs that are part of larger texts where the graph has been included to make a particular point or points. Thus, one of the major research efforts involves determining the appropriate description given a graph. We conjecture that the important things to say will be dependent on the intentions that led to the selection of that particular graphical device, as well as (visual) features of the data displayed in the graph itself.

Several systems exist that attempt to allow a user to "visualize" a graphic. For instance, [Meijer, 1992] describes a system that converts a video image into a "soundscape". Others describe a transformation of information possible by taking advantage of the underlying structure of the information [Barry et al., 1994]. Notable projects include ASTER [Raman, 1997] and several projects from the Science Access Project at Oregon State [Bargi-Rangin et al., 1996], [Gardner, 1998], [Walsh & Gardner, 2001], [Walsh et al., 2001]. These projects rely on meta-document information to help interpret the graphical components in the intended way. A great deal of work also comes from the Multimodal Interaction Group (part of the Glasgow Interactive Systems Group [Brewster, 1994]). Much of this work involves the application of earcons [Blattner et al., 1989] and sonically-enhanced widgets. This work and others (e.g., [Kurze et al., 1995], [Blenkhorn and Evans, 1995] [Kennel, 1996]) relies on a human to "translate'" the graph into a form where the earcons and widgets can be used.

While these systems are clearly extremely valuable, their basic focus is on graphical elements and their rendering in an alternative medium. In our work, rather than enabling the user to reason on a "reproduced" graphical image, we concentrate on describing the important content the graph is intended to convey. We take the view that the graph was produced in order to get across particular information. Thus one can view the summary we wish to generate as an alternative mode of getting across the important informational content of the graph.

Most previous work in the area of Natural Language Generation surrounding graphical presentation has centered on generating multi-media explanation (e.g., [McKeown et al., 1992], [Wahlster et al., 1993], [Arens & Hovy, 1990]). In particular, that work attempts to generate a presentation that contains both text and graphics. In contrast, our work is aimed at providing a textual explanation of a human-generated graphic.

Two projects that do deal with the kind of data of interest here are (1) the caption generation system associated with the SAGE project [Mittal et al., 1998] and (2) the PostGraphe system [Fasciano & Lapalme, 2000], [Fasciano & Lapalme, 1996], [Lapalme, 1999]. The caption generation system is concerned with generating a caption that enables the user to understand how to read the data in the graphic (i.e., how the graphic expresses the data). The system is thus driven by an identified set of "graph complexities" that the caption explains to the user. Rather than graph complexities, our proposed work must identify and convey aspects of the data depicted in the graph that are important to the user. Thus, a large portion of our work involves identifying features that convey importance.

The PostGraphe system [Fasciano & Lapalme, 2000], [Fasciano & Lapalme, 1996], [Lapalme, 1999] generates statistical reports that integrate both graphics and text (accompanying caption). The primary input is data in a tabular form, an indication of the types of values in the columns of the table, and the intentions that the user wishes to convey. Examples of intentions handled by PostGraphe include: present a variable, compare variables or sets of variables, and show the evolution of a variable with respect to another variable. A user may have multiple intentions to be achieved in a single report. PostGraphe uses a schema-based planning mechanism [McKeown, 1985] to plan both the graphics and the text. The graphical schema is chosen primarily on the basis of the input intentions using information about the kinds of graphics that are most effective in achieving those intentions. While the kind of information included in the PostGraphe captions are of the type we would hope to generate, there are significant differences between the input and goals of the two projects. A significant portion of our work concerns determining what these goals should be given a previously generated graphic. For example, since we don't have the input intentions, we must deduce these given the graph itself.

Note that the caption associated with the line graph we are attempting to summarize is also potentially available to our system. However, we have not yet considered how the caption may be useful. In this paper we concentrate on deducing the goals of the graphic through the properties of the graphic itself. Future work will consider how the caption might affect what is included in the explanation.

3. System components

A well-known adage is that a picture is worth a thousand words. This is because a picture captures a great deal of information, some of it only tangential to the main purpose of the picture. Our goal is to communicate only the most significant and relevant information, while retaining the ability to convey additional detail if requested by the user. In addition, the designer of a visual display has some underlying purpose, which can affect how the display is created. For example, a pie chart emphasizes comparisons while a line graph brings out trends.

Our overall methodology for summarizing a graph consists of the following components:

1. Extract the basic components of the graph, identify their relationship to one another, and classify the screen image into one of the classes of graphs being handled using techniques as described in [St. Amant & Riedl, 2000], [Futrelle et al., 1995], and [Futrelle, 1999].

2. Hypothesize possible intentions of the graph designer, based on characteristics of the type of graphical display selected. For example, both a scatter plot and a pie chart can be used to graph how an entity (such as government income) is divided up among several categories (such as social welfare, military spending, etc.); however, a

graphical designer will choose a pie chart if the intent is to convey the relative distributions as opposed to their absolute amounts.

3. Once we have determined the goals of the graphic designer, we still must decide what propositions should be expressed in the summary. There are likely to be many interesting things within a graphic, but we wish to select the few most emphasized ones for summarization. These are identified in our architecture through the use of a set of features where the features that we look for are dependent on the overall characteristics of the graphic (e.g., the features looked for in a line graph will be quite different from those in a pie chart) and the intentions determined in the previous phase. The feature use and importance values for features was motivated by a study in which human subjects were asked to describe line graphs. The use of these features will produce a set of propositions along with a rating of each indicating how important they are.

4. Construct a coherent message that conveys the most significant propositions conveyed by the graphical display and translate the message into English sentences. For generating the actual sentences we turn to FUF/SURGE [Elhadad & Robin, 1996]. This is a unification-based natural language sentence generation system that is able to translate between formal specifications of English sentences into English sentences.

The following sections concentrate on the subproblems of identifying the intentions of the graph and selecting relevant features to include in the summary.

4. Recognizing intention

A great deal of research is being conducted focusing on the automatic generation of multimedia presentations, such as that of the SAGE visualization group (e.g., [Kerpedjiev & Roth, 2000], [Kerpedjiev et al., 1998], [Green et al., 1998]). They propose that speech act theory can be extended to cover graphical presentations. A graphic can bring about an intended change in the reader's cognitive state in the same way that text can.

If we accept this claim, then we have a basis for trying to summarize a given graphic. There are well-researched and well-documented results concerning the construction of effective graphical presentations, and the means by which humans decode them. If we assume:

1. the author is being cooperative and efficient: the author is including information pertinent to the presentation, and avoiding extraneous information that might confuse or hide the intended information

2. the author is competent: the author is aware of the conventions for effective graphic design

then we are able to extract from the graphic a set of intentions the graphic is likely intended to convey.

The goal of our intention recognizer is the inverse of the graph design process: namely, we wish to use the graphic display as evidence to hypothesize the communicative intentions of the graph's author. This is a plan recognition problem [Carberry, 1990]. Typically, both planning and plan recognition systems are provided with operators that describe how an agent could accomplish different goals; these operators can include constraints, preconditions, subgoals and subactions, and effects of actions. Planners start with a high-level goal, select an operator that achieves that goal, and then use other operators to flesh out the details that describe how to satisfy preconditions and subgoals. On the other hand, a plan recognition system is provided with evidence (here an action performed by an agent such as the use of a particular graphical device), and chains backwards on the operators to deduce one or more high-level goals that might have led the agent to perform the observed action as part of an overall plan.

To cast intention recognition from graphic displays as a plan recognition problem, we must represent the knowledge that might have led the graphic designer to construct the observed graphic. We will construct two types of operators, goal operators and task operators. Goal operators will specify the tasks that must be supported in order to achieve various high-level intentions. Task operators will be of two types: cognitive and perceptual. Cognitive operators will specify the tasks that must be performed in order to accomplish cognitive tasks such as performing comparisons or summary computations. Perceptual operators will specify the tasks that must be performed in order to accomplish perceptual tasks such as looking up the value of a particular attribute.

The task of this module is one of plan recognition. This module has as input the graphical display and any accompanying discourse context, and must reason backwards to identify the author's intentions. In our architecture, input to the plan recognizer will be the graph type (pie chart, line graph, etc.) identified by the visual extraction module. The plan recognizer will also have access to an examination module that can answer questions about the graph and aid the plan recognizer in determining whether operator constraints are satisfied. For example, the plan recognizer might need to know whether an attribute could be a primary key. From this evidence, the system will chain backwards to identify perceptual and cognitive tasks that are supported by the graph and from these hypothesize the plausible intentions of the graph designer. In most cases, multiple intentions will be hypothesized. These will be ordered according to how well the requisite cognitive and perceptual tasks are enabled.

5. Identifying important features

Once the overall intention is recognized (e.g., get the reader to recognize a trend in the data), the system must determine what should be said in order to convey the trend to the user. To determine how this might be done, we conducted an experiment where we asked human subjects to produce summaries of some graphs that we provided. The graphs in the experiment were line graphs whose intention was arguably to get the reader to recognize a trend. An analysis of the resulting summaries led us to conjecture the use of features of graphical objects to help select important propositions to include in the summary. We define a graph feature as an attribute of some subset of the components of the graph. Such features can vary from local features that apply to a single point in the graph to more global features that apply to a segment of the graph or even the entire graph. In the case of line graphs, a local feature might be a maximum value at a point and a global feature might be resemblance of a portion of a graph to a parabola. The relevant features will be different for different kinds of graphs (e.g., line graph versus pie chart). In studying our generated summaries and a number of collected line graphs from popular media such as USA Today and Newsweek, our analysis suggests several relevant features that we hypothesize will be useful in generating summaries:

- Extrema: the extreme values of a line graph are often significant. These are point values, and include the initial value, global minimum, global maximum, and final value
- Slope: slope refers to the rate of change of a line segment (or sequence of line segments viewed as a single straight line from the starting point of the initial segment to the terminal point of the last segment). Slope is relevant because it indicates how quickly the data is changing.
- Variance: variance refers to the raggedness, or jumpiness, of a sequence of line segments. A sequence of line segments that form a straight line would have zero variance. Variance may be significant since it indicates the stability or consistency of changes in the data.
- Analogy: analogy refers to a concept that a portion of the data resembles. For example, a sequence of line segments may overall resemble a parabola or sine wave. If the user is familiar with the selected concept, comparison to it provides a compact method of enabling the user to easily visualize the data in his mind.
- Cycles: a cycle is a repeated pattern in the line graph. As with analogies, cycles provide compact methods of characterizing graphical information.

Further experimentation needs to be done to identify a complete set of features. However, this set is a starting point that can be added to as further analysis is done. We define significance of a feature to be the strength of the feature in the graph. For example, the more ragged a sequence of line segments, the more significance will be assigned to the variance feature. We hypothesize that significance of a feature cannot be determined in isolation. For example, the significance of a feature will be affected by the domain of the graph, with some features of greater importance in some domains than in others. In addition, significance will also be affected by the portion of the graph that the feature spans. For example, large variance over a small segment of a graph will have less significance than large variance over a longer segment of the graph. We propose to evaluate our corpora of ranked features and explanations produced by human subjects to identify what aspects of the graph and its domain should be taken into account in estimating significance. The result of reasoning on these features will be a set of propositions -- one for each feature whose significance is great enough to include in the summary.These propositions will be organized and eventually rendered as text by the remaining system components.

6. Conclusion

We have set forth a strategy for creating a summary for a graph. What we have presented is a method that has the potential for paralleling human generated summaries. Future work might consider how to combine this basic methodology with world knowledge, user preferences, and consideration of the context of the presentation. Such knowledge will ultimately be necessary for completely natural sounding, natural language summarization of any given graphic.

References

[Arens & Hovy, 1990] Arens, Y., Hovy, E. H. (1990). How to describe what? Towards a theory of modality utilization. In *Proc. of the 12 Conference of the Cognitive Science Society*. Cognitive Science Society.

[Bargi-Rangin et al., 1996] Bargi-Rangin, H., Barry, W. A., Gardner, J. A., Lundquist, R., Preddy, M., Salinas, N. (1996). Scientific reading and writing by blind people - technologies of the future. In *Proceedings of the 1996 CSUN Conference on Technology and Persons with Disabilities*, Los Angeles, CA.

[Barry et al., 1994] Barry, W. A., Gardner, J. A., Raman, T. V. (1994). Accessibility to scientific information by the blind: Dotsplus and ASTER could make it easy. In *Proceedings of the 1994 CSUN Conference on Technology and Persons with Disabilities*, Los Angeles, CA.

[Blattner et al., 1989] Blattner, M., Sumikawa, D., Greenberg, R. M. (1989). Earcons and icons: Their structure and common design principles. *Human Computer Interaction*, 4(1): 11--44.

[Blenkhorn & Evans, 1995] Blenkhorn, P. Evans, D. G. (1995). Access to graphical information for blind people through speech and touch. In *Proc. of the Second TIDE Congress*, pages 298--301, Paris.

[Brewster, 1994] Brewster, S. A. (1994). Providing a structured method for integrating non-speech audio into human-computer interfaces. PhD thesis, University of York, UK.

[Carberry, 1990 Carberry, S. (1990). *Plan Recognition in Natural Language Dialogue*. ACL-MIT Press Series on Natural Language Processing. MIT Press, Cambridge, Massachusetts.

[Elhadad & Robin, 1996] Elhadad, M., Robin, J. (1996). An overview of surge: a re-usable comprehensive syntactic realization component. In *Proceedings of the Eighth International Generation Workshop (INLG96) (demonstration session)*, Herstmonceux Castle, Sussex, UK.

[Fasciano & Lapalme, 1996] Fasciano, M., Lapalme, G. (1996). PostGraphe: A system for the generation of statistical graphics and text. In *Proceedings of the Eighth International Generation Workshop (INLG96)*, pages 51-60, Herstmonceux Castle, Sussex, UK.

[Fasciano & Lapalme, 2000] Fasciano, M., Lapalme, G. (2000). Intentions in the coordinated generation of graphics and text from tabular data. *Knowledge and Information Systems*.

[Futrelle, 1999] Futrelle, R. (1999). Ambiguity in visual language theory and its role in diagram parsing. In *Proceedings of the IEEE Symposium on Visual Languages*, pages 172-175.

[Futrelle et al., 1995] Futrelle, R. P., Zhang, X., Sekiya, Y. (1995). Corpus linguistics for establishing the natural language content of digital library documents. In Adam, N., Bhargava, B., & Yesha, Y., editors, *Digital Libraries. Current Issues*, pages 165-180. Springer-Verlag.

[Gardner, 1998] Gardner, J. A. (1998). The quest for access to science by people with print impairments. *Computer Mediated Communication*, 5(1).

[Green et al., 1998] Green, N., Carenini, G., Kerpedjiev, S., Roth, S., Moore, J. (1998). A media-independent content language for integrated text and graphics generation. In *Proc. of the Workshop on Content Visualization and Intermedia Representations (CVIR'98)* of COLING '98 and ACL'98, Montreal, Canada.

[Kennel, 1996] Kennel, A. R. (1996). Audiograf: A diagram-reader for the blind. In *Proc. of the Second Annual ACM Conference on Assistive Technologies*, pages 51-56, Vancouver, British Columbia, Canada. ASSETS '96.

[Kerpedjiev et al., 1998] Kerpedjiev, S., Carenini, G., Green, N., Moore, J., Roth, S. (1998). Saying it in graphics: from intentions to visualizations. In *Proc. of Information Visualization*, Research Triangle Park, NC. IEEE.

[Kerpedjiev & Roth, 2000] Kerpedjiev, S., Roth, S. (2000). Mapping communicative goals into conceptual tasks to generate graphics in discourse. In *Proc. of Intelligent User Interfaces 2000*, New Orleans, Louisiana.

[Kurze et al., 1995] Kurze, M., Perie, H., Morley, S., Deconinck, F., Strothotte, T. (1995). New approaches for accessing different classes of graphics by blind people. In *Proc. of the Second TIDE Congress*, pp. 268-271, Paris.

[Lapalme, 1999] Lapalme, G. (1999). Comparing PostGraphe with the rags architecture. In *Proc. RAGS Workshop -- Reference Architectures for Generation Systems*, Edinburgh, Scotland, UK.

[McKeown, 1985 McKeown, K. (1985). *Text generation: Using discourse strategies and focus constraints to generate natural language text*. Cambridge University Press, Cambridge UK.

[McKeown et al., 1992] McKeown, K., Feiner, S., Robin, J., Seligmann, D., Tanenblatt, M. (1992). Generating cross-references for multimedia explanation. In *Proc. of the 1992 National Conference on Artificial Intelligence*, pp. 9-16, San Jose, CA. AAAI.

[Meijer, 1992] Meijer, P. B. L. (1992). An experimental system for auditory image representations. *IEEE Transactions on Biomedical Engineering*, 39(2): 291-300.

[Mittal et al., 1998] Mittal, V. O., Moore, J. D., Carenini, G., Roth, S. (1998). Describing complex charts in natural language; a caption generation system. *Computational Linguistics*, 34(3): 431--468.

[Raman, 1997] Raman, T. (1997). *Auditory User Interfaces*. Kluwer Academic Publishers, Boston.

[St. Amant & Riedl, 2000] St. Amant, R., Riedl, M. (2000). A perception/action substrate for cognitive modeling. In *HCI. International Journal of Human-Computer Studies*. To appear.

[Wahlster et al., 1993] Wahlster, W., Andre, E., Finkler, W., Profitlich, H. J., Rist, T. (1993). Plan-based integration of natural language and graphics generation. *Artificial Intelligence*, 63:387-427.

[Walsh & Gardner, 2001] Walsh, P., Gardner, J. A. (2001). Tiger, a new age of tactile text and graphics. In *Proceedings of the 2001 CSUN Conference on Technology and Persons with Disabilities*, Los Angeles, CA.

[Walsh et al., 2001] Walsh, P., Lundquist, R., Gardner, J. A., Sahyun, S. (2001). The audio-accessible graphing calculator. In *Proc. of the 2001 CSUN Conference on Technology and Persons with Disabilities*, Los Angeles.

[Zhang & Krishnamoorthy, 1994] Zhang, S. H., Krishnamoorthy, M. (1994). Audio formatting of a graph. In *Proc. of the 1st Annual ACM Conference on Assistive Technologies*, pages 148-156, Los Angeles, CA. ASSETS.

Remembering how to use the Internet -
An investigation into the effectiveness of VoiceHelp for older adults

Wesley Morrissey, Mary Zajicek

Speech Project
School of Computing & Mathematical Sciences
Oxford Brookes University, Oxford OX3 0BP, U.K.

Abstract

This paper describes an experiment to assess the effectiveness of voice output (messages) to communicate help to older adults whilst using BrookesTalk [1], a speaking web browser originally designed for blind and visually impaired people. The messages used to provide help to the user are referred to as 'VoiceHelp'. The experiment described is part of a programme of research that was initiated by findings connected with developing BrookesTalk. It was found that it had a very poor uptake amongst older adult visually impaired users, with 82% unable to master the technology (Zajicek & Hall, 2000). This work explores possible reasons for this finding and seeks to determine potential solutions.

1. Introduction

A pilot study showed that older adult users encountered a number of problems whilst using the browser either with or without VoiceHelp messages. In particular, it highlighted an inability to remember the choices that had been output in the form of messages. Although clear deficits in memory efficiency and retrieval success are present in the older population, studies have shown that older adults are still capable of learning (West, 1995).

The objective of this experiment was to establish clear and memorable instructions, in the form of VoiceHelp, to aid older adults in using the browser.

2. Motivation for investigating VoiceHelp

With normal ageing sensory input channels (sight, hearing, touch) show a gradual decline making the use of a standard graphical user interface more difficult. The use of a solely graphical user interface can put quite a strain on the visual channel, so an auditory interface can reduce the load on the users visual system (Mountford & Gaver, 1990). VoiceHelp appeared to be a useful way to present structured information to older adult users, particularly in cases where visual acuity had shown a decline.

In the pilot study, synthetic speech output was employed for presenting the help information to the users. Although no formal analysis had been undertaken, observational analysis clearly showed that the messages were too long, providing too much functionality and thus they did not improve the user's performance. If messages are created without care then they may be ineffective or even hinder performance. Many studies have been conducted to investigate the changes in short-term memory in older adults, notably age related decrements may be present in working memory functioning with higher-level cognitive functions such as comprehension of difficult text or messages, and the ability to draw inferences (Cohen, 1988; Zack & Hasher, 1988). However, for the purpose of this experiment it was important to use messages contextually relevant to the application.

This experiment tried to answer three main questions. The first was: Are VoiceHelp messages a good method of communicating information to older adult users? The second was: What is the optimum length (in terms of chunks of information) of VoiceHelp messages to be effective? This was investigated by looking at the recall rates of the messages. The third question was: Are function keys or lettered keys more memorable?

3. Experiment

An experiment was devised to test whether older adult subjects could recall the information presented in a way that was pertinent to using the browser.

3.1 Subjects

Experimental subjects: Thirty subjects, two groups of 15, were used from Age Concern Day Centres in Oxfordshire. They were randomly allocated to one of two groups. None of them had used the Internet or indeed a computer. The average age was 81.74 for the group assigned to the shorter messages and 80.74 for the group assigned to the longer messages.

Control subjects: Thirty subjects, two groups of 15, were used from various sources (students and non-students). They were randomly allocated to one of two groups. All had some experience of using the Internet or a computer. The average age was 27.6 for the group assigned to the shorter messages and 32.5 for the group assigned to the longer messages.

3.2 Messages used

The VoiceHelp messages consisted of pre-recorded natural speech messages, as work carried out by Luce and Pissoni (Luce & Pissoni, 1983; Luce et al, 1983) had shown that synthetic speech puts extra load on working memory, particularly in the case of older adults. The subjects undertook a pre-experimental test first. This test acted as a control between the message groups and accustomed the subjects to what would be expected of them in the experimental tests.

3.3 Pre-experimental test message

The pre-experimental test message was intended as the first message users would actually hear if they were using BrookesTalk with VoiceHelp. It was aimed to put users at ease and make them feel in control as they are given commands to stop and start the VoiceHelp, as well as the facility to repeat the message if they wished to.

3.4 Experimental test messages

The experimental test messages are shown below. Both the short and the long versions of message 1 convey information about the status of the application, which in this case indicates there is no page loaded. Given this, the user is then provided with the functions available for the status of the application. The shorter message gives a choice between entering an Internet address and carrying out a search, using two function keys. The longer message provides two additional function keys to change the browser settings and to hear more details. Each message ends with the instruction on how to repeat the message.

Short experimental test message 1
This is the main menu
There is currently no page loaded:
To enter an internet address, press F1
To perform an internet search, Press F2
To repeat these instructions press R.

Long experimental test message 1
There is currently no page loaded:
To enter the URL of a page, press F1
To perform an internet search, Press F2
To change the settings of the browser, press F7
To hear more details about options available to you, Press F8
To repeat these instructions press R.

Message 2 conveys the status of the application (page loaded). Once a page is loaded there is a great deal more that the user could potentially do. However, the functions that were thought to be most useful were chosen to prevent the messages from becoming too lengthy. Both messages then gave the instruction to be carried out if the user wished to read the page. For the longer version this involved two keystrokes (as would have been employed by the original version of BrookesTalk) 'Press return and then press the right arrow key'. The shorter version simply required the user to 'Press the right arrow key'. The shorter message related information about just one Function key to follow the links. The longer version involved the use of four Function keys to: follow the links; hear the headings; hear a summary; or hear an abstract. The shorter message also gave an option to return to the main menu, using the letter key 'M'. Both messages ended with the option to repeat the message if required.

Short experimental test message 2
You have a page loaded:
To read the current page, press the right arrow key.
To follow the links on this page, press F4
Go back to the main menu, press M.
To repeat these instructions, press R.

Long experimental test message 2
You have a page loaded:
To read the current page, press return and right arrow key.
To hear the headings on this page, press F3
To follow the links on this page, press F4
To hear the page summary, press F11
To hear the page abstract, press F12
To repeat these instructions, press R.

702

4. Experimental design and procedure

An independent-subjects design was used, involving two groups of 15. One group heard the shorter messages, whilst the other group heard the longer messages (the pre-experimental message was the same for both groups). The experiment was run in exactly the same way for the control group of younger adults. Immediately after hearing a message, each subject was presented with an orally administered questionnaire, with multiple-choice answers.

Training: The subjects were not given any training prior to the experiment as to what operations the various lettered and Function keys facilitated, as the experiment was designed to be a 'worst case' evaluation of the messages output. If the messages resulted in effective recall under these conditions, then in a situation where incremental structured training was given the resulting performance should improve.

5. Experimental hypotheses

The main aim of this experiment was to discover if VoiceHelp messages could convey memorable information in the form of natural speech output to support older adults to use BrookesTalk. If they could not then there would be little point in using them and alternative methods would have to be devised. The main hypothesis was that VoiceHelp would be effective at communicating help with the recall rates of shorter messages being higher than that for longer messages. The length of message for the purpose of this experiment was measured in terms of chunks of information, represented by the number of options available to the user.

6. Results and discussion

6.1 Results of pre-experimental test for older adults

The subjects in the group listening to the shorter and longer messages recorded mean correct scores of 4.14 and 3.94 respectively (out of a maximum of 5). A t-test showed that there was no significant difference in the means of the two groups in the pre-experimental test, $p = 0.34$ at 5% level of significance.

6.2 Results of pre-experimental test for younger adults (Control Group)

The younger adult groups also showed that there was no significant difference ($p=0.28$) between the two message groups. However, the younger adult groups scored higher than the older adult groups with a mean of 4.94 for the short messages and 4.87 for the longer messages. This result gave a statistically significant difference in means between the younger and older groups for the longer messages ($p=0.02$) and the shorter messages ($p=0.001$).

6.3 Results of experimental test for older adults

The next stage of the analyses involved taking the total scores for each subject for the experimental test. The scores were converted into a percentage as the subjects in the longer message group had more available options than the subjects in the shorter message group.

The group listening to the shorter messages scored 54.8% on average, whilst the group listening to the longer messages scored 26.1% on average, showing a statistically significant ($p=0.001$) increase in performance.

6.4 Results of experimental test for younger adults (Control Group)

Again, the younger control group showed no significant difference ($p=0.31$) in performance between message groups. The group listening to the shorter messages scored 91.93% and the group listening to the longer messages scored 90.13%.

6.5 Further analysis of Experimental Test results for older adults

Further analysis was conducted on the tests taking results from comparable questions.

This analysis showed that older adults undertaking the task with shorter messages scored higher. A t-test was performed on these percentage scores ($p=0.003$). The group listening to the shorter messages showed a statistically significant increase in performance, showing a 23.7% better performance.

The two groups listened to experimental message 1. The short message consisted of four chunks of information; The longer message consisted of five chunks of information, as for the shorter message, but with an extra function key option (to hear more options).

Is there a page loaded? The graph shows that neither group recalled whether a page was loaded or not, however, the graph showing the responses after message 2 show that nearly half the subjects in the short message group picked up the page status as did just over a quarter of the long message group, having this information brought to attention

The graph below shows the questions elicited from Message 1 for comparison.

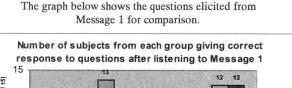
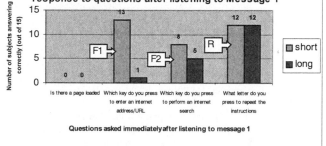

through the questionnaire after message 1 appeared to make it more salient.

Which key do you press to enter an internet address? Which key do you press to enter the URL of a page? Thirteen out of fifteen of the short message group correctly recalled the Function Key F1. However the long message group were presented with the information using the label 'URL' rather than 'internet address' and all but one failed to recall this. This low score supports observations from the pilot, where technical jargon leads to confusion.

Which key do you press to perform an internet search? Just over half the short message groups correctly recalled the Function Key F2, as did one-third of the long message group.

What letter do you press to repeat the instructions? Twelve out of fifteen in both groups recalled the correct answer. This result may be affected by the 'recency effect' since it was the last item heard, however, 'recency' is generally negated by distracting information and the preceding questions would have acted as such. This result may indicate that using lettered keys may convey more meaningful information than Function Keys. This trend is also demonstrated in the graph below, which shows a breakdown of the number of subjects who answered correctly after listening to Message 2.

The two groups listened to experimental message 2. The short message consisted of five chunks of information; the longer message consisted of eight chunks of information.

Is there a page loaded? As mentioned above, more of the subjects picked up this chunk of information and recalled it correctly.

How do you read the current page? Nearly half of the short message group recalled their answer correctly 'right arrow key', but none of the long message

The graph below shows the questions elicited from Message 2 for comparison.

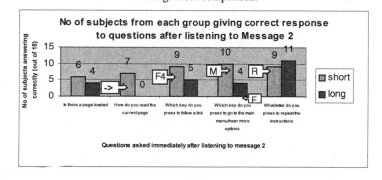

group gave the required the response, 'return and right arrow key'. This result supports observations in the pilot, which suggested that the use of multiple keys was confusing.

Which key do you press to follow a link? Nearly two-thirds of the short message group recalled the Function Key F4 correctly, whilst one-third of the long message group recalled F4 correctly.

Which key do you press to go to the main menu? Which key do you press to hear more option? For the short message group the required response was the letter 'M', and two thirds of the group answered this correctly. The long message group required the response F8 to hear more options. This correct response rate for this question was less than one-third, indicating that output of too many chunks of information is detrimental to recall, particularly non-meaningful items, such as F8, F11 and so on.

704

What letter do you press to repeat the instructions? The scoring for both groups was again high for this question with the eleven of the long message group and nine of the short message group scoring correctly.

7. Conclusions

Some general conclusions can be drawn from this experiment. The overall recall of shorter messages (65%) was better than that for longer messages (40%) for older adults. The overall recall for shorter messages for younger adults (94%) was only slightly better than that for the younger adults recalling longer messages (92%).

The condition for the experiment was the 'worst case' scenario as no training was given to the older adult subjects for them to become familiar with the functions available. Given these conditions they would be expected to perform better in terms of recall with more practice and increased familiarity of the terms used. These results suggest that VoiceHelp could be successfully employed to aid older adults in their use of the browser provided shorter messages were used.

From the graphical evidence it is apparent that the use of a higher number of function keys leads to a decrement in recall of this information, however, recall of the lettered keys remained similar for both groups regardless of the number of items in the messages or the number of function keys mentioned. The messages demonstrated that the information chunks were largely retained when one or two Function keys per message were output, however, the use of more than two Function keys in one message leads to a decline in recall and confusion.

8. Guidelines

From the results of this experiment some guidelines have been drawn up for use when creating VoiceHelp messages for the browser for older adult users.

The length of the message should be kept as short as possible, without losing structure and meaning.

In order to be in line with other applications, F1 should be used as a help key rather than as a function of the application.

The use of lettered keys appears to aid recall, since they have a pneumonic association with the operation to be carried out.

The use of more than one key to carry out an operation should be avoided.

The number of function keys used in the messages should be kept to a minimum, a maximum of two per message. The number of function keys reflects the extent of the functionality of the browser. Since a small number of function keys reflects a low functionality browser, then it would be advised that the functionality of the browser was kept to a minimum.

As the user gains confidence and familiarity with the browser, additional functionality could be implemented. This is supported by a study conducted by Czaja et al (1990) that concluded that older people are willing and able to use computer communication systems if the system is simple to use, features are added incrementally and they are provided with a supportive environment.

References
1   www.brookes.ac.uk/speech

Cohen, G. (1988) Age differences in memory for texts: production deficiency or processing limitations? In L.L. Light & D.M. Burke (eds) *Language, Memory and Aging.* Cambridge University Press, pp 171-190.

Czaja, S., Clark, C., Weber, R., Nachbar, D., (1990) Computer communication among older adults, Proc. Of the Human Factors and Ergonomics Society 33rd Annual Meeting, pp 146 – 148

Luce, P.A., Pisoni, D.B. (1983). Capacity-demanding encoding of synthetic speech in serial-ordered recall. Research on Speech Perception, No.9, Speech Research Laboratory, Indiana University, Bloomington, IN.

Luce, P.A., Feustel, T.C., Pisoni, D.B. (1983). Capacity demands in short-term memory for synthetic and natural word lists. Human Factors, 25, 17-32.

Mountford, S.J., Gaver, W. (1990). Talking and listening to computers. In B. Laurel (Ed.), *The art of human-computer interface design,* pp. 319-334. Reading, Massachusetts: Addison-Wesley.

West, R.L., (1995) Compensatory strategies for age-associated memory impairment

Zacks, R.T., Hasher, L. (1988) Capacity theory and the processing of inferences. In L.L. Light & D.M. Burke (eds) *Language, Memory and Aging.* Cambridge University Press, pp 154-170.

Zajicek, M., Hall, S., (2000) Solutions for elderly visually impaired people using the Internet, In proc *A New Age for Old Age*, BSG 2000, Oxford.

Usability of Transport web sites by elderly travellers

Annie Pauzié

INRETS/LESCOT
National Institute of Research on Transport and Safety
Laboratory Ergonomics and Cognitive Sciences in Transport
25 avenue François Mitterrand, 69675, Bron, France.
pauzie@inrets.fr

Abstract

According to available statistical data, the cybersenior population is growing. To ensure that this population is not excluded from the information society, it is important to take into consideration their characteristics, needs and requirements when developing web sites. In this framework, and because it has been shown that mobility promotes mental and physical health, a specific care has to be devoted to web sites giving information on transport modes.

1. Background

Population using transport has a wide heterogeneity in terms of functional capacities and cultural background. It is so a great challenge for the ergonomists to integrate this heterogeneity in order to make accessible and usable the transport environment for all.

Nowadays, the important improvement of the transport facilities lies in the area of information and communication features. Indeed, through the technological evolution, transport operators are more and more able to give accurate, personalised and real time information to travellers.

In addition to the numerous transport services that are going to be available on portable devices such as Personal Digital Assistant and mobile phone, the development of transport web sites offers new opportunities to inform on "door to door" multimodal trip at the city level or at the European level. This information can be personalised by a variety of criteria according to user's preferences and needs. Furthermore, the search is conducted before the trip, with no time constraint. These two aspects make this media of communication of a great interest for elderly population, in order to support them in the preparation of their trip. In this framework, it is then important to check if the design of these sites, the modalities of dialogue and the proposed functions are fitting with the elderly specificity.

2. The web and internet use by seniors

A commonly belief is that older people are resistant to change and unwilling to interact with high tech products. Nevertheless, available data do not confirm this stereotype, older people are not technophobic, the demonstrated existing relation is between computer anxiety and experience, rather than anxiety and age (Charness, Schumann & Boritz, 1992). In fact, the decision to accept or to reject the computer use is mainly determined by the interaction of factors such as perceived needs, sociocultural influences and the design of the technology (Gitlin, 1995).

Indeed, concerning the use of internet and web sites, the number of baby boomers and seniors online is the fastest growing Internet population, in US as well as in Europe and in Japan.

A comparison survey between US and Japan use of internet by individuals age 50 and over indicated that average hours of internet use per week was 16,2 in US and 15,7 in Japan (SeniorNet, June 2000, http://www.seniornet.org/). The age of the cybersenior can be quite high: in the previous survey, 85 respondents were between 75-79, 24 between 80-84 and 7 over 85. In France, there is a web site devoted specifically for nonagenarian (http://www.chez.com/nona).

One of the main motivation for this part of the population to be connected is " to stay in touch with friends and relative" followed by " research on various topics" (SeniorNet survey, April 2000, http://www.seniornet.org/).

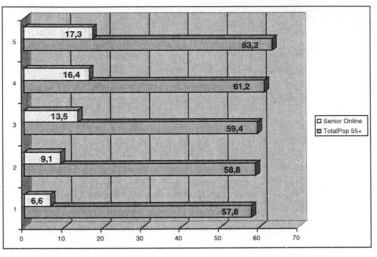

Source eMarketer   www.eMarketer.com

*Increase of the Seniors Online Population in comparison with the Senior Total Population
over 55 years old in millions of persons from 1999 to 2003 in USA.*

3. Functional abilities of elderly and HCI

In this framework, we need to understand the implications of the age-related changes in the functional abilities for the design and the functions proposed by information technologies.

A gradual decrease in perceptual performance can be noticeable from the sixtieth year, in terms of presbyacusis (decrease in high frequency sensitivity for hearing) and presbyiopia (visual accommodation reduction).

For auditory mode of communication, some studies have identified specific preferences in relation with age. For example, people between 70-75 years old used programs generating a variety of sound effects ; they found the interface more difficult to use in comparison with the silent mode, possibly as they were overwhelmed by the multimedia effects (Nielsen & Schaefer, 1993). The consequences of a reduced audio spectrum may be deleterious to speech perception (Jerger, 1973). In this framework, there are projects to develop speech signal processing techniques to compensate typical hearing losses encountered by elderly (Faulkner, Fourcin & Moore,1990 ; Faulkner, Ball & Fourcin, 1990). Furthermore, the comparison between human speaker and text-to-speech synthesizer indicates that the comprehension of synthetic speech increased short term memory demand, with a poorer performance for older people (Al-Awar Smither, 1993). Nevertheless, computer display can happen to be an improvement if compared to traditional way of information. Indeed, participants average 71 years old prefer reading computer interface rather than a classical printed leaf-let, with an even higher performance and preference score for a multimedia versus a text-only presentation when considering the computer display on the screen (Ogozalek. 1994). This study illustrated that cyberseniors can have great benefits of multimedia interactive information systems.

The mode of dialogue with the computer has been designed by young adults for young adults. Older people are more likely to suffer from arthrisis which make devices such as keyboards and light pen difficult to use for this group (Czaja, 1997). Nevertheless, the keyboard as user interface would be more effective use than a mouse (Ellis et al., 1991), as older people make a greater number of mouse errors than younger people because they have specific difficulties with small target (Charness & al., 1995). Individuals with hand trouble have particular difficulty using the mouse, analysis of videotape showed that the difficulty was about to learn to control the find positioning aspects of movement and the authors recommended the touch screen. According to this brief overview, it seems that future efforts should be devoted to improve devi   allowing dialogue with computer, with mouse and keyboard specifically designed for old hands.

An other solution is brought by the improvement in the vocal recognition accuracy and reliability. Indeed, in the Ogozalek & Praag study (1986), it appeared that even if voice input does not improve performance in comparison with keyboard, this way of communication is greatly preferred by all the participants whatever their age. And older users appear to be more enthusiastic, through preference ratings, behavioral observations and post-experimental debriefing (Ogozalek & Van Praag, 1986).

In the software design, some features can be specifically preferred by old users, because their cognitive specificity such as memory deficiencies. For example, it has been shown that there is a differentiated performance in relation to age according to the mode of display like on-screen menu, functions keys and pull-down menus : older people perform quickly and with fewer errors while using the pull-down menu, as this mode would reduce the memory demands (Joyce, 1989). An other example is concerning the e-mail use, where old participants sometimes had difficulty remembering the address name (Czaja, 1997). In the same study, providing on-screen reminders of basic commands, like "press enter after entering the address" was helpful for the elderly.

Design interface which requires the lighter working memory demand would be suitable for old people, but finally would be suitable for the entire population of users (Charness et al., 1992). This is confirmed by the content of the general ergonomic guidelines set up for older user (Charness & Bosman, 1990). The difference is that if these guidelines are not respected, the young user can manage to overcome the difficulty, rather than the older can give up the process.

4. Multimodal transport information web sites

Concerning the content and the interface of transport web sites, the users needs are highly diversified due to the wide heterogeneity of the travellers population, in terms of functional abilities, motivations of the trip and cultural background. The goal of the ergonomic approach is to take into account as much as possible of this diversity to be included in the HCI specifications. The elderly population, which is also very heterogeneous in terms of functional abilities, has the common specificity to present a general slowness, in terms of motor, perceptive and cognitive abilities, because of a general weakening and because the need to check information and to seek for confirmation. Ageing process is also often linked with the fear to get lost in unfamiliar trips.

Whatever the age of the traveller, it is always more convenient to have access in advance to precise information about the planned trip (time schedule, availability of public transport, type of road networks,...), in order to take decision, but if this possibility is a pleasant option for a young traveller, it could be considered in some cases for the old traveller as the element which makes him decide to travel or not (Pauzié, 1999a). These data point out the importance of an easy access to the information about transport for the elderly population, with part of their mobility relying on well designed transport web sites.

European surveys allow to identify the functional specifications of transport information systems after an investigation of the requirements of the elderly population: it concerns accessibility modalities, walking distance for pedestrians, criteria of choice for transport modalities ... (Pauzié, 1999b).

An investigation on elderly persons using various existing transport web sites based upon trip search scenario will allow:
  – to support the design of interface and mode of dialogue for computer in general, according to elderly needs and requirements
  – to support the development of web site in general, with recommendations based upon cyberseniors specificity
  – to support the development of transport web site, by identifying the more relevant functions and preferences to display to fulfil the requirements of the old travellers.

5. Conclusion

In our society, almost everybody needs technology for normal daily functioning, and people of the third age are no exception to this. In 1987, an investigation showed that the elderly do not use computers very extensively because they see them as too complicated, or because some of the physical impairments of advancing age make using standard systems too difficult for them (Tobias, 1987). Fortunately, since then, computers use has been drastically simplified and the cybersenior population is growing.

Researches in ergonomic have adopted the "Design for all" concept, where the systems design is targeting all the population by taking into consideration specificity of some users category, such as elderly, in order to be convenient for the wider part of the population. In addition to this approach, in some cases, it can be relevant to consider the " Gerontechnology" concept, consisting in providing technical options developed especially for elderly people, with the assumption that elderly citizens may have purposes and abilities different from younger adults that technology may help to fulfil and that requires specific development (Graafmans & Brouwers, 1989). Whatever the approach, we reach a point where it is crucial for the elderly population to be considered by designer with a specific care, in order to avoid to exclude them as potential users.

References

Al-Awar Smither J., 1993, Short Term Memory Demands in Processing Synthetic Speech by Old and Young Adults, Behaviour and Information Technology, 12, 6, 330-335.

Charness N., Bosman E.A., Elliot R.G., 1995, Senior friendly input devices: Is the pen mightier than the mouse ? 103 Annual Convention of the American Psychological Association. New York, August.

Charness N., Schumann C.E., Boritz G.A., 1992, Training older adults in word processing: Effects of age, training technique and computer anxiety, International Journal of Ageing and Technology, 5, 79-106.

Charness N., Bosman E.A, 1990, Human factors and design for older adults. In J.E. Birren and K.W. Schaie (eds.) Handbook of the Psychology of Aging (3rd ed), 446-464, New York: Academic Press.

Czaja S.J., 1997, Computer technology and the older adult, in Handbook of Human-Computer Interaction, Helander & col. (ed.), 797-813.

Ellis L.B.M., Joo H. and Gross C.R.., 1991, Use of a computer-based health risk appraisal by older adults, Journal of Family Practice, 33, 390-394.

Faulkner A., Fourcin A.J.& Moore,B.C.J.,1990, Psychoacoustic aspects of speech pattern coding for the deaf, Acto OtoRhinoLaryngologica, suppl.469, 172-180.

Faulkner A., Ball V., Fourcin A.J., 1990, Compound speech pattern information as an aid to Lipreading, Speech, Hearing and Language, University College London, Dept. Of Phonetic and linguistics, Working Progress, 4, 65-80.

Gitlin, L.N., 1995, Why older people accept or reject assistive technology, Generations, 19, 41-46.

Graafmans J. A. M., Brouwers T., 1989, Gerontechnology, The Modeling of Normal Aging: Environments, Products, and Technologies for the Aging Population, Proceedings of the Human Factors Society 33rd Annual Meeting, 1, 187-190.

Jerger J., 1973, Audiological findings in ageing, Advances in OtoRhinoLaryngology, 20, 115-124.

Joyce B.J., 1989, Identifying differences in learning to use a text-editor: the role of menu structure and learner characteristics. Master's thesis, State University of New York at Buffalo.

Nielsen J., Schaefer L., 1993, Sound Effects as an Interface Element for Older Users Computing for Older User, Behaviour and Information Technology, 12, 4, 208-215.

Ogozalek V.Z., Van Praag J., 1986, Comparison of Elderly and Younger Users on Keyboard and Voice Input Computer-Based Composition Tasks Voice Enhancement, Proceedings of ACM CHI'86 Conference on Human Factors in Computing Systems, 205-211.

Ogozalek V.Z, 1994, A Comparison of the Use of Text and Multimedia Interfaces to Provide Information to the Elderly Multimedia in Use, Proceedings of ACM CHI'94 Conference on Human Factors in Computing Systems, 1, 65-71.

Pauzié A., 1999a, From user needs to system specifications of PT and multimodal information systems,., report Infopolis 2 project, DG XIII.

Pauzié A., 1999b, Results of the ergonomic evaluation of different versions of the mock-ups, "Design of multimodal information systems", CODE project conference, Lisbon, 9 & 10 December.

Tobias C.L., 1987, Computers and the Elderly: A Review of the Literature and Directions for Future Research Age Research on Skill Acquisition, Assessment, and Change and Applications to Design, Proceedings of the Human Factors Society 31st Annual Meeting, 866-870.

A simple interaction scheme for quick exploration of long videos

Denis Payet, Henri Betaille, Marc Nanard, Jocelyne Nanard

LIRMM, Univ. Montpellier, 161, rue Ada, 34392 Montpellier cedex 5, France
(payet, betaille, nanard)@lirmm.fr

Abstract

Exploring a document consists in quickly getting an overview of its contents, with the ability to access any relevant part at any level of detail. Designed in a perspective of universal access the Video Explorer interface lets the user control both the focus in the data set and the local level of detail of the overview. The interaction scheme and the implementation architecture take care of response time issues. The technique is not limited to video exploring but also helps explore other kinds of data (e.g.: large still images, genetic sequences, algorithm traces...).

1. Introduction

Estimation about networking services in the next decade points out the increasing use of video document exchanges for private use as well as for professional tasks. Universal access to video documents on the web is an important challenge for stepping towards information society. Although digital video and video on computers seem to be quite familiar, a deeper analysis assesses that exploring distant video is not yet universally accessible (Stephanidis, 2000). In this paper, we focus on a technique that makes it more universally usable (Shneiderman, 1999) and accessible to explore videos located on distant servers. This technique, initially developed for professional use, provides universal usage to several other kinds of information. We focus on the notion of exploration, a very universal notion, which differs from the notion of reading or of searching.

2. Exploring as a universal access feature

Exploring is a natural and primitive notion for humankind. It is even more primitive than pointing, which however has proved to be the base of an extremely important paradigm for human computer interaction (Engelbart, 1967). A baby explores its environment long before pointing to objects! Unfortunately, computer interaction metaphors pay not enough attention to the need for exploring (CBSSE, 1995, chap. 7). On the one hand, computers provide alternative solutions such as "search" and all the associated information retrieval techniques. On the other hand, zooming interfaces focus more on visualization techniques than on interaction.

2.1 Exploring versus reading or searching

Exploring is distinct from reading or searching. Exploring consists of a recursive process in which a human gets extremely quickly an overview of a given set of information and goes on exploring subsets or oversets of the already explored information. For instance, if someone is attracted by a title or by an image on a magazine cover, she explores the magazine before buying it. She does not *read* it but scans the set of pages very quickly to get an overview of the contents, focuses on some pages and explores them without reading, or even explores a few sentences or images. Enabling exploration is an important marketing criteria. Therefore, exploration tools have been already developed for new media. For instance, in supermarkets, one may explore Audio CD before buying them. A machine reads the bar-code on the disc label in order to access its contents in a database; the user interactively drives the exploration just by moving the finger on the title list displayed on a screen. Remark that she does not 'listen' to the music but 'explores' the disc to get a feeling on it. The direct manipulation paradigm is well suited to support the exploring activity, since it enforces the notion of *engagement* (Hutchins et al., 1986).

2.2 Exploring video

Like exploring a book or a room, the purpose of exploring a video is to quickly get an idea of its contents, but with the ability to access any relevant part at the desired level of detail. Therefore, exploring a video cannot be reduced to playing some sequences nor to looking at an image album. Typically, exploring a video consists in looking overviews of sequences at variable levels of detail. Engagement supposes that the user fully controls the segment size and position, and the level of details at which the overview is presented. Today, video is one of the most difficult media to explore, especially on the Web. Although most video players have a slider to seek the video content, this artifact does not really provide support for easy exploration. Let us consider a one hour video record about a young baby, on CD. While playing it, one may notice an especially pleasant smile of the baby and want to picture it. Unfortunately, the slider cannot select the desired picture directly! Why? Just because the slider width in the video player window cannot provide direct access to more than a few hundred frames, whereas the one hour video is 90000 frames long! In this example, more than 10 seconds separate two adjacent pixel positions in the slider: the baby has a long time for smiling! Such a situation rarely occurs in a so extreme manner for home videos since most of them are very short, but it is a real problem when exploring long documents.

Exploring video on the Web is even the worst. It is far from being fast, natural, and easy, due to constraints on data transmission rate. Video players are not relevant at all to this service: they have been designed for playing, not for exploring. Some players rely on locally stored video. They support immediate seeking, but, on the web, their use is restricted to very short movies since downloading the whole file is needed before playing. Such long delays are incompatible with an exploration activity. Conversely, streaming enables to play a video before receiving its whole contents. Streaming is well suited for looking at live video, but it does not provide with real time exploration features since seeking requires interaction between the client station and the server. Practically, today no one can explore at will a large video located on a distant site in an efficient manner, like universal usability criteria define it.

3. Design rationale of a video explorer

The design follows the stepping suggested in (Stary, 2000).

3.1 Task analysis

Engagement in exploring sequential information requires the user have full control on two independent variables:
- the current *focus*, which should be moved forward and backward,
- the current *level of detail* of the exploration at the current focus.

One can remark that the focus has a rather continuous evolution law, whilst the desired level of detail changes rather discontinuously. Typically, one scans a magazine continuously by turning pages, but stops on a page for observing it in more details. Since the amount of information a human can receive in a given duration is limited, the level of detail is directly related to another notion, the *range* of the focus. For instance, when turning pages, the level of detail is very low and the range is the whole magazine. When stopping on a page, the level of detail becomes higher and the range shrinks to the page. The range notion is important since it allows the implementation to limit the amount of information to deliver: the user does not need the whole magazine when exploring a single paragraph.

The *focus*, the *level of detail* and the *focus range* are general notions which apply to any kind of exploration, and make sense for video exploration. They will be considered further as three variables to be controlled at the semantic level (Norman, 1986). They may be related to Shneiderman principle for information visualization "Overview first, zoom and filter, then details on demand".

3.2 Interface metaphors for information exploring

In a universal access and usability approach, the interface metaphors must be adapted to a wide range of users and support a wide set of usage (see http://www.universalusability.org/). In this aim, three major interaction metaphors and their associated interaction style have been studied for a better coverage of use situations. The first interaction metaphor is oriented towards a clear understanding of the interaction scheme at the task level: it clearly separates the controlled variables. The second one is oriented towards a better feedback of the focus and range within the full document and enables direct manipulation of these notions. The third one is dedicated to expert users and put the stress on the directness and continuity of control: a 2D gesture is used to jointly control the focus and the level of detail, thus enabling a very short articulatory distance (in the sense of Hutchins) of the exploration control.

3.2.1 Separating the controlled variables

In the first metaphor, the focus is controlled as a continuous variable by a slider, whilst the level of detail is discontinuously changed by two buttons which increase or decrease it. The focus slider width is mapped to the focus range, thus enabling very precise selection when the range is small. To access part of the document which are outside of the current range, the user needs first to widen the range by lowering the level of detail. Since the variables are clearly separated, and their action well identified and easily understandable, this interface is better suited for novice users who need to understand the effect of each command and are not used to mix elementary actions into a single command. Furthermore the paradigm is very close to those of video players.

3.2.2   Mapping the controlled variables onto a 2D surface

With the second metaphor, the two controlled variables are mapped to a rectangular surface called the exploration pad. On the x axis, the width of the pad represents the whole document; On the y axis, the height represents the precision level. Clicking anywhere in the pad assigns both variables in a single interaction. A visual feedback of the focus and of its range is directly represented on the pad as a line whose length represents the focus range. During mouse pressed interaction, the feedback is moved accordingly, enabling the user both to precisely locate the focus and to catch the relationship between detail and range. To optimize the pad surface, the relationship between the y location and the range is a $k/2^y$ law: Moving vertically by one unit divides the range by two. It offers more space to select precise exploration whilst little space is dedicated to large views. This non-linearity better fits the natural semantics of the precision level, and is better accepted by the user than a linear law. The exploration pad enables direct selection and seems quite easy to understand, even for naive users. Nevertheless, it suffers from the same drawback as video players: the x coordinate is mapped onto the whole document. As a consequence, the upper line of the pad, which selects the highest level of precision, cannot select more focus locations in the document than the pad width in pixels. Therefore, the pad needs to be associated with a slider to seek precisely into the selected range. Consequently, its use is less intuitive for novices than it could seem at first sight. The focus variable is jointly handled by two controls: one for roughly setting it within the whole document with the pad, the other for precisely setting the current focus within the focus range, with the slider.

3.2.3   Using 2D motion to precisely and jointly set the controlled variables

The drawback of the pad metaphor where the focus assignment is split on two interactors can easily be overridden by changing the semantics of the 2D mapping. In this metaphor, a single slider is used to jointly control two parameters: the main one (the focus location) is expressed along the axis of the slider, and the minor one (detail level) by the distance between the pointer and the slider while dragging. The slider width is dynamically mapped onto the current range, rather than onto the document width. The joint y motion dynamically changes the current level of detail and thus the associated range accordingly. During a 'mouse pressed' motion, the current x location is interpreted according to the current range which can be adjusted at will by the y motion, thereby putting any focus in the whole document accessible at any level of detail within the reach of a single continuous interaction (Nanard, 2001).

Let us suppose the user is currently exploring a given segment of the document. Two cases are to be considered:

- She wishes to explore a new focus which is in the current focus range. She just has to move the pointer horizontally to reach the new focus and then to move it vertically to select the appropriate level of detail.
- She wishes to explore a focus which is outside the current focus range. She just has to first move the pointer vertically down to a lower level of detail in order to extend the current range until it includes the desired focus, then to move horizontally for reaching the new focus, and then to move vertically to the desired level of detail.

The x and y motions are freely combined within a continuous interaction. For instance, as and when the y motion increases the level of detail and reduces the range, the user refines the target focus, thus taking advantage of the increased precision on the x axis. To make motion precise, the Alt and Shift alteration keys can be combined with the mouse motion to temporarily constrain horizontal or vertical motion, as usually done in most drawing software.

This interface metaphor is both simple once understood and very efficient to use since it enables to directly reach any part of the document at any level of detail in a single interaction. Nevertheless, it is better suited for expert users who are already familiar with complex pointer motion and get the feeling of 'riding' the explorer.

3.2.4   Plasticity of the interface and feedback

Practically, the three metaphors are integrated in the VideoExplorer interface to add flexibility to it (Calvary et al., 2000). We have observed users behaving opportunistically in choosing one of the interaction schemes, according to the work context. Redundancy of commands helps to better fit to the task and to increase accessibility. Redundancy does not add ambiguity when the metaphors are not contradictory and when a clear and consistent feedback acts as glue between them. It helps the user to understand the effects of her actions. In the VideoExplorer, general feedback of the focus location and of its range within the whole document is suitable so that the user keeps a spatial awareness of the exploration, exactly like she has when browsing a book in her hand. A small ribbon like feedback area is located at the top of the explorer. Its width always represents the whole document length. A colored area shows the location of the current focus range within the whole document. A red line in it locates the current focus within its range. This feedback is continuously updated according to the pointer position.

3.3   Client server architecture for exploring on the Web

Exploring distant documents, especially video documents on the web, relies on a client-server architecture. This section deals with the distribution of functions among the server and the client, with respect to ergonomic constraints. A precise balance of the activities handled by both is necessary to get a smooth behavior of an explorer.

The main constraint we have to take care of is response time. Exploration, by its nature, requires very short response time, otherwise it is no longer an exploration. By luck, the task analysis has pointed out two distinct types of activity: adjusting the focus, which is continuous, and changing the level of detail, which cares less for discontinuity since it corresponds to a change in the user's activity. It follows two consequences:

- exploring within the focus range must necessarily be fully done on the client side in order to suppress any transmission time in the interaction loop to enable a quick and thus enables smooth exploration,
- changing level of detail implies a server reaction to provide the client with new data for exploring the new focus range at the desired level. Since the response time needs to remain low, it is strictly necessary to transmit only relevant information, i.e. only the information that the user is really able to access while exploring the focus range. As a very simple example, in a video, it is irrelevant to transmit more frames than the slider can select. As noticed in section 3.1, a change of detail level breaks the activity, thereby a short delay is accepted.

The role of the server is to elaborate the overview of the selected range and to deliver it to the client. According to the desired level of detail, some overviews can be prepared once for all in the server, whilst others are computed on the fly. For instance, each video can be associated in the server with a ready to use summary from which lowest level overviews are prepared, whilst overviews at higher levels of detail are produced by sampling frames in the selected segments of the full video. In any case, the size of an overview always remains low and constant, since a high precision exploration concerns a shorter range, and a low precision a larger one. Furthermore the elaboration of an overview is requested only when the user expresses a change in the level of detail by releasing the mouse.

In the VideoExplorer, a set of 100 frames described as low quality JPEG images is transmitted as overview. In order to reduce the latency delay between the selection of a new range and its exploration, the transmission of the new overview is done asynchronously and progressively so that the client may start exploring the new focus before it is fully downloaded. The closest frame to the focus is transmitted first and frames from both sides are alternately transmitted to quickly provide the client with samples in the selected range. A colored indicator located on the cursor tells whether or not the selected frame is updated yet. A cache mechanism on the client side provides with jokers images for the missing frames until they arrive. One may remark that the first exploration of a document starts with a low level overview of the whole document which, by the way, provides with such jokers for any part of the document. Thereby, when the client explores new areas of the document, the explorer client can instantaneously deliver approximate images and refine them as and when the proper ones are received from the server.

In order to consider 'exploring' as a general interaction paradigm (Shneiderman, 1999), the notion of *overview* must remain abstract, thus is not limited to video. In the VideoExplorer the protocol between the client and the server only defines which segment is concerned, but lets the overview elaboration strategy remain at the charge of the server. In order to exemplify this, we present in the Application section two examples of exploring activities. In one of them, overviews of documents which are not at all videos are prepared by the server to be explored like videos.

4. Applications

4.1 The video explorer for INA

In the context of a research and development project aiming at providing new services for video documents in digital libraries, we are developing an environment for experts who study audiovisual archives. The French 'Institut National de l'Audiovisuel' (INA) is an institution in charge of storing archives of national radio and TV channels. It is now one of the largest repository of audio-video archives (Auffret, 2000) with more than one and a half billion of hours and has already started to turn into digital format a part of its collections. Basically, INA serves as a patrimonial archive library. Archives are mainly accessed by specialists for instance by TV producers who insert archive segments within a production, and also by domain experts such as historians, sociologists, economists, who study audiovisual archives for research purposes. Till now, they needed either to go to INA's building to look at video tapes or to buy copies to work in their own institutions. The purpose of the VideoExplorer is to provide them with a more universally usable tool directly available on a private Web portal. Typically, such users query the archive base to search for sequences, but they also need to 'explore' the answers to decide whether or not they are relevant. They need just exploring, not playing the whole set of answers. The purpose is to choose some relevant segment or even to extract images from the video. The constraints are to do it quickly for economical reasons, but in a simple, natural and easy way for ergonomic reasons. Usability is the key to success.

This first application of the VideoExplorer greatly improves the accessibility and usability of video archives. Although its use is currently restricted to end users who are allowed to access archives, the implemented mechanisms can easily be extended to any kind of video documents, and offer a more universal exploration facility. For instance, TV on demand services could let users explore video before viewing them. Since only a very small part of the document is transmitted, this could solve a part of copyright problems associated to previews.

4.2 Algorithms trace exploration

Algorithm trace exploration is one of the most surprising applications of the VideoExplorer. It is typically concerned with universal access. Understanding algorithms still is a privilege of a very small part of world's population, exactly like calculus was the privilege of some erudites a few centuries ago. Today, distant learning puts at the reach of anyone the possibility of studying computer sciences and algorithmic. As a consequence, providing efficient tools for making algorithm understanding simpler is an important challenge for universal accessibility. Visualization of algorithm traces is often used to help understand some algorithms. But most visualizations just 'play' the algorithm and list or picture what happens, but offer no means to really 'explore' the run, i.e. to have an overlook at the whole trace and to focus on some details, and even to go back to already seen steps for a deeper analysis. With the video explorer approach, the server runs the algorithm to be explored and elaborates an overview of the selected range of steps, at the desired level of detail. At low resolution, it keeps only semantically rich information associated to macroscopic properties of the data structures. At high resolution, it provides with more detailed information. The overview is exported in a XML format whose DTD is specific to the abstractions involved in the algorithm. On the player side, the overview is transformed into an animated picture, which can be explored back and forth like a video. The meaningful abstractions used in the algorithm are exhibited in the overview to be enhanced on the display. See online a demo of *graph shortest path algorithm* at: (http://lirmm.fr/~payet/these/evad).

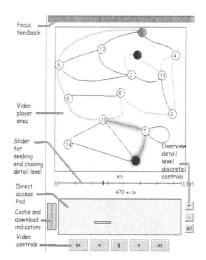

Figure 1: Video Explorer exploring the traces of a *shortest path algorithm* like a video.

5. Discussion and conclusion

Although searching is known as an efficient way to quickly find information in very large data sets, at the end, the human always have to explore the retrieved data, i.e. to quickly look at them to catch what is really important. The video explorer contributes to universal access on two points. First, it implements exploration as a universal mechanism to access video as well as many other kinds of data (e.g.: large still images, genetic sequences, algorithms traces...). Second, it provides a set of combinable interaction schemes, which makes it quite universally usable by different kinds of users or for different context of exploration. Controlled experiments are under development.

References

Auffret G. (2000) *Structuration de documents audiovisuels et publication électronique*, PhD Thesis, Université de Technologie de Compiègne, France.

Calvary, G., Coutaz, J., Thevenin, T. (2000). Embedding Plasticity in the Development Process of Interactive Systems. *6th ERCIM Workshop "User Interfaces for All"*, CNR-IROE, Florence, Italy.

CBSSE (1995). Information Access and Usability in *Emerging Needs and Opportunities for Human Factors Research*. Commission on Behavioral and Social Sciences and Education. National Academy Press.

Engelbart, D.C. *et al.* (1967). Display-Selection Techniques for Text Manipulation. *IEEE Transactions on Human Factors in Electronics*, Vol. HFE-8, No. 1, March 1967, 5-15.

Hutchins, E.L, Hollan, J.D., & Norman, D.A.. (1986). Direct Manipulation Interfaces, in *User Centered System Design*. S. Draper & D. Norman (edts). Lawrence Erlbaum Associates, 88-124.

Nanard, M., Nanard, J., Payet, D. (2001). Design rationale of a video explorer. ACM Conf. CHI'2000.

Norman, D.A. (1986). Cognitive Engineering, in *User Centered System Design*. S. Draper & D. Norman (edts). Lawrence Erlbaum Associates.

Shneiderman, B. (1999). Pushing Human-Computer Interaction Research to Empower Every Citizen, Technical Research Report 99-77, Institute For Systems Research, University of Maryland; http://www.isr.umd.edu.

Stary, C. (2000). A structured Contextual Approach to Design for All. 6th ERCIM Workshop "User Interfaces for All", CNR-IROE, Florence, Italy.

Stephanidis, C. (2000). From Adaptations to User Interfaces for All. 6th ERCIM Workshop "User Interfaces for All", CNR-IROE, Florence, Italy.

How much does compliance with the W3C Web Content Accessibility Guidelines improve web site usability for the blind and vision impaired? A Human Factors Perspective

Robert Pedlow, Mathew Mirabella, Casey Chow

Human Factors Research Group
Telstra Research Laboratories
770 Blackburn Road Clayton
Victoria 3168, Australia

Abstract

This paper reports on two studies that investigated 1.) Accessibility problems detected by different accessibility assessment methods and 2.) The extent to which designing web sites to comply with the World Wide Web Consortium (W3C) Web content accessibility guidelines (WCAG), delivers improved usability for blind or vision impaired users. The first study compared blind, partially sighted, and normally sighted users ability to carry out three tasks (involving registration, web based e-mail, and finding and playing a music track) using a large corporate web site (site_anon.com[1]). Some of the problems encountered by blind and visually impaired users in completing the scenarios included: problems completing the registration form using a screen-reader; difficulty locating the message in a web based e-mail application; and problems with low contrast for vision impaired users. The pages were also assessed for WCAG compliance using Bobby and by an expert review of the HTML code. The results showed that they did not meet the single A compliance level of the W3C accessibility guidelines. Some of the guidelines these pages did not meet included: providing alternative text for all images; providing a linear text alternative for tables laying out text in column format; and associate labels with their form controls. The sighted comparison group encountered few or no problems in completing all of the scenarios and were also able to navigate through web pages in very much shorter times than were the blind and visually impaired users. The results showed that some types of non-compliance with WCAG violations were detected by automated reviews using Bobby, expert reviews of the code, and by the blind and vision impaired users interacting with the pages using assistive software. The second study compared in a controlled experiment, blind and normally sighted users ability to complete a version of the site_anon.com registration form that was modified to be AAA compliant with the W3C WCAG, with their ability to complete a non-compliant version of the same form. The results showed that it is possible to substantially improve the usability of pages for blind users while having no effect on their usability or visual presentation for sighted users. Based on the findings of the two studies we conclude that to ensure WCAG compliance it is necessary to use a range of assessment techniques including review by disabled users who are skilled in the use of assistive technology. We further conclude that it is possible to significantly improve usability for blind users without compromising either the visual presentation of web pages or usability for sighted users.

1. Overall introduction

The release by the World Wide Web Consortium (W3C)'s Web Accessibility Initiative (WAI) of the Web Content Accessibility Guidelines (WCAG) in May 1999 represented a critical step forward for the development of accessible web content (Dardailler, Brewer and Jacobs, 2000). Reflecting the perceived importance of accessibility, national governments around the world have increasingly moved to encourage and or require web sites to meet the W3C guidelines. In Australia a number of recent events have raised the profile of web accessibility including a statement by the Human Rights and Equal Opportunity Commission (HREOC) that web sites that comply with the W3C WCAG will be regarded as meeting the requirements under the Australian Commonwealth Government Disability Discrimination Act (DDA) 1992 (Innes, 2000). For the current study, accessibility was defined as an attribute of design specifically compliance with the W3C WCAG. Usability was defined for this research as "the extent to which product can be used by a defined user population to achieve specified goals with effectiveness, efficiency and satisfaction in a specified context of use " ISO 9241:11 (1997). The research investigated the effect of changing the accessibility of web pages on the usability of the pages for users with visual impairment and normally sighted users.

Specifically in the pilot study we investigated the types of issues detected by the different accessibility assessment techniques and in the experimental study the amount of usability improvement provided for users with disability by re-design of existing content to make it comply with the WCAG.

2. Pilot study

2.1 Introduction

The W3C WCAG (1990) in the conformance section recommends the use of automated assessment tools, manual review, testing how sites perform with different browsers and browser configurations and involvement of people with disability in design and testing of sites. For application of the guidelines in an industry setting, given the relative time and costs of the different assessment methods, it is essential to determine the relative effectiveness of these different methods in detection of page accessibility guideline violations. Thus this study assessed the accessibility of several selected components of site_anon.com using three different assessment methods: automated review with Bobby, expert review by researchers with skills in HTML coding and testing with blind and partially sighted users accessing the site with their preferred assistive technology and a comparison group of sighted users. We hypothesised that the different assessment techniques would identify different types of accessibility problems that would all impact on the overall accessibility of the pages.

2.2 Method

2.2.1 Subjects and procedure

The subjects who took part in the testing comprised 14 totally blind and 15 partially sighted users who were all clients of the Royal Victorian Institute for the Blind and a comparison group of 14 normally sighted participants. The blind participants all used screen readers to access the computer while the vision impaired group comprised seven voice and seven large print users. The testing for blind and vision impaired participants was carried out at the Royal Victorian Institute for the Blind and for comparison group subjects at the Human Factors Research Group usability laboratory. The participants were asked to complete 3 basic task scenarios with the site_anon.com site which were: Join site_anon.com, Check their e-mail and Locate a particular music track (Bob Dylan's Like a rolling stone) in the audio archives and listen to it.

2.3 Results and Discussion

2.3.1 User testing results

Table 1 below shows the number of subjects in each group who successfully completed each of the three scenarios.

Table 1: Numbers of blind, partially sighted and sighted users who completed each scenario

	Tasks		
Participants	Join site_anon.com	Read my e-mail	Find audio archives track
Blind (n=14)	3	7	0
Partially sighted (n=15)	3	7	2
Sighted (n=15)	13	11	10

Since the three groups had comparable levels of computer and Internet experience the differences between these groups in the ability to interact with the computer and the Internet in carrying out the test scenarios are unlikely to reflect differences in computer or Internet experience. However, it should be noted that some of the issues encountered by blind and vision impaired participants did appear to be caused by these participants level of knowledge of the adaptive technology (eg. JAWS) rather than issues in the design of the web site itself. Some of the specific problems identified from user testing were:

- Blind and partially sighted users had great difficulty locating the join link on the home page and took much longer to do so than normally sighted users.
- Blind users in particular had difficulty completing the registration form and in most cases were unable to do so.
- Blind and partially sighted users had difficulty locating the message link on the webmail page - did not comply with guideline 13.1 on identification of link targets
- Finding the audio archive track was very difficult for partially sighted users due to low contrast for one link. - did not comply with guideline 2.2 on provision of sufficient contrast.
- As indicated some of these problems were due to non compliance with specific accessibility guidelines in the design of the pages while the others reflected a combination of design issues.

2.3.2 Bobby and expert assessment results

The HTML code of all the site_anon.com web pages involved in the scenarios used in the testing was further reviewed for compliance with the WCAG with the Bobby analysis tool and a manual review by TRL researchers with skills in HTML. The W3C define three levels of conformance with the content accessibility guidelines: A - all priority 1 checkpoints satisfied, AA - all priority 1 and 2 checkpoints satisfied and AAA - all priority 1 2 and 3 checkpoints satisfied. None of the pages evaluated met the single A level of conformance. The highest frequency accessibility problems identified by Bobby were:

14% (902/6115) Consider adding keyboard shortcuts to frequently used links;

12% (745/6115) If any of the images on this page convey important information beyond what is in each images alternative text, add descriptive (D) links;

12% (745/6115) If any of the images on this page convey important information beyond what is in each images alternative text, add a LONGDESC attribute. It is notable that these are all issues that are fairly easy to conform to from an HTML coding perspective.

2.3.3 Discussion

The findings from this study supported our hypothesis that different WCAG violations would be identified by expert and automated reviews of the web page code compared to user testing. In the course of the user testing and page review we examined only a small fraction of the total set of pages making up site_anon.com. This suggests that for present at least the steps involved in making sites accessible must include testing/ design reviews of at least key elements of the site by disabled users accessing the site with assistive technology. Such reviews need not necessarily be in the form of formal usability testing with large numbers disabled users. Indeed we suggest that it may potentially be more effective to have heuristic reviews by small numbers of "high-end" users with disability who are skilled in the use of asssistive technology.

3. Experimental study

3.1 Introduction

The objective of this study was to determine the amount of improvement in usability that would be delivered to vision impaired users by re-designing an existing set of non-compliant web pages, namely that pages in the registration scenario in the pilot study, to be fully compliant with the accessibility guidelines. In addition for this study we wanted to see if we could provide substantially improved accessibility for blind users while having little or no impact on the visual presentation of the site. A further issue explored in this study was the effect of the re-design changes required to provide accessibility on the usability of the site for normally sighted users. We hypothesised that the design changes required to make the registration form compliant with the accessibility guidelines would provide substantially improved usability for blind users while having no negative impact on usability for sighted users.

3.2 Method - Subjects, design and procedure

A total of 21 blind subjects took part in the testing. These subjects were all voice users with minimal levels of functional vision who were clients or staff of the Royal Victorian Institute for the Blind and had experience of using the computer and using a screen reader to access the Internet. The testing for blind users took place at the Royal Victorian Institute for the Blind. A comparison group of 18 sighted subjects completed the same tasks as the blind subjects at the Human Factors Research Group usability. All subjects completed both the modified and original versions of the join form. The design was balanced for order with random allocation of participants to order of presentation. The web pages were presented on a local disk. Users were told that they would complete the same scenario i.e. joining site_anon.com using two different versions of the web pages.

3.2.1 Materials

Two versions of the join site_anon.com page were created for the testing. One was a WCAG non-compliant version of the page. The second was redesigned to be as far as possible WCAG compliant while retaining the same look and feel as the original page. Both versions of the form were presented with the original unmodified home page. To create the redesigned page the reviews described below were carried out and modifications made. All priority 1, 2 & 3 issues identified with Bobby 3.2 and by expert review were corrected except for the following. In order to fully comply with the priority 2 checkpoints, two other issues required fixing, namely; that long lists of select forms were to be grouped into a hierarchy and that relative positioning (% values) be used instead of absolute positioning (pixel values). Only browser compatibility issues for Netscape 5.x and I.E 5.x were corrected. Based on a review of videotapes from the pilot study the following changes were made. Additional texts describing certain confusing input forms were added. Some of these texts were hidden using the 'style' attribute's visibility attribute. In particular, at the first name, init, last name input text boxes and also the phone number input text boxes. More descriptive text at the "click here" links was added and the title of the submit button - 'next', was changed to - 'submit button'. The document was passed through the W3C website's HTML validator. Except for the following all HTML validation issues identified were corrected: misplacement of <P>, <BLOCKQUOTE>, <TABLE>, <TD> and <TR> tags, further some of these tags were not closed. The document was also run through the W3C's HTML Tidy program which gave warnings regarding some unclosed tags (nb these were not corrected). The issues listed for all reviews that were not changed would have altered the visual presentation of the page. Overall this approach was considered to substantially meet the recommendation provided in the W3C web content accessibility guidelines for verifying conformance (Web Content Accessibility Guidelines, 1999).

3.3 Results and discussion

The results shown in Table 2 below report the numbers of blind and sighted subjects who correctly completed the scenario by page (i.e. whether the page was WCAG compliant) and order of presentation. The results show that there appears to be a small advantage for blind users accessing the WCAG compliant page regardless of whether this was the first or second page they encountered in testing although the difference is not statistically significant. The results also show that for sighted users all of the scenarios were completed with no difference as a result of WCAG compliance.

Table 2: Numbers of blind and sighted users who completed each scenario and usability ratings

Participants	CORRECT COMPLETION		USABILITY RATING	
	Presentation		Presentation	
	First	Second	First	Second
Blind				
non compliant page	5*	6#	2.1 (0.9)	2.8 (1.1)
compliant page	8#	8*	3.3 (1.2)	3.7 (1.3)
Sighted				
non compliant page	9*	9*	4.0 (0.7)*	3.7 (1.1)
compliant page	9*	9*	3.8 (1.2)	3.9 (0.6)*

Notes. Figures in ()'s show cell standard deviations; * cell n=9; # cell n=12

The usability ratings given by subjects were on a scale from 1 very difficult to use, 2 difficult to use 3 neither easy nor difficult 4 easy to use 5 very easy to use. The results show that blind users rated the compliant page as more

718

significantly usable than the non-compliant page on the first presentation. For the sighted users there were no significant differences between all four conditions. It is notable that the blind users' rating of the WCAG compliant page fall in the same range as the sighted users ratings. The results indicate that by redesigning the registration page to be WCAG compliant we have improved the usability of the page for blind users while having effectively no impact on the usability for sighted users.

4. Overall conclusions

The findings from the pilot study show the importance of using a range of assessment techniques in order to ensure WCAG compliance. The findings from the experimental study suggest that it is possible to improve the usability of web site registration pages for the blind while having no impact on the usability or visual presentation of the site for sighted users. We have not thus far investigated the effect of WCAG compliance on web-site usability for other disability groups (e.g. motor-impaired) however we would also expect some level of benefit for these groups. From our review of the literature we have not discovered any other published research looking at the impact of WCAG compliance on web-site usability. Overall the findings from this research provide some initial evidence that it is possible to make web sites accessible and usable for disabled users with no negative impact on non-disabled users.

Note
1. Site_anon.com is a fictitious name.

Acknowledgments.
The authors wish to formally acknowledge the co-operation and support for this work given by the Royal Victorian Institute for the Blind. The support of Distributed Systems Technology Centre DSTC who provided the graduate studentship for Casey Chow is acknowledged. The intellectual contribution to this paper by Dr Liz Bednall and Dr Peter Ostojic, both members of the TRL Human Factors Research Group, are also acknowledged. The permission of Paul Kirton, Director of Research, Telstra Research Laboratories to publish this research is acknowledged.

References
Dardailler, D. Brewer, J. and Jacobs, I. (2001). Making the web accessible. In Constantine Stephanidis (Ed.), User Interfaces for All: Concepts, Methods and Tools (pp 571-588).Mahwah, New Jersey: Lawrence Erlbaum Associates.
Innes, Graham (2000). Personal communication at the International Web Accessibility Summit Melbourne Australia November 15-16, 2000.
ISO 9241:11 Ergonomic requirements for office work with visual display terminals (VDTs) - Part 11: Guidance on usability. Sweden: ISO.
Web Content Accessibility Guidelines (1999) - http://www.w3.org/TR/1999/WAI-WEBCONTENT-19990505/

Information coverage –
Incrementally satisfying a searcher's information need

Erik Proper

Ordina Institute
Groningenweg 6, 2803 PV Gouda
The Netherlands
E.Proper@acm.org

Theo van der Weide

University of Nijmegen
Toernooiveld 1, 6525 ED Nijmegen
The Netherlands
TvdW@cs.kun.nl

1. Introduction

The Internet has become the virtual reality of mankind – a world that we shape without many of the imperfections of reality. We can jump to literally every place in no time and reach every resource anywhere anytime. In particular this last promise of *information at your fingertips* is under siege. The growing complexity of information space overwhelms the wired consumer and the vast increase in information is outpacing the improvement of retrieval tools.

Traditional information retrieval systems aim to satisfy a searcher's information need by matching some explicit formulation of the searcher's information need to the set of available information carriers (such as documents, web-pages, etc.) and then returning the set of information carriers that best match the formulation of the searcher's information need. From the perspective of the searcher, this traditional approach has two serious drawbacks:

Need formulation: Searchers are presumed to have a very clear understanding of their information need, even though it is not likely that they will be able to articulate their precise information needs in terms of a query language. In the case of the Internet, this becomes even more apparent as the collection of available information is sheer endless.

Need satisfaction: Even when the entire set of returned information carriers is indeed relevant to the searcher's information need, they are still required to manually wade through the result sets in search of the right combination of information carriers to cover their information need.

In traditional information retrieval systems, searchers are not provided with an advice on an effective order to best read (a selection of) the information carriers to cover their information need. The word 'effectively' may refer to time, financial costs, low overhead, etc.

Other drawbacks have to do with the quality of document characterization and limitations of matching algorithms. On the next few pages we report on ongoing research. This research extends on work as reported in [Proper and Bruza, 1999; Weide et al, 1998]. The current state of the research is that we have (formally) defined the information coverage problem area in the style of [Proper and Bruza, 1999]. As a next step, we are developing strategies and algorithms to actually develop an (idealised) information portal.

2. Information coverage paradigm

In our research, we take the perspective that searchers are not just looking for a set of relevant information carriers, but really expect an information retrieval system to help them in covering their information need in an effective way. It is our belief that an information retrieval system should really play the role of an information portal as illustrated by the information coverage paradigm as shown in figure 1.

On the left hand side there we find a searcher who is in need of information. Facing the searcher, on the right hand side, we find a collection of available information carriers. This may be a limited collection of information carriers that are available in some library, but could also be a sheer endless set of carriers, such as all information available via the Internet.

In the middle we find an *information portal*. The information portal aims to satisfy the searcher's information need. It tries to do so not only by communicating with the searcher in terms of queries and result sets. An information portal should really communicate with searchers on two levels, rather than one.

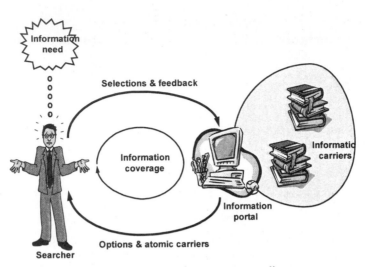

Figure 1: The information coverage paradigm

This first level is the *clarification level*, focusing on the clarification and discovery of the searcher's actual information needs. At this level of communication, the *need formulation* issue as discussed above plays a crucial role. The communication between the searcher and the information portal should ideally be seen as an interactive dialogue in which the information portal tries to clarify the precise information need while incrementally satisfying this need. The concept of *query by navigation* as discussed in, e.g. [Bruza and Weide, 1990] and [Bruza and Weide, 1992], is an example of such an approach.

The second level is the *information level*, which is concerned with the actual information aimed to fulfil the searcher's information need. At this communication level, the *need satisfaction* issue is a major challenge. Communication with respect to the actual information the searcher is looking for involves the information portal providing the searcher with information carriers, and some form of (relevance) feedback from the searcher to the information portal. This feedback may pertain to the informational content, as well as aspects such as the medium used, length, genre, etc. Ideally, the feedback from the searcher to the information portal should be as complete as possible with regards to the perceived relevance to the searcher's information needs. The more feedback the information portal receives from the searcher with regards to the relevance of information carriers, the better the information portal will be able to tune the set of selected information carriers to the actual information need.

The combination of both levels of communication, with the intention of covering a searcher's information need, is what we refer to as *information coverage*. In our research we aim to develop a theory for this information coverage process, where we take the functionality of an *idealised* information portal as a starting point. The ensuing challenge is then to identify an as-large-as-possible subset of functionality of the idealised information portal that can actually be realised.

3. A modest view on searchers

Since we are ultimately aiming for an *automated* information portal, we have, even at the level of the idealised information portal, made some pragmatic assumptions with regard to cognitive and psychological aspects of searchers. These assumptions focus on the goals with which searchers turn to the system as well as the state of mind of the searchers themselves.

For an (idealised) information portal, it is imperative to understand the goal and expectations with which a searcher turns to the system. We presume there to be two different levels for defining the goal of searcher.

The information need with which a searcher turns to the information portal is the searcher's *cognitive goal*. In addition to the cognitive goal, a searcher is likely to have some *operational goal* as well. This goal relates to the tasks the searcher has to perform. Tasks, which have led the searcher to turn to the system in the hope to gain new knowledge relevant to the task.

While, as discussed above, it may not be an easy task for searchers to express their cognitive goal, it will be less hard for them to express their operational goal. Especially when this can be done using some pre-defined terminology in the context of the searcher's task description. A context that, for example, may be provided by a business process model or a workflow model. Operational goals can have different forms, e.g. task profiles of employees, student goals when attending courses, etc.

Finally, searchers are assumed to be 'objects' that may be in different states; the searcher states. In each of these states the searcher will be in some mood. A mood that is likely to influence the searcher's ability to take up new information from information carriers. Furthermore, in each state the searcher is viewed as having some (active) body of knowledge at their disposal; their knowledge in that particular state. In general, the intention of searchers, when turning to an information portal to satisfy their needs will be to achieve a particular change in their mood and/or to acquire additional knowledge.

We realise that the above assumptions leave out numerous intricate aspects of human cognition and psychology. However, we believe that our limitations would still yield an information portal that would be more attuned to the needs of searchers than existing approaches offer. Using the above modest view on searchers, our model of an idealised information portal may already seem to be too ambitious to automate.

4. A theory for demand and supply of information

The theory for information coverage uses the metaphor of infons [Barwise, 1989; Proper and Bruza, 1999] as a semantic base. Infons are used as a representation form of elementary information particles.

In general, the intention of searchers, when turning to an information portal to satisfy their needs, will be to achieve a particular change in their mood and/or to acquire additional knowledge. Information carriers are the vehicles by which the searcher may satisfy these intentions. Experiencing (e.g. reading, listening or viewing) an information carrier will bring a searcher in a different state. A number of assumptions describe the nature of the state change function. For example, it is assumed that experiencing an information carrier does not lead to a loss of pre-existing knowledge. What information the searcher actually absorbs, depends on the mood of that searcher.

Depending on how information from information carriers is absorbed, a number of typical searcher profiles can be defined. An example is a maximal absorber, being a searcher who will absorb all information available from an information carrier on the first time they experience this carrier.

In our research, we initially restrict ourselves to the need for knowledge, which is effectuated as a need for infons. This need is communicated to the system in a searcher-independent style, for example as a query. The actual demand of the searcher thus depends on the current state of that searcher. When matching this demand for information with the information supply from the information carriers available, a classification can be made. For example, a distinction can be made between the new information supplied by some information carrier, the overhead information that would be supplied by some information carrier, etc.

In small search tasks, a stable need for information may be assumed. However, in the general case, there may be a demand drift. For example, when reading a document on a certain topic, this may raise a further interest in some aspect of the original information need. Furthermore, the structure of compound information carriers is to be taken into account as well.

5. Cost models

When a collection of alternative information carriers is available to a searcher, some mechanism is needed to determine the relative relevance of these carriers to the searcher's needs. One factor in determining this relevance is the topical relevance of the information carrier with respect to the searcher's information need. This is, however, not the only factor involved. Other factors, such as:

- the reliability of the information provided by an information carrier,
- the amount of money that may have to be paid to obtain the carrier,
- the cognitive load the searcher must endure in reading the information carrier,
- the amount of time needed to gain access to the information carrier,

may have to be considered as well. To be able to rank the available information carriers, while taking such factors into account, a price/performance ratio is introduced. The better the price/performance ratio of an information carrier, the more preferred the carrier is presumed to be. The actual price/performance ratio associated with a set of information carriers is based on some underlying cost model. As 'electronic commerce' has increasingly gained attention, the need for such models becomes more pressing [Wondergem et al, 1998].

A complicating factor is the fact that the price/performance ratio cannot be calculated for one particular information carrier in isolation. Some of the factors involved in computing the price/performance ratio do not behave monotonous with respect to sequences of information carriers. An example of such a factor is the purchase price of an information carrier. When obtaining more information carriers from the same supplier, subsequent information carriers may, for instance, be obtained at a lower price then when they were purchased in isolation. To illustrate the effects of non-monotonic behaviour, consider the following example:

> Suppose chapters A-1 and A-2 are from book A and chapter B-1 is from book B, while A-1 and B-1 are similar in content. Let us also presume book A is priced at € 40.- while book B is priced at € 30.- and that the searcher does not yet own either other the books.

> If the information portal would have to chose from either A-1 and B-1 to relinquish the searcher's information need, the portal is likely to opt for B-1 as this requires the purchase of book B, which is cheaper then book A.

> However, if A-2 is also needed to cover the searcher's information need, then first purchasing book B might not be such a good idea after al, as it is still required to also purchase book A.

One might argue that in the age of E-Commerce, the above example is antiquated as one may choose to use micro-payments to purchase specific chapters of even smaller pieces of relevant information. However, even in the case of micro-payments, the purchase price may still behave non-monotonous. It is not unlikely that content providers will try and increase the loyalty of their clientele by offering price reduction based on a customer's volume of consumption. In that case, it may be wiser to purchase chapters A-1 and A-2 using a single content provider, rather than purchasing B-1 and A-2 from different content providers.

6. Further research

It should be the intentions of any information portal to find those information carriers that best correspond to the information requirements of a particular searcher, within the specified cost restrictions. This goes beyond the traditional retrieval task, as it also requires the system to monitor searcher behaviour, such as some approximation of the searcher's state (covering both mood and knowledge).

The incremental model has been introduced [Weide et al, 1998] to model partial satisfaction of the information need as an effect of supplying an information carrier to the searcher. Extending this model with a cost function provides the opportunity to solve simple versions of the information discovery problem. A typical application is to find a minimal cost set of information carriers that completely cover some information request.

The incremental model can be introduced within the context of the model presented in the previous sections. Currently we are looking for strategies to map the requirements of the 'idealised information portal' as discussed above on to the concrete operators as introduced in the incremental model.

References

[Barwise, 1989] Barwise, J. (1989). The Situation in Logic, CSLI Lecture Notes.

[Bruza and Weide, 1990] Bruza, P. and Weide, T. v.d. (1990). Two Level Hypermedia - An Improved Architecture for Hypertext. In Tjoa, A. and Wagner, R., editors, *Proceedings of the Data Base and Expert System Applications Conference (DEXA 90)*, pages 76-83, Vienna, Austria. Springer-Verlag.

[Bruza and Weide, 1992] Bruza, P. and Weide, T. v.d. (1992). Stratified Hypermedia Structures for Information Disclosure. *The Computer Journal*, 35(3):208-220.

[Proper and Bruza, 1999] Proper, H.A. and Bruza, P.D (1999)., What is Information Discovery About?, *Journal of the American Society for Information Science*, 50(9):737-750.

[Weide et al., 1998] Weide, T. v.d., Huibers, T., and Bommel, P. v. (1998). The Incremental Searcher Satisfaction Model for Information Retrieval, *The Computer Journal*, 41(5):311-318.

[Wondergem et al, 1998] Wondergem, B.C.M, Bommel. P. v., Huibers, T.W.C. and Weide, T. v.d. (1998). Opportunities for Electronic Commerce in Information Discovery. In F. Griffel, T. Tu, and W. Lamersdorf, editors, *Proceedings of the International IFIP/GI Working Conference on Trends in Distributed Systems for Electronic Commerce*, TREC 98, pages 126-136, Hamburg, Germany.

A Different Approach to Real Web Accessibility

António Ramires Fernandes, Fernando Mário Martins, Hugo Paredes, Jorge Pereira

Universidade do Minho, Portugal
Email: sim@di.uminho.pt

Abstract

Accessibility is sometimes taken for granted, however for people with certain disabilities there are still a number of obstacles between them and the Information Society. At a time where the Information Society is becoming more and more important it is imperative to make sure that no one is left out. The web is becoming the centerpiece of this new information age; therefore if the sentence "Information Society for All" is to have any meaning then the web must be accessible to everyone. Nevertheless the web is turning into a showcase for designers and technology experts. Although this is pushing many people to the web it does drive some sectors away. People with specific disabilities, namely the blind community, find it harder and harder to read common web pages. This paper reports on a work on a talking browser showing some features that either are not yet widely implemented or, in some cases, have never been reported. Furthermore a new approach is presented, based on information extraction scripts for template urls, to provide real accessibility for blind users. This latter approach, although restricted to a small subset of the web by its nature, aims at giving blind users further views of a web page which will provide them with a clean extraction of the main information on a web page.

1. Accessibility today

When browsing the web today, we're amazed at the graphical quality and the richness of content with Flash animations, Java Applets, and other fancy multimedia plug-ins. It is very common to see HTML being used as a design language, instead of a content language. HTML tables are frequently used for layout instead of data. The design industry is clearly enjoying this new media. That's OK, we've got enough bandwidth for this, and in the future bandwidth will keep increasing so there are no technical problems in doing this. As a matter of fact we believe it's great. This approach drives people to the web making it grow. Unfortunately it also drives people away from the web.

Blind people are at a severe risk of being left out of the information society since the extraction of information from web pages can be extremely hard. Does this mean that we shouldn't use rich data types, or even simple images in a web page? We don't think that trying to discourage users and developers to use these data types is the way to go. In our understanding this would only create a wider gap between the blind community and the Information Society.

Recently, in 1999, the W3C released a first version of the accessibility guidelines [1]. This document defines a set of guidelines for web page construction. The guidelines are built around the idea that web designers can still use all that technology provides, including Flash animations, Java, 3D, and other rich media types. The main purpose of this document is to provide the web page designer with suitable, and of course accessible, ways of making that content available to users with disabilities. The guidelines are a step in the right direction towards having an information society for all. Nevertheless its implementation requires goodwill, and maybe even funding, from the content publishers.

The guidelines by W3C specify how the pages should be built in order to be accessible; the next step is to get a tool to actually read the page. Two different approaches are available today: screen readers and talking browsers. In here we will focus mainly on the talking browsers. This latter approach has seen some good examples as, for instance, the IBM Home Page Reader [2], Brookes Talk [3], and pwWebSpeak [4]. The talking browsers have a demanding task on their hands: to convert into meaningful text a HTML document regardless of how compliant it is with the W3C guidelines.

The screen reader is only concerned with the look of the document. If the document's graphical layout is clear then a good screen reader will be able to read the page. However the screen reader has no access to the document's structure, and this is where the talking browsers do have some advantage. If the W3C guidelines are followed then the talking browser ought to do a better job than the screen reader. Furthermore the talking browser can take advantage of the knowledge provided by the documents structure to present the user with different views of the document. There is one disadvantage though, if the structure of the document does not reflect its visual appearance then the talking browser can be misled, and provide an interpretation of the document which was not intended.

The issue is no longer to find a technology to read a web page. The real issue is reading it in a reasonable amount of time, allowing the blind user to grasp the information at a speed comparable to that of a sighted user.

There are still problems which have to be solved for talking browsers, namely how to deal with plugins, Java and JavaScript. In Earl et. al. [5], a review of two talking browsers, IBM HPR and pwWebSpeak, both manufacturers acknowledge this point (HPR release 3.0 claims to reads pages with JavaScript). Perhaps the ideal tool will be a merge between a talking browser and a screen reader.

In section 2 we'll describe our talking browser, the AB browser, focusing mainly on the features that we believe are either not yet widely implemented, or are new to accessibility tools. Section 3 explores a new approach for providing real accessibility for specific sites.

2. The audio browser: standard issues

The Audio Browser, AB hereafter, is a talking browser developed at the University of Minho under the support of the research program CITE IV. Although there are already a few good talking browsers, there was none at the time that spoke standard Portuguese (IBM is planning on launching a Brazilian Portuguese version in March 2001). Furthermore we believed that we could explore some directions yet unseen in other similar products. While we don't intend to establish a direct overall comparison to other similar products, we do think that launching new features is required to keep the ball rolling. In this paper we give a more comprehensive treatment to the features that are not common in other talking browsers, as opposed to common or basic features.

One of the main disadvantages a blind user has when accessing a web page, is the inability to quickly find and/or move to a position in a document. The ability to see gives the user the power to locate the desired information in a quick glance. For instance, the second table, the e-mail of the Webmaster, or the list of anchors in a long document with a table of contents. Providing, as much as possible, a fraction of that power to disabled users quickly became our main goal, giving the user the required content as fast as possible. Although we're clearly a long way from reaching our goal we believe that we have implemented some uncommon, yet useful, features for talking browsers.

In AB the web page is presented in three different ways: audio for blind users; visually as in a standard browser; and in a large hi-contrast text box for those with some usable vision. In this way, our browser allows for different users to share the web experience together. This feature is desirable for full inclusion of the blind community in the Information Society and as far as we know only the latest version of HPR shares this feature.

In AB four views of the web page are provided: the document itself, the forms of the document, the links, and the tables. Each view will be explained in detail in the next subsections.

The browser also has common features such as history of links visited and bookmarks. When book marking a document, we give the user the possibility to store not only the page's address, but also the current position in the document. This is a useful feature in long pages since the user can stop at the middle of a page and later, in a future visit to the page, restart from the same position.

2.1 The links view

When the user swaps to the links view, it is presented with the full set of links in the document divided in six categories:
- Anchors to the current page
- Internal links, i.e. links within the same domain
- External links, i.e. links to other domains

- Internal files, i.e. files downloadable from the current domain
- External files, i.e. files downloadable from other domains
- E-mail list

We believe that this classification narrows down the search to locate a link. The categories are presented initially and the user can then navigate between them with up and down arrow keys, pressing [return] opens up a category. Navigation within a category is done in a similar fashion with the up and down arrow keys. When the user is upon a link it has the option of having the document's text read, immediately before and after the link, to provide a context to the user. After the text is read the user can stay on the document or return to the previous location in the links list. This is mainly useful when web pages have very small or even meaningless link text, as in for example "…please press here", where the link is the word "here".

The list of anchors in the current page is extremely useful because it can provide an easy navigation in those long documents that have a table of contents. In this case the list of anchors is almost the table of contents.

Separating between internal and external links helps the user to navigate in the results page of a search engine like Google. In this case the internal links provide the user with the search engine options whereas the external links are a quick shortcut to the results. The advantages of the e-mail list become obvious when, for instance, the user is looking for the technical support e-mail.

2.2 The tables view

In this view the tables of the current web page are provided in a list, similar to the links view. The user can then select a table and jump to it in the document. Therefore, if the user knows that the football scores are at the end of the page in the second table, the user can jump there with a small number of keystrokes. As in the links list the user can listen to the text immediately before and after the table with a single key press.

As opposed to links, tables have no identifiers so the table view is of little use if you're not familiar with the structure of the page. On the other hand if you already know which table you want to look at, then this is a quick way of getting to those scores.

Table navigation is done with the arrow keys. Pressing the modifier keys provides further options like reading the first/last non-empty cell of a row/column. There is also the possibility of navigating inside the table without loosing the current position. If the user presses Alt and the arrow keys, navigation is performed as usual with the arrow keys, but upon releasing the Alt key the user is sent back to the previous position.

2.3 The forms view

The forms view displays the forms in the web page. As in the other views the idea is to provide the user with a quick shortcut to information. For instance in a search engine, the forms view gives direct access to the input query. Again this feature aims at providing yet another shortcut to the information.

2.4 The document view

This is the main view, where the whole web page is read. Three modes are available to the user, in each a different perspective of the web page is given. Each mode was built based on our experience during the browser's implementation. We soon realized that there was no unique way if reading all web pages, in a clear way. Furthermore some users will want to have full knowledge of the pages structure whereas others will just want to have the text read to them.

In order to cater for these situations and requests from the users we do provide different ways of reading a web page. One mode gives the user total knowledge of the web page structure, including all the tags, their relation and hierarchy. Navigation is done in the hierarchical structure of the web page using the arrow keys. However this mode adds too much extra text to the page's text, therefore getting the actual text of the page can be cumbersome.

In order to provide a smoother reading, but still have all the information available we also provide a mode where the tags text is eliminated. The text is presented in a hierarchical tree based on the headers (if available). Tables are linearized and the reading is row based. Nevertheless no information is lost because the user can still jump to the tables view and have a non-linearized version of the tables.

The third mode available is for those users who just want a piece of text on the web page. For instance, in a news web site, when jumping to a page for a particular piece of news, it is common to have the links to other sections on the left side. In a talking browser these implies that before reading the actual text, all the links must be read because they appear first in the html code. This third mode removes all the stand-alone links from the text, i.e. a link that has no text before or after the link tag will be excluded from the reading. In practice this generally implies that all navigation links tend to disappear and the text that is read is just the main text on the web page.

In the example of a news web site almost only the text for the piece of news is read. Note that the user can still jump to the links section. However, in some sites using JavaScript some extra text might also be read, therefore delaying access to the information.

At any time the user can switch between the different modes using the keyboard, and select the mode that suits him/her better.

As in the tables view the user can also navigate without loosing the actual position. When pressing Alt together with the arrow keys a temporary cursor is used to move in the document, and the cursor item will be read. Upon the release of the Alt key, the user is sent back to the previous position.

3. Using scripts for template URLs

Accessibility has two different meanings for blind and sighted users. These latter users have the ability with a quick glance at a web page to identify the locations where the information is present. Sighted users can go directly to the relevant information avoiding paying any attention to the list of links on the left of the page, or the banner at the top. Blind users, on the other hand, have to cope with all the text of a web page when using a talking browser or a screen reader. In a typical news page, full of navigation elements, links to sponsors, and to many other entries, the blind user must either have all this text read, or do a bit of navigation in the web page, before the tool gets to the actual information.

Even with the different views we already provide in our browser there are pages where the main text of the page is not readily available to the user. For instance, we've tried with pages of a local newspaper that uses JavaScript and this causes a lot of garbage to be presented to the user.

Since site design is very diverse it is impossible to cater for all cases. On the other hand it is extremely easy to define a script to extract the information from a page, or even a set of pages from a particular site. The ability to incorporate scripting in a talking browser, gives the user the best of both worlds. The user can still travel freely on the web, and for a restricted number of selected sites the user is presented with the information right away, without any extras. The fact that this feature is incorporated in a talking browser allows the user to switch to the default views, without scripting; therefore the extras are not lost. Scripting provides the user with a new view of the web page.

Scripts can be written for template URLs, i.e., the URL is defined using wildcards in a similar fashion to how a list of folder or filenames is specified. When the talking browser detects a URL for which there is a script the document is presented using the script to extract information. An audio message is provided to the user in order to let him/her know that a script has parsed the page's content. Using a single keystroke the user can change to the default view, i.e. the full web page.

Writing a single script is a very simple process. However to get to a significant number of sites, time is required to implement all those scripts. Basically this is a never-ending process, where scripts are added in a continuous process. Furthermore web sites do design updates and this may require a script update. With time a large database of scripts can be build, covering the major news sites, search engines, and sites with relevant specific information for blind users.

Because of the dynamic nature of the scripts database we needed a way of keeping the previously installed talking browsers up to date. The solution we found is based on a central server that contains the database. When a talking browser is activated, a connection to the server is made where it is checked if there were any updates to the database, either new or modified scripts. If this is the case then the talking browser will download the new and updated scripts, and store them locally.

4. Conclusions and future work

There's still a lot of work to be done before the blind community can fully experience the power of the Internet. The graphical nature of the web implies that they will always be outsiders to some extent. Nevertheless this is no reason to quit, quite the opposite, these problems drive us to improve our browser.

While we believe that we've provided the blind user with a good set of document views, we realize that there are still a considerable number of pages where unwanted text is read to the user. One direction of research to cope with this problem is the study of heuristics to provide new views of a web page.

We also are looking into ways of building a script editor that could be used by the blind user. This may be just a dream, but if we could make it come true then the blind user could define his/her own view of specific web pages. Different users could have different views of the same web page, and they would even be able to share those views. Obviously this feature would be only for power users, still we think that this is a step in the right direction towards increasing the level of accessibility for blind users.

Another issue is multi language support. Right now our browser only speaks in Portuguese, but its implementation was designed having this goal on mind. Furthermore, the interface and pre defined messages can also be translated with very little effort, so having English, Spanish, French, and other versions is a straightforward process.

Acknowledgements
The research reported in here was supported by SNRIP: The National Secretariat of Rehabilitation and Integration for the Disabled under program CITE IV.

References
[1] *Web Content Accessibility Guidelines 1.0.* [On-line]. Available: http://www.w3.org/TR/WAI-WEBCONTENT/
[2] *IBM – Home Page Reader.* [On-line]. Available: http://www-3.ibm.com/able/hpr.html
[3] *Brookes Talk.* [On-line]. Available: http://www.brookes.ac.uk/schools/cms/research/speech/info.htm
[4] *pwWebSpeak.* [On-line]. Available: http://www.prodworks.com
[5] Earl, C., Leventhal, J., Wehberg, K. *A review of IBM Home Page Reader and pwWebSpeak.* [On-line]. Available: http://www.onlinejournal.net/afb/AW/1999/1/0/prod_eval2.html
[6] Zajicek, M., Powell, C., Reeves, C. (1998).*A Web Navigation Tool for the Blind.* 3rd ACM/SIGAPH on Assistive Technologies, California.

Developing Guidelines for Designing Usable Web Pages for Older Chinese Adults

Pei-Luen Patrick Rau, Jia-Wen Shiu*

Department of Management Information Systems, Chung Yuan Christian University,
Chunli 320, Taiwan
* Institute of Communication Studies, National Chiao Tung University, Hsinchu 300, Taiwan

Abstract

This research developed and examined guidelines for designing web pages for older Chinese adults. The past findings on how aging affects ability in general and computer skills were reviewed. These guidelines were developed and arranged into six categories: layout, character, color, hyperlink, object, and picture. Based on these design guidelines, a heuristic evaluation was conducted by three usability specialists (young adults) for two Chinese web sites providing on-line news. Three older users participated in a think aloud test for the same web sites. The results of the heuristic evaluation and the thinking aloud test were compared and discussed.

1. Introduction

Taiwan has a rapidly graying population and increasingly ubiquitous Internet use. A shifting age ratio combined with the spread of the Internet means that the importance of older users and the importance of interfaces for older users have increased considerably. In general, the literature on aging and skill acquisition indicates that older adults have more difficulty acquiring new skills than young people and that they achieve lower levels of performance. Further, there is substantial evidence that there are age-related declines in most component processes of cognition including attention processes, working memory, information processing speed, encoding and retrieval processes in memory and discourse comprehension (Park, 1992). Essentially, we need to understand the implications of age-related changes in functional abilities for the design and implementation of Web interface.

There are several relevant cognitive, perceptual and motor changes that occur with increased age (Czaja, 1997). Older adults take more time to perform most motor and cognitive tasks in general. In screen design, this means that fast-moving objects should be avoided. Morrell and Echt (1997) suggested that Helvetica improved reading performance, they also suggested colors schemes, font thickness, physical spacing, justification, and line width for older users. Research has shown dark text on a light background to be more efficient for older adults (Tobias, 1987). Snyder (1988) noted that negative contrast might minimize the distraction from glare, which older adults are particularly sensitive to. The proper organization and amount of information on a screen are important because of the age-related decline in visual search skills and selective attention (Ellis & Kumiawan, 2000). Highlighting information and using perceptual organization, such as grouping, would help older adults to access the necessary information more effectively (Czaja, 1997).

Hwang (Hwang, Wang, & Her, 1988) found that a Chinese character size of 24x24 dots (4.85mm), line spacing of 15 dots (3mm), and character spacing of 3 dots (0.61mm) or 6 dots (1.21mm), would improve performance without increasing visual fatigue. For older Chinese adults, the character size, line spacing, and character spacing of Chinese words may need to be larger. She found that there was no significant display format (vertical and horizontal) effect on any performance measure.

2. Design guidelines

Having reviewed the literature on interface design issues, we offer general guidelines with respect to designing Web interfaces to accommodate the needs of older Chinese users. (Czaja, 1997; Ellis & Kumiawan, 2000; Hwang, Wang, & Her, 1988; Mead, Spaulding, Sit, Meyer, & Walker, 1997; Morrell & Echt, 1997; Morris, 1994; Snyder, 1988; Tobias, 1987). We classified these guidelines into six areas: character, color, layout, hyperlink, object, and picture. The guidelines are briefly discussed as follows:

• Character

"Provide both simplified and traditional Chinese versions." It takes a lot of effort for older adults to accommodate different Chinese character systems.
"Use Min font for Chinese." Most Chinese newspapers and web content providers utilize min font.
"Make character spacing as large as possible." Character size and spacing are important, because older adults have some difficulty reading the screen.

- Layout
"Keep the web layout as simple as possible." Older adults have difficulty processing complex or confusing information and are more likely to experience interference form irrelevant information (Plude & Hoyer, 1985). Designers should keep web layout as simple as possible.
"Use attention-attracting and grouping techniques." A lot of information must put into homepages, designers should follow "Gestalt laws" to fit human perception.

- Hyperlink
"Make sure that links are placed where they are easy to see." The location of Links is very important. Designers should place links on the top or the left of the pages. Links should be large and have space around them to prevent accidental selection.

- Object
"Minimize the amount of scrolling users must do". Older adults have difficulty using the mouse for dragging, so web pages should be short.

- Picture
"Avoid decorative animation and pictures, wallpaper patterns and flashing text." Designs should use only simple, highly relevant graphics and avoid decorative animation and pictures. The speed of animated pictures and texts should be slowed down.

3. Heuristic evaluation

- Participants: three usability specialists (young adults who were 24, 25, and 27 years old). All of these three evaluators were graduate students majoring in user interface design or communication technology. These evaluators had taken at least one graduate level course on usability engineering or user interface design. Two out of the three evaluators were female and the other was male.

- Apparatus: Each evaluator used a Pentium II personal computer with a 17" CRT monitor. The presentation mode was controlled at 800*600. The browser was Microsoft Internet Explore Version 4.0 or above.

- Materials: Two Chinese web sites providing on line news services. The first was the "People net" (http://www.people.com.cn), the second was the "Chinatimes.com" (http://www.chinatimes.com.tw). These sites were selected based on their enormous reputation and page views. The former was presented in simplified Chinese by default and the later was presented in traditional Chinese. A traditional Chinese version of People net was utilized in the evaluation because all of the participants were in Taiwan. The traditional Chinese version of People net was exactly the same as the default simplified version of People net except for the character system.

- Procedure: three usability specialists evaluated the home pages of People net and Chinatimes.com for one hour individually. During the evaluation, the three evaluators were separated to avoid discussion. The evaluators were asked to record all of the usability problems associated with the provided guidelines. After the evaluation, the results were collected and analyzed by the authors.

- Results: 47 usability problems were detected. Layout was the category associated with the most usability problems among the six categories.

4. Think aloud

To determine the actual "in use" problems and to compare the differences in guideline implementation between young adults and older adults, we designed a think aloud test.

- Participants: three older adults (53, 55, and 57 years old). None of the three older adults had experience browsing the two web sites. Two out of the three older adults were female and the other was male.

- Apparatus: as in heuristic evaluation.

- Materials: as in heuristic evaluation.

- Procedure: Participants were first instructed about the test and allowed to practice. Each participant was then asked to browse the homepages of the two web sites and to think aloud about what he/she was doing for one hour. During the test, the experimenter was prompting and listening for clues about how each participant was dealing with the home page and recording all of the results for further analysis.

- Results: The three participants rarely felt comfortable browsing the pages. Characters and the picture elicited the most concern among the six guideline categories.

5. Summary of design recommendations

The results from the heuristic evaluation and think aloud test are summarized in Table 1. Guideline significance ratings were determined according to the test results. The heuristic evaluation was based on the score for each guideline as the number of associated usability problems found by the three usability specialists. During the think aloud test the experimenter recorded all of the usability problems found by the three older users. After the test, all of the identified usability problems were analyzed and attributed to one of the six categories

Table 1 Rating of Significance of Guidelines for Designing Usable Web Pages for Older Chinese Adults

Priority	Category of Guidelines	Older Adults	Usability Specialists (Young Adults)	Examples of Usability Problems
1	Character	8	5	• Small characters • Too many characters in each line • Small line space
2	Picture	7	9	• Too many animations • Too many flashing effects • Big banners
3	Hyperlink	5	9	• Unclear hyperlinks • Too many hyperlinks • Lack of navigation aid
4	Layout	3	15	• Lack of grouping techniques • Complicated layout
5	Object	3	4	• Long pages • Too many objects
6	Color	0	5	• Sharp colors

As shown in the table, older adults considered the characters and picture the most significant categories among the six categories. However, usability specialists (young adults) considered layout the most significant category among all. This difference implied that user interface designers, who are mostly young adults, have difficulty recognizing usability problems for older adults, even on the basis of design guidelines. Both older and young adults did not consider the Hyperlink category very significant. The possible reason is due to the test design, which focused on home page design rather than web site information architecture.

6. Conclusion

The general agreement with the literature lends support to the notion that our findings are firm and general. We think that Web page designers can improve the usability of Chinese web pages by applying the guidelines proposed in this paper. However, a wider sample and variety of sites must be evaluated to refine and produce a more complete set of guidelines. In addition, other tests must be conducted to determine which guidelines have a more significant effect on the usability of a page. This research examined only the visual decline and cognitive impairment issues for older Internet users. More research is needed with regard to the I/O mechanism of older adults, cognitive overload and the tendency to get lost in hypertext surroundings.

References

Czaja, S. J. (1997) Computer technology and the older adult. In M. E. Helander, T. K. Landauer, & P. Prabhu (Eds.), *Handbook of Human-Computer Interaction* (2nd ed., pp.797-812). New York: Elsevier.

Desurvire, H. W., Kondziela, J. M., & Atwood, M. E. (1992). What is gained and lost when using evaluation methods other than empirical testing. In Monk, A., Diaper, D., & Harrison, M. D. (Eds.), *People and Computers VII*, Cambridge University Press, Cambridge, U.K. 89-102.

Ellis, R. D., Kurniawan, S. H. (2000). Increasing the usability of online information for older users: a case study in participatory design. *International Journal of Human-Computer Interaction*, 12(2), 263-276.

Hwang, S. L., Wang, M. Y., & Her. C. C. (1988). An experimental study of Chinese information displays on VDTs. *Human Factors*, 30(4), 461-471.

Mead, S. E., Spaulding, V. A., Sit, R. A., Meyer, B., & Walker, N. (1997). Effects of age and training on world wide web navigation strategies. *In Proceedings of the Human Factors and Ergonomics Society 41st annual meeting* (pp. 152-156). Santa Monica, CA: Human Factors and Ergonomics Society.

Morrell, R.W., & Echt, K. V. (1997). Design written instructions for older adults: Learning to use computers. In A. D. Fisk & W. A. Rogers (Eds.), *Handbook of human factors and the older adult* (pp. 335-361). New York: Academic.

Morris, J. M. (1994). User interface design for older adults. *Interacting With Computers*, 6, 373-393.

Park, D. C. (1992). Applied cognitive aging research. In F.I.M. Craik and T.A. Salthouse (Eds.), The handbook of aging and cognition (pp.449-494). Hillsdale NJ: Lawerence Erlbaum Associates.

Plude, D. J. & Hoyer, W. J. (1985). Attention and performance: identifying and localizing age deficits. In N. Charness (Eds.), Aging and human performance (pp.47-99). New York: John Wiley.

Nielsen, J. & Mack, R. L. (1993). *Usability Inspection Methods*. New York: John Wiley.

Snyder, H. L. (1988). Image quality. In M. Helander (Eds.), *Handbook of human-computer interaction* (pp.437-474). Amsterdam: Elsevier.

Tobias, C. L. (1987). Computers and the elderly. A review of the literature and directions for further research. *In Proceedings of the Human Factors Society 31st annual meeting* (pp. 866-870). Santa Monica, CA: Human Factors and Ergonomics Society.

Multimodal spatial querying: what people sketch and talk about

Isolde Schlaisich, Max J. Egenhofer

National Center for Geographic Information Analysis
Department of Spatial Information Science and Engineering
University of Maine, Orono, ME 04469-5711, USA
{isolde,max}@spatial.maine.edu

Abstract

People want to use geographic information systems (GISs) while spending little time for learning how to use the system. This trend increases the need for easy-to-learn and intuitive user interfaces. In order to make GISs easier to use for untrained users, we investigate an interaction method that mimics the natural communication between people and allows users to formulate queries. In human-human interaction people often communicate about space by talking and simultaneously drawing a freehand sketch; therefore, this combination of modalities is attractive for describing spatial scenes when querying a GIS. To assess the usefulness of this approach, we have conducted a human-subject experiment in which subjects were asked to draw spatial queries and talk at the same time. We analyzed what modality subjects preferred when they had the choice of sketching and talking. We found that subjects gave complementary information through the two modalities, indicating that sketching cannot replace the verbal descriptions, and vice versa. We also found redundant information in both modalities, which provides a solid foundation for integrating elements from the two modalities.

1. Introduction

Future geographic information systems (GISs) must be easier to use for everyday users who do not have extensive training in the use of computer systems. GISs are widely used to query and analyze spatial data in such applications as car navigation, tourist information, and real estate management. While the community of interested GIS users is growing rapidly, the current systems are often too difficult to use or require too long a time to learn them. State-of-the-art GISs offer a WIMP (windows, icons, menus, pointers) interface where users often struggle to find their way through the available menus or type the commands they wish to execute. People are, however, more familiar with other approaches to communicate about space in everyday life, such as speech, drawings, gestures, and writing.

We are investigating a *Sketch-and-Talk* user interface, which combines the input of spoken language and drawn sketches to build an effective query method. This kind of interface is expected to lead to a broader acceptance of GIS and suit the demand for a wide-ranging access to computer systems (Stephanidis 2001). For this goal we conducted an experiment to investigate how people communicate about space, simulating the interaction with a *Sketch-and-Talk* GIS. We recorded the responses of twelve subjects while they were asked to respond to four spatial queries. The analysis of the drawings and the corresponding text showed that all subjects used both interaction modalities. The spatial objects in the queries are at times stated in only one modality and at times in both, generating complementary and redundant information. More than half of the spatial objects were mentioned in both modalities, giving further evidence that one modality alone would be insufficient.

The remainder of the paper is structured as follows: Section 2 briefly discusses interaction methods with GISs. Section 3 describes the experiment to determine what people sketch and to what they refer with voice annotations. Section 4 analyzes the experiment, followed by our conclusions in Section 5.

2. Interaction with geographic information systems

GISs have a large audience of users due to the attractive use of spatial access and spatial analysis in a variety of application areas (Longley *et al.* 1999). Since many potential GIS users often lack specific training in the use of GISs, the need for more intuitive user interfaces arises. One viable approach to the design of GISs and their user interfaces is the consideration of people's common-sense knowledge of the geographic world (Egenhofer and Mark 1995).

New interaction methods that deviate from the traditional ones are asked for in order to make GISs more usable (Egenhofer and Kuhn 1999; Bass 2001). These interaction methods include talking, drawing, and gesturing. Sketching as a spatial query language offers a fairly natural way of expressing spatial information (Egenhofer 1996a), as was demonstrated with a recent implementation of Spatial-Query-by-Sketch (Blaser and Egenhofer 2000). It enables the users to draw a spatial configuration of the scene they are looking for and search for the most similar settings in a spatial database. Sketching alone, however, is limiting and a more efficient interaction could be achieved by offering multiple modalities (Oviatt and Cohen 2000).

In human-human interaction people typically use spoken language or written comments to augment sketches (Egenhofer 1996b). The two interaction modalities are especially useful for GISs, because some types of spatial information are better expressed by drawing a sketch, while others are better conveyed with words. Particularly spatial relations and metric information are drawn faster and with less ambiguity than if they are given verbally. For example, it is more precise to describe the spatial configuration of a house close to the northwest shore of a lake by drawing a sketch than by explaining the same details with words. On the other hand, attributes of an entity, such as the address of a building, are better stated in spoken or written language. Sketching and talking are most powerful in a spatial query language if they are used together. For instance, a person who describes the way from her workplace to her house by drawing a sketch that shows her house, the streets surrounding it, and crucial landmarks, could also simultaneously use spoken language to add information about the name of the streets. If such an intuitive interaction style were available with a GIS, it would enable more casual users to access and analyze more spatial information, because this combination mimics the way people often communicate about space. Better use of spatial information will enable people to make better-informed decisions in their everyday lives.

3. Experiment

To explore how people use sketches and verbal annotation, we simulated a *Sketch-and-Talk* environment and asked subjects to state various queries by drawing on a touch-sensitive screen and talking at the same time. The sketch was entered through the Spatial-Query-by-Sketch interface (Blaser and Egenhofer 2000), while the spoken words were recorded with a tape recorder.

Six male and six female subjects in the age range of 20-29 years participated in the experiment. The instruction for the first question (Query 0) was to interact with an information system by sketching the way from the subject's house to the post office. Subjects were not told that they could use voice input as well. The reason for the first query was to allow subjects to familiarize themselves with the sketch system. After that, they were told to imagine that the system could process spoken language as well. Their task was now to formulate another three queries by sketching and talking. The particular instructions were: "Communicate with the system to retrieve the place where you grew up and went to school" (Query 1), "Look for your dream vacation resort" (Query 2), and "Look for a campus that has a setting that is similar to the University of Maine" (Query 3). Figure 1 shows an example of one subject's sketch, describing her way from home to school and the accompanying text.

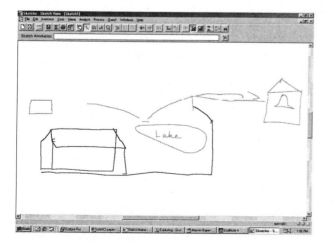

"I lived with my grandparents on a farm. It took me approximately 10 to 15 minutes to school, because we lived on a hill. I lived off of a major route and then I would have to change over to another route. I would have to go around a lake, which would be here, and go up another road, ascend another hill, and there was my school. It was situated so that, if you looked out across the road from the farm, you could actually overlook the lake and the school."

Figure 1. Sketch of subject 3, describing her way to school and the corresponding text.

4. Analysis

For the analysis of the experiment, the spoken text was typed into a text file and cleaned up from noise. To compare the content of the sketch with the text, we counted in each sketch the number of recognizable objects drawn and in each text the words as well as the number of objects (i.e., nouns) used in the text. Examples of spatial objects that can be found in Figure 1 are buildings, streets, and a lake. The twelve subjects sketched 428 objects in queries 1-3, while they referred to 314 objects in the spoken parts of queries 1-3.

Table 1. Average number of objects per sketch, and words and objects per annotation.

	Query 0	Query 1	Query 2	Query 3	Average
Number of objects in sketch	6	10	9	17	11
Number of words in text	–	156	120	188	155
Number of objects in text	–	10	7	11	9

Table 1 shows the average number (arithmetic mean) of the objects that were found in the four sketches and the average number of objects in all sketches of the experiment in the second row. In each of the 48 sketches, subjects drew on average eleven objects. This number is relatively small, confirming Blaser's (1997) finding that a person draws a sketch of average complexity with 5 to 25 objects.

The number of words in the text is approximately 16 times higher than the number of objects. The comparison of drawn objects versus verbally mentioned objects shows that on average up to 50% more objects were drawn than talked about.

4.1 Choice of modality

The experiment showed that people embrace the offer to sketch and talk as a method of interaction with a GIS. Some subjects were talking more, while others were sketching more, but none did completely ignore one of the modalities. All subjects used both modalities in at least one of their queries. Ten of the twelve subjects used both modalities for each query. One subject only talked to formulate query 2, while another subject used only sketching in queries 2 and 3.

Five out of the twelve subjects (42%) already talked during the first query, although they had no information about the possibility of voice input. They did not expect the system to process the words. This voluntary use of sketching and verbal annotations is another indicator that the simultaneous use of sketching and talking is a natural interaction style for spatial querying.

4.2 Complementarity

Users have a choice to formulate different parts of the query with the modality that best fits their mental model. This type of interaction may lead to simple requests, with fewer mistakes and without complicated formulations. At the same time, some parts of a query may be found only in the sketch, while others may be expressed only verbally; therefore, the two modalities promote a complementary query formulation. Through a single interaction modality, this information would be either lost or users would have to reformulate the query using the only available modality. Table 2 shows that 46% of the 428 objects that appeared in the sketches only exist in the sketched part of the queries. On the other hand, the 83 objects that are only mentioned while talking comprise only 26% of the 314 objects in the spoken part. These results indicate that objects mentioned in the verbal description are more likely to have a graphical counterpart, while almost half of the drawn objects lack a verbal annotation.

Table 2. Number of objects that appear only in a sketch, only in the verbal part, and in both.

Objects in sketch only	Objects in sketch and talk	Objects in talk only
197	231	83

4.3   Redundancy

In the same way the two modalities enable complementarity, they offer users to describe the same content redundantly through either sketching or talking. Redundancy may be an indication for the importance of a piece of a query. Redundancy also provides an important glue when integrating the sketch and the spoken request into a single query.

Subjects often referred redundantly to an object in the sketch and in the verbal annotation. For example, in Figure 1 the objects farm, lake, school, and routes where drawn in the sketch as well as stated in the text. Table 2 shows that 231 of the objects are referred to in both modalities, which is 54% of all sketched objects and 74% of all objects in the verbal part. This indicates that the subjects took advantage of the possibility of simultaneous input of sketches and voice, and produced a remarkable quantity of redundant information.

5.   Conclusions

Simpler and more intuitive user interfaces are needed to give inexperienced users access to GISs. The combined use of sketching and talking mimics the way people typically communicate about space, therefore, these two modalities is a natural way to formulate questions about space. We expected that inexperienced GIS users are more at ease to combine speech and sketches to convey their knowledge of space than to use a single mode. Also the use of the two modalities—speech and drawing—is especially suitable for communication about spatial topics (Egenhofer 1996b; Oviatt 1996). Although the emphasis of this paper is on providing easy-to-use interfaces for inexperienced users, the approach of *Sketch-and-Talk* can also facilitate the routine work of experienced GIS users. The natural way of interaction will make them at ease with the use of the system and might improve their performance as well.

We conducted an experiment to demonstrate that people make use of both modalities when formulating spatial queries. We found significant complementary and redundant information in the spoken and sketched parts of the queries. From a user-interface point of view, these results demonstrate that both modalities are needed for a natural formulation of spatial queries. From a query processing point of view the information about redundancy forms a solid basis for query processing, as it enables to integrate the sketched and the spoken query parts.

Acknowledgments

This work was partially supported by a grant from the National Science Foundation under grant number IRI-9613646. Max Egenhofer's research is further supported by NSF grants IRI-9613646, IIS-9970123, and EPS-9983432; the National Institute of Environmental Health Sciences, NIH, under grant number 1 R 01 ES09816-01; the National Imagery and Mapping Agency under grant number NMA202-97-1-1023 and NMA201-00-1-2009, and by Lockheed Martin M&DS.

References

L. Bass (2001) Interaction Technologies: Beyond the Desktop. In: C. Stephanidis (Ed.), *User Interfaces for All: Concepts, Methods, and Tools*, pp. 81-95. Mahwah, NJ, Lawrence Erlbaum Associates.

A. Blaser (1997) *User Interaction in a Sketch-Based GIS User Interface*. Technical Report, National Center for Geographic Information and Analysis, University of Maine, Orono, ME.

A. Blaser and M. Egenhofer (2000) A Visual Tool for Querying Geographic Databases. In: V. Di Gesù, S. Levialdi, and L. Tarantini (Eds), *AVI2000—Advanced Visual Databases*, Salerno, Italy, pp. 211-216.

M. Egenhofer (1996a) Spatial-Query-by-Sketch. In: M. Burnett and W. Citrin (Eds), *VL '96: IEEE Symposium on Visual Languages*, Boulder, CO, pp. 60-67, IEEE Press.

M. Egenhofer (1996b) Multi-Modal Spatial Querying. In: M.-J. Kraak and M. Molenaar (Eds), *Advances in GIS Research II, Seventh International Symposium on Spatial Data Handling*, pp. 12B.1-15. Delft, The Netherlands.

M. Egenhofer and W. Kuhn (1999) Interacting with GIS. In: P. Longley, M. Goodchild, D. Maguire, and D. Rhind (Eds), *Geographical Information Systems: Principles, Techniques, Applications, and Management*, Vol. 1, pp. 401-412. New York, John Wiley & Sons.

M. Egenhofer and D. Mark (1995) Naive Geography. In: A. Frank and W. Kuhn (Eds). *Spatial Information Theory*. Lecture Notes in Computer Science 988, pp. 1-15. Berlin, Springer-Verlag.

736

P. Longley, M. Goodchild, D. Maguire, and D. Rhind, Eds. (1999) *Geographical Information Systems: Principles, Techniques, Applications, and Management.* New York, John Wiley & Sons.

S. Oviatt (1996) Multimodal Interfaces for Dynamic Interactive Maps, *Conference on Human Factors in Computing Systems - CHI'96,* pp. 95-102, ACM Press, New York.

S. Oviatt and P. Cohen (2000) Multimodal Interfaces That Process What Comes Naturally. *Communications of the ACM* 43(3): 45-53.

C. Stephanidis (2001) User Interfaces for All: New Perspectives into Human-Computer Interaction. In: C. Stephanidis (Ed.), *User Interfaces for All: Concepts, Methods, and Tools,* pp. 3-17. Mahwah, NJ, Lawrence Erlbaum Associates.

A Simple Approach to Web-Site Usability Testing

Mikael B. Skov, Jan Stage

Department of Computer Science, Aalborg University
Fredrik Bajers Vej 7, DK-9220 Aalborg East, Denmark

Abstract

Universal access requires dissemination of usability engineering competence to a wider audience. This article reports from an empirical study of web-site usability testing conducted by a diverse group of people with no formal education in software development or usability testing. 36 teams of four to eight first-year university students with an interest but no education in information technology were trained in a simple approach to web-site usability testing that can be taught in one week. It is concluded that basic usability testing skills can be developed in a week.. The article describes how they applied this approach for planning, conducting, and interpreting a usability evaluation of the same web-site. The student teams gained competence in defining good task assignments and ability to express the problems they found. They were less successful when it came to interpretation and analytical skills. They found quite few problems, and they seemed to lack an understanding of the characteristics that makes a problem list applicable.

1. Introduction

Universal access relates to the fundamental relation between use and technology. We are on the move towards a society where the fusion of emerging information technologies will facilitate new or improved patterns of work and interaction. In this way, information technology may impose radical changes to our daily life in a variety of ways. These changes will create new demands that are summarized in the requirement for information technology that is available and accessible to anyone, anywhere, anytime. The World Wide Web represents a significant move towards the ideal of universal access. Today, the Web makes substantial amounts of information available to anyone with a computer that is connected to the Internet. We are still far from the ideal, but the Web has definitely brought us a step forward.

The Web has also imposed drawbacks. The present work practices in web-site development seem to largely ignore the body of knowledge and experience that has been established in the disciplines of software engineering, human-computer interaction, and usability engineering, cf. (Sullivan & Matson 2000). Web developers experience a strong push for speed because users of web sites rapidly change preferences and patterns of use, and this makes customers and management ask for development cycles that are considerably shorter than in traditional software development (Anderson 2000, Broadbent & Cara 2000). Many web sites are designed and implemented in fast-paced projects by multidisciplinary teams that involve such diverse professions as information architects, Web developers, graphic designers, brand and content strategists, etc. Such teams are usually not familiar with established knowledge on human-computer interaction (Braiterman et al. 2000). Moreover, the strong limitation in terms of price and development time effectively prohibits usability testing in the classical sense, conducted by experienced testers in sophisticated laboratories as illustrated in cf. (Fath *et.al.* 1994, Rohn 1994, Rubin 1994).

There are a broad variety of suggestions that aim at overcoming the challenges and time pressure that are typical for web-site development. These include iterative testing to spur creativity and inform the design process (Braiterman et al. 2000), user-centered research to improve the designers' understanding of the prospective users (Anderson 2000), and life stories as a means to surface implicit requirements (Broadbent & Cara 2000). The focus in all of these proposals is on analysis and design. The implied lack of focus on usability issues and practical skills with usability testing reflects a potential barrier for universal access of information on the Web, cf. (Sullivan & Matson 2000).

This article reports from an empirical study of the potential for supporting universal access through dissemination of fundamental usability engineering skills. Our aim is to teach a simple approach to usability testing to people with an interest in information technology but without formal education in software development or usability engineering, and to do it in less than a week.

2. Usability testing approach

The simple approach to usability testing was developed through a course that was part of a curriculum for the first year at Aalborg University, Denmark. The overall purpose of the course was to teach and train students in fundamentals of computerized systems with a particular emphasis on usability issues. The course included ten class meetings, each lasting four hours that was divided between two hours of class lectures and two hours of exercises in smaller teams. The ten class meetings compromised the following topics: #1 introduction and computer networks; #2 usability issues: guidelines and measurement; #3 usability testing: think-aloud protocol; #4 usability testing: questionnaires; #5 computer architecture; #6 usability testing: data analysis; #7 usability issues: techniques and documentation; #8 web-site: usability; #9 web-site: orientation and navigation; and #10 web-site: visual design. Thus all class meetings, except number one and five, addressed aspects of usability and web-sites. The purpose of the exercises was to practice selected techniques from the lectures. In the first four class meetings, the exercises made the students conduct small usability pilot tests in order to train and practice their practical skills. The last six exercises were devoted to conducting a more realistic usability test of a web-site.

A number of techniques for usability testing were presented in the course. The first of two primary techniques was the think-aloud protocol, which is a technique where test subjects are encouraged to think aloud while solving a set of tasks by means of the system that is tested, cf. Nielsen (1993). The second technique is based on questionnaires that test subjects fill in after completing each task and after completion of the entire test, cf. Spool (1999). Additional techniques such as interviewing, heuristic inspection, etc. were also presented to the students.

The tangible product of a usability evaluation is a usability report, that identifies usability problems of the product, system, or web-site in question. It was suggested that a usability report should consist of an executive summary (1 page), description of the approach applied (2 pages), results of the evaluation (5-6 pages), and a discussion of methodology (1 page). It was also emphasized that the problems identified should be categorized, at least in terms of major and minor usability problems. In addition, a report should include all data material collected such as log-files, tasks for test subjects, questionnaires etc. A prototypical example of a usability report was given to the students for inspiration.

3. Empirical study

We have made an empirical study of the usability approach that was taught to the students. This study involved 36 teams of first year university students who used the approach to conduct a usability evaluation of the email services at the Hotmail web-site. The 36 teams consisted of 234 students in total, of which 129 acted as test subjects, from such diverse educations as architecture and design, informatics, planning and environment and chartered surveyor. These studies are all part of a natural science or engineering program at Aalborg University. Figure 1 describes the 36 teams that participated in the study.

Team size Average	Team size Min / Max	Number of test subjects Average	Number of test subjects Min / Max	Age of test subjects Average	Age of test subjects Min / Max
6.5	4 / 8	3.6	2 / 5	21,2	19 / 30

Figure 1: Team sizes and test subject data of the teams that participated in the empirical study.

Each student team was required to apply at least one of the two primary techniques, and they were allowed to supplement this with other techniques according to their own choice. With both techniques, the team should among themselves choose a test monitor and a number of loggers (they were recommended to include two loggers), who should examine the system, design task assignments for the test subjects, and prepare the test, cf. (Rubin 1994). The rest of each team acted as test subjects, and the web-site used for testing was kept secret to them.

Each team was given a very specific two-page scenario stating that they should conduct a usability test of the Hotmail web-site (www.hotmail.com). The scenario included a list of features that emphasized the parts of Hotmail they were supposed to test. Each usability test session was planned to last approximately one hour. Due to the pedagogical approach of the university, each team has its own office. Most teams conducted the tests in this office, which was equipped with a personal computer and Internet access. After the test, the entire team worked together on the analysis and identification of usability problems and produced the usability report.

The usability reports were the primary source of data for our empirical study. All reports were analyzed, evaluated, and marked by both authors of this paper. First, we worked individually and marked each report in terms of 17 different factors. The markings were made on a scale of 1 to 5, with 5 being the best. Second, all reports and evaluations were compared and a final evaluation on each factor was negotiated. In this article, we focus on the following five factors: #1 the planning and conduction of the evaluation, #2 the quality of the task assignments, #3 the clarity and quality of the problems listed in the report, #4 the practical relevance of these problems, and #5 the number and relevance of the usability problems identified. In order to provide a basis for comparison, we have also employed usability reports produced by teams from eight professional laboratories, cf. (Molich 1999). These teams have evaluated the same web-site according to the scenario mentioned above, and their reports were analyzed, evaluated, and marked through the same procedure as the student reports.

The specific conditions of this study limit its validity in a number of ways. First, the environment in which the tests were conducted was in many cases not optimal for a usability test session. In some cases, the students were faced with slow Internet access that influenced the results. Second, motivation and stress factors could prove important in this study. None of the teams volunteered for the course (and the study) and none of them received any payment or other kind of compensation; all teams participated in the course because it was a mandatory part of their curriculum. This implies that students did not have the same kinds of incentives for conducting the usability test sessions as people in a professional usability laboratory. Finally, the demographics of the test subjects are not varied with respect to age and education. Most test subjects a female or a male of approximately 21 years of age with approximately the same school background and recently started on a design-oriented education. The main difference is the different curricula they follow.

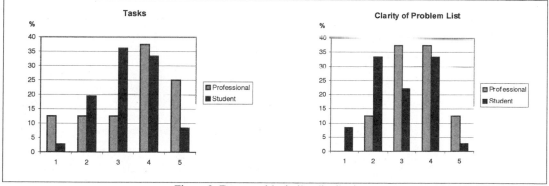

Figure 2: Factors with similar distributions

4. Usability test results

The results of the study are illustrated in figure 2, 3 and 4 where factors with similar and different distributions are shown. The first of these factors is the task factor where the distribution of markings is illustrated on the left hand side of figure 2. This factor measures the relevance of the tasks, the number of tasks, and the extent to which they cover the areas specified in the scenario. The student teams cover all five elements of the scale with a few at the top and at the bottom. The average is 3.3 for all student teams, which is a little above the middle. The professional laboratories score almost the same result, with distribution across the whole scale and an average of 3.5. This is by no means impressive for the professionals; thus the comparable result produced by the students is rather due to a general low quality of the tasks defined by the professionals.

The second factor with a broad distribution is the clarity of the problem list, cf. the right hand side of figure 2. This factor measures how well each problem is described, explained, and illustrated and how easy it is to gain an overview of the complete list of problems. Here, the student teams are distributed mainly around the middle of the scale with a few at the top and at the bottom. Their average is 2.9, just below the middle. Compared to the professional laboratories, they are also doing quite good. The professionals are distributed from 2 to 5 with an average of 3.5. Again, this is not impressive for professional laboratories.

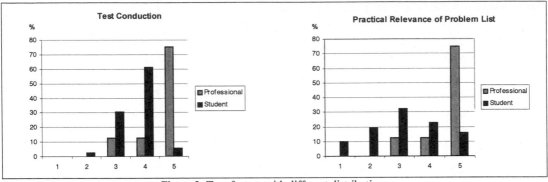

Figure 3: Two factors with different distributions

The results also reveal factors with a more biased distribution. The first of these is the conducting of tests, which reflects how well the tests were planned, organized, and carried out, cf. The left hand side of figure 3. The majority of student teams score 4, which indicates well-conducted tests with a couple of problematic characteristics. The average on 3.7 also reflects the general quality of the test processes. The professional laboratories score an average of 4.6 on this factor, and 6 out of 8 score the top mark. This is as it should be expected because experience will tend to raise this factor.

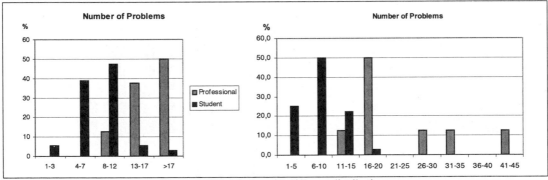

Figure 4: Third factor with different distributions

The second factor that exhibits a difference is the practical relevance of the problem list, cf. The right hand side of figure 3. The student teams are almost evenly distributed on the five marks of the scale, and their average is 3.2. Yet when we compare these to the professional laboratories, there is a clear difference. The professionals score an average of 4.6 where 6 out of 8 laboratories score the top mark. This difference can partly be explained from the experience of the professionals in expressing problems in a way that make them relevant to their customers. Another source may be that the course has focused too little on discussing the nature of a problem; it has not been treated specifically with examples of relevant and irrelevant problems.

A key aim in usability testing is to uncover and identify usability problems, and this is the third factor in this study with differences, cf. figure 4. The student teams are on average able to find 7.9 problems. They find between 1 and 19 problems with half of the teams finding between 6 and 10 problems. Thus the distribution seems to be reasonable, but compared to the professionals there is a clear difference. The average for the professionals is 23.0 problems identified; only one of them scores in the same group as a considerable number of student teams; that is between 11 and 15 problems. Only one student team identified a number of problems that is comparable to the professional laboratories.

5. Conclusions

Significant usability problems in web-site interfaces are a barrier for universal access. Thus a low level of skills in usability engineering is likely to prohibit moves towards the ideal of universal access. This article has described a simple approach to usability testing that aims at supporting the ideal of universal access. The core aim of this approach is to quickly teach fundamental usability skills to people without any formal education in software development and usability engineering. Whether this approach is practical has been explored through a large empirical study where 36 student teams have both learned and applied the approach within 40 hours or a week's work.

The student teams gained competence in two important areas. They were able to define good tasks for the test subjects, and they were able to express the problems they found in a clear and straightforward manner. Overall, this reflects competence in planning and writing. The students were less successful when it came to the identification of problems, which is the main purpose of a usability test. Most of the teams found too few problems. It was also difficult for them to express the problems found in a manner that would be relevant to a practicing software developer.

The simple approach to usability testing did provide the students with fundamental skills in usability engineering. Thus it is possible to have usability work conducted by people with primary occupations and competencies that are far away from software development and usability engineering. We see the approach as a valuable contribution to universal access as emphasized here: "Organizations and individuals stuck in the hierarchies and rigidity of the past will not foster what it takes to be successful in the age of creativity, the age of the user, and the age of the Internet economy"(Anderson 2000).

References

Anderson, R. I. (2000). Making an E-Business Conceptualization and Design Process More "User"-Centered. interactions, 7, 4 (July–August), 27-30.

Braiterman, J., Verhage, S., & Choo, R. (2000). Designing with Users in Internet Time. interactions, 7, 5 (September-October), 23-27.

Broadbent, S. & Cara, F. (2000). A Narrative Approach to User Requirements for Web Design. interactions, 7, 6 (November–December), 31-35.

Fath, J. L., Mann, T. L., & Holzman, T. G. (1994). A Practical Guide to Using Software Usability Labs: Lessons Learned at IBM. Behaviour & Information Technology, 13, 1 & 2, 25-35.

Molich, R. (1999). Comparative Usability Evaluation - CUE-2 Reports. http://www.dialogdesign.dk/cue.html

Nielsen, J. (1993). Usability Engineering. Morgan Kaufmann Publishers, Inc., San Francisco, California.

Rohn, J. A. (1994). The Usability Engineering Laboratories at Sun Microsystems. Behaviour & Information Technology, 13, 1 & 2, 25-35

Rubin, J. (1994). Handbook of Usability Testing – How to Plan, Design, and Conduct Effective Tests. John Wiley & Sons, Inc., New York.

Spool, J. M., Scanlon, T., Schroeder, W., Snyder, C., & DeAngelo T. (1999) Web Site Usability – A Designer's Guide. Morgan Kaufmann Publishers, Inc., San Francisco, California.

Sullivan, T. & Matson, R. (2000). Barriers to Use: Usability and Content Accessibility on the Web's Most Popular Sites. In: Proceedings of Conference on Universal Usability, November 16-17, Washington, ACM, 139-144.

Ubiquitous Multimedia Services with XML

Petri Vuorimaa, Juha Vierinen, Jussi Teirikangas

Telecommunications Software and Multimedia Laboratory
Helsinki University of Technology, Finland, P.O. Box 5400, FI-02015 HUT
email Petri.Vuorimaa@hut.fi

Abstract

New multimedia capable consumer devices like 3rd generation mobile phones, portable DVD players, and digital television set-top-boxes are becoming more common. Although, the devices have very different sizes and purposes, they have two common features. First, they are all connected to Internet, and thus most of services run on top of a www browser. Second, the new devices are aimed for mass markets, and thus they should be as easy to use as possible. Unfortunately, no design conventions or general usability guidelines exist, so far. In this paper, we study next generation multimedia devices. As a test environment, we use the X-Smiles XML browser. We show how a multimedia ice hockey service can be used on different kinds of multimedia devices.

1. Introduction

Extensible Markup Language (XML) (Bray et al., 2000) is the next generation www markup language. The main advantage of XML is that the content is separated from the presentation. The content is modelled with XML, while the user interface is defined with an associated Extensible Stylesheet Language (XSL) (Adler et al., 2000) file and ECMAScripts (ECMA-262, 1998). The services are run on top of a XML browser. Vuorimaa, Ropponen, and von Knorring (2000) have developed an XML browser called X-Smiles. The browser has been implemented using Java language. It supports many XML related specifications, e.g., XSL Formatting Objects (XSL FO) (Adler et al., 2000), Synchronised Multimedia Integration Language (SMIL) (Hoschka et al., 1998), Scalable Vector Graphics (SVG) (Ferraiolo et al., 2000). Vierinen and Vuorimaa (2001) have implemented X-Smiles Graphical User Interfaces (GUI) for laptop PC and digital television. We are currently implementing GUIs for different kinds of mobile devices.

Our idea is that all consumer devices connected to the Internet could include the X-Smiles browser. Each device has its special GUI, though. Then, the multimedia services coded with XML can be fitted to different kinds of devices simply by changing the associated XSL stylesheet. This creates a ubiquitous access to the multimedia based www services.

In this paper, we describe how the above described environment can be realised. In Chapter 2, we present the X-Smiles XML browser. In Chapter 3, we introduce a demonstration multimedia service. Next, we describe the different GUI versions of the X-Smiles browser. Finally, conclusions are given in Chapter 5.

2. X-Smiles XML browser

The X-Smiles is intended for embedded devices, e.g., PDAs, mobile phones, and digital television set-top-boxes. The idea is that the next generation multimedia devices would have a small operating system (e.g., mobile Linux) and a Java virtual machine. The X-Smiles browser runs directly on top of these. If all multimedia services are implemented using XML then no other system software is needed. Thus, the size of the system software is small. Sivaraman and Vuorimaa (2000) have estimated that the total size of the software is about 20 MB. Thus, the software system can be fitted into various consumer devices.

2.1 Architecture

The architecture of the X-Smiles browser is depicted in Fig. 1. The *XML Parser* parses the XML documents. The *XSL Processor* processes the document according to the related XSL stylesheet. The output is stored into a data structure, which can be accessed via the *Document Object Model (DOM)* (Apparao et al., 1998) *Interface*. The *ECMAScript Interpreter* runs the scripts contained in the XSL stylesheet. The processed documents can be displayed by different *Markup Language Functional Components (MLFC)*. So far, we have implemented MLFCs for source view, document element tree, XSL FO, SMIL, and SVG.

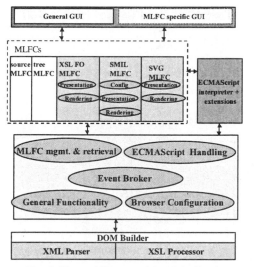

Fig. 1: The architecture of the X-Smiles browser

The basic graphical user interface (GUI) resembles a typical PC user interface, as shown in Fig. 5. The GUI is an independent component connected to the browser using the so-called *bridge* pattern (Gamma, Helm, Johnson, and Vlissides, 1994). Thus, the basic GUI can be easily switched to a special GUI. This can be done in real-time, and thus the XML based content service can be tested with different kinds of GUIs.

The core of the browser ties the different modules together. The *MLFC management and retrieval* unit takes care of loading the appropriate MLFC, which can also be retrieved over the network, although this feature is not used much at the moment. The *ECMAScript Handling* unit co-ordinates the operation of the ECMAScript interpreter. The *Event Broker* unit distributes the events. The *General Functionality* unit controls the GUI, document history, etc. Finally, the *Browser Configuration* unit is responsible for management of the different features of the browser such as the parser and stylesheet processor, home page, etc.

2.2 Basic Operation

When the X-Smiles browser reads a new XML document, it has to do certain steps to process it. First, the browser reads the document source. The processing is done using any XML parser, which uses the Simple API for XML (SAX). The parser processes the XML file node by node and creates different SAX events. The DOM implementation works together with the parser. It takes the SAX events as input and constructs the DOM model of the document accordingly. The next step is to read the XSL stylesheet. First, the browser determines, whether a stylesheet is attached to the XML document. If a stylesheet is associated with the XML file, the browser reads the stylesheet and creates a DOM data structure. Since the XSL stylesheet is also a valid XML document, it can be processed in the same way as the original XML document. The next step is to actually process the stylesheet. The browser calls the XSL processor with the stylesheet and document as parameters. Both are DOM trees as described above. The XSL processor creates a new DOM tree from the XML document based on the instructions given in the XSL stylesheet.

2.3 MLFCs

After the final DOM tree has been constructed, it has to be rendered. Since there are different XML languages available, different rendering modules have to be used. In X-Smiles, this is achieved by using different MLFCs. Each MLFC knows how to render one specific type of XML dialect. So far, we have implemented a MLFC for SMIL, XML FO, and SVG. The SMIL MLFC was the first true MLFC that was developed. The SMIL MLFC supports most of the features of the SMIL version 1.0. SMIL 2.0 is not supported, yet. The Java Media Framework (JMF) (Gordon and Talley, 1999) extended with custom JMF players developed in the project, is used for rendering and synchronisation of both continuous (e.g., audio and video) and static (e.g., text and images) media elements. The XSL FO MLFC is based on the Apache XML FOP Project (http://xml.apache.org/fop/). However, FOP cannot be used in the browser as such, as it provides a way to preview a XSL FO document, but allows no means of user interaction – a very basic feature in any browser. Thus, we have extended the FOP to add user interaction using links and forms.

The SVG MLFC can render vector graphic documents. It uses a modified version of CSIRO SVG Toolkit (http://sis.cmis.csiro.au/svg/), which supports most of the SVG features. There are still several things missing or only partially implemented, since the toolkit is not yet complete. In addition, we have basic MLFCs, which can display source code and tree view. The source code MLFC is useful for debugging. The content of the original XML document, an associated XSL stylesheet, as well as the resulting source code of the processed document can be viewed. The tree view presents the DOM structure of the processed document in a tree-like format, thus providing an easier way to examine the structure of the document.

2.4 ECMAScript Interpreter

The current ECMAScript interpreter used in the X-Smiles is Rhino (http://www.mozilla.org/rhino/), a full Java implementation of the ECMAScript scripting language. ECMAScript provides service developers a convenient method to implement interactivity. The scripts can manipulate the original XML document via the DOM API. Scripts may also call

744

library functions. The scripts are defined in CDATA sections embedded in the XML document or in the XSL stylesheet. The scripts are run either when documents are loaded or in response to user or timer events. For example, a user action, such as clicking a mouse button on an image, might create an event, which in turn activates a function defined using ECMAScript.

3. A multimedia service demonstration

Marttila and Vuorimaa (2000) have developed three multimedia services to test the X-Smiles browser. The first one is a movie theatre service, which allows the user to view information about on-going movies, look for available theatres, and make reservation for tickets. The service contains limited amount of multimedia elements as movie trailers. The second one is distance education service, which contains lecture slides and videos. The videos and lectures and synchronised, and thus the service contains more multimedia. Finally, Tennis multimedia presentation contains text, sound, images, videos, and animations, which are tightly synchronised. Thus, the service is the most demanding. The first two services have been implemented using XML and XSL, while the third service has been implemented using SMIL. The results show that XML can be used as an *interchange format* for mobile multimedia services.

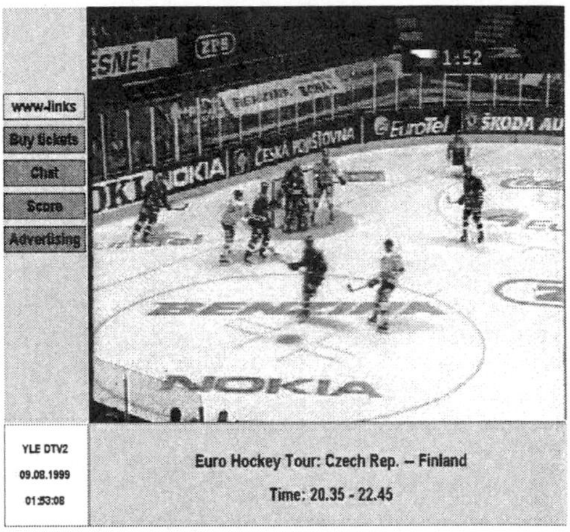

In this study, we chose to use a fourth type of service: sports program. Peng and Vuorimaa (2000) have initially implemented the service for digital television, as shown in Fig. 2. Therefore, the service is based on existing sports program format. Most of the time the viewers use the service to watch the on-going ice

Fig. 2: A screenshot of the original interactive sports Java application for digital television

hockey game. During the breaks of the game – or when the game is boring – the viewer can activate the interactive part of the service by pushing a special button on the remote controller. Then, the viewer can access related www links, buy tickets for the forthcoming games, chat with other viewers, view statistics of the game, participate in sponsor contest, etc. The original GUI is composed of four areas. The main menu is on the upper left side of the screen. The viewer can select the different parts of the service with the 'up' and 'down' buttons of the remote control. The selection is done with the 'ok' button. The lower left corner shows the channel logo, time, and date. The lower right side of the screen is the information area. The content can be scrolled with the 'up' and 'down' buttons. The user can return back to the main menu with the 'back' button of the remote controller. Finally, the on-going ice hockey game is shown in the upper right part of the screen. The content of the service is composed of streaming video and textual information. These two main elements are loosely integrated. Thus, the content can be coded as two separate streams: data and video. The data can be presented as tree, where the main menu is on the top level, and the data of each part of the service is presented as sub tree. Thus, the data can be presented as one XML file. An example of the XML content is shown in Fig. 3.

The main advantage of XML is that it is neutral, since it does not contain any information on how the content should be shown to the user. This information about the GUI is coded in a separate XSL stylesheet, which contains also the related ECMAScripts. The task of the XSL stylesheet is to transform the original XML content so that it also contains formatting information. In the following examples, the formatting information is coded using XSL FO. The transformation is done by using the pattern matching rules of the XSL file. Each rule selects a part of the original XML content and attaches the formatting information to the output, as shown in Fig. 4.

A single XML and XSL file pair can be used to generate several screens of the GUI. Another alternative is to have a separate XSL file for each screen of the GUI. In addition to the formatting information, each GUI screen contains also definitions of how the interactivity is implemented. This information is usually coded as event handlers, which process various keyboard and mouse inputs. In the case of XML, the event handlers are coded with ECMAScript.

```
:?xml version="1.0"
    encoding="UTF-8"?>
:?xml-stylesheet type="text/xsl"
    href="./hockeyDTV.xsl"?>
:!DOCTYPE hockeygui SYSTEM
    "hockeygui.dtd">
<hockeygui>
  <mainmenu>
    <item href="./stats.xml">
      Statistics
    </item>
    <item href="./links.xml">
      WWW Links
    </item>
    <item href="./chat.xml">
      Chat
    </item>
    <item href="./betting.xml">
      Betting
    </item>
  </mainmenu>
  <videostream href="http://
    www.hockey.fi/tvstream.ra"/>
</hockeygui>
```

```
<xsl:template match="link">
  <fo:block color="white"
      font-size="22pt" space-
      after.optimum="3pt">
    <fo:basic-link external-
        destination="{@href}">
      <xsl:value-of select="."/>
    </fo:basic-link>
  </fo:block>
</xsl:template>
```

Fig. 3: An example of the XML data content

Fig. 4: XSL transformation used to define the GUI

4. Results

The objective of this paper is to study how the above-described service can be used in different kinds of multimedia devices. Thus, we developed three versions of the service: PC, mobile, and digital television. The first version of the service is intended for laptop PCs. It acts also as a basic template for the two more specialised versions: mobile and digital television user interface.

All three versions are described in more detail in the following examples. The examples have been implemented using the X-Smiles browser. The target devices (i.e., mobile PDA and digital television) have been implemented as virtual prototypes of the real devices. Thus, all the experiments have been done with a laptop PC, which has a touch display and remote control attached to it. Thus, the GUI can be controlled either with mouse, remote control, or the buttons of the virtual prototype.

Fig. 5: The PC version of the ice hockey service

4.1 Laptop PC

The laptop PC version of the service is shown in Fig. 5. The form of the user interface is similar to the original digital television implementation (cf. Fig. 2). The main difference is that the user interface is controlled with a mouse. Text input is done with keyboard. Thus, the amount of event handlers is limited and most of the interactivity is implemented by using hyperlinks.

4.2 Mobile PDA

The mobile PDA version of the service is shown in Fig. 6. The most limiting factor of the GUI is the size of the screen. Thus, much smaller resolution of the video has to be used. The user interface is controlled either with stylus (i.e., touch screen of the device) or the buttons of the virtual prototype. In practise, most of the interactivity can be implemented with hyperlinks, but special event handlers are also required for buttons. The main task of the event handlers is to control the location of highlight point, and switch the screen when 'ok' or 'back' buttons are pushed.

746

4.3 Digital Television

The digital television version of the service is shown in Fig. 7. The user interface is very similar to the original example (cf. Fig. 2). The user interface is controlled by a remote control. Thus, special event handlers are required for button input. In practise, the event handlers are similar to event handlers of the mobile PDA device.

5. Conclusions

In this paper, we have shown how multimedia services can be adapted to different kinds of devices. The results show that XML is a convenient interchange format for the future multimedia services. The content can be coded in neutral format and adapted to various services by switching the related XSL file. In practise, the modifications required to each device version were small. The basic function of the GUIs were the same. The different video resolutions had the biggest effect. Changes to the formatting were limited. The mobile PDA and digital television devices require also changes to the input processing. Fortunately, the event handlers were similar. As a conclusion, we expect that XML can effectively be used for adapting the content for multidevice services. The basic PC user interface can be used as template for other devices. This can be done easily especially, if the user interface has simple menu structure. The usability aspects of such GUI should be studied further, though. In the above examples, the input and output devices were rather traditional. At the moment, we are also experimenting on using multimodal user interfaces. Voice control and output is especially interesting in mobile services. To study the use of voice, Teppo and Vuorimaa (2001) have integrated VoiceXML functionality to X-Smiles browser.

Fig. 6: The PDA version of the ice hockey service

Fig. 7: The digital television version of the service

Acknowledgement
The X-Smiles browser was originally developed in a student software project during 1998-99. After that, the development has continued in the GO-MM project. We would like to thank all the participants. The X-Smiles XML browser is available at http://www.x-smiles.org.

References

Adler, S. et al. (2000). Extensible stylesheet language (XSL) version 1.0. *W3C Working Draft*.

Apparao, V. et al. (1998). Document object model (DOM) level 1 specification - version 1.0. *W3C Recommendation*.

Bray, T. et al. (2000). Extensible markup language (XML) 1.0 (2nd ed.). *W3C Recommendation*.

ECMA-262. (1998). ECMAScript language specification. *European Computer Manufacturers Association (ECMA)*.

Ferraiolo, J. et al. (2000). Scalable vector graphics (SVG) 1.0 specification. *W3C Candidate Recommendation*.

Gamma, E., Helm, R., Johnson, R., Vlissides, J. (1994). *Design Patterns: Elements of Reusable Software*, Addison-Wesley.

Gordon, R., Talley, S. (1999). *Essential JMF: Java™ Media Framework*, Prentice Hall.

Hoschka, P. et al. (1998). Synchronized multimedia integration language (SMIL) 1.0 specification. *W3C Recommendation*.

Marttila, O., Vuorimaa, P. (2000). XML based mobile services. *8th Int. Conf. in Central Europe on Computer Graphics, Visualization, and Interactive Digital Media, WSCG'2000*, Czech Republic.

Peng, C., Vuorimaa, P. (2000). Development of java user interface for digital television. *8th Int. Conf. in Central Europe on Computer Graphics, Visualization, and Interactive Digital Media, WSCG'2000*, Czech Republic.

Sivaraman, G., Vuorimaa, P. (2000) Compact windowing system for mobile devices. *2nd Int. Symp. Mobile Multimedia Systems & Applications, MMSA2000*, Delft, The Netherlands, 134-141.

Teppo, A., Vuorimaa, P. (2001). Speech interface implementation for XML browser. accepted to *2001 Int. Conf. Auditory Display*, Espoo, Finland.

Vierinen, J., Vuorimaa, P. (2001). A browser user interface for digital television. *9th Int. Conf. in Central Europe on Computer Graphics, Visualization and Computer Vision, WSCG'2001*, Czech Republic.

Vuorimaa, P., Ropponen, T.,, von Knorring, N. (2000). X-Smiles XML browser. *2nd Int. Workshop Networked Appliances, IWNA'2000*, New Brunswick, NJ, USA.

Universal Web Approach to Web Contents for a Company Web Site

Kazuhiko Yamazaki

Human Interface Design Department, IBM Japan Ltd.,
242-0001, 1623-14 Shimotsuruma, Yamato, Kanagawa, Japan
kazkaz@jp.ibm.com

Abstract

The purpose of this study is to explore a design methodology for a company Web site from a universal design viewpoint. This paper describes a proposal for a universal Web site design approach, which consists of universal browser, universal content, and universal transcoding. This paper focuses on the universal content and how it includes various adaptations for items such as the layout, images, terminology, user interface, voice output, and a consistent Web structure. To develop a Web site based on this approach, a design method is proposed using XML, and our description includes design templates for each user segment and a unified database with user information. After the approach was defined, experiments were conducted for a company Web site to evaluate the proposed design approach and philosophy. These results indicate that the proposed approach and methodology can help to make a company Web site with a consistent design perspective.

1. Introduction

This paper focuses on a design method suitable for Web page design that is usable by all people based on a universal design approach. Universal design means the design of products and environments usable by all people, to the greatest extent possible, without the need for adaptation or specialized designs [1-4]. All people include all ages, all cultures, all locations, and all disabled people, but universal design does not mean that all people will be able to use all products or services. Some severely disabled individuals will still need specific modifications [5]. The goal of universal design is not just to eliminate physical barriers but also to eliminate mental barriers, considering economic factors such as availability, cost, and price.

Recently, the rapid spread of the Web helps people to get information and send information easily. It also helps many people to participate in social activities. For example, people with physical impairments and senior citizens have been able to easily get information by browsing the Web. Without accessing Web, there is some information that is not easy to find elsewhere. For example, recently students who are looking for recruiting information are required to access the Internet to get information—paper versions of some of the materials are unavailable. Considering this situation, it is important to adopt a universal design approach for Web access. To adopt universal design principles for the Web [6-9], several studies have been performed and approaches have been developed, such as special software on users' PCs, making design guidelines [10], and studying transcoding software on the server [11].

In reality, it is not easy to adopt a universal design approach to the Web because universal design means to cover a wide range of users and Web content consists of self-contained pages. It is very difficult for Web designers to find out answers to questions such as: What are the design requirements, how are ideas and designs evaluated; and where is a good place to provide adaptations—at the users' browsers, at the server side, or in advance from the designers' side. The situation is very complicated, making it hard for Web designers to adopt universal design principles. Many designers are waiting for concrete proposals of simple unifying concepts leading to the adoption of universal design principles for the Web. This report proposes such a universal Web approach to provide universal design principles for the Web, including "universal browser", "universal transcoding", and "universal content". For this paper, the author focuses on universal content and a detailed proposal for the implement of suitable standards.

2. Universal web

2.1 Proposal for a universal web

The author proposes a structure for a universal Web that will be usable by all people. As shown in Figure Fig.1, the proposed universal Web design has three basic concepts to make the Web usable by all people: universal browser, universal transcoding, and universal content. The universal browser part of the approach involves adaptations at the user side by installing special software—user-adapted browsers that can mask some differences. The universal content part of the approach involves adaptations at the content side by the designer. The universal transcoding part involves a solution between the users and the designers, where the user-required customizations are inserted by transcoding software on the servers. All three concepts are not strictly required to make a universal Web, and if even one of the three approaches is sufficiently flexible in adapting content, other approaches may not be required.

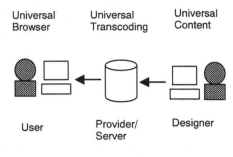

Fig. 1. Structure of the universal Web

The universal browser concept is to adapt the Web content for all people by using software on the client PCs which users are using to access the Web. This software adapts normal Web content so those users are able to access the Web easilyWeb . For example, voice-reading software is one example of these approaches [10]. Universal transcoding is a concept to adapt the Web content for all people by automatically making modifications on a server. Without any special preparation by the users or designer, the users are able to access the Web in a way that is adapted to each user group. Universal content is an approach to adapt the Web for all people from the designer side. The designer prepares several Web pages for each user group. Without any special preparation, users are able to access Web content that is adapted to various user groups according to their abilities and interests.

2.2 Target User

The goal of making the Web accessible for all people is not easy to implement. To clarify the user profiles, the author selected the following eight user groups as representative Web users—satisfying the needs of these eight user groups should satisfy the requirements for a universally accessible Web.

- Novices: People who are beginners to accessing the Web and need support or help.
- Experienced people: People who are able to access the Web without any support and help. Most existing Web pages are categorized as targeted for these users.
- Elementary school students: Young children are interesting in the graphic information and have limited knowledge.
- High school or junior high school students: Students at the high school or junior high school levels are primarily interested in information which is related to their various interests.
- The elderly: Older people may have many of the same limitations and may benefit from the accessibility features and functions provided to accommodate people with disabilities.
- People with physical impairments: People with motor or mobility impairments require switches, latches, and controls that are easy to access and manipulate, diskettes and media that are easy to insert and remove, and alternate input/output capabilities.
- People with visual impairments: People who are blind or visually impaired need audio or tactile access to information from the computer. They also need the ability to navigate screens using only a keyboard, audio device, or tactile device.
- People with hearing impairments: People who are deaf or hard of hearing require visual representations of auditory information that the computer or workstation provides.

3. Universal content

3.1 Adaptation by universal content

This study focuses on universal content for user adaptation. The user adaptation involves design activities to adapt for each user group from general information and basic material. These design activities include several kinds of adaptations as follows:

- Image adaptation: Images are closely related to users' feelings. For example, to change graphics or to change background and text colors is one of the ways for image adaptation. For example, to adapt for students, a designer changes to modern and dynamic graphics from more general graphics, and to adapt scenarios for older people, a designer changes to special graphics, which are easy for the elderly to understand.
- Layout adaptation: Layout is related to information priority, ease of use, and the general feeling of a page. A Web designer provides layouts for user groups to make it easy to find the content which each user is interested in.
- Interface adaptation: Interface adaptation includes designs for links, buttons, and text sizes, which are very closely related to user integration and ease of use for each user group. For example, to adapt content for the elderly, a designer changes to larger text from general text sizes, including larger buttons (which are made as graphic objects).
- Content adaptation: Content adaptation is to change the content itself for each user group. For example, to adapt for students, the content for news is selected from the students' interest categories such as music, movie, sports, fashion, etc.
- Terminology adaptation: Text on the Web is changed to adapt for each user group using specialized dictionaries. For example, to adapt for children, technical words are changed to general words more easily understood.
- Structural adaptation: Structural adaptation is to change the intended user linkage for each user group. For example, to adapt for PC novices, a designer might change to a simple step-by-step flow of links to make the pages easier to access.
- Audio adaptation: Audio adaptation is to change audio information for each user group. This includes both audio data and adaptations for reading software. For example, to adapt for people with hearing impairments, a designer can change pages to eliminate the sounds and add visual tokens instead. Using audio to adapt for people with visual impairments, a designer can change pages to eliminate large graphics and add detailed text descriptions for graphics.

Table 1. Relationships between User Adaptations and User groups
(A means close relationship between user and method of user adaptation,
B means moderate relationship, C means limited relationship)

User Adaptation	Novice	Experi-enced	Element-ary	High school	Elderly	Physical Impair-ment	Visual Impair-ment	Hearing Impair-ment
Images	A	B	A	A	A	C	C	C
Layout	B	C	B	B	A	A	B	C
Interface	A	B	B	B	A	A	B	B
Content	A	B	A	A	A	C	C	C
Terminology	A	B	A	B	A	B	B	B
Structure	A	B	A	B	A	B	A	B
Audio	C	C	C	C	B	C	A	C

3.2 Realizing universal content

Based on the user groups as shown in Section **2.2**, and the user adaptations as shown in Section **3.1**, the author studied design approaches for universal content. As shown in Table 1, for each user group, several user adaptations are related. For example, image adaptation is closely related to adapting for novices, students, and the elderly. This table was created from discussions with experienced Web designers based on their experiences. Based on this table, designers should use several adaptations to design each page so that it is adapted for each user group. However, it is not practical for each designer to make eight pages instead of the single current page. To realize universal content, a more convenient way is required, and XML and a user characteristic database offer one approach.

There are two ways for users to access adapted pages: user selection and automated selection by client software. User selection is when users select one of the user groups on the entry pages by themselves. Automated selection uses client software that selects one of the user groups automatically based on a user profile, which each user has already input. For automated selection, users need client software, so user selection is a more practical way to access adapted pages.

3.3 Proposed structure of universal content

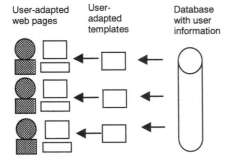

User-adapted web pages   User-adapted templates   Database with user information

Figure 2. Structure of universal content

The author focuses on using XML technology in order to create universal content. As shown in Fig.2, the proposed structure of universal content consists of user-adapted Web pages, user-adapted templates based on XML technology, and a database with user information. If the designer prepares eight templates, one for each user group, and links the content with user information using the database, eight adapted Web pages can be prepared automatically. For this purpose, a special server is not required. For the database, user information about the content and about each object such as a text and a graphics, needs to be prepared, and based on this information, selected information is combined with selected templates for each user group. As shown in Table 2, the relationships between user groups and methods for user adaptation were studied. This table was created from discussion with an experienced Web designer. User-adapted templates can contribute to user adaptations from the layout, interface, and structural viewpoints. The database with user information contributes to user adaptation from the image, content, and terminology viewpoints. Audio adaptations are related to both the templates and database.

Table 2. Methods for user adaptation
(A means close relationship, B means moderate relationship, C means limited relationship)

User Adaptation	Adaptation by template	Adaptation by database
Image	B/ change image data, change color	A/select adapted image file
Layout	A/ change basic layout	C/
Interface	A/ change interface such as link, button	C/ select adapted interface data
Contents	C/	A/ select adapted text
Terminology	B/ template with adapted text	A/ contents with adapted text
Structure	A/	C/
Audio	B/ audio information on template	B/ audio information on content

3.4 Experiment

To evaluate the proposal for universal content, the author prepared a detailed structure for the top page of a company Web site based on the proposed structure for universal content. An experienced Web designer created eight Web page versions for the top page of the company Web site by using XML technology.

Examples appear in Figs. 3 and 4, showing how this page is adapted for senior citizens using the template for the elderly and the database with information about that group. The layout is updated from the template with news in a large font on right side instead of another large graphic. The interface is adapted using the template to have bigger buttons and link areas. The actual content is adapted using the database and news items are picked for senior citizens. The terminology selected according to the template is adapted for the elderly to make it easy to understand. The structure is adapted according to the template to eliminate a second layer within the page. For audio, to support reading software, several tags and descriptions were added as recommended by the template.

The result of this experiment, the universal Web approach, offers many possibilities to use XML technology with adaptive templates and a database with user information to easily make user-adapted Web pages.

4. Discussion

The author has proposed a design concept for the Web based on a universal design approach for all users, and proposed a universal Web structure. In addition, a method to structure the universal content with XML technology was proposed. After creating the proposal, eight versions of the top page for a company Web site were created. The preliminary work indicates that the universal content approach helps designers to make several adapted Web pages easily. As a next step, author is planning to prepare real Web content to evaluate the performance of the XML system and its usability and effectiveness for each of the user groups, as well as to learn more about the design process itself.

Fig 3. Web for experienced people   Fig 4. Web for the elderly

Acknowledgements

For help with this paper, I would like to thank Dr. Kotaro Hirano of Kobe Design University and Manabu Sasajima of IBM Japan.

References

[1]   http://www.design.ncsu.edu/cud/, The Center for Universal Design, NC State University, 1997.
[2]   Kose S., *Design for the 21st Century*, Toshibunka, 1998.
[3]   http://www.jp.ibm.com/lead/0719.html, Universal design approach to products (in Japanese), IBM Japan Ltd., 2000.
[4]   Yamazaki K., Universal design approach to portable computers (in Japanese), Technical report Vol. 2000 No. 81, pp. 33-38, Information Processing Society of Japan, Tokyo Japan, August 2000.
[5]   Covington G.A., Hannah B., *Access by Design*, VAN Nostrand Reinhold, 1997.
[6]   Stephanidis C., *User Interfaces for All*, Lawrence Erlbaum Associates, 2001.
[7]   Shibata E., Yamazaki K., Yamazaki M., Design approach to company Web site from universal design viewpoint −1 (in Japanese), *Human Interface Symposium*, Human Interface Society of Japan, 1998.
[8]   Yamazaki M., Shibata E., Yamazaki K., Design approach to company Web site from universal design viewpoint −2 (in Japanese), *Human Interface Symposium*, Human Interface Society of Japan, Osaka, Japan, October 1999.
[9]   Yamazaki K., Yamazaki M., Shibata E., Design approach to company Web site from universal design view point −3 (in Japanese), *Human Interface Symposium*, Human Interface Society of Japan, Tsukuba, Japan, September 2000
[10] IBM Corporation, IBM accessibility guidelines. Available at http://www.austin.ibm.com/sns/guidelines.htm, 1999.
[11] Asakawa C., A study of nonvisual Web access using, voice input and output (in Japanese). *Technical report of IPSJ LSP-17-10*, pp. 53-58, Yamanashi Japan, 1997.

PART 8

APPLICATIONS

HCI applications for professional driver seats and their impact to driver's health and efficiency

Angelos Amditis[1], Evangelos Bekiaris[2], Simon Sartor[3]

1. Institute of Communication and Computer Systems (ICCS)
National Technical University of Athens
9, Iroon Politechniou str., 15773 Zografou , Athens, Greece
Tel: +301 7722398, Fax: +301 7723557, e-mail: angelos@esd.ece.ntua.gr

2. Aristotle University of Thessaloniki, Transport Research Laboratory,
61, Ermou st., 54623 Thessaloniki, Greece
Tel: +3031 241078, Fax: +3031 256037, e-mail: trnspcon@compulink.gr

3. ISRINGHAUSEN GmbH & Co.KG, An der Bega 58, 32657 Lemgo, Germany
Tel.: +495261 210 481, Fax.: +495261 210 310, e-mail: Simon.Sartor@isri.de

Abstract

Low frequency (4 to 6 Hz) cyclic motions, like those caused by a vehicle's tires hitting the road, can put the human body into resonance. Just one hour of seated vibration exposure can cause muscle fatigue, weaken the soft tissue and make a worker more susceptible to back injury. Truck drivers, agricultural machinery operators, travelling sales representatives, subway operators, tractor drivers and construction vehicle operators are common victims of back problems. Truck drivers are four times more likely to have a herniated disk and tractor drivers with more than 700 driving hours per year have 61% pathologic changes in the spine. Thus, one million back injuries occur per year in USA alone, costing around $90 billion, whereas there are 5,4 million Americans with low back pain disabilities. To face this serious problem, SAFEGUARD project is co-funded by EU (DG RESEARCH), as part of Quality of Life program, aiming at integrating assistive technology in ergonomic seat design with modern sensor and in-vehicle informatics technology, to develop a new generation of adaptable, ergonomic and yet affordable heavy vehicle seats. To reach this goal, a 3D anthropometrical FEM and VR model of professional driver is built, a new modular seat assessment methodology is developed and several VR mock-ups are realized and tested in 10 Pilots Europe-wide.
One of the key elements of the new seat characteristics is that it will include innovative user interface elements (display screens, sounds, haptic elements) to control the various seat functionalities. Furthermore, the seat will be autonomously adaptable to key driver's characteristics and even to the traffic scenario, based upon stochastic optimization algorithms.
Following the "design for all" approach, SAFEGUARD Consortium bases the seat requirements and design on the needs not only of healthy but also of drivers with back problems. For this reason E&D users will participate also in the project tests to guarantee the optimum performance of the system for all users. Furthermore, the developed know-how is expected to have significant secondary applications in the field of ergonomic chairs and wheelchairs for mobility impaired people, and especially seat adaptations for E&D drivers and travelers.

1. Introduction

Sitting motionless for long periods of time as well as experiencing vibrations while driving lead to increased back injuries among professional drivers, like truckers, fork-lift operators, tractor drivers, taxi drivers and others. The lumbar spine in the seated human flattens as compared to standing, resulting in a 65% increase in disc pressure [Anderson et al.]. Vibrations in the range of 4-6 Hz increase the risk of disk herniation three-fold in people who spend more than half their workday driving [Pope et al.]. Other studies indicate that static seated posture and vehicular vibration lead to fatigue as result of increased postural muscle use.

Occupations that involve driving automobiles, motorcycles, buses, tractors, trucks and heavy construction and agriculture machinery are present more frequently in patients with **low back pain** than in a reference group without back pain. Low back pain is the leading major cause of industrial disability in those younger than 45 years [Brendstrup et al., Boshuizen et al.].

Agricultural tractors and earth moving machinery in particular are responsible for some of the most common, prolonged and severe occupational vibration exposures among civilians. Recognition of the potential hazards has resulted in the need to work towards the development of standards concerned with both the vibration transmitted by the seats of these vehicles and the vibration exposure of the vehicle occupants. Although international standards have been defined in the past for agricultural wheeled tractors and field machinery, with regards to measurement of whole body vibration of the operator (Standard 5008 – 1979) and the measurement of the transmitted vibration (Standard 7096 - 1982), more work is required for the definition and evaluation of a method for the measurement, evaluation and provision of an acceptance level for the whole-body vibration transmitted through the seat to an operator during simulated vibration. For each of four classes of earth moving machinery (including new technology combine harvesters, scrapers, loaders, crawlers and graders) an acceleration power spectral density function needs to be redefined. Several investigators reveal that long-term exposure to whole-body vibrations can induce degenerative changes in the lumbar spine. However, there exists little consensus on the various risk factors and the magnitude of their causal impact on back pain for the workers of agricultural machinery.

Currently on-line production of car-vans-tractors does not take into account the above particular needs of the professional drivers with respect to short-term effects to their driving ability as well as to long-term effects to their health. Hence, space allocation, ergonomics of primary and secondary control elements, inappropriate task identification and vibration related discomfort leads to drowsiness, fatigue, inappropriate driving behaviour but also repetitive strain syndromes, back and shoulder pain.

Seat design can play an important role in how much postural musculature is recruited and ensuing effects of static seating and vibration. Both health and safety issues are involved in this intricate consideration. A smart seat design requires prior data regarding the balance of muscle fatigue and alertness decrement. The lack of sufficient data may hold back vital improvements in proper seat design. Furthermore, the design of a new and reliable methodology for vehicle seat assessment in terms of short and long-term impacts to driver's health as well as driving behaviour will help professionals to choose the best options for their work characteristics and remain healthy throughout their working life. Improvement of performance and vibration related pain would decrease the overall cost of sustaining risk free driving, and increase the living standard of the people who drive. In parallel, embattling fatigue and drowsiness through comfort enhancement will contribute to the enhancement of road safety.

To face the above problem, a project called **SAFEGUARD** has been established, co-funded by DG Research within the Quality of Life programme. This project involves a pan-European Consortium of 12 participants from 7 European countries, encompassing, among others, car manufacturers (Fiat), seat manufacturers (IsringHausen), bioengineering laboratories (KUL, COAT), ergonomers (IAO, ICCS) and specialised institutes in developing technological aids for alleviating mobility problems and disabilities (AUTh).

Among the objectives of SAFE-GUARD the following should be mentioned:
- To establish a clear aetiological classification and identification of the occupational hazards and performance limitation factors that relate to the agricultural workers and professional drivers.
- To develop a new reliable, cost-effective and modular methodology to assess the health impact of vehicle seats to professional drivers of different types of agricultural vehicles as well as tractors, trucks, small vans and cars.
- To develop an optimum concept seating system, with advanced comfort, easily adaptable to the user's specific anthropometrical needs, for drivers of combine harvesters and trucks.
- To develop an adaptable seat profile and UI that will equally well serve young and older users and will support the needs of those that already face back pain problems.
- To establish guidelines with regard to car-van-tractor-agricultural machinery vibration related problems of the human spine and upper body and draft standards on how to protect the driver.
- To provide a designer's checklist to promote the application of those guidelines to the maximum extent to different types of existing vehicles.

2. Problems in current UIs of professional drivers seats

A professional seat offering optimum comfort is certain to be a rather complex one in terms of support functionalities. The following list provides an overview of the possible seat functionalities to be controlled by the driver. The selected properties presented below reflect many years of experience of SAFEGUARD partners and are currently under thorough revision during the user needs analysis of the project.

Truck Seat
– vertical shock absorber adjustment
– horizontal suspension system (mechanical), on/off
– integrated Pneumatic-System to fit backrest to the individual shape of the body, inflation or deflation of various air chambers
– heating or climate system in cushion and backrest
– massage or rotary passive motion systems
– swivel
– anthropometric adjustments:
    o horizontal adjustment
    o armrests, infinitely adjustable and tilting
    o cushion depth adjustment (x-dir)
    o seat height adjustment
    o backrest adjustment
    o seat tilt adjustment
    o head restraint adjustment, x-,z-dir
    o shoulder adjustment, x-dir

Agri-Seats for tractor and harvester
– vertical shock absorber adjustment
– horizontal suspension system (mechanical), on/off
– integrated Pneumatic-System to fit backrest to the individual shape of the body, inflation or deflation of various air chambers
– heating or climate system in cushion and backrest
– swivel
– anthropometric adjustments:
    o horizontal adjustment
    o armrests, infinitely adjustable and tilting
    o cushion depth adjustment (x-dir)
    o seat height adjustment
    o backrest adjustment
    o seat tilt adjustment

To perform all these functionalities by knobs and levers is feasible and is indeed the norm today. However, such a seat is designed for few "ideal" professionals, since:
• Elderly users (and in the agriculture business of today a lot of machinery operators are well over 50) would lack the memory capacity or the hand dexterity to operate such a complex UI. In effect, they would use only the standard configuration of the seat, thus reducing their comfort and endangering their health.
• People suffering already from back pain and/or spinal cord disorders would either need a specialised (much more expensive) seat or would run the risk of being injured by a sudden (erroneous) re-adjustment of the seat, while working.
• People with disabilities (i.e. FIAT displayed in 1999 a tractor model fitted with lift for wheelchair users and another, adapted for one hand operation, within its relevant AUTONOMY program) would require specialised UIs.

To avert these problems, as well as to optimise the UI functionality for all users, SAFEGUARD considers an innovative UI development process, that takes the above user groups well into account.

3. Towards a new concept of designing professional seat interfaces

The innovation brought by SAFEGUARD in the UI design of professional seat consists of three complementary approaches.

3.1 An inclusive Virtual Model

A number of seat models will be built and optimised using Virtual Reality (VR) s/w. Such models will be parametrically tested not only with the "mean user" model but also with virtual models of specific user categories (i.e. E&D), to take their ergonomical requirements (in terms of fast twitch / slow twitch fibre ration, physiological cross-sectional area, resting length, force-length and force-velocity curves, maximal voluntary contraction, etc.). Furthermore, the particular problems of people suffering already from back pain problems will be considered, facilitated by a relevant epidemiological and aetiological study within the project. "Risky" tasks in terms of UI operations and seat adjustments will be recognised and considered at the UI design level.

3.2 An optimum and highly adaptable UI prototype

Alternative means of simplifying the UI, to make it user-friendly for all users types, will be considered. The following figure 1 shows the possible inclusion of a small screen (or even touch-screen) for providing visual feedback and guidance.

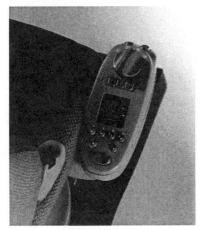

Figure 1: Alternative design of seat UI

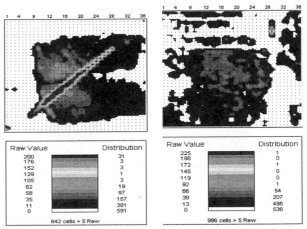

Figure 2: Buttocks pressure distribution after prolonged sitting (2 hours) of a patient with spastic diplegia with and without active seat pressure regulation (courtesy of TRANSWHEEL Consortium)

A number of conceptual drawings will be performed to propose UIs that can be easily adapted to different users cohorts needs (i.e. to operate with horizontal or vertical steering, used by some E&D). Nevertheless, the ultimate goal is to result in an "intelligent" seat, that is a seat that can be self-adaptable to the different environmental conditions, such as type of road, vehicle, driver's weight and driving style, driver's specific characteristics (i.e. disability, preferences, etc.), temperature and humidity, level of road-induced vibrations, etc.

Indeed, the employment of advanced pressure sensors and seat pressure regulators within SAFEGUARD seat may support an uneven seat profile, that would be optimal for particular drivers cohorts. When we measure the pressure at the buttocks of the user, as an indicator of his/her seating comfort using a specialized sensor (X-sensor), where red colour denotes areas with high pressure and blue colour areas with low pressure, it is obvious than an active seat pressure regulation enhances the seat pressure distribution and minimizes its peaks, as shown in Figure 2. This is especially important for people with low limb disabilities that in many cases lack the appropriate seat pressure feeling and thus do not change by themselves their seating position to relieve it.

3.3 Testing with all types of users

Both the truck and agricultural seat prototypes will be tested in a series of pilots in 10 different locations Europewide, including laboratory (vibrations, safety, virtual environment) and on-site (test track, road) tests. During these tests users of different profiles (including elderly and people with back pain problems) will test the seat concepts and prototypes. Even more innovative is the fact that users will test the seat performance, functionalities and UI under different scenarios of use, including fatigue and prolonged driving in a relevant Pilot, trying to correlate seating comfort levels of the driver with focused attention, specific task performance and stress parameters.

4. Conclusions

The total cost of low back pain to the United States economy is around $90 billion a year. One million back injuries occur per year and low back pain disables 5.4 million Americans; 100 million work days are lost each year, and low back pain accounts for 20 % of all work related injuries [Jelcic, 1995]. Tractor drivers with more than 700 tractor driving hours per year were found to have 61 % pathologic changes of the spine ; of those with 700 to 1200 hours, 68 % were affected ; and of those with greater than 1200 hours, 94 % were affected. Low back pain syndromes parallel pathologic changes. It is therefore obvious that the development of a new generation of comfortable and safe vehicle seats is a prerequisite to disability prevention for the professional driver. Such technologies seem to be viable, as they target the very large professional driver markets and are integrated into a product that has already a substantial cost (i.e. heavy vehicle seat).

On the other hand, to guarantee that such a seat would not only prevent but also cure relevant disabilities, as well as it is user-friendly and easy to use by all types of users, the needs of elderly and disabled users have been considered in all steps of HCI development within the project, from UI and seat functionalities conceptualisation to prototype development and testing. Into the bargain, a major secondary effect will be the use of such systems for particular types of elderly and disabled drivers as well as wheelchair users (since a wheelchair is also a "vehicle" and the road-imposed vibrations to it have been modelled within TRANSWHEEL DE-3013 project). By taking into consideration also the needs of E&D drivers, integrating the needs of elderly professional drivers (over 50) as well as data of wheelchair users (from TRANSWHEEL project), SAFEGUARD aims to offer its products as helpful solutions also for the rehabilitation market. By the followed innovative approach SAFEGUARD truck / agricultural machinery seat will be most probably the first serie in the mass market of a seat family, truly designed for all professional drivers.

References

Boshuizen, H.C., Bongers, P.M., Hulshof, C.T.J. (1992). Self-reported back pain in fork-lift truck and freight-container tractor drivers exposed to whole-body vibration. *Spine*, 17, 59-65.

Brendstrup, T., Biering-Sorensen, F. (1987). Effect of fork-lift truck driving on low-back trouble. *Scandinavian Journal of Work, Environment and Health*, 13, 445-452.

Foley, D.J., Wallace, R.B., Eberhard, J., (1995). Risk factors for motor vehicle crashes among older drivers in a rural community. *J Am Geriatr Soc*, 1995 Jul; 43(7), 776-81.

Gyi, D.E., Porter, JM (1998). Musculoskeletal problems and driving in police officers. *Occup Med (Lond)*, 1998 Apr, 48(3), 153-60.

Hildebrandt, V.H. (1995). Back pain in the working population: prevalence rates in Dutch trades and professions. *Ergonomics*, 1995 Jun, 38(6), 1283-98

Hostens, I., Papaioannou, G., Ramon, H. (1999). Static pressure distribution tests on seats of mobile agricultural machinery. *17th Congress of the International Society of Biomechanics*, 8-13 August, Calgary, Canada.

Jelcic, I. (1995). Improper body posture of bus drivers. *Arh Hig Rada Toksikol*, 1995 Mar, 46 (1), 89-93

Krause, N., Ragland, D.R., Greiner, B.A., Fisher, J.M., Holman, B.L., Selvin, S. (1997). Physical workload and ergonomic factors associated with prevalence of back and neck pain in urban transit operators. *Spine* 1997 Sep 15, 22(18), 2117.

Langauer-Lewowicka, H., Harazin, B., Brzozowska, I., Szlapa, P. (1999). Evaluation of health risk in machine operators exposed to whole body vibration. *Med Pr* 1996, 47(2), 97-106.

Lee, M.S., Park, J.H., Park, K.S. (1996). Reality and Human Performance in a Virtual World. *International Journal of Industrial Ergonomics*, 18 (2/3), 187.

Lings, S., Leboeuf-Yde, C (1998). Whole body vibrations and low back pain. *Ugeskr Laeger*, 1998 Jul 13,160(29), 4298-301.

Pope, M.H., Magnusson, M., Wilder, D.G. (1998). Kappa Delta Award. Low back pain and whole body vibration. *Clin Orthop*, 1998 Sep, (354):241

Adaptation of interactive courseware

Margherita Antona[1], Anthony Savidis[1], Constantine Stephanidis[1,2]

[1] Institute of Computer Science
Foundation for Research and Technology - Hellas
Science and Technology Park of Crete
GR-71110, Heraklion, Crete, Greece
Email: cs@ics.forth.gr

[2] Department of Computer Science, University of Crete

Abstract

This paper focuses on the notion of content and interaction adaptation in web-based courseware. The adopted perspective is that adaptation should continuously deliver, at any time during the use of the system, the most appropriate interactive learning content to each individual learner. To this purpose, the paper briefly discusses the concept of adaptation in educational applications from the wider perspective of universal access to Information Society Technologies, and discusses some aspects of a proposed architectural framework suitable for supporting content and interaction adaptation in such a context.

1. Introduction

Long distance computer-based instruction, and in particular web-based instruction, is progressively acquiring a fundamental role in shaping new inclusive forms of education in the emerging Information Society, and is attracting considerable attention at both government and industrial level [see, e.g., European Commission, 2000; Kerrey et al., 2000; Hodgins, 2000]. The advent of the World Wide Web (WWW) and of multimedia technologies has marked a turning point in the way computer-based education is conceived, designed, delivered and experienced. The WWW enables the wide distribution of educational material in which content is presented using a wide variety of media (hypertext, graphics, animation, audio, video, etc). Web-based learning involves an increasing variety of target learner groups, including, for example, adults, children in secondary education, people in geographically dispersed areas, (re-)trainees, disabled and homebound, etc., in an increasingly wider variety of educational activities, including pre-university and university courses delivered on the web, training in industry, life-long learning, learning on demand, and many others. As a consequence, distance learning is evolving from a form of education in which the learner and the teacher are physically separated, to the provision of "whatever educational opportunities to anybody, anywhere and at anytime, characterised by a great diversity in practice" [Spodick, 1995, p.1]. The role of technology, in this context, is not only limited to providing more efficient, simple and cost-effective solutions for overcoming distance and time in education, but also to create and make widely available new forms of learning, exploiting the capability of dynamic electronic content to be manipulated, structured, retrieved and presented in a variety of ways. In this respect, one of the most promising characteristics of Web-based learning is the concrete possibility of providing adaptation and individualisation of learning resources.

2. Adaptation in interactive learning system

In the application domain of education, adaptation has been investigated in a number of recent research efforts, mainly falling into the category of Intelligent Tutoring Systems and Adaptive Hypermedia. The goal of Intelligent Tutoring Systems is to use the knowledge about domains, learners, and teaching strategies to support flexible individualized learning and tutoring [Brusilovsky, 1999; Murray,1999]. Adaptive hypermedia systems apply different forms of user models to adapt the content and the links of hypermedia pages to the user, and have found in education one of their main applications [Brusilovsky, 1996; De Bra, 1999]. In these efforts, several adaptation techniques have been developed, the most relevant of which are curriculum sequencing [Vassileva, 1997;

Huebscher, 2000; Stern and Park Woolf, 2000], adaptive navigation [Calvi & De Bra, 1997; Weber and Specht, 1997], and adaptive presentation [Brusilovsky et al., 1998].

On the other hand, in recent years, the concept of adaptation has been investigated in the broader perspective of providing built-in accessibility and high interaction quality in applications and services in the emerging Information Society [Stephanidis, 2001a; Stephanidis, 2001b]. Adaptation characterises software products that automatically adapt themselves according to the individual attributes of users (e.g., mental / motor / sensory characteristics, preferences), and to the particular context of use (e.g., hardware and software platform, environment of use). In this respect, adaptation concerns the interactive behaviour of applications and services as well as the content. In Human-Computer Interaction, self-adaptation of a system's interactive behaviour have been proposed as a framework for providing accessibility and high interaction quality to all potential users, on the basis of each user's individual abilities, requirements and preferences, as well as of the context in which interaction takes place and of the adopted interaction technology [Stephanidis, 2001b]. Therefore, adaptation implies the capability, on the part of the system, of capturing and representing knowledge concerning alternative instantiations suitable for different users, contexts, purposes, etc, as well as of reasoning about those alternatives to arrive at adaptation decisions. Furthermore, adaptation implies the capability of assembling, coherently presenting, and managing at run-time the appropriate alternatives for the current user(s), purpose(s) of use and context [Savidis and Stephanidis, 2001].

From the point of view of interactive courseware, all the types of adaptation mentioned above are relevant. In a learner-centred perspective, content should adapt, as a minimum requirement, to the learner's previous knowledge, individual progress, learning goals and learning style. Content adaptation should be applied each time a particular course topic is to be entered, so that the learner is initially presented with the appropriate material, as well as during learning sessions, so that the subsequently presented material is selected on the basis of learner specific requirements at run-time. Interaction, on the other hand, should be adapted, as a minimum requirement, to the learner's computing experience.This implies that both course structure and learning sessions are flexible and can be assembled on the fly and modified at run-time as necessary, and that the learners' characteristics, the educational material, the course structure and the course delivery process are appropriately modelled, so that the best match between their characteristics can be established at any point in time during the use of the system.

3. Learning system architecture

This section discusses a proposed architecture for learner-based content and interaction adaptation in a courseware environment comprising both authoring and delivering facilities, and in which concrete learning object instances are physically distributed. Adaptation is considered to be based on learner information, course information, learning objects, and process information. A high-level representation of an architecture for the core of such a system is depicted in Figure 1.

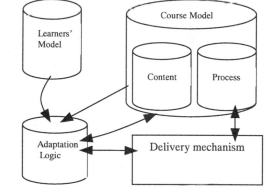

Figure 1. A high-level architecture for adaptive courseware

3.1 Learner Model

A learner model contains explicitly modelled assumptions that represent the characteristics of learners that are relevant to the system. Several techniques are commonly used for learners' modelling in Intelligent Tutoring Systems, such as fixed stereotypes, overlay models, or combinations of the above. A number of sources of information may be used to construct a learner model. The system acquires data about the user and infers learner characteristics from this data. The validity of the assumptions depends on the technique used to acquire the information. Automatic modelling may be unreliable, and therefore collaborative and cooperative modelling is frequently chosen [Conlan and Wade, 2000].

The properties chosen to represent the user should be pertinent to the potential adaptation by the system. Learner characteristics frequently modelled are the user's previous knowledge, goals and objectives, cognitive style, learning style, maturity, general ability, confidence, motivation, preferences and background [Specht & Weber, 1996]. For interaction adaptation, knowledge about relevant learners' characteristics is also necessary, including. sensory, motor and cognitive abilities, expertise in the use of computers, of the web, and of the particular system, knowledge about specific interface components, colour and media preferences, preferences in interaction and navigation styles, etc.

In order to ensure wide applicability and reusability, the learner's model should be general enough to address different categories of knowledge domains, learning purposes and training processes. For example, it should be possible to create learner profiles for formal classroom education, professional training, informal learning, etc. This implies the need of differentiating the type of information to be modeled according to its relevance for the specific case. For instance, in classroom-based training, evaluation records may need to be maintained, while in professional training, organizational-role specific information will need to be managed. To achieve such a flexibility, the learner model should be based on meta-data taking into account currently proposed standards [IEEE-LTSC, 2000b, IMS, 2001a], and encompass appropriate abstract attributes, enabling concrete learner models to be instantiated. As intensive use of modeling is required, models should be easy to create, consult and up-date. A possible technique for model information input is the provision of alternative templates for specific domains. Query facilities for existing profiles should also be provided.

3.2 Course Model

Besides information about learners, appropriate models of the data to be delivered in a course and of the delivery process are necessary. Content information should be logically separated from course information. This enables a course to be delivered in different form by controlling semantic parameters. The content model identifies: (i) the elementary constituent units of content; and (ii) the structure in which elementary constituents are organised. In recent standardisation efforts, units of content are referred to as 'learning objects' [IEEE-LTSC, 2000a]. Such a unit can be a paragraph or other piece of text, an image, a video clip, etc. The unit internal structure should not be relevant for the model. The knowledge domain can be described in the content model in terms of concepts (or terms) and relationships between concepts. Concepts are used for defining prerequisites providing meaningful paths through the information [De Bra, 1999]. The granularity at which the content is stored affects the level of content adaptation that may be achieved, as well as the reuse potential [Conlan and Wade, 2000].

In the proposed approach, content can manifest itself in different forms, and, therefore, learning objects should not have a singular mapping to content resource items, and should not be physically attached to content. Rather they should be abstract, i.e., capable of linking to different content items. This can be accomplished by associating learning objects to content via resource identifiers, as opposed to embedding content within learning objects, also supporting a greater degree of reusability. Alternative templates can be supplied for different meta-data categories. For interaction adaptation, learning objects should also include information concerning their interaction characteristics (e.g. color, font, interface objects, required interaction techniques), to be matched with learners' interaction requirements and preferences. Current proposed standard specification can be considered as a comprehensive modelling of learning objects, aiming to be domain-free [IEEE-LTSC, 2000a]. However, due to intended generality, such a specification offers a taxonomy of semantic attributes which do not allow for domain-specific concepts. Therefore, domain specific templates should also be defined.

The structure of a course may vary depending on many factors. For instance, educational material for a course may be structured as a hypermedia book, a tutorial, a training session, a seminar, etc. Course structures are usually specified on the basis of a hierarchical organisation model, in which the most primitive constituent components are the learning objects [IMS, 2001b]. Different templates can be provided for different type of courses. An important requirement on course templates is that they cater for alternatives to be made available to different users in different situations, and provide information for selecting amongst alternatives, i.e., they define multiple courses in one hierarchical structure, depending on the learner's level, purpose of learning, background, acquired prerequisites, etc. Learning objects' templates and course templates should appropriately link, in order to enable the implementation of facilities such as, for example, search, evaluation, selection, and acquisition of learning objects, etc. The process model should also provide information for course management at run-time.

3.3 Adaptation Logic

In Adaptive Hypermedia and Intelligent Tutoring Systems, adaptation decision-making is often embodied in the "pedagogical knowledge" of the system and is implemented through rule-based approaches (e.g., Vassileva, 1997; De Bra et al., 1999). Alternative approaches include Bayesian networks for calulating probabilities (Henze, 2000), fuzzy logic (Nkambou, 1997), and neural network and connectionist methos (Mullier et al., 1999).

In the proposed architecture, the adaptation logic is the component in charge of providing the reasoning mechanism required for accomplishing both interaction and course adaptation, on the basis of the relevant characteristics captured in the models. The reasoning process is assumed to take place both at the beginning of and during each learning session, and to drive to decisions concerning: (i) the learning material to be displayed, and (ii) how it should be assembled. In other words, the adaptation logic provides a decision mechanism for dynamically selecting, among the multiple instances of course components and learning objects available, the one(s) appropriate for presentation at a given point in a learning session of a specific user. In this respect, the adaptation logic combines the information provided in all models in a coherent whole. Although general adaptation criteria can be defined, it is unlikely that all adaptation cases for a variety of learners' characteristics and courses can be captured at a general level. Therefore, adaptation criteria should be editable, i.e., course authors should be allowed (and supported in) defining their own adaptation logic on the basis of the knowledge domains, learners and purposes their course is designed for.

3.4 Delivery Mechanism

Apart from deciding which learning objects should be supplied to learners, the physical form of such objects should be retrieved and delivered appropriately. Delivery may imply the presence of mechanisms to assemble a physical structure that is not only "acceptable", but also high-quality. For instance, assume that the learning objects correspond to independent paragraphs in a document, while the necessary physical data are supplied in different locations, and may contain both text and images. The assembly of those items into a single document should be carried out in a way leading to a high-quality visual document structure. Additionally, the delivery platform should be able to handle the retrieval and display of such physical data (i.e., text and images in their particular format). The content assembly process can be based, for example, on domain-specific HTML templates.

In web-based learning systems, a clear separation between the content and the user interface is not easily performed. The reason is that content is also interactive, supplying all kinds of interaction objects and techniques which are normally classified as belonging to the interfacing layer. Therefore, two types of user interface can be identified: the delivery interface, and the content interface. The former provides some standard facilities such as history control, navigation, and communication, while the latter provides content-specific interaction methods depending on the physical delivery of a learning object (i.e., embedded in form). Consequently, to accomplish user interface adaptation, both levels of interfacing need to be affected. The delivery interface can realise adaptation according to information on learner's interaction requirements and preferences. While the delivery interface is chosen at the beginning of a session, the content interface is chosen at run-time in case interaction alternatives exist.

4. Conclusions

This paper has discussed some important aspects of a proposed architectural framework suitable for supporting content and interaction adaptation in web-based courseware. A wide range of research issues emerges from the above analysis. To mention only some examples, the granularity at which content should be broken down and analysed constitutes a very important aspect, likely to highly affect the adaptation process and the necessary information. Appropriate granularity levels are likely to vary for different knowledge domains and media, and different modelling solutions are required to be supported. Another important issue concerns the source of knowledge for continuous learner modelling during system use. It is likely that completely automated assessment of the user progresses is not a viable solution, and that human assessment is required at some point in the process. However, on-line monitoring of dynamic user characteristics, which may change in the time-frame of a session, is also required. Therefore, learner characteristics should be carefully investigated with respect to when and how their values and history need to be up-dated.

References

Brusilovsky, P. (1996). Methods and techniques for adaptive hypermedia. User Modelling and User Adapted Interaction. *Special issue on Adaptive Hypertext and Hypermedia*. v. 9, n. 2-3, 87-129.

Brusilovsky, P. (1999). Adaptive and Intelligent Technologies for Web-based Education. In: C. Rollinger & C. Peylo (eds.) *Künstliche Intelligenz, Special Issue on Intelligent Systems and Teleteaching*, 4, 19-25.

Brusilovsky, P., Eklund, J., Schwarz, E. (1998). Web-based education for all: A tool for development adaptive courseware. *Computer Networks and ISDN Systems*, 30, 291-300.

Calvi, L., De Bra, P. (1997). Improving the Usability of Hypertext Courseware through Adaptive Linking. *Proceedings of the Eighth ACM Conference on Hypertext*, Washington DC, pp. 224-225.

764

Conlan, O., Wade V. (2000). Novel Components for supporting Adaptivity in Education Systems - Model-based Integration Approach. In *ACM MM 2000 Electronic Proceedings*. [On-line]. Available at: http://www.acm.org/sigs/sigmm/MM2000/ep/toc.html

European Commission. (2000). *e-Learning – design tomorrow's education. COM(2000) 318 final. Report from the Commission to the Council and European Parliament*. [On-line]. Available at: http://europa.eu.int/comm/education/elearning/comen.pdf.

De Bra, P. (1999). Design Issues in Adaptive Web-Site Development. *Proceedings of the 2nd Workshop on Adaptive Systems and User Modeling on the World Wide Web*, Toronto and Banff, Canada, pp. 29-39.

De Bra, P., Houben, G.-J., Wu, H. (1999). AHAM: A Dexter-based Reference Model for Adaptive Hypermedia. *Proceedings of the ACM Conference on Hypertext and Hypermedia*, Darmstadt, Germany, pp. 147-156.

Henze, N. (2000). *Adaptive Hyperbooks: Adaptation for Project-based Learning Resources*. PhD Thesis. University of Hannover.

Hodgins, W. (2000). *Into the Future. A Vision Paper*. White paper commissioned by the Commission on Technology & Adult Learning of the American Society for Training & Development (ASTD) and National Governors. [On-line]. Available at: http://www.learnativity.com/into_the_future2000.html.

Huebscher, R. (2000). Logically Optimal Curriculum Sequences for Adaptive Hypermedia Systems. In P. Brusilovsky, O. Stock, C. Strapparava (Eds.), *AH2000*, LNCS 1892, Springer-Verlag, pp. 121–132.

IEEE-LTSC (2000a). *Draft Standard for Learning Object Metadata. IEEE P1484.12/D5.0* . [On-line]. Available at: http://ltsc.ieee.org/wg12/index.html

IEEE-LTSC (2000b). *Draft Standard for Learning Technology — Public and Private Information (PAPI) for Learners (PAPI Learner). IEEE P1484.2/D7, 2000-11-28*. [On-line]. Available at: http://edutool.com/papi/

IMS (2001a). *IMS Learner Information Packaging Information Model Specification. Version 1.0. Public Draft Specification*. [On-line]. Available at: http://www.imsproject.org/profiles/lipinfo01.html

IMS (2001b). *IMS Content Packaging Information Model Version 1.1. Public Draft Specification*. [On-line]. Available at: http://www.imsproject.org/content/packaging/cpinfo11.html

Kerrey B., Abraham, P.S., Bailey, G., Brown, R.W., Enzi, M.B., Gage, J., King, D.R., Pfund, N., Isakson, J., Arkatov, A., Bingaman, J., Collins, S.R., Fattah, C., Gowen, R.J., McGinn, F., Winston, D., (2000). *The Power of the Internet for Learning: Moving from Promise to Practice*. Report of the Web-based Education Commission to the President and Congress of the United States. [On-line]. Available at: http://interact.hpcnet.org/webcommission/index.htm.

Mullier, D.J, Hobbs D.J., Moore D.J. (1999). A Hybrid Semantic/Connectionist Approach to Adaptivity in Educational Hypermedia Systems. In *Proceedings of ED-MEDIA'99*, Seattle, WA.

Murray, T. (1999). Authoring Intelligent Tutoring Systems: An analysis of the state of the art. *International Journal of Artificial Intelligence in Education*, Vol. 10, pp. 98-129.

Nkambou, R. (1997). Using Fuzzy logic in ITS-course generation. In: *Tools with Artificial Intelligence*, IEEE press, pp. 190-194.

Savidis, A., Stephanidis, C. (2001). The Unified User Interface Software Architecture. In C. Stephanidis (ed.) *User Interfaces for All – Concepts, Methods and Tools* (pp. 389-415). Mahwah, NJ: LEA (ISBN 0-8058-2967-9, 760 pages).

Specht, M., Weber, G. (1996) Episodic Adaptation in Learning Environments. *Proceedings of EuroAIED 1996*, pp 171-177.

Spodick, E.F. (1995). The Evolution of Distance Learning, Presentation, Hong Kong University of Science & Technology Library, August 1995. [On-line]. Available at: http://sqzm14.ust.hk/distance/evolution-distance-learning.htm

Stephanidis, C. (2001a). Adaptive techniques for Universal Access. *User Modelling and User Adapted Interaction*, 11 (1-2): 159-179.

Stephanidis C. (2001b). User Interfaces for All: New perspectives into HCI. In C. Stephanidis (ed.) *User Interfaces for All – Concepts, Methods and Tools* (pp. 3-17). Mahwah, NJ: LEA (ISBN 0-8058-2967-9, 760 pages).

Stern, M. K., Park Woolf, B. (2000). Adaptive Content in an Online Lecture System. In P. Brusilovsky, O. Stock, C. Strapparava (Eds.): *Adaptive Hypermedia and Adaptive Web-Based Systems. Proceedings of AH 2000*, Trento, Italy, August 2000, p. 227 ff.

Vassileva, J. (1997). Dynamic Course Generation on the WWW. In: Boulay, B. & Mizoguchi, R. (eds.) *Artificial Intelligence in Education: Knowledge and Media in Learning Systems*. IOS, Amsterdam, 498-505.

Weber, G. and Specht, M. (1997). User Modeling and Adaptive Navigation Support in WWW-based Tutoring Systems. In A. Jameson, C. Paris & C. Tasso (Eds.) *User Modeling: Proceedings of the 6th International Conference 97, Springer*. [On-line]. Available at: http://citeseer.nj.nec.com/weber97user.html

Accessibility of government services - A macroergonomics perspective

Albert G. Arnold, David J.M. Beentjes

Centre of Expertise Foundation
Consultants for Informatisation of Governmental Institutions
2517 JS The Hague
The Netherlands
b.arnold@hec.nl

Abstract

In this contribution a macroergonomics perspective to the accessibility of Dutch governmental services is taken. In general, the accessibility or responsiveness of the Dutch local and national governmental institutions is still problematic. It is argued that the accessibility can only be improved if one takes into account organisational, cultural and political issues. An important step to more accessible service provision is the establishment of an integrated front office. At an integrated front office citizens can acquire a broad range of services within a specific area. In this contribution three cases are presented concerning governmental and non-governmental actors, who try to establish such an integrated front office. From the cases it becomes clear that the problems encountered are largely due to organisational, political and cultural aspects. In order to overcome these problems a more citizen-oriented attitude needs to be adopted by all parties involved starting with the politicians and top administrators. More specific, these responsible persons should create the necessary conditions for co-operation and stimulate the change processes by putting the citizen in a central role.

1. Introduction

Governmental institutions play an important role in the creation of the information society for all. Apart from their role as legislator and rule enforcer, they constitute a large service-providing organisation, which is important in the life of all citizens. This contribution is on the accessibility of the Dutch governmental institutions. However, we believe that the concept of 'accessibility' does not really reflect the kind of services a government (national or local) should render. The concept has a rather passive connotation, instead governments should become responsive or even pro-active.

In this article we are taking a so-called *macroergonomics* perspective. The goal of macroergonomics specialists is to optimise workplace productivity by recognising the effects of systemic interactions between individuals and environmental, technological, and interpersonal variables (see also Hendrick, 1987). In line with the macroergonomics approach, our opinion is that universal access to governmental services heavily depends on organisational and cultural issues. Aspects of technology and interfaces are important as well, but become opportune in the later stages of establishing an integrated front office.

In the past years a number of initiatives have been taken in the Netherlands to make the Dutch local and national government more accessible (e.g. http://www.vanboxtel.nl, http://www.ol2000.nl, http://www.overheid.nl and http://www.gemnet.nl). Despite these initiatives, an anthology (Vervest et al., 2001) on the digital government responsiveness on the internet showed a rather gloomy picture concerning governmental accessibility. Dutch governmental institutions are generally inward facing organisations that mainly rely on one way communication with citizens. This gloomy picture is supported by other studies too (e.g. Deloite & Touche, 2000; Ministerie van BZK,1999; Pro Active International, 2000).

The http://www.ol2000.nl initiative is aimed at improving accessibility, by the integration of services of local and national institutions and to offer these services to the citizens via one front office. In our opinion, the development of such physically or digitally integrated front offices is a necessary and important step in the process of becoming more responsive. Such an office makes governmental services more accessible to the public, because citizens have to deal with only one office for example for the procurement of a social security benefit. There are two options for an integrated front office. The first option is to bring together various service providers at one location – the front office, but still render the actual services in a separate and sequential way. In this situation the client does not have to go to different locations, but still has to deal with more than one civil servant. The second option is more sophisticated in its degree of integration. The various services of the different donor organisations are rendered by one civil servant.

In this contribution three cases are presented of governmental and non-governmental actors try to establish an integrated front office.

2. Cases

The cases have been studied in a research conducted at the Faculty of Computer Science at the University of Twente. The objective of this research was to make an inventory of the problems encountered by governmental institutions during the process of service integration (Beentjes, 2000).

2.1 Centre of Work and Income

The Centre of Work and Income (CWI) was realised through co-operation between the local Social Security Office, the Governmental Employment Agency, and the Employment Insurance Administration (GAK). The objective of a CWI is to get as many people as possible to work or to supply an unemployment benefit. The reason to start an integrated front office was caused by the political climate of the Local Security Office and the Governmental Employment Agency. Politicians were not very satisfied with the performances of both institutions. Feelings of loyalty towards the others constituted the driving forces for the third party. Moreover, legislation is on its way to realise more CWIs in the Netherlands.

The way of operation of a CWI should be client-oriented, efficient and effective. Traditionally the clients of the CWI are send from one organisation to the other in order to get a job or a benefit. In the new centres all services are brought under one roof. A complete integration of services was intended in this case. A number of problems were encountered during the installation and operation of the Centre of Work and Income:

- The ambition level of this project was rather high. The goal was full integration of all services. During the integration process this goal appeared to be unattainable due to differences in complexity and completion time of the various tasks involved.

- The partners involved in the CWI were too much focussed on their own individual interest. As a consequence they drew back employees. These employees were more or less forced to concentrate on the tasks traditionally fulfilled by their own organisation. The Local Social Security Office experienced a backlog in the processing of social security applications. In the front office the employees of the other donor organisations were increasingly helping out their colleagues of the Social Security Office. However, the Dutch government pays the Employment Insurance Administration on the basis of the number of applications processed. A number of their employees was called back to the own organisation in order to increase the productivity and secure funding.

- Basically, the partners involved underestimated the consequences of this far-reaching project. Especially, they did not understand the differences between for example culture, ambitions and way of working of the other partners.

- The partners were not willing to make their own back office organisation inferior to the integrated front office.

2.2 Municipal real estate office

The objective of this project was to provide the private sector in the Amsterdam region with real estate information through a regional website. The partners in this project were four municipalities. The integration of real estate information provision means a substantial improvement of service to local entrepreneurs. For example, if a potential buyer of an office building wants to know the condition of the soil at the location of the building, he needs to visit three different institutions. After the integration of the real estate information the buyer only needs to visit one electronic front office. The information offered covers real estate information, documentary information, and contact information. The four municipalities applied the same functional design of their real estate information, which should make the information exchange and integration relatively easy. The four partners automatically delivered their information from their back offices to the front office. Despite these favourable conditions, the services offered were not fully meeting the real estate information needs of industry in the region due to a number of causes:

- A proper inventory and analysis of the information needs of the potential clients in the region was not made at the beginning of the project. After some time when the need patterns of the clients were known, it appeared that the information required was not available in digital format.

- The necessary changes in the back offices to meet the information needs of the clients were underestimated and were never realised for two reasons. First, the costs of digitising information (e.g. maps) were rather high and were not anticipated. Second, the organisational effort needed was great. Within each municipality there were many stakeholders. Furthermore, the municipalities had strong reservations to publish specific information, for example the land prices. In other words, it was difficult to reach consensus among parties.

- The digital real estate office was an initiative of civil servants and was not directly supported by their managers or administrators. This situation has been a frustrating factor and severely lowered the ambition level.

2.3   Integrated emergency room

The police, fire department and ambulance service wanted to integrate their emergency rooms. The main motivation for integration was the rising costs of technology and also the fact that in some emergency cases multi-disciplinary action is indicated. Each partner on itself could not raise enough money to implement the next generation of digital communication networks. The intention was to bring together the geographically dispersed emergency rooms and to fully integrate all of the various tasks of the three partners. The implementation of the integrated emergency room has taken much longer than expected due to the following problems:

- The main objection against the integrated emergency room came from the ambulance services. They thought that the integration of services would mean deterioration of the quality of their own services. Their conviction was that firemen and policemen are unable to make adequate decisions concerning incoming medical emergency calls.

- Political pressure heavily influenced the attitude of the police and fire departments. Local politicians were eager to show that they were prepared for big catastrophes. An integrated emergency room would provide facilities to cope with such a situation. However, the question is whether an emergency room needs to be based on catastrophes, which hardly occur or on daily accidents which require, most of the time, the service of only one organisation.

- The needs of the public concerning emergency services have not been studied. And therefore the possible opinions of the public have been more or less misused by the various partners to make their own argumentation stronger.

3.   Discussion

The cases presented in this contribution make it clear that a macroergonomics perspective is indicated in order to improve the accessibility of Dutch governmental services. It is by no means the technology or the interfaces, which are decisive. Rather the organisational, cultural and political issues appear to be much more important. The following influential issues can be distinguished (see also Bekkers (1998)):

- *The superior role of the back office*
  The donor organisations are in fact setting the rules for their front office employees. The front office cannot take care of her own management. Also the rules of the back office still hold for the front office employees as well as their clients.

- *The incapability of the back office*
  The back offices are not always capable to deliver the information required by citizens. This is especially the case when the service needs of the citizens are studied in depth.

- *The lack of co-operation*
  The willingness to co-operate within the own organisation but also with partners from outside is often not very strong. In order to create accessible services intensive co-operation between all parties

involved is necessary. In particular, co-operation with people from outside – public and private – the own organisation is necessary. The cultural differences of people with different backgrounds should be taken into account more seriously.

- *The attitude of politicians and administrators*
  Politicians are often oriented towards direct profits, which can be gained from the electorate. They do not sufficiently see the importance of an accessible government. An accessible and responsive government supports the citizen's feeling of involvement and thereby provides the government her legitimatisation.
  The administrators often underestimate the effort, which is needed for the change process. And they are not creating favourable conditions for co-operation, for example the financial thresholds for co-operation in terms of their budgets are generally not taken away. Furthermore, in line with politicians, they also do not realise how ICT (information and communication technology) could improve the accessibility of their services.

4. Conclusions

The main conclusion is that the solutions to these accessibility problems can only be found if one takes a macroergonomics approach. All the parties involved in making governmental services more accessible must start to adopt a client friendly attitude (Vervest & Dunn, 2000) starting with the politicians and top administrators. More specific, the persons who are responsible should create the necessary conditions for co-operation and stimulate the change processes by putting the citizen in a central role.

References

Beentjes, D.J.M. (2000). *Tussen de overheid. Een exploratief onderzoek naar interorganisatorische geïntegreerde dienstverlening.* Twente: Universiteit Twente, Faculteit Bedrijfsinformatietechnologie, master thesis.

Bekkers, V. (1998). *Grenzeloze overheid. Over informatisering en grensveranderingen in het openbaar bestuur.* Alphen aan den Rijn: Samson.

Deloite & Touche (2000). *Overheid oNLine. Trendonderzoek naar gemeenten en provincies op internet.* Rotterdam: Deloite & Touche.

Hendrick, H. (1987). Macroergonomics. In: P. Hancock (Ed.) *Human Factors in Psychology.* New York: North-Holland.

Ministerie van BZK (1999). *Internet Monitor Overheidswebsites.* 's-Gravenhage: Minsterie van BZK.

Pro Active International (2000). *Project Gemeenschappelijk Webbeleid.* Amsterdam: Pro Active International.

Vervest, P., A.G. Arnold, L.Meuleman & M. Wolters (2001). *Toelichting op het pamflet. Een overheid voor ieder-één.* Utrecht: Postbus 52, www.postbus52.com.

In-Vehicle Trip Information for All

Wilhelm Bauer[1], Harald Widlroither[1], Evangelia Portouli[2]

1. Fraunhofer-Institute for Industrial Engineering IAO, Nobelstrasse 12, D-70569 Stuttgart, Germany, Tel.: +49-711-970-2105, fax: +49-711-970-2083
E-mail: Harald.Widlroither@iao.fhg.de

2. TRD S.A., 387, Mesogeion Ave., 15341 Ag. Paraskevi, Greece,
Tel.: +30-1-65477975, fax: +30-1-6549922, E-mail: trnspcon@compulink.gr

Abstract

In a co-operation between INARTE (TR 4014) and TELSCAN (TR1108) European projects additional usability tests of the IN-ARTE system were performed with a group of 10 elderly drivers. The subjects had to fulfil several interaction tasks with a virtual prototype of the IN-ARTE HMI. The system functions and the HMI were introduced to the subjects in a defined sequence. The user behaviour was observed and interviews were performed targeting at the evaluation of acceptance and usability of the IN-ARTE functions and HMI. Comparisons could be made with a group of younger drivers, who were involved in former IN-ARTE experiment with the same design. During the tests the experimentators found qualitative differences in behaviour between elderly (required more explanation, longer habituation phase) and other drivers. However the evaluation of the results concerning usability and acceptance of the IN-ARTE system did not show significant differences between the two groups. The expressed needs concerning the HMI were not different. Based on these tests it cannot be concluded that the IN-ARTE concept and HMI requires special adaptations to the need of elderly drivers, but could be concluded that adapting the IN-ARTE interface to match the extended behavioural needs of elderly drivers results in a better interface for all.

1. Introduction

The aim of IN-ARTE project was to develop an integrated autonomous on-board system to be able to build an extended view of the environment in front of the vehicle, integrating signals from anti-collision radar, road recognition CCD sensor, digital road map, and navigation system in order to guide and warn the driver through an optimum HMI in a series of rural areas related traffic tasks, such as intersection handling, speed selection while negotiating curves, obstacle detection, etc. (1).
The TELSCAN project has provided assistance regarding its system development so as the needs of elderly users are met. The assistance should guarantee that the special needs of the user group are taken into consideration in the design process of the HMI. More specifically, this assistance has included additional simulator experiments at Fraunhofer IAO with a virtual prototype of the relevant in-vehicle HMI to assess and define how elderly drivers can benefit best from the IN-ARTE application. In this context Fraunhofer IAO has performed a limited number of user tests with subjects that represented the user group above 50.

2. Methodological Approach

Based on the specification of functions (1,2,3) and the application of general HMI guidelines (4,5,6,7) a virtual prototype of the IN-ARTE HMI has been developed with ALTIA Design. In ALTIA Design the HMI is modelled in a graphical environment on the computer. Controls, displays and to a certain degree environmental conditions can be animated to demonstrate possible user actions and system output. With a control editor it can be determined how and when components of the animated model communicate with one another. Thus simulations of the system behaviour and therefore of human-machine interaction can be realised, communicated, studied, evaluated and trained. The virtual ALTIA prototypes are visual and acoustic representations of the hardware and software of the HMI and can be run on multimedia computers. The visual representation is realised by a virtual display terminal. The virtual control actuators can be operated by a pointing and clicking device, e. g. a mouse. The demonstrator for user tests focus primarily on acceptance issues, barriers in the dialogue (comprehension) and use of redundant information.

770

This version included a tracking task representing lane keeping during the experiments. Figure 1 shows the components that were included in this virtual HMI demonstrator.

No.	Identifier
1	Steering wheel as input device for tracking task
2	'Repeat' Button
3	Brake-Pedal
4	Accelerator Pedal
5	LCD-Display
6	Traffic Scenario
7	Warning lights in dashboard and mirror device
8	Tracking task
9	Speedometer

Figure 1: Components of virtual demonstrator

Figure 2 displays the virtual demonstrator/simulator for user tests with a typical test scenario: The driver is warned before an intersection by a warning light in the mirror and speech output. In addition redundant display information is given.

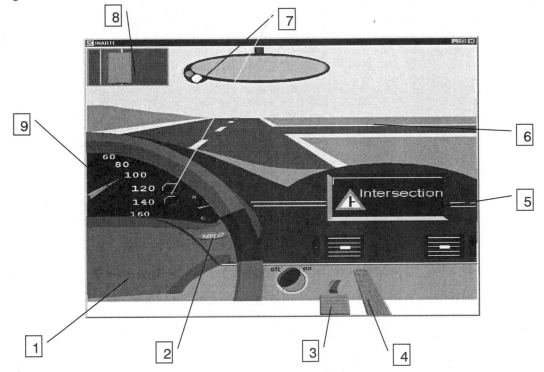

Figure 2: Virtual HMI simulation

Methods of direct open observation, with enclosed standardised written questionnaires, were chosen for this user test. This method allows a particular economical evaluation. "Direct observation" means that the behaviour of the user is directly observed by the test operator during the test. At the observation, the operating behaviour and mistakes made by the user are noted. By this means, one can gather conclusions for necessary changes within the system. In order to obtain information about the thoughts of the subjects it is necessary that the subjects think aloud throughout the test. This procedure is very helpful when explaining many of the subject's operating behaviour. Additionally, the operation can also be queried throughout the test.

A written questionnaire (in German) for the experimenter and the subject complements the observations. In order to recognise behavioural tendencies which give information on required changes, only a relatively small number of subjects are needed as this does not deal with research experiments, rather with a development lead prototype evaluation. Similarly to the oriented evaluation and optimisation, only one design is examined within the tests.

Figure 3 shows the set-up of the test and the components of the simulator. To get more realistic proportions of the dashboard and the road scenes, the simulation was projected with a video beamer. The input device for the subjects is a mouse, the test operator uses the keyboard to control all functions of the HMI and the several road scenes. Warning sounds and speech output is realised via sound card and loudspeakers, because the timing of sound output is controlled by ALTIA software.

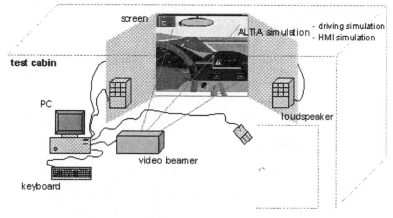

Figure 3: Test set-up

3. Summary of test results

When the test results were analysed the answers of the user questionnaire and the results of the observations are used for the final conclusions. All together 10 subjects completed the test, 6 men and 4 women. One subject had experience with several navigation systems and the use of a additional display. 4 subjects are driving more than 10 000 km per year. The number of subjects is not sufficient to create a representative statistical collective, but important tendencies and predictions can be detected with this approach.

Considering speech output, tests lead to the result, that short messages were preferred. But it is supposed, that in the real traffic situation the comprehensibility of the speech is much more difficult because of the environmental influence, like noisy vehicles or conversation of passengers inside the car, and the increasing mental workload of a real driving situation Therefore it is necessary to prepare the driver that a speech message will be presented. This can be achieved either by an extended speech with some additional words, which does not give any further information, or a suitable tone will fulfil this task. The simulation tests lead to the result, that a tone preparing the driver for a speech message is useful to attract driver's attention. The selected warning tones were accepted very well for attracting driver's attention.

Elderly people had greater problems in noticing visual messages, either icons or text. They often considered this as disturbing and distractive. Icons were sometimes only noticed as some changes in the display, but their meaning often remained unclear. This may be attributed to their reduced field of peripheral vision and reduced adaptation capabilities.

Elderly people considered speech output as the most effective way to inform the driver about any events, because there is no need to distract driver's visual attention from the street, like a display does. In this context, single words were accepted best, because the information is reduced to the essential and therefore easier and faster to understand.

Regarding the use of an additional warning light to indicate danger, there was a very high acceptance by the majority of the users and by all elderly drivers of a mirror device solution, because it is situated in the peripheral field of vision and perceived well.

Regarding the use of warning tones there were a lot of opinions expressed. Elderly users preferred most mild tones for events occurring quite often, while for rare / dangerous events, also quite penetrating tones can be used or have to be put in.

Based on the results of the trials, the following points can be outlined, to achieve an HMI acceptable by all users. Text has to correspond to the following guidelines to be easily read by elderly and disabled:

- Easy-to–read-letters without serifs

- Distance between the letters 1/6 – 1/2 of a letter's height, depending of it's shape (tall letters like 'I' require more distance for good readability).
- Text only in capitals should be avoided, rather a mixture is to prefer. For single words, plain capitals are suitable.

A good combination of colours for elderly and disabled is the use of background-foreground-colours blue-yellow for displaying text, the icons express their warning character through red frames, also on blue background. Speech messages should be short and should refer only to significant warnings. An ideal warning sound should have a frequency of 1000 Hz and rectangular waveform with a duration of 0.8 s. To emphasise the warning character, this tone is superposed in rectangular form with a chord and switched of and on once every 0.167 s. Events occurring with a higher frequency require more mild tones, e.g. 'gong'-similar for a good acceptance.

When the above were applied to the simulated HMI of IN-ARTE, the users' acceptance of the system enhanced, not only among the E&D but also among the "standard" users.

4. Conclusions

The IN-ARTE system is generally accepted by elderly drivers. The tendency to use the system is absolutely clear. Only the automatic intervention in certain situations, which are not considered as acutely dangerous e.g. reducing speed before intersections, is not fully accepted. Main reason for that is the slight confidence in the system, i.e. the reliability is not assumed very high.

Comparing the previous acceptance questionnaire with the results after the tests, it must be stated, that user acceptance declined after the test, but is still positive. The virtual IN-ARTE HMI as it was implemented for the user tests met with the expectations of the subjects of the tests. The feature automatic intervention of the system could not be tested with the virtual simulator, but the approach pre- and post-acceptance questionnaire made the subjects conscious about this functionality and they could estimate the consequences of using such a system.

Although tests (excluding the questionnaires) were performed in English, subjects stated that the system warnings, especially the speech messages, fulfilled their purpose to the full extent. Like any new product the user must adapt himself, and after a certain time he's used to get information from a display and perhaps will notice the direction of warning sounds.

Generally the implemented system warnings have no disturbing, annoying or scaring effects to the driver, only some elderly people felt scared about the warning sounds and uttered, that in really dangerous situations maybe warning sounds could increase the confusion already generated by certain events. As the annoyance of a warning system highly depends on the frequency of events, this subject must be carefully considered in additional tests.

From the above it is clear that in order to develop a user-friendly HMI for all drivers, there is a need to:

- Follow the "design for all" concept in developing in-vehicle HMI, so that we do not exclude parts of the driving population from using it and in order to offer better solutions for all.
- Provide easily adaptable solutions, both in terms of installation location (because of the co-existence of driving aids for E&D drivers) and the thresholds (i.e. volume, contrast, reaction time allowed).
- Test newly developed aids with E&D drivers and not with the standard middle-aged or even young male test driver, as is usual the case.

If the above considerations are taken into account throughout the design process, the resulting application will not only gain recognition from the E&D sector of the Market but will surely appeal as more attractive to all drivers.

References

1. Saroldi, A., Lilli, F. (1999). *System Function Definition*. EU project IN-ARTE (TR 4014), Del. 3.1
2. Portouli, V., Bekiaris, E. (1999). *Driver Warning Strategies*. EU project IN-ARTE (TR 4014), Int. Del. 4.1
3. Portouli, V., Bekiaris, E. (1999). *Emergency detection and handling strategies*. EU project IN-ARTE (TR 4014), Int. Del. 4.1
4. Fairbanks, Rollin J.; Fahey, Sara E.; Wieville, W. (1995). *Research on Vehicle-based Driver Status / Performance Monitoring: Seventh Semi-Annual Research Report*. U.S. Department of Transportation.
5. Handikappinstitutet Sweden et al. (1997). *Environmental Control Systems, Part I: Nordic Requirements*. Institutes of disabled Persons in Sweden, Denmark, Iceland, Norway and Finland, Tampere
6. Lerner, N.D., Kotwal, B.M., Loyns, R.D., Gardner-Bonneau, D.J. (1996). *Preliminary Human Factors Guidelines for Crash Avoidance Devices (NHTSA DOT HS 808 342)*. Task Force HMI
7. *European Statement of Principles on Human Machine Interface for In-Vehicle Information and Communication Systems* (1996).

A Multicultural Perspective on Digital Government Usability

Shirley A. Becker

Computer Science & Software Engineering
Florida Institute of Technology
150 West University Blvd.
Melbourne, Florida 32901
becker@cs.fit.edu

Frances Crespo

Software Engineering
Florida Institute of Technology
150 West University Blvd.
Melbourne, Florida 32901
fcrespo@email.com

1. Internet opportunities

The original intention of the World Wide Web was to make information accessible to anyone regardless of cultural considerations and geographic location. From a government perspective, the Internet provides an unprecedented opportunity for accessibility to each and every individual in the US. Elmagarmid & McIver (2001, p. 32) point out that e-government has the potential to profoundly influence a citizen's concepts of civil and political interactions with government. Government web site usability issues, inclusive of multicultural accessibility, need to be addressed in order for this to happen.

Government web sites, in terms of cultural diversity, must provide for effective information dissemination for multi-directional communication. Most of the US government web sites are English-based making it difficult if not impossible for use by non-English speaking individuals. Though some government web sites provide multilingual support (e.g. Social Security Online (www.ssa.gov)), too often, there are communication roadblocks associated with their use. In addition, many of these sites are focused on one-way communication such that information may be presented in several languages, but offer little support via a two-way interaction between the individual and the government.

Much of web development to date has focused on building local web sites. Localized web sites focus on a particular design in terms of language, text layout, symbols, dates, currency, numbers, and other usability elements in order to target a local audience. Unfortunately, there is little guidance on how to address the development of highly adaptable web sites in order to meet the needs of a multicultural audience.

In this paper, we describe web site usability from a multicultural perspective. This is an extension of our ongoing work in web site usability and internationalization. Section 2 provides an overview of web site usability factors as they relate to government web sites. Section 3 illustrates usability issues associated with existing government, and Section 4 concludes the paper with future research directions.

2. Multicultural web site usability

Powell (2000) formally describes web usability as allowing the user to manipulate the site's features in order to accomplish a particular goal. Nielsen (2000), Schneiderman (1998), and others have similar definitions of web usability in terms of taking into account user satisfaction to achieve usage goals. From a multicultural perspective, web usability means the design and information content of a web site must be adaptable to accommodate the needs of a diverse audience (Tuominen, 1998).

The perception of multicultural usability is impacted not only by culture but also by user characteristics such as gender, age, educational level, and technology skills, among others. As such, it is critical to profile users in order to gain insight as to what usability means in terms of design layout, information content, performance, navigation, and other factors. User profile attributes that should be taken into account are presented in Table 1.

Table 1: Web Usability Profile

User Profile Attribute:	Potential Impact on Usability
Age	Visual and audio impairments and technology experience
Gender	Needs further study
Computer Skills	Typing & keyboard skills and use of the web and email services.
Impairments	Visual, hearing, and physical impairments that might visual, audio, and physical use of the internet
Culture	Symbols, colors, date, time, units of measure, address, and currency formats
Language	Syntax, semantics, and textual flow
Environmental Attributes:	Potential Impact on Usability
Modem Speed	Low modem speeds should be taken into account when using graphics and animation. Performance statistics may be gathered for lowest, moderate, and high-speed access.
Browser	Web pages display differently in the various browsers and browser versions. These differences need to be taken into consideration, as they will impact usability.
Monitor Size	The size of the monitor will impact usability especially in terms of design layout, navigation, and information content.
Usability Goals	
The usability goals, associated with a particular web site, are needed to determine what usability issues might impact consumer understandability and ease-of-use. A usability goal, for example, might be: ● Informative – provides information about services, resources, and other links. This typically requires minimal bi-directional communication except for email. ● Interactive – provides bi-directional communication in that forms may be downloaded by the consumer. The consumer may send email, ask questions, or obtain online support from the web site.	

In addition to user profile attributes, it is important to identify the environmental attributes that impact the consumer. Modem speed is typically an issue, in terms of performance usability, as graphics, animation, and audio impact download time. Browser and monitor size are also factors especially with the introduction of hand-held devices.

A usability assessment model was developed to identify and measure eleven usability factors that impact an individual's online experience. These usability factors include: *design layout, design consistency, design standards, navigation, security, reliability, customer service, personalization, accessibility, performance, and reliability* (refer to Becker & Mottay (2000) for a brief description of the model). Several of the usability factors, as they relate to multicultural support of government web sites, are presented below.

Design Layout is the visual presentation of the web page in terms of background color, white space, horizontal and vertical scrolling, font size and color, text flow, and others. The design layout impacts understandability and ease-of-use. Page layout may be influenced by cultural differences in usability, such as the meaning of a particular color. Often, for example, a web developer assumes that the color red is universal in its meaning of "warning" or "error." In certain cultures, this is not the case. Another usability issue is text layout, which is a critical usability issue in supporting "Bidi" or bi-directional languages. Not only does the text have to flow in a right-to-left direction, the layout order of supporting objects (e.g., "okay" and "cancel" buttons) needs to be considered (Dallet, 1999).

Navigation refers to the navigational schema of a web site in terms of search paths and traversal mechanisms. Navigational simplicity is promoted through the effective use of links, frames, buttons, and text. Navigational considerations, from a multicultural perspective, include ready access to other languages on a page. This may require the use of symbols and the multiple language tags/links for easy page traversal. Navigational symbols should be universally meaningful and represent navigational flow of the overall web site. Navigational errors messages should be presented in the currently used language. Figure 1 illustrates navigational issues associated with several government web sites.

Design Consistency is the consistent location and content of web objects within and across web pages. From a multicultural perspective, the look and feel of a web page should be predictable to support understandability and ease-of-use. Figure 2 shows a Spanish version of a government web page that has a design layout that is different than its English version. The Spanish version, for example, has a greater number of graphics that don't add significant value in terms of semantic support, link traversal, or symbolic notation.

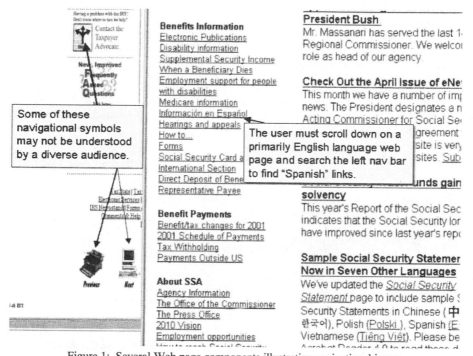

Figure 1: Several Web page components illustrating navigational issues

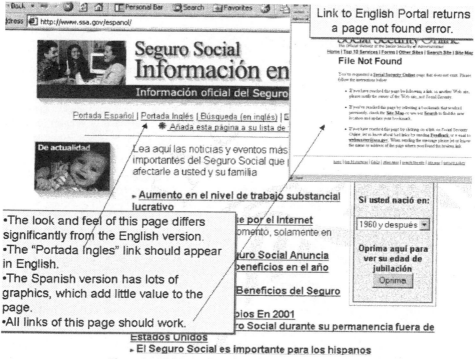

Figure 2: Web page illustration of design layout issues

Information Content includes meaningful error messages, prompts, button labels, textual descriptions, and help, among others. From a multicultural perspective, information that is translated from one language to another should be grammatically correct, not archaic, and not localized (jargon). Figures 3 and 4 show examples of government web pages whereby information content could be improved. Figure 3 shows an English language error box that is associated with the Spanish version of this web page. Figure 4 shows two web pages; the top page has inconsistencies in regard to English and Spanish textual presentation. The bottom web page provides an informal warning, in Spanish, regarding the usability of the page to which this message is linked. Though not shown here, the linked page is difficult to use because of its mix of English and Spanish.

Figure 3: Information content: Error message box is in English

3. Web Usability Transformation Rules

Our current research efforts are focused on developing a set of usability transformation rules that would provide guidance on the transformation of one web object into one or more web objects. The goal of this initial work is to understand cultural "usability" requirements or constraints, which would impact the transformation of a web object into another web object. The set of rules would focus on object transformation and not translation because of the number of usability attributes that might need to be taken into consideration. The basis of support for multicultural web sites is quite complex and cannot be viewed as a direct translation of textual content from one language to another.

The Bidi languages, for example, require a set of usability constraints to be taken into consideration during language translation. A set of usability transformation rules would have to include text layout (right-to-left), web object alignment as the textual display will impact the alignment of other web objects on the page, conditions placed on other web objects (e.g., message boxes with buttons), special characters and symbols, font size and type, as well as, other usability issues. A transformation algorithm would be used to transform one web object into another using these usability transformation rules.

Our research is XML (Extensible Markup Language)-based as it supports the standardization and customization of multicultural web sites. XML schemas and other supporting technologies provide the basis for web page, object, and attribute transformations based on a user profile. Multicultural data such as, language, color use, text layout, font style and size, symbols, etc., would be encapsulated in the XML code. Thus, any web site making use of this XML technology would have a standard means of personalizing a web page from various cultural perspectives.

4. Conclusion

This work is in the early stages of development, as multicultural support for government web sites has only recently been recognized as an important issue. In fact, the web page examples, shown in this paper, were taken from the government web sites that offer multi-language support. Many government web pages offer no or little support for non-English speaking web users.

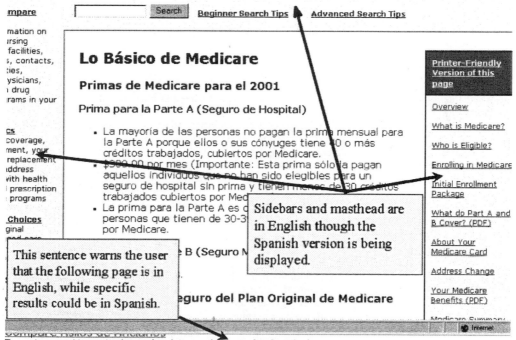

Figure 4: Web pages with mixed English and Spanish text.

References

Becker, S., F. Mottay. (2001). A Global Perspective of Web Site Usability. *IEEE Software*, 18 (1), pp. 54-61.

Dallet, N. (1999, October). *DHTML Localization on the Windows Update Site DHTM*. Microsoft Internet Developer.

Elmagarmid, K., W. McIver, Jr. (2001). The Ongoing March Toward Digital Government. *IEEE Computer*, 34 (2), pp. 32-38.

Nielsen, J. (2000). The Unbearable Lightness of Web Design. *Software Design & Usability* (pp. 50-76). Copenhagen Business School Press, Copenhagen, Denmark..

Powell, T. (2000). *Web Design: The Complete Reference*. Osborne McGraw-Hill, Berkeley, CA.

Schneiderman, B. (1998) *Designing the User Interface: Strategies for Effective Human-Computer Interaction*. Addison-Wesley Longman, Reading, Mass.

Tuominen, T. (1998). Going Global: Not for the Halfhearted. *Site Builder Network Magazine*, http://msdn.microsoft.com/workshop/management/intl/internatln.asp (Jan 8).

Subjectivity and Cultural Conventions: the role of the Tutor Agent in the Global Network

Paolo Bussotti[a], Gianluca Vannuccini[a], Davide Calenda[b], Franco Pirri[a], Dino Giuli[a]

[a]Electronics & Telecommunications Department, University of Florence,
Via S. Marta 3, 50139 Firenze, Italy
[b]Political and Social Science Department, University of Florence,
Via Valori 9, 50132 Firenze, Italy

Abstract

Proper exploitation of subjectivity by all types of users must be considered as a key requirement for the evolution from the Internet to the Global Network. A new structure of the logical Network architecture has thus been devised. Such an architecture can be devised while conceiving it as two integrated parts. One is the subjectivity engine, namely the Subjectivizing Gene Imprinter (SGI); the other is the Actuative Stack, conceived of as a conventional multi-layered architecture, suitably re-structured and linked to the SGI. The latter, which cognitively stores and elaborates user needs both for the user and for his/her relationships in the Network, dynamically supplies and updates "assistance modules" to the Actuative Stack, which runs them. This paper focuses on the SGI sub-system and its way of interpreting user needs - both from the point of view of direct human-machine natural language interaction and that of the monitoring of his/her activity - and translates them into the dynamic modelling of adequate subjectivity-based assistance modules for the Actuative Stack (i.e. the "gene imprinting" on the stack).

1.   Introduction

As the Internet is rapidly becoming the most important human activity field, both commercially, bureaucratically as well as in the socially-oriented field, the concept of the Global Network is nowadays embracing different cultural contexts worldwide. Unfortunately, the dream of a global network for a substantial improvement in the quality of life is contradicted by the reality of a network that was developed for diverse and simpler purposes; it is becoming more and more apparent that the architecture of this Network is overloaded by its current usage. The evident lack of contractual symmetry in E-commerce, the continuous risk of privacy violation and frauds are just a few examples which reveal the barely adequate nature of the current Internet network as a candidate to evolve into a true Global Network. It is expected that only a deep architectural intervention – at least at the Application level for compatibility reasons - could ameliorate the situation, and the most urgent aspect is the condition of the final user, in the sense of providing a "network depth" to the user, a representation of his/her subjectivity that may condition all his/her Network activities and some form of individual and collective assistance from the Network itself (Giuli, 2001a, 2001b, 2001c). This work has evident links with Human-Computer Interaction (HCI) research (Stephanidis, 2001), especially in the context of intelligent assistance to the user, as it can be made possible by the user-subjectivity model exploitation. As is well known, in fact, one of the major goals of Human-Machine Interaction has always been the development of systems that could effectively *assist* the human operator in his/her tasks, in order to fill the apparent cultural gap among, for instance, an *operator's manual* functional description of a system and what he/she really needs to know when problems occur. Such systems, known as Intelligent Assistant Systems (IAS) (Boy, 1991) have assumed the paradigm that *man should be a part of the design for overall effectiveness*. The current approach abandons the presumed *objectivity* of nature (and man's ability to know it objectively) and adopts epistemological paradigms (Winograd et al., 1986) based on *cognitive semantics, philosophical hermeneutics* and *theories of belief*. The meaningful power of natural language (NL) and the light it unconsciously casts on subjectivity is, according to us, a major instrument for deep human cognitive modelling: in these works we have extended the idea of machine-assistance to the very heterogeneous context of the Network, and we intend to try and show how a NL-based interaction can lead user communities to create and socially recognise cognitive meanings (and dynamically share them on the net), as well as allowing a user tutor-agent to pragmatically extract a reliable dynamic profile of the user by means of dialog.

The present paper reports on the specific results of current research within the framework programme called "B.E.S.T. beyond Internet"(Giuli, 2001a), while a general outline of such research is provided by a companion paper

(Giuli, 2001c). Focus on further specific and related subjects, thus completing the review of such research, is made in other companion papers (Giuli, 2001d; Calenda & Giuli, 2001; Vannuccini et al., 2001; Pettenati & Giuli, 2001).

2.   An agent for tutoring the final user

2.1   The system architecture and the Tutor Agent

All the traits summarised and structured through the metaphor of the *subjectivising gene* (Giuli, 2001b, 2001b) represent wide functional and inter-dependent problematic areas in which the knowledge from the subject (i.e. from its User Profile (Vannuccini et al., 2001)), as well as the accepted cultural and social competence in the Network - including that from end-user communities - can fruitfully give the user that *depth of dimension* as regards the Network, whenever dynamically and adequately used for the aim of (1) configuring a living and operating environment which is better suited to his/her subjectivity and (2) providing him/her with more awareness about objects and actors (Giuli, 2001d) envisaged as populating the Network. When engaging in relationships with other actors, either for commercial and/or socially-oriented purposes (e.g. thematic communities, free information services), or when delivering-receiving information contents, the user could benefit from the proposed system which is an extremely flexible *cognitive support* and possesses a wider competence than the user himself (i.e. a user-shared competence) which is capable of capitalising user needs, goals, plans both from his/her explicit requests and/or by inferring them from verbal interaction and from the monitoring of his/her performative/pathemic attitudes during his/her use of the Network. The proposed meta-model of the logic Network architecture derives from a typical stratified architecture for the sequential resolution of different-level communication characteristics. As is well known, the ISO-OSI stack (or the TCP/IP stack) provides internetworking by resolving, layer-by-layer, all formal heterogeneity of diversely originated information fluxes. Actually, OSI layers solve a set of ordered specific heterogeneity problems by means of static solutions, which fit the requisites of implementation-independence and effectiveness. The presented case deals with additional problems by envisaging, at various levels, the possible introduction of subjectivity conditioning aspects: for example, one can request the personalisation of an application interface, as well as the personalisation of the format presentation of contents. Moreover, a relational profile may introduce a different treatment of communication, based on previously stipulated agreements among the interlocutors and stored user-beliefs together with degrees of confidence in them. If the system is enabled to understand what the user attributes as being important in a communication, it can automatically attribute adequate levels of privacy to the outcoming flux (stream colouring and/or encryption), as well as select subjectively-retained important incoming fluxes with respect to the other fluxes. Finally, when addressing a Network service the user can be allowed to use natural language (NL) taking advantage of the collective localisation and connotation (i.e. as regards reputation) of services.

An evident aspect of imprinting the fluxes at all levels by means of the subjectivising gene is that, given their inherent nature, user beliefs and competencies are subject to evolve in quite an unpredictable way (just as with collective competence, even if the latter evolve more slowly). This implies that the envisaged solutions are dynamic, polymorphic (i.e. potentially providing a wide variety of strategies), and subject to unpredictable updating. An effective approach – based on object-oriented modelling – could be the definition and use of *strategic modules* to be run by the stack modules, which, in this context, are referred to as *assistance modules*. Therefore, a cognitive sub-system must be introduced which interacts with the information stack in order to dynamically deliver user knowledge-based effective solutions to the latter in the form of assistance modules. Thus, a partially interconnected architectural solution is required, as in figure 1. The multi-layered structure, namely the Actuative Stack, integrates the SGI while interacting with it, thus, implementing the overall new logical Network architecture, re-conceived and re-structured so as to allow the subjectivity-gene imprinting. While this paper focuses on the SGI, further insights as regards the Actuative Stack (AS), are given in the companion paper (Vannuccini et al., 2001). An *agent*, named the Tutor Agent (TA), is proposed as the abstract model which

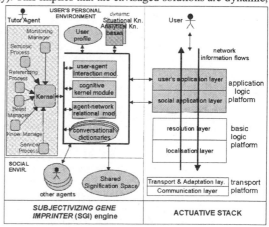

Figure 1: The SGI engine within a logical architecture for the Global Network

deals with cognitive and instrumental (i.e. updating of assistance module) aspects, and has as its *goal* the "imprinting of the subjectivizing gene" in all user activities or, in other terms, in adapting all types of Network activities initiated by him/her, to the user's needs. This can be achieved by inserting the agent in the so-called Subjectivizing Gene Imprinter (SGI) sub-system, which provides it with the ability to perceive and interpret user needs (both on a linguistic interaction basis, as well as by means of user action-monitoring), and at the same time, enabling it to access user-shared community competence. Indeed, its "imprinting" is actualised in the competence-based updating of assistance modules to the various stack layers within the actualising sub-system.

2.2 An overview of the Subjectivizing Gene Imprinter sub-system

As illustrated in figure 1, the SGI sub-system is modelled as a *core* processing unit (the Tutor Agent), which controls functional modules (FM) (i.e. the User-Agent Interaction Module (UAIM), the Cognitive Kernel Module (CKM) and the Agent-Network Relational Module (ANRM)), and uses their functions to access knowledge repositories such as the User Profile/s (UP), the user community encyclopaedic Shared Signification System (SSS), and portions of it – tailored (and dynamically updated from the SSS when required) on the knowledge about the user (or about other Network actors the user is engaged with) - represented by the Conversational Dictionaries (CDs). Knowledge is, then, stored and compiled from the analytical to the situational form by the CKM functions which operates on the dynamic knowledge basis (DKB). For instance, the kind of knowledge from the UP and the SSS might probably be analytical (e.g. frames, scripts) as well as basic operating knowledge in the DKB; however, the form of knowledge which is embedded in each assistance module is presumably situational, mostly for effectiveness. Moreover, the Cognitive Kernel Module has adequate methods for the delivery of assistance modules to the requesting stack layer, together with other more strictly cognitive functions. It is worth noticing how the Tutor Agent is articulated in concurrent and co-ordinated sub-process modules (or lower-level agents), which actively and reactively use the FM functions for the overall activity. Among these sub processes, the Service Process (ta-SP) which deals with the control of all the operations (i.e. not strictly cognitive) which allows the subsystem to carry out its tasks and interact with the Actuative Stack in an effective manner; the Monitoring Manager (ta-MM) which handles all conventionalised exception-signals from the monitoring of user activity on the terminal; and, the three processes which are responsible for cognitive processing, namely the Semiosic Process (ta-SS), the Belief Manager (ta-BF) and the Knowledge Manager (ta-KM). In addition, the Referencing Process (ta-RP), which actually references (in this case) Network objects, is often initiated by semiosis (i.e. the process of sign recognising and interpreting, often conditioned by the context and the circumstances in which the significant expression is perceived), though the processes are neatly separated: for example, the fact that a person can tell lies - which could be checked as being "plausible" or not by *referencing* real world objects – does not at all disturb the process of comprehension as regards the liar's utterance, i.e. *semiosis*. The Semiosic Process is, in turn, composed by three co-ordinated sub-process, namely the Syntactic, Semantic and Pragmatic Sub-processes. The access by the Actuative Stack towards the SGI sub-system is evidently a feedback reaction, and it is performed by the assistance modules which are embedded and active within each layer. In fact, together with their main function of conditioning the communication fluxes within their layer, assistance modules also monitor the results of their operations, according to pre-established - and cognitively generated - indicators of the felicitous outcome of their actions, from the user-centred viewpoint. In the case of exceptions, assistance modules back-signal the anomalous situation – through formalised messages - and wait for a decision from the SGI sub-system, while at the same time periodically informing the SGI sub-system about their views of the user activity status. In other situations, it can happen that it is the SGI sub-system itself which forces the updating of assistance modules, on the basis of direct communication with the tutored user, be (s)he, either, explicitly requesting a change or, implying inferences from his/her utterances.

2.3 Subjectivity-oriented Network assistance: the need for a natural language in HCI

A crucial point in designing an intelligent assistance system is, of course, that of human-machine communication or, from a semiotic viewpoint, the human-machine capability of sharing concepts, expressions and meanings. Indeed, human-machine communication allows the system to dynamically evolve its knowledge about the user, and helps it to try to stay tuned to the user's real needs. As a matter of fact, recent research has highlighted how, nowadays, the postulated global human signification code reveals itself as an intricate network of sub-codes - in the synchronic dimension - which is subjected to unpredictable variations in the diachronic dimension (Eco, 1975; 1998). In our scheme the SSS is both fed and utilised by users referring to it: the ability to "condense" similar meanings from different user contributions as well as to "branch" context-based parts of them - so that a Tutor Agent could provide and retrieve user-contestualized information - is implemented to approximate the dynamic behaviour of human semantics. Together with *semantic competence* – that is, by definition, culturally *conventionalised* (although within a multitude of contextual domains), the system also requires a *conversational competence*, which is the capability to understand speech acts; to memorise - and to "hyper-encode" - whole phrases previously pronounced; to hazard "ad hoc" inferences, on *supposed* references to those conversational contexts as well as on ellipsis; to understand user idiosyncracies; to fill the discourse using a wide variety of presupposition; to create new codes (e.g. neologisms); to recognise subjective possible *timic polarisation* beneath the surface of his/her *modalised* utterances. The problem of communication is apparently less critical for those assistant systems working in a well-defined activity domain, and which can rely on already defined and tested operating procedures. Since an operator, even if assisted, is expected to develop a competence in his/her application domain, as well as to learn how to master some technical languages of his/her machine, a convergence towards a human-machine compromise code is generally welcomed. On the contrary, the present paper, deals with assistance in the very wide and non homogeneous domain represented by the Network. The Network user is not generally a Network operator but, more probably, a simple "navigator": therefore (s)he should be assisted (tutored) in the most effective way, assuming as a design hypothesis that (s)he is completely unskilled and unaware of the several problems that (s)he can encounter browsing the Network. This implies the most radically user-centred choice: if the system assistance wants to effectively deal with all typologies of users, it must allow users to express themselves in their natural language, casting light on their most real (and perhaps unconscious) needs. Natural language could be an extremely important key to the problem, also keeping in mind Searle's "expressibility principle" (Searle & Searle, 1970), according to which any speech act (including negotiations, promises, assertions, requests, referencing to real-world things) can be expressed by language.

3.   Understanding and knowledge for assistance

3.1   Dynamic Knowledge Bases, Natural Language and Theory of Belief

A Tutor Agent requires a minimum conversational dictionary for interacting with the user and other Network actors, i.e. a portion of the shared encyclopaedia possibly tailored on the activity and interest contexts of the target actor. More specifically, for what concerns the tutored-user, this can be done on the basis of background profile data, explicitly inserted by the user - which are reliable as far as they represent a relatively "objective" user's knowledge about himself. They are used by the Semiosic Process as representative expressions of the user's activity and intellectual contexts, in order to extract an adequate dictionary from the Shared Signification Space. In case there is no matching, the semiosic process may try to: (i) identify synonymous expressions by browsing the SSS, (ii) ask the user about synonymous expressions or, as a last resort, (iii) ask the user to define (possibly in natural language) the mentioned context: this will be inserted in the local dictionary, both for new words and meanings, as well as for new contextual meaning of already existing words. The qualitative level of conversation with the user - and, as a result, the capability of effective assistance for vaguely stated user problems – is destined to increase with the knowledge (or the belief framework) the system develops about the user's needs and with the refining of its conversational competence. User needs should be allowed to be expressed in natural-language speech acts, which can either express directly a user's belief ,or provide a basis for system "beliefs interpretation" from the utterance. The interpretation of monitoring messages about the user's activity is a further information source about the user. Hence, a space of "system beliefs about its environment" is developed, incrementally, including, a set of system beliefs about user beliefs. When dealing with the relational profile of the user, a further belief nesting is required, as the system has to manage a space which also includes all system beliefs about user beliefs about other people's beliefs. In order to manage this complex belief space, the Tutor Agent provides a specialised sub-process, the Belief Manager, which in collaboration with the Semiosic Process, is capable of inference of various types, including deductions (which are no longer "true", but "plausible" in the belief-based intensional domain), inductions, abductions and all types of presuppositions. In this way, together with keeping track of previous conversations and elaborating contextual conversational patterns (e.g. on the basis of stereotypical conversational pattern), a *conversational competence* is approached, up to a certain point. A belief space needs to be formalised, but it is held to be important to enable the system to maintain the relation among natural language expressions and formalised internal language: from a radical point of view, every formalisation should be made dynamically and contextually. This is also required when dealing with conceptual categories and personal axiological investments: in fact, although some conventionalised categories can certainly be defined, the process of categorisation as well as the semantic and axiological opposition of members has been proven to be very dynamic, and is often limited to the conversational context. The system should, then, be enable to create categories in the conversational context, possibly starting from standardised ones and elaborating them by pragmatic interaction with the user.

3.2   A User Assistance – oriented Artificial Semiosis

The process of semiosis (and its syntactic, semantic and pragmatic collaborating sub-processes) has been defined for humans as the process – started by the perception of an expression that is considered to be significant - for accessing the meaning of the referred cultural concept through the theoretically unlimited exchange of significant expressions - or *interpretants* : if a concept is a *cultural unit*, derived from a cultural segmentation of the experienced world, as well as from abstraction and categorisation activities, it is assumed to be able to be represented through a set of properties - or distinguishing traits; each property, however expressed (e.g. by a text, an image etc.), requires a further signification of other concepts, and so on *ad infinitum*. Indeed, during semiosis, humans add new elements and structural links, over and above, the pregressed knowledge, and the process may stop when cognitive satisfaction is met. If an encyclopaedic support is provided to a system (as in this case), a semiosis can be emulated by a proper system process, and it is the key for approaching natural language (together with conversational competence and belief management). However, the *condition of satisfaction* for the system must be related to the goals of this sort of *artificial semiosis*, which are sub-goals of the main purpose of assistance. These may concern interpreting user's enunciations, with the aim of deciding whether (independently from the formulation of the enunciate) it introduces some new belief and/or reinforces/weakens some already stored user's beliefs, or whether it allows him/her to express subjective frames and meanings; interpreting commands through intelligently locating describing/operating frames, and act consequently; providing a meta-linguistic function (a help) to the user for words he does not know; interpreting user judgements to feed the SSS, in order to describe new code elements, as well as provide a mechanism for their acceptance in the semantic space (which independently, condense and/or contextually separate meaning sources); perform autonomous (transparent) belief maintenance and other actions. Form enunciates from concepts and communicate to the user (i.e. asking for explanations or confirmation during pragmatic activities), or to other subjects for the user's benefit.

4.   Conclusions

A new logical network architecture has been proposed in order to deal with user subjectivity requirements. A cognitive part (i.e. the Subjectivizing Gene Imprinter) employs intelligent assistance, based on user need interpretation and semantic competence, and compiles that analytical knowledge into situational knowledge to be embedded into "assistant modules", so as to be delivered/updated and run on the other part of the architecture, which is an actuative multi-layered stack (i.e.the Actuative Stack). This paper focuses on the SGI engine and the Tutor Agent (its processing core), and proposes a natural-language-based approach to knowledge dynamic feeding as well as to user needs interpretation. The approach is based on the process of semiosis (i.e. co-ordinated syntactic, semantic and pragmatic processes) and on the transformation from analytical to synthetic knowledge. This work is an outline of the global approach, and further research is necessary for actual integrated results.

References

Boy, G.A. (1991). Intelligent Assistant Systems. *Knowledge-Based Systems*, Vol. 6. London: Academic Press.
Calenda, D., Giuli, D. (2001). Subjects, subjectivity and privacy in the Global Network. In *proceedings of UAHCI 2001 Conference*, Louisiana, New Orleans (present volume).
Eco, U. (1975) (Ed.) *Trattato di Semiotica Generale*. Bompiani, Italia.
Eco, U. (1998) (Ed). *I Limiti dell'Interpretazione*. Bompiani, Italia.
Giuli, D. (2001a, February). *Rationale and objectives of the B.E.S.T. beyond Internet framework research program*. Dept. of Electronics and Telecommunications, University of Florence, Internal report DET-1-01.
Giuli, D. (2001b, February). *Subjective information and the subjectivizing gene for a new structure of the instrumental relation and interaction processes in the Global Network*. Dept. of Electronics and Telecommunications, University of Florence, DET-2-01 (in Italian).
Giuli, D. (2001c, February). *Subjectivity, citizenship and Public Administrations in the evolution towards the Global Network*. Dept. of Electronics and Telecommunications, University of Florence, DET-3-01 (in Italian).
Giuli, D. (2001d). From Individual to Technology towards the Global Network. In *proceedings of UAHCI 2001 Conference*, Louisiana, New Orleans (present volume).
Pettenati, M.C., Giuli, D. (2001). Human Subjectivity Profiling Factors for the Global Network. In *proceedings of UAHCI 2001 Conference*, Louisiana, New Orleans (present volume).
Searle, P.G., Searle, J.R. (1970, December) *Speech Acts: an assay of philosophy of language*. Paperback.
Stephanidis, C. (Ed.). (2001). *User Interfaces for All - concepts, methods, and tools*. Mahwah, NJ: LEA (ISBN 0-8058-2967-9).
Vannuccini, G., Bussotti, P., Pettenati, M.C., Pirri, F., Giuli, D. (2001). *A new Multi-layer Approach for the Global Network architecture*. In *proceedings of UAHCI 2001 Conference*, Louisiana, New Orleans (present volume).
Winograd, T., Flores, F. (1986). *Understanding Computers and Cognition*. USA: Ablex Publishing Corporation.

Virtual Environments - Improving accessibility to learning?

Sue Cobb[1], Helen Neale[2], David Stewart[3]

[1] Research Manager - [2]Special Needs VE Researcher
Virtual Reality Applications Research Team (VIRART)
University of Nottingham
University Park, Nottingham, NG7 2RD, UK
Fax: +44(0)115 8466771
www.virart.nottingham.ac.uk
[1]sue.cobb@nottingham.ac.uk, [2]helen.neale@nottingham.ac.uk

[3]Head Teacher
Shepherd School
Harvey Rd off Beechdale Rd
Nottingham, NG8 3BB, UK
Fax: +44 (0)115 9153264
www.shepherdschool.org.uk
[3]info@shepherdschool.org.uk

Abstract

The Virtual Reality Applications Research Team (VIRART) has been developing desktop Virtual Environments for Special Needs Education since 1991. Many of these projects have been completed with support and collaboration from Shepherd School in Nottingham, a large day school catering for students aged 3-19 with severe or profound learning disabilities. This paper will discuss the utility of these virtual environment applications as a teaching medium in schools. We will describe the virtual environments, why they were built and what the teaching purpose was. We will then discuss issues surrounding usage of completed applications; this will also touch upon issues relating to cognitive and physical barriers to student access to virtual environments. Finally, we will discuss the wider issues surrounding the relationship between researchers, educators and students in establishing ground rules for successful partnership in working together to produce useful and usable tools for education.

1. Introduction

The Virtual Reality Applications Research Team (VIRART) at the University of Nottingham has been examining the use of virtual reality technology and building virtual environments (VEs) since 1991. During this time VIRART's research has investigated the potential for Virtual Reality (VR) in Industry (Wilson, Cobb, D'Cruz, & Eastgate, 1996), and education (Cobb, Neale, Crosier, & Wilson, 2001) and has assessed usability and guidelines for VR use (Nichols et al., 2000). Of all these, the application area to which VR has received most interest and uptake is education, and in particular special needs education. This is possibly due to the foresight and motivation of special education teachers who have collaborated in VR research projects and certainly this will have a huge influence on uptake. It is also possible that the interest in VR for special needs education is indicative of some unique potential that this technology may provide access to learning, not provided by other teaching media.

In 1991 teachers from the Shepherd School viewed existing, albeit fairly primitive, demonstrations of desktop virtual environments. They recognised that this technology could allow students to gain experiences not otherwise available to them. For example, it could be possible to bring inaccessible places, such as mountains, into the classroom, to offer new experiences to students such as skiing or driving a car. Logically, if it was possible to replicate real world places in a virtual environment, then it may also possible to build virtual learning environments to support practical and social skills development. The opportunity for self-directed, interactive exploration of a virtual environment offered the facility for experiential learning by allowing the student to explore the VE at their own pace and interact with the VE in real time (Bricken, 1991). Augmented learning and self-paced discovery suggested that VR may offer potential as a suitable teaching medium in special needs education (Middleton, 1992; Powers & Darrow, 1994).

The staff from the Shepherd School then embarked upon a long-term relationship with VIRART in which we attempted to harness the potential of virtual reality technology; to construct virtual learning environments that could be applied as a teaching facility within the special needs classroom. Ten years on we have conducted and completed a number of research projects aimed at developing and evaluating virtual learning environments. Some of these projects are described as case studies in section 3. During this period there has been much learning on both sides about the practicalities of establishing true and meaningful partnerships between researchers, school staff and students. This paper will examine the implementation of these case studies and reflect upon lessons learned.

2. Virtual Environment Applications

2.1 Introduction

Virtual Learning Environments (VLEs) are computer-generated 3D interactive environments. These can be viewed on any standard Pentium PC and allow users to move freely around using a navigation device such as a joystick, and interact with virtual objects using a mouse or keyboard. Because it uses standard PC hardware, it can be used in schools across the UK with minimal additional investment. All of the virtual environments described in this paper were built using the Superscape virtual reality toolkit (VRT). These required the user to purchase a license for the Superscape Visualiser software to view the virtual environments. Snapshots of some of these virtual environments are shown in the picture gallery at the end of this paper.

2.2 Experiential Environments

A number of VEs were built to allow the user to experience different situations. The Virtual Skiing environment allowed users to slaalom down a ski slope; Virtual Town allowed users to drive a car, walk or fly a helicopter; and Virtual Bowling allowed users to position and play bowls (Brown, Kerr, & Wilson, 1997). These virtual environments had no objectives to support cognitive learning but were intended to offer students control over situations which may otherwise be unavailable to them such as skiing or driving. As a first investigation of the use of virtual environments for special needs users, our research objective here was to assess whether VR technology was at all appropriate for special needs education.

As well as enabling the VR development team to explore the capabilities of the VR systems they were learning to use, this project allowed the informal assessment of how well these VEs would be received by users with severe learning disabilities. Informal observations were extremely positive. Many of the students could use the VEs using the standard computer interfaces, and enjoyed the experience. Staff at the school were also excited by the opportunities for using VR in the classroom and saw the potential of this technology for future applications aimed at supporting learning objectives. For pupils with Profound and Multiple Learning Disabilities (PMLD) these worlds could enhance the establishment of virtual rooms where environments could be experienced in a controlled manner.

2.3 Language Development

The Makaton language of symbols and signs (Walker, 1985) is used as a communication medium in the Shepherd School. A programme of VEs were built to encourage the association of an object in a 3D interactive environment with a Makaton symbol and sign. This was done by allowing the user to move around the VE on the right side of the screen, explore and interact with the objects represented by the Makaton symbol to gain an understanding of their meaning and function A virtual manikin performed the associated Makaton hand sign when interacted with and a Makaton communication symbol was permanently displayed on the left side of the screen (Brown, Cobb, & Eastgate, 1995). Unfortunately, the funding for this project did not allow for an iterative process of evaluation and re-development to allow for end-user needs, and the VE was only evaluated (in terms of its usability rather than educational effectiveness) several years after its development.

The use of the Makaton programme of VEs was evaluated in 1997 within the Shepherd School (Neale, Brown, Cobb, & Wilson, 1999). Teachers liked the VE and their students (aged between 7 and 9, all with severe learning disabilities) could use it. However, they commented that some of the signs made by the manikin were unclear which may lead to incorrect learning and misinterpretation. Unfortunately, the effectiveness of these environments as a teaching aid for communication has never been tested.

2.4 Life Skills Training Environments

The enthusiasm amongst staff and students for the VEs described above, led to the initiation of a set of experiential environments that allowed the practise of important life skills. A house, with a functional kitchen and a shop were developed. Within the house the user could navigate around and interact with some objects and within its kitchen a user could boil a kettle and make a cup of coffee. In the shop, items could be selected from the shelves and then paid for at the checkout. Some studies were conducted to assess the usefulness of these VEs and it was found that a group of users who practised shopping skills in a VE were better at these skills in a real shop than those who had received no training (Standen, Cromby, & Brown, 1998).

As an attempt to provide an holistic learning experience, these individual learning environments were further developed and integrated into a Virtual City in which a user was engaged in making choices and carrying out procedural tasks associated with dressing, washing, cooking, shopping, catching the bus, ordering food in a café and dealing with emergency situations (Brown, Neale, Cobb, & Reynolds, 1999; Cobb, Neale, & Reynolds, 1998; Neale et al., 1999). This programme was developed with a consortium of groups who provide services to people with a learning (Brown, Kerr, & Bayon, 1998; Meakin et al., 1998). A user-centred development process was carried out and the VEs were refined after user and expert reviews and an in depth evaluation study. This study also found some instances of skill learning and transfer from VEs to the real world (Brown et al., 1999; Cobb et al., 1998; Neale et al., 1999).

2.5   General Outcomes

There have been some encouraging outcomes from the projects detailed above. We have found that Virtual Environments may provide a motivating and enjoyable media to be used for skill learning. We have also found that many users with severe learning disabilities are able to access VEs using standard input devices, such as the joystick and mouse. However, barriers still exist as some groups of users are unable to manipulate them effectively. It is estimated that as many as 40% of the students at Shepherd School are prevented from using virtual learning environments purely because the input devices are inappropriate for them.

There are two main barriers to accessibility of virtual environments; physical ability to manipulate the input devices effectively and cognitive ability to understand how to interact with the virtual environment. These are compounded when the input devices interface to the virtual environment through abstract metaphors. For example, using a mouse to position a cursor over a virtual object and clicking the button to lift the object up. There are two main ways in which we are trying to address this problem:

* ❖ reducing cognitive workloads by simplifying interaction requirements within the VE software (Neale, Cobb, & Wilson, 2000)
* ❖ providing wider access to students with physical disabilities by building direct manipulation input devices (Cobb, Starmer et al., 2001).

Users frequently need support to use the programmes. In school this may mean that a teacher or teaching assistant will be required to sit alongside a student to help them get through the program. This carries obvious restrictions in terms of personnel resources.

The potential of VR for learning has been identified by some of the studies described above. It has been suggested that VR can be useful for learning as it lends itself to a constructivist style of learning (Crosier, 2000; Winn, 1993). VEs allow users to directly interact with objects or people in the environment to explore cause and effect. This active learning is done at a pace determined by the individual and activities may be repeated as required. By representing complex tasks within a realistic environment this may aid transfer to the real world. An examination of how VIRART's environments support exploratory learning styles is reported in Neale et al. (1999).

Over the 10 years of developing VE applications for users with special needs, our approach to involving end-users and teachers in the development process has changed. Although many of the ideas for the earliest VEs developed originated from teacher suggestions, more recent applications, such as the Virtual City, have been built with much more involvement of end users (Neale, Cobb, & Wilson, 2001). Some user-group members expressed that their involvement in this process boosted their confidence and self-esteem (Meakin et al., 1998). One of the activities used to get participants to generate ideas and make decisions about which ideas they wanted to see included in a VE was by creating, as a group, picture based storyboards. Another positive spin-off from the project was that this method has since been implemented in classroom decision-making sessions.

3.   Implementation in the classroom

Despite all of the promise and excitement surrounding development if these applications, we find that they are still not widely used in the school. In part, this is due to the nature of working with innovative technology; the virtual reality developer platform used to create these virtual environments has itself developed over the last ten years. As a direct consequence many of the first virtual environments we created cannot be run on newer versions of the visualiser. For a while this problem was solved by installing different programmes each with corresponding versions of the visualiser on separate computers. However, local developments in the school IT infrastructure shifting to an intranet system meant that this was no longer possible. As a result only the newest version of the visualiser is accessible and so early virtual environments can no longer be used.

This being said, we cannot lay all of the blame for unused programmes on progression of technology itself. Many reasons for the lack of actual use of such VR programmes in schools are related to implementation. For example, the provision of teacher training, technical support and user manuals that integrate the programme with teaching/curriculum areas will have a huge affect on whether the systems are used. Some of our projects did include teacher training and documentation for basic system use, enabling the teachers from the classes worked with to use the programmes on their own. However, in projects where there was little or no little helpful documentation about using the VEs, teachers not directly involved in the study are unable to use the application.

Problems of continuity exist between the research organisation and the user site. Student projects usually last for one year, and after this time the student leaves the University. Versions of software are updated and need installing on school systems. Teachers may also leave, and if training is not available for new members of staff then it is likely that programmes will not be used. These problems may be addressed by improving the documentation that comes with each product – (which visualiser it works with, how it is used). Better communication between the development team and technical support at the school would also help. Within such projects the issues of implementation should be addressed from the very first stages of project management. This activity should be included in project planning, ideally involving teachers in the writing of user manuals.

4. Conclusions and further work

Research involving VIRART and the Shepherd School in the development of VE applications for people with special needs has progressed substantially over the last 10 years. We started in 1991, working with special needs educators to brainstorm ideas for possible applications, and have now developed and evaluated a series of applications designed for a variety of different learning aims, which are described in section 2. At the same time there have been vast advancements in the technology with which these environments are used; in terms of the processing power and affordability of hardware, and the capabilities and realism afforded by VR development platforms.

VEs have been shown to be effective learning tools, which may be used to support real world skill development. For profoundly disabled students the environments can provide stimulating and creative experiences, supporting a multi-sensory approach to learning. Projects have been developed with contributions from teachers and end-users at the school, and the process of involving these user groups has resulted in the inclusion of relevant and appropriate content and a feeling of ownership amongst all of those involved in this process. However, so far we have faced a number of barriers to the successful use of VR in schools. Many of the reasons for this are due to insufficient considerations at the start of the project given to requirements for implementation. The user-centred development process needs to be extended to consider not just the learning needs of the users, but also the requirements of the school and staff in the eventual implementation of the VE. A number of activities and resources to support this implementation may need to be provided, including: teacher training, VE programme manuals - for installation and use, a closer relationship between technical developers at the University and technical support staff at the school. It may also be important to specify early on in the project how this relationship will change after the project has finished, and how any further support will be provided.

From our experiences so far we have been greatly encouraged by the enthusiasm and support from teachers and students from the Shepherd School for VE use and development. Both partners have had to learn from the other, as VIRART researchers have learned about the needs of teachers and people with learning disabilities, staff and students have learned about the process of developing VE products. Successive development projects have facilitated increasing levels of collaboration from both partners, as we have learnt how to work together and consider the needs and limitations of the other partner. We hope to continue to work together to further improve accessibility to learning through VE development and use. Good practice in the process of developing virtual learning environments is equally applicable to other user groups with special educational needs in mainstream education and the wider community. VIRART are currently collaborating on projects to teach social skills to people with Autistic Spectrum Disorders (Parsons et al., 2000) and to support rehabilitation in recovery from stroke (Hilton, Cobb, & Pridmore, 2000).

Acknowledgements

This paper has commented upon the experiences of a research group working in partnership with a special needs school over a period of ten years. Many people have been involved in the projects conducted and funding has been provided via a variety of sources. It is impossible to acknowledge individuals by name and so we would like to thank all researchers, VE developers, School staff, users, students and community support groups collectively. Without huge efforts and enthusiasm from these people we would not have got as far as we did.

References

Bricken, M. (1991). Virtual Reality learning environments: potentials and challenges.*Computer Graphics, 25*(3), 178-184.

Brown, D.J., Cobb, S.V., Eastgate, R. (1995). Learning in Virtual Environments (LIVE). In R.A. Earnshaw, J. A. Vince, H. Jones (Eds.), *Virtual Reality Applications* (pp. 245-252). London: Academic Press.

Brown, D.J., Kerr, S.J., Bayon, V. (1998). The development of the virtual city: a user centred approach. Paper presented at the *2nd European Conference on Disability, Virtual Reality and Associated Technologies*, Skovde, Sweden.

Brown, D.J., Kerr, S.J., Wilson, J. R. (1997). Virtual Environments in Special-Needs Education. *Communications of the ACM, 40*(8), 72-75.

Brown, D.J., Neale, H.R., Cobb, S.V.G., Reynolds, H. (1999). Development and evaluation of the virtual city. *International Journal of Virtual Reality, 4*(1), 28-41.

Cobb, S.V., Neale, H. R., Crosier, J. K., Wilson, J. R. (2001). Use of Virtual Environments in Education. In K. Stanney (Ed.), *Virtual Environment Technology Handbook.*: Lawrence Erlbaum Associates, Inc.

Cobb, S.V., Neale, H.R., Reynolds, H. (1998). Evaluation of virtual learning environments. Paper presented at the *2nd European Conference on Disability, Virtual Reality and Associated Technologies*, Skovde, Sweden.

Cobb, S.V., Tymms, S., Cobb, R.C., Pridmore, T., Webster, D. (2001). Interfacing tactile input devices to a 3D virtual environment for users with special needs. Paper to be presented at the *9th International Conference on Human-Computer Interaction (HCI2001)*, New Orleans, 5th-10th August.

Crosier, J.K. (2000). *Virtual environments for science education: a school-based development.* PhD Thesis, University of Nottingham.

Hilton, D., Cobb, S.V., Pridmore, T. (2000). Virtual reality and stroke assessment: Therapists perspectives. Paper presented at the *3rd International Conference on Disability, Virtual Reality and Associated Technologies*, Sardinia, 23-25th September.

Meakin, L., Wilkins, L., Gent, C., Brown, S., Moreledge, D., Gretton, C., Carlisle, M., McClean, C., Scott, J., Constance, J., Mallett, A. (1998). User group involvement in the development of a virtual city. Paper presented at the *2nd European Conference on Disability, Virtual Reality and Associated Technologies*, Skovde, Sweden.

Middleton, T. (1992). Applications of virtual reality to learning.*Interactive Learning International, 8*, 253-257.

Neale, H.R., Brown, D.J., Cobb, S.V.G., Wilson, J.R. (1999). Structured evaluation of Virtual Environments for special needs education.*Presence: teleoperators and virtual environments, 8*(3), 264-282.

Neale, H.R., Cobb, S.V., Wilson, J.R. (2000). Designing Virtual Learning Environments for People with Learning Disabilities: usability issues. Paper presented at the *ICDVRAT, The International Conference of Disability, Virtual Reality and Associated Technologies*, Sardinia.

Neale, H.R., Cobb, S.V., Wilson, J.R. (2001). Involving users with learning disabilities in virtual environment design. Paper to be presented at the *1st International Conference on Universal Access in Human-Computer Interaction (UAHCI)*, New Orleans, 5th-10th August

Nichols, S.C., Ramsay, A., Cobb, S.V., Neale, H.R., D'Cruz, M., Wilson, J.R. (2000). *Incidence of virtual reality induced symptoms and effects (VRISE) in desktop and projection screen display systems* (274/2000): Health and Safety Executive.

Parsons, S., Beardon, L., Neale, H.R., Reynard, G., Eastgate, R., Wilson, J.R., Cobb, S.V., Benford, S., Mitchell, P., Hopkins, E. (2000). Development of social skills amongst adults with Asperger's Syndrome using virtual environments. Paper presented at the *International Conference on Disability, Virtual Reality and Associated Technologies*, Sardinia, 23-25th September.

Powers, D.A., Darrow, M. (1994). Special education and virtual reality: challenges and possibilities.*Journal of research on computing in education, 27*(1), 111-121.

Standen, P.J., Cromby, J.J., Brown, D.J. (1998). Playing for real.*Mental Health Care, 11/12*, 412-415.

Walker, M. (1985). *Makaton Vocabulay Development Project, Fourth Edition.*: Obtainable from 31 Firwood Drive, Camberley, Surrey, UK.

Wilson, J.R., Cobb, S.V., D'Cruz, M., Eastgate, R. (1996). *Virtual Reality for Industrial Applications: Opportunities and Limitations.* Nottingham: Nottingham University Press.

Winn, W. (1993). *A conceptual basis for educational applications of VR* (TR-93-9): Human Interface Technology Lab.

Picture Gallery:

Virtual City   Virtual skiing   Virtual bowling   Virtual supermarket   Virtual kitchen

In-vehicle telematic systems HMI for elderly drivers

Alessandro Coda, Sergio Damiani, Roberto Montanari

Fiat Research Centre - Strada Torino, 50; 10043, Orbassano, Torino, Italy
a.coda@crf.it, s.damiani@crf.it, r.montanari@crf.it

1. Summary

Telematic functions are growing up dramatically in the automotive field. This is expected to be a big advantage for all drivers in general and particularly for the elderly ones. At the meanwhile, these drivers will have to cope with a huge amount of information. To guarantee that driver's workload will be maintained within tolerable levels and will not provoke any safety's risk, a user-centred approach has to be adopted in the in-vehicle **Human Machine Interfaces** (HMI) design.

The article will explain the steps and the results derived from the adoption of a user-centered methodology approach in the design of the on-vehicle telematic HMI (starting from the user needs identification until the final product development). As already mentioned, the main focus is on the **elderly drivers,** who are becoming to be considered, at least within the automotive field, one of the most important customer target. Therefore, all **cognitive** and **physical attitudes** will be taken into account. It will be also explained how these attitudes are essential for the definition of the **telematic functions** that will be implemented into the cars and in which way the best interactions between driver and in-vehicle HMIs can be designed.

This article is structured in fourth parts. In the first one, a state of the art of the telematic opportunities related to the automotive domain will be introduced. Special attention will be given to the impact of these functions in the on-board information systems. In the second part, the user-centered approach adopted in Fiat Research Centre will be explained in details. Some examples where this methodology has been adopted fruitfully (for instance in the concept car NEA) will be reported as well.

Nevertheless, the project cycle requirements oncoming from the user-centered approach and the physical and cognitive attitudes which belong to the world of the elderly people has to be merged. This is the objective of the third part of the paper. Aware that this is a really innovative field, we reported in the last part of the paper some open items about the ways to build up an in-vehicle telematic HMI which is expected to be usable for the elderly drivers.

2. The impact of the telematic functions on the automotive domain

Thanks to the telematic revolution, a consistent amount of new functions are arriving in the automotive domain and are expecting to change the driver behaviour and the people mobility. At the meanwhile, this opportunities are increasing the information and functions available in the car. Therefore, the designers who are building up the future dashboard have to harmonise these functions with the driving constraints. In particular, they have to avoid any improvement in the driver's distraction due to the complexity of the on-board information systems.

These risks can be avoided in two ways. The first one, that will be detailed in the next paragraph, consists in the design of an easiest human machine interface (for instance, reducing the number of the input keys, simplifying the complexity of the menu, supporting the interaction with a voice recognition system). The other one is based on the introduction of an intelligent layer (the so called dialogue manager) which selects the best interaction according to the external car conditions (traffic and environments) with the driver and the user's mental workload. Both approaches (that are starting to be merged) have to consider that telematic providers could offer a huge amount of new opportunities during the vehicle life's cycle. That's why these interfaces should be able to implement new services and functions when and if available.

Basically, among the main innovative telematic functions has to be considered: the telephone system (actually based in Europe on the GSM technology); the advanced telephone functions like WAP; the emergency calls (in particular for mechanical, medical and security support); the GPS' based route guidance and support; the satellite antitheft; the travel and traffic information; the distance car diagnosis (the so-called telediagnosis) and the floating cars management. Besides, new functions are becoming to be available or they will be quite soon even if in an experimental version as well. For instance: the on-vehicle Internet based information system; the long range GPRS and UMTS platforms; the short range Bluetooth interface.

The drivers will interact with other innovative functions that are not actually included among the telematic ones. That's why, according to the scientific literature, some authors are starting to group the car's functions in several sets. Montanari, Andreone & Damiani (2000) propose, on the basis of the EU funded project COMUNICAR[1] findings, four groups:

- Telematic (GSM, GPS, etc.);
- Multimedia and entertainment (DVD, MP3, etc.);
- Comfort (climate control, etc.);
- Advanced Driver Assistant Systems (preventive safety functions like longitudinal, lateral and rear warnings, collision avoidance, etc.).

A real improvement in the quality of the drive is expected by the **integration** of these groups. Nevertheless, this integration has to be designed on the basis of the specific interactional requirements of every single group. For instance, the multimedia functions are not so critical for the driving as the collision avoidance ones. That means two different strategies in the HMI design procedure. Namely: the possibility to share the output between driver and passengers as well as a not distractive easy-to-use device for the multimedia functions; a capture-attention display able to suggest the right manoeuvre to the driver in case of dangerous situations for the collision avoidance display (Carrea, Deregibus & Montanari, 2000).

3. User-centered methodology for the in-vehicle telematic Human Machine Interface

To guarantee the design of the best HMI in terms of usability, a methodology based on the user expectations should be follow. This is the so-called "**user-centered approach**" (Norman and Draper 1986; Nielsen 1992; Norman 1999)

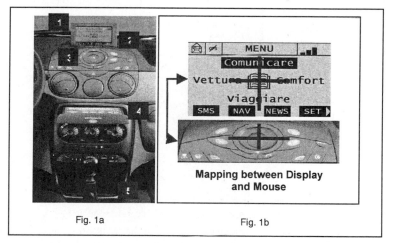

Fig. 1a   Fig. 1b

Figure 1 - Infocar project HMI

and it is devoted to build up all the design steps around the final users. In particular several activities are carrying on like: *test sessions* (where the usability of the prototype is tested and redesigned according to the findings), *focus groups* and/or *brainstormings* (where the user's ideas are encouraged and collected) and *surveys* on *users' samples* (to define the user expectations preliminarily and to forecast the info-system market impact).

Following the methodology explained below, the Fiat Research Centre developed and is designing some on-board telematic information systems. The first one is called "**Infocar**" and it is based on a on-board architecture where GSM and GPS modules are merged (fig.1). The HMI designed for this system in 1998 has been built on a simple user's interaction scheme (fig.1a). This means: (1) - a visual Liquid Crystal Display located in the centre of the dashboard to be visible both for the driver and the frontal passenger; (2) - some multifunctional buttons related to the LCD (as in the cellular phone, they are soft keys to activate quickly the system's functions); (3) - a mouse behind the LCD to manage the visual interface as a whole and few specific keys (no more than 6) to activate set ups like on-

---

[1] See www.comunicar-eu.org for details

line help, exit, main page, etc.; (4) - a keyboard to be used when the car is stopped; (5) - three keys for the emergency calls; (6) - a voice recognition system. The HMI has been designed following a central symmetry layout displayed as the mouse. All the main items can be found and selected by the driver without any scrolling (fig. 1b).

The second generation of the telematic in-vehicle information systems developed by Fiat Research Centre are mainly represented by the Lancia concept car **NEA**[2]. The strong concept which underlies the NEA project is the design of a sort of driver's friend to help and support its human partner during the driving task. Aware of this aim, a great number of **safety** (ACC, collision warning, emergency brake, etc.), **telematic** (GSM, GPS, traffic information) and **entertainment** applications (DAB, MP3, etc.) have been implemented. The scope is to assure the possibility for the driver to be informed and protected by any potential dangerous situation and to be continuously connected to the external world as it could be done from home or from the office.

To reach this scope, an innovative communicational scheme has been designed (fig.2). It is structured in two areas: *the first one* (1) is mainly dedicated to the primary information for the driver (namely the safety and vehicle functions); *the second one* (2) is devoted to the telematic services. Because of the difference between the information available in the two areas, the whole HMI has been designed according to the following criteria: the HMI for the primary functions is thought to get driver's attention suggesting the right manoeuvre to face critical situations (therefore, all the information are immediate, unambiguous and impressive); the HMI for telematic services has to give additional information reducing driver's distraction. This second set of functions is managed by the driver using a **touch screen**. Therefore, any physical input devices (neither buttons, knobs nor key) has been included in the dashboard. For the telematic area, a structure of the menu where is possible to see contemporarily several hierarchic levels of information has been provided to avoid any scrolling (see Deregibus, Montanari & Bianco 2000 for details on HMI structure).

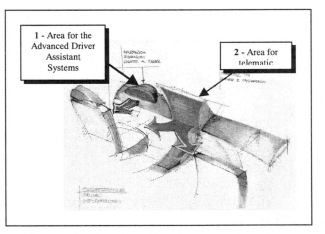

Figure 2 - Lancia NEA dashboard communication scheme

While the main usability problems is starting to be reduced thanks to the results of the projects mentioned above, a new generation of telematic in-vehicle interface are studying now. They are based (as announced in paragraph 2) on an **intelligent layer** able to manage the users' interaction accordingly to the car external conditions and the drivers' mental workload (Montanari, Andreone & Damiani 2000; Weathley, Rembonski, Gardner, Hurwitz, MacTawish & Gardner 2000; Baligand, Simoen, Bellet, Boudy, Bruyas et al. 2000). First vehicles which will show this interface concept are expected by the beginning of 2002.

4. How telematics could aid the elderly driver

By 2025 elderly people (namely people over 60 years old) will increase in Europe from 38.9% to 49% and in U.S. from 16.5% to 25%. That means a big change in the next two decades social structure. Nevertheless, this change does not mean what the common opinion thinks about the elderly people. In fact, elderly people are not mainly poor, disable or institutionalised. On the contrary, the improvements in the health treatments (80% of the elderly people has a good health) as well as in the education (the number of older adults with four years of college has doubled in

---

[2] See www.lancianea.com for details

last 20 years) are expanding the opportunities of this social group. One of the main consequence is a bigger mobility demand. Some analysis forecast that the vehicle miles travelled (VMH) by the over 65 aged driver will be 35% increased by 2025.

At the meanwhile, elderly people have to cope with some disadvantages. In particular, their perception and attention are reduced as well as the strength and flexibility. To support these physical constraints, car makers are augmenting car's accessibility and optimising the vehicle ergonomics features. On the other side, the reduction in the cognitive capabilities has to be taken into consideration not only during the traditional on-board information system design (namely, the font size of the instruments, the contrast between background and foreground colours, etc.), but also in the integration of the new telematic functions. In particular, these new services will affect positively the quality of the elderly people mobility. For instance, the supporting oncoming from an emergency aid could save ones life; the use of an easy navigation system could help a person who has problem to find the final destination, especially in an unknown place. Nevertheless, even if the telematic revolution could represents a big opportunity for the elderly people to increase the quality of their drive, the human machine interfaces of these devices should be designed according to the perception and attention attitudes of this target.

Several models have been conceived to represent the cognitive interpretation of a stimulus, especially if it is a question of information. One of this model to evaluate the interactions of a user with an informative input has been reported as follow. It was firstly proposed by Wickens in 1993 (Wickens 1993) and it is here integrated with remarks emerged from the current activities in Fiat Research Center. The result is a sort of **guidelines framework** on which the first prototypes of the on-vehicle telematic HMI - especially thought for elderly users - are based.

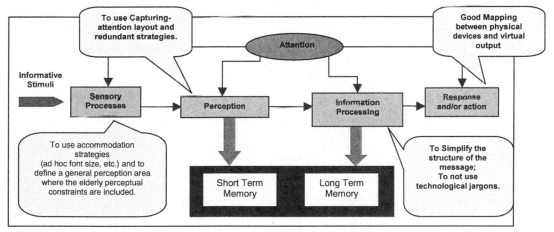

Figure 3 - Cognitive processing model adapted by Wickens 1993

The model starts to explain the cognitive process that takes place when an informative stimulus occurs. In the on-vehicle telematic information system domain, **informative stimulus** could be considered the output of the visual display that is communicating to a driver an incoming call. It has been proved that one of the main difficulties for the elderly people regards the visual outputs decoding. According to the researches, night vision capability is deteriorated after 40 years while the sensitivity to glare is dramatically increased. The 90% of the population over 50 age use bifocals and the useful-field-of-vision is diminished.

Some **accommodation strategies** could grant the right perception of a stimulus by the users. For instance, the size of the font should be adapted to the visual capability of a mean elderly person while an auditory signal should be perceptible for user who have senile hearing troubles. Nevertheless, these outputs must not to be annoying for the others. Therefore, it is becoming necessary to define a **general perception area** where the common senile constraints are included.

Given that some elderly people could have a reduction in the short term memory "storage" capacity, the perception of a message should be supported through an **output layout** particularly **capable to capture the driver attention** as well as **redundant strategies**. In this second case, a message can be repeated or a multimodal solution, where the visual and the acoustical outputs have been provided at the same time, could be adopted.

It is known that a stimulus is perceived as a **meaning message** if it is recognised by the previous user knowledge (Eco 1975, 1997) or it is cultural compliant with the context of use and perception (Gibson 1979). Even if the familiarity of the elderly people with the new information technologies is increasing (according to the surveys, the fastest growing segment of computer and web users are the over 50 people), it is really important to **simplify** as much as possible the **syntactical** and **semantic structure** of the **message** and to not use **difficult technological jargon** in the outputs.

Finally, if the message is well understood, the corresponding **actions** can be done properly. Nevertheless, if the devices are not easy to be handle because they are not ergonomical, some troubles could occur. Therefore, it has to be paid attention to the physical devices that manage the interaction between user and HMI. In the already mentioned Infocar project (see paragraph 3), **the mouse has the same structure of the virtual HMI** (on the basis of the Norman theory (Norman 1988) related to a right mapping between physical devices and cognitive artefacts). The usability tests done in Fiat Research Centre with a sample of elderly users confirmed this rule demonstrating a good level of usability of this system as well as a fast learning period.

5. Conclusions: telematic HMI for elderly people within the universal design approach

Aware that talking about a car for elderly people is neither advantageous in economic terms, nor convenient in respect to the social impact of these models (a boundary between aged drivers and the others could derive from this proposal), some authors are introducing the concept of universal design. It means to design a car for every kind of driver where the physical and cognitive attitudes of the elderly people are included.

In respect to the telematic services and on-vehicle HMI, the guidelines framework proposed in the previous paragraph should be merged with some items that could be considered the *"open conclusions"* of this paper. First of all, the HMI have to compensate and to cooperate with the declining capabilities of the aged users. Secondly, a special effort during the HMI design should be put on a seamless transition from existing systems to the new ones. Finally, the telematic systems should remain open in two directions: from the system to the infrastructure, because of new functions and services can be provided after the car's purchasing; from the system to the users, because of the interaction could be personalised and it could be dependable (according to the next on-board information system generation) on the external car condition and the level of activities of the driver in the driving task.

All these items are intended to be the main challenges and research directions to merge the human machine interface design focused on the elderly users with the universal design approach. Each of the conclusions reported below explain, in fact, not only an approach useful to design specific information systems but also the working framework in which all the interface design activities (especially related to the automotive field) should be included.

References

Baligand, B., Simoen, A., Bellet, Th., Boudy, J., Bruyas, J. H. et al. (2000). *Driving Attentional Demands in Extreme Situations: New Technologies to Managing Fatigue in Complex and Monotonous Driving*, Technical Paper on ITS 2000 World Conference , Torino (Italy).

Carrea, P., Deregibus, E., Montanari, R. (2000). *The invisible hand: how to design the best interaction between Advanced Driver Assistant Systems and Car users*. Convergence 2000, Paper 2000-01-C034, International Congress on Transportation and Electronics, Dearborn, Michigan (US).

Deregibus, E., Montanari, R., Bianco, R. (2000). *TRIPMATE: HMI for a "Connected care" vehicle*, Technical Paper on ITS 2000 World Conference , Torino (Italy).

Eco, U. (1997). *I limiti dell'interpretazione*. Milano: Bompiani.

Gibson, J.J. (1979). *The ecological approach to visual perception*. Boston: Houghton Mifflin College.

Montanari, R. Andreone, L., Damiani, S. (2000). *Comunicar: multimedia communication inside car*. Technical Paper on ITS 2000 World Conference , Torino (Italy).

Norman, D.A., Draper, S. W. (1986). *User Centered System Design*. Hillsdale, NJ: Erlbaum.

Norman, D.A. (1988). *The psychology of everyday things*. New York: Basic Books.

Weathley, D., Rembonski, D., Gardner, J., Hurwitz, J., MacTawish T., Gardner, M. (2000). *Driver performance improvement thought the Driver Advocate: a research initiative towards Automotive Safety*. Convergence 2000, Paper 2000-01-C075, International Congress on Transportation and Electronics, Dearborn, Michigan (US).

Wickens, C. (1993). *Psychology*. Boston: Houghton Mifflin College.

Easily Access to Technical Information of the Old Construction Handbooks By A Wbi System
M.I.C.R.A. - Manuale Informatizzato per la Codifica della Regola d'Arte

Rossella Corrao, Antonio De Vecchi, Simona Colajanni

D.P.C.E., University of Palermo, viale delle Scienze, 90128, Palermo, Italy
Tel.: +39 91234147/153/148, Fax: +39 91488562
e-mail: {corrao,simcola}@dpce.ing.unipa.it; devecchi@unipa.it

Abstract

The paper reports on the first results of a research we are carrying out at the Department of Project and Civil Engineering (D.P.C.E.) of the Faculty of Building Engineering in Palermo. The aim of the research is to create an electronic knowledge repository based on the information about the building construction rules reported in the handbooks edited since the 18th century and normally stored in different libraries around Europe. The system M.I.C.R.A. (Manuale Informatizzato per la Codifica della Regola d'Arte) is a WBI System able to allow different kind of users to easily find the information stored in the old construction handbooks and to immediately compare them with each other. Specific tools have been designed to support different study activities. Actually, the system is able to support study activities performed by experts or university students: these two users categories require specific tools able to satisfy their particular needs. The paper reports on the project of these tools, the design strategies and the specific "research criteria" we have adopted to improve cognitive access to the system. In particular we focus on the different kind of access that the system is able to allow users and on the interface of the system that was aimed at reducing the cognitive overload of the system users.

1. Introduction and design strategies

In the field of Architecture and Building Construction is increasing the tendency to search for information in the old construction handbooks to more easily find the best solutions to the recovery of ancient buildings. This approach allows a better development and a more careful conservation treatment of the ancient buildings (De Vecchi et al., 2001). These handbooks are stored in different libraries around Europe and for this reason they are not easily accessible by the users interested in having information about the old buildings construction. Furthermore they are not in commerce anymore and because of their state of preservation they can't be consulted by a wide range of people.

In the old construction handbooks the information is organized by subject. Texts and images are stored in different part of the handbooks preventing users from making a clear view of the building technologies (Guenzi, 1981).

The difficulty derived from the large quantity of information that is possible to extract from the old construction handbooks, the different way to face matters by the authors of the handbooks and the impossibility to easily manage this information have led to the definition of a new kind of handbook by using the New Information Technologies.

This handbook is able to allow different kind of users (from experts in the fields of Architecture and Building Construction to university students) to have easily and immediately access to the repository of constructive rules brought by the different authors of the old construction handbooks.

The underlying rationale of this system goes beyond the principles established for paper textbooks, which are organised by subjects. Actually, the information stored in the old construction handbooks has been organized in the system M.I.C.R.A.-Manuale Informatizzato per la Codifica della Regola d'Arte[1] considering buildings as technological system made up of a set of 'technical components' characterised by their own structural and spatial continuity. Therefore for a ready reference and an easy contextualization, the information has been systematised by subdividing each technical component (floor, ceiling, roof, curtain walls, etc.) into functional elements and/or layers[2].

This approach enables the user to understand how every *f. e. and/or l.* are made up and built; it also allow to overcome the limits due to the information structure of the old construction handbooks.

---

[1] Electronic Handbook to codify the "workmanlike".

[2] From now on you will read *f. e. and/or l.* instead of functional elements and/or layers.

794

Furthermore, this codification of data obtained by the division in *f. e. and/or l.*, makes the information taken from different handbooks homogeneous and, consequently, easier to mutual compare.

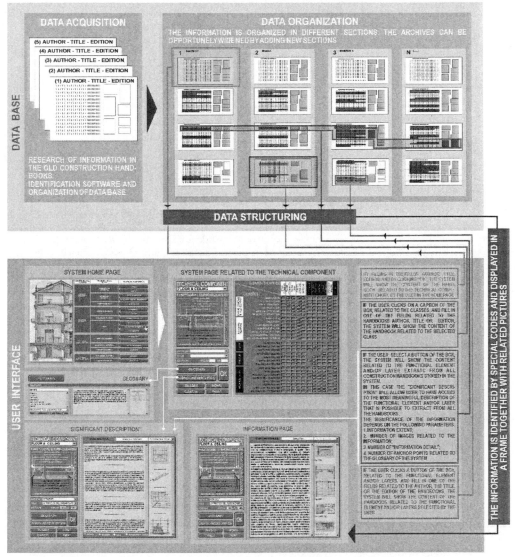

Fig. 1: The structure of the system M.I.C.R.A.: organization user interface and data

2. User interface

The user can search for specific information on *f. e. and/or l.* simply by retrieving excerpts that are identified by special codes and displayed in the system pages together with relevant images.

Various access modes are available: the user can retrieve the required information either directly, by selecting specially provided query boxes, or by refining search methods.

With such a system the user can obtain specific search results more or less rapidly, according to the amount of data input at the beginning of the query. The Access to the system is provided through a home page showing the available *f. e. and/or l.* and an image represents a section of an ancient building *(Fig. 2)*.

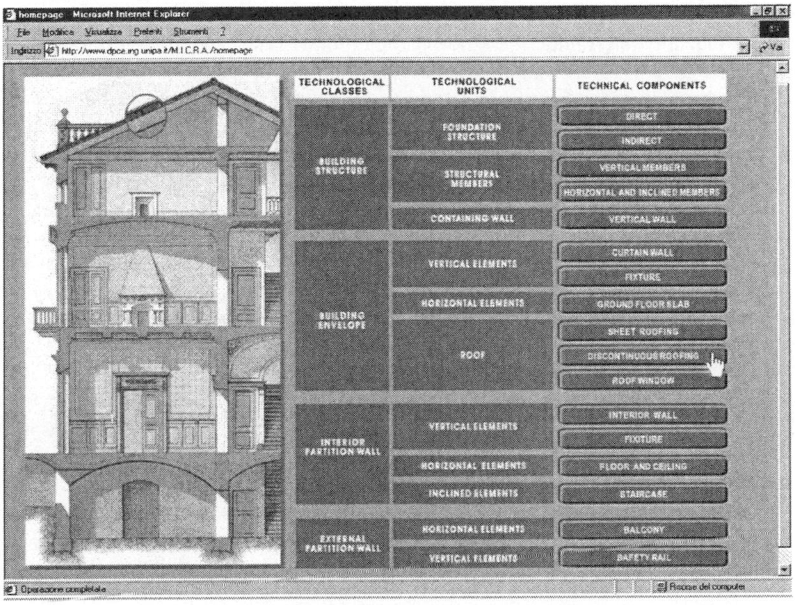

Fig. 2: The home page of the system

The active buttons on the home page are related to some technical components of the ancient building. They allow, if touch lightly, to underline on the image the technical components selected by the user.

By clicking the buttons it is possible to enter the opening page that shows a drawing in which it is possible to identify the elements or functional layers related to the information stored in the system.

The user can have access either to sets of descriptions showing the features of the *f. e. and/or l.* sought or to single descriptions simply by using the specific tools foreseen in the system and related to refined search criteria.

These tools are organized in one frame of the system pages.

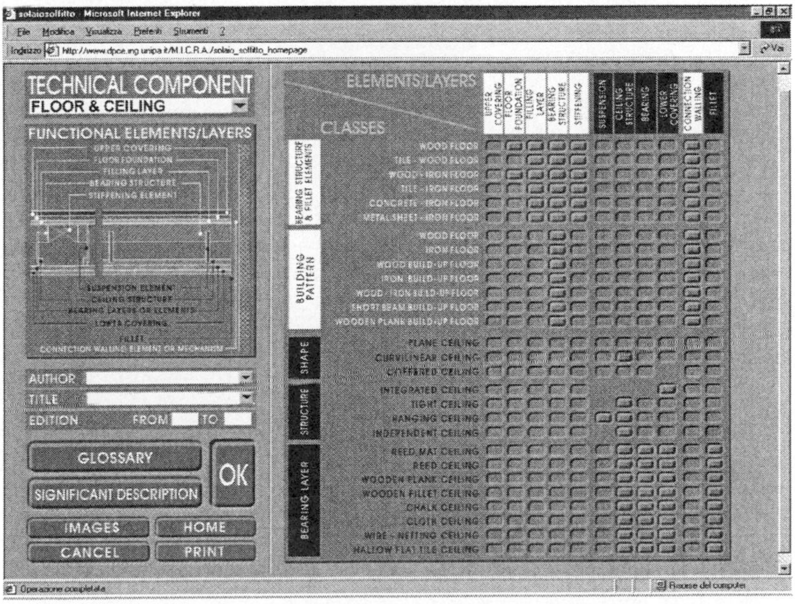

Fig. 3: One of the system pages related to the technical element "solaio-soffitto"

Among these tools there are:

- Query boxes to identify the Author, the Title and the Edition of the handbooks put into the system;
- a button to have access to the "Significant description" of *f. e. and/or l.* selected by the user. This tool allows users to have immediate access to the most complete description of the *f. e. and/or l.* automatically selected by the system on the basis of particular criterions.
- a button to have access to the "Glossary" of the system. By using this tool is possible to know the meaning of particular words that is possible to find in the texts of the old construction handbooks;
- a button (Images) to activate the research of the information stored in the system by leading to the images related to the text.

In the same side of the system page is possible to find a drawing of the technical components on which the user can have further information. This drawing shows each *f. e. and/or l.* linked to the Glossary. In another frame of the system page there is a box made up of buttons referred to *f. e. and/or l.*. Each column in the box refers to *f. e. and/or l.* related to the technological classes[3] that show the possible configuration of the technical element on which the system is organized. Actually, the Information on *f. e. and/or l* can be more accurate by relating them to a set of *t. c.*. Depending on the user's needs, information may be accessed through these different *t. c.*; at the same time different building techniques may be highlighted. The box contains all the *f. e. and/or l* arranged according to the *t. c.*.

3. Access to the information of the system M.I.C.R.A.

The system M.I.C.R.A. can be used to perform research activity according to the users special needs. In particular, the system is able to support study activities performed by experts or university students: these two users categories require specific tools able to reach particular goals more or less rapidly . So, specific tools have been designed to support these different study activities.

The user can program his research through the use of the query boxes above mentioned and shown on the left side of the system page. These tools are always active even if the page is changed by the user.

The user can have direct access to the description given by a chosen author by entering data into the query boxes and selecting a button on the box. In this case the information will be shown together with significant images, in another frame on the right side of the system page *(Fig. 4)*. This research mode can be used, in particular, by the expert.

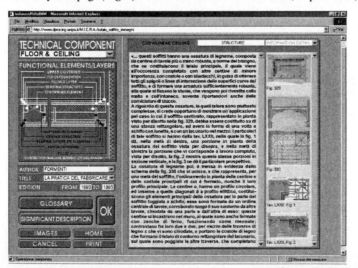

Fig. 4: One of the system pages showing the information related to the technical component "solaio-soffitto". Text and images are shown according to the research criterions specified by the user

If the users specify all the information related to the "Author", "Title" and "Edition" in the query boxes, they will be able to have access to the whole content of the handbook related to the technical element.

---

[3] From now on you will read *t.c.*. instead of technological classes.

If the users know only one of the three research criteria above mentioned the system automatically shows a list of information related to the other two research criteria allowing users to choose the one more interesting for them.
If the users select one button of the box and don't enter data into the query boxes related to the author's name, the title of the handbook or the edition, the system will show the information related to the selected functional element and/or layer by extracting these information from all the handbooks stored in the system. This research mode can be used, in particular, by the university students.

3.1 Advanced research tools: "Glossary", "Significant description" and "Images" access

The user can search for descriptions of unusual or specialized terms that are found in the handbooks by selecting the "Glossary" button. For each description the system will provide, if any, one o more significant images taken from the handbooks.
The "Significant Description" tool gives access to the most meaningful description of the *f. e. and/or l.* chosen by the user. The information is retrieved on the basis of specific parameters according to the user's requirements (number of words, images, links, specifications, etc.) (Gudivada et al., 1997). For each parameter the system will display data regarding the same functional element and/or layer and ranked for relevance.
Moreover it is possible to activate a research for images that are organised in a specific section of the system. It is possible to have access trough the button "Images".
The possibility to have access to the information by using images allows the not much expert user, to have more easily access to the textual information stored in the system. Actually the images represent, in particular for students, a simple way to understand the problem related to technical elements.
In short, depending on the amount of data input at the beginning of the query and according to their special needs, users can achieve specified goals with effectiveness, efficiency and satisfaction simply by using the tools above described. In this way we tried to consider different abilities, preferences and requirements of users in the variety of contexts of use.

4. Potential of the system and future development

The potentialities of comparison of the system represent a main starting point for the activation of didactic strategies based on the method of the case-based learning. For this reason we focused on the use of technologies related to the net. These result particularly useful because the recent languages used to structure the information on the Web allow to get an easy organization of data. By using these kind of "languages" is possible to simplify the management of data and to activate effective algorithms of research (Corrao et al., 1998).
As a consequence, the system have been designed to be used from computers located in different geographical contexts, simply connected to Internet but a stand alone version of the same system will be developed in the way to make it also available from PC not connected to the net. In this kind of system the information research is an application not limited to the simple search of key words, but it is also possible to consider the use of synonyms, the variation of the technical terminology, the comparison between documents related to similar matters: with the aim to resolve these kind of problems and to support users research activities we are considering the opportunity to introduce techniques of Artificial Intelligence, and particularly the Neural Nets. By using these techniques the system will be able to control the "meaning" of the information and to show them in a particular order. In such a way users can have access to the information more useful to their research activity.

References
De Vecchi, A., Corrao, R., Colajanni, S. (2001) Una normativa finalizzata al recupero di elementi costruttivi della tradizione locale. AA VV *Norme per l'edilizia storica diffusa. Quale spazio per la tutela, Workshop Proceedings*, Ancona.
Gudivada, V.N., Raghavan, V.V., Grosky, W.I., Kasanagottu, R. (1997). Information Retrieval on the Worl Wide Web. *IEEE Internet Computing*.
Corrao, R., Fulantelli, G. (1998). Cognitive accessibility to information on the Web: insights from a system for teaching and learning Architecture through the Net. *AA VV, Towards an Accesible Web, Proceedings of the 4th ERCIM Workshop User Interfaces for All*, Långholmen-Stockolm. [On-line]. Available: http://ui4all.ics.forth.gr/UI4ALL-98/corrao.pdf
Guenzi, Carlo (1981). *L'arte di fabbricare. Manuali in Italia 1750-1950*, Milano, BE-MA Editrice.

Accessible software for early maths education: focusing on content, strategies, interface

Giuliana Dettori[a], Michela Ott[b]

[a] Istituto per la Matematica Applicata, C.N.R., Genova, Italy

[b] Istituto per le Tecnologie Didattiche, C.N.R., Genova, Italy

dettori@ima.ge.cnr.it   ott@itd.ge.cnr.it

Abstract

Students presenting learning disabilities may have trouble accessing some educational tools, including software. In this paper we analyse what accessibility might mean when referred to content, strategies and interface of educational software. We point out how a very precise analysis of these components should be made in order to allow learning-disabled pupils to make good use of them. Based on this discussion, we conclude that the accessibility of a software application does not only depend on its characteristics, but also on how it is used in relation with each pupil's learning objectives and current cognitive level, as well as the whole educational setting.

1. Introduction

In the wide range of disabilities affecting school students in the developmental age, learning difficulties are statistically significant (CCLP, 1998). If not correctly tackled from the beginning, they can evolve into real handicaps, leading to early school dropout and difficult integration into the labor world. Learning-disabled pupils obviously need particular attention and care, but do they also need special tools? In this paper we try to answer this question, as regards educational software. We therefore discuss which features should be considered, and from what point of view, when deciding to use a software tool in a remediation program for learning-disabled pupils.

Effectiveness and functionality are certainly major objectives (Hines, 1994) to pursue when choosing software products to use with such children; this means choosing tools which, if properly applied, can actually contribute to attaining the proposed educational objectives. Effectiveness and functionality, however, seem to be strictly connected with accessibility: no significant educational goal may be reached if we rely on tools (no matter how powerful, rich and appealing) which prove difficult to use for these particular pupils, and that result in overlooking their specific, individual needs from the point of view of cognitive abilities (Simon & Halford, 1995), emotional/behavioural attitudes, competence already developed.

We are currently carrying out a long-term experimental project on the use of educational software in mathematics education, with elementary school children (2nd to 5th grades) that have specific learning difficulties (underachievers but not cognitively impaired). In this framework, the software has been used both for functional diagnosis and as a remediation tool. This experience led us to reconsider the issue of software accessibility for learning-disabled pupils; we analysed more closely than is usually done in the literature the nature of the main aspects of educational software tools, that is, content, educational strategy and interface.

These considerations might prove useful not only for mathematics, but also for the other basic abilities, such as learning to correctly read and write, whose lack marks from the beginning pupils who are unable to keep pace with the rest of their class. The experience devluped in the above-mentioned project, together with previous experiences in the fields of dislexia or disgraphia (Morchio et al, 1989), suggest us the consideration that with learning-disabled sudents it is particularly necessary to make a very precise analysis of content, educational strategy and interface, since even small shades of differences can result into cognitive obstacles for these particular students.

2. Software content

When a superficial analysis of a software tool is performed, the *content* appears easy to determine, simply by examining the user's manual or the set of exercises proposed. However, this is totally misleading since the cognitive task underlying the solution of arithmetic exercises actually differs according to how they are formulated.

Let us consider, for example, the case of programs that drilling pupils in addition operations. An exercise requiring computation of the sum of two numbers (e.g., 4+3= ...) aims to develop reckoning abilities, and suggests a view of the operation as a procedural process. If the result of the sum is to be chosen from a (small) set of possible results (e.g., 4+3= <5; 7; 2; 10>), the exercise lends itself rather to the development of evaluation capabilities (or lets the pupil guess rather than reason!). If an addition is to be performed using three given numbers (e.g., write an operation using the following symbols: 4, 7, 3, =, +) or choosing three numbers from a given set, then the accent is rather on emphasizing the relational character of addition. If the result is shown and one of the addends is missing (e.g., 4+?=7), the relational value of addition can be exploited to introduce the concept of difference. Though they all concern the same operation, these exercises actually have a different knowledge content and present different conceptual difficulties. The resulting impact on students' learning and cognitive development will differ depending on which of these exercises they tackle. This may not be obvious to a non-mathematician tutor, nor is it usually explained explicitly in the user's manual –(and we wonder if it is always clear even to most software designers!). Though the relational view of the operation usually proves to be more difficult to grasp than the procedural one, whether these different views of sums are equally accessible to a pupil strongly depends on his/her personal cognitive characteristics and on the way they are introduced. A careful mentor, paying attention to the different levels of understanding of the different forms of operation, can focus more precisely on each pupil's cognitive gaps (Fuchs et al., 1989).

Fig.1: The *Rooster* exercise in the "MathMax" software package (Dainamics)

Moreover, not all exercises concerning operations on numbers are actually focused on reckoning abilities. An interesting example is that shown in Figure1, where the user has to arrange the numbers by shifting each row left or right, so that the numbers in the column under the rooster sum up as assigned. In this exercise, making computations is secondary; what is actually exercised is the capability to make plans in order to reach the expected result. It is not surprising, then, that some children in our experimental group who were able to handle such simple sums without difficulty found it particularly arduous. On the other hand, learning to plan a solution strategy is an important part of mathematics learning, since it is necessary for developing problem solving abilities: we were not surprised, then, to find that the same pupils who found difficult in working out a solution strategy for the above exercise or in learning one from their mentor turned out to be poor problem solvers. Hence, it is also important to consider what kind of *solution strategies* a program allows or supports, and by means of what kind of tools. Though strategies are often built-in features of software tools, teachers should not feel compelled to accept them passively, since it is possible to have some control through interaction with the pupils by guiding them to make their reasoning processes explicit.

A third important component of mathematical content knowledge is the construction of meanings for concepts and symbols (Carey, 1991). It must be clear to the teacher that being able to apply the computation technique correctly for some operation does not imply understanding its meaning, and this is witnessed by the number of pupils who have learned to perform operations but still try to guess which one needs to be applied when solving a problem. None of the above mentioned exercises concerning sums offers any tool to build the meaning of the addition operation, which can be constructed and understood only in a problem solving (i.e. applicative) environment, possibly initially disregarding the use of arithmetical symbols in order to avoid conceptual overloading (Carreter, 1988).

Hence, from the point of view of software content, accessibility essentially means proposing exercises and activities concerning the very shades of knowledge that the pupil has been prepared - and is ready - to learn or to rehearse, bearing in mind that by themselves learning-disabled pupils are hardly able to make connections between different aspects of the same mathematical object , and understanding that a slight difference in point of view, if not suitably addressed, can hinder the pupil from making use of an apparently valid learning experience.

3. Educational strategy

Educational strategy refers to *how* a software tool leads the user to acquire some content knowledge. It is deeply embedded in the design of each educational program and is a key point to be checked in order to make any educational software effective with respect to some educational objective (Dreyfus, 1993).

Most software tools currently available for early maths education follow a drill-and-practice approach, that is, offer a predetermined set of exercises which consist in answering a question or solving some operation, but do not involve the construction or exploration of meanings. Such programs can help pupils to become fast and fluent in performing some operations but have little impact on their understanding of the concepts at play if meanings are not previously constructed and supported by means of suitable explanations and activities. Other programs, on the other hand, show a hands-on, constructivist approach (Norman, 1996), or make widespread use of representations as support to concept acquisition, or offer communication tools to favor learning as a shared process within a group of pupils. What makes one or other approach preferable is usually the nature of the educational objective (Hammond, 1992): simulation-based activities lend themselves to exploration and understanding of meanings, while repetitive ones aim to fix and automate purely algorithmic skills such as summing up numbers or memorizing multiplication tables.

Recent trends in mathematics education research tend to privilege educational strategies based on personal construction and discovery (Nardi, 1996), rather than the drill-and-practise one, which nevertheless is the type most often implemented by commercial software tools simply because it can be more easily automated.

In our experience, however, we have noticed that some repetitive tasks can prove to be as challenging as explorative ones for learning-disabled children; working out a solving strategy to tackle simple computations, or building a mental model to support table memorization, can turn out to be an exploration task as well. The same exercise, suitably introduced by the teacher, can be equally used to make a child grow in his/her mathematical thinking or to fix things (which is not necessarily a simple task for these particular pupils!). What makes the difference is whether such activity interacts with the child's proximal development zone (Vygotsky, 1986), that is, it is ready to be understood under teacher's guidance, or has already been understood and its knowledge must be consolidated, and this cannot be determined a priori but only evaluated by the mentor through careful observation of the child's attitude and behaviour.

Hence, we can say that a software is accessible from the point of view of the embedded educational strategy if the child is in a condition to make use of it to grow cognitively. As with the case of content, strategy accessibility depends not only on the software itself, but also on the use that the teacher is planning, or able, to make of it with respect to the educational objective and the current cognitive needs of the pupil.

4. Interface

The *interface* determines to what measure the users are put in a condition to make good use of the above-mentioned components, in that it concerns how the content is structured and presented. It not only determines how appealing, understandable and easy to use a program will be, but can also support thinking development, e.g. by means of a variety of meaningful representations, or by allowing a stimulating and non-constraining dialogue, hence possibly proving to be a concrete tool of cognitive growth. It is natural then to wonder what accessibility means as concerns the interface.

Though every pupil obviously has his/her own difficulties and cognitive style, the drawbacks which are usually shared by all these children are: scarcity of attention and memory, difficulty in understanding assignments, difficulty in conceptualizing, that is, building suitable mental representations to support mathematical operations as well as organizing the acquired knowledge at metacognitive level by recognizing analogies and connecting related concepts. Hence, based on a broad analysis of the literature and on our own experience, we think that interfaces accessible to learning-disabled children should present in wider measure those characteristics of consistency, essentiality and clarity which are anyway desirable for educational software.

Therefore, features which can hinder learning appear to be:
- poor logical organization, which makes the connection between learned concepts or abilities difficult;
- overly "nice" graphical presentation, rich in gadgets devoid of any meaning functional to the solution activity, which distracts, rather than helps, pupils perform the cognitive task at hand: even apparently easy tasks often require a big mental effort of these children, who welcome any possibility to switch from attentive work to play;
- proposing representations which are too symbolic to be easily understood, or purely descriptive, and useless for supporting an operative understanding of concepts;
- assigning tasks which require too much time to be solved, since the attention span of learning-disabled children is usually limited;
- joining too many concepts or abilities to be learned in one exercise, which results in an excessive cognitive effort;
- requiring to work fast, or not giving children enough time or chances to work out a correct answer, which makes them lose pace and prevents them from learning, since organizing knowledge into mental representations and retrieving it requires time, especially for children who are unable to do so instinctively.

Some features, on the other hand, appear to ease learning:
- assigning tasks which make sense in relation with the life experience of the pupil; one of the main difficulties with learning mathematics is the centrality of abstraction, which increases more and more with the level of complexity; adding more abstraction at task level can be too much for these children;
- making use of different representation registers (Duval, 1995) for the same assignment (such as symbols, words, graphics, sound), in order to convey the same information in different, connected ways;
- giving meaningful feedback for every answer; this gratifies the pupil when the answer is correct, and helps pinpoint problems when it is not.

Finally, it ought to be remembered that no software is so versatile as to *really* adapt its functioning to the needs of every pupil, hence a tutor's intervention is always necessary to introduce, explain, graduate and combine tasks. This is even more necessary with pupils who are hardly able to adapt their cognitive activity to a program's functioning.

5. Conclusions

Students presenting learning disabilities may have trouble in accessing some educational tools, including software. Accessibility in this case is to be understood as linked not to physical but to cognitive barriers, since we are not dealing with a physical problem. However, similarly to what happens with material situations, software should be considered accessible when pupils are in a condition to make good use of it.

When a software tool is used in a remedial program for learning-disabled pupils, attention should not be limited to the interface, since software accessibility is also strictly connected to the content and educational strategies embedded in the product. These three aspects are strictly interconnected and the issue of making each of them accessible cannot be suitably tackled independently of the others. The accessibility of content and educational strategy of software tools does not depend only on their peculiar characteristics, but also on the use that the teacher is planning, or is able, to make of them with respect to the educational objectives and the current cognitive needs of the pupil.

Acknowledgements
The experience mentioned has been carried out within the SVITA projectsinvolving CNR (IMA and ITD), ASL3 Genovese and Provveditorato agli Studi di Genova.

802

References

Carey, D.A. (1991). Number sentence: linking addition and subtraction word problems and symbol. *Journal for Research in Mathematics Education*, 22 (4) 266-280.

Carraher, T. (1988). Street mathematics and school mathematics. *Proceedings of PME-XII*, Veszprem, Hungray, I, 1-23.

CCLP-Coordinated Campaign for Learning Disabilities (1998). Tools for Living with Learning Disabi-lities. Retrieved. Jan 30, 2001 from: http://www.ldonline.org/ld_indepth/technology/tools.html

Dreyfus, T. (1993). Didactic design of computer-based learning environments. In C. Keitel and K. Ruthven (Eds.), *Learning from computers: Mathematics education and technology* (pp.101-130). NATO ASI Series F, N.121, Springer Verlag.

Duval, R. (1995). *Sémiosis et pensée humaine*. Bern, Switzerland: Peter Lang.

Fuchs L.S., Fuchs D., Hamlett C.L. (1989). Computers and curriculum-based measurement: Effects of teacher feedback systems, *School Psychology Review*, 18, 112-125.

Hammond, N. (1992). Learning with Hypertext: problems, principles and prospects. In: McKnight et al. (Eds.). *Hypertext: A Psychological Perspective*. Chichester: Ellis Horwood.

Hines, S. (1994). Creating Value for Students with Learning Difficulties. *Journal of Educat. Television*, 2, 79-91.

Morchio, B., Ott, M., Pesenti, E. (1989). The Italian Language: Developmental Reading and Writing Problems. In P.G. Aaron & R. Malatesha Joshi (eds.), *Reading and Writing Disorders in Different Orthographic Systems* (pp. 143-163). NATO ASI Series D, N.52, Springer Verlag.

Nardi, B.A. (Ed.). (1996). *Context and consciousness*. The MIT Press.

Norman, D.A., Spohrer, J.C. (1996). Learner centred education. *Communications of the A. C. M.*, 39 (4), 24-27.

Simon T.J., Halford G.S. (1995). *Developing Cognitive Competence: New Approaches to Process Modeling*. Hillsdale, New Jersey: Lawrence Erlbaum Associates.

Vygotsky, L. (1986). *Thought and language*. The MIT Press.

Anyone, anywhere access to community-oriented services

Pier Luigi Emiliani

Institute of Research on Electromagnetic Waves "Nello Carrara",
National Research Council
Via Panciatichi, 64, 50127 Firenze, Italy
Tel: +39-055-431090 - Fax: +39-055-410893 - Email: ple@iroe.fi.cnr.it

Abstract

This position paper describes and discusses PALIO[1], a new project recently funded by the European Commission's Information Society Technologies (IST) Programme. PALIO builds on previous European research and technological development efforts to provide a novel understanding of anyone, anywhere access to community-oriented services.

1. Introduction

In recent years, there have been several efforts in the direction of mainstreaming accessibility issues to provide universal access to general-purpose community-oriented services. This trend, which is likely to continue, is indicative of the technological advances which have taken place over the years and the increasing appreciation of the need to address diversity in the Information Society as a design challenge.

In this position paper, we aim to report on experiences towards universal access to interactive applications and services, which have been accumulated over the years from participation in European Commission supported collaborative research and development activities. In particular, we aim to show how novel concepts about universal access have progressively moved from formation to realization. This will be attempted by drawing upon early efforts to provide user interface software tools for user interfaces for all (Stephanidis, 2001), then reflecting upon their application in a large scale project, namely AVANTI (Stephanidis et al., 2001; Stephanidis et al., 1998) and finally, considering their relevance and influence on the PALIO project.

2. Universal Access

The notion of universal access to the Information Society (Stephanidis et al., 1998; 1999) is rooted on the concept of universal design, as it has evolved over the years. Universal design refers to the conscious effort to consider and take account of the widest possible range of end user requirements throughout the development life-cycle of product or service (Story, 1998). In recent years, it has been applied in interior and workplace design (Mueller, 1998), housing (Mace, 1998) and landscapes (Story, 1998).

In the context of Human Computer Interaction (HCI), design for all implies a proactive approach towards products and environments that can be accessible and usable by the broadest possible end-user population, without the need for additional adaptations or specialized (re-) design (Stephanidis & Emiliani, 1999).

There have been several efforts in the form of technical research and development projects, which have aimed to provide insights towards new user interface development frameworks and architectures that account (explicitly or implicitly) for several issues related to accessibility and interaction quality. Examples include the European Commission funded project TIDE-ACCESS TP1001[2] (ACCESS Technical Annex, 1993), the Japanese FRIEND21

---

1 The partners of the IST-PALIO project (IST-1999-20656) are: ASSIOMA S.p.A.(Italy) - Prime Contractor; CNR-IROE (Italy); Comune di Firenze (Italy); FORTH-ICS (Greece); GMD (Germany); Telecom Italia Mobile S.p.A. (Italy); University of Sienna (Italy); Comune di Siena (Italy); MA Systems and Control Ltd (UK); FORTHnet (Greece).

2 The partners of the TIDE-ACCESS (TP 1001) consortium are: CNR-IROE (Italy) - Prime contractor; ICS-FORTH (Greece); University of Hertforshire (UK); University of Athens (Greece); NAWH (Finland); VTT (Finland); Hereward College (UK); RNIB (UK); Seleco (Italy); MA Systems & Control (UK); PIKOMED (Finland).

804

initiative (Institute for Personalised Information Environment, 1995) and more recently the AVANTI Project[3] (Bini & Emiliani, 1997). Many of the concepts introduced by these projects, were subsequently taken up by industrial initiatives in an attempt to comply to legislative clauses or as a result of genuine interest towards increased access to interactive systems and services. For instance, the notions of *abstraction, platform independence* and *interoperability* (Akoumianakis et al., 2000) as well as the notion *of pluggable-look and feel* are now being considered in both Microsoft's Active X technologies and SunSoft's accessibility initiatives.

3. Universal Access to Community-Oriented Services

Building upon the results of earlier projects, such as ACCESS and AVANTI, we are now pursuing the universal access challenge at another level. Specifically, the PALIO project sets out to address the issue of anyone and anywhere access to community-wide services. This is an extension of previous efforts, as it accommodates a broader perspective on adaptation and covers a wider range of interactive encounters beyond desktop access. In what follows, we will briefly overview how this project addresses the issue of universal access and how it advances the current state of affairs by considering novel types of adaptation based on context and situation awareness.

3.1 The PALIO system

PALIO is a project, which is funded by the EC's IST Programme. The main challenge of the PALIO project is the creation of an open system for accessing and retrieving information without constraints and limitations (imposed by space, time, access technology, etc.). Therefore, the system should be modular and capable of interoperating with other existing information systems. In this scenario, mobile communication systems will play an essential role, because they enable access to services from anywhere and at anytime. One important aspect of the PALIO system will be the support of a wide range of communication technologies (mobile or wired) to access services. In particular, it will be possible for a user equipped either with a common cellular phone or an advanced WAP phone to access services wherever he/she is; referring to Figure 1, the *Augmented Virtual City* (AVC) centre will adapt the presentation of information to the different access technologies.

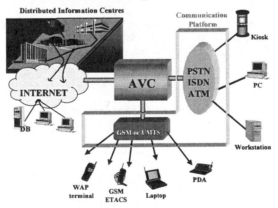

Figure 1: PALIO architecture

The PALIO system envisages the adaptation of both the information content and the way in which it is presented to the user, as a function of user characteristics (e.g., abilities, needs, requirements, interests); user location with the use of different modalities and granularities of the information contents; context of use; the current status of interaction (and previous history); and finally used technology (e.g., communication technology, terminal characteristics, special peripherals).

The PALIO information system consists of the following main elements (see Figure 1):

---

[3] The partners of the ACTS-AVANTI AC042 consortium are: ALCATEL Italia, Siette division (Italy) - Prime Contractor; IROE-CNR (Italy); ICS-FORTH (Greece); GMD (Germany), VTT (Finland); University of Sienna (Italy), TECO Systems (Italy); STUDIO ADR (Italy); MA Systems and Control (UK).

1. **A communication platform** including all network interfaces to inter-operate with both wired and wireless networks;
2. **The AVC centre** which is composed of the main adaptation components, a service control centre, and the communication layers from and to the user terminals and the information services;
3. **Distributed Information Centres** in the territory, which provide a set of primary information services.

The AVC centre is the architectural unit, which manages diversity and implements the mechanisms for universal access. The AVC will be perceived by users as a system, which groups together all information and services that are available in the city. It will serve as an augmented, virtual facilitation point from which different types of information and services can be accessed. The context- and location- awareness, as well as the adaptation capabilities of the AVC, will enable users to experience their interaction with services as a form of 'contextually grounded' dialogue, e.g., the system always knows the user's location and can correctly infer what is 'near' the user, without the user having to explicitly provide information to the system.

The envisaged main building blocks of the AVC are depicted in Figure 2, and can be broadly categorised (according to their role within the project) into the adaptation infrastructure; the service control centre; the software communication layer to and from the user terminals; the distributed information services and their integration infrastructure. What is of interest to this conference is the adaptation infrastructure, which is briefly elaborated below.

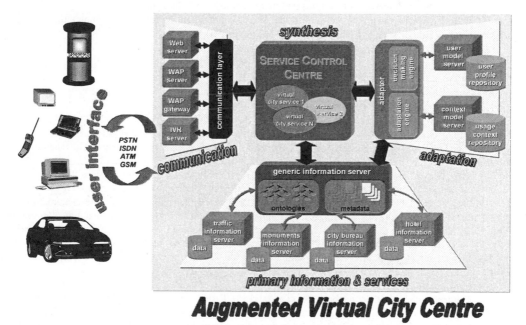

Figure 2: AVC centre architecture

3.2 Adaptation infrastructure

The Adaptation Infrastructure will be responsible for content and interface adaptation in the PALIO System. It will be composed of the Adapter, the User Model Server and the Context Model Server.

The **User Model Server** (UMS) will integrate and manage information concerning user characteristics (e.g., interests, interaction style, disabilities), in the forms of both individual user profiles constructed during interactive sessions, and user stereotypes for groups of users that share a number of characteristics. The UMS will contribute to the provision of support for adaptation at the information content and presentation levels, by making user-related information available to the Adapter. User-related information will be provided directly by the user. However, the system will be designed so as to be functional even in the case where users do not provide the required information when requested for it (e.g., through interactive question-answer sessions). Furthermore, the system will formulate assumptions about the user at run-time, on the basis of interaction behaviour. The UMS will retain individual user profiles in the User Profile Repository. When permitted by the user, these profiles will be maintained between sessions. The latter case (i.e., having profiles that contain information from multiple interactive session of the user with the system) is expected to have particular added value in the use of the system, as some of the characteristics and interaction patterns of users can only be observed over a longer period of time, and not within the usually limited temporal constraints of a single session.

The **Context Model Server** (CMS) will assemble the context profiles, using information retrieved from the **Usage Context Repository**, and will inform the adapter about the current context. Context related information includes, for example, the user location, the characteristics of the terminal and network used to access the system, etc. It should be noted that, in order to collect information about the current context of use, the CMS will communicate (directly or indirectly) with other components of the PALIO system, which will be the primary carriers of such information. These first-level data collected by the CMS will then undergo further analysis, with the intention to identify and characterise the current context of use. Similarly to the UMS, the CMS will be able to respond to queries made by the Adapter regarding the context, as well as to relay notifications to the Adapter about important modifications to the current context of use, which may necessitate the triggering of specific adaptations.

The **Adapter** will constitute the basic adaptation component of the system. It will integrate information concerning the user, the context of use, the access environment and the interaction history, and will adapt the information content and presentation accordingly. Adaptations will be performed on the basis of:

- user interests (when available in the User Model Server and/or resulting from the ongoing interaction);
- user characteristics (when available in the User Model Server);
- user behaviour during interaction (provided by the User Model Server);
- type of telecommunication technology and terminal (provided by the Context Model Server);
- current service state and the session history (provided by the Service Control Centre);
- location of the user in the city (provided by the relevant information services, e.g., GPS);
- type of input request;
- etc.

The Adapter will itself comprise two main modules, the **Decision Making Engine** (DME) and the **Adaptation Engine** (AE). The DME will be responsible for deciding upon the need for adaptations, based on: (a) the information available about the user, the context of use, the access terminal, etc., and (b) a knowledge corpus that will relate specific (types of) adaptations with such pieces of information. Combining the two, the DME will attempt to make decisions about the most appropriate adaptation for any particular setting and user / technology combination addressed by the project. The AE will be complementary to the DME and will undertake the instantiation of the decisions communicated to it by the DME. The decoupling of reaching adaptation decisions and effecting them (resulting from the presence of the DMS and the AE as two distinct functional entities), bestows the PALIO system with high levels of flexibility, as new (types of) adaptations can be introduced into the system very easily, while the rationale for arriving at an adaptation decision and the functional steps required to carry it out can be varied separately.

It should be noted that one of the very important properties of the described architecture is the fact that there is continuous monitoring of the interaction between users and the PALIO system, so that modifications in the user behaviour, the context of use, or the status of communication facilities and devices can be identified, interpreted, and used to trigger the appropriate type of adaptations (e.g., inclusion of links to relevant information, spatial or temporal restructuring of information elements, modification of presentation parameters such font, colour, voice gender, volume, etc.)

4. Concluding remarks

We have briefly reviewed how PALIO, a European Commission funded collaborative project under the IST Programme, seeks to advance a novel understanding of universal access in community-oriented services. PALIO constitutes a substantial extension over previous efforts on universal access, since it introduces and explicitly accounts for novel types of adaptation and new interactive encounters beyond the desktop. Accordingly it pursues an architectural model of interaction which is expected to be of wider applicability in service communities other than tourism.

Acknowledgements

The author is the Technical Manager of the PALIO project and wishes to acknowledge and express his sincere appreciation to all members of the consortium.

References

ACCESS Technical Annex (1993). *Development Platform for Unified Access to Enabling Environments*. Technical Annex (TIDE TP 1001), The ACCESS Consortium.

Akoumianakis, D., Savidis, A., Stephanidis, C. (2000). Encapsulating intelligent interactive behaviour in unified user interface artifacts. *Interacting with Computers, Special Issue on The Realities of Intelligent Interface Technology*, 12 (4), 383-408.

Bini A., Emiliani P.L. (1997). Information about Mobility Issues: the ACTS AVANTI Project. In *AAATE Conference*, Porto Carras, Tessaloniki (pp. 85-88).

Institute for Personalised Information Environment (1995). *FRIEND21 Human Interface Architecture Guidelines*. Tokyo, Japan.

Mace, R. (1998). Universal Design in Housing. *Assistive Technology*, 10, 21-28.

Mueller, J. (1998). Assistive Technology and Universal Design in the Workplace. *Assistive Technology*, 10, 37-43.

Stephanidis, C. (Ed.). (2001). *User Interfaces for All - Concepts, Methods, and Tools*. Mahwah, NJ: Lawrence Erlbaum Associates (ISBN 0-8058-2967-9, 760 pages).

Stephanidis, C., Emiliani, P-L. (1999). Connecting to the information society: a European perspective. *Technology and Disability* Journal, 10(1), 21-44.

Stephanidis, C., Paramythis, A., Sfyrakis, M., Savidis, A. (2001). A Case Study in Unified User Interface Development: The AVANTI Web Browser. In C. Stephanidis, (Ed.), *User Interfaces for All - Concepts, Methods and Tools* (pp. 525-568). Mahwah, NJ: Lawrence Erlbaum Associates. (ISBN 0-8058-2967-9, 760 pages).

Stephanidis, C., Paramythis, A., Sfyrakis, M., Stergiou, A., Maou, N., Leventis, A., Paparoulis, G., & Karagiannidis, C. (1998). Adaptable and Adaptive User Interfaces for Disabled Users in the AVANTI Project. In S. Trigila, A. Mullery, M. Campolargo, H. Vanderstraeten & M. Mampaey (Eds.), *Intelligence in Services and Networks: Technology for Ubiquitous Telecommunications Services - Proceedings of the 5th International Conference on Intelligence in Services and Networks (IS&N '98)*, Antwerp, Belgium (pp. 153-166). Berlin: Springer, Lecture Notes in Computer Science, 1430.

Stephanidis C. (Ed.), Salvendy, G., Akoumianakis, D., Bevan, N., Brewer, J., Emiliani, P.L., Galetsas, A., Haataja, S., Iakovidis, I., Jacko, J., Jenkins, P., Karshmer, A., Korn, P., Marcus, A., Murphy, H., Stary, C., Vanderheiden, G., Weber, G., & Ziegler, J. (1998). Toward an Information Society for All: An International R&D Agenda. *International Journal of Human-Computer Interaction*, 10(2), 107-134.

Stephanidis, C. (Ed.), Salvendy, G., Akoumianakis, D., Arnold, A., Bevan, N., Dardailler, D., Emiliani, P.L., Iakovidis, I., Jenkins, P., Karshmer, A., Korn, P., Marcus, A., Murphy, H., Oppermann, C., Stary, C., Tamura, H., Tscheligi, M., Ueda, H., Weber, G., & Ziegler, J. (1999). Toward an Information Society for All: HCI challenges and R&D recommendations. *International Journal of Human-Computer Interaction*, 11(1), 1-28.

Story, M.F. (1998). Maximising Usability: The Principles of Universal Design. The *Assistive Technology Journal*, 10 (1), 4-12.

Evaluating the usability of Brazilian Government web sites

Elza Maria Ferraz Barboza[a]*
Eny Marcelino de Almeida Nunes[b]
Nathália Kneipp Sena[c]

[abc] Brazilian Institute for Information on Science and Technology (IBICT),
SAS Quadra 5, Lote 6, Bloco H, CEP 70070-914, Brasília, DF - Brazil

Abstract

Web sites of Brazilian federal government have been analyzed and evaluated, particularly the sites of the ministries that integrate the Brazilian Society Program. A checklist based on ergonomic principles was established. Evaluation criteria were clustered into four large main headings: scope and purpose, content, graphics and multimedia design, workability. The results point out that the governmental agencies should improve their web sites aiming at a better adequacy to ergonomic standards.

1. Introduction

The present paper proposes criteria to analyze and evaluate the web sites of the Brazilian Government in terms of adequacy to ergonomic standards, especially their usability. The sites that offer information services were selected from the "Information Society Report" published by the Information Society Work Group which is part of the Ministry of Science and Technology. That online content brings public sectors closer to citizens and businesses leading to better public sector services. Some lack of standardization has been observed after analyzing each site and comparing them. A list of criteria was elaborated. Positive and negative aspects of each site are presented not forgetting the limitations imposed by the subjectivity and idiosyncrasies inherent to the chosen method of analysis.

2. Methodology

This study is a small part of a much larger project whose goal is the standardization of governmental web sites. In this first approach the heuristic evaluation (analytical approach) was adopted.

Two main premises have been considered in order to elaborate a checklist for evaluating the sites:

    a) Page type
        Governmental web pages have been categorized as informational ones. They have the purpose of presenting factual information.
    b) The quality criteria which is to be defined according to the objectives and goals of the provider.

The existing literature related to the topic was analized and the work was based on authors like Shneiderman, Nielsen, A.G. Smith, McMurdo and on guidelines from institutions such as IBM, ISO (Standard ISO 9241, parts 10, 11, 12) and University of Yale as well. It is well worth mentioning that those sources focus on general site building instead of a particular type of page. Then some items were adapted and some were add observing their adequacy or not to web pages of the Brazilian government.

The items (39) were established and grouped into four general requirements mentioned below.

1) **Scope and purpose** (3 items): this issue verifies the intended coverage of the source the information it contains and what is the purpose of the site and its audience. If it is clearly stated and whether it is successful in addressing its stated topics leaving nothing significant information out.

2) **Content** (12 items): this criterion relates to the organization of information and its structure in the site. The sub-items verifies whether the information is free from grammatical, spelling and other typographical errors the quality of its writing; the content is clear in its presentation and the text is easy to read and existing links to other information resources.

3) **Graphics and multimedia design** (13 items): this criterion point out the lay-out of the pages verifying if the font is clear, simple and easily readable and related with the surface that it occupies; the organization in the heading composition is outstanding the titles according to interest of content and is displayed in harmonious way with the blank spaces and images; whether the icons clearly represent what is intended and if any multimedia content is incorporated appropriately.

4) **Workability** (11 items): this criterion relates how easily and reliably the users can access the site and obtain information or material from it. The sub-criteria relate to the degree of user-friendliness, easy of connectivity and they can be effectively accessed.

The last items of Content, Graphics and multimedia design, and Workability requirements (n. 15, 28 and 29, see Annex 1) despite being considered important, as they are related to kind of professionals involved in sites elaboration, they were not appraised in this stage of the work. So it was given a punctuation to 36 items only.

The sites were analyzed individually by each evaluator. A value has been assigned (using one-point scale) to each item (y = 1 positive score; n = 1 negative score) and the cumulative average was summed up. The results obtained by each professional in relation to a particular site were compared in order to verify if their analysis were matching. Then an average score was calculated and it was verified which site stands out as the best/worst in matching or not the ergonomic principles.

3. Findings

The investigation we described in this paper has given us an accurate view of the ergonomic aspects found on federal government web pages and the general results can be seen in Annex 1 . Some lack of standardization has been observed after analyzing each site and comparing them. The results are depicted in Table 1.

Table 1: Scores obtained by the sites

Ministries	Scope		Content		Graphics		Workability		Total	
	Y	N	Y	N	Y	N	Y	N	Y	N
A	2	1	8	3	6	6	6	4	22	14
B	2	1	6	5	10	2	8	2	26	10
C	2	1	5	6	6	6	5	5	18	18
D	2	1	5	6	10	2	6	4	23	13
E	2	1	5	6	9	3	6	4	22	14
F	2	1	6	5	8	4	6	4	22	14
G	2	1	5	6	5	7	6	4	18	18
H	2	1	6	5	6	6	2	8	16	20
I	2	1	4	7	8	4	6	4	20	16

In relation to the requirement "Scope", the sites did not include the Ministry mission. This item was considered important because it gives to the user an idea of the activities developed by the institution. All them distinguished their programs, services and products, delivering useful information to the users while citizens.

Some sub-items of "Content" requirement were not filled out satisfactorily, as it can be observed in Annex 1. The best result was obtained by the site of the Ministry A in relation to the content analysis (72.7%) and that of the Ministry I obtained the worst result (36.4%).

It was demonstrated that it is not a standard procedure the mention of sources of information and the mention of the professional who elaborate the contents. Besides, date and up-dating aren't mentioned in the sites. It was noticed the absence of postal address, telephone and fax number. So the user doesn't have a chance of interaction.

In the "Graphics and multimedia design" requirement, the best result was obtained by the Ministries B and D, with 83.3% both, followed by the Ministry E with a positive percentage of 75%. The site of Ministry was that presented a less satisfactory performance (41,7%) in this item.

810

In the "Workability" requirement, the Ministry B had the best result with 80% addressed of the criteria. The site that presented a less desirable performance was the Ministry H with only 20%. It was observed that the site which had the higher average of answers in accordance with the established criteria was that of the Ministry B (26), and that of the Ministry H was that whose discrepancies reached the highest average (20). A graphic showing the general performance (Y/N) obtained by each site is visualized in Figure 1.

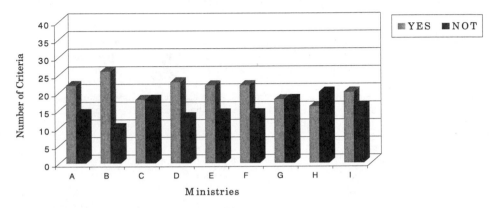

Figure 1. Values obtained by each Ministry

The Ministries C and G accomplished the criteria in 50%. The Ministry H fulfilled the requirements in 44.4%, that means less than half of the items required. The total result obtained by the Ministries was compared and it can be visualized in Figure 2.

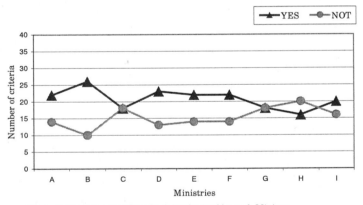

Figure 2. Comparative table of results obtained by each Ministry

Summarizing, eight sites need a revision concerned to "Content" requirement (B, C, D, E, F, G, H, and I). Four of them should give attention to the "Graphics and Multimedia Design" (A, C, G and H). We still suggest to Ministry I that a greater emphasis must be laid on the item "Content", the Ministry G on "Graphics and multimedia design" and Ministry H on "Workability" issue as we can see in Table 2.

The most part of the web sites have filled out positively a percentage above 50% of the evaluation criteria. It was concluded that the sites need a general revision in order to be addressed to the ergonomic criteria. The improvement of a strategy for enhancement of characteristics of the services that is peculiar to each one is needed as well. This analysis will serve as a baseline for standardization of governmental web pages. Comparison with other page domains regarding their peculiarities (.com; .org etc) will also be possible.

Emphasizing that the design characteristics have influenced on the overall effectiveness of the sites, we still suggest as a conclusion of this first phase of project, the implementation of these criteria for standardization of visual and informational presentation of the Brazilian Government web sites, and its extension to sites in the level of state and municipal instance.

Table 2: Percentage of items in agreement with ergonomic criteria

Ministries	Scope	Content	Graphics	Workability	Total
A	66.7	72.7	50.0	60.0	61.1
B	66.7	54.5	83.3	80.0	72.2
C	66.7	45.4	50.0	50.0	50.0
D	66.7	45.4	83.3	60.0	63.9
E	66.7	45.4	75.0	60.0	61.1
F	66.7	54.5	66.7	60.0	61.1
G	66.7	45.4	41.7	60.0	50.0
H	66.7	54.5	50.0	20.0	44.4
I	66.7	36.4	66.7	60.0	55.5

References

Alexander, J.E. & Tate, M.A. *Checklist for an informational web page.* [On-line]. Available at: http://www2.widener.edu/wolfgram-memorial-Library/inform.htm

Barboza, E.M.F., Nunes, E.M. de A. & Sena, N.K. (1999). *Análise dos web sites governamentais [Analysis of Governmental Web Sites].* Brasília, IBICT/Sociedade da Informação.

Brazil. Federal Government (1999). *Manual of trademark use.* Brasília.

European Communities. European Comission (1999). *Public sector information: key resource for Europe (Green Paper on Public Sector Information in the Information Society).* Luxembourg.

IBM. *Guide to web style: quick reference.* [On-line]. Available at: http://www.sun.com/styleguide/tables

Instone, K. *Site usability evaluation.* [On-line]. Available at: http://style.webview.com/wr/pub97/10/10/usability/index.html

ISO. (1996). Ergonomic requirements for office work with visual display terminals (VDTs). Part 10:Dialogue principles. Genève. (ISO 9241-10:1996).

(1998). Part 11: Guidance on usability. Genève. (ISO 9241-11:1998).

(1998). Part 12: Presentation of information. Genève. (ISO 9241-12:1998).

Lynch, P.J. & Horton, S. (1997). The Yale manual style. New Heaven, Yale University. [On-line]. Available at: http://info.med.yale.edu/caim/manual/

McMurdo, G. (1998) Evaluating web information and design. *Journal of Information Science,* 24(3), pp.192-204.

Nielsen, J. How users read on the web. [On-line]. Available at: http://www.useit.com/alertbox/9710a.html

Ten mistakes in web design. From the World Wide Web: http://useit.com/alertbox/9605.html

Nunes, E.M. de A. (1999). *Aplicação de métodos de avaliação ergonômica em interface de sistemas de recuperação de informação em bases de dados [Application of ergonomic evaluation methods on information retrieval system interfaces of databases].* Brasília. xi, 14p. Dissertação (Mestrado em Ciência da Informação). Universidade de Brasília.

Ribeiro, M. (1998). *Planejamento visual gráfico [Graphic visual planning].* 7 ed. rev. e atualizada. Brasília, Linha Gráfica Editora. 500 p. Inclui ilustrações.

Shneiderman, B. (1998). *Designing the user interface: strategies goes effective human-computer interaction.* 3.ed. Reading, Ma., Addison-Wesley. 638p.

Smith, A.G. *Testing the surf: criteria goes evaluating Internet information resources.* [On-line]. Available at: http://info.lib.uh.edu/pr/v8/n3/smit8n3.html

ANNEX 1

Comparative Table

EVALUATION CRITERIA	A	B	C	D	E	F	G	H	I
SCOPE									
1. Institutional goal clearly enunciated on the site's first page.	❀								❀
2. Headings that emphasize programs, services and products offered by the Ministry.	❀	❀	❀	❀	❀	❀	❀	❀	❀
3. Useful to citizens on the area covered by the Ministry.	❀	❀	❀	❀	❀	❀	❀	❀	❀
CONTENT									
4. To show sources of presented information is a standard procedure.		❀		❀					
5. Index for site content.		❀				❀	❀	❀	
6. Link to frequently asked questions (FAQ) submitted to the Ministry.	❀	❀							
7. Good quality text showing appropriate style of hypertext syntax.	❀			❀	❀			❀	❀
8. Name of the person in charge with content elaboration is mentioned.	❀						❀	❀	
9. Webmaster's e-mail address.	❀		❀		❀	❀			
10. Mailing address, telephone and fax numbers at the Ministry.	❀		❀						
11. Pages are dated and up-dating frequency is shown.	❀								
12. Links to other information sources on the subjects covered by the institution.	❀	❀	❀	❀	❀	❀	❀	❀	❀
13. Content reflects established partnerships among the Ministries and announces their common programs/actions.	❀	❀	❀	❀	❀	❀	❀	❀	❀
14. Information about initiatives related to "Avança Brasil" [Advance Brazil] program.									❀
15. *Content is elaborated by information and/or communication professional.*									
GRAPHICS AND MULTIMEDIA DESIGN									
16. Harmonious, interesting lay-out.	❀	❀	❀	❀	❀	❀	❀	❀	❀
17. Number and type of colors are addressed to ergonomic recommendations.		❀	❀	❀	❀	❀	❀	❀	❀
18. Fonts obey ergonomic recommendations.		❀	❀	❀	❀	❀	❀		
19. Federal government's logo is shown on pages.	❀	❀	❀						
20. Ministry's logo is shown on pages.	❀	❀	❀	❀	❀	❀	❀	❀	❀
21. Lateral, top, bottom frames help to see how site was indexed.	❀	❀	❀	❀	❀	❀	❀	❀	❀
22. Images favor reading.	❀				❀				
23. There is a relationship between icons and content.	❀	❀	❀	❀	❀	❀	❀	❀	❀
24. Pictures transmit meaningful information.		❀	❀						
25. Multimedia resources fit site's objectives.									
26. The use of banners or any kind of animated image with marketing purposes is avoided.				❀		❀		❀	❀
27. Text and image areas are designed for good visualization at different screen sizes.	❀	❀		❀			❀		❀
28. *Graphic project is made by graphic or web designer.*									
WORKABILITY									
29. Total size of image files in home page does not exceed 100KB, thus minimizing loading time.		❀	❀	❀	❀	❀	❀		❀
30. It always maintains a return link to first page.		❀	❀	❀	❀	❀	❀		
31. Designed to load in the latest generation of computers as well as in older machines.	❀	❀		❀		❀	❀		❀
32. There is a search mechanism.		❀	❀				❀		
33. There is a site map.		❀					❀		
34. Site's indexing and general arrangement help navigation throughout it.	❀	❀		❀	❀	❀	❀		❀
35. Features such as forms are provided. They work properly and add value to the site.	❀						❀		
36. Documents disseminated by the Ministry are available for downloading.	❀	❀	❀	❀	❀	❀	❀	❀	❀
37. Textual URL showing correlation between words/acronyms and the name of the institution.	❀	❀	❀	❀	❀	❀	❀	❀	❀
38. Pages' HTML source code includes meta tags providing site description and keywords for search engine robots.								❀	
39. *Webmaster is a computer science and/or information technology professional.*									
Total	22	26	18	23	22	22	18	16	20

Achieving universal access for European e-government applications -
A case study of two European projects in the field of e-voting and smartcards

Antonis Galetsas,[1]   Stéphan Brunessaux,[2]   Stefan Hoernschemeyer[3]

[1] European Commission, Belgium
[2] MATRA Systems & Information, France,
[3] Zuendel & Partner, Germany

Abstract

This paper will report the current research on innovative systems contributing to the achievement of the Information Society for all. A basic prerequisite for achieving this aim is the provision of "Universal access", which in this context signifies the right of all citizens to obtain access to community-wide pool of information and services.
The work of two EC supported research projects will be described. These are called CYBERVOTE and FASME.
The project CYBERVOTE aims at providing, for all citizens, a reliable and trustworthy eVoting system. The human interface of the system is of particular importance for the success of the overall system. Its characteristics and design decisions will be reported.
The project FASME aims at providing a JAVA card based system, which will be able to handle the transfer of citizens' administrative data in a reliable, secure and transparent way. The functions of the human interface of the system will be reported.
In addition to the research work of the above two projects a review will be made on HCI and UA issues as they apply to the areas of eVoting and the handling of administrative data, with smart cards, on a European level.

1. Introduction

Electronic communications, and in particular Internet and mobile technologies, are bringing about enormous changes in every facet of our lives and opening new possibilities for individuals and business. Not least amongst these changes is the potential to give every citizen wider access to, and participation in, the democratic process.
European Union research programmes have addressed the use of electronic data exchange by public administrations since 1992. Over that period, the European Commission has supported over 100 shared-cost RTD actions with an EU contribution totalling more than 200 M.EUR. The subject matter has covered all aspects of government and public administration and the entities that interact with them.
The challenge now facing the European Union RTD is to assist the public sector to adapt to the changes brought about by the information society. The guiding principle of EU funded research actions up to now has been **"access for all,"** for example supporting multi-lingualism , mobility, special needs, etc.
This paper will report on RTD work done in the area of eGovernment, and in particular addressing issues of UA for eVoting and administrative data transfers. In both cases, HCI issues are crucial for the overall success of the projects. Thus special attention is given to the analysis of the user requirements, and user acceptance of the prototype.

2. eGovernment

E-government refers to the delivery of information and services through the Internet or through other digital means. Many governmental units are adopting the digital revolution and are putting online a wide range of materials, from publications and databases to actual government services, for the use of citizens.
A survey conducted in the USA (Assessing E-Government, 2000) during the summer 2000, revealed that the e-government revolution has fallen short of its potential. Government websites are not making full use of available technology, and there are problems in terms of access and democratic outreach. E-government officials need to work to improve citizen access to online informative and services. Amongst the more important findings of the research are:

  ➢ *Only 5 percent of government websites show some form of security policy and 7 percent have a privacy policy*

> *15 percent of government websites offer some form of disability access,*
> *4 percent offer foreign language translation features on their websites*
> *22 percent of government websites offer at least one online service*
> *There is a need for more consistent and standard design across government websites.*

The European Commission has identified the above and other obstacles to the realisation of the Information Society in Europe, and designed the Information Society Technologies (IST) research programme, aiming at contributing towards their resolution.

3. The European Project "CYBERVOTE"

3.1 General and UA project Objectives

The CyberVote project, "an innovative cyber voting system for Internet terminals and mobile phones", is a research and development project partially funded by the European Commission under the IST programme. This project commenced on 1 September 2000 and will finish on 1 March 2003.

CyberVote aims at contributing to the development of European democracy by providing all citizens with a modern electronic voting system. It is expected that the use of such a system will increase the overall participation of European citizens in all kinds of elections. More specifically, it should increase the participation of the young, physically handicapped, elderly, immigrants and socially excluded people.

The goal of the CyberVote project is to develop and demonstrate a user-friendly on-line voting system integrating a highly secure and verifiable Internet voting protocol, and designed to be used at local, regional, national or European elections.

This system will allow voters to cast their vote using Internet terminals such as PCs, handheld devices and mobile phones. It will rely upon an innovative voting protocol, designed within the project, that uses advanced cryptographic tools. This protocol will ensure authentication of the voters, integrity and privacy of their vote when sending it over the Internet and during the vote counting and auditing processes.

The project will evaluate to what extent on-line voting influences the voter participation. CyberVote should improve the voting process for all voters, but examples of citizens who should particularly benefit from CyberVote include people with limited mobility (the disabled, the ill, hospital patients, the elderly etc.), people travelling during the election day, and expatriates. The system will satisfy their requirements by allowing them to cast their vote without the need to go to their usual polling station. However, voters will still have the option to vote via the usual paper procedure.

This system will be tested in 2003 during trial elections that will be held in Germany, France and Sweden. These trials will involve more than 3000 voters and will allow full assessment of the system before any potential product launch. Additional information on the project can be found on http://www.eucybervote.org.

3.2 Approach

The CyberVote consortium is currently analysing the user requirements from both the voter's viewpoint and the administration's viewpoint. This task will be completed by March 2001. Then, an ergonomics plan will be defined during the second half of 2001.

This plan will ensure the acceptability (user-friendliness, usability, usefulness and trustworthiness) of the CyberVote system by all users (voters and administrations). Special attention will be given on how to combine a user-friendly approach with security controls and on how to present the information on the small displays of mobile phones. The Web Content Accessibility Guidelines issued by the W3C (Web Content Accessibility Guidelines 1.0, 1999) will be considered to see to what extent CyberVote can be made accessible to people with disabilities.

One of the key issues is the optimal use of the platform display with respect to the language used. The various possible types of voting interaction will be identified together with the approach to carry them (use of natural language, light NL or coded language). The voting vocabulary will then be identified. The size of the words to use and their complexity will be analysed for each target language. The logic of the voting dialog will be defined for the whole voting procedure. The dialog will then be tested with each target language. This process will be integrated on the various target platforms (mobile phones, smart phones, PDA and PCs). The objective will be to ensure that the voting procedure and the dialog remain consistent whatever platform or language used.

Several tests will be carried out to test their approach. A first trial simulation consisting mainly of animations of user displays will be conducted by spring 2002 in order to get an early feedback of the overall functioning of the system.

Large scale pilot applications involving thousands of people will be conducted in late 2002 – early 2003 with the cities of Issy-les-Moulineaux (France), Bremen (Germany) and Kista/Stockholm (Sweden).

Before carrying out these tests, the consortium will define the composition of the testing population. It is expected to include various profiles of citizens (young, elderly, lower class, upper class, ill, computer addicts, immigrates, etc.). A procedure will be defined which will consist of a cross sample of people testing the system, followed by the results being organised and analysed. This assessment will include measures of the behaviour of the voters such as moves of the mouse when using a PC, uses of selection keys and roller (when using a mobile or a smart phone), number of error messages displayed, duration of the voting session and so on. In addition to that, the trial voters will be asked to fill in an assessment grid, which will allow them to rank the trial on different acceptability criteria. All these results will be compiled and analysed in order to ensure the largest use of the system.

3.3 Expected results

The very first results will be available in spring 2002 while the final results will only be available in April 2003. The objective is that the trial applications show a satisfaction rate of 80% from both the administration project end-users and the trial voters.

By overcoming the barriers imposed by time and distance, the CyberVote system will propose an alternative to traditional voting that will allow people with special needs to reintegrate themselves in political life. Among these people, disabled and elderly people are sometimes resigned or discouraged by the physical efforts they have to undertake to participate in polls and elections. Immigrates are sometimes kept outside of the political life simply because they cannot always afford to go back to their own countries to participate in elections.

3.4 Future work

This research project will not address all categories of the population. However, should these trials prove successful it is expected that commercial implementation of the CyberVote system will follow in 2003. Then additional work should be carried out to ensure a universal access by all categories of the population. For instance, the combined use of a touch screen, text-to-speech synthesisers and headphones or dynamic Braille display could be investigated to propose a solution for blind people.

4. The European Project "FASME"

4.1 General and UA project Objectives

The FASME project, "Facilitating Administrative Services for Mobile Europeans" is an RTD project, which intends to show the feasibility of a Javacard as a means of providing data for administrative documents. All technical developments are driven by the actual needs of all potential Javacard users. The user needs are being identified by workshops where the audience includes citizens and civil servants from different countries and backgrounds.

An important feature of the prototype system is its ability to be used by different groups of users e.g. the elderly, people with special needs etc. After some years of research work in the area of universal access (UA), the time has come to start implementing the research results in the emerging information society. This is manifested by national legislative actions in the different member states throughout Europe.

4.2 Approach

In order to make the Javacard system comply with the user requirements three main sources of input were applied; literature research, involvement of experts during interviews and workshops, involvement of end users during interviews and workshops. The results delivered by these sources were implemented during the technical development of the Javacard prototype.

The most relevant literature sources available to the project were the ISO-guidelines, documentation of earlier projects in the e-government sector where the FASME partners have taken part and guidelines on the development of graphical user interfaces for people with special needs and for those with visual impairment.

Information researched here was applied to the development of the Graphical User Interface (GUI) prototype.

A first GUI prototype was presented to end users to help in assessing their needs.

To precisely analyse the user requirements, citizens and civil servants have been grouped into clusters:

- Moving Citizens
- Commercial Companies dealing with support for administrative services (including voluntary or charity organisations which provide social support to citizens)
- Clerks in the front-offices of municipalities
- Clerks in the back-office on the management-level within the municipalities

These user groups have been interviewed and were invited to test the different versions of the GUI and of the prototype during the development phase. The reactions and opinions of the users have been documented concerning both the handling of the GUI and the interaction with the prototype.

4.3 First results

The development of the FASME-system has produced two user-interfaces:
1. The citizen GUI and the biometric sensor
2. The FASME-Javacard

4.3.1 Citizen Graphical User Interface (GUI)

The GUI for the citizen will be used for both kiosk applications and Internet access to the system. The GUI will therefore have to fulfil the requirements defined by the different legal regulations ensuring universal access to e-government services for user groups, the requirements defined by experiments with users and the common requirements of the different EU-member states.

The following features were included:
- A consistent GUI with a general dark yellow layout ensuring a common branding policy.
- Automatic choice of style and layout. According to the profiled card information, the system can also recognise the requirements of people with visual impairments and it displays the screen according to these requirements.
- Large, grey click buttons which are tagged by bold-lettered descriptions and commonly known icons such as the stop sign.
- Descriptive "alt" tags[1] to indicate the meaning of the buttons (not available on touchscreen).
- Display of a process bar at the top of the screen for orientation indicating the current step the user currently has to perform.
- Display of a progress bar during data transaction.
- Choice of language: The system reads the preferred language from the card in which the user wants the information presented, but it also offers the choice of other languages which the user can set as his preferred language.
- Every step is explained concisely to avoid an information flood. Additional information though is available on request for every step.
- The biometric sensor as part of the system is required for secure identification of the cardholder. The system guides the user through the identification process.

4.3.2 The Javacard

The Javacard as a means of identification and transportation of personal data will need to have a visually simple readable style and clear structures. The citizen knowing that this card stores a huge amount of personal data has to trust the security of the card. To maintain this trust the card is equipped with external security features like a hologram, guilloches (small impregnated lines) and is manufactured with secure printing procedures. As the card has to be issued by an official body national guidelines will have to be followed. This leads to the situation that in the different member states different data sets are allowed to be included on cards or special requirements are made to display the data, for example:
- No information may be given in Finland concerning the gender and in the Netherlands no information on the religion of the cardholder

---

[1] Alt-tags (or Tool Tip Text) are small texts appearing when the user moves the mouse over a picture or button. Short information texts can be displayed in this way without crowding the screen with information excessively.

–   The display of a photo on the card is under discussion due to privacy reasons in some European member states.
–   Different letter type sets have to be used in Greece and in the future member states in Eastern Europe after an expected EUExpansion to the East.

The Javacard offers as new technical functions multifunctionality, flexible updates by renewing and adding cardlets, integration of security related components (Crypto-Processor, Biosensor). Further by a Secure Card Extension (SCE) developed during the project the diskspace has been extended to a Secure Card Server. (Maibaum & Cap, 2000).

4.4   Future work

Validation of user requirements
To validate the results of the FASME project interviews, questionnaires and user observations are planned. According to the results of these activities the usability of the developed GUI will be tested and improvements applied. Also there will be the possibility to use the GUI analysis tools such as WAMMI *'Website Analysis and Measurement Inventory'* and MUMMS *'Measuring the usability of multimedia systems'* to measure the user's emotional responses, the learnability and the efficiency of the GUI.

5.   General UA Issues for eGovernment Applications

Universal Access for eGovernment applications is more important now and in the future than in the past. This is because the number of online services offered to the citizens and companies will increase exponentially. However, no human-computer interaction can match the sophistication of face to face interaction. This will continue to be true for the near future and result in the ever increasing exclusion of more and more people from the possibility of access to publicly available information and services. In addition, economical constrains will also limit the access to the available information, because of the unavailability of the required infrastructure.

As a consequence of the above all eGovernment applications should be designed for the widest possible number of user groups. This is a real challenge taking into account the difficulty defining the different user groups for each application plus the additional development costs required to include such features.

Legislation, of course, will make certain UA features compulsory for the eServices of the future. Thus, more RTD work will be required to find out how such features could be implemented in the most cost-effective way.

In the two projects described above, the main UA features adopted are:
>   **Multilanguage capabilities:** Considered to be the absolute minimum requirement for any application designed for an integrated Europe.
>   **Mobility:** The applications do not depend, in general, on the geographical/space position of the user. Thus they provide access to many user groups which are not addressed by conventional applications.
>   **Special needs groups:** Provisions are made for the subsequent integration of special features **Quality in Use for ALL:** ISO 13407 provides guidance on achieving quality in use by incorporating user-centred design activities throughout the life cycle of interactive computer systems

References
Assessing E-Government (2000, September). *The Internet, Democracy, and Service Delivery by State and Federal Governments*. Darrell M. West, Brown University Providence, RI 02912.
Web Content Accessibility Guidelines 1.0, W3C Recommendation (1999, 5-May)
      http://www.w3.org/TR/WAI-WEBCONTENT/
Maibaum, N., Cap, C.H. (2000). Javacards as Ubiquitious, Mobile and Multiservice Cards. *1st PACT 2000 Workshop on Ubiquitous Computing, Philadelphia, USA.*

Open Adaptive Hypermedia:
An approach to adaptive information presentation on the Web

Nicola Henze

Institut für Technische Informatik, Abt. Rechnergestützte Wissensverarbeitung
and Learning Lab Lower Saxony, University of Hannover
Appelstr. 4, D-30167 Hannover
henze@kbs.uni-hannover.de

Abstract

In this paper we will propose our adaptive hyperbook approach for building open, adaptive hypermedia systems. We will describe the underlying document indexing approach which facilitates the adaptation. The indexing approach is minimal in the sense that it requires only a content description (keywords) of a document. The user modeling component is capable of generating reading sequences, retrieve and annotate documents, choosing suitable examples, etc. on base of this content description by using an knowledge model (incorporated as a Bayesian Network) of the application domain.

1. Introduction

Since the emergence of the World Wide Web (WWW) in 1991 (Berners-Lee,1991), the value of information has got a new dimension. Nowadays, millions of computers are connected via the internet, humans can collect information from nearly anywhere in the world, visit virtual galleries, go shopping at virtual market places, check their bank accounts or take a look at the actual situation at the south pole {http:bat.phys.unsw.edu.au/~aasto/}.

This enormous amount of information is also a chance for experience and learning. But effectively selecting information from the internet is still a hot research topic as the effectiveness of search machines increase with the precision of the query. The information contained in the internet is often useless for exploring or learning, as learners need guidance to build up a mental model of the area they are working on before being able to make sufficiently exact queries.

Promising approaches in research come from the area of adaptive hypermedia systems (Brusilovsky, 1996). Adaptive hypermedia systems combine hypermedia systems with intelligent tutoring systems. The aim of these systems is to personalize hypermedia systems to the individual users. Thus, each user has an individual view and individual navigational possibilities for working with the hypermedia system.

A contribution to adaptive hypermedia systems are adaptive hyperbooks (Henze, 2000) which personalize the access to information to the particular needs of users. They give users the ability to define their own learning goals, propose next reasonable learning steps to take, support project-based learning, give alternative views, and they can be extended by documents written by the learners. Adaptive hyperbooks have been developed in the KBS hyperbook project at the Institut für Rechnergestützte Wissensverarbeitung, whose English name (Knowledge Based Systems, KBS) gave the name for the project. Adaptive hyperbooks are information repositories for accessing distributed information. A main focus of hyperbooks is the extendibility of the system in respect to the World Wide Web. To create open adaptive hypermedia systems, the indexing approach chosen in KBS allows to treat each information unit equally independent of its origin. Thus, HTML pages from the World Wide Web can be integrated and adapted to a particular user's needs in the same way as documents stored in the hyperbook's library.

2. Adaptation tasks of hyperbooks

One of the main goals of student modeling in educational hypermedia is student guidance (Brusilovsky, 1996). Students have learning goals and previous knowledge which should be reflected by the hyperbook for adapting the content or the link structure of the hyper-document. For our KBS hyperbook system we follow a constructivist pedagogic approach, building on project based learning, group work, and discussions (Henze and Nejdl, 1997). Such a project-based learning environment leads to particular requirements for adaptation, in order to adapt the project resources presented in a set of hypermedia documents to the student's goals (for a specific project) and to the student's knowledge. It has to support the student learner by implementing the following adaptation functionality:

- Adaptive Information Resources: give the students appropriate information while performing their projects, by annotating necessary project resources depending on current student knowledge.
- Adaptive Navigational Structure: annotate the navigational structure in order to give the student additional information about appropriate material to explore or to learn next.
- Adaptive Trail Generation: provide guidance by generating a sequential trail through some part of the hyperbook, depending on the student's goals.
- Adaptive Project Selection: provide suitable projects depending on the student's goals and knowledge.
- Adaptive Goal Selection: propose suitable learning goals depending on the particular student's knowledge.

The user modeling component has to fulfill various tasks. On the one hand it has to enable the above stated adaptation functionality. On the other hand, it has to enable further adaptation functionality which depends on the openness of the KBS hyperbook approach (Henze and Nejdl, 2000): Information resources located anywhere in the WWW should be included in the curriculum of the student's work with the hyperbook, explanations and examples can origin from the hyperbook's libraries or from any other location in the WWW.

3. Enabling adaptation: Indexing documents

The connection between the KBS hyperbook system and the user modeling component is based on indexing any kind of information resources. The index concepts are called knowledge items (KIs). Knowledge items are similar to the domain model concepts used in (Brusilovsky and Schwarz, 1997) or the knowledge units in (Desmarais and Maluf, 1996).

3.1 Modeling the user's knowledge

The knowledge of a user is modeled as a knowledge vector. Each component of the vector is a conditional probability, describing the system's estimation that a user U has knowledge on topic KI – on base of all observations E the system has about U :

Definition 1: Knowledge Vector (KV(U))

$$KV(U) = (\ P(\ KI_1|\ E)\ ,\ P(\ KI_2|\ E)\ ,\ \ldots,\ P(\ KI_n|\ E)\ )$$

where $KI_1, \ldots, KI_n$ are the knowledge items of the application domain and E denotes the evidence the system monitors about U's work with the hyperbook. Observations about the student's work with the hyperbook are stored for each KI. Thus, the KIs are, on the one hand, concepts describing the application domain of a book, on the other hand , they are random variables with four discrete values coding the knowledge grades expert, advanced, beginner and newcomer.

The evidence we obtain about the student's work with the hyperbook changes with the time. Normally, the student's knowledge increases while working with the hyperbook, although lack of knowledge is equally taken as evidence. Since every kind of observation about a student is collected as evidence, the knowledge vector gives - at each time - a snapshot of the student's current knowledge.

3.2 Indexing information: HTML pages, examples, projects

Each information resource is indexed by some set of knowledge items describing the content of the resource. These resources can be general HTML pages, examples, projects, etc. The origin of an information resource is not relevant for indexing, only the content defines the index.

Definition 2: (Content Map) Let $S \neq \emptyset$ be the set of all *KI*s, and let H be a set of HTML pages. Then

$$I : H \rightarrow P(S) \setminus \{\emptyset\}$$

is the content map, which gives for each information resource in H the index of this resource, e.g. the set of *KI*s describing its content. $P(S)$ denotes the power set of S. To identify the index of an information resource we can scan the text for keywords or phrases. Actually, the indexing is done by the author of an information resource by hand.

4. Discussion

We have described the use of knowledge items for indexing all kinds of information, belonging to the hyperbook, or located anywhere in the WWW. The use of an indexing concept in student and user modeling in this way is new. Most approaches model dependencies like prerequisites or outcomes directly with the information resources themselves (Brusilovsky, 1999).

We separate knowledge and information, as we model learning dependencies solely on the set of *KI*s of a hyperbook. The connection between the student modeling component and the hyperbook system is the content map (definition 2), which maps each information resource to a set of s.

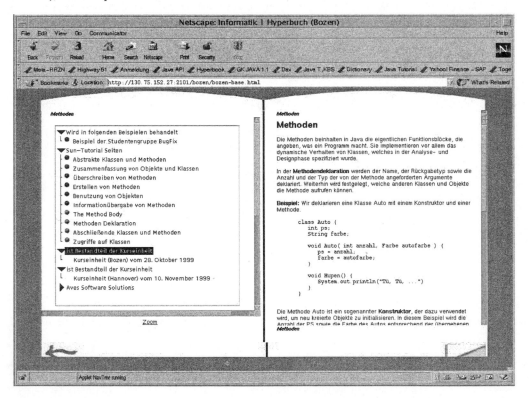

Figure 1: Hyperbook Document "Methoden" (Methods) with links to examples, Sun Units
and to two lectures where this document has been used

This separation is advantageous in many aspects. As the KBS hyperbook system allows different authors to write parts of a book, they become independent from the work of others: They can write (and index) their information entries without caring about the other content of the hyperbook. The KBS hyperbook system is an open hypermedia system, allowing to include information resources located anywhere in the WWW. As all information resources are equal in the sense that they only need to be indexed for being integrated in a particular hyperbook, this openness is enabled by the indexing concept, too. In addition, all kinds of information resources from arbitrary origins are fully integrated and adapted to the student's needs: We can propose programming examples in the WWW, generate reading sequences which contain material of the hyperbook library and the WWW, calculate the educational state of HTML pages in the WWW according to the student's actual knowledge state, etc. In fig.1 we can see an information page about Methods in Java. For this page links to examples, to the Java Sun Tutorial and to corresponding lectures have been generated in real-time and have been annotated with additional reading suggestions: A red ball indicates non-recommended information, a green ball indicates recommended information, and a white ball marks already known information.

In addition, the use of a separate knowledge model makes the hyperbook system robust against changes. If we add additional information pages or change contents, we only have to (re-)index these pages accordingly. No further work has to be spent on updating other material, as it would be necessary if knowledge, and thus reading or learning dependencies, would have been coded in the material itself.

The chosen way of implementation enables us to apply different inference mechanisms to the student modeling component. The inference technique we currently use for the KBS hyperbook systems is Bayesian inference (Henze, 2000).

5. Conclusion

We have proposed an approach for building open adaptive hypermedia systems on the Web. These open systems contribute to universal access as they allow to adapt information located anywhere in the internet to a particular user's needs, goals and knowledge. Further work will concentrate on simplifying the discussed indexing approach by applying information retrieval techniques for (semi-) automatic indexing and retrieval of documents in the Web.

References

Berners-Lee, T. (1991). *World Wide Web: An illustrated seminar*. Held as an Online Seminar in 1991. [On-line]. Available: http://www.w3.org/pub/WWW/Talks/General.html.

Brusilovsky, P. (1996). Methods and techniques of adaptive hypermedia. *User Modeling and User Adapted Interaction*, 6(2-3), 87-129.

Brusilovsky, P., Schwarz, E. (1997). User as student: Towards an adaptive interface for advanced web-based applications. In *Proceedings of the Sixth International Conference on User Modeling, UM97*, Sardinia, Italy.

Brusilovsky, P. (1999). Adaptive and Intelligent Technologies for Web-based Education. In Special Issue on *Intelligent Systems and Tele-Teaching* (Künstliche Intelligenz), 4, 19-25.

Desmarais, M.C., Maluf, A. (1996). User-expertise modeling with empirically derived probabilistic implication networks. *User Modeling and User Adapted Interaction*, 5, 283-315.

Henze, N. (2000). *Adaptive Hyperbooks: Adaptation for Project-Based Learning Resources*. PhD thesis, University of Hannover, Germany. [On-line]. Availabale: http://edok01.tib.uni-hannover.de/edoks/e002/313646791.pdf.

Henze, N., Nejdl, W. (1997). A web-based learning environment: Applying constructivist teaching concepts in virtual learning environments. In *IFIP 3.3 and 3.6 Joint Working Conference: The Virtual Campus: Trends for Higher Education and Training*, Madrid, Nov. 1997.

Henze, N., Nejdl, W. (2000). Extendible adaptive hypermedia courseware: Integrating different courses and web material. In *Proceedings of the International Conference on Adaptive Hypermedia and Adaptive Web-Based Systems (AH 2000)*, Trento, Italy.

Universal access in Health Telematics
Some factors that influence the design of Health Telematics Services

Ilias Iakovidis[1]

European Commission, DGXIII
Rue de la Loi 200, B-1160 Brussels, Belgium
Email: Ilias.IAKOVIDIS@cec.eu.int

Abstract

This paper is concerned with the issue of end user acceptance of Health Telematics products and services from the perspective of universal access and Human-Computer Interaction (HCI). Specifically, the paper aims to correlate acceptance and benefit with detailed and thorough knowledge about the target users. Such knowledge may relate to individual user characteristics, preferences, literacy, point of access, as well as to the hardware, software and bandwidth. To cope with variation in the above set of user parameters means a strong commitment to universal access, which can be attained by inducing universal deign principles into user-centred design practices.

1. Introduction

Today's health care has a threefold objective. First, it provides information, services and support to help people remain healthy. Second, it makes quick, effective and convenient treatment, care and support available to help those who fall ill. Finally, it provides services to enable people who are ill or vulnerable to live their life with maximum dignity and independence. All objectives are increasingly supported by technological solutions in the field of Health Telematics, based on an appropriate health information infrastructure. Despite ongoing progress, there is a substantial concern regarding the acceptance by, and the benefit of, these solutions to the intended user groups (Iakovidis, 1998a; 1998b) In this context that the notion of *Universal Access,* becomes critically important for ensuring social acceptability of Health Telematics products and services. Universal access implies the accessibility and usability of Information Society Technologies (IST) by anyone, anywhere, anytime. Its aim is to enable equitable access and active participation of potentially all citizens in existing and emerging computer-mediated human activities (Stephanidis et al., 1998). *Universal Access* becomes predominantly an issue of design, and the important issue arises of how is it possible to design systems that permit systematic and cost-effective approaches to accommodating all users. *Design for all* in the Information Society has been defined as the conscious and systematic effort to proactively apply principles, methods and tools, in order to develop IST products and services that are accessible and usable by all citizens, thus avoiding the need for a posteriori adaptations, or specialised design (Stephanidis et al., 1998). The field of HCI plays a critical role towards facilitating Universal Access, as citizens in the Information Society experience technology through their contact with the user interface of interactive products, applications and services. In addition to HCI, there are at least two other important levels of concern: the telecommunications infrastructure, and the digital content of applications and services.

In the field of Healthcare Telematics, the issues of Universal Access and Universal Design have recently started being systematically investigated. For example, the IS4ALL (Information Society for All) EC-funded Thematic Network aims to to provide the European and Health Telematics industry with a comprehensive code of practice on how to appropriate the benefits of universal design. This paper is concerned with the question of end user acceptance of Health Telematics products and services, focusing mainly on HCI issues and, to a lesser extent, on issues concerning the digital content of products and services. Specifically, extending previous work (Iakovidis, 1997; 1999a, 1999b), the paper aims to correlate acceptance and benefit with detailed and thorough knowledge about the target users. It is claimed that unless design and development of Health Telematics applications and services follow a user-centred protocol, a serious risk to obtain sub-optimal results arises. The cornerstone of a user-centred design protocol is based on the notion of knowing the user. Traditional user-centred design offers a wide collection of techniques, which facilitate a tight evaluation feedback loop, to allow early identification of design deficiencies and their correction when the cost is not a prohibitive factor. Nevertheless, in Health Telematics there are very few

---

[1] The views developed in this paper are that of the author and do not reflect necessarily the position of the European Commission.

studies focusing on what developers need to know about their target users, how to collect the required data and how to use the collected data to improve design. This paper is an attempt to shed light into this complex issue, with the intention to help developers when designing health telematics services.

2. Parameters that shape design in Health Telematics

Several user-related factors need to be accounted for when developers decide on Health Telematics application content, interfaces, media, and design approaches. In many cases, the explicit involvement of members of the target audience during the development phases is deemed as essential to a successful product. Below, some of the user-related issues, which may be used to drive design efforts, are described.

Individual Characteristics

Individual characteristics, such as age, gender, disability, race and ethnicity, cultural factors, and socioeconomic status, may influence health-information-seeking behavior, and can account for differences in the amount and type of health-related information and support that individuals seek. Some people do not seek much information or support, while others who do may encounter serious barriers to the use of Health Telematics applications (Eng et al., 1998). Willingness to use health information technology may also be an important consideration in designing IHC applications. Individual characteristics and preferences can be accommodated by "tailoring" content and interfaces. Ensuring accessibility of an application's interface is essential to reach users with physical disabilities (WWW Consortium, 1998). Tailored information has been found to be more effective in providing consumer information (Brennan et al., 1998) and is preferred by patients (Jimison, Fagan et al., 1992). Some users, however, may resist over-customized applications that are too narrowly focused.

Individual Preferences

The concept of individual preferences is important for Health Telematics applications that focus on health decision-making (Mulley, 1989). Although patients need information about the quality-of-life associated with the medical outcomes of possible decisions, reliable assessment of individual preferences and risk attitudes for clinical outcomes are probably weak links in clinical decision-making. Recent efforts to explore the use of computers in communication about health outcomes and in assessment of patient preferences for various health outcomes have started to address these issues (Goldstein et al., 1994). Information on patient preferences is important for tailoring information to patients and for providing decision support (Jimison, 1997). In addition to differences in preferences for health outcomes, patients differ in the degree to which they choose to be involved in decision-making. Age (younger more than older persons), gender (women more than men), and education level (higher-educated more than less-educated persons) are generally strong predictors of desire to be involved in medical decisions. In addition, there is a greater desire to be involved in decisions in health areas that generally require less medical expertise, such as a knee injury, than those that are more complex, such as cancer.

Literacy

An individual's reading ability impacts on application design. The problem of low literacy skills is widespread in many countries in the world. In the United States (Baker et al., 1996), about one of every five adults reads at or below the fifth grade level. Only about half of people examined comprehended written health education materials and average reading levels were well below what is needed to understand standard health brochures. Additionally, an analysis of medical information on Web sites showed that, on average, materials were written at a 10th grade reading level, which is not comprehensible to the majority of people (Graber et al., 1999). Lack of health literacy may also be an acute problem among the elderly. A person who has completed a certain grade level should not be assumed to be able to read at that level. Generally, health materials should be written at least three grade levels lower than the average educational level of the target population (Jubelirer & Linton, 1994). There is a danger, however, that excessively simplifying materials may reduce the value of the program to more educated users. Interactive media can help in this situation because they can be used to accommodate a range of users with varying levels of health and technology literacy. Text characteristics, organization and clarity also impact on comprehension and retention of material. To address shortcomings in reading literacy, multimedia techniques can be used to facilitate comprehension. Information can be conveyed through video, audio, and graphics, instead of, or in addition to, text. Additionally, presenting material in multiple languages would increase comprehension for non-native English speakers.

Point-of-Access

The ideal point of access for many of the functions of Health Telematics is the home[2]. The home, as a context of use, offers the advantage that it allows the user to access the application at any time of day, in privacy and comfort. However, some applications can function effectively in more public settings, such as schools and work-sites. For example, shared decision-making applications, because of their typical one-time-use nature, may function effectively through these and other access sites, including libraries and health professional offices. Disease coping and behavior change applications that offer both information and emotional support, however, need to be immediately accessible at any time. Hence, to be used effectively, they need to be available in the home and/or portable. In addition, access to the Internet at work and home expands the availability of online employer-sponsored wellness programs that traditionally were only available at worksites. Public access points need to be selected with a thorough understanding of the target audiences and will depend on the type of application and the relationship between intended users and the setting. For instance, many underserved populations harbor a distrust of certain institutions that might otherwise be appropriate candidates for delivering IHC applications. If government-sponsored sites, such as clinics and public buildings, are not trusted by these populations, then alternate settings, such as community centers and places of worship, may need to be considered.

Hardware, Software, and Bandwidth
The capability and performance characteristics of the hardware and software and communication pathways used by target audiences to access Health Telematics applications and services are important considerations for design. The functions and content of the application should be matched to the level of technology available to typical users. For example, integrating extensive graphics or full motion video into an application that is intended for groups of users who do not typically have computer graphics accelerators or large bandwidth access is counterproductive. The quality of a user's experience when accessing an interactive application via a slow dial-in modem versus a much faster ISDN connection is so different as to almost render them as different programs.

3. Aiming for universal access
Given the above non-exhaustive landscape of health care user parameters, the question posed is how to cope with such variation and diversity when designing systems, which may be accessed by different users, including health professionals, managers and patients. A classical example of such an application is the Electronic Health Record. Existing implementations are far from the desired quality level (Iakovidis, 1998a). Clearly, there are numerous technical challenges as well as non-technical ones (Stephanidis et al., 1998; 1999).

3.1 The technical challenge
On the technical level, much relates to the capacity of existing systems and services to realise suitable adaptations. Adaptation is a technique that has long been used in the field of HCI, and comes in two primary types, namely adaptability and adaptivity (Stephanidis, 2001). Adaptability implies changes to the interactive system prior to the initiation of an interactive episode, and is primarily concerned with tailoring the system to be accessible by the target user groups. Adaptivity involves dynamic updates of the dialogue on the basis of inferences derived from the system's user model and from monitoring the end user's interaction patterns. In recent efforts funded by the EC, the scope of adaptability and adaptivity have been revisited with the intention to provide the basis for universal access (e.g., Stephanidis & Emiliani, 1999; Stephanidis, ed., 2001). These efforts have demonstrated that by taking into account additional aspects, such as for example the context of use and the technological infrastructure, the system can implement a wider range of adaptable and adaptive behaviours, which constitute the basis for universal access. However, universal access is not merely a technical challenge.

3.2 Policy aspects
There are additional policy factors that may have an impact on feasibility, viability and economic efficacy of universal access solutions in Health Telematics. Examples include health data and information standards, network security issues, and legislative actions at the federal and state levels, addressing issues such as medical information privacy, confidentiality, trust and security. It is important to note that such policy issues are already undergoing extensive review in the context of expanded use of telecommunications and computer technologies. However, such revisions are being undertaken in the light of cost savings and / or quality improvement. While each of these targets can be attained through various pathways, the importance of universal access in the long run needs to be stressed.

---

[2] There have been several applications of advanced technologies to facilitate home-based care; for example see Balas & Iakovidis, 1999.

The commitment to universal access is not only due to the demographic changes, which are evident in our society, but also due to the emergence of the Information Society, in which access to information becomes the ultimate threshold for participation and inclusion, as well as sustainable development and economic prosperity.

3.3   Appropriating the benefits

Universal access in Health Telematics may help reduce health disparities through its potential for promoting health, preventing disease, and supporting clinical care for all. Recent data indicates that the profile of Internet users may be becoming more representative of the general population (PRCP&P, 1999), but the poor and others who have preventable health problems and lack health insurance coverage are unlikely to have access to such technologies (Eng et al., 1998). Enhancing access to health information and support may promote more efficient use of services, reduce the total costs of illness, and help avert preventable health conditions that disproportionately impact lower income populations. Although data on the impact of Health Telematics on underserved populations are limited, some studies suggest that it can improve health knowledge, attitudes, and cognitive functioning (Carroll et al., 1996), enhance emotional well-being (Gustafson et al., 1993), and reduce utilization of health services without impacting health (Gustafson et al., 1999). If these effects can be consistently replicated, substantial improvements in public health and health care cost savings can be realised. In addition, as reliance on online health information and support resources become more common for routine functions such as making appointments and communicating with health professionals, universal access to Health Telematics services becomes an increasingly essential component of emerging health information infrastructures.

4.   Concluding remarks

Clearly, to attain universal access in Health Telematics involves a concentrated effort from all parties concerned. This is now appreciated by the European Commission's research policy and it is expected to be explicitly accommodated in its 6th Framework Programme. Barriers to be removed include those related to technology infrastructure access and those associated with the characteristics of nonusers and the information and applications themselves. Certain populations also have difficulty accessing online health resources because most health telematics applications are designed primarily for educated, literate, and able-bodied audiences. Many people have inadequate skills in science, technology, or reading literacy; cannot understand or use health information, have a physical disability, or cannot communicate in English (WWW Consortium, 1998). Underserved populations are keenly interested in using technology including the Internet (Hoffman & Novak, 1998). Studies show that, with appropriate training, many underserved groups including low-income families, residents of inner cities and rural areas, disabled and elderly people, racial/ethnic minorities and drug users can successfully use technology to address health concerns. Studies suggest that low-income consumers are savvy about persuasive marketing communications (Alwitt & Donley, 1996), want independent information when purchasing a range of products (Mogelonsky, 1994), and, thus, can critically evaluate information. Providing universal access will require a collaborative effort among a wide variety of stakeholders at all levels (Milio, 1996). EC's research policy is explicitly focused on collaborative, interdisciplinary and human-centered research and development. Moreover, it is now realised that without external intervention, market forces are unlikely to address the needs of those without access. While universal access at home is ultimately desired, for the near term, until home access is universally available and affordable, universal access may necessitate a combination of private (i.e., home) and public (e.g., schools, libraries, public buildings, post offices, shopping malls, community centers, health care facilities, places of worship) access points (Eng et al., 1998).

References

Alwitt, L.F., Donley, T.D. (1996). *The Low-Income Consumer: Adjusting the Balance of Exchange.* Thousand Oaks, CA: Sage Publications.

Baker, D.W., Parker, R.M., Williams, M.V., Pitkin, K., Parikh, N.S., Coates, W., et al. (1996). The health care experience of patients with low literacy. *Arch Fam Med.* 5, 329-334.

Balas, A., Iakovidis, I. (1999). Distance technologies for patient monitoring. BMJ Vol. 319, [On line]. Available at: http://www.bmj.com/cgi/content/full/319/7220/1309

Brennan, P.F., Caldwell, B., Moore, S.M., Sreenath, S., Jones. J. (1998). Designing HeartCare: custom computerized home care for patients recovering from CABG surgery. *J Amer Med Informatics Assoc.* (suppl), 381-385.

Carroll, J.M., Stein, C., Byron, M., Dutram, K. (1996). Using interactive multimedia to deliver nutrition education to Maine's WIC clients. *J Nutrition Educ.* 28, 19-25.

826

Eng, T.R., Maxfield, A., Patrick, K., Deering, M.J., Ratzan, S., Gustafson, D. (1998). Access to health information and support: a public highway or a private road? *JAMA*. 280, 1371-1375.

Goldstein, M.K., Clarke, A.E., Michelson, D., Garber, A.M., Bergen, M.R., Lenert, L.A. (1994). Developing and testing a multimedia presentation of a health-state description. *Med Decis Making*. 14, 336-344.

Graber, M.A., Roller, C.M., Kaeble, B. (1999). Readability levels of patient education material on the World Wide Web. *J Fam Pract*. 48, 58-61.

Gustafson, D.H., Hawkins, R., Boberg, E.W., Pingree, S., Serlin, R., Graziano, F., et al. (1999). Impact of patient centered computer-based health information and support system. *Am J Prev Med*. 16, 1-9.

Gustafson, D.H., Wise, M., McTavish, F., Taylor, J.O., Wolberg, W., Stewart. J., et al. (1993). Development and pilot evaluation of a computer-based support system for women with breast cancer. *J Psychosocial Oncology*. 11, 69-93.

Hoffman, D.L., Novak , T.P. (1998). Bridging the racial divide on the Internet. *Science*. 280, 390-391.

Iakovidis, I. (1997). Interacting with Electronic Healthcare Records in the Information Society. In M.J. Smith, G. Salvendy, R.J. Koubek (eds), *Design of Computing Systems: Social and Ergonomic Considerations, Proceedings of the 7th HCI International Conference*, San Francisco, USA (pp. 811-814). Amsterdam: Elsevier.

Iakovidis, I. (1998a). Towards Personal Health Record: Current situation, obstacles and trends in implementation of Electronic Healthcare Records in Europe. *Intern. J. of Medical Informatics*, 52(123), 105-117.

Iakovidis, I. (1998b). Convincing cases: important component of user acceptance. In *"User acceptance of Health Telematics Applications: looking for convincing cases", Studies in Health Technology and Informatics* (vol 56, pp. 8-14). IOS Press.

Iakovidis, I. (1999a). User Acceptance in Health Telematics: an HCI perspective. In H-J Bullinger & J. Ziegler (Eds), *Human-Computer Interaction: Communication, Cooperation and Application Design, Proceedings of the 8th HCI International Conference*, Germany (pp 863-867). Mahwah, NJ: LEA.

Iakovidis, I. (1999b). Learning from past mistakes. In *Proceedings of the European Conference on Electronic Health Records (EUROREC'99)*, (pp. 4-9).

Jimison, H,B. (1997). Patient-specific interfaces to health and decision-making information. In: R. Street, M. Gold, T. Manning (eds.) *Health Promotion and Interactive Technology: Theoretical Applications and Future Directions* (pp. 141-155). Mahwah, NJ: LEA

Jimison, H.B., Fagan, L.M., Shachter, R.D., Shortliff,e E.H. (1992). Patient-specific explanation in models of chronic disease. *Artif Intell Med*. 4, 191- 205.

Jubelirer, S.J., Linton, J.C. (1994). Reading versus comprehension: implications for patient education and consent in an outpatient oncology clinic. *J Cancer Educ*. 9, 26-29.

Milio, N. (1996). Electronic networks, community intermediaries, and the public's health. *Bull Med Libr Assoc*. 84, 223-228.

Mogelonsky, M. (1994). Poor and unschooled, but a smart shopper. *AmDemographics*. 16, 14-15.

Mulley, A.G. (1989). Assessing patients' utilities: can the ends justify the means? *Med Care*. 27, S269-28l.

PRCP&P (The Pew Research Center for the People & the Press) (1999). *Online Newcomers More Middle-Brow, Less Work-Oriented: The Internet News Audience Goes Ordinary*. Washington, DC: The Pew Research Center for the People & the Press; 1999. [On-line]. Available at: http://www.people-press.org/tech98sum.htm.

Stephanidis, C. (2001). Adaptive techniques for Universal Access. *User Modelling and User Adapted Interaction International Journal, 10th Anniversary Issue*, 11(1-2), 159-179.

Stephanidis, C. (Ed.). (2001). *User interfaces for all: concepts, methods and tools*. Mahwah, NJ: LEA. (ISBN 0-8058-2967-9).

Stephanidis, C., Emiliani, P-L. (1999). Connecting to the information society: a European perspective. *Technology and Disability*, 10(1), 21-44.

Stephanidis, C., Salvendy, G., et al. (1998). Toward an Information Society for All: An International R&D Agenda. *International Journal of Human-Computer Interaction*, 10(2), 107-134.

Stephanidis, C., Salvendy, G., et al. (1999). Toward an Information Society for All: HCI challenges and R&D Recommendations. *International Journal of Human-Computer Interaction*, 11(1), 1-28.

World Wide Web Consortium. *Web Accessibility Initiative (WAI)*. (1998) URL: http://www.w3.org/WAI.

Knowledge and media engineering for distance education

Huberta Kritzenberger, Michael Herczeg

Institute for Multimedia and Interactive Systems, University of Luebeck
Seelandstrasse 1a, D-23569 Luebeck

Abstract

Designing media for learning is a difficult process and any deficiencies in the authoring process are likely to be reflected in the student's educational experiences. Course sequencing has became an important research issue for educational hypermedia, which among others resulted in standardization issues of learning metadata. However, the complexity of the design process has often been overlooked. A big problem is the necessity to start from a text-based tradition of teaching and learning, which is used to design knowledge in a fixed way and in a hierarchical and linear structure of knowledge management. Educational media will only efficiently support the learning process, if the conceptual design models of the members of the design team fit together and if these conceptual models take into consideration the tasks and needs of different learners in the information society. Vital aspects of the conceptual models of the designers on course design are concerned with user- and task analysis, which in the context of web-based courses is at least to some extent a didactic conception of the media.

1. Introduction

Educational media, for example web-based courses, are used by students with many different goals and levels of knowledge (Kritzenberger and Herczeg, 2000). There is a need for web-based courses that allow to modify and adapt parameters accordingly and to tailor courses flexibly to the needs of different user groups. On the other hand it is obvious that educational media are produced in a text-based tradition of teaching and learning. Starting from this tradition means that content authors produce hierarchical, linear text-structures of courses. However, a linear printed book does not adapt to the needs of different users, e.g. with respect to different levels of difficulty, learning goals, learning strategy, or media preferences. These original "book documents" are in subsequent production stages transferred into hypertext, enriched with hypertext and multimedia functionality, as well as with interactive elements. The later issues are normally not done by the content authors themselves, but by other experts involved in the production process, e.g. concept makers, multimedia producers, pedagogues, software ergonomic experts, etc.. They cooperate as distributed design team, which is separated in time and place.

The text-based tradition of producing courses compels the authors to decide on a structure too early in the authoring process. The result in this kind of text-based production process suffers from severe problems. Among these problems, the hypermedia documents tend to become linear and do not model the (true) semantic structure of the domain itself. This problem is vital in systems which do not provide dynamic links, because the author has to provide the correct links for every possible user of that hypermedia system. Course sequencing became an important research issue in the last years. It is the goal of the sequencing approach to generate a lesson for a target group, e.g. for students, which is capable to be tailored to the needs of that group. One result of the sequencing approach is the work on standardization issues of learning metadata, such as Instructional Management Systems (IMS) or the efforts of IEEE's Learning Technology Standard Committee (LTSC).

Furthermore, not only the semantic structure has to be taken into consideration during the production process but also the different didactic media conceptions, which includes user- and task-analysis (Herczeg, 1994) with respect to learners and the learning process.

2. Knowledge engineering in the didactic field for task- and user-adequate educational media

It has been noted for the process of authoring hypermedia systems, that it is a complex one, which has often been overlooked by hypermedia designers (cf. Theng and Thimbley, 1998; Conklin, 1987). From the practice of the development of educational courses it is obvious that the authoring process of educational hypermedia can be described in several stages (see following sections), where the development is distributed over different people contributing to the development process with different kinds of expertise and different conceptual models. These people of the design team are normally distributed over time and over place. Each of theses experts develop his or her own conceptual model and didactic knowledge is represented in different ways and qualities. One of the main problems in the production process of didactic media is the didactic transformation of teaching and learning contents and goals. By many practitioners, this problem is reduced to a question of collecting and structuring of content.

This paper is based on experience in two national projects. The one project is called "Virtual University of Applied Sciences" www.vfh.de (period of duration 1998-2003). It aims at establishing a location independent university with a curriculum for computer science of multimedia systems and for business engineering (Bachelor, Master). The authors of this paper are involved in the production of web-based courses, in the design of user-adequate learning spaces and in the support of the design process. Their focus is on usability recommendations and quality management for the course material during the development process. Other aspects like teaching strategies, learning processes or technical issues concerning the course production are supervised by other dedicated consulting groups within this project. The other project "Distance Education in Medical Computer Science" (started in January 1999) aims at providing a complete course of studies for the specialization of students in medical computer science. The course of studies is offered at a virtual university (Hagen, Germany). Our responsibility is to transfer the linear text documents (mostly MS-Word format) into hypermedia networks and multimedia courses (Kritzenberger and Herczeg, 2001).

2.1 Stage 1: Conversion from linear text documents to HTML-documents

Content authors normally start writing the courses as linear text documents. This way of producing course material is a quite normal way of teachers' knowledge organization, as people seem to be not good at writing in a non-hierarchical fashion. Extensive research on that suggests that readers form mental hierarchical representations of texts have been cited (Charney, 1987; van Dijk and Kintsch, 1983). However, this compels users to decide on a structure too early in the authoring process.

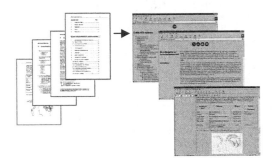

Figure 1: Conversion from linear text-documents to HTML-document

Content authors write linear text documents, which are transferred into web-based courses (HTML-documents) not by the content authors but for example by media producers. Normally, a conversion tool is used for this task, which reflects directly the document markup in order to preserve the hierarchical structure in the first-order hypertext (Rada, 1992). The most significant first-order links are those connecting outline headings. The linear hypertext is similar to Trigg's (Trigg, 1988) idea of paths authored into hypermedia systems. That is nodes are encountered one after the other, guiding the learners through the hyper-space. However, as the hypertext can be used in different ways from the original text, specific media aspects have to be considered in design, e.g. how to give the learner orientation in the information space.

The conception of the text documents by the content authors implicitly covers a didactic planning, which can partly be deduced by analysis of the sequential structure of the instructional texts and elements. In further stages of course production, these considerations on didactics have to be transformed into a proper media conception.

2.2 Stage 2: Conception of the hypermedia document

Although our common way of organizing information in a text is through hierarchies, it is not the proper way of structuring domain knowledge for hyper-spaces. Therefore, in the second stage, the text structure is re-organized the domain knowledge into a hypertext with elementary knowledge units and their relationships, for which a variety of notations exist (Murray, 1998). Among them are ER-Model (Verdejo et al., 2000) and semantic networks (Fischer and Steinmetz, 2000), that is as a directed graph in which concepts are represented as nodes and relations between concepts are represented as typed links (Conklin, 1987). IEEE Learning Technology Architecture (LTSC) proposes a set of knowledge library (knowledge base) which is responsible for the sequencing of a lesson, while the actual compilation of the lesson is performed in the delivery component. A set of semantic relations (e.g. super concept, part-of, problem-solution, instance, causes etc.) is used, which are stored as metadata to describe how concepts relate to other concepts. A similar approach is used in Multibook (Steinacker et al., 1999; Fischer and Steinmetz, 2000). The sequencing of a lesson is then a filter to select a specific structure to be presented to the learner. In this conception the student's learning process is modeled as navigational alternatives over the structure of semantic relations between the concepts (knowledge base). Conceptual relations may for example be typed as pre-condition, post-condition, invariant, satisfy condition, derive as activities etc.

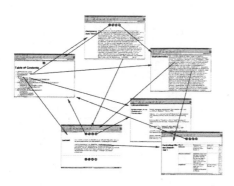

Figure 2: Hypermedia Course

At this point of modeling of the learning environment an inherent problem comes up. In many cases there are not sufficient kinds of relational types, in order to characterize the relation between two knowledge elements in a proper way. Rich typologies of relations are borrowed from discourse analysis models (e.g. Rhetorical Structure Theory) (Mann and Thompson, 1987) in order to type complex relationships between concepts. Another problem in this stage of the production of web-based courses arises from a didactical point of view for the conception of media. Semantic networks are models for the representation of propositional knowledge. The reduction of didactic modeling on this kind of representation and to reduce analysis on it means to neglects other kinds of knowledge of which didactic analysis should also be aware of (Kerrcs, 1998).

2.3 Stage 3: Multimedia course

In the current practice of the design process of web-based courses, the hypertexts are enhanced with time-based multimedia like audio, video or animations etc.. This is regarded as being helpful for a more precise presentation for some kinds of facts and a more concrete way explaining processes, which would otherwise be too complicated to be explained properly and therefore hardly be understood by the learners.

For example can simulations help to demonstrate how time-based actions follow each other in complex processes. However, the question on supporting the learning process with multimedia remains unsolved and the senses are prone to be overloaded. Unfortunately, practice shows that the conception and design of media is often used as secondary question in didactic models for the production stage of multimedia courses. With the growing significance of media in teaching and learning with web-based courses in distance education, a didactic planning model centered around the media aspect is necessary.

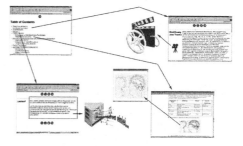

Figure 3: Multimedia Course

Propositional knowledge representation in semantic networks is not sufficient for the proper representation of the goals of teaching and learning activities. In didactic thinking there are further goals describing for example what quality of knowledge acquisition in the learning process should be achieved. It is helpful to analyze the kinds of knowledge (Tennyson and Rasch, 1988) on which learning content is based in order to deduce consequences for presentation:

- declarative knowledge: abstract ⟷ concrete
- procedural knowledge: meta-cognitive ⟷ domain specific
- contextual knowledge: anchored ⟷ analog (scripts, mental images, cognitive maps...)

Only declarative knowledge is properly structured and represented with propositions and relations between them, whereas procedural knowledge is rule-based knowledge ("if... then..."). Contextual knowledge is bound to experiences of concrete situations, which allow another kind of access to knowledge structures and can hardly be described with propositions of production rules. But here is a chance to analyze the domain and to find content which is of the kind and to model it with digital media.

2.4   Stage 4: Interactive computer-based training

Interactivity means a quality of action and reaction of user and system and vice-versa, where the focus is on internal cognitive operations of the learner.

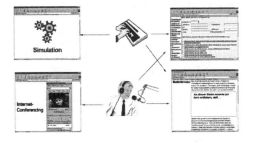

Figure 4: Interactive computer-based training

For the didactic conception of interaction with computer-based training program, the main focus has to be on the learner's cognitive operations. Also information processing procedures of the learners have to be considered. This considerations go beyond the models of knowledge representation and look at the learning procedures itself, that is how knowledge is acquired by the learner. It implies that different stages of acquisition of expertise have to be considered in the design, e.g. the compilation of knowledge. Another aspect is, that the design should inspire the learners to various kinds of alternative learning activities.

2.5   Stage 5: Computer-Supported Cooperative Learning (CSCL)

The last stage in the development of web-courses for lifelong learning is the integration into a tele-cooperation and communication environment, for example into a virtual university. A special advantage of this constellation is the combination between information and communication component. The CSCL approach is based on the following premises: that learning and skill acquisition is facilitated by generative problem solving, collaborative work and use of multiple cases.

As collaborative work is central to learning, students are expected to assignment in groups in order to articulate and reflect their thinking and subsequent transfer. Furthermore, the availability and use of multiple cases during problem solving facilitates learning new knowledge and transfer of previous solutions to the current problem (Kolodner, 1993). A shared electronic environment has to seamlessly integrate a full variety of functions and has to tie together tools that students will use while solving problems and collaborating.

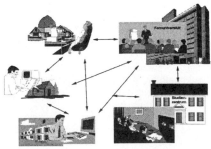

Figure 5: CSCL

3. Conclusions

User- and task adequate learning environments to be used for lifelong learning for all, need didactic media conception with transformation through different stages of production, where different media constellations are in the center of conceptual models of the developers. The problem in development is, that the expert knowledge is distributed over the different members of a design team. The conclusion lies near that the quality of educational media will grow, if this knowledge, that is the conceptual models of the developers, is better managed. A user- and task-based framework for analysis and a proper tool for knowledge management will be helpful to solve this problem (Kritzenberger, Hartwig, and Herczeg, 2001).

References

Charney, D.: Comprehending non-linear text: The role of discourse cues and reading strategies. In: Proceedings of Hypertext 1987 Conference. Charlotte, N.C., November 13-15. ACM, N.Y., 1987, 109-120

Conklin, J.: Hypertext. An Introduction and Survey. IEEE Computer, September 1987, pp. 17-41

Fischer, S. and Steinmetz, R.: Automatic Creation of Exercises in Adaptive Hypermedia Learning Systems. In: Proceedings of Hypertext 2000. San Antonio, Texas. ACM, 2000, 49-55

Herczeg, M.: Software-Ergonomie. Addison-Wesley-Longman and Oldenbourg-Verlag, 1994

IEEE Learning Technology Standards Committee (LTSC). IEEE P1484.12 Learning Objects Metadata Working Group. URL: http://ltsc.ieee.org/wg12

Instructional Management Systems (IMS). URL: http://www.imsproject.org

Kerres, M.: Multimediale und telemediale Lernumgebungen. Konzeption und Entwicklung. München, Wien: Oldenburg Verlag, 1998

Kolodner, J.: Case-Based Reasoning. Morgan Kaufmann, San Mateo, CA, 1993

Kritzenberger, H. and Herczeg, M.: Completing Design Concepts for Lifelong Learning. In: Proceedings of ED-MEDIA 2000. World Conference on Educational Hypermedia, Multimedia and Telecommunications. 26th June – 1st July 2000, Montréal, Canada. AACE: Association for the Advancement of Computing in Education, 2000, 1374-1375

Kritzenberger, H. and Herczeg, M.: Approaches to Quality Management for Developing WBT-Courses. In: Proceedings ICDE-World 2001. 20th Conference of the International Council for Open and Distance Education. Düsseldorf, 1st- 5th April, 2001

Kritzenberger, H.; Hartwig, R.; and Herczeg, M.: Scenario-Based Design for Flexible Hypermedia Learning Environments. In: Proceedings of ED-MEDIA 2001. International Conference on Educational Multimedia, Hypermedia and Telecommunications. AACE: Association of the Advancement of Computing in Education, 2001 (to appear)

Mann, W.C.; Thompson, S.A.: Rhetorical Structure Theory: A Theory of Text Organization, Technical Report RS-87-190, USC/Information Science Institute, 1987

Murray, T.: Authoring knowledge based tutors: Tools for content, instructional strategy, student model, and interface design. Journal of Learning Sciences 1998, 7(1), 5-64

Rada, Roy: Converting a Textbook to Hypertext. ACM Transactions on Information Systems 1992, 10(3), 294-315

Steinacker, A.; Seeberg, C.; Reichenberger, K.; Fischer, S.; and Steinmetz, R.: Dynamically Generated Tables of Contents as Guided Tours in Adaptive Hypermedia Systems. In: Proceedings of ED-MEDIA 1999. International Conference on Educational Hypermedia, Multimedia and Telecommunications. Seattle, Washington. AACE: Association for the Advancement of Computing in Education, 1999, 640-645

Tennyson, R.D.; Rasch, M.: Linking Cognitive Learning Theory to Instructional Prescriptions. Instructional Science 1988, 17, 369-385

Theng, Y.; Thimbley, H.: Addressing Design and Usability Issues in Hypertext and on the World Wide Web by Re-Examining the "Lost in Hyperspace" Problem. Journal of Universal Computer Science, 1992, 4(11), 839-855

Trigg, R.: Guided Tours and Table Tops. ACM Transactions on Office Information Systems 1988, 6(4), 233-247

Van Dijk, T.; Kintsch, W.: Strategies of text comprehension. New York: Academic Press, 1983

Verdejo, M.F.; Rodríguez-Artacho, M.; Mayorga, J.I.; Calero, M.Y.: Creating Web-Based Scenarios to Support Distance Learners. In: Franklin, S.; Strenski, E. (Eds.): Building University Electronic Educational Environments. Boston, Dordrecht, London: Kluwer Academic Publishers, 2000, 142-153

Bridges for learning – language training and interaction over a distance

Magnus Magnusson, John Sören Pettersson

Karlstad University
Division for Educational Science, Department of Special Education
Division for Information technology, Department of Information Systems
S-651 88 KARLSTAD, Sverige/ Sweden
Tel: +46 54 700 17 14 Fax: +46 54 83 23 12
Magnus.Magnusson@kau.se
John_Soren.Pettersson@kau.se

Abstract

Three projects are described briefly as examples of ongoing work with multimedial distance communication in the form of different types of videophone technology in therapeutical and educational use in Sweden. Project one examplifies language therapy for adults with aphasia. Project two examplifies professional networking among speech therapists and project three exmplifies social interaction among pupils in primary schools.

1. Background

The most common application in the Information Technology/ IT-field today, is clearly the PC or the personal computer. The grandchild of the large and impersonal mainframe-systems of the 1960's has become a sort of cybernetic mental prostheses to individuals and organizations, expanding and supporting certain mental or cognitive procesesses in a way which has given birth to ideas about the human machine or the technified human being, meaning that there might seem to be no clear borderline between a human being and the computer as a tool.

The PC as a tool is very hard to define or to classify. Most tools in human history of technology have been linear and uni-dimensional, meaning that you are supposed to use the tool in one or possibly a few general situations with very specific aims and content and that the result of using the tool can be seen directly. The typical tool has also been non-representative, that is, the tool has never symbolized anything. This doesn't mean that tools have never been used as symbols but the symbol-content has always been unintentional in relation to the intended usage of the tool. One example is the usage of a pen to symbolize literature, writing, being literate etc. The PC, however is multifunctional, indirect and representative which makes it very difficult to define and generalize according to traditional terminology. Humanity has always developed tools.

Tools can be used directly or indirectly as a means to an end. A direct usage is for instance when you use a monkey-wrench to open the pipes of a sink. An indirect usage is when you use a book or an information video to learn how to use the monkey wrench. There, the book or the video is a tool for learning to use the tool for fixing the sink. In modern society we are totally dependent on an indefinite number of tools which are helping us in more or less complicated structures to manage or even survive in modern society. A very important type of tool in modern society is the learning (or teaching) tool since society is so complicated that we need to learn how to do things constantly. The learning/teaching situation can be adapted according to people's needs through the means of different types of tools. The best tools become integrated with the user in a way that it is advisable to use the word transparency to define the relation.

With modern Information-Technology-based tools, learning and teaching can be performed over a distance, that is, teaching, learning, training and supervision can be performed without moving the participants, just moving the messages between them, using some sort of medium and different formats of the messages. Thereby it seems to become easier to raise the productivity of education in terms of number of teaching hours for every pupil and it suddenly becomes possible to analyze education in terms of producing something where the level of productivity and efficience can be raised.

Distance education as a concept has been used as mostly in relation with learning and teaching within a traditional educational framework. However, in the field of therapy and clinically oriented training and treatment, distance learning could also be a possibility. In Sweden, several field projects have investigated methods for giving language training or therapy to different groups of people with language and speech disabilities, notably persons with Aphasia or mental retardation. In those projects the technology used has been different types of off the shelf videophone technologies and the aim has been to create a sort of experienced virtual therapeutical or educational multimedial "rooms" over the ISDN-network or Internet. Therapists as well as clients have taken part in short term and long term projects and some results will be described, concentrating on considerations for the future.

2. Research

At Karlstad University several projects have been investigating the usage of different types of videophone technology, for children as well as for adults, in a training or educational situation. The paper will present some of these projects, including a project on language training for people with Aphasia using ISDN-technology and a WebCam-project for children at a primary school.

The authors represent different perspectives and disciplines. Magnusson is a speech pathologist with a doctoral degree in Pedagogics, working at the moment with distance education related problems for adults and adolescents with speech and language disabilities. Pettersson has a doctoral degree in General Linguistics and is working within a framework project HumanIT to develop general IT-ideas. He is concentrating on learning processes of children, using computers.

We will focus on three projects or field trials:
1. Language training for people with Aphasia
2. Professional use of speech pathologists of videotelephony
3. WebCam-based communication for young pupils at school

2.1 Project 1

In the first project an 18-month field trial involving 31 persons with Aphasia indicate that participants easily forget the presence of the intermediate technology, even if the quality of the transmission is limited. It also indicates that participants experience higher quality of communication and also positive effects of training.

2.2 Project 2

In the second project, professionals experience wide possibilities to enlarge their professional competence and their possibilities of developing new methods of therapy of value for their clients. In both projects, participants experience the technology as functioning, trustworthy, available and undramatical. The experience of difficulty or not to learn to handle the technology is clearly related to the participants' knowledge of computer usage in general.

Seemingly, videophone technology could be described as a good tool, both for learning about things as well as for manging something and as a means to an end. A tool which easily becomes transparent in the sense that the user takes it as granted and it remains unseen as long as it functions well. It supports the mutual feeling of the users interacting with each other of being together in a sort of virtual and common room. A result of this feeling is that the interaction becomes more unfettered and "natural" in the sense that the interactors do not feel the presence of any "filter" which could be experienced as something to interfere with the interaction. A good example of this is a report on a study of the lifeworld experiences of the participants of project one (Magnusson, 2000). All the data in that study have been amassed through videophone dialogues.

The technology used has been an integrated off-the-shelf PC-based videophone system (Intel Pro-Share) where it has been possible to use the PC-based videophone as a common computer at the same time as it has been used as a videophone. This means that it has been possible for the users to share a common programme or application at the same time as they have been participating in a videophone-call. The experience of sitting in a virtual or even common room "room" has thereby been heightened and is comparable to a feeling of (physical) togetherness. The programmes used have all been off-the-shelf and no special development has been needed in the projects. The"adaptions" have all been built upon availability of human support. Assistants and supervisors have been available all the time, especially early in project one. The assistants themselves have partly been people with language disability which has included an element of teacher training into the projects.

2.3 Project 3

In the third project, results are not available in the same degree since the project is recently started. However, as the work reported above indicates, which has been carried out by the first author, it is possible to obtain a "full" or at least meaningful communication through the videophone system thanks to the possibility to aid spoken dialog with gestures as well as text messages. In this project, the second author has had his own reasons for attempting a webcam project for young school children.

The objective of this project is "international" training of the children, especially when it comes to linguistic competence (including cross-linguistic strategies). The immediate goal is to develop a web-based environment for Swedish children to "play" with Norwegian children (the Swedish and Norwegian languages are closely related). Future prospect includes letting children with another mother-tongue than Swedish have access to peers in Sweden and perhaps teachers and relatives in their (or their parents') home country. It would also be desirable to aid language training even when it comes to foreign languages not spoken at home, especially so for families returning to Sweden after some years abroad.

Communication across national borders often entails great linguistic skill: knowledge of a foreign language as well as of writing. For this project the webcam is thought of bridging this double linguistic gap. Just like the videophones used in the earlier described project, today's webcam software includes the simultaneous exchange of webcam signal and other things like written chat and sketch pad (cf. Microsoft's NetMeeting). Around such things as a simple drawing pad, different plays can be developed; initially, the children in each country are taught the games in advance of the webcam sessions. A simple thing is to have one child at each site drawing a picture of each other (in NetMeeting both drawings appear simultaneous side-by-side in the same sketch pad window while each video image appear in a window close by).

It would be interesting to have special software where the children can play different language-dependent games, where for instance one child is referee and the other tries to learn a few words in the language of the first one through a memory card game. Or forming two teams with members from each site so that people will try to explain word-meaning by enactment before the webcamera or by using the sketch pad. Admittedly, webcam transmissions have no high quality of sound and image while still demanding more of the Internet connection than most modem connections allow for. However, despite the deficiencies in transfer quality, there is a distinct advantage of using a video image. The fact that the children are able to see each other in real time implies more than merely allowing them to gesticulate. Signs of passivity are being transmitted: if one part remains silent, the other part is able to detect whether the first one listens intensively or has started to do something else, or whatever the reason for the silence might be.

Lack of passive signals characterizes other Internet-mediated communications like written "chat" and "netgames". For young children the immediateness, the parallelism, of an interaction is very important for the perceived degree of importance of the interaction. After all, the ambient situation is what children are most aware of. On the other hand, games, including netgames, can provide excellent opportunities to develop "togetherness", mutual understanding, and in due course even language. But in order for this to come about via Internet games, there has to be a more fully fledged communication channel than the one provided by the games per se if we want to see younger school children being able to develop a mutual understanding via the net. The situation has to be rather like an ordinary playground where indeed children of different nationalities meet and begin playing with each other. That is where the webcam image comes into the picture of international classroom correspondence, as argued above.

For the moment, we are trying to establish contact between a grade 1-3 class in Sweden and a grade 3 class in Norway (the phrasing "trying to establish contact" reflects various die hard technical problems). A project assistant had previously worked in Norway and engaged this Swedish class in a peer-to-peer correspondence with her Norwegian class. Many of the pupils in both these classes had experienced great joy, but when the Swedish teacher left that Norwegian class, no adult had been able to sustain the correspondence. Written communication with such young children demands a deeply committed adult. But we have to admit that although we use off.the-shelf systems, the technologically related problems seem to make up for an obstacle nearly as great.

The new Norwegian class has been awaiting installation of Internet and computers, while in the Swedish class a couple of webcam-equipped computers have been installed in two different rooms to test technology and get some initial feeling for how the children react and interact. The children liked everything, even when the sound signal malfunctioned in one direction. Of course, this was quite a new experience for them. They could see each other, there was even a video window with one's own picture (to make obvious for the child/children at each machine whether they were visible for the web camera or not), and they could mysteriously draw and write on the same whiteboard window on the computer screen. They drew and withdrew their own and each others pictures. It is interesting to see a graphic "dialog" evolving like that. There are indeed netgames consisting only of a drawing chat, but we think that will not really allow for the development of support for pedagogical aims. Also, a pure drawing chat will not allow the children to meet each others as persons.

During November and December several interviews were carried out among the staff in the Swedish class, including special teachers and after-school club staff. The interview concerned if they could see any use of webcam communication within their own pedagogic work and if they had any suggestions on what activities would be suitable to this end. The results of the interviews show ambiguous feelings among the teachers towards the technology as well as different pedagogical approaches.

Especially the class teacher discussed activities in terms of different exercises or tasks which individual or groups of pupils could perform. On the other hand, one leisure time pedagogue emphasized that one has to discuss whith the children before designing activities, which probably reflects her background: she has no teacher's exam but exam for after-school club staff. She is less bound by a curriculum. The two youngest memebers of the staff were very positive to using computers, while the rest demanded that everything work before they have to deal with the equipment. One made it clear that there was simply no time for her to engage in this and she saw little use of the project. Referring to all technological problems this school unit has had long before this project started, she said she had lost all hope in computer activities.

3. Conclusion

Such reaction stands in stark contrast to the findings in the Aphasia project. Obviously, the presence of technical assistance plays a major role. This could be compared with a recent evaluation of the effect of additional computer-support resources for schools in Sweden (Birnik & Eliasson, 2000). It was found that some of the technical problem-solving had been lifted off from teachers, but it was not always the case that teachers felt they, as a consequence, had been given time to work more pedagogically with the computer in their teaching. Rather, the attitudes among teachers to information technology seemed to be a more important factor to pedagocical development.

The more positive user-reactions out of projects one and two might also be explained as a result, either of the higher quality of transmission or as a result of the participants feeling that they were participating in a pioneering venture. This might also be a result of the fact that groups with this type of disability often bear witness to a feeling of being left outside of the rest of society (Magnusson, 2000). Another reason might simply be the fact that the first two projects have been running for a long period of time.

The projects mentioned in this overview are representative of larger initiatives within which several new projects, investigating concepts like telepresence, distance education etc will be investigated.

The projects mentioned have been supported through grants from The Board of Communication Research, HumanIT at Karlstad University and the Swedsh Institute for Technical Aids.

References

Birnik, H., B. Eliasson (2000). IT-support i skolan gör datorn till pedagogiskt verktyg? Utvärdering av ett IT-rådgivarprojekt i sju svenska län. *Karlstad University Studies* 2000:05.

Magnusson, M. (2000). *Language in Life – Life in Language*. Karlstad University, Dissertation in Education.

Evaluation in the Development of a tele-medical Training System

Yehya Mohamad, Holger Tebarth[2]

[1] GMD - German National Research Centre for Information Technology
Institute for Applied Information Technology
Schloss Birlinghoven
D-53754 Sankt Augustin

[2] KaS – Klinik am Schloßgarten
Clinic for Psychiatry and Psychotherapy
Am Schloßgarten 10
D-48249 Dülmen

Abstract

Pedagogical Animated Agents that inhabit interactive training environments can exhibit lifelike behaviours and have the ability to impart and coach memory strategies in a very suitable way to children. In addition to provide problem-solving advice in response to children's activities, these agents may also be able to play a powerful motivational role. The project aims a web-based cognitive training for an economical, prolonged and controlled intervention in large numbers of neuropaediatric memory impaired patients.

1. Introduction

Recent neuropsychological research assumes, that children with cognitive disorders e.g. caused by epilepsy can benefit by special training of memory strategies [Haverkamp et al.1999]. Though up to now this kind of memory training requires a specialized coach who performs an individual and motivational support.

In the consequence we started to develop an autonomous tele-medical Training system that is primarily based on the concept of **A**nimated **P**edagogical **A**gents (**TAPA**) [Tebarth, Mohamad & Pieper 2000]. The basic structure is determined by interconnected diagnostic and training modules. Diagnostically relevant input give the starting point for the training level while the training modules themself recognize the children's performance dynamically.

An aimed interface adaptivity is based on non-trivial inferences from input information. The implementation of adaptivity is based on three main sources or methodologies [ANDRÈ E, RIST T & MÜLLER J.]:

- Pre-collected and entered data about the individual child
- Data received from input devices during a training session
- Data calculated pre and during a training session using Bayesian networks and Markov chains

To control the important influence of motivational and attentional aspects on the children's performance and training benefit the agents have to consider a lot of data that are obtained by several input-devices like e.g. eye-tracker. Responding to the risk of misinterpretation of these input data we developed for a first evaluation step some prototype scenarios that include different communication situations. The agents job about that is to verify either falsify the interpretation of these input data by so called repair questions or dialogues.

These repair questions were developed systematically with respect to the socio-psychological concept of meta-dialogue and can be expressed in formalized pattern. A logical model for the development of meta-dialogue was developed and tested.

In this regard basic concepts of the so-called "Interpersonal Perception Method" (IPM) were used, which has proven to be useful in psycho-therapeutical diagnostics of interpersonal relationships. The two basic IPM-assumptions that had to be considered are:

1. Behaviour is a function of experience.
2. Experience as well as behaviour always has a relationship to someone else or something else other than oneself.

To design an optimized and effective agent-based training environment for children with cognitive disorder, it is furthermore essential to understand how the children perceive the animated pedagogical agent with regard to affective dimensions such as encouragement, believability and clarity. So we have built an emotional model to implement all expressions of the agent (facial play, gesture, body movements and speech).

Currently we started a pre-study training of 30 mnemonically impaired children with a prototype scenario of TAPA. First results will be presented at the UAHCI 2001 Proceedings.

2. Purpose

The project's purpose is to develop a tele-medical intervention system with the following main characteristics:

- Performing an agent-interface with a high suitability for children using different input devices.
- Integration of diagnostic and training module allowing valid assessment of the cognitive starting point as well as performance changes during the training.
- General consideration of the motivational and attentional situation of the child while training with the system is necessary to achieve the expected training effect. To receive information about the current motivational situation as valid as possible different input techniques are used and crosschecked.
- The system should be able to realize an individual influencing of the child's motivation and attention by reactively changing the agents behaviour and task presentation within an adaptive system environment.
- Furthermore TAPA is to improve the children's perceiving of their self-efficacy. Due to repeated and continuous feedback by animated agents the child can learn to estimate its own performance more realistically.
- Basically the whole TAPA-system is aimed for a web-based use with an individual and autonomous training at home, which is necessary to obtain a prolonged effect in large numbers of affected children.

3. Development

The development and authoring of appropriate scenarios and story required some basic consideration. First the design of a hierarchical structure had to be built for scenario sequences and chains. Second the scenarios were interconnected to meet the requirements of both modules (diagnosis and training). Carry on here we build a model that combines individual graphics to animated scenarios using animation software.

The context story consists of different and sequential episodes involving several agents. The design of all individual graphics is performed with different software e.g. Flash 5.0, Corel Draw, Mash 5.0. The different graphics are combined to a virtual world, which helps to tell the story:

- Leading Agents (LA) introduce themself to the child and interview the child regarding learning-self-concept and estimation of own cognitive performance and abilities. The child´s answers are used for adaptivity and evaluation.
- The diagnosis module needs different scenarios and input to obtain the actual attention and motivational level of the child, this allows as well that the training session can be adapted to the child´s abilities and motivation, activation and attention.
- In the training session many tasks will be offered by different side agents (SA). The child has to solve them under the assistance of the LA. Later this kind of tasks could easily transferred to every day situations to allow using the learned knowledge in real situations [LESTER JC, STONE BA & STELLING GD].
- A `motivation´ panel (fun-o-meter) is used, on which the child could tell the system about his current level of motivation. Children with cognitive impairment have often problems with the perceiving of their motivational relevant self-efficacy.

Regarding the emotional modelling we had to define and variegate gesture, gaze and facial play of the agents. In different situations the agents have to show different emotions e.g. to be happy and to tell that in gestures to the child. So we built a model with all possible emotional states. Those emotional states should be applied in an optimized way to the current situation [JOHNSON WL & RICKEL J.].

Furthermore we had to integrate the individual agent into its environment to build a consistent and closed training context. All agents communicate to each other and to the virtual world and its objects. This allows them to react to any change in this world and to perform finally an adaptive behaviour to the child.

(Hard and software requirements: Pentium II or equivalent PC, Windows Operating System, Microsoft Agent Technology Package, Flash player, Speech Recognition Software, MSAccess, IIS 4.0 or higher or PWS.)

838

4. Evaluation

Developmental evaluation is currently performed in a preliminary study that consists in testing 15 learning disabled children, 15 epileptic children and 15 unimpaired, healthy children by the prototype scenario.

This first step helps to analyse communicational functionality as well as usability and is necessary for optimizing of meta-communication. Within the scope of the context story TAPA-prototype is able to do the following:

- Measuring of the memory performance (regarding a special meta-memory strategy)
- Realizing an adequate, for the time being performance-adapted training
- Registering the child's performance-related self-concept via agents' questions
- Registering the child's estimation of its own performance via agents' questions
- Measuring the child's fun and motivation work on the tasks via a special input panel (fun-o-meter)
- Using of different technical input devices and saving data on timescale

In addition it is necessary to investigate in which way the agents, animation effects and tasks can influence the child's attention, emotions and motivation that speak at the physiological behaviour level and how to measure them. The following parameters can give indirect physiological hints for the actual motivation: reaction speed as time between task output and answer input; frequency and intensity of sound remarks via voice recognition and sound registration; speed, rest time and target of child's eye movement registered by eye tracker and occasionally heart frequency and skin conducting ability measuring with a finger clip.

Furthermore there is a parallel behaviour observation through experts and parents regarding the child's motivation on the basis of a video documentation as well as there is a later supplementary questionnaire for child and parents that measures also the situation-related motivation.

In a further step the data collected on these three ways are statistically compared and cross-checked to examine the validity of the technical survey. If there are statistically significant correlations between the technical, observational and the psychometrical collected data, the corresponding inputs can partially used for future motivation measurement and could function as depending variable for the adaptional system.

Next to these more methodological but necessary questions the preliminary study allows to answer some more agent-related aspects:

- How does the child perceive the virtual characters (agents)? > Asking the child for special traits of the agent e.g. intentions, sympathy, believebility.
- Which aspects of the interface (a) technical: mouse, voice generator, voice recognition and (b) agent related: behaviour, kind of questions, frequency of interaction had motivating and which demotivating character for the child? Which problems occur and what should be modified?
- Which are functional and dysfunctional aspects in communication outflow e.g. did the repair questions reached their function goal in meta dialogues?
- How important is the computer experience of the child?
- To which position on a motivation scale do the child categorize the different animation events in comparison to every-day events?

Main evaluation on the second will involve a pre-/post-test design with at least n = 200 cognitive impaired and n = 100 unimpaired children. The necessary evaluation criteria for the diagnostic module are committed to concurrent validity, precision, reliability and predictive accuracy of diagnostic module via comparison with conventional (meta-)memory tests.

Training efficiency, ecological validity and individual benefit of intervention module will be performed within the pre-/post-test design and via alternative examination of every-day memory performance including a comparison to efficacy-studies on human trainers.

Next to the effectivity criteria the users' acceptance and subjective perceiving of TAPA should be evaluated.

Altogether the main study should allow to answer several questions, e.g.:

- Result of a cost benefit analysis and prognosis to the usage of agent-based tutorial systems within the educational and therapeutic area.
- Which connection exists between the intellectual efficiency of the child and the spontaneous use of meta-memory strategies on the one as well as training success on the other hand?
- Which connection exists between special disease parameters (e.g. type of epilepsy, frequency of seizures) and the spontaneous use of meta-memory strategies and training success [Dam 1990]?
- Which connections exist between self concept and estimation of the own expected performance an the one hand and spontaneous, effective application of the strategies and training success on the other hand?

- Whether meta-memory knowledge and strategy use can be improved beyond the concrete TAPA session, e.g. whether transfer training on everyday life level and modification (by consistent feedback of the Agents) can improve the self perception?
- To what extent could the different parameters determined in the preliminary study be used as motivation indicators allowing a motivating adaptivity of the system?
- Whether and to what extent the child's motivation can be influenced by appropriate adaptive motivation work of the agents?

5. Perspective

The project tries to consider many interdisciplinary aspects in its parts. On the one hand there is aimed to realize an integration of agents with a database and graphical environment as well as an evaluation of agent oriented interfaces. On the other hand we consider psychological research on meta-memory training and tele-medical systems.

This project might give a deciding contribution to the research in the aforementioned areas and could build the basic of further research especially on using new technologies and devices to achieve adaptivity of user interaction with agent oriented interfaces. The expected results could play an important role in future research on the cognitive as well on the perceptual side of the implementation of emotional models fore agent oriented user interfaces. Possibly they will help in the further development of tele-therapeutical and tele-pedagogical intervention system.

References
ANDRÈ E, RIST T & MÜLLER J. Employing AI methods to control the behaviour of animated interface agents. *Applied Artificial Intelligence,* 1999, 13: 415-448.
DAM M. Children with Epilepsy: The effect oft seizures, syndromes, and etiological factors on cognitive functioning. *Epilepsia,* 1990, 31(4): 26-29.
HAVERKAMP F, TEBARTH H, MAYER H & NOEKER M. Serielle und simultane Informationsverarbeitung bei Kindern mit symptomatischer und idiopathischer Epilepsie: Konsequenzen für eine spezifische pädagogische Förderung. *Aktuelle Neuropädiatrie '98,* 1999: 251-255.
JOHNSON WL & RICKEL J. Animated Pedagogical Agents: Face-to-Face Interaction in Interactive Learning Environments. Marina del Rey (California/USA): University of Southern California, Center for Advanced Research in Technology for Education (CARTE) 2000.
LESTER JC, STONE BA & STELLING GD. Lifelike pedagogical agents for mixed-initiative problem solving in constructivist learning environments. *User Modelling and User-Adapted Interaction,* 1999, 9: 1-44.
PIEPER M. Sociological Issues of HCI Design, In: STEPHANIDIS C. (Ed.): User Interfaces for All – Concepts, Methods, Tools. Mahwah (NJ): Lawrence Erlbaum 2000.
TEBARTH H, MOHAMAD Y & PIEPER M. Cognitive Training by Animated Pedagogical Agents (TAPA) Development of a Tele-Medical System for Memory Improvement in Children with Epilepsy. 6th ERCIM Work shop *User Interfaces for All*; CNR-IROE, Florence, Italy 25-26 October 2000.

User Interfaces for training E&D drivers

Aristotelis Naniopoulos[1], Maria Panou[2]

1. Aristotle University of Thessaloniki, Civil Engineering Department, Transport Systems
Research Group, Ermou 61, 54623 Thessaloniki, Greece
Tel : +30-31-256033, Fax : +30-31-256037, Email : naniopou@ccf.auth.gr)

2.TRUTh S.A., 18, Navarinou sq. , 54622 Thessaloniki, Greece,
Tel.: +30-31-283722, Fax: +30-31-262784, E-mail: trnspcon@compulink.gr

Abstract

People with Special Needs (PSN) encounter various problems while trying to acquire a driving license. In the majority of cases, they have to visit specialized driving schools, which are often far from their residences. Furthermore, there is not a concrete manual of instructions for best practices to follow, according to their impairment. The Greek national project ODIGO, co-funded by HORIZON programe of the European Commission, has dealt with this issue. Its aim was to provide universal access of PSN to driving training facilities, namely driving schools, software, manuals. Among others, the project has developed an educational manual and multimedia software for driving instructors, to be used in driving training of disabled as well as non-disabled trainees, so as to create a training infrastructure accessible by all. Such a generic manual and software, including specialized sessions and data for PSN presents some innovative aspects in international level.

1. Introduction

The problem of training PSN in driving has not been solved adequately. Getting a driving license is very important for PSN, as their job opportunities are closely related to their ability to transfer from and to their workplace. Unfortunately, PSN drivers are very few compared to the drivers "without special needs", due to difficulties throughout training. The procedure of acquiring a driving licence for PSN consists of the following steps :

1. Assessment of fitness to drive (locomotory diseases, other medical/functional problems).
2. Determination of adaptations required to compensate for the locomotory diseases / functional problems.
3. Acquisition of required equipment and adaptation of car.
4. Training of PSN in driving.
5. Examination.

The above steps are all dealt with efficiency in various European countries, except from the training of PSN. Indeed, there is no handbook, manual or software to support sufficiently driving instructors on how to train the PSN in driving. Therefore, if a PSN wants to learn to drive, he/she has to go to special driving schools, which are not many or they are more expensive.

A questionnaire survey has been conducted through EU countries about the reasons that restrict PSN from getting driving licenses. The conclusions of this survey, derived by the answers that PSN gave, are:

* There should not be special driving schools for PSN.
* All people, with or without special needs, must go to the same driving schools. In order to achieve this, all driving schools must be trained and equipped so as to be able to train all people.
* The training system must be adapted too, in order to comply to PSN residual abilities.

For these reasons, through ODIGO / HORIZON project a new integrated training system has been developed. This is based on a training manual, which contains about 100 driving exercises. These exercises are adequate for all ("standard") drivers. Furthermore, each exercise contains special instructions about how to teach it and where to put emphasis in case of PSN trainees.

Furthermore, a multimedia tool was developed for training all drivers, including PSN. This in fact presents in a user-friendly way, the exercises of the manual and the special instructions for PSN. It also includes a driving aids' database with aids animations, so that PSN can easily have an overview of available aids. The software includes text, pictures, video clips of various situations, as well as a database of driving aids for people with special needs with pictures, animation (successive pictures of the aid from different angles) and video clips.

2. Description of the training software for PSN and driving instructors

When the program starts, at the Selection Screen, the user has the following options.

- **Chapter Selection**: Clicking on this area, gives the user access to the Instruction manual for the driving instructors. He/she will be able select which chapter or which paragraph to view.
- **Keyword search**: This is another way to access the Instructor manual. The user can give a keyword and the program will give all relevant paragraphs or chapters of the manual relevant to this word.
- **Driving aids**: If the user clicks on this area, he/she can access in different ways the database of driving aids that is included in the manual. He/she can then get information on specific aids.
- **Help**: A help screen appears with information on the current screen. Help is also available on every screen by clicking on the ambulance.

Also, the user can go to the previous screen by clicking on the relevant button or exit the database by clicking on the button "Exit".

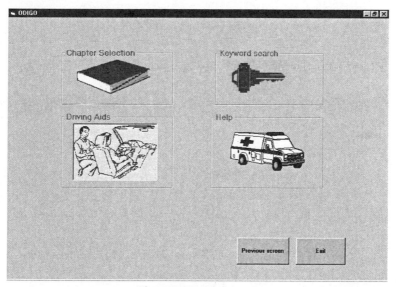

Figure 1: Selection screen

After the user selects the required exercise the relevant text appears. On the right of the screen a list with the pictures to which the text refers to is shown. By clicking on the figure caption a screen with the picture and a relevant short text appears.

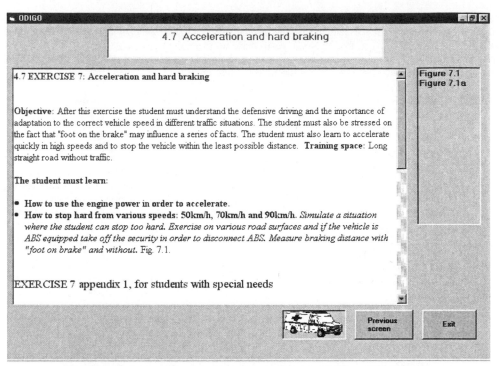

Figure 2: Text screen (for general purpose training and PSN specific training)

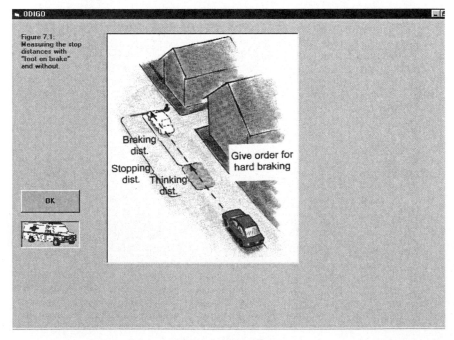

Figure 3: Picture form

The driving aids database can be accessed in different ways, such as:
- Selection from a List of Aids: A list with the aids in the database appear, from which the user can select an aid and view its main characteristics.
- Search by disability: The user can search the aids that are suitable for a particular disability
- Search by keyword: The user can enter a keyword and get a list of the relevant aids in the database.

The information on the selected aid includes the following:
- ➢ Short description of the aid
- ➢ Price (in EUROs)
- ➢ Disabilities for which it is suitable

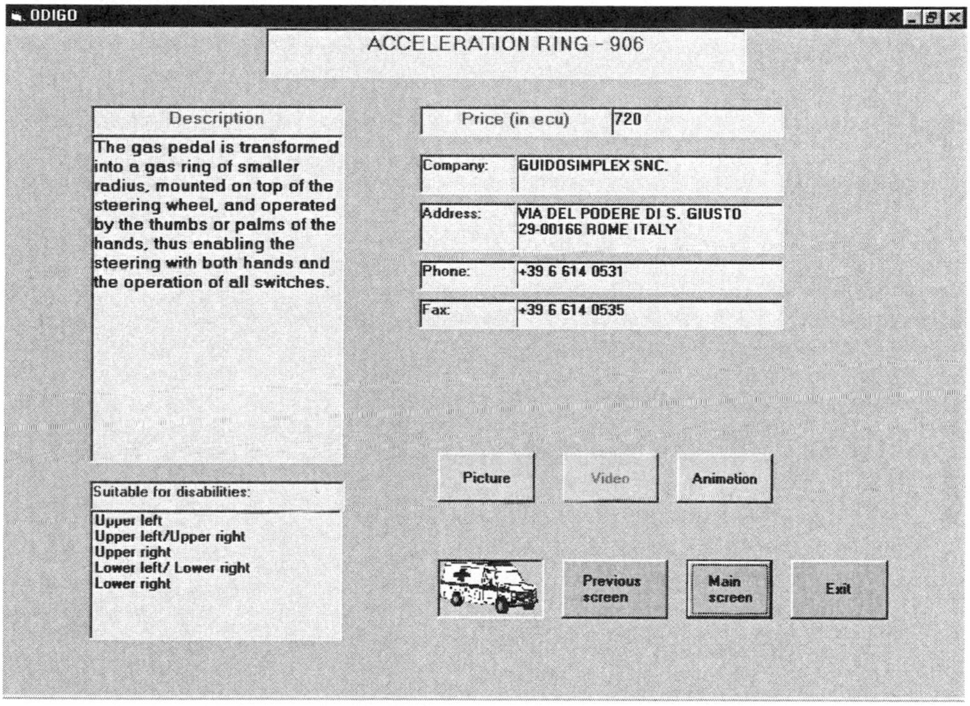

Figure 4: Information presentation on the selected aid

There are also three options from this screen:
- ➢ **Picture button**, which gives a screen with the picture of the aid.
- ➢ **Video button**, which gives a screen where a short video with demonstration of the aid is shown. The user can enlarge the video by clicking on the 'Zoom' button and put it again in its initial dimensions by clicking on it again.
- ➢ **Animation button**, which gives a screen where successive pictures of the aid from different angles are shown. The user can stop the animation by freezing on one picture and continue later by clicking on the relevant button.

Tests have been performed with driving instructors using the training software for 5 days and then using it to train their own students and volunteer PSN. Driving instructors were asked to complete an acceptance questionnaire, subjectively rating various items in a 0 to 9 scale (usability, controllability, user-friendliness, helpfulness, efficiency. Average rating were over 6 for all items except helpfulness, which was 5.28. Therefore, specific help on each button was added in addition to the existing help menu. As a result instructors were fully satisfied with the software and manual and rated it as containing very useful advice on best practices, not only for PSN but for all learner drivers.

This software is publicly available to all driving instructors and PSN and can be obtained from INIOCHOS Centre of the National Institute for the Rehabilitation of the Handicapped of Greece.

3. Conclusions

From the above work it is obvious that the application of the "design for all" principles generates the best results for everybody. Indeed, all drivers liked and found useful the compendium of advices, developed for PSN. The training s/w and database design, that was developed following the E&D interface design recommendations of ADA, was highly appreciated by all trainees who underlined that what they liked more was the "User friendly interface of the tool". Furthermore, through the use of this methodology for driver training, PSN can be integrated into the society and their quality of life is enhanced, because:

- They can get a driving course in any driving school, without being obliged to go to special driving schools for PSN.
- At usual driving schools, they can learn to drive in standard training vehicles, using only some (pre-installed) driving aids.
- Getting a driving license makes PSN feel more independent. Also, the ability of autonomous transportation enhances their possibilities of getting a job.

References

Bekiaris, E., Portouli, E, (1999, November). *Manual for PSN training in driving.* ODIGO project.

Naniopoulos, A., Bekiaris, E. (1992, May). *Critical issues and survey of past R&D work for DSN.* TELAID (V2032), Del. 1.

Naniopoulos, A., Bekiaris, E. (1993, August). *Prospective aid systems for categories of needs of DSN.* TELAID (V2032), Del. 6A.

Organisation for Economic Cooperation and Development (1985). *Traffic Safety of Elderly Road Users.* Paris.

Vanderheiden, G.C. (1994). *Application Software Design Guidelines: Increasing the Accessibility of Application Software to People with Disabilities and Older Users.* Version 1.1, June 15, 1994, the Trace R & D Center, Dept. of Industrial Engineering, University of Wisconsin – Madison.

Personalised and Contextualised Content Delivery for Mobile Users

Gregory M.P. O'Hare

Practice and Research in Intelligent Systems & Media (PRISM) Laboratory
Smart Media Institute (SMI), Department of Computer Science
University College Dublin (UCD), Belfield, Dublin 4, Ireland
Gregory.OHare@ucd.ie

Abstract

This paper addresses the issue of delivering information to mobile users, when it is needed, where it is needed and in a form relevant to the individual users technological context. We consider access issues and means by which these can be overcome. This paper considers two broad categories of mobile user those that move through a physical environment and those that move through a virtual environment. We provide examples of personalisation within these two arenas.

1. Introduction

Within this paper we seek to investigate the implications of access within the ever-changing computational landscape that exists today. We are witnessing the next major evolution of computing, the move from desktop, to palmtop, to wearable and digestable computing metaphors. Each of these metaphors presents it's own unique access problems. This paper is concerned primarily with the need for relevant information, when it is needed and in a form relevant to the individual users technological context.

Within this paper we want to focus upon this content access issue in the light of the evolving computing context that of all pervasive computing. Section 2 will justify the need for symbiotic information spaces. Section 3 presents the differing dimensions of mobility while section 4 presents examples of personalisation within the sectors of Personal Digital Assistant (PDA) and Collaborative Virtual Environments (CVEs).

2. Symbiotic Information Spaces

Visualisation research has long recognised that differing depiction models prove more appropriate for differing bodies of data or content. The choice may be determined by, the nature and form of the content in tandem to the expectations and experiences of the users.

The overall information space associated with a given problem area may be segmented into numerous different constituent spaces. Each of these may, and more probably will, be inter-related. It is the conjecture of this paper that, in order to achieve the maximum information content derivable from a given information space, they must be inextricably linked and furthermore the user must be able to seamlessly move from one such space to another. Want, Fishkin, Gujar, & Harrison (1999) blur the traditional boundaries by associating electronic services and actions in a virtual world with objects in a physical world. We consider in this paper how we might achieve a symbiotic relationship between these differing spaces or views, such that collectively they be viewed as the information landscape each being used in order to convey that content for which they are deemed most appropriate.

By way of an analogy consider the movement of a user within a 3-Dimensional (3D) virtual world. The navigation achieved through the world results in a simultaneous movement in potentially three dimensions, namely the X, Y and Z planes. In addition it is conceivable that a user navigating in such a space can be considered to be moving through associated dimensions like of course the time dimension and that of the social dimension to mention but two. We shall develop this point further in section 4 before doing so let us reflect on the varying forms of mobility.

3. Mobility

Mobility can be characterised in many ways. We define mobility to be the ability of a user/artifact to move within a given space. In particular there are two instantiations of mobility that we will consider. Firstly, the movement of a user and secondly the movement of autonomous artifacts. In the former there are two broad subcategories, those of user movement within a physical space and user movement within a virtual space. However, such spaces may not necessarily be disjoint and that whilst the user navigates one such space they may be simultaneously moving through parallel, yet fully integrated spaces. In the latter category artifacts may take the form of robots or software agents.

4. Personalisation

Personalisation is the watchword for service delivery today. No longer can users be crudely herded and coerced into categories, which all too often exhibit inappropriate levels of granularity. We shall consider personalisation within two specific sectors, firstly the mobile handheld sector and secondly in the Collaborative Virtual Environment (CVE) sector.

4.1 Personalised Mobile Services

The number of Personal Digital Assistants (PDAs) in the world has seen an explosion and in terms of universal access this movement away from the desktop activity exclusively but rather toward support for the mobile citizen presents numerous and profound challenges. In Ireland alone research as of November 2000 indicates that 5% of adults aged between 15 and 74 own a PDA (137,000) while a further 4% intend to purchase one in the very near future. Content delivery faces ever-increasing demands. These manifest themselves in the need for timely relevant information delivered in a form sensitive to the user's technological context. By context we refer to the limitations of the user device for example (screen size, colour/monochrome, interaction modality etc) and communications bandwidth.

Figure 1: The HIPS Interface

We will illustrate the need and provision of *context sensitive services* (Schilit, Adams & Want, 1994) through two examples, both located within the tourism and tourist services sector which has proven a rich field for the cultivation and deployment of such services.

The first example is drawn from the HIPS (Hyper-Interaction within Physical Spaces)[1] project (O'Rafferty, O'Grady, & O'Hare 1999), O'Grady, M.J., O'Rafferty, R.P. & O'Hare. G.M.P. (1999). HIPS is an electronic handheld tour guide that presents context sensitive information to the user. As the HIPS user approaches a location (exhibits, museums, archeological sites or cities) so they will be supplied with personalized multimedia information specific to their position and orientation. HIPS is motivated by earlier context aware tourist guides like GUIDE (Davies, Cheverst, Mitchell & Friday 1999) and CyberGuide (Long, Kooper, Abowd, & Atkeson 1996).

The visitor can receive a presentation consisting of audio and text descriptions, images and video clips of the current location. The HIPS interface is illustrated in Figure 1. HIPS allows a user to navigate a physical space and a related information space at the same time, that is, as the user explorers a location (e.g. a museum), she will be supplied with personalised multimedia presentations as new exhibits are encountered. Just as a tour-guide is expected to provide personalised and relevant information, HIPS must be aware of the users interests, expertise, and experience, it must also give location relevant information and react to how the user reacts to the presentations provided (i.e. feedback). Simple user profiling thus supports the personalisation process. The content is retrieved from a multimedia database where the tuple of location (longitude, latitude and orientation) and user preferences are used as the primary key. The former is supplied from a GPS device and electronic compass and the latter from the user profile.

[1] HIPS (Hyper-Interaction within Physical Spaces) Project was funded by the European Union Esprit Programme Long Term Research (LTR) Project No. 25574.

One of the key features of the outdoor HIPS demonstrator is that it is rapidly configurable to a new city (or town or campus). Since GPS does not have the same infrastructure setup cost unlike other localization technologies like IR our system is easily portable from one city to another. For example it was originally developed for the UCD campus, but was recently demonstrated in Siena which required only a short data configuration time.

A second project entitled AD-Me (Advertising Delivered to the Mobile E-commerce user) investigates the delivery of context sensitive advertising. This *push technology* function is driven by an underlying user profile and the user location. Again initial user profiles are used to bootstap the system and thereafter are dynamically updated based upon user migration and activity within a physical environment. The service is augmented with a range of *pull technology* functions including a *find me function* which will find a sought object that is closest to the users current position. Objects are selected from a list of standard objects including *inter alia* bancomat, taxi rank, pharmacy, train station, and police. The system addresses interoperability and delivers its service in a manner sensitive to the users technological context. Figures 2 & 3 depict the find-me taxi rank function as rendered on an html enabled device and a WML enabled device respectively. We use PHP as the mapping technology interfacing with DB2.

Figure 2: AD-Me Interface within HTML

Figure 3: AD-Me Interface in WML

4.2 Personalised Virtual Environments

If we consider Collaborative Virtual Environments (CVEs), within such CVEs the worlds are cohabited with a group of recurrent users. On going research efforts have concluded that whilst user immersion in a virtual world is often beneficial, immersion and social inclusion in a *virtual community* is crucial. Social and professional relationships can emerge as recurrent encounters are made between avatars within the virtual world. In many environments shared experience maps, in the form of shared memories exist which can significantly empower group activity. Thus as time evolves users compile photofit impressions of fellow users. Such models would encode *inter alia* the individuals perception of their colleagues in terms of such features as degree of friendliness, level of co-operation, competence. Thus we could consider users as simultaneously moving through a time dimension while moving through the social dimension.

In moving through information spaces users provide an insight as to their interests and preferences. In order to farm this information effectively one need be able to record, filter and make inferences based upon user interactions. In order to achieve this we advocate an agent based approach. One such system which embraces an agent oriented approach to the management and mediation of the user experience within CVEs is that of the ENTER system[2].

The ENTER (ENvironment which Totally Envelops the useR) (Guinan, O'Hare & Doikov2000) system seeks to provide a totally immersive 3-Dimensional e-commerce environment which is highly configurable and personalisable based on the perceived needs of the individual user. The ENTER system architecture as outlined in (Figure 4), adopts an agent oriented approach to its design and implementation. The system is broken down into

[2] ENTER (Environment which Totally Envelops the user) has been supported through a Enterprise Ireland Applied Research Grant No. ARP/96/117.

different functional components known as *agents* (Shoham, 1993; Wooldridge & Jennings 1994; O'Hare & Jennings 1996), which exhibit characteristics that conform to strong agenthood (Wooldridge & Jennings 1994). Five agents interact with each other to facilitate the realisation of ENTER, these are a *Design & Build Agent* (D&B Agent), *Presentation Agent*, *Login Agent*, *Listener Agent* and *Analysis Agent*. Other systems have also adopted an agent oriented approach to CVE management (O'Hare et al 2000; Nijholt 1999).

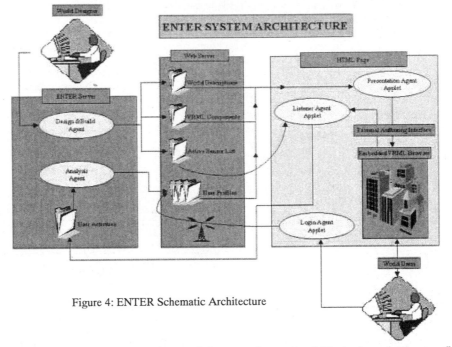

Figure 4: ENTER Schematic Architecture

Within the ENTER system one of the key thrusts of the research was the ability to dynamically reconfigure the world within which the user is situated. The classical world of retailing has long recognised the importance of product location. Dynamic world reconfiguration is delivered yielding personalised and contextualised user views. In the ENTER environment this can be achieved based upon an individual user profile where products and services most relevant to an individuals needs are assembled in the foreground while those of less relevance are pushed into the background. In the retail domain, information and service presentation based on perceived user needs are a key to market penetration.

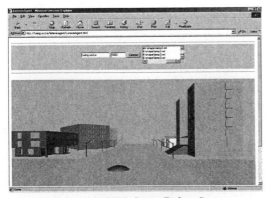

Figure 5: ENTER Scene Before Swap

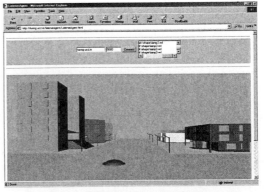

Figure 6: ENTER Scene After Swap

Additional difficulties associated with user disorientation and the dynamic reconfiguration of regions of worlds within which other shoppers are currently situated are being addressed (Guinan, O'Hare, & Doikov 2000). Reconfiguration would generally only be a local operation affecting merely the immediate foreground of the particular shopper. Indeed it would more often than not be the situation that reconfiguration can only be achieved where co-occupation did not occur.

5. Conclusion

The paper has reflected upon the need for symbiotic information spaces and how these can be simultaneous navigated. In addition it has considered the delivery of content to mobile users within two distinct contexts the physical space and the virtual space. It is our conjecture that the personalisation and contextualisation demanded in these two very different domains are in fact very similar and can be yielded by harnessing similar technologies namely location aware techniques and user profiling. We have considered the delivery of content to differing devices with restricted interaction modalities and offered illustrations from on-going research projects.

References

Davies, N., Cheverst, K., Mitchell, K., Friday, A. (1999) 'Caches in the Air': Disseminiating Tourist Information in the Guide System. *In Proc. of 2nd IEEE Workshop on Mobile Computer Systems and Applications (WMCSA '99)*. New Orleans, Louisiana, 25 - 26 February.

Guinan, T., O'Hare, G.M.P., Doikov, N. (2000). ENTER: The Personalisation and Contextualisation of 3-Dimensional Worlds. In *Proceedings of Eight Euromicro Workshop on Parallel and Distributed Processing (EURO-PDP 2000)* (pp. 142-149). IEEE Computer Society Press.

Long, S., Kooper, R., Abowd, G.D., Atkeson, C.G. (1996). Prototyping of Mobile Context-Aware Applications: The Cyberguide Case Study. In *Proc. 2nd ACM International Conference on Mobile Computing (MOBICOM)* (pp. 97-108), Rye, New York, U.S.

Nijholt, A. (1999). The Twente Virtual Theatre Environment:Agents and Interactions, TWLT 15. In *Proceedings of the Fifteenth Twente workshop on language technology*, May 19-21, Enschede, The Netherlands.

O'Grady, M.J., O'Rafferty, R.P., O'Hare. G.M.P. (1999). A Tourist-Centric Mechanism for Interacting with the Environment. *In 1st International Workshop on Managing Interactions in Smart Environments MANSE'99*, Dublin, December. Springer-Verlag Publishers.

O'Hare, G.M.P., Jennings, N.R. (Editors.) (1996). *Foundations of Distributed Artificial Intelligence*. Sixth Generation Computer Series, New York: Wiley Interscience Publishers (296 pages, ISBN 0-471-00675).

O'Hare, G.M.P., Sewell, K., Murphy, A., Delahunty, T. (2000). ECHOES: An Immersive Training Experience. In *Proceedings of International Conference on Adaptive Hypermedia and Adaptive Web-based Systems (AH2000)*. Springer Verlag.

O'Rafferty, R.P. , O'Grady, M.J., O'Hare, G.M.P. (1999). A Rapidly Configurable Location-Aware Information System for an Exterior Environment. *In Proc. International Symposium on Handheld and Ubiquitous Computing (HUC 99)*, September 27-29, Karlsruhe, Germany.

Schilit, B.N., Adams, N.I., Want, R. (1994). Context-Aware Computing Applications, In *Proceedings of the Workshop on Mobile Computing Systems and Applications*, Santa Cruz, CA, December. IEEE Computer Society.

Shoham, Y. (1993). Agent-Orientated Programming, *Artificial Intelligence* (Vol. 60, pp. 51-90). Elsevier Science Publishers.

Want, R., Fishkin, K. Gujar, A., Harrison, B. (1999). Bridging physical and virtual worlds with electronic tags. In *Proc. ACM CHI '99*, Pittsburgh, PA, May 15--20. ACM Press

Wooldridge, M., Jennings N.R (1994). Agent Theories, Architectures and Languages: A Survey. In *ECAI Workshop on Agent Theories, Architectures and Languages*, Amsterdam, The Netherlands, August, *LNAI 890*. Springer Verlag Publishers.

Contextualized Information Systems
for an Information Society for All

Reinhard Oppermann, Marcus Specht

GMD - German National Research Center for Information Technology
Institute for Applied Information Technology (FIT-ICON)
D-53754 Sankt Augustin
{_reinhard.oppermann,_marcus.specht_}@gmd.de

Abstract
The paper describes some ideas about contextualized information services based on location, user, environment, and equipment awareness. An information society where people have access to information and communication facilities wherever they are overstrains people by information overload and selection and configuration effort if selection and presentation means do not consider the context of use. An example for a location aware adaptive information system is presented and discussed and future directions for enriched multi-modal information presentation are proposed.

1. The need for contextualized information

In an information society many people are no longer users of dedicated computers but they will be information and communication recipients, senders, or processors in a wide variety of contexts and with a wide variety of devices. Several research projects and applications on adaptive information systems in the past focused on scenarios where either a static user using a desktop computer or a mobile user stopping by at kiosk systems to pick up information where taken as a basis. New developments in the field of wearable and ubiquitous computing nowadays allow for using computers in really mobile scenarios. With the growing mobile usage of computers the need for new interaction paradigms increases because classical desktop paradigms do not work for this scenario, e.g., when walking around, even with a technically very sophisticated head mounted display, classical desktop interaction for manipulating information objects are not usable. One important consequence of that development is that accessibility problems become more relevant for all users. "Fully enabled" persons are in a similar situation like disabled people, because of restricted resources for interaction with the computer system in a special context. This shows up the need for approaching interface design and information exchange in a way to look at users and their available interaction resources based on their current interaction context and also gives a new view on the developments of interaction research for disabled people.

Information will be relevant and will increase both quantitatively (amount of information) and qualitatively (importance of information for the person) for more and more people. To cope with the information flow the person needs a contextualized view on the information available and relevant in a given situation. The contextualized view is specific for a person, a task, a location and the technical infrastructure the person is equipped with like the device and the communication bandwidth. A contextualized view is no isolated snapshot of an interaction situation. A contextualized view is embedded into a process of the person's activity over time and space. Considering an individual's activity process, including the social group the individual is interacting with, allows the system

- to select the information most probably relevant for the individual in the given task,
- to present this information in the most suitable way for the given environment and the given technical equipment
- to support the most effective interaction technique to receive and enter information and commands.

If information systems consider the demands of the person continuously in the process of activities they can support the person in an effective, efficient and pleasurable way avoiding information gaps and overload.

In this paper we will discuss some ideas about contextualizing interaction and building contextualized information systems and present an example for a contextualized information system, e.g. a nomadic museum guide "hippie" (Oppermann & Specht, 1999).

2. Context aware information systems

For adapting the interaction and information to an individual user and his/her current context several models for describing the context and the person's characteristics are necessary. For describing the context of use (CoU) of an information system we need to define the parameters of context we want to consider for adaptation.

Several approaches have defined context models and described different aspects of context taken into account for context-aware systems. (Schilit, Adams, & Want, 1994) has mentioned: where you are, who you are, and what resources are nearby. (Dey & Abowd, 1999) discuss several approaches for taking into account the computing environment, the user environment, and the physical environment and distinguish primary and secondary context types. Primary context types describe the situation of an entity and are used as indices for retrieving second level types of contextual information. In most definitions of context four main dimensions of a context are considered:

Location: We consider location as a parameter that can be specified in electronic and physical space. An object or entity can have a physical position but also an electronic location described by URIs or URLs. Location-based services as one type of context aware applications (Schilit et al., 1994) can be based on a mapping between the physical presence of an object and the presentation of the corresponding electronic artefact.

Identity: The identity of a person gives access to a lot of second level type contextual information. In some context-aware applications a highly sophisticated user model holds and infers information about the user's interests, preferences, abilities, knowledge and detailed activity logs of physical space movements and electronic artefact manipulations.

Time: The time is an important dimension for describing the context. Beside the specification of time in CET format. Categorical scales as an overlay for the time dimension are mostly used in context-aware applications (e.g., working hours vs. weekend, mapping onto a calendar of a person to get information about free vs. busy hours). For nomadic information systems a process-oriented approach can be time dependent and is used in workflow systems for selecting information in mobile working scenarios.

Environment or Activity: The environment of a context describes the objects and the physical location of the current situation. In several projects approaches for modelling the objects and building taxonomies or an ontology about their interrelations are used for selecting and presenting information to a user.

From our point of view contextualized information systems should at least take these four parameters into account for adapting the current CoU of a user. In the following section we will describe how we have modeled these dimensions in the development of Hippie and how we plan to extend the system in new projects.

3. A nomadic information system for exhibition visitors

The nomadic information system contains three models to identify the CoU. A *domain model* describes and classifies which objects of the domain information are to be presented and processed. A *space model* describes the physical environment where the nomadic system is used and the location of the domain objects in the physical space. A *user model* describes the knowledge, the interests, the movement, and the personal preferences of the user. The domain model and the space model are assumed to be static, i.e. the domain objects are described and their location is identified before the systems are used. If changes occur in the environment, the domain model or the space model has to be updated explicitly. The user models are dynamic, i.e. the users' interactions with the information system and their movements in physical space are evaluated to update the user model automatically.

The nomadic information system Hippie has been developed for a cultural environment, providing information about two domains, an art exhibition and a fair. The nomadic user is supported in the course of the preparation of a visit, during the actual visit, to the evaluation of a visit by contextualizing the interaction to the current CoU.

For the preparation of a visit hippie supports the user in:
- individual information access with annotation and communication possibilities,
- awareness about events in the physical space by combinations with awareness platforms like described in (Gross & Specht, 2001). By infrared sensors connected to significant positions in the physical space (exhibits, transits) and infrared receivers and electronic compass connected to the client of the visitors,
- adaptation of information to interest, knowledge and preferences of visitors by adaptive tour proposals and content adaptations, and
- graphical understanding support of exhibits for non-experts to initiate advanced viewing.

During the visit the user is supported by:
- a high quality 3d audio soundscape,
- a high precision localization system for location aware information selection and presentation, and
- intuitive interaction facilities, like physical information points and information selection by gestures and movements.

For evaluating a visit a nomadic user is supported by:
- detailed reporting of a past visit,
- recommended facilities for extending the current knowledge and arousing interest.

Contacting tools for visitors in other CoU for cross reality discussion see (Gross & Specht, 2001). The prototype "Hippie" has been implemented for an art exhibition in the castle of Birlinghoven, the headquarter of GMD in

852

Sankt Augustin. Evaluation experiments were conducted with 60 visitors using three comparative guidance media. The results revealed that the prototype was effective to support the visitors in their knowledge acquisition about the art domain. It turned out that for user satisfaction of novices the computer handling needs improvements. This effect is due partly to the hardware device that is still too heavy and to difficult to use while roaming.

A main interest for the evaluation was the adaptive user support by the system. Assessing adaptive features of a system is a difficult task. Adaptivity evaluates the user history and can be the more valid the more data about a user can be collected. In experiments access to user data is typically limited. The time for evaluation studies is restricted. In the current study usage sessions varied from .5 to 1.5 hours with a mean of 69 minutes. To present adaptive tour proposals for the visitor during the time period of the experiment, the system can not analyse more than around 30 - 45 minutes visiting recordings to evaluate the user interests.

In the experiment conducted we mainly evaluated three adaptive features of the hippie prototype:

- *Adaptive exhibit recommendation*: Based on the localisation of the visitor and an internal model of the physical space, the system proposed objects that where close to the user.
- *Adaptive content selection*: Based on the user's previous exhibit presentations the prototype adapted the detail of information presented about an artwork to reduce redundancy of presentations and to arouse user's interest with additional information.
- *Adaptive tour proposals*: Based on the user's movements in physical space and on his/her information requests about exhibits the prototype proposed tours of exhibits that where similar to the estimated interests of the user in the user model.

60 visitors were invited to use different information media: a traditional guidebook, a simple audio guide and the Hippie system. 20 visitors used each of the three media for intensive use and the other two media for short supplementary tasks. During these tests 18 of 20 Hippie users assessed the announcement function of the system showing a new exhibit near to the user by answering 2 questions in a questionnaire. 14 of 20 Hippie users assessed the adaptivity of the system providing a personal information space by selective information presentation and responded to 2 questions of a questionnaire. 8 of 20 Hippie users used the adaptive tour function of the Hippie and responded to 4 questions of a questionnaire.

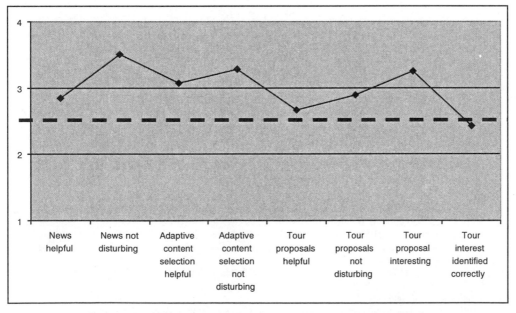

Figure 1: Assessment of adaptive features of a mobile exhibition guide Hippie

Figure 1 shows the mean values of the visitors' assessment of adaptive system features. For the assessment a Likert scale was used with four categories: 1 = completely disagree, 2= disagree, 3= agree, 4 = completely agree. The dotted line marks an assumed neutral line 2.5 where assessments are neither positive nor negative.

The results for the „New exhibit" announcements show positive support for the idea to automatically inform the visitor about exhibitions the system can give explanations about. The mean of the helpful-assessment is M=2.8 and the assessment that the „News"-announcement not being disturbing is even M=3.5.

The system takes into account the visiting history of the visitor for each exhibit. The first time a visitor gets the full information set being appropriate according to the user interest model. The second time the presentation encompasses only the basic content (name and thumbnail of the exhibit) together with access to further information content via attribute buttons. This is meant to be a way to prevent the user from information overload. The assessment of adaptivity for re-approaching an exhibit based on the already presented information during the visitor's first approach is rated highly helpful (M=3.1) and evaluated as not-disturbing (M=3.3). This feature reflects the knowledge the user is assumed to need when coming to an exhibition.

The assessment results for the adaptive tour function show that the adaptive proposal of an exhibit tour matching the personal interests of the visitor is appreciated by the subjects: it is in particular interesting (M=3.3), helpful (M=2.7) and not disturbing (M=2.9). The "Tip" was announced by a three times blinking bulb with a following report of relevant exhibits seen by the visitor constituting the hypothesis of a tour interest and a tour proposal.

The question whether the tour proposal meets the actual interests of the individual visitor was assessed critically: The value of 2.4 is less than the neutral 2.5 what means that more than half of the subjects were happy with the proposal content; the answers to this question showed the biggest variance between the subjects.

To conclude, the adaptive features of the prototype turned out to be helpful for the visitors. Announcements of relevant exhibits, adapted information content and adapted tour proposals were appreciated. More extensive evaluation of user interests by the user model is needed to enable the system to propose exhibit tours adapted to the actual art interests of the visitors.

A main criticism in expert workshops and a main result of the user evaluations was the complicated interaction with small laptop computers, table PCs, or even with wearable computers like the Xybernaut™ MAIV. Therefore for supporting a usable and intuitive interface for mobile users during a visit we try to develop a completely new kind of interface in the project LISTEN (http://listen.gmd.de). In LISTEN the user wears a lightweight headphone that displays 3d audio information and his/her movements in space will be tracked with an adequacy of 10 cm and 5 degrees. The high quality of audio material used in LISTEN and the 3D audio rendering will allow for intuitive interaction of the visitor with his/her physical environment while splitting resources in a traditional and well-known way. The visitors *look* to the exhibits and *listen* to the explanations of a guide or are embedded in an audio augmented soundscape.

The aim of CRUMPET is to implement, validate, and trial tourism-related value-added services for nomadic users (across mobile and fixed networks). In particular the use of agent technology will be evaluated (in terms of user-acceptability, performance and best-practice) as a suitable approach for fast creation of robust, scalable, seamlessly accessible nomadic services. The implementation will be based on a standards-compliant open source agent framework, extended to support nomadic applications, devices, and networks.

With these developments we are trying to realize prototypes that come closer to a real nomadic system with adequate interaction facilities for different CoU and a continuous support for a nomadic user based on a user model, a world model, and additional information about semantics of the world model in taxonomies.

4. Conclusion

For an information society to allow people to receive and process information and communicate suited to the current context of activity context aware services are needed. The context can be specified by location and activity tracking of the user evaluating the position and interaction of the user based on a model of the task or interest of the individual, the current physical environment and technical equipment. The examples shown in the paper demonstrate current solutions and future demands to enrich the contextualization of the information selection and presentation for seamless roaming. For content selection modelling of the user and the environment is crucial, for information presentation and interaction modelling of appropriate modality like visual and auditory display needs further development.

References

Dey, A. K., & Abowd, G. D. (1999). *Towards a better understanding of Context and Context-Awareness.*: College of Computing, Georgia Institute of Technology. Technical Report.

Gross, T., & Specht, M. (2001). *Awareness in Context-Aware Information Systems.* Paper presented at the Mensch & Computer 2001, Bad Honnef (Germany), 167 - 175.

Oppermann, R., & Specht, M. (1999). *A Nomadic Information System for Adaptive Exhibition Guidance.* Paper presented at the ichim. Cultural Heritage Informatics 1999, Washington, 103 - 110.

Schilit, B. N., Adams, N., & Want, R. (1994). *Context-Aware Computing Applications.* Paper presented at the Workshop on Mobile Computing Systems and Applications, Santa Cruz, CA.

Evaluation of a mobile travel information service by deaf and mobility-impaired users

Björn Peters

Graduate School of Human Machine Interaction at the Linköping Technical Institute, Sweden and Swedish National Road and Transport Research Institute - VTI, Linköping, Sweden

Abstract

A user field evaluation of a personal mobile traffic and travel information service was conducted in the Gothenburg area. The aim of the evaluation was to evaluate the quality and availability of services provided on the PROMISE homepage on Internet. Sixty subjects participated in the evaluation and 58 completed it, of these were 10 users with disabilities, 6 deaf and 4 with mobility impairments. This presentation focuses on the users with disabilities. The users were provided with one of two types of handheld terminals (communicators), NOKIA 9000 or Ericsson MC 12, with which they had access to the PROMISE services. The users were instructed how to use the terminals and how they could access the travel information services. They used the communicators for approx. 4 weeks in their daily life activities. Questionnaires were used to collect background information and the users' opinion about the services and the terminals. Communication logs were used to collect data about the actual usage. The trip planning service was especially appreciated. However, long response times and communication problems were reported. It was also revealed that some travel information was outdated or insufficient or missing. Suggestions for improvements were given. The SMS facility was much appraised by the deaf users. The screens were considered to be somewhat small and too sensitive to daylight. Some users also thought that the keys on the keyboard were too small. Finally, it was concluded that even if a number of problems were revealed in the evaluation this kind of service has a good potential to improve independence and mobility for travelers with disabilities. However, the investigated services and technology has to be further developed. It is, finally, proposed that the findings from this study combined with design guidelines presented in the TELSCAN Handbook of Design Guidelines can be used to further develop the PROMISE product in order to realize an inclusive design for all.

1. Introduction

TELSCAN (TELematic Standards and Coordination of ATT systems in relatioN to elderly and disabled travellers), was a horizontal project within the EC Telematics Application Programme, which made a number of collaborative evaluations of ITS (Intelligent Transport Systems) applications from an E&D travellers' perspective (Naniopoulos, 2000) during the fourth framework. One of these was conducted together with the PROMISE project in Gothenburg (Ojala, 1996). PROMISE developed and evaluated personal mobile traveler and traffic information services. An earlier version of the PROMISE system was evaluated by Karlsson (1995). However, travelers with disabilities were not included in that evaluation. The idea with PROMISE was to utilize the possibilities of small handheld computers combined with mobile telephone such as the NOKIA 9000. The services provided by PROMISE, in Gothenburg, were accessed via Internet connections and included services like multi/intermodal trip planning, route planning and guidance, information services like timetables, yellow pages, tourist information and weather forecast (see figure 1).

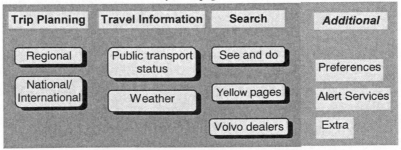

Figure 1 PROMISE services in the Gothenburg region

Alert Services utilized the SMS (Short Message System) facility and with this option the user could subscribe for e.g. weather information. The weather forecast could be sent on request to the user as short text messages. Two extra services were also available. City Navigator was a route planning (car) service for Stockholm. This service had a different user interface compared to the one for Gothenburg. Finally there was a dictionary (English/Swedish) available. There were though no services available, which were specifically designed or intended for travelers with disabilities.

2. Method

Volvo/VTD was the PROMISE partner responsible overall for the Swedish evaluation. In total, a group of 60 users were invited to participate in a user evaluation, 58 completed the evaluation (Karlsson, 1998). This presentation will only cover the ten users with disabilities. Four mobility impaired and six deaf users were included among the users as an initiative from the TELSCAN project. The users were provided with a terminal, NOKIA 9000 or Ericsson MC 12 (see figure 2) for approximately 4 weeks. Information about the services and instructions on how to use the terminals was given to the users before they received the terminals. They also received a user manual. The PROMISE services were also accessible on stationary PCs with Internet access. A questionnaire was used to collect background information such as age, disability, and travel habits. Communication logs were collected for each user during the trails. At the end the users filled out a questionnaire and were invited to participate in a group discussion.

Figure 2 Ericsson MC 12 and NOKIA 9000

3. Results

The mean age for the ten users with disabilities was 43 years. Out of the four mobility impaired users were two wheelchair users and two did not depend on any mobility aids (rheumatics). Gender distribution was equal. All of the hearing impaired was completely deaf. Nine were commuters with regular travel habits. Seven had their own mobile phone. When asked if they used their own mobile phone for trip planning, one said always, one often, one sometimes, two seldom, two never, and three did not answer. On average the users borrowed the terminals for 20 days. They were asked how much they were willing to pay for getting access to the Promise services and the average amount was approx. 160 SEK/month.

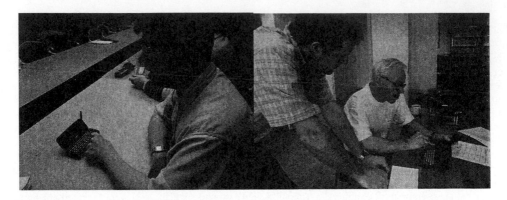

Figure 3 Training and instructions of use is provided to users

Analysis of user logs for the total group (n=58) revealed that the users overestimated their actual use of the terminal. The service most preferred was the possibility to plan a trip in advance. However, most of the services were considered in need of improvements. The users with disabilities also noted that some information was outdated or missing like specific information on which they depend, e.g. access to parking space for disabled, special transportation services. Suggestions for improvements were given. Most of the users experienced system errors, long response times and communication problems. Specifically, users with disabilities considered that this was an important disadvantage.

They were also asked about their opinion about the terminals they had used. The Ericsson system consisted of a separate terminal (hand computer) and phone while the Nokia system was one unit with both phone and terminal. Here are the results of some of the questions asked about the terminals:

I think the terminal was ......
much too small (1), too small (2), right size (6), too big (1) much too big (0)

I think the terminal was ....
much too light (1), too light (1), right weight (5), too heavy (1), much to heavy (1), missing (1)

It seems like most of the users were quite satisfied with the weight and size of the terminals. They were asked some questions about the keyboard. The results were as follows:

Table 1 Users' opinion about the keyboards

I think the keys on the keyboard were (number of users)......

	1	2	3	4	5		Missing
Too small		4	4			Too big	2
Unclear	1	1	5	1		Clear	2
Too few		1	6			Too many	3
Well placed	2	1	4	1		Badly placed	2

Most users were satisfied but there seemed to be some difficulties with the size of the keys on the keyboard (average rating = 2.5). They were also asked to give their opinion of the screen.

Table 2 Users' opinion about the screens

I think screen was (number of users)......

	1	2	3	4	5		Missing
Too small	4	2	2	1		**Too big**	1
Unclear		3	1	3		**Clear**	3
Easy to read	1	1	1	3	2	**Difficult to read**	2
Not sensitive to daylight	1	1	1	3	3	**Sensitive to daylight**	1

As can be seen from the results above most of the users were quite satisfied with the weight and size of the terminals. The keyboard was considered to be rather good but could be improved. The screens were too small, too difficult to read and too sensitive to daylight. Most users seemed to prefer a terminal in one unit as the NOKIA 9000. The deaf users considered the SMS functions specifically useful. Even if there is a lot that needs to be improved it seems like the evaluated services can provide travelers with disabilities with a support, which can make them more mobile and independent.

During the concluding meetings, as the terminals were returned, the users were also interviewed in order to capture their opinion. Here are some remarks made during these reassembly meetings.

- Promise offers great search facilities which can simplify trip planning
- Promise made me change my travel routines
- Weather (alert) information was very useful, but could be more frequently updated
- Shortest route is not always the best, especially for E&D travelers
- A print function would be helpful
- For trip planning there is no information about total trip length
- There is no logic in the combinations of street names/numbers for trip planning
- Some information was old (3 years)
- SMS/fax was very much appreciated by deaf users as it provided a way to communicate with all people (deaf or not) - this is an option that should be exploited better

They also wanted better communication links (higher speed), better connections (no disturbances), no yellow pages, lighter terminals (rheumatic), free of charge number information services, buttons (keys) require to much force for people with low force resources (rheumatics).

4.   Conclusions

The evaluation reported here must be considered as a pilot test with a few users. Thus, it is impossible to draw any definite conclusions from the results above. Furthermore, it has to be kept in mind that people with disabilities are very diverse due to differences in their impairments. However, the results can be used to indicate what needs to be done in order to really provide some useful help to those that really needs it. This type of services has a definite potential with respect to travelers with disabilities. Comparing to the previous evaluation (Karlsson, 1995) it seems like some of the problems have been resolved, better dialogue and improved hardware, however it should also be noted that important problems still remains, long response times and lacking quality of services provided. In order to make both services and hardware useful for users with disabilities it is important that they are included early in user tests so that their specific needs can be considered. An inclusive design for all should be applied. Two examples can be mentioned:

1.   an improved screen, which is not so sensitive to daylight would be beneficial to all users.
2.   the current types of keyboard were difficult to use e.g. too small for travelers with e.g. dexterity problem a problem which will increase in cold weather also in this case would an improvement make it easier for all users.

In order to achieve this aim it can be helpful to consult the TELSCAN Handbook of Design Guidelines for Usability of Systems by Elderly and Disabled Travellers (Nicolle, 1999). This handbook was compiled from the work conducted in TELSCAN and as this process ran parallel in time as this collaborative study it was not possible to fully consult these recommendations at the time when the PROMISE system was designed. This should be a task for the future. In conclusion it seems like both the traffic and travel information services tested and the hardware needs to be further improved to properly suite users with disabilities and further testing is needed. However, the potentials are however good for a system like PROMISE with respect to travelers who are elderly or disabled.

References

Karlsson, M. (1995). Krav på ett portabelt informationssystem för resenärer - Utvärdering av PROMISE-Systemet (Requirements on an portable information system for travelers – evaluation of the PROMISE System In Swedish) (1995:10). Gothenburg: Institute for Consumer Technology; Chalmers University of Technology.

Karlsson, A.-S. (1998). *Demonstration Results - Swedish Test Site*. Internal PROMISE Report.

Naniopoulos, A. (ed.) (2000). FINAL REPORT TR 1108, TELSCAN (TELematic Standards and coordination of ATT systems In relatioN to elderly and disabled travelers), Thesaloniki, Greece.

Nicolle, C., Burnett, G. (eds) (1999). *TELSCAN Handbook of Design Guidelines*. Thesaloniki, Greece. [On-line]. Available: http://hermes.civil.auth.gr/telscan/telsc.html

Ojala, T. (1996). User Needs Analysis: Disabled and Elderly People (PROMISE Report Deliverable 3.1). Tampere: Nokia Research Center.

Smart Medical Applications Repository of Tools for Informed Expert Decision (SMARTIE)

Juan J. Sancho (1), Angela M. Dunbar (1), Carlos Díaz (1), Juan Antonio Cobos (1),
Susan Clamp (2), Chris Kirke (2), Petros Papachristou (3), José Esteban Lauzán (4),
Christian Ohmann (5), Hans-Peter Eich (5), Jean-Pierre Thierry (6),
Marie Gabrielle Verdier (6), Clive Tristram (7).

(1) Department of Medical Informatics. Fundació IMIM; Barcelona. Spain.
(2) Clinical Research Unit. University of Leeds, Leeds, United Kingdom.
(3) ATKOSoft S.A., Athens. Greece.
(4) Department of Medical Informatics. Heinrich-Heine-Universitaet. Duesseldorf. Germany.
(5) SEMA Group sae. Madrid. Spain.
(6) SYMBION. Paris. France.
(7) ETS Futur Dessin. Poitiers. France.

Abstract

The aim of the SMARTIE project is to demonstrate, through a working example, the appropriate methodologies using IT technology to encapsulate clinical knowledge and distribute them under a single interface. They will take the form of a collection of multiplatform medical software applications to provide the physician with critical information to manage and inform the patient/citizen. The four key points of these applications are: 1) They will share the same basic, simple and friendly, interface. 2) They will be attached to pre-existing Clinical Information System applications and validated in real scenarios. 3) They will have a scientific content supervised by the European Medical Associations related to each particular field. 4) They will be developed with the aim to become a suite of Open Source licensed applications after the support of the project from the EU in order to be able to copyright, and distribute the interface, the attachment technology and the applications themselves.

1. Background on Computer Assisted Medical Decision Making

There is currently an apparent paradox in the medical decision support field. On the one hand visionary leaders recognise the overwhelming need to support medical decision making via electronic patient records, automated alerts and reminders, and clinical guidelines that are clearly documented, constantly maintained, and delivered just-in-time and on-point (Sancho, 1999)[1]. They view such technologies as strategic weapons in the fight to reduce waste, poor quality, and preventable expense in a health care industry that has grown worldwide to about $1 trillion per year. As but a few examples, studies show that much of the $12.4 billion spent annually on adverse drug events is preventable, 30% of all diagnostic tests may be unnecessary, 25% of all procedures may be unwarranted, 50% of all post-operative complications are preventable, and healthy lifestyles reduce lifetime medical expenses by half. [2,3,4,5]

The paradox is that despite the dramatic contributions of expert systems to other sectors and despite the continued need for medical decision support, there has been a failure of Artificial Intelligence in Medicine (AIM) to revolutionize the health care field. For example, Miller (1994) pointed out that general purpose medical Expert Systems (ESs) like Internist-QMR, Illiad, or DXplain were too broad to be useful, but that experience seemed to indicate that narrowly focused, special purpose expert systems could be effective. Only three years ago, Coiera[6] pointed out that hardly any expert systems at all have caught on, and his website that tracks AIM trends lists only 3 or 4 dozen ESs with minimal success. A closer look at these ESs reveals their success is constrained, and at times reversed after a short interval.

By repositioning the AIM vision into more of a human centred support approach and by addressing some severe infrastructure obstacles, one should reasonably expect the paradox to begin to fade, and the hoped for revolution to quietly move forward.

The repositioning needed involves envisioning AIM as a subset of human centred systems HCS (AIM). This vision should include viewing AI and ESs in the context of distributed human-computer co-operation[7], joint person-machine cognitive models[8], support and partner roles rather than as pure mentoring[9], expert error critiquing systems rather than expert systems[10], situated on-demand tutoring in the context of decision support[11], and computer-based reminding of already known items[8].

In particular, the theory of human performance is that any task, to complete it correctly, requires knowledge that loosely put ranges across cues, skills, experience, heuristics, etc. Whatever it's called, let us label the know-how used to complete a task as:

A - irrelevant knowledge that gets incorrectly added (errors of commission)

B - the portion of correct knowledge that one uses correctly

C - the portion of correct knowledge that one knows about, but forgets to include (errors of omission)

D - the portion of correct knowledge that one hasn't yet learned, missing knowledge (errors of ignorance)

This four part taxonomy is only the top level of an extensive ontology of heuristics and errors types culled from the literatures in cognitive psychology, behavioural decision making, human factors engineering, and so on[12].

The human task performer goes through their workday sometimes encountering problems for which they are highly competent and for which all the knowledge falls under type B, and they perform acceptably. At other times during the day, they may encounter situations they are less competent at and they begin to use heuristics and commit errors under the type A and C categories. Also, during the day they invariable encounter situations requiring new knowledge for which they as yet have no training, and thus they fall into the type D category. In short, to get through the day, most professionals must have:

1. *access to critics, debiasers and influencers to help them overcome Type A mis-judgment problems.*

2. *reminder and alerting agents to support their slips, lapses, forgetting and Type C heuristics.*

3. *life long, situated, on-demand learning assistance to overcome Type D knowledge half-life issues.*

4. *Ideally, there should also be a correct-Type-B knowledge capture system that supports corporate memory, case-based knowledge management, and asynchronous expertise transfer and subsequent sharing.*

Often, the technology to support (1) through (4) is absent or ineffective, and professionals obtain these items from direct contact with trusted colleagues, organisational support staff, timely informal and formal meetings, readily available reference materials, built in review and feedback processes, and the like. Increasingly often as systems and problems grow more complex, however, they "do without" and they perform at sub-optimal levels causing the organisation adverse cost and quality consequences.

To effectively deploy the four broad categories of technologies is not as simple as "turning engineers loose" on the situation. The tech push approach fails so often, as it has in AIM, from a simple misunderstanding of the need. The need here is not for a machine to take over the humans' jobs (as an expert system tries to do), nor is it for any one type of technology. Rather the need is for a plethora of approaches, roughly grouped into the four major categories: (1) to (4) listed above. Knowing what to deploy in a given organisation depends on a number of human modelling skills, and on a human cognitive task analysis that formally inventories and models when problems of type A/B/C/D occur and that maps out what forms of technologies (1,2,3, and/or 4) to deploy in support of each sub task (Sancho, 1998)[13]. Drilling down through this process for each subtask and sub-subtask is a time consuming effort that should also conform to the realities of the spiral model of software engineering.

The largest challenge affecting almost everyone in Medical Informatics today is that of **software integration and access**. As healthcare institutions evolve into networks of facilities and as people live longer, relocate more often, and aspire to ownership of their own records, overcoming the integration challenge and producing and maintaining the "virtual health record" for each citizen becomes a more urgent need with each passing year. To meet this challenge there are hundreds of organisations and thousands of individuals currently participating in dozens of committees of numerous standard-setting bodies. Most of these several dozen major standards pertain to the data layer and comprise vocabulary models, financial and clinical aspects of the patient record model, distributed healthcare messaging protocols, media representation, and others that will foster the secure yet open sharing of information between applications in the health care field.

Still, many of these standard-setting efforts seem to be floundering in the absence of a more rigorous science to base them on. As but one example, vocabulary standards need to be based not on long lists of synonyms, but on description logic and concept models of healthcare information semantics: e.g., see Rector et al. (1993)[14]. Likewise patient record and messaging standards need to be grounded in formal models of the underlying conceptualisations of the items being encoded and transmitted. A second example lies in field trials and empirical analysis (what works? what is a waste of effort? what is still missing? (Sancho, 1999)[15]. How do mediators, query agents, active databases, event handlers, smart brokers, speech act theory, etc. fit in, or not?) that synergistically combine many of the standards into an open, reusable, distributed agent architecture, such as in the HOLON project[16]. Another, not unrelated issue to the acceptance and use of patient record systems concerns the reliability of real time, spoken language transcription and understanding. These and many other issues suggest there is significant need for further scientific investigation before integration can fully mature.

While the previous challenge affects the data layer, this challenge concerns the "business logic" layer of the healthcare sector. This is the newer of the two challenges, and it is what I shall refer to as the **knowledge management** challenge. On the clinical side, evidence-based medicine and the randomised clinical trial are producing new practice guidelines at an increasing rate (Ohmann, 1999)[17]. Clinical workers increasingly will find themselves falling into the non-B portion of the A/B/C/D taxonomy, and working in the absence of HCS (AIM) and decision support will become increasingly unrealistic, and not just if care institutions continue to shift toward patient-workload gatekeepers and general practitioners (Sancho, 1998)[18]. At present, clinical institutions are woefully unprepared for this knowledge deluge.

It may very well be in the end that the early failure of AIM was less a failure of the technology itself as it was of the ways in which it was being inserted into the field. If we are judicious in merging AIM with human centred systems designs and doing this in concert with the large scale software engineering and open standards based approaches, we may very well hear clinicians and patients alike in the next quarter century wondering how healthcare ever got anywhere without the functionality AIM provides. And, if the AI is implemented in an integrated, open standards, HCS-based approach, few will even realise its there.

1.1 Background of the SMARTIE Consortium in Computer Assisted Clinical Decision Systems

The knowledge in clinical practice is represented by the knowledge that some experts have about best practice management. This knowledge has not been ever properly (universally, consistently, timely and updated) disseminated to the working clinician. The know-how about the best management practice for many diseases remains the privilege of few and the real user of this knowledge either does not apply it at all or uses outdated knowledge. The Computer Assisted Decision Support Systems (CADSS) –successors of the Expert Systems of the 80's, tried to partially solve this absurd situation without any uniform success.

Very few CADSSs have been used in routine clinical practice, although scoring systems for various measures of prognosis or severity of a condition are used. Even a dedicated palmtop calculator has been built and distributed to calculate a single severity score for ICU patients! (APACHE II score). Most of these however are done by hand. The main problem has been the integration of computer packages into routine clinical use. Some problems occur with attitudes of clinicians towards these but other significant problem has been the user interface and accessibility of systems for busy clinicians. There is not a problem with the take up of new technology in medicine if it is perceived as useful (EKGs, MRI, CT) but how these are perceived and how they fit into the clinicians' way of working is crucial to success.

In 1992, a survey was carried out by one of the partners of SMARTIE about the formal decision aids that were used by clinical gastroenterologists. All clinical members of the Deutsche Gesellschaft fur Verdauungs und Stoffwechselkrankheiten (n=584), received a questionnaire about the use, knowledge and attitudes to computerised decision aids in general. Formal decision aids were used by 56% of the participants (mainly severity scores and disease stage classifications)[19]. Although computer based formal aids (i.e.: Expert Systems) were only used by a minority, the majority of participants believed that CADSS's would become part of clinical routine and would improve education and clinical practice once their friendliness and availability were improved. In a survey of 796 clinicians, medical students and patients concerning the use of formal computer aided decision support systems over 80% of participants felt that the use of decision support systems would help improve decision making and patient care[20].

2. Overview of the project

SMARTIE is a research project funded by the European Union. Its aim is to develop a software package for use in clinical medicine which incorporates a large number of existing internationally accepted clinical algorithms and presents these in a format which makes them easy for the working clinician to use.

3. Reasons for the SMARTIE project

We believe that many clinical algorithms or "Decision Support Tools" (DSTs) currently in use would benefit from computerisation. Potential benefits include: 1) Easier and quicker to use, 2) Less chance of error and 3) Rapid access to important and specific clinical knowledge.

Some examples of potential DSTs include: The "Fine criteria for community acquired pneumonia", "An algorithm to estimate fluid volume depletion using clinical features" or "The BTS guidelines for acute asthma".

4. How will the Decision Support Tools be developed?

The literature will be searched for potential DSTs and the available evidence base for each will be compiled. Clinical experts will be asked to assess each DST on a number of specified grounds in order to judge its suitability for computerisation. The final list of selected DSTs will be developed into software applications (currently termed "SMARTIEs").

Each SMARTIE will undergo extensive structured testing under both laboratory and clinical conditions, involving technical and informatics specialists, clinical experts and user groups. Should the SMARTIE format prove successful it is envisaged that it would provide a common development process and a universal software interface for further developments in clinical computer aided decision support systems.

References
[1] Sancho JJ, Planas I, Doménech D, Martín-Baranera M, Palau J, Sanz F. IMASIS. A Multicenter Hospital Information System – Experience in Barcelona. En: Iakovidis I, Maglavera S, Trakatellis, User Acceptance of Health Telematics Applications. Looking for convincing cases. Amsterdam. IOS Press. 1988; 35-42.
[2] Einarson, T., "Drug Related Hospital Admissions," The Annals of Pharmacology, v. 27, 1993, pp. 832-40.
[3] Anon. JA, "Strategies for Quality Care and Lower Costs," Preventive Med., Spring 1995, p.23.
[4] PHS, Healthy People 2000, Washington DC: US Public Health Service, 1993.
[5] Eddy, D.M., "Clinical Decision Making: From Theory to Practice," JAMA, 270, 1990, 520-6.
[6] Coiera, E.W., "Artificial Intelligence in Medicine: The Challenges Ahead," JAMIA, 3:6, Nov/Dec 1996, 363-66.
[7] Silverman, B.G., "Computer Reminders and Alerts," IEEE Computer, v.30. n.1, Jan.1997

[8] Silverman, B.G., Murray, A., "Collaborative Human Judgment Theory: Issues for Improving Knowledge Based Systems, in Full-Sized Knowledge Based Systems Workshop Proceedings, Washington DC: GWU Tech Report, May 1990.

[9] Silverman, B.G.,"Expert Intuition and Ill-Structured Problem Solving", IEEE Trans. on Engineering Management, v. EM-32, no. 1, Feb. 1985, pp.29-32.

[10] Silverman, B.G.,"Human-Computer Collaboration", Human-Computer Interaction, v. 7, n. 2, Summer 1992b.

[11] Tuttle, M., "Medical Informatics Challenges of the 1990s: Acknowledging Change," JAMIA, 4:4, 1997, 322-5.

[12] Silverman, B.G., Critiquing Human Error: A Knowledge Based Human-Computer Collaborative Approach, London: Academic Press, 1992a.

[13] Sancho JJ, Planas I, Doménech D, Martín-Baranera M, Palau J, Sanz F. IMASIS. A Multicenter Hospital Information System – Experience in Barcelona. En: Iakovidis I, Maglavera S, Trakatellis, User Acceptance of Health Telematics Applications. Looking for convincing cases. Amsterdam. IOS Press. 1998; 35-42.

[14] Rector, A., "Using Concept Models for Representing Terminology: The GALEN Approach,"

[15] Sancho JJ, Sanz F. Learning Just-in-Time in Medical Informatics. En: Iakovidis I, Maglavera S, Trakatellis, Information Technologies for Medical Informatics Education. Amsterdam. IOS Press. 1999.

[16] Silverman, B.G., Safran, C., et al., ""HOLON: A Web-Based Framework for Fostering Guideline Applications" Fall AMIA Conference Proc., Nashville: AMIA, October 1997, pp. 374-8.

[17] Ohmann C, Eich HP, Sancho JJ, Díaz C, Faba G, Oliveri N, Clamp S, Cavanillas JM, Coello E. European and Latin-American countries associated in a networked database of outstanding guidelines in unusual clinical cases. En: Kokol P et al, eds. Medical Informatics Europe'99. Amsterdam. IOS Press. 1999; 59-63.

[18] Sancho JJ, Oliveri N, Faba G, Campos M, Clamp S, Sanz F. Using the Internet to tackle unusual clinical cases: ELCANO. En: Oliveri N, Sosa-Iudicissa M, Gamboa C, eds. Internet, Telematics and Health. IOS Press. 1997; 371-376.

[19] Ohmann.C, H.O. Formal decision aids in gastroenterology – results of a survey. Zeitschrift fur Gastroenterology 1992: 30; 558-64

[20] Clamp, S.E. Impact on and attitudes of society to computer-aided decision support systems in clinical medicine. PhD. Thesis. University of Leeds. 1995.

Human Factors in the Design of
Wireless Point-of-Care Medical Applications
in U.S. Department of Defense (DoD) Settings

G. Rufus Sessions, Jeffrey I. Roller

Telemedicine and Advanced Technology Research Center (TATRC)
USA Medical Research and Materiel Command (USAMRMC)
Fort Detrick, Frederick, MD, 21702-5012

Abstract

The U.S. Army Telemedicine & Advanced Technology Research Center (TATRC) has formed a Wireless Medical Enterprise Working Group (WMEWG) to investigate the utilization of wireless medical digital assistant devices (MDAs) in DoD settings, to facilitate management of this research program and to provide intra- and interoffice coordination's of R&D efforts. This paper discusses procedures and projects within the WMEWG that employ human factors engineering and knowledge management practices in the design and conduct of this research program, notably clinical needs assessment and usability studies designed to ensure that the devices and software systems that are developed adequately incorporate user requirements and feedback. Progress in giving health care providers wireless point of care access to a broader spectrum of medical information technologies will additionally enhance the access of their patients to these technologies, there by promoting the objectives of the universal access.

1. Introduction

The use of wireless, hand-held or Personal Digital Assistant (PDA) computer technologies to provide broad point-of-care support for medical treatment is exploding in medical communities. U.S. Department of Defense (DoD) medical R&D activities have been (Roller, et al., 1995) and are currently engaged in numerous efforts to exploit such technologies. Medical Digital Assistants (MDAs) may be defined to any small, portable and unobtrusive computing and/or telecommunications device that assists in the collection, retrieval or communication of data relevant to medical care. The U.S. Army Telemedicine & Advanced Technology Research Center (TATRC), a component of the U.S. Army Medical Research and Materiel Command (USAMRMC), is investigating the utilization of wireless MDAs in DoD settings. The TATRC Wireless Medical Enterprise Working Group (WMEWG) was formed to facilitate management of this research program and to provide intra- and interoffice coordination's of these R&D efforts. This paper discusses procedures and projects within the WMEWG that employ human factors engineering and knowledge management practices in the design and conduct of this research program.

2. Objectives

Objectives of the WME program include: (1) to explore use of wireless networking in medical settings (both within Medical Treatment Facilities [MTFs] and in field), including advanced technologies for data gathering and omni-directional transfer of vital information between the battlefield, theater operations, and distant strategic positions including CONUS (Continental United States) facilities; and (2) to develop systems that make use of MDAs as point-of-care end agents in a wireless distributed computing environment that enables access to a myriad of remotely based information systems. Prospective efforts include: concept development to exploit wireless access to an intelligent store & forward e-Mail system; integration of MDA technology into existing systems for access to results reporting and interactive data/order entry, encompassing legacy-based patient records and medical information systems in the DoD; access to remotely-based referencing information over a wireless network for clinical decision support (i.e., CD-ROM text, Internet sources, and locally resident databases); establishment of an intuitive method of information collection for nursing, medical corps, and ancillary personnel. Models created in this prototype will potentially enable an efficient and non-intrusive "behind the scenes" aggregation of data to be used for wide variety of purposes including, but not limited to, staffing needs assessment, outcomes-based appraisals, and sundry patient/provider pattern analyses so critical in an era of managed care. The hypothesis being tested is that mobile wireless MDA devices will make caregivers more efficient and knowledgeable in the detailed information that is needed at the point of care, resulting in enhanced capabilities to deliver care and better overall care for patients. These technologies also have the potential for reducing medical errors by providing more immediate access to data and support for caregivers in monitoring critical parameters and safety guidelines.

3. Clinical needs and usability assessments

The initial efforts of the TATRC working group include conducting both a clinical needs assessment for MDAs in military medical treatment facilities (MTFs) across all levels of care and a formal usability test of prototype MDA systems. Many different wireless hardware and software applications are being developed by industry. Relatively little attention, however, is being devoted to learning from the end users exactly what functions are needed or how best to insert such technological enhancements into existing clinical business practices, nor how usable these systems are in practical clinical applications. The current project is designed to conduct a thorough formal needs assessment in actual clinical settings in military medical treatment facilities (MTFs) across all levels of care and to monitor the use of wireless MDA devices in various settings.

The outcome of this project will produce: 1) Documentation of the needs of clinical users for wireless applications in MTFs across all levels of care. 2) Documentation of the parameters of usability of MDA devices. 3) Expansion of the experience base within the military medical community with the power and versatility of wireless MDAs. This will result in enhancing the ability of end users to inform the R&D community of their needs. These outcomes will provide the basic understanding of user needs that will guide R&D efforts, and will reduced the real potential for expensive "stovepipe" efforts that dilute the power of available R&D resources.

The proposed project will employ phased combinations of participant-observer, expert panel methodologies and laboratory testing to answer the following questions:

1) Which currently available, COTS wireless hardware and software combinations are most readily adopted in actual military clinical settings, both in fixed facility and field, and by what segment of caregivers (nurses, physicians, administrative personnel, etc.)? This should lead to a documented list of technologies that are efficacious and practicable in military medical settings. 2) What works and does not work in day-to-day clinical business practices? 3) What do various segments of caregivers report that they need/want from point-of-care devices that existing COTS technologies do not provide? 4) What technical barriers currently exist to restrict or prevent the expansion of local use of wireless technologies throughout the military (e.g., security, electromagnetic interference/safety, etc. issues)?

Preliminary market survey and literature review work by the Wireless Medical Enterprise Working Group at TATRC revealed that several commercial off-the-shelf (COTS) devices using both Windows CE and Palm Operating Systems, employing Wireless Applications Protocol (WAP) technologies for connectivity, provide the hardware/software base that affords the necessary characteristics of extensibility and expandability of medical applications.

Hand-held units can store large quantities of medical reference materiel, ranging from simple outlines of clinically useful information to full textbooks, to drug manuals. Hand held devices are also useful for simple word processing, database entry, and Internet applications like e mail, newsgroups, and web browsing. Useful sites accessible by wireless handhelds include the MEDLINE database, MDConsult or MedScape.

Examples of some of the clinical applications that will be evaluated on appropriate hand-held devices include: a) Ephysician, which provides important medical information, electronic prescription and lab test ordering on a hand-held device from any location. b) EPocrates, a clinical drug reference for health care professionals on a hand-held device. c) MobilePractice. Includes patient management, medical references, and current medical literature. d) Mobile-Patient Viewer. Displays real-time physiological data, retrospective e) and numerical data from a variety of medical devices and ambulatory ECG transmitters. f) PocketClinician Medical Library, which provides a variety of access to reference information. g) WardWatch, which allows patient tracking on wards. h) Various specialty-specific calculators like DoseAssist, MedCalc, Pregcalc, etc. i.) Portable electronic patient record software (ex. PatientKeeper, etc.) applications may be tested, although there will be no expectation of using "live" patient data in

866

this Phase as a substitute for other official patient record keeping. Additionally, AvantGo server to wireless device technologies will be utilized to support the kind of forms-based applications that physicians on rounds may eventually use to replacing paper-based temporary recording systems.

The methods for the needs assessment will involve a participant observer study of the use of wireless technologies followed by a conclusion phase involving a structured expert panel focused on specific questions/issues arising out of the first phase of the study. Clinicians serving as participants will be recruited from military MTFs, to include field medics. Volunteer clinicians will first be given minimal training in using the selected hardware/software configurations, and will be instructed in the types of uses appropriate to their particular clinical environments. They then will use the devices in their day-to-day work, gaining experience and recording their experiences on dairy-type data forms residing of the devices themselves. At intervals the research team will visit with each participant to interview them on their experiences, ideally at their work sites, so as to capture information relevant to their particular experiences. To the extent possible, participants will be given experience with each of the different devices and software packages selected for evaluation.

Following the completion of the initial familiarization phase of the study, expected to require 3 mos., the data will be reviewed to determine critical issues/questions that will be subjected to Expert Panel methodology for documentation of consensus by Subject Matter Experts (SMEs). Selected participants will be invited to take part in a structured Expert Panel process that will result in the generation of requirements documentation . Following their newly acquired experiences using wireless point-of-care devices in their medical practices, these clinicians will now qualify as SMEs for not only their clinical expertise but also in the technical matters in question for the requirements generation process.

Immediately following the clinical needs assessment in the field trial of Medical Digital Assistants (MDAs) we will substantially improve the understanding of user needs by evaluating the human factors of medical MDA use in the context of a formal laboratory usability test, and will combine this information with that generated in the clinical needs assessment study to develop a clear picture of MDA users needs, human factors, and a vision of potential opportunities and barriers for future development. Our benchmark usability study will address both MDA hardware and software user interface design issues, as well as systems-level issues relating to the integration of MDA technology into complex clinical infrastructures.

4. Knowledge management activities

The Working Group makes use of a variety of knowledge management tools and practices to forward the work of the group in a virtual collaborative effort. The group maintains an open web site to serve as a portal to the activities of the working group to provide access to background information about the WME effort, special documents and reports, and a web-enabled interactive database of literature relevant the research effort. This web site may be viewed at http://www.tatrc.org.

Another important tool used by the WMEWG is a closed bulletin board style web environment supporting virtual collaborations and document sharing within the working group as well as list serve e-mail connectivity. The WMEWG WebBoard environment contains the majority of the information pertinent to the work of the WMEWG and is linked to the group's web Home Page. The WebBoard serves as a virtual collaborations tool for intragroup communications and for convenient web-enabled document sharing in a manner similar to a File Transfer Protocol (FTP) site. The WebBoard provides a convenient means to publicly communicate as a group and to organize and archive threaded e-mail communications and documents. Moderated discussions focused on specific issues or subtasks of the group are held ongoingly on the WebBoard and are available for future research or browsing. Sophisticated search engine technologies will be applied to the growing body of material collected on the WebBoard to facilitate data mining and retrieval activities.

As part of its work, the WMEWG is conducting ongoing literature searches and market surveys, and has collected references to a variety of literature pertinent to the use of MDAs. This material has been entered into a web-enabled database to facilitate the collection and sharing of literature within the group. The database supports online addition and retrieval of references, abstracts and links to full text where available. Keywords and categories specific to the work of the Working Group are added to each reference, so that users can easily search and filter the database to meet specific needs for information.

Additional knowledge management activities include the creation and maintenance of web accessible documents and spreadsheets displaying summary material developed by the working group, accessible to the group on the WebBoard or on the Home web site. Examples of the types of materials available to the group include a Glossary of terms, tables of comparisons of MDA devices, availability of MDA software and standards pertinent to wireless medical applications.

5. Summary

The activities of the TATRC Wireless Medical Enterprise Working Group are designed to facilitate and provide broad support for a wide variety of R&D activities and projects investigating the use of wireless MDAs in DoD. Ultimately this program is geared towards investigating questions, the answers to which, are indispensable to defining the appropriate role of point-of-care wireless technologies in the delivery of modern military health care. Progress in giving health care providers wireless point of care access to a broader spectrum of medical information technologies will additionally enhance the access of their patients to these technologies, there by promoting the objectives of the universal access.

References
Roller, J. I., Zimnik, P. R. Calcagni, D. E. and Gomez, E. R. (1995, October). Project PROMED: The Development of a Wireless Hospital Based Network Utilizing Personal Digital Assistants. Paper presented at the 34th Annual Meeting of the Armed Forces District, The American College of Obstetricians and Gynecologists, San Diego, CA.

868

Perceptually-Seductive Technology
in Special Needs Education

Eva L. Waterworth, John A. Waterworth

Interactive Institute Tools for Creativity Studio,
Tvistevägen 47, PO Box 7964
SE-907 19 UMEÅ, Sweden
E-mail: eva.lindh.waterworth@interactiveinstitute.se
Phone: +46 90 18 51 36, Fax: +46 90 18 51 37

Abstract

We contrast a self-contained desktop VR environment (the Cave, after Plato) with an augmented reality environment (the Telescope, after Galileo) as two ways of implementing what we call Perceptually-Seductive Technology (PST). PST is a general class of sensory augmentation where information is presented in multiple modalities, in concrete forms and in 3D spaces. We believe that approaching computerised information through PST is the most promising route to open up the information society to all. In this paper we report an evaluation of two versions of the Cave, one based on a familiar place in the real world, the other a fantasy world that exists only on screen, as applied to the education of upper secondary school pupils with various physical handicaps. Although the fantasy world was more relaxing and fun to begin with, the familiar world was judged the more effective – as well as being more fun in the long run.

1. Introduction

The basis of all learning is human memory. And yet, paradoxically, the development of traditional educational technologies has tended to reduce human memorisation skills. This is not so surprising if we compare information technology with physical machines such as automobiles and earth movers. As we use our muscles less, because of the availability of these technological aids, we tend to lose some of our earlier physical strengths. Similarly with information technology; the more we depend on technological aids to hold information, the less capable we are of retaining information in memory. We claim that this loss of memory skills has also tended to reduce creativity in education, since we view learning as a creative process and creative thought as a learning process. We consider this to be especially true for people with special needs; for them, link between learning and everyday creativity is particularly obvious.

However, this analogy between information technology and mechanical technology rests on a certain view of information and information technology, where computers act as 'cognitive artefacts' (Norman, 1993) providing access to information generally represented in abstract, linguistic forms. This view of technology presents educators with a problem, which is 'how can artefacts created to serve the function of reducing mental effort be designed to encourage that very effort?' (Waterworth, 1994).

We are experimenting with recent information technologies in educational settings for physically-disabled students, in particular virtual and augmented environments, which tend to exercise perceptual faculties rather than replacing cognitive skills of memorisation and retrieval. We believe that, by presenting information in different perceptually-seductive and emotionally-stimulating ways, we can provide technological tools for the enhancement of both memory skills and creativity of people with special needs. This approach is in contrast to Norman's (1993) suggestion that (with new 'experiential' technology such as multimedia and VR) 'One can have new experiences in this manner, but not new ideas, new concepts, advances in human understanding: For these, we need the effort of reflection'(p 17). While we agree that reflection (that is, abstract, linguistic processing) is important, we are exploring ways in which *perceptually-seductive technology* (PST), incorporating techniques largely discarded since the earliest information technologies became widespread, can provoke both creative ideas and robust learning (see Waterworth and Waterworth, 2000). PST is a general class of sensory augmentation where information is presented in multiple modalities, in concrete forms and in 3D spaces. We believe that approaching computerised information through PST is the most promising route to open up the information society to all, because it capitalises on corporeal being in general, rather than specific learned skills.

We are experimenting with two main approaches to implementing PST for people with special needs, which we call the Cave (after Plato) and the Telescope (after Galileo), and which their very different assumptions about how new knowledge may be acquired. In the Cave the learner is immersed in a created and limited virtual environment that is designed to display only relevant information and is not affected by the reality outside. While looking through the Telescope, on the other hand, the learner is still part of external reality, but with the addition of technologically-augmented features. We have implemented two versions of the Cave – the Classroom Theatre and the Memory Theatre - in which information is stored and accessed by navigation around two different desktop virtual worlds.

In the remainder of this paper we describe our findings from evaluating the Classroom Theatre and the Memory Theatre, and then draw some general conclusions about PST applied to special needs education.

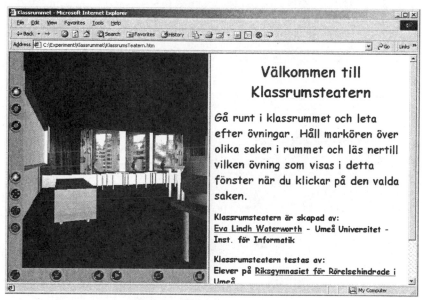

Figure 1 - The Classroom Theatre.

2. Testing the cave with handicapped users

The experiment described below was designed to contribute to two projects, one called 'The Classroom Theatre' and the other called 'The Cave or the Telescope? Mnemonics in Virtual and Augmented Reality'. Both projects are at fairly early stages in the design and development process. The main goal of the Classroom Theatre project is to set up a shared virtual arena with separate workspaces for the pupils: this also includes the design of software that supports memory and makes learning more engaging and fun. The other project, The Cave or the Telescope? Mnemonics in Virtual and Augmented Reality, on the other hand, is aimed at another target group and a slightly different aim, namely to improve everyday life for elderly people with some kind of communication problem. The goal here is to design an aid to be used for communication which at the same time assists and perhaps improves the person's memory. Both projects are based on the idea of using Perceptually Seductive Technology (PST) for people with special needs to improve their everyday creativity, their own abilities and everyday life. The formative evaluation described in this paper is the first experiment we have carried out on PST applied to special needs. The aim of the evaluation was to gain suggestions and ideas of how to improve the design of the prototypes. These improvements will then be implemented, further tested and evaluated. Consequently this is the first step involving implementation and testing, in an iterative design process.

2.1 The classroom theatre and the memory theatre

The first set of evaluations test the theoretical ideas behind PST and the use of computerised memory theatres in education (see Yates, 1992, for a fascinating survey of the history of mnemonic techniques). The approach adopted to test this is a formative evaluation where the pupils use the two different desktop memory theatres, the Classroom Theatre and the Memory Theatre, during a period of two weeks per environment. The Classroom Theatre is a simplified version of the classroom where the pupils have their lessons. One of the main ideas behind the design of the classroom is that it is supposed to be familiar to the pupils. The Memory Theatre on the other hand is design as a typical Shakespearean theatre in the same style as Robert Fludd's memory theatre (see Yates, 1992). In this case the idea is that the environment is not something that is real and known to the pupil, it is more or less a fantasy world. Both environments are evaluated by students and then compared to each other in terms of their usefulness in education for pupils with special needs.

Both environments run in web-browsers, which have plug-ins that to support both VRLM-code, and Flash-applications. Each is presented as a web-page consisting of two frames, one that shows the VRLM-world of the actual theatre and the other frame that shows the Flash-movie of the actual lesson (see Figure 1). In the VRML-worlds the learner "walks" around the theatre to find clues of the lesson that he or she is going to study. The lessons are hidden on different kinds of objects in the actual theatre, for example on doors, balconies, computer screens or whiteboards. When the learner points to the marker a message is shown to indicate which lesson he or she is pointing at. The film of the lesson starts in the right frame when the learner clicks on the marker of the lesson.

Figure 2 - The Memory Theatre

The lessons were based on the book *Datorkunskap* that is used in Sweden to acquire basic skills about how to handle a computer. It includes among other things lessons on Microsoft Excel™ and Microsoft PowerPoint™. The first session of the experiment was based on learning about databases in Microsoft Excel™. In this session all the information that is in the book was represented in the different films of the lessons. There was no sign of where the exercises were hidden apart from the message that was shown at the bottom of the window. The second session of the experiment was based on learning about Microsoft PowerPoint™. Here the idea was that the learner gets the general information from the book or the teacher and only exercises were presented in the films. The exercises were based on the same theme as in the book but the actual tasks were different. In this version of the theatres there is a coloured sign where a lesson is hidden unlike the first version used in the first session.

The Classroom Theatre (Figure 1) is a simplified representation of the learner's actual classroom where the lesson is taking place. It includes the main objects in the room, but all minor objects were excluded. The Memory Theatre (Figure 2) on the other hand is a fantasy world designed as a Shakespearean theatre from the Middle Ages, according to Robert Fludd's ideas of memory theatres (Yates, 1992).

2.2 Target Users and Procedure

The formative evaluation was conducted at *Riksgymnasiet för rörelsehindrade* at Dragonskolan in Umeå, Sweden (upper secondary school). The class consisted of seven pupils with different kinds of disability. Two of the pupils where absent during all the sessions because of illness, while the other five pupils participated during all sessions. The five remaining pupils were all boys aged between sixteen and seventeen. In these kinds of classes the pupils are often absent from school because of treatment or illness, but during these sessions the five pupils were present even though they did not feel quite well.

The software was examined through assessment of the pupils' opinion by administering a specially designed questionnaires where the pupils evaluated each the two different theatres and also compared the two different theatres against each other. At the first lesson of using the software the pupils were given an introduction to the structure of the software and how to use it. Half of the group were then given access to the Classroom Theatre and the other half to the Memory Theatre. After the introduction the pupils started to work with their respective theatres. After using one of the theatres over two weeks, which amounted to six one-hour sessions, the pupils were given one questionnaire to evaluate the theatre and their experience of using it. For the next following two weeks, the pupils switched theatres. After this the pupils used the same questionnaire to evaluate the second theatre and their use of it. They were also given a third questionnaire aimed at evaluating the theatres in general and also inviting comparisons between the two theatres.

3. Summary and Discussion of the Findings

One reflection from the evaluation that was remarked upon by their usual teacher is that all five students participated in all the sessions, even when feeling unwell, which is very rare. This could indicate that they found the use of the theatres very interesting and because of that they did not want to miss out. But at the same time it could be a Hawthorne effect, just that they wanted to participate because it was interesting that somebody showed them interest and that they were selected and pioneers. But it seems to us and to the teacher that they really enjoyed the sessions whatever ever the reason was.

The results suggested that the pupils preferred the Classroom Theatre to the Memory Theatre as stated before. This could be a result of the incompatibility of experience and reflection, as discussed in the Introduction to this paper. One possible reason is that the pupils felt more relaxed in the Memory Theatre than in the Classroom Theatre. In the Memory Theatre they used more attention just to experience the environment in itself since it was unfamiliar and a fantasy creation. In the Classroom Theatre on the other hand they were more reflective because the environment were familiar even though virtual, and that meant that most of their attention could be used to reflect, especially over the task itself. The Memory Theatre was initially more fun and they seemed more relaxed when using it, but at the same time they rapidly became tired of it.

All three questionnaires had an open question at the end where the pupil could add ideas for improvements to make the theatres more fun and/or effective to use. One suggestion was that the Classroom Theatre should be bigger and include more details. This would make the room more engaging and a bit more exciting, but still capitalise on familiarity.

One conclusion from this first set of experiments is that the pupils find the use of PST, and specifically VR memory theatres motivating and enjoyable. A less predictable finding was that a virtual version of a familiar room was more effective than a specially-designed memory theatre. However, the power of familiar concrete places as cues to memory has been recognised since the time of the ancient Greeks (Yates, 1992).

Another conclusion is that the environments should try to mix experiencing and reflection better. This could be achieved by creating some breaks to trigger switches between the two, for example by adding some unexpected events in the design. One example might be that when a pupil discovers the right lesson to study and he clicks on the label something unexpected happens, for example the troll, who is the instructor in the animations of the lessons, says something or does something unexpected. This event should be different for different lessons and at different times. Another suggestion is to include sound in the application.

In sum, the outcome of this set of experiments has resulted in the redesign of the actual environments according to the suggestions made, and these will become the designs of the environments in the next set of experiments.

Our next step is a comparison of our two main approaches to implementing PST, the Cave and the Telescope. The beneficial effect of familiarity we found with the present evaluation leads us to predict that the Telescope approach, based on the augmentation of a real, familiar environment, will be particularly effective in as a learning environment.

Critics may accuse PST of taking a very general view of learning, by assuming that learners share the same abilities and characteristics. But the interpretations and associations underlying perceptually-seductive learning are made by the individual learner. We aim to provide technological mediation that allows information to be explored, and acted upon, with perceptual and emotional engagement, while developing conceptual ideas. This is a step closer to real life learning, and one that opens up the possibilities of information access for all.

Acknowledgements
Artwork for The Memory Theatre was created by Linda Bergkvist. We gratefully acknowledge the assistance and cooperation of staff and students at *Riksgymnasiet för rörelsehindrade* at Dragonskolan in Umeå, Sweden, for their participation, and of Informator (formerly NUDU Läromedel AB) of Stockholm for permission to use materials from their *Datorkunskap* course book. We are also grateful for project funding from KFB (the Swedish Transport and Communications Research Board).

References
Norman, D. A (1993) *Things That Make Us Smart*. Reading, Mass., USA: Addison-Wesley.
Waterworth, J. A (1994) Review of Things That Make Us Smart by Donald A Norman, *SIGCHI Bulletin*, **26**, (4). 1994, 78-79.
Waterworth, J. A. and Waterworth, E. L. (2000) Presence and Absence in Educational VR: The Role of Perceptual Seduction in Conceptual Learning, *Themes in Education* 1:1, 7-38.
Yates F.A. (1992) *The Art of Memory*, London U.K: Pimlico.

ICT in special needs education: what are European practitioners asking ICT researchers for?

Amanda Watkins

Project Manager, European Agency for Development in Special Needs Education, Middelfart, Denmark

Abstract

The use of computer and information technological aids by a teacher working with a child who has special educational needs is one possible end-point of the research and development process the UAHCI conference is concerned with. An *information society for all* needs to consider and meet the needs of the least as well as most able of its citizens. This paper considers some of the central issues raised by ICT and special education specialists who participated in a study conducted by the European Agency for Development in Special Needs Education. They highlight a number of concerns about access to appropriate hardware, software and support for pupils with special educational needs and the teachers who work with them. This paper suggests that there needs to be more direct dialogue between ICT researchers and this group of users if the potential to apply ICT to meeting the needs of such children and a real information society for all is ever to be fulfilled.

1. Preamble

About 10% of the population of Europe has some form of recognised disability (European Commission (1999)) and it is estimated that there are 84 million pupils and students – approximately 22% or 1 in 5 of the total school aged population - who will require *special educational provision* either in a mainstream classroom, as part of a special class or within a separate institution (Eurydice (2000)). Depending on the way a child is identified and assessed in the countries of Europe, pupils with *special educational needs* (SEN) make up between 2% and 18% of the school age population (Meijer (1998)). The statement of the Salamanca World Conference (1994) suggests that in the past special education was defined in terms of a range of physical, sensory, intellectual or emotional difficulties children may present. However, it is now necessary to widen the concept of *special needs education* to include all children who, for whatever reason, are failing to benefit from school

Today, the provision of education for pupils with special needs varies across Europe according to different educational policies. Meijer (1998) distinguishes between a one-track approach (full inclusion of all pupils in the mainstream system), to a two-track approach (mainstream and segregated) and finally a multi-track approach (with intermediary provisions between mainstream and segregated systems). Despite differences in political standpoints and practical provision, all European Union countries are in agreement that meeting the educational needs of every individual pupil and student can be viewed as an important element of guaranteeing the quality of life of European citizens. In all countries, *Information and Communication Technology* (ICT) is increasingly seen as a major tool in meeting this challenge.

The OECD study Schooling of Tomorrow (in press) clearly shows how ICT is set to transform pupils' school experiences in all countries. However, the present indications are that the *Information Society for All* is far from a reality for all European school pupils. Information from the European SchoolNet (1999) shows the disparity in hardware access across countries (ranging from 7 to 150 pupils per machine in primary and 5 to 37 in secondary schools) and Internet access (between 5% and 90% of primary schools and 48% and 100% of secondary schools connected to the internet).

European wide information on ICT usage with pupils who have a range of special educational needs is limited. One of the few sources of information available is the European Agency project this paper is based upon.

2. The ICT in SNE project

The main objectives of the European Agency for Development in Special Needs Education are to work for the improvement of quality in special needs education and the creation of a long-term extended European collaboration in this field. It is this context of co-operative international work that has lead over the last three years to the completion of a number studies that look at different aspects of special education. Each study has

874

aimed to identify trends in successful functional support or patterns of provision that seem to promote the philosophy of a *school for all* as outlined in the *Charter of Luxembourg* (1996).

The *Information and Communication Technology (ICT) in Special Needs Education (SNE)*(2001) project began in 1999 and initially focussed upon an information gathering exercise, establishing a web database of relevant overview information relating to ICT policy and practice in special education situations in 15 European countries – Austria, Flemish Community of Belgium, Denmark, Finland, France, Germany, Greece, Iceland, Ireland, Luxembourg, Netherlands, Norway, Portugal, Sweden, UK. Information was collected via a network of ICT in SNE experts in each of the countries, who acted as the key contributors as well as a main target audience for the database. The information collected provides an overview for each of the participating countries, highlighting the strengths and weaknesses of the policy and infrastructure of equipment, support and access to information and training available to teachers in special education settings. An analysis of this overview information has lead to a synthesis report also available on the Agency website (as well as in print format).

Four key issues emerge from the analysis of the information that appear to be important considerations for ICT researchers whose work may be intended for an educational context:
-   What policies direct the availability and use of ICT in special education situations and how do educational (and not research or technology) policies dictate a teacher's access to hardware and software?
-   What forms of support and training are available in ICT for special education teachers and how can limitations in support and training place constraints upon the introduction of new technologies into the classroom situation?
-   What are the current and perceived future developments in the field that an ICT researcher may be interested in pursuing in their work?
-   What key information sources are available for teachers and professionals who support them and how can ICT researchers use these channels to disseminate their information in a more targeted way?
The information below is a synopsis of the project findings in relation to these issues.

3. ICT policies

National policies on ICT dictate the infrastructure of hardware and software made available to teachers and pupils. Policies also have a direct impact upon a teacher's access to training, support and information relating to ICT.

Only four countries identified that there was a National ICT Strategy that had an influence upon ICT in SNE (Dk, Ice, Po, UK). Of the 15 countries involved in the study, 14 stated that there was no specific policy on the use of ICT in SNE, but that there was an overall policy for ICT usage in education which covered special needs education. Only Portugal identified a specific policy, which is drafted as a particular element of National Disability and SEN legislation. Evaluation of ICT policies was being conducted in 6 countries (Au, Fi, Nl, No, Sw, UK).

None of the countries identified links between educational ICT policies and IST or R&D policies or initiatives. This fact seems to be a real issue as it can be argued that an opportunity to co-ordinate efforts – particularly in terms of matching practical needs with research interests – is lost.

The central point to be made within the context of the UAHCI conference is that ICT researchers and developers with an interest in the education sector as an audience for their work, may need to be familiar with the educational policies aims and constraints that dictate their potential users' work.

4. ICT support

The availability of appropriate support in implementing ICT is as important as having the appropriate hardware and software to use for many teachers. Four countries (Ice, Ire, No, UK) had recognised National Agencies for ICT in Education. Three countries (Dk, Sw, UK) offered support services that worked directly with teachers working in SNE. 11 countries (Au, Dk, Fi, Fr, Ice, Ire, Lux, No, Sw, UK and some of the Lander in Ger) provided specialist resource centres where teachers could obtain advice, materials and information. Specialist websites and on-line networks were provided in 14 countries (Au, B(Fl), Dk, Fi, Fr, Ger, Gr, Ice, Ire, Lux, Nl, No, Sw, UK). Most countries suggested that individual schools may have named staff with special expertise acting as ICT co-ordinators, but these staff were not necessarily those with SEN expertise. Most countries offer general ICT courses for teachers, but specialist training in applying ICT to meet SENs was only made available in initial teacher training in one country (Ire) and as part of in-service teacher training in seven countries (Au, Dk, Fr, Ger (variations across Lander) Gr (distance learning) Ice, UK).

Support for individual teachers in using specialist ICT can be provided at national, regional, local, school or colleague levels. Whilst this can lead to a range of flexible information, advice and practical support services it also presents problems in terms of split responsibilities of levels of government, difficulties in accessing funding and potential lack of co-ordination in provision of information services.

All contributors to the study argued that access to appropriate hardware and software to meet individual learning needs was a key factor in encouraging teachers to implement ICT in their work with pupils with SEN. However, the existence of equipment does not guarantee its implementation – teachers need to be trained and practically supported in its use, a factor which cannot be overstated enough to hard and software developers.

5. Future developments

Throughout the study, the contributors referred to very concrete and specific examples of potential developments in ICT that needed to be looked at by researchers in more depth. Specific suggestions related to four areas of ICT usage:

5.1 Hard and Software requirements
- Pedagogical robots such as Valiant Robot or Jeulin's Turtle (Fr)
- Devices like Lego Dacta or Electronics Lego (Fr)
- Development of gesture recognition devices (Po)
- Development of individual hard and software solutions for children with severe disabilities (Au)
- Development of software for diagnostic purposes (Au)
- Development of the wearable computers (Po)
- In the development of educational software, there should a clear focus upon the educational context cultural, ethnic, philosophical and psycho pedagogical (Gr, Po)

5.2 Internet Access
- Email with speech input/output (Ge)
- Browsers for pupils with severe learning difficulties, directed by only few icons (Ge)
- Improvement of networking facilities to allow more efficient co-operation of institutions (Au, Gr, Ire)
- Building an ICT network (discussion groups, list servers) in Europe between teachers working with pupils with rare conditions, for example blind-deaf pupils (No)
- Creation of a very simple system for installation of websites (Ge)

5.3 Compatibility/application issues
- Work towards common European models of practises (Fi)
- Adaptation of standard software to the needs of the children with different SENs (Au)
- Models to simplify the frames of windows programmes (Ge)
- The translation and the adaptation of European methodology to and within individual country standards (Gr)
- Co-operation in order to get a standardized storage format for text, pictures and sounds in different teaching materials and software according to the different needs of the SEN students (Sw)
- Integrated research concerning hardware and software in order to ensure compatibility (Po).

5.4 Pedagogical research
- Research into psychological and pedagogical aspects of ICT & SEN children (Gr, Ice)
- Research and development projects on the effect of ICT in the learning process (No)
- Research work concerning how ICT may help support the inclusion process of pupils with SEN (No)
- Transnational projects exploring video-conferencing in particular, with email and web-based support, to ascertain the value of international communication in special education (Fr, Ire)
- Systematic investigations into models of teacher training and support (Ice, UK)
- Information about teachers' real needs for products (Nl)
- Research into systems directly related to educational environment and its communication requirements (Po)

The participants in the project all had very clear ideas about what developments would need to be made in these four areas if the needs of children with SENs were to be better met through the application of ICT.

They were also very clear about wanting to be active partners in such developmental processes *if* they were given the opportunity to participate in projects and programmes.

6. Getting Information

Access to appropriate information is in itself a key source of support for teachers and professionals who support them. The project participants identified the following as information they would like to have access to on an international basis:

6.1 ICT policies
- Information on ICT policy, its implementation in practice and evaluation of policies in other countries (Fr, Ire, Sw)
- Comparative reports of support structures for ICT and SEN, with statistics and trends identified (Ire)

6.2 ICT usage
- Concrete examples of projects dealing with ICT in special education settings (Au, Lux)
- Practical information on latest hardware and software developments (Gr)
- Overviews of information on available hardware and software relating to particular types of special need (Gr, Nl)
- Examples of special projects, innovations and best practises of ICT (Fi, Fr, Ire, Lux, No, UK)
- International exchange of resources; comparisons of resource provision across Europe (Au, UK)
- Information on training and training resources that are available (Ire, Nl)
- Information about ICT management in schools (Nl)
- Information about integrating ICT into educational processes (Nl)
- Results and ideas from research and development projects in other countries (No)
- A database employing a fixed evaluation schema which estimates perceptions of the quality of educational software (Po)

6.3 ICT users

- Addresses of experts and institutions in the different countries (Au, Ire)
- Information about key international conferences/seminars etc (No)
- International contacts on ICT development and implementation (Nl).

Much of this information already exists at an individual country level, but there is work to be done to co-ordinate this information and make it available, not only internationally, but also to other groups of ICT users and researchers.

7. Sharing Information

The European Union *e*Europe Action Plan (2000) underlines the fact that education is the foundation of Information Society, since it has an impact on the life of every citizen. The EC Communication *Towards a European Research Area* (2000) argues that there is a real need to improve co-ordination between research and industry, encouraging trans-European research and sharing of knowledge between the research and business worlds.

The argument of this paper is that there is a real need for educators to be included in this sharing of knowledge and experience – in particular, there is a need to include special educators who represent the diverse community of learners with special needs. Class teachers – and the specialist staff who support them – are all asking for more readily available and applicable IT to use with a wider range of pupils who have special educational needs. However, at present, their very specific voices as ICT users – and representatives of pupils as direct ICT users – are not being heard by the researchers who are maybe not only interested in what they have to say, but may also be in a position to already provide them with some of the solutions they are looking for.

The current project suggests that a form of "vicious circle" is in evidence to a certain extent: appropriate hard and software solutions are not seen as being readily available to teachers, therefore these teachers are not motivated to investigate using any ICT solutions. Teachers do not develop their ICT competence and knowledge base and are therefore not linked into an ICT culture where they can contribute to debates. A lack of teachers' voices in ICT debates means they do not have as much impact on possible R&D developments as they could usefully have. No input into debates from a classroom practitioner perspective means specialist appropriate solutions are not developed and so the circle turns again.

Whilst access to appropriate ICT solutions is not the only factor in encouraging teachers to use ICT with pupils with SENs – policies and subsequent ICT infrastructure and support as outlined above being equally important components – it is a contributing factor and one which can be addressed in a number of ways at an international level using ICT solutions.

The next phase of the ICT in SNE project will be to develop the existing database in order to provide more specific information on ICT policies and examples of policy implementation in the different countries. It will also address the issue of identifying and providing contact information on key people and organisations for ICT and SEN in the participating countries.

These developments however do not address the main issue behind this paper – how can the contact between the two communities of special educators and ICT research and development workers be facilitated for mutual benefit?

As a long-term development to the European Agency's work in this field, it has itself made the leap and developed working partnerships with ICT/IST researchers, notably ICS FORTH in Greece. Within the framework of an EC funded project, a thematic network of special education practitioners and researchers and ICT/IST researchers and developers is being established under the title Special Educational Needs – Information Society Technology – Network (**SEN-IST-NET**). The aim of this project is to establish the conditions – both technological mechanisms and information resources – to open the lines of communication between teachers as "ordinary users" of technology and researchers in order to make access to the "right" sort of technology a more likely proposition for more children with special educational needs.

Via the establishment of a web portal with communication facilities within and between different ICT communities and the development of multilingual resource banks of information relating to ICT and SEN it is intended to facilitate the sort of dialogue between individuals and groups that leads to long term networking and real developments in understanding about a very specific ICT user group's needs.

Despite the practical – and sometimes political – issues relating to the front line application of technology in classrooms, special educators across Europe are ready and willing to be more involved in the process of developing the technological solutions they may some day use with pupils with special educational needs. At the moment, they are not sure whether their ideas will be listened to by ICT researchers or how to make their voices heard. Our experience in the European Agency is that ICT/IST researchers are ready and willing to listen to what is being said by these special educators. If the dialogue between these two groups can be facilitated to a greater extent, then a real step forward will have been made towards achieving a genuine information society that includes all citizens.

References

European Agency for Development in Special Needs Education, Edited by Meijer, C.J.W. (1998) *Integration in Europe: Trends in 14 European Countries,* Middelfart, Denmark

European Agency from Development in Special Needs Education, Edited by Watkins, A (2001) *Information and Communication Technology (ICT) in Special Needs Education (SNE),* Middelfart, Denmark

European Commission Communication (1999) *Towards a Barrier-free Europe for People with Disabilities, a Roadmap to the Achievement of Greater Community Added Value.* Brussels, Belgium

European Commission (2000) *eEurope 2002 - An Information Society For All* Prepared by the Council and the European Commission for the Feira European Council

European Commission Communication (2000) *Towards a European Research Area* Brussels, Belgium

European SchoolNet (1999) *ICT developments in the European Union* Brussels, Belgium

EURYDICE (2000) *Key Data on Education in Europe* Luxembourg

Helios Study Partners (1996) *Charter of Luxembourg* DGXXII, Education and Youth, Brussels, Belgium

OECD (in press) *Schooling for Tomorrow Initiative* Paris, France

UNESCO (1994) *World Conference on Special Needs Education: Access and Quality* Salamanca, Spain

Information and Communication Technology (ICT) in Special Needs Education (SNE) web database http://www.european-agency.org/ict_sen_db/index.html

More information on the ICT in SNE project or the SEN-IST-NET can be obtained from Amanda Watkins aw@european-agency.org

878

Clinical Usability Engineering for Computer Assisted Surgery

Andreas Zimolong[1], Klaus Radermacher[1], Bernard Zimolong[2], Guenther Rau[1]

1 Helmholtz-Institute for Biomedical Engineering, Aachen
2 Work and Organizational Psychology, Ruhr-Universität Bochum

Abstract

For the broad range of systems for computer assisted surgery a design process is proposed which is based on the iterative design approach of ISO 13407 and considers the design of a medical work system according to ISO 6385. It is based on thorough task analysis of the conventional medical procedure and aims to derive user requirements and system specification from this input. As a high-level design goal clinical usability is established, which expands the notion of usability from ISO 9241 to include learnability and reliability. A key element in the design approach is system evaluation already in early stages of the development life cycle, in order to detect design deficiencies and to feed back the results into the design process. Examples of guideline- and user-based evaluation will be given. Results from a comparative testing of different input devices for an intraoperative planning task are also presented, which indicate an advantage of the touchscreen for the given task.

1. Introduction

Systems for computer assisted surgery (CAS systems) are designed for the support the surgeon to improve the outcome of a surgery intervention. In general, two different approaches can be discriminated: with robotic systems, part of the surgery is autonomously carried out by the robot, while the surgeon takes over supervisory control (cp. Sheridan 1987). Surgical robots by now can be found in clinical routine for a number of different interventions, e.g. total hip replacement or total knee replacement (Delp 1998). In a broad study involving more than 1000 cases of total hip replacement, Bargar et al. (1998) showed that a radiographically superior implant fit and positioning was obtained. However, in the same study no significant difference could be found when comparing the subjective patient's ratings on the medical outcome of the intervention.

In contrast to the robotic systems which perform part of the surgical intervention autonomously, passive systems do not move the surgical tool but provide the surgeon with information relating to position and orientation of the surgical tool in relation to the surgical site. Movement of the tool is performed by the surgeon, whereby the human operator is an integral part of the control loop (*manual control*, cp. Sheridan). System with visual or haptic feedback are available, however systems in clinical routine provide visual feedback and include application for spine surgery (Merloz 1998), and cup placement in total hip replacement (DiGioia 1998). Other passive CAS systems make use of intraoperative acquired data such as fluoroscopic images of the spine using conventional C-arm technology (Nolte 2000). In the approach developed by Radermacher et al. (1998) no additional computerized devices are needed intraoperatively. Instead, preoperatively manufactured individual templates customized to the patient's bone are and used intra-operatively as guiding means after the surgeon has positioned the template formclose on bone.

Summarizing it can be noted that there exists a great variety of CAS systems which do not only differ in the way they assist the surgeon during surgery (actively or passively) but as well in the kind of information output and input they process. The later has a strong influence upon the way the planning of the intervention is performed: for systems which make use of preoperatively acquired data such as x-ray computer tomography (CT), the planning is performed preoperatively, in an office-like environment. This is a completely different context of use in comparison to the operating theatre environment. Systems using intraoperative acquired data require an intraoperative planning, as a consequence the sterile working conditions of the operating room need to be especially considered. The sterile working surgeon e.g. needs sterile devices for interaction with the technical equipment, whereas standard input devices are in general not sterilizable. Thus interaction must be performed by using special devices or through sterile barriers such as drape and plastic tubes.

A further important characteristic of CAS systems is their temporal restricted use: working with the system makes up only a very limited part of the overall work spectrum of the involved health care professionals. The frequency of use can range between once per day and once a month or even less. One reason for this is that CAS systems are build only for a specific kind of disease treatment, the less frequent this kind of treatment, the less often the system is used. Even though in future times frequency of use may rise as systems' range of use broadens up, working with CAS system will make up only small parts of the duties of health care professionals.

2. Clinical usability

Most CAS systems do not integrate as an additional tool into the operating theatre, but change completely the way parts of surgery is performed. Using these systems implies additional task steps such as draping of the equipment and registration of image information, task steps which do not concern treatment of the patient, but might in fact influence its outcome. A survey performed by Radermacher using the AET workload measurement and analysis approach for different orthopedic interventions revealed that the surgical task is already characterized by high time demands and a high amount of mental demands concerning acquisition and processing of multimodal information (Radermacher 1999). Resources for additional cognitive tasks are limited, an increase of workload might increase the probability for human error. When designing the CAS work systems it is therefore mandatory to control and minimize the overall workload, to optimize usability of the system. In the following five high-level design goals for CAS systems are presented, which are summarized by the concept of *clinical usability*.

Effectiveness, efficiency and user satisfaction are fundamental design goals which are derived directly from the definition of usability of the ISO 9241. In this sense, effectiveness describes the completeness by which the task goal is reached (e.g. medical outcome measured by clinical scores), and efficiency the relation of resources spent to reach that goal in relation to effectiveness (e.g. operating time in relation to medical outcome), (ISO 9241). With the introduction of new CAS systems into the operating theatre, these design criteria need to be demonstrated against a "gold standard" such as the conventional procedure. This extends the conventional clinical study design, which usually concentrates on measures of clinical outcome and complications.

Learnability. CAS "... technology is only one facet of surgery, it must not predominate other facets ...." (Merloz 1997), which applies to training as well. Health care professionals are requested to continuous medical education, additional technical training is thus tolerated only to a very limited extent, especially if workload is high. The systems thus needs to be self-descriptive and easy to learn. Technical tasks which stem from using the computer system (e.g. saving of data files) should be avoided or automated as much as possible.

System-Reliability. The Medical Devices Directive (MDD, DIRECTIVE 93/42/EEC concerning medical devices, 14th June 1993) requires reliability of CAS systems: "Devices incorporating electronic programmable systems must be designed to ensure the repcatability, reliability and performance of these systems according to the intended use. ..." (MDD, Annex I - Essential Requirements, ¶ 12.1). However it does not clarify upon the relation between system design and human error. Bogner et al. (1994) attribute the cause for human error to an insufficient design of the medical work system, to systems badly adapted to human capabilities and limitations. It is thus compulsory to extend the notion of reliability to cover the whole CAS work system, i.e. not only the reliability of the technical components is considered but the probability of human errors is reduced as well.

This concept of reliability includes such high level system design goals such as error tolerance (cp. ISO 9241 part 10). As ¶ 12.1 of Annex I - Essential Requirements of MDD continues: " ... In the event of a single fault condition (in the system) appropriate means should be adopted to eliminate or reduce as far as possible consequent risks". With the extension of the systems boundaries to include the human operator, this means that the system has to be designed tolerant against e.g. user input error (single fault condition).

In the following methods will be presented which incorporate clinical usability into system design. Furthermore, examples for studies will be given which evaluate systems or system components with respect to clinical usability.

3. Method

Towards achieving the objective of high clinical usability a system design approach was established which is based on the ergonomic principles in the design of work systems (ISO 6385) and the user centered design process for interactive systems (ISO 13407). ISO 6385 identifies three aspects of work system design: Design of work space & work equipment, design of work environment and design of work process. With respect to CAS systems the design process thus comprises the hard- and software system design, the integration of the CAS system in the OR and the development of clinical protocols for the pre- and intraoperative use of the system, respectively.

ISO 13407 on the other hand concerns the system design process. It advocates a strong involvement of the prospective users into the design process and distinguishes four stages: Analysis of the context of use, identification of organizational and user requirements, system design and design evaluation. By feedback of the results from system design evaluation into the design process an iterative design process is established, which aims at continuos

880

design improvement. Evaluation by prospective users and usability experts already in early stages allows to refine the designer's understanding of the context of use, and possibly induces new or refined user and organizational requirements.

Basically, two different approaches to evaluation are considered: user- and guideline-based evaluation. Guideline-based evaluation is performed to identify flaws and inconsistencies in system design using established standards and guidelines. As the number of potential users of CAS systems is comparatively small and time constraints of the user group are high, this type of evaluation is particularly important. However, up to now there is a lack of standards for CAS systems with specific design requirements. Thus it is necessary not only to adopt high level requirements such as compatibility and controllability, but also to adopt and, if necessary, to adapt and validate evaluation criteria from literature and standards established for other domains. With this focus the ISO-Evaluator (former EVADIS, cp. Oppermann 1997) was used to evaluate the design of two different surgical planning systems. The original set of testitems (from ISO 9241) was extended by further criteria from standards relevant for CAS systems (e.g. testitems concerning compatibility, ISO 6385), and from literature (e.g. Wickens (1987): principle of the Moving Part; Sanders (1993): number of reference frames, tracking displays, etc.).

For user-based evaluation test-user use a prototype of the system to solve a given task, while being observed by the evaluator. In out test settings the test-users received a standardized introduction to the medical task, to ensure that all started with the same medical background knowledge. Training on the CAS system for operation planning was not provided, as the intention was to measure how well first-time users were able to use the system without prior training. During the trial test users were monitored using a video-based eye-tracker system for event recording, while the evaluator took notes on system and user performance. Questionnaires and interviews concluded the tests.

Further user-based evaluation was conducted to determine the suitability of different input devices for an intraoperative planning task. For this comparative study a selection of three devices was made from an extended

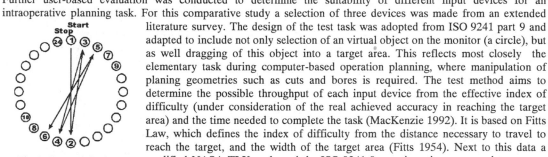

Fig. 1: Test-task for input
devices

literature survey. The design of the test task was adopted from ISO 9241 part 9 and adapted to include not only selection of an virtual object on the monitor (a circle), but as well dragging of this object into a target area. This reflects most closely the elementary task during computer-based operation planning, where manipulation of planing geometries such as cuts and bores is required. The test method aims to determine the possible throughput of each input device from the effective index of difficulty (under consideration of the real achieved accuracy in reaching the target area) and the time needed to complete the task (MacKenzie 1992). It is based on Fitts Law, which defines the index of difficulty from the distance necessary to travel to reach the target, and the width of the target area (Fitts 1954). Next to this data a modified NASA TLX scale and the ISO 9241-9 questionnaire were used to capture subjective workload and satisfaction ratings.

Basic information for system design and subsequent evaluation needs to be gathered by analysis of the context of use (cp. ISO 13407). For CAS systems this context of use is constituted by the conventional medical procedure and related settings and organizational structures in the operating room. In the framework of the EC-funded project MI[3 1] and VOEU[2] an approach for system design was established which derives user and organizational requirements from task analysis of the conventional procedure. The analysis is based on a model of human information processing such as presented by Wickens (1987) and describes each task of the medical procedure by information/ material to be perceived, the decision to be taken, and the action to be performed. This task description is supplemented by information concerning experiences of the surgeon and the medical staff on potential bottlenecks and complications of this task, their impact on the outcome of the medical intervention, and ratings on severity. Next to identification of the context of use this information is used to derive requirements on the CAS system from those bottlenecks and complications, which might be improved by using the CAS system in comparison to the conventional technique.

4. Results

The results of the guideline-based evaluation of both CAS planning systems are presented in **Fig. 1**, where those requirements full-filled by the systems are classified by dialogue principles of the ISO 9241. Only those test-items were taken into account, which were applicable to both CAS systems, about half of the entire set. This was not due

---

[1] MI[3] - Minimally Invasive Interventional Imaging (IST – 1999 – 12338)
[2] VOEU - Virtual Orthopaedic European University (IST – 1999 – 13079)

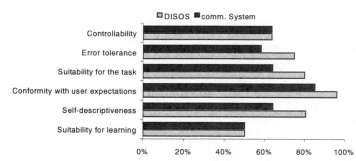

Fig. 1: Comparison of evaluation results for two different surgical planning CAS systems: passed test-items classified by dialogue principles of the ISO 9241

to the specific context of use of the systems, but due to the specific interface designs: both systems do not provide forms for alphanumeric data entry, and DISOS also does not possess a menu structure. For comparison of ergonomic design quality this graphical representation is however only of limited value, as test-items are not weighted. For identification of system deficiencies and subsequent improvement failed test-items were analyzed in detail.

A subsequent qualitative user-based evaluation of systems aspects improved according to the suggestions against the former design yielded higher usability rating. However, the user-based evaluation revealed further design deficiencies which hampered a smooth interaction with the system and which had not been identified before. This indicates that both evaluation approaches complement each other.

Further user-based evaluation concerned comparison of input-devices for an intraoperative planning task under sterile conditions. The trial was split up into two phases: in the first phase the learnability of the input device was determined. For this, the test user repeated the same task with the same parameters until a stagnation of the continuous improvement of the captured time ratings in task completion was observed (Fig. 2). The test-user was then said to have reached stable state were no significant further improvements could be expected, which concluded the learning phase. In the second phase, the difficulty of the task was stepwise increased by increasing the distance between start and target (following Fitts' Law).

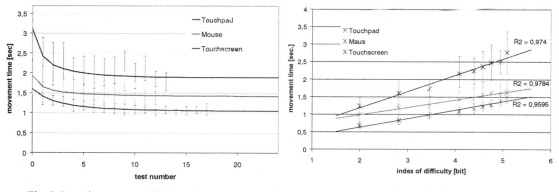

Fig. 2: Learning curves of different input devices   Fig. 3: Throughput vs. index of difficulty

15 subjects tested two or three input devices each. The results indicate that touchscreens allow the highest throughput (Fig. 3), account for fastest movement time and thus are most suitable for direct manipulation tasks. Furthermore, they were rated superior by the test users (Fig. 4). For touchpads, large improvements were achieved during the learning phase, whereas workload was rated significantly higher. Despite the fact that most test users were well accustomed to the use of computer mice, achieved throughput was less and users preferred touchscreens.

5. Conclusions

In the scope of an iterative user-centred approach to system design, user- and guideline-based evaluation provide important, complementary information on design deficiencies, which can be achieved already in early stages of the development life-cycle. They furthermore provide important input for definition of training requirements, which however should not replace an appropriate user-centered system design. Thus integration of the evaluation into the system development process is important towards achieving the goal of high clinical usability of CAS systems.

882

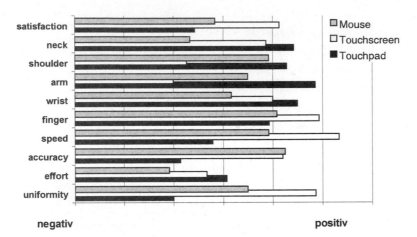

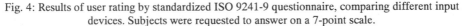

Fig. 4: Results of user rating by standardized ISO 9241-9 questionnaire, comparing different input devices. Subjects were requested to answer on a 7-point scale.

References

Bargar WL; Bauer A; Börner M (1998): *Primary and revision total hip replacement using the Robodoc system.* Clinical Orthopaedics And Related Research (354). 82-91

Delp SL; Stulberg SD; Davies B; Picard F; Leitner F. (1998): *Computer assisted knee replacement.* Clin Orthop (354). 49-56

DiGioia A.M., Jaramaz B., Blackwell M., Nikou C., Colgan B., Moddy J., Simon D.A., Morgan F., Kischell E., Aston C. (1998): HipNav: A Surgical Navigation and Measurement Tool for Placement of Acetabular Implant in Total Hip Replacement. In: A.M. DiGioia, L.-P. Nolte, S. Delp, S.D. Stulberg (Eds.): Syllabus, Computer Assisted Orthopedic Surgery Symposium (CAOS/USA '98), 1988, Pittsburgh, USA, pp.158-162

DiGioia AM, Jaramaz B, Blackwell M, Simon DA, Morgan F, Moody JE, Nikou C, Colgan BD, Aston CA, Labarca RS, Kischell E, Kanade T (1998):The Otto Aufranc Award. Image guided navigation system to measure intraoperatively acetabular implant alignment. Clinical Orthopaedics And Related Research (355). 8-22.

Fitts PM (1954): The information capacity of the human motor system in controlling the amplitude of movement. J Experimental Psychology 1954:47(6): 381-391

ISO 13407:1999 Human-centred design processes for interactive systems

ISO 6385:1981 Ergonomic principles in the design of work systems.

ISO 9241. Ergonomic requirements for office work with visual display terminals (VDTs)

MacKenzie S (1992): Fitts' Law as a Research and Design Tool in Human-Computer Interaction, Human-Computer Interaction, Vol. 7 1992. S. 91-139

Merloz P, Tonetti J, Pittet L, Coulomb M, Lavallee S, Troccaz J, Cinquin P, Sautot P (1998): Computer-assisted spine surgery. Computer Aided Surgery, 3(6). 297-305.

Nolte LP, Slomczykowski MA, Berlemann U, Strauss MJ, Hofstetter R, Schlenzka D, Laine T, Lund T (2000): A new approach to computer-aided spine surgery: fluoroscopy-based surgical navigation. European Spine Journal 9 Supplement 1. 78-88.

Oppermann, R., Reiterer, H. (1997): Software evaluation using the 9241 evaluator. Behaviour & Information Technology, 1997, vol. 16, No. 4/5, pp. 232-245.

Radermacher K, Portheine F, Anton M, Zimolong A, Kaspers G, Rau G, Staudte HW (1998): Computer assisted orthopaedic surgery with image based individual templates. Clinical Orthopaedics And Related Research (354). 28-38.

Sanders M.S., McCormick E.J. (1993): Human Factors in Engineering and Design. 7th ed. McGraw-Hill International.

Wickens, C. D. (1987): Information Processing, Decision-Making and Cognition. in G. Salvendy (ed.): Handbook of Human Factors, John Wiley and Sons Inc. N.J.

Merloz P (1997): Clinical interest of enhanced OR. Public Deliverable 30, IGOS (EU TA Program HC1026HC)

PART 9

ASSISTIVE TECHNOLOGIES

One-dimensional User Interface for Retrieving Information from the Web for the Blind

Chieko Asakawa[a], Hironobu Takagi[b]

IBM Japan, Ltd., Tokyo Research Laboratory
1623-14 Shimotsuruma Yamato-shi Kanagawa-ken, 242-8502, Japan
+81 46 273 {4633[a], 4557[b]}
{chie[a], takagih[b]}@jp.ibm.com

Abstract

Web search engines are the fundamental tools to use the Web effectively. It is, however, not easy for the blind users to access such engines' sites and get information nonvisually, because visual effects are used a lot to improve the intuitive operation for sighted users. In this paper, we will first discuss the difficulties of accessing Web search engines using voice output. Then we will describe a one-dimensional user interface for retrieving information from the Web which we developed to resolve the issues. Our system consists of three components: a one-dimensional input method, a component to create an accessible results list, and a navigation method for the list. Blind users do not need to access general Web search engines' sites, since our system internally accesses these sites to get the results.

1. Introduction

The Web has an important role as a new information resource for the blind, since they can get any kind of information all by themselves. These days, the Web has been becoming much more visual to meet a variety of requirements for sighted users. This tendency makes nonvisual Web access much harder. There are two main difficulties when blind people try to retrieve information from the Web. One is because of the two-dimensional forms used in Web search engines, which are designed for intuitive operation at a glance, and another is the difficulty of understanding the complicated layout of the results pages. Web search engines are the fundamental tools to use the Web, so they should be easily used by any user. For this purpose, there are some related approaches (Asakawa et al., 2000; SETI; Google; Babylon) to simplify web forms. However, they still use the Web forms, and are not for voice output. We therefore decided to develop a one-dimensional user interface using voice output for retrieving information from the Web.

The system consists of three components. The first component is a one-dimensional input method for a keyword search. This allows users to retrieve information about a keyword from the Web as though it was a text editing process. The second component creates an accessible results list suitable for voice output. The third component is a navigation method for the results list, using sound effects.

In this paper, we will first introduce how the blind users currently access Web search engines, and then describe an overview of our system. Finally, we will give some conclusions and plans for the future.

2. An example of the current method for nonvisual web access

Figure 1-A is a Yahoo top page, and Figure 1-B is a reading example by Home Page Reader (Asakawa, 1998), which reads a Web page sequentially in the tag order. The page is well organized for visual scanning, while using voice output it takes a long time to find the input box and submission button. Moreover, the results page layout is complicated to read aloud, while the results can be spotted in a glance. The results page also includes information unrelated to the results, such as shopping lists, banners, indexes, and so on. As shown in Figure 2, that unrelated information is read before the results appear. As a result, it takes more time to find the exact result the users want. Even after they get to the results section, the explanation for each hit is often not enough to confirm if the link should be followed, especially when using voice output, so users need to open pages anyway to find out that which is the one they want. This wastes a lot of time on irrelevant hits.

Yahoo! Mail
free from anywhere | Instant Stock Alerts
Download Yahoo! Messenger | Yahoo! Photos
Share holiday photos
online

[ ] Search advanced search

Y! Shopping − Apparel, Books, Computers, DVD/Video, Luxury, Electronics, Music, Sports and more

Shop Auctions · Classifieds · PayDirect · Shopping · Travel · Yellow Pgs · Maps Media Finance/Quotes · News · Spc

(A) Visual browser

(Start of form 1) [Map: r/a1] [Map: r/p1] [Map: r/m1] [Map: r/wn] [Map: r/i1] [Map: r/hw] Yahoo! Mail free from anywhere [Blank] Yahoo! Photos Share holiday photos online [Text:] [Search: Submit Button] advanced search Y! Shopping - Apparel , Books , Computers , DVD/Video , Luxury , Electronics , Music , Sports and more Shop Auctions -Classifieds -PayDirect -Shopping -Travel -Yellow Pgs - Maps Media Finance/Quotes -News -Sports

(Underlined words are read by female voice.)

(B) Home Page Reader

Figure 1: An example of nonvisual web access (Yahoo! top page)

[Map: google.com] [Map: yahoo.com] Help - Personalize Search Result Found about 205000 web pages for Mariners [Free Horoscopes!] Categories Web Sites Web Pages Related News Events Search Books! [Buy Books!] - MARINERS - Search Books - Search DVD/Video Inside Yahoo! Matches Shopping: Over 300 Seattle Mariners items on Yahoo! Shopping Team Info: Seattle Mariners Team Page, Calendar , Roster Community: Seattle Mariners Message Board , Clubs , Chat League Info: MLB News & Scores , Standings , Stats Auctions: Over 250 Seattle Mariners listings on Yahoo! Auctions Web Page Matches (1 - 20 of about 205000) Seattle Mariners Official Web Site Seattle Mariners Official Website -- Welcome to the official web site of the Seattle Mariners Baseball Club. This page is the entryway containing a few ... http://www.mariners.org/ [More results from www.mariners.org] Photos: Mariners ... Mariners Photos, (Page 1 of 3). Jump to page 1 2 3 ... Maradona, (5, Oct 31). Marathon, (50, Nov 7). Mariners, (7, Nov 18). Maryland, (15, Nov 19). ... http://sports.excite.com/photo/topic/sports/mariners [More results from sports.excite.com]

Figure 2: An example of nonvisual web access (Yahoo! result page)

3. Overview of the system

Figure 3 shows the system architecture. It works on the client side as a front-end processor. It is a memory resident program and can be used at any time by any application. Our system consists of three components: a one-dimensional input method, a component to create an accessible results list, and a navigation method for the list.

3.1 A one-dimensional input method

The first component is a one-dimensional input method for a keyword search. Once the system is loaded, the KeyHook Module observes Windows keyboard events in the text edit controls. When the spacebar is pressed, the module recognizes that a keyword input is complete, and it automatically retrieves information about the word from the Web. Thus an editing process replaces the input box, and pressing the spacebar is a substitute for a submission button. If this were done every time the spacebar is pressed during a text editing process, it would be too intrusive,

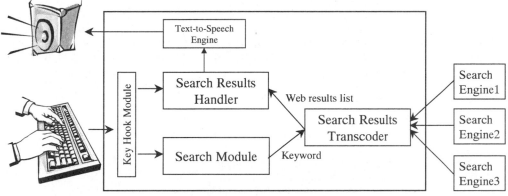

Figure 3: System architecture

Table 1: Navigation commands

Key combination	Function
ALT + CTRL + O	toggle switch for automatic/manual search mode
ALT + CTRL + W	explicit search command
ALT + CTRL + RIGHT/LEFT	move forward/backward between items
ALT + CTRL + HOME/END	jump to the top or bottom of the list
ALT + CTRL + UP/DOWN	move forward or backward between the groups
ALT + CTRL + P	speak current item position
Enter	open link
ALT + CTRL + T	speak total number of the results

so the Search module filters words based on an English parser and picks up proper nouns and unknown words as keywords. When a word like "Mariners" is input followed by the spacebar, as an example, the system starts the search. It is also possible to use this method manually. First, users need to turn off the automatic search mode by using the "ALT + CTRL + O" command (a toggle switch for on/off). Then the cursor should be placed just after the word "Mariners" using a screen reader, so the user can use the search command "ALT + CTRL + W" to start the search. In this case, the search command is the substitute for a submission button. A user also can select words using the ordinary Windows methods such as a key combination of the Shift key and the cursor movement keys, before pressing the search command.

3.2   Creation of an accessible results list

The second component creates an accessible results list suitable for voice output. The Search Results Transcoder works as an intermediary between a server and a user. When the transcoder receives a keyword from the Search Module, it searches for the keyword in the registered Web search engines, creates an accessible results page, and sends it back to the client.

There are several steps to create the list. First, it removes information unrelated to the search results from the results pages, and then consolidates all the results with their links from the different results pages into one list, called the "Web results list". This is possible because each registered engine's page structure has been analyzed in advance. Next it deletes duplicated links from the list, since several engines often return the same hit. In our Web results list, duplicated links are not included, but users can find out which engines returned the same hit, since the system saves a list of the engines' names before the results are consolidated. At this stage, it reorganizes the list. For example, the top-ranked hits from each search engine will all be moved close to the top of the consolidated results list. Finally, it inserts explanations for each link.

Usually, the explanations of the links that are included with search engine results are insufficient for complete understanding, especially using voice output, so the system downloads the destination pages in the list and picks appropriate sentences as the explanation for voice output. The system picks three sentences as the explanation of each link to give users enough information. This process is based on heuristic rules such as using a complete sentence which contains the keyword that appears on the page. Then the Web results list is ready to be sent to the client.

The following list is the Web results list, for "Mariners" as the search string.

1. Link: <u>Seattle Mariners</u> http://www.mariners.org/
    (returned by AltaVista, Direct Hit, Excite, Google, Infoseek, Internet Keyword, WebCrawler)
    Explanation:
    1) Seattle Mariners -- Official Web Site of the Seattle Mariners
    2) Welcome to the official web site of the Seattle Mariners Baseball Club.
    3) This page automatically transports users to the Flash intro of the Seattle Mariners web site.

2. Link: The Mariners' Museum http://www.mariner.org/
    (returned by Excite, Google, Infoseek)
    Explanation:
    1) An overview of Mariners' Museum research library and museum collections.
    2) The Mariners' Museum - Newport News, Virginia
    3) Copyright 1995-2001 by The Mariners' Museum

3. Link: Seattle Times: The Seattle Mariners http://www.seattletimes.com/sports/mariners/
    (returned by AltaVista, Excite, WebCrawler)
    Explanation:
    1) The Mariners signed pitcher Brian Falkenborg to a minor-league contract and invited him to spring training.
    2) Even though their player payroll is already pushing its $80 million limit, the Mariners still might consider expanding their search for run production to include Sammy Sosa.
    3) But the more likely scenario is that the Mariners will try to resuscitate the moribund talks with the San Diego Padres for third baseman Phil Nevin in spring training.

The underlined text is a title of a linked page. In the case of the first hit, "Seattle Mariners" is the title of the destination page, and it is extracted by the system along with the explanation. The title is also presented as a link. The prototype system uses seven search engines: AltaVista, Direct Hit, Excite, Google, Infoseek, Internet Keyword, and WebCrawler. The first hit was returned by all of the registered engines. The system does not read aloud all of the engines' names as written in the example list above. Instead, it will use sound logos as described below.

The second hit is not related to Seattle Mariners. It is really about a museum related to sailors. Users would be able to tell before opening the page that it is not the page they want to visit, since phrases in the explanation, such as museum research library and Newport News, Virginia would give them some hints.

The third hit seems to be an article about the Seattle Mariners. Since the explanations are picked from the article's content, users can easily decide whether it should be opened or not.

3.3   The user interface for accessing the web results list

The third component is the user interface to navigate and read through the results list. The Search Results Handler handles the navigation commands (Table 1) as well as reads the list using the various sound effects. ALT + CTRL + the right or left cursor key is used to move forward or backward between items in the list. An item includes a link and its explanations. ALT + CTRL + the Home or End key is used to jump to the top or bottom of the list. ALT + CTRL + the up or down cursor key is used to move forward or backward between the groups (see below for the definition of "group") in the list. The system can read out the number of results as well as the exact position in the list based on commands.

The system presents additional information related to the current item using sound effects to give users more information besides the voice output produced by the text-to-speech engine. We will describe two examples of the sound effects. First, the eight notes of the musical scale are used to indicate the current item's approximate position in the list. The list is divided into eight equal groups, and each group is assigned to one note. When each result in the first group is read, the "do" sound of the first group is heard. Similarly, when each result in the second group is read, the sound is "re". In this way, the approximate position in the list can be presented to a user, while they are listening to the results. The length of the note represents the current item's position in the group. Second, sound logos are used to represent each registered Web search engine. If an engine site already has one, the system uses the existing sound logo. We create sound logos for search engines that do not have their own yet. After the explanation of an item is read, the sound logos of the engines that returned the hit are played, one after another. In the case of the first hit of the example above, seven different sound logos would be played. In the case of the second and the third hits, three different sound logos would be played, creating a kind of short melody. Users can get used to the little melodies, allowing quick and intuitive recognition of the engines that returned the hits. When a user finds a relevant link in the list, the Enter key is pressed to open it. After a page is loaded, he or she can read it using their usual voice browser.

4. Conclusions

We have been testing this system with some blind users. Even those who have never accessed the Web could search for a keyword and get the results from the Web without any difficulty. Experienced users could track down the exact link they wanted much faster than by using the usual methods. Our system allows blind users to retrieve information from the Web one-dimensionally by means of an ordinary editing process, so they can use it without any stress or training, and the sound effects help users to get additional information while listening to the text information. We will continue investigating the usability of the one-dimensional user interface of this system compared with two-dimensional ordinary user interfaces for nonvisual Web access. We hope to report on concrete comparative results at the next opportunity. Our plan is to extend the features to apply the system to existing Web pages, based on our annotation-based transcoding methods (Asakawa et al., 2000) and Home Page Reader (Asakawa, 1998) experience. If Web pages were presented directly via the "Web results list", blind users would not even need to use Web browsers. They would be able to enjoy surfing the Net nonvisually using the one-dimensional interface of this system. We also plan to adopt this method to locally saved files and dictionaries, so that, it would be possible to provide users a universal one-dimensional user interface for retrieving information about a keyword from any kind of information resource.

References

Asakawa, C. (1998). User Interface of a Home Page Reader. In Proceedings of The Third International ACM Conference on Assistive Technologies ASSETS '98, pp. 149-156.

Asakawa, C, Takagi, H. (2000). Annotation-Based Transcoding for Nonvisual Web Access. In Proceedings of The Fourth International ACM Conference on Assistive Technologies ASSETS 2000, pp. 172-179.

Babylon, Babylon.com Ltd., http://www.babylon.com/

Google, http://www.google.com/

SETI: Search Engine Technology Interface, Agassa Net Technologies Inc and KIA Internet Solutions Inc, http://www.seti-search.com/

Document Reader for the Blind

Chieko Asakawa[a], Hironobu Takagi[b], Takashi Itoh[c]

IBM Japan, Ltd., Tokyo Research Laboratory
1623-14 Shimotsuruma Yamato-shi Kanagawa-ken, 242-8502, Japan
+81 46 273 {4633[a], 4557[b]}
{chie[a], takagih[b]}@jp.ibm.com

[c]IBM Japan, Ltd., Yamato Science Laboratory
1623-14 Shimotsuruma Yamato-shi Kanagawa-ken, 242-8502, Japan
+81 46 273 2636
JL03313@jp.ibm.com

Abstract

These days electronic documents are becoming visual, since authors are paying more attention to visual appearance, so electronic documents are becoming inaccessible to blind users. In this paper, we will describe the user interface of a document reader for the blind, which allow users to navigate through various types of formatted documents with a common user interface.

1. Introduction

Personal computers have played an important role in enabling the blind to access printed documents. Blind users can communicate with others by e-mail and can read and write electronic documents by using screen readers (Jaws) and other assistive technology software (Asakawa, 1998; pwWebSpeak). These days, however, electronic documents are becoming increasingly visual, with various fonts, colors, and illustrations, since authors are paying more attention to visual appearance. These visual characteristics are making electronic documents increasingly inaccessible to blind users. For example, they have a hard time reading such documents with screen readers, and need to know how to operate a variety of applications just to read documents. Consequently, they tend to convert formatted documents into plain text format. However, the logical structure of the sentences and rich text information in the original document is lost by converting formatted documents into plain text.

To solve these kinds of problems, we decided to develop a document reader for the blind. Our system consists of two components. The first component of the system provides a single user interface to navigate through documents formatted by common word processors and presentation software, using the logical structure of the sentences. A numeric keypad is used for the navigation commands. In the second component, the system is capable of navigating through documents with rich text information in addition to the logical structure of the sentences. To realize these functions, our system uses a document object model (Microsoft Developer Network) of each application. In this paper, we will describe the user interface of our document reader. Then, we will offer some conclusions and plans for future.

2. Overview of the System

Our system uses a document object model of each application. It allows external applications to access and update the contents, structure, and style of a document. So our system can extract the contents of a document using the logical structure of the sentences, rich text information and graphical elements without regard to its visual layout.
Figure 1 shows the structure of our document reader. It communicates with each application's object model through the COM interface, and it handles TTS (Text to Speech) engines through an ActiveX○,R control (Microsoft Agent). Three TTS engines are now available with the system: ViaVoice○,R Outloud, L&H, and ProTALKER○,R for Japanese. The TTS engine can be toggled by pressing a key combination on the numeric keypad when the system is active. Each component uses the numeric keypad for the navigation commands. When a Braille pin display is

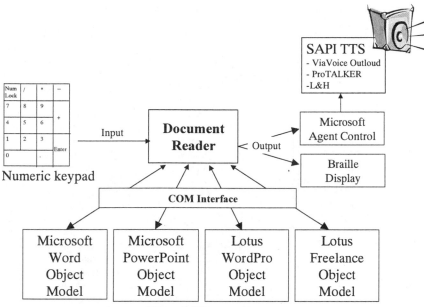

Figure 1: The system structure

connected, the system is also capable of Braille output. In this section, we will describe the user interface of the two components.

2.1 Navigating Through Various Formatted Documents with a Single User Interface

In the first component, the system provides a single user interface to navigate through documents formatted by generally used word processors and presentation software, using the logical structure of the sentences. Currently, our prototype system of this first component allows users to access Microsoft Word and Lotus Word Pro○,R as word processors and Microsoft PowerPoint○,R and Lotus Freelance Graphics○,R as presentation software. (For brevity, the names of the companies producing these programs will be omitted after this.)

Documents should be opened through the Explorer (not Internet Explorer). When a document is opened, it is listed in the document selector of the system. There are two modes while the system is active: document selector mode and document reading mode. When the minus key is pressed, the document selector becomes active. The name of any document in the selector can be pronounced by pressing the up/down cursor keys. When the Enter key is pressed, the document becomes ready for reading.

When a user presses one of the navigation keys on the numeric keypad, the system communicates with an object model and, after getting the appropriate information, sends it to a TTS engine. Figure 2 shows the navigation keys for word processors. It is based on the same basic key assignments as Home Page Reader (Asakawa, 1998), so we assigned the functions for moving in the previous or next direction by columns. The left column, keys 1, 4 and 7 are for moving backwards to the previous item. Keys 2, 5 and 8 are for reading the current position. The right column, keys 3, 6 and 9 are for moving forward to the next item.

We then, assigned the functions for moving based on the logical structure of the sentences by rows. The bottom row, keys 1, 2 and 3 are for moving between pages or headings. Keys 4, 5 and 6 are for moving between paragraphs or sentences. Finally, the top row of keys, 7, 8 and 9, are for moving between words or characters.

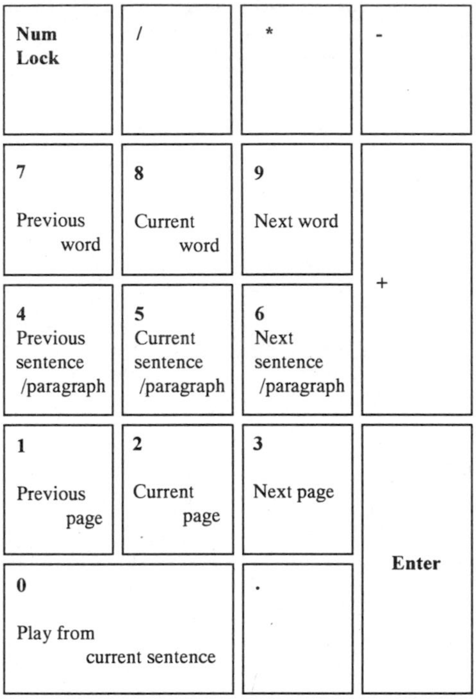

Num Lock	/	*	-
7 Previous word	8 Current word	9 Next word	+
4 Previous sentence /paragraph	5 Current sentence /paragraph	6 Next sentence /paragraph	+
1 Previous page	2 Current page	3 Next page	Enter
0 Play from current sentence	.		Enter

Figure 2: Navigation keys for word processors

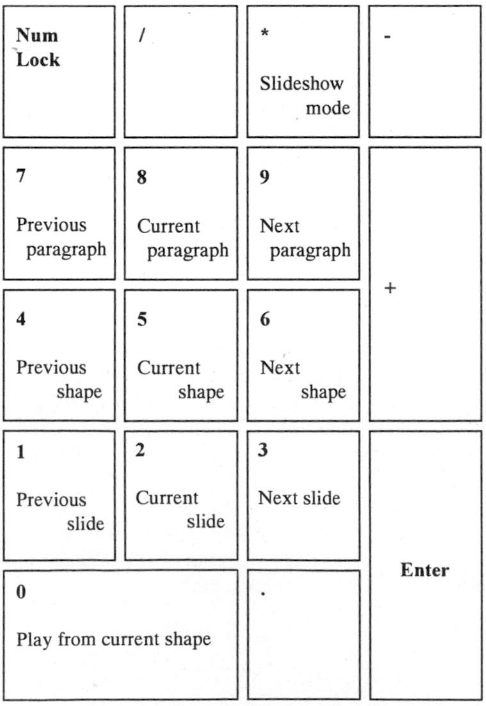

Num Lock	/	* Slideshow mode	-
7 Previous paragraph	8 Current paragraph	9 Next paragraph	+
4 Previous shape	5 Current shape	6 Next shape	+
1 Previous slide	2 Current slide	3 Next slide	Enter
0 Play from current shape	.		Enter

Figure 3: Navigation keys for presentation software

We also assigned extended functions to keys with the combination of the plus key. Pressing the plus key followed by the 1/3 key, for example, can play from the first/last page of a document. Since each application has different characteristics, each object model has different functions. Because of this problem, the system cannot provide a completely universal user interface for various formatted documents. Since there is no function for determining the current character in the Word Pro object model, for example, the system could not provide a character jump key for Word Pro. When a Word Pro document is loaded, the character jump key does not work.

Figure 3 shows the navigation keys for presentation software. The concept of the key assignment is the same as for the word processors. Presentation materials are especially difficult to read through screen readers. Our system reads all the text information contained in a slide, just as if it were a text document. When a page jump key is pressed, it reads the title of a page, so it is easy for users to follow presentations. The reading order in a slide can be toggled to Y order and Z order by pressing the plus key followed by the 2 key. Y order means that it reads from the top to the bottom on the screen. Z order means that it reads from the front most figure towards the back. Since the Freelance object model does not have functions for extracting words and sentences, the system cannot provide the word/sentence jump key for a Freelance document.

2.2 Navigating Through Documents with Rich Text Information

To focus on a representation method for rich text information using voice output, the system uses a single object model, specifically Word's to implement as many useful functions as possible for this purpose. Figure 4 includes the new keys for reading and navigating through documents with rich text information. A key combination with the enter key is used for these commands.

Pressing the enter key followed by 1 or 3 is a font related jump key, such as font size, underline, bold, italic, font face and font color. It can be toggled by pressing the enter key followed by the 2 key. Pressing the enter key followed by 4 or 6 is a paragraph and line related rich text jump key, such as alignment, style, left indent and line spacing. It can be toggled by pressing the enter key followed by the 5 key.

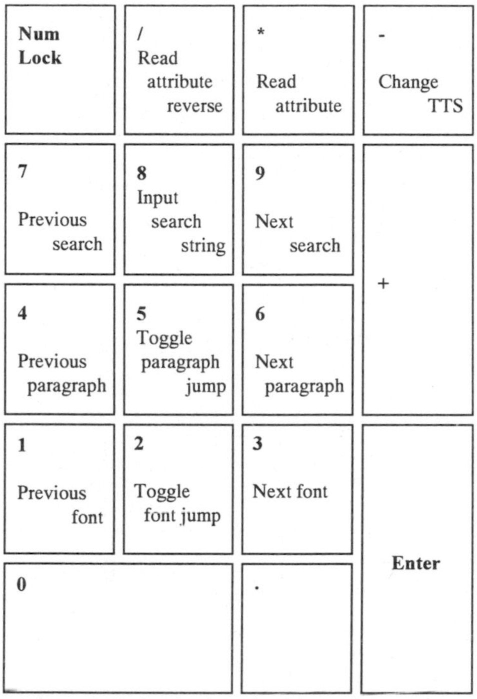

Num Lock	/ Read attribute reverse	* Read attribute	- Change TTS
7 Previous search	8 Input search string	9 Next search	+
4 Previous paragraph	5 Toggle paragraph jump	6 Next paragraph	
1 Previous font	2 Toggle font jump	3 Next font	Enter
0	.		

Figure 4: Navigation keys for rich text information

The system uses both a regular male voice and a high-pitched male voice. The regular male voice is used for the contents and the high-pitched male voice is used for attributes. The announcement of attributes can be set by pressing the plus key followed by the asterisk key.

There are three settings: no announcements, announcing only the word, "change" at the point of an attribute change, and announcements of each attribute change at the point where it is changed. For documents with many attribute shifts, it may be too verbose to read the contents with the attributes. In such cases, users can select "no announcement" since it provides a key combination for reading attributes one by one.

When this key combination is pressed--the enter key followed by the asterisk key--the system reads each attribute in effect at the current location. The system has 15 attributes that can be announced, so users have to press this key combination 15 times to hear all attributes. To avoid this, the system provides another key combination to move backwards through the attribute list--this is the enter key followed by the slash key. In addition, the asterisk key is used for reading all 15 attributes at one time. There are actually more than 15 attributes that can be extracted by the functions of the Word object model, but in the prototype system, we picked the most frequently used attributes, because it might be too confusing to deal with all of them nonvisually.

3. Evaluation

We evaluated each component of our document reader with blind and sighted subjects. Table 1 shows the task list for each component. Each task was tested using three methods: (1) with a screen reader, (2) with our system and (3) with the basic software alone (by sighted subjects). Each task was tested by three subjects. Each subject was an intermediate-level user who can operate a word processor. Before starting the evaluation test, instruction in using our system was given. Table 2 shows the results for the tasks. In the case of the Word document for Task 1, the instructions say that the answer is written in the "Evaluation" section, so it is easy to find it with our system, using the heading jump key. After reaching the section, the subjects can use the sentence jump key to find the answer. Sighted subjects can see the difference between headings and sentences just by looking. Since it was only a 2-page paper, it took them only 18 seconds even if they read through it. But in the case of Task 2, there were 8 pages, so it took 74 seconds on average, because they had to read longer. Our system allows blind subjects to jump to each bold string, so it was very easy for them to find it. All tasks took longest with a screen reader. This is because there is no effective way of navigating through a document using the logical structure of the sentences and rich text information while using a screen reader. After the evaluation test was over, some free comments were offered by the blind subjects. One user said that he has never thought the formatted documents could be read so easily and he doesn't want to convert them to plain text format any more.

Table 1: The task list for evaluation of each component

Task 1	Word documents	Open a 2-page document which has two columns. Find the paragraph which explains Task 2 in the "Evaluation" section. (Without using a string search.)
	PowerPoint slides	Open a presentation package with 20 slides. Find the page which shows the background of the system. In the page, find the number of shapes used. Then find the number of items for the background. (Answer: The 5th slide, 4 shapes and 4 items)
Task 2		Open a 8-page document which has two columns. Find the sentence which is written in bold. (It appears in the 5th page.)

Table 2: The average result of the tasks

| Software | Subject | Task 1 [sec] | | Task 2 [sec] |
		Word	PowerPoint	
Screen Reader	Blind	85	80	245
Document Reader	Blind	38	58	115
Original software	Sighted	18	30	74

4.   Conclusion

Our system allows users to read differently formatted documents with a single user interface using the numeric keypad. It presents the most frequently used functions to users, even computer novices can easily start using it. We hope that in this way, computers will be more useful and convenient for blind users. We will keep testing to determine how much visual information can be converted to nonvisual representations and how much of the converted nonvisual representations can be understood by blind users. We will then study object models for other applications such as spreadsheets, mail software and databases, to provide a consistent user interface for as many applications as possible. Currently, the functions of each object model differ from each other. In the near future, we hope to create a list of the necessary functions for each object model to give them equal accessibility. If all applications' object models are standardized, then accessibility interface applications like our Document Reader can provide truly universal user interfaces for the blind.

References

Asakawa, C. (1998). User Interface of a Home Page Reader. In Proceedings of *The Third International ACM Conference on Assistive Technologies ASSETS '98*, pp. 149-156.

JAWS, Henter-Joyce, Inc.: http://www.hj.com/

Microsoft Developer Network, Office Developer Home, http://msdn.microsoft.com/officedev/default.asp

pwWebSpeak: The Productivity Works, http://www.produworks.com/pwwebspk.htm

From a specialised writing interface created for the disabled, to a predictive interface for all: The VITIPI System

Philippe Boissière, Daniel Dours

IRIT – UPS – CNRS (UMR 5505)
118, Route de Narbonne, F-31062 Toulouse Cedex, France
{boissier, dours}@irit.fr

Abstract

VITIPI's aim is to increase speed text acquisition in all computer applications. Unlike the other systems, it doesn't display lists of unexpected words but provides letters without end-user's intervention. It takes into account previous words for prediction. VITIPI database is made up on texts corpora so that it is well adapted to user's vocabulary and grammar.

VITIPI was conceptualised for speech troubled speakers to increase speedwriting. Writing is essential in the *information society*, especially for troubled speakers who used it to communicate easily. But sometimes, it is also difficult for them to write easily and quickly, so it should be very helpful for them to use a writing assistance tool based on a word predictor treatment.

In a first step, we want to skim through different systems and various approaches, and we will try to point out the VITIPI characteristics, and justify our choices. In a second one, we will show VITIPI principles. Finally as an example, we will give VITIPI last results on meteorological sentences, and explain how it will be helpful for every inexperienced computer user.

1. Introduction

Who can imagine the place that takes the activity writing for persons with disabled or disorders speech? When the spoken mode is substituted for the writing mode, this becomes the only efficient mode of communication. For speakers having speech disorders, writing mode has to be quickly done otherwise the communication can fall down. So the problem clearly appears : How could speed writing be increased?

The performance of the speech recognition systems allows the speech mode to enter quickly text strings [Privat 00]. For users having speech troubles, this alternative communication is not usable. So you have got to use keyboards or virtual keyboards [Piwetz 95] to type your text as quickly as possible. Your keyboard, or virtual keyboard, has to be correctly adapted with special devices to the disabled user [Emiliani 96]. Nevertheless speedwriting is very low.

It should be very helpful for these disabled to have a word predictor system that writes letters (or words) on user's behalf. We shall see on the following 2.2 section that the basic principles of VITIPI system are simple: when a user is starting writing the first letters of a word, the system displays either the ending of a word or a part of it as soon as there is no ambiguity. As long as the word remains incomplete, it goes on writing as many letters as possible and automatically corrects typing or orthographic mistakes in real time, even if the word is unknown to the system.

Who can live in our *information society* without the ability of writing? Everyone is nowadays using computer to write documents, e-mails, and so on. But if you are elder or not master of keyboard, you may write very slowly and you will be disabled too! On the other hand, with new coming media like PDA (Personal Digital Assistant), mobiles, and WAP (Wireless Access Protocol) telephones, the generalised use of computer and automatic teller machine in every task life require the typing mode to interact for everyone. By the use of PDA or WAP devices based on very small keyboards/screens can introduce writing disabilities for ordinary people. So how people with special needs will be able to use these new media?

To answer to the user's needs and to consider the adaptability characteristics of a writing assistance tool, we describe the VITIPI system, as an assistant tool to write for all. In a first step, we want to skim through different systems and various approaches. In a second one, we will show VITIPI principles. Finally as an example, we will give VITIPI results on meteorological sentences, describe how it will be access on every Windows software and/or word processors, how it will be adapted to every disabled, and explain how it will be helpful for every inexperienced computer user.

2. Problematics

2.1 Existing systems

There are different AAC (Augmented and Alternative Communication) tools for writing. One class of these tools is based on abbreviations [Derouault 83] or shorthand code[1]. The limits are that the user has to learn and to memorise a large abbreviation list.

For alternative devices such as "ManyTwo[2]" with Kalido[3] box, user has to press on simultaneous keys, which is not an ergonomic way for gesture disorders. The systems [Hunnicutt 85], [Bertenstam 95] and [Maurel 00] are displaying a list of words on the screen, then the user has to select one of them. This approach is disturbing for the user because he has to read the words list (which could be often out of context) and then he could forget the sentence he wants to write. These systems don't offer function to write a word that does not belong to the list. In consequence, they don't integrate neither orthographic or typographic checking procedures. Another inconvenient concerns the introduction of new words which needs to enter grammatical attributes of words [Godbert 93], [Pasero 98a, 98b].

For severe handicapped communication, alternative communications based on iconographic language can be used like described by [Brangier 97, 00] and [Pino 00]. These are very useful to display quickly usual and daily sentences. Their limitations concern the semantic relation between the concept and the mental representation as well as the vocabulary size and the grammatical possibilities to build sentences.

At the end, we could notice that many word predictors are linked to specialised word processor. If the user wants to use a commercial editor or web browser, he can't use its predictor.

Looking outside AAC devices, we can find a kind of prediction for mobile phones. With mobile phones, each keypad number can display three or four letters. Pressing one time on "1" displays "a" two times on "1" gives "b", and so on. The Nokia 3.110[4], 7 210[5] or Siemens T9[6] Mobile phones allow you to type only once on each keypad. Each time it displays the most likely word it finds in its database. For example, if you want to type : "*easy input*", Nokia successively displays : *e, da, far, easy , easy i, easy in, easy hop, easy inpu, easy input*. It could be noticed that user may be disconcerted with all these changing suggestions. There are no predicted letters even in the ending word. Furthermore, if user makes typing or orthographic mistakes, he can not find its word.

2.2 VITIPI principles

During the specification phase of the VITIPI system, we have tried to keep in mind the running limitations of the previous systems disagreements. The VITIPI's aim is to increase the text writing speed. Unlike the others systems, it does not display lists of unexpected words but displays letters without end-user's intervention. VITIPI knowledge base is made up with user's corpus so that it is well adapted to user's vocabulary and to sentence structure (i.e. grammatical rules and semantic topics). VITIPI system integrates checking, prediction mechanisms for word known and inference for unknown words. The prediction is based on the modelling of the previous context (characters or words). When VITIPI is facing with an unknown or unexpected word containing typing errors or/and orthographic mistakes, unlike the other systems, it can predict the end of the word, thanks to inference procedures [Boissière 96]. The VITIPI system will not be linked to a devoted word processor, it will be able to run on all windows applications.

---

[1] Shorthand for windows : http://www.pcshorthand.com/
[2] Many-two : http://www.egi.fr/CMS/Download/Kalido/ManiTwo/Kalido_ManiTwo.HTML
[3] Kalido : http://www.egi.fr/CMS/RetD/Projets/Kalido/Kalido_D10.html
[4] Nokia 3.210 http://www.nokia.com/phones/3210/serious/predictive.html
[5] Nokia 7.210 http://www.nokia.com/phones/7110/phone/new/predictive.html
[6] Siemens T9 http://www.T9.com

For instance, in a running context of isolated French words (e.g.. without taking into account the word context), VITIPI was able to display *26 % of predicted letters* on a vocabulary size of 5,930 words. *72 % of typing errors* and *75 % of orthographic mistakes were corrected*. Different tests were conducted on isolated word corpora. The main conclusion can be expressed as follows: there is a relation **between the vocabulary size and the ratio prediction** (*predicted letters divided by total letters*). More the vocabulary size grows, more VITIPI's system will be faced with ambiguities, and the ratio prediction goes down. VITIPI system will have to wait for the user's letters to clear up ambiguities. Inversely, more the vocabulary size decreases, more the ratio prediction goes up. If we want to increase the ratio prediction, we have to reduce the vocabulary size without hindering the user. To reduce theses inconveniences, a solution based on the use of contextual linguistic information can be developed. It is important to outline that after a words (characters) succession, a very small set of words (characters) can be predicted. Our system uses this property to build the set of words that could be written at this very instant.

To implement these principles, various procedures and functions were built. One creates from a corpus, an adapted knowledge treelike structure called *transducer*, well adapted to realise letters prediction with isolated words. When system is faced to an unknown or misspelled word, five specialised inferences (for more details, consult [Boissière 90a, 92] were developed based on the transducer implementation. To take into account previous words for prediction, we have added a kind of *N-gram* model [Lesher 99]. Opposite of *N-gram* methodology, we don't compute the number of words, 2-gram, 3-gram, etc... The principle relies on the storage of all *N-gram* (N <= 10) in the same *transducer*: One can consider words of corpus as "*letters of an alphabet*". Sentence is a string of words. If words become "*letters*", then sentence can be seen as a "*word*". Thus, words of the corpus become an input of the transducer. So letter and word predictions are realised by a single structure. One can be feared that this structure quickly grows at an exponential rate. To avoid the increasing size of the structure, minimisation mechanism is automatically done. This procedure can operate on beginning of words or sentences. A procedure that factorises the ending of words and sentences had been developed, it does not introduce unexpected words or sentences in the system. Finally, as for isolated words, specialised inferences were developed to take into account sentences or strings of words that don't belong to corpora. All these procedures and functions were already described in various publications ([Boissière 90a, 90b, 96, 00], the lecture will refer to them for more details.

Now VITIPI is an assistant software compatible to WINDOWS API. Then, it will be considered as a virtual keyboard so that it could be used with every PCs soft (*words processing, spreadsheets, web browsers...*).

3. VITIPI Results

Our aim is to evaluate the VITIPI system when it is faced with unknown situations (*new sentences*), but also to estimate its adaptation abilities. To do this, we daily take the weather forecasting on the French web site of METEOFRANCE [7].

First day weather forecasting text was introduced in the system in order to create the first version of the transducer, which then had been minimised. This first transducer was used for simulate writing of second day weather forecasting. New words were computed with the rate of input letters user typed. First and second days weather forecasting texts were joined to build a second transducer. This second transducer was used for simulate writing of third day weather forecasting and compute the same parameters. The same process was repeated until 19th transducer was obtained and used for simulate writing of *20th* day weather forecasting. Finally, the following results were provided.

When first transducer simulates writing of second day weather forecasting, system displays *16 %* of output letters with isolated words (figure 1). By the opposite, if system takes into account *9* previous words, it displays *17.5 %* of output letters. In the same way, when *19th* transducer simulates writing of *20th* day weather forecasting, system displays *29 %* of output letters with isolated words. On the contrary, when *9* previous words are taken into account, system provides *44 %* of output letters.

---

[7] http://www.meteo.fr/temps/index.html

898

It could be feared that those results are typical to a domain-specific meteorological corpus, and they will be different with a general language. To test it, we have added a basic French lexicon that contains 5,930 useful words. It could be noticed (figure 1) that if we consider isolated words, results are 6 % lower than those obtained without basic lexicon. If we take into account 9 previous words, results obtained with basic lexicon are only *2.5 %* lower than those obtained without lexicon.

If now we use transducers to rewrite their own learning corpus, all sentences are known. When basic lexicon isn't used, results (figure 2) are *35.5 %* with isolated words, on the other hand we have got *77%* with previous words. If now basic lexicon is added, results are *8 %* lower with isolated words, but with 9 previous words they are unchanged.

Taking into account previous words is very useful fore prediction rate. We also have to outline hat if we add a basic lexicon, the results are worse with isolated words than those with previous words. It seams that system with previous words doesn't need basic lexicon.

First VITIPI version that only runs with isolated words, was tested in a specialised French secondary school [Dubus 96], and a match was made with HANDIWORD. For low level pupils (6° 5°), HANDIWORD was preferred because they don't have good mastery of writing and it is easier for them to select displayed words. By the opposite, high level students (from 4° and upper) find VITIPI better than HANDIWORD. They understand VITIPI functions and use them very well. It has been proved that VITIPI increases twice or three times student speed writing.

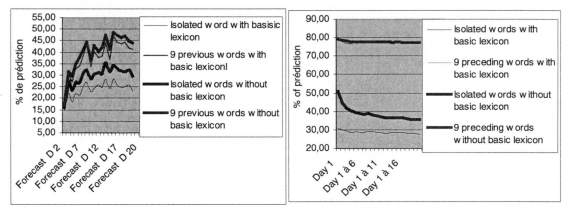

Figure 1: Part of lexicon in prediction rate (with unknown sentences)   Figure 2: Part of lexicon in prediction rate (with known sentences)

4. Conclusion

At the beginning, VITIPI project was built for disabled people. It has been explain in the introduction that this system will be very helpful for everybody. Like inclined planes, remote controls, special devices made for disabled are very helpful and useful for everyone. So we have got to continue to make special devices and to carry research on disabled and elder people.

Acknowledgments
The authors would like thanks to their colleague, N. Vigouroux for their various comments and discussions during the writing process of this paper.

References
[Bertenstam 95] Bertenstam J. Hunnicutt S. Adding Morphology to a Word Predictor. TIDE: The European Context for Assistive Technology, pp. 312-15.

[Boissière 90a]   Boissière Ph. Dours D. Writing-assistance system for disabled persons in a man-machine communication. 5th European Signal Processing Conference, pp. 1651-54.

[Boissière 90b] Boissière Ph. VITIPI : Un système auto-organisationnel pour faciliter le dialogue écrit homme-machine. Thèse d'Université, Laboratoire I.R.I.T (UPS).

[Boissière 92] Boissière Ph. Dours D. De l'aide à l'écriture pour les personnes handicapées à un système d'aide à la communication et au contrôle de l'environnement; Informatique 92 (EC2), pp. 645-56.

[Boissière 96] Boissière Ph. Dours D. VITIPI : Versatile Interpretation of Text Input by Persons with Impairments., 5th rnational Conference on Computers for Handicapped Persons , pp. 165-72.

[Boissière 00] Boissière Ph. Dours D VITIPI : Un système d'aide à l'écriture basé sur un principe d'auto-apprentissage et adapté à tous les handicaps moteurs. IFRATH Handicap 2000, pp. 81-96.

[Brangier 97] Brangier E. Pino P. Le Drezena A. Lamaziere J. Prothèse interactionnelle Pallier les déficits interactionnels des handicapés lourds avec une interface de contrôle d'environnement. LLIA N° 123 Interfaces, pp. 202-06.

[Brangier 00] Brangier E. Gronier G. Conception d'un langage iconique pour grand handicapés moteurs aphasiques. IFRATH Handicap 2.000, pp. 93-100.

[Emilliani 96] Emilliani P.L. Ekberg J. Kouropetroglou. Petrie H. Stephanidis C. The Access Project : Development platform for unified access to enabling environments, 5th national Conference on Computers for Handicapped Persons,.

[Deroualt 83] Derouault A.M. Merialdo B. Stehele J.L. Une expérience de transcription automatique sténotypie français, TSI Vol 2 N°5

[Dubus 96] Dubus N. Evaluation de l'interface intelligente d'aide à la saisie informatique, VIITIPI au lycée "Le Parc Saint-Agne" Journal d'Ergothérapie, MASSON, Mars 1996, pp. 95-100.

[Godbert 93] Goodebert E. Pasero R. La connectivité en langage naturel : Modelisation de contraintes sur le nombre. 13-IEME Conférence Internationale, Intelligence Artificielle Systèmes Experts, Langage Naturel.

[Guenthner 93] Guenthner F. Krüger-Thielmann. Pasero R. Sabatier P. Communication aids for handicapped persons. 2nd European Conference on the Advancement of Rehabilitation Technology, ECART'93.

[Hunnicutt 85] Hunnicutt S, A lexical prediction for a text-to-speech system. Report of Speech Dept Communication STOCKHOLM STL-QSPR 2-1/1985.

[Lesher 99] Lesher, G.W., Moulton, B.J., & Higginbotham, DJ (1999). Effects of ngram order and training text size on word prediction. Proceedings of the RESNA '99 Annual Conference, Arlington, pp. 52-4, VA: RESNA Press.

[Maurel 00] Maurel D. HandiAS : Aider la communication en facilitant la saisie rapide de textes. IFRATH Handicap 2000 , pp. 87-92.

[Pasero 98a] Pasero R. Sabatier P. Linguistic games for language Learning and tests, an ILLICO application. Computer-Assisted Language Learning (CALL)

[Pasero 98b] Pasero R. Sabatier P. Concurrent Processing for Sentences Analysis, Synthesis and Guided Composition. Natural Language Understanding and computational Logic, Lecture Notes in Computer Science, Springer.

[Pino 00] Pino P. Brangier E. Environnement Digital de Télécommunication : Adaptation automatique du temps de défilement aux caractéristiques et intentions de l'utilisateur. IFRATH Handicap 2000, pp. 125-30.

[Piwetz 95] Piwetz Ch. Eiffert F. Heck HH. Müller-Clostermann B. An Adjustable User Interfaces Providing Transparent Access to Application Programs for the Physically Disabled. Sigcaph Newsletter ACM PRESS N° 51, pp. 11-6.

[Privat 00] Privat R. Vigouroux N. Conception de systèmes multimodaux de consultation de serveurs d'informations par et pour les personnes âgées. Ergo-IHM 2000, Rencontres Doctorales.

Standards in Mainstream and Adaptive Interface Design Required for Efficient Switch User Access

David Colven, Andrew Lysley

ACE Centre Advisory Trust
92 Windmill Road, Oxford OX3 7DR, UK
Tel +44 1865 759800, Fax+44 1856 759810,
E-mail info@ace-centre.org.uk, www.ace-centre.org.uk

Abstract

Access to Windows and Internet environments is now almost essential for learning and employment. Many of the interfaces currently available are not universally accessible by people with disabilities. It is not reasonable to expect 'mainstream' software engineers to accommodate extreme accessing needs. However, properly adapted mainstream applications can provide disabled people with powerful tools for independence. There is therefore, a requirement for an alternative means of operating these systems. It is also no longer acceptable or necessary to use conventional, slow emulation of keyboard and mouse controls for doing this. It is important for people who may only be able to use one switch to minimum effort for maximum effect. Just the task of opening up an application and re-sizing it so that on-screen keyboard and work area are conveniently placed is considerable for a switch user.

This paper will demonstrate how such systems can be constructed for people with severe physical disabilities. At the ACE Centre we have been developing a shareware tool, SAW (Switch Access to Windows) as a design environment for switch user interfaces. This is an object oriented interface creation utility to enable switch users access all Windows functionality. Arising from our experiences of the SAW design process we propose standards that should be applied to 'mainstream' interfaces and the suggest the requirements made of the operating system which are required for such adaptations to work efficiently. We stress the design requirements, development methodology, and the role of the designer in creating alternative custom ergonomic solutions indeed, switch users can design their own interfaces. Examples of successful outcomes will be demonstrated.

For the future further European projects such as Comlink (a Java based communication and control software construction and configuration utility) have extended the functionality created in SAW to multi-lingual and multi-modal communication and control environments. WWAAC (WorldWide Augmentative and Alternative Communication) is a European project initiative to make web and email-based technology more accessible to people with communication, language and/or cognitive impairment. WWAAC will use an extended version of Comlink to expand the interface capabilities to symbol based electronic communication. The progress of these two projects and their influence on interfaces for augmentative and alternative communicators will be reported

1. Introduction

The ACE Centre has been concerned with providing tools for communication to children with physical and language difficulties since 1984. At the beginning only what now seem crude devices were available, restricting the sort of activities these children could access. These included Apple II and BBC 6502 based microcomputers, typewriters and simple communication aids. Despite these limitations many groundbreaking communication systems were developed, programs that compare well even with those available on the most modern systems.

1.1 Challenges of windowing environments

The advent of Macintosh, Acorn RiscOS and Microsoft Windows operating systems promised to open many more opportunities for our client children, the users. In particular, because these systems were multi-tasking it offered the hope that powerful mainstream applications would be available, giving these users widened horizons to a whole new world of creativity, information and communication.

Given the existence of legislation designed to protect the rights of disabled people in areas of North America and Western Europe, it is surprising that more forethought has not yet been given to alternative and appropriate access Information and Communications Technology (ICT). For example, many multimedia resources do not provide keystroke equivalents for mouse actions, buttons, 'hot' keys, menus and functions. This makes life for the disabled user very difficult. In some applications recommended standard system calls to the keyboard are not made hampering alternative access. The 'standard' user interface designs of all operating systems and mainstream applications often assume that it doesn't really matter where menus and buttons are placed on the screen so that buttons and 'hot' areas that are repeatedly used on different screens are rarely placed at consistent points where a switch driven pointer can easily target them. All in all this makes switch access to multimedia slow, laborious and even impossible.

1.2   Standard Adaptations

The accessibility options available in both Microsoft Windows and MacOS do give many users with certain types of physical, hearing and sight problems greater access to these systems and both Microsoft and Apple have given time and effort into opening up their software to many more people. However, there still remains a core of people with severe problems for whom access to keyboard and mouse calls needs wholesale re-design.

Key issues relating to effective access in ICT have thus emerged over the years since the advent of Windows 3 and before that the Macintosh and RiscOS operating systems. The first of these is the fundamental assumption that everybody can and wishes to control a graphical user interface exclusively with a mouse or trackball. This needs to be challenged.

2.   The switch users and their interfaces

The experience of the ACE Centre is focused on children and young people. 60% of the children we see have cerebral palsy, the majority of whom have severe athetosis and are switch users.

Most switch users access computers by means of scanning systems. These can be external devices but are now more commonly on-screen scanning keyboards or selection sets. The term "selection set" is used because it describes a collection of items on the screen that can be selected from. (At first they would use single switch with cause and effect programs. Over time, experience and improvements in muscle tone some children develop so that the scan can become faster and less tiring to use as a single switch.. However, this over-simplification hides a range of needs that can be helped by quite subtle differentials in switch interfacing.)

Fig 1: A switch user at work

2.1   Switch Access

Using one or two switches to do anything on a computer is very slow and laborious. This is particularly true for mouse pointer control. All actions have to be chosen from a selection set of items. A selection is made from a scan which moves over an array of items and items can be selected, scanning items contained within it, or to causing input (via keyboard or mouse buffers or system calls). In general, selection sets are constructed as layered hierarchies of objects. At the lowest level there are items which have a range of properties that control their interactions with the software environment.

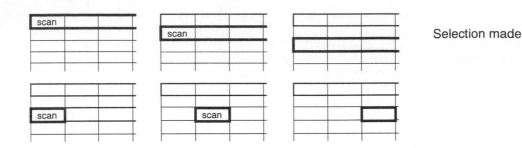

Fig 2: Switch row/column scanning

A single item in a selection set can contain orthographic and/or symbol displays, output data including keystroke and sound, operating system commands and controls that modify the interface – for example calling up a new selection set. The layout of items within sets is most important (Venkatagiri, 1999). Items themselves can contain sub-items that can form a hierarchy of selections. Both Comlink (Colven et al., 1996) and SAW are component-based development systems. The advantage of flexible component-based architectures is that flexible and complex systems can be constructed which can simplify many tasks for the user.

2.2   Switch control

Usually a single switch is used for automatic scanning (autoscan) such selections sets and two switches are used for user controlled (step) scanning although there are many variations. Single switch autoscan is the most efficient method for most users as long as they have developed good timing The scan proceeds at the scan rate, set by the system controls until a press event is received. Step scanning gives more control to the user. However, this demands that the user physically access and control two switches reliably. Full descriptions of these and other scanning options are described in detail elsewhere. (Colven et al., 1990)

2.3   The designers role

The role of the interface designer is to match the needs of the severely disabled user to the capabilities of a range of applications, taking into account their intellectual and physical abilities. This normally requires the interface of the applications themselves to be compliant with the requirements of the switch input utility.

The main requirements of the applications' interfaces in priority order are:
1. Direct keyboard shortcuts for all functions
2. All menu items to have single letter keyboard shortcuts
3. Menu access to all functions (through popup dialogues as necessary)
4. Mouse click only functions to be consistently placed on the screen
5. All mouse click 'hotspots' to be visible

2.4   Customizing the switch user's interface

Utilities like SAW (Switch Access to Windows) provide flexibility to developers of interface systems for people with a broad range of disabilities. They provide versatile mouse and keyboard interface design tools. The versatility of SAW, however, gives greater scope for producing efficient and custom designed user interfaces for those who use switches or pointing devices. The lessons learned from SAW have been incorporated in the interface designer Comlink (Comspec) (Colven et al., 1996)

2.5   Selection set design - The bond between designer and user

The relationship between the designer and the switch user in the interface design process is critical and often unique to each user. They should work in partnership to produce flexible, fast and friendly access to a variety of software and hardware environments including multimedia and the Internet. This bond between the designer and the end user is intrinsic to the heuristic and iterative evolution and successful outcome of switch interfacing. The central tenet of the SAW (Switch Access to Windows) and Comlink Projects is that designers of user interfaces and end-users should be able, with the help of simple design tools, work together to create ergonomic and easy to use interfaces. Indeed utilities like SAW can act as 'bolt-on' wizards to any application.

Looking to the future the modular Java based development system Comlink is to be used in the WWAAC project www.wwaac.org to implement modified interfaces to Web based information retrieval, E-mail and e-chat for users who are unable to use orthographic systems. These users need to access written materials by means of symbols or icons only.

3. Case study - Elaine

Elaine is 32. She has severe athetoid cerebral palsy and no speech. Elaine lives at home with her parents and attends a day centre three times a week. Although she does not enjoy paid employment, she is a writer, poet and journalist. She has used computer technology for writing since 1983. Elaine accesses her IBM laptop computer via two foot switches at a scanning rate of 0.3 seconds per scan. She makes very few errors in selection despite her rapid rate.

Her Windows interface consists of a SAW menu with two main application items, 'Word' and 'Internet'. From this she can either access a suite of Word switch selection sets or a sub-menu consisting of 'E-mail' and 'The WEB'.

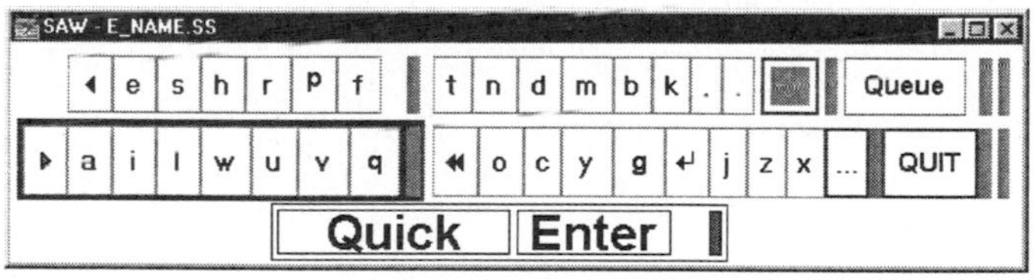

Fig 4: An ergonomic layout for addressing E-mails.

Fig 3: Set for accessing the Web

Selecting E-mail opens up a modified interface to Eudora tools enabling her to read or send her mail, write new messages or reply to old ones. Selecting the WEB on the other hand opens the door to Internet Explorer. Elaine is a big fan of Rod Stuart (her only real problem!) and has visited many of the 250 Rod Stuart sites, downloading vast quantities of information relating to her hero.

4. Case study - Sandip

Sandip is 25. He too has severe athetoid cerebral palsy and no speech. English is his second language. He has completed an economics degree. In contrast to Elaine he uses a single pull/push switch which he accesses with his hand at a scanning rate of 0.2 seconds per scan. Sandip uses a SAW menu with Microsoft Office and Internet applications. He is an avid E-mail user and often uses it to ask for technical information and training relating to SAW.

"As I find it difficult to talk and I cannot write the Internet has benefited me enormously. With E-mail, I can communicate to anyone and anywhere in the world without my speech getting in the way. ..... E-mail has given me the ability to 'say' the things how I want. I cannot keep information very confidential over the phone as they (my helpers) usually speak for me . When I have written letters it's only natural for a person to glance over my letters when they put them in envelopes. With E-mail I can send the messages with no physical help which means I keep things confidential."

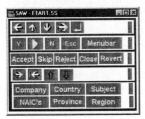

For Sandip access to the Internet is no longer a novelty, rather it is an essential tool for communicating and retrieving information. He has recently started investigating using a specially designed interface (fig 5) so that he can start work for the Financial Times World Reporter as a reviewer. A custom designed system has been created for him to carry out this task efficiently and effectively.

Fig 5

References

Venkatagiri y, H. (1999). Efficient Keyboard Layouts for Sequential Access. *AAC*, 15(2), 126-128.

Colven, D., Detheridge, T. (1990). *A Common Technology for Switch Controlled Software*. NCET ISBN 1 85379 112 1.

Colven, D., Lundälv, M. (1996). From SAW to Comspec. *Med. Eng. Physics* 19 (3), 181-186.

Hierarchical web frame navigation for the visually impaired

Tsuyoshi Ebina[a] , Teruhisa Miyake[b]

[a] Communications Systems Section, Communications Research Laboratory
4-2-1,Nukuikita-machi, Koganei, Tokyo, 184-8795, Japan
[b] Division for the Visually Impaired, Tsukuba college of Technology
4-12, Kasuga, Tsukuba, Ibaraki, 305, Japan

Abstract

This paper focuses upon the improvement of non-visual frame navigation for visually impaired people. All frame components of a web page are analyzed with regard to frame reference, and the frame structure is represented in a hierarchical frame diagram easily understood by the visually impaired. Combining frames also reduces the number of frame-switching operations required for navigation. A prototype implementation of the proposed methodology successfully analyzed 21 of 25 web pages.

1. Introduction

Web page simplification researches are being attempted (Ebina et al., 2000a, 2000b; Zajicek et al., 1998). Web frames, which unify some parts of web pages and combine them in a single view, are often used to transform a complex visual structure into a simple visual representation. The frame-based web page, however, is sometimes too complex to be understood by visually impaired people. This is because the contents of frames are laid out distributively and users have to perceive the frame structure. Another reason is that speech-based web navigation software focuses on only one frame at a time and visually impaired users have to switch focuses frequently. And because the links in a frame often update the components of other frames, navigation cannot be made easier by simply selecting a frame. This paper describes two ways to improve non-visual frame access, making it easier for the visually impaired to navigate the web.

The first improvement is a frame structure analysis that uses a web reference diagram to represent frame-based page structure. This diagram represents the relationships between frames. Some parts of the contents of different frames are related each other and thus will constitute the main theme of the web site. If some of the contents of a frame are not related to the other content on the page that frame is on, those isolated parts are not important. Perhaps they are simply advertisement icons or page title logos. The network automatically extracted by this analysis represents the hierarchy of the frames on a page page, and this hierarchical representation is one that is perceived more easily by visually impaired users. It lets them understand the frame structure without actually reading all the content on a page.

The second improvement is a mechanism that simplifies frame switching. When people browse a frame-based web page, they have to switch frame focus often. The introduced mechanism enables users to access multiple frames without switching focus.

2. Nonvisual web frame accessibility

Web accessibility standardization efforts are in progress and the World Wide Web Consortium (W3C) has presented a guideline for assuring accessibility by visually impaired people (World Wide Web Consortium, 1999). For example, the guideline directs web page designers to "Title each frame to facilitate frame identification and navigation" and gives this strategy the highest priority. Accessibility would be greatly improved if web pages satisfied this requirement, but many web sites do not have useful titles on their frames. When there are no titles on a frame, visually impaired people usually infer the frame contents from the attributes of the frame names. Any screen reader or self-voicing browser has frame navigation facilities, but their functions are limited and insufficient. This software can tell visually impaired users what the general content of a frame is but cannot follow multiple events in multiple frames. These readers and browsers can follow only one frame at a time (Paciello, 2000). The users have to infer each frame's role from the title and name they are given. Home Page Reader (Asakawa et al., 1998) reads the names and contents of the frames on the web page, and users can often infer the role of each frame from the attributes of the given name. For example, they will understand that the frame given the name "menu" will be used as a menu on the web page. But web page designers do not always give each frame an appropriate name, and page structure cannot be inferred from a set of meaningless formal labels such as 1.htm, 2.htm, and 3.htm. A page structure overview function for the visually impaired is therefore needed.

Frame-based web pages have *menu* frames, and people change views by clicking an anchor on the *menu* frame. But because users are irritated when they have to switch frames frequently, frame unification techniques are also needed.

3. Web frame diagram

The web frame diagram introduced in this section represents frame reference relationships.

First, all the components of the frames on the page are extracted. The frame components shown in Figure 1, for example, will be grouped into three frames. These frames are then transformed into a tree explanation like that shown in Figure 2. The frame in Figure 2 can easily be understood from the set of frame names, but this figure does not represent frame structure in a way that shows inter-frame reference. Inter-frame relationships as well as the set of frame names have to be represented.

The web frame diagram described in this paper produces an easy-to-understand frame structure and presents this structure into users. It is made by analyzing the link tags on all web frames and creating a web frame transition diagram. For example, the following anchor updates the contents in the frame *main*.

<a href = "http://www.crl.go.jp/topics/1.html", target = "main">

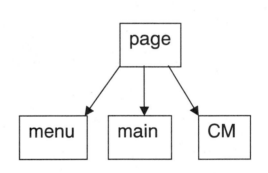

| Figure 1. Frame-based web page Explanation | Figure 2. Tree-based frame Explanation |

If the target attribute on an anchor refers to another frame the frame containing that anchor is regarded as a *menu*, which is a collection of the links on that page. Therefore the menu frame and the main frame are combined. The web frame diagram thus obtained is shown in Figure 3, which shows that the index frame works as a menu and that anchors in index frame affect the contents of the main frame. Note that no anchors in the main frame affect the index frame—except for _blank and _top, which trigger the frame reset procedure.

The web frame diagram described in this paper also eliminates isolated frames. A menu frame refers to other frames but an ad frame, for example, neither refers to other frames nor is referred to by other frames. Therefore the web frame diagram will generate the following result:

"There are three frames on the page – a main frame, a menu frame, and an ad frame. The menu frame updates the contents of the main frame. The ad frame is isolated from other frames. All frames contain _top link tags. Both the main frame and the menu frame contain _blank tags"

Visually impaired users can easily understand this message and understand the page structure before they access the contents.

4. Hierarchical web frame representation

Web speech software reads only one frame at a time, and a frame-switching operation is needed. For example, sighted users select a category from the menu frame and get relevant contents reflected on the main frame. If they want to view the contents associated with another category, they select a link from the menu frame and change their eye focus to the main frame. But visually impaired people have to switch the focused frame more often, and this frame-switching task is often so complicated that people are too frustrated to use the frame-based web page. Figure 4 illustrates an example of the repeated focus-switching operation.

The proposed frame accessibility improvement combines related frames. The menu frame refers to the main frame, so these two frames are combined. The focus is first set on the menu frame so that users can get an appropriate overview of the page. Then the focus is moved to the main frame so that users can navigate the *main* frame contents.

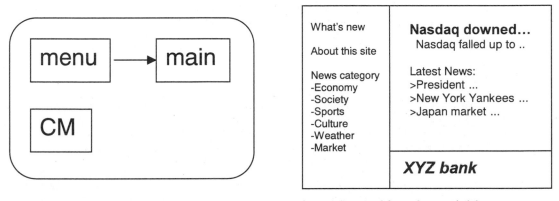

| Figure 3. Web frame diagram | Figure 4. Repeated frame focus switch between menu and main frames |

A pseudo-cursor is used in order to read contents without change cursor position. The pseudo-cursor was originally used to improve the accessibility of text-based screen access systems. It is mainly used to search text far from the cursor while editing. But the functionality of the previous pseudo-cursor is limited to text editing tasks.

Visually impaired users often have to switch focus into the menu frame simply to update *main* frame contents. Therefore the pseudo-cursor is set on the menu frame. While they are focusing on the main frame (and reading the main frame contents), they are also able to read a menu frame by using the pseudo-cursor. When they want to switch content categories, they simply select the appropriate anchor on the pseudo-cursor. This reduces the need for frame-switching and speeds navigation.

The proposed frame navigation algorithm automatically combines a menu frame and other referred frames by analyzing anchor tags. Users do not have to infer the role of each frame from the given name and the title. By using the pseudo-cursor they can select menu links while they are focusing on another frame.

5. Frame structure analysis on web sites

Frame-based pages are so widely various in structure that some of them may not improved by the proposed heuristics. We therefore investigated the characteristics of existing frame page design. Selecting 25 frame-based home pages from famous web sites and then extracting and analyzing their frame structures, we found that 15 of them contained only two frames. These home pages were of two types: those with no references between two frames, and those with hierarchical relations between the two frames. The frame components of the first type of pages were a main frame and an ad frame, isolated from each other. In this case, the frame with the larger content is extracted as the main frame. When the proposed methodology is used with this kind of frame structure, the isolated frame is eliminated.

In the web frames with a hierarchical structure, one frame is used as index while the other is used as the main frame. Here the proposed methodology improves the frame accessibility by reducing the number of frame switch operations required.

The structure of web pages with more than three frames is a mixture of the above frame structures. The most complicated of the frame-based pages we investigated was http://www.tv-asahi.co.jp/, which is composed of six frames.

The three frames *left, bangumi,* and *main* constitute a hierarchical network, and the contents of these three frames make up the main topic of the page. Unfortunately, the *main* frame itself has two menu nodes: *left* and *bangumi*. In the current prototype implementation of the proposed methodology, a menu frame is not uniquely identified. Users have to decide which of the *left* and *bangumi* elements should form the menu frame, but this problem will be improved by facility extension. The contents of three isolated frames—*top, yajiuma,* and *ad*—are not part of the main topic of this page.

Without the proposed methodology, a user would jump into the first frame, *top*, which has no link path, and would thus be isolated from the other frames. With the proposed methodology, however, visually impaired users could avoid reading the three isolated frames, which are not part of the main topic. Frame-switching operation would also be simplified, cutting down on total page access time.

Analyzing the isolated frames in these 25 web sites, we found that they were used as advertisement message icons or title logos. Thus isolated frames are relatively unimportant from the viewpoint of page contents.

Twenty-two of the sites were processed appropriately, no main topic was extracted from three sites, and no *menu* frame was extracted from the other.

6. Discussion

Frame diagnosis prior to actual reading is useful when visually impaired people access a web page. Previous page frame access systems did not analyze the frame structure, and visually impaired users had to infer page structure from the name attributes. The page analysis system described in this paper, however, generates simple inter-frame relationships that will be helpful even if the users do not actually read each web frame.

The proposed analysis is also useful while the user is reading the web page. For example, if an anchor's target attribute is _blank, the current frame structure is extended. The conventional web access software does not read target attributes, so the user may be confused about where the current focus is. The proposed analysis helps them understand the frame structure. And as noted in the previous section, the pseudo-cursor reduces the number of frame-switching operations dramatically.

7. Conclusions

This paper introduced frame accessibility improvements using a web frame diagram that translates frame structure into representation more easily understood by visually impaired people. It also identifies isolated frames not important to the content of a page. And the use of a pseudo-cursor enables users to access multiple frames even when they are not focusing on a frame. The pseudo-cursor also reduces the number of frame-switching operations required.

Twenty-one of 25 web sites were processed successfully, and menu frames were extracted from all but one of the 25 sites. Web frame accessibility needs to be further improved, and we are going to extend the proposed method to handle a greater variety of web frames.

Refrences

Asakawa, C., Itoh, T. (1998). User Interface of Home Page Reader. In *Proceedings of Assistive Technologies 98* (pp. 149-156). ACM press

Ebina, T., Igi,S., Miyake,T. (2000a). Fast Web by Using Updated Content Extraction and a Bookmark Facility. In *Proceedings of Assistive Technologies 2000* (pp. 64-71). ACM press.

Ebina, T., Igi, S., Miyake, T. (2000b). Annotation-Based Transcoding for Nonvisual Web Access. In *Proceedings of Assistive Technologies 2000* (pp. 172-179). ACM press.

Paciello, G.M.(2000). *Web Accessibility for People with Disabilities* (pp. 79). CMP books.

World Wide Web Consortium (1999, May). *Web Content Accessibility Guidelines 1.0.* [On-line]. Available: http://www.w3.org/TR/1999/WAI-WEBCONTENT-19990505/wai-pageauth.html#tech-frame-titles

Zajicek, M., Powell, C., Reeves, C. (1998). A Web Navigation Tool for the Blind. In *Proceedings of Assistive Technologies 98* (pp. 204-206). ACM press.

A Taxonomical Approach to Special Needs Design in HCI

William H. Edmondson

School of Computer Science, University of Birmingham, Edgbaston, Birmingham B15 2TT, UK
w.h.edmondson@cs.bham.ac.uk

Abstract

The paper contrasts a list-based approach to interface design with a structured, analytical, taxonomic approach. The taxonomy employed categorizes behaviour in 'orthogonal' terms: manipulation, selection, symbol mediation. It is argued that the taxonomical approach permits more coherent design responses to the needs of special users.

1. Introduction

Conventional approaches to grouping styles of interfaces seem obvious. Typically one finds lists of interfaces such as: command line, WIMP, form-fill-in, direct manipulation, speech, WYSIWYG, and etc. The HCI student finds these terms in textbooks (e.g. Shneiderman, 1998), and the user meets them in advertising material. The difficulty for the cognitive scientist and engineer is that such a listing has no obvious structure and no obvious implications. The designer gains no insights from the list. Attention to users with special needs is therefore most likely to be expressed in technical terms – modification of interface technology, provision of special hardware, tailorability, and so forth. This too is found in list format, for example in Bergman and Johnson (2000). Users are also described in list format. There are power users, experts, intermittent users, intermediate users, and novice users. This list is similarly unhelpful, and tends not to include users with special needs. But we can easily elaborate the difficulties, and in so doing we lay the groundwork for returning to the list of interface styles with an analytic perspective. The descriptive list of categories of interface styles can be transformed into a taxonomic approach to interaction with computers, which in turn unlocks discussion of the issues that face special groups of users, and allows us to address the topic of widened access to information technology.

2. Users

The need for careful analysis rather than simplistic listing is revealed when we look at the detail. The listing above arranges users in order of their level of expertise with a computer. The designer is supposed to be attentive to the needs of novices and experts, and to provide pathways for migration from one level of experience to the other. But the notion of expertise is so confused as to render the designer's task unnecessarily difficult.

2.1 System experience

Users of computer systems will have varying experience of such systems. All users will have used a system for the first time, but this might happen in childhood in some cultures, and adulthood in others, as part of instruction, or in self-directed exploration. The factors that shape this initial experience include: economic (personal and national), culture, health, physical capability, motivation and education, as well as age. The experience will become part of a user's world model of technology, but it might also be informed or guided by a lifetime's prior experiences. One difficulty, therefore, with a 'levels of expertise' approach is that where this relates to experiences with computers it is virtually impossible to categorize users. The purpose of such a categorization is to provide comparability between users as well as to guide provision of specific technologies for specific users. Even a simple measure may be inadequate – does the number of hours a child spends playing 'computer games' contribute to overall experience of computers? Does the deep cultural/domestic exposure to the MiniTel system in France mean that the French are particularly advantaged? The variability of experience/exposure identified here will continue to be encountered by the developers of systems and applications – maybe for decades to come.

2.2 Application experience

Users of specific applications, or even classes of applications, will be similarly varied. Despite the marketplace dominance of some products, users will have varied experiences with regard to, say, word processing. These differences will not just be superficial differences with respect to, for example, which keystrokes are required to achieve specific effects. Conceptions of document structure, formatting, and so forth will vary in subtle ways between products, and even between versions of products. Categorization of users will be difficult.

Additionally, of course, there is the tricky issue of assessing expertise in the case of a long-term and frequent user of a word processing application (say, Microsoft Word) who then tries to use a presentation system (say, Powerpoint) for the first time. Are there generics of application usage which become learned and exploited to speed one's encounter with a new application? How can this be possible on the basis of exposure to one application?

2.3 Domain experience

From time to time a designer will have to work with groups of users who have experience of a specific domain irrespective of any encounters with a computerized interpretation. One example might be in a machine workshop where metal parts are fashioned using lathes, drills, milling machines, etc. Increasingly such machines are computerized, and operators are less and less likely to have been through apprenticeships. This means that within a group of users of such modern machines one can find varying levels of domain experience – where the domain is not defined by the capabilities of the computer controlled machines. Likewise, experienced typesetters have a wealth of domain experience when they encounter computerized document preparation/printing systems for the first time, novice typesetters less so. In what ways, then, are such people (in)experienced?

2.4 Special needs experience

We cannot avoid the reality that for people with special needs the world of technology does not always look like successful special purpose provision. All too frequently the specialist needs user will have a wealth of experience in finding 'workarounds' for problems they encounter. They bring this specialist experience to new technology, and it is sometimes possible that radical developments in technology can temporarily disadvantage such a user by both rendering their specialist skills unusable, and posing new learning tasks.

2.5 Summary

We have seen that 'obvious' descriptive schemes may be unhelpful. The problem is that the supposedly useful categorization of users actually makes it impossible to recognize the reality and the detail both of the varied experiences of all users and the special worlds of special needs users. It is rightly and widely argued that better design helps everyone, not just the special needs user (cf. the example of kerb ramps and the discussion in Bergman and Johnson, 1995). However, designers cannot produce better design just by encouragement; they need analytical frameworks and insights. The interaction taxonomy briefly described below offers a more structured account of interaction behaviour within which the needs of special users can be understood. But as in so many cases, one has to work at insights, and sometimes the value of examples has to be dug out. To return to the example of the kerb-ramp – the problem is seen as that of providing mobility for people in wheelchairs, not on foot. Actually, the reality is that kerbs are provided to stop wheeled vehicles getting too close to pedestrians; they are intended as obstacles to wheels, not people. In fact there are many occasions when wheels need to move from one space to the other – not just the wheels of wheelchairs. The reason that the ramps turn out to be so useful is precisely the generality of the situation. However, now in Europe one is increasingly finding an adaptation made for blind users – the surface of the ramp and adjacent pedestrian walkway is being replaced with special bumpy/irregular material so that the presence of the ramp can be detected by foot. This is fine – but it can interfere with wheeled use of the ramp (wheel-chairs, child/infant mobility devices, shopping trolleys, emergency wheeled trolleys in accident situations, load-carrying trolleys, cyclists, etc.). The adaptation for the blind is not thought through – other solutions are possible, but this is not the place to discuss them.

3. Interaction behaviour

We have used the obvious example of user categories to illustrate the more general problem we need to tackle: simple listing of so-called interface styles will not yield useful generalizations. Additionally, we should be aware of the need to get at the underlying problem(s) in order to appraise solutions. However, unlike the difficulty of categorizing user experience it is possible to establish a *categorization system* for interaction. The categorization system discussed below was originally proposed by Edmondson (1993) (see also Edmondson , 2000). It is a three layer system, which is systematic and as such constitutes a taxonomical or structured system, not just a collection of names or technologies.

The notion is that there are three 'orthogonal' types of behaviour which humans use widely, and three aspects of interaction which need to be considered when people use technologies. Although the scheme needs refinement, it is offered here because it contributes to clarification of the needs of specialist users in general (as will be demonstrated). It is important that we strive for this, else provision for specialist users will simply constitute an enormous catalogue of idiosyncratic provisions. The aspects of interaction can be considered as layers – and these are *underlying, observable*, and *superficial*. The three 'orthogonal' behaviours are *manipulation, selection*, and *symbolic*. We will first summarize the classification of behaviours.

3.1 Behaviours

The scheme proposed sets out manipulation, selection and symbolic mediation as three essentially different ways of behaving with objects and artefacts to achieve changes in the world. Briefly, these are described as follows:

Manipulation. The focus here is on physical manipulation, movement (and monitoring of movement), and sensory feedback. The activity is closely coupled to the actor in time – one moves one's hand to do something and the monitoring, feedback, evaluation, etc. are directly and immediately engaged as part of the on-going process. The person's concern can be with both static and dynamic aspects of the world – and the original sense of the word manipulation, pertaining to the handling of things, is appropriate.

Selection. People select things, values, locations, etc., on the basis of known or perceived characteristics, often with reliance on recognition. Characteristics can be logically appraised: mutually exclusive properties have to be recognized as such, likewise mutually compatible properties. For example, a typeface can be bold and italic, but not Times and Courier. When one makes a selection (cognitive process) one may also have to express that selection somehow – and this involves some other activity that is not part of selecting as such. Selections may be expressed slowly, and there may be a delay between making a selection and expressing it.

Symbolic mediation. Manipulation and selection are fundamentally different from each other and both are fundamentally different from symbolically mediated behaviour. This latter comes in four flavours:

4. specification/command behaviour. In the context of HCI, this is clearly seen in the UNIX operating system, where any command issued is a specification of the desired state of the machine (and maybe details of how to get to that state).

5. description/instruction behaviour is most often seen in human dialogue when a person gives an instruction, or a description of the required outcome, to another, and that person (agent) has to bring about the desired state, perhaps interpreting rather than executing the instruction. Travel agents are good examples of such agents.

6. information is the behaviour where a person provides information to a system, or another person, in response to some requirement (security check, data entry, etc.).

7. query is the behaviour where a person queries another person or system, regardless of the motivation for so doing.

It should be noted that symbolic behaviour relates to activities/states which are not necessarily physically or temporally closely coupled to the person engaged in the behaviour. If I ask my travel agent to sort out an itinerary for me, I do not monitor their every move/call/act/negotiation/etc., I get on with something else.

7.1 Aspects of interaction

The three types of behaviour are manifest, in varying degrees, in the three aspects, or layers, of interaction – underlying, observable and superficial.

Underlying interaction. The first or deepest layer is concerned with the user's intentions or local goals. These intentions are those which motivate the observed behaviour in some obvious and direct fashion. They are closely linked to the user's conception of the task to be accomplished. One can think of the underlying layer or level as being behaviour, but something that one might achieve by 'telepathy', if such a thing were possible. Whilst obviously imprecise, this description helps the reader to focus on the cognitive precursors of behaviour. Examples of where/how the behaviour types are found underlyingly might be:

8. I might have the intention that the windows on my 'desktop' should be resized and rearranged so that they tessallate nicely (revealing contents with no wasted space). My task is to rearrange the desktop.

9. I can stand in front of the counter at a baker's shop and make a selection of which type of loaf I want to buy, and this will eventually lead to a change in the world. However, note that one's selection is not a physical act, nor an intention to act. It has to be expressed somehow, but that is not actually part of selecting.

10. I might have an intention/desire to delete a file; to give a vague instruction to my butler; to give some data to my accountant; to query my secretary about tomorrow's schedule.

Superficial interaction. It is helpful to consider this here, as it clarifies the complexities of observable interaction, considered below. Superficial interaction is, essentially, doing things with symbol systems or muscles. Talking into a computer, typing commands, moving a mouse or joystick – these are all superficial aspects of interacting. For example, if I am standing in a crowded elevator and I ask someone near the buttons to press the button for the fourth floor, I am simply using vocal symbolism to effect a button press. I could reach out and poke the button with my umbrella. The interaction is not 'symbol mediated' in one case, and 'manipulation' in the other. Indeed, pressing the button itself is only one way of several possible ways of indicating to the mechanism my desire that the elevator permit me to leave at the fourth floor. Notice that superficially there can be no selection – that is a cognitive action or process which has to be imputed to the actor or deduced from the observable behaviour.

Observable interaction. One can observe a person manipulating an object and often one can work out why they are doing so. Notice that the observation often has to be as coupled into the reality as the original manipulation. My observation of someone taking the lid off a paint tin, observed five weeks later in a video-taped review for purposes of assessing dexterity, is not quite the same as when that person is standing next to me as s/he does it. One can observe a person making a decision – they look indecisive until they look decisive. But actually the observation of interest is the expression of the selection. This may be physical, or symbolic, or both. What is important is that the circumstances of the activity – use of a menu in a restaurant, or to select font properties – need to be known to the observer for them to form a clear view that the underlying activity is selection, at least potentially. The elevator example is relevant here. My request for the fourth floor, either vocally, or via my finger on the button, looks like a selection (or expression thereof). Actually, it could as easily be a specification, if I knew where I was going and why. And similarly, my selection of 12 pt Times could actually be expression of a specification. In general, we seem to be good at interpreting the behaviours we observe – of reading them.

11. Interaction design

Taking the three types of behaviour and three layers or levels of behaving in combination yields the scheme or taxonomy we need. It is important to recognize that the task being undertaken by a person is an important part of determination of the significance of their behaviour. When the behavioural/interactional taxonomy is combined with appreciation of the task the value of the scheme becomes clear, and its utility in resolving problems for special needs users becomes obvious. This is best illustrated with examples.

11.1 Ticket machine

Consider a machine for use at airports for selling tickets to travellers. (The payment task will not be considered at this point – see below.) One might imagine, having seen something like this at an airport, that a designer would opt for a form-fill-in style of interface. A keyboard will be required for some entry fields, and there might be drop-down menu lists for other fields, and so forth. So one will need a mouse or track-pad in addition to the keyboard. It all seems pretty obvious. And for special needs users – well, it depends on their needs. A visually impaired person might want to be able to enlarge the display, or parts thereof, and a person with only one hand (either in reality, or because the other is holding tight to a child's hand), might need a special keyboard, and so forth. This approach to the perceived problem – how to get the widest possible range of users to feel comfortable with the designer's choice of interface – seems fine. We have to trust the designer picked the right interface, but this is convincing if one remembers that airline tickets look like completed forms, and the interface screen layout can be made to look like an airline ticket. For people in a foreign language situation the form-fill-in interface is good technology because labels/instructions can be provided in many languages. The taxonomical approach goes about things differently. What is any user trying to do when they buy a ticket? They are specifying a valid permit for travel (and then paying for it). The task is underlyingly specificational, and thus the designer's first assumption must be that the interface must provide for specificational input. For people able to use form-fill-in technology the role of that interface for specification seems mostly clear. Observable selection behaviour (menus) is frequently used for specification underlyingly (as when one knows one wants 12 point Times but must 'select' those characteristics from a menu). But, when it comes to special needs users we should stop thinking about modifications to the interface and instead think about what sort of interface best supports a specificational interaction. The answer is clear – symbol mediation is the general solution. It happens that this is often well served by text and menus in a form-fill-in interface, *but that is not the only way to match the task*. Voice input would serve as well, and would suit visually impaired users (assuming recognition technology can perform well enough in appropriately engineered/installed solutions). Notice that for many people the solution is to go to an agent (human) and use symbol mediated interaction with that person. This has the embedded advantage of description over specification – some vagueness is tolerated with the agent ensuring the valid specification is nonetheless delivered. For many special needs users another human being is actually a good solution, not a fundamental failure with the technology. However, when all is said and done, if a technological solution *is* required the designer must return to basics – and find an interface which supports coherence between the superficial, observable and underlying aspects of behaviour to deliver, in this case, a specificational interaction.

11.2   Payment

Payment for services/items is often arranged by machine. One dips one's credit card into a slot and types in a few numbers and the transaction is complete. Once again the technology is likely to be straightforward, and might include a form or something similar (sequences of dialogue boxes are often forms spread out over time, a source of frustration if the task has no sequence underlyingly – as is the case when specifying a ticket). Once again the generic situation has to be analyzed. The task is a mixture of specification and information (payee, amount of money and validation details). If a keypad is not suitable for special needs users, or the screen is not readable, and voice is ruled out by concerns for privacy as well as reliability, then a human intermediary is again a sensible option. This is not a technical failure, just a designed solution. There could be other designed solutions.

11.3   Mismatched interfaces

The interaction taxonomy discussed above expresses the obvious – both why it is that some interfaces are ideally suited for their task, and why some are hopelessly wrong. The three orthogonal behaviours, if mapped unambiguously from underlying to observable levels, will yield well justified interfaces. If one has a manipulation based task underlyingly – e.g. control of a vehicle – then mapping this into an observable interaction which is, say, voice input (specificational interaction), will yield a poor, even dangerous, interface. Likewise, a joystick control for the Unix operating system is unlikely to be very successful. The clarity of analysis offered for poor interfaces indicates that the taxonomical scheme should be well suited for analysis to broaden access to computers.

12.   Conclusions

HCI designers of applications for all users, including special needs users, must avoid ad hoc developments and 'fixes'. The systematic account of interaction offered above reveals the sorts of factors which designers must consider if they are to provide appropriate interfaces for tasks and applications. Additionally, the scheme reveals what needs to be done for different sorts of users, including special needs users – special needs have to be identified and defined in terms appropriate to the taxonomy. The taxonomical approach outlined above looks helpful, but it cannot deliver the right solutions automatically. It remains the case that designers will have to be trained to be considerate – the interaction taxonomy should make such training much easier. Wider access to the utility offered by computers and computer controlled artefacts *can* be 'designed in' from the start. Special needs users should be able to expect that their needs are provided for, and this now looks increasingly plausible.

References

Bergman, E., Johnson, E. (1995). Towards Accessible Human-Computer Interaction. In J. Nielsen (ed) *Advances in Human-Computer Interaction.* (Vol. 5) New Jersey: Ablex.

Bergman, E., Johnson, E. (2000). Sun Microsystems Accessibility Program - Designing for Accessibility. [On-line]. Available: http://www.sun.com/access/developers/software.guides.html

Edmondson, W. H. (1993). A Taxonomy for Human Behaviour and Human-Computer Interaction. In G. Salvendy & M.J. Smith (Eds), *Advances in Human Factors/ Ergonomics 19B: Human-Computer Interaction: Software and Hardware Interfaces,* (pp. 885-890). Amsterdam: Elsevier.

Edmondson, W.H. (2000). A Taxonomy of Users' Behaviour in Human Computer Interaction. In M.M. Taylor, F. Néel, D.G. Bouwhuis (eds), *The Structure of Multimodal Dialogue – II,* (pp. 335-348). Amsterdam: John Benjamins.

Shneiderman, B. (1998). *Designing the User Interface: Strategies for Effective Human-Computer Interaction.* (3rd ed.). Reading, Massachusetts: Addison-Wesley.

Early Experiences Using Visual Tracking for Computer Access by People with Profound Physical Disabilities

James Gips[a], Margrit Betke[b], Philip A. DiMattia[c]

[a]Computer Science Department, CSOM, Fulton Hall 460, Boston College, Chestnut Hill, MA 02467, USA, gips@bc.edu, http://www.cs.bc.edu/~gips

[b]Computer Science Department, Boston University, 111 Cummington Street, Boston, MA 02215, USA, betke@cs.bu.edu, http://www.cs.bu.edu/faculty/betke

[c]Campus School, Lynch School of Education, Campion Hall, Boston College, Chestnut Hill, MA 02467, USA, dimattia@bc.edu

Abstract

A dozen people who cannot speak and have very limited voluntary muscle control because of cerebral palsy or traumatic brain injury have tried using a new technology called the Camera Mouse to access the computer. Nine of the people were successful, in many cases being able to spell words or run commercial software or access the internet for the first time. The Camera Mouse is non-invasive and uses a standard video camera and room lighting to track slight movements of the head or thumb or toe, whatever part of the body a person can control. The Camera Mouse acts as a mouse substitute.

1. Introduction

We work with people who are excluded from access to computers and the internet and even from communication with fellow human beings. These people have severe physical disabilities either from birth or from accidents or from degenerative diseases. They are unable to speak and have very little or no voluntary muscle control. Their level of mental functioning might not be known because of their inability to communicate. Children with severe physical disabilities but fully functioning minds can go uneducated. People with severe physical disabilities often are isolated, spending hours in bed or in a wheelchair at home or in an institutional setting staring at the wall or at the television.

Computer and communications technology can make all the difference in the world for people with profound physical disabilities. Our approach has been to develop technologies that allow people to control the mouse pointer on the screen by moving their eyes or by small movements of their head or toe or thumb. If people can control the mouse pointer and issue clicks then they can use the computer to communicate with people, to run educational software, to access the internet.

We have developed two technologies to provide access to the computer for people with severe physical disabilities. The first technology, EagleEyes (Gips, Olivieri & Tecce, 1993; DiMattia, Curran & Gips, 2001), uses five electrodes placed around the eyes to allow a person to control the mouse pointer by eye movement. More recently, we have developed the Camera Mouse, a technology that uses a standard video camera to allow a person to control the mouse pointer by slight movements of the head or any other portion of the body. In Gips, Betke & Fleming (2000), we introduced the Camera Mouse technology (patent pending). Here we report on our experiences with the first dozen people, mostly people with severe cerebral palsy, in using the Camera Mouse for computer access.

Several other assistive technology products allow people to control the computer through head movements. The Tracker 2000 from Madentec (www.madentec.com) and the HeadMouse from Origin Instruments (www.orin.com) use near infrared light to track a reflective dot placed on the head. The HeadMaster 2000 from Prentke Romich (www.prentrom.com) mounts the light source on the head instead of near the monitor. The Gyro-HeadMouse from Advanced Peripheral Technologies (www.advancedperipheral.com) embeds mercury switches in a hat to measure head angles. People have benefited enormously from these products. The Camera Mouse uses existing light and has nothing placed or mounted on the person's head. We hope a product based on the Camera Mouse technology will be easier to use, less invasive, more flexible, and much less expensive and thus will be of wide benefit.

2. The Camera Mouse

The Camera Mouse uses a video camera to capture images of the part of the body being moved and uses the computer to track the movements in the images in real time. These body movements are converted into mouse pointer movements on the screen. The result is that the Camera Mouse is a completely non-invasive, easy-to-use access device. A video camera sits on top of the monitor or below the monitor pointed towards the user. The user moves his head slightly. The mouse pointer moves accordingly. As an option to the user, selections can be made using "dwell time." For example, a mouse click can be generated if the user keeps the mouse pointer at a given location for some settable time, usually 0.7 seconds. As an additional option, a double click can be generated if the mouse pointer is held in the location for a longer period of time. The Camera Mouse is a general mouse replacement system that works with commercial software, including communications software, educational software, and web browsers, and with custom-developed software.

The initial version of the Camera Mouse (Gips, Betke & Fleming, 2000) used two computers: one for the visual tracking and one for the user application program. The current version of the Camera Mouse uses a single computer. The Camera Mouse program runs in the background while the application program runs in the foreground. The Camera Mouse works by taking in 30 frames of video per second. Initially a person clicks on the feature (for example the tip of the nose or the middle of the lower lip) to be tracked. The Camera Mouse then tracks that feature by finding the portion of the next frame that most closely matches it. The process is repeated for each frame. An example of visual tracking is shown in Figure 1.

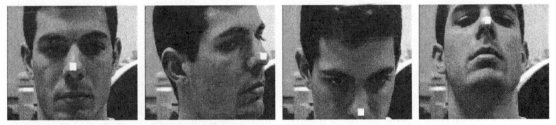

Figure 1: Automated visual tracking.

The window with the image of the user is minimized while the user runs the application program. In the current version, the tracking program is given control of the mouse pointer by pressing the CapsLock key. Control is returned to the mouse in the same way.

With the support of the Camera Mouse both commercial software and custom-developed software can be used. The Camera Mouse enables people to use Internet Explorer and Netscape Navigator to access the web. We use a wide variety of educational software. We use both custom-developed spelling programs and Clicker 4 for communication. We use both commercial and custom-developed entertainment software to develop skill control and just for fun.

3. Jordan using the Camera Mouse

The first person to have her own Camera Mouse system was Jordan – a thirty-three month old girl with severe cerebral palsy. Jordan can't talk but she can move her chin up and down a little and her head from side to side a little.

916

Here is Jordan in her wheelchair trying a system in our lab. Her mom is next to her. The camera is black and below the monitor.

Figure 2: Jordan controls the mouse pointer by moving her chin.

Jordan is moving her chin to move the white arrow (mouse pointer) to pop the bubble that contains a bird. With the help of the Camera Mouse she regularly uses this commercial program for children.

Figure 3: Jordan uses the Camera Mouse to run commercial software.

4. Results

People without physical disabilities quickly learn to control the computer through the Camera Mouse by moving their head.

We have tried the Camera Mouse with a dozen people who have severe physical disabilities. Each person cannot speak and has very little voluntary muscle control. Ten have cerebral palsy. Two have traumatic brain injury, one from an automobile accident, one from a motorcycle accident. Nine of the people are continuing to use the Camera Mouse. Six can use the Camera Mouse to spell using an onscreen keyboard system. Three of the people did not have sufficient muscle control to use the Camera Mouse. They are using EagleEyes to control the mouse pointer by moving their eyes. One of the people who was not successful was close and will be given additional opportunities to try the Camera Mouse in the future. The results are summarized in Table 1.

Table 1: Summary of results for the first dozen people with disabilities to try the Camera Mouse.

Age	Gender	Condition	Continuing to Use?	
2	M	Cerebral Palsy	Yes	Obtaining a system for home.
3	F	Cerebral Palsy	Yes	First with home system.
6	F	Cerebral Palsy	Yes	Spelled name. Obtaining a home system.
8	M	Cerebral Palsy	Yes	Spells naughty words and laughs.
11	M	Cerebral Palsy	Yes	Obtaining a home system.
14	M	Cerebral Palsy	Yes	Spells words. Obtaining a home system.
15	M	Cerebral Palsy	No	Close, but could not control reliably.
19	M	Cerebral Palsy	No	Does not have sufficient muscle control.
23	M	Traumatic Brain Injury	No	Does not have sufficient muscle control.
31	M	Traumatic Brain Injury	Yes	Spelled "TAKE OFF DAD"
37	M	Cerebral Palsy	Yes	Spelled "MERRY CHRISTMAS"
58	M	Cerebral Palsy	Yes	Spells, explores internet on home system.

Some of these people were unable to unequivocably demonstrate their level of mental functioning before trying the Camera Mouse. Often the parents were convinced of their child's intelligence and perhaps had found subtle ways to communicate within the family, but the authorities were doubters. Spelling using an onscreen keyboard with no outside intervention provides a definite proof of a level of mental functioning and a level of learning. It obviously is a critical step in providing contact and communication with the outside world and in beginning to formulate a coherent plan for education and intellectual development.

The dozen people have a high level of difficulty in controlling their head (or thumb or toe). With a traditional onscreen keyboard, where the keys are right next to each other and the keyboard fills much of the screen, they run into the Midas Touch problem described by Jacob (1991); everything the mouse pointer touches and rests on is selected. People without physical disabilities have no trouble spelling with a traditional onscreen keyboard. For people with more limited control we usually use a custom-developed two-level spelling program. The first screen shows five large buttons, each with a group of letters (ABCD, EFGH, etc.). The person clicks on the group containing the letter desired. The next screen shows a button for each letter in the selected group. There is plenty of resting area between the buttons. We are still experimenting with the optimal configuration of these screens.

The Midas Touch problem also occurs using Internet Explorer or Netscape Navigator with the Camera Mouse. As the person is reading the text on the webpage, the mouse pointer is resting somewhere. If it rests on a link for 0.7 seconds (a typical threshold for triggering a mouse click) then the link might be activated and the user is off in an unintended direction through the web. Our future work will explore better ways to navigate through the web using a technology like the Camera Mouse.

There is an important story to be told about each of the dozen individuals listed in the table. The gentleman who is 58 with cerebral palsy spends his time in bed at home in New Jersey. He had no expressive language ability and no voluntary movement below the neck. We were approached by his 80+ year old father who wanted his son to be able to communicate before he died. We set up a prototype system in the home. The gentleman is now using the Camera Mouse to communicate with his father and access a variety of software.

Here is a portion of an email sent by Jordan's mother, Cheryl. (Cheryl has given full permission to use the pictures and email.)

> We are a family of four. What makes us different is that we have a daughter who is three and has Cerebral Palsy ... we had no idea if or how much Jordan understood. To our amazement she followed every direction that was given her. Can you just imagine the joy we had watching our daughter exploring her environment and dreaming of what the possibilities were with this application in her life? The possibilities seemed endless now. Since then Jordan has become more sophisticated with this computer ... it has given her the chance to actively participate in learning what her typical developing peers are learning, it has given her a way to communicate her thoughts, it gives the school that she is attending a way to adapt the curriculum so that Jordan can participate in a REGULAR preschool, it put Jordan in a situation where people can see her

ABILITIES rather than her disabilities, and it allows a level playing field where Jordan can play from.

In our situation we were very fortunate that the Camera Mouse System was available to us ... I feel strongly that this system needs ... support so that many individuals will be given the same opportunities to learn and express themselves. This type of independence is a rare gift for someone that has any type of disability where they are unable to explore their world. The best thing that the Camera Mouse System has given us, is watching both of our daughters sharing and playing at something that brings them so much joy and giggles. That is an experience that should not be missed by other families.

Thank You,

Cheryl, Mike, Alyssa, and Jordan

Since obtaining the Camera Mouse, Jordan has transitioned from an early intervention program to a pre-school regular education program. Despite her inability to vocalize and her very limited muscle control, she is functioning at grade level and above in most areas of cognitive and social development. Our goal now is to help Jordan access the regular schoolwork that is afforded her peers.

5. Future work

Our major thrust is to improve the Camera Mouse system itself by improving the algorithm used for tracking, the interface with the program, and the cost of the camera and video capture hardware needed. We continue to work on developing new communication software. We are contemplating a better web browser to use with the Camera Mouse to mitigate the Midas Touch problem.

More ambitiously, we would like to expand our vision software to include recognition of gestures such as laughter and blinking and eye movements so that we can more fully allow people with limited physical abilities to access the computer.

We are planning a website that will serve as a "meeting place in cyberspace" for people who use the Camera Mouse and EagleEyes. The site will allow the people in England or Boston or Florida using these technologies (or, indeed, anyone) to engage in cooperative and competitive problem solving and educational development. It will allow them to chat with each other and see each other on live video. We hope it will begin to break down the isolation in these people's lives.

References

Gips, J., Olivieri, P., Tecce, J.J. (1993). Direct Control of the Computer through Electrodes Placed Around the Eyes. In M.J. Smith and G. Salvendy (eds.), *Human-Computer Interaction: Applications and Case Studies (Proceedings of the Fifth International Conference on Human Computer Interaction*, Orlando (pp. 630-635). Elsevier.

DiMattia, P., Curran, F.X., Gips, J. (2001). *An Eye Control Teaching Device for Students Without Language Expressive Capacity: EagleEyes*. Edwin Mellen Press.

Gips, J., Betke, M., Fleming, P. (2000). The Camera Mouse: Preliminary Investigation of Automated Visual Tracking for Computer Access. In *Proceedings of RESNA 2000* (pp. 98-100). RESNA Press.

Jacob, R.J.K (1991). The Use of Eye Movements in Human-Computer Interaction Techniques: What You Look At Is What You Get. *ACM Transactions on Information Systems* 9(3), 152-169.

Kinematic profiling in object location and line drawing tasks by visuo-spatial neglect subjects

Richard M. Guest, Michael C. Fairhurst, Jonathan M. Potter[*]

Electronic Engineering Laboratory, University of Kent, Canterbury, Kent, UK, CT2 7NT.
[*]Nunnery Fields Hospital, Canterbury, Kent, UK.

Abstract

Assessing the functional characteristics of a user or user group is essential in any system design, but is of critical importance in dealing with systems to support rehabilitation programmes. This paper reports on the development of a technique, based on computer analysis of essentially "pencil-and-paper" tests to detect and assess visuo-spatial neglect in CVA patients, as a precursor to developing appropriate rehabilitation programmes. Quantitative results are presented which will facilitate more accurate monitoring of rehabilitation progress.

1. Introduction

In almost any sphere of system design, facilitating user access implies an *a priori* understanding of the nature of the abilities and functional characteristics of potential system users. At the same time, computer-based systems are increasingly being used to support therapy and rehabilitation programmes, and systems for this purpose must therefore be explicitly designed to provide an appropriate degree of meaningful interaction between a (potentially) highly specialised user group (almost certainly with constraints on its interface requirements) and the system itself. This paper addresses such issues in a particular context, reporting on the use of a series of pencil and paper line drawing tasks for the detection of visuo-spatial neglect within a geriatric stroke patient. In particular, these tasks are used to study the kinematic profile of pen movements made by the test subject across the surface of an attached graphics digitisation tablet and establish any performance characteristics between the different sides of the visual field. Visuo-spatial neglect is a condition that can occur following a stroke or head injury causing a patient to fail to react to stimuli positioned within the opposite side of the visual field from the location of the lesion (Halligan et al., 1989). This can lead to problems in performing everyday tasks such as washing, dressing and eating (Edmans & Lincoln, 1990). The effects are more prevalent in patients with a lesion in the right hemisphere of the brain (Right cerebral vascular accident - RCVA) leading to a deficit in the left of the visual field. Visuo-spatial neglect is recognised as barrier to recovery following a stroke and therefore accurate assessment of the condition is critical to the selection of effective schemes of rehabilitation (Robertson, 1994).

Such an assessment, aimed at providing a greater degree of accuracy in the evaluation of rehabilitation, is ultimately likely to offer increased success in specifying effective rehabilitation programmes and therefore improved outcomes for patients. Since the effects of stroke are likely severely to impair participation in many everyday activities, this type of work has a clear and direct bearing on universal access to information and data which would otherwise be unattainable by such a group of individuals.

A standard technique for the assessment of neglect is the use of a series of pencil and paper based tests which can be used to *quantify* performance (Halligan & Robertson, 1992). These tests involve the completion or drawing of a task printed on a sheet of paper which is placed directly in front of the patient. Using a pencil or pen, typical tasks involve the cancelling of printed targets or the drawing of simple geometric shapes. The responses of these tasks are then evaluated by therapists or trained assessors. Standard assessment techniques analyse the omissions or drawing variation from normal performance. For example, in a cancellation task, standard analysis is to count the number of targets cancelled on the completed overlay (Albert, 1973). Observed response from a Right CVA neglect subject is a failure to cancel left hand side targets. In drawing tasks, RCVA subjects often omit the extreme left hand side components of a geometric shape or representational drawing (Heilman et al., 1994). This study aims to assess if the static differences which are evident in neglect performance are replicated by dynamic constructional differences according to the position within the visual field. The research reported builds on recent studies (Mattingley et al., 1994; Konczak & Karnath, 1998) by further investigating a neglect subject test response in terms of planning and drawing phases of task completion and assesses the impact of the introduction of processing tasks such as movement to a specified target amongst distractors on test performance. Velocity profiling of a straight line response with respect to the direction of drawing is also investigated in term of deviation from a normal profile.

2. Method

The set of tasks reported in this paper are a series of line drawing assessments involving visual discrimination. Using five overlays, the drawing profile tasks enable a more detailed understanding of the movement dynamics involved in drawing images. By asking the test subject to draw a line the length (or width) of the overlay, the pen is in contact with the tablet for a greater time, thus producing a clearer movement feature dynamic, such as pen velocity, whilst also allowing the extraction of a variety of kinematic-based timing measurements, such as acceleration and deceleration phase timings. Repeated movement back and forth across the width of the page can indicate mean differences as a neglect subject moves the pen in and out of the neglected field. The overlays can be divided into two distinct task groupings:

Line Drawing Tasks: The first three overlays contain a series of dots at each side of the page. The dots are 10 mm in diameter, located 57 mm from each vertical edge of the overlay and are separated vertically by 46 mm. Starting at the bottom right hand side of the overlay (selected because the right hand side is least affected by the patient's neglect), the test subject must move the pen across the page to the lowest left hand side dot. Having located this target, a movement is made back across the page to the second lowest target. This 'zig-zag' pattern is repeated until all dots have been visited, finishing in the top right hand corner.

The two other overlays contain variations of this task, introducing distractors (10 x 10 mm squares) placed amongst the circles. The test subject must discriminate between the targets and only move the pen to the circles. The order of targets and distractors is uniform on each side of the second overlay, but randomised in the third overlay.

Visual Field Square Drawing Tasks: The second set of two overlays require the test subject to join dots to form a square. The dots for this task are located 64 mm apart in a single side of the visual field. No drawing sequence is specified for this task. The first of the overlays has the dots printed in test subject's right visual field, swapping to the left visual field in overlay 5. In this way performance characteristics from each side of the visual field can be assessed.

3. Feature Extraction

Analysis is performed by assessing each individual line segment. Mean feature results can thus be obtained for groups of drawing movements, for example, all lines drawn in a leftward direction. Component timings are assessed by separately analysing the *drawing time* (time when pen is moving across the tablet surface and marking the task overlay) and *pre-movement time* (time *prior* to drawing movement when pen is removed or stationary on the tablet). In this way, measurements can be taken concerning different phases of the drawing process: the planning phase, when the test subject is observing the movement target and establishing the required pen trajectory and the drawing phase indicating motor performance and positional feedback processing of the pen. The total time for a component 'x' is therefore the sum of these two time measurements:

Time to drawn component 'x' = pre-movement time before drawing component 'x' + drawing time
of component 'x'

The separation of timing measurements has been used to analyse the responses of cancellation-based tasks (Donnelly et al., 1999) demonstrating constructional differences between left and right sided visual fields. Pen velocity across the surface of the tablet was calculated by taking the first derivative of the coordinate pair displacement against time. Third order, four coefficient polynomial modelling was used to obtain a derivative of displacement at each coordinate point (Williams, 1986). Using a constant sampling time-base, the following approximation uses displacement values of four sets of coordinates at times t-2, t-1, t+1 and t+2.

$$\frac{ds}{dt} \approx \frac{1}{12}(-s_{t+2} + 8s_{t+1} - 8s_{t-1} + s_{t-2})$$

The displacements used within the calculation can be extracted on an axis-component basis obtaining the separate horizontal and vertical velocity features or by calculating the Euclidean displacement using both the x and y components. The mean velocity is obtained by summing the velocities at individual points within the segment and dividing by the number of samples taken.

A velocity profile, plotting the obtained velocity values against time for an individual segment, can provide additional information into the kinematic aspects of drawing. Typical asymptomatic velocity response from a straight line segment produces a 'bell-shaped' profile (Plamondon, 1991). Figure 1 models this normal performance in drawing a straight line segment. The profile skew to the start of the segment indicates a shorter acceleration phase in comparison to the deceleration phase following the peak velocity. The peak velocity is the highest recorded velocity value within the segment analysis.

Several timing features can be obtained from the profile. The time to reach the peak velocity (acceleration phase) and the time from peak to zero velocity (deceleration phase) have been used as performance indication features (Mattingley, 1994). The segment drawing time is thus the sum of these two phases:

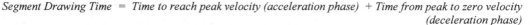

$$Segment\ Drawing\ Time\ =\ Time\ to\ reach\ peak\ velocity\ (acceleration\ phase)\ +\ Time\ from\ peak\ to\ zero\ velocity\ (deceleration\ phase)$$

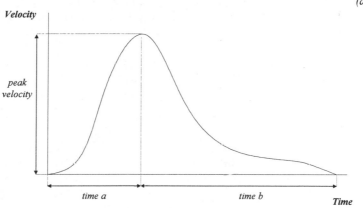

Figure 1: Velocity profile model of single line component

A velocity skew percentage is an intra-profile indicating the position in the profile in terms of total segment drawing time where the peak velocity occurs. The calculation (with reference to Figure) is made according to:

$$velocity\ skew = \left(\frac{time\ to\ reach\ peak\ velocity\ (time\ a)}{total\ drawing\ time\ of\ the\ component\ (time\ a + time\ b)} \right)$$

4. Experimental Method

Six test subjects were assessed for this study, as characterised in Table 1. The BIT score is a standard clinical assessment of neglect obtained from the Rivermead Behavioural Inattention Test (Wilson, 1987). A score of 129 or below indicates the presence of neglect.

Table 1: Characteristics of subjects in case study

Subject	Sex	Age	BIT Score
Neglect 1	Female	78	110
Neglect 2	Female	81	20
Control 1	Male	70	146
Control 2	Female	67	146
Control 3	Female	66	145
Control 4	Female	79	146

5. Results

Results are presented in Tables 2, 3 and 4. A subscript + or – indicates a response outside the bounds of normal subject performance. A '/' character indicates that a movement in the given direction was not obtained.

Table 2: Control subjects mean (and standard deviation) velocity profile results

Task	Movement	Skew (SD)	Mean Vel (mm/sec) (SD)	Max Vel (mm/sec) (SD)	Mean Pause and Move Time (sec) (SD)
Line Drawing 1	Left to Right	0.47 (0.12)	38.40 (12.36)	70.06 (24.19)	3.02 (0.88)
	Right to Left	0.31 (0.11)	28.36 (4.38)	55.54 (5.20)	1.73 (0.15)
Line Drawing 2	Left to Right	0.47 (0.2)	26.60 (6.47)	48.43 (7.37)	0.59 (0.26)
	Right to Left	0.38 (0.1)	30.09 (3.47)	52.26 (8.45)	2.1 (0.11)
Line Drawing 3	Left to Right	0.48 (0.07)	36.42 (6.37)	66.54 (10.92)	2.29 (0.32)
	Right to Left	0.39 (0.09)	29.51 (4.67)	55.35 (8.77)	1.78 (0.72)
RHS Square Drawing	Left to Right	0.30 (0.03)	21.03 (4.87)	34.10 (5.63)	1.09 (0.26)
	Right to Left	0.25 (0.03)	15.93 (5.95)	29.46 (7.48)	/
LHS Square Drawing	Left to Right	0.34 (0.17)	19.21 (6.82)	35.08 (4.09)	/
	Right to Left	0.43 (0.17)	16.07 (2.73)	28.48 (5.83)	2.04 (0.1)

Table 3: Neglect subject 1 mean (and standard deviation) velocity profile results

Task	Movement	Skew (SD)	Mean Vel (mm/sec) (SD)	Max Vel (mm/sec) (SD)	Mean Pause and Move Time (sec) (SD)
Line Drawing 1	Left to Right	0.49 (0.19)	25.05 (3.09) [-]	55.09 (6.53) [-]	4.47 (0.69) [+]
	Right to Left	/	/	/	/
Line Drawing 2	Left to Right	0.41 (0.19)	23.07 (7.12)	45.65 (11.43)	2.17 (0.33) [+]
	Right to Left	0.49 (0.15)	28.19 (7.87)	52.92 (13.00)	2.33 (0.62)
Line Drawing 3	Left to Right	0.41 (0.17)	27.38 (3.50) [-]	61.19 (22.02)	4.34 (0.04) [+]
	Right to Left	0.59 (0.20)	28.37 (11.95)	61.65 (28.18)	6.73 (0.40) [+]
RHS Square Drawing	Left to Right	0.34	10.95 [-]	24.75 [-]	1.46 (0.34) [+]
	Right to Left	0.53	14.62	33.92	/
LHS Square Drawing	Left to Right	0.25	21.11	38.58	0.17 (0.40)
	Right to Left	0.76 (0.26) [+]	12.91 (4.60)	26.79 (7.01)	/

Table 4: Neglect subject 2 mean (and standard deviation) velocity profile *results*

Task	Movement	Skew (SD)	Mean Vel (mm/sec) (SD)	Max Vel (mm/sec) (SD)	Mean Pause and Move Time (sec) (SD)
Line Drawing 1	Left to Right	/	/	/	/
	Right to Left	0.57 (0.29) [+]	4.45 (1.53) [-]	12.00 (3.44) [-]	5.12 (1.28) [+]
Line Drawing 2	Left to Right	0.30 (0.13)	3.01 (1.54) [-]	6.63 (3.98) [-]	11.26 (1.81) [+]
	Right to Left	0.32	11.25 [-]	3.73 [-]	/
Line Drawing 3	Left to Right	0.49 (0.12)	3.70 (2.26) [-]	9.61 (6.62) [-]	6.72 (1.20) [+]
	Right to Left	0.84 [+]	2.39 [-]	6.75 [-]	/
RHS Square Drawing	Left to Right	0.56	5.00 [-]	14.33 [-]	2.47 (0.25) [+]
	Right to Left	0.25 (0.49)	4.22 (3.36) [-]	13.00 (10.75) [-]	/
LHS Square Drawing	Left to Right	0.49 [+]	3.36 [-]	10.75 [-]	1.80 (0.35)
	Right to Left	/	/	/	/

6. Summary

The results from this study show that we are able to detect dynamic performance differences between neglect subjects and a control population. Using a conventional graphics digitisation tablet to capture the positional data of the pen in real time, we can note that a number of features are sensitive to the amount of neglect exhibited by a subject. Neglect Subject 1 who has moderate neglect, performs normally in an assessment of velocity profiles. The mean movement and pause times prior to drawing are however significantly larger for this test subject. Neglect Subject 2 has severe neglect (as indicated using the BIT test results). In the responses to the tests, Neglect Subject 2 failed to complete the task satisfactorily (failed to joined up dots) and this is reflected in the lower maximum and mean velocity values, although the value skew values produce results which indicate a significantly longer pen acceleration time. Again, the movement and pause times provide a sensitive feature for neglect measurement.

The line drawing tasks are part of a wider computer-assessed battery of neuropsychological tests for neglect which have been used in a clinically-based trial. The combination of accurate and objective assessment of static features and the novel dynamic measurements both aid the diagnosis of neglect but also further the understanding of the condition with respect to constructional and timing aspects of test performance. It is hoped that this information will aid the clinical diagnosis and increase the accuracy of performance measurement and hence produce a clearer indication of rehabilitation progress. Hence, the work reported here, while of immediate and obvious clinical value in rehabilitation, is also of importance to the wider field of "universal access", since it offers a novel and important perspective on issues relating to the assessment of functional characteristics of a significant clinical population.

The authors acknowledge the support of the UK South Thames NHS R&D Project Fund

References

Albert, M. L. (1973). A simple test of visual neglect. *Neurology*, 23, pp. 658-664.

Donnelly, N., Guest, R., Fairhurst, M., Potter, J., Deighton, A., Patel, M. (1999). Developing algorithms to enhance the sensitivity of cancellation test of visuo-spatial neglect. *Behavior Research Methods, Instruments and Computers*, 31(4), pp. 668-673.

Edmans, J.A., Lincoln, N. (1990). The relationship between perceptual deficits after stroke and independence in activities of daily living. *British Journal of Occupational Therapy*, 53, pp. 139-142.

Halligan, P.W., Marshall, J.C., Wade, D.T. (1989). Visuospatial neglect : Underlying factors and test sensitivity. *The Lancet*, ii, pp. 908-910.

Halligan P.W., Robertson I.H. (1992). The assessment of unilateral neglect. In Crawford, J.R., Parker, D.M., McKinlay, W.W (Eds.) *A Handbook of Neuropsychological Assessment* (pp. 151-175). Hove, UK.: Lawrence Erlbaum Associates.

Heilman, K.M., Watson, R.T., Valenstein, E. (1994). Localization of lesions in neglect and related disorders. In Kertesz, A.(Ed.), *Localization and neuroimaging in neuropsychology* (pp. 495-523). San Diego, CA: Academic Press.

Konczak, J., Karnath, H. (1998). Kinematics of goal-directed arm movements in neglect: control of hand velocity. *Brain and Cognition*, 37, pp. 387-403.

Mattingley, J.B., Phillips, J.G., Bradshaw, J.L. (1994). Impairments of movement execution in unilateral neglect: A kinematic analysis of directional bradykinesia. *Neuropsychologia*, 32, pp. 1111-1134.

Plamondon, R. (1991). On the origin of asymmetric bell-shaped velocity profiles in rapid-aimed movements. In Requin, J., Stelmach, G.E. (Eds.), *Tutorials in Motor Neuroscience* (pp. 283-295). Kluwer Academic Publishers.

Robertson, I.H. (1994). The rehabilitation of attentional and hemi-attentional disorders. In Riddoch, M.J., Humphreys, G.W. (Eds.), *Cognitive Neuropsychology and Cognitive Rehabilitation* (pp. 173-186). Hove, UK: Lawrence Erlbaum Associates.

Williams, C.S. (1986). *Designing Digital Filters*. New Jersey: Prentice-Hall.

Wilson, B., Cockburn, J., Halligan, P. (1987). *The Behavioural Inattention Test*. Fareham, Hampshire: Thames Valley Test Company.

A Study of HCI for People in Japan with Communication Disorders

Tohru Ifukube

Research Institute for Electronic Science, Hokkaido University, Sapporo 060-0812, Japan
ifukube@sense.es.hokudai.ac.jp

Abstract

Over a period of 28 years, the author has developed a basic research approach for assistive technology. As reviewed in this paper, he and his co-researchers have designed several aid devices as well as obtained many basic findings concerning human information processing. Some of the devices have been manufactured in Japan. Moreover, the technologies as well as the basic findings have been applied to construct human-centered computer interfaces. Consequently, it can be said that these newly developed computer interface technologies have led to improvements in the design of models for developing assistive communication devices.

1. Introduction

We have been carrying out the human computer interface studies and have designed several aid devices which are now or shall soon be put into practical use for people with communicative and perceptual disorders (Ifukube, 1997). Moreover, we have obtained many findings concerning both the mechanisms of perception and concept formations in the human brain. In this paper, we will refer to our research regarding the aid devices and explain how that research has been related to new information technologies such as speech recognition, speech synthesis and virtual reality. Lastly, we will introduce our research approach in the development of human-centered computer interfaces.

2. Some assistive devices for communication disorders

2.1 Tactile display for hearing and/or visual aids

Through fundamental research on auditory and tactile information processing, we developed a portable tactile voice coder (tactile vocoder) for the hearing impaired about 25 years ago. The device makes it possible to produce Japanese vowels as well as some consonants such as fricatives by touching a piezo-electric vibrator array (Ifukube, 1982). Although it has been put into practical use, this device has not been widely used because of its high price and large size.

Recently, with the help of the Japanese national government, we have designed a new tactile display for the hearing and/or visually impaired

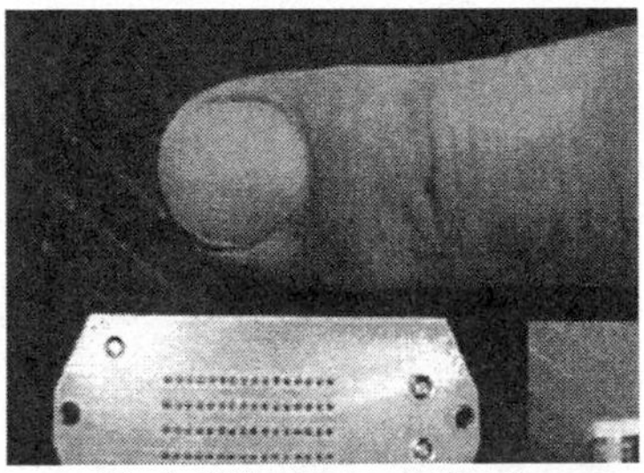

Fig.1: Vibrator array for the hearing/visually impaired

(Ifukube et al., 2000). The device includes a tactile stimulator matrix consisting of 16 rows and 4 columns and a small computer that can be worn on a person's arm like a bracelet. The tactile display is 2 cm long, 1 cm wide and 1 cm high so that it can be attached to a finger tip as shown in figure 1.

With this new tactile vocoder, speech signals picked up by a microphone are divided into 16 components by use of a digital filter. Next, each intensity of the component is converted into a unique vibratory intensity in the stimulator matrix. Our research has indicated that by using this device, the average identification rate for Japanese consonants reaches 70 % following one week of training. The device, including the tactile display and converter, will be offered at the low price of around $500 (US).

This tactile vocoder can be used together with a hearing aid, sign language, lip-reading and cochlear implants. Speech signals obtained from cellular phones and e-mail voice messages can also be converted into vibratory patterns. It could likewise enable a person to catch alarm sounds while walking on a street or even to detect emergency information broadcasts from radios.

In order to display the images drawn on a paper or computer screen, a CCD camera attached to the bottom of the stimulator matrix detects images. The density of the image is then converted into a vibratory intensity corresponding to each pixel. By scanning a whole image, the subjects are able to construct the corresponding image inside the brain. This tactile display can further be used to catch an outline or shade of images, pictures and graphs presented on a computer display with a GUI.

2.2 Voice recognition system with a see-through HMD for the hearing impaired

As an aid for individuals with acquired deafness, about 20 years ago our prototype tactile vocoder was applied to both a voice typewriter and an 8 channel cochlear implant. This particular hearing-disabled group was chosen because, unlike people who are born deaf, it is particularly difficult for those with acquired deafness to learn lip-reading and sign language. As to the cochlear implant, the author conducted the research at Stanford University where many basic findings were obtained from psycho-physical experiments involving a volunteer to whom our electrode array was implanted (Ifikube et al, 1987; Miyoshi, 1999). The technology used in our cochlear implant was also applied to an implantable tinnitus suppressor for people with tinnitus, also known as a ringing ear. (Matsushima, 1996).

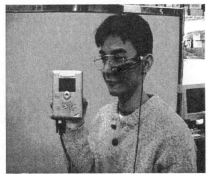

The voice typewriter was successfully designed using a micro-computer with a 32 kilobyte memory. This device can convert monosyllabic voice sounds (such as /a/ /ka/ /sa/ /ta/ /na/ and /ha/) into both Japanese letters (kana) and Chinese letters (kanji) almost in real time and show them on a display (Ifukube, 1984).

Fig.2: Speech recognition system with a see-through HMD

Although modern speech recognition results in frequent recognition errors, humans are often able to guess the correct meaning of sentences through context. By combining this guessing ability with non-verbal information (including lip-reading, facial expressions, and a speaker's gestures), the present speech technology could prove useful as a communication aid for the hearing impaired especially for the acquired deaf (Kosakai et al., 2000).

We have designed a new hearing assistive device which utilizes a portable speech recognition system combined with a see-through Head Mounted Display (HMD), as shown in figure 2. By use of this system, it was ascertained that both the non-verbal information and the guessing-through-context ability does indeed contribute to the comprehension of spoken language.

2.3 Digital hearing aid for the elderly hearing impaired

In general, elderly people who have suffered from a hearing impairment have less ability to understand spoken language, even though they can hear the speech sounds. This phenomenon seems to be due to a decrease in the recognition of auditory time patterns in the speech area of the cortex. Therefore, in cooperation with the Hitachi company, we have designed a hearing aid which can slow down speech without any pitch frequency change by using a digital signal processor, as shown in figure 3. Through our research, we have obtained evidence that this device is effective in helping the elderly sensori-neural hearing impaired to catch the meaning of rapidly spoken sentences. (Nejime et al., 1996).

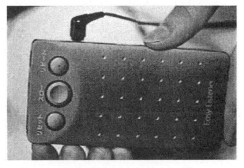

We have also proposed a digital processing method to modify tone contrast for the hearing impaired. This was designed especially for the spoken Chinese language which consists of 4

Fig.3: Digital hearing aid with a speech rate conversion function

tones that are used to determine the meaning of each character. Our research has shown the claim that a hearing aid with tone modifications is significantly effective for the hearing impaired, especially for the elderly Chinese (Lu et al., 2001).

926

We are planning to design a new digital hearing aid which has both speech rate conversion and tone modification functions.

2.4   Electric artificial larynx for people with speech production disorders

An electric artificial larynx is designed to offer substitute speech for patients who have undergone a laryngectomee, thereby losing their natural speech function. However, the electric artificial larynx makes it difficult to produce a natural sounding voice.

In order to improve the conventional electric artificial larynx, we have proposed a new method that can allow laryngectomees to control intonation by use of their respiration. This idea is based on an analytical study of a mynah bird's vocalization mechanism which can imitate the human voice, especially as regards intonation patterns and pitch fluctuations (Lu et al., 2001; Aoki et al., 1998).

The device consists of three parts. The first part is a pressure sensor that can detect exhaled air pressure produced from a stoma made by a surgical incision into the neck. The second part is a transformation circuit made of a microprocessor that can convert air pressure into a pitch frequency. The third part is an electromechanical vibrator that can be attached to the neck.

Fig.4: Manufactured electrical larynx

This type of artificial larynx has been manufactured by a company in Japan (Ifukube et al., 1999). Up to now, about 1,500 devices have already been sold. Figure 4 shows one such manufactured electrical larynx. The device also has a singing mode by which the laryngectomees are able to sing 5 songs by using melodies stored onto the memory. It is priced at $700 (US).

2.5   A GUI screen reader for Japanese blind people

We have proposed a screen reader software as a supporting tool for GUI operation by blind Japanese. The proposed software transmits information on the screen to blind users through synthesized speech. It does not require the use of pointing devices; all computer operations can be done from the keyboard. Our screen reader builds an environment in which blind people can use applications like word processors in Microsoft Windows on their own. The keyboard commands were designed not only to enable blind people to use Windows, but also to let them enjoy such advantages of Windows 95 as common interfaces among applications and multitasking operations.

In view of the above, we developed a prototype screen reader in 1996 called the 95-reader. Our screen reader for Windows is composed of several pieces of software (Watanabe et al., 1998): the hooking program, the main reading program, the speech synthesis driver, and the IME (input method editor) for blind people.

Based on the results of the evaluation done by blind users, the prototype screen reader was revised. In particular, a reading mode was added, thereby enabling the user to choose the speed, pitch, length of pauses at punctuation, accent and the kind of voice (male, female, or computer). This revised screen reader was put into practical use in 1997 and has become widespread throughout Japan. The screen reader has been improved to make it compatible with Windows 98 and 2000. Almost 80% of blind users in Japan are employing our screen reader. Figure 5 shows the package case of the screen reader.

Fig.5: A package case of the Windows 95 Screen-reader

By using the GUI screen reader, blind users will be able to utilize a LAN application – including the use of Internet mail – to enable them to send and receive documents with not only other blind people but with sighted people as well. Such a change should greatly improve their working environment.

2.6   Obstacle detection for the visual impaired

Using a microprocessor and ultrasonic devices, 20 years ago we developed a new model of a mobility aid in order to enable blind people to better perceive their surroundings. The ultrasonic devices transmit swept ultrasounds within 1 ms, while reflected ultrasounds are slowed down by a rate of 50:1, resulting in their being heard as localized audible sounds. This mobility aid is modeled after a bats' echo-location system. Our evidence shows that this device is very effective, especially at detecting small obstacles which appear in front of the head (Ifukube et al., 1991).

However, it is generally recognized that most blind people can detect obstacles without a mobility aid by using an ability known as the "obstacle sense". To better understand the mechanism of this obstacle sense, we have conducted certain psychophysical experiments of blind students. From the experimental results, we have ascertained that the reason why the blind can detect the obstacle is due to their ability to discriminate the tiny difference in sound field between themselves and an obstacle compared with when there is no obstacle present.

Fig.6: Speaker array for displaying sound spatial patterns

Furthermore, we were able to make the blind "hear virtual obstacles" by controlling the sound field produced from a speaker array which is shown in figure 6. This study has also been related to virtual reality research. In future, we intend to design another mobility aid device that will make use of the ability of the obstacle sense.

2.7   Virtual reality studies for perceptual disorders

Since 1997, our laboratory has been conducting a research project on a mixed reality (MR) system with the aim of constructing a kind of augmented reality. This is being done in cooperation with the Japanese goverment and Canon company. MR involves integrating elements from the real world with a virtual world. Information from reality and virtual reality are put together and displayed in the MR environment. Before applying the MR technologies to aids for perceptional disorders, we evaluated how the MR stimulation influenced the human body by using MR equipment including an arch screen (figure 7), a motion base and a biomedical measurement tool installed at the Sapporo research branch.

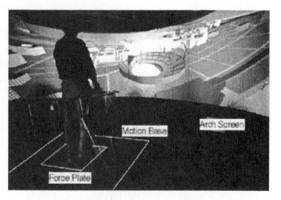

Fig.7: Arch-screen and motion base with two force plates inside the mixed reality system laboratory

From the evaluation experiments, we have obtained a principle guideline for the HMDs and 3D-displays (including their content) (Ifukube, 2001). Based on this guideline, we have determined how the MR technologies can be applied as an aid for perceptual disorders without any harmful effects.

Up to now, we have proved that MR technologies are useful in designing an auditory aid with both a speech recognition device and a see-through HMD, as mentioned in section 2.2. We have also suggested that a moving sound image added to a moving visual image is helpful for rehabilitation of the sense of balance, especially in the case of the elderly.

3. Conclusion

The basic findings concerning human sensory functions and assistive devices for the disabled will be useful for designing information devices including the virtual reality system. In the near future, the newly developed technologies of human computer interaction will be applied to the design of better models of assistive devices. This is our research approach of universal accessibility as schematically shown in figure 8.

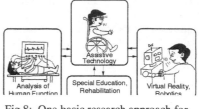

Fig.8: One basic research approach for UAHCI

Acknowledgement
The author thanks Shuichi Ino, Norihiro Uemi, Chikamune Wada, Norishige Nagai, Yoshito Nejime, Tetuya Watanabe and graduate students involving in this study for their help in running the experiments and designing the devices. Funding was supported by Grant-in-Aid from the Ministry of Education and Science (No. 10305031).

References

Aoki, N., Ifukube., T., (1998). Analysis and perception of spectral 1/f characteristics of amplitude and pitch fluctuations in normal sustained vowels. *J. Acoust. Soc. Am* 106, 423-433.

Ifukube, T., (1982). A cued tactile vocoder. In J. Raviv (ed), *Computers in Ading the Disabled* (pp. 197-215). North-Holland Publishing Co.

Ifukube, T., (1984). Design of a Voice Typewriter. *CQ Publishing Co.*. (195 pages, in Japanese),

Ifukube, T., (1997). Sound-based Assistive Technology for the Disabled. *Corona Publishing Co.*, (250 pages. in Japanese)

Ifukube, T., (2001). A guideline for the design of 3D-displays based on physiological parameters for use in an MR environment. *Proceedings of Inter, Symp. on Mixed Reality, Special Session: Final Report of the MR Project*, 116-124.

Ifukube, T., Sasaki, T., Peng, C., (1991). A blind mobility aid modeled after echolocation of bats. *IEEE Trans., BME* 38 (5), 461-465.

Ifukube, T., Uemi, N., (1999). A New Electrical Lalynx with Pitch Control Function. *Proceedings of the 2nd East Asian Conference on Phonosurgery,* 1-5

Ifukube, T., Wada, C., Ino, S., Shoji, H., Tsunashima, H., (2000). On a project for practical use of a hearing assist device using the tactile sense. *Technical Report of IECE,* WIT00-14, 13-18

Ifukube, T., White, R. L., (1987). A speech processor with lateral inhibition for an eight channel cochlear implant and its evaluation. *IEEE Trans., BME* 34 (11), 876-882

Kosakai, A., Nara, H., Ino, S., Yoshida, J., Ifukube, T., (2000). A speech recognition assist device for the hearing impaired which presents both in complete verbal information and non-verbal information. *Technical Report of IECE,* 13, 7-12

Lu, J., Uemi, N., Li, G., Ifukube, T., (2001, in press). Tone Enhancement in Mandarin Speech for Listeners with Hearing Impairment. *IEICE Transactions on Information and Systems.*

Matsushima, J., Miyoshi, S., Ifukube, T., (1996). Surgical method for implanted tinnitus suppresser. *Inter. Tinnitus Journal*, 2 (1), 21-25

Miyoshi, S., Ifukube, T., Matsushima, J., (1999). Proposal of a new method for narrowing and moving the stimulated region of cochlear implants -animal experiment and numerical analysis. *IEEE Trans., BME.,* 46 (4), 451-460

Nejime, Y., Aritsuka, T., Ifukube, T., Matsushima, J., A., (1996). Portable digital speech- rate converter for hearing impairment. *IEEE Trans., Rehabilitation Eng.,* 4 (2), 73-83

Watanabe, T., Okada, S., Ifukube, T., (1998). Development of a GUI screen reader for blind persons, *Systems and Computers in Japan.* 29, 13, 18-27.

The development of a tool to enhance communications between blind and sighted mathematicians, students and teachers:
A global translation appliance[1]

Arthur Karshmer[a], Gopal Gupta[b], Klaus Miesenberger[c], Enrico Pontelli[d], Hei-Feng Guo[e]

[a]University of South Florida
[b]University of Texas at Dallas
[c]University of Linz, Austria
[d]New Mexico State University
[e]State University of New York at Stony Brook

Abstract
Mathematics as an area of advanced study or practice is hampered by the very tool that makes it possible: the Braille based math notation. These notations suffer from several problems from the fact that they are context sensitive to the problem that there are numerous standards used in the world. The current work is an effort to automatically translate between various Braille based math notations. Then, for example, a blind mathematician in Germany can convert his Marburg math representation to Nemeth and email it to a colleague in the United States.

1. Introduction

The primary difficulty encountered by a visually-impaired student in pursuing studies in Science, Mathematics, Engineering or Technology (SMET) is how to read and write mathematics. To overcome the limited expressiveness of six dot based Braille, a plethora of notations for marking-up mathematics have been devised: these include notations such as the Nemeth Math code, the Marburg code, the French standard, the Stuttgart standard, etc.

One problem with these notations is that they are Braille-based and designed specifically for visually impaired students/teachers/researchers of mathematics, and thus are not known to sighted individuals. As a result, written technical communication between sighted individuals and visually impaired individuals is quite difficult. A seemingly simple solution for solving this problem is to build translators that automatically translate back and forth between mark-up notations used by the visually impaired (such as Nemeth [Nemeth, 1972] code and Marburg code [Marburg-Lahn, 1946]) and those used by sighted individuals such as LaTeX and MathML. Communication between visually impaired individuals is also difficult because each group uses a different notation (Marburg vs. Nemeth, for example).

Our solution to solve this problem is to build automatic translators that translate one notation for the visually impaired to another, and back. However, these translators are difficult to build due to the complex and context sensitive nature of these Braille-based mathematics mark-up notations. Recently we have had remarkable degree of success building translators from Nemeth math notation to LaTeX using an innovative approach based on logic programming and denotational semantics.

The task of building such a translator was hitherto considered impossible. In this project we propose to leverage our techniques, based on logic programming and denotational semantics to develop a general framework in which translation can be automatically achieved within a set of mark-up languages. For this project, this set will consist of the Marburg code, Nemeth code, LaTeX and MathML, but there is no reason that this set can't be expanded in the future to include more braille codes and markup languages. The tool we will develop will allow free inter-conversion from one notation to another. Our approach is based on developing a common intermediate format (CIF) for representing mathematics, and building translators using the technology of logic programming and denotational semantics for translating back and forth between the Marburg, Nemeth, LaTeX, and MathML notations and the CIF (see Figure 1)

1   Some of the work reported in this paper was supported by the U.S. National Science Foundation under project number HRD #9800209

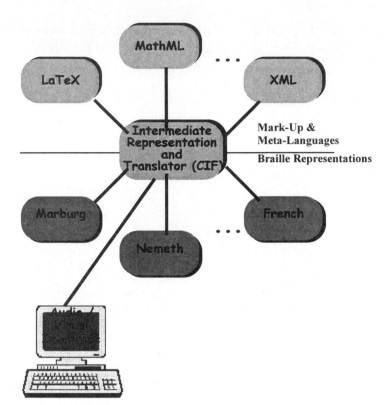

Figure 1. The Overall Structure of the Translator

The CIF will also be used to develop mark-up notation independent auditory browser for understanding of complex mathematical expressions by visually impaired users. The auditory browser will be designed to convey the structure of a mathematical expression as well as its content via speech output. The user will also have the ability to interactively navigate the expression and focus on its subparts in order to better understand the expression.

The project work will be done by an international team of computer scientists from USA and Europe who have considerably expertise in disability-related research. The final piece of the project will a sound-based tool for directing the translation process that will be designed for use by both the sighted and visually impaired user. Our team has had considerable experience in designing and building these devices.

2. The Translation Problem

A major problem with automatically translating languages designed for marking-up mathematics for the visually impaired is that they are very complex. Complex to the point that they cannot be processed automatically on the computer. This complexity is present due to two main reasons:

- These languages were designed decades ago, when very little was understood about computer processability of languages. For example, the design of the Nemeth math notation (done in 1951) pre-dates Chomsky's seminal work that led to the study of syntactic structure of languages [Chomsky, 1976]. In fact, the computer was just in the process of being developed then.
- These languages are designed to facilitate reading by blind readers. Thus, they tend to make the context in which an expression occurs explicit in the notation. This is done in order to make sure that the blind user is aware of the context at any given moment. For example, the level of exponent in the Nemeth math notation is explicitly stated for each exponent. Thus, the expression, x^{y^2+1} will be coded in (ASCII) Nemeth as

x\^\,y\^\,\^\,2\^\,+1 with two \^\, before **2** to indicate that it is in a second level exponent. This building of the context into the expressions puts notations such as Nemeth math and Marburg in the class of languages called context sensitive [Aho & Ullman, 1986]

Due to this complexity, it is very hard to construct parsers for these mark-up languages using the traditional approaches used in computer science.

Note that a parser is a software system that will check if a Nemeth or Marburg document is syntactically correct, and produce a parse tree. A parser is a first step in any automatic computer processing of Nemeth, Marburg, or any other notation for that matter. In fact, after the parsing process is over, the information in the marked up document is encapsulated in the parse tree which is used for any further processing of that document. Traditional parsing technology is designed only for those languages that are context-free and that are LALR(1) [Aho & Ullman, 1986]. Mark- up languages for visually impaired such as Nemeth or Marburg are neither.

In fact, the problem of parsing these mark-up languages was considered unsolvable until recently, when Gupta, Karshmer, and Guo proposed an approach based on logic programming and denotational semantics that resulted in the first ever parser being built for Nemeth math [Guo et al., 2000; Karshmer, Gupta, et al., 1998 a&b]. In this work we leverage this approach further to build logic programming based translators for translating

Marburg to Nemeth, Marburg to LaTeX, etc. In fact, we go a step further and propose to design a system in which the mappings between patterns of mark-up are specified at a very high- level, and a logic programming system that acts as a translator is generated automatically.

3. Conclusions

The software systems developed will greatly aid the visually impaired community in study of mathematics. The ability to translate between any of the four formats Nemeth, Marburg, LaTeX, and MathML will allow blind mathematics students to access mathematics documents coded in various notations. Thus, a student who only knows Nemeth, can also access Marburg documents. He/she can also access math expressed in LaTeX or math published on the Web using MathML by using appropriate translators.

Visually Impaired students/researchers/teachers will be able to understand complex mathematical expressions better with the use of the auditory browser via repetitive navigation and abstraction techniques.

The flexibility of the CIF will induce designers of other Braille-based math notations (such as the Stuttgart standard and the French standard) to develop translators from their notation to the CIF and back, thus making their documents accessible to other blind users. At the same time, they will make documents coded in other notations accessible to users who are only familiar with their standard.

The Global Translation project, based on technologies already developed, holds the promise of allowing visually impaired mathematicians to communicate with both sighted and visually impaired colleagues around the world. Currently this process can only be achieved with the intervention of human translators, adding time and complexity to the procedure.

The current work will automate the translation process between standard mark-up languages and numerous Braille-based math notations.With the addition of a user interface designed to be used by both sighted and visually impaired users, the project should open simplified communications between mathematicians around the world.

References
Aho, A, Ullman, J., and Sethi, R. (1986). *Principles of Compiler Design.* New York: Addison- Wesley.
Chomsky, N. (1976). *Syntactic Structures.* The Hague: Mouton.
Guo, H-F., Gupta, G., Karshmer, A., et al. (2000) ."Computer Processing of Nemeth Braille Math Notation." *Proceedings of the 7th International Conference on Computers Helping People with Special Needs (ICCHP 2000).* Vienna: OCG Press

Karshmer, A., Gupta, G. , Geiger, S., and Weaver, C., (1998). "A Framework for Translation of Braille Nemeth Code to LaTeX: The MAVIS Project." *Proceedings of the 3rd ACM Conference on Assistive Technologies.* New York: *ACM Press.*

Karshmer, A., Gupta, G., Geiger, S. and C.Weaver, (1998). "Reading and writing mathematics: The mavis project." London: *Behaviour and Information Technology*, 18(1), 2-10.

Marburg-Lahn Braille, Blindenstudienanstalt, Marburg-Lahn, Germany 1946

Miesenberger, K., Batusic, M., St\öger, B. (1997) "LABRADOOR: LaTeX-to-Braille-Door, Le LabradoorLaTeX-to-Braille-Door: une passerelle du LaTeX au Braille, www.snv.jussieu.fr/inova/ntevh/labradoor.htm

Nemeth, A. (1972). *The Nemeth Braille Code For Mathematics And Science Notation: 1972 Revision.* New York: American Printing House For The Blind.

Predictive and Highly Ambiguous Typing for a Severely Speech and Motion Impaired User

Michael Kühn, Jörn Garbe

Fachbereich Informatik
Universität Koblenz-Landau
Postfach 20 16 02, D–56016 Koblenz, Germany
kuehn@uni-koblenz.de
jgarbe@acm.org

Abstract

Typing commands is the access of choice for many experienced computer users because of the speed and flexibility frequent interactions can take place and the ease of defining and using macros. One specific group of persons who can be counted as frequent computer users are speech impaired people relying on electronic devices for augmentative and alternative communication (*AAC*).

But many speech impaired users are severely motor impaired, too. Therefore, the usage of a physical or on-screen keyboard is prohibitively costly for them with a standard keyboard layout. This paper presents a communication aid based on a highly ambiguous keyboard with word completion for communication, interaction and navigation.

1. Electronic communication for severely speech impaired users

One outstanding feature of the information society nowadays is the importance of text as communication media in e-mails, SMS, newsgroups, and webpages. Even in synchronous situations like chatting electronic devices (computers, mobile phones, hand-helds) are used to transmit textual information . The success of ubiquituous communication has partly its reasons in the advantages of written messages via SMS: 1. they may be used for a synchronous or asynchronous dialogue as wanted; 2. they can be absorbed very fast and 3. they can be displayed in private (an important point for the use in classrooms). Likewise, chats, e-mails and newsgroups have become attractive alternative channels of synchronous or asynchronous communication for a varity of virtual communities (groups of interests).

For the fraction of population affected by analphabetism, this development establishes a new barrier for participating in society. On the other hand, electronic communication could become an equally accessible communication channel for speech impared members of tomorrow's information society. However, many strongly speech impaired people have severe motion impairments, too, because of the same causation: by birth (e.g. cerebral palsy), by accident or by disease (e.g. stroke, Lou Gehrig's or multiple sclerosis). For them, typing text with a standard keyboard or the usage of a graphical input device might be impossible. Often only the operation of a single or very few switch signals via buttons, joystick, eye tracking, EEG or other sensors is possible.

There are several solutions available for substituting the physical keyboard of a computer by a virtual, on-screen keyboard with different layouts that can be operated with one or two physical switches by scanning: In the simple case of linear scanning, the user presses a switch A repeatedly to step from one key of the virtual keyboard to the next key of the scanning cycle until he reaches the desired key that is then activated by pressing a switch B. For row-column scanning, the user first steps through rows of keys with switch A, selects the wanted row by switch B and then steps through the columns of keys in this row, finally activating the desired key again with switch B. In case only one physical switch is operated, one of the switches can be substituted by self-activation after a certain delay.

Thus, the operation of a keyboard is simulated with one or two physical switches, but typing text letter by letter this way is cumbersome: in general, the typing of a word of six letters on a 32 keys keyboard by row-column scanning will take dozens of scanning steps. The achieved written communication rate ranges from less than 1 to 5 wpm (*words per minute*) — a typist's rate in comparison ranges from 10 to 20 wpm depending on her level of expertise (Darragh and Witten, 1992).

2. Ambiguous keyboards and word prediction

Instead of reducing the number of required keystrokes for a word by using an enlarged keyboard with e.g. iconic codings, Kushler (1998) proposes the use of an ambiguous keyboard with multiple letter key assignments in the way phone keys are used for the so called vanity phone numbers: every phone digit key from 2 to 9 is additionally labeled with at least three letters. Instead of memorizing digit sequences for the phone number of telephone services, it suffices to dial the phone keys labeled with the letters of e.g. the name of the provider.

If arbitrary words are coded this way many encodings will be the same for different words, i.e. the decoding function is not injective. Therefore, the user has to select the intended word in a list of word suggestions (ordererd e.g. by word frequencies) whenever the mapping between the encoding and the intended word is not one-to-one. Nevertheless, as Witten (1982) already pointed out, with nine letter keys only eight percent of 24,500 words actually are encoded ambiguously and thus need adisambiguation step. Following studies (Foulds et al., 1987) concentrated on ambiguous typing where only the next character is predicted based on n-grams and corrected if necessary, as the size of an electronic full word or even syllable dictionary would have been too limited (Arnott & Javed, 1992). The drawback of this character-based disambiguation is that the user has to pay attention to the predicted word prefix after each keystroke.

Today, the storage of large vocabulary in main memory of a standard computer is not a serious problem anymore. The advantages of an ambiguous keyboard with eight keys and word disambiguation are enumerated by Kushler (1998) for users of alternative and augmentative communication devices as follows: 1. The efficiency of an ambiguous keyboard is near to one keystroke per letter. 2. Apart from literacy, no memorization of special encodings is required. 3. Attention to the display is required only after the word has been typed. 4. A keyboard with fewer keys can have larger keys for direct selection. 5. The average time to select a key by scanning is reduced considerably. 6. Simple, linear scanning can be used efficiently to select a key. 7. Fewer keys may allow direct selection with various input devices.

Surely, the main arguments for a character-based text input with an ambiguous keyboard are that the keystroke per letter efficiency is very high and that there is no need for the user to learn any special encodings besides standard orthography. Further, if the user did a typing mistake, this error is corrected more easily as if the typing mistake is done via scanning on a 100+ keys keyboard. Finally, a character-based vocabulary can be extended very easily with new words, whereas an iconic language system has to depend on the user to memorize the icon word associations and reaches its limits, if we do not restrict ourselves to a base vocabulary of a few thousand words.

The use of an ambiguous keyboard is compatible with another technique to accelerate typing input: the prediction and completion of user input. Actually, the completing prediction of commands and other user input (e.g. Darragh and Witten, 1992) to accelerate keyboard input for motion impaired users is a well-known and much used feature of standard software like word processors or command shells nowadays. The key idea is to present the user at every step of user input a ranked list of predictions what the user is going to type next based on a statistical model of the recent input.

On top of reducing the typing costs there is another advantage of this input enhancing feature for users who do not always memorize exactly the set of admitted inputs in a certain context: The user can type the prefix of the intended input and request a list to be displayed for possible completion alternatives, thus reducing the cognitive load of memorization for orthography (Newell et al., 1995).

3. The UKO pilot study

UKO is the abbreviation for German "*Unbekanntes Kommunikationsobjekt*" (i.e. "unidentified communication object") – the name given to our communication aid by its pilot user, a fifteen-year old girl with cerebral palsy who visits a regular school. This communication aid consists of an ambiguous keyboard with four letter keys plus an "enter" and a "delete" key. The keys are scanned cyclically. A second physical switch is used to activate the word completion with the word suggestions depicted on the right side (as in Figure 1).

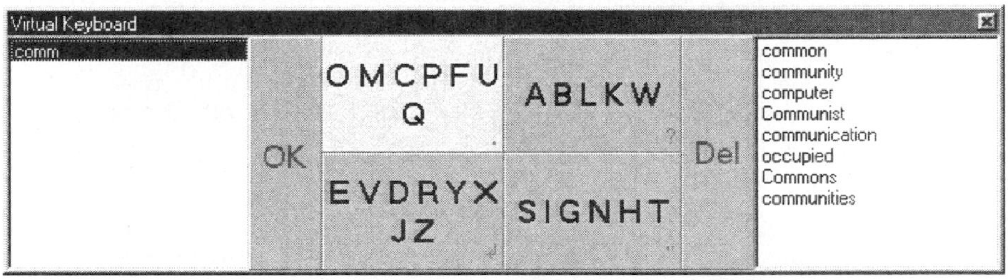

Figure 1. The UKO virtual keyboard (after selecting the first key four times)

In his diploma thesis, Garbe (2000) followed an evolutionary optimisation approach to find an ambiguous letter key assignment for a keyboard with six keys that is to be used by cyclic scanning (cf. Levine and Goodenough-Trepagnier, 1990). The UKO system includes keyboard layouts for German and English and the corresponding lexica from the CELEX lexicon database with 360,000 resp. 160,000 word form entries (Baayen et al., 1995). Using the 360,000 German word form lexicon including word frequencies information, the ambiguous word form lists do not grow longer than 34 entries.

This motivated us to look further for even smaller keyboards, and, actually, a keyboard with only three letter keys is used now, with a fourth key functioning as a quasimodal command key (Raskin, 2000) that is used in connection with one of the letter keys to invoke the "enter", "delete" and "complete" function of the UKO system. These three plus one keys will be selected directly by our pilot user who formerly has used two-switch-scanning until now. Therefore, another considerable typing speed-up is expected.

But even with linear scanning the six keys of the UKO keyboard as depicted in Figure 1, experimental results compared with an iconic communication system are encouraging: an time improvement of 18% could be achieved on a small test set of ten sentences with ten words in average, mainly since there were on average two unknown words per sentence using the iconic vocabulary, which then had to be spelled rather costly. Using the larger CELEX lexicon for the UKO system, only two of the 101 words in total were unknown for the ten sentences. The sentences with no unknown words have been entered in approximately the same time of 8.56 seconds per word resulting in word rate of 6.94 wpm. The overall achieved word rate for this sample is 5.12 wpm for the iconic system and 6.22 wpm for the UKO system.

4. Computer interaction and navigation

Shneiderman (1986) points out that for experienced "power" computer users command typing is their way of choice for computer access because of the speed and flexibility frequent interactions can take place and the ease of defining and using command macros. Certainly, many people relying on electronic devices for augmentative and alternative communication can be accounted to this group, as their need for communication leads to frequent computer usage. Using an ambiguous keyboard, commands (as well as other inputs like numbers or special characters) may be treated like words where the dictionary of possible word inputs consists of the possible command names (numbers, special characters).

This way, interaction with the communication aid as well as with an external device like a computer, a TV or another environmental control is performed as flexibly and efficiently as the written communication with the ambiguous keyboard itself. The user is not required to switch between different interaction models but stays in the framework of ambiguous word typing with completion. Finally, defining and using macros for text and commands in any arbitrary mixing gives the user of an AAC device the possibility to adapt this tool to her needs better than anybody else could do.

Two-dimensional navigation in texts on a display is a frequent task, to search for a specific location that may not be visible in the display, but also to point to a location on the screen for e.g. selecting a hyperlink. The spatial navigation to an arbitrary position on a display in general takes dozens of keystrokes if only the four cursor keys are available. Raskin (2000) highlights the advantages of an incremental search as it is offered by the text editor Emacs: In a forward or backward incremental search, the cursor jumps after each entered character of the search string to the next (or last) text location from the starting position and highlights the region of text that matches the prefix of the search string that was entered so far. If this is not the aimed text position, the user either enters more characters of the search string or jumps to the next occurrence of the prefix entered so far.

Incremental search has several advantages in comparison to delimited search and cursor navigation: 1. The search is stepwise undoable by deleting the last character of the search string prefix. 2. It is savvier with user time, starting with search as soon as the user enters one character. 3. The user only has to enter the smallest prefix of the search string really required to find the desired location. 4. Mistyped or inappropriate inputs are detected and corrected (or improved) quickly. 5. In contrast to cursor navigation, the user only says where to go, not how.

Incremental search is fully compatible with ambiguous keyboards. The text where the search takes place itself restricts the disambiguation context of the search pattern, such that the input will be less ambiguous than it would be entered e.g. as free text in a document. Of course, for browsing through an unknown context, another kind of navigation is needed. But if the user knows the text, incremental search is a very efficient and adequate way of finding and selecting a specific text position even with an highly ambiguous keyboard.

5. Conclusions

Following the idea of Witten (1982) and the argumentation of Kushler (1998), we propose the use of highly ambiguous keyboards for people with severe motion impairments to communicate and interact in the information society. We have given the arguments for ambiguous keyboards for users relying on a very limited set of input sensors. We have presented a pilot study of our communication aid and compared some test results with the word rate via scanning of a large-scale iconic keyboard. We have extended our investigation to include general computer interaction and text navigation and argue for the compatibility with highly ambiguous keyboards. This way, we gain an uniform approach to electronic communication and interaction for severely motion impaired users.

Future work will include implementation and evaluation of macros and incremental search with ambiguous keyboards. A stronger investigation of linguistic knowledge, specific context domains as well as other natural language processing techniques may lead to a further speedup of highly ambiguous text input – for a programmatic argumentation on natural language processing for alternative and augmentative communication, we refer to (Newell, Langer & Hickey, 1998).

We like to stress at this point that not only motion impaired users may benefit from highly ambiguous typing. There is the commercially very successful system "T9"®[1] from Tegic Communications that provides in several languages an easy text input for mobile phone short messages as well as for handhelds. Other competing systems use another keyboard labeling ("Octave" by e-acute[2]) or mixed unique/ambiguous letter keys ("WordWise" by Eatoni Ergonomics[3]). We do not know of another ambiguous text input system using only three or four letter keys as it is the case in our approach.

Predictive and highly ambiguous typing cannot promise to give the motion motor impaired user the same speed access as a standard keyboard or even voice input for unimpaired users does. In fact, our first 'power' user of the predictive and ambiguous typing, a 15-year old girl with athetotic cerebral palsy visiting a regular school, gains an input speed of 6 words per minute scanning with two physical switches through an on-screen keyboard with four plus two keys – approximately with the same speed she entered words with her former, icon based communication device if the words are contained in its rather restricted dictionary. But she would not miss the flexibility of just

---

[1] http://www.t9.com/
[2] http://www.e-acute.fr/
[3] http://www.eatoni.com/

writing words letter by letter without checking that the words are available and without having to remember thousands of (picture-based or other) encodings, additionally to the usual orthography.

References

Arnott, J.L, Javed, M.Y. (1992). Probabilistic character disambiguation for reduced keyboards using small text samples. *AAC Augmentative and Alternative Communication*, 8(3).

Baayen, R.H., Piepenbrock, R., Gulikers, L. (1995). *The CELEX Lexical Database (Release 2) [CD-ROM]*. Philadelphia, PA: Lnguistic Data Consortium, University of Pennsylvania.

Darragh, J.J., Witten, I.H. (1992). *The Reactive Keyboard*. Cambridge Series on Human-Computer Interaction. Cambridge University Press.

Edwards, A.D. (1995a). *Extra-Ordinary Human-Computer Interaction: Interfaces for Users with Disabilities*. Cambridge Series on Human-Computer Interaction. Cambridge University Press.

Edwards, A.D. (1995b). *Computers and people with disabilities*. Chapter 2 in: Edwards (1995a), pp. 19-43.

Foulds, R.A., Soede, M., Balkom, H. van (1987). Statistical disambiguation of multi-character keys applied to reduce motor requirements for augmentative and alternative communication. *AAC Augmentative and Alternative Communication*, 3(4).

Garbe, J. (2000). Optimizing a layout of an ambiguous keyboard using a genetic algorithm. Diploma thesis, Universität Koblenz-Landau.

Kushler, C. (1998). AAC using a reduced keyboard. *Proceedings of CSUN 98*. http://www.dinf.org/csun98/csun98_140.htm .

Levine, S.H., Goodenough-Trepagnier, C. (1990). Customised text entry devices for motor-impaired users. *Applied Ergonomics*, 21(1).

Newell, A.F., Arnott, J.L., Cairns, A.Y., Ricketts, I.W., Gregor, P. (1995). *Intelligent systems for speech and language impaired people: A portfolio of research*. Chapter 5 in: Edwards (1995a), pp. 83-101.

Newell, A., Langer S., Hickey, M. (1998). The rôle of natural language processing in alternative and augmentative communication. *Natural language engineering*, 4(1).

Raskin, J. (2000). *The Humane Interface: New Directions for Designing Interactive Systems*. Addison-Wesley.

Shneiderman, B. (1986). *Designing the User Interface: Strategies for Effective Human-Computer Interaction*. Addison-Wesley.

Witten, I. H. (1982). *Principles of Computer Speech*. Academic Press Inc., London.

Investigating the parameters of force feedback assistance for motion-impaired users in a selection task

Patrick Langdon[1], Simeon Keates[1], P. John Clarkson[1], Peter. Robinson[2]

[1] Engineering Design Centre, University of Cambridge,
Department of Engineering, Trumpington Street, Cambridge, CB2 1PZ UK

[2] Computer Laboratory, University of Cambridge, New Museums Site,
Pembroke Street, Cambridge CB2 3QG UK

Abstract

This paper reports empirical studies investigating the parameters of use of a mouse-based, haptic interaction device as an assistive aid for people with motion impairments resulting from disease or injury. Force feedback was generated within an inexpensive, mouse-based force feedback device in order to present constant directional forces, and tactual and vibration sensations corresponding to user interface events. Two experiments made a detailed investigation of specific parameters of a realistic, point-and-click GUI selection task. The results suggest that parameters obtained from temporal and frequency analysis of cursor movements can be used to modulate force feedback in user interfaces, to the benefit of motion-impaired users.

1. Introduction

Medical conditions such as Cerebral Palsy, Parkinson's disease, Muscular Dystrophy, spinal injuries and the after-effects of stroke, give rise to symptoms such as reduced strength, restricted movement and a continuum of impairments involving spasms, tremors and movement control problems. These populations display irregular, jerky movements, possibly in conjunction with slower, linked writhing (athetosis and chorea). They may also be poor at sequencing and co-ordinating movements (apraxic) or may have essential tremor with repetitive movements (dyskinesia) and spasms (rhythmic myoclonus). Motion impairments may arise from central nervous system dysfunction or be muscle spasms and dystonia arising from other neurological damage or dysfunction. Many of these impairments are prevalent in the older population. Despite these problems, computers can be beneficial to those with motion impairments (Busby, 1997). This paper reports two empirical studies investigating the interface and force feedback parameters of use of a mouse-based, haptic interaction device as an assistive aid.

1.1 Haptic feedback

The existing GUI paradigm relies principally on visual feedback, often supported by sound and with minimal involvement of haptic feedback. Haptic perception is the gathering of information about objects outside of the body through the tactile and kinaesthetic senses (Loomis and Lederman, 1986). Motion-impaired users often exhibit decreased motor control and muscle strength, but not necessarily a decreased haptic sensitivity. The sensitivity of able-bodied and motion-impaired users to haptic feedback has been demonstrated using devices such as the Phantom. However, this is an expensive research tool that is unlikely to be used routinely as a general-purpose interaction device. Force feedback technology is also available using a mouse that is, in principle, capable of generating both tactual and force feedback haptic interactions with the user as a result of its very wide range of movement generation capabilities. It has also been shown that such a device can improve interaction for able-bodied users in cursor control tasks (Dennerlain 2000). It may be possible to enrich the standard user interface with haptic textures and edges in order to signal the location of regions as the mouse passes over them. In addition, it is also possible to use force feedback within the input device to present constant directional forces, and tactual and vibration sensations corresponding to user interface events. This force feedback mode has the capability of boosting or aiding user input, in the case of muscle weakness, and damping or restraining user inputs, in the case of muscle spasm or tremor.

The relative utility of haptic feedback using devices such as the force feedback mouse has been demonstrated by the authors in work previously reported (Langdon, Keates, Clarkson, Robinson, 2000). This paper describes two experiments that extend those findings by examining the improvement in users performance in a point and-click selection task with varying parameters and also introduces a new approach to specifying the dynamic force feedback necessary to improve interaction through the analysis of user cursor movements.

2. Force feedback experiments

Trials were carried out with 6 motion-impaired users at the Papworth Trust (Cambridge, UK), a charitable organisation dedicated to the care of the motion-impaired. Vibro-tactile and force feedback were varied in tasks representative of the standard User Interface (UI). The users were sampled so as to represent a wide range of movement capabilities. Two users; PV1 and PV2, were non-ambulant and severely impaired by some combination of tremor, spasm or weakness; PV3, PV4, PV5, PV6 were more capable, three being ambulant. The users were partially representative of populations with Cerebral Palsy, Freidrich's Ataxia and Kalman-Lamming's syndrome.

2.1 Experiment 1

This experiment examined whether force feedback was discernible in a realistic UI task and whether it was effective in improving the selection performance of motion-impaired users. This task was repeated both with and without the force feedback active. Four motion-impaired users, PV2, PV3, PV4 and PV6, participated in this experiment.

Figure 1. The target display layout for experiment 1.

The performance improvement over time was also measured as an indication of learning. A 2D flattened projection map of North America was drawn on the screen and the 10 target locations were distributed as fixed city locations on this map. After the start signal was displayed, the user was required to move to, and click on, the inner circle of the target circle whose outer circle area was flashing. The outer circle sizes remained constant with the inner circles increasing in diameter for the easier conditions. A trial consisted of moving from one target city to another as it flashed. The program then lined together the targets visited by the user. The targets were always visited in the same sequential order, enhancing the learning effect over trials, as discussed below. During the force feedback assisted trials the mouse was strongly attracted to the center of the target once the outer circle was reached. During the unassisted mouse trials the interface behaved as a normal point-and-click mouse. The users performed the task using the easy, medium and hard settings, corresponding to differing sizes of targets.

Scores were substantially improved during the force feedback assisted trials, at around half the unassisted time on average. User PV2 was unable to perform the task in the unassisted mode, taking an excessive time to complete half the targets. Figure 2 show the time to complete a trial averaged across the motion-impaired users for 10 trials. There was a corresponding decrease in errors. The step increase in times between trials 5 and 6 arose because the trials were conducted over a period of 2 days, one week apart, and this corresponds to the week-long break in the trials. There is evidence of significant improvement in times over trials implying that learning occurred over the two batches of 5 trials (Three factor ANOVA F=2.73; df = 9, 27; p< 0.05), but the step increase implies that there is a potential for unlearning as well, without regular exposure.

EFFECT OF FEEDBACK, DIFFICULTY AND TRIALS ON TIME TO
COMPLETE

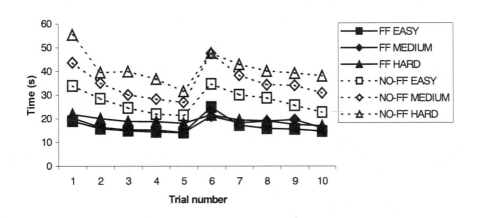

Figure 2. The times to complete a trial averaged across the motion-impaired users.

2.1   Experiment 2

This experiment was a detailed investigation of specific parameters associated with a point-and-click, selection task. Four motion-impaired users, PV1, PV2, PV3, and PV4, participated in this experiment.

The aim was to quantify the effect of the test interaction variables on the measured time taken to select the target outer circle after it was highlighted. The experiment involved the users being presented with 16 target circles arranged equidistantly around a central circle on the screen. The aim was for the users to click on each target circle in a random order determined by the software (Figure 3). The experiment was performed without any assistance, with the following conditions: (1) no assistance just a selection task; (2) pointer trails behind the cursor during selection; (3) with 66Hz, saw-tooth vibration while the cursor was over the target area; (4) colour change to yellow while the cursor was over the target area and (5) force feedback gravity wells consisting of a spring force acting in the direction of the target edges with an effect range of 400% of the target diameter (6) all effects, consisting of gravity wells, colour, vibration and pointer trails together.

Figure 3. The target button layout for experiment 2.

The gravity well effect for the force feedback gives the sensation that that there is a second, larger circle around the target that corresponds to the extent of the gravity field. Entering that outer circle causes the cursor to become subject to the gravity and it is attracted by a spring force towards the center, but whose influence stops at the target edge. The average times obtained across all the users are shown in Figure 4. The use of a gravity well around a target appeared to improve the time taken to complete the task by approximately 10-20%. The most dramatic improvements were seen for the more severely impaired users as was revealed by the comparative cursor traces of able and impaired users aiming for the target circles without and with force feedback assistance (Figure 5). However, the average times shown in Figure 4 represent considerable variability between users and are also not directly comparable with the results from experiment 1 where, for example, the gravity well extended to the center of the target.

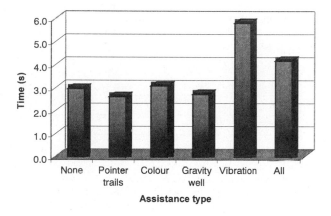

Figure 4. The average time per target, for the different types of assistance.

The addition of vibration, however, almost doubled the time to perform the task. The presence of vibration in the 'All' assistance type (a combination of all the other types) could have also adversely affected those results. These results correspond to those obtained using the Phantom with able-bodied users (Oakley 2000).

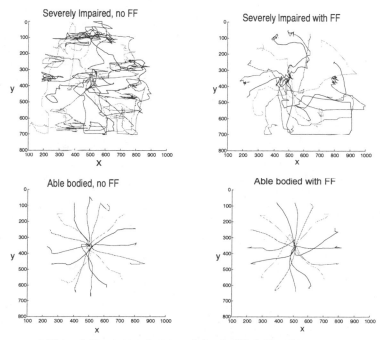

Figure 5. Cursor trace for users aiming for 20 pixel targets

3. General discussion

Overall, a strong positive effect of using force feedback to enhance interaction was observed for motion-impaired users. In particular, the times to complete the trials for the first task were reduced by 30–50% of times for unassisted interaction modes. The results for the gravity wells condition of the second task suggested a 20%-30% improvement in times in comparison with no assistance. However, this reduced effect may have been due to the gravity well attraction force ceasing at the target boundary. To examine this possibility, the experiment was repeated with the same users and method but using only the conditions: no assistance; gravity well to edge; and gravity well to center. Table 2 shows he results, to date, suggesting that the modified force feedback assistance has an impact of approximately 50% improvement over no assistance, compared with a 20% improvement for the gravity well to edge condition. These results are in agreement with both the results obtained from the first experiment and also those of the second experiment, though their overall significance awaits full statistical analysis.

Table 1. The time to select the target in seconds for two gravity-well conditions, averaged over 4 users

No Assistance	Edge Gravity-Well	Center Gravity-Well
4.8	3.4	2.5

The presence of a statistically significant learning effect over trials in experiment 1, raises the possibility of using force feedback as therapeutic training. Improvement in performance would be expected with continued use of the force feedback interface and learning may also carry over to the use of normal interfaces. However, force feedback needs to be complementary to the existing input. For example, the results suggest that specific types of vibration feedback may actually have a negative impact on specific aspects of the interaction.

4. Conclusions

A general conclusion of this work is that the availability of comparatively cheap force feedback devices means that the use of haptic interaction has the potential to become extremely important technology for enabling universal access to computers. The users tremor was significantly damped by the force feedback, an application whose value has been confirmed by others (Pledgie et. al, 1999). These experiments suggest that carrying out frequency and temporal analysis of users cursor movements in standard GUI sub-tasks will yield parameters that can be used in the modulation of force feedback to assist users. Such a modulated force-feedback may be pre-set prior to interaction or may be adapted to the user during the interaction. Two possible approaches suggest themselves: first the use of vibrational feedback aimed at enhancing GUI elements using primarily the cutaneous receptors, and secondly, the use of directional forces, both constant and transient, to influence the dynamic progress of the users' interaction through their effect on the kinaesthesic receptors.

References

Busby, G. (1997). Technology for the disabled and why it matters to you. In: *IEE Colloquium Digest Computers in the service of mankind: Helping the disabled*, Digest No. 97/117, 1997, pp 1/1-1/7.

Dennerlein, J.T., Martin, D.B., Hasser, C. (2000). Force feedback improves performance for steering and combined steering-targeting tasks. In: *Proceedings of CHI 2000* (The Hague, Netherlands), ACM Press, pp. 423-429.

Langdon,P, Keates, S., Clarkson, P.J., and Robinson, P. (2000). Using haptic feedback to enhance interaction for motion-impaired users;. In: *Proceedings of the 3rd International Conference on Disability, Virtual Reality and Associated Technologies. pp 25 – 32. Sardinia Italy*.

Loomis, I.M., Lederman, S.I. (1986). *Tactual Perception*. In: Handbook of Perception and Human Performance, Perceptual Organisation and Cognition.

Oakley, I., McGee, M.R., Brewster, S., Gray, P. (2000). Putting the feel in 'Look and Feel'. In: *Proceedings of CHI 2000* (The Hague, Netherlands), ACM Press, pp. 415-422.

Pledgie, S, Barner, K, Agrawal, S Rahman, T, (1999). Tremor suppression through force feedback. Proceedings of the International Conference on Rehabilitation Robotics, pp16 – 26. Stanford USA.

ViKI: A Virtual Keyboard Interface for the Handicapped

Blaise W. Liffick

Department of Computer Science, Millersville University
Millersville, PA 17551 USA
1.717.872.3536 - liffick@cs.millersville.edu

Abstract

Although computerized assistive devices for most handicaps are becoming more widely available, one major drawback is that users must currently provide customized interfaces for every computer system they use. In addition to specialized input and output devices, such assistive devices also frequently require customized software. This paper proposes an alternative solution, by providing a customized interface (both hardware and controlling software) on a laptop computer that is then used to control standard software on any other host computer. In this way, the user can carry their specialized interface to essentially any general computer, and can interact with any standard software supported on the host. The laptop system becomes a virtual keyboard to the host machine.

1. Introduction

As accessible as assistive devices have made computers for handicapped users of all types, there is still at least one situation in which such users of specialized interfaces are at a distinct disadvantage. Assistive devices generally allow a user to set up a one-to-one relationship between the specialized device(s) and a particular computer. While this certainly helps users to create a customized workstation that satisfies their personal needs, it is, nonetheless, allowing customized access to just one computer. There are many situations that come easily to mind where this is insufficient: laboratory settings where particular computers may have specific software or hardware needed by the user, that cannot be easily made available on their own workstation; academic settings where the user may need to use computers in more than one lab; business and industry settings where the user needs to collaborate with others on a project, and so must be able to access computers other than their own; any time the user travels but needs to access a computer.

The ViKI (**Vi**rtual **K**eyboard **I**nterface) project is an attempt to demonstrate the efficacy of using a laptop computer as a virtual keyboard to access a host computer that is running standard software. Figure 1 shows the Specialized Interface Laptop (SIL) system connected to a host computer running standard software that will be controlled by the user of SIL. SIL provides any specialized connections required for assistive devices, and any special screen display the user might need in order to interact with those devices (e.g. a scanning keyboard representation). Upon entering an input into SIL, the input is then passed on to the host computer in a form as if it were generated by a standard keyboard and mouse combination.

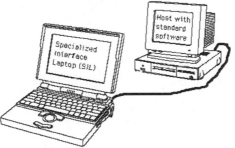

Figure 1: Specialized Interface Laptop (SIL) connected to host computer.

The hardware and software on SIL can be customized to the specific needs of the user, while the hardware and software on the host remain unchanged. In this way, the user can connect SIL to any host computer he or she likes, while maintaining his or her customized interface on a portable laptop.

2. Background

The impetus for this project comes mainly from personal experience with setting up specialized interfaces for disabled students. There have been occasional attempts on campus to set up a "handicapped workstation" in one lab or another in order to assist a particular student. This typically entails purchasing the specialized hardware and/or software that a student with a particular handicap could use to access a computer in one of the campus labs. One specific computer in the lab is designated as the special "handicapped station," which is then used almost exclusively by the one student for whom it was modified.

This has been greatly unsatisfying for a number of reasons. First, such modifications often have made the workstation unusable by non-handicapped students. This has meant that this particular station in the lab has been greatly underutilized compared to all other lab systems. Consider the need of a particular individual for a multi-modal access mechanism, such as suggested by Smith *et al* [10], where the interface might consist of several devices such as a head-mounted pointing device in conjunction with foot controls and speech recognition. Second, customizing the station for one particular handicap has often meant that the station is unusable by students with other, different handicaps, making it necessary to provide additional special workstations within the lab in order to accommodate multiple handicapped students. This is also a problem when there is more than one student in a particular class requiring accommodation, even if they could make use of the exact same hardware and software. Third, most accommodations for handicapped users are truly customized to the individual, so that any generically configured system will likely fall short of the needs of an individual user. Finally, on our campus we have numerous general-purpose labs, as well as many special-purpose labs dedicated to a particular department and/or application area. It is unrealistic to imagine that all of these labs can provide assistive devices to accommodate all handicapped students, yet we are legally (and ethically) required to provide such accommodation in order to avoid limiting a particular student's access to a specific educational experience.

Further impetus is provided by the observation of the increasing ubiquitousness of computer technology. The growing availability of automatic teller machines (ATM), personal data assistants (PDA), cellular telephony, interactive kiosks, interactive television, Internet access, etc., all points to a need for providing handicapped users with personal portable interfaces [1, 9]. The need for such "universal access" continues to increase as assistive devices continue to make handicapped users more mobile and independent [11]. Consider, further, that the number of people with handicaps continues to increase, and should probably also include the fact that people are living longer but are developing disabling conditions as they age [7, 12]).

The notion of an electronic notebook [2] suggests that non-disabled users would benefit from a portable system customized to intellectual needs. While such a device typically includes general tools such as an appointment calendar, "To Do" list, or email system, there is some indication that how the system would be used by a particular individual would make the system in some fashion desirable. At the very least, particular professional activities might require specialized software in order to accommodate, say, a physician's notebook vs. a biologist's notebook vs. a writer's notebook, etc.

Perhaps the ultimate extension of such personal devices is the wearable computer [2, 5]. In these cases the system takes on a very personal relationship to the user, and in a way becomes like an additional sense, i.e. as an additional way for the user to interact with the surrounding world. The portable nature of these systems, and the aspect of these systems being an additional means for gathering and interpreting input from the external world are quite important.

Feiner [4] has suggested that there has been a growing impetus to "decouple the personal computer and its interface..." His ultimate extension of this notion is hybrid devices that marry small, precise controls for use over small domains with less precise devices controlling access to larger virtual spaces. In our case, the small, precise controls might be our laptop device, while the larger domain is the host computer.

Specialized mobile systems are also of interest with respect to this project. Mobile information browsing [1, 5] is likely to increase substantially in the near future. The use of mapping systems in automobiles, electronic books, personal electronic assistants [3, 9], etc. all point to a trend in the use of portable electronic devices to access an increasing amount of information.

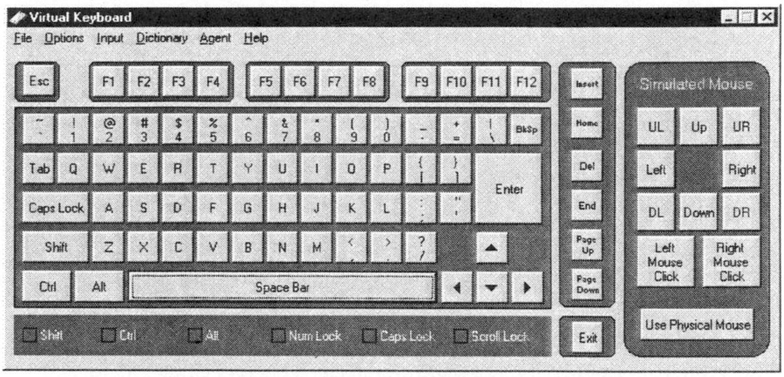

Figure 2: The Virtual Keyboard screen layout.

3. The project

The ViKI project was envisioned in three main phases. Phase 1, the main subject of this paper, developed the necessary software for the SIL system to mimic all of the interaction methods available from standard keyboard and mouse, with communication between SIL and the host system via a serial interface. Phase 2 of the project will be to replace the serial communications with a direct connection to the keyboard port of the host system, thus eliminating the need for the TSR communication program for the host. Phase 3 will be to replace the hard-wired connection between SIL and host with an infrared wireless system.

3.1 Phase 1

The SIL software runs under a Microsoft Windows® environment and was developed in Visual Basic®. It consists primarily of the virtual keyboard emulator (VKE) and software for communicating with the host system (see Figure 2). The host system is also Windows® and Visual Basic® based. The two systems connect together through their serial ports using a null modem cable.

The VKE software provides an on-screen representation of a keyboard, which can be manipulated with either the laptop's keyboard or through pointing and clicking with the laptop's pointing device (we connected an external mouse for demonstration purposes). We are not proposing this representation of the display as an actual useful interaction method, but are using it as a means to demonstrate the efficacy of the idea. Later research will explore options for providing actual interfaces that might be suitable for assisting particular handicapped users.

There was no attempt in this first phase to in any way optimize the virtual keyboard or the interaction methods of the user. As seen in Figure 2, the on-screen VKE is simply mimicking a fairly standard keyboard layout. The VKE resembles a standard keyboard and has most of the features of one. When the VKE software is run, data is read from the system registry and loads either the appropriate customized features or a set of defaults. Since this demo system has also been set up for speech recognition, the speech dictionary is also loaded at startup time.

The keyboard buttons can be clicked using the pointing device to send a keystroke equivalent to the host system. The VKE also provides a mouse simulator (the right side of Figure 2), which will move the host's mouse cursor in the indicated direction. The VKE's menu bar houses access to customizing controls, such as selecting the main input device, setting background color, mouse speed, etc (see Figure 3). Most options within the menu bar are saved in the system registry, allowing the user to customize the VKE to his or her need.

At this time, interaction with the VKE is provided through external pointing devices such as a mouse, trackball, or joystick, through the laptop's keyboard, and through a commercial voice recognition system. The pointing devices can be used to either directly control a cursor on the SIL display (in order, for instance, to point to and select by clicking a particular key on the VKE), or to remotely control the cursor on the host system.

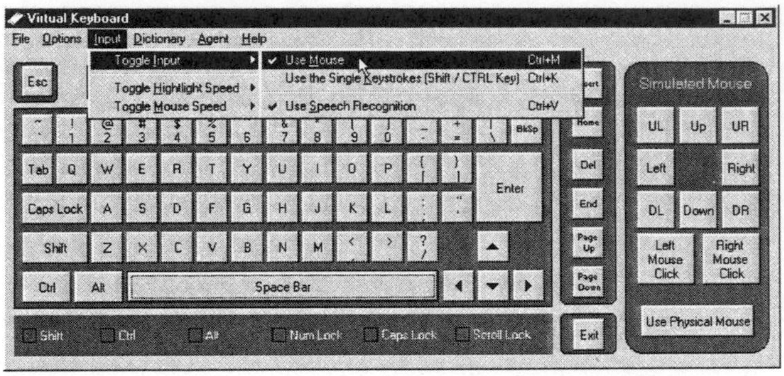

Figure 3: A look at the nested **Input** menu, containing user controls

Interaction can also be set to a scanning keyboard mode, whereby the user can select a keystroke to be sent to the host using a single-switch input. In this mode, the rows of buttons begin to highlight one by one. If the user presses the *shift* key, that row is selected and scanning continues by then highlighting one button at a time through the row. A second pressing of the *shift* key will select that one key for transmission to the host computer. After selecting one key, scanning is resumed. Scanning speed can be set through a menu control. Another control that can be set by the user is sticky keys, so that multi-key combinations can be entered one key at a time.

The SIL system can also act in a stand-alone mode while still using the VKE software. This option allows information to be sent to an active application running on the laptop. In order for the VKE to perform this action, VKE must know which application program to send data to. To accomplish this, the VKE presents a list of all applications found on the laptop's hard drive, allowing the user to select one to execute. Upon doing so, the VKE obtains the ID of the application, which allows VKE to send data to the application.

From the host system side, a small terminate-and-stay-resident (TSR) program (the Communicator) is required at this time. Communicator accepts SIL signals as input via the host's serial port, and reroutes those signals as input to some application running on the host. The Communicator must have primary control of the host system, although the application software must have also been started but suspended. Once the application has been started, Communicator is run and immediately minimized and placed in the system tray.

3.2 Phase 2

The second phase of this project will be to replace the serial connection on the host side with a direct connection to the keyboard port. The goal is to eliminate the need for the Communicator software on the host computer, so that there is no need whatsoever to modify anything on the host system. This will make the SIL truly portable (at least with respect to Windows-based systems), allowing the SIL to be connected to any compatible system in any lab on campus.

3.3 Phase 3

Most laptops today come equipped with an infrared port. This may make it possible to eliminate the need to physically plug the SIL into the host computer. We envision being able to simply carry the SIL within carrier distance of the host computer and being instantly linked. Infrared interfaces are already used as a means of transferring files between systems. In addition, wireless keyboards are widely available from a number of sources. Our hope is to replace the wireless keyboard with a laptop acting in its place.

Admittedly this idea seems to reintroduce the drawback of the Phase 1 Communicator software, i.e. that the host system must be modified in some way in order for the SIL system to communicate with the host. This is true - every system the user wishes to interact with must have an infrared interface installed. Our thinking, however, is that someday such interfaces will be installed as a standard feature not only on PC's, but also on all types of consumer products and service systems. ATM's, kiosks, etc. could all be outfitted with a simple infrared interface at a very low cost, finally making such specialized computers accessible to the handicapped. This is now becoming a reality with the introduction of the Bluetooth protocol for wireless communication of electronic devices. [8]

4. Future research

There are two areas of potential future research we are interested in pursuing once Phase 3 of ViKI is completed. First is the development of a multitude of interfaces that would be of actual use to handicapped users. While our demonstration system virtual keyboard may be of some practical use, there is at least an argument to be made for experimenting with alternative keyboard layouts. The QWERTY layout, while ubiquitous and arguably well known to most users, is probably not the most efficient layout for many users. A strictly alphabetic layout might be preferable, or even something more radical such as a circular design based on digrams (2-letter combinations) or letter frequencies. With a virtual keyboard, it should be possible to easily investigate many types of specialized layouts.

The second area for future research would be in exploring two-way communication between the SIL system and the host. The current project is one-way communication only, from SIL to host. What might be accomplished if the communication were two-way? This is of particular interest for exploring systems for the vision impaired, since such users might not be able to read the screen of a device with which they wish to interact, such as an ATM machine. Could a duplex communication system be used to allow such users to have the host system's screen read to them using the reading technology housed in their personal interface?

Acknowledgments

This material is based upon work supported by the National Science Foundation under Grant No. DUE-9551245. Any opinions, findings, and conclusions or recommendations expressed in this material are those of the author and do not necessarily reflect the views of the National Science Foundation. Special thanks to Rob Shneider for the development of the ViKI demonstration system.

References

[1] Ark, W. S. and Selker, T. A look at human interaction with pervasive computers. *IBM Systems Journal.* Vol. 38, No, 4

[2] Bass, L., Kasabach, C., Martin, R., Siewiorek, D., Smailagic, A., and Stivoric, J. (1997). The design of a wearable computer. *Proceedings of the Conference on Human Factors in Computing Systems (CHI 97),* March 1997. Atlanta, GA:139-146.

[3] Erickson, T. (1996) The design and long-term use of a personal electronic notebook: a reflective analysis. *Proceedings of the Conference on Human Factors in Computing Systems (CHI 96),* April 1996. Vancouver, British Columbia, Canada:11-18.

[4] Feiner, S. and Shamash, A. (1991) Hybrid user interfaces: breeding virtually bigger interfaces for physically smaller computers. *Proceedings of the ACM Symposium on User Interface Software and Technology,* November 1991. Hilton Head, South Carolina:9-17.

[5] Kawachiya, K. and Ishikawa, H. (1998) NaviPoint: an input device for mobile information browsing. *Factors in Computing Systems (CHI 98),* April 1998. Los Angeles, CA:1-8.

[6] Mann, S. (1996) Smart clothing: the shift to wearable computing. *Communications of the ACM,* August 1996:23-24.

[7] Miller, B., Bisdikian, C. (2000) *Bluetooth Revealed.* Prentice-Hall. 2000.

[8] Muller, M., Wharton, C., McIver, W., and Laux, L. (1997) Toward an HCI research and practice agenda based on human needs and social responsibility. *Factors in Computing Systems (CHI 97),* March 1997. Atlanta, GA:155-161.

[9] Robertson, S., Wharton, C., Ashworth, C., and Franzke, M. (1996) Dual device user interface design: PDAs and interactive television. *Proceedings of the Conference on Human Factors in Computing Systems (CHI 96),* April 1996. Vancouver, British Columbia, Canada:79-86.

[10] Smith, A., Dunaway, J., Demasco, P. Peischl, D. (1996) Multimodal input for computer access and augmentative communication. *Proceedings of the Second Annual ACM Conference on Assistive Technologies* (ASSETS '96), April 1996. Vancouver, British Columbia, Canada:80-85.

[11] Shneiderman, B., Between hope and fear: Universal access, medical records, and educational computing, *Communications of the ACM* 40, 2 (February 1997): 59-62.

[12] Warden, A., Walker, N., Bharat, K., and Hudson, S. (1997) Making computers easier for older adults to use: area cursors and sticky icons. *Factors in Computing Systems (CHI 97),* March 1997. Atlanta, GA:266-271.

Re-inventing icons: using animation as cues in icons for the visually impaired

Stephanie Ludi, Michael Wagner

Computer Science and Engineering Department, Arizona State University,
P.O. Box 875406, Tempe, Arizona 85287-5406, United States

Abstract

Icons remain in use as a means of interacting with computers. Now such Graphical User Interfaces are also being used in handheld devices and appliances, often using small displays. With these new appliances and means of accessing resources, the visually impaired find such interfaces difficult to access. For users who possess functional vision, the graphical nature of the interface is not in itself a problem. Instead the interfaces and their components need to be examined and redesigned in order to be more effective. This paper presents the potential use of animated icons as a means of increasing accessibility in general and in the case of limited display space.

1. Introduction

Icons are commonplace in Graphical User Interfaces, as small static pictographs that represent a task that can be performed or a feature of the system. These icons, in conjunction with other aspects of Graphical User Interfaces, can make technology difficult to access for the visually impaired especially as the display sizes on many appliances remain small. Where magnification software is available the cost can be prohibitive, the software may not always be compatible or portable, and the user may find the constant scrolling to be frustrating. Also, people wish to utilize their remaining sight with minimum overhead. The user should not adapt to the interface; the interface should adapt to the user. As a result, icons need to be re-invented, and adding animation to icons may be an answer.

The overall objective of the study* is to investigate how effective animation in icons can enable visually impaired computer users, those who possess functional vision, to utilize a Graphical User Interface in a similar manner as those who have no visual impairment. The use of animation in icons has been shown to increase the usefulness of icons in terms of clarification and ease of use (Alpert, 1991; Baecker, Small & Mander., 1991; Nielson, 2000) for mainstream users, but new applications of animation in the user interface may benefit the visually impaired population. By re-examining icons, they can be redesigned and updated to enable more access to computers.

2. Research questions

As the investigation of the usefulness of animated icons takes several steps, this work highlights the questions regarding the comparison of animated and static icons. Specifically, we seek to answer the following questions:

- Is there a perceived size difference between static icons and animated icons?
- If a size difference exists, is the size in animated icons smaller than that of static icons?

3. Method

3.1 Participants

The overall population of the legally blind is diverse, where vision cannot be corrected to better than 20/200 in the better eye, or if the visual field is 20 degrees or less, with correction (SSA, 2000). An additional criteria was for participants to have used a computer within the last 6 months. For the study, volunteers from the university and the community had visual acuity between 20/200 and 20/800 (inclusive). The ages ranged from the traditional college age to the elderly. About half of the participants used computers on a daily basis.

Due to the difficulty of locating participants who fit the criteria, three of the eight participants were used who fell into the criteria interval when not utilizing glasses or contact lenses. These three participants are referred to as possessing simulated legal blindness. The other five participants fit the study's legal blindness range with best correction. None of the eight adult participants used corrective aids during the study. The participants were compensated for their time and travel.

3.2 Materials and Equipment

The experiments were conducted on PCs (550 MHz Dell Optiplex GX1) with 17-inch monitors, running Microsoft Windows 98. The screen size was 300mm across and 225mm tall and a resolution was maintained at 800 pixels by 600 pixels. As a result, each pixel had height and width dimensions of .375mm. The icon sizes were manipulated and recorded according using pixels.

The icon identification program was developed using Macromedia Director 7. The program captured the size of each identified icon and saved it to a text file. The icons themselves were created using Macromedia Flash 4. The icon and software development was conducted on a PC (600 MHz Dell Dimension XPS T600) and a Macintosh G3.

3.3 Design

A 2-Forced Choice design, outlined in (Ermer-Braun, 2001), was used for the set of animated icons and static icons, where in each trial the participant selected either the target icons or the icons selected as noise (an abstract icon).

Within the icon test sets for each session, the order that the icons were displayed was randomized. The icons used for the study, presented in Figure 1, are representative of icons found in standard word-processing, email, and drawing programs.

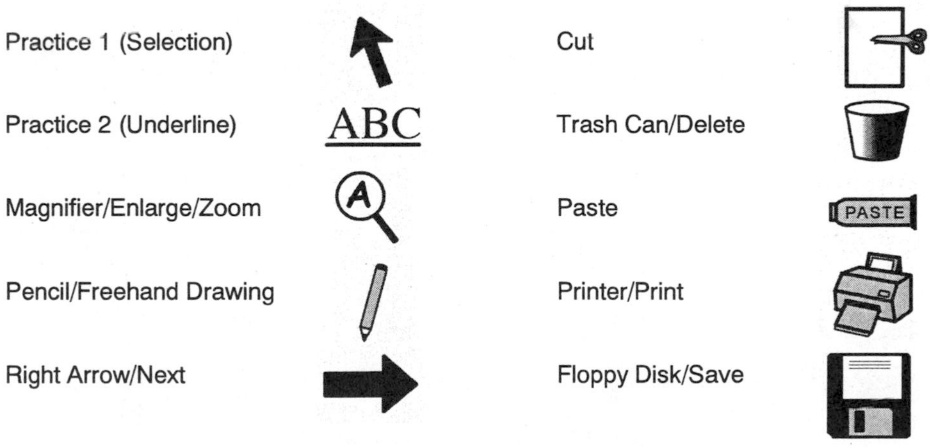

Practice 1 (Selection)

Practice 2 (Underline)

Magnifier/Enlarge/Zoom

Pencil/Freehand Drawing

Right Arrow/Next

Cut

Trash Can/Delete

Paste

Printer/Print

Floppy Disk/Save

Figure 1. The ten icons used in the study.

While the icons are meant to be commonplace, the icons were designed rather than originating from pre-existing programs. A sampling of two animated icons is presented in Figure 2.

* This study is part of a larger project funded by the National Science Foundation. (Award #: IIS-9978183).

All of the icons are in black-and-white so that color would not complicate the study. Animated icons were played at 9 frames per second, as that speed seemed to be a good balance between presenting motion fluidly though not too quickly so as for the purpose of the motion to be lost on the participant.

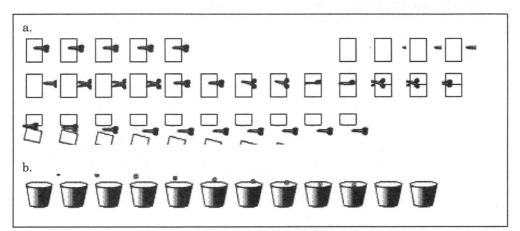

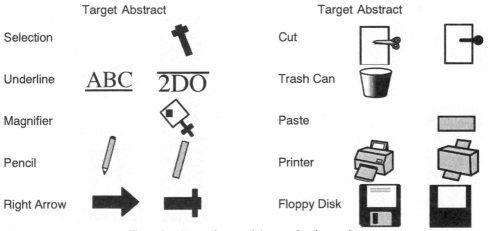

Figure 2. Sample of animated icons. a. Cut, b. Delete

During the trials, abstract icons were also used in order to provide noise for the participant. Abstract icons, discussed in (Stots, 1998), were designed to match the movement for the target icons (in the case of animated icons), but were generalized in appearance. The abstract version of each icon and the static icon it is associated with is presented in Figure 3.

Figure 3. Abstract icons and the associated target icons

3.4 Procedure

The trials were conducted with one participant at a time, sitting 19 inches away from the computer monitor. The first set of trials was randomly chosen as either the animated icon or the static icon set, and the second trial was conducted a week later using the remaining icon set. For each trial within the set, one of the 10 icons was designated as the correct icon for a task while the associated abstract icon to the target was selected as noise. The first two icons were practice icons, while the remaining 8 were for the study.

For each trial, the facilitator presented the icon's function to the participant. Next the "correct" icon and the "dummy" icon were each displayed for 10 seconds in a random order. The facilitator told the participant when the first icon was displayed and when the second icon was displayed. After the icons were displayed, the participant was required to select the number of the icon that was perceived as corresponding to the stated task. If the correct icon was selected, the program recorded the selection as a hit. The icon's width and height were increased by 2 pixels until three consecutive hits occurred. At this point, the program recorded the icon size and proceeded to the next icon in the test set.

4. Results

The first question to be addressed is whether static and animated icons are perceived at different sizes. While omitting the two practice icons from the analysis, the data can be examined by the analysis of the difference between the perceived size for the icons.

Table 1. Animated and static icon size comparisons shown as the number of trials. The Size Difference was determined by the static icon size minus the animated icon size.

Icon	Size difference < 0	Size difference = 0	Size difference > 0
Arrow Right	2	1	5
Cut	0	1	7
Delete	4	2	2
Magnifier	3	2	3
Paste	3	2	3
Pencil	1	0	7
Print	4	1	3
Save	5	0	3
Total trials	**22**	**9**	**33**

The overall size difference was calculated for each icon from each participant, where the animated icon size is subtracted from the static icon size. Then the size differences were grouped according to whether the animated icon size was larger than the static icon size (size difference < 0), the animated icon size was the same as the static icon size (size difference = 0), or if the animated icon size was smaller than the static icon size (size difference > 0). The sizes were then tabulated for each icon.

According to Table 1, in only 9 of the 66 trials (14%) are the static and animated icons perceived at the same size. Since a perceived size difference does exist between static and animated icons, the next question to examine is if the perceived size of animated icons is smaller than that of static icons. Animated icons were perceived at a smaller size than the static version (noted in Table 1) 33 times out of the 64 trials (51.6%). While the overall result does not offer much credibility for the use of animated icons, the examination of the data regarding specific icons shows that animation can make a difference. For example the Cut and Pencil icons show that the animated version was perceived at a smaller size than the static version 7 out of the 8 trials for each icon. By contrast the Delete version of the animated icon was identified at a smaller size than the static version in only 2 of the 8 trials. Clearly other factors within the icons themselves may contribute to the disparity in the data. For example, the dynamic activity of the icons, as the difference between the Delete icon and Cut icon presents. As such, a definitive answer to the second research question cannot be determined without further investigation into the icons themselves.

5. Conclusions and future work

In regards to the first research question, size differences between animated and static icons were apparent in nearly all of the trials. The answer to the second question is more complex. Of the set of trials where a size difference existed, the size of animated icons was less than that of static icons just over half of the time. Some icons had a more prominent size difference than others. Examples include an Arrow pointing right, Cut, and a Pencil/Drawing icon. However in some instances, the static icon sizes were smaller than the animated icon counterpart. The Delete and Print icons are examples where the phenomena occurred.

While quantitative data was not gathered on the participants' comments, several people noted that the animated objects caught their attention much more than the static objects. The animation had the potential be more

952

memorable than the static versions of the icons, thus allowing for an interface that is easier to learn. This potential benefit may also influence the design of interfaces with animated icons. Further study will investigate this idea.

While the potential for animated icons being used to conserve screen space exists, future work will investigate the discrepancies uncovered in this study. Since no definitive trends or icon sizes were determined, more work needs to be done. The type of animation in the icons needs to be analyzed to detect patterns of recognition. Also conducting tests in the context of use can be useful by providing a more realistic environment for potential users to work through tasks and assess the effectiveness of animated icons. When the framework of how animated icons can be clarified, more choices can exist so that the visually impaired can access technology and resources.

References
Alpert, S. (1991). Self-describing animated icons for human-computers interaction: a research note. *Behaviour and Information Technology*, 10(2), 149-152.
Baecker, R., Small, I., and Mander, R. (1991). Bring icons to life. *Proceedings of CHI '91 Conference on Human Factors in Computing Systems*, 1-6.
Ermer-Braun, E. (2001). A Method to Measure the Recognition of Animated Icons by the Legally Blind Computer User. Arizona State University, Unpublished Master's Thesis.
Nielson, J. (2000). *Designing web usability*. Indianapolis: New Riders Publishing.
Social Security Administration Disability Benefits Publication No. 05-10052 (2000, March). Washington DC: Social Security Administration. Retrieved May 5, 2000, from the World Wide Web: http://www.ssa.gov/pubs/10052.html#Blind
Stots, D. (1998). The usefulness of icons in the computer interface: effect of graphical abstraction and functional representation on experienced and novice users. *Proceedings of the Human Factors and Ergonomics Society 42nd Annual Meeting*, 453-457.

Teaching LaTeX to blind and visually impaired students at the University of Applied Sciences Giessen-Friedberg

Erdmuthe Meyer zu Bexten, Martin Jung

University of Applied Sciences Giessen-Friedberg, Department of Mathematics, Computer Science and Natural Sciences, Wiesenstrasse 14; D-35390 Giessen; Germany
erdmuthe.meyer-zu-bexten@mni.fh-giessen.de

Abstract

Studying Natural Sciences causes many problems for blind and visually impaired students as handling mathematical expressions is extremely difficult due to their disability. While simple formatted printed texts can be read easily by scanners with optical character recognition software, this fails for mathematical expressions.

There are some specially designed mathematical notations for blind and visually impaired people with various properties. All of them are exclusively used by blind or visually impaired people. Producing mathematical expressions for communication purposes is also difficult, since people without vision impairment are not trained to read these codes. In schools not all blind pupils learn the same mathematical codes. As a result even blind people cannot communicate with each other.

At the Institute for Blind and Visually Impaired Students (Zentrum für Blinde und Sehbehinderte, BliZ), we decided to use the typesetting system LaTeX for mathematical representations for our students. We have developed a concept to teach this "language" to our blind and visually impaired students in co-operation with our partners. This concept can be adapted to different levels of mathematical skill.

1. Introduction

There are 155,000 blind people in Germany and a larger number of people with serious visual impairments (estimated 550,000 people). The possibility of doing a university degree is open to them today. Varied and widely differing degree subjects are available. This progress has its roots in the development of ever more efficient and reasonably-priced computers with adapted access, as well as other modern communication technology, creating enormous possibilities for the visually impaired in particular. It is particularly in the natural sciences and technical degree subjects that mathematics plays an important role. During lectures, for example in physics, electrical engineering or chemistry, students are confronted with (mathematical) formulae again and again, causing various problems for them.

How can this problem be solved? Over a period of some years, many mathematics codes for use on the computer (like e.g. [1] and [2]) have been created especially for blind people. The aim of these special character codes is firstly to enable blind people to write mathematical formulae by means of the normal computer keyboard and secondly to print out formulae in Braille displays and thus make them readable.

BliZ offers special support for those visually impaired students studying in different faculties and is following a completely new path in using the typesetting system LaTeX [3], which is recognized and employed world-wide. This is based upon the following facts: In the German-speaking countries, there is unfortunately no uniform and compulsory computer mathematics code for schools. As our students come from many and varied school backgrounds throughout Germany, they have thus all learned different mathematics codes. The problem is further exacerbated by the fact that the codes are not compatible with one another, which constitutes a significant disadvantage for exchange of ideas and for communication. Due to this lack of orientation and the non-compulsory nature of the codes, more and more "private codes" are being invented. If necessity demands, new mathematical symbols are created "on the fly". In addition to this the mathematics codes are not particularly user-friendly, especially with regard to sighted people. Interaction and communication between the sighted and the blind is hardly possible at all. All of the above-named reasons have convinced us to teach our blind students LaTeX.

2. Mathematics Codes

The mathematics codes are an old and much-discussed problem. For many years the deliberations and discussions have been going on as to the question of which mathematics code is the "best" and how to establish it as quickly as possible. The German Society for the Blind and Visually Impaired (Deutscher Blinden-und Sehbehindertenverband e.V., DBSV) has been pressing the necessity of a uniform code system for the blind for many years. The mathematics code taught to pupils and students should not, according to the self-help group for the blind, differ according to the place of training or the type of school attended (integrated, segregated). The incompatibilities still existing between the different mathematics codes lead by definition to considerable communication problems between the users and if trainees move from one place of training to another. In the following subsections several of the commonly-used mathematics codes for the blind are presented in brief:

- Marburg mathematics code (MBS)   - Karlsruhe and Dresden ASCII - mathematics code (AMS)
- Stuttgart mathematics code (SMSB)   - Bochum mathematics code (BMPS)
- HrTeX

Detailed information about each of the codes can be found in [1]. The mathematics codes all have a corresponding representation in Braille and in addition to this as ASCII characters on the monitor.

2.1 Marburg Mathematics Code

The Marburg mathematics code (MBS) was developed in 1955 by blind mathematicians for blind mathematicians. However, at that time representation on and working with computers was not important. Other criteria (like[1]) like clarity, ease of use, compactness with regard to utilizing space on Braille paper and compatibility with traditional Braille in literature were more paramount at that time. The MBS has been used successfully for many years at many schools for the blind and also in Braille books. This code can also be used in combination with contracted Braille. Unfortunately, the MBS is not suited for interactive computer working techniques and is additionally difficult for sighted people to learn.

2.2 Karlsruhe und Dresden ASCII-Mathematics Code

At the universities of Karlsruhe and Dresden the "ASCII mathematics code for the blind" (AMS) is employed for the representation of mathematical formulae in scientific texts. This notation was developed as a part of the project "Information Technology for the Blind and Seriously Visually Impaired" at the University of Karlsruhe and reworked at the Center for the Visually Impaired at the University of Karlsruhe in co-operation with a work group "Study for the Blind and Visually Impaired" at the Technical University in Dresden. The representation of mathematical symbols on a computer is usually undertaken in graphical form, which is, however, unreadable for the blind and severely visually impaired. In order to make mathematical symbols accessible to them, they are converted into a representation by characters (AMS). The number of characters is limited to the ASCII character type set available on the computer. Texts so created are system-independent and can be read and altered using any simple editor program. It is the aim of the AMS to introduce both syntax-orientated means of representation for mathematical expressions as well as ones which accentuate the semantics of the expressions. The AMS additionally allows short forms of frequently used expressions to be employed. Both contribute to an improved readability. The disadvantages of the AMS are that, firstly, it is not based upon the traditional presentation format of the 6-point notation and, secondly, that long expressions and a multitude of brackets lead to a decreased quality in the readability of the expressions. Further information can be called up on the Internet under [2].

2.3 Stuttgart Mathematics Code

Dr. Schweikhardt (University of Stuttgart, Institute for Information Technology) developed the Stuttgart mathematics Code for the Blind (SMSB) and reworked it in 1989 in order to meet the demands of integrated lessons. The SMSB Code set as its aim a clear 1:1 ratio of Braille characters and ordinary letters.
This mathematics code is a further development of the Marburg system, but is an 8-point code and thus has a character base of 256 characters. The SMSB makes a very compact representation of expressions possible. In

addition to this it is easily transferable into other formats. The disadvantages of this mathematics code are firstly that it is very difficult to learn for both blind and sighted people and secondly that special software is needed for input and output.

2.4 Bochum Mathematics Code

In the framework of a project "Integration of Blind and severely Visually Impaired Pupils in normal Schools" at the Heinrich-von-Kleist-Gymnasium (Bochum), another mathematics code was developed. The developer of this code is Mr. Jandrik Kraeft. It was his aim to develop a code especially for pupils in which the structure of the expression also became clear. In his code, for example, the digits in a fraction in the numerator are marked differently to those in the denominator. Unfortunately this code is not easily accessible to the sighted and the representation on the computer is difficult for them to understand.

2.5 HrTeX

At the J. Kepler Universität Linz, Institute for Computer Science they developed a method for the tactile presentation of structured information encoded in LaTeX. This concept is called ``HrTeX'' – Human readable TeX [4]. HrTeX defines on the one hand the rules for the linear tactile rendering of LaTeX documents of mathematical formulae. On the other hand, it integrates a HrTeX mathematics code, an extensible object library. The objects consist of single TeX macros which represent various mathematical symbols or whole mathematical expressions.
HrTeX was developed to provide an easy to read and adaptable notation for mathematics (HrTeX code) and to be used with computers for blind persons. An another objective was to reduce the efforts of preparing literature and to improve access to documents.

3. LaTeX

The layout system of LaTeX is based upon the type-setting system TeX developed by Donald E. Knuth (Stanford University, USA) in the mid seventies. The disadvantage of the extremely efficient and extensive TeX system is that it requires considerable experience of programming to use it and especially to exhaust all of the possibilities in doing so. This means, however, that the system is therefore not available to everyone.

This significant deficit gave the American scientist Leslie Lamport [3] the idea to develop the LaTeX program package, which is a further development of TeX. LaTeX is, however, considerably more user-friendly. In addition to this, it is available for all operating systems. Not all of the necessary mathematical symbols and characters necessary to describe mathematical expressions are to be found on the keyboard.

Schriftsystem	Schreibweise
Symbolik für Sehende	$\dfrac{(a+b)^{n+1} - c}{a - b}$
Stuttgarter Schrift	▌ (a+b) ⌐ n+1 ‡ -c ■ a-b ▌
Karlsruher Schrift	((a+b)**(n+1)-c)/(a-b)
LaTeX	\frac{(a+b)^{n+1}-c}{a-b}

Figure 1: Example of mathematical code representations (see [1])

4. LaTeX versus Mathematics Codes

The mathematics codes briefly described in the previous chapter are codes specially developed for the blind with the aim of making mathematical formulae "visible" (tangible) for them. LaTeX has particular advantages in comparison to them, which are represented here in brief:

1. A great advantage of LaTeX for the blind is that it represents a standard which is well-known, widely-used and recognized throughout the world and is thus accordingly used intensively by many sighted people as well. Irrespective of which mathematics codes one decides to use, this is always an isolated solution and not particularly simple for sighted people to access.
2. LaTeX can be used both to format texts and to present and describe mathematical formulae and tables.
3. The description of complex tables and mathematical formulae is carried out in a text-based and row-based manner, so that a print-out and thus check by means of a Braille display print-out is always easily possible.
4. LaTeX is, in contrast to MS-Word, not a WYSIWYG system, as the formatting is described in text form. This leads to a much easier and more comfortable readability for the blind by means of the Braille display.
5. An exceptional feature of LaTeX is that document classes (styles) for various types of formats (e.g. letters, reports, books) already exist and are now at the disposal of the users. In addition to this, several classes of documents can be defined. Users do not have to worry about the painstaking formatting of their texts. This aspect may seem trivial to a sighted person, but for people who cannot see what their document looks like, it is of very great significance indeed.
6. The LaTeX system is available on practically all computer and operating systems.
7. There is manifold literature from starter books to specialist literature to assist new-comers to the area.
8. In an age in which electronic media have already firmly established themselves in schools, LaTeX plays an important role. Scripts and exercise sheets are often created using LaTeX and passed onto students by means of the web.

While the above-mentioned mathematics codes consist of a host of mathematical characters and written representations for differing mathematical content, LaTeX offers a simple concept which can be summarized in a few rules:

- Short forms for particular mathematical symbols are compiled according to generally understandable, mnemonic considerations.
- The prefix notation, i.e. placing the command words – like \frac oder \sqrt , in front of each expression improves readability for the blind person who reads sequentially.
- There is a host of mathematical literature written in LaTeX. The same is the case for various computer books.

Despite the many points which speak for using LaTeX, there are also some problems:

1. LaTeX is a very extensive type-setting system with a myriad of commands which have to be learnt before use. This is not necessarily easy and takes up a lot of time. A lot can naturally be achieved with very few commands, but the more complex the mathematical formulae become, the more commands are naturally necessary to present them. For this reason it is appropriate to begin with the introduction of LaTeX early in schools.
2. The more complex a mathematical formulae is, the longer its presentation in the LaTeX notation becomes. This means that the readability of mathematical expressions for the blind becomes worse and increasingly difficult in proportion to the length and complexity of the expressions themselves.
3. The type-setting system is unfortunately not taught in schools today, but only employed at university level. However, even at this level it is not always offered as a course, but has to be learned independently. This is, of course, a particularly difficult problem for the blind.

5. Possible solutions

In the last few months there have been many discussions between the visually impaired students, teachers from the BliStA (Deutsche Blindenstudienanstalt e.V./German institute for the blind), professors from our University, and LaTeX experts. In the framework of these discussions, various possible solutions with regard to learning and working with LaTeX for the visually impaired have been worked out. These possible solutions are listed here in brief:

There is a consensus of opinion that the Marburg mathematics code is not compatible for work at the computer. It is also agreed that the traditional mathematics code should continue to be used in schools. The integration of LaTeX into normal mathematics lessons is not an alternative, as less able pupils might thus have even more problems with mathematics. However, the suggestion that LaTeX be taught to those students in the 16-18 age group intending to take up studies in the natural sciences or a technical apprenticeship, met with much approval. The pupils will then not have to learn LaTeX additionally during their studies.

Graduates of the BliStA institute do not have any knowledge of LaTeX at this time. According to our current experience, it is most effective to offer a basic course in LaTeX directly before the begin of the degree course at university level. Then, parallel to the usual mathematics lectures, specialist knowledge in particular areas relevant to the lectures can be acquired later in the course.

In the future a multi-level training concept is planned by the BliStA. Beginning with training for a very simple document format in the 16-18 age group, the layout system of LaTeX can be learned step by step in tune with the increasing knowledge and demands of the pupils. Entry into the LaTeX system is made easier by the following rules:

- The extremely extensive, often very complex and therefore not readily comprehensible "header" for the formatting of the text should be avoided at the beginning.
- Macros should also be avoided, they are not absolutely necessary at the beginning.
- In addition to this, the many special functions such as italics, bold etc. are not necessary. It is paramount for the visually impaired to learn the mathematical notation in LaTeX and, when this has been completed to a satisfactory level, the other functions of LaTeX, with regard to its characteristics as a type-setting system, can be taught.
- Building blocks for formatting texts (macros) should be used in such a way that it is only necessary to insert the numbers. These building blocks are still to be developed.

6. Closing remarks

In the very first year of the BIZ the advantages of LaTeX compared to other mathematics codes has been evident. Although learning LaTeX is not easy for the blind, the advantages it brings are enough to waylay these initial difficulties, in that it can also be used by fellow students and professors, facilitating a better co-operation with and integration of the students or, indeed, making this possible for the first time. Many lecture scripts and exercise sheets as well as exams have been written in LaTeX by lecturers at our University and put at the disposal of blind students. All usual mathematics codes are in a position to describe any complex formulae, however, LaTeX offers much more than just another mathematics code. Due to the document classes or styles and the extensive possibilities for type-setting, LaTeX is much more a universal tool. Taking this into consideration, it is important to press the demand that LaTeX be taught at school level as a mathematics code, instead of other "local", privately developed alternatives. Lessons in LaTeX should, however, take into account the suggestions in the above capital on "Possible Solutions".

References

[1]   Kalina, U. (1998), *Welche Mathematikschrift für Blinde soll in den Schulen benutzt werden?*, Beiheft Nr. 5 der Zeitschrift blinde/sehbehinderte, Zeitschrift für Sehgeschädigten-Pädagogik, Heft 3, pp 78-94

[2]   Information on mathematical codes for blind:{http://elvis.inf.tu-dresden.de/asc2html/ams/h-000001.htm}

[3]   Lamport, L. (1985), *LaTeX - A Document Preparation System*, Addison-Wesley Co., Inc., Reading, MA

[4]   Batusic, M., Miesenberger, K., Stöger, B. (1985), *Access to Mathematics for the Blind - Defining HrTeX Standard*, Proceedings of the 5th International Conference on Computers Helping People with Special Needs, ICCHP'96, Linz, Austria, July 1996, pp. 609-616. Oldenbourg Wien München

A Tele-Nursing System using Virtual Locomotion Interface

Tsutomu Miyasato

ATR Media Integration & Communications Research Laboratories
2-2-2 Hikaridai, Seika-cho, Soraku-gun, Kyoto 619-0288 Japan
Tel: +81 774 95 1401, Fax: +81 774 95 1408, E-mail: miyasato@mic.atr.co.jp
Interdisciplinary Graduate School of Science and Engineering
Tokyo Institute of Technology

Abstract

This paper describes a telecommunication system that combines locomotion interfaces which we have developed up to now with a wheelchair typed moving equipment from the viewpoint of communications that allow both shared experiences and shared emotions using VR (Virtual Reality) technologies. We proposed "Tel-E-Merge"- a system in which the user can enter the environment of another person in a remote location to carry on a conversation - and are conducting research into new VR equipment aimed at the practical application of this system. In Tel-E-Merge, we expect subconscious and nonverbal interaction to be the key point to enhance the reality of the remotely located partner. This paper explains one application for making this system a reality: a "Tele-Nursing" system. The Tele-Nursing system with realistic sensations focuses on shared experiences from the perspective of non-verbal communications between the helper and the patient in the context of nursing in a remote environment.

1. Introduction

A person alone is isolated, and people need to communicate with others. In this context, telephone has become everyday necessities in our society, and an increase in the number of mobile telephones has made it possible to telephone from essentially any location in the world.

We have been working on new communication media, especially media for daily communications. Our primary consideration is assisting users who want to have a chat. For example, when a person travels alone in a foreign country or when a person views an impressive artwork alone, he or she may want to share this feeling with family members or friends. We have proposed "Tel-E-Merge" as a new communication method for such a situation, i.e., "Wishing you were here!"[1] that can not be satisfied by even a TV-phone. We coined the term Tel-E-Merge with a double meaning: "Tele-Merge" and "Tel-Emerge". More specifically, we want to make it possible to merge a remotely located person, a tele-visitor, into a tele-inviter's space through VR systems.

One form of Tel-E-Merge that we imagine is that a person (Tele-inviter) in a foreign country invites his friend (Tele-visitor), who is in a communication booth at home, to join him for a chat. In this case, the friend is invited into a scenic spot where the tele-inviter actually is, and the tele-visitor's image is superimposed on a mobile robot. At the same time, the tele-visitor can see an image of the person visiting the scenic spot (Tele-inviter) by using VR devices. In this research, we intend to shape Tel-E-Merge into a type of medium that allows conversations to take place between remotely separated persons while they walk together. In particular, our focus is on the sense of locomotion, the sense one feels while walking on the real ground.

This paper introduces a Tele-Nursing system with realistic sensations using locomotion interfaces that we have developed. The Tele-Nursing system with realistic sensations focuses on shared experiences from the perspective of non-verbal communications between the helper and the patient in the context of nursing in a remote environment.

2. Walking sensations

When people walk together, their walking motion is sometimes felt to be synchronous. Moreover, they can find out a lot of information about a place, like the sound, scale, hardness, humidity, and so on of the space by going there on foot. People usually do such a complicated task unconsciously while still paying a great deal of attention to their partner. Such an unconscious interaction while walking can be thought of as a key approach to enhancing reality or the presence of one's partner.

We have been investigating a series of development programs for communication devices that use locomotion with the Tel-E-Merge system. We have developed two types of locomotion interfaces, ATLAS (ATR Locomotion Interface for Active Self Motion)[2] and GSS (Ground Surface Simulator)[3] for applying Tel-E-Merge. ATLAS and GSS are walking simulators that enable walking in a virtual space. The development of these systems focused on creating a situation, through a virtual space, in which a user in one indoor location feels exactly as though he were walking and talking with another person in a remote location.

3. Locomotion interfaces

3.1 ATLAS [2]

In the ATLAS system, an active treadmill is installed on a motion platform with 3-axis rotation, and there is a function that freely tilts the walking surface (Fig. 1).

The notable feature of ATLAS is its function that smoothly matches human walking movements, canceling out any forward motion. That is, ATLAS can cancel out the forward motion of a person walking via an actively operating treadmill. In order to achieve this canceling-out function, a method is required to detect the walking movements of the user. Because it uses a non-contact, non-restrictive sensor, ATLAS offers the advantage of not interfering with the natural walking motion of the human user. Furthermore, by rotating or tilting the entire treadmill with the motion platform, it is possible to give a sensation of turning, or to generate or recreate the slope of a graded path.

In the ATLAS system (Fig. 1), a CCD camera fixed at the front of the treadmill ahead of the user's position records an image of the front user's feet; through image processing, walking speed is detected by obtaining a movement pattern at the front of the feet. An ultra-red filter is fixed to the CCD camera, and an ultra-red floodlight is positioned on approximately the same axis. In this way, the user only needs to affix a reflective marker to the front of his shoes; there is no need to wear or carry any other special equipment or sensors.

The treadmill used in ATLAS is a modified version of a commercially available treadmill, with a maximum belt speed of 4.0 m/s. The available walking area is 145cm (L) x 55cm (W), and the response delay after receiving speed commands is 0.1 seconds.

Figure 1. Locomotion interface: ATLAS [2]

3.2 GSS [3]

Figure. 2 shows an external view of the GSS. Because the GSS can recreate an uneven walking surface, the user can experience the sensation of walking on uneven terrain in a remote location, or on an uneven surface that has been generated virtually. The GSS has the form of a treadmill, but with the following significant features:

1) Based on walking movements and ground formation information stored in a computer, the pattern of irregularities on the ground is recreated on the walking surface of the belt.

2) As in the case of ATLAS, the user need only walk normally on the belt, and his position is constantly maintained at the belt's center. This is accomplished with measurements of the user's walking movements and automatic control of the belt speed.

With these two functions, GSS can provide the user with the sensation of walking on uneven terrain.

The size of the walking surface that recreates the irregularities is 150cm (L) x 60cm (W). Underneath the surface of the walking belt, there are six panels with corresponding actuators for the irregularity simulation function. The panels are positioned in strips (25cm wide) lying at right angles to the direction of the belt's movement. The movement of the simulated irregularities is one-dimensional, in the direction of the walking movement, and the maximum height of the irregularities is 15cm. The maximum elevation speed is 15cm/sec, so that the shape of the terrain can be simulated without delay in following the user's walking movements.

The maximum height of the irregularities that actually move up and down is 15cm, but it is possible to generate level differences of over 15cm. For example, in the case of a graded path that has a continuous slope, it is necessary to give the user the sensation that the slope has a continuous angle. This is accomplished by the following process:

(1) Before the stepping foot is set down, the elevator panel in front is raised,

(2) While the lowered foot is carried backward by the belt, the panel carrying that foot is returned to a level position,

(3) When the foot on the opposite side is raised and before it is lowered, the next panel in front is raised as before.

Using this method, it is possible to create a sensation of slope that is larger than in the case of methods that simply create a maximum elevation difference of 15cm. It is also possible to create an endless upward or downward staircase.

Figure 2. Locomotion interface: GSS [3]

4. Tele-nursing system with realistic sensations using virtual locomotion interface

The Tele-Nursing system with realistic sensations focuses on shared experiences from the viewpoint of non-verbal communications between the helper and the patient in the context of nursing in a remote environment.

We move around on foot every day without being particularly conscious of that activity. However, for people who cannot move around freely, particularly due to aging or physical disabilities, the very act of walking can be extremely difficult. For this reason, the wheelchair is a very important means of transportation that takes the place of walking [4-5]. It is said, however, that human beings are social animals; even if a person is able to move, this is not enough. A person alone is isolated, and human beings need to communicate with others [6].

The authors are trying to develop a means of communication that allows a sharing of physical sensations and emotions through Tel-E-Merge, which uses VR technologies. In this paper, we propose one practical application of the Tele-Nursing system that uses locomotion interfaces.

Along with developing the two types of locomotion interfaces described earlier, the authors are also conducting a test production of the mobile equipment, called a Tele-GSS, which becomes the user's counterpart and moves around in a remote location as a reflection of the user's movements.

The Tele-GSS is equipped with a sensor that measures the irregularities in the ground; as it moves, it sends information on the measured irregularities to the locomotion interface. In this way, as the changes in the ground are recreated on the walking surface of the GSS belt, the user on the GSS is able to experience the sensation of the irregularities in the surface of a remote location. The Tele-GSS has a TV phone function, one of the goals of which is to create a channel for conversation and at the same time take the place of the person speaking in a remote location to bring out a feeling of actually being there. Furthermore, a helper in the remote location can remotely control the Tele-GSS with a patient on it.

The following are features of the Tele-GSS:

(1) Controlled by a person (helper) in a remote location via a communication line. In this case, the speed is equivalent to the walking speed of the remote helper.

(2) The helper's voice and an image of the helper's face are projected through the display segment. At the same time, the movements of the helper turning his or her head are also recreated in the equipment.

(3) Visual and audio information from the environment surrounding the patient and the Tele-GSS are sent to the helper in a remote location.

From the point of view of the patient on the Tele-GSS, the voice of the "remote helper" can be heard from behind, and it seems as though the helper is pushing a wheelchair. It is also possible to carry on a conversation. In addition, the helper⌐s face is also visible to passers-by through the display.

Figure 3 shows the Tele-GSS controlled by a helper on the ATLAS in a remote location. The helper with an HMD (Head Mounted Display) walks on the ATLAS while watching the image sent from the Tele-GSS. In order for the remote helper to manipulate the Tele-GSS easily, it is important to create realistic sensations, as though the helper is actually pushing the Tele-GSS like a wheelchair in which the patient is sitting. In this sense, because Tele-Nursing via the Tele-GSS brings about stimuli of the dynamic physical sensation of walking, the helper can share physical experiences with the patient in a remote location. Although the connection is made through communication lines, the experience created is one that would have been difficult to achieve using existing communication methods.

By sending information regarding the movement of the Tele-GSS to the locomotion interfaces, the patient in the Tele-GSS and the helper can feel the same sensation of moving together in a virtual space.

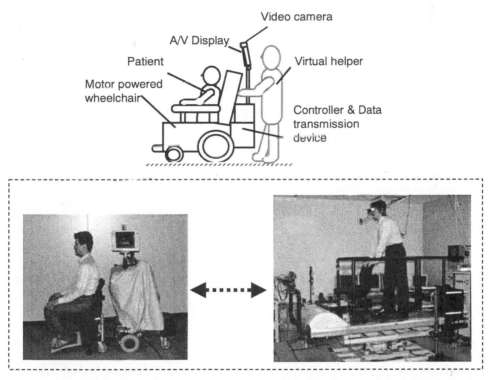

Figure 3. A patient on the Tele-GSS and a helper on the ATLAS in a remote location

5. Considerations

5.1 Methods of remote control for the Tele-GSS

If we were only facilitating operation of the Tele-GSS from a remote location, then manipulation by a mouse or joystick would be a feasible method. Since the subject is a living human being, however, the helper operating the Tele-GSS remotely would feel a psychological resistance to this method. Furthermore, the settings based on the relationship between the amount of movement in the operator hands during the remote operations and the movement of the wheelchair is a critical problem from both the perspective of safety and that of the of the remote helper's proficiency.

The goal of Tele-Nursing using the Tele-GSS, which is a wheelchair Tel-E-Merge apparatus, is communication between humans to achieve nursing operations that can be carried out easily by anyone.

5.2 Conversation between remote locations

When two people communicate, the feeling of sharing the same environment is very important. Through this sharing, each person involved can feel the temperament of the other as well as the atmosphere of the particular location, or notice things that the other person is focusing his or her attention on. This allows the speakers to grasp key points in the conversation, allowing the conversation to progress more smoothly.

If we were simply trying to facilitate a conversation between people in remote locations, then we could consider a conversation over portable telephones. With a portable telephone, however, there is no channel for non-verbal communication, which is important in smooth communications. Recently, we have seen the emergence of phones that can send images as well, allowing the users to transmit facial expressions and other types of non-verbal information; however, even with these devices there is no physical interaction when the conversation takes place via a communications line.

On the other hand, the Tele-GSS creates a situation in which the remote helper and the patient can interact through the wheelchair. A feeling of "actually being there" can be transmitted between the patient and the remote helper, via the actual body of the wheelchair. In this case, communication is not made possible simply because there is sound and images; as with such devices as "In Touch"[7], the apparatus creates a basis for non-verbal communications through the physical sensations of the helper's operations.

As explained above, by using Tele-Nursing with virtual locomotion, the sensations of the patient's wheelchair going up a hill or across uneven ground can be transmitted to a remote helper. In this way, the helper can manipulate the wheelchair as though he or she were actually with the patient.

6. Conclusion

In this paper, we presented the Tele-Nursing system as a part of the research being carried out to creating the optimal environment for communications between people using VR technologies.

The Tele-Nursing system applies a locomotion interface and remote-controlled moving equipment as means of communication. In particular, the system aims to achieve non-verbal communication through shared physical experiences between a helper and patient in the context of nursing in remote locations. The system makes it possible to stimulate dynamic physical sensations, such as the sensation of walking or the sensation of moving and thus provides a communication environment in which the persons involved can share physical experience.

References

[1] T. Miyasato and R. Nakatsu (1998). User Interface Technologies for a Virtual Communication Space. In IEEE and ATR Workshop on Computer Vision for Virtual Reality Based Human Communications, pp. 105-110. Bombay, India: IEEE

[2] H. Noma and T. Miyasato (1998). Design for Locomotion Interface in a Large Scale Virtual Environment, ATLAS: ATR Locomotion Interface for Active Self-Motion. In ASME-DSC. 64. pp. 111-118.

[3] H. Noma, T. Sugihara, and T. Miyasato (2000). Development of Ground Surface Simulator for Tel-E-Merge System. In IEEE VR2000, pp. 217-224. New Brunswick, USA: IEEE.

[4] K. Stanton, P. Sherman, M. Rohwedder, C. Fleskes, et al (1990). PSUBOT - A Voice - Controlled Wheelchair for the Handicapped. In Proc. of IEEE Midwest Symposium on Circuit and Systems. 2. pp. 669-672. : IEEE.

[5] S. Khalaf and P. Siy (1983). The Intelligent Multifunctional Robot Wheelchair. In Proc. of Robotic Intelligence and Productivity Conference, pp. 145-148.

[6] T. Miyasato (1991). VOICE-AID: Pushbutton-to-Speech System for Vocally Handicapped People. IEEE Journal of Selected Areas in Communications, 9(4), pp. 605-610.

[7] S. Brave, A. Dahley, P. Frei, and H. Ishii (1998). In Touch, In SIGGRAPH98 Conference Abstracts and Applications. p.115. Orlando, USA: ACM

The practical side of teaching the elderly visually impaired users to use the e-mail

Takuya Nishimoto, Masahiro Araki, Yasuhisa Niimi

Department of Electronics and Information Science,
Faculty of Engineering and Design,
Kyoto Institute of Technology,
Matsugasaki, Kyoto, 6068585, JAPAN
nishi@vox.dj.kit.ac.jp

Abstract

It is important for visually impaired persons to have the opportunity to take lessons specialized for the visually impaired and novice PC users. From this viewpoint, we designed and implemented the UKK system, which guides the typing lessons with recorded sound and synthesized speech. We also investigated the usability of the popular e-mail tools that the visually impaired can use, and the Bilingual Emacspeak Platform (BEP). As a result, we found BEP is powerful and it has a good conceptual model, but one of the problems is the lack of the manual browser that the novice user can use. We also propose the manual browser of BEP, which uses the VoiceXML browser we implemented. We expect the user will browse the manuals or access to the content using speech input and output, while they use the self-voicing Internet tools with keyboard.

1. Introduction

According to the survey by the Japanese Ministry of Health and Welfare in 1996, there are 305,000 visually impaired persons in Japan. Those who can use Braille among them are less than 10%. Braille is not commonly used because it is especially difficult for the elderly visually impaired persons to learn Braille. On the other hand, only 1,000 blind persons can use personal computer (PC), according to the survey.

It is important for visually impaired persons to use Internet with voice, which gives the way of real-time communication and the chance of attending social activities. In recent years, Japanese Text-to-Speech (TTS) systems were available on the standard PC with audio output, though they required additional hardware. These days, there are some screen-readers for the Japanese version of Microsoft Windows, which help the visually impaired persons to access web and read or write e-mail. They are sold at low prices and do not require any additional hardware.

There remains the problem, however, that the user cannot even setup and learn such systems without the help of a sighted person. There is very limited opportunity to take lessons specialized for the visually impaired and novice PC users. A non-profit organization SCCJ has been carrying out "the Internet school for visually impaired person" in Kyoto City since May 1999. Many people including those who live in far-away places are participating in the school.

From the practical side of teaching such software, the most serious problem is learning the touch-type operations of the keyboard, because many adults were not taught the typing skill during their education in Japan.

In the case of sighted persons, the novice PC users can find the alphabets in the keyboard and use the computer from the beginning. The blind users, however, are forced to learn the touch-typing operations before the lessons about the PC itself, as shown in Figure 1.

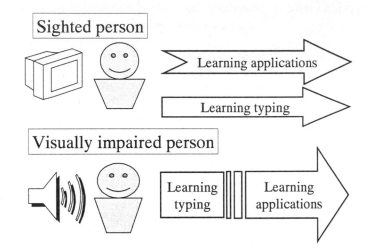

Figure 1. If the novice PC users are sighted persons, he or she can learn touch-typing
and learn applications simultaneously. However, if the person using the screen-reader
software is visually impaired, the skill of touch-typing is required before the
lessons about applications itself.

To learn the touch-typing of the keyboard without another person's help, there is a need of keyboard training tools that do not use the visual display. In Japan, there has been no such software until 2000. Therefore, we designed a system that guides the typing lessons with recorded sound and synthesized speech. This software is named "Uchikomi-kun (UKK)."

2. Development of typing lesson software

As shown in Figure 2, the training system we designed and implemented consists of the UKK player and the UKK contents. The contents are described as the plain text files. The messages in the files are outputted with text-to-speech (TTS) engine. The contents may also include recorded sound files.

We performed preliminary experiments with visually impaired persons and logged the operations of the users. As a conclusion, good results were produced in some cases, and many people said that our system is very useful.

Our evaluations also showed that the poor quality of TTS engine impedes the training, because the synthesized speech we used is sometimes inaudible, especially for the aged persons. For this reason, all the key-echo messages and some of the other messages were replaced in the recorded sound files.

The lessons may take a very long time for the novice users, so it is important for such software to attract the user's interest. From this viewpoint, we introduced an entertainment element into the contents. In our training scenario, two virtual characters appear and talk to the user. One is Uchikomi-kun, who has the synthesized voice and reads most of the guide messages. The other character is a calligraphy master whose name is Kakuunsai. His role is to encourage the user. Recorded sound files are used for the second character's voice, the music and the special effects. The story consists of 13 lessons, which correspond to the stories of travel from the east coast to the west coast of the USA. Professional musicians and narrators are participated to make the contents.

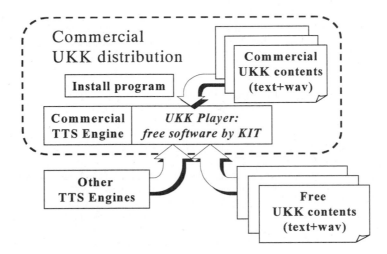

Figure 2. Configuration of UKK, the touch-typing training system we designed
and implemented. It consists of the UKK Player software and the UKK contents
including text and recorded sound files. A TTS engine is required to use the system,
UKK contents are easily modified or replaced by the users.

The unique feature of the system is that the contents are very easy to modify or be replaced by the users. We defined a simple notation for the system, so the files are easy to understand. In fact, the contents of the system were written by a person who is a visually impaired and skilled PC user, but not a software engineer. He has many experiences of teaching screen-reader and Internet software to the visually impaired users, so he could turn them to his advantage in writing the contents.

The system can perform not only the lessons for alphabet, but also the Roman-characters, which are used in the Japanese Input Method Editor (IME). The contents are written as the sequential commands, which are executed in the order, and the event-driven commands, in which the corresponding command is performed when the specific key is pushed.

The user interface is carefully designed. The complicated menu of the multiple hierarchies does not exist. All the lessons can be selected from the single menu using two-digit numbers.

We found that many users were impatient to type the key to go forward or skip some messages, so we decided to make the asynchronous input and output acceptable in our system. Such functions are equivalent to the barge in attribute defined in the VoiceXML specification, which is a description language of the dialog systems [1].

This system is jointly developed with the SCCJ (NPO). The UKK player is open-source software, which is implemented with Visual Basic 6.0 and Speech API (SAPI) 4.0. We performed the evaluations of this version in December 2000 with more than 10 persons who are aged or visually impaired. Most of the testers said that our system is very unique and useful.

We also are supported by the Social Welfare and Medical Service Corporation (WAM), which is a non-profit corporation established under the supervision of the Japanese Minister of Health and Welfare. Supported by WAM, CD-ROM version of the system will be released in March 2001.

3. Evaluation of e-mail softwares

We also investigated how to teach the e-mail software itself. From the practical viewpoint, we evaluated the usability of popular two e-mail tools on Windows system and the Bilingual Emacspeak Platform (BEP), which is currently developed as the open-source software [2]. In the evaluation, we only used the speech output and do not use the screen.

The blind users can use very few Internet software, because most systems are originally designed to use the GUI. It is also said that the software originally designed for GUI has inferior usability for blind users, to the one designed for CUI or AUI (Auditory User Interface)[3]. In our opinion, the problem is not that what kind of UI model a software is based on. The conceptual models, the consistency between the operations, the labor in operations, the burden of the short-term memory, and the design of the auditory feedback is more important for such discussion [4].

We also expect that many people will use speech recognition in the near future. It is, however, difficult for visually impaired people to use large vocabulary automatic speech recognition (LVCSR). It is because the current performance of LVCSR requires the manual correction of errors, which is very hard work without vision. We are also proposing a new application that uses LVCSR effectively [5], but we do not discuss it in this paper.

There are some self-voicing e-mail tools for Windows, but they are not easy to learn. Novice users may need a tool with simple conceptual model and the functions which are limited. It is, however, important for give them the opportunity to use more powerful environment and assist them to learn the expert usage.

The popular e-mail tools that the visually impaired can use, MM-Mail, WinBiff, and BEP (with Mew) were investigated. The results found MM-Mail and WinBiff had some inconsistency in operations, and the conceptual models are relatively complicated. BEP is powerful and it has a good conceptual model. The disadvantages of BEP are:

1. All the messages are in English, so the Japanese novice user has some trouble.
2. There are difficult to learn for the novice user.

The first problem is related to internationalization of Emacspeak itself. The second is, there is a need of a good help-file viewer for visually impaired. There is, of course, the info browser in Emacs, but it also requires some basic skill of using Emacs itself. We are focusing on the second problem currently.

4. Voice help for applications

We expect the user will browse the manuals or access to the content using speech input and output, while they use the BEP or other self-voicing Internet tools with keyboard. We can use UKK player itself as the self-voicing document browser for that purpose. However, we decided to make the browsing tool of BEP manuals based on VoiceXML 1.0. We are implementing a VoiceXML browser which runs on Windows and IBM SR/TTS engines. Using VoiceXML, finding the contents becomes much easier, because we can use speech recognition. VoiceXML also makes the maintenance of the contents very easy.

As the preliminary test, we asked a novice user of Emacs/Mew to use e-mail with BEP. The subject was a university student. He learned the operations of BEP using our VoiceXML based help browser without visual display. It took about 20 minutes to learn the usage of BEP, then he could read the e-mail using BEP.

5. Concluding remarks

Figure 3 shows our current works (solid line) and future works (dotted line). The UKK player will be merged into the future version of our VoiceXML player. The VoiceXML browser can use the server-side implementations of HTML (Web) or E-mail tools, so the novice user can learn typing and use the Internet with our VoiceXML browser alone. Other applications or environments for the professional such as BEP can be used, with the manuals provided in VoiceXML format. The tools described in this paper will be distributed at http://www.sccj.com/ and http://www-vox.dj.kit.ac.jp/nishi/.

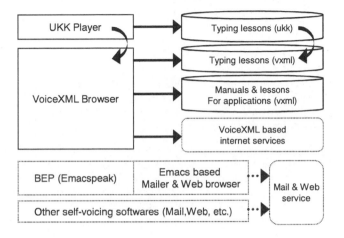

Figure 3 shows our current works (solid line) and future works (dotted line). The novice user can learn keyboard and use the Internet with our VoiceXML browser alone. Other applications can be used with the manuals provided in VoiceXML format.

References

1. http://www.w3.org/Voice/.
2. T. Watanabe, K. Inoue, M. Sakamoto, M. Kiriake, H. Shirafuji, H. Honda, T. Nishimoto, and T. Kamae, Bilingual Emacspeak and accessibility of Linux for the visually impaired, Proceedings of Linux Conference 2000 Fall (2000) 246, (in Japanese).
3. T.V. Raman, Auditory User Interfaces -- Toward the Speaking Computer --, Kluwer Academic Publishers (1997).
4. M. Zajicek, The construction of speech output to support elderly visually impaired users starting to use the internet, Proceedings of ICSLP2000, Volume I, pp.150-153, (2000).
5. T. Nishimoto, H. Yuki, T. Kawahara and Y. Niimi, An Asynchronous Virtual Meeting System for Bi-Directional Speech Dialog, Eurospeech'99, pp.2471-2474 (1999).

Bringing interactive audio documents to life on the WWW

T. V. Raman, Angel Luis Diaz

IBM Research

Abstract

The following insight "The advent of electronic documents makes information available in more than its visual form – electronic information can now be display independent!" led to the development of Audio System For Technical Readings (Aster) in the early 90's. Aster is a computing system that audio formats electronic documents to produce interactive audio documents.

At the time, the World Wide Web (WWW) was still in its infancy. Though most mathematical books and papers were authored in TeX, the marked-up source for such documents was rarely if ever published on the WWW – readers were limited to downloading static visual snapshots as embodied in DVI or PS files. As the WWW matures, we are now moving from the world of *see-only* static HTML documents comprising mostly of text and images to a world where content is encoded as richly structured XML. Rather than using the handful of markup tags prescribed by HTML, authors can now mark-up their documents using domain-specific XML-based languages to produce semantically rich content encodings.

The first example of such semantically rich content is mathematical and scientific content authored in MathML; with a single stroke, mathematical equations on the WWW that were until now *What You See Is All You Have* GIF images can be replaced by *live* mathematical content that can be displayed and manipulated in a manner best suited to the user's needs and abilities.

In this paper, we give an overview of how interactive audio documents as envisioned in Aster are beginning to be realized six years later in the present-day context of XML-based WWW combined with the availability of rich content browsers such as the IBM techexplorer Hypermedia Browser.

1. Introduction

Interactive audio documents as envisioned in Audio System For Technical Readings Aster allow the reader to:
- Browse the document by navigating its underlying structure.
- Listen to the current selection using a preferred rendering style.
- Interactively switch rendering styles to obtain multiple views.
- Interact with live content from online WWW pages and symbolic algebra systems such as Mathematica.

All of this requires suitable high-level internal representations of the underlying document structure – see Raman, 1998. High-level representations are easily constructed from richly marked up XML documents. The XML Document Object Model (XML DOM) provides a standardized tree representation for XML documents; standard XML parsing tools can produce such representations given valid document encodings along with the document DTD. Context-specific information needed for producing rich aural renderings is obtained from the domain-specific semantic tagging used to markup the document. This paper focuses explicitly on mathematics and the use of Mathematical Markup Language (MathML) in encoding such logical structure.

Using MathML in conjunction with XML-aware authoring and browsing tools, students and teachers alike can create, proof-read, and browse mathematical content using a multiplicity of modalities such as visual presentations, outline views, and aural renderings. The rest of this paper will focus explicitly on aspects of aural interaction in the context of authoring and browsing structured mathematics in an eyes-free environment.

2. Browsing structured mathematics

Browsing structured mathematics involves walking the expression tree, applying user-customizable transformations, and producing a rendering to the chosen modality. This section outlines the various XML-based standards we use in attaining these goals.

IBM techexplorer Hypermedia Browser is a cross-platform Web browser plug-in and Microsoft Windows ActiveX control that supports interactive publishing of scientific documents encoded in TeX or MathML. An important aspect of our work on techexplorer is redefining the meaning of *online publishing* of scientific material. The key requirements for scientific *e*-publishing include:

* document logical structure,
* two-dimensional math layout,
* object semantics as well as presentational markup,
* inclusion of dynamically-generated markup, and
* connectivity to a wide range of local and remote applications.

Thus, we needed to add many features beyond what is needed for traditional typeset mathematics. Further, mathematical objects was not static; in many situations, readers need to re-use the math for further computations. This means that the math must be represented in a way that describes ist meaning and not just its presentation. The result of a computation may need to be pasted back into the original document, making the document dynamic.

The first specification from the W3C HTML Math working group was the Mathematical Markup Language (MathML). MathML 1.0 was released as a W3C Recommendation in April, 1998. As the first W3C endorsed XML application, MathML is a XML vocabulary for encoding mathematics. MathML provides a much needed foundation for the inclusion of mathematical expressions in Web page and as a common encoding for scientific processors.

Each MathML element falls into one of three categories: presentation elements, content elements and interface elements. Just as titles, sections, and paragraphs capture the logical structure of a textual document, presentation elements are meant to express the syntactic structure of math notation. Content elements describe mathematical objects directly, as opposed to describing the notation which represents them.

The W3C Document Object Model (DOM) recommendation brings the structure of an XML document to life by providing a platform and language-neutral interface that allows executable code to dynamically access and update the content, structure and style of documents. User agents like techexplorer provide access to the DOM from several programming languages.

The W3C XSLT (the Extensible Stylesheet Language Transformations) and XPath (the XML Path Language) recommendations describe a tree-oriented transformation language for converting instances of XML documents from one XML vocabulary into another.

3. Authoring content

Structure-based authoring is especially conducive to an eyes-free computing environment.
This is because the author can focus on the content and leave the minutii of maintaining a structurally valid document to the authoring environment. The production step of generating a final-form representation is managed by the authoring environment. To use such environments effectively, authors also need to interactively preview their work.

The first author uses Emacs package PSGML as a structure-aware XML editor to author XML documents. PSGML} is fully speech-enabled by Emacspeak} – see Raman, 1997a; Raman, 1996; Raman,1996; Raman1997b. Given the DTD for MathML, this authoring environment provides context-sensitive menus and keyboard shortcuts for inserting MathML constructs as appropriate. Emacspeak's various navigation features permit me to quickly move through complex structures as they are being authored. Emacs' outlining features allow one to obtain outline views of complex constructs. Since the validity of the structure is continuously maintained by the authoring environment, constructs are ready to be previewed in a MathML browser as soon as they have been authored. When our present work of integrating audio formatting into IBM TechExplorer is completed, one should be able to obtain instant aural previews of mathematical content as it is being created.

Notice that as described, the user is being provided functionality comparable to what users of WYSIWYG environments have come to expect, with the additional advantage that at the end of the day, the user is left with a well-structured document – rather than just a What You See Is All You Have document, as all too often happens in the WYSIWYG world. Computer algebra systems like Wolfram Research's Mathematica and MathSoft's MathCAD provide the ability to export techexplorer ready MathML documents. Mathematica documents are live in that the mathematical expressions can be updated inside the browser and sent to a compute engine for re-evaluation. MathCAD can import an HTML/MathML document for further editing.

4. Proof-reading

MathML content can be browsed using IBM TechExplorer. This browser can either run as a stand-alone browser, or provide MathML browsing services to other applications. The browser is well-integrated with standard XML tools and provides access to the underlying document structure using standardized APIs like DOM. In addition, it is possible to integrate this browser with XML Style Language Transforms (XSLT) to interactively apply different transforms to obtain multiple views.

We use the facilities described above in conjunction with IBM ViaVoice speech technologies to produce interactive aural renderings of MathML constructs. Audio formatting to produce such aural renderings happens in several stages:

- The selected portion of the DOM is transformed using a user selectable XSLT stylesheet to produce output suitable for a text to speech (TTS) engine, and conforming to any one of possible rendering styles.
- The output of this transformation consists of text annotated with TTS commands for controlling various aspects of voice characteristic and prosody. Eventually, this will be updated to produce Speech Synthesis Markup Language (SSML).
- The annotated output stream is sent to the TTS engine to produce spoken output.

As spoken output progresses, the user can interrupt the rendering, and navigate the structure. Moving through the structure results in the system summarizing the node that is selected. Such summaries are designed to be succinct while providing sufficient contextual information; the typical summary is of the form "Context" is *type*." For example, when moving through an equation, the user hears "*Left hand side* is *summation*."

The ability to rearrange nodes in the selected portion of the DOM provides the multiple rendering functionality described in system Aster. Thus, given an expression of the form

$$(A \otimes B)^T = A^T \otimes B^T$$

the listener can have the same expression rendered using different styles to produce different *auditory views* of the same content as in:

- The transpose of the Kronecker product of A and B is equal to the Kronecker product of B transpose and A transpose.
- The transpose of quantity A Kronecker product B is equal to B transpose Kronecker product A transpose.

In general, the ability to hook up user customizable XSL transformations to the underlying DOM representation before the final aural rendering is produced allows the user to traverse the same expression in different ways to obtain a complete understanding of the underlying content. Notice that this happens implicitly in the case of visual browsing, where the eye \emph{dynamically} scans across a *static* piece of paper or visual display; these steps need to be made explicit in the case of auditory browsing, where otherwise a *passive* listener would be limited to listening to an *actively* scrolling display.

5. Conclusions

To summarize, our work demonstrates how evolving cross-platform XML standards like MathML enable interactive scientific documents via the following:

- MathML, a standard markup for math,
- DOM, a standard application programming interface for XML,
- and XSL, a standard transformation language

together bring interactive scientific documents to life on the WWW.

References

Raman, T. V. (1994). Audio System for Technical Readings, PhD thesis, Cornell University, http://cs.cornell.edu/home/raman.

Raman, T. V. (1996a). Emacspeak - direct speech access, Proc. of The Second Annual ACM Conference on Assistive Technologies (ASSETS '96)}.

Raman, T. V. (1996b). Emacspeak: A speech interface, In Tauber, M. J., Bellotti, V., Jeffries, R., Mackinlay, J. D., and Nielsen, J., editors, Proceedings of the Conference on Human Factors in Computing Systems : Common Ground, pp 66-71, New York: ACM Press.

Raman, T. V. (1997a). Auditory User Interfaces --Toward The Speaking Computer}, Kluwer Academic Publishers.

Raman, T. V. (1997b). Emacspeak: A speech-enabling interface: Moving toward auditory user interfaces, Dr. Dobb's Journal of Software Tools, 22(9):18--20, 22, 23.

Raman, T. V. (1998). Aster Audio System For Technical Readings, Lecture Notes In Computer Science. Springer Verlag.

Virtual Reality for Persons with Central Nervous System Dysfunction: Assessment and Treatment in the Information Society for All

Albert A. Rizzo[ab], J. Galen Buckwalter[b], Maria Schultheis[c], Ulrich Neuman[a], Todd Bowerly[d], Laehyun Kim[a], Marcus Thiebaux[e], Clint Chua[a]

[a]Integrated Media Systems Center, University of Southern California
3740 McClintock Ave, EEB 131, Los Angeles, California, 90089-2561, USA

[b]School of Gerontology, University of Southern California
3715 McClintock Ave, Los Angeles, California, 90089-0191, USA

[c]Kessler Medical Rehabilitation Research & Education Corp.
1199 Pleasant Valley Way, West Orange, NJ 07052, USA.

[d]Fuller Graduate School of Psychology
180 North Oakland Ave., Pasadena, CA. 91101, USA

[e]Information Sciences Institute, University of Southern California
4676 Admiralty Way, Marina del Rey, CA 90292-6695, USA

Abstract

Virtual Reality (VR) technology is increasingly being recognized as a useful tool for the study, assessment, and rehabilitation of cognitive processes and functional abilities in persons with central nervous system dysfunction. Much like an aircraft simulator serves to test and train piloting ability, virtual environments can be developed to present simulations that target human cognition and behavior in normal and impaired populations. The capacity of VR technology to create dynamic, multi-sensory, three-dimensional (3D) stimulus environments, within which all behavioral responding can be recorded, offers clinical assessment and rehabilitation options that are not available using traditional methods. The assets available and issues involved in the design of VR scenarios for use by persons with cognitive/functional impairments are considered in this article within the context of Information Society for All and Universal Access perspectives.

1. Introduction

We are experiencing the emergence of an information society, increasingly based on the production and exchange of information. As this vision unfolds, those who are able to thoughtfully design, develop and apply information technology and telecommunications (IT&T), will be in a position to drive fundamental advances for promoting human welfare. In order to maximize the potential benefits of this paradigm shift for those with special needs, it is necessary to focus efforts on the development and application of more usable and accessible IT&T. This direction fits well with the "Information Society for All" concepts that have recently been addressed in the human-computer interaction literature (Stephanidis, Salvendi Akoumianakis, Bevan, Brewer, et al., 1999). Efforts in this area support the development of IT&T that accommodates the broadest range of human abilities, skills, requirements and preferences. The potential results of such efforts could substantially redefine the assessment and rehabilitative strategies that are used in the area of disabilities, particularly with clinical populations having cognitive and functional impairments due to central nervous system (CNS) dysfunction. CNS dysfunction and its resulting impairments can occur through a variety of circumstances. The most frequent causes include traumatic brain injury (TBI), neurological disorders, developmental and learning disabilities, as well as complications from medical conditions and surgical procedures. Impairments that are common sequelae to CNS dysfunction involve processes of attention, memory, language, perceptual-motor function, spatial abilities, higher reasoning and functional abilities (i.e., instrumental activities of daily living (IADLs)). The co-occurrence of significant emotional, social, vocational, self-awareness and motor deficits further complicates the needs of CNS patients. Because of the pervasive nature of CNS dysfunction and resulting disabilities, the costs to individuals and society is significant. However, in the age of "managed care", short-term cost savings are often prioritized over long-term goals and systematic standard rehabilitation efforts are typically curtailed before positive benefits have time to emerge. Advances in more usable IT&T could have significant impact in this area by providing more cost/effective automated Virtual Reality (VR) assessment and rehabilitation tools that can supplement the necessary hands-on work of clinical professionals. Addressing Universal Access concerns in the design of these "clinical" VR applications in a manner that promotes more seamless and naturalistic interaction could also serve to drive R&D that could produce benefits for a broader range of unimpaired users.

Early innovations in IT&T have already played an important role in improving the lives of persons with disabilities. For example, early developments in IT&T have improved accessibility via HCI advances which now allow persons with tetraplegia increased control over their home environment by way of more "seamless" voice recognition computerized assistance (e.g., Platts & Fraser, 1993). Advancements in IT&T telecommunication tools have provided vocational opportunities for home-bound individuals that permit competitive work to be conducted from one's home (Flynn & Cherie-Clark, 1995) and also support increased interaction with the outside world (Post, van Asbeck, van Dijk, & Schrijvers, 1997). These applications support functional independence and provide the impetus for rehabilitation specialists to continue exploring the potential assets available with emerging IT&T systems keeping Universal Access and Information Society for All concerns in mind.

2. Virtual reality definitions and rationale for use with cns populations

One form of IT&T that is being rigorously investigated for its potential use with clinical CNS populations is virtual reality. Virtual Reality can be defined from an HCI perspective as "...a way for humans to visualize, manipulate, and interact with computers and extremely complex data." (Aukstakalnis and Blatner, 1992). In essence, VR can be viewed as an advanced form of human-computer interface that allows the user to "interact" with, and become "immersed" within a computer generated virtual environment (VE) in a more intuitive and naturalistic fashion. VR integrates real-time computer graphics, body tracking devices, visual displays, and sensory input devices to immerse a participant in a VE that changes in a natural way with head and body motion. The believability of the virtual experience or sense of presence is supported by employing such specialized technology as head-mounted displays (HMDs), tracking systems, earphones, gesture-sensing gloves, interaction/navigational devices and sometimes haptic-feedback devices. The combination of a HMD and tracking system allows the computer to generate images and sounds in any computer-modeled (virtual) scene that corresponds to what the user would see and hear from their current position if the scene were real. While HMDs are most commonly associated with VR, other methods incorporating 3D projection walls and rooms (known as CAVES), as well as basic flatscreen computer systems have been used to create interactive scenarios of value for assessment and rehabilitative purposes. Methods for navigation and interaction such as data gloves, joysticks, 3D mice, treadmills and some high-end "force feedback" mechanisms that can provide tactile feedback have also been developed. However, challenges in existing interface design still need to be addressed before the level of naturalistic VE interaction is achieved to make them truly useful, particularly with clinical CNS populations.

The rationale for VR applications in the assessment and rehabilitation of populations with CNS dysfunction is fairly straightforward. By analogy, much like an aircraft simulator serves to test and train piloting ability under a variety of systematic and controlled conditions, VEs have been developed to present simulations that target human cognitive and functional processes that are relevant for assessment and rehabilitative purposes. This work has the potential to improve our capacity to understand, measure, and treat the impairments typically found in clinical populations with CNS dysfunction as well as advance the scientific study of normal cognitive and functional/behavioral processes. The unique match between VR technology assets and the needs of various clinical application areas has been recognized by a number of authors (Rizzo, 1994; 1997; Pugnetti, 1995; Rose et al. 1997; Schultheis & Rizzo, 2001) and an encouraging body of research has emerged (Rizzo, Buckwalter and van der Zaag, 2001). What makes VR application development in this area so distinctively important is that it represents more than a simple linear extension of existing computer technology for human use. VR offers the potential to deliver systematic human testing and training environments that allow for the precise control of complex, dynamic 3D stimulus presentations, within which sophisticated behavioral recording is possible. When combining these assets within the context of functionally relevant, ecologically valid VEs, a fundamental advancement emerges in how human cognition and functional behavior can be assessed and rehabilitated. This potential was recognized early on in a visionary article ("The

Experience Society") by VR pioneer, Myron Kruegar (1993), in his prophetic statement that, "... Virtual Reality arrives at a moment when computer technology in general is moving from automating the paradigms of the past, to creating new ones for the future". (p. 163) However, before this paradigm shift can occur, advances in the design of VEs need to occur that will make them more accessible and usable.

3. Usability issues

3.1 VR assets for assessment and rehabilitation CNS applications

In order for VR applications to be efficiently developed in this area with "Information Society for All" concerns in mind, one must be apprised of *both* the assets available with the technology and the unique user characteristics that vary both between and within clinical CNS populations. Numerous assets for VR use in this area have been postulated (see Table 1) and these present exceptional possibilities for developing human cognitive performance assessment and rehabilitation scenarios.

Table 1: VR assets for cognitive/functional assessment and rehabilitation applications

1. The presentation of *ecologically* valid assessment, training and treatment scenarios.
2. Delivery of stimuli (dynamic, interactive 3D) that would be difficult to present using other means.
3. Total control and consistency of stimulus delivery.
4. The presentation of repetitive and hierarchical stimulus challenges which can be varied from simple to complex, contingent upon success.
5. The provision of "cueing" stimuli or visualization tactics (i.e., selective emphasis) designed to help guide successful performance.
6. The delivery of immediate performance feedback in a variety of forms and sensory modalities
7. The capacity to pause assessment, treatment and training for discussion and/or integration of other methods.
8. The option for self-guided practice and independent testing and training when deemed appropriate.
9. The modification of sensory presentation and response requirements based on the user's impairments (i.e., movement, hearing, and visual disorders).
10. The availability of a more naturalistic/intuitive performance record for review and analysis by the user/therapist/researcher.
11. The integration of virtual human representations (avatars) for systematic applications addressing social interaction.
12. The design of safe testing and training environments which minimize the risks due to errors.
13. The introduction of "gaming" factors into the assessment/treatment/training scenario to enhance motivation.
14. The potential to create low cost libraries of VEs that could be distributed/shared (possibly over the internet).

3.2 Clinical usability questions

When considering the design of VR scenarios for persons with CNS dysfunction, as with unimpaired populations, it is essential to have advance specification of the user assets and limitations that could potentially affect the optimal use of these systems. Users with CNS dysfunction may typically manifest a variety of deficits in such areas as basic arousal, attention processes, sensory/perceptual abilities, memory, motor skills, multi-tasking and organizational facility, among impairments in other cognitive/functional domains. Task and system requirements also need to be considered in an interactive fashion with such user characteristics. Consequently, before creation of an application in this area can commence, certain questions need to be addressed to maximize effective R&D to promote access and usability (Table 2). The questions listed in Table 2 provide an intuitive stepwise approach for deciding on how and when to use VR in this area from a "hybrid" clinical/HCI perspective.

Table 2: VR/Clinical Usability Questions.

1. Can the same objective be accomplished using a simpler approach?
2. How well do current VR assets fit the needs of the cognitive/functional approach, task or target?
3. How well does a VR approach match the characteristics of, and is usable by, the target clinical user group?
4. What is the optimal level of presence needed for clinical CNS VR applications to be usable and useful?
5. Will clinical CNS target users be able to navigate and interact within the VR in an effective manner?
6. What is the potential for side-effects due to the characteristics of different CNS clinical groups?
7. Will VR assessment results and treatment effects generalize or transfer to the "real world"?

These general usability questions address a range of issues that are commonly cited in the human factors and HCI literature and are particularly important when considering the wide variability in user skills seen with clinical CNS populations. To reach the intended goals of the "Information Society for All" perspective in this area requires interdisciplinary cooperation between usability specialists and scientists who have domain-specific knowledge of the range of cognitive/functional impairments seen with CNS clinical populations. User-centered design and evaluation methods are vital to this process throughout the application lifecycle and considering the questions presented in Table 2 is essential to support the rational development of VR systems for users with CNS dysfunction. Although it is beyond the scope of this article to address these issues in detail here, involved treatments of this area from both the HCI and Clinical Neuropsychology perspectives can be found elsewhere (Hix, Swan, Gabbard, McGee, Durbin, King, 1999; Rizzo et al, 1998,2001; Brown, Kerr and Bayon,1998). However, some of the primary observations on how attention to these questions might promote Universal Access concerns in this area can be briefly discussed.

An initial needs and requirements analysis is the first step to justifying that a given VR application can serve a useful purpose beyond what currently exists with traditional assessment/rehabilitation methodologies (Question 1). This incorporates expert knowledge as to the task components and how VR assets can better address these targets (Question 2). For example, early success has been seen in VR applications that use 3D modeling of functional environments for training wayfinding in clinical populations with topographical disorientation. In these applications, stroke patients with amnesia (Brooks et al., 2000) and children who use wheelchairs and have limited opportunities for independent exploration (Stanton et al., 2001), displayed excellent transfer of training from the virtual environments to the real environments (a rehabilitation unit and school building, respectively). These are examples of applications where unique VR assets are well matched to the needs of the application (wayfinding). Other examples of successful VR application development can be seen in VR programs that integrate user-centered input from clinical populations. This consideration of user characteristics (Question 3) has been later shown to promote usability (and learning) by these groups as demonstrated by the thoughtful work of Brown et al., (1998) incorporating input from students with severe learning disabilities in the design of VR lifeskill training scenarios. These applications were also shown to be effective using flatscreen display technology that has been commonly seen as engendering a lower level of immersion with consequent limiting of the sense of "presence". However, these applications still produced good clinical benefits and therefore underscore the need to consider what the optimal presence requirements might be for a given application. This is important for determining when reductions in system costs and complexity can occur while still supporting the production of usable and useful VR systems (Question 4).

From an HCI perspective, a primary concern involves how to design more naturalistic and intuitive tools for human interfacing with complex systems. In order for persons with cognitive impairments to be in a position to benefit from VR applications, they must be capable of learning how to navigate within the environment. Many modes of VR navigation (data-gloves, joy sticks, space balls, etc.), while easily mastered by unimpaired users, could present problems for those with cognitive difficulties. Even if patients are capable of using a VR system at a basic level, the extra non-automatic cognitive effort required to navigate may serve as a distraction and limit the assessment and rehabilitation process. In this regard, Psotka (1995) hypothesizes that facilitation of a "single egocenter" found in highly immersive *interfaces* serves to reduce "cognitive overhead" and thereby enhance information access and learning. This is an area that needs the most attention in the current state of affairs for VR applications designed for populations with CNS dysfunction (Question 5). As well, side effects of VR usage as cited in Question 6 needs to be considered and continued monitoring of ALL VR applications is necessary before consistent guidelines can be specified for safe use by CNS groups across the range of factors that could influence side effect occurrence. Finally, transfer of training and generalization tactics need to be preprogrammed into VR system design and evaluation to determine the best course for creating VEs that produce optimal real world impact for the broadest range of clinical populations (Question 7). Useful recommendations for accomplishing this can be found throughout the applied behavior analysis literature, particularly in a "classic" paper by Stokes and Baer (1977).

4. Conclusions

In line with the principles engendered in the Information Society for All, attention to these sorts of user-centered issues is seen as vital in the design, development and implementation of virtual environments for the broadest range of potential users. By investigating the factors that need to be considered in the creation of VEs for CNS clinical populations, enhancements in accessibility that are relevant for unimpaired populations could likely be advanced as well. VR, in it's idealized naturalistic and intuitive form, should in theory be well poised to promote the development of tools that manifest good universal design in such domains as: equitable use, flexibility of use, simple and intuitive use, perceptible information, tolerance for error and low physical effort. However, the challenges that exist for the design of VR systems that maximize access and usability by persons with cognitive/functional impairments are substantial. If these challenges are successfully addressed, this application area could have a significant impact on the level of standard care available to populations that are often "bypassed" by new IT&T developments. This would serve a useful human purpose for this group directly, while at the same time driving the methodology needed to truly create an Information Society for All! These issues will be detailed during the conference presentation in the context of descriptions of VR application development in the areas of assessment and rehabilitation of attention, memory and visuospatial processes.

References

S. Aukstakalnis, D. Blatner, Silicon Mirage: The Art and Science of Virtual Reality. Peachpit Press, Berkeley CA., 1992.

B.M. Brooks, J.E. McNeil, F.D. Rose, R.J. Greenwood, L.A. Attree, A.G. Leadbetter, Route learning in a case of amnesia: A preliminary investigation into the efficacy of training in a virtual environment, *Neuropsychological Rehabilitation*, 9 (1), 63-76, 1999.

D.J. Brown, S.J. Kerr, V. Bayon, *The development of the Virtual City: A user centred approach.* In P. Sharkey, D. Rose, and J. Lindstrom (eds.), Proceedings of the 2nd European Conference on Disability, Virtual Reality and Associated Techniques, Sköve, Sweden, pp. 11-16, 1998.

C.C. Flynn, M. Cherie-Clark, Rehabilitation technology: Assessment practices in vocational agencies, *Assistive Technology*, 7, 111-118, 1995.

D. Hix, J.E. Swan, J.L. Gabbard, M. McGee, J. Durbin, T. King, *User-Centered Design and Evaluation of a Real-Time Battlefield Visualization Virtual Environment*, Proceedings of the IEEE Virtual Reality 1999, pp. 96-103, 1999.

M.W. Krueger, The Experience Society, *Presence*, 2(2), 162-168, 1993.

R.G. Platts, M.H. Fraser, Assistive technology in the rehabilitation of patients with high spinal cord injury lesions, *Paraplegia*, 31, 280-287, 1993.

M.W. Post, F.W. van Asbeck, A.J. van Dijk and A.J. Schrijvers, Spinal cord injury rehabilitation: 3 functional outcomes, *Archives of Physical Medicine and Rehabilitation*, 87, (3 Suppl), S59-64, 1997.

M.T. Schultheis, A.A. Rizzo, The application of virtual reality to rehabilitation. *Rehabilitation Psychology*, 2001.

J. Psotka, Immersive training systems: Virtual reality and education and training, *Instructional Science*, 23, 405-431, 1995.

L. Pugnetti, L. Mendozzi, A. Motta, A. Cattaneo, E. Barbieri and S. Brancotti, Evaluation and retraining of adults' cognitive impairments: Which role for virtual reality technology? *Computers in Biology and Medicine,* 25(2), 213-227, 1995.

A.A. Rizzo, *Virtual reality appilications for the cognitive rehabilitation of persons with traumatic head injuries.* Proceedings of the 2nd International Conference on Virtual Reality and Persons with Disabilities (pp. 135-140), 1995.

A.A. Rizzo, J.G. Buckwalter, Virtual reality and cognitive assessment and rehabilitation: The state of the art. In G. Riva (ed.), *Psycho-neuro-physiological Assessment and Rehabilitation in Virtual Environments: Cognitive, Clinical, and Human Factors in Advanced Human Computer Interactions* (pp. 123-146), IOS Press, Amsterdam, 1997.

A.A. Rizzo, J.G. Buckwalter, C. van der Zaag, Virtual Environment Applications for Neuropsychological Assessment and Rehabilitation. K. Stanney (Ed), *Handbook of Virtual Environments*, 2001.

F.D. Rose, E.A. Attree, B.M. Brooks, Virtual environments in neuropsychological assessment and rehabilitation. In G. Riva (Ed.), *Psycho-neuro-physiological Assessment and Rehabilitation in Virtual Environments: Cognitive, Clinical, and Human Factors in Advanced Human Computer Interactions,* (pp. 147-156). IOS Press: Amsterdam, 1997.

D. Stanton, P. Wilson, N. Foreman, H. Duffy, *Virtual environments as spatial training aids for children and adults with physical disabilities,* In Sharkey, Cesarani, Pugnetti, Rizzo (Eds.) Proceedings of the 3rd International Conference on Disability, Virtual Reality, and Associated Technology, Alghero, Sardinia, Italy, September, 23-25, 2000.

C. Stephanidis, G. Salvendi, D. Akoumianakis, N. Bevan, J. Brewer, P.L. Emiliani, A. Galetsas, S. Haataja, I. Iakovidis, J.A. Jacko, P. Jenkins, A.I. Karshmer, P. Korn, A. Marcus, H.J. Murphy, S. Stary, G. Vanderheiden, G. Weber, and J. Ziegler, Toward an information society for all: An International Research and Development Agenda, *International Journal of Human-Computer Interaction*, 10 (2), 107-134, 1998.

T.F. Stokes, D.M. Baer, An implicit technology of generalization, *J App Behav Analysis,* 10, 349-367, 1977.

Adapting haptic game devices for non-visual graph rendering

Patrick Roth[a], Christoph Giess[b], Lori Petrucci[a], Thierry Pun[a]

[a]Computer Science Department
CUI, University of Geneva
CH-1211 Geneva 4, Switzerland

[b]Division Medical and Biological Informatics
Deutsches Krebsforschungszentrum, Heidelberg
Patrick.Roth@cui.unige.ch

Abstract

We describe in this paper the design of a system of graph representation for blind and visually impaired people. Our system is based on an audio and haptic force feedback (FF) interaction. We use a game force FF mouse as haptic display. The auditory representation is used to improve the kinesthetic rendering. We also compared our system with another approach based on a high quality haptic device.

1. Background

In the domain of human-computer interaction for blind users, the non-visual representation of scientific data has become a very active domain of investigation. We focus here on the rendering of graphs given by an equation $Y = f(X)$, where X is the independent variable and Y the dependant one.

Current rendering approaches are either based on auditory or on haptic rendering models. Auditory rendering involves sonification to represent the graphical data (Mansur 1985; Edwards 1993; Flowers 1995; Sayhun 1999). Most sonification methods map a graph to a tone whose pitch represents the Y coordinate and time represents the X coordinate. Experimental results however point out that users have difficulties to interpret rate of changes in the graphs (i.e., first derivative information).

To solve this problem researchers replace audition by the sense of touch, by means of haptic FF devices (Fritz 1999; Ramloll 2000). With such devices, users feel the graph by following its path on a virtual plane using either a stylus or fingertips. The main problem with this solution stems from the high cost of the haptic systems involved, such as the PHANToM apparatus (http://www.sensable.com). In contrast, this paper describes a non-visual scientific data representation system that uses auditory rendering to augment the performances of a low cost haptic FF device (WingMan FF mouse). We also report the experiment that we conducted with eight blind and visually impaired participants. In this experiment, we compared our system with the traditional haptic FF rendering method that uses high quality FF device.

2. Audio haptic designs

2.1 System design for low quality FF device

Mechanical, ergonomic design and cost issues make low cost game FF devices and more precisely the WingMan FF mouse poor in performance with respect to several factors. Sjöström (Sjöström 1999) mention three points that affect the performances of haptic such low-end devices. First, the small size of the workspace combined with the nature of the manipulandi degrades the manipulation precision (Cutkosky 1990). For such a small workspace, the haptic interface must be combined with some form of fingertip grasp rather than full hand grasp in order to preserve a reasonable level of precision.

978

A second problem comes from the limited number of degrees of freedom (DOF) enabled by the 2-DOF WingMan FF mouse. In terms of haptic visualization, the number of DOF characterizes the information bandwidth allowed by a haptic FF device. With a 3-DOF device such as the PHANToM, a graph could be mapped onto two dimensions while the third dimension might be used for additional information and so increase the information bandwidth.

The last factor comes from the force range that the FF mouse provides. As defined by Fritz (Fritz 1996), a device must be able to exert enough force to create a realistic sensation, that sensation depending on the type of grasping. Since our FF mouse exerts a maximum force of 1 N, it does not allow rendering a very realistic sensation for hand grasping.

In the sequel, we show how such poor performances can be enhanced by the combined use of audio and haptic rendering, thus enabling good representation with low-cost devices.

2.1.1 Haptic rendering

The haptic rendering of the graph is based on a virtual force that attracts the mouse cursor towards the graph. The point P_{min} on the graph with the minimum distance from the mouse cursor c is first determined (see figure 1.a). If the distance d is below a given threshold, an attraction force F is activated; F is orthogonal and directed towards the graph, as well as proportional to the distance d (see figure 1.b).

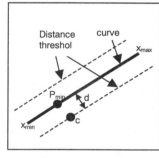 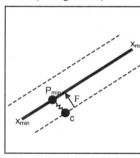 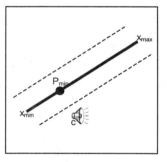

Figure 1.a. distance setting;   1.b. attraction force;   1.c. audio rendering.

In addition, to go around the lack of precision due to the limited size of the workspace, we provide zoom-in and zoom-out functions. This functionality allows the user to dynamically change the exploration interval (i.e. x_{min} and xmax) of the horizontal axis.
 at any time.

2.1.2 Auditory rendering

In order to increase the information bandwidth, we choose complement haptic exploration to work also withby the auditory channelfeedback.The audio rendering may provides additional sonic information during the haptic exploration. The This additional information relies conveyson graph properties that are:

- Minimum or maximum y-values
- Intersection point(s) between the graph and either horizontal and vertical axis
- Slope (i.e., a tone pitch that represent the first derivative value of P_{min})

Each property is represented by a different tone, making simultaneous rendering possible (Brewster 1995) so as to increase the information bandwidth.

Finally, an alarm is used in order to increase the FF sensation. When the user leaves the region around the graph specified by the distance thresholds, an auditory signal is played. This signal is characterized by its spatial location that corresponds to the position of the point c relative to P_{min} (see figure 1.c).

2.2   System design for high cost FF device

For representing the graph, the system that worked with high cost FF device used uniquely haptic as output modality. The haptic rendering approach that was implemented is identical to that used for the FF mouse (see section 2.1.1). The point P_{min} on the graph with the minimal distance to the PHANToM pointer is determined and, if below a given threshold, an attraction force is activated. This force is perpendicular to the derivation of the graph and proportional to its distance.

The x-axis and y-axis are rendered with the same technique but lower attraction forces then the graph. This allows distinguishing between graph and coordinate system. Additional forces are used to mark both ends of the selected part of the graph as well, giving the user the feeling of a "closed system".

All these forces are overlaid during the interaction with the system. When the user follows the run of the graph, the points of intersection with both axes can be felt as small leaps.

3.   Hardware configuration

The evaluation of the first system was made on a Pentium PII equipped with a SoundBlaster Live sound card and a Sennheiser headphone. For the 3D sound rendering, we used the DirectSound3D library available from Microsoft. For the kinesthetic interaction, we worked with the Logitech WingMan Force Feedback mouse (http://www.logitech.com).

For the evaluation of the second system, we used PHANToM Desktop System from SensAble Technologies. The implementation was done using SensAble's GHOST library for haptic rendering.

4.   System evaluation

We have conducted a series of experiments in order to compare the respective efficiency of the two methods described above. Each method was experimented by four participants aged between 20-45 years old. In each group, two of the participants were congenitally blind and the two others were visually impaired. At all times, the visually impaired participants were not blindfolded, but they did not have access to the computer screen. All the participants had an equivalent knowledge about scientific graphs.

Prior to the experiment, we gave each participant the same amount of time (20 minutes) to get acquainted with the system under test.

4.1   Tasks and scenarios

Each participant was given three different scientific graphs (one at the time) for evaluation:

* a linear: y = 1-x, x∈[-2;2] (see figure 2.a);
* a parabolic curve: y = x2, x∈[-2;2] (see figure 2.b);
* a periodic sinusoid: y = sin(x), x∈[-4;4] (see figure 2.c).

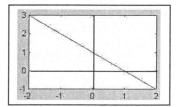 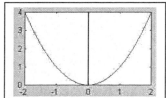 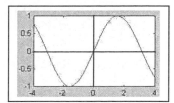

Figure 2.a. y=1-x;   2.b. y=x^2;   2.c. y=sin(x).

For every graph, after interacting with it, each participant was asked to reproduce it. The blind participants reproduced it using a tactile medium, where the visually impaired used a pencil and a sheet of paper. We asked all participants to place the horizontal and vertical axis while drawing. Our goal was to check the efficiency of each local information rendered by the encoding method that was being tested. After completing each task, we asked them several question, on the graph rendered such as the periodicity, minimum and maximum values location and continuity. At the end of the experiment, we then asked them to evaluate the system they had tested.

4.2 Results and discussion

The results of the experiment reveal that the group that used our Audio-Haptic system had problems in distinguishing curvature (see table 1 and figure 2.a). The zoom-in and zoom-out features that we implemented were not efficient enough to compensate the lack of precision due to the size of the workspace and the nature of the manipulandi. We therefore have to opt for solutions based on the modification design of our FF device. A solution would be to increase the size of the workspace. Sjöström (Sjöström 1999) found that according to his experience, the ideal workspace size should be 6 x 7 cm. Another solution would be to modify the manipulandi as a pantograph in order to provide a finger grasping.

Table 1: Percentage of participants that were able to correctly reproduce the shape of the graph, his curvature and to place the horizontal and vertical axis.

Graph	Modality	Reproduce		Axis positioning	
		Shape	Curvature	Horizontal	Vertical
1 – X	FF-Audio	100 %	100 %	100 %	100 %
	FF-PHANToM	100 %	100 %	100 %	100 %
X²	FF-Audio	100 %	50 %	100 %	100 %
	FF-PHANToM	75 %	75 %	100 %	100 %
Sin(X)	FF-Audio	75 %	0 %	75 %	100 %
	FF-PHANToM	100 %	100 %	50 %	25 %

As for periodic graphs, participants that used the second rendering method (FF-PHANToM) had difficulties in placing horizontal and vertical axes (see figure 2.b). The reason comes from the kinesthetic design that was implemented. Here, we represented in the same way the graph and his axis. Participants had problems to distinguish when the graph intersected the axis.

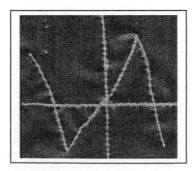 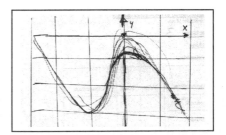

Figure 2.a. reproduced with FF-Audio;   2.b. reproduced with FF-PHANToM

A possible solution to this problem would be to use the third DOF provided by the PHaNTOM device to represent the intersections between the graph and his axis. Giving the user additional information might be a solution to overcome this problem. This could be accomplished by rendering graph and axis with different friction and/or on different layers in the three-dimensional space. Giving the user the choice which information should be haptically rendered would be an other possibility.

5. Conclusions

In this paper we have presented the design of a non-visual graph representation system that used game FF device. The results obtained during this experimentation survey prove that it is possible to increase the quality of low quality FF device by adding sonification. This enables the use of low cost FF devices for scientific data visualization, which represents an advantage on current solutions. However, such device needs to be modified (e.g., size and manipulandi) in order to compensate for the lack of rendering precision. Actually we are working on several modifications of the FF mouse such as the increase of the workspace, and the redesign of the manipulandi (i.e., finger based).

For the next step, we plan to adapt our system for other non-visual representation of pure graphical information such as diagrams, maps and drawings.

Acknowledgments
The authors are grateful to Alain Barrillier, Denis Page, and Denis Rossel for their help in the design and evaluation of the prototype.

References
Cutkosky, M.R., Howe, R.D. (1990). Human Grasp Choice and Robotic Grasp Analysis. *Dextrous Robot Hands*. S.T. Venkataraman and T. Iberall, Ed. Springer-Verlag, New York.

Brewster, S. A., Wright, P. C. and Edwards, A. D. N. (1995). Parallel earcons: Reducing the length of audio messages. *International Journal of Human-Computer Studies* **43**(2): pp. 153-176.

Edwards, A. D. N. and Stevens, R. D. (1993). Mathematical representations: Graphs, curves and formulas. in D. Burger and J.-C. Sperandio (ed.) *Non-Visual Human-Computer Interactions: Prospects for the visually handicapped*. Paris: John Libbey Eurotext. pp. 181-194.

Flowers, J.H. Hauer, T. (1995). Musical versus visual graphs: Cross modal equivalence in perception of time series data. *Human Factors*, 37, 553-559.

Fritz, J. P., Barner, K. E. (1999). Design of a Haptic Visualization System for People with Visual Impairments, *IEEE Transactions on rehabilitation engineering* vol. 7, No 3, pp. 372-384.

Fritz, J. P. (1996). Haptic Rendering Techniques for Scientific Visualization. *MSEE Thesis*, University of Delaware, USA.

Mansur, D.L., Blattner, M.M., and Joy, K.I. (1985). Sound graphs: a numerical data analysis method for the blind. *Journal of Medical Systems* 9(3), pp163- 174.

Ramloll R. et al. (2000). Constructing Sonified Haptic Line Graphs for the Blind Reader: First Steps. *4th International ACM Conference on Assistive Technologies* (ASSETS'2000, Virginia, USA, 13-15 November 2000), pp. 17-25.

Sahyun, S. C. (1999). A Comparison of Auditory Versus Visual Graphs for Use In Physics and Mathematics. *Ph.D. thesis*, Department of Physics, Oregon State University, USA.

Sjöström, C. (1999). The IT Potential of Haptics – touch access for people with disabilities. *licentiate thesis*, Center for Rehabilitation Engineering Research (CERTEC), Sweden.

An annotation editor for nonvisual web access

Takashi Sakairi, Hironobu Takagi

IBM Research, Tokyo Research Laboratory
1623-14, Shimotsuruma, Yamato, Kanagawa 242-8502, Japan

Abstract

Most Web page authors do not consider that their pages might be rendered by nonvisual devices such as voice browsers. However, it is possible to provide accessible Web pages for blind users by transcoding Web pages. A proxy server between a Web server and a Web browser can modify Web pages by using annotations. Since authoring annotations is a time-consuming task, a good authoring tool is required. This paper describes an annotation editor that eases annotation authoring for nonvisual Web access.

1. Introduction

The Web has an important role for our daily life and business. Although researchers recognize that Web pages must be accessible for all users, most Web page authors do not consider that their pages may be rendered by nonvisual devices such as voice browsers [1]. For examples, the value of an ALT attribute of an IMG element is "Click Here!", or complex nested TABLE elements are used only for multicolumn layout. Blind users can obtain little information from such Web pages by using voice browsers.

It is possible to provide accessible Web pages for blind users by transcoding Web pages [2, 3]. A proxy server between a Web server and a Web browser can modify Web pages by using annotations. If an annotation for an IMG element describes that the element is a link to an advertisement of a securities firm and the proxy server transcodes the Web page by using the annotation, a blind user can obtain more information from the modified Web page than from the original one. If an annotation for elements describes roles, such as a header, footer, main content, general index, and advertisement, and the proxy server changes the order of the elements according to their role, a blind user can obtain important information quickly.

Annotations can be made either automatically or manually. Some kinds of annotations can be generated automatically, but manually authored annotations are required for better transcoding. For example, the title of a Web page linked to an IMG element can be acquired automatically; however it is difficult to determine automatically whether the IMG element is an advertisement or not.

We are developing a transcoding proxy system for nonvisual Web access [4, 5]. Since authoring annotations is time-consuming task, a good authoring tool for annotation authors is required. We designed and developed an annotation editor to simplify annotation authoring. We assume annotations are authored by volunteers who wish to help blind people. Because the number of Web sites and Web pages is huge, many volunteers are required. Consequently ease of use is the top priority design issue.

2. An annotation editor

In our transcoding proxy system, annotations are represented by XML. Many XML editors have been developed as research prototypes and products; however, most conventional XML editors are designed for general purposes. Since such editors are not suitable for annotation authoring, we designed and implemented our own annotation editor. The main difference between an annotation editor and a general purpose XML editor is that the former is designed to handle two kinds of documents, a target document to be annotated and an annotation document.

2.1 Target web page selection

The first thing that an annotation author has to do is select a target Web page to be annotated. In usual Web browsing, we load a Web page by clicking an ANCHOR element on a currently displayed Web page or by selecting a bookmark. We can also enter a URL in a text field to load a Web page, but this is a less common case.

In our annotation editor, annotation authors can select target Web pages as in normal Web browsing. When an annotator invokes the annotation editor, it displays two windows; one is a Web browser window, and the other is a control window. Annotators browse Web pages by using the Web browser as usual until they find a target Web page to be annotated. When they push a "Start Annotation" button on the control window, they can start authoring annotations for the target Web page.

2.2 Element selection

The second thing that an annotation author has to do is to select elements on the target Web page. Elements of a Web page can be represented by a Document Object Model (DOM), and elements can be addressed by XML Path (XPath) expressions. It is possible to visualize the DOM structure of a target Web page as a tree. It is, however, difficult for annotation authors to specify desired elements in the DOM tree viewer. Some WYSIWYG HTML authoring tools use TABLE elements for laying out objects in multiple columns for visual devices, but not for specifying logical structure. Many Web pages even contain nested TABLE elements just for multicolumn layout. The DOM structure does not reflect the logical structure of the Web page, so the output of WYSIWYG HTML authoring tools tends to be complex. The number of DOM elements in a Web page of a typical news Web site may number in the hundreds, and the DOM tree may be more than ten levels deep.

Although Web authors do not worry about simplifying the DOM structure of their Web pages, they do try to make the rendered image of the Web pages as readable as possible for visual devices. For annotators, selecting DOM elements from the rendered Web pages is easier than using the DOM tree viewer.

In our annotation editor, annotation authors can select desired elements by using a Web browser. When they move the mouse pointer on the Web browser, the background color of the DOM element under the pointer changes. When they click on a DOM element, the element is selected and highlighted.

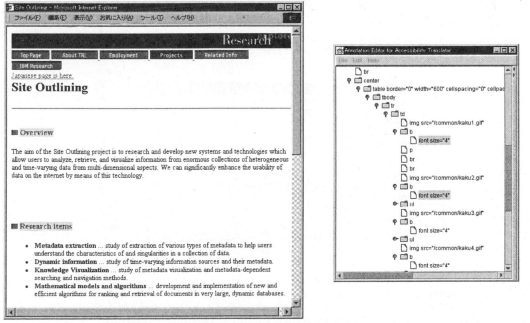

Figure 1. A Web browser (left) and a DOM tree viewer (right)

984

Although elements can be select in the Web browser, a DOM tree viewer is useful for annotation authors who understand the syntax of HTML. They can display the DOM tree viewer if they want to. If they display both the Web browser and the DOM tree viewer, they can also select elements in the DOM tree viewer. In that case, the highlighted elements of the DOM tree viewer are synchronized with the Web browser. Figure 1 shows the Web browser and the DOM tree viewer. Two elements, "Overview" and "Research items", are highlighted in the Web browser, and corresponding elements are also highlighted in the DOM tree viewer.

2.3 A description dialog box

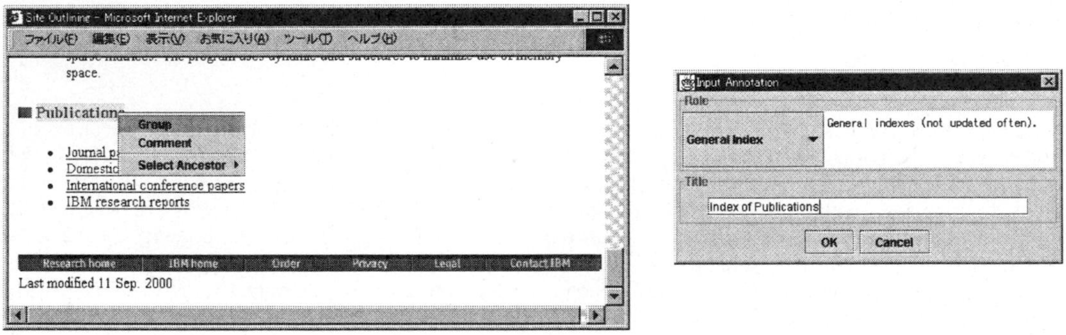

Figure 2. A popup menu on a Web browser (left) and a description dialog box (left)

After selecting elements, annotators can show a dialog box in order to create a description, which is the basic component of an annotation, by choosing a popup menu. There are two types of description: group and comment. Group descriptions are used for grouping elements based on their roles, such as a header, footer, main content, general index, and advertisement. Comment descriptions are used for assigning commentary information to elements. Figure 2 shows a popup menu and a description dialog.

2.4 An annotation table view
A general purpose XML editor displays a document as a source code view and a tree structure view, but our annotation editor does not display such views. Since the annotation syntax of our transcoding system is simple, it is not necessary to display complex structures.

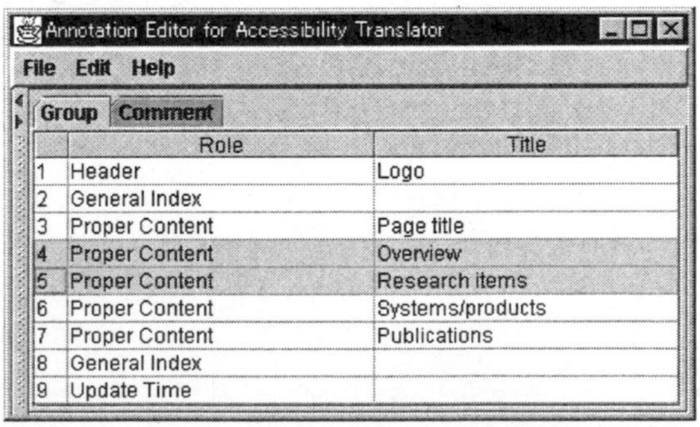

Figure 3. An annotation table view

In our annotation editor, annotation data is displayed in a table. Each row indicates one description. When annotation authors select one or more descriptions, DOM elements associated with the selected descriptions are highlighted in the Web browser and in the DOM tree viewer, so they can easily find which DOM elements have already received annotations. Figure 3 shows an annotation table view. The two types of descriptions each have their own table. Annotators can change the displayed table by selecting tabs at the top of the table. They can also change the order of descriptions in the table.

3. Implementation

The main part of our annotation editor is implemented as a Java application. A Web browser is controlled by a Java applet and a JavaScript program. The JavaScript program is used for controlling the DOM structure of a target Web page. It highlights selected elements in the Web browser, and notifies the Java applet of the annotation author's action. The Java applet communicates with the JavaScript program by using LiveConnect, and it also communicates with the main Java application by using Java Remote Method Invocation (RMI).

When an annotation author specifies a URL of a target Web page, the Java application loads the Web page, and modifies the Web page by inserting the JavaScript program and the Java applet, and then loads the modified Web page into the Web browser.

4. Conclusion and future work

We have designed and implemented an annotation editor that simplifies annotation authoring for nonvisual Web access. We plan to develop an experimental portal site for senior citizens with a portal company by using our transcoding proxy system. We will use our annotation editor to prepare annotations for the site, and will report the evaluation result of our annotation editor.

Web Accessibility Initiative (WAI) [6] of World Wide Web Consortium (W3C) published Web Content Accessibility Guideline [7]. It explains how to make Web content accessible to people with disabilities. Many tools such as Web Accessibility Versatile Evaluator) WAVE [8] and Bobby [9] help users to check Web content accessibility. We plan to add the Web content accessibility checking function to our annotation editor.

References

1. M. Rowan, P. Gregor, D. Sloan, and P, Booth, Evaluating Web Resources for Disability Access, in *Proceedings of ACM Conference on Assistive Technologies*, pp. 80-84 (2000).
2. M. Hori, G. Kondoh, K. Ono, S. Hirose, and S. Singhal, Annotation-Based Web Content Transcoding, in *Proceedings of 9th International World Wide Web Conference*, pp. 197-211 (2000).
3. W. Huang and N. Sundaresan, A Semantic Transcoding System to Adapt Web Services for Users with Disabilities, in *Proceedings of ACM Conference on Assistive Technologies*, pp. 156-163 (2000).
4. H. Takagi and C. Asakawa, Transcoding Proxy for Nonvisual Web Access, in *Proceedings of ACM Conference on Assistive Technologies*, pp. 164-171 (2000).
5. C. Asakawa and H. Takagi, Annotation-Based Transcoding for Nonvisual Web Access, in *Proceedings of ACM Conference on Assistive Technologies*, pp. 172-179 (2000).
6. Web Accessibility Initiative, World Wide Web Consortium, http://www.w3.org/WAI/.
7. World Wide Web Consortium, Web Content Accessibility Guidelines 1.0, http://www.w3.org/TR/WAI-WEBCONTENT/, (1999).
8. Kasday, L. R., A Web Accessibility Evaluation and Repair Tool with Application to Section 508 Standards, In *Proceedings of CSUN's Annual International Conference "Technology and Persons with Disabilities"*, (2001).
9. Center for Applied Special Technology, Bobby, http://www.cast.org/bobby/.

Computer based communication on and about mathematics by blind and sighted people

Waltraud Schweikhardt

Universität Stuttgart, Institut für Informatik
Breitwiesenstr. 20-22, D-70565 Stuttgart
schweikh@informatik.uni-stuttgart.de

Abstract

In this paper, we introduce requirements for written electronic communication between blind and sighted people about mathematics. We put together four groups of demands, which a mathematical notation for the blind should fulfill, and show examples for a solution. Finally, we list technical preconditions for a suitable learning and working environment.

1. Introduction

Mathematics accompanies a child from its first year of school to its last one. Mathematics is present for students at all levels of education. Many adults are confronted with mathematics during their professional time. At least kinds of arithmetic appear also in daily life. However, mathematics uses a specific notation to express terms, facts and theorems. The conventional symbols are just abbreviations as for instance "+" for "plus" or "-" for "minus". There are others to introduce a complex expression such symbols as root or integral. Conventions like a two dimensional arrangement of numbers or variables express matrices or determinants, others powers or binomial coefficients. Sighted people regardless of their cultural background are familiar with those conventions.

Mathematical notation is international; there exist very few national particularities. The special symbols can be written down with paper and pencil, but it took long until typewriters provided for mathematical characters and nevertheless, it was time-consuming to use them. Therefore, sighted mathematicians have preferred for a long time to communicate written materials by using paper and pencil.

Since modern computer based tools have become accessible and usable by more and more people, electronic communication about mathematics has become realistic. Formulas can for instance be written with LaTeX or the formula editor, which is embedded in WinWord. Blind scientists have also become used to LaTeX. LaTeX uses only characters of the ASCII-character-set, and in most countries there exists a convention to encode them in Braille. It is encoded in 8-dot-Braille and allows a braille-character for each visible character on the screen.

2. Mainstream education for blind children

Blind children in a class together with sighted boys and girls are taught facts, which are new for them as well as for the sighted. The subject matter is the same for all. New contents, but especially the related notation causes difficulties if blind pupils can only follow the explanations by listening. They urgently need a possibility to grasp also new symbols and the notation by touch. This is especially true for mathematics. It is necessary that there are equivalent symbols to show writing and representation as well as to the sighted as to the blind. Each tactile character must have one and only one visual correspondent in order to make it possible for sighted teachers and sighted classmates to understand the written product of the blind easily. As a consequence, parents, who want to help their child, really have the chance to do so. Finally, this results in a notation, which can e translated one, character or symbol at a time from tactile to visual notation and vice versa.

$$+- \; (\,) \; [\,] \; \{\,\} \; <> \; \lfloor \; \partial \cap \div \; ...$$

$$+- \; (\,) \; [\,] \; \{\,\} \; <> \; \lfloor \; \partial \cap \div \; ...$$

The above examples of SMSB (<u>S</u>tuttgarter <u>M</u>athematik<u>s</u>chrift für <u>B</u>linde), the Stuttgart Mathematics notation, are shown in „black print" as they also appear on the screen and in their corresponding braille characters in visual form.

2.1 Agreement on an international mathematical notation for the blind

Mathematics uses its own notation, which has been developed out of the mathematical language. General requirements, which concern the design of a notation for the blind, are necessary to open the door to the language and the written representation of elementary and complex expressions. The tactile notation must render the intuitive understanding, which the optical characters allow.

Since Louis Braille invented a notation for the blind at the end of the last century, braille has been based on six dots per character. In consequence, mechanical machines were developed as 6-dot-machines and have been used to write and to communicate. Also the first electronic stand-alone devices were based on six dots. These were devices like Versabraille of Telesensory Systems in Boston, Massachusetts, USA, the Braillex-device of Papenmeier in Schwerte, Germany or the Braillocord of AIB – electronic in Berlin, Germany, which were built in the seventies and used 6-dot-braille-lines for the output.

Only the late K.P. Schönherr built a braille-line to be used as a braille terminal with an 8-dot-line. In 1980 the first Schönherr-braille-display was connected to a microcomputer [6]

The arrangement and numbers of dots looks like

Each dot may be tactile or flat. This results in a set of 2 power 8 characters.

```
1 o o 4
2 o o 5
3 o o 6
7 o o 8
```

An agreement on an international mathematics notation for the blind, which is based on 8 dots, is necessary after all.

2.2 Requirements to a mathematical notation

Our requirements to a suited notation consist out of 4 theses:

Thesis 1: A mathematical notation for the blind must be readable by the finger.

This requirement means more than the demand for a tactile correspondent for each mathematical symbol. The notation must assist the reader in comprehending complex terms. Vision makes it possible to grasp the type of many mathematical expressions at a glance. Examples may be a fraction, a root or an integral. A blind reader, however, has to touch one character after the other to complete a puzzle, which finally makes clear what the content of the string is.

A reasonable approach to design a mathematical notation for the blind is to model it according to speaking out expressions unmistakable. One should aim at a writing, which allows concluding the type of a complex term already from the first character. This is, however, not necessary for expressions like 10+4 or 23x+10y or even x^4, because they are simple and short enough.

Thesis 2: The number of characters in a mathematical term should be as small as possible.

The comprehension of complex terms increases with the compactness and clearness of the notation [4]. That is the reason for the variety of mathematical symbols as + for plus, - for minus or ⌐ for root. A mathematical notation for the blind should follow the same principles and allow the same degree of abstraction as the notation of the sighted. Therefore, 64 characters, which are provided by 6-dot-braille are not enough.

Thesis 3: Tactile symbols should be understandable intuitively.

Many mathematical symbols include visual components, which facilitate understanding them. Examples are arrows, which are used in identifiers for vectors or in the writing of sequences, series, and their limits.

Symmetries in semantic concepts should be represented by tactile symmetries. Examples are all kinds of brackets and the signs for smaller and greater.

To make clear that a character or a string of characters is raised or lowered, an indication is needed. The signs should show the direction where the characters are placed and consider symmetries.

Examples: a_i, x^5, and x^{2n+5}.

In SMSB: $\alpha^{TM}\iota$ x $\prod 5$ $\xi \mid 2\nu + 5\cdot$

$$\alpha^{TM}\iota \qquad \xi\prod 5 \qquad \xi \mid 2\nu+5\cdot$$

In SMSB straight vertical arrows mean that only one raised or lowered character will follow. Diagonal arrows indicate strings. The end of such a string is marked
Straight lines in printed form should be represented by straight tactile lines.

Thesis 4: Integrated learning and working of blind and sighted students and colleagues should be supported.

Mathematical texts should be easily transformed, that means also in "real-time", from the notation of the blind to an understandable visual version and vice versa: In addition, computer programs are needed to translate them into the representation to which sighted students are used are used. SMSB-terms can be translated by a macro into the formula-field of the Microsoft Formula-Editor, what results in a graphical representation of the term [2].
It is extremely helpful if a teacher works out assignments only once for a class. It can be written in the mentioned visual form for the blind, who will read it on the braille-output-device of their computer or in the printed tactile form. The material will be translated by a computer program into the adequate representation for the sighted, who read it on the screen or take a printout.
It should be clear that it is difficult for children, who are just doing their first steps in mathematics, to learn the language of mathematics, symbols which help to shorten the written representation of expressions and a notation which differs from the methods their sighted classmates learn. In a further step in the education it will be possible to learn other notations to write down mathematical facts, theorems or proofs in addition.

2.3   Existing mathematical notations

There exist mathematical notations in 6-dot-braille as well as in 8-dot-braille. In Europe, the notation with the longest tradition is the "Internationale Mathematikschrift für Blinde" of the "Deutsche Blindenstudienanstalt" in Marburg/Lahn in Germany [3] and [5]. Another 6-dot-braille-code for mathematics is the Nemeth-Braille-Code for Mathematics and science in the United States. The size of 6-dot-cells fits well with the size of the fingertip which has to read them, but blind computer users have been used to 8-dot-braille since more than ten years in the meantime. It has been claimed already in the eighties, that 8-dot braille will be the notation or the future.

SMSB has been successfully used in Germany in mainstream education where blind students are educated together with sighted boys and girls for nearly 15 years by now. At the Universities of Karlsruhe and Dresden the ASCII-mathematical-Notation AMS (ASCII-Mathematikschrift) [10] has been used for about ten years.

3. A Computer-based learning and working environment for mathematics

The learning and working configuration will consist of a computer with a braille-output-device, probably also with speech output and a conventional keyboard as basic peripheral devices.
The software includes a dialog-manager that receives the entered characters, hands them over to an interpreter, which transforms it into the correct presentation on the output-devices, and which also returns the reaction in the adequate representation.

During the last years, different character sets have been used what caused several problems. In 1981, when we developed the first version of SMSB, we used the APL-character set. This set of the programming language APL provided already then, in the time of the 7-bit-ASCII, 256 characters. These were even well suited for mathematics, because the programming language APL uses many "mathematical" characters. Later, the necessary SMSB-characters were also mapped to ASCII-characters to enable DOS-users to work with SMSB. This was and still is the case in schools.

For SEM, the Stuttgart Editor for Mathematics, which was also called Stuttgart Emacs, we developed an SMSB-character set out of true type fonts in 1991. SEM became a German subset of the Emacs-editor.
The next generation made it necessary to adapt SMSB to Windows [9].

3.1 Keyboard-layout

It is recommended to design keyboard layouts which allows writing the characters, which are used in the mathematical notation by typing them. The keys should be overlaid according to mnemonics. This, however, results in different layouts for different languages. While a combination with key w may be "write" in English speaking countries, it could mean "Wurzel" (root) in German speaking ones. Another but far worse solution would be to enter the characters by typing their numbers in the character-set.

3.2 Requirements on user-interfaces for braille-output-devices

Software for braille-output-devices like braille-lines or braille-printers must support programs, which allow to alter, to change and to enter new, also own character-sets for the braille-lines easily. It should also be possible to bind character-sets to applications.

4. Concluding remarks

Arithmetic, geometry, and algebra are basic techniques and sciences in our culture. Everybody is taught the necessary skills like reading and writing. Thanks to Louis Braille also blind people have got access to written communication. Modern techniques shorten the local gap between people, also between the sighted and the blind. Nevertheless, it seems that the access to mathematics is only possible for blind specialists, while the sighted are introduced into the world of mathematics already in their first years of school. Thanks to the computer and special peripheral devices, the chances for blind children to gain an education, which is adequate to their mental abilities, have been raised. It became easier or it became possible at all to teach blind and sighted children and students together also in mathematics. The introduced requirements for a suited notation are however neccssary to realise equal treatment of both groups. The possibility of communication between blind and sighted mathematicians is achieved.

References
[1] BATUSIC, M., MIESENBERGER, K., STÖGER, B., „LABRADOOR – a contribution to make mathematics accessible for blind", in: Edwards, A.D.N., Arato, A., Zagler, W.L. (eds.) (1998). Computers and Assistive Technology ICCHP'98, Proceedings of the XV. IFIP World Computer Congress.
[2] CHRISTIAN, U. (1999). „Entwurf und Implementierung eines Dialogprogramms zur Umsetzung von SMSB-Termen in eine grafische Darstellung", Diploma Theses No. 1719, Universität Stuttgart.
[3] EPHESER, H., POGRANICZNA, D., BRITZ, K. (1992). „Internationale Mathematikschrift für Blinde", in: J. Hertlein, R.F.V. Witte, (ed.), Marburger Systematiken der Blindenschrift (Teil 6), Deutsche Blindenstudienanstalt, Marburg.
[4] IVERSON, K.E. (1980). „Notation as a Tool of Thoughts", in Communications of the ACM, Volume 23, Nr 8.
[5] SCHEID, F.M., WINDAU, W., ZEHME; G. (1930). „System der Mathematik- und Chemieschrift für Blinde", Marburg/Lahn.
[6] SCHWEIKHARDT, W. (1980) „A Computer Based Education System for the Blind", in: Lavington, S. H. (ed.), Information Processing 80, pp 951-954, North Holland Publishing Company.
[7] SCHWEIKHARDT, W. (1983 und 1989). „Stuttgarter Mathematikschrift für Blinde, Vorschlag für eine 8-Punkt-Mathematikschrift für Blinde", technischer Bericht an der Universität Stuttgart, Institut für Informatik.
[8] SCHWEIKHARDT, W. (1998). „Stuttgarter Mathematikschrift für Blindc", technischer Bericht an der Universität Stuttgart, Institut für Informatik.
[9] SCHWEIKHARDT, W. (1998). "8-Dot-Braille for Writing, Reading and Printing Texts which Include Mathematical Characters" in Alistair D.N. Edwards, András Arato, Wolfgang L. Zagler (Eds.), Proceedings of the XV. IFIP World Computer Congress, 31.8.98 - 4.9.98, Wien/Budapest, pp. 324-333.
[10] Studienzentrum für Sehgeschädigte, Karlsruhe: „ASCII-Mathematikschrift", 5. Auflage, 1999.

Objective and quantitative evaluation measure on ability of obstacle sense by using acoustical VR system and body movement measuring device in rehabilitation for the visually impaired

Yoshikazu Seki[a] , Kiyohide Ito[b]

[a]National Institute of Advanced Industrial Science and Technology,
1-1 Higashi, Tsukuba, Ibaraki 305-8566 Japan

[b]Future University-Hakodate,
116-2 Kamedanakano-cho, Hakodate, Hokkaido 041-8655 Japan

1. Introduction

"Obstacle sense" is an experience-based ability to detect objects that generate no sound without using visual sense or haptic perception, but using auditory sense.   It is one of the important sensory substitutions of the visually impaired. The cue that announces an object is sound field variation (sound reflection, diffraction, etc.) caused by the existence of an object.   Learning the relationship between object existence and sound field variation enables a person to master obstacle sense [1].   It is very important to construct systematic training method to master this ability through education and rehabilitation of the visually impaired. We proposed systematic auditory training to acquire the obstacle sense based on both its acoustical mechanism and acoustical VR technology in HCI '99 [2]. Now we propose the evaluation method of the training results by using the acoustical VR technology and the body movement measuring device in this paper.

2. Principle of evaluation

When we present the sound field where the moving wall exists to the subject who has mastered the obstacle sense, the subject's body sways unconsciously with the wall synchronously by perceiving its movement. This phenomenon is called "auditory kinesthesis ," and is observed only in the obstacle sense mastered subject. Thus, we can evaluate the degree of acquisition by measuring the degree of auditory kinesthesis [3]. This principle enables objective and quantitative evaluation whereas the conventional method was depend on the trainer's qualitative subject. Our evaluation system described bellow is based on this principle.

3. Proposed system

Our evaluation system consists of the acoustical VR system and the body movement measuring device shown in Figure 1. The acoustical VR system can produce the sound field where the moving wall exits by emitting the simulated reflected sounds through loudspeakers, and its detail was reported in HCI '99. The loudspeakers are arranged on a circle of 2m radius, and the subject stands at the center of the circle. The simulated reflected sound is generated by the digital delay device, and the delay time of the reflected sound is calculated by the personal computer and transferred through MIDI. The body movement measuring device is 3SPACE Fastrak made by Polhems. The 3SPACE source coil that generates magnetic field locates above the subject's head, and the sensor coil that measures its position by receiving the magnetic field is fixed the back of the subject's neck. The acoustical VR system simulates the subject by the various "virtual wall movements," and the body movement measuring device observes the subject's auditory kinesthesis.

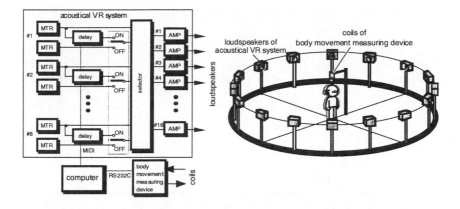

Figure 1   Our evaluation system for obstacle sense by using acoustical VR system and body movement measurement device.

4. Evaluation experiment

We performed the evaluation experiments to test our system. The subjects were 10 blind persons who reported that they have already mastered obstacle sense by conventional training and their daily experiences, and 7 sighted persons who have not mastered. The two loudspeakers of our system that located in front and back of the subject were used as the reflected and direct sound sources, respectively (Figure 2 shows the experimental setup). This arrangement supposed that the wall exists in front of the subject. The stimulus sound was the pink noise (40Hz –

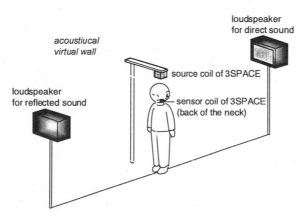

Figure 2   Experimental setup for test our system.

8kHz), and the sound pressure level was 56dB(A) at the position of subject's head. The acoustical virtual wall movement was repeated linearly in the range between 0.2m and 0.8m away from subject's head and its velocity was 0.03m/s (i.e. go-and-back period was 20s). The repetition of movement was 5 times, that is, the duration of one trial

992

was 100s. The result of experiment was obtained by averaging the 10 trial movement curves to cancel the random body motion.

The results of the experiment show that 8 of the 10 blind subjects and 1 of the 7 sighted subject swayed synchronously with the wall movement, and others did not sway (Figure 3 shows the typical results of the blind and the sighted). Furthermore, 5 of the 8 blind subjects made namely large movements. These results demonstrate effectiveness of our system by suggesting that measuring the auditory kinesthesis can apply to the evaluation of the obstacle sense. The other 2 blind subject did not show remarkable movement because they were bad at perceiving the wall in front of them.

5. Control experiment

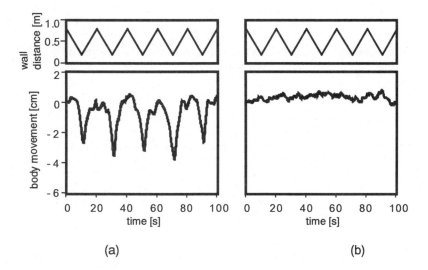

Figure 3   Typical results of (a) the blind and (b) the sighted. Upper panels show the virtual wall movements, and lower show body movements.

We performed the control experiment to test whether the subjects who showed the auditory kinesthesis in the evaluation experiment move with other acoustic variations that are not the factors of the obstacle sense. Control conditions were 3 case; loudness variation (loudness of the sounds varied periodically), sound image variation (sound image moved between front and back periodically), and pitch variation (pitch of the sounds varied periodically). The results show that the subjects did not move larger with the 3 conditions than the virtual wall movement condition (Figure 4 shows the typical results). This results suggests that the body movement is caused by only the acoustic factors of the obstacle sense.

6. Conclusions

We proposed the evaluation system of the obstacle sense for the rehabilitation of the visually impaired, by using the acoustical VR technology and the body movement measuring device, based on the principle of the auditory kinesthesis. We tested the evaluation experiment and the control experiment, and we demonstrated the possibility of our system. Our techniques will contribute to the rehabilitation of the visually impaired by shortening the training period and ensuring the evaluation of training results.

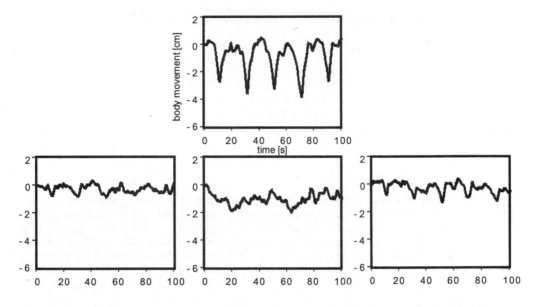

Figure 4   Typical results of the control experiment. Upper panel is the same as Figure 3 (a), lower left shows "loudness variation," middle "sound image variation," and right "pitch variation."

References

1. Seki Y. (1997). On a Cause of Detection Sensitivity Difference depending on Direction of Object in Obstacle Sense, IEEE Transactions on Rehabilitation Engineering, 5, pp. 403-405.
2. Seki Y. (1999)., Systematic Auditory Training of Obstacle Sense for the Visually Impaired by using Acoustical VR System, Human-Computer Interaction (Proceedings of HCI International '99), 2, pp. 999-1003.
3. Ito K., Seki Y.(1999). The Tentative Study on an Application of "Auditory Kinesthesis" to an Objective Measure of Obsyacle Sense (II), Proceedings of the 25th Sensory Substitution Symposium, 1-6.

994

Universal Access of DOS-like Information Systems

Norman L. Soong

Department of Computing Sciences
Villanova University
Villanova, Pennsylvania
Norman.Soong@villanova.edu

Abstract

The disk-based information infrastructure, the Disk Operating System (DOS), is a common information warehouse. The popular Graphical User Interface (GUI) provides the users with an efficient method to access the warehouse. In this paper, we point out that one may extend the essence of system navigation from visual systems like GUI to non-visual ones. This idea was derived from our experience with two experimental, operational Audible User Interface (AUI) systems, one was based on a mouse and speaker configuration and the other on keyboard and speaker. We will outline a set of universal access features that were used in GUI and AUI systems.

1. Introduction

The dominant method of information browsing in modern computing systems is the popular family of window-based GUI features, available as MS/Windows form Microsoft Corp., OpenLook from Sun Microsystems, and Motif from Hewlett Packard, etc. These systems require sophisticated hand-eye coordination that favor users of sharp eyesight, nimble wrists, and steady hands. The visual icon systems are designed by the nimble and for the nimble.
The warehouse to all GUI systems is the DOS, that was commercially introduced in 1965 as the DOS/360 from IBM, in the 1970s the Unix system from AT&T, followed in 1980 by the MS/DOS from Microsoft. While the GUI features may be designed for the visually nimble, its characteristics can be shown as inherently system independent and is suitable to support different super structures, such as the AUI, and possibly conversational, and hand-writing systems.
We suggest that behind the modern GUI system there is a universal conceptual framework that can support the direct manipulation of information objects in a DOS bedrock. In its basic form, information may be in the context of raw or processed data, system hardware attributes, DOS infrastructure, user commands, etc. We want to extrapolate the navigational essence of a DOS-based GUI system to a non-visual AUI system, to demonstrate that a set of the GUI features supports universal information access.

2. Audible user interface

The experimental AUI systems is designed to navigate a Microsoft DOS system with objects of drives, directories, sub-directories, files, and short cuts. With the system monitor disabled, the users receive output from the AUI system via a voice synthesizer and a pair of speakers. The design was implemented on two hardware configurations: mouse and speaker; and keyboard and speaker. The goal is to locate any object in the DOS system if one knows the name of the object and the path that leads to it. [1] Upon locating the object, desired operations may be carried out. If the object is a directory, it may be opened, if a text file, it may be read through the voice synthesizer, if a mail message, supposedly it may be read by an audible mail reader, and if an html-file, it may be read by an audible web browser, etc.
The intersection between the functions of the popular GUI and the experimental AUI is a set of navigation features, represented by audible icons. While the GUI icons are visible on the monitor, inaccessible to the non-visual users, the AUI ones are audible through the speakers. The DOS information warehouse and the basic notion of navigation are universal components in modern computing systems. This paper will identify these components.

2.1   Existing non-visual systems

Current development efforts on aural interface are focused on the notion of talking window systems. In 1993, the Berkeley System Inc. introduced the first commercial talking window system, OutSpoken. It was followed by the ScreenReader system from IBM Corp. [2] A system called JAWS is also popular recently. These systems can read out text information through the system speakers and some limited success to read spreadsheet applications. Later systems include WinSpeak from Germany and Windowlize from Microsoft Inc.

With the telecommunication system increasingly employing digital technology, there is a flurry of phone-based interactive voice response applications. They are suitable to process menus, lists, and forms based information and they reach everyone who can access a phone. This class of system is slow, reflecting the pre-GUI, program-centered technology. [3]

The ultimate audible interface would be the conversational mode of interaction. For a true conversational interface, one would need the support of natural language processing, far beyond the scope of currently available, program-centered, voice command systems, such as the Voice Type from IBM Corp. and the Voice Pad from Kurzweil Applied Intelligence.

2.2   Universally usable and friendly navigation features

Universally usable navigation features should support information browsing in a DOS-like environment, provide friendly, affirmative responses, and maintain operational efficiency. To navigate the DOS bedrock, one must be able to assign drives, to navigate the directory hierarchy, to locate objects, etc. These basic features should be usable to both visual and non-visual users.

Components of friendly and affirmative systems are many. One should liberally use echo to confirm the outcome of an operation and the presence of objects, provide a query feature to identify the current object, use a distinct mechanism to identify the volatile activations from the non-volatile ones. To assist the navigators who are lost in the information space, three types of query are introduced, the 'where am I' query, the 'who are you' query, and the 'what can you do for me' query. These features make the GUI system easy and pleasant to use. In this paper we discuss how in an aural system, similar features are introduced.

To provide operational efficiency, one need to provide direct object manipulation, different levels of system reset for fire escape, suggestive responses for memory easement, and critical path to reach frequently needed objects. This set of design pointers defines a minimal, user-friendly, universally accessible system. These features and much more have been studied and experimented with in a visual, GUI environment. In the next several sections, we will discuss a subset that is universally accessible.

2.3   Object address space

Modern interface design centers on the technique of direct object manipulation. [4] In general, there are two types of objects in a WYSIWYG system, the information objects such as files, directories, shortcut, etc., and the navigation and service objects such as 'open directory', 'close directory', 'change files name', etc. Every object has two ways of presenting its identity: a human-friendly name, and an anti-social system address. We define the address of a current object in a DOS tree-like file structure as:

   Addr(current.object) = I(P,N) = { The Nth object at path P } .
   I(P,N).Identity = Visual and audible name of current object .
   I(P,N).Property = Other attributes of current object .

2.4   Icon space and object space

A practical information system contains a large number of information bearing objects, e.g., the number of files, and a small user workspace, e.g., the actual Window of a GUI system. We define two abstract spaces to accommodate the presentation of information from one large space; the DOS object space, to a small one, the Icon space. We may view the Icon Space as a template over the DOS Space and the users may directly manipulate only objects appear in the Icon Space. [5,6] Two types of object may appear in the Icon Space Template, a set of service icons and some information bearing DOS objects. The popular GUI system is a visual Icon Space with the template exposing groups of DOS objects. The pull-down menu is one form of the templates. In an AUI system, audible pages serve as the template.

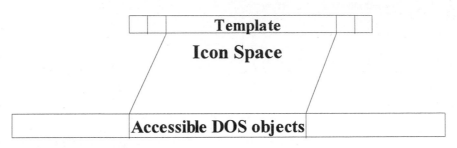

DOS Object Space

What appears next is a list of browsing features that are extracted from GUI systems and experimented with in the AUI systems to demonstrate their universal access appeal.

Address(Current.Object) = $I(P,N)$
Address(Initial.Object) = $I(P_0,0)$

Load system:	Null	→	$I(P_0,0)$	/* Defines initial system condition
Page up:	$I(P,N)$	→	$I(P+\Delta,N)$	/* Moves the template up
Page down:	$I(P,N)$	→	$I(P-\Delta,N)$	/* Moves the template down
Level up:	$I(P,N)$	→	$I(P+1,N)$	/* Opens a new directory
Level down:	$I(P,N)$	→	$I(P-1,N)$	/* Closes the current directory
Reset sys:	$I(P,N).A$	→	$I(P,0).A_0$	/* Reset current directory attributes
To.root:	$I(D,0)$	→	$I(P_0,0)$	/* Return to initial system load condition
Exit:	$I(0,0)$	→	Escape	/* Exit Window or exit AUI
Who.are.you:	$I(P,N)$	→	$I(P,N).$Identity	/* Present or pronounce its name
Do.it:	$I(P,N)$	→	Activate($I(P,N)$)	/* Execute (volatile) the current object

Interested readers may wish to find how the tasks of the features are accomplished in GUI and how there are many universal ways to perform the same task. In their AUI reincarnation, these features are represented by a single audible icon, hence 'Who are you' may be applied to a DOS object and have the filename pronounced or if applied to a navigation icon and have 'Page up' pronounced. Name pronouncement does not imply activation. 'Do it' activates the last referred object and is the only volatile icon in this group. ('Page up' is non-volatile, yet 'Page up' followed by 'Do it' is.)

3. Conclusion

In the following sections, we will briefly outline four AUI architectures that are based on the mouse and speaker configuration. Current effort focuses on a keyboard and speaker system that could be demonstrated.

3.1 The concentric icon architecture:

A major ergonomic issue in an audible icon system that is based on a mouse and speaker configuration involves the alignment of the upward orientation and the location of the center of the space. [7] A concentric design shapes its icons in concentric square rings. While traversing outward from the center, such a system does not require any directional alignment. It does need a 'Reset' feature that will reset the mouse to the center of the icon space.

The following diagram illustrates the implementation of this architecture. A visually impaired user would normally unable to see this screen display. Any slight movement of the mouse over this icon space will generate a sound to inform the user of his position in the audible space.

Each ring is a sonic icon and it is assigned a unique sonic presence. While traversing through the sonic space, different sounds are heard. "Current icon" is the one that is associated with the current mouse position. Moving the mouse to the next icon produces a different sound. In this fashion, the users are directly interacting with the objects. A left mouse click is the 'Who are you' query and a right-click activates it. The meaning of activation depends on the nature and the type of the icon. At the center is the 'return' icon and at the edge are the 'page up' and 'page down' icons. Move the mouse beyond the sonic space; one would receive a warning at the boundary followed by the dreadful silence of empty space.

3.2   The column icon architecture:

A collection of long rectangular columns is perceived to have more tolerance against the loss of upward orientation against lateral mouse movements. A column design is a more compact design; hence, more icons may be accommodated in the template.

3.3   The matrix icon architecture

A matrix design is a repetitive pattern of many identical n-by-n icon grids. In the sample implementation, each grid is patterned after a telephone or calculator pad. Being already common in existing commercial applications, this pattern minimizes user training. By packing repetitive versions of the same grid pattern (9 by 9), the slippage from one grid to another may create (hopefully) less disruptive side effects. This design proved to be less friendly than the concentric and column designs.

3.4   The Random icon architecture:

This design is based on the observation that teenagers can find their clothing in an unorganized bedroom, if enough interchangeable clothing items are present. Each mouse location (x,y) is assigned a uniform random number, R, and the projection of 'R mod n' will produce the identity of an icon, where 'n' is the number of icons in the template. Most monitors provide more than 250,000 locations, sufficient for random access. On the average, one would find the desired icon in n/2 times of trial. Such a non-design architecture may be used as the minimal standard against which one may compare other forms of conscientious efforts. Paradoxically, initial experimental evidence indicated that up to 8 to 10 audible objects the random architecture can match or out-perform the conscious ones.

3.5   Summary

We have demonstrated that some parts of the modern user interface may be universally usable, if properly designed. While the actual implementation of the interface may differ between visual and non-visual systems, we have shown that a subset of the essential navigation features could share a common design, hence universally usable. Next stage of our effort will be directed to extend the universally usable interface design beyond navigation to include object maintenance, creation, modification, deletion, etc.

References

1.   Soong, N.L., Bangarbale, G. (1997). Some Experimental Suggestions to Assist the Visually Handicapped End-Users to Navigate the Information Space, Proceedings of HCI International '97, pp.433-436.
2.   Schwerdtfeger, R.S. (1991). Making the GUI Talk, Byte, pp.118.
3.   Resnick, P., Virzi, R.A., Relief from the Audio Interface Blues: Expanding the Spectrum of Menu, List, and Form Styles. ACM Transaction on Computer-Human Interaction, 2-2, pp.145-176.
4   Shneidreman, B. (1983). Direct Manipulation: A Step Beyond Programming Languages, IEEE Computer, 16:8, pp.57.
5.   Blattner, M.M., et al. (1989) Earcons and Icons: Their Structure and Common Design Principles, Human-Computer Interaction, Vol.4, No.1, pp.11-44
6.   Bly, S. (1982). Sound and Computer Information, Ph.D. thesis, University of California at Davis.
7.   Soong, N.L., Santhanam, B. (1998). Some Cognitive Issues on the Design of Audible User Interface to Computers, Proc. of HAAMAHA '98 Conference, Hong Kong, pp.431.

Distance communication system for the blind using sound images

Masahiko Sugimoto*, Kazunori Itoh**, Michio Shimizu*

*Nagano Prefectural College, 380-8525 Japan.
**Faculty of Engineering, Shinshu University, 380-8553 Japan.

Abstract

Recently, various distance communication systems have been developed for use on the Internet. These systems can be of great use in supplying anywhere the information necessary for the blind. In this paper, we examine the web-based system, because by using this system the blind can recall the graphical patterns at any time. We show the usefulness of a virtual sound screen and tablet with tactile guide. Through experiments using this system, subjects in distant locations could send images accurately by sending several sound images. We found that it is possible for the blind to recognize graphic patterns.

1. Introduction

Recently, various distance communication systems have been developed for use on the Internet. These systems can be of great use in supplying anywhere the information necessary for the blind. However, general distance communication systems at present are very difficult for the blind to use, because they are based mainly on the transfer of visual information.

Computer user interfaces have been realized using graphical presentation that imitates a real desktop environment so as to simplify user operation by clicking on graphical objects with a pointing device. These kinds of interfaces are convenient for sighted people, but are difficult for the blind to use. In earlier systems, blind computer users could access a command-line based interface by using a screen reader which reads the text displayed on the screen and provides information via a speech synthesizer or a tactile text display. Today, screen readers and voice browsers are useful programs for obtaining information without looking at the display of a personal computer (Earl,1997), (Paramythis,1999).

Previously, we proposed a sound support system which indicates position using sound (Itoh,1990), (Itoh,1995). We also previously proposed a drawing support system using a virtual sound screen for the blind (Itoh,1997) and showed that the virtual sound screen can be used in the mouse cursor leading method (Shimizu, 1998).

In this paper, we examine the use of a virtual sound screen in a distance communication system experimentally and show the usefulness of this system. In our distance communication system, it is possible to use a general personal computer with a web-based system connected to the Internet. Therefore, without using other special hardware, the system can be used to support communication between sighted and blind people. This means that the system has a good chance of being widely utilized. As a result of many experiments, we found that it is possible for the blind to recognize graphic patterns using this system.

2. The distance communication system

The real-time system and the web-based system are well known distance communication systems. The web-based system is a method of accessing at any time the data stored in a web-server (Sakumoto,2000). We examined the web-based system, because by using this system the blind can recall the graphical patterns at any time.

Figure 1 shows a block diagram of the web-based system for the blind. A web-server is connected to the Internet, and is accessed from a client PC. A tablet (WACOM FAVO) and headphones are connected to the client PC. The communication software for the blind made using the Live Audio function of JavaScript is stored beforehand on the web-server. When blind people access the web-server using Netscape(R) Communicator, the communication software is downloaded and stored on the client PC. Pen positions on the tablet correspond to positions of sound images on the virtual sound screen. When blind people push writing positions on the tablet with the pen, position data are transported from the tablet to the PC system. The sounds can be heard by the blind through the headphones connected to a sound board in the PC.

In this system, the virtual sound screen is composed of a 9×9 matrix of sound images synthesized according to sound location. The sounds corresponding to the vertical axis of the screen use different frequencies. Low frequencies correspond to low positions and high frequencies correspond to high positions on the screen. Each display point is divided logarithmically from 500Hz to 3000Hz. The horizontal axis is expressed by changing the sound level difference between the left and right speakers of the headphones. The sound levels are set 5dB apart with a maximum difference of 20dB. The stereo signal outputs from the sound board are converted into 30ms tones interspersed with a 10ms silent interval.

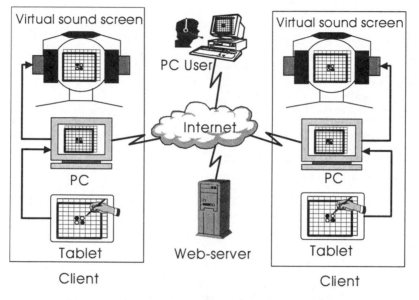

Figure 1. The Web-based system for the blind.

Figure 2. The tactile guide on the tablet.

Figure 2 is a photograph of the tactile guide on the tablet. An acrylic tactile sense guide with a 9×9 (81 point) matrix of holes on the tablet corresponds to the virtual sound screen. Blind subjects send point positions by feeling the holes and pushing a pen into the desired hole. Point positions (sounds) received can be identified by pushing the pen in various holes until the same sound is heard.

Figure 3 shows the display of the browser in our system. There are two kinds of buttons on the browser. One kind corresponds to the drawing position on the virtual sound screen and the other corresponds to function keys. Function keys correspond to the buttons on the upper part of the display in Figure 3. When the button of a function key is pushed, a voice message is output. To send point positions, the pen is pushed into the corresponding hole on the tablet, e.g., to send position 80, the pen is pushed into the top left hole on the tablet. This produces the same result as clicking button 80 with a mouse.

3. Input system experiments

The size of the operation area of the tablet was examined. The examinees were three 20-year-old sighted students wearing eye masks. Three sizes of operation area were examined: 81×81mm, 54×54mm and 27×27mm. In addition, we experimented using a tablet with and without a tactile guide. When subjects searched for the point corresponding to the received sound, the correct answer and time were measured. Figures 4 (a) and (b) show the experimental results.

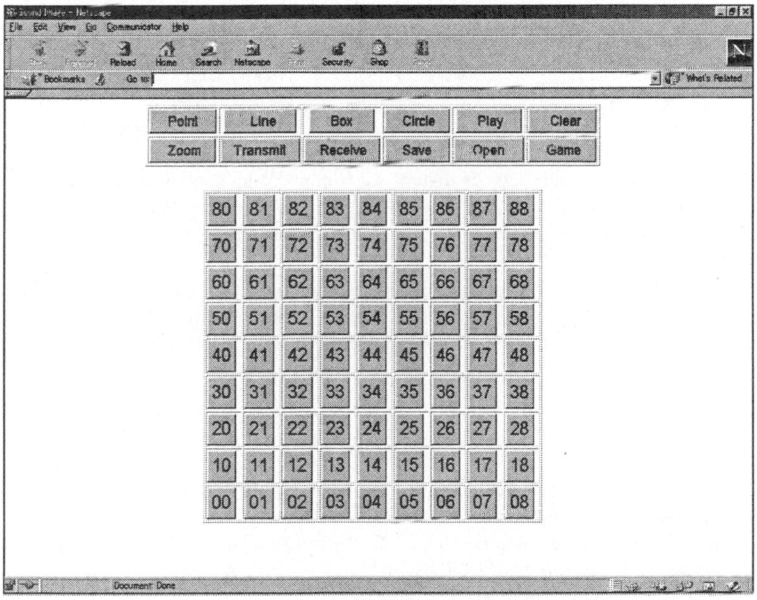

Figure 3. Display expression.

When no tactile guide was used, the accuracy was the highest for the 54×54mm guide, and the time was also the shortest. The accuracy when the tactile guide was used was higher than when no tactile guide was used. However, the accuracy falls for the 27×27mm guide, since the holes are small. Also, the time when the tactile guide was used was 50% less than when no tactile guide was used. Therefore, we found that the size of the operation area which is effective for the system is 54×54mm.

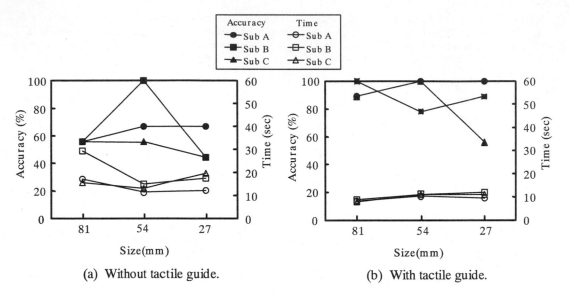

(a) Without tactile guide.

(b) With tactile guide.

Figure 4. The evaluation of the size of operational area.

4. Conclusions

Through experiments using our system, subjects in distant locations could send images accurately by sending several sound images. We found that simple patterns can be correctly recognized. We also concluded that a web-based system is more effective than a real-time system for communication of the blind using the Internet.

By using our system, the blind will be able to get information such as maps and layouts of buildings. Our system can be also applied to games such as Chess, Checkers and Othello. It also can be utilized not only for communication between the blind, but also for sighted persons. In the future, our system will be modified into a multi-point network system.

References

Earl, C.L., & Leventhal, J.D., (1997). Windows 95 access for blind or visually impaired persons: An overview, *J. of Visually Impairment & Blindness, News Service*, 91, 5-9.

Paramythis, A., Sfyrakis, M., Savidis, A., & Stephanidis, C. (1999). Non-Visual Web browsing : Lessons learned from the AVANTI case study, *Proceedings of HCI*, 812-817.

Itoh, K., & Yonezawa, Y., (1990). Support system for handwriting characters and drawing figures to the blind using feedback of sound imaging signals, *J. Microcomputer Applications*, 13, 177-183.

Itoh, K., Inagaki, Y., Yonezawa, Y., & Hashimoto M., (1995). A support system for handwriting for the blind using a virtual sound screen. *Symbiosis of Human and Artifact*, Elisevier, Amsterdam, 803-808.

Itoh, K., Shimizu, M., & Yonezawa, Y., (1997). A drawing support system for the visually impaired computer user using a virtual sound screen, *Design of Computing Systems (HCI'97)*, Elisevier, Amsterdam, 401-404.

Shimizu, M., Itoh, K., & Yonezawa, Y., (1998). Operational Helping Function of the GUI for the Visually Disabled using a virtual sound screen, *Proceedings of the ifip99 world computer congress (ICCHP '98)*, 387-394.

Sakumoto, K., Sugimoto, M., Sakurai, H., Kuramochi, H., Ishihara, M., & Shikata,Y.,(2000). The Evaluation of the Real-time Distance Learning System and the Plan of the Distance Learning System in which Real-time and Web-based System are integrated, *Proceedings of International Ergonomics Symposium 2000*, 305-308.

Page-customization allowing blind users to improve web accessibility by themselves

Hironobu Takagi[a] , Chieko Asakawa[b]

IBM Japan, Ltd., Tokyo Research Laboratory

1623-14 Shimotsuruma Yamato-shi Kanagawa-ken, 242-8502, Japan +81 46 273 {4557[a], 4633[b]}
{takagih[a], chie[b]}@jp.ibm.com

Abstract

There are several issues when blind users try to access the web, since a voice browser reads the Web contents in HTML tag order. In this paper, we first discuss these issues of nonvisual web access. Then we will describe a method that blind users themselves can use to customize Web pages for improving Web accessibility.

1. Introduction

There are several issues when blind people try to access the Web nonvisually. One major issue is the time it takes to find the main content of a page. A voice browser [1, 2, 3] reads the Web contents in the order the tags appear in the HTML file. However, there are usually banners, indexes and forms at the top of an HTML document, and the blind people have to listen to that heading information first before the main content is read. Similar issues often interfere when blind users try to find specific information on a page, since the Web page layout is designed for intuitive recognition using visual scanning. Such problems make nonvisual Web access harder. The WAI (Web Accessibility Initiative) [4] of W3C has been trying to improve Web accessibility by defining a set of guidelines for writing accessible Web pages and by promoting activities [5] to spread this information throughout the world. However, it will take a long time to let the Web authors know the importance of Web accessibility and to get their active cooperation. Handicapped users don't want to just wait and hope until the Web becomes more accessible.

Figure 1. An example of general Web page layout (IBM News page)

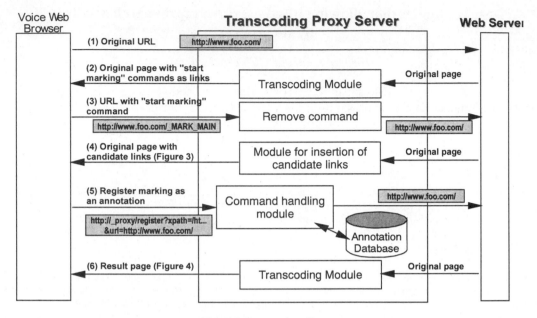

Figure 2. Transaction diagram

Therefore we decided to develop a system that allows blind users themselves to improve the Web's accessibility. This system consists of two components, a Web-based application for annotating a page while viewing it, and a server-based system for page customization based on the annotations. This customization is done by a proxy server located between the Web servers and the users' browsers (see Figure 2). The users' annotation authoring tool is a Web-based application—also added to the pages as they pass through the proxy server—, so blind users can conveniently annotate Web pages while surfing. They do not need to operate any other application program besides their familiar browser. In this paper we will first describe the inconvenience of nonvisual Web access, and then describe the overview of our system. Finally, we describe our conclusions and plans for the future.

2. Nonvisual web access issues

It is easy for sighted users to find the main content in the Figure 1. However it takes much more time for blind users to find it. Voice browsers read the text on a page in tag order. In the case of Figure 1, first the IBM logo, headers, and indexes would be read, and finally the main content is read. This took 40 seconds at the default rate (180 words per minute) of a voice browser [1]. This page is a generally used Web page layout. It means that blind users face this kind of problem almost every time they visit any Web site.

3. Overview of the system

The system consists of two components, one Web application for annotating a page with users' specified information and a proxy-server component that customizes pages based on the annotations. Users see the system as a Web-based application, which allows blind users to annotate Web pages while surfing with their familiar browser.

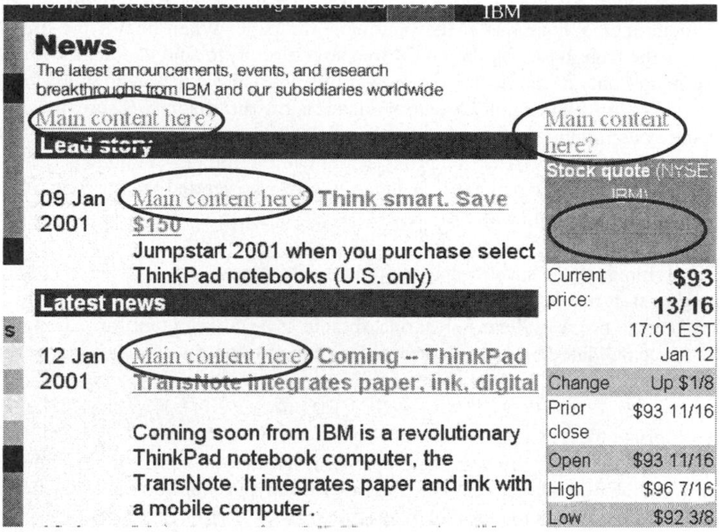

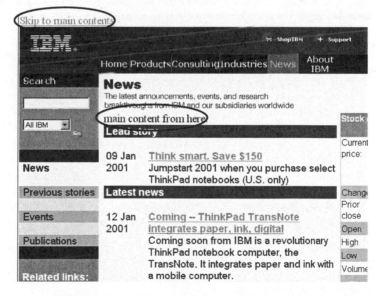

Figure 3. Page with main content candidates

Figure 4. Page with jump to main content link transcoded

3.1 Web Application For Annotation

The first component of the system provides five kinds of fundamental annotations for making nonvisual Web access easier. They are marking the starting point of the main content, marking an HTML block element as the starting point for reading, selecting an appropriate form in a page, commenting on an element, and selecting a main frame file in a frame-using page.

(1) Marking the starting point of the main content:
Figure 2 shows a transaction diagram for the system. Every page accessed through our proxy contains links for creating the abovementioned annotations at the bottom of the page. When one of the links is selected, the system analyzes the layout of the page based on the DOM tree structure [6] to find candidates for each kind of annotation. In finding the best 'main content' candidates it uses various heuristics, such as looking for a table cell with a long string (without links), or an element in a central location on the display. These are some of the most likely candidates, which the system then presents as possible starting points of the main content. These candidates actually appear as encoded links (see Figure 3) that the proxy server can interpret as commands. When the user selects one of the candidate links, the appropriate command is sent to the proxy server and the position of that element is registered in an annotation database as an XPath expression.

(2) Marking an HTML block element as the starting point for reading:
This function is used to start reading from a certain position in a page. It encodes links at the block elements, such as 'p', 'hr', 'table', 'form' and so on. A user can select one of them as a starting point for reading. This is also useful when all of the main content candidate links are incorrect. This function produces many more candidates for the starting point.

(3) Selecting an appropriate form in a page:
All but one form is eliminated from the page, and links are encoded for every form element. This avoids confusion with multiple Web forms. Optionally, a user can elect to see only an input box and a submission button for a form. Our system encodes links for every input box in the page.

(4) Commenting on an element:
This function is used to describe any element. It encodes candidate links based on our tag list for commenting, such as 'img', 'form', 'table' and so on. The list can be customized.

(5) Selecting a main frame file in a frame-using page:
This selects the main frame in a framed page. It offers encoded links for every frame element. The links for this selection only appears in a frame-using page. The goal is to avoid confusion with framed pages.

Users do not need to annotate all of the pages in a site. The system is capable of evaluating the similarity of the layout of the pages in a site based on the DOM tree structures, so some of the annotation information can be shared by all similar pages. This can be done automatically when the annotation is registered in the annotation database.

3.2 Page Customization

The second component of the system customizes pages based on the annotations. This is done as the pages pass through our proxy. The customizing features are as follows:
(1) Insertion of a 'skip to main content' link:
This function inserts a link to the main content at the top of a page (see Figure 4). When a user clicks the link, it jumps to the main content position in a page that was previously annotated by a user. In this way, the user can skip all heading information at the top of a page. When there is no annotation for the main content, it automatically uses the first candidate in the main content candidates list as the target of the "skip to main content" link.

(2) Insertion of a "skip to block element" link:
The transcoder also inserts a link at the top of a page to any predefined block elements. When more than two block elements are annotated, there would be links corresponding to each of the selected elements, such as "skip to block element 1", "skip to block element 2", and so on.

(3) Presenting only one form in the page:
When an annotated page with an appropriate form is opened, it only presents one form in the page. There is also an optional setting for a simple form, so the form will include only an input box and a submission button. This helps blind users input data into Web forms simply and quickly.

(4) Presenting commentary annotations:
This function replaces images with descriptive text (from the commentary annotations). When there is an annotated image link with a comment, the transcoder also uses that comment as the alternative text for the image link. This feature helps in describing images, and may clarify help messages.

(5) Opening a main frame:
When a main frame file for a framed page has been identified, that file is opened instead of a framed view. The other frame files defined in the framed page are converted to links at the top of the main frame page, so each frame file can be opened separately. This avoids blind users having to open each frame file to find the main frame file using Home Page Reader, or from having to find a main frame on the screen using a screen reader every time they visit such framed pages. When there is no annotation for the main frame, the system automatically looks for the main frame file based on heuristics, such as the largest file and a file containing a lot of links.

4. Conclusion

Our system allows blind users to improve the accessibility of some parts of the Web by themselves. Before that, they could only ask Web authors to improve the Web's accessibility. That is a passive approach and when authors do not choose to help with their special requirements, it causes stress and frustration. This system allows users to use the Web actively and effectively, since they can solve the problems by themselves. Our plan is to extend the range of user annotation features, as well as extending the methods for automatic layout analysis for purposes like presenting main content candidates. Our ultimate goal is to develop a transcoding system to improve the Web's accessibility by integrating annotations by users, authors and volunteers as well as to use automatic layout analysis and semantic analysis of the Web's contents.

References

1. Asakawa C., User Interface of a Home Page Reader, in Proceedings of The Third International ACM Conference on Assistive Technologies ASSETS '98 (April 1998), 149-156.
2. pwWebSpeak: The Productivity Works, http://www.prodworks.com/catalog/catalog_pwwebspeak.html
3. Jaws, Freedom Scientific, Inc., http://www.freedomsci.com/
4. Web Access Initiative (WAI): World Wide Web Consortium (W3C), http://www.w3.org/WAI/
5. Bobby, Center for Applied Special Technology (CAST), http://www.cast.org/bobby/
6. Document Object Model (DOM), World Wide Web Consortium, http://www.w3.org/DOM/

Voice and display interfaces to capture elderly and disabled persons' physical problems

Toru Takeshita, Tomoichi Takahashi, Yasunori Nagasaka, Masayoshi Ono

Graduate School of Business Administration and Information Science,
Chubu University
1200 Matsumoto-cho, Kasugai City, Aichi Pref., 487-8501, Japan

Abstract

The number of elderly and disabled persons who require nursing is increasing, and it is becoming an urgent matter to reduce the workload of the nursing people. We are conducting a research work to recognize, analyze, verify, and transmit to a database mainly physical problems spoken by persons who are nursed so that a remotely-located doctor, nurse or attendant can look at their physical records. A short natural sentence consisting of a part of the human body and the its problem (pain, fever, loss of motion, breakdown, etc.) spoken by an aged person is captured by voice recognition software and semantically analyzed using dictionaries to identify the body part and problem, which are immediately illustrated with a human-shaped 3D figure on the computer display. The problem area is shown in red or a different color depending on the type of the problem and is enlarged for better visibility. If an error in understanding the problem by the computer is recognized, then the user can correct it by choosing a right item in a menu on the screen by saying its item number or by doing an alternative.

The identified problem information is formatted (in XLM) and transmitted to a remote computer and stored into its data base. The doctor, nurse or attendant in charge can look at all or selected information stored in the database them regularly or on a time-available basis or can be alerted to the condition requiring immediate cares.

1. Introduction

An HCI interface has been built to have a wearable computer react to a spoken physical problem by changing the position and color of the problem part of the human-like 3D figure on the screen (Smailagic, and Siewiorek, 1999; Billinghurst and Starner, 1999). The information extracted from voice-entered sentences is transmitted for viewing by the nursing person(s). The interface to allow the display of the received information in an effective way at any time helps reduce the need of their attending or being located closely to the elderly or disabled all the times.

This system is designed to provide interfaces for persons who require care and for those who are in charge of care, so that the people who cannot normally access the computing facilities can benefit from them, thus increasing access in human-computer interaction.

2. System overview

This system consists of two pieces, one on distributed micro-computers and another on an Internet server machine.

It enables the elderly or disabled person to enter into a wearable computer a physical problem (a body part and its condition) spoken in a natural sentence. A word indicating urgent help and/or a short note (several words) can be appended. The wearable computer has been chosen as it must go where its user goes.

Keyboard operations are replaced by a microphone as input device and a voice recognition software engine provided by the IBM Via Voice SDK (IBM, 2000) .The captured information is converted to the XML format, and transmitted to an Internet web server which stores XML documents into its database.

On the server, an XML database system (eXcelon Release 2.0), XML Parser, ASP, etc. are used to allow the display of all or selected physical condition information in details or in summary formats as required by the user.

3. Voice recognized information extraction, verification, and transformation

When the system is up, a human-like 3D figure is displayed on the computer screen (Fig. 1). This serves as a prompt to the user to say what physical problem he has. When he has said it in a natural Japanese sentence, the color of the body part indicated by the user is changed according to its condition. For example, when the user says "left_leg", it is recognized as a lower left part of a human body. Likewise "aches" is recognized as having pain. Then the color of the left leg part of the displayed figure is changed to red so that the person can easily verify whether what he said has been correctly understood or not (Fig. 2).

An early design of this system was of a command type, and so a body part and its condition were separately pronounced and independently captured. It often occurred that they were not correctly recognized. The occurrences of this failure has been reduced by using the recognition engine provided by the Via Voice SDK. The engine determines proper words by contextual analysis. A Japanese sentence pronounced naturally is converted to a text string containing two key words (mostly in Kanji and their pronunciations).

Fig 1. Initial 3D figure   Fig 2. Problem part in red   Fig 3. Similarly pronounced words

The text string is scanned to detect those words which match words included in the body-part and condition vocabulary lists, thus extracting the body part and its condition respectively. They are immediately reflected on the 3D figure.

If the user finds an error, then he can specify in the displayed sentence by voice or with the mouse the wrong word. This causes a window containing a list of the words with similar pronunciations to pop up (Fig. 3). He tells the computer which is the correct one by saying the number assigned to it or by pointing it with the mouse. Then the wrong sentence is replaced by a right one. The system has been extended to accept sentences such as "I have a fever" and "I feel cold," which do not contain a body part

To assist novice users, synthesized voice instructions and explanations are produced. For example, pushing the help button on the right side in the pop-up window makes the computer speak "Say the number assigned to the right word."

As an "OK" signal is given, the body part and condition (as well as the urgency and a brief note) together with a time stamp are transformed to the XML format, and transmitted to the Internet web server located at the care center.

4. The contents and retrieval of XML database

The physical condition information received from distributed computers used by the aged or disabled is stored into an XMLStore of eXcelon (eXcelon Corp, 1999) and can be reviewed periodically or on a time-available basis by persons in charge of nursing such as nurses, doctors, care workers. When the degree of the urgency is high, the computer rings its bell to draw immediate attention

(1) Nursing Person

(2) Body Part

(3) Urgency

(4) Start Date

(5) Style Sheet

(6) Query Statement

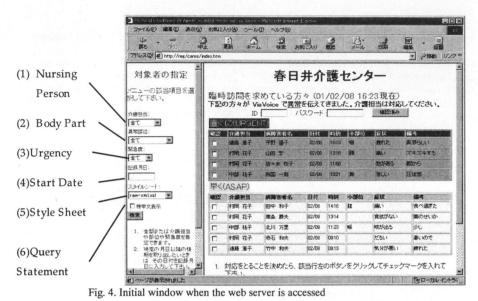

Fig. 4. Initial window when the web server is accessed

The XML document concerning each person who is registered in the system includes the following: ID No, name, sex, birthday, address, photo, current location, information on the person to be contacted in emergency, reason for nursing, information on person(s) in charge of caring for him or her, followed by a series of problem records each consisting of the body part, its condition, date and time, degree of urgency, etc.

When the server computer is accessed with the URL of this system specified, the initial window shows in a table format on the right-hand side the information on the people who have reported physical problems requiring immediate or ASAP attention (and not been taken care of yet). Each row includes the name of the nursing person in charge, the name of the person having a problem, its body part and condition (Fig. 4).

The left-most field of each row contains a check box which is to be clicked with a mouse when the person in charge confirms that an action is being taken. The next time the server is accessed, the clicked row will not appear.

The left-hand side pane of the window has the following six entry forms (horizontal boxes): (i) Person in charge of nursing, (ii) Body part having problems, (iii) Degree of urgency, (iv) Start date of displaying records, (v) Style sheet, and (vi) Check box for query statement (to be displayed or not).
The first four allows the retrieval conditions including defaults to be entered.

The specifying a style sheet allows selection of the contents and formats of the information displayed. If 'care_details.xsl' is selected, then the photo and personal information of each individual and his physical records satisfying the given retrieval conditions are shown in the descending order of dates on the screen (see Fig. 5). When 'care_table.xsl' is selected, the physical problem records grouped by the person(s) in charge of nursing are shown in the condensed tables. In either case, the user can specify the starting date of displaying stored records in order to avoid lengthy listings of outdated old records. If the 'raw-xsl' is chosen, then the information is shown in the XML format so that all the details can be displayed.

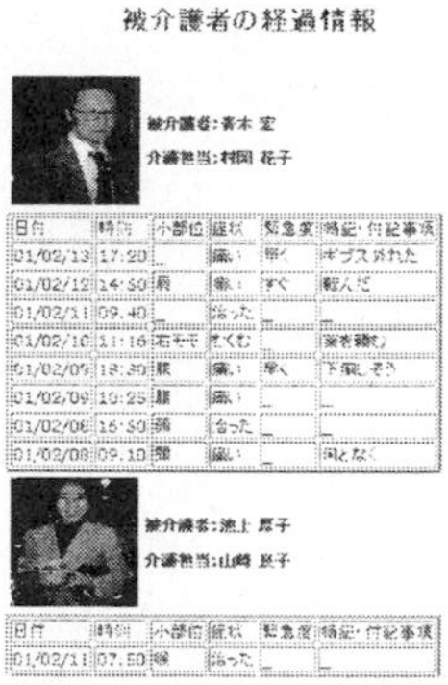

Fig. 5. Retrieved Information

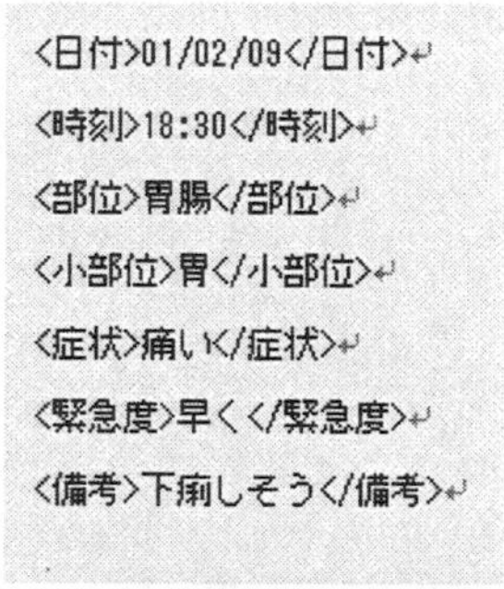

Fig. 6. A physical condition record

```
<date>01/02/09</date>
<time>18:30</time>
<part>digestive</part>
<subpart>stomach<subpart>
<condition>pain</condition>
<urgency>ASAP</urgency>
<note>may have diarrhea</note>
```

5. Conclusions

We have developed a prototype of a physical condition information system utilizing a voice recognition and XML database software to support the person(s) who are caring for elderly and/or disabled people located remotely.

 A physical problem spoken in a natural sentence is entered into a wearable computer (presently substituted by an A5-size ThinkPad). It is captured by a voice recognition engine and converted to a character string. Its contents are analyzed to extract the body part and the nature of its problem. After visual verification using a 3D figure on the screen, they are transmitted in the XML format to a remote computer.

Further study will be conducted on more proper handling of physical conditions related to the whole body or without a particular body part such as "No appetite" and "Very tired", and on the voice entry of the readings of the thermometer, pulse meter, and blood pressure gauge, and other items such as hours slept, the amount of meal taken, frequency of urination, etc.

Also, the use of VoiceXML (Lucas, 2000) which supports simple "directed" dialogs is being studied to replace the implementation of the first part of this system.

Acknowledgement

We acknowledge a grant from the High-Tech Research Center Establishment Project of Ministry of Education, Culture, Sports, Science and Technology.

References

Billinghurst, M. and Starner, T. (1999). Wearable Deveices: New Ways to Manage Information, *IEEE Computer, Vol.32. No.1* , 57-64.

eXcelon Corp (1999), *eXcelon Programmer Guide.*

IBM Developer's Corner (2000), *Via Voice Software Development Kit (SDK),*

http://www-4.ibm.com/software/speech/dev/sdk_windows.html.

Lucas, B. (2000), VoiceXML for Web-based Distributed Conversational Application, *Communications of the ACM, Vol. 43, No. 9,* 53-57.

Smailagic, A. and Siewiorek, D. (1999). User-Centered Interdisciplinary Design of Wearable Computers, *ACM Mobile Computing and Communications Review, Vol. 3, No. 3,* 43-52.

Meeting Rehabilitation Goals with a Multi-Sensory Human Computer Interface – A Case Study

Michael R. Tracey, Corinna E. Lathan

AnthroTronix, Inc.
387 Technology Drive
College Park, MD 20742
&
The Catholic University of America
Biomedical Engineering Department
Washington, DC 22206

Abstract

In this project we used play and computer-based simulation to assist in achieving functional rehabilitation goals - In this case, self-feeding. A user-centered design approach was used to develop a total system, enabling a child with a disability to experience situations and sensations, which would not be accessible to him in the physical world. A visual and vibrotactile interface immerses the user in the simulation and stimulates the user's imagination to make him feel a bat hitting the ball. Lifting the arm and activating a tilt sensor generated input signals. Improvements in range of motion were recorded with use of the system.

1. Introduction

A disability is an impairment that leads to an inability to perform an activity in the manner or within the range considered normal for a human being [1]. The application of a device, service, or strategy to aid a person with a disability in achieving an activity is known as an assistive technology. In recent years, the increased use of assistive technologies has provided many individuals with disabilities with an overall improvement in quality of life and an increase in self-esteem. However, the technical sophistication of many assistive technologies overwhelms some operators, which can lead to abandonment.

Assistive technology may also be applied to therapeutic interventions, with the intent of developing the function necessary for activity performance without the use of assistive technologies. In this respect, the assistive technology would be used in an exercise or therapy setting to improve strength, flexibility, etc. Incorporating computer technologies can also provide simulations of experiences that would be difficult, or impossible, for a child with disabilities to engage in otherwise [2].

Types of learning that can potentially be facilitated in virtual environments include motor skill learning. In this case, self-feeding was the goal for a 16-year-old young man with Cerebral Palsy. A user-centered design approach was adopted in development of the total system. Physical therapy reports for the user indicated a need to exercise the arms. This exercise may promote the motor skill development necessary to lift a utensil to the mouth, achieving the functional and therapeutic goal.

2. Tools and methods

2.1 Human centered design methodology

Human-centered system design is a method that acknowledges the human-technology interface as the key to optimizing human performance. Figure 1 shows a conceptual model that identifies and assesses the important human-centered parameters of the system. The flow of information through the interface to the technology forms a closed loop that starts and ends with the human operator.

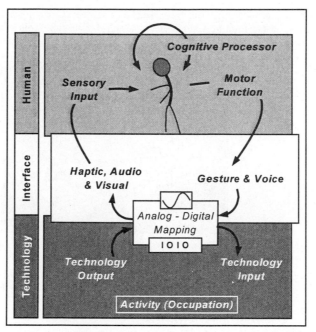

Figure 1: Information Flow

At the top of Figure 1, a simple model of the human processing system is shown. This is an idealized model of human information-processing capabilities where the user is described in terms of three processors: perceptual, cognitive, and motor [3, 4]. Sensory information is gathered from the interface, which is visual, tactile, and auditory in this application. This sensory information flows into working memory through the perceptual processor where it can then be accessed by the cognitive processor to decide if and how to act on it. Finally, the motor processor executes any actions.

The information flow through the perceptual processor to working memory is considered to be a "low-capacity channel" [4]. In other words, there is a limited ability of a person to pay attention to all of the sensory input available or to process all the available information. From the point of view of human-centered design, the key is not to overload channel capacity, but to give the optimal information for successful human performance. A good system design will take into account operator capabilities. Presenting the operator with multi-modal information to act upon will enhance the experience in the computer-based setting and provide a more realistic sense of immersion [5].

2.2 System design

A baseball simulation running on a Pentium Class PC was configured to present a virtual environment on a PC monitor. The game was configured to recognize keystrokes for commands, such as "swing bat" and "pitch". Working with the physical therapist, customized accelerometers were embedded within a wrist brace and a small vibro-tactile device was constructed from an unbalanced dc motor, and fastened to the top of the brace just over the knuckles. Raising or lowering the arm activated the accelerometers and was mapped to game input signals.

With the design centered on the operator, all input signals from the customized sensors were sent to a TNG-3b (Mindtel, LLC) data acquisition box. The TNG-3b interfaces with the computer through the RS232 serial port. Freely downloadable software called NeatTools [6] is used to map operator-input signals, and gestures. Inputs are then converted into game commands that correspond to the arm movement. The arm-gesture activity to signal a bat swing was the same motion prescribed as therapy to learn how to lift a utensil. We focused on the motion that the operator was capable of, and was prescribed by the therapist, rather than dictating the input gesture. This provided a truly human-centered system that is easily customizable, inexpensive, and technically straightforward.

Vibrotactile motor control signals are sent out from NeatTools via the PC parallel port. When the bat is swung, the player feels the bat move in the form of a slight tactile feedback sensation. This feeling is reinforced with visual feedback from the screen. An image of the operator using the system at home is presented in Figure 2.

Figure 2: Using the System at Home.

The home-based PC was also outfitted with a USB digital video camera. Using MS NetMeeting and NeatTools based socket communications we can videoconference to check sensor locations and calibrate thresholds remotely over the Internet and play "on-line".

2.3   Universal design

In designing this system great effort was taken to apply the principles of universal design, as defined by Rehabilitation Engineering Research Center (RERC) on Universal Design and the Built Environment [7]. A universal design ensures products are functional for all people, which is especially important when designing for operators with impairments. Each operator is unique and has different interface requirements.
The gesture-based interface can accommodate a wide range of individual preferences (equitable) and abilities and is useful to people with a wide range of disabilities (flexible). The interface is simple and intuitive to use. It is based on inexpensive, commercially available software and is not game or platform dependent. Information from the system presented to the operator is multi-modal and easily perceived. Tolerance for error may be set relatively high by adjusting the thresholds. The physical effort required to operate the system is very low and adaptable and the system is unobtrusive and approachable. The size, posture or mobility requirements of the operator will not affect the usability of the system.

3. Results

Prior to intervention with our system, clinical measurements were taken by the operator's physical therapist. The introduction of this intervention was meant to compliment botulinum toxin injections. Botulinum toxin is injected into an overactive, or spastic, muscle to hinder the transmission of acetylcholine. This effectively decreases spasticity and allows for increased range of motion. The effects of botulinum toxin are relatively short lived. Exercise therapy will help to maintain the benefits of botulinum toxin [8].

Two baseline measures of strength and range of motion were taken prior to previous botulinum toxin injections. Over a six to eight week period following baseline treatments, range of motion returned to pre botulinum toxin injection measures. With the combined botulinum toxin injection and this intervention, enhanced range of motion measurements were sustained over a ten-week period. The system was used at least three times per week, with one session per week occurring "on-line". Table 1 provides range of motion and strength measures for the weeks prior to and post intervention.

Table 1: Range of Motion and Strength Measures

Weeks Prior to (-) & Post (+) Intervention	Range of Motion (passive)	Strength	Range of Motion (subject able to actively match passive range)
-24	65°	Data not available.	-90°
-18*	100°	7 lbs.	-80°
-6	85°	8 lbs.	-90°
0*	90°	8 lbs.	-90°
+10	95°	8 lbs.	-85°

* = Botulinum toxin injections

4. Conclusions

Computer based play and simple, customized, unobtrusive sensors can mask exercise and facilitate an exciting environment in which to strengthen motor skills. Future work should focus on better integration of the simulation with feedback devices. For example, the vibrotactile information could differentiate a hit from a missed pitch.

We have developed and implemented a successful human computer interface for therapeutic play. The system has also achieved functional goals. One of the main reasons for assistive technology abandonment is an overwhelming complexity in the technology. Our novel video tele-conferencing system allows troubleshooting and provides assistance in the set-up and maintenance, from a distance. This allows the therapist to retrieve performance data and keeps the operator, health care professionals and rehabilitation engineer involved in all stages from design and development to practice.

Acknowledgments

This research was supported in part by the US Army Medical Research Acquisition Activity, MCMR, Assistive Technology and Neuroscience Research Center (ATNRC), grant # DAMD 17-00-1-0056, located at the National Rehabilitation Hospital, and the National Institute on Disability and Rehabilitation Research, U.S. Dept. Education grant #H133E980025. All opinions are those of grantees.

References

1. Kirchner, C., *People with Disabilities: A Population in Flux*, in *Rehabilitation Engineering*, R.V. Smith, Leslie, J.H., Editor. 1990, CRC Press: Boca Raton, FL. p. 3-28.
2. Deitz, J.C., Swinth, Y., *Accessing Play Through Assistive Technology*, in *Play in Occupational Therapy for Children*, L.D. Parham, Fazio, L.S., Editor. 1997, Mosby: St. Louis, MO. p. 219-232.
3. Cook, A. and S. Hussey, *Assistive Technologies: Principles and Practice*. 1996, St. Louis: .Mosby-Year Book.
4. Card, S., T. Moran, and A. Newell, *The Psychology of Human-Computer Interaction*. 1983, Hillsdale: Lawrence Erlbaum Associates.
5. Burdea, G., P. Richard, and P. Coiffet, *Multimodality Virtual Reality: Input-Output Devices, Systems Integration, and Human Factors*. International Journal of Human-Computer Interaction: Special Issue on Human-Virtual Environment Interaction, 1996. **8**(1): p. 5-24.
6. Lipson, E., Warner, DJ & Chang, YJ. *Universal interfacing system for interactive technologies in telemedicine, disabilities, rehabilitation, and education*. in *Medicine Meets Virtual Reality:7*. 1999: Ios Press.
7. NC State University Center For Universal Design, *The Principles of Universal Design, Version 2.0*. 1997, RERC on Universal Design and the Built Environment.
8. NIH, *Clinical Use of Botulinum Toxin*, in *Consens Statement Online*. 1990, NIH: Bethesda, MD. p. 1-20.

A Barrier Free Systems for the Next Generation

Kazuo Tsuchiya[a], Yoshihiro Tachibana[b], Midori Shoji[a], Shinji Iizuka[a]

[a]Yamato Lab. IBM Japan Ltd., 1623-14 Shimotsuruma, Yamato, Kanagawa, 242-8502, Japan

[b]IBM Japan, Ltd. 19-21 Nihonbashi Hakozaki-chou, Chuo-ku, Tokyo, 103-8510, Japan

Abstract

This paper proposes a barrier free approach to working with personal computers. Through support work for customers wanting to use assistive technologies with personal computers, it was found that a large portion of the customer help requests involved information about the assistive technologies themselves. They are lacking the fundamental information on what features to choose and how to use them. We present a new approach to guiding users and support personnel who may not have much experience and knowledge about the possibilities so they can choose and install appropriate assistive technologies.

1. Introduction

"Barrier Free" and "Universal Design" are now keywords. In the computer industries, software and hardware designers also have to consider accessibility in their products. There are a number of ways to adapt or adjust personal computers and information kiosks for people with disabilities. Most of them use Windows User Assistance Help parameters and other assistive technology applications. However, what parameters or what assistive technology (AT) applications needs to be used for individuals with disabilities is not an easy-to-answer question. Who knows these parameters and solutions? Where we can find someone who will set up computers for people with disabilities? The demands for adapting computer interfaces or parameters for individuals' preferences are quite high, while the people working to meet those demands are in short supply. Moreover, a personal computer setup suitable for one person is not guaranteed to work for another person.

The background of this project was the finding that Japan lacks a large number of personnel, industries, or institutions that can provide analyses of services to people with disabilities in terms of computer accessibility issues. IBM Japan's Accessibility Center recognizes these social needs through its activities as one of the core institutions that provides assistive technology products, information, and support services. The Telecommunications and Advancement Organization of Japan has sponsored research by IBM Japan to create a barrier free system for the next generation that is easy to use for all.

2. The objective of the Barrier Free System Project

The objective of the Barrier Free System Project is to research what kind of system will be helpful for the people with disabilities to obtain information on computer accessibility. In addition, the system should include an easy-to-use function to setup Windows-based computers or information kiosks using Windows-based platforms.

The Barrier Free System Project is a four-year project running from 1998 through 2001. We did a comparative study on barrier free systems in Japan and the US in 1998. In 1999, we developed a prototype in order to get comments from professionals in this field. Given their suggestions and comments on the prototype, we continued improving the prototype design from the aspects of usability and practicality throughout the year 2000. The prototype will be examined in 2001 in schools, rehabilitation institutions, regional centers, and if possible, in some workplaces. This system is intended to be used by relatively inexperienced users and support personnel.

3. Three main features of the Barrier Free System

Our Barrier Free System has three main components, the Assessment Manger, an IC card, and the Adaptation Manager.

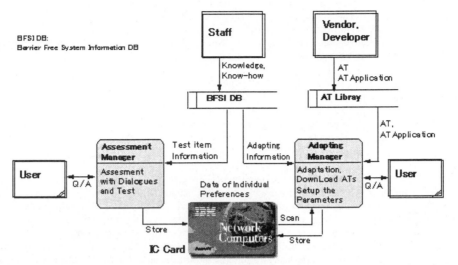

Figure 1. Three main components: the Assessment Manager, the IC Card, and the Adaptation Manager.

Assessment Manager

There are several approaches to assessing requirements and preferences of a user with disabilities. Our survey indicted that Japan lacks personnel who have experience and knowledge to customize a personal computer to an individual

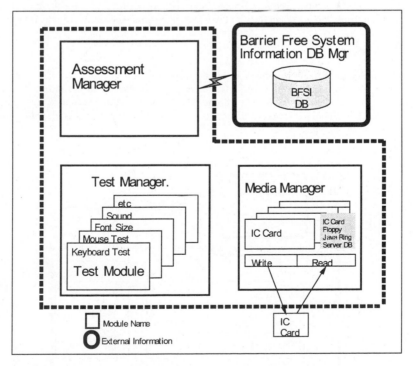

Three approaches are:
- Using information of the type of disabilities and their degrees;

- Having experience in choosing appropriate ATs; or
- Gathering information through a dialogue between the user and a personal computer.

The third approach has the advantage that the person who is supporting the assessment needs less experience and knowledge compared to the first two methods.

The design of the dialogue calls for considering several alternatives:
- Asking about the type and the degree of disabilities;
- Asking about the type of AT the user would like to use;
- Asking about the type of home appliances the user currently uses; and
- Trying some simple tasks on the personal computer.

The forth type of dialogue is considered to be especially feasible.

The Assessment Manager provides three assessment menus for tests are administrated by the Test Manager: A standard course, a quick course, and a customized course. The standard course consists of tests of all input and output methods, using heuristic dialogues and tests of various devices. Input methods tests include the keyboard, mouse, and onscreen keyboard. Output methods tests cover a screen reading program, sound output, and screen enlargement capabilities. The quick course aims to reduce the number of test items by providing some sub-menus based on the classification of disabilities, such as mobility disabilities, hearing disabilities, and visual disabilities. Some people do not have to go through the standard full assessment method. The customized course is designed for professionals in institutions such as rehabilitation centers and schools, where someone is able to select the tests for the individuals and tailor the course.

IC Card

The project has chosen to use an IC card with no contacts as the information storage media. There are two major reasons: the high potential of IC card for future general use, and its intrinsic accessibility in terms of lightness and requiring no complicated user actions. The user of the system is only required to place the IC card on the IC card reader tray. The user's information including the set of parameters is instantaneously read into the system.

Adaptation Manager

Adaptation in our Barrier Free System is defined as when a person with disabilities is able to automatically set up appropriate ATs on a personal computer. The Adaptation Manager handles all of the information in the IC card. According to the recorded information in the IC card, the Adaptation Manager configures computers by adjusting the parameters of the AT applications in the computers or by downloading required AT applications from an AT library and then adjusting the parameters.

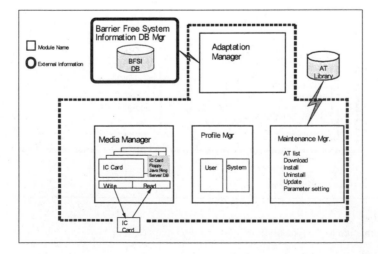

4. Conclusions

The Assessment Manager makes it possible for someone who is not an expert to assess and choose ATs and parameters easily regardless of their skill or experience. When a new AT becomes available in the market, a new test module to handle the new AT can be developed and distributed to the client systems by referencing the Barrier Free System Information Data Base. The Adaptation Manager will install new a AT by downloading it from an AT library, and the changes of the system's configuration are controlled by the BFSIDB. The results of assessments and the preferences of the user are stored in the IC card and personal computers can be adapted to the user automatically when the IC card is presented. This will reduce the effort required to customize a personal computer when it is shared with others.

References

Mizuta, et al. (1999). A Study on the Barrier Free System for the next Generation, Research and Development Report for 1998, Telecommunications and Advancement Organization of Japan, Tokyo, pp. 27-31. In Japanese.

A multimedia editor for mathematical documents

Gerhard Weber

Harz University of Applied Studies and Research
Friedrichstr. 57, 38855 Wernigerode, Germany
gweber@hs-harz.de

Abstract

MathsML is the latest development for mark-up of web-enabled applications. We have developed a multimedia editor for editing mathematical equations written in MathsML and text using standard techniques for accessibility by students and/or teachers who are blind.

1. Introduction

Mathematics is a scientific discipline students experience as both a source of knowledge and a challenge for acquiring mathematical skills. As the World Wide Web, has mathematics a universal approach, it's notation can be read in almost any language and culture. But unlike browsing the Web is "doing mathematics" with computers still only possible for a small group of professionals and academics.

Techniques for editing mathematical documents on computers are as diverse as the users who create single equations, solve problems by transforming equations, or document a calculation. In this paper we concentrate on students solving mathematical assignments and teachers preparing those assignments. By explicitly addressing users who are blind, we will demonstrate the use of multimedia techniques for doing mathematics. For use by both blind and sighted users, each group requires a user interface to create, modify, or simply to read equations and text. Combinations of acoustic media with GUIs can possibly increase the effectiveness of computer based mathematics for all users.

An approach for universal design of an interactive system for mathematical equations is needed as documents are prepared both for non-visual and visual use. Exchange of comments on text as well as on equations and terms facilitates the process of learning mathematics considerably and is a standard didactic principle in teaching mathematics. As a pre-requisite, only through the use of appropriate data formats it can be attempted to allow coherence between visual and non-visual presentation of complete documents or their elements.

2. Mathematical documents

Mathematical documents stored in computers have developed in several generations which can be described as

- *first generation document*s: proprietary data formats for desktop publishing which do not allow electronic exchange,
- *second generation document*s: non-standardised mark-up languages such as LaTeX and a variety of SGML-based document type definitions (DTD, see Sydow, 1994) which allow submission to publishers, and
- *third generation document*s: standardized mark-up languages such as XML and the MathsML DTD which have been defined to allow electronic exchange with all readers .

Editing using second generation mark-up languages for mathematical documents has allowed to become independend from particular printers. Recent developments allow to translate from LaTeX documents with sophisticated mathematical contents to the industry standard portable document format (PDF) directly (Goossens, Rahtz, 1999).

Unlike any other approach is LaTeX open for extensions. Authors are free to create new commands in order to extend the mark-up language to avoid repetitive use of formatting commands. For example can a new binary operator be based on visually overlapped glyphs.

This advantage turns into an disadvantage since exchange of mathematical contents using non-standard mark-up information is difficult. For example, semantic information about each operator within an equations is needed to transform equations within symbolic processors such as Mathematica or Maple. Up to now, several incompatible sets of LaTeX macros have been developed for their use within symbolic processors.

Third generation documents can carry information such as the number of operands an operator takes and hence can be exchanged between typesetter and symbolic processor. MathsML 2.0 recently has received the status of a recommendation by the W3C and was designed to be a third generation mark-up language. MathsML assists no method for extending the mark-up language but provides both a set of formatting tags as well as a set of mark-up tags for semantic interpretation of operands and operators. The following table 1 compares different MathsML representations of the term

$$\[x^{3} + 5x - 8 = 0 \]$$

Table 1: Different versions of MathsML markup for \[x^{3} + 5x - 8 = 0 \]

formatting tags (version 1)	formatting tags (version 2)	interpretation tags

`<msup>`
`  <mi> x </ mi>`
`  <mn>3</mn>`
`</ msup>`
`<mo>+</ mo>`
`<mn>5</mn>`
`<mi>x</mi>`
`<mo>-</mo>`
`<mn>8</mn>`
`<mo>=</ mo>`
`<mn>0</mn>`
	`<mrow>`
	`  <mrow>`
	`    <msup>`
	`      <mi>x</mi>`
	`      <mn>3</mn>`
	`    </ msup>`
	`    <mo> + </ mo>`
	`  <mrow>`
	`    <mn>5</mn>`
	`    <mo>&InvisibleTimes;</ mo>`
	`    <mi>x</mi>`
	`  </ mrow>`
	`  <mo>-</mo>`
	`  <mn>8</mn>`
	`</ mrow>`
	`<mo>=</ mo>`
	`<mn>0</mn>`
	`</ mrow>`
		`<reln>`
		`<eq/>`
		`<apply>`
		`<minus/>`
		`<apply>`
		`<plus/>`
		`<apply> <power/>`
		`<ci>x</ci> <cn>3</cn>`
		`</apply>`
		`<times/>`
		`<cn>5</cn>`
		`<ci>x</ci>`
		`</apply>`
		`</apply>`
		`<cn>8</cn>`
		`</apply>`
		`<cn>0</cn>`
		`</reln>`

Neither LaTeX nor MathsML were designed to be read by students and teachers. We have therefor extended a commercially available converter and typesetter accepting both MathsML and LaTeX (IBM Techexplorer, see Goossens and Rahtz, 1999). Our extension is based on file transfer and a converter program. The converter uses a Java-based XSLT processor. It is driven by an XSL file covering a subset of MathsML tags (see Figure 1).

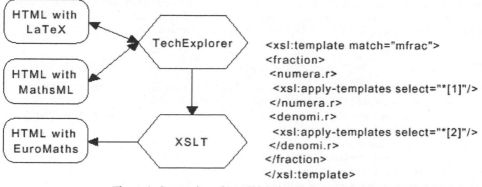

```
<xsl:template match="mfrac">
<fraction>
 <numera.r>
  <xsl:apply-templates select="*[1]"/>
 </numera.r>
 <denomi.r>
  <xsl:apply-templates select="*[2]"/>
 </denomi.r>
</fraction>
</xsl:template>
```

Figure 1: Conversion of LateX/MathsML to EuroMaths using XSLT

The result of this conversion is a document which consists of a simplified form of HTML with embedded equations. Equations are in this file are described using the EuroMath mark-up language. Within the MATHS project structure described by this mark-up method have shown to be suitable for students (Maths, 1997).

Various limitations in automatic transcription have shown that a subset of the EuroMath DTD can be used at the moment. For example is the end of an integral not clearly marked in MathsML output generated by TechExplorer (version 2.0). But polynoms, fractions, roots and nested terms put into parentheses produce sufficient mark-up.

Functional testing has to be based on manipulations of equations. In the following we will discuss therefor a editor accepting the generated file and which is able to generate input for our converter. A round-trip therefor consists of creating a web page with LaTeX mathematics, converting it to an input file for an editor, editing text and equations and saving modifications. The saved file can be exported again into LaTeX. If this is comparable to the original term then the test is passed.

3. Developing mathematical documents

Development of proofs and other transformations through humans by use of interactive systems has developed in several steps:
- *first generation editing*: symbolic representation programming languages require progra mming-like skills to use the computer for interactive development of equations and mathematical terms.
- *second generation editing*: repetitive use of cut and paste operations for text-only linear character strings.
- *third generation editing*: WYSIWYG Systems allow to develop equations by means of interaction objects such as menus, toolbars, and even handwriting (see Suzuki, 1999).
- *forth generation editing*: tutoring systems which develop a user model and accompany the user (student, teacher, lecturer, etc.) by means of an intelligent expert in an adaptive manner.

Consistency among interactive systems has not yet been developed. There is no industry standard interactive equation editor as commercial and non-commercial systems (e.g. W3C's Amaya) implement various of above aspects differently.

Non-visual presentation of mathematics is required if users cannot see equations either due to lack of vision or due to constraints such as lack of display space. Both auditory (Stevens and Edwards 1994; Raman, 1994) and Braille based presentation (Schweikhardt, 2000) are suitable alternatives. In general, multimedia presentation of mathematical equations combined with mult imodal input can assist all users in learning mathematics in a better way.

An interactive WYSIWYG editor for interaction via auditory output as well as braille based input and output has been developed within the MATHS project. One of the strengths of this system is the semantically oriented EuroMath DTD which allows synchronisation between different output techniques for each entity carrying mark-up information (MATHS, 1997). As no standard approach for access to interactive systems through assistive devices was available the MATHS workstation is a hardware dependent implementation. In the following we discuss a new approach to make it more independent of particular hardware for speech and braille.

4. M3 Editor

We have developed the M3 (multimedia mathematics) protoypical editor as a new interactive system on the basis of rendering techniques used within the MATHS workstation. Requirements for M3 are:
- printing documents for desktop publishing using a wide range of printers and electronic file formats (PDF),
- multimedia capabilities of the MATHS workstations,
- accessible through industry standard assistive devices and screen readers,
- synchronisation of different media for collaborative use between sighted and blind people,
- integration with MathsML-based documents, and
- open for transcription into different mathematical braille notations.

In order to print documents, we rely on typesetting capability of TechExplorer. Since this is a plug-in for Netscape, printing, file save and file open are standard operations provided by the browser.

To ensure multimedia capabilities the software architecture of M3 forsees a layered approach (see figure 2):
- a web page written in HTML and Euromath DTD mark-up is parsed,
- menus control document appearance and browsing of equations and text.
- a graphical user interface both visualizes text only
- a renderer transcribes mathematical terms into braille or spoken output , and
- a screen reader and assistive devices access the graphical user interface.

As a novel approach to the design of GUIs for accessible editors we have visualized the concept of synchronising a focussing interaction object with spoken output through a textfield. This field is writeable and readable as long as it contains text. The editor recognizes any equations and makes the textfield read-only. If the user moves on to text the textfield will accept input from a keyboard again.

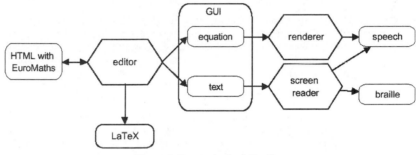

Figure 2: Layers in the M3 editor

By separation of text from equations within the graphical user interface it becomes possible to pass implicitely control of the user interface between a screen reader and a specific renderer for mathematical terms. Figure 3 shows a snapshot of the M3 editor prototype.

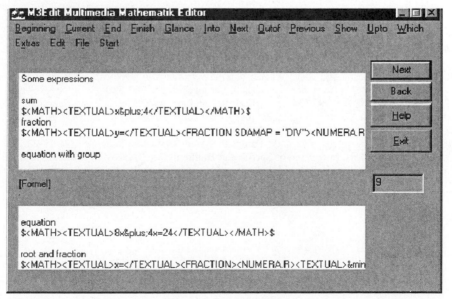

Figure 3: The M3 Editor

Only if there is a caret or selection being displayed then the screen reader becomes active. Making the textfield read-only requires the screen reader to be practically silent and the editor can generate speech output while the user is browsing the structure of an equation.

Input for this editor is menu-driven. A verb such as "next" combined with an object such as "fraction" forms the command "Alt-NF" consisting of two keys and a modifier key. The number of terms (and implicitly tags) is limited and motivated by semantic structures. The phrases for verbs and objects have been selected to fit naturally with unique characters of the alphabet. A configuration file allows to localize the menu to different languages.

At the point of this writing only browsing through an equation is implemented. Input of new terms and their manipulation will be based on the same menu technique. As has been shown in the Maths workstation braille-based input is a possible extension as well.

5. Conclusions

MathsML, a recommendation for mathematical document mark-up by the World Wide Web Consortium, provides the basis for "doing mathematics" with an editing program, but it provides not the basic structure for a multimedia user interface. A conversion to a more simple structure such as the EuroMath DTD is useful for classroom use. The graphical user interface of a multimedia editor should separate equations from text to allow also access through standard assistive devices and screen readers.

Acknowledgements

This work has been partially funded by the Human Language Technologies unit under the IST Programme by the Commission of the European Union (Project IST-2000-27513). We want to thank also Sebastian Breit, Ine Langer, Michael Menz, Robert Nitschke, Tino Reichardt, and Kurt Weimann for their help in implementing the M3 editor.

References
Goossens, M. and Rahtz, S. (1999). *The LaTeX Web Companion*, Reading, Mass.: Addison-Wesley.
Maths (1997). Final Report, available from F.H.Papenmeier, Talweg 2, 58840 Schwerte, Germany.
Raman, T.V. (1994). *Audio Systems for Technical Reading*, Ph.D. thesis, Department of Computer Science, Cornell University, NY, USA.
Schweikhardt, W. (2000). Requirements on a mathematical notation for the blind, in Vollmar, R. und Wagner, R. (eds.) *Computers Helping People with Special Needs ICCHP 2000*, Wien: Österreichische Computer Gesellschaft, 661-670.
Stevens, R. D. and Edwards, A. D. N (1994). Mathtalk: The design of an interface for reading algebra using speech, in W. L. Zagler, G. Busby and R. R. Wagner (eds), *Computers for Handicapped Persons: Proceedings of ICCHP'94*, Lecture Notes in Computer Science 860, Berlin: Springer, 313-320.
Suzuki, M.; Fukuda, R.; Ohtake, N. (2000). Optical Recognition and Braille Transcription of Mathematical Documents, in Vollmar, R. und Wagner, R. (eds.) *Computers Helping People with Special Needs ICCHP 2000*, Wien: Österreichische Computer Gesellschaft, 711-718.
von Sydow, B. (1994). Editing mathematics in the Euromath System, *Euromath Bulletin, 1* (2), 17-23.
Weber, G. Stevens, R.D. (1996). Integration of Speech and Braille in the MATHS Workstation, in Klaus, J.; Auff, E.; Kremser, W.; Zagler, W. (eds.) *Interdiscplinary Aspects on Computers Helping People with Special Needs*, ICCHP'96, (July, 17-19, 1996, Linz), Wien: Oldenbourg, 617-626.

Improving access for elderly and severely disabled persons: a hybrid adaptive and generic interface

Linda White*, Jenny Jerrams-Smith*, David Heathcote**

*University of Portsmouth, Department of Information Systems,
Portsmouth PO1 3HE, UK

** Bournemouth University, Department of Design, Engineering and Computing
Bournemouth BH12 5BB, UK

Abstract

The current work promotes Universal Access by developing software that enables very elderly and/or severely disabled people to participate in the Information Society. In addition to providing software suitable for elderly and disabled people in general, adaptive software has been developed which meets the individual requirements of some special needs users.

Software was developed which would support users with severe cognitive deficits. The trial task used in this study was chosen to be motivating to both the elderly and the young users alike.

The adaptive aspect of the interface was developed to meet the requirements of the physically disabled as identified at an earlier stage of the empirical investigation. Analysis of the results produced a classification of the relevant deficits of these users and of the resultant interface requirements for people with either specific physical disabilities or groups of disabilities. Software was developed that would provide the necessary adaptivity based on this classification.

1. Introduction

The work described in this paper promotes Universal Access by developing software that enables very elderly and/or severely disabled people to participate in the Information Society. Although such participation is likely to improve the quality of life for many (Anderson, Bikson, Law and Mitchell, 1995), it must be noted that Stoll (1995) suggests that use of the Internet can cause social isolation. However, people who are disabled often experience social isolation, as do many elderly people because their friends die or their family members move away. Indeed, a recent survey of 1,091 adults aged 65 and over (Mori, 2000) found that 12% feel trapped within their own home and 2% have gone for at least one week without speaking to neighbours, friends or family. Older people with both low income and long-term illness were almost three times as likely to feel isolated as older people in general. There is therefore considerable value in the development of software that helps to provide such people with new interests and to support their access to friends and relatives (Katz and Aspden, 1997). Indeed, increased social contact could make their lives both happier and healthier (Cohen and Wills, 1985).

We believe that our work is applicable to both disabled and elderly people because people often become disabled in various ways as their age increases. Two types of task need to be supported in order to approach universal access for people who are disabled and/or elderly. Firstly, tasks involved in helping people to learn how to use their computer, and secondly, tasks involved in supporting people in using their computer in the long term.

2. Interface functionality: supporting all disabled and elderly users

Our work indicates that two types of supportive software should be developed, and thus a hybrid interface was developed that includes both generic and adaptive components. The generic aspects of the interface consist of software that will alleviate some common problems that are eventually likely to be experienced by many elderly people and by younger people who have some cognitive impairment. We shall consider two examples of such problems, which occur in elderly people as their cognitive ability degrades. These are reduction in working memory capacity and reduction in attention, that is, the ability to remain focussed on a specific task. Our work has indicated (Jerrams-Smith, Heathcote and White, 1999) that when working memory capacity is reduced, people become progressively less able to process information in parallel (or carry out tasks in parallel). We have therefore developed an interface to our software that requires only serial processing of tasks. This will support people both as they learn to use their computer and in the long term. As far as attention problems are concerned, it is particularly important to support people while they are learning how to use their computer. We therefore decided to provide an attention-catching environment that both includes the tasks that they need to learn and also tests their capabilities. Our research indicated that elderly people are usually extremely interested in their own history and that of their family, and will often talk at length about these topics even when taking part in an empirical study. We therefore developed an environment known as "My Story" which encourages people to record information about themselves and their family.

3. Interface functionality: supporting specific needs of individuals

In addition to providing software suitable for elderly and disabled people in general, the hybrid interface also includes adaptive software that attempts to meet the individual physical requirements of some special needs users. These requirements were identified at an earlier stage of the empirical investigation. Analysis of the results produced a classification of relevant deficits of users with physical or cognitive impairments. We have subsequently developed a classification of the disabilities and of the resultant interface requirements for people with either specific physical or cognitive disabilities or groups of disabilities. The physical categories were extended from those of Edwards (1995). Based on this classification, knowledge-based software that provides the necessary adaptivity has been developed. It diagnoses the physical problems of the individual user and prescribes the adaptation which would best suit that user. For instance, it can diagnose when the user responds very slowly, keeping a key pressed down too long (causing unwanted repeated letters), and can prescribe an alteration to the 'key repeat rate' to compensate for this (White, 2000).

4. A classification of relevant deficits of users with physical or cognitive impairments

Moderate physical disabilities causing difficulties in using standard input devices
P1:   Unable to perform simultaneous keyboard actions such as shift, control, Alt.etc.
P2:   Cannot respond quickly (e.g. keys held down too long).
P3:   Cannot use standard mouse or other pointing devices which use fine motor movement.
P6:   Have trouble hitting the required key and accidentally touching adjacent keys.

Severe physical disabilities preventing the use of standard input devices
SP1:   Very limited motor movement requiring switch facilities
SP2:   Need to connect special input devices or interfaces (sip-puff, eyegaze, etc.).

Visual impairments causing difficulties in using standard input and output devices
V1:   Screen display is too small to see.
V2:   Colour blind persons cannot see information presented through some colours.

Visual impairments that make it impossible to use standard input and output devices
B1:   Need electronic access to information displayed on the screen in order to use special non-vision display substitutes.
B2:   Do not have eye-hand coordination required for mouse tasks, touch screens, etc.

Hearing difficulties
H1:   Cannot hear auditory warnings or other indications or output (including speech).
H2:   Cannot hear sounds at normal volume.

Speech disabilities, or no speech capability

S1:   Unable to speak clearly enough to use standard speech recognition systems

S2:   Unable to speak clearly enough to record personal instructions

Cognitive difficulties with reading textual information

CAV1:   Problems reading words on the screen.

CAV2:   Unable to comprehend complex sentences / passages with a high fog index.

CAV3:   Trouble locating target words on screen.

Cognitive difficulties that cause problems when using a computer

CAL1:   Unable to use a standard keyboard input device.

CAL2:   Cannot understand the way a cursor can be moved on the screen.

CAL3:   Choosing between alternatives is problematical.

CAL4:   Has difficulty in understanding screen displays.

CAL5:   Unable to comprehend the meaning of sound / speech output.

CAL6:   Tasks often left unfinished.

5. The Information Technology Safety-net (ITS), a software prototype

The above classification has been incorporated into software components in a system known as the Information Technology Safety-net (ITS). It provides the hybrid generic and adaptive interface and includes the following modules :

VICI - a virtual community interface, that replaces the usual windows interface. This module is intended to enable users to interact with the ITS system and with their carers, friends and family;

My Story - a set of adaptive modules based on a books metaphor. It is part of VICI and uses four books that both test the user's physical and cognitive capabilities and support users' learning. The books consist of typical tasks that users need to master in order to use their computer effectively and they help users to learn to perform these tasks. In addition, the books are intended to be motivational in that they are based on activities likely to interest the users. They include : "This is my Life", to which the user adds information about him/herself; "Family Tree" to which the user adds family details; an address book; and a medical appointments book;

The user modeller - gains initial information about the user from a questionnaire and from simple game-like tests and uses this to build the individual user model;

A set/reset module - sets or resets the computer control panels in response to the contents of the user model;

The adaptivity engine - a knowledge based system that consults the user model in order to offer suitable interfaces for users to test; and

The interface designer - monitors the user of the "My Story" books. It makes changes to the user model when the user selects an interface preference. At the end of a session the user can revert to the original user model or save the new one for future use.

6. The Future

ITS is currently being tested in empirical studies with users. Ease of learning to use the system and ability to use it efficiently and effectively are being tested and compared with performance of users who are not supported by ITS. Quantitative measures include time taken to complete specified tasks and number of errors (of a variety of types).

References

Anderson, R. H., Bikson,T. K., Law, S. A. & Mitchell, B. M. (1995). Universal access to e-mail: Feasibility and societal implications. Rand Corporation. Santa Monica, CA..

Cohen, S. and Wills, T.A. (1985). Stress, social support and the buffering hypothesis. *Psychological Bulletin*, 98, 310-357.

Edwards, A.D.N. (1995) Computers and people with disabilities. In A.D.N. Edwards (Ed) Extra-Ordinary Human Computer Interaction. Cambridge University Press. Cambridge, UK.

Jerrams-Smith, J., Heathcote, D., & White, L. (1999). Working Memory Span as a Usability Factor in a Virtual Community of Elderly People. *Proceedings UKAIS Conference*, University of York, UK

Katz, J.E. and Aspden, P. (1997) A nation of strangers? *Communications of the ACM* 40 (12) 81-86

Mori (2000) A survey of 1,091 adults. Commissioned by Help the Aged, UK.

Stoll, C. Silicon Snake Oil. New York, NY. Doubleday.

White, L (2000). Usability issues and users with a range of specific needs. *Proceedings of HCI 2000 Conference: Volume 2*. Sunderland, p39-40.

Sound News: An Audio Browsing Tool for the Blind

Cliff Williams

Biomedical Eng, Science & Health Systems
Drexel University
Philadelphia, PA 19104
cwilli@drexel.edu

Marilyn Tremaine

Center for Advanced Information
Processing, Rutgers University
Piscataway, NJ 08854
tremaine@cs.princeton.edu

Abstract

Sound News is an audio and tactile interface that allows eyes-free browsing of newspapers. It works by mapping categories of text captured from a web page onto a small inexpensive touch pad that has been segregated into tracks by the addition of physical dividers glued on the touch pad. To navigate the interface, a user runs a finger across one of the tracks. The user hears spoken categories of the text (e.g., newspaper sections) and non-speech audio cues that indicate boundaries of a category. If a particular category is found interesting, a user can then zoom into more information on that category by pressing a zoom in key on the side of the touch pad. The same track will then speak newspaper subsections in the category chosen. A user can zoom in until a newspaper article of interest is found and then listen to this article. Zooming out is accomplished by pressing a zoom out key on the other side of the touch pad.

The two additional tracks in the interface help a blind user make appropriate settings for the interface, e.g., speech speed, volume, etc. and to control the reading, e.g., spell out a word, go back a sentence, etc. Usability studies run on both blind and sighted users indicate that the interface is readily learnable and also suggest areas needing redesign especially in the reading-control area. The interface provides an eyes-free mechanism for accessing any material that is organized hierarchically, e.g., books, directories, pull down menus, radio stations and newspapers via a small touch track and three keys. It's simplicity makes it much easier for users to browse available options, converting a GUI design into a T(touch)UI design.

1. Introduction

Today's newspaper is a well-designed document, based around a multi-layered hierarchical structure of sections and articles. Sections group articles based on content, and sub-sections further refine this grouping. Other visual cues besides section headings and newspaper location such as story length, heading size, and breakout captions indicate the importance and salience of news to readers. Although newspapers are now carried on the web and thus, available to blind readers with text-to-speech or Braille reader support, the visual cues that help the sighted reader "sample" the news through browsing are not available.

This paper presents a system that allows blind readers to browse the newspaper with the same advantages and strategies as sighted readers. It is based on an interaction style we call Audio Browser which allows a user to explore any structured textual content through the use of a touch pad and audio feedback. We call our system, Sound News.

2. The sound news design

The Sound News system is intended to be a software environment and information service for use on touch and text-to-speech-enabled handheld computers. We envision users of Sound News downloading a web version of the newspaper from participating content providers to their personal data assistant. Using the touch pad on the PDA, the user can then browse the news and listen to selected stories via computer-synthesized speech. One of our design goals is to make Sound News universally usable, that is, marketable to all users who wish eyes-free access to the news, e.g., individuals who drive to work daily. Without this design goal, products made for blind users do not generate enough market share to make their purchase price accessible.

Complete realization of this idea will involve the research and development of three distinct components. First is the design of the Sound News user interface. Second is the development of an information infrastructure for transferring web-based information to the Sound News environment. The final component is the porting of the first two to a handheld platform. The scope of this paper is limited to the first component, design and development of the Sound News user interface.

Because one of the design goals was an interface that was useful for sighted as well as blind users, any needed user feedback could not require significant learning, e.g., the memorization of multiple different sounds representing system actions. This limited our output to audio rather than tactile feedback and, in particular, to speech. However, we reasoned that the inclusion of secondary non-speech audio sounds could be learned through constant usage and eventually substitute for the slow method of serial speech feedback. For our input mechanism, we were also constrained by the amount of effort a sighted user would be willing to exert to learn input commands. Assuming that such a user would not be willing to learn a variety of special commands or keystrokes when they are accustomed to 'pointing and clicking' in a graphical user interface, we looked for a design that would support similar 'pointing and clicking' with audio menus. We also decided to eliminate the keyboard. Although visually impaired users might be motivated to learn a set of special keys, our sighted users would not. We had to come up with an input method that involved as few keystrokes as possible. This left us with a dilemma because keys are useful discrete input devices for blind people.

We selected a touch pad as our input device. We could map available spoken menu options to positions on the touch pad and allow them to be selected by using the one or two keys on the side of the touch pad. The touch pad would also give us a limited size physical space with boundaries and directionality that a blind person could use for orientation. We selected the EZ-Pointe Touch pad produced by PC Concepts as our input device. This model was chosen because of the application-programming interface (API) provided by Synaptics that allowed us to convert the relative positioning behavior of the touch pad to absolute positioning. This touch pad is also small enough to eventually be integrated into a Personal Data Assistant. We modified the touch pad by taping two of the buttons together and gluing two paper clips to the touch pad space to separate it into three equal tracks. The device is illustrated in Figure 1. The modifications gave us three buttons and three tracks that we could program for user input.

The three tracks on our input device were used as follows. The top track is the browsing track. It contains the categories of the newspaper to be browsed. This is laid out in equally spaced sectors with the number of sectors equal to the number of information categories that are being presented. Dwelling a finger on any sector on the track initiates a spoken description of the category. The middle track is used for controlling the fineness of the speech presentation, that is, a sentence at a time, a word at a time or even single letters for those times when the spelling of a name or address is needed. The bottom track allows the user to set speech parameters such as volume, speed and voice.

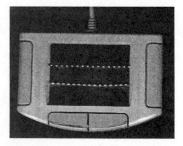

Figure 1. The modified touch pad

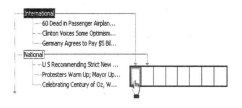

Figure 2. Mapping of a hierarchy level to touchpad sectors

A user browses the newspaper by running a finger back and forth on the top track of the touch pad. The audio display on this track varies depending on the user's actions. Initially, the track contains the top level categories of the newspaper, e.g., "international," "national," "sports." etc. As the user runs a finger over the categories, each is spoken (displayed). If a desired category is found, the "zoom-in" button on the right hand side of the touch pad is pressed. This changes the auditory display of the top track to the set of subsections in that category. Navigation in and out of sections is based on a "zoom" metaphor, with the left button zooming in and the right button zooming out. Figure 2 illustrates a display of the top categories of the newspaper. If the "zoom-in" key is pressed while the finger is resting on or has last rested on the international category, the headlines under that category will now be the displayed information that is given when the user runs a finger over the top track.

When a sector is entered, the information for the corresponding article or sub-section is displayed. For articles, this information includes the article headline, followed by the article's position in the current section. If the article was the third of five items in the section, its position would be displayed as "Item 3 of 5". This provides logical position information in addition to the spatial position users are able to perceive from their contact point on the touch pad. The article's byline follows its position. For sections, the section name is displayed followed by its position as described above. Following this is a summary of the section's contents. For example, a user might hear "International section; item one of nine; eight articles; two sub-sections."

Text navigation is accomplished through the middle touch pad track and the right and left buttons. The units used for text navigation (spell-out, word, sentence, paragraph, and article) are mapped to sectors across the middle track just as newspaper items are mapped to sectors in the first track. Users begin the text review process by browsing across the second track until they locate the desired unit. The last "touched" unit becomes the active text navigation unit. The user can then move forward and back in the text by pressing the left and right buttons respectively. Output can be stopped at any time by pressing the middle button at the bottom of the touch pad.

Sound attributes of the audio output are controlled through the bottom track and the side buttons. To change an attribute's value, the user must first browse across the track and find the desired attribute. As in the previous section, the last "touched" attribute becomes the active one. Next, the left and right buttons are used to increase or decrease the value of the attribute respectively. When one of the buttons is pressed, the system speaks the attribute's new value, e.g., "240 words per minute" for the speech rate.

A variety of non-speech audio cues are added to give additional feedback to the user. When a user presses the zoom-in key, a quick "whooshing" sound is played. The reverse of that sound is played when zooming out. A muffled, bumping sound is used to signify actions that are not valid for the current context. In this design, these situations are 1) zooming in on an article, and 2) zooming out from a top-level section. As the user moves a finger over the sectors, a quick sharp clicking sound is used for each boundary crossing. Finally, as an article is being displayed, a series of short notes rising the musical scale is played to indicate progress in the story. First, one note is played, then when the story is one quarter done, two notes are played until the entire eight notes or full octave is played near the end of the story.

3. A usability evaluation of sound news

A usability study was carried out to examine our design premises. In particular, we wanted to determine if both blind and sighted users could navigate a newspaper, finding and listening to stories of interest while skipping others of no interest. We also wanted to assess how easy it was to learn the functions needed to operate Sound News and whether our non-speech audio cues aided the users.

Participants in the study consisted of five visually impaired subjects and five fully sighted subjects. All subjects were using the system for the first time. Seven of the subjects were male and three were female Five had no visual disability, one was partially sighted, and four were completely blind. All but two of the subjects had significant experience with computers.

We downloaded and massaged 104 full-text articles (approximately 10,000 words) from the July 23, 2000 edition of the online New York Times for our study's corpus. Each user session began with an introduction to the Sound News project and a brief, interactive spoken tutorial on Sound News. Subjects were asked to perform 10 representative tasks with Sound News and then answer a set of questions. Browsing tasks asked the subject to find articles under different conditions. Environmental control tasks asked the subject to alter some attribute of the sound environment. Text navigation tasks asked the subject to move backwards or forward in an article's body by different textual units.

Subjects were asked four sets of questions. The first set consisted of ten true/false questions designed to examine the user's mental model of the system. The next set asked the user to explain in their own words how to accomplish a set of tasks with the system. The third set posed twelve statements that the subject evaluated on a 5-point scale (with 1=Strongly Agree and 5=Strongly Disagree). The last set asked three general questions on overall attitudes towards the system.

All subjects exhibited browsing behavior and were able to navigate with no problems. When asked to select four articles of interest, no subject picked all four articles from the same section. Five subjects, four sighted and one blind, picked at least one article from a sub-section. Three of these subjects' articles, two sighted and one blind, were in a sub-section at the very last level. All subjects were able to relocate articles they had previously browsed. Average navigation scores for finding articles were 1.96 for sighted users and 1.46 for blind users with a score of 1.0 equal to optimal navigation and 2.0 equal to double the number of steps needed to find the article.

Subjects used a serial browsing strategy browsing from left to right over each sector until the target item was found. Four subjects also used a "jump" strategy moving immediately to a position near the target item, often jumping directly to it. All subjects began searching for the article by first moving to the outer level of the paper even when already positioned at the correct sub-section.

The biggest problem encountered by subjects was navigating the text by grammatical units. Seven of the ten users, four sighted and three blind, had significant trouble using the text navigation feature to listen to a story. Problems centered around two common mistakes. Five users, three sighted and two blind, tried to use the left and right buttons to move in the text before first selecting a unit from the middle track. The other troubled users would remember to select a unit from the middle track, but would then move back to the top track and try to use the right and left buttons. All users were able to perform the sound environment change tasks without difficulty.

On the True/False questions, only one question was consistently missed: "The articles of a section are organized according to importance." All sighted subjects also missed the statement, "The set of notes I hear in the background when I am listening to a story indicate my progress through the story" and only three blind subjects reported even hearing the background notes while using the system. Answers to the Action Recall questions were generally accurate, although frequently vague or incomplete. The evaluation statement responses were generally positive about the system expressing neither fatigue nor confusion with any aspect of the system.

Answers to the final set of general questions were positive. Users, especially blind users, were very receptive to the system. When asked what they liked about the system, blind users seemed very pleased with what they were able to accomplish while using the system. One user commented, *"It's been so long since I've read a newspaper."* Three users commented on the spatial mapping between newspaper items and the touch pad. A blind user noted that he likes how *"it's not serial."* A sighted user commented, *"I liked that there was a geometric layout where I could go back to where I was on the slider."* Two sighted users commented that the physical interaction was a bit tiring due to the layout of the touch pad and the three tracks, specifically the top track. One blind user thought the touch pad was too sensitive. Finally, a single user noted that the text navigation scheme was a bit confusing.

4. Related work

Designing interfaces for blind users requires insight beyond the realization that a non-visual interface will be required. Accommodations must be made for how blind users are accustomed to navigating and browsing environments both real and virtual. There is a significant body of research describing interfaces designed for blind users. Our work has built upon these previous efforts.

The GUIB (Textual and Graphical User Interfaces for Blind People) Project provides blind users access to graphical user interfaces (GUI's) (Weber, 1993). GUIB uses a touch pad to allow users to scan large features of a GUI display such as windows and icons. Characters within these units are mapped to a Braille output device. Sound News loses this visual spatial orientation but overcomes two practical issues. First, the GUIB system assumes the user is able to read Braille. Second, the system uses a special input/output device, which significantly increases the cost of the end product. Like GUIB, the Mercator project (Mynatt & Weber, 1994) also seeks to provide blind users access to GUI interfaces. However, Mercator differs in several significant ways. First, Mercator is an auditory interface, relying on a standard keyboard for input and a combination of synthesized speech and non-speech audio for output. Second, Mercator is focused on mapping the GUI interface at the level of interface objects, rather than at the pixel level. Mercator constructs its mapping of graphical interfaces by examining the user controls present in the interface and organizing them into a hierarchical structure. Sound News borrows this use of a hierarchical structure for navigation.

PC-Access (Ramstein et al., 1996) is another system for providing GUI access to blind users. As with GUIB, PC-Access seeks to preserve the direct manipulation style of interaction. There are two versions of the system, differing in mode of input. The first version uses an absolute-position mouse coupled with a large drawing tablet. The second uses a custom haptic device, the Pantograph. Both rely on auditory output in the form of synthesized speech and non-speech sounds. SoundNews borrows the absolute positioning but does not seek to preserve the visual format of the display. Martial and Dufresne (1993) present another direct manipulation-based GUI adaptation, the Audicon. The Audicon's input device is a mouse operating on a touch tablet with user-adjustable grooved grid marks that are easily felt by the fingers. Feedback is given through synthesized speech and non-speech audio. Audicon most closely parallels the design of Sound News except that Sound News gains efficiencies by not imitating the visual space.

The DAISY Playback Software is a system for playing digital talking books on PC's using a standard keyboard (Morley, 1998). Sound News borrows the text navigation functions and several of the non-speech audio cues featured in this system. DAISY is keyboard based but both systems are concerned with presenting large, continuous blocks of content organized into hierarchical structures. The ACCESS hypermedia system is designed to provide blind accessibility to hypermedia material. The system allows users to read large bodies of hypermedia material in a manner similar to the DAISY Playback Software. Although synthesized speech was the primary mode of output, the system was also capable of presenting digitized audio clips, digitized audio with sound effects, musical nodes, and tactile pictures (Morley *et al.*, 1999). This system was also keyboard based and did not have the small size and portability envisioned in the Sound News design.

5. Conclusions

This research has yielded several useful findings. First, universal usability, at least in terms of blind and sighted users, is attainable in an interface such as Sound News. In fact, the design constraints of universal usability improved the design for visually impaired users. It is worth noting, however, that sighted users were noticeably less enthusiastic about the system than were blind users. We also show that using an inherently spatial device, such as a touch pad, to represent a logical structure is possible. The usability study demonstrated that non-speech audio cues work best when they are given as direct feedback to a user's motor behavior, e.g., the clicking sounds used to differentiate sectors on the touch pad were sounded only as a user moved a finger past a boundary. In contrast, the sounds used as audio progress markers were not mapped to their underlying meaning.

We also achieved the goal of building a system that is both simple and inexpensive. The touchpad cost less than $30 US dollars, and all of the software support technologies are available via free downloads from the web. Several smaller goals were not achieved. The text navigation scheme needed for supporting auditory reading is too confusing. We also need to develop a better method for communicating reading progress.

Acknowledgements

We would like to thank David Goldfield from Abilitech Corporation and George Conway, Director of Human Services for the Philadelphia Associated Services for the Blind for their contributions to this work. We are grateful to the Calhoun Endowment from the Drexel University School of Biomedical Engineering, Science and Health Systems for financial support.

References

Martial, O., Dufresne, A. (1993). Audicon: easy access to graphical user interfaces for blind persons—designing for and with people. *Proceedings of the Fifth International Conference on Human-Computer Interaction*, pp. 808-813.

Morley, S. (1998). Digital talking books on a PC: a usability evaluation of the prototype DAISY playback software. *Third Annual ACM Conference on Assistive Technologies*, pp.157-164.

Morley, S., Petrie, H., O'Neill, A-M., McNally, P. (1999). Auditory navigation in hyperspace: design and evaluation of a non-visual hypermedia system for blind users. *Behaviour & Information Technology*, 18:1, 18-26.

Mynatt, E. D., Weber, G. (1994). Nonvisual presentation of graphical user interfaces: contrasting two approaches. *Proceedings of ACM CHI'94 Conference on Human Factors in Computing Systems, pp.* 166-172.

Ramstein, C., Martial, O., Defresne, A., Carignan, M., Chassé, Mabilleau, P. (1996). Touching and hearing GUI's: design issues for the PC-Access system. *Second Annual ACM Conference on Assistive Technologies*, 2-9.

Weber, G. (1993). Adapting direct manipulation for blind users. *Proceedings of ACM INTERCHI'93 Conference on Human Factors in Computing Systems, Adjunct Proceedings*, pp. 21-22.

CULTURAL, LEGAL, ETHICAL AND SOCIAL ISSUES

ELAI: An Interface "Model"
Towards Fostering Universalization of Interfaces

M. V. Ananthakrishnan

Tata Consultancy Services, Nortel Bldg., Raheja Estate, Borivili(E)
Mumbai 400 066 (INDIA)

Abstract

Universal interfaces that are all pervading across languages and cultures are as elusive as ever. Near solutions have been attempted through similar layouts using colours, icons etc. in keeping with local cultures. Hofstede has talked about five dimensions of culture that influence interfaces. But, the move towards universalisation of interfaces can only be possible through selection/design of proper icons and metaphors. The current paper talks about an ergonomically laid-out application interface (ELAI) that could possibly give cues to the development of the "elusive" design. The paper details out (a) the features of the "interface", (b) the positioning of items on the interface and (c) the ergonomic features involving human capabilities.

1. Introduction

The Universal Interface.... What is it after all? Easy to define... one that would be usable for any lay user form any corner of the world. "Unfortunately, only a few researches in the field have focussed on the social aspects of the user modeling to analyse and understand the relationship between user behaviour and motivation" (Inaba, 1996).

Globalisation of business, the mixed "advent" of e-business, the mind-boggling advances in software and hardware and the "intrusion" of information technology into every aspect of human life and well-being (health, wealth and character)... all are looking forward to the panacea... in the form of an universal interface. HCI, which acts as the intermediary between the Application Developer and the Interface Designer, takes into account all the aspects viz., physical, psychological, cultural, social and even aesthetic aspects in total interface design.

Thus, with all the where-with-all, is the universal interface still elusive? Apparently, it is so. So, "talking" interfaces.... interfaces that contain icons and metaphors, could possibly be the panacea to this haunting issue of developing universal interfaces.

2. Review of select studies and current issues

According to Staples (1993), " the relatively brief history of graphical user interfaces has already been marked by an increasing tendency toward 'realism' in the representation of interface entities".

"Interaction metaphors will be closer to reality, providing means of interaction similar to the users' everyday tasks. Multi-modal interaction will be used to diversity in requirements due to the different culture of the user" (Stephanidis et al., 1995). Continuing, Stephanidis et al.(1995) mentions that "metaphors from the real world have been used to model the application in two-dimensional space, using 'virtual' objects".

Russo (1993) brings out vividly the issue of problems confronting 'internationalisation' of interfaces. He warns about the wide variations and levels of acceptance in virtually every feature of the interface, be it the form of writing or colours or even the perceptions of icons/metaphors. Cultural appropriateness relates to how cultural differences have an impact on people and their behaviour. Future research will include data collection modes such as interviews, direct observation of user behaviour and focus groups... and a research project to evaluate users' understanding of interface metaphors across several cultures.

Yeo (1996) talks about overt and covert factors with reference to the 'internationalise or not' decision on software. According to him, overt factors are the tangible ones viz., date, time, telephone number, address formats, order sequence, reading and writing direction and currency, to name a few. The covert factors, on the other hand, are intangible and would depend on culture or special knowledge. Typical covert factors would be colour, icons, metaphors and mental models. Therefore, any attempt to internationalise through icons, metaphors etc. could be 'suicidal'.

The design of a Cultural User Interface (CUI) (Yeo, 1996) necessarily calls for the application to be separated into the functional component and a user interface component. A recent example is the Alpha UIMS "that allows an application to be separated into three components encompassing the application, display technology and user interface" (Klien, 1995).

Looking at the brighter side of the move toward the internationalisation of interfaces, Griffiths et al. (1995) observes: "by doing so, we are also preserving a wider variety of perspectives, potential insights and solutions to the world's problems".

3. Hofstede's dimensions of culture

Hofstede (1997), the Dutch anthropologist, carried out an extensive study on IBM employees across 53 countries, the analysis of which resulted in his epoch-making identification of the five dimensions of culture viz., Power Distance, Collectivism vs. Individualism, Femininity vs. Masculinity, Uncertainty Avoidance and Long-term vs. Short-term orientation. So, do these have a role to play in the design of universalisation of interfaces? Yes and no. Can an interface ignore the five dimensions of culture and be accessible to all? ELAI could possibly open up ideas for such attempts.

4. ELAI

Elai is the acronym for Ergonomically Laid-out Application Interface. Elai, in Tamil (the language of the state of Tamilnadu in South India), stands for a leaf, the banana leaf in this specific metaphor. Further, the social scenario used here is the typical wedding feast, served after the marriage rituals are completed. The guests squat on the floor, in parallel rows back to back, and a banana leaf is placed in front of each one of them, with the tip of the leaf pointing to the left-hand side of the guests.

Figure 1: A typical wedding feast scenario – a guest squatting in front of a banana leaf

Figure 1 shows a typical guest squatting on the floor. With the banana leaf placed in front of her, the various items are served on the leaf

a) in a definite sequential order
b) in particular positions on the leaf

The guests are expected to "refrain" from commencing their eating until the principal item (rice) is served. Once the rice is served, followed by a small amount of clarified butter (called "nei" in Tamil), the guests offer their thanksgiving to God and then begin to eat. Figure 2 shows a leaf with the various items served at this stage. The analysis of the leaf, with its various "divisions" viz., and eating(working) area, is shown in Figure 3.

5. The interface: an analytical study

5.1 The interface per se

A close study of Figure 3 will bring out the reasons for the selection of such a leaf and shape. The leaf is biodegradable and can be disposed after use, without polluting the environment. In fact, even all the eatables served are all biodegradable.

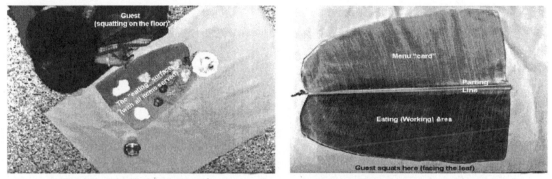

Figure 2: The fully-laid-out leaf (as seen by the guest)   Figure 3: The principal 'divisions' of the leaf

The rib of the leaf is the "parting" line between the "working" area and the menu "card". This is analogous to
a) the Menu Card that one gets when eating in a restaurant
b) the Typical Windows interface, where the various "tools" are available for use on the "development" area

The various items are served in "sampling" quantities to help the guest taste each item and decide on an ABC classification of the items as "like most", "like" and "don't like". This help in avoiding wastage. Further, it is akin to the Western "buffet" system except that in this case, the "items" move while the guests are stationery. In fact, each item or a group of similar items is handled by individual "servers", who continuously move around on a round-robin basis.

All the features highlighted above are the ones that go on to make the interface "talk". As a result, "unfamiliar" guests (the ones who possibly are not familiar with the local customs, a fact that is not uncommon in cosmopolitan cities like Mumbai, Kolkata, New Delhi, Chennai and Bangalore in India), have to just observe their more "familiar" neighbours and "ape" them without any difficulty. Hence the emphasis on the universality of such an interface.

5.2 The interface : a scientific analysis

Figure 4 gives a close-up of the fully laid out leaf, just prior to the guest commencing his/her eating. Each item has a specific position of the leaf, besides the broader classification mentioned earlier in Figure 3. The unique positions are decided by the
a) access sequence, as programmed by the "culture" factors
b) access based on need

1040

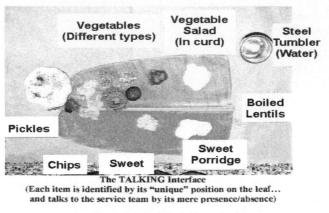

Vegetables
(Different types)

Vegetable
Salad
(In curd)

Steel
Tumbler
(Water)

Boiled
Lentils

Pickles

Sweet
Porridge

Chips   Sweet

The TALKING Interface
(Each item is identified by its "unique" position on the leaf...
and talks to the service team by its mere presence/absence)

Figure 4: The fully laid-out leaf

5.2.1 The "culture" factors

The culture factors, although so termed, have scientific reasons. For example,
a)   the sweet porridge on the right hand bottom corner (of the leaf)
b)   the pickle on the left hand top corner
c)   the tumbler of drinking water on the right hand top of the leaf

At this stage it would not be out of place to mention that the guests necessarily use their right hand for eating. The left hand in invariably used (i) for picking up the tumbler and/or (ii) for non-verbal communication with the service personnel e.g., saying "some more", "enough", "No, thank you", "the neighbour wants" etc. Therefore, to ensure a "sweet" commencement to the feast, the sweet porridge is placed on the right hand bottom corner of the leaf... a position where the hand will naturally go first, ergonomically speaking!

The pickle, normally hot and spicy, is located on the left-hand top corner. This is done to restrict access to it till the logical requirement arises, else one would end up "screaming" after the "pickle" starter!. The water tumbler is placed on the right hand top corner (with respect to the guest)..... to make sure that the guest does not access it unless it is absolutely necessary (say, a hiccup or a temporary "choke").... Else he/she would land up with more of liquid than solid!

5.2.2 Access based on need

The primary area of action is essentially the central part of the lower half of the leaf (figure 3). As such, all items are distributed uniformly around this "area". Further, the guest is free to "taste" the samples before going in for refills. However, this becomes evident when the "treat" of the feast, the sweet porridge, is served in large helpings. The guest(s) clear up as much of the "working" area to facilitate more "helpings". Using the GUI language, the various items are reduced to "icons" and the current application area is made "full" screen!

5.3   The power of the universal interface

The interface metaphor discussed in this paper is unique in that it takes care of the total communication between the guests and the "service" team. It has no boundaries and even a "stranger" guest (one new to the culture) is at home. The stranger has to just seat himself besides a guest (who is aware of the system) and "ape" him right through the process... and that too successfully!

The power is vividly illustrated through the Figures 5 and 6. Figure 5 shows certain areas of the leaf as empty i.e., the items there have been consumed and are possibly awaiting a refill! The service team, on its round-robin tour, observes this.... And lo and behold!, someone arrives with the item and refills (Figure 6). In addition, the guest, can even indicate this refill request by pointing out the item (without even caring to know its name!) on the leaf. In no time, a service team member will arrive with the choice item!

The "TALKING" Interface
(empty spaces on the leaf 'talk' to the 'service' team...)

Figure 5: The empty spaces 'talk'

The "TALKING" Interface
(...inviting the 'service' team to serve the 'exhausted' items, if desired by the guest)

Figure 6: The system's reply ... Refill on!

6. ELAI as an universal interface "prototyping model"

This typical social function (marriage feast) in India has been selected because it illustrates the existence of universal interfaces as also the relevance of ergonomics in layout and design. All this is possible because of a new feature in Interface Design, which the author introduces viz., TALKABILITY ... a characteristic as important as the familiar USABILITY. Putting it more crisply, any universal interface will have to heavily depend on "natural" interfaces and possibly move away from the too familiar but "painful" Windows (replete with buttons, standard icons and the like!).

This example should possibly trigger off the thinking in terms of global interfaces/icons that would help people use software/applications across countries, cultures and societies.

Acknowledgements

The author wishes to thank the Tata Consultancy Services for the opportunities provide to handle the GUI lectures as well as responsible for development of computer-based training software. These responsibilities have indeed motivated the author to try this approach in introducing "natural" interfaces and sustain it with the continued contributions from the programme attendees.

References

Hofstede G. (1997). *Cultures and Organisations : Software in the Mind*. New York: McGraw-Hill

Inaba M. (1996). *User Modeling based on Cultural Theory, a project in the Computer Science Department of the University of Hawaii*. Manoa.

Klein D. (1995). Developing Applications with the Alpha UIMS, *Interactions*, 2(4)

Russo P, & Boor S. (1993). How fluent is your interface?, *INTERTECH' 93: Human Factors in Computing Systems*.

Staples L. (1993). Representation in Virtual Space : Visual Convention inn the Graphical User Interface. *INTERTACH' 93, Human Factors in Computing Systems*.

Stephanidis C, & Sfyrakis M. (1995). *Telecommunications for all* (Ed: P R W Roe), ECSE-EC-EAEC, Brussels.

Yeo A. (1996). Worldwide CHI: Cultural User Interfaces, A Silver Lining in Cultural Diversity, *SIGHI*, 28(3).w

The legislative impact in Australia on universal access in telecomunications

Gunela Astbrink

GSA Information Consultants
15/12 Woodstock Avenue, Taringa Qld 4068, Australia
Ph. +61 7 3876 0880
Email: g.astbrink@gsa.com.au

Abstract

This paper will provide an overview of the impact that legislation has had on the universal access to telecommunications products and services for people with disabilities in Australia. While there is a long way to go before universal access is achieved, some successes will be highlighted in this paper. Legislation to be discussed includes the Disability Discrimination Act 1992 and the Telecommunications Act 1997. Discrimination cases brought before the Human Rights and Equal Opportunity Commission which have influenced access will be analysed. The importance of consumer involvement and representation to government and industry in achieving better access will be described. The work being done by industry associations will also be featured with regard to accessible products and web sites. The paper concludes with a note of warning that, without clear government and industry policies, based on legislative and consumer input, there is always a threat that accessibility will deteriorate.

1. Introduction

The provision of improved access to telecommunications for people with disabilities in Australia has been on the disability political agenda for the past 20 years. It is instructive that it was in 1981, in the International Year of Disabled People, that the first documented attempts took place to research the accessibility requirements to telecommunications for people with disabilities. (Wilson & Keating, 1981) This report concentrated mainly on detailing different types of disability and the particular equipment required to meet their needs. It was not until 1988 that an academic study by Newell highlighted the social aspects of disability and that functional requirements by people with disabilities for telecommunications products should be considered rather than focussing specifically on certain types of disability. (Newell, 1988)

However, it has been only in the past five years that significant changes in legislation has started the move towards universal access in telecommunications. While current legislation, the resultant regulatory framework and the activities of government and industry do not by any means reflect what people with disabilities would define as an information society for all, the steps taken are still significant.

This paper will outline the successes by discussing the impact of legislation especially the Disability Discrimination Act 1992 and the Telecommunications Act 1997. It will concentrate on how particular types of telecommunications technology have been made more accessible through the efforts of consumer input into the reform process.

However, it is important to note that the aim is not only to provide specialised equipment but also to move towards a system where the majority of people with disabilities will have easier access to a range of products and services. In other words, products and services are designed for a larger proportion of the population which includes people with disabilities.

2. Legislation

The Disability Discrimination Act 1992 (DDA) and the Telecommunications Act 1997 are inextricably linked in a number of ways. Complaints of discrimination under the DDA influenced the insertion of particular clauses in the Telecommunications Act. The Telecommunications Act also referred back to the DDA with reference to the supply of services to people with disabilities.

2.1 Disability Discrimination Act 1992

The impact of the Disability Discrimination Act 1992 (DDA) was to be significant for improved access to telecommunications. The DDA is broad-based legislation with few references to telecommunications. Complaints of discrimination lodged with the Human Rights and Equal Commission are mediated and if not successful, legal decisions are handed down. Most notable was the Scott vs. Telstra case in 1995 where a Deaf person stated that he was being discriminated against due to Telstra failing to supply a text telephone for him to make use of the telecommunications network. In a landmark decision, the Human Rights and Equal Opportunity Commission found that Scott had been discriminated against, leading to Telstra, Australia's major carrier, initiating a text telephone provision program. (Bourk, 1998) This has benefitted Deaf, hearing and speech impaired people. What is just as interesting is the increased usage of the network and subsequent income to the carriers once there were more users with text telephones.

More recently, in 2000, was the Maguire v. SOCOG case. The Sydney Organising Committee for the Olympic Games (SOCOG) was found to have discriminated against Bruce Maguire who is blind, that after repeated requests, it continued to provide a web site for the Sydney Olympic Games of which key parts were not accessible for a blind person using screen reading software (HREOC, 2000). While the web site was not changed in time for the Olympic Games, this particular case is the first in Australia and amongst the few internationally which clearly states that discrimination has occurred if a web site is not accessible.

It is preferable for progress to be made other than through discrimination cases. To guide this process, the Disability Discrimination Act encourages organisations, both government and non-government, to develop a Disability Action Plan. An Action Plan provides details on how an organisation will improve accessibility and includes an implementation plan and timeframe. An advantage for an organisation in having an Action Plan is that it decreases the likelihood that a discrimination case can be brought against it if it has clearly shown that plans are underway to implement particular accessibility strategies. Telstra was the first major corporation in Australia to lodge a Disability Action Plan with the Human Rights and Equal Opportunity Commission. In 2000, Telstra lodged its second Action Plan after evaluation of its first Plan had been completed (Telstra, 1999). Just previously, Cable & Wireless Optus lodged its first Disability Action Plan (Cable & Wireless Optus, 1999). This means that Australia's two largest carriers have a set of guidelines in place on improving accessibility. However, there are many issues outstanding which still need specific consumer input and consultation.

2.2 Telecommunications Act 1997

Just as significant in moving towards an information society for all has been the Telecommunications Act 1997 and the Telecommunications (Consumer Protection and Service Standards) Act 1999. This legislation is very specific and was enacted as a result of increasing liberalisation and competition in the telecommunications market. Of relevance are the sections on Universal Service Obligations, the National Relay Service and Industry Development Plans as well as the Telecommunications (Equipment for the Disabled) Regulations 1998.

It is important to note that, under the Universal Service Obligations, the supply of a standard telephone service does not limit itself to a voice-based service but includes customer equipment needed by people with disabilities. It refers to the Disability Discrimination Act as a compliance measure. The Scott vs Telstra case would have had a distinct influence on this inclusion in the Telecommunications Act.

The concept of universal service can be portrayed in many different ways. An influential report by the Consumers' Telecommunications Network (Wilson & Goggin, 1993) stated that there are five elements to universal service:
" 1. universal geographical availability
   2. universal accessibility
   3. universal affordability
   4. universal technological standard
   5. universal telecommunications and participation in society" (p. 27)

This report recommended that universal service be recognised as a social and economic objective and that it go beyond the more traditional concept of universal service usually defined as merely geographical availability. This has only been achieved to a very limited degree.

The universal service provider has, until the end of 2000, been Telstra. However, under amendments to the legislation, regional universal service providers will operate in three pilot regions in Australia based on a tendering process. Universal service providers are obliged to offer a disability equipment program and consult with consumers about the equipment to be supplied as specified by the Telecommunications (Equipment for the Disabled) Regulations 1998. Telstra has been providing disability equipment but, with a number of different carriers and the advent of regional universal service providers, changes are expected in the method of provision of equipment.

The National Relay Service for text telephone users is also carefully outlined in the Act. This is a twenty-four hour service funded by the federal government and run by an independent organisation, the Australian Communication Exchange (ACE) but with carefully constructed performance measures. The government regulator, the Australian Communications Authority oversees the National Relay Service and conducts an annual Consumer Forum to ensure that ACE is delivering the service as required.

Many new carrier licenses have been granted over the past few years. Amongst the conditions for the granting of a license is the provision of an industry development plan under the Act. This specifically mentions the requirement for the reporting of a carrier's relationships in connection with the production and supply of equipment for use by people with disabilities and the reporting of research and development to address the needs of people with disabilities. There is much more that could be done in this area as research and development impacting on people with disabilities has been quite limited over many years.

The Telecommunications Act states that consumer representation may be funded by the federal government to ensure that consumers are informed about telecommunications developments and that consumer interests are clearly presented to government and industry. This has become particularly important with the deregulation of the market and the many changes occurring in the provision of products and services. Three organisations are currently funded to represent residential consumers (Consumers' Telecommunications Network), small business consumers (SETEL) and people with disabilities (TEDICORE). Smaller amounts of funding have also been allocated to peak disability organisations to cover costs of representation activities.

3.   Towards universal access

A number of measures have been taken by industry to accommodate accessibility in the development of products and services largely as a result of their legislative requirements.

The Australian Communications Industry Forum (ACIF), the industry self-regulatory organisation, has established a Disability Advisory Body comprising representatives from peak consumer-based disability organisations. This Body provides input on industry codes and standards under development and their potential impact on people with disabilities.

The federal government through the Australian Communications Authority requested that ACIF develop a Disability Standard for telecommunications equipment to ensure improved accessibility. Industry and consumer representatives spent eighteen months debating various accessible telephone features for inclusion under the Standard. This Standard would be binding under the Act and thus have a major impact on the import and manufacture of equipment in Australia. Industry was concerned, as usual, about the potential additional cost of adding accessible features to equipment. Unfortunately, there was little understanding of the general benefits to many segments of the population of incorporating accessible features in equipment. However, the Standard does state that there should be a raised mark on the "5" keypad for people with vision impairments and that telephones have inductive coupling for people with hearing impairments.

The Australian Telecommunications Industry Association (ATIA) has, together with consumers, set up the Disability-Industry Partnership. This informal group comprises representatives from manufacturers and distributors as well as disability organisations. Its current project is the development of a web-based database which will act as a public compliance instrument for companies' products in order to meet the requirements of the Disability Standard. However, the database will extend to list products which meet a specially-developed set of accessibility criteria. These criteria have been developed by consumer representatives based on international accessibility guidelines. It is believed that this type of initiative is unique internationally. It is hoped that by using the accessibility criteria, companies will import, develop and promote more products which have accessible features.

The federal government, in 2000, instituted its Government Online strategy. This states that all federal government information of a public nature shall be available through the internet. One section of the strategy specifically states that all web sites shall meet the W3C guidelines on Web Content Accessibility. Interestingly, the rationale for this was that it would benefit people with disabilities as well as people in rural areas who had slow connections. People in rural areas therefore should not have to cope with graphics and plug-ins which reduced the speed with which they could access the information. Thus, while the W3C guidelines were principally designed for improving access to people with disabilities, they can also be beneficial to other user groups. This clearly indicates that improving access to people with disabilities improves access generally.

4. Inclusive design

Inclusive or universal design can be defined as helping: "to simplify life for everyone by making products, communications, and the built environment more usable by more people at little or no extra cost." (Center for Universal Design, http://www.design.ncsu.edu/cud) There are two parts to universal design. The first is to design products which are flexible enough, as is commercially practical, to be directly used by people with the widest range of abilities. The second is to design products so that they are compatible with the broadest range of assistive technologies for people who can not use or efficiently access the products directly. However, there still needs to be a recognition that, in some cases, provision should be made for people with disabilities to access additional equipment and services. This has been the case with the cochlear car implant adaptor. The importance of inclusive design is that a product is designed in a flexible manner to accommodate more people than previously.

The Trace Center based at the University of Wisconsin-Madison in the U.S. has conducted a large amount of work with industry to develop inclusive design from a theory into practice. Many examples are given of people who are not disabled but may need the same functions to use a particular product as people with disabilities. For example, people may wish to use a mobile phone menu whilst driving and thus require voice output. A mobile phone may be used in a noisy environment and thus require enhanced volume. These functions benefit vision and hearing impaired people respectively.

In Australia, seminars on inclusive design have been held with Telstra and a telephone designer. One positive result is the establishment by Telstra of an Accessibility Centre within the Human Factors team at the Telstra Research Laboratories. This Centre will advise and support Telstra staff in the development of accessible online services. There have also been discussions with major IT research institutions to incorporate inclusive design into the product design cycle. This is a long-term process but it can have significant results if products become more generally accessible not only to people with disabilities but the population as a whole.

5. Consumer consultation

Consumer consultation and involvement is vital for the achievement of better access to products and services. Through the Australian legislative process, telecommunications companies and in particular the universal service provider are required to have formal avenues in place for consumer consultation. Telstra organises twice-yearly Disability Forums and has set up a consumer advisory group for its disability equipment program. This is based on requirements under the universal service obligations of the Telecommunications Act.

As well, Cable & Wireless Optus has established a more broadly-based Consumer Liaison Forum comprising consumer representatives from a wide variety of consumer organisations.

Again, on a broader basis, Telstra has its Consumer Consultative Council with representatives from a wide range of organisations representing ethnic communities, isolated children in remote regions of Australia, financial counsellors, older people, women, people with disabilities, Indigenous people, internet users, small business consumers and young people. The Council is co-chaired by a senior manager of Telstra and a consumer representative. There is a separate secretariat to support the work of Telstra and one to support the work of consumers in the development of policy through the Council.

Goggin & Newell (2000) state, with regard to consumer consultation, "As with any relationship, the nature of that benefit will always depend upon people willing to get together, acknowledge difference, and yet also to find a common ground." (p. 22)

6. Conclusion

The impact of legislation has enabled more active consumer representation on vital issues which in turn has led to more regulatory steps taken to ensure that people with disabilities have improved opportunities of achieving better access to products and services.

However, a note of caution is needed as progress can be quickly reversed if there is an overarching change in policy and attitude by government, regulators and industry. It is therefore vital to continue using legislation to help achieve universal access to telecommunications in the future.

References

Astbrink, G. (1995). An overview of telecommunications and disability in Australia. In *HFT '95, 15th International Symposium on Human Factors in Telecommunications*. Telecom Australia: Melbourne.

Bergman, E. & Johnson, E. (1995). Towards accessible human-computer interaction. In *Advances in Human-Computer Interaction*. Vol. 5. Ablex Publishing: New Jersey.

Bourk, M. (1998). *Universal service and people with disabilities: An analysis of telecommunications policy making from 1975-1997*. Unpublished master's thesis. University of Canberra: Canberra.

Brandt, Å. (1995). *Telephones for all: Nordic design guidelines*. Nordic Committee on Disability: Århus.

Cable & Wireless Optus (1999). *Disability Action Plan: Achieving access for all*. Cable & Wireless Optus: Melbourne.

Gill, J. (1999). *Telecommunications - Guidelines for accessibility*. COST 219bis: London.

Goggin, G. & Newell, C. (2000). Telstra and consumers: The TCCC and the future. *Telecommunications Journal of Australia,* 50 (3), pp. 19-22.

Human Rights and Equal Opportunity Commission (2000). *Bruce Lindsay Maguire v. Sydney Organising Committee for the Olympic Games (Respondent)*. <http://scaleplus.law.gov.au/html/ddadec/0/2000/0/DD000200.htm> Access date: 6.2.00

Newell, C. (1988). *Australian telecommunications and disabled people*. Unpublished MA (Hons.) thesis. Department of Science and Technology Studies, University of Wollongong: Wollongong.

Telstra (1999). *Telstra's second Disability Action Plan 1999-2001*. Telstra: Melbourne.

Wilson, D. & Keating, C. (1981). Disabled people and telecommunications. In *Telecommunications Journal of Australia* 31 (1), pp. 36-43.

Wilson, I. & Goggin, G. (1993). *Reforming universal service: The future of consumer access and equity in Australian telecommunications*. Consumers' Telecommunications Network: Sydney.

Subjects, Subjectivity and Privacy in the Global Network

Davide Calenda[a], Dino Giuli[b]

[a]Political and Social Science Department, University of Florence
Via Valori 9, 50132, Firenze, Italy
[b]Electronics & Telecommunications Dept, University of Florence
Via S. Marta 3, 50139, Firenze, Italy

Abstract
With this paper we want to stress the main critical elements of the current Network evolution trend, which influence the way through which final users can exploit their subjectivity in Network interaction and relations. Specific attention is devoted to the Network-subject[*] behaviour related to final user privacy. We will stress how relationships generally result undermined by asymmetries in the information management, which condition the building of trusted and stable relationships among the subjects. Proper solutions aimed at re-balancing power among Network subjects, according to final user subjectivity, must take into account the dynamism that properties- requirements - such as negotiation and self-determination can offer to the entire information process related to Network interaction and relations.

1. Introduction

Subjectivity is treated here as the end-user requisite and possibility to consciously and autonomously express and represent his attitudes, values, preferences and agreements in relations with others Network subjects. It thus concerns the degree of control that the end-user has on information processes, related to his/her activities carried out throughout the Net. Subjectivity thus becomes a requisite to be exploited in the Network, so as to support the symmetric evolution of its architecture, with respect to the possible relations and interactions which may occur among users, be they final users or providers. In this paper three main dimensions of subjectivity are stressed. i) User-identity representation in cyberspace, concerning endogenous subjective requisites and elements that structurally configure relationships in the Network. ii) Personalized interaction in the Network that concerns both user and provider attitudes, types of behaviour and requisites. iii) Power in information management as regards the influence of technology, and design as well as subject role and behaviour in structuring hierarchies of power and interests in Network interaction. We therefore envisage, in an ideal Network environment, that end-user should be able to negotiate with autonomy, in a consensual and effective way, finalising his/her actions so as to establish personalized and understandable relations and interactions, since this appears to be a requisite for a balanced, and hopefully a subsequent stable, development towards the global Network (Giuli, 2001a, 2001b, 2001d). In this paper, we try to analyse the above-mentioned dimensions, considering privacy as a basic issue, since it conditions the structural relations among the three aforementioned aspects. We will focus on approaches adopted by different subjects with respect to privacy regulation, trying to highlight critical elements of debate that are closely connected to the exploitation of user-subjectivity in the Network.
The present paper reports on specific results of the current research framework within B.E.S.T. Beyond Internet open framework program (Giuli, 2001a). While a general outline of such research is provided by a companion paper (Giuli, 2001d), so as to complete the overall picture of the reviews available on such research, focus on further specific and related subjects is made in other companion papers (Pettenati et al., 2001; Bussotti et al., 2001; Vannuccini et al., 2001).

2. End user representation in a new frame

In the early stages of the Internet, the dynamics of the Network interaction process were more readily understandable. Users often interacted with each other mostly in an informal way, by just sharing technological and intellectual competencies. At that stage, users at that time, were mainly Network professionals, and were directly concerned with its evolution. Experiences were mainly topic-oriented, especially for research purposes and in order to support communities which generally ensured the individual's satisfaction through highly professional Network support. Communities made it possible to aggregate and articulate interests according to representations that users managed with respect to specific services. Trust and reputation were important factors, as regards both individuals and services, that spread throughout cyberspace. Their diffusion, as constitutive relational elements, was reinforced by the possibility to recognise Network user identities which were usually strongly connected to that of the real world. The first virtual community experiences, show, for instance, that user's mail domains were mainly institutional, and individuals were developing interpersonal relations, just revealing themselves as real known subjects, rather than hiding behind virtual

---

[*] The term "subject/s" is used interchangeably with "individual/s" and providers throughout this paper.

unknown identities as often happens today[*]. Final user identity representations were thus quite stable, and though deception often affected searchers in cyberspace, reciprocity in interaction ensured a more structured relationship. In those contexts, the personalization of relations depended mainly on the extent of the user's emotional involvement. Network tools, rules and functions were more directly understandable than at present, also because it was the end users who were normally involved in their definition and design.

With the widespread diffusion of Network and telematic applications many changes have occurred, and some cleavages, concerning both the subjective and instrumental dimension of network interactions, are rising. Asymmetries and weaknesses, currently emerging in interaction among subjects, reflect the spread of new behaviours and interests that are addressing the way which user subjectivity is conceived and treated. The role that subjective and personal sphere play in Network development, should be understood both as regards cyberspace and user differentiation, which are related with the widespread diffusion of human activities through the network. At the same time, cyberspace seems to be moving toward a more formal and structured space, which is gradually being governed by new paradigms. Such a trend is reinforced by the widespread spawning of commercial providers in the Web-context, as well as by the widespread growth of connected applications, which are influencing the overall architectural design of cyberspace[**]. The management of personal information, specially related with commercial purposes, is even more important for the success of interaction and relation modes, determined by user attitudes, status, preference, privacy, and expectations. For the end-user, the management of personal information together with its representation, bears a strong relevance, given his/her expectation that subjective requirements be actually met. Subjective requirements should now be defined also taking into account the emerging limits as well as the tendencies, both of current technological evolution as well as subject behaviour in the Network. As for providers, the management of such processes currently represents a means to gain user trust in order to achieve stable relations, be they aimed at commercial, informational, institutional or political objectives.

3. Limits and trends for privacy in the current explotation of user subjectivity

Approaches and solutions concerning privacy issues - together with their limits and trends - constitute critical factors which affect the current policies adopted by Network providers as regards user subjectivity. These issues become fundamental once the systemic deployment of human subjectivity profiling is considered as a basic requirement for the provision of symmetry among Network subjects (Giuli, 2001d). Though debate also concerning government control of individuals and society, which radically connotes the issue as being political as well as philosophical, in this paper we principally address the implications on the relationship between the end-user and the service providers. On-line privacy specifically concerns the collection, treatment and use of the personal data which the user discloses directly or indirectly to providers when accessing their services. Problems related to the practices adopted by Internet providers with regard to personal data are well analysed in the literature[*]. We can emphasise that privacy problems revolve around three main issues: i) the provider's approach, through which user personal data requested is not strictly related to the delivery of a specific service (data creep practices); ii) data collection using invisible modalities which make use of spy technologies such as cookies, web bug, etc., iii) use of data for different purposes to those indicated but without the user's previous and/or informed consent. The entire issue of problems related to personal data collection started to become widespread, with the invasive onslaught of commercial providers to the Network, thus influencing both the design architecture and the policies adopted in order to make user behaviour more *searchable* and *trackable* (Lessing, 1999). In order to safeguard users, solutions have therefore been developed, mainly on the basis of two main approaches: Law-Regulation and Self-Regulation[***].

The Law regulation approach normally interprets privacy first and foremost as a right that must be defended in cyberspace just as in real life, through the regulation of both behaviour and technologies. Although its effectiveness has

---

[*] The present trend is also due to the spread of *consumer oriented services*, such as AOL, Yahoo! etc., which provide users with possible anonymous mail domains (Smith et al., 1999).

[**] As of January 23, 2001, there were 35,158,339 domains registered worldwide, with 21,243,588 being ".com" domains on. (http://www.DomainStats.com/01/23/01).

[***] Law-Regulation approach is mainly adopted in the EU, Canada and Australia. Other countries, especially those in Latin America, are presently compiling detailed laws on privacy. Self-Regulation is mainly empowered in the USA where private subjects, such as consumer associations and companies, are called on to guarantee privacy by implementing coding technology and regulating behaviours. A lot of conflict seems to exist between the states that support the two different approaches, but a convergence of ideas seems to be emerging among international organisations. See the *Safe Harbour* agreement adopted in 1999 by EC and USA.

[*] Up-dated information on publications about privacy issues. On-line:
http://www.privacyfoundation.org/resources.html.

still to be proven, Regulation by Law still seems insufficient, in our opinion, to support subjectivity management, in the manner in which we define it throughout this paper. A structural problem is due to the fact that, in reality, personal data interchange already represents the constituent elements of the Network as an individual-oriented space which is increasingly becoming economically viable. A further problem is represented by the fact that the Internet is profoundly changing the space dimension of the privacy right effectiveness, making it more dependent on the variability and mutability of the frameworks within which users interact. However, we must also stress that an important conceptual evolution on privacy on the Network, will be achieved, once jurisdictions, such as that in Germany and other EU states recognize it as the right of the user/individual to self-determine his/her own informational flow process. The concept of self-determination identifies the electronic user's identity as an important social and political right for individuals. Furthermore, self-determination indirectly refers to other concepts – requisites - such as autonomy, negotiability and power, which encompass subjectivity*. Concerning the user autonomy in the self-determination of his/her personal data, it mainly depends on off-line individual cultural and social backgrounds (Pettenati et al., 2001). Yet, autonomy also depends on interaction modes; in other words, it depends on how much autonomy space is provided by other subjects to users during interactions and relations. Providers are acting as if user autonomy could be safeguarded simply by offering customization options to users, or by informing them on privacy and behaviour policies adopted, in an effort to win their trust. Nonetheless this approach fundamentally ignores the dynamic and contractual nature of relations based on subjectivity: autonomy should really be negotiated, if it is to be effective. At present, the effectiveness of user negotiation power, depends more on both provider behaviour and the technology available, rather than on individual decisions; the main space for negotiation that the user has during interaction, is the possibility to move from one site to another if subjective preferences are not matched. This possibility reflects the opportunity that an individual has in a traditional mass-oriented space - such as in mass media broadcasting - by channel zapping. It also highlights a development limit for the present Network environment in favour of a more individual-oriented space for interaction and relations. User power to negotiate individual preferences based on values, privacy, attitudes, and needs, represents an optimal situation that should be fully supported by the Global Network of the future. In the current situation, the final user appears to us as being the weak partner in the relationship with provider, while the latter, continuously increases his power by supporting customized interaction through instrumental and behaviour strategies - user-friendly interfaces, ambiguous personal data collection, profiling etc. The analysis of the relevant treats in the use of such instruments and strategies undoubtedly leads to the evidence of the relational asymmetry between the user and provider over Internet.

4. Relation asymmetries in customization policy

Many Internet service providers are struggling to offer personalized features, in an effort to reduce network user unpredictability, thus aiming at winning user friendship and trust. Commercial providers are currently the ones most interested in personalization, i.e. customization, thus orienting interaction modes by users – be them treated as single users or categories - in accordance with the services and products to be offered. Success in this activity depends mainly on the quantity and quality of the personal data collected**. Generally speaking, apart from the specific problems related both to privacy and self regulatory practice leaded by providers, such trends reflect the way subjectivity is currently perceived, and, hence, treated. Conceptual and technological instruments, identified as the means that, at present, are used to provide personalized interaction, leave little space to user self-awareness and/or the individual's ability to infer knowledge on the related information treatment process***. Commercial providers generally want to *type-set* the user, from an economic point of view, instead of trying to understand *who* the user really is. The main goal of the providers is to discriminate user behaviour in order to discriminate the market. In other words, user interaction with a service provider generally means that the user wanders in a framework in which rules and categories related to the personalization are yet to be defined by the provider. It is, to-date, extremely rare for a user to join the Network and be represented by a profile which (s)he, as an individual, has previously defined. How the personalization is currently

---

* Privacy issues, therefore, include a multiple aspects related to structural and subjective variables (Pettenati et al., 2001). With respect to such emerging complexity, Lawerence Lessing campaigns for a regulation of code that would make it more difficult to search and monitor behavior and personal data. He interprets privacy as a user's exclusive property, which can be accessed by other subjects only by negotiation (Lessing, 1999).

** Personal information interchange is a core trend in the New Economy phase which is characterized by the transition toward a global market where goods and services are increasingly related to individual cultural, social and experiential needs. As Jeremy Rifking points out, individual life acquires economic value (Life Time Value) in the New Economy paradigm, thus, implying the increased value of the control of personal information (Rifkin, 2000).

*** Marvin Minsky treating of machine's design according to user subjectivity stress the importance of self-reflection requisite implying that "our machine must keep records that describe the acts and the thinking they've recently done so they'll be able to reflect on the results of what they tried to do. This is surely an important part of what people attribute to consciousness" (Minsky, 2000).

designed and operated, support the thesis that profiling policy currently performed - especially by commercial providers - tends to implement *normalization* rather than supporting the *diversity* - and privacy - associated with user subjectivity and needs. Consequently, the constituent phase of the interaction and relation process is affected by grounding asymmetries (Giuli, 2001d). Nevertheless, user self-classification with respect to access to specific services and features, represents a key issue to strengthen individual subjectivity and privacy; this is perceived as being especially critical in applications such as electronic citizenship, finance and e-commerce due to the implications they introduce in the private sphere[*]. Empowerment of self-classification implies a high degree of the individual's self-awareness, and it represents a means through which the user could define and control his/her representative *views* in the Network (Pettenati et al., 2001). At present, limitations on such a possibility, lead many users to join the interaction environment, while keeping their own identities secret and using, instead, either false ones or using anonymous services (Oppliger, 2000). Partial solutions, such as costuming services related to user personal-profile management, are provided on the Internet especially by super providers such as Hotmail or Yahoo!. *Passport* service, which is offered by Hotmail for example, gives the user the possibility to define his/her own profile according to predefined choices, including the disclosure of personal data (demographic, financial, etc.). Through *Passport*, a user personal-profile is maintained and stored in Hotmail central database, and it can be disclosed by the single user to other sites that accept *Passport* service. *Passport* is used to support access to services, and it is mainly designed to increase trust and privacy in e-commerce applications. Its diffusion mainly depends on the trust that users place in the provider, especially as far as privacy policy and provider reputation is concerned. The definition of procedures and rules by which the end-user can share the user personal-profile with the provider according to the privacy preference, are still critical issues[**].

Proper solutions should instead help the user in understanding - by means of an instrumental support, even exploiting the available shared-knowledge space - what relational and application views need to be used to adequately place users in different relational and application frameworks during Network interaction (Pettenati et al., 2001). This approach implies that the personal machine should be able to learn from the individual, how to represent subjectivity in specific frames, also by consulting his/her personal profile, which should be exclusively maintained by user or shared with trusted subjects. This requires the individual to be able of dynamically contributing to define - and formalize - his/her identity and subjectivity through a continuous dialogue with the machine and the entire Network space. This means that solutions currently proposed to reinforce the personal information management in the interaction through the Network, should be evaluated on the basis of their ability to solve dialogue gaps among individuals, technology and whole Network resources.

5. The role played by network social subjects in supporting privacy and subjectivity

Privacy solutions proposed by network subjects tend to self-regulate behaviour and technology, and are strictly associated with negotiation and autonomy issues. Social subjects play an important function, i.e. virtual communities and intermediary organizations such as Trust Organizations and consumer associations, whose scope is to support the user in his/her taking control over interaction in which personal data are exchanged. The Virtual Communities that act as Computer-Supported Social Networks (CSSN's), support users in solving many kinds of problems, from the social to the technological ones, and keep them informed about critical issues such as privacy, security and rights on the Internet. Apart of ideal leaded, many virtual communities - that are generally non-profit - pragmatically aim at the single user in structuring his/her own interaction in the Network. Users can thus orient action towards short or medium-term goals. The success of such communities in supporting social organizations in the Network depends first and foremost on the amount of shared interests among members. If the community is well organized, structured and linked, it can offer a social space to the single participant where he can experiment the access and the management of many resources which are useful in moving around the Network. Though doubts have been advanced by some critics respect to the degree of exclusiveness of many communities[***], the expectations about the growth and spread of virtual communities, which are also built by commercial providers in the marketplace, together with the increase in new spontaneous participation spaces throughout the Network, could enable final users to strengthen their position of power in Network interactions. Support could be provided to user by enabling information and technological instruments to allow communities to play

---

[*] Stefan Brands has well developed auto-classification concept in application connected to citizenship and business (Brands, 2000).

[**] Directive members of Yahoo!, treating of personalization policy, stated that the treatment and maintenance of end-user personal data according to privacy and security implies a change of companies' mentality. Personal data "should be guarded just as much as the most secret of trade secrets" (Manber et al., 2000).

[***] The principal factors involved in the concept of exclusiveness include: language (both from a linguistic as well as idiosyncratic point of view) and the competency which is necessary for the newcomer to become a member. These communities often offer membership at a cultural cost which many newcomers may not be in a position to support, hence, such an approach tends to inhibit the potential participation of users in this type of cyber-social space.

their envisaged Network role, as well as by implementing an active political role of users in such communities. The mobilization and the articulation of user sensitivity through communities also means the growth and spread of collaborative mechanisms on the Network; the latter constitutes a useful means by which to orient the user social-action as well as the exploitation of subjectivity. Trust Organizations, such as *BBBonline* and *Trust'e*, also support users by acting as trust intermediaries between service providers and users. Both these organizations enforce privacy policies and promote negotiation by posting sites with *Web Seals* or providing them with privacy programs. Such organizations are normally non-profit consortiums and deal mainly with applications such as e-commerce and finance where trust, privacy and contractual rights are perceived, not only as critical, but also imperative. The effectiveness of such self-regulation instruments depends principally on subject discretion in accepting program, as well as leading rules (Cavoukian et al.). Nonetheless, this actually represents a step forward towards a socio-dynamic (and socio-technological) approach to the issue. User subjectivity must play a major role in future, as regards negotiation power in the rule- definition game. User-subjective preferences could probably be expressed in a direct manner through the negotiation of the interaction modes, or in a indirect way by expressing value judgements on provider behaviour by means of the use of reputation mechanisms, claims etc (Zacharia et al., 2000).

6. Conclusions

Proper enabling technology should, thus, help users even to increase their self-awareness and power during Network interaction. Network-social subjects should be considered and promoted as necessary intermediary Network subjects, on account of their ability to articulate user interests, participate and make representations using collaborative mechanisms. For users, this would further represent an important shared-knowledge space being made available, while also providing then with trusted network subjects. The proper handling of privacy, by means of new enabling technology and operational modes, is a basic requirement for the above objectives. Users should be provided with adequate resources so that they can personally manage the information process as well as provide suitable personal profile representation in Network relationships, in accordance with their own specific subjective requisites. This would represent a means to help overcome the current asymmetry which exists among Network subjects.

References

Giuli, D. (2001a, February). *Background and objectives of the B.E.S.T. beyond Internet framework research program.* Dept. of Electronics and Telecommunications, University of Florence, Internal report DET-1-01.

Giuli, D. (2001b, February). *Subjective information and the subjectivizing gene for a new structure of the instrumental relation and interaction processes in the Global Network.* Dept. of Electronics and Telecommunications, University of Florence, DET-2-01.

Giuli, D. (2001c, February). *Subjectivity, citizenship and Public Administrations in the evolution towards the Global Network.* Dept. of Electronics and Telecommunications, University of Florence, DET-3-01 (in Italian).

Giuli, D. (2001d) From Individual to Technology towards the Global Network. In *Proceedings of UAHCI 2001 Conference*, Louisiana, New Orleans (present volume).

Pettenati, M.C., Giuli (2001). Human Subjective and Relational Profiling Factors in the Global Network. In *Proceedings of UAHCI 2001 Conference*, Louisiana, New Orleans (present volume).

Bussotti, P., Vannuccini, G., Calenda, D., Pirri, F., Giuli, D. (2001). Subjectivity and Cultural Conventions: the Role of the Tutoring Agent in the Global Network. In *Proceedings of UAHCI 2001 Conference*, Louisiana, New Orleans (present volume).

Vannuccini, G., Bussotti, P., Pettenati, M.C., Pirri, F., Giuli, D. (2001). A new Multi-layer Approach for the Global Network architecture. . In *Proceedings of UAHCI 2001 Conference*, Louisiana, New Orleans (present volume).

Smith, M.A., Kollok, P. (1999). *Communities in cyberspace.* New York: Routledge.

Lessing, L. (1999). *Code and other Laws of Cyberspace* (157). New York: Basic Books.

Rifkin, J. (2000). *The age of access: the new culture of hyper-capitalism, where all of life is a paid-for experience.* New York: J.P. Tarcher/Putnam.

Minsky, M. (2000). Commonsense - Based Interfaces. *Communication of the ACM*, Vol. 43, No. 8, 70, 67-73.

Brands S. (2000). *Rethinking Public Key Infrastructures and Digital Certificates – Building in Privacy.* The MIT Press.

Oppliger, R. (2000). Privacy protection and anonymity services for the World Wide Web (WWW). *Future Generation Computer Systems* n.16, 379–39.

Manber, U., Patel, A., Robinson, J. (2000). Experience with Personalisation on Yahoo!. *Communication of the ACM*, Vol. 43, No. 8, 37, 35-39.

Cavoukian, A., Crompton, M. *Web Seals: A Review of Online Privacy Program.* [On-line]. Available: http://www.ipc.on.ca/english/pubpres/sum_pap/papers/seals.htm

Zacharia, G., Moukas, A., Maes, P. (2000). Collaborative reputation mechanisms for electronic marketplaces. *Decision Support Systems*, No. 29, 371–388.

Bridging the Digital Divide:
Case Study of Anti-Exclusion Measures in Ireland

Alexis Donnelly

Department of Computer Science, Trinity College Dublin, Ireland
Alexis.Donnelly@cs.tcd.ie

Abstract

This paper outlines some measures taken to bridge the digital divide in Ireland. The Irish case is interesting due to the existing economic and social conditions – growing social inequality and weak levels of social support, but increasing resources to address them. Ireland's recent economic prosperity offers the possibility of seriously addressing these inequalities as never before. The paper examines some recent policy initiatives in Ireland in the light of several principles that can be used by marginalized groups in addressing exclusion from the emerging information society.

1.   Background

Despite spectacular "celtic tiger"economic performance, a recent report from Ireland's Economic and Social Research Institute indicates that Irish society remains deeply divided and social inequalities continue to widen (Economic and Social Research Institute [ESRI], 2000). National budgets have tended to distribute wealth unequally despite a National Anti-Poverty Strategy and a Combat Poverty Agency to advise government. Ireland spends least on social supports relative to GDP of all countries in the EU. However, the performance of the economy and consequent increased resources provides Ireland an unprecedented opportunity to address social inequalities, including those that arise in the information society.

The Irish Information Society Commission was set up to advise government on measures to prepare Ireland for the emerging world wide information society. While the Commission has a broad remit and a largely commercial focus, some account is being taken of social inclusion in its recommendations. This presentation examines these recommendations and some relevant initiatives in Ireland to bridge the emerging digital divide. Many social inequalities are mirrored in the emerging information society (Kaye, 2000; Policy Action Team 15, 2000; National Telecommunications and Information Administration, 1999). Unless explicit action is taken to address inequalities emerging in the information society, many marginalised groups will be further excluded.

This paper takes the perspective that social divisions progress from separation, to tension and eventually strife and that it is better to promote full participation and inclusion rather than pay the costs of exclusion. This presentation takes the view that an information society that fails to take corrective measures for inclusion is not sustainable. To quote Lincoln, "a house divided against itself cannot stand".

2.   Principles and commentary

In the discussion that follows below, we present some general principles that have been found useful in the Irish context for addressing aspects of the digital divide. The principles were developed by the author (Donnelly, 2000) and were inspired by discussions in the Connected Communities Advisory Group of the Irish Information Society Commission. This group was comprised of academic experts and representatives of the community and voluntary sector working with various socially excluded groups in Irish society. Some of the principles later appeared in the Commission's Third Annual Report (Information Society Commission [ISC], 2000). This paper includes discussion of these principles in the light of more recent Irish and European policy developments.

2.1   Accompaniment principle

Accompany the marginalised group in its struggle for inclusion; do not foist technology upon the group blindly.
If ICTs can be used in a tailored way to support the marginalised group's struggle for inclusion, they are perceived as relevant and the excluded group is motivated to acquire necessary technical skills and mastery. The metaphor of accompaniment on a journey seems appropriate here. In a mutual learning process technical experts acquire knowledge about the group's needs in order to design and deploy ICTs appropriately. The group also learns more about the technology in a way that is deeper than straightforward training – the group acquires technical mastery. The key is to learn the agenda of the marginalised group and work with it (Moggeridge, Plant, and Kamm, 2000).

The Third ISC report recommends clearly identifying the benefits of participation in the information society and making it "relevant to people's lives" (ISC, 2000, p. 64). It also advocates "develop[ing] IT access initiatives through partnership, focus[ing] on shared discovery" (p. 64). Though the report is notably silent about targets and funding, a fund of £2.5 m (Euro 3.17 m) has been established under the Department of Public Enterprise to support adoption of ICTs by the community and voluntary sector (Department of Public Enterprise [DPE], 2001). A recent White Paper on the Voluntary and Community Sector (Department of Social, Community and Family Affairs [DSCFA], 2000) is noticeably non-committal on funding of these needs, or initiatives to support them.

2.2   The Three Ts: Technology, training and technical support

Provide the three Ts in roughly equal proportions, but adapted to local need.
Preliminary findings indicate that inadequate training and technical support can doom an ICT project with marginalised groups to failure (Employment Initiative in Ireland, 1998). When financial resources are scarce, training and technical support are sometimes the first items to be cut from the budget. Training may have to be provided in a non-standard way (for example with groups with physical impairments).

The ISC report recommends providing "technology, training and technical support tailored to local need" (2000, p. 64), though it is noticeably silent on targets, standards and funding sources. However, it remains to be seen whether the Department of Public Enterprise fund will prove useful in supporting this tailored delivery.

2.3   Continual observation principle

Evaluate and research the project throughout its life.
In order to ensure that a project is addressing its social goals, evaluation during the project is vital (Moggeridge et. al., 2000). This provides opportunities for corrective action and supports the mutual learning process. In a wider social context, research to determine the extent and nature of need helps to target resources effectively.

The Third ISC report recommends that "on-going evaluation forms an integral part of all state-funded projects promoting access to ICTs and that best practice is disseminated" (2000, p. 67). Under the heading of research in this sector (p. 72), the report recommends a qualitative analysis of the extent of social exclusion in the information society. A *quantitative* picture would be more useful in targeting resources efficiently. This might be done using the national census or the Quarterly Household Survey, as has been done in the United States (Kaye, 2000). The ISC report recommends that the newly formed National Disability Authority should carry out research on the needs of people with disabilities in the information society, that national programmes of ICT research should include socio-economic research and that technological research should address social needs (p. 67). The report is imaginative in suggesting mainstreaming of research on social exclusion into major national funding programmes and drawing the attention of technology researchers towards social needs, but does not suggest targets, standards or funding. The EU Fifth Framework Programme is one example of a research programme that attempts some social targeting.

2.4   Highlighting principle

Highlight models of good practice.
There is an old Irish proverb - 'mol an óige agus tiochfaidh sé' - which roughly translated means "praise youth and it will blossom". Drawing attention to models of good practice clearly helps to replicate them. However, they also

serve a less widely known motivational function: they inspire similar groups (sometimes in another region or country) to undertake similar initiatives.

The Third ISC report adopts this principle in its recommendations regarding on-going evaluation and "that best practice is disseminated" (p. 67). The report also highlights some innovative Irish projects using ICTs to address issues of social exclusion (p. 70). The Department of Public Enterprise initiative in this area (DPE, 2001) is a welcome development in funding streams for such projects. However, the relative silence of the Department of Social Community and Family Affairs on this issue (DSCFA, 2000) gives cause for concern. A scheme of annual awards for exemplary ICT initiatives among marginalised groups would be a useful highlighting mechanism.

2.5   Universal principle

Apply Universal Design principles universally.
Universal design (UD) is the design of products, environments and communications to be useable by all people, regardless of impairments, to the greatest extent possible without the need for adaptation or special design. The barriers faced by people with disabilities are often addressed by application of UD principles and especially so when applied from the start of a project. Requirements ignored at the start of a project are more expensive or impossible to incorporate later. Unfortunately the UD approach is "more honoured in the breach than in the observance". Delay or selective application amounts to violation of UD principles. The principles are universal, every group can (and should) contribute and the resulting benefits accrue to all. Successful mainstream technologies such as the telephone, the typewriter, the starter motor, the radio pager and speech recognition had their origins in assistive technology. UD is not just a "special needs" issue.

The e-Europe initiative has been criticised for confining discussion of UD principles to a forum involving people with disabilities (Seymour, 2000; European Disability Forum [EDF], 2000). The discussion of the e-Europe project was launched with a discussion document in PDF, a format inaccessible to people with visual impairments. However the Third ISC report was made simultaneously available in HTML and steps were taken to improve the accessibility of the ISC website. Nevertheless, discussion of UD principles was confined largely to the Connected Communities Advisory Group (CCAG) of the ISC and there was no interaction with the other advisory groups (Connected Government, Infrastructure, Legal Issues and lastly Lifelong Learning). The Third ISC report (ISC, 2000)makes several recommendations regarding application of UD principles:
- promotion of UD as a benefit for all, including people with disabilities (p. 66).
- promotion of the WAI guidelines to the private sector (p. 72). It is recommended that legislative measures requiring accessibility be adopted if voluntary compliance does not work. It is also recommended that forthcoming e-commerce legislation include accessibility requirements (p. 75)
- adoption by government of UD principles and the WAI guidelines, particularly in procurement policies (p. 87). Public service tenders should require WAI-compliance with immediate effect and government websites should be level AA WAI-compliant by the end of 2001. The report honestly admits that earlier government web-publishing guidelines (Interdepartmental Group, 1999) including accessibility have not been adhered to (p. 85).
- promotion of better design of e-commerce websites (p. 92). It is noted that accessibility and ease of use (components of design excellence) give commercial advantage in the competitive e-commerce marketplace.
- funding of research on the social aspects of the Information Society within mainstream research and development programmes (p. 97).

While the above recommendations demonstrate some integrated strategic thinking on the part of the ISC and development of its strategy from the earlier IT Access for All report (ISC, 2000a) (that did not mention WAI guidelines) some opportunities for policy development were not taken. The laissez-faire approach in relation to accessibility of public sector websites may continue the proliferation of inaccessible sites that will require later expensive repair. Current equality legislation in Ireland unfortunately allows a service provider to evade access requirements unless the cost of access is other than nominal. The laissez-faire approach is also inconsistent with the later recommendation for access requirements in e-commerce legislation. The potential for applying UD principles to forthcoming e-commerce legislation and including accessibility requirements does not appear in the legislative section of the Networked Economy chapter. Similarly, no requirements for accessibility appear among recommendations concerning infrastructure developments in that section of the same chapter. There are no recommendations for the telecommunications regulator regarding ensuring accessibility. Relevant areas include interoperability of SMS services of competing service providers, accessibility of directory services to customers with disabilities and accessibility of equipment (telephones, public internet access points and information kiosks). Finally, the chapter on Lifelong learning includes no measures regarding the application of UD principles to its recommendations despite the fact that unemployment rates among people with disabilities are 70-80% and these groups are often excluded from mainstream education.

2.6   Urgency principle

Take informed action as soon as possible.
The urgency for action to address social exclusion is clear: failure to act means that divisions will persist and, in the case of the information society, barriers will remain and divisions widen. When a recommendation is made, specific and measurable targets should be set and adequate resources should be allocated. In many cases history shows that is cheaper to have an excluded group participating equally in the labour force than to pay the costs of supporting them (or of not supporting them) on the margins of society. In relation to the e-Europe Initiative, people with disabilities express dissatisfaction with absence of measurable targets and deadlines and with the pace of change, they offer valuable suggestions on possible courses of action Seymour 2001; EDF 2000).

The third ISC report is very weak on setting specific targets and deadlines in most areas. Most indicators of progress mentioned in the ISC's Benchmarking Report (ISC, 2000b] refer to commercial and infrastructural measures (with the exception of the percentage of Community and Voluntary groups using the internet). The suggested source for this data (Department of Social Community and Family Affairs) may ignore those groups active in health and education areas. The percentage of the adult population with literacy problems is mentioned among the knowledge and skills indicators. In Ireland, some 50% of the adult population were found to have literacy problems sufficiently serious to prevent them participating fully in social and economic life (OECD 1999). The Irish government has recently announced a major national adult literacy and numeracy programme costing £73.6 m (93.4m Euros) covering the period 2000-6. There is no suggestion to measure the percentage of public or private sector websites complying with WAI Guidelines.

3.   Discussion and conclusions

The principles discussed in the previous section have highlighted issues that appear to arise in other national contexts. The availability of internet connections of sufficient bandwidth and at sufficiently low cost are not being deployed equally throughout society and corrective action must be taken.

Financial, literacy and technological awareness barriers must be addressed in a serious, targeted manner if current social exclusions are not to be carried over into the information society. Initiatives that recycle older PCs to disadvantaged groups are a welcome example of overcoming financial barriers, but it may be difficult to provide technical support for some equipment. Also some groups, such as visually impaired surfers, often require the latest equipment to work with their assistive technology. New methods of providing internet access via digital TV, set-top boxes and internet kiosks have received relatively little attention from the universal design perspective, except perhaps recently in some countries with strong human rights / accessibility legislation. It appears that stringent legal requirements are necessary to concentrate manufacturers' minds, although the lucrative "elder market" is attracting the interest of some manufacturers.

The eEurope initiative, while it attempts to raise awareness of universal design, is notably silent on legislative measures that would complement such a move.   Universal Design principles have been applied during the development of W3C standards in the Web Accessibility Initiative (WAI). The WAI content guidelines have received some publicity and are being adopted by governments as an accessibility standard. The eEurope initiative is targeting the lowest (A) level of compliance by April 2001, thought the AA level is considered the minimum required for equitable access. Far more serious is the slow response of manufacturers of authoring software to produce tools that produce WAI-compliant HTML. Stringent procurement policies by large government and other bodies would help to speed the production of such tools. The WAI standards have somewhat unfairly been criticised as leaning too much towards visual impairment at the expense of other impairments, despite the fact that the former group is perhaps most disadvantaged in a largely visual environment. The necessary technical nature of the WAI content guidelines has been cited as a barrier to their uptake by web designers who are generally not familiar with the details of HTML. The provision of further introductory, awareness-raising documents by WAI is a welcome move to address this gap in the literature.

This presentation has examined recent measures taken in Ireland to bridge emerging inequalities in the information society. These developments are discussed with reference to principles developed by the author (Donnelly, 2000) inspired by discussions among an advisory group to the Irish Information Society Commission. Although some general principles have been set forth in the latest report from this Commission (ISC, 2000) and some welcome initiatives have been taken, further ambitious initiatives including deadlines and targets are required if the widening social divisions in Ireland's information society are to be addressed and reversed.

References

Department of Public Enterprise. (2001). *Announcement of the CAIT Initiative (Community Application of Information Technology)*. Press Release, 18 January. (See http://www.irlgov.ie/tec/press01/jan18th01.htm).

Department of Social, Community and Family Affairs. (2000). *Supporting Voluntary Activity: A White Paper on a Framework for Supporting Voluntary Activity and for Developing the Relationship between the State and the Community and Voluntary Sector*. Dublin, Government Publications Office. (also available at http://www.dscfa.ie/dept/reports/volact.htm).

Donnelly, A. (2000). Towards an Inclusive Information Society: Some Principles from the Margins. *Proceedings of the Sixth ERCIM Workshop on User Interfaces for All*. (pp 229-240). Florence, Italy: IROE. December.

European Disability Forum. (2000). *Mainstreaming e-Participation of Disabled People in the eEurope Action Plan: EDF Response to the European Commission e-Europe Initiative*. (Doc EDF 00/07). Brussels, Belgium: EDF.

Employment Initiative in Ireland. (1998). *Getting Connected: Social Inclusion in the Information Society*. Dublin, Ireland: WRC Economic Consultants.

Economic and Social Research Institute. (2000). *Bust to Boom: The Irish Experience of Growth and Inequality*. Dublin.

OECD. 1999. *International Adult Literacy Survey*. Available from Statistics Canada. (http://www.nald.ca/nls/ials)

Information Society Commission. (2000). *Information Society Ireland: Third Report of Ireland's Information Society Commission*. Dublin: Government Publications Office. Also available at http://www.isc.ie.

Information Society Commission. (2000a). *IT Access for All*. Dublin: State Apartments, Dublin Castle, March. Also available at http://www.isc.ie.

Information Society Commission. (2000b). *Benchmarking Ireland as an Information Society*. Dublin: State Apartments, Dublin Castle, May. Also available at http://www.isc.ie

Interdepartmental Group. (1999). *Recommended Guidelines for Public Sector Organisations*. (Report, ISBN 0-7076-6275-3). Dublin, Ireland: Department of the Taoiseach, November. Also available at: www.irlgov.ie/taoiseach/publication/webpg/guidelines.htm

Kaye, H.S. (2000). Computer and Internet Use Among People with Disabilities. (Disability Statistics Report No. 13). Washington, DC: U.S. Department of Education, National Institute on Disability and Rehabilitation Research. (Also available at http://www.dsc.ucsf.edu).

Moggeridge, A., Plant, N. and Kamm, R. (2000, September). Academics and the Voluntary Sector Learning Together: Formative Evaluation in Community Information Systems Projects. Paper presented at *The Sixth Researching the Voluntary Sector Conference*, University of Birmingham, of the National Council of Voluntary Organisations.

National Telecommunications and Information Administration. (1999). *Falling through the Net: Defining the Digital Divide. A Report on the Telecommunications and Information Technology Gap in the United States*. National Telecommunications and Information Administration. Also available at www.ntia.doc.gov/ntiahome/digitaldivide

Policy Action Team 15. (2000). *Closing the Digital Divide: Information and Communication Technologies in Deprived Areas*. London, England: Department of Trade and Industry.

Seymour, K. (2000). *eEurope: an Information Society for All: Response from the Royal National Institute for the Blind, UK*. London: RNIB. Also available at http://www.rnib.org.uk/digital/eeurope.htm

An Evaluation Perspective: Access to Telecommunications in Europe

Jan Ekberg, Erkki Kemppainen

National Research and Development Centre for Welfare and Health, (STAKES)
Box 220, FIN-00531 Helsinki, Finland

Abstract

In this paper we try to look at telecommunications services from the point of view of equality. In the light of this criterion, interpreted as availability and accessibility, we then analyse some actions taken in the legislation area. Then we look at some developments in telecommunications services in Europe in order to find out if equality has been improved.

1. Introduction

The key sense of the term "evaluation" refers to the process of determining the merit, worth or value of something. A fundamental element of evaluation is setting a criterion in relation to what is evaluated. The basic logic of evaluation consists of establishing criteria of merit, constructing standards, measuring performance and comparing with standards, and finally synthesising data into a judgement of merit or worth. (Scriven 1991; Rossi & Freeman 1999.)

Our perspective is that of equality. We can interpret availability and accessibility as evaluation criteria for telecommunications equality. The PROMISE project (Cullen, 1998) presented five A's (availability, accessibility, affordability, awareness, appropriateness) which all in a sense could be used as evaluation criteria but we will restrict ourselves to only the ones mentioned.

But, why equality, and why availability and accessibility?

2. The motivation of the evaluation criterion

Law is a strong basis for a criterion. Equality is a fundamental right. Equality and the prohibition of discrimination have been enacted into law in various ways in different countries in Europe. At a European level the Treaty of Amsterdam explicitly includes the prohibition of discrimination on grounds of disability, for example. It is included also in the Charter of Fundamental Rights of the European Union. The Treaty of Amsterdam provides the basis for a crucial leap forward to promote equal rights for people with disabilities at EU level (Commission of the European Communities, 2000).

It is important to see how equality is interpreted and presented in specific legislation. Recent telecommunications Directives have some requirements for services but they require further action from the Commission and the Member States. The Directives are relevant for the provision of text telephony services or directory services, for example. These examples can be seen as an interpretation of equality and are thus also criteria for the evaluation of the access to telecommunication.

The *Directive on the application of open network provision (ONP) to voice telephony and on universal service for telecommunications in a competitive environment (98/10/EC)* defines the universal service in Article 2 as a defined minimum set of services of specified quality which is available to all users independent of their geographical location and, in the light of specific national conditions, at an affordable price.

Article 8 in this Directive concerns specific measures for disabled users and users with special social needs. It says that Member States shall, where appropriate, take specific measures to ensure equal access to and affordability of fixed public telephone services, including directory services for disabled users and users with special social needs. According to the Preamble, specific measures for disabled users could include, as appropriate:

- making available public text telephones or equivalent measures for deaf, hearing or speech impaired people,
- providing services such as directory enquiry services free of charge or equivalent measures for blind or partially sighted people, and
- providing itemised bills in alternative formats on request for blind or partially sighted people.

The other important new Directive is the *Directive on radio equipment telecommunications terminal equipment and the mutual recognition of their conformity (99/5/EC)*. This relates to essential requirements as defined in Article 3. In accordance with the procedure laid down in Article 15, the Commission may decide that apparatus within certain equipment classes or apparatus of particular types shall be so constructed that it supports certain features in order to facilitate its use by people with a disability.

3. Measuring performance: the situation of some services

Now we should set the standard against which we measure availability and accessibility. We could name some services as standards: text telephones, directory services etc. The situation of telecommunications services has been sporadically studied recently in the project COST219bis. For the time being it is difficult to define the degree of accessibility. In this context we can say only whether a certain service exists or does not exist. However, the accessibility of some services has been improved.

There are text telephony and relay services available in European countries, but the situation and the organisation of services varies. The rapid technological development has created problems for directory services in the new legal and market environment, but ways to arrange better services are being sought in some countries.

The availability of appropriate accessible terminals and services has also been analysed in order to evaluate how telecommunications accessibility has improved during the last decade. The goal is to evaluate if equality has improved in real life.

From the equality point of view it can be argued that disabled people to-day encounter a much more accessible ICT environment than 10 years ago.

Relay services for deaf people are much more versatile than earlier and can serve also people using signing or using Bliss. Communication between different text telephone protocols can be handled automatically.

The widespread use of mobile SMS have provided hearing impaired people and part of deaf people with a communication channel that can be used by a majority of people.

Hearing aids have developed giving better quality and background noise reduction. The interference between mobiles and hearing aids still exist but can in some cases be remedied.

People needing speech recognition in order for instance to control their computers or mobiles can to-day find very useful solution.

The transition from line based operating systems towards graphical interfaces created a lot of accessibility problems for blind people some 20 years ago. The Windows systems that were mostly very difficult for visually disabled people to use have now been much more accessible due to screen readers, large font, speech output etc. Some features are available in ordinary personal computers or can easily be installed. Scanners have become affordable and the popularity of electronic post makes it easy to send scanned material to a friend for analysis.

Internet services like the Web electronic mail and file transfer have provided mobility impaired people with new options for a more equal participation in the ICT community.

Decision makers have start to realise that Internet services are becoming very important and that they have to be made available and accessible to the citizens. The *e*Europe initiative has as a goal to ensure that public Web services become accessible.

4. Synthesis: achievements, problems and possibilities for success

There exist some legal measures, which could be used to improve the situation. National measures could affect the availability of directory services for blind persons. The measures of the European Commission could affect the accessibility of terminal equipment.

The eEurope initiative has focuses, among other things, on the participation for all and accessibility. It is agreed that the European Commission and the Member States adopt the Web Accessibility Initiative (WAI)[1] Guidelines for public web sites by the end 2001 (Council of the European Union and Commission of the European Communities, 2000).

A problem is created with the new multimedia break through. The requirement that information should be presented in optional modes is not always fulfilled. Not to talk about the requirement to have the information always also in textual form as for instance tags describing a picture or what is shown in a video clip.

Still there are problems. All who would need accessible ICT for participation on equal footing in society cannot afford the devices and services. Interface devices especially meant to facilitate life of disabled users are normally expensive because they are only manufactured in small series.

The best solution would be that all equipment that are manufactured in large numbers should be design in such a way that as many as possible could directly use them. Design for All would help accessibility come true.

In the course of technological development there appears new barriers. What is all the time needed is awareness raising as well as research and development work.

More information about accessibility and links to other sources is found at:
COST 219 ("Telecommunications, Access for disabled people and elderly" with Web address: http://www.stakes.fi/cost219) or
INCLUDE ("Inclusion of disabled people and elderly in telematics": http://www.stakes.fi/include) or
PROMISE ("Promoting an Information Society for Everyone": http://www.stakes.fi/promise).

References

Commission of the European Communities (2000). *Towards a Barrier Free Europe for People with Disabilities*. Communication from the Commission to the Council, the European Parliament, the Economic and Social Committee and the Committee of the Regions. COM(2000) 284 final. Brussels, 12.05.2000.

Council of the European Union and Commission of the European Communities. *eEurope 2002* (2000). *An Information Society For All*. Action Plan prepared by the Council and the European Commission for the Feira European Council 19-20 June 2000. Brussels, 14.6.2000.

Directive 99/5/EC of the European Parliament and of the Council of 9 March 1999 on radio equipment and telecommunications terminal equipment and the mutual recognition of their conformity (1999). OJ L 91, 7.4.1999, p. 10.

Directive 98/10/EC of the European Parliament and of the Council of 26 February 1998 on the application of open network provision (ONP) to voice telephony and on universal service for telecommunications in a competitive environment (1998). OJ L 101, 1.4.1998, p. 24.

Cullen, K. (1998). *The Promise of the Information Society*. Gummerus Printing, Jyväskylä, Finland 1998 (ISBN: 951-33-0550-3)

Rossi, P., Freeman, H., Lipsey, M. (1999). *Evaluation. A systematic approach*. Sixth Edition. Thousand Oaks: SAGE Publications.

Scriven, M. (1991). *Evaluation Thesaurus*. Fourth Edition. Newbury Park: SAGE Publications.

---

[1] http://www.w3.org/WAI/

From the Individual to Technology towards the Global Network

Dino Giuli

Dipartimento di Elettronica e Telecomunicazioni, University of Florence, Firenze, Italy
Via S. Marta 3, 50139, email: giuli@diefi.det.unifi.it

Abstract

New enabling technology is needed to overcome the current limits of Internet to cope with the human subjective requirements arising from the widespread use of the Network. A new and systematic approach is, therefore, needed in the search for solutions, moving from the individual to technology along the research and the development path towards the Global Network. The systematic exploitation of human subjectivity is a key factor in this framework. A general analysis is carried out in this paper, which aims at orienting that research line, as well as at collocating some related and specific results achieved in the meantime.

1. Introduction

The pervasiveness of solutions and applications continuously made available by Information and Communication Technologies is leading people – i.e. those who live in the most developed countries – into a new era, in which potentially human activities, in almost all their aspects, can be carried out using the Network, adopting a combination of architectures, languages, protocols, terminals and applications. Although theoretically, technological solutions are already available to support the process of becoming citizens of a global and inter-networked social and economic world, there are, undoubtedly, factors which can slow down, and can decrease the force of this trend. Some main factors have been identified as being critical to the pervasive use of information technology in everyday life and they can mostly be associated back to the origins of the prototype of the Global Network, i.e. the Internet [7,10]. If the Network has to be used to support *formal relations as well as interactions*, the need arises to make the entire system evolve and adapt in order to cope with the new needs and requirements related to specific, as well as new, applications in an ever expanding social-technical environment. Hereafter, we use the term Global Network, to refer to the interconnected instrumental resources, which operate by means of information and communications technologies, and are accessed for the purposes of information, interaction, communication and knowledge. In addition, they are made ubiquitously available for access by individuals, in accordance with their abilities, capabilities and needs. The Global Network - which acts as an instrumental support within the complexity of human non-material activities, and, subsequently, leads to the effective development of material activities - is, thus, definitively aimed at supporting the widespread improvement of the quality of life. In this sense, the Global Network could have a strong growing impact and could be better applied in all contexts related to human activities. (execution of citizens' rights and duties, cultural and educational applications, work activities, spare time activities and hobbies, assistance & personal services, finance & business, management of life & its environment, etc.). While this can certainly bring a substantial improvement to the quality of life, we must also be aware that the pervasiveness of the Global Network in different areas of human life has relevant effects on the individual's way of life, work and relations, thereby transforming the human and subjective sphere, making cultures uniform, by modifying human perception, emotion and cognition mechanisms, etc. To this extent and given such a use, the Global Network is essentially an *interaction environment* which supports direct and/or indirect *relations among people*. Therefore, information about subjects interacting in this environment, is crucial to develop the capability of the system to adapt and match user needs and requirements, as well as to support the exploitation of their natural, human, individual and social role in the Network. To date, the Internet, as instance of a social creation for relationships and interactions, does not really meet the requirements of symmetric and balanced availability, and neither does it allow the control and exploitation of subjective information. In the sequel, we describe some elements to highlight the current asymmetry of the relations between subjects interacting through the Network, in order to define important pre-requisites of a more symmetric interaction. To this end, a basic pre-requisite is the introduction of a complete human subjective profile which can be represented, managed and controlled by network users themselves, to support more actions of self-awareness and interactions on the Global Network. This analysis moves along a basic research path, going from the individual to technology, towards a Global Network. This path accounts for the human subjectivity which is at the very origin of the process, as well as for the instrumental operational Network solutions, which can be structurally tailored to subjective factors, as a means to identify the main need and scope so as to obtain an effective evolution towards the Global Network [8]. A lot of fundamental analyze have been developed to date, and constitute a relevant grounding as to the solution approaches here presented and envisaged [1,2,3,4,5,6].

To this end, interdisciplinary research directions are, therefore, needed and here highlighted. A common set of guidelines is also provided on specific aspects, which are dealt with, in companion papers [10,11,12,13] which the reader should refer to for a deeper analysis. Such guidelines and insights, also summarize major current achievements to date, within the research framework program named **B.E.S.T. beyond Internet** (Bridging Economy and Society with Technology **beyond Internet**), which has been promoted within the interdisciplinary Ph.D. program "Telematics and Information Society" at the University of Firenze [7].

2. Towards a more symmetric interaction

The current manner of interaction on the Global Network, implements an intrinsic *asymmetric manner* of relations between the subjects – be they final users and/or providers – with a clear imbalance between the information available to and manageable by each partner [10]. No answer has been given to-date, to support individuals in the *representation or, even more so for the management of the user's own private information,* or again, in order to facilitate user interaction and the subjective exploitation of a much more *conscious end-user role* on the Global Network. Enabling symmetry thus implies a balanced positioning of different subjects and actors within the Network, even as regards the availability of instrumental subsidies which could eventually enable subjective adaptation, as well as direct control over the modes and scopes of Network relations and interactions. In order to obtain a more balanced interaction, we, hereafter, identify some general pre-requisites which should be matched within an ideal Network environment. First and foremost, individuals should be able to *express their subjective information,* maintain *control* and have the ability to *manage* it – using appropriate instruments and applications. Secondly, there should be intrinsic mechanisms enabling *trust* and *privacy protection* during Network relations and interactions, thus promoting adequate institutional and operational activities on the part of trusting subjects, to be accredited for the role of intermediary among Network end-users. The capability to have a balanced relationship, is, therefore, closely associated with (i) the *instrumental elaboration capability*, of personal information made available to and manageable by end users, to assist them even at their terminals (ii) the direct availability of *proper profiling information* for all Network subjects (iii) the *communication capability* made available to end-users.

3. User Subjective Profile

Based on the above pre-requisites, we need to develop a way to systematically *represent the user, in particular the individual,* as a real complex subject, characterized by his or her social background, personality, context, situation etc., accounting also for all the elements related to personal relations with the social – global to local – environment, including the network. To this extent, we introduced a user *subjective profile* as a *complex set of digital information, related to the subjectivity of the user, employed to allow and facilitate personalization and negotiation during Network interaction according to user interests and needs – actively expressed and represented by the same user, as well as automatically produced or inferred.* The subjective profile, aims at describing, in a systematic and dynamic way, the individual profile which is composed of Background and Foreground information accounting respectively for the consolidated motivation and behavior, as well as for the contingent factors – related to the specific time and place – which describe the user and his/her possible modes of interaction on the Network [8,11]. These elements, if properly described and managed, should then give the users new possibilities for greater awareness together with balanced actions with other subjects within the Global Network.

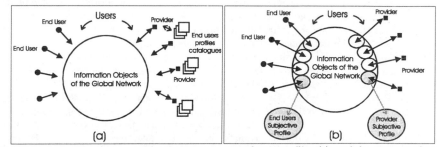

Figure 1: The present asymmetric manner for the management of user profiles (a), and the prospected symmetric one allowed by the systematic introduction of user-subjective profiles in the Network (b).

Figure 1(a) presents an unbalanced situation for the handling of user subjective profiles. Providers usually build strategic databases of end user profiles, and thus have the very economically valid possibility to provide customized offers. To date, end users have, therefore, no direct control and/or management of their personal data, once they submit them during network transactions, or while they are simply navigating on the Internet [10]. Alternatively, Figure 1(b) shows the prospective way of carrying out a more symmetric interaction: users – be they end users or providers –over the Network, using their subjective profile as a means of intermediation, which in turn supplies them, not only with the possibility of having a more tailored answer to their specific needs of the moment, but also with the possibility to show only the personal data they decide to make available to others for the purpose of a specific interaction on the Network, thus, increasing their negotiation power. Further insight on subjectivity, privacy and asymmetry, as well as that related to human subjectivity and relation profiling factors, is provided, according to the above approach, in the companion papers [10,11] respectively, to which the reader is referred.

4. Influencing technologies to match individual needs

Information and Communication Technologies, do their utmost to offer new solutions for new, even newly created, needs – from communications technologies, to applications - in a form, which is certainly not systemic, thus making the inheritance and amplification of problems and imbalances somehow possible, at different levels of the Network interaction and relation process. We would then ideally like, to find *helps* matching the above described needs, so as to help users in their actions to determine interaction on the Network, in a system which is becoming more and more complex and difficult to manage. These *helps* should be able to exploit the important wealth of *subjective knowledge* and to promote *self-awareness* in actions and interactions, as well as exploit *collective intelligence and knowledge* (considered as an investment capital), which derives from relations and interactions among Network subjects. The natural consequences of this reasoning, include the idea of thinking about the instrumental processes in which relations and interactions are grounded, as being systematically influenced and conditioned by subjective factors, in order to best match and adapt user needs – where *needs* is intended in its wider sense (individual and relational) as referred to throughout this paper.

When thinking of relations and interactions among subjects as processes which involve phases as well as the bilateral exchanges of information, we can try to imagine a mechanism by which processes themselves – in terms of acts, and objects – are conditioned and determined, by being based on subjective information which offers profiles of the interacting parties, as shown in Figure 1. Establishing and carrying on relations, imply agreeing upon some ways of acting – for instance on the way of transmitting and receiving information by each user side, but its success also implies accounting for private information concerning the parties – for instance the way and purposes of autonomously treating the results of the interaction by each user. We, thus, perceive the need to have subjective informative elements – both for common and individual spaces – which condition the interaction-relation processes in the network, at each step as well as at more intense levels, all along their development and dynamic evolution.

4.1 The *subjectivizing gene* in instrumental environment development

Based upon the above analysis, we proceed by addressing the factors which identify the *imprint* of each user, as those being crucial for the definition of the basic structure of a proper instrumental network environment which truly matches subjective needs, while allowing balanced operations and roles among the network interacting subjects. Such an imprint can be conceptually represented through the metaphor of the *subjectivizing gene* [8]. The *subjective profile*, is of course one main component of the *gene*, but the latter also includes other important factors both from the human and from the network perspectives.

Figure 2 represents the *subjectivizing gene*, associated with a specific user (gene-imprinter). Such a representation specifically refers to an end user. The elements in the core of the gene, without the information added by the elements in the crown, would just represent the ordinary process typically performed during network activities – i.e. without the elicitation of any subjective factor. The connection between the core and the information contexts of the crown, utterly changes the process driven by the former, while, at the same time, subjectivizing it.

The elements pertaining to the "crown" of the subjectivizing gene are hereafter described:

- *Human User Background Profile;* this is the information which describes the personality of the end user, as regards his/her interaction and relation, both in the real and on the networked world [11].
- *Human User Foreground Profile;* this is the information which describes specific spur-of-the-moment subjective user factors, related to the specific moment and place, which can at times condition his/her relations and interactions both in the real and in the networked world. These factors also account for the instrumental characteristics of the terminal, at present, available on the user side. Together with the Background Profile, these factors concur to define an instrumental representation of user self-awareness: the individual conscience, as relevant for interaction in the Network environment [11].

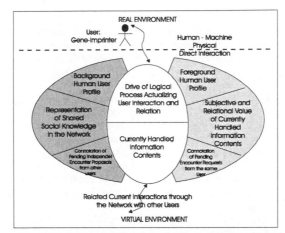

Figure 2: *Subjectivizing Gene*: subjective user-imprint

- *Representation of the Shared Knowledge;* this is the information which allows representation of the collective intelligence and common-sense shared by a community of Network users; it can also be produced in relation to specific interests, scopes and types of relationships in the community. We implicitly assume that it is possible to exploit it as a set of instruments, together with the user profile, in order to match both the subjective requisites which derive from the need to facilitate one's physical and cognitive interaction mechanisms, and the trust requirements for user relations on the Network.
- *Subjective and Relational Value of the Handled Information Contents;* these values describe factors of the currently handled information contents, once they have a subjective and a relational value. These subjective and relational values can define attributes related to privacy and copyright, sharing rights, etc.
- *Connotation of Pending Independent "Encounter Proposals" from Other Users;* this information concerns the subjective connotations of incoming solicitations which are "encounter proposals" directed to the user (gene-imprinter), through network interactions which are independently performed by other users.
- *Connotation of Pending "Encounter Requests" from the Same User;* this has the same value as the previous element, except that these solicitations are self-expressed by the user.

The elements belonging to the center of the subjectivizing gene are hereafter described, and they define the operational context of the gene itself:

- *User Human Direct Physical Interaction;* this describes the physical actions which are carried out directly by the person, as his/her commands together with instrumental communications produced at the terminal side. Such actions are connected to the physical perception process displayed to the user, at the terminal side, as well as by the cognitive process contextually fed within the user himself. The structure and implementation of network interaction and relation processes, in the different ways permitted by the explication of the subjectivizing gene, are also intended to facilitate such physical and cognitive interaction processes by the user.
- *Related Current Interactions via the Network with Other Users;* this information connotes current interactions with other users which influence the current instrumental interaction and relation process of the user at hand, affecting all the other operational and information contexts of the gene, as needed.
- *Drive of the Logical Process Acting User Interaction and Relation;* this is the drive of the logical process (typical of the operating software), which actualizes user relations and interactions, through the elaboration of information in different informational contexts pertaining to the gene. This process uses the necessary communication and processing resources purposely available for that specific purpose in the different components of the physical and logical platform of the Network (including the user terminals).
- *Currently Handled Information Contents;* these are the information contents, from objects or subjects, which are being treated in the current logical process, for the production, transmission, or transformation of the information itself.

With regard to the temporal projection of the gene, we notice that the gene can account for the evolution of subjectivity factors, thus enabling gene adaptation to the evolution of the user's personality on the network.

5. Conclusions

The present development trend of the Global Network does not appear balanced, and thus, is unstable. It is mainly conditioned by economic factors, while the individual readiness to access and use the Network must, indeed, meet the individual's desire to extend human subjective space, for the purpose of information, communication, interaction and knowledge, within a social-technical environment.

Thus, the systematic and balanced exploitation of subjectivity, together with its protection, appear to be priority conditions for a personally tunable and balanced, hence stable, development of the Global Network. This is also to be considered as a means to achieve compatibility with different economical, social and cultural contexts. At the same time, we are aware that simple regulatory actions are insufficient and, indeed they could block the development of a wider environment capable of supporting and sustaining human activities throughout the Network. For these reasons, we must cope primarily with the structural technological factors which are currently dominating, but, at the same time, also limiting, proper developments in the manner proposed above. Meanwhile, new technology is continuously being made available and, in the majority of cases, it is currently moving in the direction of some main lines of development such as: the improvement of the exploitation and control of subjective factors, which concern the end user for the support, adaptation and customization of functionality and services, from the view-point of the providers; the rapid development of mobile communication technologies; the interoperability of networks for the integrated transport of information, ever faster broadband networks; etc. Despite these elements, the direction towards a Global Network which truly responds to user needs, must be grounded on some fundamental new action lines which have not yet been addressed sufficiently in a systemic way. Basic objectives must still be pursued and achieved: these include: the development and diffusion of a new Network culture, which promotes self-awareness and increases end user capability and self-control; a rapid transformation of goals and functions within the Public Administration which needs to take place in order to support the widespread use of technology capable of fulfilling social requirements, facilitating both the growth of trust by individuals using the Network, and the increase of guarantee of their protection; economic subjects need to pursue stable and balanced social-economic objectives on the network; a synergic interdisciplinary research needs to be promoted in order to overcome the structural and technological limits of the network. To this end, the meaning of user subjectivity has to be clearly defined and the role it has to play in the network has to be fully understood, and more intensely pursued. In the latter sense, this paper, together with the related and specific contributions made by companion papers [10,11,12,13], is an attempt to provide some guidelines and insights, which should help in outlining a proper re-structuring and development of the instrumental and operational body of the network, in the direction of the Global Network.

References

1. Winograd, T., Flores, F. (1986). *Understanding Computers and Cognition*. USA: Ablex Publishing Corporation.
2. Minsky, M. (1988, March). *The Society of Mind*. Simon & Schuster.
3. Rifkin, J. (2000). *The Age of Access: The New Culture of Hypercapitalism Where All of Life Is a Paid-For Experience*. J P Tarcher; ISBN March 27.
4. Negroponte, N. (1995). *Being Digital*. Knopf (January).
5. Friedman, T.L. (2000). *The Lexus and the Olive Tree: Understanding Globalization*. Anchor, edition May.
6. Communications of the ACM (2000, August). Special Issues on *Personalization*, Vol.43, No.8.
7. Giuli, D. (2001, February). *Background and objectives of the B.E.S.T. beyond Internet framework research program*. Dept. of Electronics and Telecommunications, University of Florence, Internal report DET-1-01.
8. Giuli, D. (2001, February). *Subjective information and the subjectivizing gene for a new structure of the instrumental relation and interaction processes in the Global Network*. Dept. of Electronics and Telecommunications, University of Florence, Internal Report DET-2-01 (in Italian).
9. Giuli, D. (2001, February). *Subjectivity, citizenship and Public Administrations in the evolution towards the Global Network*. Dept. of Electronics and Telecommunications, University of Florence, Internal Report DET-3-01 (in Italian)
10. Calenda, D., Giuli, D. (2001). Subjects, subjectivity and privacy in the Global Network. In *Proceedings of UAHCI 2001 Conference*, New Orleans, Louisiana (present volume).
11. Pettenati, M.C., Giuli, D. (2001). Human Subjectivity and Relation Profiling Factors for the Global Network. In *Proceedings of UAHCI 2001 Conference*, New Orleans, Louisiana (present volume).
12. Bussotti, P., Vannuccini, G., Calenda, D., Pirri, F., Giuli, D. (2001). Subjectivity and Cultural Conventions: the Role of the Tutoring Agent in the Global Network. In *Proceedings of UAHCI 2001 Conference*, New Orleans, Louisiana (present volume).
13. Vannuccini, G., Bussotti, P., Pettenati, M.C., Pirri, F., Giuli, D. (2001). Towards a new Multi-layer Approach for the Global Network architecture. In *Proceedings of UAHCI 2001 Conference*, New Orleans, Louisiana (present volume).

User Interfaces for the Productive Information Society

Wolf Göhring

Institute for Autonomous intelligent Systems AiS, GMD — Forschungszentrum
Informationstechnik GmbH, D-53754 Sankt Augustin, Germany, wolf.goehring@gmd.de

Abstract

A scenario where people are organizing life and work by means of technical networks and information explains the notion of "Productive Information Society". People become at the same time white and blue collar workers and consumers of their collaboratively produced goods. The Web mediates in a new way the acquisition of emerging technologies and of products, and it will influence tools and user interfaces thoroughly. Vice versa good tools and good user interfaces facilitate that usage. Issues in computer science arise from collisions between plan and reality, from concurrency, from a growing information jungle, and from psycho-physiological limitations of users. Societal issues stem from different skills, from economic competition, and from existing industrial relations.

1. Motivation

How could information technology (IT) be used by everyone for everyone's needs? Which means could help to do so? Which user interfaces would be suited for this task? Which issues and problems could arise? To find answers let us have a look on daily life. Today our life is "organized" in a specific way: Most of us are working for a salary which enables us to buy things and services produced by other people remaining anonymous and that operate in a similar way. Things and services are presented on a market where we do our choice along needs and preferences. But IT enables a deeper and broader discussion on products and services, it enables a better planning and the connection between plan, real production processes *and* consumption. We illustrate this development in a scenario where people form a productive information society. We derive issues whose solutions will influence tools and user interfaces where the "user" may be each of us in every day's situations involving spare time activities as well as those on the job. The scenario may yield ideas for benchmarkings for tools and user interfaces.

On the road to productive information society there also arise societal issues which have been discussed in more detail in [3] and in [4}. In [3] we also consider effects on sustainability.

2. Entering a scenario

"The 21st century has to be the age of a new integration of labour and living, ecology and economy, technology and culture. Architecture and urban planning are key factors in this scheme." [5]. This "age of a new integration" in mind, imagine you are an owner of a family's house and you would like to install yourself a solar thermic system. You would search for information on such systems, you would discuss the layout with some skilled and experienced people, you would visit running systems in reality *and on the Web*. You would plan the work to be done, you would buy the material and finally you would install the system with the help of some friends and craftsmen. In each stage of your work you would use IT. Based on IT, industrial, artisanal and private work and the respective consumption may be combined in a new way. Computer support simply means that one views some aspects of the system at a display or on print-outs and that one handles this information. One makes comparisons of different running systems and with further information from the Web one finds better answers to technical and cost questions not only concerning the system as a whole but also components found on the market. Entering such information into the Web the treated system is "mobile" from its beginning: One can view "life" presentations with stored as well as with current operation data. One "works" with it from anywhere: in a bureau for planning and simulation and at the construction site, during installation, maintenance, repair, reconstruction or demolition. People will link up these stages in the life cycle of a system and they will show the combined effects of their labor on nature, society and individuals. Today, workmen and craftsmen, small and medium-sized enterprises interface between the consumers and customers at one side and the distributors and manufacturers on the other side. They also interface to the public educational and advisory sector. The Web augments the traditional means of communication and cooperation of these social actors. Professional information is not only accessible to specialists but also to everyone surfing on the Web. Everybody may use technical data, data of running systems and planning tools found on the Web. Via I/O-sharing one communicates on-line and in real time with remote partners discussing technical details. Schools and universities use this new access to the productive world. During the education of consultants, planners, craftsmen, workmen, trainees, and retrainees one considers "virtual" systems, varies their configurations, compares them with real ones which have been found on the Web, and one plans new ones using Web tools. Action groups of unemployed people as well as people coming from local exchange trading systems (LETS) use Web support to participate in productive discussions about concrete, technical themes. Negotiating and processing a thing people talk about intended and produced changes. Traditionally they remember the discussions in their mind or write and draw it on paper in order to work consistently. Now, they also deal with electronic descriptions which may be produced manually or automatically by technical processes. Sending information to the Web, individual or regional experience is *systematically* opened for all. Traditional isolations may be overcome since computer supported information and communication complete what is passed by word and written on paper. But we are far away from having a complete and consistent database. It is more like a growing "information jungle" [2]. Nevertheless, the information helps to treat thoroughly and to balance the substantial dimension of man's life, to evaluate not only intended effects or those that lie nearby but also a lot of non-intended side-effects. People will find better answers to the standard question "Do we want this or someting else?" Using freely and efficiently the information on the Web as the very best database people may discuss *all* aspects of their life and may organize *all* activities needed to live. In this *productive* process *labor is reflected in products and in information* and vice versa *information is reflected in labor and in products,* in short: the social actors of the scenario form a *productive information society.*

3. Technical Issues and user interfaces

The scenario will not have been realized even tomorrow. It will take a long time during which one will be confronted with technical, social, economical and political issues. One will meet challenges of the following spatial, temporal, functional, and dynamical aspects, and to master them one needs appropriate tools with excellent user interfaces:

- Spatial aspects: People of different skills, qualifications, and positions and coming from different regions collaborate closer on the basis of technically networked information, communication, and systems. They want to perceive better the effects of their labor onto nature and onto themselves.
- Temporal aspects: The extraction of raw materials may leave traces in nature for decades or for ever. Also decades may pass from the planning until the demolition of a system. The ordering of parts, their production and delivery take days, weeks, or months. Components running in a network may communicate within seconds or milliseconds.
- Functional aspects: The access to the appropriate data, methods, and to the documentation of the various systems makes it possible to take care of resources and to elaborate innovative things efficiently. These products contain components of different functions which are connected mechanically and driven electrically, and which cooperate in an electronic network.
- Dynamical aspects: The world of products, their data and their environment are subject to permanent change. Existing systems are rebuilt, spare parts will be provided for a long time, new systems will contain new components. A system will be recycled when it is taken out of service.

In the same period in which computer power is doubling the growth of industrial productivity is far less than 10%. Most of this growth is due to organizational changes in the enterprises and to innovations in classical engineering areas and *not* to progress in computer science. (This and the consequences are discussed in [1].) In applying computer science we are confronted with the permanent collisions between plan and reality, the growing jungle of information, the marrying and divorcing of people, things and worlds, and last but not least, the physiological und psychological limitations of individuals in operating a complex and technically structured world.

These tasks meet with usability and with ergonomic requirements of computer systems. We need a good idea of "user interfaces for all". Tools and interfaces should help people to transform data from the world into knowlegde about the world, and help people to organize activities as sketched in the scenario. The interaction with computers always has an aspect of communication with some people. Thus, user interfaces represent in their totality a type of language if they are really suited for a productive information society.

3.1 Collisions between plan and reality

Concurrent activities and diverging interests generate unforeseen differences between plan and reality which are reflected in the resulting data. Information should be realistic and complete with respect to the purpose for which one wants to use it. But it is impossible to describe facts completely. Information will never be applicable without any unexpected side-effect thus always having an element of desinformation. The dynamic solution lies in the reciprocal aptitude in using *and* in the reciprocal dependence on using (foreign) information turning it by this way into knowledge. An objective idea of the quality of the virtual world is found if it is combined with reality: The description of a component proves to be true if a technical system together with the built-in component operates as expected.

- (Distributed) Data have to reflect reliably their objects for decades during their life cycles. The data have to migrate reliably over generations of hardware and software.

- The growing number of data, though remaining incomplete, and the complexity of planning force the use of heuristics whose quality has to be founded theoretically.
- We need technical support to contact programmers, data originators, designers, producers and other people or to go into the details of a program or the history of data if programming errors occur or if information seems to be obscure.
- We need tools *usable by everybody* for modeling, planning and documentation of plans *and* of reality. The models have to accept algorithmic relations between the components and to facilitate corresponding simulations. We need to get an (virtual) idea of the quality of the used data, information, and programs.

Briefly: A plan and the reality which the plan reflects should be verified each against the other. Existing contradictions have to become tractable.

3.2 The growing jungle of information

In the future we will have much more mechanized information and computerized means to manage information as they are available today. This results in a kind of an information jungle which hardly seems to be used rationally though more and better information is a prerequisite for a rational behavior. We need *efficient* solutions for the following aspects:

- One has to be able to combine the lot of pieces of information in order to view the effects of work and to make the information useful for interrelated behavior of individuals.
- Intelligent autonomous agents should enrich existing information with new one. They should structure and maintain the information, delete old one, and exploit the wealth of information in showing essentials and in extracting new information from correlated data.
- In the information jungle people have to find the needed fixed points, measures and methods in order to recognize the essentials of the information and in order to work accurately, productively, cooperatively, fairly and well protected from losses and errors.

Briefly: He who hammers a 2 1/2 inch nail with three hits he wants to work in a similarly smart manner with the intelligent agent in the laptop positioned beside his hammer drill.

3.3 Marrying and divorcing people, components and worlds

In order to handle something people bring together their competences using the technique of synchronous I/O-sharing, for example. Work finished they search for other partners in order to handle something else. The computer supported collaborative work on a temporarily common object changes the basis of information continuously. For the sake of technical documentation and the recycling of materials it has to be continued and to be archived. If data have no more meaning for existing things they have to be recognized as obsolete, and they have to be given up. Generally, partners, projects, information and technical conditions change and are situated in different and changing places giving raise to the questions about synchrony, concurrency, conflict and confusion.

We need *efficient* solutions for the following aspects of this problem:

- Intelligent agents should install and de-install software on existing hardware, and handle the migration of existing software and data to new hardware and operating systems. A general infrastructure has to provide systems for planning, documentation, verification, information retrieval, I/O-sharing during teleworking and cooperation.
- The different aspects and the concurrency of the work of the different partners have to become transparent such that the partners can secure the desired consistency and synchrony within the treated virtual and real worlds.
- One has to be able to make visible or to fade out aspects of a virtual object which are of different relevance.
- Synchronous I/O-sharing should enable remote partners to communicate on-line and in real time on the details of their common (virtual) object thus enriching the traditional process of asking, answering and demonstrating work steps.

Briefly: The Web tools have to be designed such that people may come together for a temporarily common work on virtual and real objects minimizing conflict and confusion.

3.4 Limitations of individuals in operating a complex world

The use of IT consumes time and power of the users each endowed with limited capabilities. Some of these limits are well known in physiology and in psychology. IT-tools are usefull only when their user interfaces are optimal within these human limitations. The optimum may be related to the time difference if one does the task with *and* otherwise without support of the specific tools, always including the specific time of recreation.

- How much, how often, and what kind of information do people really use to organize their daily life including their labor?
- Which piece of this information could be supported by IT, how should it be presented and in which way could it be handled?
- In which way are individuals limited to handle computer supported information?

Briefly: Which IT-support is actually a support?

4. Societal issues

In the scenario we have seen a re-integration of design and work, production and consumption, education and daily life leading to issues in social, economical and political respect intertwined with each other. Societal barriers define the extent to which people really can use the new technology. But, if things are technically possible people want to use them and they try to remove existing barriers. Let us summarize some of the societal issues and questions.

4.1 Development of a new universality of the individuals

The scenario shows how people could attain a new universality since everybody has access to the complete information presented on the Web. Yet different skills as defined by the job (white and blue collar work e. g.), by age, education and social tradition may determine different extents to which people can really access the information and contribute to its augmentation.

- What is an adequate concept of information society that comprises and combines the labor of blue and white collar workers as well as the consumers?
- Which socially defined obstacles are hindering sections of population from engaging in productive information society and which conditions are suited to support these sections in their engagement?
- What is an adequate concept of education in a productive information society?

4.2 Goods instead of merchandise

A merchandise is a product that becomes a merchandise by being produced not for self-consumption but for consumption by others. Only after exchange the other can consume the acquired thing for which he or she gave an abstract equivalent, namely money. Exchange and pricing take place under competition. Yet competition contradicts to the economy of the scenario where in relation to accessible machinery, resources, and needs people become able to determine the concrete labor each of them *will* do. There, they define their *concrete equivalent* contribution such that an abstract equivalent of their contribution becomes inapplicable.

- Which obstacles due to competition and which means to overcome obstacles exist?
- How could the exchange of products based on an abstract equivalent be replaced by agreements on concrete equivalent contributions in producing something common?

4.3 A legal framework for the productive information society

In the scenario everybody may intervene to work and to argue without any hindrance. The only regulative in the Web-based cooperation should be that one has to deliver an agreed service in return. This dependency may be organized as a mutual one since everybody has access to all information and, if productive information society has developed, to the whole productive inventory of society. At this stage people can leave ARISTOTLE's point of view: "In order to maintain society nature decided that there are leaders and retinue." (Politics, Book I, 1252a.). People can realize the first article of the Universal Declaration of Human Rights (10th December 1948): "*All* human beings are born free and equal in dignity and rights."

- Which re-definition of industrial relations would be required to behave on really equal rights in the depicted scenario?
- Which political and legal framework would encourage this development in the productive world?

5. Finding Solutions

An experimental approach with case studies where one implements some elements of the scenario, would be adequate to meet aforementioned issues and to find good solutions. A case study in this field should have *three central aspects*, namely a realistic productive aim, an analysis of the usage of the lot of software tools, and, thirdly, an analysis of the arising societal problems.

5.1 The productive aim

For example, the productive aim could be the installation of some solar thermic systems on some existing private or public buildings (schools, youth clubs, self-organized meeting centers e. g.). Some scholars from secondary schools, some apprentices, students from universities for sciences and applied sciences, other young, but jobless people together with teachers, journeymen, masters and professors could participate. Such a team of about 15 people would be very heterogeneous with respect to skills and ambitions. It would have to virtually rebuild some real buildings for the planning process, to design some solar thermic systems including 3-dimensional virtual installations in virtualized buildings. The team should compare alternatives and optimize the solution, derive components, costs, and the consumption of natural resources, and consider ways for a recycling of the system. Finally, the team should select a design, build the system and present all its work on the Web.

5.2 The usage of software tools

Secondly one would have to analyze the usage of the software tools and information and to study the role of user interfaces in this productive process. Take for example the virtual installation of the pipes in the virtual building: How can one find a suited tool in the information jungle of the Web? How long does it take to understand its different features and its usage? How can one remember the way to use it? How can one introduce actual data and modify them? Does the tool support the synchronization of concurrent and remote work, for example through I/O sharing? How can one use the tool when constructing the pipe in the building? Is it practicable to archive and to use data for a long period, even if the team has dispersed? How can one combine this tool and the needed data with others? How many time do manual operations take, the waiting for software action, the mental recognition of data, remembering the way of usage?

5.3 Societal problems

The third aspect of the study concerns societal questions. For example: Which differences result when people endowed with different skills and aptitudes (e. g. blue and white collar workers, jobless people, non-professionals, age, education) use tools and information? Which type of relations do the participants establish within the team and which concept of "justice" guides them? In which way do proprietary or free software products influence the elaboration of the productive aim? With which people inside enterprises is it possible to discuss features of needed parts? Which information is not yet existent on the Web or not accessible because lack of access rights? Does regulations influence the way the productive aim can be achieved? The challenges and issues from chapters 3 and 4 yield a raster of categories to differentiate such questions and to classify effectiveness and efficiency of the user interfaces, and respective problems. They also allow to study in more detail how society encourages people to go on to productive information society. Answers may be found through more or less detailed reports given by the participants, through interviews and through observations during work. All this yields an empirical basis of technical and social user needs and of the quality of user interfaces with respect to the productive aim in the case study which stands for a small slice of the productive information society. In cooperation with the participants one can derive optimizations. More studies of this type yield a basis for more general solutions, methods, and principles. One may also derive actual obstacles and limits for the productive information society as defined by skills, economy, industrial relations, and regulations.

6. Final remarks

Based on the idea of the procductive information society we sketched a re-integration of design and work, production and consumption, education and daily life. This gives raise to issues in technology but also in society intertwined with each other. The technical state of the art and societal barriers define the extent to which people actually can use the new technology. Barriers may lie in the division into blue and white collar work, in the production of goods as merchandise and in the legal framework concerning industrial relations (details in [3] and in [4]). Generally, labor is reflected socially by the work steps that fit together, technically by the parts forming an object and informationally by the technically supported descriptions of things and the accompanying communication. These three aspects of labor are closely interrelated, though not without contradictions. The Web helps to organize productive labor and to combine it with consumption resulting into more freedom and equality. In a coming productive information society the Web will mediate the acquisition of technology and it will reduce alienation of work. The division of labor into white and blue collar work tends to be removed. Appropiate tools, user interfaces, industrial relations, and regulations may enable *really all peopie* to participate in this process which should have been illustrated in the scenario. It could encourage to develop suitable conditions.

References

1. Brödner, P. (1997). *Der überlistete Odysseus. Über das zerrüttete Verhältnis von Menschen und Maschinen*. In: edition sigma rainer bohn verlag, Berlin.
2. Göhring, W. (1999). *Informationsurwald*. In: Marxistische Blätter 37, 6/99, 61—67.
3. Göhring, W. (1999, November). *The productive information society: a basis for sustainability*. GMD Report 72, Sankt Augustin (Germany).
4. Göhring, W. (2001). The productive information society as an alternative for the future. *Proposal for the 1st international conference on social computing (ICSC)*, Bremen.
5. Scheer, H. (1998). Chairman's Address. In *5th European Conference on Solar Energy in Architecture and Urban Planning*, May 27—30, Bonn, Book of Abstracts, p. iii.

Social Work Student Meets Computer –
A Curriculum for integrated education

Sigrun Goll

Evangelische Fachhochschule Hannover
(University of Applied Sciences)
Blumhardtstr. 2, 30625 Hannover, Germany

Abstract

Education concepts for areas such as social informatics can formulate general aims and contents; concrete ideas however are subject to constant change. Sustainable concepts should of course be orientated towards working practice, but at the same time not only be seen from the view of the organization or client involved. They should deal with the explicit needs of the social worker. The role or the needs of the social worker are of special relevance as frustration and work overload due to problematic working conditions can have serious consequences.

1. Introduction

The use of Information Technology (IT) in the social area is no longer a theoretical question but the long discussed advantages and disadvantages have been decided on. Applications of various kinds can be submitted online, and indeed many circulars will not be received without an e-mail address. Without online statistics and billing, social facilities are not able to claim any money at all.

In line with the requirements of working practice, there is a clear need to pass on the necessary competence to students of social work. The fact that the computer is used in almost all social facilities, however, relates neither to the quality improvements of social work, nor to the benefits the social workers draw from this fact.

It is often claimed that the true advantage of the computer is that it reduces administrative work, enabling the client to become the center of attention again. This opinion does not correspond to reality, because actually time is often not gained and therefore cannot be distributed. The general workload has grown so vastly because of the withdrawal of the welfare state simultaneous to an increase in poverty. Therefore, the administrative part of the work can no longer be dealt with without using the computer.

In addition, it is a widespread myth that the use of the computer per se leads to a reduction of the work load. Moreover, a number of new tasks have been added to the social worker's area of responsibility (computer aided evaluation, comprehensive obtaining of information, public relations work etc.).

These remarks should make clear that at least three aspects need to be considered when debating the use of IT within the area of social work. The benefits for both the facility and the client as well as the requirements regarding the working conditions of the social worker should be included within a curricular debate. IT and its consequences are currently leading to a comprehensive change in working conditions.

2. A Curriculum for social informatics

A Curriculum for a discipline such as the field of social services, which is so heavily influenced by socio-political regulations, necessarily includes aspects which are only transferable to other states with certain restrictions.

But in Germany too no kind of agreement will be reached in the near future on a general Curriculum, as two opinions exist concerning the education of social workers in the departments of social services. The discipline of social work is traditionally seen as a mixture of various disciplines (psychology, educational science, sociology, law etc.). Critics see the reason here for problems within the education of social workers, because they have difficulties developing a *professional identity* which is crucial for *professionalism*.

An interdisciplinary educational concept was designed at the Evangelischen Fachhochschule Hannover (University of Applied Sciences) that involved a binding framework for the lecturers with their varying professions.

This also gives an exotic discipline the chance to integrate itself instead of having to consider itself yet another "add-on" discipline. From the perspective of the interdisciplinary educational concept, a supplementary concept is evolved for social informatics.

2.1   Aims of a program for education in social informatics

The educational aims of all lectures on social services and IT are to establish competence amongst the students which enable them to make use of IT in a responsible manner whilst planning and carrying out the following:
- Independent acquisition of knowledge of general and special application programs
- Competence in judging the programs concerning suitability, software ergonomics and universal access
- Competence in formulating requirements towards the programs used within the social services
- Competence in integrating IT within specific fields of work
- Competence in judging the social implications of the use of IT

2.2   Integration within an interdisciplinary educational concept

2.2.1   Interdisciplinary educational area I: Social conditions (Social differentiation and inequalities)

a) What role do social conditions play in the use of IT – from the perspective of the clients?
When a field such as IT is considered to be very important – from an economic point of view – by the political and economical decision-makers, then demands such as "IT for All" are formulated, i.e. not only for those who can afford it or for those who don't have accessibility problems.
Although competence in dealing with computers is on the one hand comparable to the ability to read and write, a vast difference exists for the clients of the social workers as the material resources must be available in order to gain this competence.
Will it be the case that many members of our society will be deprived of their chance to take part in this technological advance due to their lack of competence and key qualifications which are required for success in the job and career market? What can be done in order to prevent this development?
b) What effects does IT have on the working conditions of the social worker?
Permanent renewals of hardware and software due to ever shorter development processes serve to downgrade the technical equipment along with the know-how required for its use. One could argue that a social worker is not responsible for technical maintenance – but reality is often different, however. Due to the financial situation of social facilities, technically qualified personnel are often lacking. Therefore social workers are often forced to take on their work.
Reports of working experience are of particular relevance to students in their educational phase, enabling them to form a picture of working reality in order to be able to evaluate the possibilities created by the use of the computer. Statements such as "We expected the computer would save work and not lead to additional work!" are typical for areas of work which have only recently been influenced by IT. Ignorance regarding the requirements arising from the use of IT, coupled with a lack of funds, often lead to problematic working conditions.

2.2.2   Interdisciplinary educational area II: Institutions and organizations (Stagnation and change)

Whilst using IT within social organizations and institutions, two aspects are of particular significance:
1. In the past, the introduction of IT was often not planned methodically – making use of systematic and central project management – at least not within smaller social institutions – but is more likely to be the result of particular efforts by individual employees or heads of department, making use of – decentralized – software programs in order to improve their working conditions.
2. The market for specific programs within the social service area is – at least in Germany – strongly competitive, as the number of customers for these programs is comparatively small. When selecting products, aspects such as quality and suitability of the product should be given special attention, especially with regard to the technical support (training, hotline) offered by the suppliers.

In general, the competence of a social worker includes the ability to establish conditions which enable an acceptable change in organizational and professional procedures for everyone involved. One important aspect here is a professional project management in which the requirements are established in co-operation with the users.
The working conditions include both the institutional procedures and the contacts between social workers and clients. One example is the design of a consultation with clients: What part should the use of a computer play here?

Should details resulting from the consultation be entered in the computer during or after the conversation? Will the client be informed of what will happen to the data, can he raise objections etc.?

Specific software, designed for the work within social institutes and organizations, includes documentation systems (adaptable for different areas of work) which enable electronic files to be kept. The use of such programs can be learned during lectures, while typical working situations can be simulated, for example, using role-play. Benefit can be drawn from both positive and negative experiences undergone by social workers in practice; these can be examined in interviews and debates. The continual dialogue between working practice and education is especially important here.

2.2.3 Interdisciplinary educational area III: Clients and target groups of social work (Life situations and ways of life)

Target group – specific use of IT

Whilst working with different target groups of social work, both specific programs and standard programs can be made use of. In work with children and adolescents, for example, computer games or Internet services can be utilized. Part of the education relates here to legal aspects such as youth protection and also educational science aspects such as youth violence.

For disabled or elderly people, IT can be of specific use; the area of Assistive Technology however currently plays a minor role in the education of social workers in Germany. For social workers, specialized knowledge of selected working aids forms a lesser aspect of their education than the question of "what is possible and what is available?". First of all, the interdisciplinary aspects of legal, psychological and educational aspects play an important role. The area of Assistive Technology is characterized by legal regulations which are constantly changing and by technological developments which offer a lasting increase of support. It is therefore logical to consider giving these aspects broader attention within future education.

2.2.4 Interdisciplinary educational area IV: strategies, competence and legitimization within the area of social services

Competence in dealing with IT

Part of this area of education is the acknowledgement of elementary knowledge of software for routine work, such as administrative tasks and acquisition of information. The aim of acquisition of software knowledge should be to develop strategies to support personnel in teaching themselves how to handle other programs.

Furthermore, practical competence includes the professional handling of programs especially useful for the social area, e.g. the information system REHADAT (Database for Rehabilitation) or consultancy advice programs for welfare aid. Virtual lectures will increase for students of social work too. An Internet-based continuation of education within the area of social work should be provided in order to enable students to gain experience and knowledge within the area of virtual communication.

3. Conclusion

The education of social workers within the area of social technology must not be limited to learning the use of application programs. The list of requirements for this education results from the various requirements of those involved: social facilities and organizations, social workers and clients. Social workers must be aware of all these requirements in order to be able to analyze their field of work and to plan and support the creation of a suitable working environment. An education program must offer support here in order to educate competent social workers.

References

van Lieshout, H., Roosenboom, P.(1995). From Teaching Computer Technology to Social Informatics. *New Technology in the Human Services*, 8 (2), 6-10

Hennig, T. (1999). Digitales Flickwerk? Erfahrungen aus einem Netzwerkprojekt im Rahmen der Hilfen zur Arbeit. *Sozialmagazin*, 24 (7-8), 46-52

Institut der Deutschen Wirtschaft (2001). Informationssystem zur beruflichen Rehabilitation http://www.rehadat.de

Wendt, W. R. (Ed.). (2000). Sozialinformatik: Stand und Perspektiven, Baden-Baden: Nomos Verl.-Ges.

Steyaert, J., Gould, N. (1999). Social services, social work, and information management: some European perspectives European Journal of Social Work 2 (2), 165-175

Information society and competencies:
Challenges for higher educational system and companies

Eila Järvenpää

Helsinki University of Technology, Department of Industrial Engineering and Management, TAI
Research Centre, P.O. Box 9500, 02015 HUT Finland
Tel +358-9-451 3653, Fax +358-9-4513665, email Eila.Jarvenpaa@hut.fi

Stina Immonen

Helsinki University of Technology, Department of Industrial Engineering and Management, TAI
Research Centre, P.O. Box 9500, 02015 HUT Finland, Stina.Immonen@hut.fi

Abstract

Development of ICT (Information and Communication Technologies), globalization of companies and industries
and efforts to develop societies towards information society have provided national educational systems with new
challenges. Wireless communication, ICT convergence, and media convergence are typical features of information
society. Majority of employees works in information and knowledge work, most of them using ICT as their tool.
The need for education models that are in line with the rapidly changing environment created by ICT is
overwhelming. This article will focus on three examples of competence development where universities and
business life are closely collaborating for mastering the demands of evolving information society. Competence
development will focus on needs of large companies, small and medium sized enterprises and new information
society professionals.

1. Introduction

Information society is the vision for future society, and today especially many Western countries have a lot of
characteristics of information society. Information and communication technologies (ICT) have a central role in
information society as an industry, a tool, a communication media and as the content of work. ICT industries are the
most rapidly growing industries in Western countries; the same trend is to be identified globally. Wireless
communication, ICT convergence, and media convergence are typical features of information society. Majority of
employees works in information and knowledge work, most of them using ICT as their tool. Most ICT companies
operate globally, both big multi-national companies with incremental globalization, and so-called born-global
companies operating globally from the beginning of their business. Figure 1 illustrates interrelations between
technology, work, business and competencies that together shape our information society.

Figure 1. Competence development in the focus of evolving information society.

In competence development, information society and globalization mean new challenges for higher education and for companies. Keil et al. (2001) state that the need for management education models that are in line with the rapidly changing environment created by ICT is overwhelming. They continue that the success and the competitiveness of the companies are based deeply on their employees, their motivation, skills and competencies. This challenges the future educational systems that will differ fundamentally from those we have been used during 1990's. First report on OECD Growth project (2000) states that the shares of better educated individuals both in the working-age population and among those employed have shown continued increase during past decades. This has been a long-term tendency and it does not itself explain why some countries have in terms of economics grown faster in the 1990s compared with the 1980s. However, a good education level may have helped some countries to cope better with the technological change. Countries that were successful in raising their GNP in the 1990s seemed to experience parallel rise in education levels among the employed and the working-age population.

In Finland during years 1989-1996, the share of individuals with higher education increased nearly 6 percentage points among those employed. At the same time, the share of higher educated people among working-age population raised only 3,5 percentage points. This is at least partly the result of weak employment among those who are less educated. This can also indicate that work life requires more, higher educated work force than before and competition over skilled employees is getting harder. For education system this might mean a challenge to increase amount of tertiary level education.

High education level in work life and innovative competence development seem to be key factors for success in the modern knowledge economy. It is not only a question of raising the overall education level but the implementation of new ICT means also comprehensive changes in competence requirements within many professions (Keil et al. 2001). The relationship between competence requirements and technological development can be described as a life-cycle of competencies (ibid.) Some competencies meet the end of their life-cycle when new technologies are adopted whereas new competence requirements are created at the same time. This article will focus on three examples of competence development where universities and business life are closely collaborating for mastering the demands of evolving information society. First two examples deal with possibilities of managing knowledge intensity and competence development in large and small and medium sized companies. Third example describes a multidisciplinary training program through which university proactively develops competencies that are needed as mediators between new information technology, business and society. Each of these examples will address what kind of competencies companies especially are searching for and how universities as higher education providers are answering to this demand.

2. Universities structuring and creating knowledge for industry and business

Universities have to provide their students with skills and competencies needed in global business in different industries and professions. Companies need well-educated and motivated people with best capabilities and competencies. For companies the challenge is to guarantee continuous competence development of their employees. Traditionally, universities have conducted their activities, teaching and research, without active co-operation in industry. Companies have trained their employees by in-house training or using established extension education institutions.

Coping with challenges of information society and globalization universities and companies need close and active co-operation. There are several ways for co-operation in teaching and research as well. For example, assignments, project works and theses can be conducted in companies, experts from industry can participate in university teaching and even give their expertise in curriculum development, and so on. Here we will discuss about two examples where co-operation between universities and companies was the platform for company competence development. The first example deals with PhD program for managers and experts in industry. The second example deals with competence development of SMEs (Small and Medium Sized companies).

2.1 Industrial PhD Program for large knowledge intensive companies

Industrial PhD Program is a full doctoral program at Executive School of Industrial Management at Helsinki University of Technology taking nominally three years. It ends with a PhD dissertation research. The program is intented to industrial candidates, who already have at least 3-5 years' experience in industry. The PhD dissertation research will focus on a real and important issue pertinent to the case company (i.e. the employer of the candidate) and its business environment. Each candidate will get a personal tutor, a professor, who will guide and tutor the research work of the candidate, and is committed to the research of the candidate.

The PhD program was developed together with university professors and top management of leading companies mainly in ICT industries. The goal of the program is to provide the PhD students with scientific knowledge and skills needed in solving complex problems in industry and business. The requirements of the program are equivalent to other PhD programs the university provides. The content of the program was focused on selected research areas relevant to industrial needs, including production, organizational behavior and business strategy. The topics of PhD dissertations deal with research problems that have both scientific and practical utilities. Among several students the research problem and/or data are from their company.

The service concept of the PhD program consists of the following elements. Thematic seminars provide the theoretical studies for PhD degree. Teaching methods consist of interactive seminars, writing essays and giving presentations. In dissertation seminars the students get advising, and feedback from professors and other students. During the entire PhD research individual supervising and advising are provided for students. Moreover, most students have also a supervisor in their own company. Professors, student and company supervisor meet several times during the PhD studies to advice and support the student and to share knowledge between the industry and the university. The PhD program helps the students also in time management. The program provides most of the teaching material in electronic format, communication about the program, seminars, etc. is regular and focused, and seminars are arranged on convenient times for students. Besides the PhD degrees the program has tightened the co-operation between the university and the program, and provided better understanding for the university about different industries. At the same time, the companies have understood the relevance and usefulness of scientific knowledge also in everyday business.

2.2   Knowledge dissemination and networking programs among SMEs

The other example deals with competence development of SMEs. Globalisation and new knowledge economy have affected also smaller enterprises' business environment in a significant way. Companies search for competitive advantage, for example, from strategic partnerships and networks. These new collaboration forms enhance firms' possibilities to compete in speed and flexibility. However, emphasis on business to business and customer care relationships sets new kinds of requirements for small and medium sized enterprises. Business, company and occupation specific, as well as personal competencies within these enterprises must be updated. Competence development requires both financial and work force resources that usually are scarce in SMEs. Typically, in Western countries most of the enterprises are small ones. For example, about 94 % of Finnish companies have 10 or less employees. To keep also SMEs on the right track along information society's development, universities should indicate more interest and motivation towards SMEs' reality.

Some steps have already been taken. European Social Fund programs of European Comission, for example promoting expertise and innovation network in Finland aims to increase collaboration between universities, large and small companies. Networking helps small companies to combine technology strategy, business strategy, and human resource strategy in a way that increase company's possibilities to survive in the global competition. However, only gradually research and university institutes have started to show their interest towards projects focusing on SMEs.

About half of the 80 projects developing centres of expertise carried out in Finland during years 1995-1998 have dealt either with business skills, entrepreneurship, export, company networks or demand - supply chain management development among SMEs in collaboration with universities. One third of the projects have concentrated on developing key individuals' competencies. Rest of the projects has operated more on strategic level creating regional or business area based networks to support expertise dissemination between universities and companies. About 40 projects of developing new teaching and learning methods for SMEs' personnel have either developed computer aided learning environments or produced teaching materials to be supplied in internet. Also in these projects, universities and other training organisations have been the main content providers.

Typical project organisers come from established training or consulting organisations that are used to operate with projects. However, research institutes or universities have not been very keen to start development projects with

SMEs although they would have the expertise needed. Projects are not supposed to be research-oriented but very practical. Maybe this is one factor that eliminates universities from participating actively in promoting expertise in SMEs.

In information society the co-operation between universities and companies is a very important way in developing competencies, both for universities and companies. Our both examples indicate the usefulness of the co-operation. However, more goal-directed efforts, learning experiences, and knowledge sharing are needed to develop the co-operation and to increase understanding about the importance, benefits and ways of the co-operation.

3. Universities creating future competencies for information society

Universities are also looking ways to answer in a proactive manner to future competence needs. Wide diffusion of information and communication technology to all sectors of the society has brought to surface a need of new kind of multidisciplinary competence. Our third example describes an engineering degree program, the Information Networks Master's program at Helsinki University of Technology. It was started in 1999 to combine information and communication technology, social sciences, and industrial engineering and management together in order to educate engineers who can act as links between developers and end-users of modern technology. The Master's program has 5 majors. A student can specialize herself either to content production, to human – information system interface, to entrepreneurship and business management of knowledge economy firms, to media technology and interactive digital media or to communications in organizations. Each of these majors aims to lower the barrier between technology designers and consumers. For example, applications of information and communication technology have still very often usability challenges to be solved. From technology user's point of view the critical point is more often the interface between several systems and the human. So, the scope must be much wider than a single human-computer interaction. This is why the studies will contain besides mathematics and computer science and engineering also sociology, philosophy, aesthetics and communication.

The Master's program is at its very beginning so no real impacts can be evaluated yet. The intake of students is 35 per year. Contrary to other Master's program at the Helsinki University of Technology applicants will have as entrance examination a "learning-from-text" test and an aptitude test. Examination in mathematics is a joint test with other Master's degree programs. In the program, the teaching is mainly based on so-called problem –based learning. The students conduct a majority of their assignments in companies, getting familiar with real life problems the companies have e.g., in computer science, communication, or business development.

In Spring 2000 there were 292 applicants of which about 43 % were women. Of the accepted students the gender ratio was approximately the same. On average, the share of women is only around 20 % in other Master's programs at Helsinki University of Technology. Chemical Technology is a distinct exception of this pattern. There the share of female students is about 50 %. However, this new study program of information networks seems to attract also female students. In the area of technology this is a welcomed concomitant of competence development for the information society.

4. Conclusions

The three examples of universities' interplay with business and industry have illustrated the significance higher education has in development of information society. On macro-level, whole nations might need new competence when structures in industrial and business life change rapidly and globalization increases multi-culturalism in societies. On meso-level, industries and companies within them encounter global competition more easily than before and companies need to be prepared for this competition. Competence development by networking can be the only possibility for smaller companies to utilize for example the research and development work done in universities and research centers. On micro-level, also individuals need new competencies to cope with the rapid evolution of their complex information environment.

Large companies, especially in the area of information and communication technology and global business seems to be interested in competencies universities have traditionally transferred to their researchers. Growing interest towards industrial PhD programs is one indication of this. Companies want their key employees to have skills of analytical thinking, restructuring, problem solving and using scientific methods even if they are not working as researchers in the company. This might be an indication of how complex the business environment is experienced even in large companies.

Small and medium sized companies (SME) are in many countries the basis of national well-being since they are so numerous and significant employers in societies. However, these companies have lesser resources to develop their competence. For example, ICT sector has many examples that outstanding technological competence does not guarantee successful business by itself. Several software and internet firms' business flaws could have been avoided if these companies had sufficient business competence as well. Another kind of pressure for the SMEs is that often they are compelled to share the risks of larger companies as their subcontractor or contract manufacturer. Networking with larger and possibly with international companies requires often continuous updating of production and quality competence. To reach this status SMEs often require support from different kinds of development programs where the costs are even shared with several partners. Universities could act here as significant networking promoters by applying public governmental or EU funding for projects where SMEs competencies are developed.

On individual level, universities are platforms for competence development that allows totally new kind of learning forms and learning contents to emerge. Since the interplay between universities and industrial and business life is getting more intense, more students from work life is attending Master's level and PhD level teaching. The platform universities are providing for all these individuals for learning must be dynamic and accommodating to the discontinuities that the future with a great certainty will have. Universities have also an important task in attracting more women in the field of technology. Applications of information and communication technology might be a new kind of forerunner to get more women involved to designing and developing technologies needed in information society.

References

Keil, T., Eloranta, E., Holmström, J., Järvenpää, E., Takala, M., Autio, E. and Hawk, D. (2001) *Information and communication technology driven business transformation. A call for research*. Computers in Industry, forthcoming.

OECD (2000). *Is there a new economy? First report on the OECD Growth project.*

Designing Trust for a Universal Audience: A Multicultural Study on the Formation of Trust in the Internet in the Nordic Countries

Kristiina Karvonen

Department of Computer Science, Helsinki University of Technology
P.O.Box 9700 HUT, Finland

Abstract

This paper presents a multicultural comparison on users' perception on computer security issues in Finland, Sweden, and Iceland. Special emphasis is given to the notion of trust: how trusting relations are formed both in an electronic environment, namely, the Internet, as well as outside it, in the real world. A set of user interviews, together with walkthroughs of existing Web sites of some e-commerce service providers, were conducted in all these three countries, in order to track down how cultural variation might affect users' perception of computer security, as well as their trusting behaviour. The outcome of the study is an understanding that cultural variation indeed plays a major role in trust-forming, and that the online behaviour of the users representing these three Nordic countries was to some extent similar, but at some points quite dissimilar to one another.

1. Introduction

The rise and fall of e-commerce depends on finding ways to ensure customers that online transactions of money are in fact secure. Internet was not created with security in mind, and has not provided for maximum security till now. However, things are changing, as the Internet society has grown more security-aware. Quite recently, the universal audience interested in the services that e-commerce can offer, has learned to demand some signs of trustworthiness from the service provider to guarantee that their privacy is not at risk. These makings of trustworthiness are, however, likely to vary from one culture to another. What creates trust in a user in one culture may not do the same for a user coming from another culture with a different cultural background. However, Internet has a truly global user base, and it is, thus, important to study the makings of trust in different cultures, in order to be able to provide for trustworthiness for everyone - either through a universal design for all, or through localised design versions of the same service. Whichever design approach we choose, the decisive factor behind the suggested solution should be that it acknowledges and addresses the cultural factors as fully as possible.

This paper presents results from ongoing studies on what promotes trustworthiness towards Web services in Finland, Sweden and Iceland, conducted in 1999-2000. The methodology to gather the research data was to use a combination of user interviews together with evaluations of existing Web services. In all, 35 users from the three countries took part in the user investigation: 20 from Finland, 10 from Sweden, and 5 from Iceland. The areas considered relevant for online behaviour as regards trust were studied, in order to get insight into the makings of the feeling of trust in general, both in and outside of the Web, and how this knowledge might be applicaple to behaviour in security-prone use situations online. These areas include the attitude towards and use of computers, phones, Internet, e-mail, passwords, and e-commerce, as well as bankcards, credit cards, and ATMs. Also, the sources of information the decision to trust was based on, was studied: who users trust in general, do they trust their bank, do they trust their local newspaper, and where do they gather their information from - from hearsay, rumours, or from various media? And, how does this information affect their behaviour in the Web?

Among the research questions that this study, in its part, tries to answer, are the following:

- How can we investigate into the effects of culture in understanding attitudes towards computer security (especially trust issues)?
- How should we define "culture" in this context? Can national borders be used as cultural borders as well?
- How weighty are cultural considerations for the overall understanding of computer security issues (and, more specifically, trust)?

- Does such a global environment, as the Internet, bring forth its own culture that can override the effects that cultural background otherwise would have on users perceptions of it? Is the Internet strong enough as a global medium to have a culture of its own, perceived in a similar manner regardless of the cultural background of the perveiver?

The above problem areas have served as a starting point for our enquiry, and we have tried to answer these questions to some extent through our studies. The last item in the list demands for a further explanation. It is important to understand that the Internet might have a culture of its own that overrides other, culture-specific expectations users might have. For example, the Web has been described as a low-trust society (Nielsen 1999, March). What this means that in the Web, we currently have a *culture of untrustworthiness*, rather than one of trustworthiness: it is natural not to trust but to suspect every service provider and information that is encountered in the Web. This kind expectations that have to do with the specific nature of the medium we are dealing with, must also be acknowledged and analysed, and taken into account, in creating culturally-adapted interfaces for, say, services, inside this medium.

As results, our paper points out in what various ways the makings of trust, needed to be willing to conduct money transactions online, for example, differ a great deal, depending on the cultural background of the customers using these services. Even in such neighbouring countries as the Nordic countries, the differences among attitudes towards what is trustworthy and what is not, seems to vary a great deal. Also, what is considered private, is a different thing in Iceland than it is in Finland and Sweden. All these differences are listed and analysed. The paper ends with a list of suggestions for a successful design for creating trustworthiness in the Web for all its users.

The rest of the paper is organised as follows: First, we will have a short overlook on previous studies on the impact of culture in usability issues. Then, we will have a look at the results of the studies conducted in Finland and in Sweden. Next, we will describe the results of the Icelandic user studies, and compare them with the results of the previous studies, especially focusing on the cultural differences among the three countries. We will conclude our presentation by design recommendations and suggestions for further research on culture-sensitive design approach for Web design.

2. Previous work on studying cultural variation in usability

Culture plays an important part in how people perceive and interpret information (Del Galdo and Nielsen 1997; Fernandes 1995). This is also true of how people perceive and interpret information about computer security. Security for users includes issues such as feeling secure, feeling private and feeling trustful (Karvonen 1999). In our studies, we have placed special emphasis on the notion of *trust*: how trust is formed, both in and outside the Internet, and on basis of what do people decide to trust someone or something? Anyhow, trusting is basically a social phenomenon (Seligman 1997) that inevitably is affected by cultural factors.

Recognising the impact of cultural background for usability issues is a corner-stone for creating truly universal design - which is something that we need in the era of such global communications and exchanges that the new media, especially the Internet, have provided us with. The importance of cultural effects is well known in the HCI field (e.g. Del Galdo and Nielsen 1997; Fernandes 1995; Shneiderman 2000; Järvenpää and Tractinsky 1999), but still many cultural issues remain unresolved. Small wonder, for the effects of culture on users' perception of usability issues is far too wide and undefined to be easy to grasp, far less to solve here and now. First problem rises already when we try to define *culture*, or, *cultural impact*. How are we going to manage to do something that has been tried over and over again by historians, sociologists, anthropologists, just to name a few, for centuries?

To suit our purposes of studying cultural variation in trust-forming behaviour, it might be best not to try to resolve the problem of how to define culture in general, but to find a definition that will be accurate enough for such a study. A solution that easily comes to mind is to draw cultural borders over national borders - that is, to treat various nationalities as culturally homogenous entities. Common sense already tells us that in doing this we will lose a whole lot of information about cultural variety within a country that more often than not are made of people with various ethnic and cultural backgrounds. For some countries, such as USA, for example, such an approach would be most inappropriate due to the huge cultural variety within that single nation. Also, by using national borders as defining borders for culture we cannot take into account the various subcultures, and other minorities inside these nations that may have their very own but strong cultural rules and likes and dislikes. However, we must choose to research on cultural effects in some way, and national borders do give us at least some kind of, even if coarse, cultural border that we can investigate. This is exactly what we have done in our studies: we have treated the Finnish, Swedish, and Icelandic users as culturally homogeneous groups that differ from each other culturally, at least to some extent. Now we will have a look on the results of these studies each in turn.

3. Acting wary and reserved: the results of the Finnish user studies

The results of the Finnish user studies have been fully reported in (Karvonen 1999), so here we will settle for drawing a general outline of the results of these studies.

The Finnish user studies were conducted as part of the TeSSA (short for Telecommunications Software Security Architecture) project at Helsinki University of Technology. In TeSSA, The main result has been the creation of a security architecture for untrusted networks such as the Internet. The architecture has been implemented using authorisation certificates, and therefore it supports anonymity, delegation, and dynamic distributed policy management. Usability was stressed as an essential ingredient of the design of this software architecture, and the user studies were an attempt to better understand the current amount of knowledge and understanding that an "average user" has on computer security issues.

The study methodology was mainly using ethnographic techniques, including both interviewing the users, together with both non-participatory and participatory observation of either going through a mock-up UI for an online movie service, or existing Web services. The users were graduate or post-graduate students at a university, aged 22 to 30 years, 8 male and 10 female. No background information about computer security was expected, just a basic knowledge of computers and the Internet. The study was strongly affected by the (ECommerce Trust Study 1999), on which the structure or the Web site walkthroughs were based. In the user interviews, the users were asked about various areas that seemed relevant to behaviour in security-prone situations. These included use of e-mail, use of credit cards, use of online banking services, use of ATMs and use of e-commerce.

The user studies showed that the current level of understanding of computer security was far from satisfactory. The trust decisions made about online purchasing were based on miscellaneous information gathered from friends, colleagues, and newspapers, rather than on use manuals or the security information provided by the service provider. The users were more ready to trust a service that existed also in the real world, for example, the online services of their bank. Users were also rather ready to trade-off with their trust level: if they really needed something that they only could get hold of through a Web retailer, they were ready to take the risk and give away their credit card number to the service provider. For example, the use of e-mail had become such an essential part of these users' daily communication that even though they were not one hundred percent sure that their e-mail was completely private, they were ready to take the risk of someone eavesdropping them, since staying in touch with other people was considered so important. In general it can be said that people were making decisions to trust on basis of the opinions of the people that were close to them, or whom they trusted to have technological expertise. This was so to some extent because information about security features was considered to be hard to get hold of, and hard to understand.

In Web site design, Finnish users valued simplicity of design. Simple design was considered to make the transaction process more easy to follow and to control. This result is, in a way, no surprise in many ways: firstly, the so-called "usability purists" have long promoted simple design as key to attracting customers (see, e.g., Nielsen 1999). Secondly, Finland is known for its taste for simplicity, and purity in design - small wonder then that this preference is reflected in the preferred Web site design as well.

4. Relaxed and careless: results of the Swedish user interviews

The Swedish user studies were conducted in Sweden in spring 2000. There were 10 users in all, all with a university degree, aged 34 to 56 years, 5 male and 5 female. Going through the mock-up UI was dropped; instead, only the user interviews and the Web site walkthroughs were repeated. The interview structure was the same as in Finland, but the Web sites were changed for ones that were best known among Swedish users. The outcome of the Swedish user study has been previously reported in more detail in (Karvonen et.al.2000).

The Swedish users were more relaxed and careless compared with their Finnish counterparts in their overall attitude towards computer security and conducting e-commerce. The Swedish users were more indulged in conducting transactions online, and their wishes about the quality of these services was more refined, and had more to do with the details of the transaction process - they were no longer wondering whether or not to do transactions online at all, unlike the Finnish users. Also, the Swedish users expressed a certain carelessness in their trust decisions, as compared with the Finns: the Swedes acknowledged that in most cases, the decision to trust a service provider enough to indulge in conducting e-commerce was based on an intuition that was mostly an emotional matter. The service seemed "pleasant" and "trustworthy", so they decided to trust it. This difference between the "more rational" Finns and "more emotional" Swedes is a cultural difference: in the Finnish culture, emotions are usually hidden, and rational thinking is greatly appreciated, whereas in Sweden, emotions can be expressed more freely, and they are given more importance also in decision-making than in Finland.

However, in preferred Web site design the Swedes expressed similar preferences as the Finns, favouring, again, simple design. Clarity ("klarhet" in Swedish) was the word used for design that was considered most trustworthy. The wish for clear design was especially strong for the transaction process, where all the unnecessary fancy features were found undesirable and distracting by most users. So at least in this respect the Finnish and the Swedish users expressed similar preferences across cultural borders.

5. The village of trust: results of the Icelandic user interviews

In fall 2000, the user studies were repeated in Iceland. This time there were only 5 users aged 21 to 37 years, of which 4 were university students and one had a university degree. There were four female and one male user. Again, the user interviews were the same, but the Web site walkthroughs were changed. The Web sites chosen for the walkthroughs were taken from (Cheskin Research 2000). The users went through two most trusted and untrusted sites according to this study.

The outcome of the user interviews with the Icelandic users was very different from the Finnish and Swedish studies. Iceland is a small and remote island with a small population of less than 300 000 people in all. What this means is that to a great extent, the Icelanders have remained "villagers": most people know each other, if not directly, then through a middleman, and this is evident in all trust structures. It also means that news travel fast in Iceland: if someone breaks the rules, everybody will soon know it. Take an exemplary excerpt from one interview:

> "Someone broke into a Webpage here, the prime court webpage pictures were changed, with the Prime Minister with sunglasses on, this was in all the news, and then people got a little scared, and they took some pics out. Maybe we are a bit naïve about these things, our society is small, and we are not so fixed on rules, and so on, we trust each other. In Copenhagen [where the user had been staying] they use the rules all the time, why don't they just trust what I'm saying."

A common way to find out about the trustworthiness of a Web service provider that was previously unknown for the user was to pick up the phone and call a friend to ask if he or she had already used that service. However, even if such human networks provided some basis for accurate decision-making online for the Icelanders, they were according to most users alarmingly trusting towards complete strangers as well. Take another excerpt from the user interviews:

> "We are not so worried about that [about giving away one's credit card number online] either, also over the phone to salesmen you can give your credit card number. There was a misuse case - some lady gave it to a guy over the phone and it was then misused. It was in the news, so I stopped doing that... I discussed what had happened with some friends."

The borders of privacy are very wide and flexible in Iceland, as compared with Finland and Sweden. For example, on basis of a mobile phone number, it is easy to find a lot of information of the owner of that phone number on Icelandic Web sites, including the person's name, address, how many other people live in the same address and their names, social security number, and in what bank the person has a bank account in. In Finland and Sweden, most if not all of this information is considered very private, and is usually not very easy to find from the Web. This is especially the case with the social security number, which is considered to confidential information that should be kept private in Finland and Sweden at least. So we can easily see that the privacy and trusting attitudes are very different indeed between Icelanders on one hand and Swedes and Finns on the other.

However, in Web site design the Icelanders again expressed similar preferences as the other Nordic users. Simplicity and clarity were, again, preferred. According to the testimony of most Icelandic users, the Icelanders were very interested in design, and aesthetics played a great role in their house decorating, for example. This was considered to be so due to the severe climate of Iceland that forced people to stay inside for most of the winter time - it was important to make the home as cosy and beautiful as possible. This was emphasized by the fact that the Icelanders preferred inviting people to their homes, rather than meeting in restaurants: the house was decorated to please the visitors as well. So, it might be expected that aesthetic design might be a big factor in evaluating Web sites as well. However, simplicity proved to be more important also among the Icelandic users.

1082

6. Conclusions

As we have just seen, cultural variance is a reality in comprehending security-related issues both online and offline. In this study, the Finnish users were most suspicious and wary about Web retailers, and demanded a high level of trust in order to be willing to trust an online service provider enough to be willing to conduct transactions of money online. The Icelandic users, however, were the most trusting, transferring their traditional trusting behaviour from real world also to online environments, demanding very low level of trust (or, having a very high level of *initial* trust) in order to be willing to conduct transactions online. The Swedish users seem to fall somewhere inbetween, acting more carelessly than their Finnish counterparts but more rationally or suspiciously than the Icelandic users. On basis of the results of these user studies we can safely claim that cultural differences do indeed matter for online behaviour, and the trust decisions are based on different ingredients among users from different cultures.

However, even if a difference in cultural backgrounds can explain differences in expressed online behaviour, for example, in readiness and willingness to trust others, be they other people or services on the Web, culture is not always a defining factor for all similarities or dissimilarities in users' behaviour, be it online or not. A good example of this fact is the finding that all the users in our studies preferred simple design, when they were wondering whether or not to trust a service provider on the Web. It is thus clear that if we want to create design that is as culturally-inclusive as possible, i.e. in the case of trust, the kind of design that promotes the feeling of trustworthiness in all our users, be they from whichever culture, we need to understand what features of, say, the Web design are culturally-dependent and which are not. In this study, simplicity of design was a feature of the design that was universal at least across three national borders, perhaps expressing a cultural feature that can be applied to all Nordic countries - a true feature of a joint *Nordic* culture. Further research may show, whether this preference for simplicity has an even wider scope - whether it is a truly *universal* design preference.

Acknowledgements
This research is based on a previous work conducted jointly with Lucas Cardholm and Stefan Karlsson from Prevas AB, Sweden. The user interviews conducted in Iceland would not have been possible without the kind assistance of Ingimar Fridriksson and the kind personnel of the Reykjavik University, Iceland.

References
Cardholm, L (1999). Building Trust in an Electronic Environment. *Proceedings of the Fourth Nordic Workshop on Secure IT Systems (Nordsec'99)*, November 1-2, 1999, Kista, Sweden, pp. 5-20.
Cheskin Research (2000, July). Trust in the Wired Americas. http://www.cheskin.com/think/studies/trust2.html
Del Galdo, E. and Nielsen, J. (eds.) (1997). International User Interfaces. Wiley and Sons, Inc., New York.
ECommerce Trust Study (1999, January). Joint Research Project by Cheskin Research and Studio Archetype/Sapient. http://www.studioarchetype.com/cheskin
Fernandes, T. (1995). Global Interface Design. AP Professional, San Diego.
Järvenpää, S.L, Tractinsky, N. (1999). Consumer Trust in an Internet Store: A Cross-Cultural Validation. *Journal of Computer-Mediated Communication*, Vol.5 (2), December 1999, http://www.ascusc.org/icmc/
Karvonen, K (1999). Creating Trust. *Proceedings of the Fourth Nordic Workshop on Secure IT Systems (Nordsec'99)*, November 1-2, 1999, Kista, Sweden, pp. 21-36.
Karvonen, K et.al (2000). Cultures of Trust: A Cross-Cultural Study on the Formation of Trust in an Electronic Environment. Proceedings of the Fifth Nordic Workshop on Secure IT Systems, NordSec 2000, 12-13 October, 2000, Reykjavik, Iceland, pp. 89-100.
Nielsen, J. (1999). Designing Web Usability. The Practice of Simplicity. New Riders Publishing, Indiana.
Nielsen, J. (1999, March 7). Trust of Bust: Communicating Trustworthiness in Web Design. Jacob Nielsen's Alertbox. http://www.useit.com/alertbox/990307.htm
Seligman, A.B. (1997). *The Problem of Trust*. Princeton University Press, New Jersey.
Shneiderman, B. (2000). Universal Usability. *Communications of the ACM*, May 2000, Vol.43, No.5, p. 85-91.

Universal usability for web sites: current trends in the U.S. Law

*Jonathan Lazar**
Department of Computer and Information Sciences
Towson University
8000 York Road
Towson, MD 21252
Internet: jlazar@towson.edu

Libby Kumin
Department of Speech-Language Pathology
Loyola College
Baltimore, MD 21210

Shawn Wolsey
School of Law
University of Baltimore
Baltimore, MD 21201

Abstract

Federal legislation in the United States mandates that accommodations be made to provide physical access to buildings and facilities for individuals with disabilities. A new issue is the accessibility of web sites for people with disabilities. Many people with disabilities access content on the web, using adaptive technology such as screen readers. Universal usability is the practice of designing information systems that are accessible by different populations with different needs. Web sites need to be designed so that they are universally usable by all users, including users with disabilities. The law is currently not clear about whether private companies must make their web sites accessible, however, U.S. Federal Government web sites will soon be required to be accessible. This paper will discuss the concept of universal usability for web sites, how to design for universal usability, how users with disabilities access the web, and will then examine the current legal status of web site accessibility for both private companies and the federal government.

1. Introduction

Web usability is the study of designing web sites that are easy to use. An increasing concern is designing information systems and web sites that all users can access-this is known as universal usability. When web sites are designed, it needs to be understood that there will be a wide variety of users accessing the web site. Some of these users are likely to include those with disabilities such as visual impairment. Certain design features, and lack of attention to detail, may hinder the user who wants to or needs to view a web site without graphics. This becomes a bigger problem when considering that some people use screen readers, which read the text on the screen through speech synthesis. Although there are many different aspects of universal usability (age, education, connection speed, browser compatibility), the fact remains that most users do not currently have legal standing to sue in the United States because of universal usability. Users are not able to sue if, for instance, they have an older browser or if they have a slow connection speed. However, in many cases, users with disabilities do have the right to sue if they are disenfranchised. Therefore, this paper focuses on the legal aspects of universal usability as it relates to users with disabilities.

In the United States, The Americans With Disabilities Act (ADA), along with case law, requires that accommodations be made for people with disabilities in public places. For instance, physical changes such as ramps and curb cuts must be provided so that those in wheelchairs can enter and access public buildings. The idea behind the ADA is that those with

---

* All correspondence should be with the first author

disabilities should have the same opportunities and access to public places as everyone else. These rules may soon apply to designing web sites that are accessible to all. In addition, by June 21, 2001 (exactly 6 months after the date that the final guidelines were released), U.S. Federal Government web sites will be required to conform to these universal usability standards, due to Section 508 of the Rehabilitation Act. How can web sites be made accessible? What implications will this have on web design? What consequences can be expected for web sites that are not in compliance? These are some of the questions that this paper will attempt to address.

2. Universal usability for web sites

Any type of information system that is built needs to meet the needs of the users (Norman and Draper, 1986). A web site is a type of information system, and therefore must be built to meet the functionality and usability needs of the targeted user population (Lazar, 2001). Unlike traditional information systems, web sites can be viewed under thousands of possible technology conditions, including different browsers (Internet Explorer, Netscape Navigator, Lynx), different browser versions (3.0, 4.0), different platforms (Windows, Linux, Mac), and different download speeds (28.8, 56k, DSL). Universal usability is the study of how to develop information systems that are easy to use in a wide range of conditions and user populations (Shneiderman, 2000).

Because web sites can be viewed from so many different technological environments, they need to be designed for flexibility. For instance, users may be accessing a web site using an older version of a web browser. Users may be accessing a web site from a text-based browser, such as Lynx. It is unwise to design a web site that can only be viewed properly by those using a specific version of a specific web browser. For instance, a web site that can only be viewed properly using Internet Explorer 5.0 will most likely be an unsuccessful web site. It can be tempting to build a web site that uses the most up-to-date technology, however, web designers need to remember that web sites should be backward-compatible, so that the web site is accessible to those using older browser versions or slow connection speeds. Many users do not have the technological expertise to upgrade browser versions, or their local network does not allow them to modify application files (Lazar, 2001). In addition, many users may be accessing the web using slower connection speeds, due to cost. Other users, who cannot afford computers of their own, may be accessing the web from schools, libraries, or community technology centers, which can provide free access, but may have time limits on usage, and may not be able to afford the most expensive connections (such as DSL, cable modem, or T1). A web site needs to be designed with all of these users in mind. For instance, figure 1 shows the Washington Post web site as it appears in a graphical browser. Figure 2 shows that same web site as it appears in a textual browser. The web site is still usable in a text-only setting.

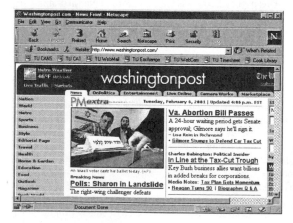

Figure 1. A web site as it appears in a graphical browser

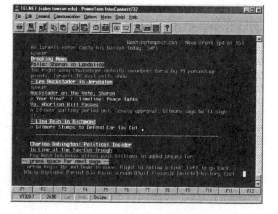

Figure 2. A web site as it appears in a textual

3. Designing for universal usability

The World Wide Web Consortium has created a set of guidelines for ensuring accessibility. These guidelines are known as the Web Accessibility Initiative (WAI), and are available at http://www.w3.org/WAI . These guidelines for web design can help to ensure that web sites can be viewed properly from many different environments, including by those users who may have a disability. Users who have a disability may need special adaptive devices, such as speech synthesizers and Braille displays, however, these users still must be considered when designing web sites. One of the most important guidelines for universal usability of web sites is to include textual equivalents for graphical or auditory content. For any graphical images that are included on a web page, a textual representation should be included, for those who are viewing web pages using graphical browsers with the graphics turned off, or textual browsers such as Lynx. In addition, web site navigation should be provided using text, so that those using a text browser can easily navigate through the web site (Lazar, 2001). Text is the most portable because it can easily be processed by speech synthesizers, Braille displays, and other adaptive devices. Graphics are displayed within a web page using the following HTML:

<img src="imagename.jpg">

where:
img is the HTML tag to display an image
src="imagename.jpg" indicates that the source file for the image is named imagename.jpg

Alternative text that will display when the graphic does not display can be included using the alt attribute in the following manner:
<img src="imagename.jpg" alt="a picture of Bill Clinton">

Textual equivalents of graphical images should always be included when designing web pages. Not only do some people use text-based browsers, but some use graphical browsers and turn off the graphics to improve the download time. In addition, when users are viewing web pages with a graphical browser, the *alt* text will appear as a "bubble" when the user places the mouse pointer over the graphic. There is simply no reason not to include <alt> information for all graphics. If a web site cannot be designed without heavy use of graphics, then it might be a good idea to design a parallel, but equivalent, text-based web site that offers the same content in a non-graphical form. Web sites such as The Seattle Times http://www.seattletimes.com have such a "parallel" web site. This non-graphical web site can also be used by those who want a quicker download time. Web sites can and must be designed to be accessible by all.

4. Web access for those with disabilities

Web designers sometimes assume that people with disabilities cannot access web sites. This is absolutely not the case. Users with disabilities may use adaptive devices, such as screen reader software, which will read the text on the screen through a speech synthesizer. Software such as Dolphin Software's *Hal* (http://www.dolphinusa.com), IBM's Home Page Reader (http://www.ibm.com), and AccessAbility's JAWS, allow users with vision impairment to still partake of the content available on the World Wide Web. Although users with disabilities can access web sites, this is only true when web sites are designed to be accessible. Although accessibility for those with disabilities should always be an important goal, a large percentage of web sites are still designed in an inaccessible manner (Rowan, et. al. 2000). This might be part of the reason why users with disabilities use the web far less frequently than the general U.S. population (Crockett, 2000). The statistic that Internet access for those with disabilities is half of the U.S. average, was recently cited by President George W. Bush, in a new policy initiative that will increase accessibility to technology for those with disabilities (Bush, 2001).

Those who use screen readers have the same needs as 1) those who browse with text browsers and 2) those who use graphical browsers with the graphics turned off to download quicker. To ignore these large user populations is simply bad design. It seems like good business sense for companies to create accessible web sites, as users with disabilities are likely to purchase items online (Tedeschi, 2001). One of the barriers to re-designing web sites for accessibility is the cost (Tedeschi, 2001). To assist in designing accessible web sites, a web site called Bobby allows web designers to test their web pages for common accessibility problems. The web site is located at http://www.cast.org/bobby/. Rowan et. al. (2000) describe other methods of evaluating web resources to meet the needs of users with disabilities. In addition, if it is expected that web sites will have large numbers of users with disabilities, then users with disabilities should be included in the design process for the web site. More information about techniques for including user input in the design process is available in the book *User-Centered Web Development* (Lazar, 2001).

5. The Americans with Disabilities Act and the Rehabilitation Act

The Americans With Disabilities Act (ADA) prohibits discrimination on the basis of disability in: 1) employment, 2) public accommodations, 3) commercial facilities, 4) state and local government, 5) transportation, and 6) telecommunications. The ADA does not specifically name all of the disability areas covered, but uses a more general definition, that an individual with a disability is defined as a person who has a mental or physical condition that substantially limits one or more major life activities, has a history or record of such an impairment, or is perceived by others as having such an impairment. Title I of the ADA relates to employment in companies having more than 15 employees, where the employer must make "reasonable accommodations" to enable a qualified employee with a disability to perform essential job functions. Title II of the ADA relates to activities of state and local government, which includes access to information. Title III relates to public accommodation, and includes provisions for ensuring "effective communication with people with hearing, vision, or speech disabilities." This may be accomplished through the use of assistive listening devices, videotext displays, taped texts, and Braille or large-print materials. Title IV of the ADA specifically addresses telecommunication relay services, and requires telephone companies to establish 24-hour telecommunication relay services, as well as adaptive telecommunication devices. The ADA does not, however, specifically address web sites.

While the ADA relates to almost all organizations, there are more specific rules for the U.S. Federal government. The Rehabilitation Act of 1973 was amended in August 1998, when President Clinton signed the Workforce Investment Act of 1998. Section 508 of the Rehabilitation Act, which relates to technology access for those with disabilities, was amended to require that all federal agencies only purchase new equipment that is accessible to people with disabilities. In addition, all federal web sites must be modified to provide accessibility to those with disabilities, by providing textual equivalents. These rules take effect by the end of June, 2001. That means that all Federal Web Sites must be accessible by that date. This ruling may also apply to downloadable files in Adobe PDF or MS-PowerPoint formats, because text is presented as a graphical image, which is not readable by those who rely on textual equivalents. Although these Section 508 rulings apply to the U.S. Federal Government and agencies, they do not apply to private organizations. For more information about Section 508, go to http://www.section508.gov/.

6. Legal status of private businesses

While Section 508 applies to the U.S. Federal Government and requires accessibility, accessibility rules for private companies are unclear. It has not been firmly established that the ADA applies to the web sites of private companies. A number of companies have had claims filed against them for having web sites that are non-accessible, but these suits have been settled out of court. For example, America Online was sued by the National Federation for the Blind for having an inaccessible web site (Tedeschi, 2001).

The National Federation of the Blind, as well as numerous plaintiffs, filed a claim against America Online(AOL) stating that AOL internet service is subject to the Americans with Disabilities Act (ADA) 42 United States Code (USC) §§ 12101 et seq. and that AOL did not comply with the ADA.. Through a settlement agreement, AOL has adapted and followed the guidelines from Section 508 of the Rehabilitation Act (29 USC 794d 2000). There are two major guidelines for Section 508. The first is a requirement that organizations clearly define their electronic and information technology policy (29 USC 794d (a)(2)(A)(i)). AOL has done this by defining its corporate commitment to accessibility and posted that policy on its website. The second requirement is listing the technical and functional performance criteria necessary to accomplish the first guideline (29 USC 794d (a)(2)(A)(ii)). AOL accomplished this by creating the "America Online Accessibility Checklist" to guide and educate its employees in making their service available. AOL had entered agreements with screen reader technology companies to develop products to be compatible with future updates of AOL. AOL agreed to release a new version of AOL, called AOL 6.0, which would be compatible with screen reader assistive technology, and which was released 4-8 months after settlement. The final provision in Section 508 of the Rehabilitation Act allows for alternative means for accessibility if the agency is unduly burdened and unable to comply with the Act. The National Federation of the Blind and AOL both concede that there are certain exceptions that will pose unusual technological difficulties. AOL has agreed to find reasonable alternatives consistent with their accessibility policy.

What does this mean for the future? While Section 508 of the Rehabilitation Act does not apply to private corporations, it can be seen by this settlement agreement that corporations may be challenged, and that they may use the

AOL case and Section 508 as a model for their web site accessibility. There have been no cases that have come before the court pertaining to the issue of decreased accessibility to web sites violating the Americans with Disabilities Act. While the Rehabilitation Act does not apply, other sections of the ADA (28 Code of Federal Regulations (CFR) 36.303 (2001) and 28 C.F.R. 35.160 (2001)) may apply to private corporations. The specific requirements are explained in 28 C.F.R 36.303 and it states that "public accommodation shall take steps that may be necessary to ensure that no individual with a disability is excluded or treated differently because of lack of auxiliary aids and services." The types of auxiliary aids and services include: "qualified readers, taped texts, audio recordings, large print materials, Braille materials, or other effective methods of making visually delivered materials available to individuals with visual impairments" ((28 C.F.R 36.303(b)(2) (2001)). Communications are explained in 28 C.F.R. 35.160 (2001) and it states that a public entity shall take appropriate steps to ensure that communications with applicants, participants, and members of the public with disabilities are as effective as communications with others (28 C.F.R. 35.160(a) (2001)). The public entity "shall furnish appropriate auxiliary aids and services where necessary to afford an individual with a disability equal opportunity" (28 C.F.R. 35.160(b)(1) (2001)). Taken together, these statutes mean that public corporations will need to provide equal opportunities for individuals with disabilities in all aspects of life, including the World Wide Web. It is expected that when a case such as this comes before the court, Section 508 will be used as a model for comparison of what is necessary for universal usability of a website.

7. Summary

Web sites should be designed to provide access for all users. Since passage of the Americans with Disabilities Act, physical access to buildings and facilities has been recognized as an important issue. The concept of physical access is now being expanded to include information access. By the end of June, 2001, all U.S. Federal Government web sites will need to be accessible to those with disabilities. It is expected that this trend will continue, and either through case law or statutory law, web sites for private organizations will be required to be universally usable by those with a variety of special needs.

References

Bush, G. (2001). Fulfilling America's promise to Americans with disabilities. Available at: http://www.whitehouse.gov/news/freedominitiative/freedominitiative.html

Crockett, R. (2000). Helping the disabled navigate the net. *Business Week*, April 3, 2000.

Lazar, J. (2001). *User-Centered Web Development*. Sudbury, MA: Jones and Bartlett Publishers.

Norman, D. and Draper, S. (1986). *User-Centered System Design*. Hillsdale, NJ: Lawrence Erlbaum Associates.

Rowan, M., Gregor, P., Sloan, D., Booth, P. (2000). Evaluating web resources for disability access. *Proceedings of the 2000 ACM ASSETS Conference*, 80-84.

Shneiderman, B. (2000). Universal Usability: Pushing Human-Computer Interaction Research to Empower Every Citizen. *Communications of the ACM, 43*(5), 84-91.

Tedeschi, B. (2001). Advocates of People with Disabilities Take Online Stores to Task. *The New York Times*, January 1, 2001.

An investigation into attitudes to, and experience of, Internet shopping

Martin Maguire, Laura-Jo Pearce

HUSAT Research Institute, Loughborough University,
Loughborough, Leics, LE11 1RG, UK
m.c.maguire@lboro.ac.uk , l.pearce2@ntlworld.com

Abstract

This paper reports on a survey of Internet users to study their pre-conceptions and experiences of Internet shopping. It looks at the characteristics of shopping sites that are most important to users and reports their comments on current sites. It asks how long users will spend looking at a site before moving on or buying something, and whether Internet shopping should be more like visiting a real shop. A study is also reported where a small sample of inexperienced users of varying ages, were asked to perform a supermarket shopping task. The results of both studies have implications for the design of Internet shopping if they are to support inclusiveness.

1. Introduction

The Internet offers a convenient means of shopping for all members of society. After initial enthusiasm for Internet-based shopping, a number of high profile failures have shown that it is unlikely to take over from traditional shopping in the near future. It is much more likely that Internet shopping will find a niche alongside traditional shopping by enriching the shopping experience for those who have the means and skills to make use of it. Potential benefits include: purchasing from abroad (e.g. music from the USA), obtaining cheaper prices on standard goods (music, CDs, books, travel tickets), purchasing by people with little time or a limited ability to get to shops, customising goods online (e.g. cars or computers) and access to specialised products (e.g. types of cheese, paintings, antiques). Internet usage is growing in popularity but the increase is uneven across age groups. Although two-thirds of 16 to 34 year olds use the World-Wide-Web, only 14% of the over 55s go online (Hartley-Brewer, 2001). This may be partly due to a lack of computer experience, confidence and motivation. One implication is that people who are elderly and possibly disabled, who could gain most from the convenience of online shopping, will fail to realise the benefits.

Researchers and marketeers are interested in indentifying the major barriers and facilitators for online shopping. It has been found that security of the Internet for monetary transactions remains a major barrier to the take-up of online shopping as shown in a survey of 1,095 online customers, carried out by NFO Interactive (Usability Lab, 2000):

Table 1: Barriers to online shopping (NFO Interactive)

Major barriers to shopping online	% of sample
Concerns about security	31%
Prefer to go to shops	21%
Not got round to it	14%
Like to see things before buying	11%
Concerns about data protection	7%

Consumer confidence is perhaps undermined by the fact that while published articles argue that Internet shopping is generally safe (e.g. Baguley, 1999), examples of security failure still appear. In one case, for example, a building society was forced to take its Internet-based share-dealing service off-line after a technical glitch allowed customers access to the accounts of other individuals (Hinde, 1999). Such stories generate negative attitudes towards using Internet shopping which discourages people from using such services in great numbers.

2. Objective of paper

This paper reports on a survey of Internet shopping and a study of inexperienced users performing an online supermarket shopping task. The objective was to investigate people's feelings about online shopping (as expressed in the survey) and the major usage problems (as identified in the experimental work) that designers should address if shopping services are to be inclusive and used more widely in the future - particularly by people who are less technically experienced.

3. Survey results

A survey was carried out among 56 members of the public who use computers, ranging in age from 18 to 70. The sample was drawn mainly from university staff and researchers, but also included members of the public with whom HUSAT had contact with. It was found that 57% of the sample use a computer at home and at work, as opposed to 36% who used a computer at work only. This provides a reminder those who learn to use a computer at work may be encouraged to purchase one for home use. Conversely, those who do not use computers as part of their work (or who have retired without using them) are less likely to develop their skills and may be encouraged less to acquire a computer for use at home. Of the survey sample, 50 of the 56 users (89%) had Internet access. Internet shoppers were initially asked to give comments (in free form) on the positive and negative aspects of Internet shopping. Table 2 below gives a summary of responses, categorised under various headings:

Table 2: Survey - comments on different aspects of online shopping

Positive and negative aspects of online shopping	
Category	**Summary of comments**
Access/convenience	Respondents liked the fact that they could shop without leaving home and outside normal shopping hours. They liked to read about goods online before making purchase decision.
Range/Availability of goods	Websites offer 'things not available elsewhere' or 'more unusual gifts'. Some commented that sites did not present the full range of a retailer's goods, stock levels, and when goods available.
Information and presentation of goods	Some (8%) felt that information on goods was comprehensive and up to date, while others (14%) thought there was insufficient information given. They also disliked the fact that on some sites you did not actually see the goods.
Price of goods	14% of respondents commented that prices were reasonable, good value or a lot cheaper than the in the high street. Just 4% found the Web useful for price comparison.
Responsiveness	Even split between those who felt online shopping was time saving (14%), and those who felt it was slow/complex (18%).
Security	Several expressed concern about the security of the Internet.
Ease of use, Navigation, Finding goods	Even split between those who felt that items were easy to find (20%) and those who did not (20%). Some thought the Internet facilitated browsing through goods, while others found it 'harder to browse aimlessly'. One stated "I feel as if I might miss something that I would to notice in a store".
Ordering of goods, Feedback	Comments on feedback provided between websites on the progress of a purchase varied between good and poor.
Delivery of goods	There were both positive and negative comments about delivery times, delivery costs and damage to goods. One person liked ease with which goods could be returned/replaced.
Payment	Some found buying online easy, while others found it awkward to check out and pay for products. Another stated that the range of card payment options was poor.
Technical issues	10% of users experienced technical problems related to unreliable sites, broken links, connection problems etc. One felt: "...at mercy of workstation. If the system freezes/crashes mid-transaction how can you know if sale has gone through?'.

To determine the most important potential benefits of Internet shopping, users were also asked to place in rank order, a set of 6 potential benefits. Table 3 below presents the average ranks (from 1 to 6). It can be seen that convenience and variety of goods are the main motivating factors behind Internet shopping.

Table 3: Ratings of importance of benefits of online shopping

Benefits of online shopping (in order of importance)	Average rank
Convenience	4.8
Wide selection of goods	3.7
More information about goods	3.3
Finding goods at lower prices	3.2
Ability to compare prices at different shops	2.9
Ability to visit shops never seen before	2.6

Users were asked how long they would spend looking at an Internet shopping site before giving up and moving on. It was found that the majority of respondents (82%) would not spend more than 10 minutes looking for an item on an Internet site. This highlights the importance of being able to locate goods rapidly. Of those who has performed Internet shopping, 60% felt that they had found it to be a straightforward process while 14% thought it not. The problems included:

- Slow and sometimes asks irrelevant questions.
- Normally very poor search facility.
- No trust in card transaction.
- Often there is no reassurance of what have done, i.e. whether transaction has gone through.
- Get annoyed about pages failing to load or loading slowly.
- Item never arrived on time, ended up cancelling order!

Respondents were asked what was the main thing about a shopping site that would either persuade them to buy from it. The most important motivating factors were: low price of goods, security of transaction; followed by: good company reputation, clear instructions, ease of navigation and buying online, comprehensive information about product, quality of service. Other factors mentioned by individuals were: immediate availability of product, speed of access, and own bank details already present on the system

Respondents were asked if they would like an online shop to be more like a real shop. It was found that of those who responded (47), just 9% stated that they would like an online shop to be like a real shop, compared with 53% who did not, while 38% thought it should to some extent. Comments made in relation to this question were as follows:

- "Less hassle on the Internet."
- "Will never really compete in the clothes market but durable goods (electrical goods etc.), are standard and the Internet is likely to be more popular as you don't need to see much of the product."
- "Would like all items in shop to be available via website which isn't always the case."
- "Some sites have very little categorisation so cannot just browse."
- "If it offered the same amount of security of your information and bank details."
- "Better categorisation of books (as in a real shop)."
- "Needs to be more interactive in terms of information exchange."
- "Need more information about the products."
- "Creation of a 3D virtual stores is too time consuming and there is insufficient band width. Buyers have to spend too long looking for products."
- "A simple picture of each item is sufficient, hi-tech 3D stuff does not impress me."
- "Details regarding size, dimensions or material can be lacking - enhance with 3D images or scale indicator."

4. User trials of supermarket shopping

In order to investigate how to design effective Internet shopping for inexperienced users, a set of user trials were performed with middle aged and older people who had minimal or no experience of computers and the Internet. They were asked to perform a grocery shopping task via one of two UK supermarket websites (Tesco and Sainsbury). The aim was to determine how easily inexperienced users would be able to carry out Internet shopping. Table 4 below summarises observations from the 6 user sessions. Alongside these observations, design needs identified and expanded in section 5.

Table 4: Observations of inexperienced users of online supermarket shopping

Summary reports of shopping trials with inexperienced users		
Subject	Observations	Design needs
Joan - aged 80 Retired. Never used computer	Problems using mouse and scroll bar. Needed explanation about how to start and support throughout. Major problem was knowing what 'aisle' each product in. Felt like a 'dunce' afterwards although she was happy to do task.	Basic PC help. Introduction to site. Help finding items.
Jill - aged 50-54 Shop assistant Never used the Internet	Needed advice to begin with to work out how to use the system. Missed prompts such as at the bottom of the list if there are further pages. Also missed tab to click on to show the next page. Once had chosen the item, user wasn't going to add it to the trolley, but saw the tab and asked what it meant. Main problem was locating the items and understanding icons. Thought there should be more help and explanation in words. Would have found it more enjoyable to go to real shop.	Introduction site. Clear screen layout and controls. Introduction to site. Help finding items. Images of items.
Helen - aged 22, Nurse. Minimal use of Internet	Double clicked rather than single clicking which caused problems as system tried to do everything twice. Proceeded without asking much advice but took a while to understand the concept of aisles and shelves which caused errors. Most mistakes were in remembering to add to the trolley. Had worked in same supermarket so knew which location most products would come under. Thought the system simple to use.	Basic help with PC/Internet use. Introduction to site. Reminders/ prompts.
Krista - aged 41, Housewife. Minimal use of Internet.	Understood the concept straightaway but struggled to locate items. Happy to browse aisles to see how site worked. Asked lots of questions and had a keen interest in shopping site uses. Thought would use herself for everyday and bulky items. Had a good attempt at additional actions e.g. specifying ripeness.	Help finding items.
Roy - aged 57 Casual postman. Minimal computer usage.	Completely unsure as to how to start so needed explanation. Finding products was difficult and got annoyed that for some products had to go through 10 pages to find each one. Distracted by special offers and missed the product because it was highlighted in red after 'visiting' it. So busy working out how to do the special offer, forgot to add it to trolley. Same happened when specifying weight of the apples. Thought Internet shopping interesting and could see its uses but would never use it as enjoyed normal shopping too much.	Introduction to site. Help finding items. Clear screen layout and controls. Reminders and prompts.
Maureen - aged 50-54 Reservations co-ordinator. Minimal computer usage	Slow start with experimenter support paid off as she sped up and ended up doing the shop quicker than most. Wary of doing anything without asking and wanted evaluator's confirmation that was not making any mistakes. Thought items should be located more easily. Thought service useful to elderly users and parents with young children but not for herself.	Introduction to site. Reminders and prompts. Help finding items.

5. Guidelines derived from the user trials

Following the user trials, a number of design guidelines were identified that, if applied, could assist inexperienced users of shopping sites and promote inclusive design.

5.1 Basic help with PC/Internet use

Inexperienced users may still be learning how to operate a PC and to use the Internet. Ideally a brief tutorial, prompt card, booklet or video should be available providing basic instructions.

5.2 Ease of registration

If too much personal information is required to be entered on forms, potential customers will find this off-putting. It should be clear which fields are mandatory and in what format registration details are to be entered. If the user makes an error filling in a form, the system should make it easy to correct or to supply the missing data.

5.3 Introduction to shopping site and process

New users will need an introduction to a shopping site so a series of animated preview screens will give them more confidence to start shopping. It is also important to explain the shopping concept and process i.e. viewing goods, finding more information about them, specifying the details (ripeness, size or amount), putting items into the shopping basket, checking out, storing a standard shopping list, etc.

5.4 Prompts and reminders

To support the shopping process, considering providing a pop-up window that guides the user at each stage (similar to a Microsoft 'Wizard') or help that they can call up to offer context related assistance. Such a 'flash-up' facility could be in the form of a shop assistant guiding new users through the initial stages of shopping. It could be settable at different levels, for instance, to pop up if the user seems to be struggling, or simply to be available on request. Special offer or promotional 'flash ups' could be provided for specific shelves.

5.5 Clear screen layout and controls

To make shopping pages easy to view and understand, the site should lead the eye to the items to buy. Shopping options should be presented using large clear text or the user should be allowed to set the text size directly or through their own browser. Item selection and navigation controls should be clearly distinguished from the background. If possible, all products within a certain section should be listed on one page. For visually impaired users, colours should be employed which contrast well. Redundant coding of items should also be used. Care should be taken that the use of corporate colours do not present a garish appearance.

5.6 Provide images of items

New users will feel reassured if they can see images of items on the screen to select from. However to provide images of all items may be too demanding of bandwidth and site maintenance. It may thus be helpful to provide images to indicate groups of items e.g. cereals, fruit and vegetables, meat, dairy products etc.

5.7 Help with finding items

As when entering any new shop, novice users will need to learn the layout of the electronic shop and to be able to locate items within logical groupings within the website. Interaction techniques can be used to support finding items, including: a high level map of the virtual shop which the user can click onto to see more detail i.e. specific goods; a 'look ahead' option where the user can 'mouse over' a general category to see the specific items; a simple search facility; and a flat link structure enabling users to find the items they want with the fewest number of clicks.

5.8 Flexibility

Novice shoppers should be supported by providing clear system instructions and feedback, while experienced shoppers should be provided with shortcuts to jump directly items frequently purchased or a standard shopping list to modify. The shopping service should also be flexible in allowing the user to easily check what items are in their shopping basket, to take them out again and to change them (e.g. for a different colour or size).

5.9 Check that site is designed for inclusion

Consider recommendations for visual, hearing and mobility impaired persons using, for example, guidance provided at: www.cast.org (Center for Applied Special Technology), www.w3.org/WAI (web accessibility initiative), www.rnib.org.uk/digital (Royal National Institute for the Blind site), www.trace.wisc.edu (TRACE centre guidelines), and www.microsoft.com/enable (Microsoft's guidelines on accessibility).

6. Conclusion

In order to perform shopping tasks successfully, users have to acquire a range of skills and knowledge. The survey and user trials highlight issues for future designers to consider in order to make online shopping accessible to a wider range of citizens in the future. Some initial design guidelines are presented but further research is needed to validate them and provide more specific recommendations.

Acknowledgement

The authors are grateful to Colette Nicolle, at HUSAT, for encouraging this paper to be written and for commenting on the initial draft. Thanks also to Samantha Wain for compiling the survey data.

References

Baguley, R. (1999). Go shopping now!, *Internet Magazine*, 37-40. April 1999.
Hartley-Brewer, J. (2001). Free Net access for all plan, *Sunday Express*, 38. 28th January 2001.
Hinde, S. (1999). We're afraid the 'e' stands for 'easily cheated', *Sunday Express*, 2. 28th November 1999.
Usability Lab (2000). Convenience stores, *PC Magazine*, London: Ziff-Davis, 9, 1, 162-177. January 2000.

Sustainable information environments and informed sustainability

Michael Paetau, Michael Pieper

GMD – German National Research Center for Information Technology
Schloss Birlinghoven
D-53754 Sankt Augustin
+49 2241 14 - {2625, 2018}
{michael.paetau, michael.pieper}@gmd.de

Abstract

Sustainable development is linked very strongly with the development of a global information society. A really sustainable development of our future societies is only possible if factors of the environment, economy, social organization and institutions are regarded as linked with each other and dependent on each other. This integrated view represents in many ways an enormous challenge. It means the step from isolated topics, local information and small groups or even individuals forward to a networking of the topics with global information as well as with individuals and groups. Sustainability can only be reached if all participating parties have access to the right information from distributed sources, if they can analyze and adequately use the information - and furthermore, if they can organize their cooperation and networking in an effective way.

The vision is to enable ubiquitous and equal access to the relevant information for all users concerning cost, technical equipment and competence, and to provide them with user interfaces and interactive analysis and planning tools which can be adapted to their needs and competence. Only then, they can effectively participate in debates in open or closed discussion rounds which are moderated with the help of mediation methods and supported through computer systems and networks.

The search for information treasures (data mining), statistical analysis algorithms, the modeling of complex data relationships, the visualization of search and analysis results, moderated topic-oriented discussion forums, geographical information systems and virtual organizations are the most important technologies and research areas which are brought together here.

1. Introduction

The relation between both, Sustainable Development on the one, and Global Information Society on the other hand, is not an automatic process. It requires a very closely tied interweaving design process. The numbers of internet connections mean nothing for the question, if we are on a way towards a sustainable future or not. And it is not evident to say »the more Computers we have, the more sustainable is our society«. Statements like this, are speaking more about myths than realistic potentials. Technology for a sustainable development must be designed following the needs of different application fields of sustainable development activities. »Design is the interaction of understanding and creation« was said by *Winograd & Flores* in their famous book »Understanding Computers and Cognition« - more than ten years ago (Winograd & Flores, 1968). Thus, designing technologies for a sustainable development needs a deep understanding of the application fields, in which various actors are operating. And it needs transdisciplinary knowledge and research.

There are two aspects of the current debate, addressing the relationship between information technology and sustainable development. *First*, ICT (Information & Communication Technologies) should be given a substantial support to the crucial points of the contemporary global change: for example the protection and sustainable use of *biological diversity* or the emission of greenhouse gases.

But Sustainable development has a *second aspect*. The possibility for the realization of the vision is coupled to the key-question of the so-called knowledge society. Sustainable Development cannot be achieved without fundamental modifications in our way of life and without a considerable change to the predominant production and consumption patterns. Relating to this, the world society has an immense control problem. And if we understand control in a cybernetic way as self-control and self-organization, it is evident, that society - understanding as world society - has a lack of knowledge. Without any doubt, there exists a lot of distributed knowledge of various actors and social systems on a variety of problems and suggestions for problem solving. But: How is it possible to make distributed knowledge available for the society in the whole?

Expressed in a sociological term: It is a question for the emergence of knowledge in complex social systems. Emergent societal knowledge is not to manage like in an institutional organization. Information-Management or Knowledge Management in the context of sustainable development has to consider the autonomy of the engaged participants. Already the interweaving of globality and locality of problems and problem solving activities - local causes and global effects -- shows clearly this dilemma. Complex societies cannot manage and control themselves like an organization with hierarchical structures, more or less well defined causal linkings of occurrences in the forms of decisions and sanctions, rules and techniques etc.

In opposite to pre-modern situations, societies are no longer characterized by a kind of homogenous structure or even a hierarchical order which guarantees for its unity. But instead, we observe an increasing particularization and autonomy of different social systems. This structural change of society has been described by a new sociological key-term: Network Society. Network Society – that has to be emphasised - is not a technical concept, but a social one. This term refers to the autonomy and self-organization of social actors and social systems (Pieper, 2001). But the realization of this concept is highly dependent of an adequate information and communication infrastructure.

In this respect there is a second challenge for ICT-Research to support the vision of Informed Sustainabilty. In what extent ICT can help to answer the unsolved question: how is sustainability possible? Can ICT help society to overcome its fragmentation? Expressed more concrete: how to apply information and communication technology to help overcome the egoistic, short-sighted and uncoordinated behavior of autonomus stakeholders. In what extent it can facilitate effective discussions to build a viable consensus for concerted action and can help to move from conceptual dicussions to effective sustainable development practice.

2. Sustainable development: from problems to strategies, from strategies to action

At the World Conference on Environment and Development in Rio de Janeiro in 1992, "sustainable development" was defined - following the Brundland Commission - as the principle that *current generations should meet their needs without compromising the ability of future generations to meet theirs*. This requires finding ways to promote social and economic development which do not sacrifice ecological and material resources. This definition implicates responsibility for the coming generation (called as inter-generational equity), but includes the demand for a human existence of the current generation (called as intra-generational equity).

And besides the aspect of securing the natural basis for life in the long term, it includes the aspect of maintaining the opportunities for development and capacity to act. Regarding the philosophical discourse we can speak of human or societal autonomy. A society can be called autonomous, if it can act rational and on the basis of own decisions instead of deterministic influences from outside, from it s natural or societal environment. Consequently, the concept of sustainability includes an ecological and a human - respectively societal dimension. In the first years after the Rio-Conference in 1992 the discussion on Sustainable Development was centered on the problem-description in environment and society, the search for consequences and causing factors and on arrangements for threshold-values of Non-Sustainability.

In different fields, especially of the environmental dimension, fundamental rules (so-called Management Rules) have been developed to define the minimum conditions necessary not to compromise the achievement of these goals. These sustainability rules are intended to serve not only as guides for the further operationalization of the concept but also as criteria to help to identify sustainable and non-sustainable development paths in various countries, activity fields and economic sectors. Quality targets, standards, action targets are steps to concretize the goals for sustainability activities.

During this period a controversal discussion arose, whether it is possible to declare such general rules and threshold-values not only for environmental problems, but also for the economic and social dimension of sustainability. In a broad discourse the question was discussed, which general goals in terms of global development can be deduced from the principles of intra-generational and inter-generational equity. This point will be exemplified below with regard to the activity field of Information and Communication.

Beforehand attention should be directed to another focus: For transferring the insights on problems into strategies of sustainability it is necessary to identify the contributions, which a particular field of societal activities and needs have on environmental pollution or social problems. For doing that, in the last years a distinction of society in several societal activity fields was advised.

- Mobility
- Constructing & Housing
- Agriculture & Nutrition
- Information & Communication

- Leisure & Tourism
- Textiles & Clothing
- Health,
- etc.

In all these activity fields particular problems of non-sustainability have been identified. If we take the example of *Agriculture & Nutrition* the following problems are named in the appropriate literature:

- emission of greenhouse gases
- eutrophication of rivers and lakes
- acidification of soil, rivers and lakes,
- soil degradation and loss (especially through erosion),

- the introduction of human and ecologically toxic substances (especially plant protection agents),
- the change of landscape
- declining of biological diversity.

On this background, the structure of the interrelation of problem identification, deducing of strategies and concretizing targets can be described more concrete. To discover in which way the potential of ICT can be used, it is necessary to translate the different action targets into activities of Information Processing.

In terms of information processing we can describe a circle of different activities dealing with global problems and sustainable development strategies. It begins with sensing and monitoring of environmental and societal problems, continues with processing of data, transformation of data into information, archiving and modeling. For processing with regard to the contents visualization and presentation in problem solving groups, analyzing and development of concrete action targets.

Following this connection between substantial activities and information processing shall be exemplified regarding two problems: first the problem of the imminent loss of biological diversity, and second the problem of the social conditions, especially the forms of communication and institutional innovations to overcome these problems and to start realizing sustainable development in practice.

2.1   Example I: Bio-diversity

The protection of biological diversity is of major importance to the conservation and development of systems which are essential for the life of the biosphere but also to human health, food supply and social development.

The Convention of Biological Diversity adopted in Rio de Janeiro leads to a number of challenges for computer science with regard to information processing and corresponding methods and technologies; e.g.

- Clearing-House Mechanism

- Making a comprehensive biodiversity survey
- Assessing and evaluating genetic erosion
    - gathering, description, archiving, aggreating, analyzing, modelling, visualization

- Developing a global biogeography
    - data mining, exploratory spatial data analysis, geographic information systems

- Overcoming societal-institutional obstacles
- Enabling transcultural discourse
    -communication, collaboration, mediation systems, Seamless Multimedia Learning Environments (SMILE), data warehousing, distributed databases, text mining, meta-data handling,, internet agents for information retrieval, web-based dissemination

- Developing methods for early recognition
    - recognizing causal complexes (Syndrom analysis), sensing, monitoring, simulation

The international clearing house mechanism is of central importance. A multimedia information and communication technology based infrastructure has to be developed to build up a global transcultural knowledge network. On the one hand, data are to be acquired and gathered, knowledge about species variety is to be made available by creating databases and this knowledge is to be used for information interchange and the coordination of planning processes.

2.2 Example II: Knowledge networks

In addition to the question how to put the guiding vision of sustainable development into concrete action targets, the question about the social conditions has gained increasing importance during the last years. Because sustainable development cannot be achieved without modifications in our way of life and without a considerable change to the predominant production and consumption patterns, the question arises: To what extent societies are able to accomplish such a serious self- change? The demand for technical and institutional innovations has to be seen in this context.

Modern societies have an immanent problem for controlling their change to a sustainable future. Innovative information and communication technology can help to overcome this problem, but it can also aggravate it. Network Society is the term, which is representing this ambiguity. Network – to emphazise it again - is not a technical concept, it is a social concept. It describes a special form of social relationships, which is characterised by

- a high degree of autonomous and self-organised activities of social actors or social systems,
- cooperation of loosely coupled social units,
- distributed knowledge.

First of all, the vision of Network-Society is replacing traditional concepts of strictly coupled and more *formal* structured society, in which - for instance - the political system was seen as able to penetrate all other parts of society. In opposite to that, Network-Society has no center, from which it could be controlled. Information and Communication will not be achieved in a hierarchical and sequential way. It is expected, that the advantage of a network-structured society could be a faster and more *flexible* reaction of particular social systems to changes in society and environment.

But until now, it has not been clarified, to what extent this loosely coupling of society can take place without undermining its own identity and stability. Especially it is not clear, in which way distributed knowledge of autonomous actors and systems can be transformed into emergent knowledge of society. Loose couplings have many advantages as well as disadvantages. They improve the particular capabilities of autonomous social systems with respect to the whole society. They produce an internal tension between integrative and centrifugal forces. This tendency could lead to some important changes in respect to the perception and reflection of environmental and social problems, especially it could lead to an increase of particularisation of problem-perception *and* problem-solving. Distributed knowledge increases the complexity in planning and decision making processes. Especially problems with global causality and effects are difficult to manage.

In a modern society two principles of evolution are in a tension: On one hand, the principle of differentiation, with an increasing specialization and thematic contraction of different social units. This will lead to the effect of an arising *inter-dependency* of the single units. On the other hand, the principle of operational closure with an increase of particular autonomy of social units. This will correspond with a circular information processing: One selectively takes notice only of that information, which supports already stabilised preoccupations. This will lead to a rise of *in-dependency* of the single units.

To overcome these difficulties, it is necessary to provide an infrastructure for coordinating autonomous social actors. This so-called knowledge network can be a platform for divergent interests in a pluralistic and polycentric society. In a world-wide context it has to become a transcultural knowledge network, which is the second vision of Informed Sustainability in a globally networked Information Society.

3. Conclusions

The structural change of society which has been pointed out, can have positive and negative effects. On the one hand the knowledge distributed in society may better be used than in a hierarchical way. An example may be the knowledge of the No-Nuke-Protest-Movement in the 70s and the early 80s. That time the public administration did not take this particular knowledge seriously into account. The result was a lack of resonance and reflectivity about environmental - and social - problems. This is an example, that in case of formal and hierarchically organized information-processing social knowledge gets lost.

Accordingly the Agenda 21 demands a wide integration and participation of citizen into the process of sustainable development (United Nations, 1999, 2000; German Bundestag, 1998). But, on the other hand - as outlined above - distributed knowledge increases complexity in planning and decision making processes. Especially problems with global causes and effects are difficult to manage. The question is: How is it possible to make distributed knowledge available to all and how can society deliberate on it?

ICT can play an important role to handle these problems. Two different stages of societal action and corresponding knowledge processing can be distinguished:

1. Observing environmental and social problems and institutional activities responding on them.
2. Conceptualization and realization of prospective strategies for Sustainable Development.

Both aspects require different technologies and methods. Observing problems require technologies for sensing and monitoring, processing, modeling, archiving to presentation, visualization and manipulation of information. In the context of conceptualization and realization of strategies, systems for mediation, planning and decision making are in the foreground.

Another point is the question, whether ICT can help to structure a world wide environmental regime, which is not organized within the borders of a centralised a single state, but could help to mange global problems. Transcultural Knowledge Networks, which have been pointed out, may be an answer, in which the potentials of Information Society Technologies can evolve.

References

German Bundestag (1998). The Concept of Sustainability. From Vision to Reality. Final Report submitted by the 13th German Bundestag's Equete Commission on the "Protection of Humanity and the Environment - Objectives and General Conditions of Sustainable Development. Bonn: Deutscher Bundestag

Pieper, Michael (2001). Sociological Issues in HCI Design. In C. Stephanidis (Ed.), User Interfaces for All - Concepts, Methods, Tools. Mahwah - New York – London: Lawrence Erlbaum Ass.

United Nations (1999). Agenda 21 - Global Programme of Action on Sustainable Development. United Nations Division for Sustainable Development: http://www.un.org/esa/sustdev/agenda21text.htm

United Nations (2000). *"Our Common Future," The Report of the World Commission on Environment and Development Report.* United Nations Division for Sustainable Development: http://www.un.org/esa/sustdev/agreed.htm

Community on a Usenet Group

Holly Patterson-McNeill

Texas A&M University-Corpus Christi
6300 Ocean Drive
Corpus Christi, Texas, USA

Abstract

The practices, activities and social structures of computer-mediated groups are emergent and not pre-determined. This study provides an analysis of the impact of computer-mediated communication (CMC) on the social behavior of a Usenet newsgroup called "alt.good.morning" (AGM). AGM is one of thousands of discussion groups on Usenet. AGM deliberately generates its own contexts and activities within an environment that includes the computer, Usenet, the people and their purposes for interaction. AGM is a vibrant community composed of people from around the world, the computer and the activities of the community.

1. Introduction

Alt.good.morning, a newsgroup distributed through Usenet, began in the winter of 1992-1993 as a small group of posters in Europe wishing each other "good morning" and has grown to over 80 messages a day from people all over the world. The conversation of AGM is routine and mundane; there is no special topic or purpose to bind its participants. Yet, it has created a vibrant culture, populated by people of all ages, from around the world, and from all walks of life. AGM is a group of distinct individuals with a shared history, with distinct behaviors and with expectations for a shared future through their interactions on AGM and through the affordances of CMC. They form a community (Traweek, 1992). The evidence that AGM is a community is found in its everyday discursive practice, in its creation of a unique place in Usenet and in its importance in the lives of the participants.

In AGM, CMC and the computer are not passive tools but active participants in the interaction, what Latour calls actants (Latour, 1987). The computer is an essential component of the AGM community; it cannot be separated from the activities of the newsgroup. The computer is both transparent and visible (Lave & Wenger, 1991) in these activities. When invisible, the computer no longer exists as a separate artifact in the activity or the delivery tool but becomes an extension of the person. After learning to use various software (Web browser, editor systems, newsreader, email, and talkers, a system similar to chat) participants are no longer aware of the existence of the computer. The computer becomes as invisible as a windowpane. The computer becomes visible when problems arise. But the invisibility of the computer does not negate its participation in the interactions. The computer and its software actively structure and enable the interactions.

This paper provides a brief description of Usenet interactions, a background description of AGM and a description of how AGM has formed a community.

2. Usenet posts

Usenet is a global exchange of messages sorted into newsgroups. These networked groups coalesce because of shared interests, not because of shared physical location. The most popular newsgroups are the social ones (Baym, 1994).

A Usenet message is called an article or a post; sending an article to a newsgroup is called posting. Newsgroup participants post articles from their private computer accounts and they read posts using specialized software, either newsreaders or Web browsers. Yet, most messages are not saved. This contributes to the perceived ephemerality of the communication.

Usenet posters are geographically distributed and their articles are received minutes, hours or even days after they have posted. Posters do not know when their messages will be read or who will be reading their messages. Because most of the readers do not post, they are given the special label of lurkers. A poster sits in front of the computer screen alone, seemingly invisible and anonymous. In addition, messages do not arrive in topical order, a single thread (group of messages with the same subject) may take days to complete and multiple threads are ongoing simultaneously. Reading messages is not the same experience for every person in the newsgroup. Because of the asynchronous nature and differences in delivery times of messages to some sites, messages and their follow-ups are read in different sequences. This isolation, disorganization and lack of consistency contrast with the connectedness felt when reading articles by people in "your" newsgroup.

Newsreaders, the software used to read and post articles to a newsgroup, are mediating tools. They format posts and allow posting similar to email tools. Thus, posts have headers, a message body with a quotation system and signature files. Additional organizational tools include kill files and threading facilities. Experienced participants use the organizational tools provided by newsreaders to structure their activities on Usenet.

Headers provide social contextualization about authors and their messages which helps create and maintain internal organization. Depending on the software used when the message is sent and what options are set for reading messages, a participant may see a rich or full header with detailed information about the post or some subset of those lines. A rich header includes the message number, the route through the Usenet sites, the sender of the message, the newsgroup(s) to which it has been sent, the subject, when the post was sent, the organization of origin, and the message length. The identity of the sender and the subject of the post provide the most important information.

The message body contains an original message or a response inserted in a previous message at relevant points. Embedding your remarks within a previous message is essential to maintaining conversational coherence. The original message provides contextualization by refreshing the memories of those who have previously read the post or by displaying the post for those who have not yet seen it. The quotation system, marking the previous message with angle brackets, provides the reader with a sense of turn-taking evident in spoken conversation. In addition, the angle bracket signals that the current poster is not accountable for the content of the quoted text but is accountable for the accuracy of the quotation (Baym, 1994).

Another common feature of a post is the signature (sig) file that is appended to the end of messages. Some systems automatically append the sig file to every message sent, thus providing a regular and consistent mechanism associating text with the author. The sig file is highly individualized with ASCII art, a favorite quotation, address(es) and telephone number(s). Distinctive sig files carry more identifying potential than names and often provide the most salient cues in learning to discriminate between individuals (Baym, 1995).

3. Alt.good.morning

Since AGM is a Usenet newsgroup, what holds for Usenet and newsreaders holds for AGM. The features of Usenet and newreaders affect the participation on AGM, yet the group has found ways to use those features to build community. AGM is defined by the posts and by the individuals who post. The average length of an original post is about 24 lines or one screen long. Follow-ups are a bit longer because they include some of the original post embedded in them. Approximately 75% of all posts are follow-ups. Over half of the posts include a sig file. Therefore the posts are highly contextualized.

AGMers post regularly to the newsgroup. There is an explicit norm that states that you should post once a day. Some post more often than that while others post several times a week. In the six-weeks of the detailed study of posting patterns, 4,047 AGM posts were received from 203 different AGMers. (There were a total of 9,839 posts, including crossposts and spam, from 427 different people. The extra posts and posters are not included in the analysis because most of their posts are filtered through the use of kill files.) Of the AGMers, twenty-eight people contributed 40 or more messages, fifteen women and thirteen men. Six of these contributed more than 100 messages, three women and three men. What is evident is that AGM has an active group of posters.

But who are these people? Using questionnaire responses from 112 AGMers, a profile of the group can be built. Over half (54%) of the responses were from the United States and the other half fairly equally divided among the United Kingdom (15%), Canada (13%) and twelve other countries (16%). AGMers range in age from twelve to over

1100

seventy-one. Most are in their twenties (42%), university aged, which is not surprising considering ease of computer access at most universities. Of the respondents, 57% are women and 43% are men. The group is well educated and approximately one-third are students.

AGMers love to communicate. They use posting and email, talker, telephone, snail mail, and face-to-face meetings. They talk about AGM with other AGMers and with people outside of AGM. They use the words "family" and "friends" when talking about AGM. As one poster said, "These people and their lives are important to me ... I like to feel that they feel the same. Giving up AGM would be losing a whole family of friends." The group is diverse enough to be interesting to many different people from around the world and homogeneous enough to engender feelings of deep friendship.

Except for acknowledging the global nature of the group, AGMers seem not to be affected by the spatio-temporal separation of Usenet participation. Perhaps this is because they use a spectrum of communication media. AGM uses the affordances of Usenet provided specifically by newsreader software to create a group of friends. They are people of all ages and all walks of life. The access AGM from school, work and home. They communicate with each other often and with a variety of media. They strongly identify with the group and express feelings of deep involvement with and even addiction to the newsgroup.

4. Community

Because it exists in the social space of Usenet and depends on CMC for its interactions, AGM is a virtual community. Nevertheless, AGM displays all the characteristics of a robust community. It contains a stable group whose presence provides a comfortable familiarity. The consistent and frequent messages from heavy posters provide a basis for the AGM voice. The posts on AGM reflect the common ideals and experiences of the group. They build and share a common history and anticipate a shared future. They have created a distinct culture, even though Usenet works against the formation of a community with its text only communication, the overwhelming number of people and messages, the asynchronous distributed system and the parquetry of messages available at any one time to any one reader (Baym, 1994).

Individual identity must be continually recreated and reinforced because of the constant arrival of newbies (new to the group). Without a physical presence, a poster must create an identity and a presence using the tools available. Posters use the Usenet structures of header, message and signature to facilitate that process. Unique names and distinctive writing styles contribute social presence cues that assist in building an identity. Much of the communication on AGM is self-disclosing. Through these stories, they create complex pictures of themselves as real people. In addition, AGMers take on specific roles in the newsgroup, including greeter, organizer of the current IRL (in real life) meeting, and maintainer of Web pages.

Together with establishing individuality, AGM posts provide a group voice and a source of group identity. When a message is embedded within another's post, a single message possesses multiple authors. Your post is transformed from "yours" to "ours". The list of frequently asked questions (FAQ) and the AGM Web page build group identity. In addition, the participants speak as members of the group and refer to themselves as members of the group. Tangible objects also create a group identity on AGM. Photos, postcard exchanges and official AGM T-shirts produce a team spirit. Thursday has become AGM day when AGMers from around the world wear the AGM T-shirts. When you wear your AGM T-shirt, you know that other people all over the world are sharing in this small act of camaraderie.

Group identity is also built out of a shared history. The FAQ with its bits and pieces of history and trivia is a starting point for the shared history of AGM. The relationships that have grown among the AGMers provide a source of group history. There are multiple romances among AGM pairs, most are international romances and many ending in marriage. Their stories are a popular part of the shared history of AGM. The first large IRL meeting occurred in Arizona (AZAGM); the following year it was in the United Kingdom (UKAGM). It has also been held in Canada and is currently being planned for West Virginia. Telling and retelling of events during an IRL meeting also constitute part of the shared history.

The expectations for a shared future include participating in the local and international IRL meetings as well as birthday celebrations. You can add your name to the birthday list in anticipation of receiving your share of AGM email from around the world. This shared future is based on friendships forged from shared ideals.

The ideals of the group are expressed in the FAQ, in the posts, and in the actual behavior on the newsgroup. One explicit "unofficial rule of AGM" is no flames. Flames are not answered on AGM; they are ignored. These norms of behavior are negotiated practices, that is, the rules of conduct are found in explicit postings and in modeled social behavior of the newsgroup. Newbies adapt their behaviors to align with these norms. The lack of flaming and the friendliness are the two most distinguishing characteristics of AGM.

The use of language on AGM is another distinguishing trait of this community and provides identifying markers for participants on AGM. These uses include the special vocabulary that has evolved, the electronic paralanguage adapted for indicating tone of voice and the general style of writing on AGM. Commonly used acronyms such as fyi, lol, and imho are used. Other acronyms and words are specific to AGM--USP for user specific posting, greebings for good morning greetings, welback for welcome back, thwap for virtual physical contact, and klem for hugs. Emoticons, intentional misspellings, capitalization for emphasis, grammatical markers, embedded words and repetition of words are common paralinguistic features common to AGM communication. These familiar strategies not only provide coherence and involvement but also provide a sharing of communication conventions and discourse (Tannen, 1989).

5. Conclusion

Despite the challenges of Usenet, AGM has built a robust community. It can be described as a global coffee shop, a back fence over which neighbors converse, or a global support session. AGM has built into the community the possibility of fact-to-face interaction through multiple backchannels of communication and IRL meetings. The newsgroup is the visible portion of this community; it is the exoskeleton of the community, a boundary between the world and its essence, the binding framework that gives structure and unity to the group. The software that supports this group provides good usability and flexibility so that these people can interact and perform their tasks (communication) intuitively and easily (Preece, 2000). Usenet was not designed to develop community but combined with other channels of communication and the creativity of people, it enables and empowers community building.

References

Baym, N. K. (1994). Communication, Interpretation and Relationship: A Study of a Computer-Mediated Fan Community. Unpublished doctoral dissertation, University of Illinois at Urbana-Champaign.

Baym, N. K. (1995). From Practice to Culture on Usenet. In S. L. Star (Ed.), *The Cultures of Computing*. (pp. 29-51) Oxford, UK: Blackwell Publishers.

Latour, B. (1987). *Science in Action: How to Follow Scientists and Engineers through Society*. Cambridge, MA: Harvard University Press.

Lave, J. & Wenger, E. (1991). *Situated Learning: Legitimate Peripheral Participation*. Cambridge, UK: Cambridge University Press.

Preece, J. (2000). *Online Communities: Designing Usability, Supporting Sociability*. Chichester, UK: John Wiley & Sons, Ltd.

Tannen, D. (1989). *Talking Voices: Repetition, Dialogue, and Imagery in Conversational Discourse*. New York, NY: Cambridge University Press.

Traweek, S. (1992). Border Crossings: Narrative Strategies in Science Studies and among Physicists in Tsukuba Science City, Japan. In A. Pickering (Ed.) *Science as Practice and Culture* Chicago, IL: University of Chicago Press.

Human Subjectivity and Relation Profiling Factors in The Global Network

Maria Chiara Pettenati, Dino Giuli

Electronics and Telecommunications Department, University of Florence
Via S. Marta 3, 50139 Florence (IT)

Abstract

This paper is intended to analyze the network user profile, in order to derive a new systemic approach for an effective instrumental exploitation of user subjectivity and relational factors in the Global Network. Proper representation and use of user profile are expected to have important repercussions because of the opportunity to offer the user more personalized network interaction, as well as of the possibility to sustain the establishment of a new user-centered network mediating system. This can account for subjectivity, relation and collectivity factors, being capable of conditioning network interactions and relations through a proper exploitation of such factors.

1. Introduction

When considering available network applications and solutions trying to cope with issues such as *personalization* and *privacy,* we notice that they currently do not offer users the possibility to *unitarily manage and dispose of their personal data and profile* [9]. Hence, they often incur in the opposite effect and offer providers fertile ground for user categorization and profiling procedures which not only violate the basic privacy principles, but also increase asymmetry in interaction and limit end user negotiation capability [4,5]. Symmetry in interaction and increased self-awareness – and, thus, negotiation capability – have therefore to be pursued by starting to reconsider *all aspects of subjectivity* as being fundamental for the network development, towards a new environment which is expected to be potentially capable of sustaining the whole range of human activities. This paper is intended to give a conceptual representation of the main elements describing both the *User Subjective Profile* and the associated network relational mechanism, with respect to network interaction and relation, the major goal being to allow their instrumental exploitation for the development of an effective end user-centered network interaction environment. The present paper reports on specific results of the current research framework program *B.E.S.T.* (**B**ridging **E**conomy & **S**ociety with **T**echnology) *beyond Internet* [1]. While a general outline of such research is provided by the paper [4], the focus on further specific and related subjects is provided in other companion papers [5,6,7] which complete the overall picture of the available reviews which report on such research.

2. The subjective profile

There is no doubt that pretending to digitally represent humans with their complexity and differentiation, is a very challenging goal, nevertheless, the need to find a common principle - grounding on subjectivity - which should consolidate the evolution of technology, is more and more urging. This principle has no reason to be, if not that of aiming to represent a set of knowledge concepts, theories, information and beliefs such as the way instruments represent user subjectivity which are partially made available directly by the user, and partially inferred or detected with appropriate techniques and methods [7]. We therefore introduce the concept of *subjective profile*, as a *complex set of digital information, related to the subjectivity of the user, employed to allow and facilitate personalization and negotiation during Network interaction according to user interests and needs – actively expressed and represented by the same user, as well as automatically produced or inferred.*
The subjective profile can be adopted to describe any type of user, i.e. end user or provider. Our approach generally accounts for both descriptions, however, the following analysis of the user's profiles will focus only on end user subjective profile, as the one which is most critical for our purposes. The subjective profile should not only be capable of being instrumentally represented i.e. "representable", but it should also be effectively produced, managed, adapted, used, controlled and safeguarded, by the user himself, and/or by someone else acting on the user's behalf [2]. If the digital representation of the user profile had such characteristics, subjects of the network environment – each represented and acting through the respective subjective profile – would acquire balanced power and capabilities. This environment could therefore evolve in order to offer the wider adaptability range to different user

subjectivity, thus being capable of supporting user *personalized* applications – intended in the deeper sense. A step further in this reasoning process, is to define more precisely the components of the subjective profile in such a way as to be suitable for data handling in the network environment. We conceptualize the personal profile as being composed of two main elements [2]:
– *Background Profile*; the information describing user personality, or settled behaviors and attitudes
– *Foreground Profile;* the information describing individual factors related to the specific moment.
These components conceptually distinguish two informative contexts, evolving in time, which differentiate between them thanks to their state of consolidation and because they are respectively the *object* of user self-awareness, and the *subject* of user action.

2.1   End user subjective Background profile

This component of the subjective profile defines stable individual traits, which are considered to be relevant for the representation and inference of user potential behavior towards network activities – be they the activation of new relations or the adaptation of services and interaction modes. The elements pertaining to the Background, can be grouped into two main contexts: those related to the *real world* and those related to the *network world*, which characterize traits which are, respectively, generally distinctive in the real world (but also relevant for the modeling of the user as a network actor), and specifically distinctive i.e. a sign of user interaction in the network environment. Some of the elements which can be included in both the above groups, are summarized in Table 1.

Table 1: End User Background Profile

a) Real World Background Elements	b) Network World Background Elements
- *Personality*; as the framework of the strictly individual concerns, capable of conferring distinction to the individual - *Public Identity* - *Recurrent Places;* such as home, work, etc. - *Wealth and Business;* the economic profile - *Biography*; such as education, religion, profession, political orientation, relational profile – social and private, etc. - *Preferences*; such as those expressed accessing specific goods or services - *Typical temporal engagement in activities* - *Conscious interests and needs*	- *Instrumental configuration of available access terminals*; the type and characteristics of available terminals - *Conditioning or limitations in instrumental interaction*; such as those due to disabilities or due to age, integrity, etc. - *Degree of instrument mastering* - *Specific aims and scope of network interaction* - *Related network subjects*; subjects with whom the user is able or willing to establish network relations - *Explicit interests and need for network interaction* - *Intensity of network activity*; such as time spent on the network, related costs, etc. - *User expressed background subjective pre-requisites for network operation* - *Network user personality*; network relational profile, network user identities for specific activities, etc.

One element of the Network Background Profile - the *network personality* and, consequently, the *network relational profile* - is particularly relevant to the instrumental representation of the network user profile and will, therefore, be analyzed in more detail in the following paragraphs.

2.2   End user subjective Foreground profile

These are the contingent elements which influence the person's behavior in the network, as for his/her relations/scopes/modes, in each specific moment/place/context/situation; currently influencing the inclusion or exclusion of network activities. In addition, it determines the adaptation for the technical-functional configuration of the network access and interaction with the current user subjective condition, on the basis of his/her possibility/willingness to activate specific network interaction and relation processes, together with his/her possibility/availability to intervene in analogous processes solicited or requested by other network subjects. The production of the foreground factors can be self-determined by the user, or inferred by the background profile, as well as being instrumentally produced by inference or monitoring, as happens through the monitoring of the state of the user and the state and type of network activity. As for the Background, the Foreground profile can further be divided in terms of its *real world* and *network world* components as exemplified in Table 2.

Table 2: End User Foreground Profile

a) Real World Foreground Elements	b) Network World Foreground Elements
- *Current activity performed outside the network;* - *New incumbent activity superimposed by the will of the user;* - *Psychological, physical, emotional state;* - *Current place of activity;* - *Agenda, Personal diary etc.*	- *Technical-functional characteristics of currently used terminal ;* - *Instrumental monitored data on current activity;* - *Monitoring data on end user psychological, physical, emotional state;* - *Instrumental data on end user current geographic position;* - *User expressed foreground subjective pre-requites for network operation*

It is worth pointing out that the process of building and updating the whole subjective profile, and its progressive growth is a recurrent process in which Background and Foreground influence each other, moving to Background some Foreground components and vice versa: the first evolves to incorporate elements which can be considered to be stable and the object of self-awareness – while the latter need to be continuously updated, since they vary and change in relation to the context – personal, environmental, relational, etc. Their changing over time thus influence the evolution of the individual *network personality.*

3.   Relation and Interaction Processes

Humans, as social beings, grow up in the real world developing and increasing experience as regards the different ways to interact and relate to other individuals or groups, in a manner which hopefully responds to their needs and personal attitudes, thus shaping their complex identities in different interaction and relations processes. Interaction and relation are interconnected processes because the first can prelude the latter, but their very nature is different. In *interaction*, there is reciprocal influence, or action and/or reaction, due to a basic need to establish a relationship. *Relation* is instead a process in which actions and manifestations make a specific relationship operational, thus following an evolution in time of a series of interactions. We would, therefore, like to be able to use information on *user subjective profile* and on his/her *relational experience* in the real and network world, in order to support the end user in performing a successful interaction. This is mainly intended to allow both (i) the promotion and exploitation of user *self-awareness,* so that he/she can become an actor on the network, and (ii) the exploitation of a *collective knowledge* [7], in order to allow easier interaction, reducing indetermination and discontent in interactions and relations.

3.1   Relation as an active process

In order to conceptualize these aspects, we would like to think about network relations in terms of *process*, either coming from the end users, or directed to the end user. The *request of encounter,* (be the user active or passive in requesting it, or be it requested by other subjects), is the *trigger for* the activation of a relation process which can be expressed at various levels of specification, thus leading to various possibilities for the development of the relationship: the actual occurrence of the encounter, the consequent activation of a new relationship, the continuation of an existing relationship or, alternatively, none of the above. In any case there must be agreement as regards the establishment of relations. Fig. 1 shows the cycle of interaction/relation phases, with connections and actions required to pass from one phase of the network-based relation process, to another.

Stages of relation in the network can thus be represented as: (i) *encounter*; the meeting between end users, or user-provider, which can either be solicited externally (from the real or network world), or spontaneous, to start an interaction and possibly to activate a stable relation (ii) *activation*; once the compatibility of subjective factors is met, the peers activate a relation process (iii) *continuation*; the relation is in this phase carried out using a definition of subjective interaction parameters, and can eventually

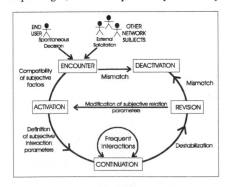

Figure 1: The network-based relation process

be consolidated and refined through recurrent interactions (iv) *revision;* if a destabilization in interaction occurs, a revision is necessary to modify the subjective interaction parameters to satisfy peers, or to lead to the interruption of the relation process (v) *deactivation;* this is the conclusion of the interaction due to the inability to define new parameters to satisfy both parts (experienced mismatches are stored to be taken into account in possible new encounters between the involved subjects).

3.2 The relational profile

The relational profile is a fundamental part of the user Background subjective profile [2], which can be structured as reported in Fig. 2. It aims at describing the *relations* – where relations thus means, therefore, a complexity of agreements, typical exchange of data, actions and reactions, between the end user and other network subjects. Information elements belonging to the relational profile can be *private*, i.e. used exclusively by the end user (related to the private and subjective handling of the relation and its consequences, such as the personal advantages due to a participation to a virtual seminar), or *shared* among interacting partners to successfully carry out the relation (such as the language agreed for the communication). The relational profile can be conceptualized in a *part* which is *common* to the network subjects with whom the end user has established or has the possibility or willingness to establish relations - this part contains information which should be exclusively available to the user (*private*) - and a *part* which is *specific*, with respect to each interacting subject (Subject 1 ™ Subject N in Fig. 2a) with whom a relation is undertaken; this information can be both of a *private* or *shared* use type.

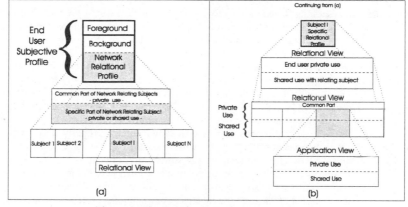

Figure 2: The Relational Profile of the End User

All the information pertaining to the relation between end user and Subject *i* (Figure 2a), is described in its *relational view*. Each relational view, can again be *private* of the end user or *shared* with Subject *i* to define modes and frameworks for possible and admissible interactions. Each relational view can be further seen as the union of many *application views* (again *private* or *shared*), introduced to differentiate information concerning activation and management of the specific interaction with Subject *i* with respect to the context and scope of the application. The application view specification, can therefore include elements such as the implicit field of human activity, typology of network application, specific scope, technical-functional and operative modes to support that specific interaction. The interaction between user - Subject *i*, described through the relational profile thus determines the whole range of possibilities for admissible application views, which are identified by specific *indicators*.

3.3 Relations among Subjects through Network Resources

End users can establish inter-personal relationships among themselves and with providers. Independently of the kind of partner, the end user executes the relation through the intermediation of *network resources* (information, contents and other resources needed, and services). These include resources made available by partners, which are also inherently responsible for the quality of such resources. Figure 3a shows the current network relation situation, in which user and provider interact directly through Network Objects; here relation is only implicit. Fig. 3b, shows, instead, the new prospected situation, in which users interact between themselves through the *subjective profile*, allowing capability for the explicit determination of relations to really match subjectivity requirements and support the adaptation of relations and interaction to various subjectivity requirements, thus, implying a balance in negotiation capability. Analogous considerations apply also for relation between the end users, as for instance is implied by interpersonal communication.

1106

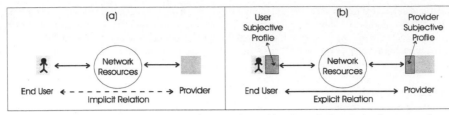

Figure 3: Interaction and relation between end user and provider through the supporting network resources

An important consequence of what has been said up to now, is that Network Resources themselves should be managed and represented with respect to related subjective factors. This mechanism should be applied for all types of information and processes implied in the relation, once referred to their contents - information to be used - or flows related to the transaction, which can contain subjective constraints related to both the social contexts and applications in which they are involved [6]. Therefore, they should include subjective meta-information such as limiting conditions or specifying factors to handle the relation respecting subjectivity requirements.

4. Conclusion

In this paper we aimed at describing *human subjectivity factors* which have to be taken into account for the development of a Global Network which truly responds to individual needs as well as aspirations and hopes to provide universal access. We have, therefore, introduced the *user subjective profile*, trying to describe its elements – Foreground and Background – so as to offer a starting point towards its integration into the network instrumental process. The key role of the network *relational profile*, within the human profile, has been pointed also out. An important remark should be made at this point, as regards the importance of collectivity for the proper exploitation of subjectivity [7]. Representing individuals with their subjective profiles, can also enable the exploitation and building of a wealth of knowledge, meanings, understandings. This can be valued and exploited in providing the user with the proper network relation. Information which derives from the description and representation of the human subjective profile, as well as from the shared social knowledge, if properly described, developed and handled, can contribute to the development of the *human subjectivizing gene*: the subjectivity driver providing the human imprint on the network, and which could suitably and structurally influence the network evolution towards the Global Network in all its aspects [4]. If the gene structure properly influences the growth of the information and relational body of the network, then the Global Network can actually become a dynamically reconfigurable and adaptable relation environment centered on the user and inherently balancing user roles in the network.

References

1. D. Giuli "Background and objectives of the *B.E.S.T. beyond Internet* framework research program" , Dept. of Electronics and Telecommunications, University of Florence, Internal Report DET-1-01, February 2001.
2. D. Giuli "Subjective information and the subjectivizing gene for a new structure of the instrumental relation and interaction processes in the Global Network", Dept. of Electronics and Telecommunications, University of Florence, Internal Report DET-2-01, February 2001 (in Italian).
3. D. Giuli "Subjectivity, citizenship and Public Administrations in the evolution towards the Global Network", Dept. of Electronics and Telecommunications, University of Florence, Internal Report DET-3-01, February 2001 (in Italian).
4. D. Giuli " From Individual to Technology towards the Global Network" HCI International 2001 9th International Conference on Human-Computer Interaction. August 5-10, 2001 New Orleans Louisiana.
5. D. Calenda, D. Giuli "Subjects, subjectivity and privacy in the Global Network". HCI International 2001 9th International Conference on Human-Computer Interaction. August 5-10, 2001 New Orleans Louisiana..
6. G. Vannuccini, P. Bussotti, M.C. Pettenati, F. Pirri, D. Giuli "A new Multi-layer Approach for the Global Network architecture". HCI International 2001 9th International Conference on Human-Computer Interaction. August 5-10, 2001 New Orleans Louisiana.
7. P. Bussotti, G. Vannuccini, D. Calenda, F. Pirri, D. Giuli "Subjectivity and Cultural Conventions: the Role of the Tutoring Agent in the Global Network". HCI International 2001 9th International Conference on Human-Computer Interaction. August 5-10, 2001 New Orleans Louisiana.
8. T. Winograd, F.Flores, "Understanding Computers and Cognition", 1986, Ablex Publishing Corporation, USA.
9. Communications of the ACM, August 2000, Vol.43, No.8 Special Issues on *Personalization*.

Creating public information environments that strengthen citizen-government relationships
Building TIES for a better society

Gary Strong,
Susan Brummel Turnbull,
Karl Hebenstreit

US National Science Foundation
4201 Wilson Blvd., Arlington, VA 22230
US General Services Administration
18TH and F ST NW, Wash. DC 20405

Abstract

The paper identifies critical issues for the shaping of the emerging Information Society, which extend beyond technology. The need for an all-inclusive Information Society is presented and the dangers resulting from the lack of a suitable policy are identified. Several new initiatives are described to facilitate progress toward goals.

1. Background

One of the most profound effects of information technology (including telephone, television, computing, and networking) has been the creation of a new basis for contractual social interaction (activities of exchange, payment, evaluation, institutional advancement, even government). The established basis for contractual social interaction in recent history has been money, or traditional capital. Contractual social interaction, i.e., that outside of family and friendships has been valued on the basis of the amount of money involved, either presently or in the future. This singular basis for contracted activity has recently given way to another, possibly even more important one, that of information.

While most information has a monetary value, it can have independent value. Some knowledge changes rapidly, particularly that which is newest, and information sources that contribute to such knowledge can be invaluable in monetary terms. Information technology has reached the point of development where access to rapidly changing information has become common for some. Learning-on-demand and other just-in-time practices are adaptations to this global, technological achievement. Now that it is possible, it is quite likely that all contractual social interaction, including government, will adapt to the value of such information capital by evolving forms that exploit the value of real-time information access.

It is yet unclear how society will be shaped in the future after adapting to the value of information as a basis for contractual social interaction, but several potential dangers can be avoided. The classing of society has occurred throughout history on several bases, including religious and monetary. The avoidance of these forms of social classification is a recognized struggle in democracies worldwide. Social stratification based on differential information access, on the other hand could become a serious issue for global human development if allowed to develop without recognition of its dangers and the development of safeguards to protect against them. A democratic approach to information access in the modern world is possible with responsible, need-driven investment in information technology research.

2. Access and Independence

Early experience has shown how to create the technical aspects of access to information through a variety of communication media and formats. Some approaches require advanced, intelligent technology, such as when spoken languages must be recognized by machines. Sophisticated technologies, such as the latter, may not be required for information access by some users, but can make all the difference for others due to language differences or disability. Now the societal aspects of information access as a new prerequisite to independence in society need to be understood. Just as financial capital and speaking the native language of a country have historically been significant prerequisites for being an independent citizen, information access will increasingly serve a similar gate-keeping function. The stakes are high for all. On one hand, experience has already shown that for many people in the world, such as those with disabilities or in isolated settings, information access through sound policy can provide both independence and societal contributions not previously attainable. On the other hand, everyone is at risk for becoming disadvantaged (and overwhelmed) by the blinding rate of changing information. Only those who find a way to attune to the world's information productively are likely to achieve their goals as independent and contributing members of society.

3. Goals

There are several goals that can be identified for exploration:

3.1. The creation of a Technology Infrastructure Extension Services (T.I.E.S.) to bring the universities to all the people and facilitate the process of technological adoption and eventual inclusion in the information society for not only K-12 educational institutions, but also to community and business groups outside the silicon valleys and dominions. The expertise and infrastructure at many universities is already distributed and of sufficient size to enable the creation of such TIES, reminiscent of the successful US Department of Agriculture Cooperative Extension Services of the past 85 years that played a major role in the US agricultural industry. Furthermore, such activities would greatly enhance the capability of universities to reinvent themselves to become more effective knowledge providers to their citizens.

3.2. The avoidance of serious socioeconomic fallout in progress toward a global information infrastructure that was experienced in the US during the Interstate Highway System build-out and characterized by:

creation of physical divisions in cities across economic zones reinforcing boundaries and further contributing to isolation; and

creation of fast urban exits that encouraged wholesale bailout of middle classes to the suburbs, leaving the poorer classes in the cities with insufficient tax base to support them.

3.3. The contextualization of economic fallout, such as described above, is a symptom of a larger, unsolved problem, which is **insufficient addressing of citizen needs in infrastructure development**. In recognition of this problem, the National Research Council (1997) recommended that the United States undertake psychological, sociological, and historical studies to determine the needs of every US citizen in the context of the National Information Infrastructure and thus to provide guidance to technologists concerning what needs to be created and what will not work for real users. In other words, there needs to be a more user-centered government with active citizen participation in policies and protocols whose outcome is likely to lead to the reinvention of society.

4. Initiatives

There are hundreds of thousands of websites, conferences, and programs aimed at addressing universal access and the digital divide, including 29 bills in the 106th legislative session of the US Congress. Four broad digital opportunities were identified by the US Senate:
1) broaden access to information;
2) provide workers and teachers with information technology training;
3) promote innovative online content and software applications that will improve commerce, education, and quality of life; and
4) help provide information and communications technology to under-served communities.

Low-cost Internet access by families at home is a broadly supported, near term opportunity, due to falling prices and service agreements. Some Federal agencies are working on the "technology push" aspects of universal information access by funding research on the capabilities needed by individuals and research on the societal implications of IT. Some agencies are funding schools, libraries and community technology centers to offer free, public access. Other agencies, such as the General Services Administration (GSA) are better suited to the "market pull" side of the equation. GSA can help Federal agency customers identify how to integrate both proven and new options for public service delivery. This will be accomplished by understanding the risks of social stratification and opportunities for mitigating closed research, development, and deployment circles that contribute to social stratification.
The role of GSA is three-fold:

1.   To share US Federal government best practice from GSA's Section 508 leadership responsibilities in assisting agencies meet requirements for accessible workplaces and electronic services. Other institutions and government bodies can learn from this evolving Federal government practice.
2.   To explore state, local, non-profit and private sector practices relevant to sustainability of open research, development and deployment practices of IT.
3.   To learn how to build a citizen-centric infrastructure by conducting Universal Access Expedition workshops to accelerate shared learning and multi-sector partnerships designed to sustain empowered citizens and their communities.

5.   Conclusion

Information capital now mimics and mixes with financial capital as a basis for contractual social interaction among individuals. We have a new, renewable resource available for building a better society. A citizen-centric public infrastructure must be created to tap this infinite resource, in a manner that enables all people to rise above the noise of difference and contribute toward a balanced, steady course to a reintegrated and reinvigorated society.

Reference
Computer Science and Telecommunications Board, National Research Council (1977) . More Than Screen Deep: Toward Every-Citizen Interfaces to the Nation's Information Infrastructure.Wash., DC: National Academy Press.

Evaluation and training on accessibility guidelines

Carlos A. Velasco

GMD - German National Research Center for Information Technology
Institute for Applied Information Technology (FIT.HEB)
Schloß Birlinghoven, D53757 Sankt Augustin (Germany)
Carlos.Velasco-Nunez@gmd.de

Abstract

The arrival of the Information Society is accompanied by a whole set of new technologies that could present access barriers for people with special needs. To tackle these accessibility issues, several clusters of guidelines have been issued by different companies and international organizations. However, the uptake of these guidelines in mainstream products is very slow. This paper discusses several issues concerning the evaluation and training process on accessibility guidelines, and how to include users with special needs in the design-loop.

1. Introduction

The **Information Society** poses to designers and developers many challenges that must be faced to make it accessible to every user independently of her needs and requirements. They must tackle with a wide variety of requirements, frequently expressed as a set of guidelines aimed to cover the widest spectrum of user needs. If these guidelines are ignored or fail to take-up, the Information Society will be split into two sectors: those able to access the information, and those who cannot access it. Therefore, to avoid this form of «Digital Divide,» it is necessary to evaluate the guidelines, and to train designers -or more adequately, implementers- on them. This paper tries to introduce end-users with special needs within the evaluation-training loop. The interaction with end-users is of uttermost importance because it allows to evaluate whether the design criteria meet their needs, and up to what extent. It is also obvious that there is not a wide pool of end-users with the technical knowledge and skills to give designers a useful feedback, thus training users is a necessary intermediate step.

2. The training process

Guidelines concerning accessibility are published by specialized groups with a deep knowledge of special needs requirements. These guidelines are highly technical and descriptive, trying to cope with different circumstances and environments. However, the take-up of these guidelines -for instance, concerning software products or web development- is slow, because their language is sometimes not adequate to their intended audience: designers. The guidelines' «end-users» - designers or implementers in the rest of the paper- have very different backgrounds: software engineering, graphic design, or even a self-didactic background. Therefore, the dissemination process must be associated with training and publication of practical handbooks on guidelines' implementation. Another practical issue, frequently ignored, is that this training process can provide useful feedback to modify and improve guidelines. To take our approach one step further, the inclusion of end-users will lead to better results. These two processes are described in the following subsections. The final aim, as it can be seen in Figure 1

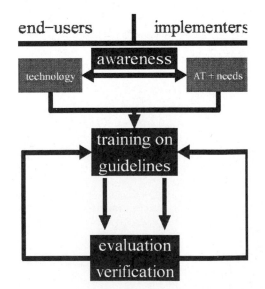

Figure 1: The global training-evaluation process for design guidelines.

is to obtain a richer set of modifications and implementation techniques.

2.1   Training end-users

The training of end-users for the evaluation of computer interfaces is a costly exercise that not every company can afford, but the benefits of this approach can lead to largely improved products, with a better market impact. The basic aim is create a meaningful amount of skilled users to produce useful feedback. The critical points, from the end-users' standpoint, are to facilitate to them:

(A) Tools to evaluate their degree of satisfaction with the performance of the interface resulting from the guidelines. This can be accomplished by comparing different interfaces (including exercises on how guidelines improve the interface for a given inaccessible element); using different input methods (different assistive technologies); and using different output methods (visual, oral, etc.).

(B) State of the art reviews in the field of computer interfaces, to make them aware of the technology limitations.

(C) Tools to provide useful feedback to publishers and implementers of guidelines.

Some authors [1] have reported difficulties to go beyond point (A), but emphasize the benefits of going further in the training process. It is necessary to remark that the design of this training phase must be carefully considered. There are two major concerns when designing the end-users' training:

i.   Ensure that all the documentation is provided in an accessible format (being HTML or text-only the «minimum» common denominator). The major problem that can be found is a negative attitude from the end-user towards the implementer or designer which arises from the fact that certain products place big hurdles towards access. The key point of any training programme is to be able to turn this attitude into useful criticism.

ii. Do not let the user focus on her own specific needs, but also on those of others, to facilitate a wider vision and to meet a broader spectrum of demands. It is important to set up a large test-bed of users to rely on.

2.2   Training designers and implementers

As mentioned earlier, there is little attention paid in the literature to training designers or implementers of guidelines. This may lead to the failure of the implementation phase of any guideline. This section tackles some of the issues that must be dealt with when planning such type of training. Our approach is based on providing to implementers with several elements that can increase their **awareness** about different interface design issues.

(A) Design for all or Universal Design

The first challenge faced is to select an appropriate term between the existing choices (Design for All, Accessible Design, or Universal Design) to define our objectives. For the first term, there are many definitions [2,3,4]. However, introducing concepts such as designing for the broader average can be vague enough to mislead guidelines implementers who are not familiar with assistive technologies. Generally, implementers consider Universal Design a highly sophisticated and not cost-effective design process by which their products will reach a wider hypothetical market.

Simplifying the message can be useful to attract designers' attention. More appropriate terms seem to be Universal Design, or «good user-based design,» which aim to address the needs of a broadest range of users by adding **customization** to the resulting interface [5,6]. Figure 2 shows a graphical interpretation of the above concept by considering a typical users spectrum. This spectrum has two elements: age and ability. It shows that common design techniques, targeted to the average user, leave out of the loop different groups of users. By applying these guidelines, the pool of potential users widens. It must be stressed that these guidelines benefit not only to people with special needs, but to users accessing the information in different contexts. The training must include as well classical communication techniques for interface design [7].

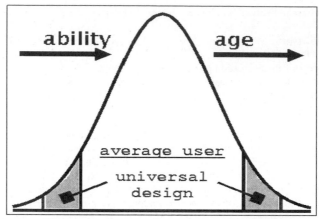

Figure 2: Typical users spectrum, showing the increase of users' pool by the adoption of Universal Design.

(B) «How-To» examples

A second step in the training of implementers or designers is to make them aware of what issues and challenges a person with special needs must face dealing with the product under consideration. Since direct interaction with the end-user is not always feasible, alternatives must be sought. Feasible alternatives are to display available multimedia material demonstrating the use of computers by people with special needs, and to host practical sessions with assistive technologies for computer access (alternative input devices, auxiliary devices, speech recognition software, speech synthesizers, and other software tools).

(C) Commercial and legal issues

It is advisable to discuss some of the figures related to the potential market for people with special needs. Some authors [8,9] estimate that by year 2020, there will be 32 million of persons with special needs in the EU. The market can be even bigger because of the progressive aging population in our countries. Those figures do not include, as mentioned earlier, persons accessing to the information in different contexts.

The legal framework in which designers operate must be discussed as well. This framework is much more important in the USA, where the publication of the **Americans with Disabilities Act** (1990) created a scenario where accessibility became a legal requirement. Further extensions of ADA were Section 255 of the **Telecommunications Act** (1996), and the most recent amendment of Section 508 (2000) of the **Rehabilitation Act** (1986). The European situation differs greatly from our American counterparts. However, some changes are expected with the **e-Europe** initiative of the European Commission.

(D) Myths

The process of creating awareness goes hand-in-hand with the process of destroying common myths. Among the most important we should mention:

- **Cost-effectiveness of accessible design**

  It is a very common belief that using accessible design guidelines is expensive because of its complexity. To eradicate this belief it is important to show some simple examples. A typical exercise is to demonstrate how the artless inclusion of meaningful ALT-tags for the images in a web page improves its accessibility.

  From the commercial standpoint, there is an unavoidable expense linked with the training process of the company's designers. However, this «one-time» investment can be recovered by the increase in the users-base (see (C)). Additionally, if not faced now, it can increase significantly the cost of development of the product or service in the long run, because its retro-fitting forced by legal reasons becomes more expensive.

- **Design appealing and lose of functionality**

  It is typical to link accessibility with unappealing software applications or text-only web sites without any multimedia content. The trainer must stress that guidelines are never about forbidding the inclusion of materials but on providing alternative tools and methods.

  This discussion can be taken a step further: guidelines implementation avoids double versions of products or services (software, web pages, etc.). Double versions are costly and user acceptability is compromised as well. It is not uncommon for a product to be rejected because of an external appearance far different from that of mainstream products. Furthermore, the idea of a mainstream product being used by a wider audience is commercially appealing. The challenge is to develop a common-looking product, which can also be functional to other end-users with special needs.

- **Accessible design aims to solve problems that should be addressed by assistive technologies**
  In the last few years assistive technology has provided a set of powerful tools that allow disabled and older people access to information and services. However, there is something that no technology can do: guess the intention of the designer. Human-computer interaction is the link between a product and the user. The implemented interface must be able to provide information -by visual or by other means- so that the end-user is able to interact with it. The same happens with technology; it can only provide whatever information is given. For example, if the printing functionality of your software application is only accessible from an icon in a toolbar, and not from the application's menu, how is the screen-reader able to figure out that your good-looking icon is for printing the open document?

3. The verification/evaluation process

The next step is to establish an interchange phase. There are always occasions when this extensive approach is not feasible, however, the interaction of the end-user with the product is always desirable [10,11,12] in the design of mainstream products such as software, Internet sites, interfaces for household equipment, etc. This is where guidelines must come into play, and trainers of guidelines implementers must play several roles simultaneously.

We strongly recommend interaction of end-users with the final product or service whenever feasible. Products addressed to multi-cultural audiences must follow similar procedures [13]. A verification/evaluation process should contain the following steps:

i. **Implement a design guideline**: Implementers should introduce a design guideline for the product or service under consideration. Whenever feasible, end-users must have access to both versions of the product (with or without the guideline implemented) and must be informed about its expected performance.

ii. **Test the resulting functionality and usability of the modification**: End-users must verify and test whether the «inclusive» version (the one with the guideline implemented) is allowing them to access the complete set of capabilities of the product or service as described to them previously. We must recall that end-users must be trained and must have access to appropriate tools to perform this task efficiently.

iii. **Exchange phase**: An exchange forum between end-users and implementers, assisted by means of a definition of common languages in order to facilitate bi-directional communication, must be established. This forum will validate whether any given guideline is fulfilling the expected results from the point of view of implementers and end-users or whether any given guideline must be modified, and how, in order to achieve the aims it was intended for. A positive disposition to collaborate between manufacturers and end-users will benefit both sides. Massive evaluations of guidelines must arise as a result of a strong cooperation between industry and organizations of end-users [14].

4. Conclusions

The training, verification and evaluation of guidelines are difficult tasks. There are no magic recipes, which can solve problems instantaneously. This paper provides to the reader some practical resources. The ideal situation will arise when interaction between end-users and implementers can take place. However, a successful interaction requires willingness to listen and to take into account the other party's demands.

The lack of this interaction does not imply that inclusive design guidelines are impossible to apply. We have shown some methods that will promote successful training of the implementers of guidelines. The objective must always be to collect experiences and to make them available to the community. There is an open field of research on adequate tools and methods to facilitate interaction with end-users, which must be explored. This will lead certainly to successful implementation of guidelines in mainstream products and services.

References

[1] Brandt, Å., and Gjøderum, J. (1995). Older and disabled users participation in standardisation work is possible and valuable. In *The European Context for Assistive Technology (Proceedings of the Second TIDE Congress)*, Placencia-Porrero, I., and Puig de la Bellacasa, R. (eds.). IOS Press (Amsterdam), pp. 51-54.

[2] Janssen, H. T. J., and van der Vegt, H. (1998). Commercial Design for All. In *Improving the Quality of Life for the European Citizen (Proceedings of the Third TIDE Congress)*, Placencia-Porrero, I., and Ballabio, E. (eds.). IOS Press (Amsterdam), pp. 84-87.

[3] Sandhu, J. S. (1998). What is Design for All. In *Improving the Quality of Life for the European Citizen (Proceedings of the Third TIDE Congress)*, Placencia-Porrero, I., and Ballabio, E. (eds.). IOS Press (Amsterdam), pp. 88-91.

[4] USER Consortium (1996). *USERfit, a practical handbook on user centred design for Assistive Technology.* ECSC-EC-EAEC (Brussels-Luxembourg).

[5] Story, M. F. (1998). Maximising Usability: The Principles of Universal Design. *The Assistive Technology Journal*, 10(1), pp. 4-12.

[6] Stephanidis, C., Salvendy, G., Akoumianakis, D., Arnold, A., Bevan, N., Dardailler, D., Emiliani, P. L., Iakovidis, I., Jenkins, P., Karshmer, A., Korn, P., Marcus, A., Murphy, H., Oppermann, C., Stary, C., Tamura, H., Tscheligi, M., Ueda, H., Weber, G., and Ziegler, J. (1999). Toward an Information Society for All: HCI challenges and R&D recommendations. *International Journal of Human-Computer Interaction*, 11(1), pp. 1-28.

[7] Mullet, K., and Sano, D. (1995). *Designing Visual Interfaces: Communication Oriented Techniques.* SunSoft Press - Prentice Hall (Mountain View, California).

[8] EUROSTAT (1992). *Rapid Reports, Population and social conditions: Disabled People Statistics.* Commission of the European Communities (Luxembourg). ISSN 10160205.

[9] Carruthers, S., Humphreys, A., and Sandhu, J. S. (1993). The market for R.T. in Europe: a demographic study of need. In *Rehabilitation Technology (Proceedings of the First TIDE Congress)*, Ballabio, E., Placencia-Porrero, I., and Puig de la Bellacasa R. (eds.). IOS Press (Amsterdam), pp. 158-163.

[10] Furugren, B., and Lundman, M. (1995). New Roles for disabled users in technology application work. In *The European Context for Assistive Technology (Proceedings of the Second TIDE Congress)*, Placencia-Porrero, I., and Puig de la Bellacasa R. (eds.). IOS Press (Amsterdam), pp. 33-37.

[11] Pascoe, J., Pain, H., McLellan, D. L., Jackson, S., and Ballinger, C. (1995). Involving users in the evaluation of assistive equipment. In *The European Context for Assistive Technology (Proceedings of the Second TIDE Congress)*, Placencia-Porrero, I., and Puig de la Bellacasa R. (eds.). IOS Press (Amsterdam), pp. 114-117.

[12] Velasco, C. A. (1998). The Information Society disAbilities Challenge (ISdAC): Paving the way for the active end-user. In *Improving the Quality of Life for the European Citizen (Proceedings of the Third TIDE Congress)*, Placencia-Porrero, I., and Ballabio, E. (eds.). IOS Press (Amsterdam), pp. 473-477.

[13] Tahkokallio, P. (1998). Though Other Eyes. From knowledge to understanding. In *Improving the Quality of Life for the European Citizen (Proceedings of the Third TIDE Congress)*, Placencia-Porrero, I., and Ballabio, E. (eds.). IOS Press (Amsterdam), pp. 63-66.

[14] Velasco, C. A., and Verelst, T. (1999). Raising awareness among designers of accessibility issues. *Interfaces* (British HCI Group) 41, pp. 6-8.

Author Index

1120

Wiberg, M. 450
Widlroither, H. 769
Wiecha, C. 210
Williams, C. 1029
Wilson, J. 506
Wilson, M.D. 286
Winckler, M. 160
Wolsey, S. 1083

Yamaoka, T. 55
Yamauchi, S. 392

Yamazaki, K. 747

Zajicek, M. 454, 700
Zaphiris, P. 496, 540
Zarikas, V. 127
Ziegler, J. 640
Zimolong, A. 878
Zimolong, B. 878
Zülch, G. 645

Subject Index